A Textbook of Head and Neck Anatomy

Copyright © B. K. B. Berkovitz, B. J. Moxham, 1988
Published by Wolfe Publishing Ltd, 1988
Printed by Grafos, Arte Sobre Papel, Barcelona, Spain

Library of Congress Cataloging-in-Publication Data

Berkovitz, B. K. B.
　A textbook of head and neck anatomy.

　Includes index.
　1. Head—Anatomy. 2. Neck—Anatomy. I. Moxham, B. J.
II. Title [DNLM: 1. Head—anatomy & histology—atlases.
WF. 17 B513t]
QM535.B46 1988　　6111'.91　　87-29431
ISBN 0-8151-0729-3

This book is copyrighted in England and may not be reproduced by any means in whole or part.
Distributed in North America, Canada, Hawaii and Puerto Rico by Year Book Medical
Publishers, Inc. by arrangement with Wolfe Publishing Ltd.

A Textbook of

Head and Neck Anatomy

B.K.B. Berkovitz[*]
Department of Anatomy
University of Bristol

B.J. Moxham
Department of Anatomy
University of Bristol

The section on the central nervous system was written by
M.W. Brown
Department of Anatomy
University of Bristol

Illustrated by
Jack Furnival

[*]Current address: Department of Anatomy,
King's College, London

YEAR BOOK MEDICAL PUBLISHERS, INC.

Preface

Major advances in medicine appear almost daily. The medical curriculum perforce must constantly change, so that new subject matter is regularly introduced alongside the traditional disciplines. In many schools, anatomy teaching has changed radically, and this is reflected in some modern textbooks which sacrifice detail to provide a simplified and general account of the structure of the human body. Paradoxically, all this is taking place against a background of increasing specialisation. We believe, therefore, that there is a need for textbooks which are restricted to specific regions of the body and which have detailed coverage. A volume dealing with the anatomy of the head and neck might be welcome for two reasons. First, this region is especially complex. Second, the head and neck is the concern of a variety of specialists: neurologists, ear nose and throat surgeons, ophthalmologists, maxillofacial surgeons, and oral and dental surgeons.

A proper appreciation of anatomy relies not merely upon the assimilation of a mass of facts, but upon an awareness of the three-dimensional disposition of structures. Thus, anatomy should be regarded as essentially a visual subject that cannot be mastered simply by reading a text. Although most textbooks of anatomy make extensive use of diagrams, the illustration usually remains subordinate to the text. Our textbook was written with the purpose of reversing this hierarchy, and is to be regarded as an adjunct to *A Colour Atlas of Head and Neck Anatomy* by McMinn, Hutchings and Logan. Indeed, figure references in the margins of our text relate to pictures of dissections in the Atlas and enable the reader to use both books simultaneously. The inclusion of numerous line drawings in the textbook furthers our aim of emphasising the visual aspects of anatomy in two ways. First, they have a didactic purpose where the use of dissections alone may complicate the understanding of a topic. Second, they illustrate areas not amenable to dissection.

We have endeavoured to follow the terminology and the order of topics used in *A Colour Atlas of Head and Neck Anatomy*. There are, however, two notable exceptions with regard to the order of topics. Within the first chapter, the description of the extracranial appearance of the skull has been rearranged so that the less complicated surfaces are considered first. In the last chapter, the description of the central nervous system has also been rearranged for pedagogic purposes, beginning at the spinal cord and working through to the cerebrum. This approach has the added advantage of enabling us to adopt the modern practice of considering neuroanatomy functionally as well as topographically.

Although anatomy is generally taught as a preclinical subject, many clinical subjects cannot be fully appreciated without it. To emphasise the relevance of anatomy to the clinic, we have provided numerous case histories which require some anatomical information for their elucidation. We hope the case histories will entertain as well as instruct, for we believe this to be an effective way of motivating students in the undoubtedly difficult task of learning anatomy. Furthermore, they ensure that the student is not merely a passive recipient of information, but must be involved in some aspects of problem solving.

We wish to thank the many friends and colleagues who helped with this book. We owe a considerable debt to Dr Malcolm Brown for writing the chapter on the central nervous system. We also gratefully appreciate the work of the illustrator, Jack Furnival. In addition to his obvious skill, he brought commendable enthusiasm to the task. The case histories were written in collaboration with several of our clinical colleagues at Bristol: M. Aldoori, C. Bevan, P.A. Bloom, R.J. Canter, M.V. Griffiths, J.E. Harcourt, S. Hickey, J. Hill, T.R. Magee, K.J. Nicpon, N.K. Ragge, R. Redmond, A.K. Robson, P.M.J. Scott, W.E. Sponsel, and R.M. Tillman. Photographic assistance was provided by D. Telling and J. Long, and secretarial assistance by E. Wheatley. Last, but by no means least, we acknowledge the help of our wives, Sylvia Berkovitz and Ruth Moxham, whose support extended well beyond the most tangible manifestations of typing and proof reading.

B.K.B. Berkovitz
B.J. Moxham
Bristol, 1988.

How to use this book

This textbook should be used with *A Colour Atlas of Head and Neck Anatomy* by McMinn, Hutchings and Logan. Figure references to this Atlas are provided in the margins of our text to enable the reader to use both books simultaneously. The system of referencing adopted is generally a two-figure system. The first figure refers to the page of the Colour Atlas on which the relevant dissection is illustrated. The second figure provides an identification number overlying a specified structure. Where occasionally a single figure reference is used, this is a page reference only, and implies either that a general appreciation of the dissection is required and/or that the page depicts a major dissection for which many other figure references will soon follow. To provide an example: figure references 26A,3; 14 indicate that on page 26 (picture A) of the Colour Atlas is found a view of the infratemporal region of the skull displaying the sphenopalatine foramen (label 3), and that a general lateral view of the skull is found on page 14.

Contents

THE SKULL 10
The extracranial appearance of the skull 12
    The norma verticalis 12
    The norma occipitalis 13
    The norma frontalis 13
    The norma lateralis 15
    The norma basalis 17
The intracranial appearance of the skull 21
    The anterior cranial fossa 21
    The middle cranial fossa 22
    The posterior cranial fossa 23

THE BONES OF THE SKULL 25
    The bones of the vault of the skull 25
    The bones of the base of the skull 29
    The bones of the face 35
    The bones of the jaws 40

SKULL BONE ARTICULATIONS 46
    The orbital and nasal apertures 46
    The orbit 46
    The nasal cavity 47
    The base of the skull 48
    The pterygopalatine fossa 48
Summary sheet – The skull 50

THE FETAL SKULL 51
Summary sheet – The fetal skull 60

THE CERVICAL VERTEBRAE 61
    The joints of the cervical vertebrae 64

THE HYOID BONE 68
Summary sheets – The bones and joints of the neck 69
CASE HISTORIES 1: THE SKULL AND CERVICAL VERTEBRAE 71

THE NECK 82
    The general arrangement of structures in the neck 82
    Surface markings of the neck 83
    The cutaneous innervation of the neck 84
    Superficial structures of the neck 85
        The anterior triangle of the neck 86
        The posterior triangle of the neck 90
    Deep structures of the neck 93
        The fascia and tissue spaces of the neck 93
        The great vessels and the nerves of the neck 99
        The viscera of the neck 118
        The musculoskeletal compartment of the neck 120
        The root of the neck 128
        The embryology of the neck 134
Summary sheets – The neck 136
CASE HISTORIES 2: THE NECK 139

THE FACE 147
    Surface markings of the face 147
Superficial structures of the face 151
    The muscles of facial expression 151
    The nerves of the face 157
    The arteries of the face 161
    The veins of the face 163
    The lymphatics of the face 164
    The parotid gland 164
    The embryology of the face 168
Deep structures of the face 171
    The muscles of mastication 171
    The temporomandibular joint 172
    The infratemporal fossa 174
Summary sheets – The face 182

THE SCALP 188
    The layers of the scalp 188
    The cutaneous innervation of the scalp 189
    The arteries of the scalp 189
    The veins of the scalp 189
CASE HISTORIES 3: THE FACE AND SCALP 190

THE ORBIT 198
    The osteology of the orbit 198
    The contents of the orbit 199
    The eye 200

The eyelids	209
The lacrimal apparatus	213
The extra-ocular muscles	215
The nerves within the orbit	221
The arteries within the orbit	226
The veins within the orbit	228
The embryology of the eye	228
Summary sheets – The eye and orbit	230
CASE HISTORIES 4: THE ORBIT	234
THE NOSE	242
The external nose	242
The nasal cavity	243
The paranasal air sinuses	249
The frontal air sinuses	249
The ethmoidal air sinuses	250
The sphenoidal air sinuses	251
The maxillary air sinuses	252
The pterygopalatine fossa	253
The embryology of the nasal cavity and the paranasal air sinuses	258
Summary sheets – The nose	261
CASE HISTORIES 5: THE NOSE	266
THE MOUTH, PALATE AND PHARYNX	272
The mouth and palate	272
The teeth	277
The oral mucosa	288
The salivary glands	289
The oral musculature	291
The innervation of the oro-dental tissues	297
The blood supply of the oro-dental tissues	304
The blood supply of other oral structures	305
The venous drainage of the oro-dental tissues	306
The lymphatic drainage of the mouth	306
The tissue spaces around the jaws	307
The pharynx	310
The nasopharynx	310
The oropharynx	311
The laryngopharynx	312
The wall of the pharynx	312
The innervation of the pharynx	318
The vasculature of the pharynx	318
Swallowing	318
The embryology of the mouth, palate and pharynx	319
Summary sheets – The mouth and pharynx	326
THE LARYNX	332
The internal anatomy of the larynx	332
The skeleton of the larynx	334
The mucosa of the larynx	338
The muscles of the larynx	338
The blood supply of the larynx	342
The innervation of the larynx	343
Speech	343
The embryology of the larynx	344
Summary sheet – The larynx	347
CASE HISTORIES 6: THE MOUTH, PALATE, PHARYNX AND LARYNX	348
THE EAR	362
The external ear	362
The middle ear	366
The internal ear	375
The embryology of the ear	379
Summary sheets – The ear	381
CASE HISTORIES 7: THE EAR	384
THE CRANIAL CAVITY	390
The meninges	390
The contents of the anterior cranial fossa	398
The contents of the middle cranial fossa	398
The contents of the posterior cranial fossa	401
THE CENTRAL NERVOUS SYSTEM	403
The morphology and characteristics of nervous tissue	403
Neurones	403
Neuroglia	414
Introduction to the topography of the central nervous system	417
Subdivisions of the central nervous system	417
Gross topography of the central nervous system	418
The embryology of the central nervous system	422
The spinal cord and spinal nerves	427

The spinal nerves	427
The autonomic nervous system	437
The gross anatomy of the spinal cord	444
The spinal meninges	445
The internal structure of the spinal cord	446
The brainstem and cranial nerves	459
The external topography of the brainstem	460
The internal topography of the brainstem	463
The cranial nerves and their nuclei	464
Ascending sensory pathways	488
Motor nuclei and descending pathways	500
Cerebellar connections and precerebellar nuclei	502
The reticular formation of the brainstem	504
The cerebellum	513
The external appearance of the cerebellum	514
The internal organisation of the cerebellum	517
The diencephalon	523
The gross topography of the diencephalon	523
The hypothalamus	524
The subthalamus	528
The thalamus	529
The epithalamus	534
The cerebral hemispheres	535
The external topography of the cerebral hemispheres	535
The cerebral cortex	540
The internal structure of the cerebral hemispheres	562
The choroid plexuses and the cerebrospinal fluid	572
The choroid plexuses	572
The circulation of the cerebrospinal fluid	574
The vasculature of the central nervous system	575
The arterial supply to the brain	577
The arterial supply to the spinal cord	582
The venous drainage of the brain	582
The venous drainage of the spinal cord	584
The blood–brain barrier	584
The circumventricular organs	585
The general organisation of sensory and motor pathways	586
Summary sheets – The central nervous system	588
CASE HISTORIES 8: THE BRAIN AND CRANIAL CAVITY	608
INDEX	632

The skull

The skull is the bony skeleton of the head and is the most complex osseous structure in the body. It protects the brain, the organs of special sense and the cranial parts of the respiratory and digestive systems. The skull also provides attachments for many of the muscles of the head and neck.

Although often thought of as a single bone, the skull is composed of 28 separate bones (Table 1). Many of these bones are flat bones, consisting of two thin plates of compact bone enclosing a narrow layer of cancellous bone. In terms of shape, however, the bones are far from flat and can show pronounced curvatures. The term diploë is used to describe the cancellous bone within the flat bones of the skull.

In order to make the skull easier to understand, two major subdivisions have been proposed. First, one can subdivide the skull into cranium and mandible. This subdivision is based upon the fact that, whereas most of the bones of the skull articulate by relatively fixed joints, the mandible is easily detached. The cranium may then itself be subdivided into a number of regions, including:

- The cranial vault: The upper, dome-like part of the skull (including the skullcap or calvaria).
- The cranial base: The inferior surface of the skull extracranially, and the floor of the cranial cavity intracranially.
- The facial skeleton: The face (including the orbital cavities and the nasal fossae).
- The jaws: The tooth-bearing bones.
- The acoustic cavities: The ear.
- The cranial cavity: The interior of the skull housing the brain.

Second, one can subdivide the skull into neurocranium and viscerocranium. The neurocranium is defined as that part of the skull that houses and protects the brain and the organs of special sense. The viscerocranium is that region associated with the cranial parts of the respiratory and digestive tracts. In many mammals, the neurocranium and the viscerocranium are reasonably distinct and the viscerocranium is particularly prominent. In man, there is no elongated snout or muzzle. Instead, the face is wide, flat and vertical. Furthermore, the neurocranium is extraordinarily large. A particularly important feature concerns the base of the cranium. The floor of the cranial cavity in most mammals is relatively flat and lies some way from the viscerocranium. In man, however, it shows a marked curvature and lies immediately above the viscerocranium (Figure 1). The neurocranium and viscerocranium of the human skull are thus not readily separable.

The evolutionary changes leading to the shape of the human skull cannot be fully explained. It has been proposed by some that the development of a large brain has been all-important, whereas others invoke the change to an upright stance. Certainly, the form of the human skull can be related to the change to bipedal locomotion. The absence of a snout has been beneficial to the development of stereoscopic vision. In addition, as the upper limbs have taken on functions other than locomotion, there is less need for the jaws to be organs of offence or defence. Indeed, the human jaws appear gracile and the ridges on the skull for the jaw muscles are very much reduced. The spinal cord in quadrupeds leaves the cranium at a foramen (the foramen magnum) located on the posterior aspect of the skull. The foramen magnum in man is positioned underneath the skull, and thus the spinal cord has become vertically orientated.

The changes in the skull wrought by the enlarged brain are no less significant. In particular, the well-developed frontal lobes of the cerebrum are associated with the

TABLE 1 Bones of the skull.

NAME	NUMBER	PRIMARY LOCATION	SHORT DESCRIPTION
ETHMOID	1	Nasal and orbital cavities of face	T-shaped. Processes form superior and middle conchae of lateral wall of nasal cavities
FRONTAL	1	Cranial vault	Forms forehead and roof of orbital cavities
INFERIOR CONCHA	2	Nasal cavity of face	Projects from lateral wall of nasal cavity
INCUS	2	Acoustic cavity	Shaped like an anvil
LACRIMAL	2	Orbital cavity of face	Situated on medial wall of orbital cavity. Related to lacrimal sac
MALLEUS	2	Acoustic cavity	Shaped like a hammer
MANDIBLE	1	Jaws	Forms lower jaw
MAXILLA	2	Jaws	Forms upper jaw. Also contributes to nasal and orbital cavities
NASAL	2	Face	Forms bridge of nose
OCCIPITAL	1	Cranial vault	Forms back of head. Also contributes to cranial base
PALATINE	2	Nasal cavity of face	L-shaped. Contributes to lateral wall of nose and hard palate
PARIETAL	2	Cranial vault	Forms mid-portion of cranial vault
SPHENOID	1	Cranial base	Butterfly-shaped. Also contributes to orbital and nasal cavities and lateral sides of skull
STAPES	2	Acoustic cavity	Stirrup-shaped
TEMPORAL	2	Cranial base	Also contributes to lateral sides of skull
VOMER	1	Nasal cavity of face	Contributes to nasal septum
ZYGOMATIC	2	Face	Forms cheek bone

development of a vertical forehead that bulges above the face. The orbits also undergo a forward rotation, making them vertically aligned and facing forwards. Because of this, the roof of the orbit becomes related to the floor of the cranial cavity, and the floor of the orbit comes to lie in the upper part of the face. As the orbits encroach on the midline, the root of the nose becomes much thinner.

There have also been changes in the location of the olfactory structures. The nasomaxillary complex becomes related to the anterior cranial fossa and the olfactory mucosa is then situated in the roof of the nose and not on the posterior walls. Many of these features are illustrated in Figure 1.

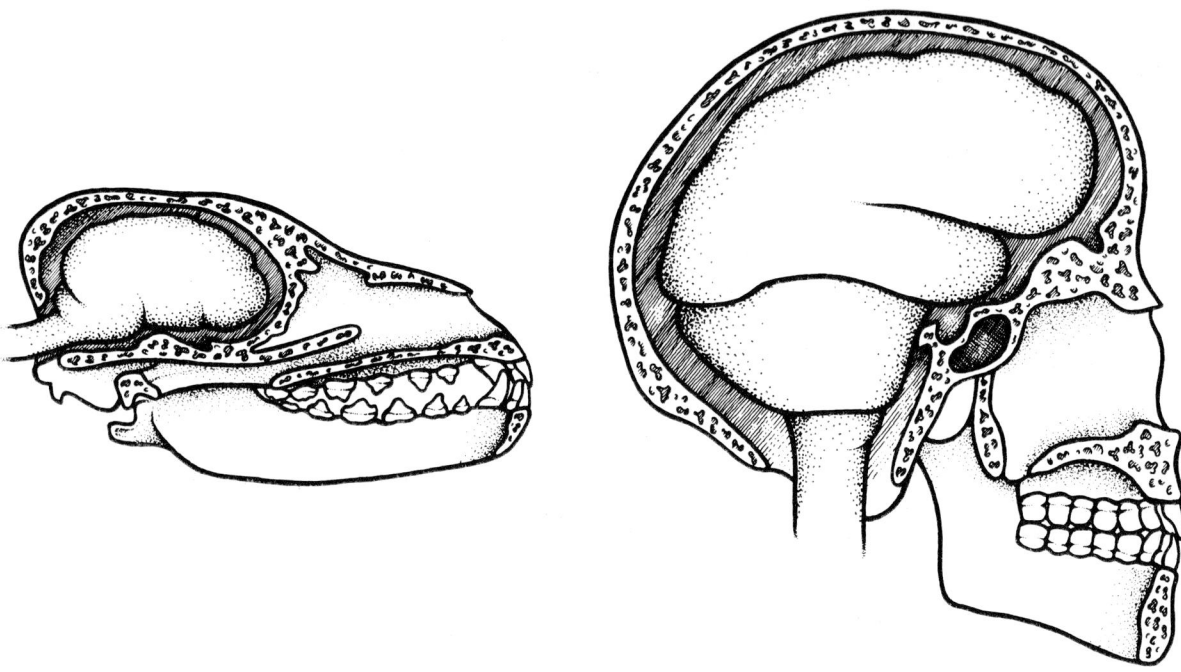

Figure 1 Comparison between the typical mammalian (dog) and human skulls. *Note the differences in a) the cranial capacity, b) the shape and position of the cranial base, c) the size and position of the viscerocranium, d) the position of the foramen magnum, e) the position and orientation of the olfactory region. Thus, the human skull has a large neurocranium, the cranial base is curved and extends over the viscerocranium anteriorly, it shows no snout and the facial skeleton is orientated vertically, the foramen magnum lies beneath the skull and the cribriform plate and olfactory area are not in the posterior wall of the nose but in its roof.*

The extracranial appearance of the skull

The following views of the exterior of the skull will be described:

- The norma verticalis — the skull seen from above.
- The norma occipitalis — the skull seen from behind.
- The norma frontalis — the skull seen from the front.
- The norma lateralis — the skull seen from the side.
- The norma basalis — the skull seen from below.

They are described in this order so that the less complex regions are considered first.

20A

THE NORMA VERTICALIS

This view is so named because the most superior point of the skull is called the vertex. The region observed is the skullcap or calvaria.

20,9
20,1

The calvaria is approximately oval in shape, the anteroposterior dimension being the greater. It is usually wider posteriorly than anteriorly. It is comprised of four bones separated by three prominent sutures. Anteriorly is found the squamous part of the frontal bone. Posteriorly is the squamous part of the occipital bone. Between the frontal and occipital bones lie the two parietal bones. The suture between the frontal bone and the parietal bones is called the coronal suture. It is this suture that gives its name to the coronal plane of the body. The midline suture between the parietal bones is the sagittal suture and this gives its name to the sagittal plane. The junction of the coronal and sagittal sutures is termed the bregma. The bregma corresponds to the anterior fontanelle ('soft spot') on the fetal skull. The suture dividing the occipital bone from the parietal bones is the lambdoid suture. The point of meeting of the lambdoid and sagittal sutures is called the lambda. This site marks the position of the posterior fontanelle on the fetal skull.

The calvaria is otherwise rather featureless. The region of maximum convexity of the parietal bone is called the parietal tuberosity. Close to the tuberosity run the superior and inferior temporal lines, though these lines are best seen in the norma lateralis. Parietal foramina may be found on either side of the sagittal suture. They transmit emissary veins (veins that link the intracranial and extracranial venous systems) from the superior sagittal sinus within the cranium. Sometimes terminal branches of the occipital arteries also pass through the parietal foramina.

THE NORMA OCCIPITALIS

Viewed from behind, the occipital bone is prominent — hence the term norma occipitalis. The lambdoid suture is also conspicuous, being seen in its entirety. A common variation is the presence of islands of bone within the suture. These sutural bones arise from separate centres of ossification, but they have no clinical significance. Inferiorly, the lambdoid suture meets the occipitomastoid and the parietomastoid sutures. These sutures lie above and behind the mastoid process of the temporal bone. The temporal bones, though most clearly seen on the lateral views of the skull, just appear as the mastoid processes to form the inferolateral parts of the back of the skull. The superolateral parts are formed by the parietal bones.

A marked feature at the back of the skull is the external occipital protuberance. It appears on the occipital bone in the midline as either a ridge or a distinct process. Extending laterally from the protuberance are two ridges called the superior nuchal lines. These lines finish above the mastoid processes. Inferior nuchal lines run parallel to, and below the superior nuchal lines. Above the superior nuchal lines may be seen the supreme nuchal lines. The external occipital protuberance and the nuchal lines are associated with muscle attachments. The supreme nuchal lines afford attachment to the epicranial aponeurosis of the scalp (see page 188). The roughened appearance of the occipital bone between the nuchal lines is also caused by muscle attachments.

The muscles attached to the skull in the occipital region are:

Muscle	Attached to:
Longissimus capitis	Superior nuchal line
Occipital belly of occipitofrontalis	Superior nuchal line
Semispinalis capitis	Between superior and inferior nuchal lines
Splenius capitis	Superior nuchal line
Sternocleidomastoid	Mastoid process and superior nuchal line
Obliquus capitis superior	Between superior and inferior nuchal lines
Trapezius	External occipital protuberance and superior nuchal line

THE NORMA FRONTALIS

Most of the features seen on the front of the skull relate to the face. In particular, there are four apertures associated with the facial skeleton: the two orbital apertures, the anterior nasal aperture (the piriform aperture), and the oral aperture between the jaws. The osteology of the orbital and nasal apertures and cavities is considered with the skull articulations (see pages 46 to 48).

10,1
10,2
10,33; 10,32
10,34; 10,35
10,6; 10,5

The upper part of the facial skeleton is formed by the frontal bone and is related to the forehead. Above the bridge of the nose lies a slight elevation called the glabella. This part of the frontal bone joins the nasal bones and the frontal processes of the maxillary bones at the frontonasal and frontomaxillary sutures. At the superior rim of each orbit are found the supra-orbital foramen (or notch) and the frontal notch. These transmit the supra-orbital and supratrochlear nerves (and accompanying vessels) from the orbit on to the forehead. Laterally, the zygomatic processes of the frontal bone join the cheek bones (zygomatic bones) at the frontozygomatic sutures.

10,25

10,13

The central part of the face is occupied by the maxillary bones. Each bone contributes not only to the upper jaw, but also to the nasal aperture, the bridge of the nose, the floor of an orbital cavity and the bones of the cheek. Beneath the inferior rim of each orbit lies the infra-orbital foramen. Through this foramen the infra-orbital nerve (from the maxillary division of the trigeminal nerve) and accompanying vessels pass on to the face. At the inferior margin of the nasal aperture in the midline lies a projection called the anterior nasal spine. The maxillary bones meet at the intermaxillary suture.

10,12

10,18

10,15
10,17
10,16

The lower part of the face is formed by the body of the mandible. In the midline is the prominence of the chin, the mental protuberance. In line with the supra-orbital and infra-orbital foramina lies the mental foramen. Through this foramen pass the mental nerve (from the mandibular division of the trigeminal nerve) and accompanying vessels.

12

The muscles attached to the front of the skull are:

Muscle	Attached to:
Buccinator	Maxillary and mandibular buccal alveolar plates in region of molars
Corrugator supercilii	Frontal bone
Depressor anguli oris	Mandible below mental foramen
Depressor labii inferioris	Mandible between chin and mental foramen
Depressor septi	Maxilla below nasal aperture
Levator anguli oris	Maxilla below the infra-orbital foramen
Levator labii superioris	Inferior rim of orbit above infra-orbital foramen
Levator labii superioris alaeque nasi	Frontal process of maxilla
Masseter	Zygomatic arch and lateral surface of ramus of mandible
Mentalis	Incisive fossa of mandible
Nasalis	Maxilla close to nasal aperture
Orbicularis oculi	Nasal part of the frontal bone, frontal process of maxilla and crest of lacrimal bone
Platysma	Inferior border of body of mandible
Procerus	Nasal bone
Temporalis	Temporal fossa and coronoid process and anterior border of ramus
Zygomaticus major	Zygomatic bone
Zygomaticus minor	Zygomatic bone

THE NORMA LATERALIS

The skull, viewed from the side, can be subdivided into three zones. Anteriorly is the face, and posteriorly is the occipital region (these have already been described). The intermediate zone shows two fossae: the temporal and infratemporal fossae. The boundary between the fossae is the zygomatic arch. The calvaria superiorly has been described with the norma verticalis.

The temporal fossa is so named because it is related to the temple of the head. The fossa is bounded inferiorly by the zygomatic arch; superiorly and posteriorly by the temporal lines on the calvaria; and anteriorly by the frontal process of the zygomatic bone. It continues beneath the zygomatic arch into the infratemporal fossa. The temporal lines often present anteriorly as distinct ridges, but become much less prominent as they arch across the parietal bone. Indeed, the superior line usually disappears posteriorly. On the other hand, the inferior temporal line becomes distinct once more as it curves down the squamous part of the temporal bone, forming a supramastoid crest at the base of the mastoid process. The superior temporal line gives attachment to the temporal fascia (see page 172). The inferior temporal line provides attachment for the temporalis muscle (see page 171).

The floor of the temporal fossa is formed by the frontal, sphenoid (greater wing), parietal and temporal (squamous part) bones. These four bones meet at an area called the pterion, where there is an H-shaped junction of sutures. This is an important landmark on the side of the skull. It overlies the middle meningeal vessels intracranially and corresponds to the sphenoidal fontanelle on the neonatal skull.

The suture between the temporal and parietal bones is called the squamosal suture. The sphenosquamosal suture lies between the greater wing of the sphenoid and the squamous part of the temporal bone.

The infratemporal fossa has the following bony boundaries: the ramus of the mandible laterally; the lateral pterygoid plate of the sphenoid bone medially; the infratemporal surface of the greater wing of the sphenoid superiorly; and the maxilla anteriorly. Beneath the zygomatic arch, the infratemporal fossa communicates with the temporal fossa. Between the lateral pterygoid plate and the maxilla lies the pterygomaxillary fissure. This fissure marks the site where the infratemporal fossa communicates with the pterygopalatine fossa (see pages 48 to 49 for the osteology of this fossa).

The lateral surface of the ramus of the mandible should also be briefly described at this point. The ramus is a plate of bone projecting upwards from the back of the body of the mandible. Most of its lateral surface provides attachment for the masseter muscle. Two prominent processes are seen superiorly, the coronoid and condylar processes. The coronoid process is the site for the insertion of the temporalis muscle. The condylar process articulates with the mandibular fossa of the temporal bone at the temporomandibular synovial joint. Between the two processes is the mandibular notch. The angle of the mandible is the region where the inferior and posterior borders of the ramus meet.

The zygomatic arch stands clear of the rest of the skull, the gap being where the temporal and infratemporal fossae communicate. Whereas the bones of the cheek comprise the zygomatic bone and the zygomatic processes of the frontal, maxillary and temporal bones, the zygomatic arch is a term restricted to that part formed by the temporal process of the zygomatic bone and the zygomatic process of the temporal bone. These processes meet at the zygomaticotemporal suture. The suture between the frontal process of the zygomatic bone and the zygomatic process of the temporal bone is called the frontozygomatic suture. The zygomaticomaxillary suture marks the union of the maxillary margin of the zygomatic bone and the zygomatic process of the maxillary bone. The zygomatic bone also joins the sphenoid bone, at the sphenozygomatic suture. As the zygomatic process of the temporal bone passes posteriorly, it becomes associated with the mandibular fossa and the supramastoid crest.

14,27; 53C,21 53D,26; 23,56	The upper border of the zygomatic arch serves as an attachment for the temporal fascia. The inferior border and the deep surface provide attachment for the masseter muscle. A small foramen, the zygomaticofacial foramen, lies on the outer surface of the zygomatic bone. Another foramen, the zygomaticotemporal foramen, is situated on the inner surface. These foramina transmit nerves and vessels of the same name on to the face.
14,20 14,19	The temporal bone is a prominent structure on the lateral aspect of the skull. As mentioned, its squamous part lies in the floor of the temporal fossa and its zygomatic process contributes to the bones of the cheek. Additional features found are the mandibular fossa and its articular tubercle, the tympanic plate and external acoustic meatus, and the mastoid and styloid processes.
26A,14 14,29 26A,13; 26A,16	The mandibular fossa has also been called the glenoid fossa. It is the part of the temporomandibular joint into which the condylar process of the mandible articulates. It is bounded in front by the articular tubercle and behind by the tympanic plate. Occasionally, there is a postglenoid tubercle. The articular tubercle is important functionally as it provides a surface down which the mandibular condyle glides during mandibular movements. The tubercle also marks the site of attachment of the lateral ligament of the temporomandibular joint (see page 173).
14,14 14,16 14,15	The tympanic part of the temporal bone contributes most of the margin of the external acoustic meatus, the squamous part forming the upper margin and the upper part of the posterior margin. The margin is roughened to provide an attachment for the cartilaginous part of the meatus. Above and behind the meatus lies a small depression, the suprameatal triangle, which is related to the lateral wall of the mastoid antrum (see page 369).
14,13 14,11; 14,10 14,12 74B,20 22,26	The mastoid process is the large prominence located immediately behind the external acoustic meatus. It is the site of attachment of a prominent muscle of the neck, the sternocleidomastoid muscle. Above the process lies the supramastoid crest which is continuous with the inferior temporal line. The mastoid process articulates with the parietal and occipital bones at the parietomastoid and occipitomastoid sutures. The junction of these sutures with the lambdoid suture is called the asterion. This corresponds to the mastoid fontanelle in the neonatal skull. A mastoid foramen may be found near the occipitomastoid suture. This foramen transmits an emissary vein from the sigmoid sinus.
14,18	The styloid process is a long, slender process that emerges from the base of the skull in front of the mastoid process. It projects downwards and forwards towards the mandible. The base of the styloid process is formed by the tympanic part of the temporal bone. The process gives attachment to several muscles and ligaments.
16	The muscles attached to the lateral side of the skull are:

Muscle	Attached to:
Buccinator	Maxillary and mandibular buccal alveolar plates in region of molars
Corrugator supercilii	Frontal bone
Depressor anguli oris	Mandible below mental foramen
Depressor labii inferioris	Mandible between chin and mental foramen
Depressor septi	Maxilla below nasal aperture
Levator anguli oris	Maxilla below infra-orbital foramen
Levator labii superioris	Inferior rim of orbit above infra-orbital foramen
Levator labii superioris alaeque nasi	Frontal process of maxilla

Muscle	Attached to:
Masseter	Zygomatic arch, lateral surface of mandibular ramus
Mentalis	Incisive fossa of mandible
Nasalis	Maxilla close to nasal aperture
Occipital belly of occipitofrontalis	Superior nuchal line
Orbicularis oculi	Nasal part of frontal bone, frontal process of maxilla, crest of lacrimal bone
Platysma	Inferior border of body of mandible
Procerus	Nasal bone
Sternocleidomastoid	Mastoid process, superior nuchal line
Temporalis	Temporal fossa, coronoid process of mandible and anterior border of ramus
Zygomaticus major	Zygomatic bone
Zygomaticus minor	Zygomatic bone

THE NORMA BASALIS

The inferior surface of the cranium is very irregular and presents the most complex of the surfaces of the skull. This complexity is increased by the fact that this region has many of the foramina through which structures enter and leave the cranial cavity. The region can be simplified by subdividing it into three zones. The anterior zone comprises the hard palate and the dentition of the upper jaw. The posterior zone lies behind a transverse plane drawn just in front of the foramen magnum. The intermediate zone is occupied mainly by the base of the sphenoid bone, the petrous processes of the temporal bones and the basilar part of the occipital bone. Whereas the intermediate and posterior zones are directly related to the cranial cavity (the middle and posterior cranial fossae), the anterior zone is related to the roof of the mouth and is some distance from the anterior cranial fossa.

The hard palate is formed by the two palatine processes of the maxillary bones and the two horizontal plates of the palatine bones. It is bounded anteriorly and laterally by the alveolus of the upper jaw, which supports the teeth.

A cross-shaped set of sutures traverses the palate. Running anteroposteriorly and dividing the palate into right and left halves is the median palatine suture. This suture is continuous with the intermaxillary suture between the maxillary central incisor teeth. Behind the central incisors, the junction between the palatine processes of the maxillary bones is incomplete, thus forming the incisive fossa. Incisive foramina (two lateral or one anterior and one posterior) pass into this fossa and transmit the nasopalatine nerves and the terminal parts of the greater palatine vessels. Running transversely across the palate between the maxillary and the palatine bones is the transverse palatine suture. This suture is incomplete on each side and forms the greater palatine foramina. Through the greater palatine foramen pass the greater palatine nerve and vessels. Behind the foramen lie one or more lesser palatine foramina, through which pass the lesser palatine nerves and vessels.

The posterior borders of the horizontal plates of the palatine bones are concave and in the midline form a sharp ridge of bone, the posterior nasal spine. To the posterior edge of the hard palate is attached the fibrous aponeurosis of the soft palate, which is formed by the tendons of the tensor veli palatini muscles.

The shape of the hard palate varies but is often dome-shaped. The arching of the palate occurs both anteroposteriorly and from side to side.

Above the hard palate are the nasal fossae, separated in the midline by the nasal septum. The posterior part of the septum is formed by the vomer. This bone lies on the body of the sphenoid. The two posterior nasal apertures (choanae) are located where the nasal fossae end. The lateral wall of each aperture beneath the hard palate is formed by the perpendicular plate of the palatine bone. A small canal called the palatovaginal canal (see page 32) is found in this region. This transmits a nerve and an artery to the nasopharynx (the pharyngeal branch of the pterygopalatine ganglion and an accompanying branch from the maxillary artery). Another canal, the vomerovaginal canal, may sometimes be found leading into the anterior end of the palatovaginal canal. It transmits a small artery (the pharyngeal branch of the sphenopalatine artery).

A prominent feature of the posterior zone of the cranial base is the foramen magnum. Associated with this foramen are the occipital condyles, the hypoglossal canals (anterior condylar canals) and the condylar canals (posterior condylar canals). Lateral to the foramen magnum are the jugular foramina. Other features of this part of the skull are the mastoid and styloid processes of the temporal bones, the stylomastoid foramina, the mastoid notches and the squamous part of the occipital bone up to the external occipital protuberance and the superior nuchal lines.

The foramen magnum is the largest foramen of the skull. Through it the cranial cavity (the posterior cranial fossa) and the vertebral canal communicate. The major structures passing through the foramen are the medulla oblongata of the brain stem, the vertebral arteries and the spinal accessory nerves. Anteriorly lies the apical ligament of the dens (an upwardly directed bony process from the second cervical vertebra) and the membrana tectoria near the atlanto-occipital joint (see Figure 14, page 66). Behind is the medulla oblongata and its covering meninges. Structures that pass through the foramen magnum with the medulla are: the vertebral arteries; the anterior and posterior spinal arteries; the spinal parts of the accessory nerves; and the meningeal branches of upper cervical nerves. The anterior margin of the foramen magnum provides attachment for the anterior atlanto-occipital membrane, and the posterior margin provides attachment for the posterior atlanto-occipital membrane (see Figure 14, page 66).

The occipital condyles lie near the anterior margin of the foramen magnum. They are facets for articulation with the vertebral column at the atlanto-occipital joints. They display marked curvature in all planes. Within each condyle is the hypoglossal canal. This communicates with the posterior cranial fossa and transmits the hypoglossal nerve. It also transmits the meningeal branch of the ascending pharyngeal artery (a branch of the external carotid artery) and an emissary vein (from the basilar plexus). Behind each condyle is a depression called the condylar fossa. The condylar canal passes into this fossa and transmits an emissary vein from the sigmoid sinus.

The jugular foramen is an irregular foramen situated lateral to the occipital condyle. It is really a large fissure formed between the jugular process of the occipital bone and the jugular fossa of the petrous part of the temporal bone. Anteriorly, the inferior petrosal sinus passes through the foramen. Midway, the foramen transmits the glossopharyngeal, vagus and accessory nerves. Posteriorly lies the internal jugular vein. A mastoid canaliculus runs through the lateral wall of the jugular fossa. This transmits the auricular branch of the vagus nerve. On the ridge between the jugular fossa and the opening of the carotid canal is the canaliculus for the tympanic nerve (a branch of the glossopharyngeal nerve to the cavity of the middle ear).

Between the mastoid process and the root of the styloid process is the stylomastoid foramen. Through this foramen emerges the facial nerve before it enters the parotid gland. Also passing through is an artery, the stylomastoid branch of the posterior auricular artery (a branch of the external carotid). Medial to the mastoid process is the mastoid notch. This is the site of attachment of the posterior belly of the digastric muscle. Medial to the notch is a groove in which runs the occipital artery (a branch of the external carotid).

The region of the occipital bone between the foramen magnum and the inferior nuchal line provides attachment for the rectus capitis posterior major and minor muscles. Between the inferior and superior nuchal lines are attached the semispinalis capitis and the obliquus capitis superior muscles. The superior nuchal line is the site of attachment of the trapezius, sternocleidomastoid and splenius capitis muscles.

22,29

22,27

The intermediate zone of the cranial base is essentially composed of four osseous structures. Anteriorly lies the body of the sphenoid bone, and posteriorly the basilar part of the occipital bone. Where these meet is a primary cartilaginous joint called the spheno-occipital synchondrosis. This joint is important for growth of the skull in an anteroposterior direction and it does not ossify until about 20 years of age. The intermediate zone is completed by the petrous processes of the two temporal bones. These are found passing from the lateral sides of the base of the skull towards the spheno-occipital synchondrosis. A petrous process meets the basilar part of the occipital bone at the petro-occipital suture. This suture is deficient posteriorly where the jugular foramen is situated. Between the petrous process and the infratemporal surface of the greater wing of the sphenoid is the sphenopetrosal synchondrosis and the groove for the auditory tube. The apex of the petrous process does not meet the spheno-occipital joint. Consequently, a large fissure is present, called the foramen lacerum. The intermediate zone is related to the middle cranial fossa and the anterior wall of the posterior cranial fossa.

68,14
68,4
69,19

68,7; 22,45

69,17
69,16
69,20
68,7; 22,45
69,18; 22,46

The intermediate zone displays a considerable number of fissures and foramina. Already mentioned is the foramen lacerum. Despite its size, the foramen does not transmit any large structures. Its upper part is related to the internal opening of the carotid canal. Thus, the internal carotid artery and its accompanying venous and sympathetic plexuses cross over the foramen lacerum on its intracranial aspect. The lower part of the foramen lacerum is filled with cartilage. Within the foramen, the greater petrosal branch of the facial nerve and the deep petrosal nerve from the carotid sympathetic plexus join to form the nerve of the pterygoid canal (see page 257). Indeed, the pterygoid canal can be seen on the base of the skull at the anterior margin of the foramen lacerum above and between the pterygoid plates of the sphenoid bone. The pterygoid canal leads into the pterygopalatine fossa and contains not only the nerve of the pterygoid canal but also accompanying blood vessels.

22,46
50E,44

Lateral to the foramen lacerum and passing through the infratemporal surface of the greater wing of the sphenoid are the foramen ovale and the foramen spinosum. The foramen ovale communicates with the middle cranial fossa and contains the mandibular division of the trigeminal nerve (and also the accessory meningeal artery from the maxillary artery). The foramen spinosum lies anterior to the spine of the sphenoid (hence its name) and posterior to the foramen ovale. It also communicates with the middle cranial fossa. It transmits the middle meningeal vessels and the meningeal branch of the mandibular nerve (nervus spinosus).

22,44

22,43

Anterior to the foramen ovale a small foramen is sometimes found, called the sphenoidal emissary foramen (of Vesalius). This contains an emissary vein linking the pterygoid venous plexus in the infratemporal fossa with the cavernous sinus in the middle cranial fossa.

29,52

Behind the foramen lacerum and within the petrous part of the temporal bone is the carotid canal through which passes the internal carotid artery.

22,36

Other features of the intermediate zone are: the pterygoid plates, pterygoid hamulus and scaphoid fossa; the mandibular fossa and its articular tubercle; the petrosquamous, petrotympanic and squamotympanic fissures; the spine of the sphenoid; and the pharyngeal tubercle on the basilar part of the occipital bone.

The pterygoid plates are processes of the sphenoid bone. There are two plates, the lateral and the medial pterygoid plates. They are located immediately behind the maxillary third molar tooth and are important for the attachment of muscles. The space between the plates is called the pterygoid fossa. At its base is a depression for

22,12; 22,14

22,13	the attachment of the tensor veli palatini muscle, which is called the scaphoid fossa. Anteriorly, the two plates are fused, except for a narrow gap (the pterygoid notch) which is filled by the pyramidal process of the palatine bone. The medial pterygoid plate has a hook-shaped process called the pterygoid hamulus, which projects behind the posterior border of the hard palate. The tensor veli palatini muscle twists around the hamulus before inserting into the soft palate. Also attached to the hamulus is a fibrous band called the pterygomandibular raphe. The other attachment of the raphe is the mandible. The raphe is important for providing the origins of two muscles (the buccinator in the cheek and the superior constrictor of the pharynx).
22,10	
22,11	
22,18	The mandibular fossa was briefly described with the norma lateralis. When viewed on the cranial base, the fossa appears as a thin-walled depression and the articular tubercle is now seen as a distinct ridge anterior to the fossa. Three fissures can be distinguished behind the mandibular fossa. The squamotympanic fissure extends from the spine of the sphenoid, between the mandibular fossa and the tympanic plate of the temporal bone, and up the anterior margin of the external acoustic meatus. Within this fissure can be seen a thin wedge of bone, which is the inferior margin of the tegmen tympani (part of the petrous part of the temporal bone, see page 34). This divides the squamotympanic fissure into two, the petrotympanic and petrosquamous fissures. The petrotympanic fissure transmits the chorda tympani branch of the facial nerve from the skull into the infratemporal fossa.
22,17	
22,39; 50D,9	
22,40; 50D,32	
22,38; 50D,42	
22,41; 50D,41	
22,42	The spine of the sphenoid is located medial to the mandibular fossa and posterior to the foramen spinosum. It varies considerably in size. The spine is the site of attachment of the sphenomandibular ligament.
22,47	The pharyngeal tubercle is found centrally on the basilar part of the occipital bone. It marks the site of attachment of the highest fibres of the superior constrictor muscle of the pharynx and of the pharyngeal raphe.
24	The muscles attached to the base of the cranium extracranially are:

Muscle	Attached to:
Digastric	Mastoid notch
Lateral pterygoid	Lateral side of lateral pterygoid plate and infratemporal surface of greater wing of sphenoid
Levator veli palatini	Petrous part of temporal bone
Longissimus capitis	Superior nuchal line
Longus capitis	Basilar part of occipital bone
Medial pterygoid	Medial side of lateral pterygoid plate, tuberosity of maxilla
Musculus uvulae	Posterior margin of hard palate in midline
Occipital belly of occipitofrontalis	Superior nuchal line
Palatopharyngeus	Posterior margin of hard palate laterally
Rectus capitis anterior	Basilar part of occipital bone
Rectus capitis lateralis	Jugular process of occipital bone
Rectus capitis posterior major and minor	Below inferior nuchal line
Semispinalis capitis	Between superior and inferior nuchal lines
Splenius capitis	Superior nuchal line
Sternocleidomastoid	Mastoid process, superior nuchal line
Styloglossus	Styloid process
Stylohyoid	Styloid process

Muscle	Attached to:
Stylopharyngeus	Styloid process
Superior constrictor	Medial pterygoid plate and pharyngeal tubercle
Obliquus capitis superior	Between superior and inferior nuchal lines
Tensor veli palatini	Scaphoid fossa, spine of sphenoid
Trapezius	External occipital protuberance, superior nuchal line

The intracranial appearance of the skull

The cranial cavity of the skull accommodates the brain and associated structures. A detailed description of the contents of the cranial cavity is given on pages 390 to 402.

The internal surface of the calvaria shows many of the features already described for the norma verticalis (see pages 12 and 13). However, the sutures tend to be less distinct because their gradual obliteration with age begins on the intracranial surface. Additional features seen intracranially include the frontal crest, some grooves for vascular structures and some depressions associated with arachnoid granulations. 20B

The frontal crest is a prominent projection into the cranial cavity anteriorly. It provides attachment for a sheet of meninges, the falx cerebri, which passes between the two cerebral hemispheres of the brain. Running from the frontal crest and across the calvaria in the midline is a groove for the superior sagittal venous sinus. On each side of this groove may be found the depressions for the arachnoid granulations (see page 574). Deep grooves for middle meningeal vessels are often found on the parietal bones. 20B,16 20B,14 20B,18 20B,15

A median sagittal section of the skull highlights the division of the cranial cavity into three distinct fossae: the anterior, middle and posterior cranial fossae. The three fossae have a marked step-like appearance, such that the floor of the anterior cranial fossa is at the highest level and the floor of the posterior fossa is lowest. 30

THE ANTERIOR CRANIAL FOSSA

28,64

The floor of the anterior cranial fossa is formed by the frontal bone (orbital plates), the ethmoid bone (cribriform plates and crista galli) and the sphenoid bone (lesser wings and jugum). Unlike the other cranial fossae, it does not directly communicate with the inferior surface of the cranium but instead is related to the roofs of the orbits and the nasal fossae. Two sutures divide the sphenoid from the other bones: the frontosphenoid suture and the spheno-ethmoidal suture. The cribriform plate of the ethmoid bone fills a gap (the ethmoidal notch) between the medial ends of the orbital parts of the frontal bone and is depressed below the level of the rest of the floor. Extending upwards from the cribriform plate is a process called the crista galli. This serves as a point of attachment of the falx cerebri. 28,10 28,7; 28,9 28,20; 28,11 28,9 38B,17 28,7

There are three openings in the floor of the anterior cranial fossa: two cribriform plates and a foramen caecum. The cribriform plates transmit olfactory nerves into the roof of the nose. The foramen caecum lies immediately in front of the crista galli. It occa-occasionally allows the passage of an emissary vein linking the superior sagittal venous sinus to the veins in the nose. The anterior ethmoidal nerve (a branch of the ophthalmic division of the trigeminal nerve from the orbit) enters the cranial cavity where the cribriform plate meets the orbital part of the frontal bone and passes into the roof of the nose by a small foramen to the side of the crista galli; indeed, the nerve 28,9 28,6 28,8 grooves the crista galli. Accompanying the nerve are the anterior ethmoidal vessels. Posterior ethmoidal vessels also run through the cribriform plate.

THE MIDDLE CRANIAL FOSSA

28; 67

28,14
28,21; 28,24
28,28

The floor of the middle cranial fossa is formed by the body of the sphenoid bone centrally and the greater wings of the sphenoid and the squamous and petrous parts of the temporal bones laterally. The middle cranial fossa is directly related extracranially to the intermediate zone of the cranial base (see page 19).

28,14

28,13
28,12

28,19
28,15; 28,17
28,16; 28,18

In the midline, the prominent structure is the pituitary fossa (sella turcica). This fossa is situated in the upper surface of the body of the sphenoid bone and lies above the sphenoidal sinuses. The anterior slope of the pituitary fossa has an elevation called the tuberculum sellae. In front of the tuberculum sellae is the prechiasmatic groove, which is associated with the optic chiasma (see page 400) and which leads into the optic canals. The pituitary fossa is bounded posteriorly by a plate of bone called the dorsum sellae. Lateral to the pituitary fossa is a groove (the carotid groove) for the internal carotid artery. Anterior and posterior clinoid processes occupy the 'four corners' of the pituitary fossa. These provide sites of attachment for a sheet of the meninges called the diaphragma sellae, which roofs over the pituitary fossa.

The regions lateral to the pituitary fossa provide deep depressions for the temporal lobes of the brain. Each region is related to the apex of the orbit anteriorly, the temporal fossa laterally and the infratemporal fossa inferiorly. With the exception of the optic canals, the lateral regions of the middle cranial fossa have all the foramina.

The openings of the middle cranial fossa on each side are: the optic canal, superior orbital fissure, foramen rotundum, foramen ovale, foramen spinosum, emissary sphenoidal foramen (of Vesalius), and the foramen lacerum.

28,19

The optic canal links the central area of the middle cranial fossa with the apex of the orbit. It transmits the optic nerve and the ophthalmic artery (a branch of the internal carotid artery).

28,51; 29B,51

The superior orbital fissure lies between the greater and lesser wings of the sphenoid bone. It is located on the anterior wall of the middle cranial fossa and links the fossa with the apex of the orbit. The fissure transmits many structures. The nerves passing through it are the oculomotor, trochlear and abducent nerves, the lacrimal, frontal and nasociliary branches of the ophthalmic division of the trigeminal nerve, and filaments from the internal carotid plexus (sympathetic). It also transmits the ophthalmic veins, and the orbital branch of the middle meningeal artery and the recurrent branch of the lacrimal artery.

28,50; 29B,50

28,48

The foramen rotundum lies within the greater wing of the sphenoid. It allows communication between the lateral part of the middle cranial fossa and the pterygopalatine fossa. Passing through it is the maxillary division of the trigeminal nerve. The foramen ovale is also present in the greater wing of the sphenoid, but it links the middle cranial fossa to the infratemporal fossa. The major structure passing through it is the mandibular division of the trigeminal nerve. In addition, there is the lesser petrosal branch of the glossopharyngeal nerve, the accessory meningeal branch of the maxillary artery, and an emissary vein from the cavernous venous sinus to the pterygoid venous plexus in the infratemporal fossa.

28,47

29C,52

The foramen spinosum lies just behind the foramen ovale. It transmits the meningeal branch of the mandibular division of the trigeminal nerve and the middle meningeal vessels. In front of, and medial to, the foramen ovale is the sphenoidal emissary foramen (of Vesalius). Both the sphenoidal emissary foramen and foramen spinosum link the middle cranial fossa and the infratemporal fossa. The sphenoidal emissary foramen is often absent but, when present, it transmits an emissary vein from the cavernous sinus to the pterygoid plexus.

28,49

The foramen lacerum lies at the junction between the apex of the petrous process, the sphenoid bone and the basilar part of the occipital bone. Structures associated with the foramen are the internal carotid artery (entering from behind and emerging above), the greater petrosal nerve and the deep petrosal nerve (which join to form the nerve of the pterygoid canal), a meningeal branch of the ascending pharyngeal artery (a branch of the external carotid artery), and emissary veins linking the cavernous sinus and pterygoid venous plexus.

Other features seen on the floor of the middle cranial fossa are the trigeminal impression, hiatuses and grooves for the greater and lesser petrosal nerves, the petrous ridge and arcuate eminence, and grooves for the middle meningeal vessels and the superior petrosal venous sinuses. With the exception of the grooves for the middle meningeal vessels, these features are associated with the petrous part of the temporal bone.

The trigeminal impression is a shallow fossa situated behind the foramen lacerum. It indicates the site of the ganglion of the trigeminal nerve. — 28,45

The greater and lesser petrosal nerves arise within the temporal bone in the region of the middle ear. They emerge on to the floor of the middle cranial fossa through hiatuses, the hiatus for the lesser petrosal nerve lying lateral to that of the greater petrosal nerve. The groove for the greater petrosal nerve runs forwards from the hiatus to the foramen lacerum. The groove for the lesser petrosal nerve runs forwards from the hiatus towards the foramen ovale. — 28,22; 28,23

The petrous ridge marks the boundary between the middle and posterior cranial fossae. A distinct bulge called the arcuate eminence lies on the petrous ridge. This indicates the position of the anterior (superior) semicircular canal of the internal ear. Running along the ridge is the groove for the superior petrosal venous sinus. This sinus links the cavernous and the sigmoid sinuses. There is a conspicuous groove for the middle meningeal vessels. This groove runs across the floor of the middle cranial fossa, from the foramen spinosum, and up on to the lateral wall of the fossa where it may divide into a groove for the frontal branches of the vessels and a groove for the parietal branches. — 28,27; 50B,26; 50B,31; 28,46

THE POSTERIOR CRANIAL FOSSA — 28; 67

The posterior cranial fossa is the largest of the cranial fossae. It contains the hind brain (the cerebellum posteriorly and the pons and medulla oblongata anteriorly). The floor and posterior wall of the posterior cranial fossa are formed mainly by the occipital bone (lateral and lower squamous parts). The anterior wall of the fossa leading up to the middle cranial fossa is formed by the basilar part of the occipital bone, the temporal bones (petrous and mastoid parts) and the sphenoid bone (dorsum sellae and posterior part of the body). The region corresponds extracranially with the posterior zone of the cranial base (see page 18). — 28,35; 28,42; 67,10; 28,15

The foramen magnum is the most prominent structure in the floor of the posterior cranial fossa. Passing through the foramen are the medulla oblongata, the apical ligament of the dens of the second cervical vertebra, the membrana tectoria of the atlanto-occipital joint, the vertebral and spinal arteries, the spinal parts of the accessory nerves, and the meningeal branches of the cervical spinal nerves. The hypoglossal canals (anterior condylar canals) and condylar canals (posterior condylar canals) lie close to the foramen magnum. The hypoglossal canal transmits the hypoglossal nerve (and its recurrent branch), the meningeal branch of the ascending pharyngeal artery and an emissary vein linking the basilar plexus intracranially with the internal jugular vein extracranially. The condylar canal carries an emissary vein between the sigmoid sinus and the occipital veins, and a meningeal branch of the occipital artery. — 28,40; 28,41

Other features found in the floor of the fossa are the internal occipital protuberance and crest, and grooves for some of the dural venous sinuses. The internal occipital protuberance lies at the confluence of some venous sinuses. Extending down from the protuberance to the foramen magnum is the internal occipital crest. This crest gives attachment to the falx cerebelli (a layer of meninges passing between the two hemispheres of the cerebellum of the brain). Grooves, which are often very prominent, are — 28,38; 28,39; 28,36

28,32; 28,37   found for the transverse, sigmoid, and superior sagittal sinuses. The occipital sinus may groove the internal occipital crest. The margins of the grooves for the transverse sinuses provide attachment for the tentorium cerebelli (a layer of meninges that passes between the occipital lobes and the cerebellum of the brain).

28,42
67,18
67,19; 28,31
67,18
30,14
50B,24

That part of the fossa in front of the foramen magnum formed by the basilar part of the occipital bone and by the sphenoid bone is called the clivus. Between the clivus and each petrous process of a temporal bone is the petro-occipital fissure. This fissure is occupied in life by a sliver of cartilage. The posterior end of the fissure is widened to form the jugular foramen. Passing through the jugular foramen is the internal jugular vein as it continues from the sigmoid sinus. In addition, it transmits the glossopharyngeal, vagus and accessory nerves, the inferior petrosal sinus and a meningeal branch of the occipital artery. The inferior petrosal sinus runs from the cavernous sinus to the jugular foramen in a groove closely related to the petro-occipital fissure. The posterior surface of the petrous part of the temporal bone shows an internal acoustic meatus for the passage of the facial and vestibulocochlear nerves into the ear. Labyrinthine vessels also pass through this meatus. Behind the opening of the internal acoustic meatus is the opening for the aqueduct of the vestibule of the ear (see pages 375, 378).

The bones of the skull

The 28 bones of the skull are grouped as follows:

- The bones of the vault of the skull.
- The bones of the base of the skull.
- The bones of the face.
- The bones of the jaws.
- The bones of the ear.

The bones in each of these categories are listed in Table 1.

The bones of the ear are described with the rest of the anatomy of the ear (see page 370).

THE BONES OF THE VAULT OF THE SKULL

The frontal bone, the two parietal bones and the occipital bone comprise the vault of the skull.

The frontal bone

38; 39

The frontal bone is a single bone located at the front of the vault of the skull. It has three parts: squamous, nasal and orbital. Within the bone are two cavities called the frontal air sinuses.

The squamous part of the frontal bone forms the major portion of the bone. Externally, it is related to the forehead and is considerably convex. Indeed, the most prominent bulges are called the frontal tuberosities. The tuberosities are most apparent on the skulls of the young. The posterior border of the squamous part is markedly serrated for articulation with the parietal bones at the coronal suture. The squamous part inferiorly forms the supra-orbital margins where it meets the orbital parts of the frontal bone. Just above the supra-orbital margins are the curved ridges of the superciliary arches. These meet above the nose to form another ridge called the glabella. The superciliary arches are larger in the male.

38,1; 39,1
38A,3
38,2
38,5
38A,7
38A,9

The lateral two-thirds of each supra-orbital margin is sharp, whereas the medial third is rounded. At the junction is found the supra-orbital notch or foramen, which transmits the supra-orbital nerve and vessels from the orbit on to the forehead. A small frontal notch or foramen may be found just medial to the supra-orbital notch and, when present, transmits the supratrochlear nerve and vessels on to the forehead.

38,6
38A,8

The supra-orbital margin projects laterally as the zygomatic process of the frontal bone. This process articulates with the frontal process of the zygomatic bone. A ridge extends backwards from the zygomatic process to divide into the anterior parts of the superior and inferior temporal lines. The area of the frontal bone below the temporal lines is the temporal surface. This surface forms the anterosuperior portion of the temporal fossa.

38,4
38D,20; 38D,21
38D,22

The nasal part of the frontal bone forms a small portion of the roof of the nose. It lies in the midline between the supra-orbital margins. The serrated free margin of the nasal part articulates with the nasal bones, the frontal processes of the maxillary bones, and the lacrimal bones. A small, thin plate of bone projects downwards in the midline as the nasal spine. This spine makes a minor contribution to the nasal septum, articulating in front with the nasal bone and behind with the perpendicular plate of the ethmoid bone.

38A,10
38A,11

The orbital parts of the frontal bone consist of two plates that lie horizontally to form the roofs of the orbits. They also contribute to the floor of the anterior cranial fossa.

38B,12; 39E,12

	Each plate meets the squamous part of the frontal bone at the supra-orbital margin.
38B,17	The orbital plates are separated from each other by the ethmoidal notch, and only
38C	occasionally are they joined posteriorly. The ethmoidal notch houses the cribriform plate of the ethmoid.

Each orbital plate presents two surfaces: the orbital (external) surface and the cranial (internal) surface. The orbital surface is concave. In the anterolateral corner is a fossa in which lies the lacrimal gland. In the anteromedial corner may be found a depression or tubercle, the trochlear fovea (or tubercle). This marks the site of attachment of the fibrocartilaginous pulley (trochlea) through which passes the superior oblique muscle of the eye. The posterior margin of the orbital plate is serrated for articulation with the lesser wing of the sphenoid bone.

(margins: 38B,12; 38B,13; 38C,14)

The margins of the ethmoidal notch articulate with the lateral masses (labyrinths) of the ethmoid bone and thereby complete the roofs of the ethmoidal air cells. Impressions of the air cells may therefore be seen on this margin. Grooves related to the anterior and posterior ethmoidal nerves and vessels can also be seen on this margin. These grooves become canals with articulation of the ethmoidal labyrinth. The frontal sinuses can be seen around the anterior portion of the ethmoidal notch.

(margins: 38B,19; 38B,15; 38B,16; 38B,18)

The internal surface of the frontal bone shows relatively few features. The squamous part is concave and laterally shows grooves related to the middle meningeal vessels. In the midline, just above the front end of the ethmoidal notch, is a ridge called the frontal crest. At the base of this crest is a groove that becomes the foramen caecum on articulation with the cribriform plate of the ethmoid. The foramen caecum may be blind-ended or may contain an emissary vein passing between the roof of the nose and the superior sagittal sinus. On following the frontal crest as it passes upwards and backwards, it gives way to a groove for the superior sagittal sinus. This sinus lies within the attached margins of the falx cerebri. Small depressions on each side of the groove for the superior sagittal sinus are related to arachnoid granulations.

(margins: 39E; 39E,1; 39E,23; 39E,24; 20B,14; 20B,18)

The muscles that gain attachment to the frontal bone are:

- Corrugator supercilii
- Orbital part of orbicularis oculi
- Temporalis

Ossification. The frontal bone develops intramembranously. Two primary centres of ossification appear in the eighth week of intra-uterine life. Each is situated in the region of the superciliary arch. A frontal or metopic suture initially divides the frontal bone in the midline. Although this suture usually disappears by the age of eight years, it may persist into adult life.

(margin: 39F,25)

The parietal bones

The two parietal bones form the bulk of the vault of the skull behind the frontal bone.

(margin: 52)

Each bone is quadrilateral in shape and presents as a curved plate. The four corners of the bone are referred to as the angles, and each is named after the bone with which it articulates. Thus, there are frontal (anterosuperior), occipital (posterosuperior), sphenoidal (antero-inferior) and mastoid (postero-inferior) angles. The frontal angle relates to the bregma, the occipital angle to the lambda, the sphenoidal angle to the pterion and the mastoid angle to the asterion.

(margins: 52,3; 52,1; 52,5; 52,7)

All the margins of the parietal bone are serrated. The sagittal (superior) margin is the longest and shows the deepest serrations. Here, the two parietal bones meet to form the sagittal suture. A small foramen, the parietal foramen, is sometimes found close to the sagittal margin. It transmits an emissary vein (joining veins on the scalp with the superior sagittal sinus) and sometimes a terminal branch of the occipital artery (a branch of the external carotid). The frontal (anterior) margin meets the frontal bone at the coronal suture. The occipital (posterior) margin articulates with the occipital

(margins: 52,2; 52A,12; 52,4; 52,8)

bone to form the lambdoid suture. The squamous (inferior) margin is the shortest margin of the parietal bone. It can be subdivided into three parts. Its central portion is concave and bevelled (at the expense of its external surface) where it meets the squamous part of the temporal bone. There is a short, thin, anterior portion (also bevelled at the expense of the external surface), which articulates with the greater wing of the sphenoid. The posterior part of the squamous margin is thicker and unites with the mastoid process of the temporal bone. 52,6

The external surface of the parietal bone is essentially featureless. It is convex and shows a protuberance near its centre called the parietal tuberosity. The superior and inferior temporal lines curve in an anteroposterior direction along its middle. 52A
52A,11
52A,9; 52A,10

The concave internal surface of the parietal bone shows grooves related to the middle meningeal vessels, which pass upwards and backwards. A groove is present along the sagittal margin for the superior sagittal sinus. Small depressions may be seen near the groove for the sagittal sinus, which are related to arachnoid granulations. A groove is also found in the region of the mastoid angle for the transverse sinus as it bends downwards to become the sigmoid sinus. 52B
52B,15
52B,13
52B,14

The parietal bone gives attachment to the temporalis muscle at and below the inferior temporal line. The superior temporal line gives attachment to the temporal fascia.

Ossification. The parietal bone develops intramembranously from two primary centres of ossification. These appear during the seventh week of intra-uterine life in the region of the parietal tuberosity.

The occipital bone 44; 45

This is a single bone that lies at the back of the vault of the skull. Its inferior portion curves forwards and contains the foramen magnum and the occipital condyles for articulation with the atlas (first cervical vertebra). The occipital bone is divided into four parts according to their relationship with the foramen magnum. The greater part of the bone lies above the foramen magnum and is called the squamous part. That lying in front of the foramen magnum is the basilar part. On each side of the foramen magnum are the lateral (or condylar) parts.

The squamous part of the occipital bone articulates with the parietal and temporal bones. The upper portion of the squamous part has the most serrated margin and articulates with the parietal bones at the lambdoid suture. The lower portion of the squamous part unites with the mastoid process of the temporal bone at the occipitomastoid suture. The region that lies at the lambda is called the superior angle. The part that lies at the asterion is called the lateral angle. 44,6
45,16
45,21
45,20

The external surface of the squamous part of the occipital bone is convex and is divided into two regions. The upper half is relatively smooth and the lower half presents a roughened appearance because of the attachment of muscles. In the midline, partly delineating these two areas, is a projection called the external occipital protuberance. The central point of this protuberance is the inion. From this protuberance, a thin ridge called the external occipital crest passes downwards to the foramen magnum. This crest gives attachment to the ligamentum nuchae, a fibroelastic band that runs up the back of the cervical vertebral column (see page 65). Three lines may be seen running across the roughened region of the squamous part. Two lines originate from the external occipital protuberance. The upper line is the supreme nuchal line. This gives attachment to the epicranial aponeurosis of the scalp (see page 188). The lower line is the superior nuchal line. Running from the external occipital crest, parallel to and below the superior nuchal line, is the inferior nuchal line. 44A,1; 45,1
44A,5
44A,2
44A,3
44A,4

The internal surface of the squamous part is concave and is divided into four fossae by a cross-shaped pattern of ridges and grooves. At the centre of the cross lies the internal occipital protuberance. A groove housing the superior sagittal sinus passes upwards from this protuberance. Passing laterally from the internal occipital protuberance are 44B
44B,27
44B,28

44B,18	grooves related to the transverse sinuses. The right transverse groove is usually continuous with the groove for the superior sagittal sinus, whereas the left transverse sinus is continuous with the straight sinus (see page 395). Finally, the internal occipital
44B,26	crest runs vertically downwards from the internal occipital protuberance before bifurcating near the foramen magnum. The upper fossae delineated by this system of
44B,17	ridges and grooves are called the cerebral fossae. They are approximately triangular in shape and are related to the occipital lobes of the brain. The lower fossae are the
44B,19	cerebellar fossae. They are related to the inferior surface of the cerebellum of the hind brain. They are larger than the cerebral fossae and are rectangular in shape.
	The ridges and grooves on the internal surface of the squamous part of the occipital bone also give attachment to layers of the meninges of the brain. To the margins of
44B,28	the groove containing the superior sagittal sinus is attached the falx cerebri, which separates the cerebral hemispheres of the brain. To the margins of the grooves for the
44B,18	transverse sinuses is attached the tentorium cerebelli, which lies between the cerebral hemispheres and the cerebellum of the brain. The internal occipital crest affords
44B,26	attachment to the falx cerebelli, which lies between the two hemispheres of the cerebellum.
44,14	**The basilar part of the occipital bone** lies in front of the foramen magnum. It is fused anteriorly with the body of the sphenoid, though up until about the age of 20
69,19	years this is the site of the spheno-occipital synchondrosis. At its sides, the basilar part
69,17	articulates with the petrous part of the temporal bone at the petro-occipital suture.
44A,15	Externally, the basilar part exhibits a small pharyngeal tubercle. To this tubercle is attached the pharyngeal raphe and the uppermost fibres of the superior constrictor muscle of the pharynx. The internal surface of the bone is grooved, forming the
66,10	clivus. Against the clivus lies part of the brain stem (the lower part of the pons and
44B,25	the medulla oblongata). Laterally, a groove for the inferior petrosal sinus may be seen.
44,7	**The lateral parts of the occipital bone** are also called the condylar parts because they
44A,12	have the occipital condyles that articulate with the vertebral column at the atlanto-occipital joints. Each lateral part articulates with the mastoid part of a temporal bone at the occipitomastoid suture.
	The external surface of the lateral part shows the following features: an occipital condyle, a condylar fossa with a condylar canal (posterior condylar canal), and a jugular process.
44A,12	Each occipital condyle articulates with a superior articular facet on the atlas vertebra. It is oval in outline and convex in all planes. The long axis of the condyle is directed anteroposteriorly and slightly medially, while the articular surface faces laterally. Each
44,11; 45,11	condyle contains the hypoglossal canal (anterior condylar canal). The main structure passing through this canal is the hypoglossal nerve. In addition, the canal transmits the recurrent meningeal branch of the nerve and sometimes a meningeal branch of the ascending pharyngeal artery (a branch of the external carotid), and an emissary vein linking the basilar plexus to the internal jugular vein.
44A,8	Immediately behind each condyle is a small depression named the condylar fossa. The superior articular facet of the atlas slides backwards into this fossa when the head is
44A,9	fully extended. The condylar fossa may be traversed by the condylar canal (posterior condylar canal). This canal allows passage of an emissary vein (linking the sigmoid sinus with the occipital veins) and a meningeal branch of the occipital artery (a branch of the external carotid).
44A,10	Lateral to the posterior part of the occipital condyle is a projection called the jugular
44B,23	process. This process is indented at the jugular notch and forms the posterior boundary of the jugular foramen.
44B,24	A bulge on the internal surface of the lateral part of the occipital bone, the jugular
44B,25	tubercle, lies above the hypoglossal canal. The tubercle may be grooved by the inferior
44B,22	petrosal sinus. Behind this is another groove, the groove for the sigmoid sinus. This

marks the site where the sigmoid sinus passes through the jugular foramen to become the internal jugular vein.

The foramen magnum is the largest foramen in the skull. The major structures passing through it are the medulla oblongata of the brain stem, the vertebral arteries and the spinal accessory nerves. Around the margins of the foramen are ligaments associated with the joints connecting the vertebral column to the base of the skull (see pages 64 to 67).

44,13

The following muscles are attached to the occipital bone:

- Longus capitis
- Occipital belly of occipitofrontalis
- Rectus capitis anterior
- Rectus capitis lateralis
- Rectus capitis posterior major and minor
- Splenius capitis
- Sternocleidomastoid
- Superior constrictor
- Obliquus capitis superior
- Trapezius

Ossification. The upper half of the squamous part of the occipital bone (interparietal part) develops intramembranously. The rest of the occipital bone develops endochondrally. Four centres of ossification appear in the squamous part. Each lateral part of the occipital bone has its own centre of ossification. There is also a centre for the basilar part. These centres appear between the sixth and eighth week of intra-uterine life. The occipital bone becomes a single bone at about six years of age.

THE BONES OF THE BASE OF THE SKULL

The bones comprising the floor of the cranial cavity are the ethmoid, frontal, sphenoid, temporal, and occipital bones. The occipital and frontal bones have already been described in connection with the bones of the vault of the skull. The ethmoid bone is considered with the face, as it contributes significantly to the orbital and nasal cavities.

The sphenoid bone

42; 43

This single bone forms much of the anterior part of the intermediate zone of the cranial base and most of the floor of the middle cranial fossa.

The sphenoid bone is often described as being butterfly-shaped. It consists of a central body and three paired processes. Extending laterally from the body are the lesser wings in front and the greater wings behind. Extending downwards from the junction of the body and greater wings are the pterygoid processes.

Anteriorly, the sphenoid bone articulates with the frontal and ethmoid bones above, and the vomer and palatine bones below. Laterally, it meets the zygomatic, parietal and temporal bones. Posteriorly, it joins the occipital bone.

There are usually two air sinuses within the body of the sphenoid bone.

The body of the sphenoid bone occupies a central position within the middle cranial fossa. The superior surface is smooth and saddle-shaped; hence the term sella turcica (Turkish saddle). The depression of the sella turcica is the pituitary fossa. The pituitary fossa is delineated anteriorly by a horizontal ridge called the tuberculum sellae. The lateral margins of the ridge are sometimes elevated to form middle clinoid processes. Above the tuberculum sellae lies the prechiasmatic groove. This leads laterally into the optic canal. The chiasma of the optic nerves is related to this surface. In front of

42,14; 43,14

42C,38
42C,37

42C,36
42C,34

42C,35	the prechiasmatic groove is the jugum. This is related to the gyri recti of the frontal lobe of the brain and to the olfactory tracts. A vertical plate of bone called the dorsum sellae forms the posterior boundary of the pituitary fossa. The free lateral margins of the dorsum sellae project as the posterior clinoid processes. The posterior surface of the dorsum sellae is concave and forms part of the clivus on which is situated the upper part of the pons.
42C,27	
42C,26	
43D,16	The inferior surface of the body of the sphenoid presents a midline ridge called the rostrum. This ridge fits into a groove on the upper border of the vomer bone.
42A,21	The anterior surface of the body of the sphenoid contains the apertures of the sphenoidal sinuses, which drain into the spheno-ethmoidal recesses in the nose. The sphenoidal sinuses are partitioned by a thin bony septum. The sinuses are bounded antero-inferiorly by two curved plates of bone, the sphenoidal conchae. These project inferiorly to form the sphenoidal crest, which articulates with the perpendicular plate of the ethmoid. The conchae also overlap on to the inferior surface of the body of the sphenoid to articulate medially with the alae of the vomer. The superolateral corner of the anterior surface meets the back of the labyrinth of the ethmoid bone and completes the posterior ethmoidal air cells.
42A,20	
42A,15	
69,19	Before the age of 20 years, the posterior surface of the body of the sphenoid is separated from the basilar part of the occipital bone by the spheno-occipital synchondrosis. This primary cartilaginous joint is a main contributor to forward growth of the face. After the age of 20 years, the synchondrosis disappears and there is then complete bony union.
42C,28	The side of the body of the sphenoid is related to the cavernous sinus. At the posterolateral region of the body (where the greater wing is attached) is located the carotid groove. In this groove runs the internal carotid artery. The groove is bounded laterally by a sharp projection called the sphenoid lingula.
42C,29	
42,1	**The lesser wings of the sphenoid bone** are two thin, triangular plates of bone projecting from the anterolateral corner of the superior surface of the body of the sphenoid. They are separated from the greater wings below by the superior orbital fissures. The main structures passing through the fissure are the oculomotor, trochlear, ophthalmic and abducent nerves and the ophthalmic veins.
42,2	
42,34	The base of each lesser wing is attached to the body by two roots that pass on either side of the optic canal. Through this canal run the optic nerve and ophthalmic artery. The anterior (frontal) margin of the lesser wing is serrated for articulation with the frontal bone. The posterior margin delineates the anterior and middle cranial fossae. It is smooth and projects medially as the anterior clinoid process. The upper surface of the lesser wing contributes to the anterior cranial fossa, the lower surface to the posterior part of the roof of the orbit.
64B	
42C,25; 42C,26	Two folds of meninges are attached to the anterior and posterior clinoid processes. The diaphragma sellae roofs over the pituitary fossa with an opening in its centre for the pituitary stalk. The tentorium cerebelli has its free border attached to the anterior clinoid processes and its attached border fixed to the posterior clinoid processes.
42; 43	**The greater wings of the sphenoid bone** extend outwards and forwards from the sides of the posterior part of the body. The anteroposterior dimensions are greater than the mediolateral ones.
42C,22	The anterior margin shows a roughened triangular portion medially (the frontal margin) for articulation with the frontal bone, and a smooth concave surface laterally (parietal margin) where it meets the parietal bone at the pterion. The lateral margin (squamous margin) is serrated for articulation with the squamous part of the petrous bone. The posterior margin articulates with the petrous part of the temporal bone.
42C,24	
42C,32	
	The external surface of the greater wing can be subdivided into lateral and anterior surfaces.

The lateral surface is further subdivided into upper and lower parts by a horizontal ridge called the infratemporal crest. The upper part is seen in the floor of the temporal fossa and is called the temporal surface. It is concave from front to back and contributes to the anterior part of the temporal fossa. There is a vertical crest anteriorly, which forms its zygomatic margin. This margin articulates with the zygomatic bone. The lower part below the infratemporal crest is almost horizontal and forms the roof of the infratemporal fossa. The foramen ovale and the foramen spinosum lie towards the back of this surface. Behind the foramen spinosum, the bone is raised to form the spine of the sphenoid (to which is attached the sphenomandibular ligament). The posterior margin here is grooved and is related to the cartilaginous component of the auditory tube running from the middle ear to the nasopharynx.

The anterior part of the greater wing is mainly formed by the orbital surface of the sphenoid. This part presents as a flat plate, which is quadrilateral and which comprises the posterior half of the lateral wall of the orbit. The orbital surface faces forwards and inwards. Its sloping posterior margin is a boundary of the superior orbital fissure. It may exhibit a small tubercle that gives attachment to the common tendinous ring associated with the recti muscles of the eye (see page 215). The inferior margin of the orbital surface is smooth and bounds the inferior orbital fissure posterolaterally. The region immediately below the superior orbital fissure is called the maxillary surface. This surface faces the maxillary bone and forms part of the posterior wall of the pterygopalatine fossa. Emerging on to the maxillary surface is the foramen rotundum.

The internal (cerebral) surface of each greater wing is concave and contributes to the floor of the middle cranial fossa. It is related to the temporal lobe of the brain.

There are three foramina in the greater wing of the sphenoid bone. The foramen rotundum is, as its name suggests, round in outline. It transmits the maxillary division of the trigeminal nerve from the middle cranial fossa into the pterygopalatine fossa. Just in front of the posterior margin of the greater wing is an oval-shaped opening appropriately called the foramen ovale. It allows the passage between the middle cranial fossa and the infratemporal fossa of the mandibular division of the trigeminal nerve, the lesser petrosal branch of the glossopharyngeal nerve, the accessory meningeal branch of the maxillary artery, and some emissary veins. Behind the foramen ovale lies the foramen spinosum. This transmits the middle meningeal vessels and the meningeal branch of the mandibular division of the trigeminal nerve.

Two additional foramina may be found in the greater wing. The sphenoidal emissary foramen (of Vesalius) is present in about 40% of skulls. It is situated medial to the foramen ovale and allows the passage of emissary veins linking the cavernous sinus and the pterygoid venous plexus in the infratemporal fossa. An occasional petrosal (innominate) foramen lies medial to the foramen spinosum and transmits the lesser petrosal nerve when this does not pass through the foramen ovale.

The pterygoid processes of the sphenoid bone arise from the inferior surface of the body where the greater wings are attached. Each process consists of two vertically aligned plates, the medial and lateral pterygoid plates. The plates are joined along their anterior margins except inferiorly where there is a small notch, the pterygoid notch. This notch receives the pyramidal process of the palatine bone. Posteriorly, the pterygoid plates diverge and enclose a space called the pterygoid fossa. A depression, the scaphoid fossa, is present between the pterygoid plates at their base.

The lateral surface of the lateral pterygoid plate forms part of the medial boundary of the infratemporal fossa. Its anterior margin bounds the pterygomaxillary fissure. The medial surface of the medial pterygoid plate bounds the lateral wall of the posterior nasal aperture (choana).

The medial pterygoid plate ends in a small, curved process called the pterygoid hamulus. At the base of the hamulus is a groove that is related to the tendon of the tensor veli palatini muscle.

The base of the pterygoid process is traversed by the pterygoid canal. The opening of the canal lies at the base of the medial pterygoid plate, just below the carotid sulcus. The pterygoid canal emerges below and medial to the foramen rotundum. Like the foramen rotundum, the pterygoid canal opens into the pterygopalatine fossa through its posterior wall. Through this canal pass the nerve and artery of the pterygoid canal.

From the base of the pterygoid process, a small plate of bone projects medially over the inferior surface of the body of the sphenoid bone to articulate with the ala of the vomer. This is the vaginal process. A groove on the undersurface of the vaginal process is converted into the palatovaginal canal by articulation with the upper surface of the sphenoidal process of the palatine bone. The pharyngeal branch of the pterygopalatine ganglion and the pharyngeal branch of the sphenopalatine artery pass through this canal.

Sometimes a canal is present between the upper surface of the vaginal process of the sphenoid bone and the ala of the vomer. This canal is called the vomerovaginal canal and anteriorly joins the palatovaginal canal. It transmits the pharyngeal branch of the sphenopalatine artery.

The following muscles are attached to the sphenoid bone:

- Lateral pterygoid
- Medial pterygoid
- Superior constrictor
- Temporalis
- Tensor veli palatini

Ossification. The sphenoid bone ossifies both endochondrally and intramembranously. The body, the lesser wings, and the roots of the greater wings ossify endochondrally. The pterygoid process and the bulk of the greater wing ossify intramembranously.

Six primary centres of ossification appear in a region corresponding to the body of the sphenoid in front of the tuberculum sellae and the lesser wings (the presphenoid component). Eight centres are found in the remaining regions (the postsphenoid component). These centres appear between the eighth week and fifth month of fetal life. The separate parts ossify to become a single bone by the first year.

The temporal bones

There are two temporal bones. Each bone contributes to the base and to the lower lateral aspect of the skull. It consists of four parts: the squamous, tympanic and petrous parts, and the styloid process. The mastoid process is here considered to be a component of the petrous part. The temporal bone also contains the auditory and vestibular systems (these are considered with the ear, see pages 362 to 380). The mastoid part has a variable number of air cells that open into the tympanic cavity.

The squamous part of the temporal bone can be divided into three regions: a temporal portion, a zygomatic process and a mandibular fossa (glenoid fossa).

The temporal portion consists of a thin, vertical plate of bone. Its external surface is gently convex and forms part of the temporal fossa. Posteriorly, the external surface shows a groove related to the middle temporal branch of the superficial temporal artery (a branch of the external carotid). The internal (cerebral) surface of the temporal portion is concave. It shows ridges related to the temporal lobe of the brain and grooves related to middle meningeal vessels. At the junction of the temporal portion of the squamous part of the temporal bone with the petrous portion of the bone there is usually evidence of a petrosquamous fissure. Rarely, a more conspicuous groove is present at this site and indicates the position of a petrosquamous venous sinus that opens posteriorly into the transverse sinus. The petrosquamous sinus communicates by an emissary vein with the external jugular vein. The upper margin of the temporal portion is bevelled (at the expense of the cerebral surface) for articulation with the parietal bone. Antero-inferiorly, the margin is thicker for articulation with the greater wing of the sphenoid.

The zygomatic process projects outwards and forwards from the lowermost part of the external surface of the temporal portion. In front, it articulates with the temporal process of the zygomatic bone to complete the zygomatic arch. Posteriorly, the upper border of the zygomatic process continues above and behind the external acoustic meatus as a ridge. This ridge passes above the mastoid process and is called the supramastoid crest. The crest gives attachment to the temporal fascia. Between the crest and the upper border of the external acoustic meatus is often found a triangular depression termed the suprameatal pit or triangle (a landmark for operative procedures on the mastoid antrum). The anterior part of the pit may bear a small projection, the suprameatal spine. The inferior surface of the zygomatic process shows a ridge called the articular tubercle. Behind the tubercle, the zygomatic process is concave where it forms the lateral margin of the mandibular fossa. In front of the external acoustic meatus, the zygomatic process terminates inferiorly as the postglenoid tubercle. This tubercle is, however, not always present.

The mandibular fossa is situated at the cranial base. It lies beneath and medial to the zygomatic process and immediately anterior to the external acoustic meatus. It articulates with the head of the condyle to form the temporomandibular joint. The mandibular fossa is an oval depression. It is bounded anteriorly by the articular tubercle. Posteriorly, the fossa is separated from the tympanic part of the temporal bone by the squamotympanic fissure. Medially, a projection from the petrous part of the temporal bone comes to lie within this fissure to divide it into the petrosquamous and the petrotympanic fissures.

The tympanic part of the temporal bone is a semicircular plate of bone that forms the anterior, inferior and posterior boundaries of the bony part of the external acoustic meatus. It is demarcated from the mastoid process behind by the tympanomastoid fissure. The anterior wall of the tympanic part bounds the mandibular fossa posteriorly. The inferior border of the tympanic part is sharp and is called the tympanic crest. The tympanic crest overlaps the styloid process and thus forms the sheath of the styloid process. The lateral margin of the tympanic part is roughened for attachment of the cartilaginous part of the external acoustic meatus.

The petrous part of the temporal bone is the solid wedge of bone that forms most of the posterior and inferior portions of the temporal bone. Internally, it lies at the boundary between the middle and posterior cranial fossae. It can be subdivided into the petrous part proper and the mastoid part. An alternative term is the petromastoid part of the temporal bone.

The petrous part proper is pyramidal in shape with an apex, a base, two internal surfaces (anterior and posterior), and an external surface (inferior). The apex has a broad, flattened surface, which articulates with the posterolateral corner of the body of the sphenoid and forms a margin for the foramen lacerum. The apex has the opening for the carotid canal, which allows the passage of the internal carotid artery towards the cavernous sinus. The base of the petrous part proper is continuous with the mastoid and squamous parts of the temporal bone, the union with the squamous part being marked by the petrosquamous fissure.

The internal (cerebral) aspect of the petrous part proper has a ridge running along the superior margin on which is a groove for the superior petrosal sinus. The ridge subdivides the internal surface into anterior and posterior surfaces. In front of the ridge is the anterior surface, which forms the posterior boundary of the middle cranial fossa and is related to the temporal lobe of the brain. Behind the ridge is the posterior surface, which constitutes the anterior portion of the posterior cranial fossa and is related to the cerebellum.

50C,36
50C,37

50,26

50C,32

50D,32
50C,34; 50C,35

The anterior surface slopes downwards and outwards. Its medial end, adjacent to the apex, is marked by a depression called the trigeminal impression. The trigeminal ganglion is situated here. Beneath the trigeminal impression is the carotid canal. By the side of the trigeminal impression is a slight concavity overlying the internal acoustic meatus. Beyond this, and about halfway along the length of the anterior surface, is a ridge called the arcuate eminence. This ridge is produced by the underlying anterior (superior) semicircular canal. Between the arcuate eminence and the petrosquamous fissure is the tegmen tympani. This thin plate of bone forms the roof of the mastoid antrum and the tympanic cavity and also contributes to the auditory tube and the canal for the tensor tympani muscle. A portion of the tegmen tympani is seen between the squamous and tympanic parts of the temporal bone at the mandibular fossa. Two small foramina may be seen in the tegmen tympani near the arcuate eminence. These foramina lead into grooves running forwards and medially towards the sphenoid bone. The medially positioned groove runs to the foramen lacerum and transmits the greater petrosal nerve. The laterally positioned groove runs towards the foramen ovale and transmits the lesser petrosal nerve.

50B,23
50B,29
50B,27
50B,25

50B,24

50B,21

The posterior surface slopes downwards and slightly medially. At its medial end, close to the apex, there is a groove for the inferior petrosal sinus. About halfway along the surface is the opening of the internal acoustic meatus. Just behind this opening, and below the level of the arcuate eminence, is a shallow fossa called the subarcuate fossa. Behind and below the fossa is a narrow slit, which is the external opening of the aqueduct of the vestibule (see page 375). The inferior margin of the posterior surface is separated in part from the occipital bone by the jugular foramen, while the posterolateral margin is related to the groove for the sigmoid sinus.

50D,43
50D,47

50D,48

50D,44
50D,45

50D,28; 50B,28

50B,28

The inferior surface of the petrous part proper contributes to the external surface of the cranial base. Its posterior third shows a pronounced depression, the jugular fossa. This fossa is related to the upper part of the internal jugular vein. On the lateral wall of the jugular fossa is found the mastoid canaliculus through which passes the auricular branch of the vagus nerve. Immediately in front of the jugular fossa is the opening of the carotid canal for transmission of the internal carotid artery. The ridge separating the opening of the carotid canal and the jugular fossa shows the tympanic canaliculus for passage of the tympanic branch of the glossopharyngeal nerve (also the inferior tympanic branch of the ascending pharyngeal artery). A triangular depression is seen close to the medial margin of the jugular fossa and directly below the opening of the internal acoustic meatus. In this depression lies the external opening of the cochlear canaliculus (see page 376). The medial margin of the inferior surface of the petrous part proper forms the upper boundary of the jugular foramen. This margin may be notched (the jugular notch) below the level of the internal acoustic meatus. Through the jugular foramen run the internal jugular vein and the glossopharyngeal, vagus and accessory nerves. In front of the carotid canal, the bone is roughened for attachment of the levator veli palatini muscle. The lateral margin of the inferior surface of the petrous part proper gives attachment to the cartilaginous part of the auditory tube.

50,16
50F,39
50F,38
50F,22

50B,21
50B,22

The mastoid part of the temporal bone lies below the squamous part and behind the tympanic part. Most of it is composed of the prominent mastoid process. Medial to the process is a deep groove, the mastoid notch. Medial to the mastoid notch is a small groove related to the occipital artery. The lateral surface of the mastoid part of the temporal bone is roughened for muscle attachments. A mastoid foramen is sometimes found near the posterior border. It gives passage to an emissary vein (linking the sigmoid sinus and the occipital veins) and to a meningeal branch from the occipital artery. The main feature on the internal surface of the mastoid is a deep groove related to the sigmoid sinus. The internal opening of the mastoid foramen is seen close to the groove of the sigmoid sinus. The thick upper border of the mastoid articulates with the mastoid angle of the parietal bone. The posterior border articulates with the squamous part of the occipital bone.

50,13

The styloid process of the temporal bone is an elongated, narrow projection of bone. It passes downwards and forwards from the base of the temporal bone, where it lies between the tympanic part and the posterior border of the jugular foramen. Between the styloid and mastoid processes lies the stylomastoid foramen. The facial nerve and the stylomastoid branch of the posterior auricular artery pass through this foramen.

50D,40

The following muscles are attached to the temporal bone:

- Digastric (posterior belly)
- Levator veli palatini
- Longissimus capitis
- Masseter
- Occipital belly of occipitofrontalis
- Splenius capitis
- Sternocleidomastoid
- Styloglossus
- Stylohyoid
- Stylopharyngeus
- Temporalis

Ossification. The temporal bone develops both intramembranously and endochondrally. The squamous and tympanic parts develop intramembranously. Each part ossifies from a single centre that appears between the second and third months of intrauterine life. The petrous part (including the mastoid portion) and the styloid process develop endochondrally. The petrous part is ossified from four centres appearing at about the sixth month of intra-uterine life. The styloid process develops from the cartilage of the second branchial arch and ossifies from two centres, one appearing before birth, the other shortly after birth. The individual parts of the temporal bone usually unite by the end of the first year. However, the distal part of the styloid process does not unite with its proximal component until at least puberty.

THE BONES OF THE FACE

This section includes not only the bones seen on the norma frontalis of the skull, but also some of the bones that contribute to the nasal and orbital cavities. Two of the bones in this region have already been described: the frontal bone with the vault of the skull, and the sphenoid bone with the base of the skull. The jaws (maxillary bones and mandible) are considered in the subsequent section.

The zygomatic bones

53C; 53D

The two zygomatic bones form the skeleton of the cheeks. Each consists of a trapezoidal body with two processes projecting from the upper and lower corners of the posterior margin (the frontal and temporal processes respectively). An orbital plate projects inwards from the upper margin.

The lateral surface of the body of the zygomatic bone is smooth and slightly convex. Its medial or maxillary margin articulates with the zygomatic process of the maxillary bone. The inferior border is free and provides attachment for the masseter muscle. The posterior or temporal margin is notched and inferiorly it projects as the temporal process, whose serrated margin unites with the zygomatic process of the temporal bone to form the zygomatic arch. Superiorly, the temporal margin continues upwards as the posterior border of the frontal process, giving attachment to the temporal fascia. The frontal process itself articulates with the zygomatic process of the frontal bone. The superior or orbital margin of the zygomatic bone is concave and contributes to both the inferior and lateral margins of the orbit.

53C,19
53C,20

53C,17
53C,18

53C,16
53C,22

Projecting inwards from the orbital margin is a shelf of bone that forms an orbital plate. Its upper or orbital surface is concave and forms the anterior part of the lateral wall of the orbit. The superior margin of the orbital plate at the frontal process

53D,23

53D,16

articulates with the zygomatic process of the frontal bone. Just below this margin, there is usually a small tubercle. This is known as the marginal tubercle (Whitnall's tubercle) and to it are attached the lateral palpebral raphe from the orbicularis oculi muscle, and the lateral palpebral ligament. The orbital plate also articulates with the greater wing of the sphenoid bone at the sphenoidal margin.

The entire surface of the zygomatic bone viewed medially is called the temporal surface. It can be subdivided into two regions. Anteriorly, it shows a roughened area for articulation with the zygomatic process of the maxilla. Posteriorly, the temporal surface includes the lower surface of the orbital plate and the temporal surface of the temporal process. This posterior region is smooth and forms the anterior boundary of the temporal fossa.

The zygomatic bone has three foramina. On the orbital surface of the orbital plate is found the zygomatico-orbital foramen. Through this pass the zygomaticofacial and zygomaticotemporal branches of the maxillary division of the trigeminal nerve. The zygomaticofacial nerve then runs through a canal and emerges on to the face at the zygomaticofacial foramen on the lateral surface of the zygomatic bone. The zygomaticotemporal nerve passes through the zygomaticotemporal foramen on the temporal surface of the bone.

The muscles attached to the zygomatic bone are:

- Levator labii superioris
- Masseter
- Temporalis
- Zygomaticus major
- Zygomaticus minor

Ossification. The bone develops intramembranously from a centre that appears during the eighth week of intra-uterine life.

The ethmoid bone

The ethmoid is a single bone that makes a significant contribution to the middle third of the face. It forms parts of the nasal septum, roof and lateral wall of the nose, and a considerable part of the medial wall of the orbital cavity. In addition, the ethmoid makes a small contribution to the floor of the anterior cranial fossa. Four parts can be distinguished: the perpendicular plate, the cribriform plate and two ethmoidal labyrinths.

The perpendicular plate of the ethmoid is a thin, quadrilateral plate of bone that descends vertically in the midline from the cribriform plate to form the upper part of the nasal septum. In front, it articulates with the nasal bone above, and the septal cartilage below. Behind, it meets the sphenoidal crest of the body of the sphenoid. Below and behind, the perpendicular plate joins the vomer bone.

The cribriform plate of the ethmoid forms the upper surface of the ethmoid bone. It lies horizontally, filling the ethmoid notch of the frontal bone. It forms the roof of the nose and derives its name from the fact that it is penetrated by numerous foramina containing branches of the olfactory nerves. Projecting vertically upwards in the midline is a triangular plate of bone, the crista galli. From the anterior border of the crista galli arise two small, wing-like plates of bone called the alae. When articulated with the frontal bone, the alae contribute to an opening called the foramen caecum. An emissary vein may pass through the opening, linking the superior sagittal sinus with veins in the nose. The falx cerebri is attached to the posterior border of the crista galli. The anterior ethmoidal nerves pass into the nose through slits lying on each side of the crista galli.

The ethmoidal labyrinths (lateral masses) hang vertically downwards from the lateral margins of the cribriform plate. Each labyrinth consists of a network of air cells separated by bony trabeculae and bounded by a lateral plate (orbital plate) and a medial surface (nasal surface). The air cells are, however, incompletely surrounded by bone and become roofed over by adjacent bones. The upper surface of the ethmoidal labyrinth articulates with the frontal bone, and the posterior surface is covered by the sphenoidal concha and the orbital process of the palatine bone.

The orbital plate of the labyrinth is a smooth, quadrilateral plate of bone. Anteriorly, the exposed air cells are covered over by the lacrimal bone and by the frontal process of the maxilla. Below, the orbital plate articulates with the maxillary bone. Above, there are grooves for the anterior and posterior ethmoidal nerves and vessels, which are converted into foramina by articulation with the frontal bone. However, the posterior groove and foramen are frequently absent.

The medial surface of the ethmoidal labyrinth forms part of the lateral wall of the nose. Two thin, scroll-like plates of bone hang down from it. The upper plate is the superior nasal concha, the lower is the middle nasal concha. The middle concha is larger and extends more anteriorly. The groove between the conchae is called the superior meatus. The posterior ethmoidal air cells open into this. The region below the attached margin of the middle nasal concha curves upwards and forwards to form a channel, called the infundibulum, in the middle meatus of the nose. The anterior ethmoidal air cells open here. A swelling of the bone in this region, called the ethmoidal bulla, overlies the middle ethmoidal air cells. An opening in or over the bulla is the site of drainage for these air cells. A thin, curved plate of bone runs lateral to the anterior part of the middle nasal concha. This is the uncinate process. The region between this process and the ethmoidal bulla above corresponds to the hiatus semilunaris (see page 245).

Ossification. The ethmoid bone develops endochondrally, and there are three centres of ossification. A centre for each ethmoidal labyrinth appears at about the fifth month of intra-uterine life. The centre for the perpendicular plate and crista galli appears after birth. The three parts become united at about two years of age.

The palatine bones

Each palatine bone is roughly L-shaped. It has a perpendicular plate that forms part of the lateral wall of the nose, and a horizontal plate that forms the posterior one-third of the hard palate. A pyramidal process arises posteriorly at the junction between the two plates.

The perpendicular plate of the palatine bone is rectangular (the vertical dimension being about twice that of the anteroposterior one). The upper border presents a notch, the sphenopalatine notch. In front of the notch lies the orbital process. Behind the notch is the sphenoidal process. In articulating with the inferior surface of the body of the sphenoid, the notch is converted into the sphenopalatine foramen. This foramen transmits the nasopalatine and posterior superior nasal nerves and the sphenopalatine vessels from the pterygopalatine fossa into the nasal cavity. The remaining part of the perpendicular plate has a lateral (maxillary) surface and a medial (nasal) surface.

The orbital process projects upwards and outwards from the upper border of the perpendicular plate. Its smooth upper (orbital) surface forms the most posterior portion of the floor of the orbit. Its lateral surface forms part of the medial wall of the pterygopalatine fossa and is closely related to the maxillary nerve. From the medial side, the orbital process is hollowed out to contain an air cell. The air cell may communicate with the posterior ethmoidal air cells or with the sphenoidal sinus. The orbital process articulates with the maxillary, ethmoid and sphenoid bones.

The sphenoidal process projects upwards and inwards from the upper border of the perpendicular plate to articulate with the body of the sphenoid, the medial pterygoid plate, and the ala of the vomer.

48B,10	The lateral (maxillary) surface of the perpendicular plate is roughened inferiorly for articulation with the maxilla (at the nasal surface behind the opening of the maxillary sinus). The upper part is smooth and contributes to the medial wall of the
48B,11	pterygopalatine fossa. The greater palatine groove is found running down the posterior margin of the lateral surface. The groove is converted into a canal (the greater palatine canal) when the bone is articulated with the maxilla. This canal transmits both greater and lesser palatine nerves and vessels. At the lower end of the
48B,7	anterior border of the perpendicular plate is a projection called the maxillary process. This covers the postero-inferior portion of the hiatus of the maxillary sinus. The posterior border of the perpendicular plate articulates with the medial pterygoid plate.
48A,5 48A,6	The medial (nasal) surface of the perpendicular plate is smooth and shows near its middle a horizontal ridge, the conchal crest. This crest articulates with the inferior nasal concha. The concavity below the conchal crest forms part of the inferior meatus of the lateral wall of the nose. The concavity above the crest contributes to the middle
48A,4	meatus and extends upwards to the region of the sphenopalatine notch where another horizontal ridge, the ethmoid crest, is situated. To this crest is attached the middle nasal concha of the ethmoid bone. The small region lying above the ethmoid crest contributes to the superior meatus of the nose.
48,3 48E,14	**The horizontal plate of the palatine bone** is quadrilateral (the mediolateral dimension being greater than the anteroposterior one). The upper (nasal) surface is slightly concave and forms the posterior third of the floor of the nose. The lower
48F,15	(palatal) surface forms the posterior third of the hard palate and, near its posterior border, may exhibit a ridge called the palatal crest, which helps to attach the tendon of the tensor veli palatini muscle. The medial border of the horizontal plate is thickened and articulates with the horizontal process of the opposite side at the median palatine
48,13	suture. The upper margin of the medial border forms the nasal crest. This articulates with the vomer. The anterior border of the horizontal plate articulates with the palatine process of the maxilla at the transverse palatine suture. The posterior border of the horizontal plate is smooth and concave. Medially, it contributes to the
48E,17	posterior nasal spine. To this spine is attached the musculus uvulae. The lateral border of the horizontal plate is attached to the inferior border of the perpendicular
48F,11	plate. At this junction is seen the lower margin of the greater palatine groove.
48,9	**The pyramidal process of the palatine bone** projects backwards, outwards and slightly downwards from the posterior surface of the palatine bone at the junction between the perpendicular and horizontal plates. It articulates in the space between the maxillary tuberosity and the pterygoid notch of the pterygoid process. The
48F,16	inferior surface of the pyramidal process shows one or more openings, the lesser palatine foramina. These are continuous above with the greater palatine groove and transmit the lesser palatine nerves and vessels.

The muscles attached to the palatine bone are:

- Medial pterygoid (superficial head)
- Musculus uvulae
- Palatopharyngeus
- Tensor veli palatini

Ossification. The bone is ossified intramembranously from a centre in the perpendicular plate. This centre appears during the eighth week of intra-uterine life.

42F; 42G	**The vomer bone**
	This single bone is a thin plate shaped like a ploughshare. It forms the postero-inferior portion of the nasal septum. Its superior border shows a groove lying between two
42,41	wing-like projections or alae. The rostrum of the sphenoid bone fits into this groove. When the ala of the vomer meets the vaginal process and body of the sphenoid, it
23,49	completes the vomerovaginal canal. Through this occasional canal passes a pharyngeal

branch of the sphenopalatine artery. The inferior border of the vomer lies anterior to the superior border and fits into the nasal crest running along the middle of the upper surface of the hard palate. The anterior border of the vomer slopes downwards and forwards. Inferiorly, it is grooved and receives the inferior border of the cartilaginous nasal septum. Superiorly, it joins with the perpendicular plate of the ethmoid. The posterior border of the vomer is free and does not articulate with other structures. It is slightly concave and slopes downwards and forwards to form a prominent midline ridge between the two posterior nasal apertures. Each side of the vomer presents a groove running downwards and forwards, which is related to the nasopalatine nerve and vessels.

42F,42

42F,43

Ossification. The vomer ossifies intramembranously from two centres appearing at about the eighth week of intra-uterine life. A cartilage initially intervenes between the two centres of ossification but is eventually resorbed. The two bony lamellae unite to form a single bone (apart from the grooved anterior and superior margins).

The nasal bones

46G; 46H

The two nasal bones form the bridge of the nose. Each nasal bone is quadrilateral (being longer than it is wide). The upper border articulates with the nasal part of the frontal bone. The lower border is sharp and notched and forms the upper boundary of the anterior nasal aperture. The lower border gives attachment to the lateral nasal cartilage. The lateral border of the nasal bone meets the frontal process of the maxilla. The medial border meets its fellow in the midline and is thickened above to form a vertical crest (this makes a small contribution to the nasal septum). It therefore articulates with the septal cartilage and with the perpendicular plate of the ethmoid.

The smooth outer surface of the nasal bone is convex, but is concave near the upper border. The internal surface is concave throughout and in its upper part shows a groove related to the external nasal branch of the anterior ethmoidal nerve.

46G
46H

The procerus muscle is attached to the external surface of the nasal bone.

Ossification. The nasal bone ossifies intramembranously from a single centre appearing at about the third month of intra-uterine life.

The inferior nasal conchae

48G–J

Each concha consists of a curved plate of bone attached to the lateral wall of the nose. The posterior end of the bone is more pointed than the anterior end. Its longer lower border lies free within the nasal cavity, is thickened and often curves inwards. The upper border serves to attach the bone to the lateral wall of the nose, articulating with four different bones. Its anterior slope is attached to the conchal crest of the maxilla. Its posterior slope articulates with the conchal crest of the perpendicular plate of the palatine bone. The middle third of the upper border is more or less horizontal and shows two small processes: a lacrimal process anteriorly and an ethmoid process posteriorly. The lacrimal process meets the descending process of the lacrimal bone and helps to complete the nasolacrimal canal. The ethmoid process articulates with the uncinate process of the ethmoid. From the middle portion of the upper border, a thin, triangular plate of bone (the maxillary process) projects downwards and partially covers the hiatus of the maxillary sinus. The maxillary process articulates with the maxilla and with the maxillary process of the palatine bone. The medial surface of the inferior nasal concha is convex and shows pits and grooves related to blood vessels. The lateral surface is smoother and concave.

48,22

48H,19
48G,21
48H,23; 48J,23
48G,20
48H,24

Ossification. The bone ossifies endochondrally from a single centre appearing at about the fifth month of intra-uterine life.

The lacrimal bones

46J; 46K

Each of these two bones is small, thin and rectangular. It lies in the anterior part of the medial wall of the orbit. In front, the lacrimal bone articulates with the frontal process of the maxilla. Behind, it meets the orbital plate of the ethmoid. Above, it joins the frontal bone and below, the maxilla. The lateral surface is divided into two areas by a vertical crest called the posterior lacrimal crest. This crest ends below in a hook-shaped process called the lacrimal hamulus. The area in front of the crest is concave and, together with the adjacent area of the frontal process of the maxilla, forms the fossa for the lacrimal sac. The area behind the posterior lacrimal crest forms part of the medial wall of the orbit. The upper part of the medial (nasal) surface of the lacrimal bone covers the anterior ethmoidal air cells. The lower part contributes to the lateral wall of the nose in the middle meatus and narrows the opening of the maxillary sinus. A descending process continues inferiorly to articulate with the inferior nasal concha and adjacent part of the maxilla to complete the nasolacrimal duct.

The lacrimal part of the orbicularis oculi muscle arises from the lateral surface of the lacrimal bone.

Ossification. The bone is ossified intramembranously from a centre appearing at about the third month of intra-uterine life.

THE BONES OF THE JAWS

The jaws are the tooth-bearing bones. They comprise three bones. The two maxillary bones form the upper jaw, whereas the lower jaw is a single bone, the mandible.

The maxillary bones

The maxillary bones support the teeth of the upper jaw and contribute much of the skeleton of the upper face. Each bone consists of a body and four processes: the frontal, zygomatic, alveolar and palatine processes. The body contains the maxillary air sinus.

The maxillary bones show many articulations. On the face, the maxillary bones articulate with each other, and with the nasal bones, the nasal cartilages and the frontal bone. Laterally, they articulate with the zygomatic bones. Each maxillary bone also joins with the vomer, the septal cartilage, the lacrimal bone, the ethmoid bone and the inferior nasal concha to contribute to the skeleton of the nasal fossa and the orbit.

The body of the maxilla is essentially pyramidal in shape. Its base faces the nasal cavity and its apex is located within the zygomatic process. It presents four main surfaces: anterior (facial), posterior (infratemporal), medial (nasal) and superior (orbital) surfaces.

The anterior surface of the body of the maxilla forms the skeleton of the anterior part of the cheek. The anterior surface is demarcated from the posterior surface by the zygomatic process and by a ridge (the zygomatico-alveolar or jugal crest) running from the zygomatic process to the socket of the first molar tooth. A ridge called the canine eminence overlies the root of the canine tooth. It separates the anterior surface into two concave areas: a shallow, incisive fossa in front and a deeper, canine fossa behind. The anterior surface is delineated from the orbital surface by the infra-orbital margin. Below this margin lies the infra-orbital foramen through which pass the infra-orbital branch of the maxillary nerve and the infra-orbital vessels.

The posterior surface of the body of the maxilla forms the anterior wall of the infratemporal fossa. Its lower, posterior extremity presents a convexity called the maxillary tuberosity. Immediately above the tuberosity, the surface of the bone is roughened for articulation of the pyramidal process of the palatine bone. In front of the tuberosity may be seen several small foramina associated with the posterior superior alveolar nerves and vessels (supplying the posterior maxillary teeth). Above the tuberosity, the posterior surface forms the anterior boundary of the pterygopalatine fossa.

The medial (nasal) surface of the body of the maxilla forms part of the lateral wall of the nose. The upper posterior part of this surface shows the large opening (hiatus) leading into the maxillary sinus. In the recent state, the hiatus is reduced in size by the articulation of the palatine bone, the uncinate process of the ethmoid, the inferior nasal concha and the lacrimal bone, and by the overlying mucosa. One or two shallow depressions may be found above and behind the hiatus. These help to complete some of the ethmoidal air cells. Behind and below the maxillary sinus lies the greater palatine groove. This is converted into a canal carrying the greater palatine nerve and vessels by the perpendicular plate of the palatine bone. In front of the maxillary sinus lies a deep vertical groove. This is called the lacrimal groove which, with the lacrimal bone, forms the nasolacrimal canal. Anterior to the nasolacrimal canal is an obliquely running ridge called the conchal crest. This articulates with the inferior nasal concha. The concave area above the crest therefore bounds the middle meatus of the nasal cavity, whereas the concave area below bounds the inferior meatus.

The superior (orbital) surface of the body of the maxilla forms much of the floor of the orbit. It is triangular in shape. A groove, the infra-orbital groove, is present midway along the posterior border of this surface and, running forwards and medially, eventually forms the infra-orbital canal. This canal exits at the infra-orbital foramen and transmits the infra-orbital nerve and vessels. The nerve is a branch of the maxillary division of the trigeminal. The maxillary nerve itself also grooves the bone at the junction between the superior and posterior surfaces of the body of the maxilla. The rounded, posterior border of the orbital surface forms the anterior boundary of the inferior orbital fissure. The medial border of the orbital surface presents anteriorly the lacrimal groove which is bounded in front by the anterior lacrimal crest. The anterior border of the orbital surface forms the medial part of the inferior margin of the orbit.

The frontal process of the maxilla projects upwards from the body and is situated between the nasal bone in front and the lacrimal bone behind. Its lateral surface is divided into two areas by the vertically-running anterior lacrimal crest. Behind the crest, the bone is grooved for the nasolacrimal canal. To the crest itself is attached the medial palpebral ligament (see page 212). The medial surface of the frontal process shows an obliquely-running ethmoidal crest. To this is attached the middle nasal concha. The bone above the crest helps to enclose the anterior ethmoidal air cells.

The zygomatic process of the maxilla projects laterally from the body. It is pyramidal in shape and has a roughened superolateral surface for articulation with the zygomatic bone. Its posterior surface is smooth and concave and forms part of the anterior boundary of the infratemporal fossa.

The palatine process of the maxilla extends horizontally from the medial surface of the maxilla where the body meets the alveolar process. Posteriorly, the boundary between the palatine and alveolar processes is sharp. Anteriorly, the boundary is less well defined. The medial edge of the palatine process is for articulation with the opposite palatine process at the median palatine suture. Behind the central incisors, the medial edge shows evidence of the incisive fossa. Through this pass the nasopalatine nerves and branches of the greater palatine vessels. An incisive canal runs through the palatine process in the region of the incisive fossa. The medial edge is thickened on its nasal surface to contribute to the nasal crest (a structure that articulates with the inferior border of the vomer). The crest is noticeably thickened anteriorly and is prolonged into the anterior nasal spine. Both the nasal and oral surfaces of the palatine process are concave. Unlike the nasal surface, the oral surface of the palatine process is rough and irregular, forming palatine grooves and spines. The posterior edge of the palatine process articulates with the horizontal plate of a palatine bone at the transverse palatine suture.

The alveolar process of the maxilla extends inferiorly from the body of the maxilla and supports the teeth within bony sockets. Each maxilla can contain a full quadrant

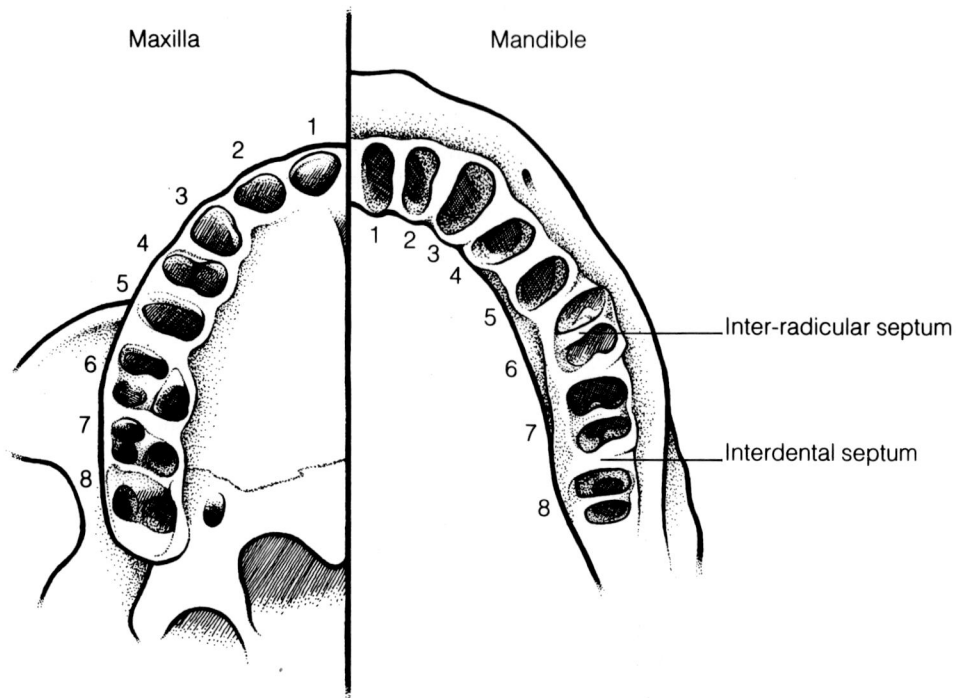

Figure 2 *The alveolar processes of the maxilla and mandible illustrating the tooth sockets. The positions of the teeth are indicated according to the Zsigmondy system (i.e. 1 to 8 for the permanent teeth in each jaw quadrant).*

of eight permanent teeth or five deciduous teeth. The form of the alveolus is related to the functional demands put upon the teeth. For example, when the teeth are extracted, the alveolus resorbs. Essentially, the alveolar process consists of two parallel plates of cortical bone (the buccal and palatal alveolar plates) between which lie the sockets of individual teeth (Figure 2). Between the sockets are inter-alveolar or interdental septa. The floor of the socket has been termed the fundus. Its rim is called the alveolar crest. The form and depth of each socket is defined by the form and length of the root it supports. The sockets thus show considerable variation. In multirooted teeth, the sockets are divided by inter-radicular septa. The apical regions of the sockets of anterior teeth are closely related to the nasal fossae, whereas those of posterior teeth are related to the maxillary sinus. In the midline, the alveolar processes of the maxillary bones meet at the intermaxillary suture. At the upper end of the suture lies a sharp bony projection called the anterior nasal spine. Adjacent to the nasal spine, and forming the floor of the anterior nasal aperture, is a notch called the nasal notch.

The following muscles are attached to the maxilla:
- Buccinator
- Depressor septi
- Levator anguli oris
- Levator labii superioris
- Nasalis
- Orbicularis oculi
- Orbicularis oris

Ossification. The maxilla ossifies in membrane from a single centre of ossification in the body. This appears at about the seventh week of intra-uterine life. It had been thought that the incisor-bearing part of the maxilla had a separate centre of ossification — hence the term 'premaxilla'. However, recent evidence indicates that ossification spreads from the body of the maxilla into its incisor-bearing component.

The mandible

The mandible consists of a horizontal, horseshoe-shaped body and two vertical rami. The body of the mandible supports the mandibular teeth within the alveolar process. The rami of the mandible articulate with the temporal bones at the temporomandibular joints.

The body of the mandible has important features on both the lateral and medial surfaces.

The lateral surface of the body of the mandible shows a ridge in the upper part of the midline, which represents the site of the mandibular symphysis (see page 52). Close to the inferior margin of the body lies a distinct prominence, the mental protuberance. On each side of the protuberance are the mental tubercles. The mental protuberance and tubercles together comprise the chin. Above the mental protuberance lies a shallow depression called the incisive fossa. Behind this fossa, a canine eminence overlies the root of the mandibular canine tooth. In the region of the premolar teeth is found the mental foramen. The mental branches of the inferior alveolar nerve and vessels pass on to the face through this foramen. During growth of the mandible, when the prominence of the chin develops, the opening of the mental foramen alters in direction from facing forwards to facing upwards and backwards. Rarely, there may be multiple mental foramina. The alveolus forms the superior margin of the body of the mandible. The junction of the alveolus and ramus is demarcated by a ridge, the external oblique line. This ridge is continuous with the anterior border of the ramus and passes downwards and forwards across the body of the mandible to terminate below the mental foramen.

The medial surface of the body of the mandible has two shallow depressions close to the midline on its inferior border. These are the digastric fossae, providing sites for the attachment of the anterior bellies of the digastric muscles. Above these fossae are the genial tubercles (mental spines). There are generally two inferior and two superior tubercles. They mark the sites of attachment of the geniohyoid and genioglossus muscles. Across the medial surface of the body of the mandible is a prominent ridge called the mylohyoid line (internal oblique line). To this is attached the mylohyoid muscle. The ridge arises between the genial tubercles and the digastric fossa and increases in prominence as it passes backwards and upwards, to end on the anterior surface of the ramus. The surface of the mandible above and in front of the mylohyoid line presents a shallow depression in which lies the sublingual salivary gland. The depression is therefore called the sublingual fossa. The shallow concavity below the mylohyoid line is the submandibular fossa in which lies the superficial portion of the submandibular salivary gland. At the posterior end of the mylohyoid line is attached the pterygomandibular raphe.

The alveolar process of the mandible continues upwards from the body. It consists of buccal and lingual alveolar plates joined by interdental and inter-radicular septa (Figure 2). Near the second and third molar teeth, the external oblique line is superimposed upon the buccal alveolar plate. The form and depth of the tooth sockets is related to the morphology of the roots of the mandibular teeth and to functional demands.

The ramus of the mandible meets the body of the mandible at an obtuse angle. The region where the inferior margin of the ramus meets the posterior margin is called the angle of the mandible. This area provides attachment for the masseter and medial pterygoid muscles and for the stylomandibular ligament. Superiorly are located the coronoid and condylar processes. These are separated by the mandibular notch. The lateral surface of the ramus is relatively featureless. It presents a surface for the attachment of the masseter muscle. In the centre of the medial surface of the ramus lies the mandibular foramen through which the inferior alveolar nerve and vessels pass into the mandibular canal. A bony process called the lingula extends from the anterosuperior surface of the foramen and gives attachment to the sphenomandibular ligament. A groove, the mylohyoid groove, runs down from the postero-inferior surface of the mandibular foramen. Below and behind the mylohyoid groove, the medial surface of the ramus is roughened around the angle for the attachment of the medial pterygoid muscle. Running down from the tip of the coronoid process is a ridge called the temporal crest. This extends down to the bone just behind the third molar tooth. The triangular depression between the temporal crest and the anterior border of the ramus is called the retromolar fossa.

The coronoid process lies anterior to the condylar process. It is a triangular plate of bone that gives attachment to the temporalis muscle.

The condylar process varies considerably in terms of both shape and size. Its broad articular head joins the ramus through a thin bony projection called the neck of the condyle. The anteroposterior dimension of the condylar head is approximately half the mediolateral dimension. The long axis of the condyle is not, however, at right angles to the ramus but diverges posteriorly from a strictly coronal plane such that the long axes of the two condyles, if extended medially, would meet to form an obtuse angle of approximately 150° at the anterior border of the foramen magnum. The convex anterior and superior surfaces of the head of the condyle are the articular surfaces. The posterior surface of the head of the condyle is broad and flat. A small depression, the pterygoid fovea, is a site of attachment of the lateral pterygoid muscle. It is situated on the anterior part of the neck of the condyle.

The mandibular canal begins at the mandibular foramen and passes initially downwards and forwards in the ramus. It runs horizontally below the molar teeth in the body of the mandible. Near the premolar teeth, the canal bifurcates into incisive and mental canals. The narrow incisive canal continues forwards towards the midline beneath the incisor teeth. The mental canal runs upwards, outwards and backwards to open on to the face at the mental foramen.

Many muscles are attached to the mandible and comprise:

- Muscles of mastication: Lateral pterygoid
                                Masseter
                                Medial pterygoid
                                Temporalis
- Muscles of facial expression: Buccinator
                                Depressor anguli oris
                                Depressor labii inferioris
                                Mentalis
                                Platysma
- Suprahyoid muscles: Digastric
                                Geniohyoid
                                Mylohyoid
- Muscles of tongue and pharynx: Genioglossus
                                 Superior constrictor

Ossification. The mandible ossifies in membrane from two centres that appear close to the site of the mental foramen during the sixth week of intra-uterine life. The two halves of the mandible are initially separated by a midline symphysis menti. With the disappearance of the symphysis at about two years of age, the mandible becomes a single bone.

The mandible ossifies around the cartilages of the first branchial arches (Meckel's cartilages). Each cartilage undergoes resorption, though its remnants form the sphenomandibular ligament and the lingula.

The development of the mandible is unusual in that, after the bone has been mapped out in membrane, secondary cartilages appear at the coronoid process, the condylar process and the symphysis menti. The cartilage of the coronoid process disappears before birth. However, the cartilage of the condyle remains until about the age of 20 years. It is thought by some to play a predominant role in the growth of the mandible. However, others doubt this, as surgical removal of the condyle does not appear significantly to affect the subsequent growth of the mandible.

Skull bone articulations

The skull is usually considered either in terms of a fully articulated structure, or in terms of the individual bones. A further appreciation of this most complex of osseous structures can be obtained by considering how the individual bones comprising a region articulate. This is of particular importance for understanding the osteology of the orbit, nose, cranial fossae and pterygopalatine fossa.

54; 55 **THE ORBITAL AND NASAL APERTURES**

54 The orbital and nasal apertures of the facial skeleton are defined by the frontal, maxillary, nasal and zygomatic bones.

55,3
55,7; 55,4
55,8
55,13

The orbital aperture (aditus of the orbit) is bounded above by the supra-orbital margin of the frontal bone, laterally by the zygomatic bone and the zygomatic process of the frontal bone, below by the zygomatic bone and the maxilla, and medially by the frontal bone and the anterior lacrimal crest of the frontal process of the maxilla.

55,12
55,14

The anterior nasal aperture (piriform aperture) is bounded mainly by the maxillary bones. The nasal bones form the superior margin of the aperture.

32A; 56–59 **THE ORBIT**

56A The orbital cavity is pyramidal in shape, with the apex pointing posteriorly. It has a roof, a floor, and medial and lateral walls. The bones that comprise the orbital cavity are the ethmoid, frontal, lacrimal, maxillary, palatine, sphenoid and zygomatic bones.

56C,1
56C,2

The roof of the orbit is formed mainly by the orbital part of the frontal bone. The lesser wing of the sphenoid bone is situated posteriorly near the apex.

58,1; 58,6
58,10

The floor of the orbit comprises of the orbital surfaces of the maxillary and zygomatic bones, with the orbital process of the palatine bone near the apex.

59,13
59,18
59,20
32A,22

The medial wall of the orbit begins at the anterior lacrimal crest of the frontal process of the maxilla. Behind this lies the lacrimal bone. Most of the medial wall behind the lacrimal bone is formed by the orbital plate of the ethmoid bone. The body of the sphenoid bone contributes to the medial wall posteriorly. Between the anterior lacrimal crest of the maxilla and the posterior lacrimal crest of the lacrimal bone is the

59,22 fossa for the lacrimal sac.

57,12
57,14

The lateral wall of the orbit is formed by the orbital surface of the greater wing of the sphenoid bone and by the orbital surface of the zygomatic bone.

There are several prominent foramina and fissures within the orbit.

32A,2; 32A,1 At the superior orbital margin are the supra-orbital notch (or foramen) and the frontal notch (or foramen). These transmit respectively the supra-orbital and the supra-trochlear nerves and vessels.

32A,15 In the floor of the orbit is found the infra-orbital groove and the infra-orbital canal for the infra-orbital nerve and vessels.

32A,24
32A,27; 32A,28

At the medial wall is situated the opening of the nasolacrimal canal, and the anterior and posterior ethmoidal foramina for the anterior and posterior ethmoidal nerves and vessels. The nasolacrimal canal is located antero-inferiorly, close to the orbital margin. The ethmoidal foramina lie at the junction with the roof of the orbital cavity.

122D,41 At the lateral wall is the zygomatico-orbital foramen (occasionally foramina). Through this foramen pass the zygomatic branches of the maxillary division of the trigeminal nerve and accompanying vessels.

Near the apex of the orbital cavity are the optic canal, the superior orbital fissure and the inferior orbital fissure. The optic canal lies within the lesser wing of the sphenoid and transmits the optic nerve and ophthalmic artery. The superior orbital fissure lies between the greater and lesser wings of the sphenoid at the junction of the roof and lateral wall of the orbit. It transmits the oculomotor, trochlear, ophthalmic and abducent nerves, together with the ophthalmic veins. The inferior orbital fissure lies at the junction of the lateral wall and floor of the orbit. Through this fissure pass the infra-orbital and zygomatic branches of the maxillary division of the trigeminal nerve and accompanying vessels.

THE NASAL CAVITY

The nasal cavity is divided into two nasal fossae by a nasal septum. The external nose has a skeleton that is mainly cartilaginous (see page 242). Each nasal fossa has a roof, a floor, a lateral wall and a medial wall.

The medial wall is formed by the nasal septum. It comprises the septal cartilage anteriorly, the perpendicular plate of the ethmoid bone postero-superiorly and the vomer postero-inferiorly. At the base of the nasal septum is the nasal crest, formed by the maxillary and palatine bones.

The bones that comprise the remaining walls of the nasal fossa are the ethmoid, frontal, lacrimal, maxillary, nasal, palatine, sphenoid, and vomer bones and the bone of the inferior concha.

The roof of the nasal fossa is formed centrally by the cribriform plate of the ethmoid bone. Anteriorly lies the nasal bone and the nasal spine of the frontal bone. Posteriorly is located the body of the sphenoid overlapped by the ala of the vomer and the sphenoidal process of the palatine bone. The cribriform plate transmits olfactory nerves into the roof of the nasal cavity and also the anterior ethmoidal nerve and vessels.

The floor of the nasal fossa is formed by the palatine process of the maxilla anteriorly and the horizontal plate of the palatine bone posteriorly.

The lateral wall of the nasal fossa consists mainly of the medial surface of the maxilla, with the large maxillary hiatus being reduced in size by the overlapping of the lacrimal and ethmoid bones above, the palatine bone behind and the inferior concha below. From the lateral wall project the three nasal conchae. The superior and middle conchae are part of the labyrinth of the ethmoid bone. The inferior concha is a separate bone. The region above and behind the superior nasal concha is called the spheno-ethmoidal recess. The region between the superior and middle nasal conchae is the superior meatus. Between the middle and inferior conchae is located the middle meatus. Below the inferior nasal concha is the inferior meatus. The nasolacrimal canal on the lateral wall of the nasal fossa is formed by the lacrimal groove of the maxilla articulating with the descending process of the lacrimal bone and the lacrimal process of the inferior concha. The sphenopalatine foramen, which links the nasal cavity with the pterygopalatine fossa, lies high up in the posterior part of the lateral wall. It is formed by the notch between the orbital and sphenoidal processes of the perpendicular plate of the palatine bone articulating with the body of the sphenoid bone. This foramen transmits the nasopalatine and posterior superior nasal nerves and the sphenopalatine vessels.

The lateral wall of the nose is noted for being the site of drainage of the paranasal air sinuses. The sphenoidal sinus drains into the spheno-ethmoidal recess. The posterior ethmoidal air cells drain into the superior meatus. The frontal and maxillary air sinuses and the anterior and middle ethmoidal air cells drain into the middle meatus. The opening of the nasolacrimal canal lies in the inferior meatus.

The posterior nasal apertures (choanae) link the nasal fossae with the nasopharynx. The bones that contribute to this region are the palatine, sphenoid and vomer bones. The posterior border of the vomer separates the two posterior nasal apertures. Each

72,12	aperture is bounded below by the posterior border of the horizontal plate of the palatine bone, laterally by the medial pterygoid plate, and above by the body and vaginal process of the sphenoid bone and the ala of the vomer. A groove on the lower surface of the vaginal process of the sphenoid bone is converted into the palatovaginal canal by articulation with the upper surface of the sphenoidal process of the palatine bone. This canal transmits the pharyngeal branch of the pterygopalatine ganglion and the pharyngeal branch of the maxillary artery. The vomerovaginal canal lies between the upper surface of the vaginal process of the sphenoid bone and the ala of the vomer. The pharyngeal branch of the sphenopalatine artery passes through this occasional canal.
72,7	
72,4; 72,2	
72A,5	
72B,15	
72B,14	

THE BASE OF THE SKULL

64–69

The cranial base has already been described in detail (see pages 17 to 24).

Internally lie the anterior, middle and posterior cranial fossae.

64,1	The floor of the anterior cranial fossa is formed by the orbital parts of the frontal bone, the cribriform plate and crista galli of the ethmoid bone, and the lesser wings and jugum of the sphenoid bone.
64,5; 64,4	
64,8; 64,7	
67,1	The middle cranial fossa consists of a central part formed by the body of the sphenoid bone, and right and left lateral parts, each formed by the greater wing of the sphenoid and the squamous and petrous parts of the temporal bone.
67,3	
67,6; 67,7	
28,42; 28,35	The posterior cranial fossa is formed by the basilar, lateral and squamous parts of the occipital bone, by the petrous parts of the temporal bones, by the mastoid angles of the parietal bones, and by the dorsum sellae and posterior part of the body of the sphenoid bone.
28,28; 28,34	
28,15	
68,69	The external surface of the base of the skull can be subdivided into three zones: the anterior, intermediate and posterior zones.
22,2	The anterior zone comprises the hard palate and is formed by the palatine processes of the maxillae and the horizontal plates of the palatine bones.
22,6	
68,14	The intermediate and posterior zones are formed by the sphenoid bone anteriorly, the occipital bone posteriorly, and the petrous parts of the temporal bones laterally.
68,4; 68,7	

The numerous foramina associated with the cranial base are described on pages 17 to 24.

THE PTERYGOPALATINE FOSSA

70

70,9	The pterygopalatine fossa lies between the infratemporal (posterior) surface of the maxilla and the pterygoid process of the sphenoid bone. It communicates with the infratemporal fossa through the pterygomaxillary fissure. The anterior wall of the fossa is the infratemporal surface of the maxilla. The posterior wall of the fossa is the pterygoid process. The medial wall is formed by the perpendicular plate of the palatine bone. The lateral wall shows the pterygomaxillary fissure. The pyramidal process of the palatine bone is situated inferiorly and articulates with the tuberosity of the maxilla. It fills the triangular gap between the lower ends of the medial and lateral pterygoid plates.
70,2	
70B,11	
70,9	
70,2; 70A,7	
70B,11	
70,8	
70B,11	The pterygomaxillary fissure transmits the maxillary artery from the infratemporal fossa, the posterior superior alveolar branches of the maxillary division of the trigeminal nerve, and the sphenopalatine veins. The fissure continues above with the posterior end of the inferior orbital fissure in the floor of the orbit. Passing through the inferior orbital fissure from the pterygopalatine fossa are the infra-orbital and zygomatic branches of the maxillary nerve, the orbital branches of the pterygopalatine ganglion and the infra-orbital blood vessels.
42A,19	Entering the pterygopalatine fossa posteriorly are the foramen rotundum from the middle cranial fossa and the pterygoid canal from the region of the foramen lacerum at
42A,18	

the base of the skull. The maxillary division of the trigeminal nerve passes through the foramen rotundum. The pterygoid canal transmits the nerve of the pterygoid canal (greater petrosal and deep petrosal nerves, see page 257) and an accompanying artery.

On the medial wall of the pterygopalatine fossa lies the sphenopalatine foramen. This foramen communicates with the nasal cavity. It transmits the nasopalatine and posterior superior nasal nerves (from the pterygopalatine ganglion) and the sphenopalatine vessels.

70,6
60A,5

At the base of the pterygopalatine fossa, the greater and lesser palatine nerves (from the pterygopalatine ganglion), together with accompanying vessels, pass into the hard palate to emerge at the greater and lesser palatine foramina.

The full contents of the pterygopalatine fossa are described on pages 253 to 258.

SUMMARY SHEET: THE SKULL

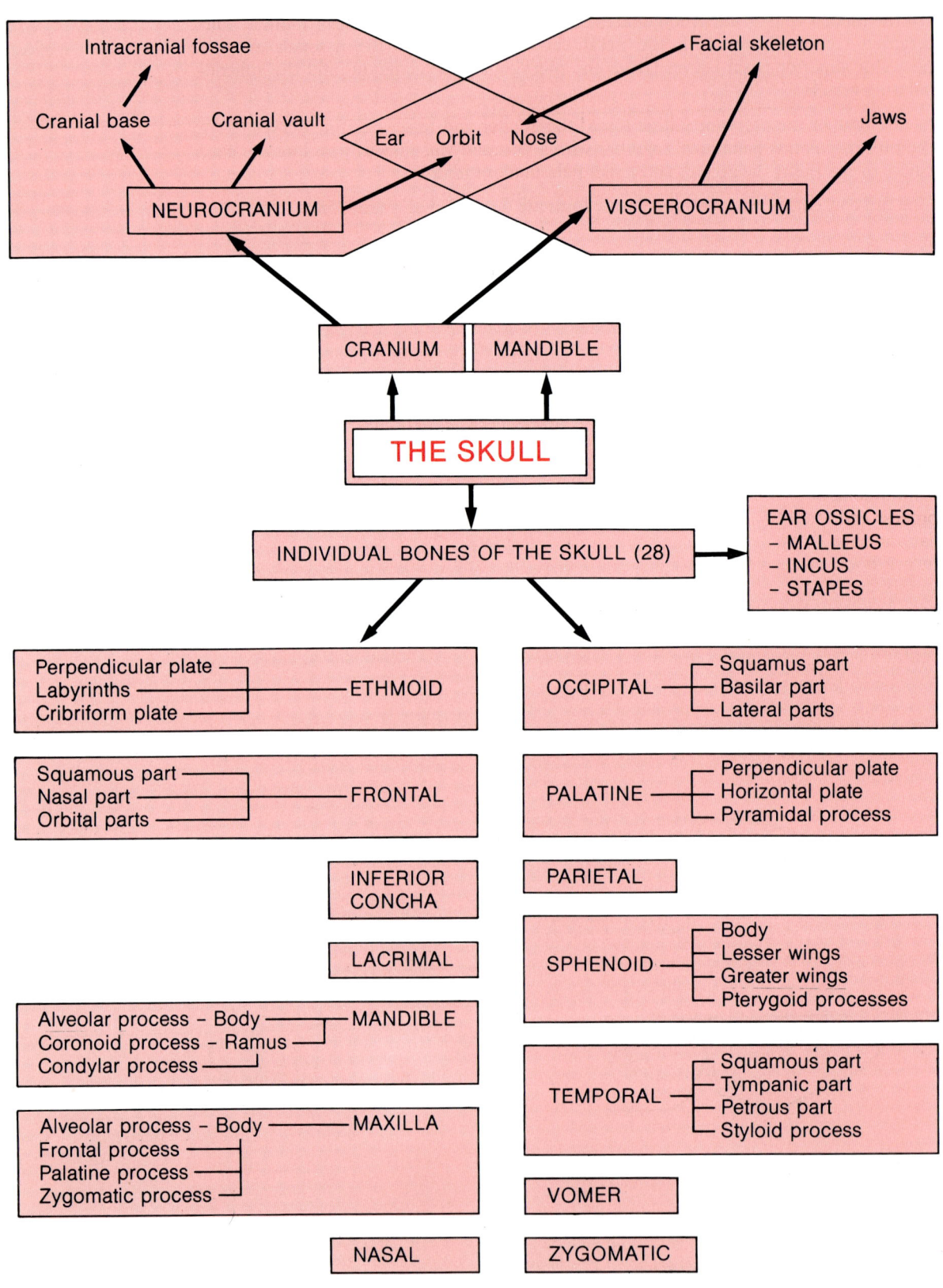

The fetal skull

The skull of a full-term fetus contains the same individual bones as the adult skull. There are, however, significant differences with respect to the proportions of the skull, the size and shape of the bones, and the way the bones articulate.

74; 75

Considering the proportions of the fetal skull, the neurocranium is much larger than the viscerocranium. Indeed, the ratio of the neurocranium to the viscerocranium is 8:1 for the fetal skull and only 3:1 for the adult skull. Thus, whereas the neurocranium expands much faster during the fetal period, the face grows predominantly in the years after birth (Figure 3). This relates to the fact that early in development the brain causes the neurocranium to expand, but later, structures such as the teeth and the jaw musculature are responsible for significant growth of the face. There is thus differential growth in the transformation of the fetal to the adult skull.

The individual bones of the fetal skull are obviously smaller than their adult counterparts, the exceptions being the ear ossicles (malleus, incus and stapes) which have almost reached adult size by birth. In addition, many of the bones have slightly different shapes, mainly because of the different proportions of their constituent parts. The bones comprising the cranial vault are more curved and the frontal and parietal tuberosities are especially prominent. The superciliary arches and the glabella on the frontal bone have not developed. The temporal bone has only a rudimentary mastoid process and the stylomastoid foramen is therefore superficial. The mandibular fossa is flat and there is no articular eminence. The external acoustic meatus is short and the future bony part is unossified. Consequently, the tympanic membrane is relatively superficial. For the ethmoid bone, only the labyrinths are ossified at birth. The fetal mandible (see Figure 10) has no mental protuberance and has a body that is large compared with the ramus. Most of the body is composed of the alveolar process containing the developing teeth. Each maxillary bone also consists mainly of its alveolar process. The palate is shallow.

75,16; 75,17

74B,23

74B,24

Concerning the articulations of the fetal skull, the prominent feature is the presence of fontanelles. These are fibrous membranes that fill in deficiencies between the bones of

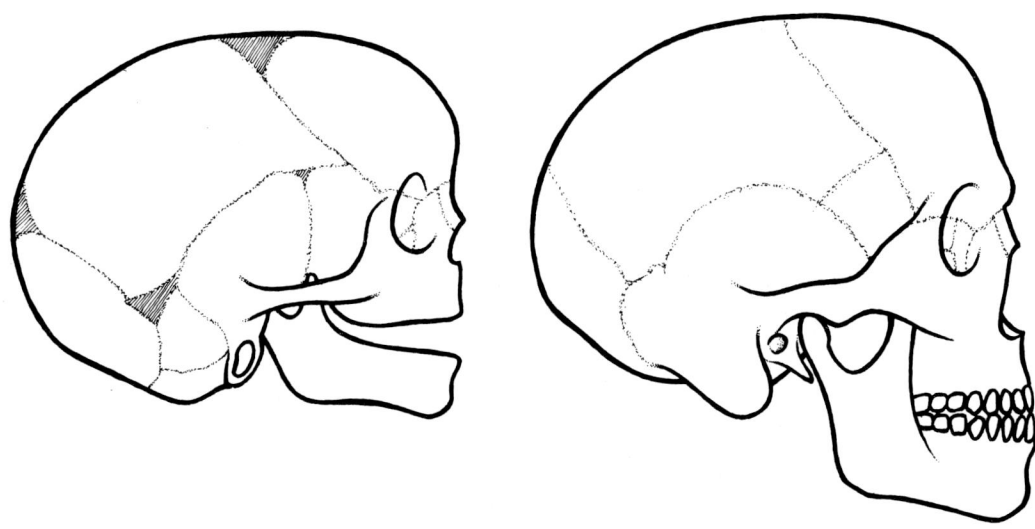

Figure 3 Comparison of the neonatal and adult skull to indicate the changes in cranial proportions. *Note particularly the greater ratio of the neurocranium to the viscerocranium in the neonate.*

74A,1; 75,1

20A,10
74C,30; 75,30
20A,2
74B,27; 74B,20
14,21
14,12

the vault of the skull. The fontanelles permit some sliding of the bones of the cranium during passage of the head through the birth canal. Any distortion of the skull occurring at birth usually disappears within a week. There are six fontanelles and these can be located with reference to the angles of the parietal bones. At the top of the cranium are the anterior and posterior fontanelles. The anterior fontanelle is the largest of all the fontanelles and is diamond-shaped. It lies between the frontal bone and the parietal bones and corresponds with the bregma of the adult skull. The posterior fontanelle is small and triangular. It corresponds with the lambda, being situated between the occipital bone and the parietal bones. On each side of the skull are found a sphenoidal (anterolateral) fontanelle and a mastoid (posterolateral) fontanelle. Both are small and irregular in shape. The sphenoidal fontanelle corresponds to the pterion of the adult skull. The mastoid fontanelle corresponds to the asterion. The posterior and sphenoidal fontanelles close within three months of birth. The mastoid fontanelle closes at 12 months. The anterior fontanelle is the last to close, at about 18 months.

The sutures on the fetal skull are smooth and wide. There are also sutures and other joints that are usually absent on the adult skull. These exist where parts of a bone have not yet fused to form a single bone.

74A,2; 75,2

Dividing the frontal bone down the middle of the forehead is the frontal or metopic suture. Its presence is responsible for the large size and diamond shape of the anterior fontanelle. The frontal suture usually disappears by the age of seven years.

74A,10

Also in the midline, separating the two halves of the mandible, is the mandibular symphysis (symphysis menti). The mandible becomes a single bone by the age of two years.

The occipital bone at birth is divided into four parts (Figure 4) corresponding to the squamous, basilar and lateral parts of the adult bone. Complete fusion takes place by the age of six years. At the junction of the basilar part of the occipital bone and the body of the sphenoid bone is the spheno-occipital synchondrosis (Figure 4). This joint

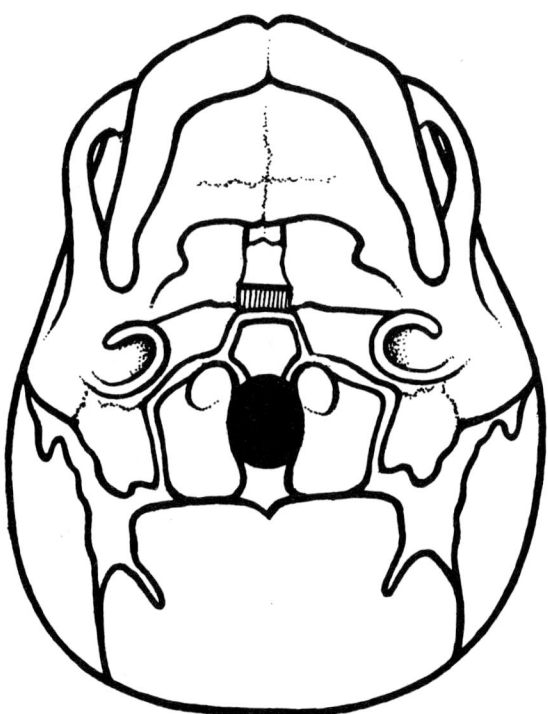

Figure 4 *Appearance of the external surface of the base of the neonatal skull.* Note that the occipital bone consists of four separate parts surrounding the foramen magnum. The hatched area represents the spheno-occipital synchondrosis.

is important for growth of the skull in the anteroposterior plane. It closes at about 20 years of age.

The sphenoid bone at birth is in three parts: a central part (the body and lesser wings) and two lateral components (the greater wing and pterygoid process on each side). These unite by the end of the first year.

The temporal bone at birth has three parts. These correspond to the petrous part, the squamous/tympanic part, and the styloid process of the adult bone. Union begins at birth and is complete by the end of the first year. The tip of the styloid process does not, however, fuse with the rest of the bone until at least puberty.

The ethmoid bone is also in three parts: the ethmoid labyrinths laterally and the cribriform and perpendicular plates centrally. Only the labyrinths have begun to ossify by birth. The three parts of the ethmoid fuse to form a single bone at about the age of two years.

The air sinuses within the fetal skull are poorly developed. The maxillary sinus is rudimentary at birth, although it is identifiable radiographically. The sphenoidal sinuses and the ethmoid air cells may also have reached a sufficient size to be of clinical significance. The frontal sinus, however, does not invade the frontal bone until the age of two years.

Concerning the origins of the individual bones of the skull, some develop intramembranously and others endochondrally (Figure 5). A few bones develop both intramembranously and endochondrally. The bones of the vault of the cranium (frontal bone, parietal bone, part of the squamous part of the occipital bone, and the squamous part of the temporal bone) develop in the membrane surrounding the developing brain. The bones at the base of the cranium (the ethmoid bone, the sphenoid bone excluding the pterygoid processes and part of the greater wings, the

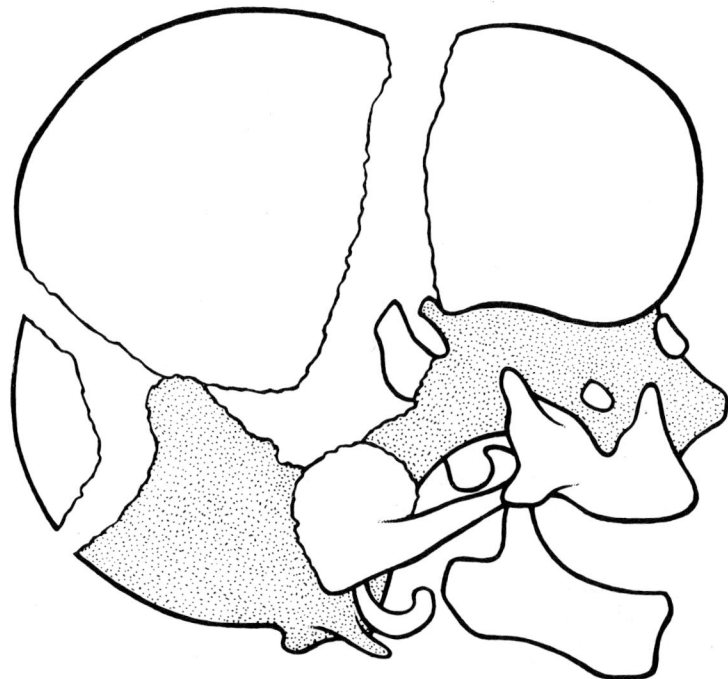

Figure 5 *Fetal skull at 16 weeks of intra-uterine life, showing the main developmental subdivisions of the skull, namely the cranial vault, cranial base and face. The bones of the cranial base (stippled) ossify endochondrally. The bones of the cranial vault and face (unshaded) ossify intramembranously.*

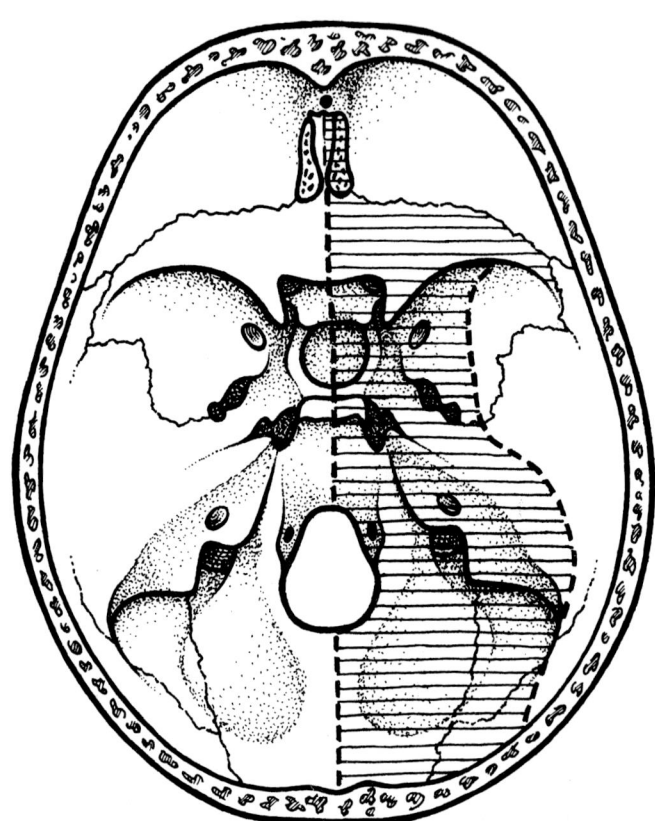

Figure 6 *Intracranial view of the base of the skull. The bones that develop endochondrally (hatched on one side) constitute the chondrocranium and are the ethmoid, sphenoid (body, lesser wings and roots of greater wings), temporal (petro-mastoid), and occipital (excluding the superior portion of the squamous part).*

petrous part of the temporal bone, and most of the occipital bone) develop endochondrally. This part of the developing skull has been termed the chondrocranium (Figure 6). The remaining bones lie in the facial region and develop intramembranously, except the inferior nasal concha which forms endochondrally. Ossification in all three regions has begun by the eighth week of intra-uterine life, although most centres in the chondrocranium appear later.

There are three main methods by which the skull grows: sutural growth, cartilaginous growth, and bone remodelling.

Growth at the sutures occurs passively as the sutures themselves do not provide a growth force but are subjected to a growth force from elsewhere (e.g. the expanding brain and eye). As the bones on either side of the suture become separated, they maintain their new positions as new bone is deposited at the margins of the suture. Sutural growth is particularly important in the first few years after birth. Some sutures eventually become obliterated with age. This usually first occurs internally.

Although cartilaginous growth is found in many regions of the fetal skull, it is limited postnatally to three principal sites: the spheno-occipital synchondrosis and the cartilages within the condyles of the mandible. The spheno-occipital synchondrosis is a primary cartilage, appearing at the outset of development as part of the chondrocranium. The condylar cartilages are secondary cartilages, having differentiated after the condyles had already started to ossify in membrane. Growth at the spheno-occipital synchondrosis and the mandibular condyles continues until about the age of 20 years. Thereafter, the cartilages become obliterated. A small amount of cartilagi-

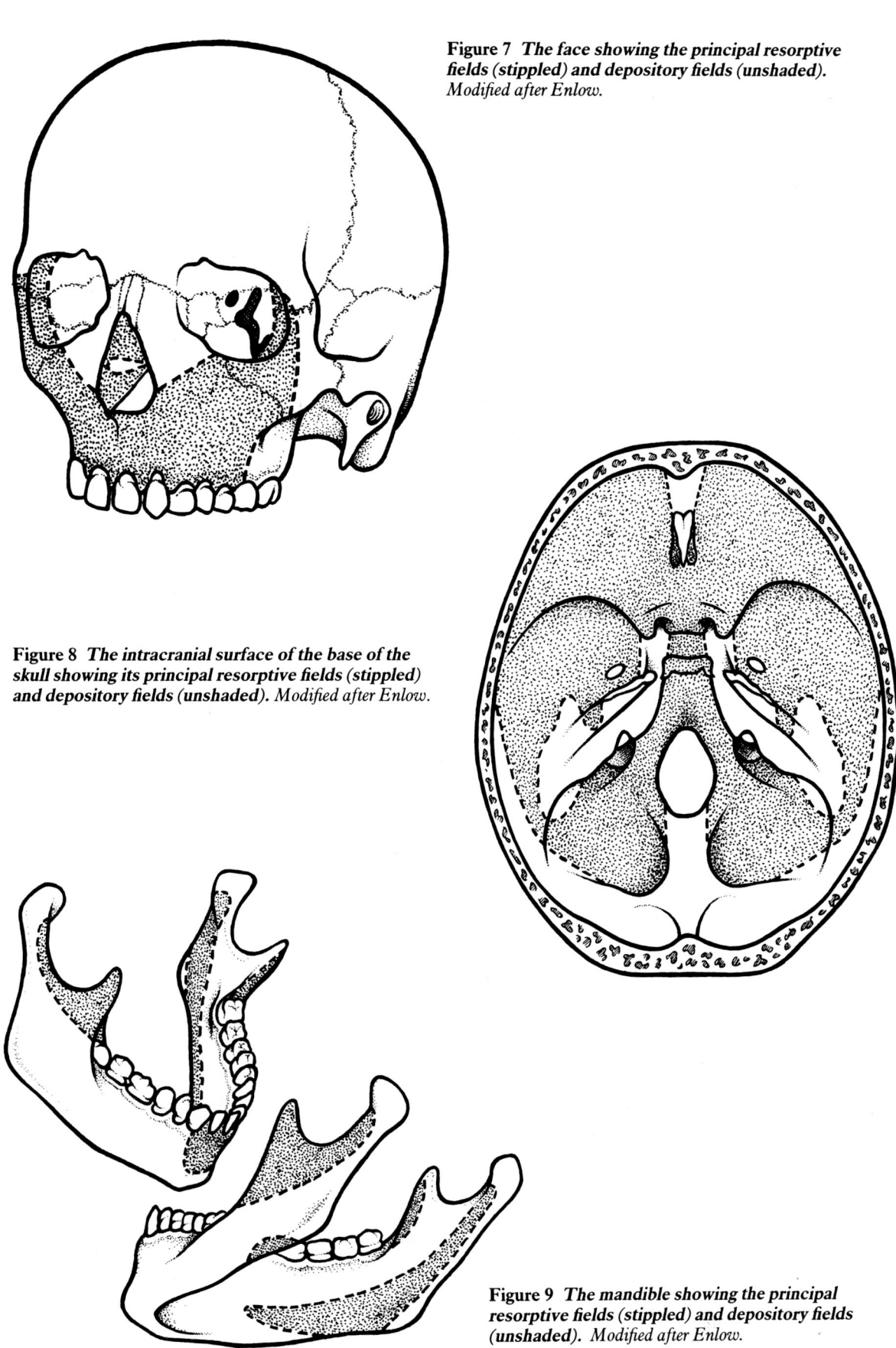

Figure 7 *The face showing the principal resorptive fields (stippled) and depository fields (unshaded).* Modified after Enlow.

Figure 8 *The intracranial surface of the base of the skull showing its principal resorptive fields (stippled) and depository fields (unshaded).* Modified after Enlow.

Figure 9 *The mandible showing the principal resorptive fields (stippled) and depository fields (unshaded).* Modified after Enlow.

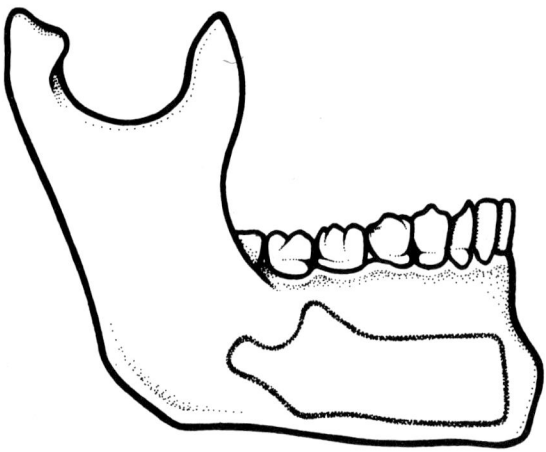

Figure 10 *Superimposed neonatal and adult mandibles.* *In the neonatal mandible, the chin is absent, the body is comparatively larger, and the angle of the mandible is more obtuse. With growth, the ramus is relocated posteriorly.*

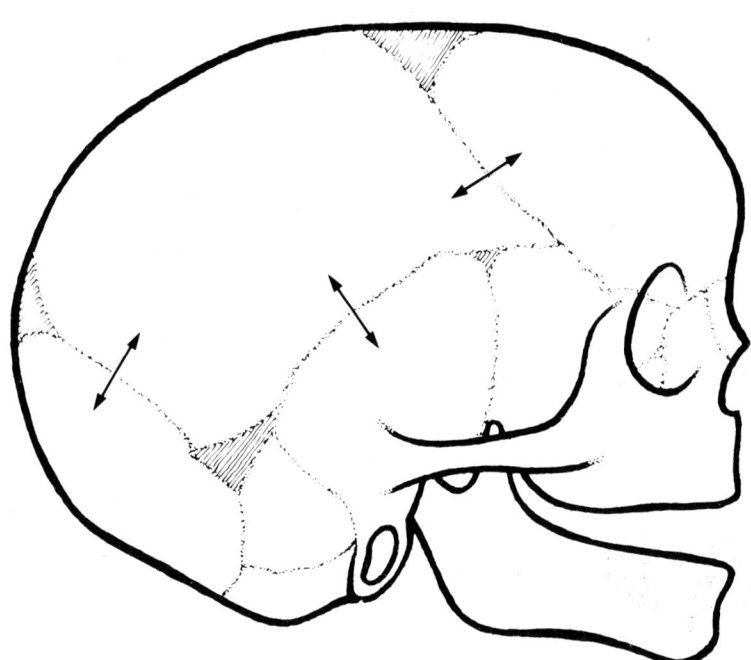

Figure 11 *The main sites and directions of sutural growth in the cranial vault of the fetal skull.* *Not shown are the frontal and sagittal sutures lying in the midline, which allow for growth in width.*

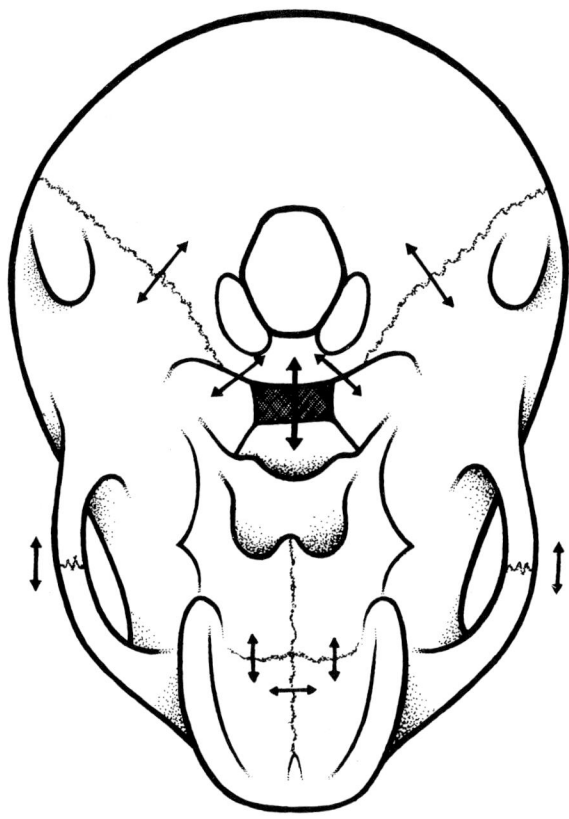

Figure 12 *The external surface of the cranial base* showing the main sites and directions of growth at the sutures. Also illustrated is the anteroposterior growth at the spheno-occipital synchondrosis (hatched). Modified after Sperber.

nous growth also occurs postnatally in the bones of the chondrocranium, which do not fully ossify until infancy (e.g. the ethmoid bone). Secondary cartilages also exist in the symphysis menti. These may contribute to the growth of the mandible but are obliterated by the age of two years.

The bones of the skull remodel at their surfaces and this involves both bone deposition and resorption. This is the predominant method by which the skull grows once sutural growth has diminished. The principal sites of this type of growth are shown in Figures 7, 8 and 9. Bone remodelling not only increases the size of a bone but can also change its shape and position. For example, because the ramus of the mandible shows deposition posteriorly and resorption anteriorly, the ramus is relocated posteriorly. The remodelling also produces a change in shape, the angle between the ramus and the body becoming less obtuse in the adult (Figure 10).

The forces responsible for growth of the skull are generated not by the bones themselves but by expansion of adjacent tissues (e.g. growth of an internal organ). The cranial vault grows primarily by sutural growth as a response to the expansion of the brain (Figure 11). Anteroposterior growth occurs at the coronal and lambdoid sutures. The vault increases in width mainly at the frontal and sagittal sutures. It increases in height along the squamosal suture. This sutural growth is accompanied by surface deposition and remodelling. The cranium reaches almost adult size at about the age of 10 years. Growth of the base of the cranium occurs by cartilaginous and sutural growth (Figure 12). Growth of the skull in an anteroposterior direction takes place at the spheno-occipital synchondrosis by cartilaginous growth. Sutural growth occurs at the occipitomastoid, palatine, and temporozygomatic sutures. In

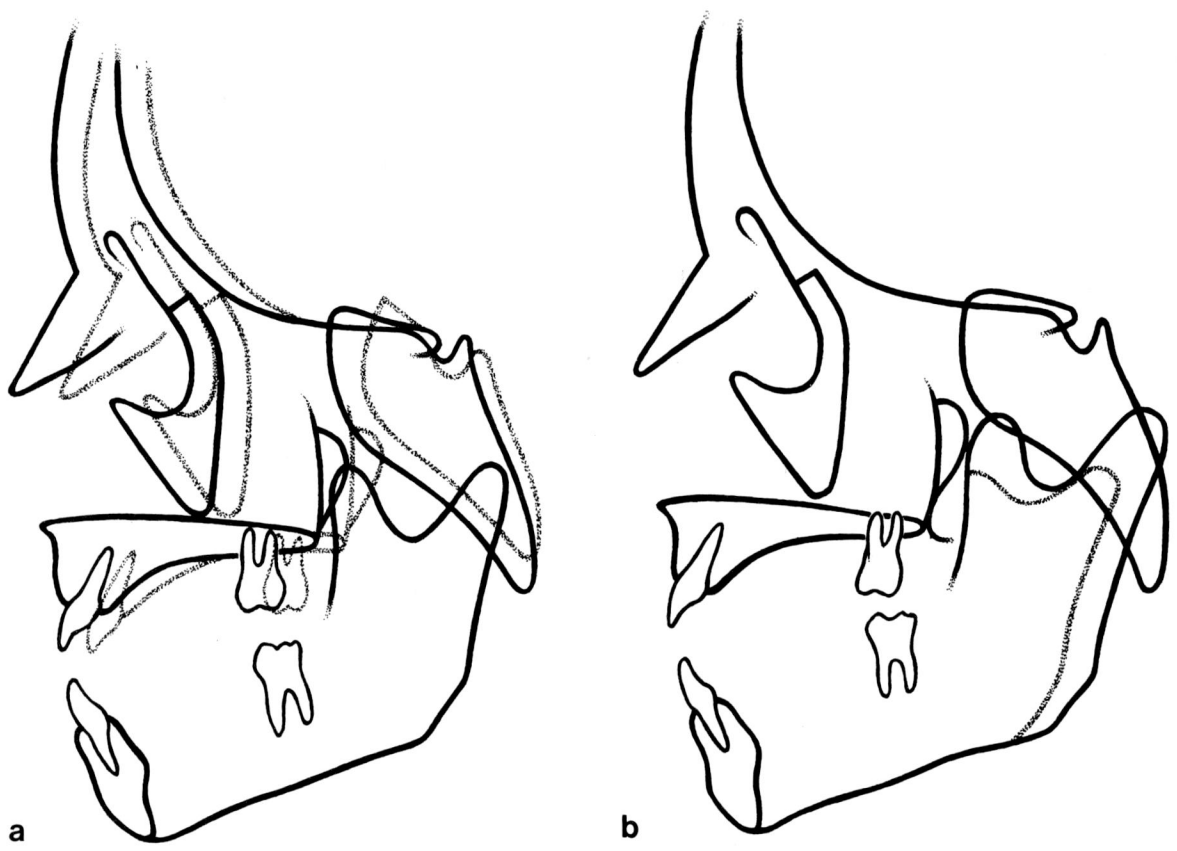

Figure 13 a) *Cephalometric tracings of the face between the ages of nine years (broken lines) and adulthood (solid lines) showing the amount of growth and the anterior direction of displacement of the nasomaxillary complex and of the anterior cranial fossa.* **b)** *Cephalometric tracings of the face between the ages of nine years (broken line) and adulthood (solid lines) showing the enlargement of the mandible by growth at the condyle and by bone deposition along the posterior border of the ramus. Modified after Enlow.*

addition, bone deposition and resorption deepen the floors of the cranial fossae (Figure 8).

Growth of the face occurs principally through sutural growth and bone remodelling. The forces responsible for sutural growth are mainly generated by growth of the temporal and frontal lobes of the brain, by expansion of the orbital contents, and by growth of the tongue. Proliferation of the cartilage of the nasal septum has also been implicated, although it has been reported that deficiencies in the septum may not affect facial growth. The direction of facial growth is shown in Figure 13. The mandible grows by means of bone remodelling, and by growth at the secondary condylar cartilages. However, there is some controversy concerning the role of the condylar cartilages. It is claimed that continued proliferation of this cartilage is primarily responsible for the increase in both mandibular length and height of the ramus. Alternatively, it has been suggested that proliferation of the condylar cartilage is a response to growth. This view is supported by experiments showing that mandibular growth is unaffected by removal of the condyle, providing normal mandibular function is maintained.

There have been several studies (including tissue culture studies) that show that bones can develop and maintain their morphology only with an adequate degree of function. Those elements that provide the functional stimulus are called functional matrices. The bony elements upon which the matrices act are called skeletal units. Even a single bone may consist of a number of contiguous skeletal units, each associated with one or more soft tissue functional matrices. For example, in the mandible the coronoid process forms a skeletal unit acted upon by the temporalis muscle. Sectioning of the temporalis during early mandibular development results in atrophy or complete absence of the coronoid process on the adult mandible. Similarly, the alveolar process is influenced by the teeth (the alveolus resorbing in the edentulous state), the condyle by the lateral pterygoid muscle, the ramus by the medial pterygoid and masseter muscles and the body of the mandible by the neurovascular bundle. Thus, because many factors can affect the development of a single bone, the development and growth of the whole skull, which comprises 28 individual bones, is understandably very complex. This complexity is compounded by the difficulty of assigning fixed points in the growing skull, as everything is moving in relation to everything else.

SUMMARY SHEET: THE FETAL SKULL

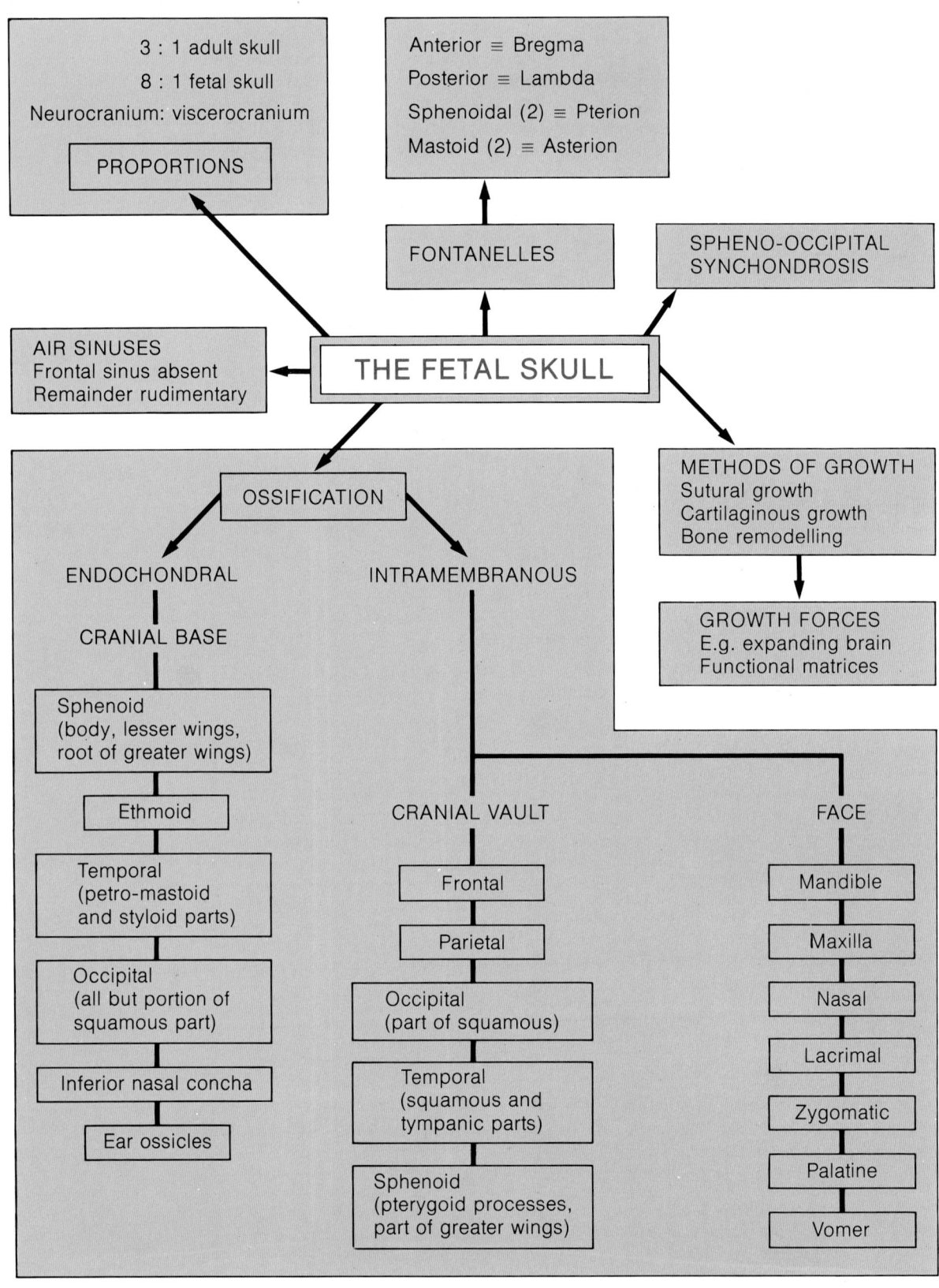

The cervical vertebrae

The vertebral column forms that part of the skeleton known to the layman as the backbone or spine. Above, it supports the skull; below, it gives attachment to the pelvic girdle, which supports the lower limbs. To its thoracic component are attached the ribs. This part of the vertebral column is also associated with the pectoral girdle and the upper limbs. In addition to its obvious role in posture and locomotion, the vertebral column surrounds, and thus protects, the spinal cord.

The vertebral column comprises 33 bones, the vertebrae. There are seven cervical, twelve thoracic, five lumbar, five sacral and four coccygeal vertebrae. Whereas the cervical, thoracic and lumbar vertebrae are separate bones, the sacral and coccygeal vertebrae are fused. Between the bodies of individual vertebrae are the intervertebral discs.

The vertebral column exhibits four curvatures when viewed from the side. The cervical and lumbar curvatures are convex anteriorly. The thoracic and sacral curvatures are concave anteriorly. In the fetus, however, the vertebral column has a single curvature which is concave anteriorly. This is referred to as the primary curvature and is present throughout life in the thoracic and sacral regions. The development of the cervical curvature is associated with the ability of the baby to hold its head upright. The lumbar curvature develops later with standing and walking.

Of the seven cervical vertebrae, the lower five appear similar, although the seventh cervical vertebra has some distinctive features. The first cervical vertebra (atlas) and the second cervical vertebra (axis) show specialisations related to the articulation of the vertebral column with the skull.

The typical cervical vertebra

The typical cervical vertebra (representing the third to the sixth cervical vertebrae) comprises two main components: a body anteriorly and a vertebral arch posteriorly. These components surround the vertebral foramen, which houses the spinal cord.

80

200B,30

The body is small compared with the body of a vertebra in other regions. It is nearly cylindrical, but is flattened anteroposteriorly. The anterior surface is convex, with depressions on either side for the attachments of the longus colli muscles. The posterior surface is flattened and forms the anterior wall of the vertebral foramen. This surface may exhibit one or more foramina related to basivertebral veins. The superior and inferior surfaces of the body are concave. The peripheries of these surfaces are smooth but centrally they are roughened. This difference relates to a difference in embryological origin (from the annular epiphysis peripherally and from the centrum centrally). The lower border of the vertebra anteriorly overlaps the upper border of the vertebra below. A lip or uncus is present at each side of the upper surface posteriorly.

80,9
80C,9
108,54
80A,12

80A,9; 80B,9

80,8

The vertebral arch can be subdivided into several components. The arch is attached to the posterolateral aspects of the body by the pedicles. These are directed backwards and outwards. The superior and inferior borders of the pedicles are notched. When adjacent vertebrae are articulated, the notches form intervertebral foramina for the passage of spinal nerves. Continuous with the pedicles are the laminae. These are thin, narrow plates of bone, which are directed backwards and medially to fuse in the midline. Associated with the vertebral arch are seven processes: four articular processes, two transverse processes, and a spine.

80,4
80E,14; 80E,15
83D,13; 200C,52
80,2

There are two superior and two inferior articular processes. They project from the junctions between the pedicles and laminae. The articular facets of the superior processes face backwards and upwards, whereas the facets of the inferior processes face downwards and forwards.

80,3; 80,13

The transverse processes are small. They project laterally and are situated just anterior to the articular processes. The most characteristic feature of the transverse process of a cervical vertebra is the presence of the foramen transversarium. This foramen is circular and transmits the vertebral vessels. The foramen subdivides the transverse process into three parts: anterior and posterior roots joined laterally by a costotransverse bar. The free extremities of the anterior and posterior roots project as the anterior and posterior tubercles. The anterior tubercle is particularly prominent in the sixth cervical vertebra. It is called the carotid tubercle, as the adjacent internal carotid artery can be compressed against it. Within the groove between the anterior and posterior tubercles lies the ventral ramus of a cervical spinal nerve. The anterior tubercle and the costotransverse bar comprise the costal element of the vertebra. Occasionally, this may be sufficiently well-developed to produce a small cervical rib.

The spine is short and projects backwards in the midline where the two laminae meet. It is bifid, and each projection terminates as a tubercle. The spines and laminae of the fifth and sixth cervical vertebrae are directed downwards as well as backwards.

The vertebral foramen cervically is triangular in outline and larger than in other regions of the vertebral column. When vertebrae are articulated to form the vertebral column, the serial vertebral foramina constitute the vertebral canal.

The seventh cervical vertebra

This vertebra has the largest spine of any cervical vertebra. The spine is directed downwards and, as it is readily palpated, the vertebra has been called the vertebra prominens. The spine is not bifid, but is thickened and terminates in a single tubercle. The transverse processes are larger than those of the other cervical vertebrae. The foramen transversarium is not circular but oval. The main vertebral vessels do not pass through this foramen. Instead, it usually transmits an accessory vertebral vein. The posterior tubercle of the transverse process is particularly prominent and gives attachment to the suprapleural membrane. If there is a cervical rib, it is almost always associated with the seventh cervical vertebra.

The first cervical vertebra

The first cervical vertebra is also called the atlas. It is the vertebra that articulates with the skull. Unlike the other vertebrae, it does not have a body, this being incorporated into the second cervical vertebra as the dens. Also there is no spine. The atlas takes the form of a thin ring of bone with anterior and posterior arches. Between the arches are situated the lateral masses. These show the articular facets and the transverse processes.

The anterior arch of the atlas is only slightly curved. There is an anterior tubercle at the midpoint of its anterior surface. On its posterior surface is an oval facet for articulation with the dens.

The posterior arch of the atlas has a pronounced curvature. There is a posterior tubercle at the midpoint of its posterior surface. The superior surface of the posterior arch shows grooves just behind the lateral masses. These grooves are related to the vertebral arteries as they wind around the lateral masses before entering the cranial cavity through the foramen magnum. The dorsal rami of the first cervical nerves also lie in these grooves.

Each lateral mass has articular facets on both the superior and inferior surfaces. The superior articular facet is oval and concave. It faces upwards and inwards to articulate with an occipital condyle. The inferior articular facet is flat. It faces downwards and inwards and articulates with the facet of a superior articular process of the second cervical vertebra. The posterior surface of the lateral mass is grooved above the level of the posterior arch by the vertebral artery as it passes backwards and inwards from the foramen transversarium. A small tubercle may be present on the inner surface of the lateral mass. This is the site of attachment of the transverse ligament of the atlas.

The transverse processes of the atlas are large and they project further laterally than all but the seventh cervical vertebra. Each process ends in a single, prominent tubercle. This tubercle is homologous with the posterior tubercle of the other cervical vertebrae. The foramen transversarium is slightly oval.

The second cervical vertebra

The second cervical vertebra is also called the axis. It can be distinguished from the other cervical vertebrae by the presence of a tooth-like process called the dens (odontoid process). The dens projects upwards in the midline from the anterior surface of the body. It articulates with the facet on the posterior surface of the anterior arch of the atlas. Rotation of the head occurs at this joint.

The body of the axis resembles the bodies of the other cervical vertebrae. The anterior overlap of the inferior border, and the depressions associated with the longus colli muscles, are especially prominent. The dens originates from the middle of the upper surface of the body of the axis. The tip of the dens is called the apex and is the site of attachment of the apical ligament. The anterior surface of the dens bears a convex facet that articulates with the facet on the anterior arch of the atlas. The lateral surfaces slope downwards and outwards and give attachment to alar ligaments. The posterior surface of the dens, near its connection with the body, is grooved by the transverse ligament of the atlas.

The pedicles and laminae of the axis are thicker than those of the other cervical vertebrae. There are no superior vertebral notches on the pedicles. The laminae meet posteriorly in a relatively broad, bifid spine. Unlike the lower cervical vertebrae, the articular processes of the axis do not lie in the same plane. The superior articular processes are located anterior to the inferior processes. The superior articular facets are larger than those of the lower cervical vertebrae and they point in a different direction, facing upwards and slightly outwards. In addition, they are situated anterior and medial to the foramina transversaria. Because of the absence of superior vertebral notches, the superior articular processes extend over the lateral surfaces of the body of the axis towards the base of the dens. The inferior articular processes are similar to those of the lower cervical vertebrae.

The transverse processes of the axis are small. Each ends laterally in a single tubercle (homologous with the posterior tubercle on a lower cervical vertebra). Because the vertebral vessels pass outwards as well as upwards to reach the foramen transversarium of the atlas, the foramen transversarium in the axis is directed upwards and outwards.

The following muscles are attached to the cervical vertebrae (from above downwards):

- Rectus capitis anterior
- Rectus capitis lateralis
- Rectus capitis posterior minor
- Obliquus capitis superior
- Scalenus medius
- Splenius cervicis
- Levator scapulae
- Intertransverse
- Rectus capitis posterior major
- Semispinalis cervicis
- Spinalis cervicis
- Semispinalis thoracis
- Multifidis
- Interspinalis
- Scalenus anterior
- Scalenus posterior
- Longus capitis

- Longus colli
- Longissimus cervicis and capitis
- Iliocostalis cervicis
- Trapezius
- Rhomboideus minor
- Serratus posterior superior
- Scalenus minimus
- Levatores costarum

Ossification. The cervical vertebrae develop endochondrally.

The third to the sixth cervical vertebrae each have three primary centres of ossification. There is a centre on each side of the vertebral arch; this appears at the tenth week of intra-uterine life. A centre for the body of the vertebra appears at the sixteenth week of intra-uterine life. The centres in the vertebral arch unite during the first year and meet with the centre in the body at about the third year. Secondary centres appear at the tips of the spinous processes, at the transverse processes and at the periphery of the superior and inferior surfaces of the body. Complete fusion of all parts occurs by the twenty-fifth year.

The seventh cervical vertebra often has additional centres of ossification for the costal elements.

The atlas ossifies from three centres. A centre appears in each lateral mass at the seventh week of intra-uterine life. The centre for the anterior arch appears at the end of the first year. A single bone is formed at about the seventh year.

The axis has the same three primary centres of ossification as the lower cervical vertebrae. There are two additional centres for the dens. These are evident at about the sixth month of intra-uterine life. A small centre is also found at the apex of the dens during the second year.

THE JOINTS OF THE CERVICAL VERTEBRAE

Many of these joints are illustrated in Figure 14.

The joints between the vertebral bodies

With the exception of the atlas, which lacks a body, the bodies of the cervical vertebrae are joined by secondary cartilaginous joints (symphyses). The intervening fibrocartilaginous discs are called the intervertebral discs. These discs increase the stability and strength of the vertebral column and act as shock absorbers. They are thicker anteriorly, conforming to the curvature of the cervical part of the spine. Each disc has a central gelatinous mass called the nucleus pulposus and a peripheral fibrous ring called the anulus fibrosus (Figure 14). The anulus fibrosus is attached to the smooth peripheral region of the vertebral body. The nucleus pulposus joins with the roughened central region, which is lined by hyaline cartilage.

The joints are strengthened by anterior and posterior longitudinal ligaments. These run in the midline along the anterior and posterior surfaces of the bodies of the vertebrae. They are attached to the intervertebral discs. The anterior longitudinal ligament is stronger and is attached above to the anterior tubercle of the atlas and to the basilar part of the occipital bone. The posterior longitudinal ligament runs inside the vertebral canal and is attached above to the body of the axis.

The joints between the vertebral arches

The articular facets of adjacent vertebrae are united by synovial joints. Concerning the ligaments associated with the vertebral arches, passing between adjacent laminae is the ligamentum flavum. This is composed primarily of elastic fibres. Whereas the tips of the spinous processes in other parts of the vertebral column are joined by

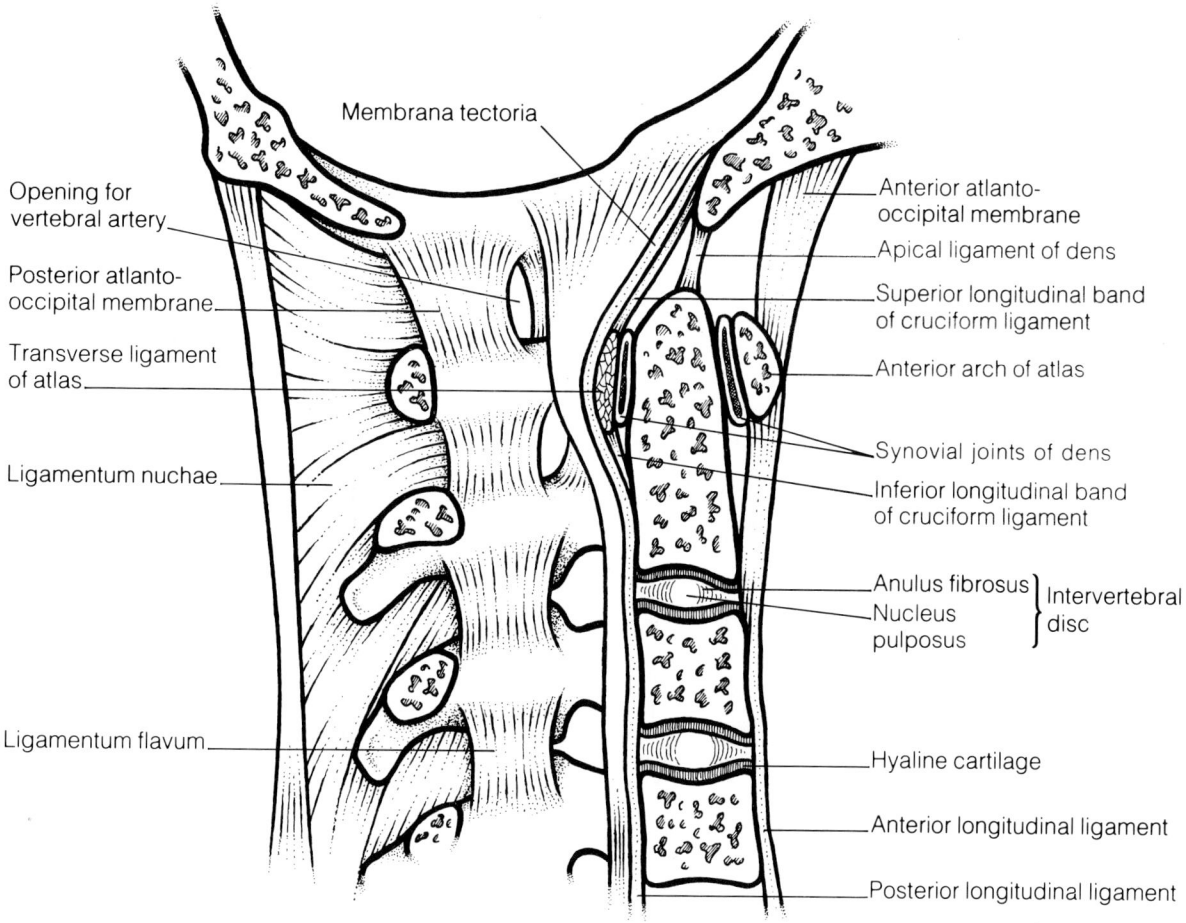

Figure 14 *Median sagittal section through the base of the skull and the upper cervical vertebrae to illustrate some of the vertebral joints and ligaments.*

interspinous and supraspinous ligaments, these are poorly developed in the cervical region and the ligamentum nuchae is present. This is a triangular fibro-elastic septum which lies between the postvertebral muscles in the midline. Its posterior border extends from the external occipital protuberance to the spine of the seventh cervical vertebra. Its anterior border is attached to the spines of the cervical vertebrae and to the posterior tubercle of the atlas. Its superior border is attached to the external occipital crest.

The atlanto-axial joints

These joints differ from the joints between the other cervical vertebrae because of the absence of a body for the atlas and the presence of the dens on the axis. There is no intervertebral disc. In front, the body of the axis is joined to the anterior tubercle of the atlas by the anterior longitudinal ligament. The inferior articular facets of the atlas articulate with the superior articular facets of the axis at the lateral atlanto-axial joints.

Another synovial joint is present in the midline between the articular facet on the anterior surface of the dens of the axis and the articular facet on the posterior surface of the anterior arch of the atlas. This joint is the median atlanto-axial joint. The dens is held in position by the transverse ligament of the atlas. This ligament is attached on either side of the dens to the small tubercles on the lateral masses of the atlas. The posterior articular surface of the dens forms a synovial joint with the cartilage-covered, anterior surface of the transverse ligament. Sometimes this joint is continuous

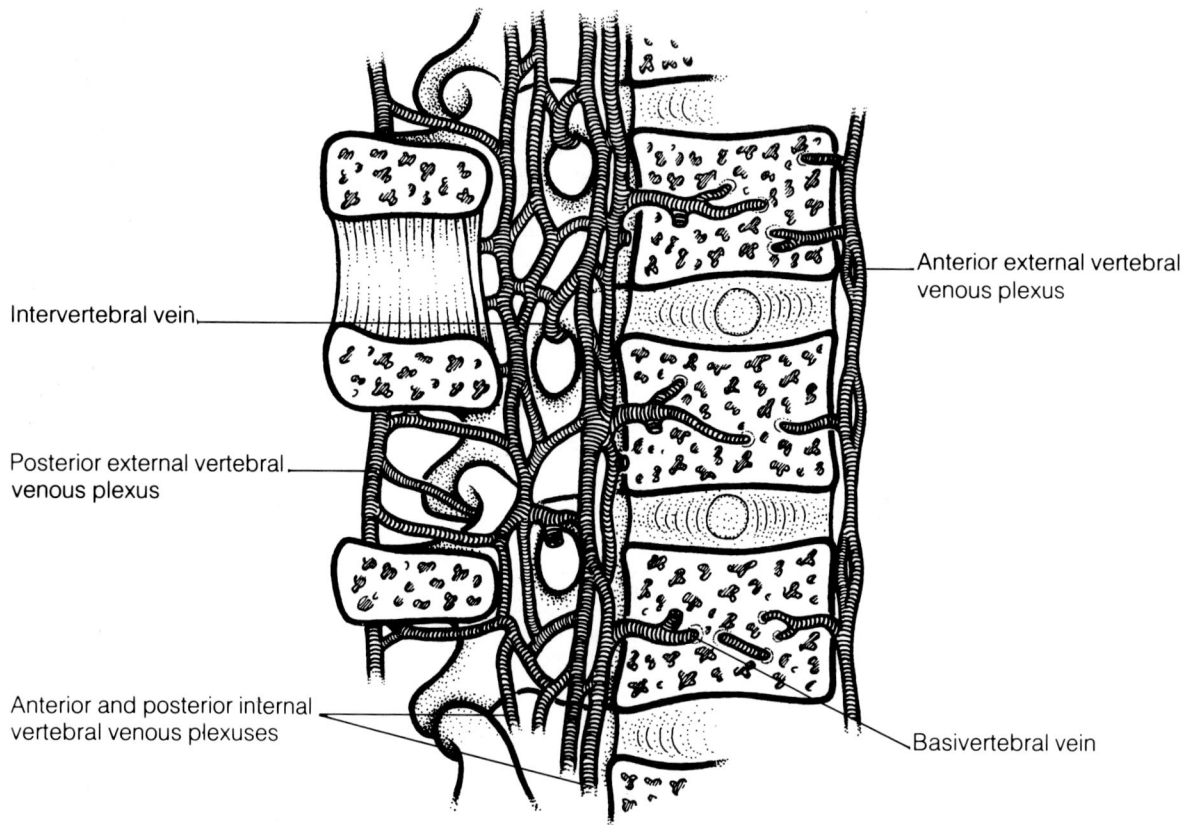

Figure 15 *The veins associated with the cervical vertebrae.*

158,40; 200B,22	with the joint cavity of one of the lateral atlanto-axial joints. Vertical extensions from the transverse ligament pass upwards and downwards in the midline. The upward extension is called the superior longitudinal band and is attached to the inner surface of the anterior margin of the foramen magnum. The downward extension is the inferior longitudinal band. This is attached to the posterior surface of the body of the axis. Because of the cross-shaped appearance of the transverse ligament and its vertical extensions, it is referred to as the cruciform ligament. Viewed posteriorly, the cruciform ligament is covered by an upwards continuation of the posterior longitudinal ligament, which is called the membrana tectoria. This membrane is attached to the inner surface of the anterior margin of the foramen magnum.
200B,27	
158,39; 200B,28	
158,41	The dens is further stabilised by three ligaments that run to the occipital bone. These are the apical ligament and the two alar ligaments. As the name indicates, the apical ligament arises from the apex of the dens. It passes vertically upwards to be attached to the anterior margin of the foramen magnum (in front of the superior longitudinal band of the cruciform ligament). Each alar ligament arises from the side of the upper part of the dens. It passes upwards and outwards to be inserted on to the medial surface of the occipital condyle.
158,37; 200B,24	

The atlanto-occipital joints

198F,73; 200B,41 158,42	The superior articular facets on the lateral masses of the atlas articulate with the occipital condyles at the atlanto-occipital joints. The upper border of the anterior arch of the atlas is connected to the anterior margin of the foramen magnum by the anterior atlanto-occipital membrane. An upward extension from the anterior longitudinal ligament of the vertebral column runs in front of the membrane to the basilar part of the occipital bone. A posterior atlanto-occipital membrane passes between the upper
158,28; 200A,19	

border of the posterior arch of the atlas and the posterior border of the foramen magnum. Between the lateral margins of the membrane and the lateral masses of the atlas pass the vertebral arteries and the first cervical nerves.

Vasculature of the cervical vertebral column. The cervical vertebrae derive their arterial supply from the vertebral arteries, with spinal branches passing through the intervertebral foramina to enter the vertebral canal.

The veins associated with the cervical vertebrae (Figure 15) are situated in two rich plexuses, one external and one internal. The plexuses freely intercommunicate, both locally and with other parts of the vertebral column, to form a series of venous rings at the level of each vertebra and particularly around the foramen magnum. The external vertebral venous plexus is located on both the anterior surface of the bodies (anterior external vertebral plexus) and on the posterior surfaces of the vertebral arches (posterior external vertebral plexus). The internal vertebral venous plexus lies between the meninges surrounding the spinal cord and the wall of the vertebral canal. It forms a complete ring around the vertebral canal and can be subdivided into a pair of anterior and a pair of posterior plexuses. Passing out from foramina on the posterior surfaces of the bodies of the vertebrae are basivertebral veins. These drain into the anterior part of the internal vertebral venous plexus. The basivertebral veins are also connected to the anterior external plexus by veins passing through the vertebral bodies. Intervertebral veins passing through the intervertebral foramina link the vertebral plexuses to veins in other regions. The vertebral veins are valveless.

The hyoid bone

152A; 152B
136,27

The hyoid bone is situated in the upper part of the front of the neck, just posterior to and a little below the inferior border of the chin. It lies between the third and fourth cervical vertebrae. The bone is horseshoe-shaped and consists of a central body spanning the midline, with greater and lesser horns on each side.

152A,3

The body of the hyoid bone is quadrilateral. It is curved so that its anterior surface appears convex when viewed from the front. A vertical median ridge is frequently present on the anterior surface. Transverse ridges may also be found. The posterior surface is smooth and concave.

152A,1

The greater horns of the hyoid bone project backwards from the lateral margins of the body. At their posterior tips, they end in tubercles.

152A,2

The lesser horns of the hyoid bone are very small. They project upwards at the junctions between the body and the greater horns. The union of the lesser horns with the rest of the hyoid bone may be osseous. In most instances, however, they are connected by fibrous tissue (occasionally by synovial joints).

152A; 152B

The hyoid bone is maintained in its position by means of the considerable number of muscles, ligaments and membranes attached to it. The muscles attached to the hyoid bone are:

- Genioglossus
- Geniohyoid
- Hyoglossus (and chondroglossus)
- Middle constrictor of pharynx
- Mylohyoid
- Omohyoid
- Sternohyoid
- Stylohyoid
- Thyrohyoid

152A,4; 142A,30
152B,14; 154C,29
152B,16
152B,15; 156C,35

The hyoid bone is suspended from the styloid processes of the temporal bones by the stylohyoid ligaments. These ligaments are attached to the lesser horns. The thyrohyoid membrane passes from the superior border of the thyroid cartilage of the larynx to the upper and posterior surface of the hyoid bone. A bursa intervenes between the membrane and the bone. There is also a hyo-epiglottic ligament which connects with the anterior surface of the epiglottis.

Ossification. The cranial part of the body of the hyoid bone and the lesser horns are derived embryologically from the cartilage of the second branchial arch. The caudal part of the body and the greater horns arise from the cartilage of the third branchial arch. The method of ossification is therefore endochondral. There are six centres of ossification, two for the body and one for each horn. The first centres appear in the greater horns just before birth. Those in the body appear at the time of birth. The centres in the lesser horns do not appear until puberty.

SUMMARY SHEET: THE BONES OF THE NECK

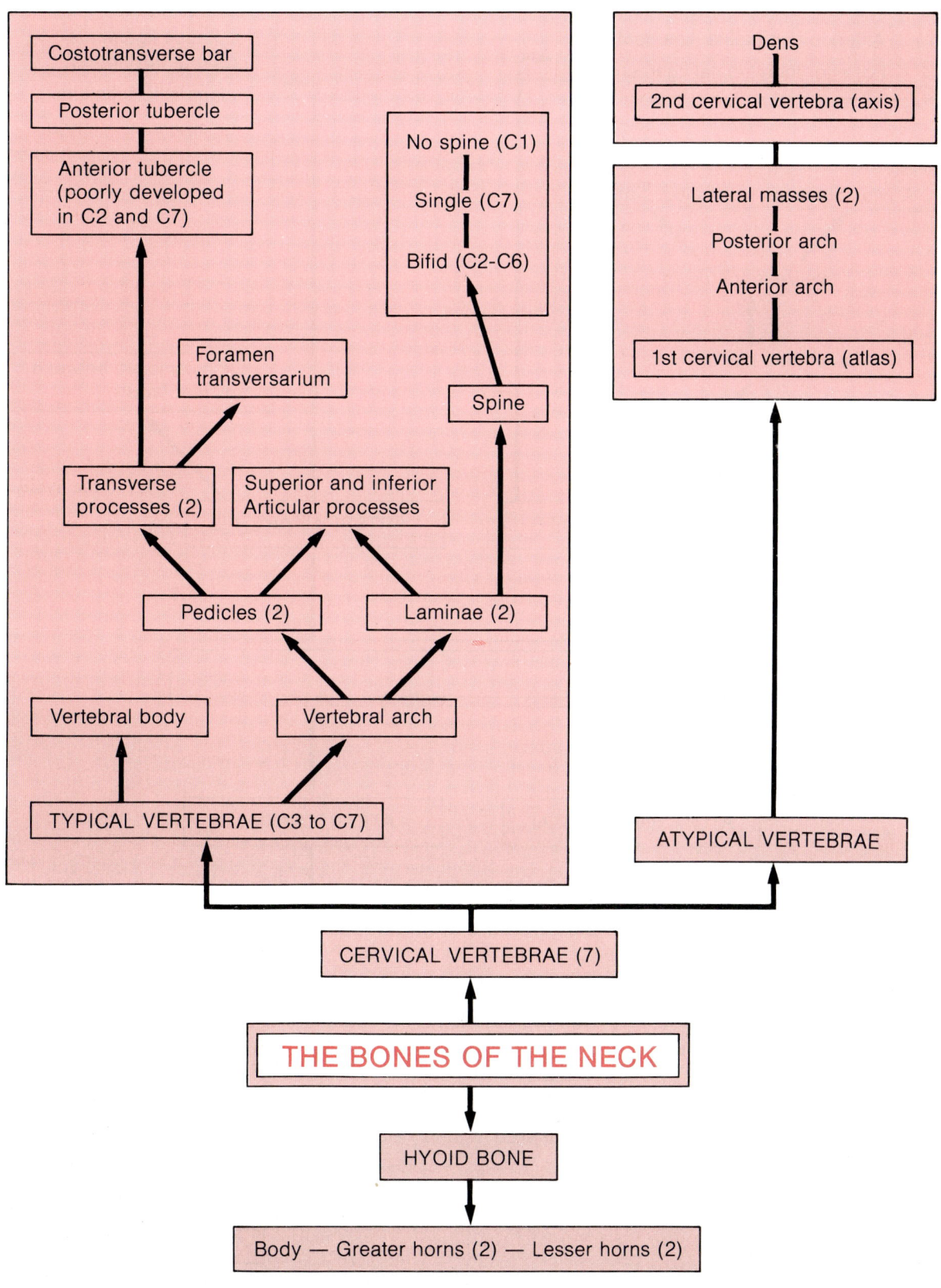

SUMMARY SHEET: JOINTS OF THE CERVICAL VERTEBRAE

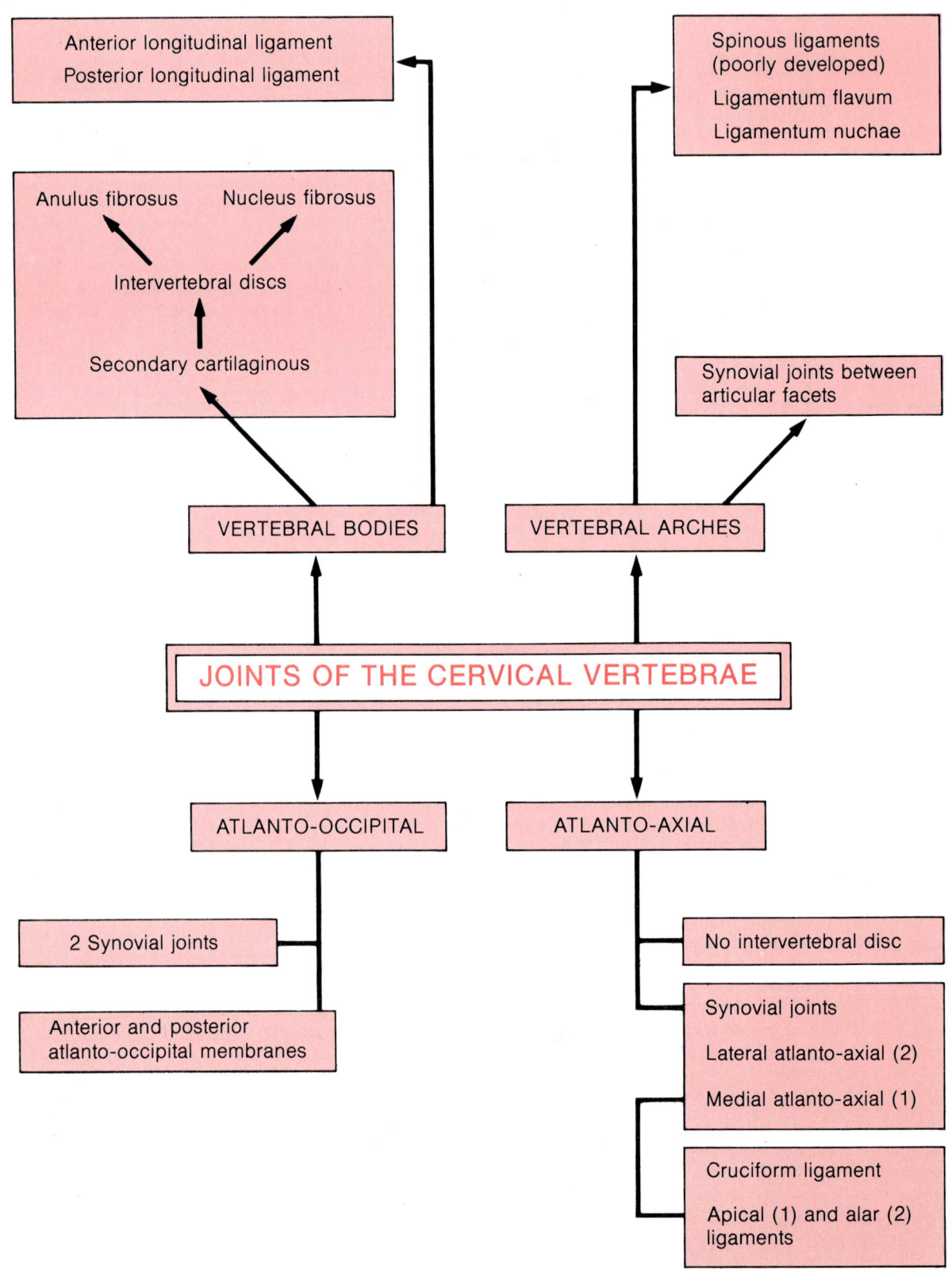

Case histories 1
The skull and cervical vertebrae

A teenager was taken to hospital having been punched on the chin during a fight. On arrival, he was conscious and aware of his surroundings. He complained of a very painful jaw and a headache. He also stated that his teeth did not fit together properly and that his lower lip was numb on the right side. Examination revealed a painful bruise on the chin, tenderness over the right angle of the jaw and blood oozing from the gum behind the right third molar tooth. The floor of the mouth was also swollen and bruised. When the patient was asked to close his jaws, a gross malocclusion of the teeth was observed and the procedure was described as painful. What has occurred to explain these findings?

The patient has a fractured mandible on the right side. This bone is the most frequently fractured facial bone. As it is curved, a blow to the lower jaw may not only produce a fracture at the point of impact but often produces further fractures, which may be unilateral or bilateral. Common sites of fracture are the coronoid and condylar processes, the alveolar process and the angle.

In the case described here, the patient has a fracture of the body of the mandible near the chin at the site of the initial blow, and a second fracture running across the ramus from just behind the third molar tooth to the angle. The bleeding behind the molar and the swelling of the floor of the mouth results from haemorrhage from the fracture sites. The malocclusion is caused by displacement of the mobile portion of the body of the mandible between the two fracture lines. The amount of displacement depends upon the pull of the muscles attached to the bone fragments and upon the direction of the fracture lines. In this case, the masseter and medial pterygoid muscles are displacing the ramus upwards, whereas the muscles around the floor of the mouth (e.g. the digastric and mylohyoid muscles) are pulling the region around the chin downwards. The numbness of the lower lip is caused by damage to the inferior alveolar nerve.

A football player presented to the accident and emergency department on a Monday morning after sustaining an injury to his face the previous Saturday. During a scramble for the ball, his left cheek hit an opponent's knee. He managed to finish the game and to enjoy the after-match celebrations. However, the subsequent pain from his cheek kept him awake at night. He also complained that occasionally he had double vision (diplopia). When he blew his nose, he felt his left eye puff up. On examination, it was noted that the patient had a depressed left cheek. There was also some redness on the lateral aspect of his left eye caused by subconjunctival haemorrhage. The lateral corner of the eye was unusually lower than the medial corner, giving the gap between the eyelids (palpebral fissure) a downward and lateral tilt. When the patient was asked to look upwards and outwards, he complained of diplopia in the affected left eye. What is the cause of the patient's symptoms?

The patient has a displaced left zygomatic bone. The fracture runs through the zygomatic arch, the zygomaticofrontal suture, and along the floor of the orbit through the suture between the zygomatic bone and the maxilla. The detached zygomatic bone is mobile and displaced downwards and inwards. This explains the depression of the cheek and the lowered lateral corner of the eye (the eyelids being attached to the zygomatic bone). The subconjunctival haemorrhage is caused by bleeding from the fracture at the zygomaticofrontal suture tracking forwards around the eyeball, but behind the conjunctiva. Diplopia when the patient is asked to look upwards and outwards indicates damage to the inferior oblique muscle of the eye. This muscle is probably trapped in the rough edges of the fractured floor of the orbit. The fracture has also involved the maxillary air sinus so that when the patient blows his nose, air passes through the floor of the orbit and causes some movement of the orbital contents.

A student fell off her bicycle and was thrown a considerable distance. She sustained serious injuries to her face, which appeared badly swollen and bruised. At the hospital, the puffiness around her eyes visibly increased while she was being examined. The patient was conscious and answered questions coherently, but was spitting out blood. In addition, swallowing and breathing were difficult. The attending medical officer placed the patient on her right side in the recovery position. He then grasped the patient's upper incisor teeth and adjacent alveolar bone and, pulling them forwards, succeeded in alleviating the breathing problems. Examination revealed numbness over the cheeks, upper lip and upper anterior teeth. The maxillae were loose and the soft palate was swollen. When the patient's molar teeth were placed in occlusion, the anterior teeth did not meet (anterior open bite). What was the main facial injury? Explain the signs and symptoms.

The patient has suffered a fracture of the face. This fracture is classified as a Le Fort II (Figure 16). Le Fort fractures are all bilateral, they extend through the pterygoid plates and the pterygopalatine fossae, and they exhibit mobility of some portion of the face. In the Le Fort II fracture, the fracture line passes through the bridge of the nose, through the medial wall of each orbit and on to the floor (involving the orbital surface,

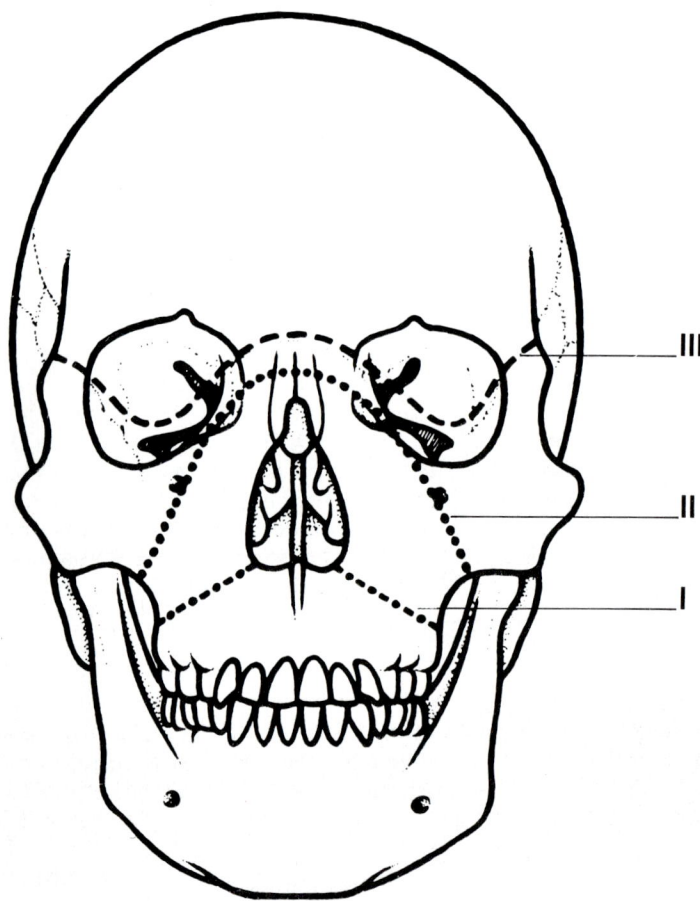

Figure 16 *The Le Fort classification of fractures of the face.*

infra-orbital canal and the zygomatic process of the maxillary bone). The fracture line passes downwards across the posterior walls of the maxillary bones and through the pterygoid processes. The maxillary sinuses are therefore involved. Thus, the complete central portion of the face is displaced, including the palate and the alveolar processes with the teeth.

The sensory loss to the cheeks, upper lip and anterior teeth results from damage to the infra-orbital branches of the maxillary nerves. The swollen soft palate is caused by the fracture line running through the pterygoid processes. This results in bleeding into the adjacent muscles of the soft palate, notably the levator and tensor veli palatini muscles.

The mobile maxillae tend to fall downwards and backwards, along the line of the fracture. This causes the difficulty in breathing and swallowing, and also the premature contact of the posterior teeth producing the anterior open bite.

Two other common fractures of the upper face are classified as Le Fort I and III (Figure 16). In the Le Fort I fracture, only the tooth-bearing parts of the maxillae are displaced, the fracture line passing through the alveolus, the base of the nasal septum and the pterygoid processes. In the Le Fort III fracture, the entire face below the nasion and the zygomaticofrontal sutures is mobile. The fracture extends backwards through the bridge of the nose, through the maxillary, lacrimal and ethmoid bones, along the superior orbital fissure, through the greater wing of the sphenoid and along the zygomaticomaxillary suture. Damage to the cribriform plate of the ethmoid bone in Le Fort II and III fractures may result in leakage of cerebrospinal fluid through the nose (rhinorrhoea) and in loss of smell because of damage to the olfactory nerves.

A construction worker sustained a bad fall on a building site. He had been unconscious for about five minutes after the accident but, by the time he reached hospital, was fully orientated and alert. On arrival at hospital, the patient was bleeding from the ear. Skull radiographs were taken but there were no obvious signs of a fracture. In view of the initial loss of consciousness, the patient was detained for observation. There was a continuous oozing of fluid from the ear. The colour of the fluid gradually changed from red to a straw colour.

The patient has suffered a fracture of the base of the skull. Diagnosis of this type of fracture is usually clinical, as the injuries are often not evident on radiographs. Continuous loss of fluid through the ear (in the absence of any other obvious cause) is pathognomonic of a basal skull fracture. This type of fracture usually involves the petrous part of the temporal bone along its longitudinal axis and is often accompanied by a tear in the tympanic membrane, visible with an auriscope. (Transverse fractures of the petrous part of the temporal bone are less common. They are not usually associated with direct damage to the tympanic membrane and are accompanied by symptoms of damage to the internal ear, such as deafness, tinnitus or vertigo.) If the bleeding from the ear is solely the result of a damaged blood vessel, it would be expected to stop quickly. Prolonged loss of fluid from the ear with the colour change is strongly indicative that cerebrospinal fluid is mixed with the blood. Indeed, cerebrospinal fluid may inhibit the coagulation of blood. Other signs that may be associated with fractures of the cranial base are facial nerve palsy, bruising around the mastoid process (Battle's sign) and around both eyes ('panda eyes').

A labourer walked too close to his workmate who was breaking a paving stone with a sledge hammer. The hammer hit him on the left side of his forehead and he fell unconscious. He was rushed to hospital but soon regained consciousness just as the doctor began his examination. The doctor observed a large bruise (haematoma) over the left side of the forehead. The patient was drowsy but fully aware of his surroundings. However, he was totally blind in his left eye. No other signs or symptoms were present in this eye. Radiographs of the skull were requested. What might they reveal?

The radiographs revealed a fracture across the base of the anterior cranial fossa. The line of the fracture coursed down the left side of the frontal bone and across the orbital plate as far as the tuberculum sellae of the sphenoid bone. The reason for the loss of sight in the left eye was damage to the optic nerve as it passes through the optic canal in the sphenoid bone. The patient was told that his sight might be permanently lost. However, sight can recover if optic nerve palsy is caused by temporary compression of the nerve resulting from swelling and/or bleeding around the optic canal.

After his car skidded on ice and overturned, a 39-year-old motorist was rushed to hospital, conscious but with fractured limbs. The man was able to talk to the hospital staff, although he was in considerable pain. While his fractures were splinted, the patient became suddenly drowsy. Further examination revealed a bruise on his left temple and weakness of his left facial musculature. A radiograph of the skull showed a fracture running vertically down the squamous part of the temporal bone (a particularly thin region) and across the base of the middle cranial fossa. Further re-examination showed his left pupil to be dilated compared with the right. The patient's condition rapidly deteriorated and he was barely conscious. An emergency computerised axial tomography (CT) scan was done. What might be happening?

The CT scan indicated a collection of blood around the temporal lobe of the left cerebral hemisphere. The skull fracture had caused the rupture of the left middle meningeal artery within the middle cranial fossa. Consequently, blood escaped between the dura and the bone to produce an extradural haematoma. Blood also passed outwards through the fracture, causing the external bruising. The extradural haematoma compressed the temporal lobe of the brain.

The facial palsy could have been produced by the pressure of the blood clot on the motor cortex of the brain. Alternatively, the fracture might have involved the floor of the middle cranial fossa and directly damaged the left facial nerve in the internal acoustic meatus or within the facial canal of the middle ear. The pupillary dilation is caused by the oculomotor nerve being pressed against the edge of the tentorium cerebelli and disrupting fibres to the sphincter pupillae. The patient was treated by a neurosurgeon, who drilled a small hole through the left temporal bone and removed the blood clot. If the pressure had not been so relieved, displacement of the brainstem at the opening of the tentorium cerebelli could have forced the crus cerebri of the opposite side against the rim of the tentorium cerebelli. This would result in a left-sided paralysis of the body (hemiplegia). Further compression of the brainstem produces decerebrate rigidity and fixed dilation of both pupils, and eventually death as the vital centres are compressed.

A husband and wife had a domestic quarrel which eventually resulted in the husband being admitted to hospital. He had been knocked out by a blow from a blunt instrument. He soon regained consciousness and was observed by the attending casualty officer to have a ragged laceration over the right occipital region. Beneath this was a large, bruised swelling. A quick neurological examination revealed no gross abnormality. Radiographically, however, a fracture of the right side of the occipital bone was evident, with a fragment of bone 2 cm by 3 cm being depressed by 1.5 cm. The casualty officer immediately re-examined the patient and found a defect in the patient's vision called a left homonymous hemianopia. This is a loss of the left visual field in both eyes. How would you account for the problems with vision?

The depressed skull fracture of the occipital bone can cause pressure on the occipital lobe of the cerebral hemisphere, where the visual cortex is situated. In this case, the right occipital lobe is affected, producing the visual problems. The right visual cortex deals with information from the left halves of the visual fields of both eyes (see page 550).

A 60-year-old accountant was just about to descend a ladder after trimming the ivy on the outside of his house when his foot slipped and he fell nearly 4 m to the ground, hitting the back of his head against a concrete path. According to his wife, he was not rendered unconscious. He did not seek medical advice, but rested quietly at home for three days until the initial headache, dizziness and nausea wore off. He then returned to work feeling much better. Five weeks later, he was urged by his employers to see his doctor, his performance at work having deteriorated. He was now slow and apathetic, although nothing more specific could be stated. Indeed, the doctor could find nothing wrong with the man and referred the patient to a consultant neurologist. A CT scan was undertaken: this showed a large collection of blood over the left cerebral hemisphere of the brain. The patient was immediately operated on and 80 ml of blood was aspirated from his subdural space.

This man had sustained a chronic subdural haematoma secondary to a head injury. The injury caused the brain to jolt quickly, rupturing several of the small superior cerebral veins passing through the potential subdural space (between the arachnoid and dura of the meninges). The subdural space increases with age, which makes the vessels more prone to injury. The damaged veins bleed very slowly, allowing blood to collect in the subdural space. The symptoms of apathy, etc., are a consequence of pressure on the cerebral cortex.

As a 50-year-old female patient walked into his surgery, the doctor took one look at her and began to write "Skull radiograph please" on a request form. The patient explained that many years ago her husband had suggested she visit her doctor as he thought she needed a tonic and seemed to be ageing prematurely. However, she hated bothering doctors and had not followed his advice until now. On examining the facial features, the doctor noted that she had a protruding lower jaw, a large nose, prominent beetling brows and an enlarged tongue. Such features were not evident 10 years ago as judged from photographs. The patient's hands were large and coarse and her feet were bursting from her sensible walking shoes. From such signs the doctor immediately suspected the patient had acromegaly (an over-production of growth hormone) caused by a pituitary tumour. Why did the doctor want a skull radiograph taken, and what might be the purpose of examining the patient's eyes?

The doctor requested the radiograph to see if there was any enlargement of the pituitary fossa, which may be caused by a tumour of the pituitary gland. Enlargement of the tumour may depress the floor of the sella turcica (giving the appearance of a double floor) and narrow the sphenoidal air sinus. Lateral pressure may also lead to erosion of the clinoid processes.

Examination of the eyes may reveal a loss of the temporal sides of the visual fields of both eyes (bitemporal hemianopia). This is caused by the tumour pressing against the optic chiasma, preventing conduction in the nerve fibres from both nasal regions of the retina (which cross over in the chiasma).

As the underlying pathological condition first appeared after the patient's epiphyses had fused, the excess of growth hormone did not increase the length of her long bones, but caused enlargement of her jaw (possibly by reactivating the condylar cartilages), frontal sinus and distal phalanges. The thickening of the subcutaneous tissues of the scalp, lips, tongue, face, hands and feet is caused by the accumulation of ground substance.

A 50-year-old engineer had worked for many years in a noisy factory where ear protectors were mandatory. However, during a routine hearing test undertaken by the company's medical officer, he mentioned that he had recently noticed he could not hear as well in his right ear. The medical officer first confirmed that the patient had indeed worn the ear protectors; he then performed hearing tests with a tuning

fork and found that the engineer had a profound neural deafness in his right ear. The company's health centre was well equipped and skull radiographs were taken. The radiographs showed that the internal acoustic meatus was enlarged on the right side. Examination of the patient's face revealed a slight palsy on the right side. What is the anatomical relationship that links the deafness to the facial palsy?

An enlarged internal acoustic meatus is often caused by a benign tumour attached to the vestibulocochlear nerve, called an acoustic neuroma (or neurofibroma). The tumour arises from the Schwann cells of the myelin sheaths. The neurological sequelae to an acoustic neuroma are: initially decreased vestibular and labyrinthine function caused by pressure (though this is often unnoticed by the patient), gradual deafness, and sometimes a facial palsy as the facial nerve, which runs in the meatus with the vestibulocochlear nerve, may be stretched over the tumour.

A 42-year-old professor of chemistry visited his doctor complaining that he was losing his sense of smell (anosmia). He had been aware for some time that he was losing his appreciation of the smell of food and wine. Furthermore, he claimed that at work the chemicals did not seem to smell as strongly. The final episode that drove him to seek medical help was when one of the porters complained that the professor had filled the laboratory with ammonia without being aware of it. The doctor first confirmed that there had not been a recent history of a cold or of trauma. Examination showed that the rest of the patient's cranial nerves were functioning normally. The doctor arranged for skull radiographs to be taken. They revealed a localised radio-opaque mass at the cribriform plate in the anterior cranial fossa.

The radio-opaque mass was diagnosed as a calcified meningioma. This is an intracranial tumour that arises from the arachnoid layer of the meninges, but which gains a very firm attachment to the bone from the dura. The professor's loss of sense of smell was caused by the meningioma pressing against the olfactory nerves. A tumour in this region can also affect the optic or oculomotor nerves. The most common site for this tumour is in a parasagittal position, where pressure can cause a progressive hemiplegia by damage to the motor cortex.

A 75-year-old pensioner from Lancashire (England) visited his doctor and complained that he was getting progressively deaf. The doctor could not help noticing that the hat the patient was wearing seemed to be far too small for him. The patient assured the doctor that the hat had fitted him a few years ago when it was bought. Radiographs of the patient's skull were taken and showed a thickening of the bones of the cranial vault and of the walls of the maxillary sinuses. There was also a patchy reduction in bone density.

The radiographic findings indicate that the patient has Paget's disease. This is a benign condition of bone resulting in excessive bone formation. When it affects the skull, it results in an enlargement of the vault (which is why the hat did not fit) and increased thickness of the walls of the maxillary sinuses. The patient's deafness could also be related to the disease, as excessive bone formation may have constricted the internal acoustic meatus and caused nerve deafness. Tinnitus may also be a symptom. Alternatively, there may be conductive deafness resulting from involvement of the malleus or incus. Although the facial nerve also occupies the internal acoustic meatus, it is rarely involved in Paget's disease. Lancashire is the world epicentre for Paget's disease, which may be caused by a latent virus infection. It is associated with an increased incidence of osteosarcoma (a malignant bone tumour).

Achondroplasia is a genetic disorder characterised by a deficiency of growth in primary cartilage. Affected individuals have very short limbs because the epiphyseal growth plates fuse much earlier than normal. Would the condition be expected to produce any changes in the skull?

The bones of the cranial vault and the mandible develop intramembranously and will be unaffected in achondroplasia. However, the most important site for cartilaginous growth in a child's skull is the spheno-occipital synchondrosis. In achondroplastic dwarfs, the synchondrosis fuses early. The result will be a lack of forward growth of the upper part of the face, giving a flattened, moon-shaped appearance to the facial profile.

A young mother had a long and difficult labour which necessitated a forceps delivery. She became rather concerned when she noticed a few days later that the right side of her baby's face appeared to droop and that the baby was unable to close her right eye properly and dribbled constantly from the right side of her mouth. What do you think has happened to give rise to these signs?

Occasionally, during a forceps delivery, the forceps become incorrectly positioned around the baby's head. The blade of the forceps may then compress the facial nerve as it emerges on to the face from the stylomastoid foramen. In the newborn, the nerve is particularly vulnerable at this site because, as the mastoid process has not yet developed, it lies in a superficial position. Damage to the facial nerve as it appears on the face leads to paralysis of the muscles of facial expression on that side.

During childbirth, the obstetrician carries out a vaginal examination to determine the orientation of the baby's head. What anatomical features on the baby's head help him to do this?

The anterior and posterior fontanelles are palpable in the midline of the cranial vault. They can be distinguished from each other during palpation by their overall shape and by the number of sutures leading away from them. The larger anterior fontanelle is diamond-shaped and has four sutures leading away from it (frontal, sagittal and two coronal sutures). The posterior fontanelle is triangular and has only three sutures leading away from it (sagittal and two lambdoid sutures). In the best position for delivery (vertex presentation), both fontanelles should be palpable, with the anterior fontanelle in front. Reversal of these positions indicates rotation of the baby within the uterus and this may result in a more difficult labour.

The anterior fontanelle of the neonate is compressible. This property is of clinical significance, as is illustrated in the following case. A woman went into labour four weeks prematurely. This was her fifth child. Events proceeded so rapidly that the baby was born on the way to hospital without a midwife in attendance. The mother stated that the baby fell a short distance on to the floor of the ambulance. Two weeks later, a doctor was called to see the baby who was unconscious and unresponsive. Palpation revealed a tense anterior fontanelle.

The combination of a rapid delivery and trauma can lead to the rupture of subependymal vessels beneath the lateral ventricles of the brain. The intracerebral haemorrhage blocks the normal flow of cerebrospinal fluid. The build-up of pressure within the cranium is responsible for the tense anterior fontanelle (and may also lead to an increase in the size of the cranium). This is a form of hydrocephaly.

Compression of the brain is responsible for the deterioration in the baby's condition. The condition can be alleviated by draining some of the cerebrospinal fluid. The procedure involves inserting a needle through the lateral aspect of the anterior fontanelle (to avoid the superior sagittal sinus in the midline) and into the lateral ventricle.

A rugby football player injured his neck when a scrum collapsed. As he fell, his neck was hyperflexed and took his full weight and that of his team mates pushing behind. He left the field of play and was taken to an accident and emergency department. He complained of a severe pain about halfway down his neck. All movements of his neck were painful. He also said that he had 'pins and needles' in his left hand. On examination, he had very little movement in his neck because of pain and muscle spasm. It was very tender over the spinous process in the region of the fifth cervical vertebra. His left arm showed no weakness in any myotome, but there was decreased sensation with pin-pricks and light touch along the radial (thumb) border of his forearm and hand. No other abnormality was found.

Radiographs showed that he had sustained a stable anterior wedge fracture of the vertebral body of the fifth cervical vertebra, the body being crushed to a wedge shape with the thin portion anteriorly. The paraesthesia and decreased sensation in his arm were all in the dermatome of the sixth cervical nerve. This suggests that some of the swelling around the crush fracture was irritating the sixth cervical nerve root as it leaves the vertebral column between the neural arches of the fifth and sixth cervical vertebrae. He was treated with a supportive collar and eventually made a full recovery.

A labourer was walking underneath some scaffolding when a concrete block fell on to his head. Fortunately, he was wearing a hard hat and was not knocked out. He did, however, have a very painful neck. When he got to hospital, he was seen supporting the weight of his head in his hands. He told the casualty officer that if he did not hold his head up then the pain in his neck extended up the back of his head. On examination, the patient would not move his head at all because of pain. There was some sensory loss, which was confined to an area over his occiput. There was no motor dysfunction. The consulting doctor thought that the patient had fractured his atlas. Why?

When the concrete block hit this man's head, the severe downward pressure of the occipital condyles on the articular facets of the atlas caused it to burst apart and to break into four pieces (a quadripartite fracture). The four pieces are: the anterior arch, the posterior arch, and the two lateral masses bearing the articular facets for the occipital condyles. The second cervical nerve divides into ventral and dorsal rami just behind the atlanto-axial joint. The dorsal ramus gives off the greater occipital nerve, which passes between the atlas and the axis to supply the skin of the occiput up to the vertex. The first cervical nerve has no cutaneous component. With fracture of the atlas, the greater occipital nerve is compressed, thereby giving severe referred pain over the occiput. Only 50% of patients with this injury survive without significant neurological involvement.

A youth riding a moped through the countryside lost control and went through a hedge. Unfortunately, a branch from a tree caught him just below the chin. A few minutes later, a passing motorist saw him in the hedge unconscious and went to his aid. He felt the motorcyclist's pulse and, finding him alive, pulled him from the hedge. The helper then went back to the road and flagged down a passing car to get help. When he returned to the scene of the accident, the youth was dead.

The motorcyclist had sustained an unstable fracture of his cervical spine. When he was moved, this action transected (or completed the transection) of his spinal cord, thereby killing him. The golden rule is that, if a cervical injury cannot be excluded, the patient should not be moved until proper equipment and expertise is at hand. The postmortem showed that the young man's cord had been transected at the level of second/third cervical vertebra. Cutting the cord at this level paralyses the body from the neck down; i.e. arms, legs and intercostal muscles, etc. Most importantly, the phrenic nerve (root value C3,4,5) is damaged and all respiratory movements cease. Radiographs showed that the pedicles of the second cervical vertebra had been fractured. This, along with disruption of the anterior and posterior longitudinal ligaments, had allowed the body of the second cervical vertebra to move with respect to the third vertebra, thus crushing the spinal cord. This particular injury, simultaneous extension and distraction of the neck causing fracture of the pedicles of the second cervical vertebra and cord transection, is the mechanism of death by hanging.

An elderly man visited his doctor complaining of a small, gradually enlarging bony lump on the top of his forehead. On taking a general history from the patient, the doctor found that the patient had progressively increasing difficulty passing urine, and that on occasions he had passed blood. Subsequent tests revealed that the patient had a malignant tumour (carcinoma) involving the prostate gland. Can you relate this diagnosis to the presence of the lump on his forehead?

The lump on the patient's forehead was a secondary deposit from the primary lesion in the prostate. The plexus of veins around the vertebrae (the vertebral valveless veins of Batson) form a link between the pelvic veins and the intracranial venous sinuses. This communication makes possible the spread of malignancy and infection between these distant sites.

A middle-aged lady, severely deformed by rheumatoid arthritis, attended her dentist complaining of toothache. The dentist decided to extract a tooth and the patient requested that this be done under general rather than local anaesthesia. The dentist agreed to this. During the ensuing operation, the tooth proved difficult to extract. After considerable effort, which resulted in movement of the patient's head, the tooth was finally removed. The patient died shortly afterwards without recovering from the anaesthetic.

Rheumatoid arthritis is a generalised disease characterised by, among other things, a weakening of ligaments and subsequent joint deformity. In this case, the cruciform ligament of the dens was weakened, and rough manipulation of the head during tooth extraction resulted in subluxation of the dens. The medulla oblongata immediately posterior to the dens was irreversibly damaged. Prevention of this accident involves previous clinical examination and investigation.

Clinical examination requires investigation of the ophthalmic division of the trigeminal nerve. The fibres of this nerve descend in the lateral part of the medulla oblongata to reach the lower extremities of the nucleus of the spinal tract of the trigeminal nerve (see page 480). Minor misalignment of the dens may therefore first disrupt these fibres. This will be characterised by loss of pin-prick sensation over the cutaneous distribution of the ophthalmic nerve. An accurate index of this is the corneal reflex. Touching the cornea with a fine wisp of cotton wool normally results in rapid blinking of both eyes, because of initial stimulation of the ophthalmic nerve. Loss of this reflex in a rheumatic patient is indicative of a weakened cruciform ligament.

Clinical investigation requires lateral radiographs of the cervical spine, preferably with the head flexed and extended. This will indicate the stability of the dens. However, this clinical investigation is not without risk.

A middle-aged bank manageress visited her doctor's surgery complaining of pain in her right arm. She had noticed initially that, while working in the garden, she occasionally felt an ache when having to exert her strength. However, her present reason for seeking medical advice was that she was unable to paint a ceiling in her home using her right hand without feeling excruciating pain in her arm within a few minutes of starting. She also noticed that her hand was pale and her fingers felt numb after attempting to paint. On examination, the doctor found that the patient's right arm was indeed paler and colder than the left. Her brachial and radial pulses were normal in the left arm but weak and feeble in the right. The doctor ordered a radiograph of her cervical spine. What might be the purpose of this?

The doctor suspected that the patient might be suffering from symptoms associated with a cervical rib. A cervical rib occurs in approximately 0.5% of people. It may be unilateral or bilateral, and is usually asymptomatic. In this case, the patient's cervical rib extended from the seventh right cervical vertebra to the first rib just before it joined the manubrium. Instead of running over the first rib, the right subclavian artery has to pass upwards over the higher cervical rib. This has produced a kink in the vessel, making it stenotic and interfering with normal blood flow. The pain in the woman's arm is ischaemic pain from the muscles, which is exacerbated by exercise. Treatment may necessitate excision of the cervical rib.

A 57-year-old nurse reported to her doctor with a history of neck pain over the past six months. She complained that the pain extended over her left shoulder and all the way down the outside of her left arm. The pain was continuous but fluctuated in intensity. She was also having problems in her job because of the difficulty of lifting objects with her left hand. On examination, the patient was found to have a stiff neck with tenderness over the lower third of the cervical spine. Pain was felt at the limit of all movements. The patient had a firm grip, but the biceps muscle was noticeably weak in the left upper arm, the circumference here being 3 cm less than that of the right upper arm. There was a diminished left biceps reflex jerk. In addition, there was markedly diminished sensation along the radial border of the left forearm. Radiographs of her cervical spine showed cervical spondylosis, a condition characterised by degeneration of the intervertebral discs and by the formation of bony outgrowths (osteophytes) running along the anterior surface of the vertebral canal. How might this condition be related to the patient's symptoms?

The patient's pain, the paraesthesia, the decreased power, and the diminished biceps reflex are the result of compression of a spinal nerve root as it passes out of the vertebral canal through an intervertebral foramen. Such compression is caused by osteophytes growing from the body of a vertebra into the intervertebral foramen. The area of numbness in this patient is related to the dermatome supplied by the sixth cervical spinal nerve. The biceps muscle is supplied by the fifth and sixth cervical nerves. This indicates that the sixth cervical nerve is being compressed in the intervertebral foramen between the fifth and sixth cervical vertebrae.

The neck

THE GENERAL ARRANGEMENT OF STRUCTURES IN THE NECK

The neck extends from the base of the cranium and the inferior border of the mandible to the thoracic inlet.

A typical transverse section through the neck is shown in Figure 17.

Superficially, the neck is surrounded by a sleeve of fascia known as the investing layer of deep cervical fascia. The fascia encloses the trapezius and sternocleidomastoid muscles. At the front of the neck, the platysma muscles lie between the fascia and the skin; the infrahyoid muscles (strap muscles) lie behind the fascia.

Posteriorly, there is a musculoskeletal compartment that consists of the vertebral column and the muscles immediately surrounding it. This is enclosed by a sleeve of fascia that is prominent in front of the vertebral bodies and is therefore called the prevertebral fascia.

There is also a visceral compartment in the neck which, depending on the level, contains the pharynx, the oesophagus, the larynx and the trachea. This compartment is enclosed by fascia which, because it is prominent in front of the trachea, is called the pretracheal fascia. The thyroid gland lies on the trachea at the root of the neck.

On each side of the neck, beneath the sternocleidomastoid muscle, lies a neuro-vascular compartment — the carotid sheath. Within the sheath lie the common carotid and internal carotid arteries, the internal jugular vein, and the vagus nerve. Detailed information concerning the fascia of the neck is given on pages 93 to 99.

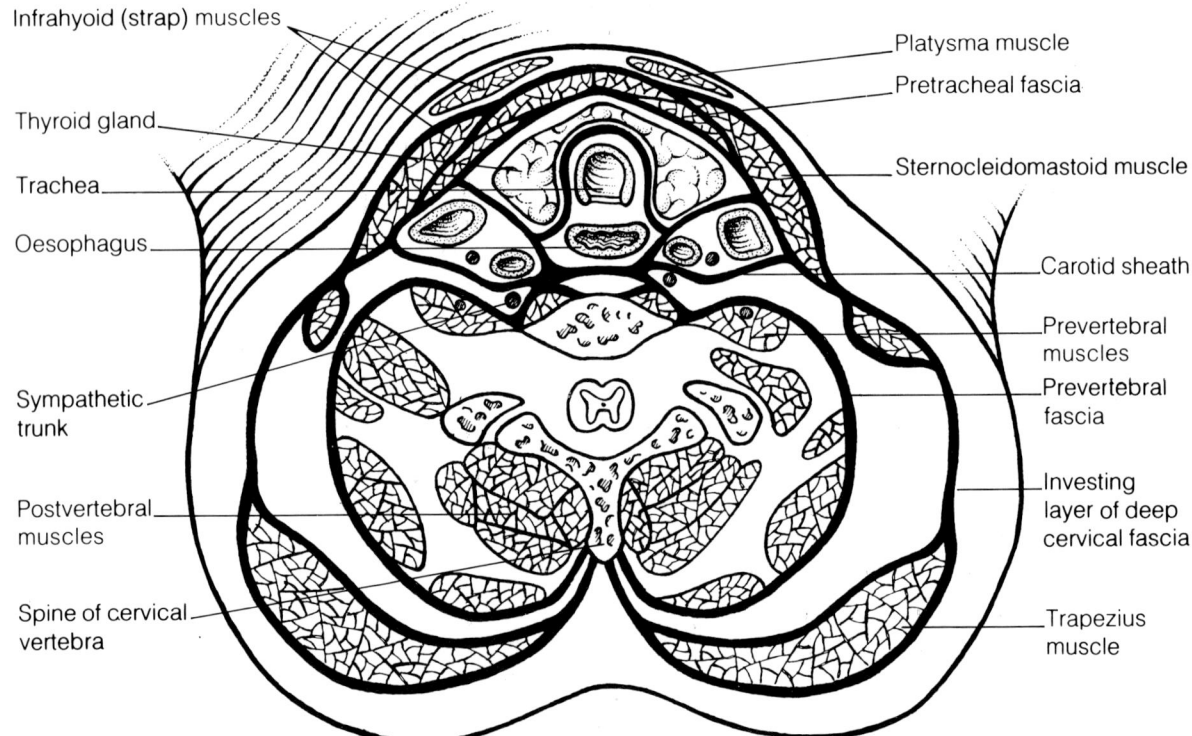

Figure 17 *Transverse section of the neck at the level of the seventh cervical vertebra to indicate the general arrangement of cervical structures.*

SURFACE MARKINGS OF THE NECK

Superiorly, ridges indicating the mastoid process and the mandible (the angle and inferior border) can be clearly seen. At the back of the head in the midline lies a prominence called the external occipital protuberance. Below and in front of the mastoid process, the tip of the transverse process of the atlas can be palpated using deep pressure. Behind the angle of the mandible lies a point indicating the lowest part of the parotid gland. The submandibular gland is situated below the inferior border of the mandible and above the hyoid bone.

Inferiorly are ridges indicating the manubrium of the sternum, the clavicle and the scapula. A depression immediately above the superior border of the manubrium (the jugular notch) is called the suprasternal fossa. It is bounded on each side by prominent ridges produced by the sternal heads of the sternocleidomastoid muscles. A supraclavicular fossa lies above the middle third of the clavicle. It is bounded medially by the clavicular head of the sternocleidomastoid muscle, and behind and laterally by the lateral margin of the trapezius muscle. If the shoulders are shrugged, the margin produced by the trapezius becomes prominent. Within the supraclavicular fossa are landmarks indicating the inferior belly of the omohyoid muscle and the upper trunk of the brachial plexus. Between the sternal and clavicular heads of the sternocleidomastoid muscle lies a depression that is sometimes referred to as the lesser supraclavicular fossa. This fossa marks the site of the sternoclavicular joint and the union between the internal jugular and subclavian veins to form the brachiocephalic vein.

Anteriorly, the thyroid cartilage of the larynx presents a laryngeal prominence ('Adam's apple') which is most conspicuous in the adult male. The upper border of the thyroid cartilage is at the level of the lower part of the fourth cervical vertebra in males, but is slightly higher in females and in children. This is the level at which the common carotid artery bifurcates. The arch of the cricoid cartilage of the larynx can be palpated at the level of the sixth cervical vertebra. At about this level, the recurrent laryngeal nerve enters the larynx, the vertebral artery enters the foramen transversarium of the sixth cervical vertebra, the cervical sympathetic trunk presents the middle cervical ganglion, the pharynx becomes the oesophagus, and the larynx leads into the trachea. The trachea passes down from the cricoid cartilage and enters the thorax behind the jugular notch of the manubrium. The thyroid gland is closely related to the trachea in the root of the neck, the isthmus joining the two lobes of the thyroid lying 1.5 cm below the cricoid cartilage. The hyoid bone is situated above the thyroid cartilage of the larynx and below the inferior border of the mandible, at the level of the upper part of the fourth cervical vertebra. Its horseshoe shape is easily discerned, the body lying in the midline and the two greater horns extending laterally towards the anterior borders of the sternocleidomastoid muscles below the angle of the mandible. At about this level, the lingual and occipital arteries arise from the external carotid artery.

Laterally, the sternocleidomastoid muscle can be seen running from the manubrium sterni and the clavicle up to the mastoid process. It is particularly prominent when it is activated by turning the head to the opposite side against resistance. The muscle is crossed superficially by the external jugular vein. The course of the vein is indicated by a line running from a point 2.5 cm below and behind the angle of the mandible to a point on the clavicle lateral to the attachment of the sternocleidomastoid muscle. The carotid pulse can be felt below the angle of the mandible and in front of the anterior border of the sternocleidomastoid muscle.

Posteriorly, the spines of the upper thoracic vertebrae and the seventh cervical vertebra are prominent. The remaining cervical vertebrae are indistinct, the spines being covered by the ligamentum nuchae which separates the postvertebral musculature on both sides. A groove runs down the midline of the neck. This is particularly marked in the young. The upper limit of the neck muscles is indicated by the external occipital protuberance and the superior nuchal lines on the back of the skull.

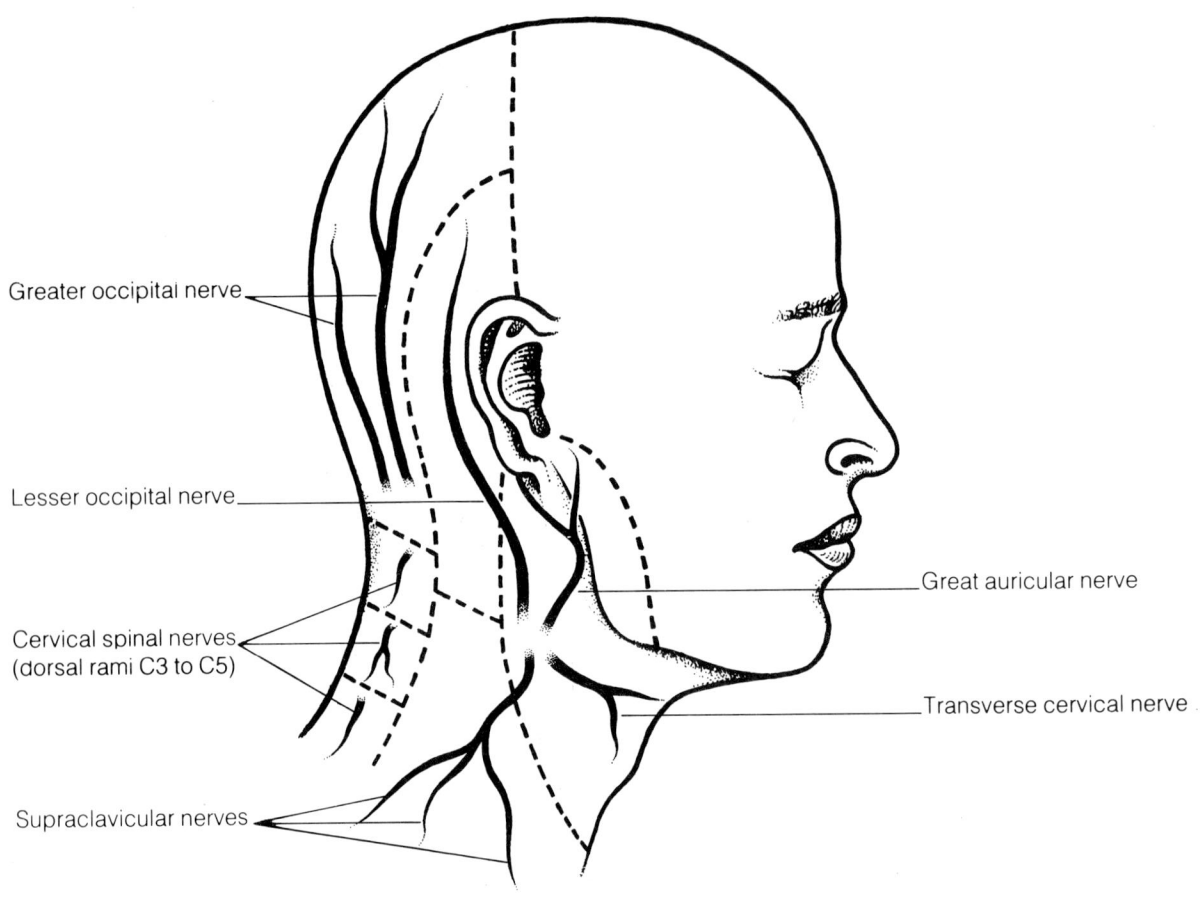

Figure 18 *Cutaneous innervation of the neck.*

THE CUTANEOUS INNERVATION OF THE NECK

The skin of the neck is innervated by branches of cervical spinal nerves, via both dorsal and ventral rami (Figure 18). The dorsal rami supply skin over the back of the neck and scalp. The ventral rami supply skin covering the lateral and anterior portions of the neck and even extend on to the face over the angle of the mandible.

The dorsal rami of the first, sixth, seventh and eighth cervical nerves have no cutaneous distribution. From the medial branch of the dorsal ramus of the second cervical nerve comes the greater occipital nerve. This pierces the trapezius muscle close to its attachment on to the superior nuchal line of the occiput and then ascends to supply the skin over the occipital part of the scalp up to the vertex of the skull. The medial branches of the dorsal rami of the third, fourth and fifth cervical nerves also pierce the trapezius to supply skin over the back of the neck in a serial manner.

The ventral rami of the second, third and fourth cervical nerves supply cutaneous branches via the cervical plexus. This plexus is located deep to the sternocleidomastoid muscle and supplies both motor and sensory branches to structures in the neck (see pages 113 to 115, Figure 37). The cutaneous branches from the plexus are the lesser occipital, the great auricular, the transverse cervical, and the supraclavicular nerves. All four nerves appear from beneath the sternocleidomastoid muscle at its posterior margin.

The lesser occipital nerve takes fibres mainly from the second cervical nerve, although fibres from the third cervical nerve may sometimes contribute. It ascends along the posterior margin of the sternocleidomastoid muscle to supply the scalp above and behind the ear, and a small area on the cranial surface of the auricle.

94,6; 99,10

The great auricular nerve receives fibres from the second and third cervical nerves. It runs up the superficial surface of the sternocleidomastoid muscle towards the ear. It supplies the skin overlying the mastoid process (the mastoid branch), much of the auricle (auricular branches), and the parotid region and the angle of the mandible (facial branches).

94,4

The transverse cervical nerve also takes fibres from the second and third cervical nerves. It crosses the sternocleidomastoid muscle horizontally to supply skin overlying the anterior part of the neck from the mandible to the sternum.

94,15

The supraclavicular nerves receive fibres from the third and fourth cervical nerves. Initially, it is a single nerve. This passes downwards towards the clavicle, where it divides into three branches — medial, intermediate and lateral supraclavicular nerves. These nerves supply skin at the root of the neck and over the upper part of the thorax.

94,12

Superficial structures of the neck

Immediately beneath the skin of the neck lie the investing layer of deep cervical fascia (see page 94) and the platysma muscle.

92

The platysma muscle

92,2; 112,23

This muscle belongs to the group of muscles comprising 'the muscles of facial expression' (see page 151). It appears as a thin, broad sheet running up the front of the neck from the root of the neck towards the mandible. However, it varies considerably in extent and may even be absent.

Attachments. The platysma muscle arises from the superficial fascia of the upper part of the thorax. It then runs across the clavicle and up the neck to insert into the lower border of the body of the mandible, the skin of the lower part of the face, and the musculature around the angle and lower part of the mouth.

12,16; 16,16

Innervation. The cervical branch of the facial nerve innervates the platysma muscle.

112,37

Vasculature. The muscle receives its blood supply from the facial artery (submental branch), and the thyrocervical trunk (the suprascapular artery).

Actions. Platysma is said to be the muscle associated with the expression of horror or surprise. It wrinkles the skin of the neck in an oblique direction, it draws down the lower lip and angle of the mouth, and slightly depresses the mandible. The muscle is also active during deep sudden inspiration.

Just visible beneath the platysma muscle are the external jugular and anterior jugular veins (see page 105).

92,4; 92,3

On removing the platysma muscle, the prominent structure seen is the sternocleidomastoid muscle.

94

The sternocleidomastoid muscle

94,2; 96,13
98,8; 112,40

The sternocleidomastoid muscle lies on the side of the neck. It is enclosed by the investing layer of deep cervical fascia which splits to pass round it. The muscle is an important landmark in the neck because many of the structures seen in both superficial and deep dissections can be directly related to it.

The sternocleidomastoid muscle forms the boundary that demarcates the regions named the anterior and posterior triangles of the neck (Figure 19). On its superficial surface lie the external jugular vein, the superficial cervical chain of lymph nodes and, emerging from behind its posterior margin, the cutaneous nerves from the cervical plexus. At the root of the neck, sternocleidomastoid is covered by the platysma muscle. Near its insertion, it is covered by the parotid gland. Beneath the sternocleidomastoid muscle are the great vessels of the neck (accompanied by the vagus nerve) within the carotid sheath. The deep cervical chain of lymph nodes, the cervical plexus, the ansa cervicalis, the upper part of the brachial plexus, and the phrenic nerve are also located beneath the muscle. Near its origin, the sternocleidomastoid muscle overlaps the infrahyoid (strap) musculature.

Attachments. The sternocleidomastoid muscle arises by two heads. The sternal head is tendinous and attached to the anterior surface of the manubrium sterni. The clavicular head is a wide muscular head arising from the upper surface of the medial third of the clavicle. The two heads merge and the muscle passes upwards, laterally and posteriorly to insert on to the lateral surface of the mastoid process of the temporal bone and the adjacent part of the superior nuchal line.

Innervation. The spinal part of the accessory nerve supplies the muscle and passes through or deep to it to emerge into the posterior triangle of the neck on its way to the trapezius muscle. The sensory innervation concerned with proprioception is associated with the cervical plexus. Even its motor supply may be derived in part from this plexus.

Vasculature. Sternocleidomastoid receives its blood supply from branches of the superior thyroid, occipital, posterior auricular and suprascapular arteries.

Actions. These vary according to whether one or both sternocleidomastoid muscles are activated. When one muscle acts, the head is tipped towards the shoulder on the same side and is rotated to direct the face towards the opposite side. When the muscles act together, the head is moved forwards. The muscles may be involved in forced expiration.

The superficial regions of the neck anteriorly and laterally are divided for descriptive purposes into anterior and posterior triangles (Figure 19). The structure delineating these triangles is the sternocleidomastoid muscle. The suboccipital triangle at the back of the neck is considered with the deep dissection of the neck (see pages 126 to 128).

THE ANTERIOR TRIANGLE OF THE NECK

This region is bounded posteriorly by the anterior margin of the sternocleidomastoid muscle, anteriorly by the median line of the neck, and superiorly by the inferior border of the mandible and the mastoid process. The sternum marks the apex of the triangle.

The anterior triangle can be subdivided in two ways. First, there are the suprahyoid and infrahyoid areas above and below the hyoid bone. Second, the passage of the digastric and omohyoid muscles across the anterior triangle defines the digastric, submental, muscular and carotid triangles (Figure 19).

The digastric triangle is also referred to as the submandibular triangle. Indeed, the most prominent structure within it is the submandibular salivary gland. The digastric triangle is bounded by the anterior and posterior bellies of the digastric muscle and by the lower border of the mandible. Its floor is formed by the mylohyoid, hyoglossus, and middle constrictor muscles. The digastric triangle contains the submandibular gland and lymph nodes, the facial, submental and mylohyoid blood vessels, and the mylohyoid, hypoglossal and glossopharyngeal nerves. The glossopharyngeal nerve lies

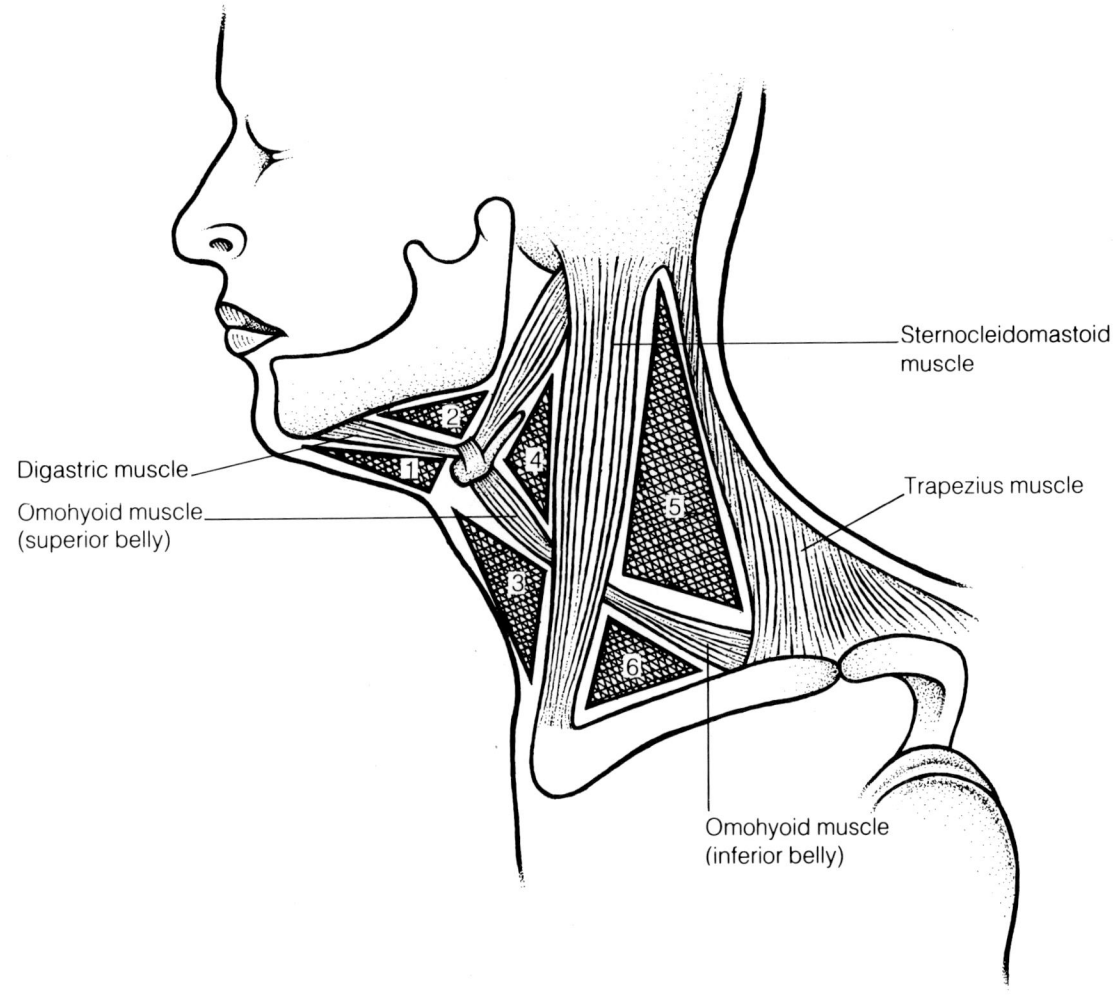

Figure 19 *The triangles of the neck. The anterior triangle between the sternocleidomastoid muscle and the midline of the neck is subdivided into the submental triangle (1), the digastric triangle (2), the muscular triangle (3), and the carotid triangle (4). The posterior triangle between the sternocleidomastoid and trapezius muscles is subdivided into the occipital triangle (5), and the supraclavicular triangle (6).*

on the stylopharyngeus muscle. The carotid sheath and the lower part of the parotid gland just appear in the posterior region of the triangle.   144A,12

The submental triangles lie above the hyoid bone, between the two anterior bellies of the digastric muscles as they approach the chin. The floor is formed by the mylohyoid muscle. The submental triangles contain the submental lymph nodes and the anterior jugular veins.

The structures within the digastric and submental triangles are considered later with the floor of the mouth (see pages 294 to 295).

The digastric muscle   102A

Attachments. The digastric muscle consists of an anterior belly and a posterior belly connected by an intermediate tendon. The posterior belly arises from the mastoid (or digastric) notch immediately behind the mastoid process of the temporal bone. The posterior belly passes downwards and forwards towards the hyoid bone, where it becomes the digastric tendon. The tendon passes through the insertion of the stylohyoid muscle and is attached to the greater horn of the hyoid bone by a fibrous

102A,44
24,15

102A,11
102A,16

loop. The anterior belly of the digastric muscle is attached to the digastric fossa on the inferior border of the mandible beneath the chin and runs downwards and backwards to the digastric tendon.

Innervation. Because the posterior belly of the digastric develops from the second branchial arch of the fetus, its innervation is derived from the facial nerve (the digastric branch). The anterior belly develops from the first branchial arch and so receives its motor supply from the mandibular division of the trigeminal nerve (the mylohyoid branch).

Vasculature. The posterior belly obtains its arterial blood supply from the posterior auricular and occipital arteries. The anterior belly receives its blood supply from the facial artery (submental branch).

Actions. Although the digastric muscle helps to raise the hyoid bone and the base of the tongue and is involved in maintaining the stability of the hyoid bone, its prime function is to assist in depressing and retracting the mandible.

Accompanying the posterior belly of the digastric is the stylohyoid muscle.

The stylohyoid muscle

Attachments. The muscle arises from the posterior surface of the base of the styloid process. It passes downwards and forwards with the posterior belly of the digastric to insert into the body of the hyoid bone at the junction with the greater horn (just above the attachment of omohyoid). Near its insertion it splits to envelop the digastric tendon.

Innervation. In common with the posterior belly of the digastric, the stylohyoid muscle develops from the second branchial arch. Consequently, it also is innervated by the facial nerve (digastric branch).

Vasculature. It receives its blood supply from the facial, posterior auricular, and occipital arteries.

Actions. It elevates and draws backwards the hyoid bone and therefore the floor of the mouth and the base of the tongue.

The digastric and stylohyoid muscles can be classified with the mylohyoid, geniohyoid, hyoglossus and genioglossus muscles as the suprahyoid musculature. These other muscles are described in relation to the floor of the mouth (see pages 293, 295). The infrahyoid musculature lies within the muscular triangle of the neck.

The muscular triangle is bounded by the median line of the neck from the sternum to the hyoid bone, by the superior belly of the omohyoid muscle, and by the anterior margin of the sternocleidomastoid muscle near its origin.

The muscular triangle contains the four infrahyoid muscles: the omohyoid, sternohyoid, sternothyroid and thyrohyoid muscles (Figure 20). The omohyoid and sternohyoid muscles are superficial; the sternothyroid and thyrohyoid muscles are deep. Because of their shape, these muscles are sometimes referred to as the strap muscles of the neck. As a group, they act to fix or to depress the hyoid bone and to elevate or to depress the larynx during swallowing.

The omohyoid muscle

Attachments. This muscle has superior and inferior bellies joined by an intermediate tendon. The superior belly of the omohyoid is attached to the lower border of the body of the hyoid bone. The inferior belly is attached to the upper border of the scapula (medial to the scapular notch). The superior belly runs down the anterior triangle of the neck, whereas the inferior belly runs across the posterior triangle. Beneath the sternocleidomastoid muscle, the two bellies are connected by the intermediate tendon. Occasionally, however, there is no tendinous intersection. The intermediate tendon is attached to the clavicle and the first rib by a fibrous band

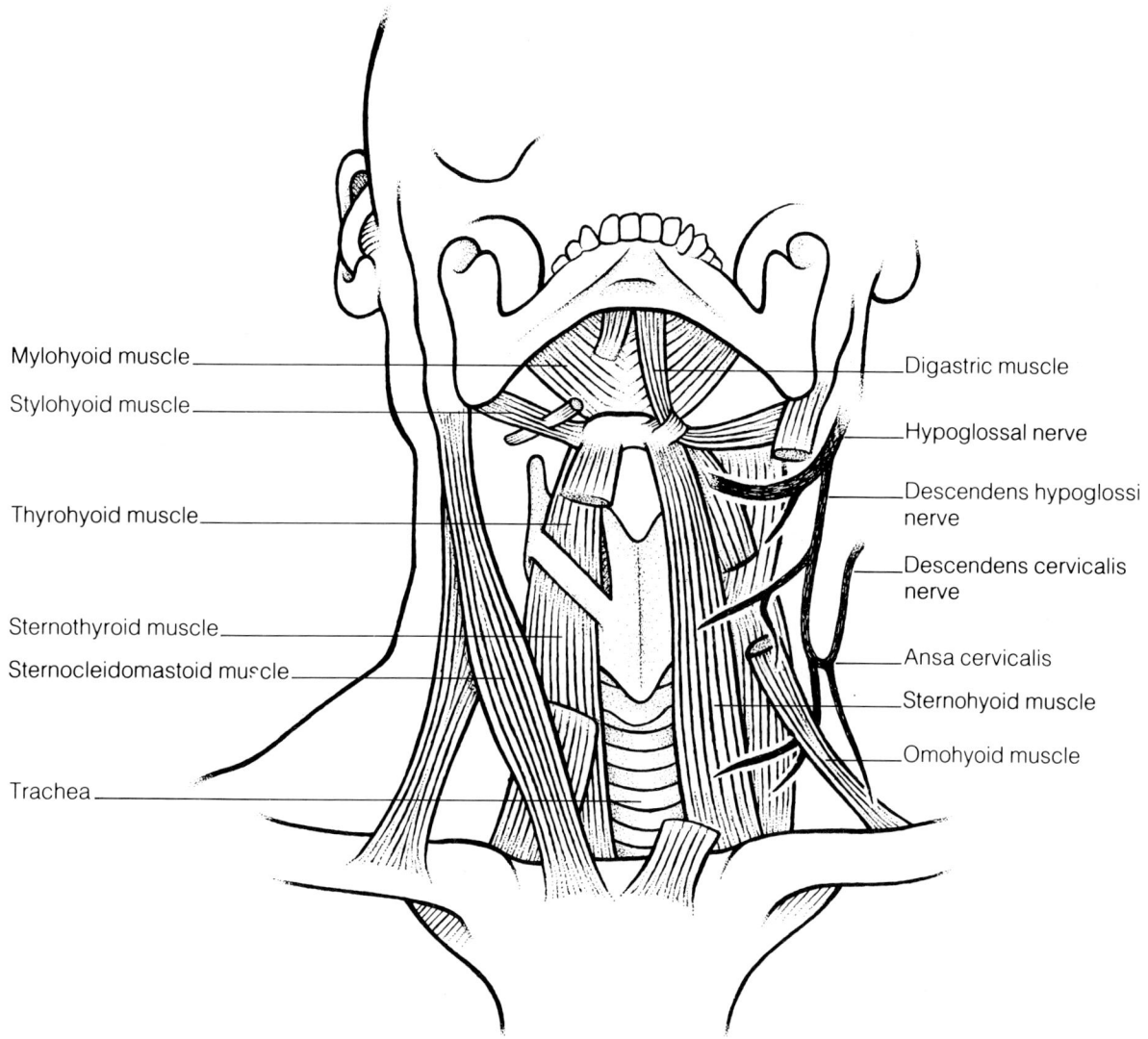

Figure 20 Muscles in the anterior triangle of the neck, *including the infrahyoid (strap) musculature (omohyoid, sternohyoid, sternothyroid and thyrohyoid muscles). Also illustrated is the innervation of the infrahyoid muscles, the ansa cervicalis.*

derived from the deep cervical fascia. It is because of this fascial sling that the 'angulation' of the muscle is maintained.

Innervation. Both the superior and inferior bellies are supplied by branches from the ansa cervicalis (see page 115). However, the superior belly receives its innervation via the superior root (descendens hypoglossi) of the ansa.

100,47

Vasculature. Branches from the lingual and superior thyroid arteries supply the muscle.

Actions. The omohyoid muscle depresses the hyoid bone after it has been elevated. It also aids depression of the larynx.

The sternohyoid muscle

96,24; 100,49

Attachments. This muscle originates mainly from the posterior surface of the manubrium sterni, but also arises from the head of the clavicle. It passes upwards and slightly medially to insert on to the inferior border of the body of the hyoid bone.

84F,29; 88B,11
152A,8

Innervation. The ansa cervicalis supplies the sternohyoid muscle.

Vasculature. Sternohyoid receives its blood supply from branches of the superior thyroid and lingual arteries.

Actions. The muscle depresses the hyoid bone after it has been elevated during swallowing.

The sternothyroid muscle

Attachments. Sternothyroid arises from the posterior surface of the manubrium sterni, below the attachment of the sternohyoid muscle. It passes upwards to insert on to the oblique line of the thyroid cartilage of the larynx.

Innervation. It is supplied by branches from the ansa cervicalis.

Vasculature. The blood supply of the muscle is derived from the superior thyroid artery.

Actions. From an elevated position during swallowing, the muscle depresses the larynx.

The thyrohyoid muscle

Attachments. This muscle can be regarded as a continuation of the sternothyroid muscle. It passes upwards from the oblique line of the thyroid cartilage to the lower border of the body and the greater horn of the hyoid bone.

Innervation. Unlike the other infrahyoid muscles, the thyrohyoid muscle is not innervated by the ansa cervicalis. In common with the geniohyoid muscle, it is supplied by fibres from the first cervical nerve which pass with the hypoglossal nerve.

Vasculature. The arterial supply to the thyrohyoid muscle is derived from the superior thyroid artery.

Actions. The muscle depresses the hyoid bone or raises the larynx.

The carotid triangle is bounded by the upper part of the sternocleidomastoid muscle, the posterior belly of the digastric muscle and the superior belly of the omohyoid muscle. The floor of the triangle is formed by the thyrohyoid and hyoglossus muscles and by the inferior and middle constrictor muscles of the pharynx. The carotid triangle is so named because it contains the bifurcation of the common carotid artery and the superior thyroid, ascending pharyngeal, lingual, facial, and occipital branches of the external carotid artery. The veins in the triangle correspond to these branches. Also present are the hypoglossal nerve, the superior root of the ansa cervicalis (descendens hypoglossi) and the internal and external branches of the superior laryngeal nerve.

THE POSTERIOR TRIANGLE OF THE NECK

The posterior triangle is poorly named, as the region lies on the lateral side of the neck. Furthermore, although diagrams of the triangle portray it as a flat region, in reality it 'spirals' around the side of the neck.

The posterior triangle is bounded in front by the posterior margin of the sternocleidomastoid muscle, behind by the anterior margin of the trapezius muscle and below by the middle third of the clavicle (Figure 19). The apex of the triangle is the mastoid process of the temporal bone. The floor of the triangle is formed by the prevertebral fascia, which overlies semispinalis capitis, splenius capitis, levator scapulae and the scalenus posterior and medius muscles (these vertebral muscles are described on pages 123 to 126). The inferior belly of the omohyoid muscle crosses the posterior triangle and subdivides it into occipital and supraclavicular triangles (Figure 19). The roof of the posterior triangle is formed by the investing layer of deep cervical fascia.

The posterior triangle contains many important structures, particularly near its base.

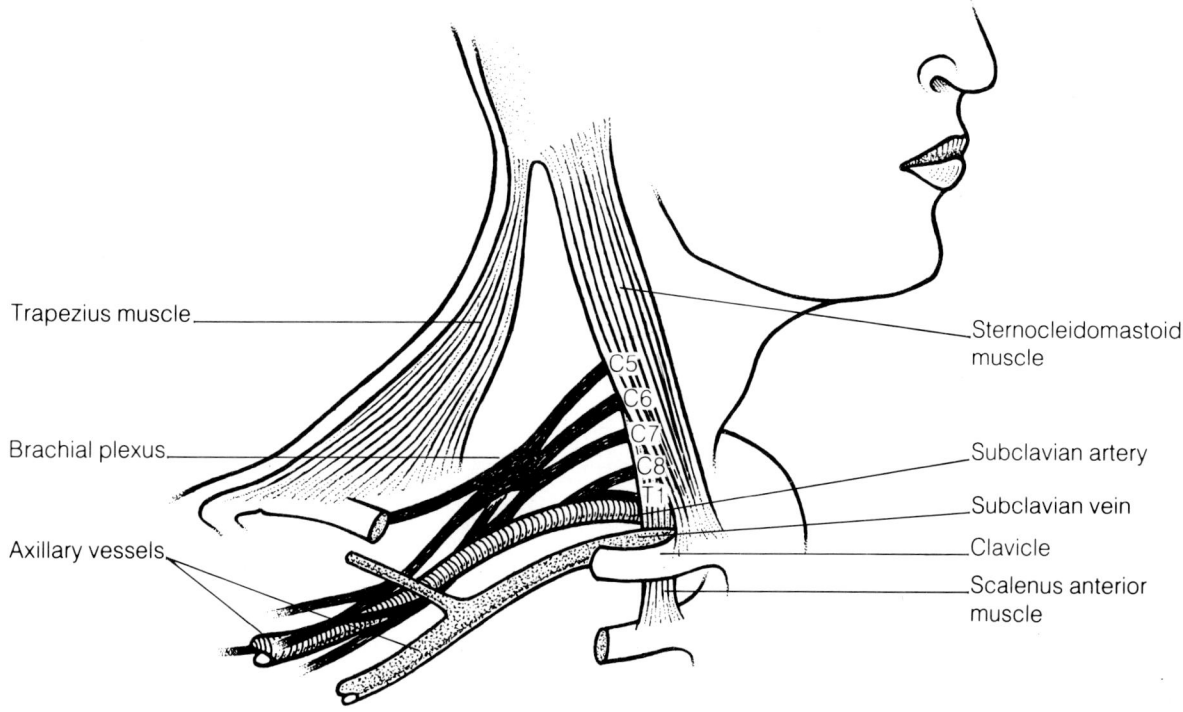

Figure 21 *Structures related to the supraclavicular triangle of the neck.*

The occipital triangle contains the occipital artery, the cutaneous branches from the cervical plexus, and the accessory nerve. The occipital artery emerges from under the splenius capitis muscle at the apex of the triangle and crosses semispinalis to penetrate the trapezius muscle. The lesser occipital nerve from the cervical plexus runs up the posterior border of the sternocleidomastoid muscle. The great auricular and transverse cervical nerves only just enter the posterior triangle, turning sharply around the posterior border of the sternocleidomastoid muscle. The supraclavicular nerves emerge from beneath sternocleidomastoid and pass across and down the posterior triangle towards the clavicle. The accessory nerve (spinal part) also emerges from the sternocleidomastoid into the posterior triangle. It runs obliquely downwards on the levator scapulae muscle to enter trapezius. The nerve is separated from levator scapulae by a distinct and dense layer of fascia.

99,37; 98,16
98,13; 99,37
99,11; 99,38
99,14; 98,10
98,8; 98,9
98,33
98,16

98,13
98,12
98,14

The supraclavicular triangle contains the subclavian artery, part of the brachial plexus, the termination of the external jugular vein, and the superficial cervical (transverse cervical) and suprascapular vessels. The subclavian artery in the posterior triangle is the third part of the artery after it has emerged from behind the scalenus anterior muscle (see page 122). The subclavian vein is not usually seen in the triangle, being behind the clavicle. The brachial plexus lies partly behind and partly above the subclavian artery (Figure 21). The superficial cervical and suprascapular vessels run across the superior margin of the clavicle. The external jugular vein is described on page 105.

104,25; 98,19
98,7; 98,21
98,25
104,25; 104,28
104,24
104,26
98,21; 98,25
98,26

The trapezius muscle (Figure 22)

98,14

The trapezius muscle forms the posterior margin of the posterior triangle, although it is best considered as a muscle of the back. Because the trapezius shares its innervation with the sternocleidomastoid muscle, these two muscles are thought by some anatomists to be two parts of a single muscle that 'splits' to reveal the posterior triangle of the neck.

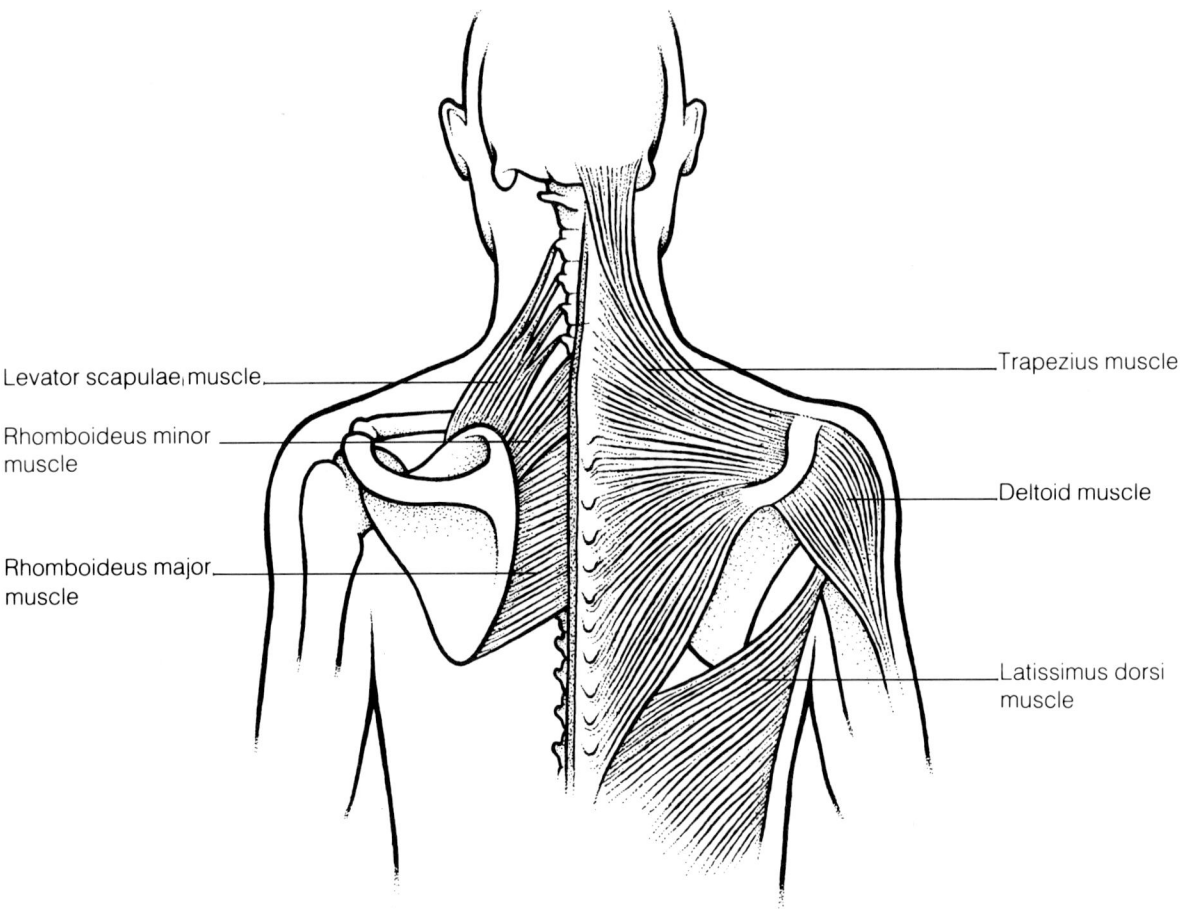

Figure 22 Superficial muscles of the back, and the back of the neck. *On the left side, the trapezius has been removed.*

24,20
200A,4; 82B,8

88,4

98,13
98,15

Attachments. Trapezius has an extensive origin from the midline of the back. It arises from the external occipital protuberance and the superior nuchal line at the back of the skull, and from the ligamentum nuchae, the spine of the seventh cervical vertebra and the spines of all the thoracic vertebrae. The superior fibres pass downwards, the middle fibres horizontally and the inferior fibres upwards to converge on the shoulder (the spine and acromion of the scapula) and the lateral third of the clavicle.

Innervation. It is supplied by the spinal part of the accessory nerve. It also receives branches from the cervical plexus, which may be related to proprioception.

Vasculature. Its arterial supply is derived from the superficial cervical and the dorsal scapular arteries.

Actions. Trapezius is involved in controlling the position of the scapula, but it can also elevate, rotate and retract the scapula when acting with other muscles. When the scapula is fixed, trapezius can move the head backwards and laterally.

Deep structures of the neck

This is a complex region, often difficult to understand. However, it may be more easily understood by considering the disposition of structures in transverse section. This enables a division of the region into four compartments (Figure 23):

- Two neurovascular compartments (i.e. carotid sheaths) containing many of the great vessels and nerves of the neck.
- A visceral compartment comprising the upper alimentary and respiratory passages, and some important glands.
- A musculoskeletal compartment around the cervical vertebral column.

In addition, deep dissection of the neck involves consideration of the root of the neck at the inlet of the thorax.

The four compartments comprising the deep region of the neck are related such that the musculoskeletal compartment provides a 'platform' posteriorly on which lie the neurovascular and visceral compartments. The compartments are bounded and defined by fascial layers.

THE FASCIA AND TISSUE SPACES OF THE NECK

There has been much controversy concerning the definition of fascia. Some anatomists have applied the term to all layers of connective tissue that interconnect neighbouring structures. Others prefer to restrict the term to distinct membranous layers that can be incised and sutured. Such a definition has obvious clinical significance. Applying this definition to the fascia of the neck, three layers can be identified: the investing layer of deep cervical fascia, the prevertebral fascia, and the pretracheal fascia. Nevertheless, the carotid sheath is one non-membranous layer usually described with the cervical fascia.

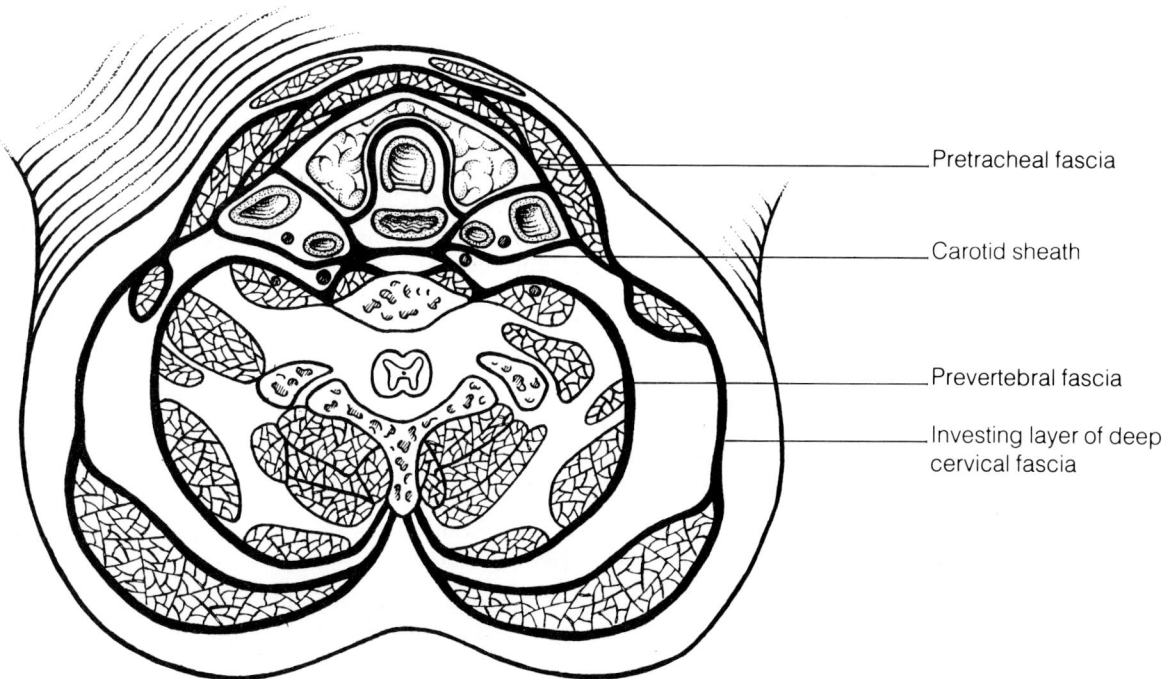

Figure 23 *Transverse section of the neck below the hyoid bone to show the main layers of the cervical fascia.*

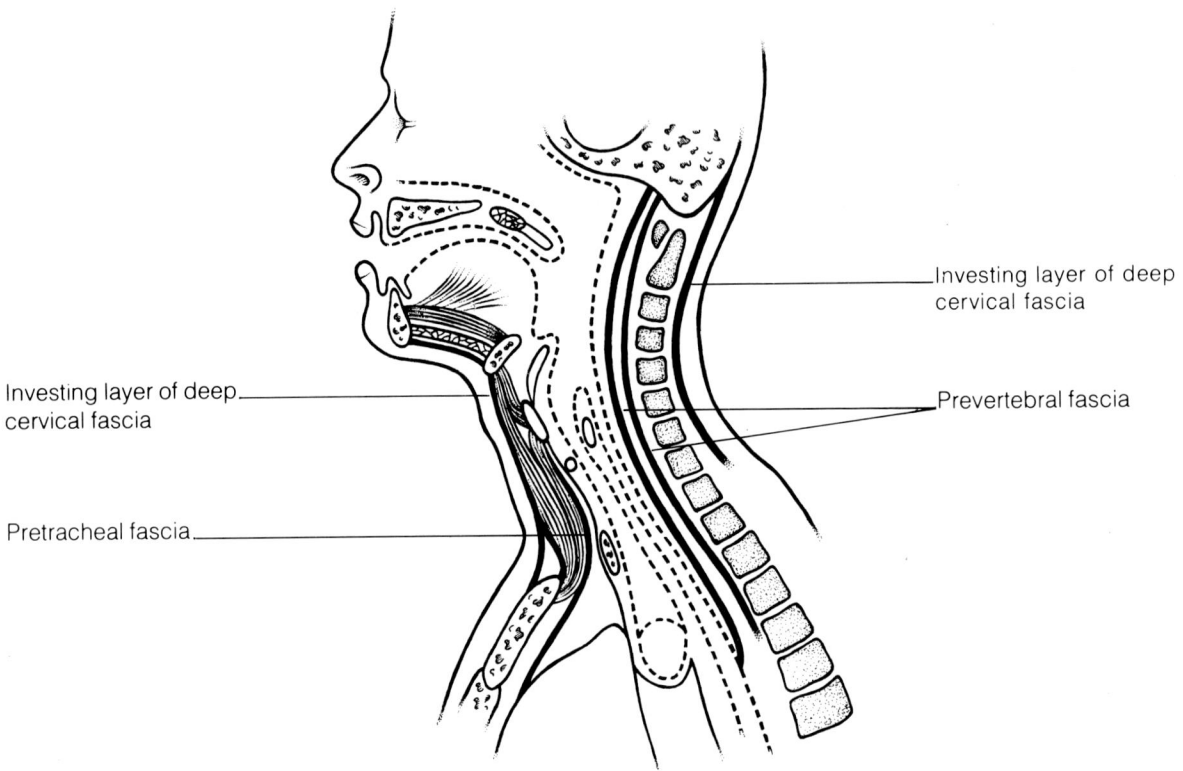

Figure 24 *Diagram to illustrate the main layers of the cervical fascia in longitudinal section.*

The general arrangement of the cervical fascia is shown in Figures 23 and 24.

The investing layer of deep cervical fascia

This fascia is located just beneath the skin and is the most superficial of the true fascial layers of the neck. It completely encircles the neck like a surgical collar. At the front of the neck, however, the fascia is situated internal to the platysma muscles. As the fascia approaches the trapezius and sternocleidomastoid muscles, it splits to enclose them. The attachments of the investing layer of deep cervical fascia are:

- Posteriorly: The spines of the cervical vertebrae and the ligamentum nuchae.
- Superiorly: The external occipital protuberance and the superior nuchal lines at the back of the skull, the tip of the mastoid process, the lower border of the zygomatic arch, and the lower border of the mandible from the angle to the chin.
- Anteriorly: The chin, the body of the hyoid bone, and the manubrium sterni.
- Inferiorly: The sternum (where it splits into superficial and deep layers with the suprasternal space between), the clavicle (where it also splits into superficial and deep layers between the attachments of the trapezius and sternocleidomastoid muscles), and the acromion of the scapula.

Note that the attachments superiorly and inferiorly are related to the attachments of the trapezius and sternocleidomastoid muscles.

Between the mastoid process and the angle of the mandible, the investing layer of deep cervical fascia splits into two layers to enclose the parotid gland, forming the parotid fascia. The superficial layer of the parotid fascia is attached to the tip of the mastoid process, the lower border of the cartilaginous part of the external acoustic meatus, and the lower border of the zygomatic arch. The deep layer extends along the base of the skull from the tip of the mastoid process towards the opening of the carotid canal, where it merges with the fascia around the internal carotid artery. Part of the deep layer extends between the styloid process and the angle of the mandible as the stylomandibular ligament.

36C,17

The prevertebral fascia

94,5

This is the fascia that encloses the cervical part of the vertebral column and the prevertebral and postvertebral muscles (i.e. the musculoskeletal compartment). It is termed the prevertebral fascia because it is particularly prominent in front of the vertebral column. Indeed, here there may be two distinct layers. The prevertebral fascia forms the fascial floor of the posterior triangle of the neck. Its attachments are:

- Superiorly: The base of the skull in front of the longus capitis and rectus capitis lateralis muscles.
- Inferiorly: It extends into the thorax to merge with the anterior longitudinal ligament of the third thoracic vertebra at the lower limit of the longus cervicis muscle.

As the fascia passes around the musculoskeletal compartment of the neck, it attaches to the transverse and spinous processes of each cervical vertebra and to the ligamentum nuchae. However, the fascia becomes indistinct posteriorly and often merges with the investing layer of deep cervical fascia.

In the root of the neck, the prevertebral fascia covers the scalene muscles. As the subclavian artery and the nerves from the brachial plexus emerge from behind the scalenus anterior muscle, the prevertebral fascia forms the axillary sheath. This sheath invests the subclavian and axillary arteries, but not the veins.

The prevertebral fascia provides a base upon which the pharynx, oesophagus and other cervical structures glide during swallowing and neck movements, undisturbed by movements of the prevertebral muscles.

The pretracheal fascia

This surrounds the viscera of the neck, but is named pretracheal because it is particularly prominent in front of the trachea. The attachments of the pretracheal fascia are:

- Superiorly: The larynx, being limited by the overlying infrahyoid musculature.
- Inferiorly: It extends into the superior mediastinum of the thorax along the great vessels to merge with the fibrous pericardium.

The pretracheal fascia merges laterally with the investing layer of deep cervical fascia and with the connective tissues comprising the carotid sheath.

The fascia forms a sheath around the thyroid gland.

The carotid sheath

Some anatomists recognise a distinct, but thin, fascial layer surrounding the carotid arteries (the common and internal carotids, but not the external carotid), the internal jugular vein, and the vagus nerve. The sheath is said to be attached superiorly to the base of the skull and inferiorly it merges with the connective tissue around the arch of the aorta. However, most reports suggest that the sheath may be nothing more than the merging of the adjacent investing cervical fascia, prevertebral fascia and pretracheal fascia.

100,18; 100,15
100,14; 102B,28

Figure 25 shows the relationships of structures in and around the carotid sheath.

The fascial layers of the neck define a number of tissue spaces. These spaces must be regarded only as 'potential spaces' and not as true anatomical entities. This is because in the healthy person the tissues are closely applied to each other, or are filled with

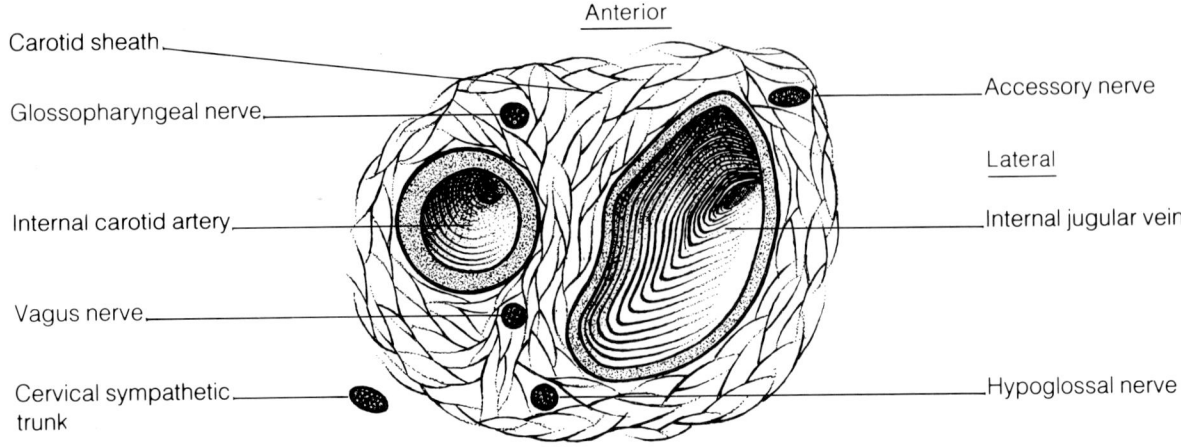

Figure 25 *Section through the carotid sheath* just below the base of the skull.

relatively loose connective tissue. Where there is pathological involvement, particularly with inflammation, this may spread through the tissue spaces, usually taking the line of least resistance. Conversely, the fascia can confine the spread of inflammation.

To aid description of the tissue spaces in the neck, we can distinguish spaces above and below the hyoid bone.

Above the hyoid bone are located the submandibular and submental spaces beneath the inferior border of the mandible, the pharyngeal spaces, and the prevertebral space near the base of the skull.

The submandibular and submental tissue spaces

The submandibular space lies on the cervical surface of the mylohyoid muscle, between the anterior and posterior bellies of the digastric muscle. It is continuous with the sublingual tissue space in the floor of the mouth (see page 308) around the posterior free edge of the mylohyoid muscle. The submental space is situated below the chin and between the anterior bellies of the digastric muscles. There are effectively no barriers between the two submandibular spaces and the submental space. Consequently, infection can readily spread right across the neck below the inferior border of the mandible.

The pharyngeal tissue spaces

The spaces related to the pharynx can be subdivided into the peripharyngeal and the intrapharyngeal spaces.

The anterior part of the peripharyngeal space is made up of the submandibular and submental spaces already described. The posterior peripharyngeal space is called the retropharyngeal space. Laterally are the parapharyngeal spaces.

The retropharyngeal space is the area of loose connective tissue lying behind the pharynx and in front of the prevertebral fascia. It extends upwards to the base of the skull and downwards to the retrovisceral space in the infrahyoid part of the neck.

Each parapharyngeal space passes laterally around the pharynx and is continuous with the retropharyngeal space. Unlike the retropharyngeal space, however, it is a space that is restricted to the suprahyoid region. It contains loose connective tissue and is bounded medially by the pharynx and laterally by the pterygoid muscles (here being part of the infratemporal fossa), and the sheath of the parotid gland. Superiorly, it is bounded by the base of the skull. Inferiorly, it does not extend right down the neck but is limited by the suprahyoid structures, particularly the sheath of the submandibular gland. The parapharyngeal space is particularly prone to receiving infections spreading from the jaws and teeth. The subsequent spread of infection from the parapharyngeal space is into the retropharyngeal space.

An intrapharyngeal space potentially exists between the inner surface of the constrictor muscles of the pharynx and the pharyngeal mucosa. Infections at this site are either restricted locally or spread through the pharynx into the retropharyngeal or parapharyngeal spaces. An important part of the intrapharyngeal space is the peritonsillar space. This lies around the palatine tonsil between the pillars of the fauces. Infections here (quinsy) usually spread up or down the intrapharyngeal space, or through the pharynx into the parapharyngeal space.

The tissue spaces above the hyoid bone are discussed further with the spaces around the jaws (see pages 307 to 309). The prevertebral space is described with the tissue spaces in the infrahyoid region of the neck.

Below the hyoid bone are located the pretracheal and retrovisceral tissue spaces in the visceral compartment of the neck, the prevertebral space in front of the vertebral column, and a space associated with the carotid sheath. Although a tissue space may be expected between the pretracheal fascia and the investing layer of deep cervical fascia, the area is occupied by the infrahyoid (strap) musculature.

The pretracheal and retrovisceral tissue spaces (Figures 26 and 27)

The pretracheal tissue space lies behind the pretracheal fascia and the infrahyoid (strap) muscles, and in front of the anterior wall of the oesophagus. The space thus immediately surrounds the trachea. It is bounded superiorly by the attachments of the

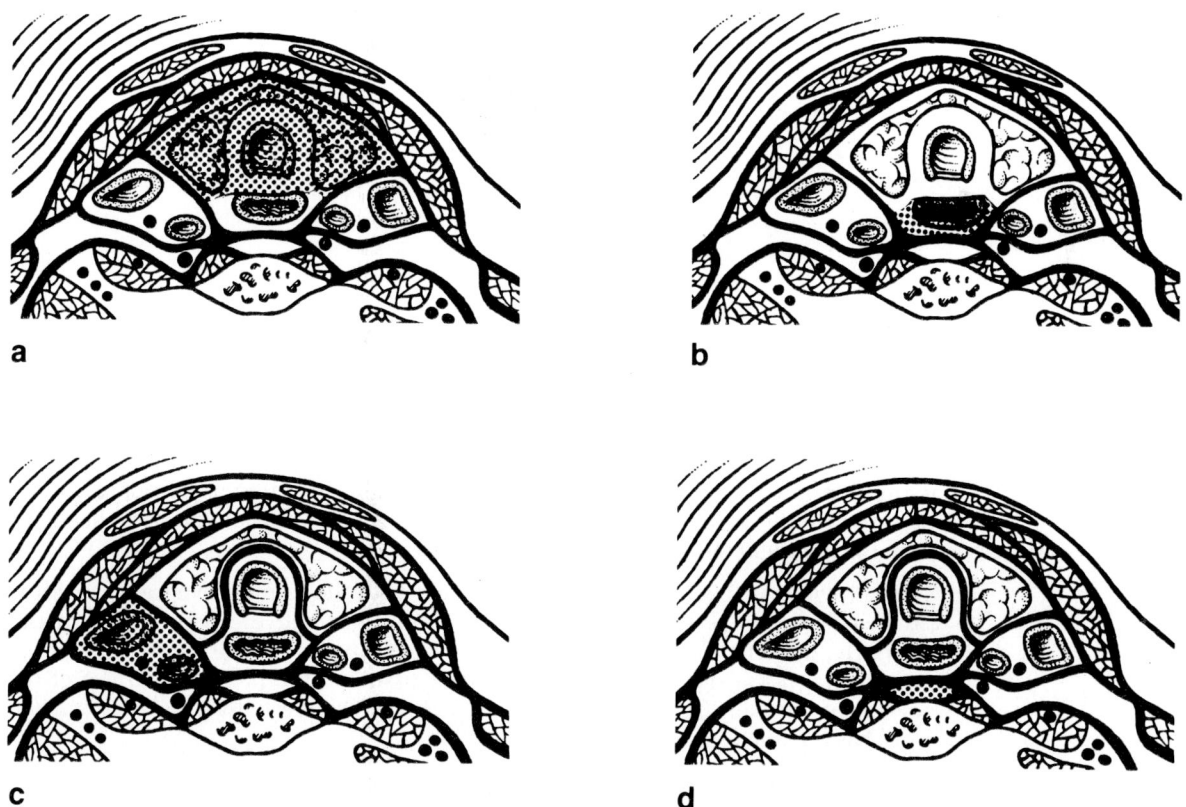

Figure 26 *The tissue spaces of the neck below the hyoid bone, indicated by stippled regions.* a) *Pretracheal tissue space.* b) *Retrovisceral tissue space.* c) *Tissue space associated with the carotid sheath.* d) *Prevertebral tissue space.*

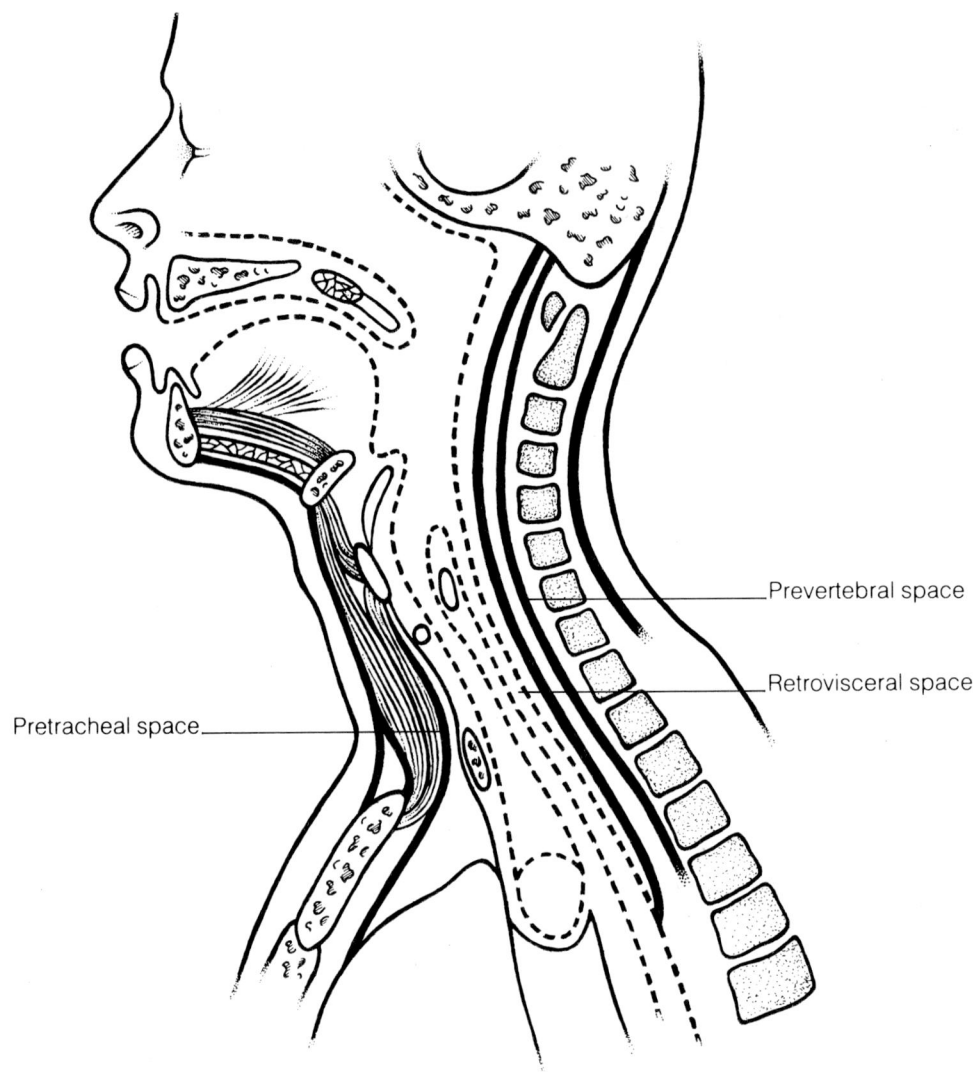

Figure 27 *Three of the tissue spaces of the neck seen in longitudinal section* (*the pretracheal space, the retrovisceral space, and the prevertebral space*). *Note that in this case the retrovisceral space is limited inferiorly by the merging of a layer of the prevertebral fascia with the oesophagus.*

infrahyoid muscles to the thyroid cartilage of the larynx. Inferiorly, it extends down into the anterior portion of the superior mediastinum. Infection usually spreads into the pretracheal space either by perforating the anterior wall of the oesophagus or from the retrovisceral space.

The retrovisceral space is continuous superiorly with the retropharyngeal space. It is situated between the posterior wall of the oesophagus and the prevertebral fascia. Inferiorly, the retrovisceral space extends into the superior mediastinum. Should the prevertebral fascia merge with the connective tissue on the posterior surface of the oesophagus (usually at the level of the fourth thoracic vertebra), the retrovisceral space has a distinct inferior boundary.

The prevertebral tissue space (Figures 26 and 27)

The prevertebral tissue space has been variously described. To some anatomists it is the potential space lying between the prevertebral fascia and the vertebral column. Others believe it to be the space between the two layers comprising the prevertebral fascia. Because the space is closed above, below and laterally, infection usually spreads into it through its fascial walls from the retrovisceral area. Inferiorly, the space extends into the posterior mediastinum.

The carotid sheath is, in reality, a layer of loose connective tissue demarcated by adjacent portions of the investing layer of deep cervical fascia, the pretracheal fascia, and the prevertebral fascia. Nevertheless, there is a potential space into which infections from the visceral spaces may track (Figure 26). However, it has been reported that infections around the carotid sheath are restricted because superiorly (near the hyoid bone) and inferiorly (near the root of the neck) the connective tissues adhere to the vessels.

THE GREAT VESSELS AND THE NERVES OF THE NECK

When the sternocleidomastoid muscle is removed, the major vessels and nerves in and around the carotid sheath are displayed. Within the carotid sheath are the common carotid and internal carotid arteries, the internal jugular vein, and the vagus nerve (Figure 25).

100
100,18
100,15; 100,14
102,28

THE CAROTID ARTERIES

There are three carotid arteries on each side of the neck: the common carotid, internal carotid, and external carotid arteries.

The carotid arteries provide the major source of blood to the head and neck. Additional arteries arise from branches of the subclavian artery in the neck (the vertebral artery, the thyrocervical trunk, and the costocervical trunk). These are described with the root of the neck (see pages 130 to 132).

The common carotid artery

100; 104

The common carotid arteries differ on the right and left sides with respect to their origins. On the right, the common carotid arises from the brachiocephalic artery as it passes behind the sternoclavicular joint. On the left, the common carotid comes directly from the arch of the aorta in the superior mediastinum. The right common carotid has therefore only a cervical part, whereas the left common carotid has cervical and thoracic parts. Both arteries terminate by bifurcating into the internal and external carotid arteries at the level of the upper border of the thyroid cartilage of the larynx.

104,6; 104,17
108,31

102,37
102,40; 102,21

The common carotid arteries usually have no branches.

Near its bifurcation, the common carotid shows two specialised organs, the carotid body and the carotid sinus. These relay information concerning the pressure and chemical composition of the arterial blood. The carotid body is a small, reddish-brown structure situated behind the bifurcation. It functions as a chemoreceptor. The carotid sinus is usually seen as a dilation of the lower end of the internal carotid. It functions as a baroreceptor. The main innervation to both the carotid body and the carotid sinus is derived from the carotid branch(es) of the glossopharyngeal nerve. The cervical sympathetic trunk and the vagus nerve also contribute.

100,15; 146,16

Some of the relationships of the common carotid artery are illustrated in Figure 17. In addition, the artery is crossed by the superior belly of the omohyoid, the superior and middle thyroid veins and the anterior jugular vein.

100,50; 104,4
102B,64

The internal carotid artery

102; 104

Following the bifurcation of the common carotid, the internal carotid continues up the neck within the carotid sheath. It leaves the neck by passing through the carotid canal at the base of the cranium and terminates intracranially.

104,2

22,36; 168,6

Within the neck, the internal carotid artery usually has no branches. The branches within the carotid canal are the caroticotympanic and pterygoid arteries.

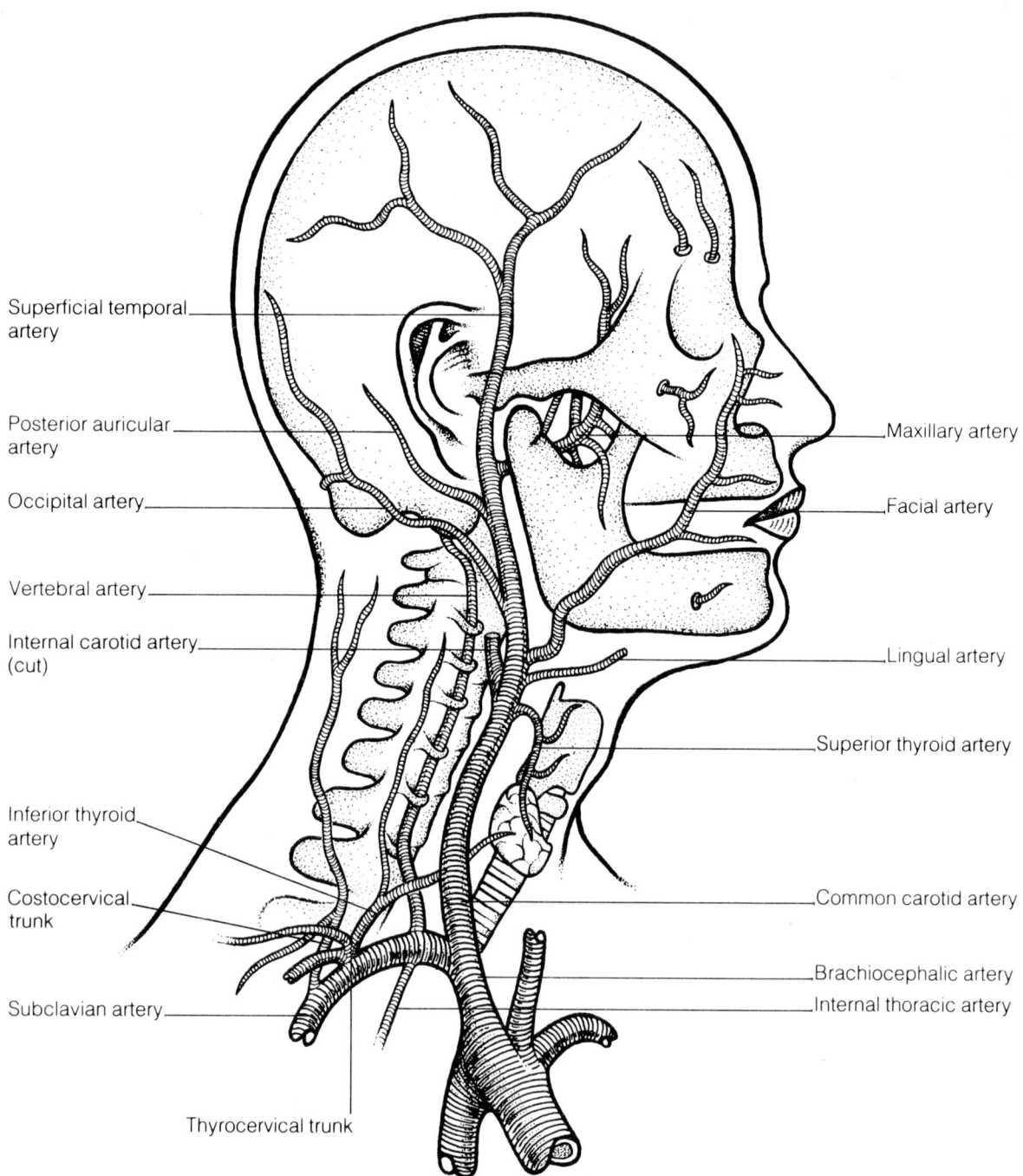

Figure 28 *The arteries of the right side of the neck* (excluding the ascending pharyngeal artery).

100

100,16
100,6; 102A,44
102A,40; 102A,1
142A,24; 142A,19

The external carotid artery

This is the component of the carotid system that lies outside the carotid sheath. From the bifurcation of the common carotid, the external carotid passes upwards, runs behind the posterior belly of the digastric muscle, crosses the styloglossus and stylopharyngeus muscles and passes into the parotid gland to divide into its terminal branches behind the condylar process of the mandible. The external carotid has eight branches (Figure 28):

Branches:
- Superior thyroid artery
- Ascending pharyngeal artery
- Lingual artery
- Facial artery
- Occipital artery
- Posterior auricular artery
- Maxillary artery
- Superficial temporal artery

The superior thyroid artery is the first branch of the external carotid artery. It arises just below the greater horn of the hyoid bone. It runs downwards on the surface of the inferior constrictor muscle of the pharynx to the apex of the lobe of the thyroid gland. The external branch of the superior laryngeal nerve lies medial and usually behind the artery. Apart from its obvious supply to the upper part of the thyroid gland, the superior thyroid artery has branches supplying the sternocleidomastoid muscle, some of the infrahyoid (strap) muscles (infrahyoid branch), and the larynx (superior laryngeal and cricothyroid branches).

100,22; 104,4
100,21; 102,22
100,23; 102A,23
154A,28

The ascending pharyngeal artery is the smallest branch of the external carotid. It arises from the posterior surface of the carotid artery, passing vertically upwards towards the pharynx. It terminates near the base of the skull. Its main branches are pharyngeal (to the muscles of the pharynx and the palate), inferior tympanic (to the middle ear), and meningeal (branches passing through the hypoglossal canal and the foramen lacerum). It also supplies some prevertebral muscles.

116C,50; 146A,3
108,2; 146A,3
108,3

The lingual artery arises from the anterior surface of the external carotid, just above the superior thyroid artery. Near the tip of the greater horn of the hyoid bone, the artery shows a distinct loop which is related to the hypoglossal nerve. The lingual artery passes through the carotid triangle of the neck and then beneath the hyoglossus muscle to enter the tongue. Here it terminates as the deep lingual artery. Other branches are the infrahyoid artery (to the omohyoid and sternohyoid muscles) and the sublingual artery (to the sublingual gland and the floor of the mouth).

100,11
100,12
154,1; 154,3
142A,34; 138,4
138,9

The facial artery also arises from the anterior surface of the external carotid. It passes forwards and upwards, deep to the digastric muscle, to enter the digastric triangle of the neck. Here, it comes into contact with the submandibular gland and weaves in and out beneath the inferior border of the mandible before crossing on to the face at the anterior edge of the masseter muscle. Its subsequent course and distribution are described in relation to the face (see pages 161 to 162). In the neck, it gives off the following branches: ascending palatine artery, tonsillar artery, branches to the submandibular gland, and the submental artery to the suprahyoid muscles.

100,3
100,6
102A,4; 140D,75
102A,2
116C,31; 144A,11
102A,7

The occipital artery (Figure 29) arises from the posterior surface of the external carotid, at the same level as the facial artery. It runs backwards across the carotid sheath where the hypoglossal nerve hooks around it. The occipital artery then runs upwards and backwards towards the suboccipital region of the scalp. It has branches supplying the sternocleidomastoid, digastric and stylohyoid muscles, the muscles of the suboccipital triangle, some of the prevertebral and postvertebral muscles, the mastoid air cells and the auricle. There is also a stylomastoid artery (in 66% of cases), a descending branch that anastamoses with the superficial cervical, vertebral and deep cervical arteries, and meningeal branches that pass through the jugular and mastoid foramina and condylar canal. Occipital branches are the terminal parts to the scalp.

116C,52; 142A,22
116C,52
99,37; 200A,2
116C,52

The posterior auricular artery also arises from the posterior surface of the external carotid, but at the level of the upper border of the digastric muscle. It passes through

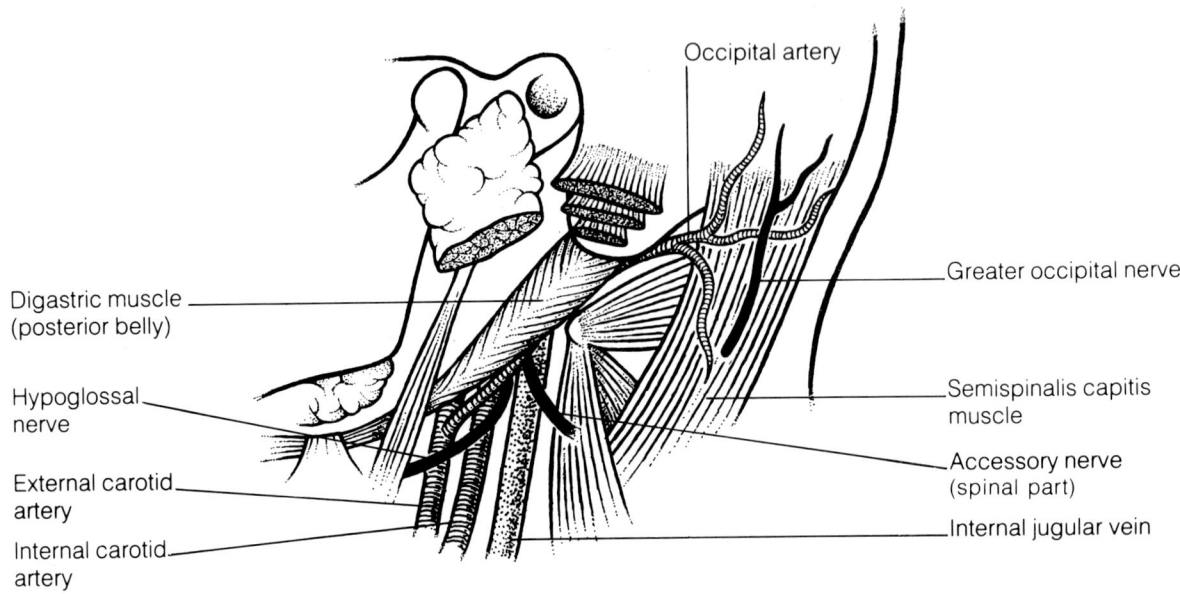

Figure 29 *The course of the occipital artery.*

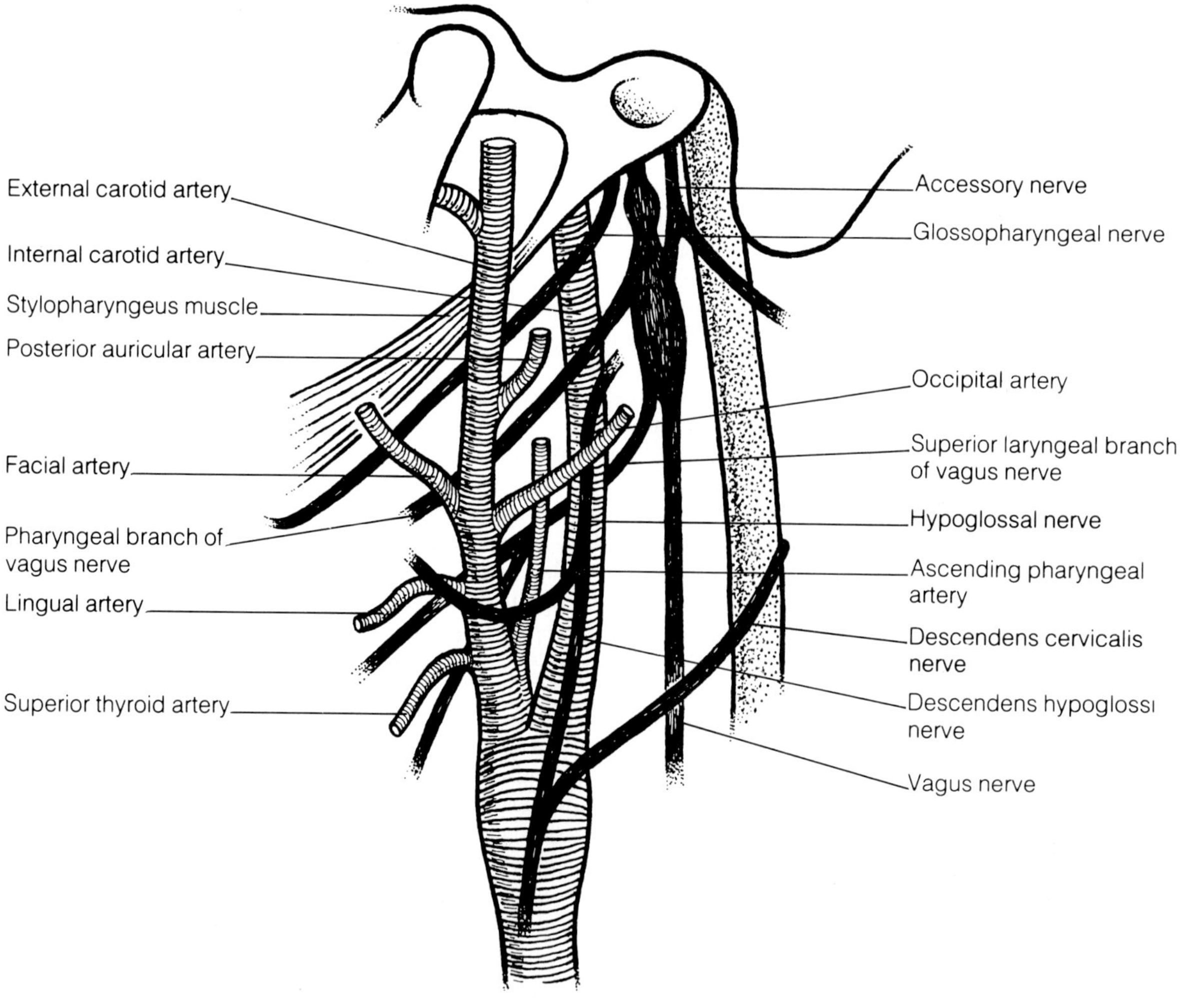

Figure 30 *The relationships of the nerves to the external and internal carotid arteries.*

the apical part of the parotid gland, along the styloid process to a groove between the auricle of the ear and the mastoid process. It gives muscular branches to the digastric, stylohyoid and sternocleidomastoid muscles, glandular branches to the parotid, a stylomastoid artery (in 33% of cases, see occipital artery, above), an auricular branch and an occipital branch to the scalp. 144A,59

The terminal branches of the external carotid artery, **the superficial temporal and maxillary arteries**, arise in the parotid gland. The superficial temporal artery is described in relation to the face (see page 162) and the maxillary artery in relation to the infratemporal fossa (see page 180).

There are many important relationships of structures, particularly nerves, with the carotid arteries. Some of the major ones are illustrated in Figure 30.

THE JUGULAR VEINS

The veins in the neck show considerable variation, in terms both of size and connections. Most of the veins in the head and neck drain into the two major venous structures in the neck — the internal and external jugular veins (Figure 31).

The internal jugular vein

This vessel is found in the deep part of the neck, running with the carotid system of arteries in the carotid sheath. It thus lies deep to the sternocleidomastoid muscle. The vein extends from the base of the skull down to the thoracic inlet where it terminates by joining with the subclavian vein to form the brachiocephalic vein. At the base of the skull, the internal jugular vein emerges through the jugular foramen as a continuation of the sigmoid dural sinus.

100
100,14

108,28; 108,47
22,35
168,21

Near the origin of the internal jugular vein is a dilation, the superior bulb. Another dilation is found near its termination, the inferior bulb. Just above the inferior bulb is a pair of valves.

Tributaries:
- Inferior petrosal dural sinus
- Facial vein
- Lingual veins
- Pharyngeal veins
- Superior and middle thyroid veins

The inferior petrosal sinus leaves the skull through the jugular foramen and joins the internal jugular at the superior bulb. 168,17

The facial vein crosses the inferior border of the mandible to meet the anterior branch of the retromandibular vein as it emerges from the parotid gland. Some anatomists call the vessel the common facial vein beyond this point. The vein then passes across the lingual artery, the hypoglossal nerve and the external and internal carotids to enter the internal jugular at the level of the greater horn of the hyoid bone. 96,11

The lingual veins are variable, but they usually follow two routes. The dorsal lingual vein drains the dorsum of the tongue and passes with the lingual artery deep to the hyoglossus muscle. It joins the internal jugular vein at the level of the greater horn of the hyoid. The deep lingual vein is seen through the mucosa of the ventral surface of the tongue. It joins the sublingual vein to become the vein accompanying the hypoglossal nerve on the hyoglossus muscle. This terminates either by joining the lingual vein or by draining directly into the internal jugular vein. 100,13

The pharyngeal veins arise from the pharyngeal plexus of veins. They usually drain into the internal jugular vein but may also pass into the inferior thyroid veins. Tributaries include meningeal vessels and the vein passing through the pterygoid canal. 146A,33

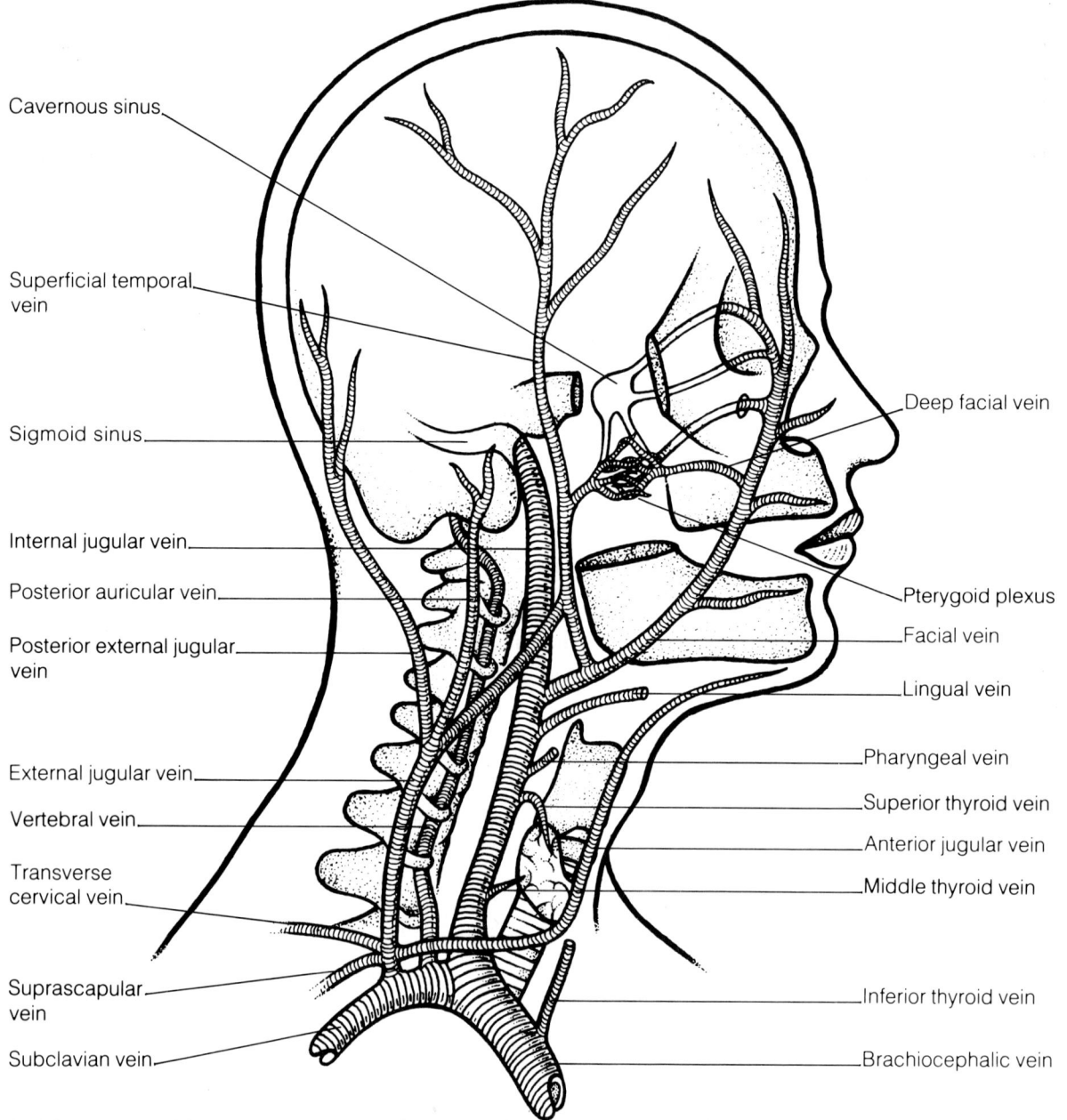

Figure 31 *The veins of the neck.*

102A,24; 104,4	**The superior thyroid vein** accompanies the superior thyroid artery. It receives not only veins from the thyroid gland but also the superior laryngeal and cricothyroid veins associated with the larynx.
102B,64; 104,7	**The middle thyroid vein** receives blood from the lower part of the thyroid gland.
92–98	**The external jugular vein**
92,4; 94,3	This vein lies superficially in the neck, on the lateral surface of the sternocleidomastoid muscle. It passes from its origin near the apex of the parotid salivary gland down to its
98,7; 98,20	termination just in front of the scalenus anterior muscle in the posterior triangle.

The external jugular vein arises by the confluence of the posterior branch of the retromandibular vein and the posterior auricular vein. It drains into the subclavian vein as it crosses scalenus anterior. The external jugular has two sets of valves, one pair where it drains into the subclavian vein and another pair 4 cm above the clavicle. Close to its origin may be found a communicating vessel with the internal jugular vein.

98,5
98,6

On the sternocleidomastoid muscle, the external jugular lies between the investing layer of deep cervical fascia and the platysma muscle. It passes into the posterior triangle after piercing the deep cervical fascia.

Tributaries:

- Occipital vein
- Posterior external jugular vein
- Superficial cervical vein
- Suprascapular vein
- Anterior jugular vein

The occipital vein is very variable. It may also be a tributary of the internal jugular vein, the posterior auricular vein, or the deep cervical and vertebral veins. It begins at the back of the scalp, where it may be joined by a vein draining the diploë in the occipital bone. Emissary veins connect the occipital vein to the intracranial venous sinuses via the mastoid and parietal foramina and through the condylar canal and occipital protuberances.

99,34; 112,45

The posterior external jugular vein also arises in the occipital region. It drains the back of the upper part of the neck and terminates at the middle of the external jugular.

The superficial cervical and suprascapular veins accompany the arteries of the same name. They may drain directly into the subclavian vein.

98,17; 106,33

The anterior jugular vein returns blood from the front of the neck. It usually arises by the confluence of veins in the submandibular region, but may also receive veins from the retromandibular, facial or parotid veins. It passes down the neck, just to one side of the midline. Above the sternum, the two anterior jugular veins are united by the jugular arch. The anterior jugular vein then curves laterally beneath the sternocleidomastoid muscle to drain into the external jugular or subclavian veins.

94,14

THE NERVES OF THE NECK

Within the carotid sheath runs the vagus nerve. Related to the carotid sheath near the base of the skull are the glossopharyngeal, accessory and hypoglossal nerves. Lying behind the carotid sheath and in front of the prevertebral fascia is the cervical sympathetic trunk. Deep to the internal jugular vein and in front of the scalenus medius and levator scapulae muscles (at the level of the first four cervical vertebrae) lies the cervical plexus of nerves. Associated with the cervical plexus is the ansa cervicalis. The brachial plexus for the arm lies in the deep part of the posterior triangle, in the root of the neck. Both the cervical and brachial plexuses are derived from the ventral rami of cervical spinal nerves. The cutaneous contributions of the dorsal rami have already been described (see page 84).

The glossopharyngeal nerve (Figure 32)

116C,49

This is the ninth cranial nerve. It emerges from the medulla oblongata of the brain stem as three or four rootlets. These rootlets are found in a groove between the olive and the inferior cerebellar peduncle. At this site, the glossopharyngeal nerve lies

188A,15; 198F,45

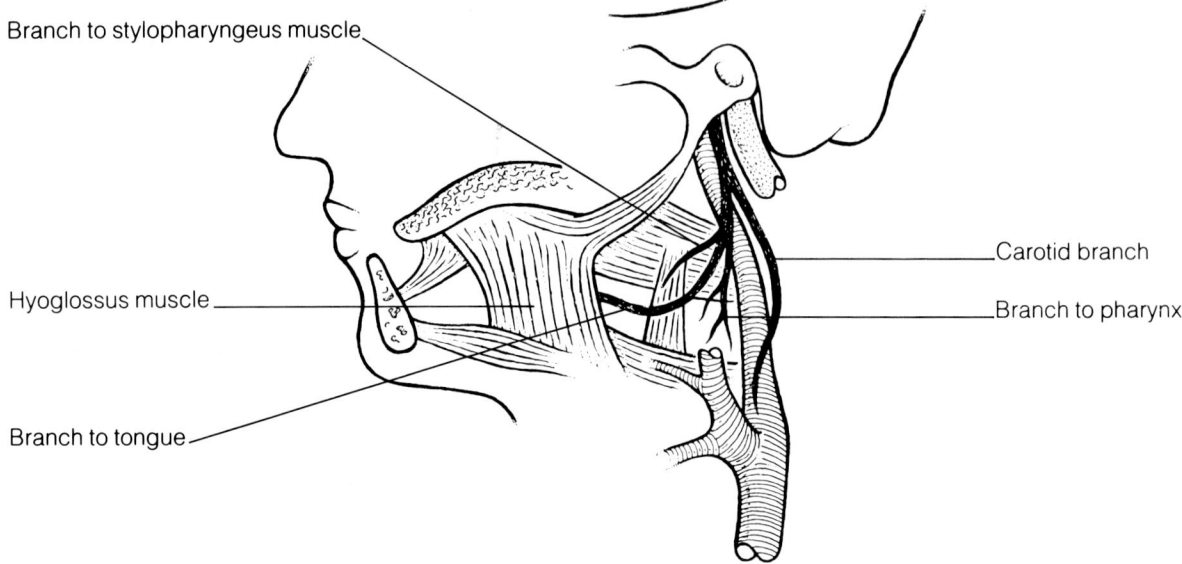

Figure 32 *The glossopharyngeal nerve.*

above the rootlets of the vagus nerve. The glossopharyngeal nerve has sensory, motor and parasympathetic fibres.

28,31; 146A,6

108,8

116C,49

144A,12; 146A,6

The glossopharyngeal nerve leaves the skull through the central part of the jugular foramen. Within the foramen, the nerve shows the superior and inferior ganglia. Below the foramen, the glossopharyngeal nerve is located anterior to the vagus and accessory nerves, passing between the internal jugular vein and the internal carotid artery. It then runs anteriorly between the internal and external carotid arteries (Figure 30) and on to the stylopharyngeus muscle. Winding around this muscle, it passes between the superior and middle constrictor muscles of the pharynx to be distributed to the tonsil, pharynx and tongue.

Branches:
- Tympanic nerve
- Lesser petrosal nerve
- Carotid branch
- Pharyngeal branches
- Stylopharyngeus (muscular) branch
- Tonsillar branches
- Lingual branches

50D,45

The tympanic nerve arises from the inferior ganglion. It passes upwards through the tympanic canaliculus to reach the middle ear cavity. Here, it contributes to the tympanic plexus which is found on the promontory of the medial wall of the tympanic cavity. This plexus provides sensory fibres to the mucosa of the tympanic cavity, the auditory tube and the mastoid air cells. From the plexus arises the lesser petrosal nerve.

28,23; 166B,53

The lesser petrosal nerve contains preganglionic parasympathetic fibres which relay through the otic ganglion to the parotid salivary gland. The nerve passes from the tympanic plexus, through the anterior wall of the tympanic cavity and on to the floor of the middle cranial fossa. It then emerges through the foramen ovale to join the otic ganglion in the infratemporal fossa (see page 179).

The carotid branch(es) arises just below the skull, as the glossopharyngeal crosses the internal carotid artery. It then passes between the internal and the external carotid arteries to the carotid sinus and the carotid body. During its course it is joined by the carotid branch of the vagus nerve.

The pharyngeal branches contribute to the pharyngeal plexus on the middle constrictor muscle (the other components of this plexus being from the sympathetic trunk and the pharyngeal branch of the vagus). The glossopharyngeal contribution to the plexus is sensory to the pharynx.

146A,12

The stylopharyngeus branch supplies the stylopharyngeus muscle (the nerve and muscle being associated embryologically with the third branchial arch).

The tonsillar branches supply the palatine tonsil. They form a plexus with the lesser palatine nerve. Branches from the plexus are distributed to the soft palate.

There are two **lingual branches** of the glossopharyngeal nerve. One branch supplies the region around the sulcus terminalis of the tongue, including the circumvallate papillae. The other branch supplies the posterior third of the tongue. The lingual branches are concerned with both taste perception and general sensation.

The vagus nerve (Figure 33)

102,28; 146,5

The vagus nerve is the tenth cranial nerve. It has the most extensive distribution of any of the cranial nerves and contains sensory, motor and parasympathetic fibres.

The vagus emerges from the brainstem at the medulla oblongata, between the olive and the inferior cerebellar peduncle. It exits the cranium through the jugular foramen with the glossopharyngeal and accessory nerves.

188A,15; 198F,45
28,31
146A,5

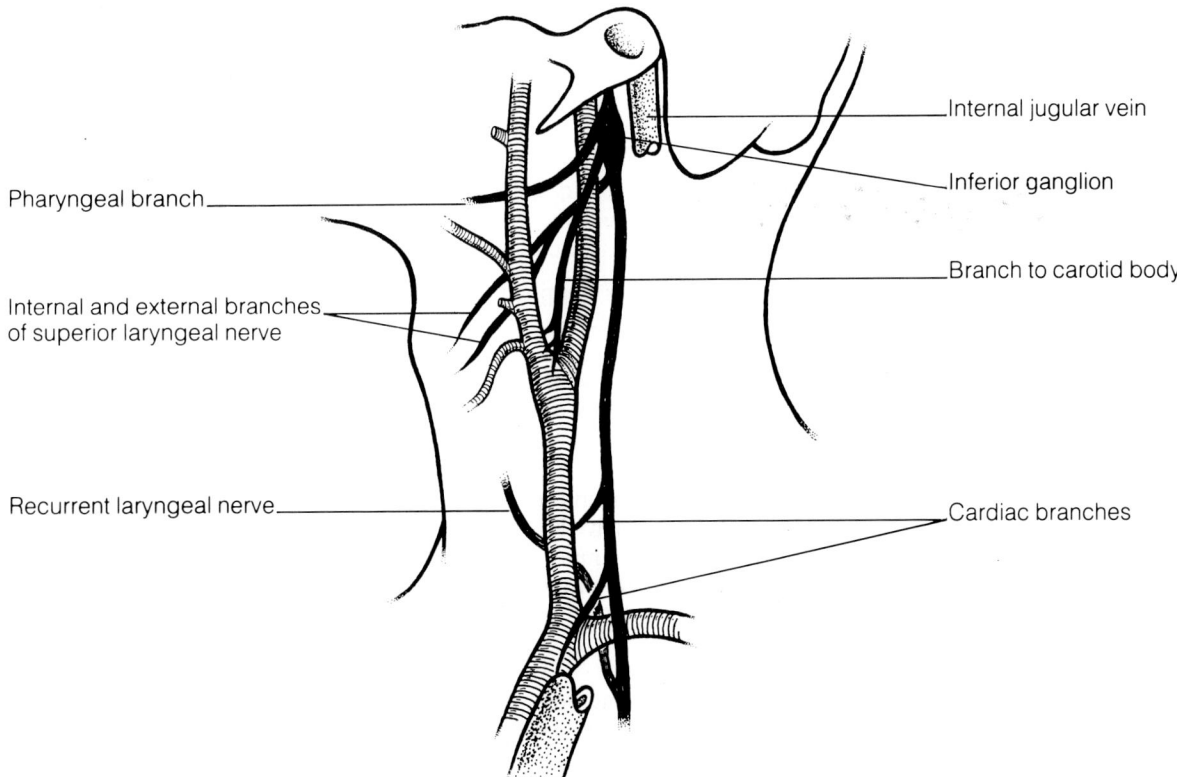

Figure 33 *The right vagus nerve.*

108,7; 146A,9

102,28
106,2; 108,6 + 33

The vagus nerve has two ganglia, the superior and inferior ganglia. The superior ganglion lies within the jugular foramen. The inferior ganglion is situated just below.

Just below the inferior ganglion, the vagus is joined by the cranial part of the accessory nerve. The vagus then passes downwards within the carotid sheath and enters the thorax at the root of the neck.

The vagus nerves in the neck differ in one important respect; namely the origins of the recurrent laryngeal nerves.

Branches:
- Meningeal branch
- Auricular branch
- Pharyngeal branch
- Branches to the carotid body
- Superior laryngeal nerve
- Recurrent laryngeal (right) nerve
- Cardiac branches

The meningeal branch(es) arises from the superior ganglion in the jugular fossa. It supplies dura in the posterior cranial fossa. There is some evidence that this nerve is not truly a branch of the vagus but is derived from upper cervical nerves and/or the superior cervical sympathetic ganglion.

50D,48
50A,15

The auricular branch also arises from the superior ganglion. It enters the temporal bone via the mastoid canaliculus on the lateral wall of the jugular fossa. It then passes out through the tympanomastoid fissure and divides into two branches. One branch joins the posterior auricular branch of the facial nerve, the other contributes to the innervation of the skin of the auricle, external acoustic meatus and tympanic membrane (see pages 363, 364, 365 and Figure 145).

146A,13

The pharyngeal branch is, in fact, derived from the cranial part of the accessory nerve. It runs from the inferior ganglion of the vagus, between the internal and external carotid arteries (Figure 30), and towards the middle constrictor of the pharynx. There it forms the pharyngeal plexus with branches from the sympathetic trunk, and the glossopharyngeal and external laryngeal nerves. The pharyngeal nerve is the main motor nerve to the muscles of the pharynx and palate.

146A,14

Although the carotid body is supplied mainly by the glossopharyngeal nerve, the vagus nerve can also contribute.

144A,54; 146A,15

96,35; 144A,31
146A,18; 154A,30
100,23; 144A,39
146A,20; 154A,16

The superior laryngeal nerve also arises from the inferior ganglion. It then passes deep to both the internal and external carotid arteries (Figure 30) on its way to the larynx. It divides into internal and external branches. The internal branch passes between the middle and inferior constrictor muscles to supply sensation to the larynx. The external branch runs down on the inferior constrictor muscle (with the superior thyroid artery) to supply the cricothyroid muscle of the larynx.

106B,27; 108,43

The right recurrent laryngeal nerve arises in the root of the neck. It leaves the vagus in front of the subclavian artery, loops below and behind the artery and then ascends towards the larynx.

102B,62; 108,43

144A,45; 144A,47
154,23

The left recurrent laryngeal nerve arises in the thorax, as the vagus passes across the arch of the aorta. Both recurrent laryngeal nerves reach the larynx by passing upwards in grooves between the trachea and the oesophagus (with the inferior thyroid arteries). They pass beneath the inferior borders of the inferior constrictor muscles to supply the mucosa of the larynx and most of the intrinsic muscles.

Figure 34 *The accessory nerve. Note that its cranial root joins the inferior ganglion of the vagus nerve.*

Usually two or three **cardiac branches** emanate from the vagus nerve in the neck. They run downwards and medially into the thorax, terminating at the deep part of the cardiac plexus.

The accessory nerve (Figure 34)   94,9; 146A,7

This is the eleventh cranial nerve. It consists of two distinct parts: the cranial accessory, and the spinal accessory nerves.

The cranial part of the accessory nerve is a motor nerve that emerges from the medulla oblongata between the olive and the inferior cerebellar peduncle. It joins the spinal part of the accessory at the jugular foramen. Once through the jugular foramen, the cranial and spinal parts again separate. The cranial part then joins the vagus nerve, eventually to be distributed in the pharyngeal branch of the vagus to the pharyngeal and palatine musculature. Some of its fibres also run with the recurrent laryngeal nerve and the cardiac branches of the vagus. Because of its close association with the vagus, some anatomists consider the cranial part of the accessory nerve to be a part of the vagus and not a separate cranial nerve.

188A,15; 198F,45
168,18; 168,19

146A,13

The spinal part of the accessory nerve is also a motor nerve, although there may be some sensory fibres. It is derived from the upper five segments of the cervical spinal cord. A series of rootlets emerges from the cord between the dorsal and ventral roots of the upper cervical nerves. They join to form a nerve trunk which passes intracranially through the foramen magnum. At the jugular foramen, the spinal and cranial parts of the accessory nerve unite but soon separate on leaving the cranium. The spinal part of the accessory nerve then crosses the internal jugular vein (usually on its lateral surface) and runs obliquely downwards and backwards to reach the upper part of the sternocleidomastoid muscle. It passes into the substance of this muscle and subsequently enters the posterior triangle of the neck. It crosses the posterior triangle on the levator scapulae muscle before entering the trapezius muscle. The spinal part of the accessory nerve provides the motor supply of the sternocleidomastoid and trapezius muscles.

198F,47

198F,45
168,18; 168,19
108,9
144A,58

94,9; 98,13

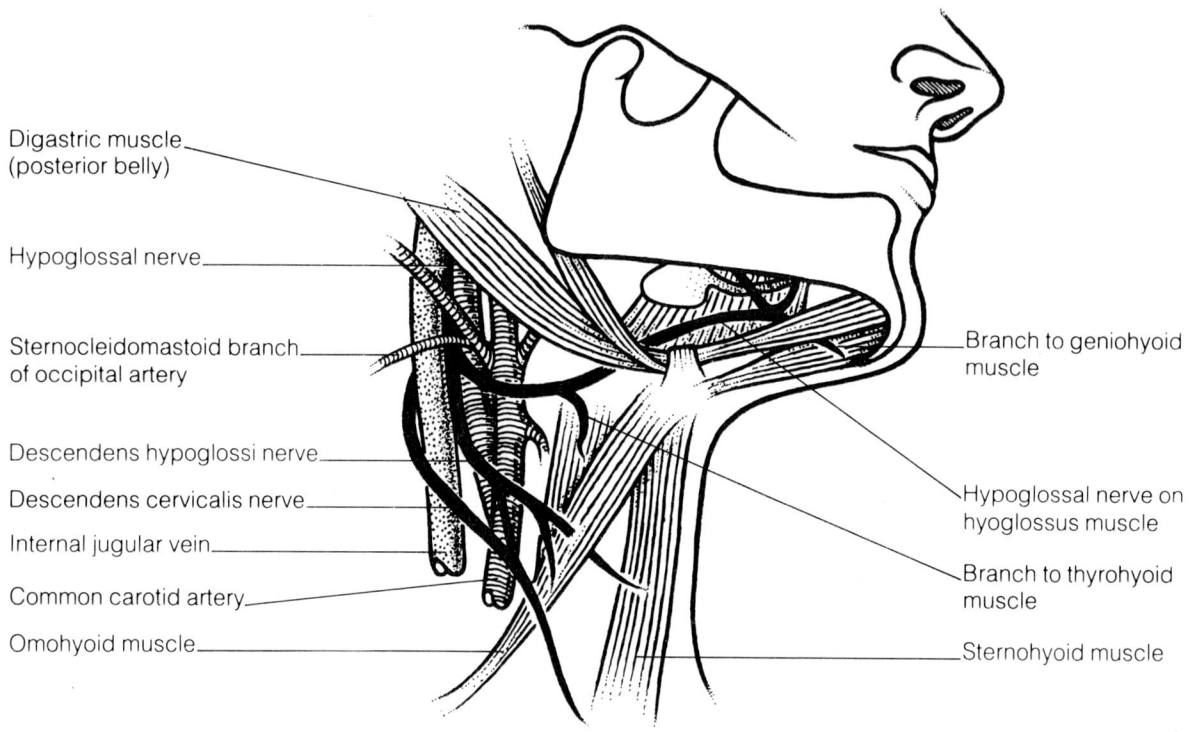

Figure 35 *The hypoglossal nerve.*

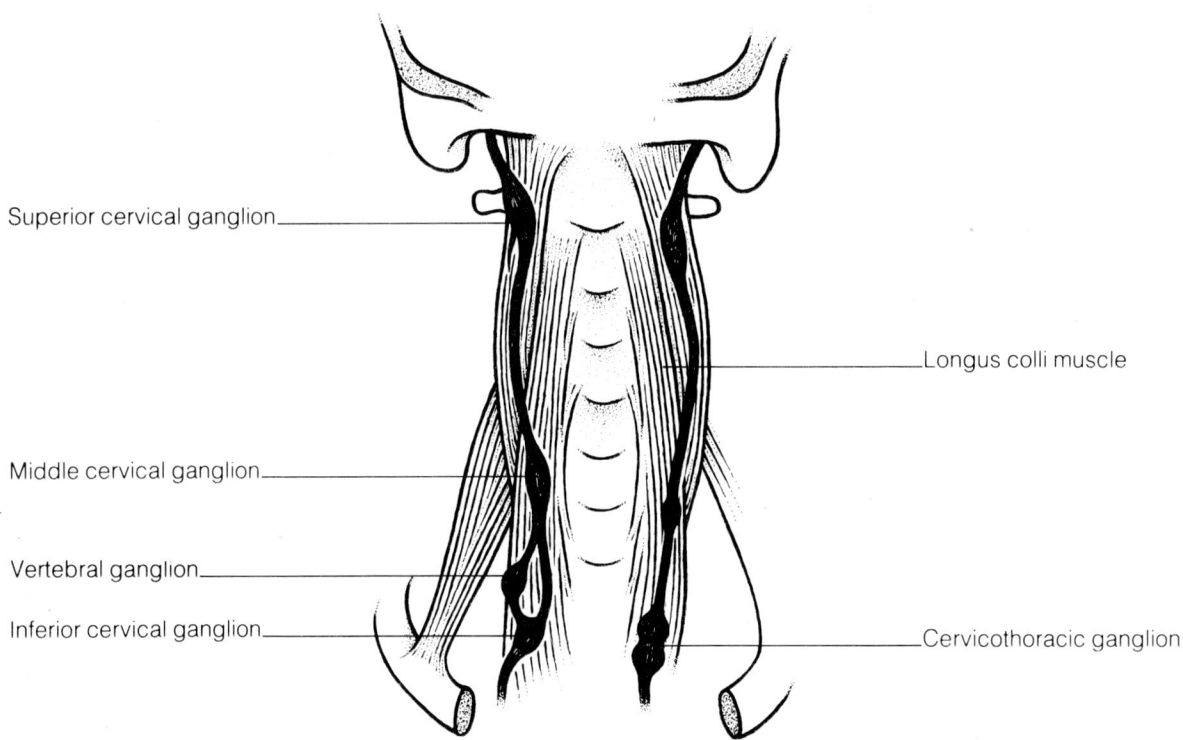

Figure 36 *The cervical sympathetic trunk.*

The hypoglossal nerve (Figure 35)

This is the twelfth cranial nerve and is a motor nerve supplying the musculature of the tongue. It originates as a series of rootlets on the medulla oblongata, between the pyramid and the olive.

The hypoglossal nerve runs through the hypoglossal canal of the occipital bone and emerges deep to the carotid sheath. It then passes downwards and, under cover of the posterior belly of the digastric muscle, outwards between the internal jugular vein and the internal carotid artery. Subsequently, it runs forwards across the vagus nerve and the external and internal carotid arteries. Indeed, it loops around the occipital artery near its origin (at its sternocleidomastoid branch). Continuing forwards, it passes below the submandibular salivary gland, on to the hyoglossus muscle, to be distributed to the muscles of the tongue (see pages 291 to 294 and Figure 113).

Like most cranial nerves, the hypoglossal nerve has connecting branches with other cranial and cervical spinal nerves and with the sympathetic system. An important connection is with the ventral ramus of the first cervical nerve.

Branches:
- Meningeal branch
- Upper root of ansa cervicalis
- Muscular branches to thyrohyoid and geniohyoid
- Muscular branches to the tongue

The meningeal branch is probably derived from those upper cervical and sympathetic fibres that communicate with the hypoglossal. It appears as the hypoglossal nerve emerges through its canal in the occipital bone. It mainly supplies the dura in the posterior cranial fossa.

The upper root of the ansa cervicalis is also derived from the ventral ramus of the first cervical nerve. This branch first appears as the hypoglossal nerve loops around the occipital artery. It passes down on the carotid sheath covering the carotid arteries, and is joined by the lower root of the ansa cervicalis from the cervical plexus to form the ansa cervicalis. The upper root of the ansa cervicalis gives a branch to the superior belly of the omohyoid muscle.

The muscular branches supplying the thyrohyoid and geniohyoid muscles are also derived from the first cervical spinal nerve. The nerve to thyrohyoid arises as the hypoglossal nerve reaches the hyoglossus muscle. The nerve to geniohyoid is given off in the floor of the mouth, above the mylohyoid muscle.

The branches to the tongue musculature are the only true branches of the hypoglossal nerve. They are distributed to the intrinsic muscles of the tongue and to the styloglossus, hyoglossus and genioglossus muscles.

The cervical sympathetic trunk (Figure 36)

The sympathetic outflow for all parts of the body is derived principally from the thoracic spinal cord (segments T1 to L2). These preganglionic fibres then pass into the sympathetic trunk via the spinal nerves as white rami communicantes. Here, they may synapse at a ganglion or they pass up or down the sympathetic trunk to a ganglion at a different level. In this manner the cervical part of the sympathetic trunk receives its preganglionic fibres from the upper thoracic nerves.

The cervical sympathetic trunk exhibits a variable number of ganglia (usually between two and four). The ganglia are designated according to their position (superior, middle, inferior, etc.).

The superior cervical ganglion lies at the level of the second and third cervical vertebrae. It is situated behind the carotid sheath on the longus capitis muscle (a prevertebral muscle). It is the largest of the cervical sympathetic ganglia and is believed to represent the coalescence of four ganglia that correspond with the upper four cervical spinal nerves.

The branches from the superior cervical ganglion are variable, but can be broadly classified into lateral, medial and anterior groups.

The lateral branches include the grey rami communicantes to the upper four cervical spinal nerves. In addition, there are branches that communicate with some of the cranial nerves: to the inferior ganglion of the glossopharyngeal nerve, to both ganglia of the vagus nerve, and to the hypoglossal nerve. The nerve that joins the glossopharyngeal and vagus nerves is called the jugular nerve. The lateral branches of the superior cervical ganglion also include nerves to the superior jugular bulb and to the meninges of the posterior cranial fossa.

There are two medial branches of the superior cervical sympathetic ganglion. There is a laryngopharyngeal branch, which supplies the carotid body and the pharyngeal plexus, and a cardiac branch.

The anterior branches pass on to the common and external carotid arteries to form plexuses. In addition to supplying the blood vessels, the plexus around the facial branch of the external carotid provides the sympathetic supply to the submandibular parasympathetic ganglion (see page 304). The plexus around the middle meningeal artery (a branch of the maxillary artery from the external carotid) serves the otic parasympathetic ganglion (see page 179).

108,5

166C,62

Emerging above the superior ganglion is the internal carotid nerve. This nerve may be thought of as the cranial part of the sympathetic system. It passes with the internal carotid artery into the carotid canal. Within the canal, it forms the internal carotid plexus around the internal carotid artery.

166C,62

The internal carotid plexus can be divided into two parts: lateral and medial.

142A,13
142A,6

The lateral part gives branches that communicate with the trigeminal and abducent cranial nerves. Superior and inferior caroticotympanic nerves traverse the posterior wall of the carotid canal to communicate with the tympanic branch of the glossopharyngeal nerve. An important branch is the deep petrosal nerve. This nerve is destined for the pterygopalatine ganglion (see page 257). It passes through the foramen lacerum and, joining the greater petrosal branch of the facial nerve, becomes the nerve of the pterygoid canal.

The medial part of the internal carotid plexus supplies the internal carotid artery itself, and communicates with the oculomotor, trochlear, trigeminal (ophthalmic division), and abducent cranial nerves. Branches also pass through the superior orbital fissure to the ciliary ganglion in the orbit (see page 225). These branches subsequently run with the short ciliary nerves to be distributed to the blood vessels of the eyeball. The fibres to the dilator pupillae travel by a different course (via the ophthalmic, nasociliary and then the long ciliary nerves). The terminal branches of the internal carotid plexus form plexuses around the ophthalmic artery and the anterior and middle cerebral arteries of the brain, passing eventually to the pia mater.

108,41; 144A,48

108,42

The middle cervical ganglion is usually situated at the level of the sixth cervical vertebra. It is the smallest cervical ganglion and is occasionally absent. It may fuse with the superior cervical ganglion. The middle ganglion lies close to the inferior thyroid artery just before it enters the gland. Some claim that it represents the coalescence of two ganglia that correspond with the fifth and sixth cervical segments.

106B,34; 106B,28

Branches from the middle cervical ganglion communicate with the fifth and sixth cervical spinal nerves (sometimes also the fourth and seventh). Two distinct cords pass down to the inferior cervical/cervicothoracic sympathetic ganglion. The anterior cord loops in front of and below the subclavian artery as the ansa subclavia. The posterior cord encloses the vertebral artery. The middle cervical ganglion also sends branches to the thyroid gland (along the inferior thyroid artery), to the heart via its cardiac branches, and to the trachea and oesophagus.

An occasional ganglion known as the vertebral ganglion may be found on the front of the vertebral artery. It can be considered as either a low middle cervical ganglion or as a detached part of the inferior ganglion. When present, it gives rise to the ansa subclavia.

The inferior cervical ganglion often combines with the first thoracic ganglion to form the cervicothoracic ganglion (stellate ganglion). The inferior cervical ganglion (or upper end of the cervicothoracic ganglion) is situated just posterior to the vertebral artery. The lower end of a cervicothoracic ganglion lies behind the subclavian artery on the first thoracic vertebra.

Branches from the inferior cervical ganglion pass to the seventh and eighth cervical nerves and to the first thoracic nerve. There are also cardiac branches and fibres that form plexuses around the subclavian artery and its derivatives. Around the vertebral artery is a plexus that continues up into the skull. This plexus eventually meets the plexus around the internal carotid artery. Some anatomists believe this to be the main intracranial extension of the sympathetic system.

The cervical plexus (Figure 37)

This plexus lies deep to the sternocleidomastoid muscle and the internal jugular vein, and in front of the scalenus medius and levator scapulae muscles. It is formed by the ventral rami of the upper four cervical spinal nerves (i.e. C1, C2, C3, C4).

The cervical plexus contains both sensory and motor fibres. In addition, grey rami communicantes near the origins of the ventral rami of the cervical nerves supply sympathetic fibres.

Branches from the cervical plexus are distributed to some of the muscles of the neck, to the diaphragm, and to much of the skin of the back of the head, the neck, and the chest around the thoracic inlet. The cutaneous nerves are superficial, the muscular branches deep.

The cutaneous nerves from the cervical plexus are the lesser occipital (C2), great auricular (C2, C3), transverse cervical (C2, C3), and supraclavicular (C3, C4). These nerves are described on page 85.

The deep (mainly motor) branches can be subdivided into those that pass medially and those that pass laterally.

The medial branches supply the following muscles:

- Longus capitis (C1, C2, C3)
- Longus colli (C2, C3, C4)
- Rectus capitis anterior (C1, C2)
- Rectus capitis lateralis (C1)

Other medial branches are the inferior root of the ansa cervicalis (C2, C3) and the phrenic nerve (C3, C4, C5). Some branches also communicate with the hypoglossal and vagus nerves, and the sympathetic trunk.

The lateral branches supply the following muscles:

- Levator scapulae (C3, C4)
- Scalenus medius (C3, C4)
- Sternocleidomastoid (C2)
- Trapezius (C3, C4)

There is also a communicating branch to the accessory nerve (C2, C3, C4).

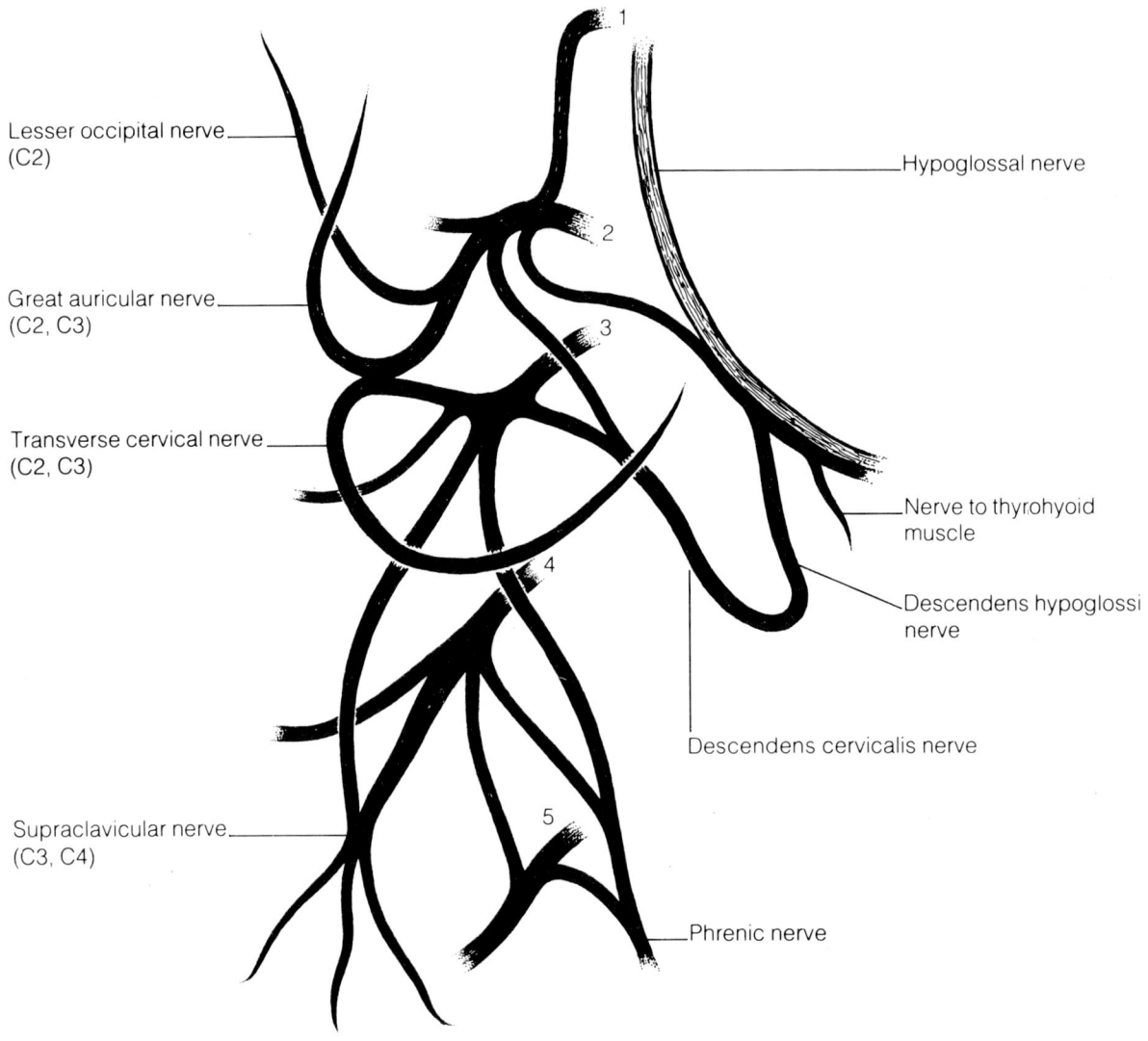

Figure 37 *The cervical plexus.*

100,36; 104,29   **The phrenic nerve** arises from the cervical plexus and usually takes fibres from the third, fourth and fifth cervical nerves (mainly from the fourth). It provides the motor nerve supply to the diaphragm.

108,24   The phrenic nerve in the neck passes downwards and medially on the superficial surface of the scalenus anterior muscle. Here, it lies under cover of a layer of the prevertebral fascia. As it passes through the thoracic inlet, it runs behind the subclavian vein and in front of the subclavian artery and its internal thoracic branch.

104,23   Some of the roots may not join the main nerve trunk until just before leaving the neck. Such roots are called accessory phrenic nerves.

The phrenic nerve contains not only motor fibres but also proprioceptive fibres to the diaphragm, and sensory fibres to the pleura and pericardium. Sympathetic fibres may join the phrenic nerve from cervical sympathetic ganglia.

100,17   **The branches to the hypoglossal nerve** from the cervical plexus arise mainly from the first cervical spinal nerve. These fibres leave the hypoglossal nerve as four distinct nerves: the meningeal branch of the hypoglossal nerve, the superior root of the ansa

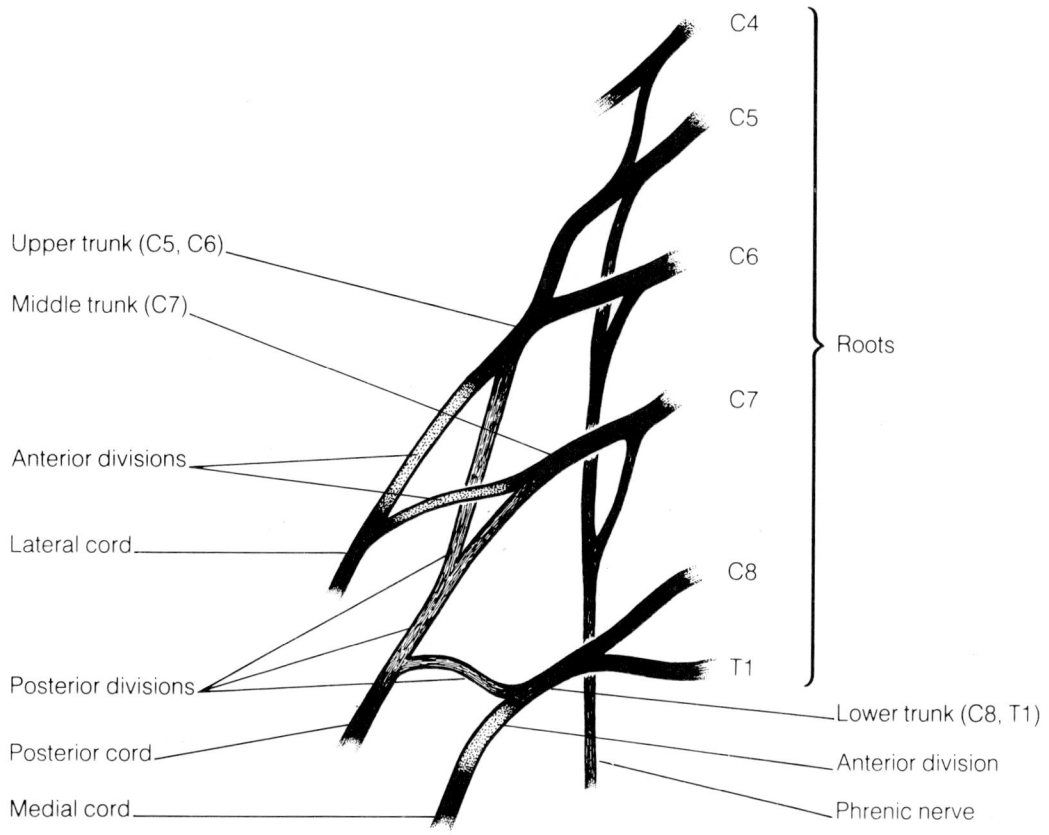

Figure 38 *The brachial plexus.*

cervicalis, and the motor nerves to the thyrohyoid and the geniohyoid muscles. Some C1 fibres also travel with the vagus nerve, and indeed may form its meningeal branch.

100,9; 144A,24

The ansa cervicalis (Figures 20, 35 and 37) is a nerve plexus located in front of the common carotid artery. It is formed by the union of two nerve trunks: the superior root of the ansa cervicalis from the hypoglossal nerve (conveying C1 fibres) and the inferior root of the ansa cervicalis from the cervical plexus (conveying C2 and C3 fibres). The superior root is also called the descendens hypoglossi, indicating its path from the hypoglossal nerve as it crosses the external carotid artery. The inferior root may also be termed the descendens cervicalis. It usually appears lateral to the internal jugular vein, crossing the vein to join the superior root. Occasionally, the inferior root may run medial to the internal jugular vein.

100,47

100,17
100,35

The ansa cervicalis supplies all the infrahyoid muscles (see Figure 20) with the exception of the thyrohyoid muscle. The innervation of the superior belly of the omohyoid muscle is often given off from the superior root just before it reaches the ansa cervicalis.

The brachial plexus (Figure 38)

98–103

This plexus lies in the deep part of the posterior triangle of the neck, between the clavicle and the lower part of the posterior border of the sternocleidomastoid muscle. It emerges between the scalenus anterior and scalenus medius muscles to pass between the clavicle and first rib, around the axillary artery and into the upper limb. The relationships of structures to the plexus in the root of the neck is illustrated in Figure 21.

98,19

The brachial plexus is formed by the ventral rami of the fourth to the eighth cervical nerves and by most of the ventral ramus of the first thoracic nerve. The plexus provides the innervation for structures in the upper limb. The general arrangement (trunks, divisions, cords, etc.) and branches of the plexus are shown in Figure 38.

The branches arising from the brachial plexus above the clavicle in the neck are:

- Nerves to the scalene and longus colli muscles (C5, C6, C7, C8)
- Communicating branch to the phrenic nerve (C5)
- Dorsal scapular nerve to the rhomboid muscles (C5)
- Long thoracic nerve to serratus anterior muscle (C5, C6, C7)
- Nerve to the subclavius (C5, C6)
- Suprascapular nerve to the supraspinatus and infraspinatus muscles and to the shoulder joint (C5, C6)

Thus, the branches are mainly motor.

THE LYMPHATICS OF THE NECK (Figure 39)

The lymphatic system is very variable. It comprises plexuses of small vessels (the lymph capillaries) which run into small masses of lymphoid tissue (the lymph nodes).

The lymph nodes in the neck can be categorised into superficial and deep lymph nodes. The superficial lymph nodes are the submental, submandibular, anterior cervical and superficial cervical nodes. The deep lymph nodes are the infrahyoid, prelaryngeal, pretracheal and paratracheal, retropharyngeal, and deep cervical nodes.

The submental nodes lie beneath the chin, on the mylohyoid muscle and between the anterior bellies of the digastric muscles. They receive lymph vessels from the anterior part of the mandible, the lower lip and the tip of the tongue (see Figure 120). Vessels from the submental nodes drain into either the jugulo-omohyoid group of the deep cervical nodes near the root of the neck or into the submandibular nodes.

The submandibular nodes are situated close to, or within, the submandibular salivary glands. The nodes receive vessels from many parts of the oral cavity (see Figure 120) and from the submental and, when present, the buccal and lingual nodes. Efferent vessels drain into the deep cervical nodes.

The anterior cervical nodes lie along the anterior jugular veins at the front of the neck. They drain lymph from the front of the neck below the hyoid bone. Vessels from the nodes pass to any of the deep lymph nodes in the neck.

The superficial cervical nodes are associated with the external jugular vein on the superficial surface of the sternocleidomastoid muscle. They receive vessels from around the lower part of the ear, the floor of the external acoustic meatus, the apical part of the parotid gland and the region around the angle of the mandible. The efferent vessels pass to the deep cervical nodes.

The infrahyoid nodes lie beneath the investing layer of deep cervical fascia on the thyrohyoid membrane. Lymph vessels from the region of the epiglottis pass to these nodes. Vessels pass from this group to the deep cervical chain.

The prelaryngeal nodes lie on the anterior cricothyroid ligament and the cricovocal membrane of the larynx. The **pretracheal nodes** lie on the trachea close to the inferior thyroid veins. The **paratracheal nodes** are found in association with the recurrent laryngeal nerves, between the trachea and the oesophagus. The **retropharyngeal nodes** are situated between the back of the larynx and the prevertebral fascia. The prelaryngeal and pretracheal nodes receive vessels from the larynx below the vocal

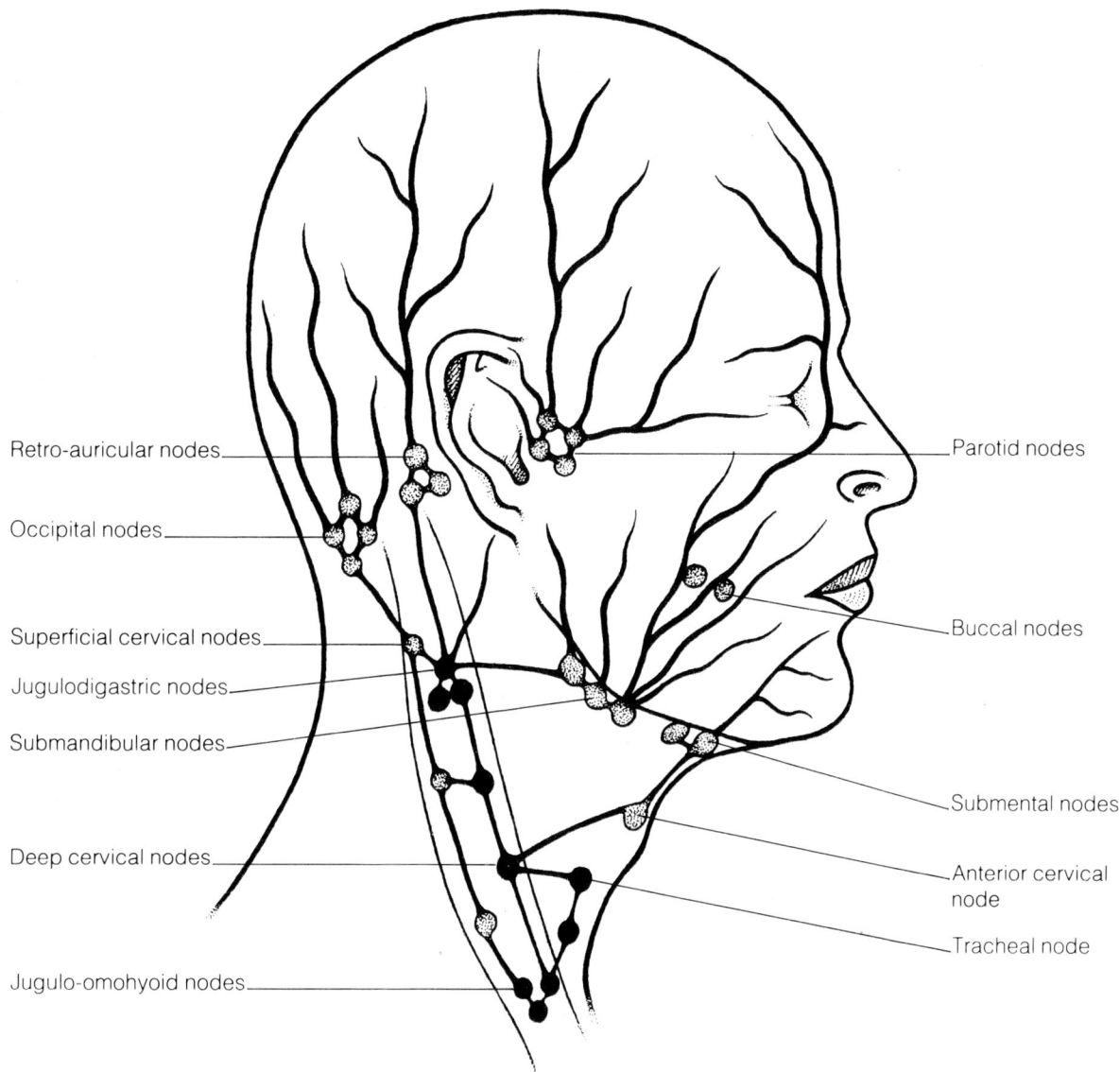

Figure 39 *The lymphatics of the neck. The nodes with the stippled shading are the superficial nodes; the nodes with the solid black shading are deep nodes.*

fold and from the trachea. The thyroid gland sends vessels to the prelaryngeal and tracheal nodes. The pharynx and oesophagus send vessels to the paratracheal and retropharyngeal nodes. Efferent vessels from all these nodes pass to the deep cervical chain of lymph nodes.

The deep cervical lymph nodes lie along the carotid sheath, deep to the sternocleidomastoid muscle. A prominent group of these nodes superiorly is located close to the digastric muscle, and is consequently designated the jugulodigastric group of lymph nodes. Inferiorly, another prominent group lies close to the omohyoid muscle, the jugulo-omohyoid group. The upper deep cervical nodes (including the jugulodigastric nodes) receive lymph vessels from the tongue, most of the nose, the air sinuses, the ear, the tonsil, the larynx above the vocal fold, the submandibular and parotid salivary glands, and from the retro-auricular and superficial lymph nodes. The lower deep cervical nodes receive vessels from all the other deep nodes of the neck, and from the submental, anterior and superficial cervical, and occipital lymph nodes.

From the deep cervical nodes, lymph is collected into the jugular trunk. The left jugular trunk drains into the thoracic duct, although it may pass directly into the subclavian or internal jugular veins. The right jugular trunk drains directly into the right brachiocephalic vein at its origin or into the right lymphatic duct (this duct collects lymph from the right arm and the right half of the thorax and also drains into the right brachiocephalic vein).

THE VISCERA OF THE NECK

The visceral structures of the neck include the larynx and the trachea, the pharynx and the oesophagus, and the thyroid, parathyroid and cervical thymus glands.

The pharynx and larynx are described on pages 310 to 318 and 332 to 346.

The trachea

The trachea is part of the respiratory system and is a tube composed of cartilages and membranes.

The trachea begins at the lower border of the cricoid cartilage of the larynx, at the level of the sixth cervical vertebra. It extends down the neck along the median plane and through the thoracic inlet. It ends in the superior mediastinum by dividing into the right and left principal bronchi.

The patency of the trachea is maintained by a series of C-shaped rings of hyaline cartilage, which are incomplete posteriorly. The cartilages are united by fibro-elastic membranes. Posteriorly, across the gap of each cartilage, is a thin coat of unstriated muscle (the trachealis muscle). Contraction of this muscle narrows the lumen of the trachea.

The relationships between the trachea and other cervical structures (see Figure 17) are of clinical importance, particularly as a tracheostomy is not an uncommon procedure. Anteriorly is situated the thyroid gland, the thyroid isthmus being at the level of the second to the fourth tracheal cartilages. Below the thyroid gland are the inferior thyroid veins, tracheal lymph nodes and occasionally a thyroidea ima artery. All these structures are covered by the infrahyoid muscles. Lateral to the trachea are the lobes of the thyroid gland and the carotid sheaths. Posteriorly, the trachea lies on the oesophagus. Posterolaterally are the recurrent laryngeal nerves. These lie in grooves between the sides of the trachea and the oesophagus.

The main blood supply to the cervical part of the trachea is derived from the inferior thyroid arteries. The veins drain into the brachiocephalic veins via the inferior thyroid plexus. Lymphatic vessels drain into the pretracheal and paratracheal nodes.

The nerve supply to the trachea arises from the vagi, recurrent laryngeal nerves, and sympathetic trunks.

The oesophagus

The oesophagus is the alimentary tube connecting the laryngopharynx to the stomach. It begins at the lower border of the cricoid cartilage. Here, the cricopharyngeal part of the inferior constrictor muscle of the pharynx acts as a sphincter.

The cervical part of the oesophagus takes a curved course down the median plane of the neck. It lies between the vertebral column and the trachea (see Figure 17). Anterolaterally are situated the lobes of the thyroid gland, the recurrent laryngeal nerves, the inferior thyroid arteries and the carotid sheaths.

The muscles within the cervical part of the oesophagus are arranged in inner and outer layers. The inner layer is circular and the outer layer is longitudinal. The circular musculature is continuous with the inferior constrictor of the pharynx.

The blood supply of the oesophagus in the neck is derived mainly from the inferior thyroid arteries. The veins drain into the brachiocephalic veins. Lymphatic vessels pass into retropharyngeal, paratracheal, or deep cervical lymph nodes.

The cervical part of the oesophagus is innervated by the recurrent laryngeal nerves and by the sympathetic plexus around the inferior thyroid artery.

The thyroid gland

102–106; 154A

The thyroid gland is an endocrine gland situated in the front of the neck. It lies on the trachea and just above the thoracic inlet. The gland is closely related to the thyroid cartilage of the larynx, and extends from the level of the fifth cervical vertebra to the first thoracic vertebra.

104,8

154A,20

The gland consists of right and left lobes joined by an isthmus. The isthmus lies just below the cricoid cartilage on the second to the fourth tracheal rings. The isthmus is occasionally absent.

104,8; 104,11
105,42; 105,46

Each lobe of the thyroid gland is conical or pear-shaped and has a narrow apex and a broad base. The apex lies beneath the oblique line of the thyroid cartilage. The base lies at about the level of the fourth tracheal ring.

A process of thyroid tissue, called the pyramidal process, often projects upwards from the isthmus. There may also be a fibrous or fibromuscular structure called the levator glandulae thyroidae. This passes from the body of the hyoid bone to the isthmus or pyramidal process of the thyroid gland. When muscular, some anatomists believe the levator glandulae thyroidae to be a derivative of the infrahyoid musculature. It can be innervated either through the ansa cervicalis or through the vagus nerve (usually the superior laryngeal branch).

The thyroid gland is invested by a capsule of connective tissue. This capsule has been variously described. Most accounts state that the gland has its own delicate perithyroid sheath which lies on the surface of the gland and which sends septa between the lobules. There is also a surrounding layer of pretracheal fascia (see Figure 17). The gland is maintained in position by this fascia and by ligamentous bands (the lateral ligaments) which attach the gland on each side to the cricoid cartilage of the larynx.

Accessory (or ectopic) thyroid tissue can be found throughout the neck and even within the tongue. This reflects the fact that the thyroid gland develops at the tongue and migrates down the neck to its adult position.

The thyroid gland has many important relationships in the neck, and these have been described with the trachea and oesophagus, and in Figure 17. In addition, the parathyroid glands lie posteriorly.

105,41; 105,44

The thyroid gland has a very rich blood supply derived from four main arteries: the two superior thyroid arteries and the two inferior thyroid arteries. There are many anastomoses between these vessels both ipsilaterally and contralaterally. The superior thyroid artery is the first branch of the external carotid artery. It descends to the apex of the lobe of the thyroid gland with the external branch of the superior laryngeal nerve. The superior thyroid artery pierces the thyroid fascia and then divides into anterior and posterior branches. The anterior branch supplies the anterior surface of the gland, the posterior branch supplies the lateral and medial surfaces. The inferior thyroid artery arises from the thyrocervical trunk of the subclavian artery in the root of the neck. As it approaches the base of the thyroid gland, the artery divides into superior (ascending) and inferior thyroid branches. These supply the inferior and posterior surfaces of the gland. The superior branch also supplies the parathyroid glands. The relationship between the inferior thyroid artery and the recurrent laryngeal nerve has clinical importance. The nerve initially lies in front of the artery but, near the thyroid gland, it usually passes behind the left inferior thyroid artery but may remain in front of the right inferior thyroid artery. An occasional artery to the thyroid gland is the thyroidea ima artery. This may arise either from the brachiocephalic artery, the right common carotid, or the arch of the aorta.

102,22; 104,4
154A,27; 154A,16
105,40

102B,51; 102B,54
104,35; 104,34
105,43

154A,25; 154A,23
102B,51; 102B,62

The venous drainage of the thyroid gland is usually via superior, middle, and inferior thyroid veins. The superior thyroid vein emerges from the upper part of the gland and runs with the superior thyroid artery towards the carotid sheath. It drains into the

102A,24,104,4

internal jugular vein. The middle thyroid vein collects blood from the lower part of the gland. It emerges from the lateral surface of the gland and drains into the internal jugular vein. The inferior thyroid vein forms a plexus with the vein on the opposite side. This plexus is located below the thyroid gland and in front of the trachea. From the plexus, the left vein descends into the thorax to terminate at the left brachiocephalic vein. The right inferior thyroid vein drains into the right brachiocephalic vein. Alternatively, there may be a common trunk draining into the left brachiocephalic vein.

The lymphatics from the gland usually pass to the prelaryngeal and the tracheal nodes. They may also pass directly into the deep cervical nodes or into the thoracic duct or the right lymph duct.

The innervation of the thyroid gland is derived from the cervical sympathetic trunk.

The parathyroid glands

There are usually four parathyroid glands, two on each side. They are small, spherical, endocrine glands which are situated at the back of the thyroid gland. To indicate their relative positions, the glands are designated the superior and inferior parathyroid glands. Alternatively, they have been called parathyroid 3 and parathyroid 4 to indicate their embryological origins from the branchial pouches; the superior parathyroid glands are derived from the fourth pouches, the inferior glands from the third pouches.

The superior parathyroid glands are said to lie near the middle of the lobes of the thyroid gland. The positions of the inferior parathyroid glands are more variable. Indeed, evidence from dissection and surgery suggests that both sets of glands are so variable in location that a meaningful description of site is not possible. This also means that their relationship to the fascia surrounding the thyroid gland is variable. Nevertheless, the parathyroid glands are usually described as lying between the posterior surface of the thyroid gland and the thin capsule of the perithyroid sheath (lying within the pretracheal fascia).

The parathyroid glands receive a rich blood supply, usually from the inferior thyroid arteries. The veins drain into the thyroid veins. The lymphatic vessels pass with those from the thyroid gland. The nerve supply is derived from the cervical sympathetic trunks.

The thymus gland

This gland has an important role in the development of the lymphoid system. It is situated mainly in the thorax, although there are cervical extensions that may reach up in the midline as far as the base of the thyroid gland. The thymus gland has a flattened, bilobed appearance. Its size varies considerably with age. It increases in size until puberty but progressively diminishes thereafter, eventually to be replaced by fat.

Accessory thymus tissue may be found in the neck. The inferior parathyroid glands may connect with the thymus by means of prominent strands of connective tissue. This may reflect the fact that both the thymus and the inferior parathyroid glands develop from the third branchial pouch.

The blood supply to the thymus gland is derived from the internal thoracic and inferior thyroid arteries. The venous drainage is via the internal thoracic, inferior thyroid, and left brachiocephalic veins. The lymphatic vessels drain into nodes within the thorax. The innervation of the thymus gland arises from the sympathetic system and the vagus nerves.

THE MUSCULOSKELETAL COMPARTMENT OF THE NECK

This compartment is composed of the cervical vertebral column (see page 61), and the prevertebral muscles and postvertebral muscles. The region is ensheathed by layers of the prevertebral fascia (see page 95).

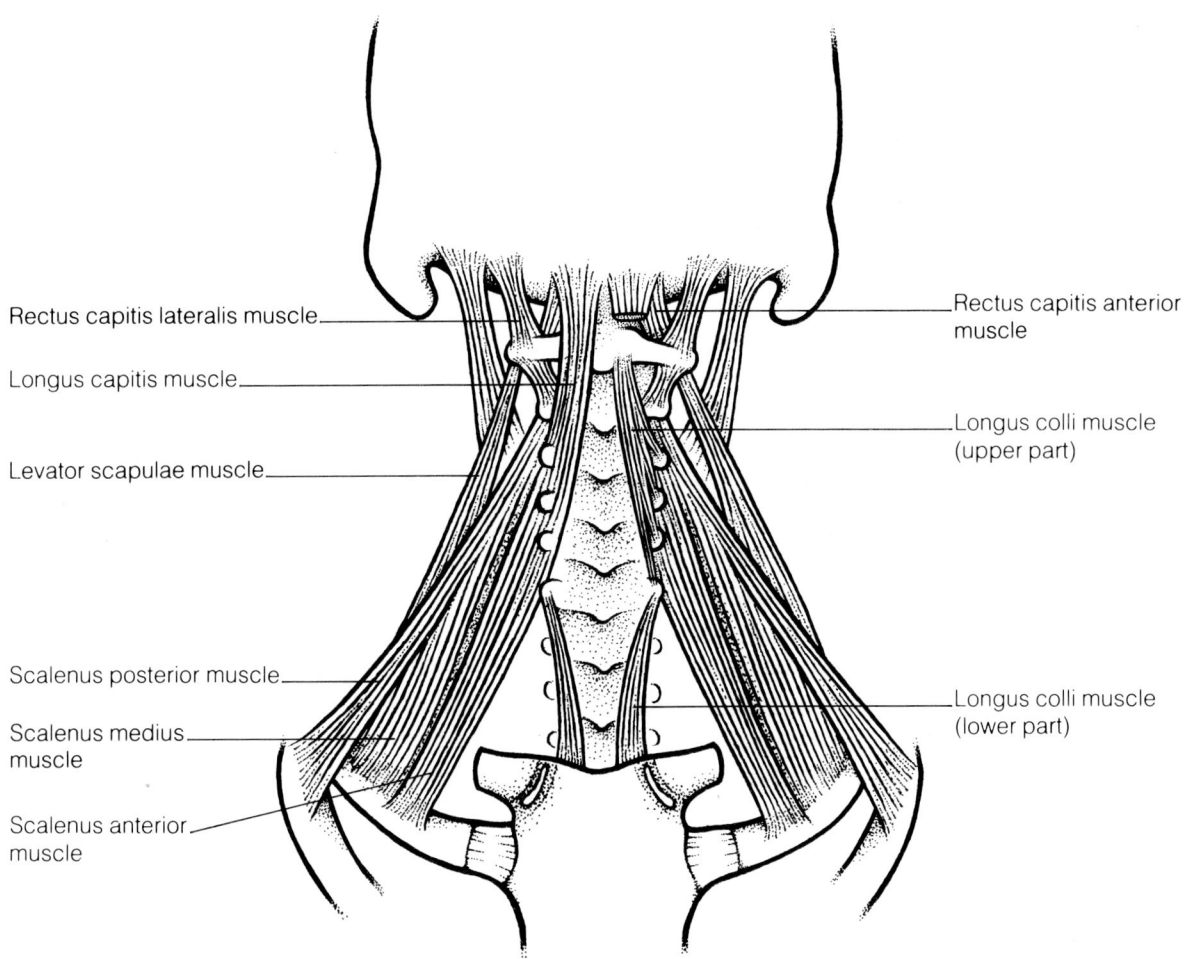

Figure 40 *The prevertebral muscles.*

THE PREVERTEBRAL MUSCLES (Figure 40)

These muscles lie in front of the bodies of the cervical vertebrae and deep to the prevertebral fascia. All are essentially flexors of the head and neck. The anterior prevertebral muscles are longus colli, longus capitis, rectus capitis anterior and rectus capitis lateralis. The lateral prevertebral muscles are the scalene muscles (anterior, medius, posterior, and minimus) and levator scapulae.

Longus colli

This muscle lies on the anterior surface of the vertebral column and extends from the first cervical vertebra to the third thoracic vertebra. It has three parts: superior oblique, inferior oblique and vertical.

Attachments. The superior oblique part arises from the transverse processes (anterior tubercles) of the third to the fifth cervical vertebrae. It passes upwards and medially to insert on to the anterior arch (anterior tubercle) of the first cervical vertebra.

The inferior oblique part originates from the bodies of the first to the third thoracic vertebrae. It passes up to the transverse processes (anterior tubercles) of the fifth and sixth cervical vertebrae.

The vertical part arises from the bodies of the upper three thoracic vertebrae and the lower three cervical vertebrae. It inserts on to the bodies of the second to the fourth cervical vertebrae.

Innervation. The motor supply of longus colli is derived from the ventral rami of the second to the eighth cervical spinal nerves.

Vasculature. Its arterial blood supply is derived from the ascending pharyngeal, inferior thyroid (ascending cervical branch) and vertebral arteries.

Actions. The longus colli muscles flex and assist in rotating the head and neck. Acting singly, the cervical vertebral column is flexed laterally.

Longus capitis

Attachments. This muscle arises from the transverse processes (anterior tubercles) of the third to the sixth cervical vertebrae. It inserts on to the basilar part of the occipital bone at the cranial base.

Innervation. Its motor innervation comes from the ventral rami of the first to the third cervical spinal nerves.

Vasculature. The blood supply to longus capitis is similar to that of longus colli, coming from the ascending pharyngeal, inferior thyroid and vertebral arteries.

Actions. It flexes the head.

Rectus capitis anterior

Attachments. This muscle lies behind longus capitis. It originates from the lateral mass of the first cervical vertebra and inserts on to the basilar part of the occipital bone.

Innervation. It is supplied by the ventral rami of the first and second cervical spinal nerves.

Vasculature. Its blood supply is derived from the vertebral and ascending pharyngeal arteries.

Actions. Rectus capitis anterior flexes the head.

Rectus capitis lateralis

Attachments. This muscle passes from the upper surface of the transverse process of the first cervical vertebra to the inferior surface of the jugular process of the occipital bone.

Innervation. Its motor supply is derived from the ventral rami of the first and second cervical spinal nerves.

Vasculature. The vertebral, occipital and ascending pharyngeal arteries contribute to the muscle's blood supply.

Actions. It flexes the head laterally.

The scalene muscles are flexors and rotators of the vertebral column. They extend obliquely from the transverse processes of the cervical vertebrae to the ribs. They provide important landmarks in the root of the neck (see page 129).

Scalenus anterior

Attachments. Scalenus anterior lies behind the sternocleidomastoid muscle. It arises from the transverse processes (anterior tubercles) of the third to the sixth cervical vertebrae. It descends almost vertically to insert on to the scalene tubercle and ridge on the upper surface of the first rib.

Innervation. The muscle is innervated by the ventral rami of the fourth to the sixth cervical spinal nerves.

Vasculature. The ascending cervical branch of the inferior thyroid artery is the main source of the muscle's blood supply.

Actions. Acting from above, scalenus anterior elevates the first rib. Acting from below, it flexes the cervical part of the vertebral column anteriorly and laterally, and also rotates the vertebral column.

Scalenus medius

Attachments. This is the largest of the scalene muscles. It arises from the transverse processes (posterior tubercles) of the lower six cervical vertebrae. It inserts on to the upper surface of the first rib, behind the groove for the subclavian artery.

Innervation. It receives its motor supply from the ventral rami of the third to the eighth cervical spinal nerves.

Vasculature. Scalenus medius receives its blood supply from the inferior thyroid artery (ascending cervical branch).

Actions. Acting from above, it elevates the first rib. Acting from below, it flexes the cervical vertebral column to the same side.

Scalenus posterior

Attachments. This is the smallest of the scalene muscles. It is also the deepest and forms part of the floor of the posterior triangle with scalenus medius. It is often difficult to separate from scalenus medius, and is consequently sometimes considered as part of this muscle. Scalenus posterior arises from the transverse processes (posterior tubercles) of the fourth to the sixth cervical vertebrae. It inserts on to the outer surface of the second rib, behind the attachment of the serratus anterior muscle.

Innervation. The motor innervation comes from branches of the ventral rami of the lower three cervical spinal nerves.

Vasculature. Both the ascending cervical branch of the inferior thyroid artery and the superficial cervical artery supply scalenus posterior.

Actions. Acting from above, it elevates the second rib. Acting from below, it flexes the lower part of the cervical vertebral column to the same side.

Scalenus minimus

This muscle is found in about 66% of people. It arises from the anterior tubercle of the sixth or seventh cervical vertebra. It inserts on to the first rib and into the suprapleural membrane which covers the apex of the lung in the root of the neck.

Levator scapulae

Attachments. The muscle takes origin from the transverse processes (posterior tubercles) of the first to the fourth cervical vertebrae. It descends obliquely across the floor of the posterior triangle of the neck to insert on to the medial edge of the scapula (between the superior angle and the root of the spine).

Innervation. The innervation is derived partly from the third and fourth cervical nerves directly, and partly from the fifth cervical nerve indirectly through the dorsal scapular nerve.

Vasculature. The blood supply comes mainly from the superficial cervical and inferior thyroid (ascending cervical) arteries. Near the attachments to the cervical vertebral column, it is supplied by branches from the vertebral artery.

Actions. As its name suggests, the levator scapulae elevates the scapula. This is accomplished when acting with the trapezius muscle. Levator scapulae also acts in association with other muscles to control the position of the scapula when the upper limb is in active use. Should the scapula be fixed, the muscle can incline the neck to the same side.

THE POSTVERTEBRAL MUSCLES

These muscles lie deep to the trapezius muscle at the back of the head and neck and behind the vertebral column. They can be subdivided into three layers: a superficial layer, a middle layer, and a deep layer.

The superficial layer consists of the splenius cervicis and capitis muscles. The middle layer comprises parts of erector spinae (iliocostalis cervicis, longissimus cervicis and capitis, spinalis cervicis and capitis). The deep layer incorporates the semispinalis cervicis and capitis muscles, and some deep slips of muscles belonging to the multifidus, rotatores, interspinalis and intertransversarii groups. The fibres in the superficial layer pass upwards and outwards. The fibres in the middle layer run parallel to the vertebral column. Most of the deep muscles lie in the groove between the spines and the transverse processes of the cervical vertebrae.

Behind the first and second cervical vertebrae are the muscles comprising the suboccipital triangle.

Splenius cervicis and capitis

Attachments. Splenius cervicis takes origin from the spines of the third to the sixth thoracic vertebrae. It inserts into the transverse processes (posterior tubercles) of the upper three or four cervical vertebrae. Splenius capitis passes over splenius cervicis. It arises from the ligamentum nuchae and spines of the seventh cervical and upper three or four thoracic vertebrae. The muscle inserts into the mastoid process and superior nuchal line on the cranium.

Innervation. The nerve supply is from lateral branches of the dorsal rami of the middle cervical spinal nerves.

Vasculature. The blood supply is derived from the occipital and superficial cervical arteries.

Actions. Acting on both sides, they extend the head backwards. Acting on one side only, they pull the head to one side with slight rotation.

The erector spinae group of muscles

This complex group of muscles is illustrated in Figure 41. In the neck, the group can be subdivided into a lateral layer (iliocostalis cervicis), an intermediate layer (longissimus cervicis and capitis), and a medial layer (spinalis cervicis and capitis).

Attachments. Iliocostalis cervicis extends from the angles of the third to the sixth ribs up to the transverse processes (posterior tubercles) of the fourth to the sixth cervical vertebrae.

Longissimus cervicis arises from the transverse processes of the upper four thoracic vertebrae. The insertion is into the transverse processes (posterior tubercles) of the second to the sixth cervical vertebrae. Longissimus capitis extends from transverse processes of the upper four thoracic vertebrae and the articular processes of the lower four cervical vertebrae to the posterior margin of the mastoid process of the temporal bone.

Spinalis cervicis takes origin from the spines of the first and second thoracic and of the seventh cervical vertebrae. It inserts on to the spine of the second cervical vertebra. Spinalis capitis arises from the transverse processes of the seventh cervical to the seventh thoracic vertebrae, and from the articular processes of the fourth to the sixth cervical vertebrae. The muscle passes upwards to insert on the occipital bone between the superior and inferior nuchal lines. Near its insertion, it blends with the semispinalis capitis muscle.

Innervation. The muscles of erector spinae are innervated by dorsal rami of spinal nerves.

Vasculature. They receive their blood supply from the occipital artery, costocervical trunk, and intercostal arteries.

Actions. The muscles are involved in extension, lateral flexion, and rotation of the head and neck.

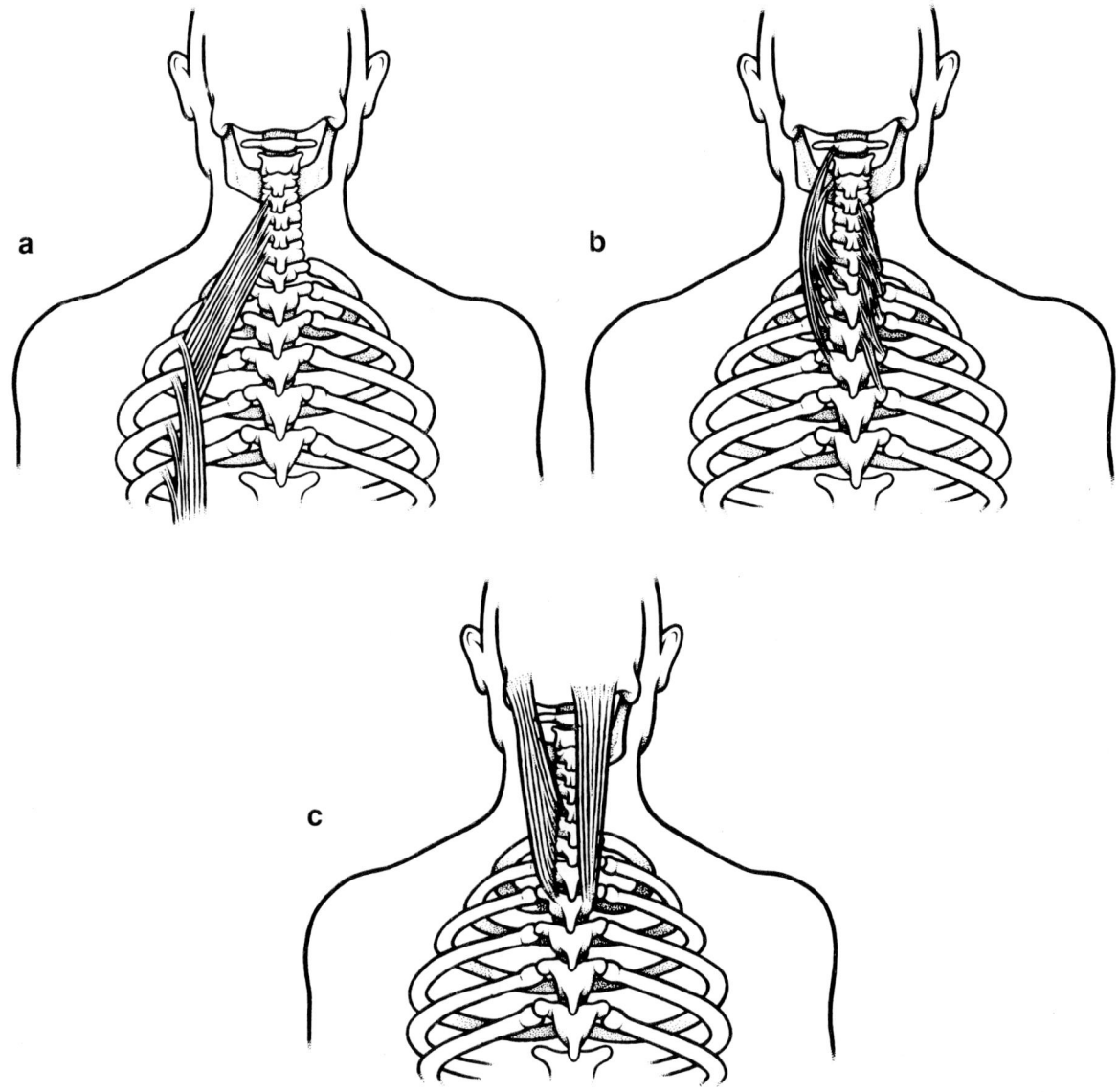

Figure 41 *Diagram illustrating some of the deep postvertebral muscles of the neck.* **a**) *Iliocostalis.* **b**) *Longissimus cervicis (left), semispinalis cervicis (right).* **c**) *Longissimus capitis (left), semispinalis capitis (right).*

Semispinalis cervicis and capitis (Figure 41)

Attachments. Semispinalis cervicis takes origin from the transverse processes of the upper six thoracic vertebrae and from the articular processes of the lower four cervical vertebrae. It passes upwards and medially to insert on to the spines of the second to the fifth cervical vertebrae. Semispinalis capitis arises from the transverse processes of the upper six thoracic vertebrae and the seventh cervical vertebra, and also from the articular processes of the fourth to the sixth cervical vertebrae. It passes upwards to insert on to the occipital bone, between the superior and inferior nuchal lines.

Innervation. The muscles are supplied by dorsal rami of the cervical spinal nerves.

Vasculature. The arterial blood supply is derived from the occipital artery, deep cervical branch of the costocervical trunk and from muscular branches of posterior intercostal arteries.

Actions. Semispinalis cervicis extends and rotates the cervical region of the vertebral

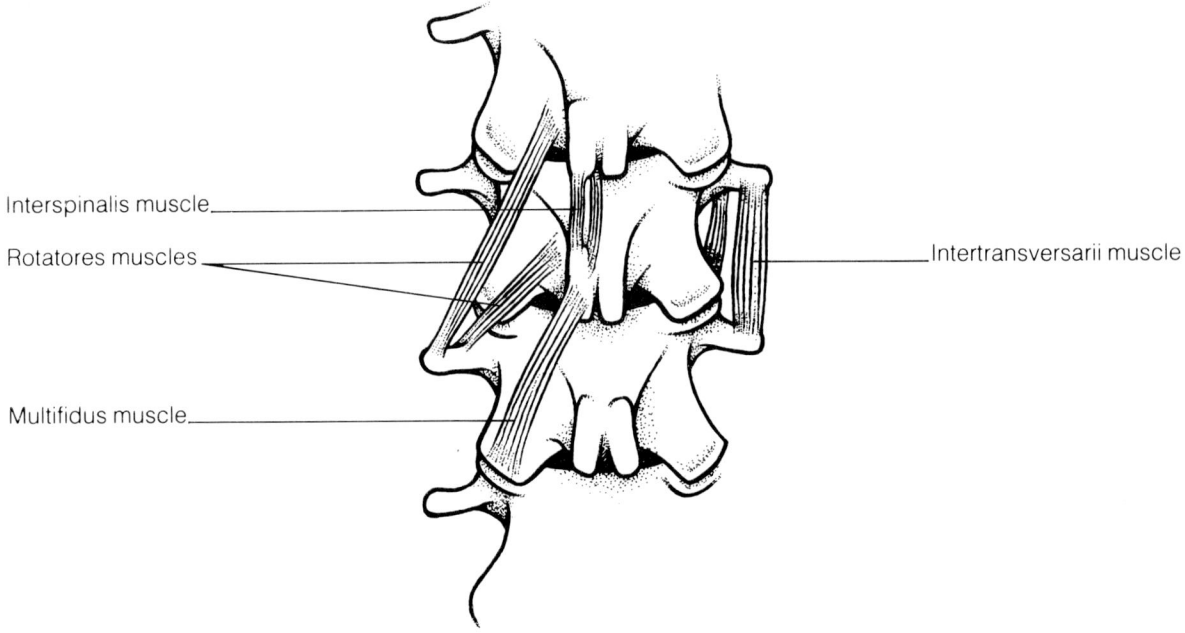

Figure 42 *The multifidus, rotatores, interspinales and intertransversarii muscles* at the back of the vertebral column.

column. Semispinalis capitis extends the head and turns the face towards the opposite side.

The multifidus, rotatores, interspinales and intertransversarii muscles are small slips of muscle that link adjacent vertebrae (Figure 42). They are involved in extension, lateral flexion and rotation of the vertebral column.

THE SUBOCCIPITAL TRIANGLE (Figure 43)

This is a region at the back of the neck that is immediately related to the first cervical vertebra. It lies under cover of the trapezius, splenius capitis, semispinalis capitis and longissimus capitis muscles.

The suboccipital triangle is bounded by the following muscles: rectus capitis posterior major (above and medially), obliquus capitis superior (above and laterally), and obliquus capitis inferior (below and laterally).

The floor of the triangle is formed by the posterior arch of the first cervical vertebra and by the posterior atlanto-occipital membrane of the atlanto-occipital joint.

Running across the floor of the triangle are the vertebral artery and the dorsal ramus of the first cervical spinal nerve (the suboccipital nerve) (Figure 43b). The artery passes horizontally behind the lateral mass of the atlas and then upwards and inwards to enter the cranial cavity through the foramen magnum. The suboccipital nerve runs between the artery and the posterior arch of the atlas to supply the suboccipital muscles.

Within the roof of the triangle are the medial branch of the dorsal ramus of the second cervical spinal nerve (the greater occipital nerve), and the occipital artery (Figure 43a). The greater occipital nerve pierces the trapezius muscle close to its attachment on to the superior nuchal line of the occiput. It then ascends to supply the skin over the occipital part of the scalp up to the vertex of the skull. The occipital artery runs with the greater occipital nerve, passing deep and then medial to the nerve.

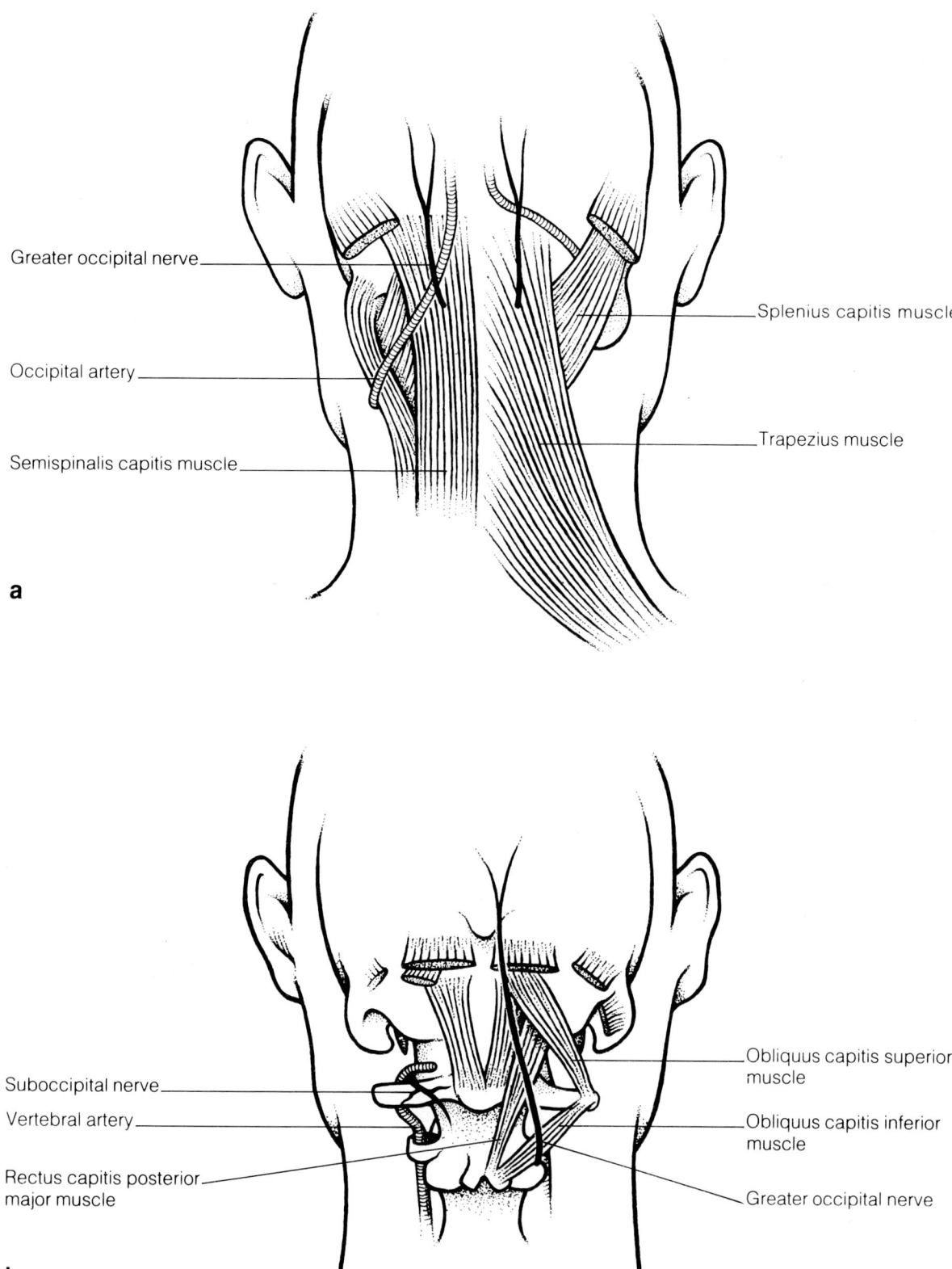

Figure 43 *The suboccipital region.* a) *Superficial dissection.* b) *Deep dissection showing the suboccipital triangle and its contents.*

Rectus capitis posterior major

Attachments. This muscle extends from the spine of the second cervical vertebra to the occipital bone below the inferior nuchal line.

Innervation. The motor supply comes from the suboccipital nerve.

Vasculature. The arterial blood supply is derived from the vertebral and the occipital arteries.

Action. The muscle is involved in extension, lateral flexion and rotation of the head.

Obliquus capitis superior (superior oblique)

Attachments. It originates from the transverse process of the first cervical vertebra and inserts at the occipital bone between the superior and inferior nuchal lines.

Innervation. It is innervated by the suboccipital nerve.

Vasculature. The blood supply comes from the vertebral and occipital arteries.

Actions. The superior oblique muscle is involved in extension and lateral rotation of the head.

Obliquus capitis inferior (inferior oblique)

Attachments. The muscle passes upwards and laterally from the spine of the second cervical vertebra to the transverse process of the first cervical vertebra.

Innervation. It is supplied by branches from the dorsal rami of the first and second cervical spinal nerves.

Vasculature. The arteries supplying the muscle are branches of the vertebral and occipital arteries.

Actions. The muscle rotates the first cervical vertebra and the skull around the dens of the second cervical vertebra.

Between the two rectus capitis posterior major muscles lie the rectus capitis posterior minor muscles.

Rectus capitis posterior minor

Attachments. This muscle takes origin from the tubercle on the posterior arch of the first cervical vertebra. It passes upwards to insert on to the occipital bone, below the inferior nuchal line and close to the midline.

Innervation. Its motor supply comes from the suboccipital nerve.

Vasculature. Both the vertebral and occipital arteries send branches to supply this muscle.

Actions. It extends the head.

THE ROOT OF THE NECK

The root of the neck is the junctional region between the neck and the thorax.

The thoracic inlet itself is narrow, and is bounded anteriorly by the manubrium sterni, posteriorly by the first thoracic vertebra, and laterally by the first ribs and their costal cartilages. The apices of the lungs occupy the lateral parts of the thoracic inlet. The major structures passing between the neck and thorax lie in the mid-region between the lungs.

The contents of the root of the neck include:

- The brachiocephalic artery and the brachiocephalic veins
- The common carotid arteries } within the carotid sheaths
- The internal jugular veins

- The subclavian arteries and veins
- The thoracic and right lymphatic ducts
- The vagus, recurrent laryngeal and phrenic nerves
- The sympathetic trunks
- The brachial plexuses
- The oesophagus and trachea

A key landmark in the root of the neck is the scalenus anterior muscle which attaches on to the first rib (see page 122). Below and in front of scalenus anterior lies the subclavian vein as it runs across the first rib and into the upper limb. Passing down the superficial surface of the muscle is the phrenic nerve (see page 114). Occasionally, the roots of the phrenic nerve may be separate and are referred to as accessory phrenic nerves. Also in front of scalenus anterior is the lateral part of the internal jugular vein, and the inferior thyroid, superficial cervical and suprascapular vessels. The vagus nerve and the common carotid artery lie between the scalenus anterior muscle and the trachea. Behind the scalenus anterior muscle are the subclavian artery, the ventral rami of the nerves comprising the brachial plexus, and the apex of the lung.

104,28; 106,4
108,26
104,24; 108,28
104,29; 108,24
104,23
104,1; 102,51
104,37; 104,27
104,31; 104,6
104,25
104,26

The brachiocephalic artery

104,17; 108,46

The major arteries passing from the thorax into the neck are different on either side. On the right, the major artery is the brachiocephalic artery. This divides behind the right sternoclavicular joint into the right subclavian and the right common carotid arteries. On the left, however, the left subclavian and the left common carotid arteries originate directly from the arch of the aorta in the superior mediastinum of the thorax. The brachiocephalic artery occasionally provides a thyroidea ima branch to the thyroid gland.

108,48; 108,49
108,32; 108,31

The brachiocephalic veins

104,16; 104,18

These major veins lie anteriorly in the root of the neck. Each brachiocephalic vein is formed behind the medial end of the clavicle by the union of the subclavian and internal jugular veins. The subsequent course of each brachiocephalic vein varies. The right brachiocephalic vein passes vertically downwards into the mediastinum and is located anterolateral to the brachiocephalic artery. The left brachiocephalic vein runs obliquely across the midline, behind the manubrium sterni, to join the right brachiocephalic vein on the right side of the superior mediastinum. In its course, the left brachiocephalic vein crosses the left subclavian artery, the left common carotid artery, and the brachiocephalic artery.

108,30; 108,28
108,35
104,18; 108,47
108,46
104,16; 108,30
108,32; 108,31
108,46

Tributaries:

- Vertebral vein
- Accessory vertebral vein
- Internal thoracic vein
- Inferior thyroid vein
- First posterior intercostal vein

In addition, the left brachiocephalic vein receives the left superior intercostal vein.

The vertebral vein arises in the suboccipital triangle at the back of the head. It initially appears as a plexus formed by the confluence of vessels from the internal vertebral plexuses (see page 67) and from the deep muscles of the upper part of the neck. This plexus descends around the vertebral artery in the foramina transversaria of the upper five cervical vertebrae. A single vein is formed, which passes through the foramen transversarium of the sixth cervical vertebra. This vein runs down in front of the vertebral artery to open into the brachiocephalic vein.

104,30

106A,19; 108,34

The vertebral vein communicates with the sigmoid sinus (through the condylar canal) and with the occipital vein. Its tributaries are the anterior vertebral vein, the deep cervical vein and sometimes the first posterior intercostal vein.

The anterior vertebral vein originates from a plexus around the transverse processes of the upper cervical vertebrae. The vein accompanies the ascending cervical artery, lying between the longus capitis and scalenus anterior muscles. The anterior vertebral vein drains into the vertebral vein close to its termination into the brachiocephalic vein.

The deep cervical vein accompanies the deep cervical artery. Like the vertebral vein, it commences from the venous plexus in the suboccipital region. The deep cervical vein passes down between the transverse process of the seventh cervical vertebra and the neck of the first rib, and ends in the vertebral vein (or in the brachiocephalic vein).

The accessory vertebral vein also arises from the plexus around the vertebral artery. It passes through the foramen transversarium of the seventh cervical vertebra to terminate at the brachiocephalic vein.

104,19

The internal thoracic vein drains the anterior wall of the thorax. It appears in the root of the neck from behind the first rib (just lateral to the sternum). The vein crosses the apex of the lung for a short distance to drain into the brachiocephalic vein.

105,45; 102B,60

104,12

The inferior thyroid vein initially arises as a plexus from the lower part of the thyroid gland. From this plexus, right and left inferior thyroid veins pass downwards to drain separately into their respective brachiocephalic vein. Alternatively, they may unite to form a common vein which enters the left brachiocephalic vein.

The first posterior intercostal vein drains the posterior part of the first intercostal space. It is a tributary of either the brachiocephalic vein or the vertebral vein.

The left superior intercostal vein receives the second and third (sometimes fourth) posterior intercostal veins from the left side of the body. It passes between the left vagus and phrenic nerves to reach the left brachiocephalic vein. The right superior intercostal vein drains into the azygos vein in the thorax.

104–108

The subclavian artery

104,25; 106,11
108,48
104,22; 84A,7

The subclavian artery is one of the principal features in the root of the neck. As mentioned earlier, its origin differs according to which side it is on. The subclavian artery runs into the root of the neck beside the trachea and then passes behind the scalenus anterior muscle. The artery then continues across the apex of the lung, on to the first rib (between the rib and the clavicle), and runs into the upper limb as the axillary artery. Its branches are widely distributed — to the neck, head and brain, to the upper limb, and to the thorax. The subclavian artery can be divided into three parts according to its relationship to the scalenus anterior muscle: proximal (first part), deep (second part), and distal (third part) to the muscle.

104,28; 106,4

104,6; 104,33
104,24
106B,2; 106B,27
106A,14; 108,37
104,26

Concerning the relationships of the subclavian artery, the first part lies behind the carotid sheath and is crossed by a loop of the sympathetic trunk (the ansa subclavia). Behind the artery lies the apex of the lung. Below and in front is the subclavian vein. On the right side, the vagus nerve gives off the recurrent laryngeal nerve which then passes up beneath the artery. On the left, the thoracic duct crosses in front of the subclavian artery. The third part of the subclavian artery is related to the brachial plexus.

Branches:
- Vertebral artery
- Internal thoracic artery

- Thyrocervical trunk:
    Inferior thyroid artery
    Superficial cervical artery
    Suprascapular artery
- Costocervical trunk:
    Deep cervical artery
    Superior intercostal artery

Most of the branches of the subclavian artery arise from its first part. However, the left costocervical trunk arises from the second part of the left subclavian artery.

The vertebral artery arises from the upper surface of the subclavian artery. It passes upwards, medially and backwards to reach the foramen transversarium of the sixth cervical vertebra (N.B. the foramen for the seventh cervical vertebra does not transmit the vertebral artery). In this part of its course, the artery lies behind the common carotid and inferior thyroid arteries, and the vertebral vein. The left vertebral artery is also crossed by the thoracic duct. The vertebral artery then passes upwards, through the foramina transversaria of the remaining cervical vertebrae, and in front of the spinal nerves, to reach the lateral mass of the atlas. It arches around the posterior surface of the lateral mass within the suboccipital triangle. Indeed, the artery lies within a groove on the upper surface of the posterior arch of the atlas, accompanied by the suboccipital nerve.

$108,39$

$108,38; 108,34$
$108,37$
$200B,39; 200B,38$

$200A,16; 200A,18$
$200B,39; 200B,42$
$200A,17$

The vertebral artery then pierces the meninges and runs upwards through the foramen magnum. On the ventral surface of the medulla oblongata, the vertebral arteries join to form the basilar artery. The intracranial course of the vertebral arteries and the basilar artery are described on pages 577 to 580.

$198F,48; 198F,49$
$168,25; 184,24$
$168,26; 184,26$

Extracranially, the vertebral artery gives off spinal and muscular branches. The spinal branches are arranged segmentally and supply the spinal cord (plus meningeal layers) and the cervical vertebrae. The muscular branches are found primarily in the suboccipital region.

The internal thoracic artery originates from the lower surface of the subclavian artery. It runs downwards into the thorax, behind the costal cartilages. The artery provides branches to the intercostal muscles. It has no branches in the neck.

$106,15$
$104,20; 108,29$

The thyrocervical trunk is short and lies close to the scalenus anterior muscle. It gives rise to three branches: the inferior thyroid, superficial cervical and suprascapular arteries.

$100,44; 102B,52$
$104,34; 106,20$
$108,38$

The inferior thyroid artery runs upwards on the medial surface of the scalenus anterior muscle. It then passes medially between the carotid sheath and the vertebral artery, finally descending to reach the inferior part of the thyroid gland where its glandular branches supply the inferior and posterior surfaces. Near the thyroid gland, the inferior thyroid artery is related to the recurrent laryngeal nerve (see pages 108 and 119). The artery has branches that supply the oesophagus, trachea and the lower part of the pharynx. Branches also supply prevertebral and infrahyoid muscles.

$102B,51; 104,35$
$108,42; 154A,25$
$102B,62; 154A,23$
$108,40; 106B,35$

An ascending cervical branch arises from the upper part of the inferior thyroid artery and passes up the neck close to the transverse processes of the cervical vertebrae. It supplies adjacent muscles, gives branches to the spinal cord, and anastomoses with branches from neighbouring vessels (e.g. vertebral, occipital, and ascending pharyngeal arteries).

$102,29; 106,6$

The inferior laryngeal branch of the inferior thyroid artery passes upwards (with the recurrent laryngeal nerve) beneath the inferior constrictor muscle of the pharynx to supply structures within the larynx.

$144A,46; 154A,22$

100,39; 102,31 104,37; 106,8	**The superficial cervical artery** crosses the posterior triangle to supply the trapezius muscle. To reach this muscle, the artery crosses the phrenic nerve and brachial plexus.
100,41; 104,27 106,10; 108,27	**The suprascapular artery** runs downwards and laterally, across the phrenic nerve, the brachial plexus and the third part of the subclavian artery, before passing to the upper border of the scapula. In the neck, it gives branches supplying the sternocleidomastoid and platysma muscles.

The costocervical trunk originates from the back of the subclavian artery. It passes backwards towards the neck of the first rib, where it branches into the deep cervical and superior intercostal arteries.

The deep cervical artery runs upwards within the postvertebral muscles (between semispinalis cervicis and semispinalis capitis). It contributes to the blood supply of the postvertebral musculature.

The superior intercostal artery descends into the thorax to give rise to the first posterior intercostal artery.

108,52 — Variations in the origin and branches of the superficial cervical and suprascapular arteries are common. Furthermore, a branch termed the **dorsal scapular artery** often arises from the third part of the subclavian artery. It passes deep to levator scapulae to reach the shoulder, where it contributes to the supply of the trapezius muscle.

104–108

The subclavian vein

104,24
84A,9; 104,22
104,28
104,25; 104,21
108,28; 108,35
108,30

The subclavian vein begins at the outer border of the first rib as a continuation of the axillary vein from the upper limb. It runs across the upper surface of the first rib, within a groove. Here, the vein lies below, and in front of, the scalenus anterior muscle and the subclavian artery. Behind the subclavian vein is the apex of the lung. The vein passes behind the sternoclavicular joint, where it meets the internal jugular vein to form the brachiocephalic vein. The external jugular vein drains into the subclavian vein just lateral to the scalenus anterior muscle. The right subclavian vein receives the right lymphatic duct near its junction with the right internal jugular vein. The left subclavian vein receives the thoracic duct.

108,51
108,37

100–108

The thoracic duct

This is the main collecting duct for the lymphatics of the body. It is associated with all regions except the right side of the head, neck, thorax and arm. The thoracic duct enters the neck through the thoracic inlet, between the oesophagus and the left pleura, and behind the left common carotid artery and the left vagus nerve. It then runs between the left common carotid and left subclavian arteries, and in front of the left vertebral artery and thyrocervical trunk, to enter the left subclavian vein. The thoracic duct in the root of the neck receives the left jugular trunk (draining the left side of the head and neck) and the left subclavian trunk (draining the left arm). Valves guard the entrance of the thoracic duct into the subclavian vein.

108,37

106A,14

108,36
106A,12

106,28; 108,51
104,32
106B,31
106B,30

The right lymphatic duct

This duct receives the right jugular trunk (draining the right side of the head and neck) and the right subclavian trunk (draining the right arm and right side of the thorax via a mediastinal trunk). It passes near the medial border of the scalenus anterior muscle to drain into the right subclavian vein. Valves are found where the right lymphatic duct joins the subclavian vein.

The vagus, recurrent laryngeal and phrenic nerves, the sympathetic trunk, and the brachial plexus are described between pages 107 and 116. The trachea and the oesophagus are described with the viscera of the neck (page 118).

The apex of the lung and the suprapleural membrane

The apex of the lung is rounded and extends into the root of the neck to a level about 2.5 cm above the medial third of the clavicle. It is covered by the cervical pleura and the suprapleural membrane. The lung apex is crossed by the subclavian vessels. Laterally, it is related to the scalenus medius muscle. Behind is the sympathetic trunk and the cervicothoracic ganglion, whilst medially lie the great vessels and the trachea and oesophagus. The suprapleural membrane is a thin fascial sheet that covers the apex of the lung, thereby strengthening the cervical pleura. Anteriorly, the supra-

104,21

104,24; 104,25
104,39
104,17; 104,13

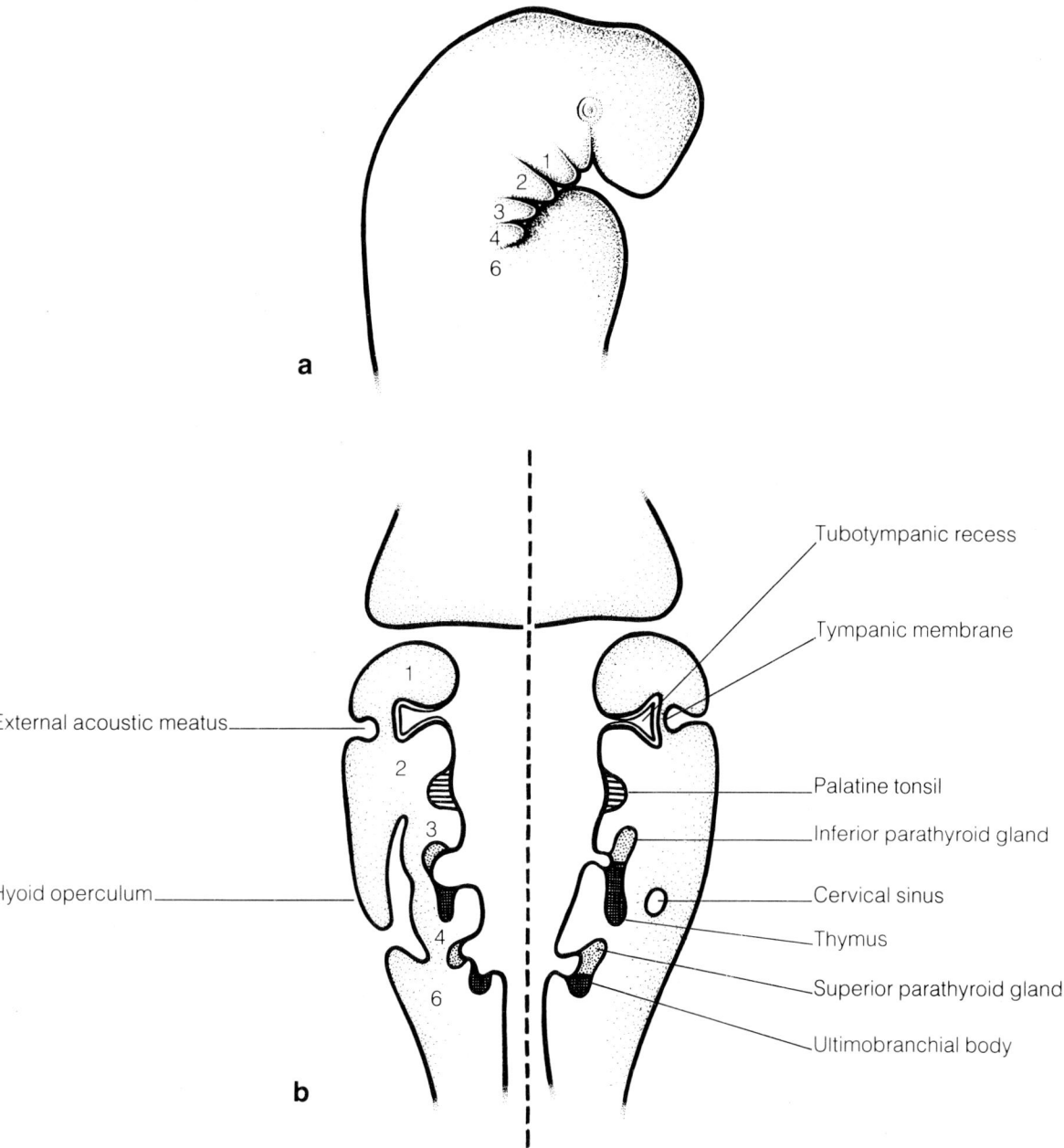

Figure 44 *Developing branchial arch system.* **a**) *Lateral view (four weeks in utero) to demonstrate the branchial arches. The fifth arch is transitory.* **b**) *Coronal section showing the development of the branchial clefts and pouches. The left side shows the hyoid operculum growing down from the second arch. The right side shows the completed growth of the hyoid operculum, enclosing the cervical sinus. The numbers 1 to 6 in both* **a**) *and* **b**) *indicate the location of the branchial arches.*

pleural membrane is attached to the inner border of the first rib. Posteriorly, it is attached to the transverse process of the seventh cervical vertebra.

THE EMBRYOLOGY OF THE NECK (Figure 44)

The neck develops as a series of swellings called branchial arches. These arise from the proliferation and migration of mesenchyme. Some consider that a significant portion of the mesenchyme is of neural crest origin. Six branchial arches are produced initially, but the fifth arch does not contribute significantly to further development.

Each branchial arch is supplied by a particular cranial nerve which innervates the structures derived from that arch.

The branchial arches are covered externally by ectoderm and are separated by grooves called the branchial clefts. The first branchial cleft subsequently forms the external acoustic meatus. The remaining clefts disappear during development, being covered by a downgrowth (the hyoid operculum) from the second arch. This results in the enclosure of an epithelial-lined region, the cervical sinus. The sinus usually disappears.

The mesenchyme within each arch is supported by a cartilaginous component which may or may not ossify before contributing to definitive skeletal components. Myotomes present in each arch also give rise to muscles which, though they may migrate from their initial site of origin, retain the nerve of their respective arch.

The internal surfaces of the branchial arches are lined by endoderm. The arches are separated internally by grooves called the branchial pouches. The endoderm of each branchial arch also contributes to adult structures, though the tonsil and thymus become secondarily invaded by lymphocytes.

TABLE 2 The principal derivatives of the branchial arches.

BRANCHIAL ARCH	NERVE SUPPLY	SKELETAL DERIVATIVES	MUSCULAR DERIVATIVES	DERIVATIVES OF BRANCHIAL POUCHES
FIRST	Trigeminal nerve	Malleus, incus, anterior ligament of malleus, spine of sphenoid, sphenomandibular ligament, lingula of mandible, mental ossicles	Muscles of mastication (masseter, temporalis, medial and lateral pterygoids) anterior belly of digastric, mylohyoid, tensor tympani, tensor veli palatini	Auditory tube and tympanic cavity
SECOND	Facial nerve	Stapes, styloid process, stylohyoid ligament, lesser horn of hyoid bone, upper part of body of hyoid bone	Muscles of facial expression, stapedius, stylohyoid, posterior belly of digastric, buccinator	Palatine tonsil
THIRD	Glossopharyngeal nerve	Greater horn of hyoid bone, lower part of body of hyoid bone	Stylopharyngeus	Inferior parathyroid glands and thymus gland
FOURTH	Vagus nerve (superior laryngeal branch)	Thyroid cartilage, epiglottis	Cricothyroid	Superior parathyroid glands
FIFTH —	Does not develop significantly			Ultimobranchial body (calcitonin-secreting cells in thyroid gland)
SIXTH	Vagus nerve (recurrent laryngeal branch)	Cricoid cartilage, arytenoid cartilages	All intrinsic muscles of larynx (excluding cricothyroid) and muscles of pharynx	

The derivatives of the branchial arches are shown in Table 2.

The thyroid gland arises during the fourth week of intra-uterine life as an endodermal thickening on the ventral surface of the branchial arches. Its development is initially related to the developing tongue (see page 321), the site of origin of the gland corresponding to the foramen caecum of the adult tongue. The gland sinks into the substance of the tongue and then descends in the midline of the neck in front of the hyoid bone. It reaches its eventual position on the trachea by the seventh week. During its descent, the thyroid gland is connected to the tongue by a duct called the thyroglossal duct. This duct usually disappears, but may persist caudally to give rise to the pyramidal process of the gland.

SUMMARY SHEET: SUPERFICIAL STRUCTURES OF THE NECK

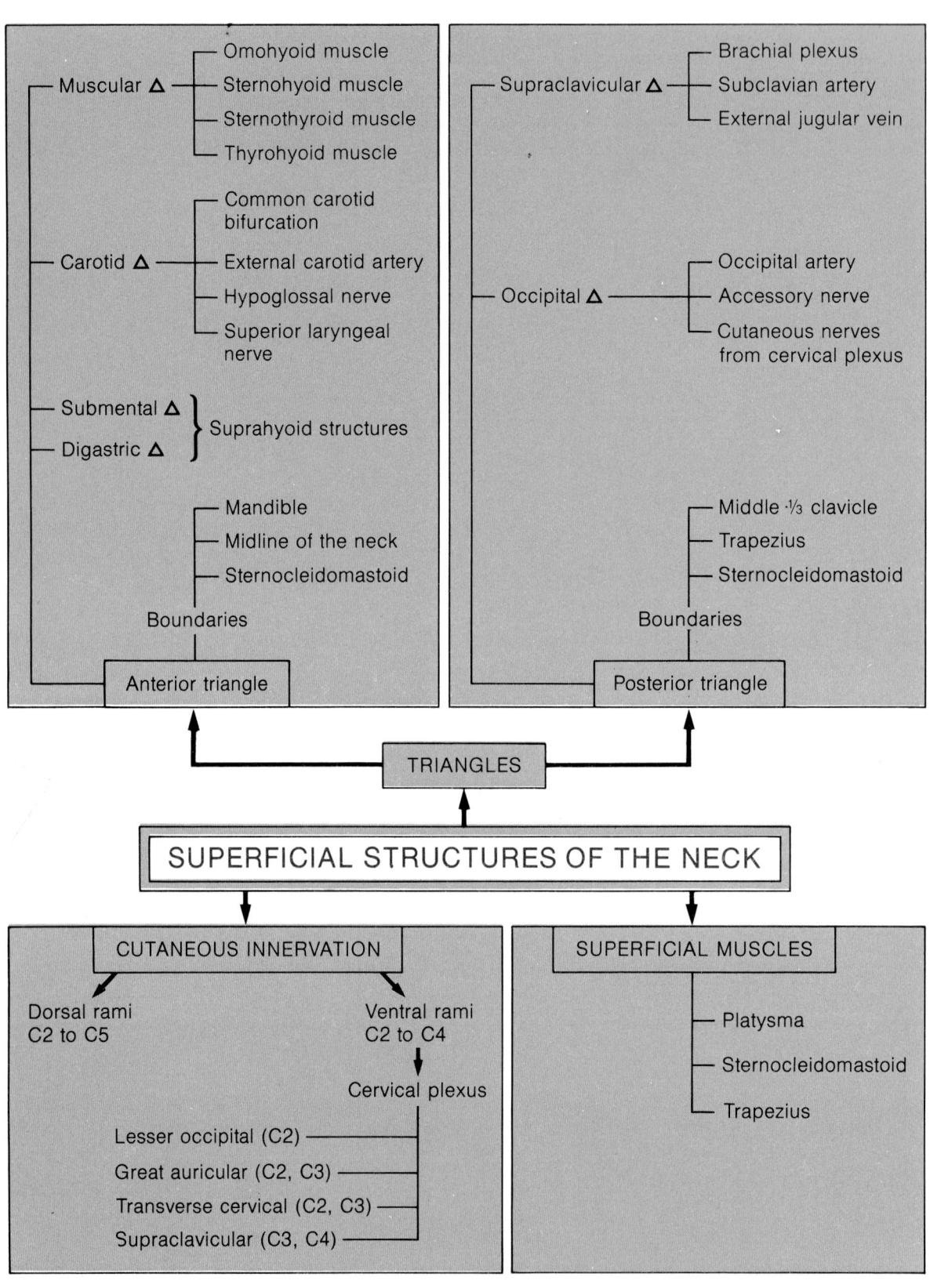

SUMMARY SHEET: DEEP STRUCTURES OF THE NECK 1

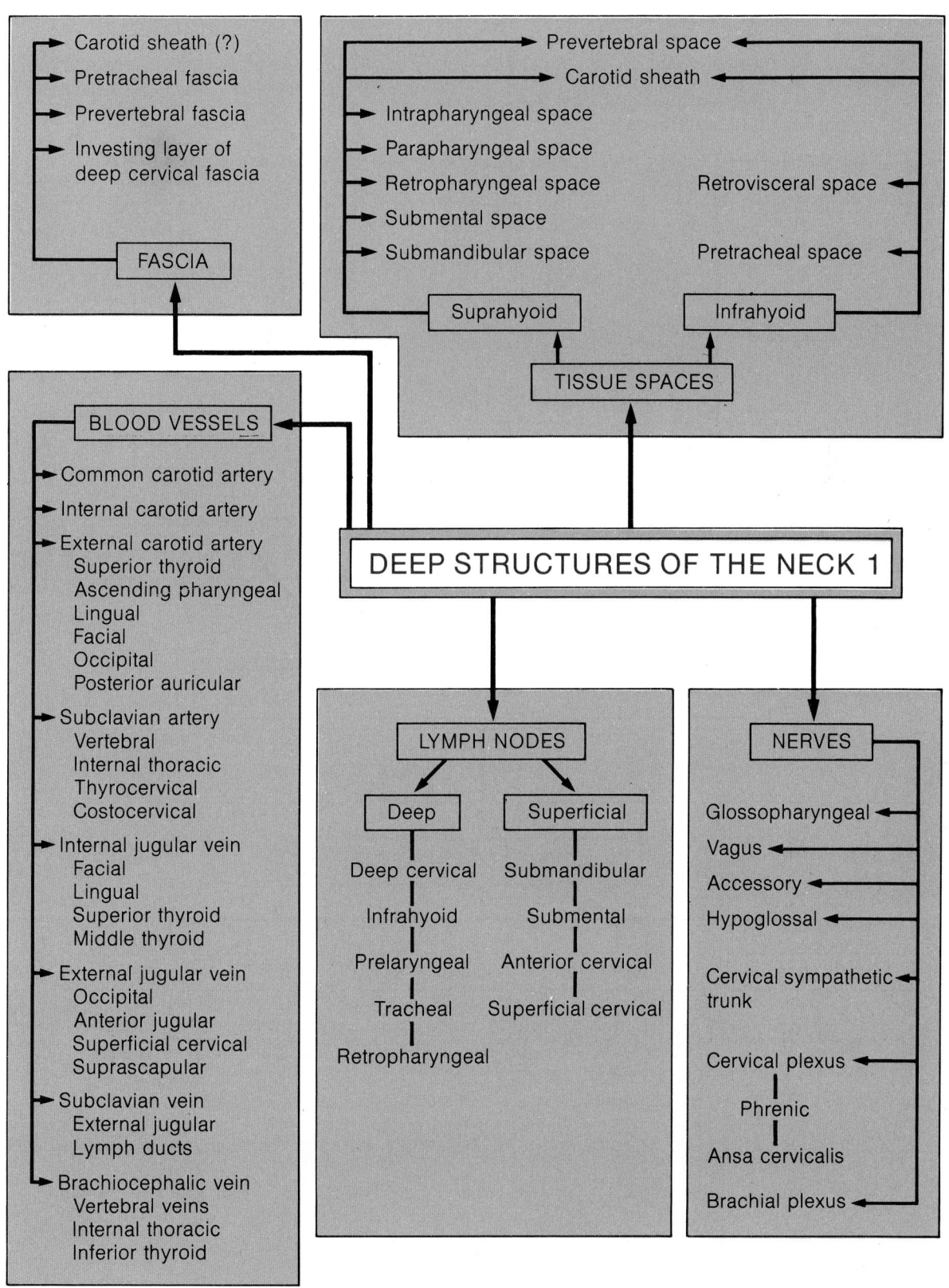

SUMMARY SHEET: DEEP STRUCTURES OF THE NECK 2

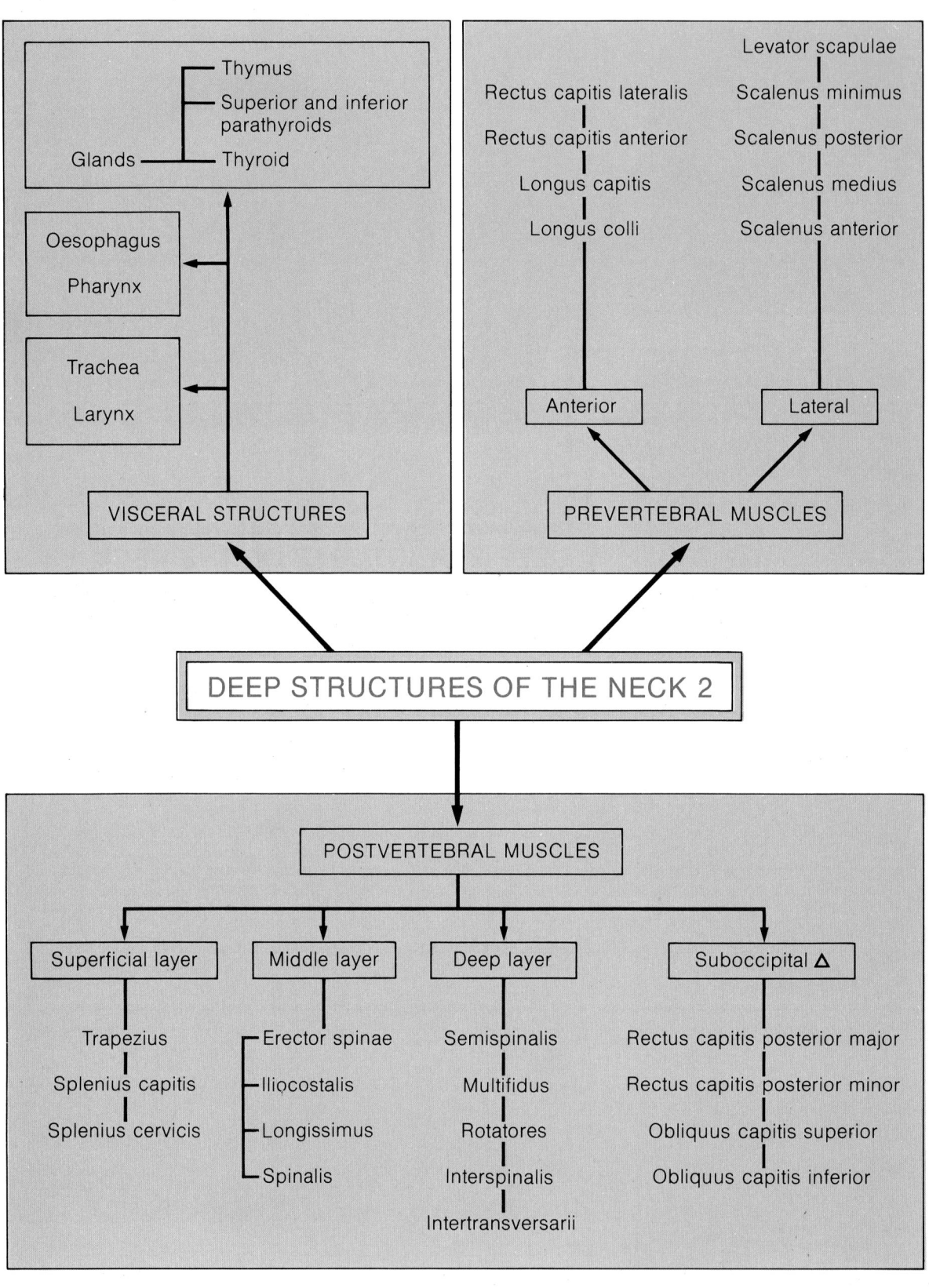

Case histories 2
The neck

A paediatrician was asked to see a child who, since birth, had had a small, fluctuant swelling on the anterior border of her left sternocleidomastoid muscle, just beneath the angle of the jaw. It had never caused the child any problems, but the mother was rather anxious. The paediatrician was able to give reassurance, as the child had a harmless congenital defect associated with retained elements of the branchial arches.

The child has a branchial cyst. The second branchial arch of the embryo overgrows the lower arches, resulting in an enclosure of an epithelial-lined region (the cervical sinus). The epithelium usually disappears, but remnants may form a branchial cyst on the lateral aspect of the neck, anterior to the sternocleidomastoid muscle. The cyst may open on to the surface, forming a branchial fistula. The fistula may also extend internally between the external and internal carotid arteries to open into the pharynx (in the region of the tonsil). It is possible that infected lymph nodes form cysts that can be misdiagnosed as branchial cysts.

A 58-year-old woman developed a painless swelling on the right side of her neck, just below the angle of the mandible. Several months later, while stretching to replace a light bulb, she felt a sudden pain in the region of the swelling. Over the next few hours the swelling increased in size and became pulsatile. She now experienced difficulty in swallowing and in moving her tongue. Her tongue deviated to the right on protrusion. A dilation (aneurysm) of the internal carotid artery, caused by a weakness in the wall of the vessel, was diagnosed by her doctor. How do you account for the symptoms?

The activity of stretching led to the escape of blood through the wall of the aneurysm and into the adjacent tissues. This caused pain and rapid enlargement of the swelling. The swelling was pulsatile, indicating its arterial origin. The last four cranial nerves (glossopharyngeal, vagus, accessory and hypoglossal nerves) and the cervical sympathetic trunk pass close to the internal carotid artery near the base of the skull and may be affected by any abnormal growth in the area. In the case described here, the signs indicate that the aneurysm complex had compressed the right pharyngeal branch of the vagus which is responsible for innervating most of the muscles of the palate and pharynx. This caused the difficulty in swallowing. There was also interference with the right hypoglossal nerve, affecting the tongue musculature. When the tongue was protruded, it deviated towards the affected side because of the unopposed activity of the left genioglossus muscle.

A casualty officer was called to see a patient who had developed palpitations of the heart (supraventricular tachycardia). An attending medical student was surprised to see the doctor press over the left side of the patient's neck in the region of the anterior border of the sternocleidomastoid muscle (at the level of the upper border of the thyroid cartilage of the larynx). What do you think the casualty officer was doing?

The casualty officer was compressing the patient's carotid sinus. This is located in the wall of the internal carotid artery just distal to the bifurcation of the common carotid artery. The pressure stimulates the baroreceptors (stretch receptors) in the sinus. Nerve fibres from the baroreceptors pass with the glossopharyngeal nerve to the medulla oblongata of the brainstem. The vasomotor and cardio-inhibitory centres in this part of the brain are responsible for increasing vagal output to the heart. Thus pressure over the sinus produces a slowing of the heart (bradycardia) and a drop in blood pressure.

A young man visited hospital because of a swelling in his neck which he had noticed for the previous four weeks. The doctor observed a non-pulsatile, soft, smooth, well-circumscribed swelling in the region of the thyroid gland, near the midline. The doctor suspected that the swelling was either a thyroglossal cyst close to the thyroid gland, or a thyroid goitre (endogenous enlargement of thyroid tissue). The two conditions can be distinguished clinically on the basis of differences in their attachments to surrounding structures.

Epithelial remnants of the embryonic thyroglossal duct may occur at any position along the descent pathway of the developing thyroid gland. Such remnants may become cystic. Thyroglossal cysts are situated near the midline and are usually attached to the tongue musculature by strands of fibrous tissue. Thus, when the tongue is protruded, thyroglossal cysts move upwards. In the case of a thyroid goitre, the enveloping pretracheal fascia prevents the gland moving upwards on protrusion of the tongue. The gland will be elevated during swallowing, however, as the pretracheal fascia binds the thyroid gland to the larynx.

A dentist was extracting a tooth under local anaesthesia when the crown broke and a fragment was inhaled by the patient, who began to choke. The dentist tried without success to dislodge the fragment. The patient stopped breathing, became cyanotic and collapsed. The dentist realised that an alternative pathway for getting air into the lungs had to be provided immediately. How might a knowledge of anatomy help him to undertake this emergency surgical procedure?

The site of obstruction is usually at the vocal cords. Consequently, an airway may be obtained by making an artificial opening in the lower part of the larynx (laryngotomy) or in the upper part of the trachea (tracheotomy).

For a laryngotomy, an incision is made through the anterior cricothyroid ligament. This procedure is suitable in an emergency as the airway is superficial at the point of entry and no major structures are interposed. Just beneath the laryngeal prominence is a slight depression between the lower border of the thyroid cartilage and the upper border of the cricoid cartilage. This depression marks the site of the anterior cricothyroid ligament. An incision can be made through this ligament and a tube of some sort inserted (even the outer case of a ballpoint pen). A laryngotomy is most readily accomplished if the neck is extended (e.g. by placing a pillow under the neck), as the larynx is then brought close to the surface.

If the procedure is not an emergency and can be undertaken in hospital, a tracheostomy is preferred as there is no likelihood of damage to the vocal cords and the site of entry provides a more stable position for the air tube. For a tracheostomy, a vertical or transverse incision can be used. The vertical incision is a midline incision from the cricoid cartilage downwards. The incision passes between the anterior jugular veins. The infrahyoid muscles are then separated vertically and retracted laterally. The pretracheal fascia covering the trachea and the thyroid isthmus is thus exposed. The thyroid isthmus (located usually at about the level of the third tracheal ring) can be retracted or, if necessary, surgically divided. An opening can now be made into the exposed trachea (usually beween the second and third rings) and a tube inserted.

A woman was admitted to a general surgical ward for the removal of a large swelling in the left lobe of her thyroid gland. During the preoperative investi-

gations, she was routinely referred to the ENT (ear, nose and throat) department to have her vocal cords examined. Why?

The nerves innervating the muscles of the larynx are the recurrent and external laryngeal nerves. These nerves run close to the thyroid gland and can be damaged during surgery. However, they may already have been affected by the disease process and it is therefore important to examine the vocal cords preoperatively to determine whether this is the case.

A middle-aged woman presented to her doctor with a six-month history of increasing difficulties with breathing (dyspnoea) and with swallowing (dysphagia). She had also developed stridor (high-pitched, noisy respiration) and the veins in her neck were dilated. There was a fullness in the lower part of the front of her neck, which elevated on swallowing. This woman has developed a retrosternal swelling of her thyroid gland (goitre). How do you explain the symptoms?

The attachment of the sternothyroid muscles to the thyroid cartilage effectively binds the thyroid gland to the larynx and limits upward expansion of the gland. However, there is no limitation to downward expansion behind the sternum. As the goitre came to occupy much of the thoracic inlet (which is narrow and has bony margins), it caused compression of adjacent structures. Compression of the trachea produced the dyspnoea and stridor. Compression of the oesophagus caused the dysphagia. The anterior jugular veins became dilated because they were compressed against the manubrium sterni. Radiographic examination aids the diagnosis of a retrosternal goitre by revealing displacement of the trachea.

A middle-aged man underwent a subtotal thyroidectomy for a hyperactive, enlarged thyroid gland. He recovered well from the operation, but five days later complained of tingling and numbness in his face, fingers and toes. This became worse, and he developed cramps and spasm of his hands. What is the likely explanation for these symptoms?

The patient has developed hypoparathyroid tetany, a rare complication (less than 1%) of subtotal thyroidectomy. The four parathyroid glands are closely related to the posterior surface of the thyroid gland but may not always be readily identified. Tracing their blood supply from the inferior thyroid artery may help to locate them. At operation, some parathyroid tissue may be removed or the blood supply interfered with. The 'carpo-pedal' spasm and the tingling and numbness result from hypo-calcaemia. This causes nerves to become hyperexcitable, and tapping the skin near the angle of the mandible may trigger off involuntary contraction of the facial musculature (Chvostek's sign).

A worried mother took her 18-month-old daughter to the doctor because she had a 'wry neck' (torticollis). The child's head was tilted to the left and her face was directed upwards and to the right. The doctor learnt that the child's birth had been difficult and that forceps had been required for the delivery. Indeed, the torticollis was caused by the difficult birth. Which muscle in the neck do you think has been damaged?

The sternocleidomastoid muscle acting singularly tips the head towards the shoulder on the same side and rotates the face to the opposite side. The muscle on the left side was stretched during the forceps delivery with subsequent haemorrhage into the muscle. The haematoma thus formed was subsequently invaded by fibrous tissue which contracted and so shortened the muscle. This pulled the mastoid process down towards the sternoclavicular joint of the same side, thus producing the 'wry neck' deformity. Similar signs can occur suddenly in adults as a result of muscular spasm. The cause of this is uncertain.

Following an emergency call to a patient suffering from a cardiac arrest, a consultant arrived to find his junior doctor already present but unable to gain peripheral venous access via the veins in the cubital fossa. This was necessary in order to administer fluids and resuscitative drugs. What alternative route in the neck may provide rapid venous access?

The most rapid venous access in the neck is through the external jugular vein. This vein lies superficially on the sternocleidomastoid muscle. The patient is placed in the 'head down' position to allow the vein to fill and become easily recognisable. After cannulating the external jugular vein, it may subsequently be necessary to establish a central venous line into the superior vena cava and the right atrium. A route from the external jugular vein is not the route of choice: the vein is constricted as it passes through the investing layer of deep cervical fascia in the roof of the posterior triangle, and it bends sharply to join the subclavian vein. A more satisfactory route is via the internal jugular, brachiocephalic or subclavian veins.

The external jugular vein is also an indicator of venous pressure. When a patient lies supine with the head on a pillow, the normal level of blood pressure makes the vein visible for about one-third of the way up the neck. As the patient sits up, the vein gradually becomes indistinct. With conditions leading to raised venous pressure (e.g. obstruction of the superior vena cava), the external jugular vein becomes more conspicuous.

A casualty officer had to insert a pressure monitor into the subclavian vein of a patient. This is accomplished by introducing a venous catheter via a needle inserted through the skin. While this procedure was being attempted, the patient complained of a sudden, sharp pain in his chest and increased difficulty with breathing. The patient's condition rapidly deteriorated. A pneumothorax was diagnosed. How might this have happened?

A pneumothorax develops when air is introduced into a pleural cavity. The lung collapses as a result. Pneumothorax is a complication of subclavian vein puncture because of the proximity of the vein to the pleural cavity. Because of the obliquity of the thoracic inlet, the suprapleural membrane rises about 2.5 cm above the midpoint of the clavicle. The subclavian vein grooves the anterosuperior surface of the suprapleural membrane, and so is close to the parietal pleura. The subclavian vein is usually cannulated by an infraclavicular approach. A pneumothorax may result from missing the vein completely, or from passing the needle right through the vein and into the pleural space.

A medical student in Africa was asked to examine a child who had a large swelling in his neck. The swelling was superficial and situated at the anterior border of the right sternocleidomastoid muscle. It was fluctuant and painless. The student observed that the right palatine tonsil was enlarged, with a purulent exudate from the tonsillar crypts. The child was known to be drinking unpasteurised milk. What might the swelling in the child's neck be caused by?

This child has tubercular cervical lymphadenitis and the swelling in the neck is termed a 'cold abscess'. The organisms were present in the unpasteurised milk and they gained entrance to the body via the palatine tonsil. Infection initially spread from the tonsil via the lymphatics to the jugulodigastric nodes and then to the rest of the deep cervical lymph nodes lying alongside the internal jugular vein. As a result of infection, the lymph nodes broke down and coalesced, a process called caseation. At this stage, the pus would have been confined beneath the investing layer of the deep cervical fascia. However, it eventually eroded through the fascia to form a superficial fluctuant mass or 'cold abscess'.

A child was referred to hospital, obviously very unwell and with a high temperature. She was also 'off her food', as swallowing was difficult and all neck movements painful. A blood culture was positive, showing the presence of pyogenic organisms. A radiograph showed evidence of a lesion within a cervical vertebra and a soft tissue shadow extending down to the region of the third thoracic vertebra. What might be wrong with this child?

The child has a pyogenic osteomyelitis involving a cervical vertebra. Pus has left the vertebra anteriorly to come to lie in the prevertebral tissue space. This space is limited anteriorly by the prevertebral fascia, which ends at the third thoracic vertebra by blending with the periosteum. This explains the radiological and the general clinical findings. As the potential prevertebral space is limited anteriorly and inferiorly, the pus is localised behind the oesophagus, causing dysphagia.

A doctor examined a man who had been involved in a motorbike accident six months previously. The patient's left arm was hanging limply by his side. The arm was medially rotated and pronated and there was considerable wasting of the biceps and deltoid muscles. The doctor also noted a loss of sensation down the lateral side of the patient's arm. What injury do you think the patient sustained?

This patient has damaged part of his brachial plexus, giving rise to an Erb–Duchenne palsy. When he was thrown from his motorbike he landed awkwardly, so that his head was displaced excessively to the right and his left shoulder depressed. The roots of the brachial plexus derived from the fifth and sixth cervical nerves were torn. This resulted in paralysis and wasting of a number of arm muscles supplied by these nerves (i.e. the supraspinatus, infraspinatus, subclavius, biceps brachii, brachialis, coracobrachialis, deltoid, and teres minor muscles). The limb hung down by his side and was medially rotated because of the unopposed action of the subscapularis muscle. The forearm was pronated because of the loss of the action of biceps. The loss of sensation down the lateral side of the arm is related to the fact that the dermatomes here are associated with the fifth and sixth cervical nerves.

A teenager presented to the casualty department with a knife wound in his neck. The wound, which was situated about 8 cm below the right mastoid process, was sutured once bleeding was controlled. One week later, when the patient returned to have the stitches removed, he complained of being unable to shrug his right shoulder and of having difficulty in raising his right arm above shoulder level. However, he was able to turn his head normally from side to side. What structure has been damaged to produce these symptoms?

The knife passed into the posterior triangle of the neck. The main structure damaged was the spinal part of the accessory nerve which runs almost vertically downwards on the levator scapulae muscle. This part of the accessory nerve supplies the trapezius muscle and the resulting denervation accounts for the patient's inability to elevate the scapula and clavicle (as in shrugging the shoulders). Acting with the serratus anterior muscle, the trapezius muscle is usually involved in elevating the arm above the head by rotating the scapula in an anterior direction. This ability was also lost in the case described here. The spinal accessory nerve is also at risk when diseased cervical lymph nodes are surgically removed. The nerve can be distinguished from branches of the cervical plexus by the fact that it emerges into the posterior triangle from within the substance of the sternocleidomastoid muscle rather than deep to it.

A young man passed out whilst cleaning his shoes very vigorously. He was diagnosed as having subclavian steal syndrome. The underlying cause of this condition is a congenital narrowing (stenosis) of a subclavian artery proximal to the origin of its vertebral branch. Bearing in mind the connections of the subclavian artery, can you suggest the pathway whereby blood reaches the affected arm, albeit with a consequent loss of consciousness?

At rest, sufficient blood reaches the affected arm. During exercise, however, the subclavian artery is unable to supply sufficient blood and there is a reversal of blood flow in its vertebral artery to bypass the stenosis. Blood supplying the arm now comes from two sources: first from its own subclavian artery, and second from the contralateral subclavian artery via its vertebral branch. In the latter situation, blood passes from the contralateral vertebral artery to the circle of Willis and down the vertebral artery on the affected side. Blood is therefore diverted from the brain to supply the arm, causing the patient to faint. The term 'steal' is used to denote that blood is 'taken away' from one side of the body to the other.

A 65-year-old man went to his doctor with a firm, painless swelling in his neck. It was situated beneath the angle of his jaw, just behind the anterior border of the left sternocleidomastoid muscle. The lump had been present for many months. Cursory clinical examination revealed little else. A biopsy showed that the mass was a lymph node that had been invaded by a malignant tumour. Where could the primary lesion be?

The involved node belongs to the jugulodigastric group of nodes. This group is part of the chain of deep cervical lymph nodes located alongside the internal jugular vein. Almost all lymph from the head and neck ultimately drains into this chain of nodes. It is therefore essential to examine thoroughly all parts of the head and neck (including pharynx, larynx, nose and ears) when an enlarged deep cervical node is found.

A newborn baby appeared quite healthy, though saliva dribbled from the corners of her mouth. On her first feed, however, she started to choke and became cyanosed. Oesophageal atresia was diagnosed.

The respiratory system develops as a ventral diverticulum (the respiratory diverticulum) from the lower part of the developing foregut. With subsequent growth and the development of an oesophagotracheal septum, continuity between the respiratory and alimentary systems is lost (except at the pharynx) and the trachea becomes completely separated from the oesophagus. However, congenital defects may arise. The most common is oesophageal atresia. In the predominant form, the oesophagus is in two separate portions. The upper portion is blind-ended and therefore swallowing is not possible. This accounts for the dribbling of saliva. When the baby is fed, fluid will flow over into the larynx and trachea, causing the baby to choke and to become cyanosed. The diagnosis can be supported by the fact that a fine-bore nasogastric tube cannot be advanced more than a few centimetres into the oesophagus. The lower portion of the oesophagus is connected to the trachea (oesophagotracheal fistula). This may allow for the regurgitation of gastric juices through the fistula. Unless treatment is immediately undertaken, a rapidly fatal pneumonia will develop.

An elderly man went to his doctor complaining of progressive difficulty in swallowing (dysphagia). He also complained of having bad breath (halitosis) and of occasionally regurgitating his food. Several hard lumps could be palpated on the side of the patient's neck and a soft, reducible swelling was found just behind the right side of the thyroid cartilage. Radiographic examination after a barium meal revealed a mass which was later confirmed as a carcinoma in the cervical part of his oesophagus. Explain the signs and symptoms.

The hard lumps were enlarged lymph nodes, indicating spread of the tumour from the oesophagus to the postero-inferior group of deep cervical lymph nodes. The carcinoma led to difficulty in swallowing and to a build-up in pressure above a narrowing of the oesophagus. This pressure caused the mucosa to protrude through the weakest point of the pharynx, giving rise to the soft reducible swelling in the neck. Such a swelling is called a pharyngeal diverticulum. This kind of diverticulum is often associated with the inferior constrictor muscle of the pharynx (between the cricopharyngeal and thyropharyngeal parts). The diverticulum extended into the posterior triangle, where it was palpable. Decomposition of food lodged in the diverticulum led to the halitosis. The diverticulum was also responsible for the regurgitation of food.

Oesophageal carcinoma may spread through the wall of the oesophagus to involve closely related structures. One of these structures is the recurrent laryngeal nerve. In such cases, the patient may present with hoarseness of the voice because of paralysis of some of the intrinsic muscles of the larynx.

The face

The face lies anterior to the auricles and extends from the hair margin to the chin. Its main functions are related to the presence of many of the organs of special sense and to two prominent openings, the oral and nasal apertures, which mark the entrances to the digestive and respiratory tracts. The face also plays an important role in communication, being concerned with both speech and facial expression. Furthermore, each individual is recognised by the idiosyncratic variations in the shape of the face.

SURFACE MARKINGS OF THE FACE

110; 118A

The shape of the face relies not only upon the form of the facial skeleton (see pages 13 and 14) but also upon the disposition of the soft tissues. The skin creases depend upon the arrangement of the underlying facial muscles, and within the cheek there may be a prominent mass of fat (the buccal pad of fat).

112,27; 140A,17

The essential surface anatomy of the face is familiar. For descriptive convenience, the face is usually subdivided into: the forehead, the temporal region, the orbital region, the external nose, the cheek, the oral region and chin.

The forehead extends from the hair margin to the eyebrows and is a region common to both the face and the scalp. It is relatively featureless. On each side, about half-way up, the forehead sometimes appears bossed (corresponding to the frontal tuberosities of the frontal bone). There are a variable number of long, transverse creases in the skin. These become particularly prominent when expressing surprise or fright. Between the eyebrows lies a small elevation called the glabella. Vertical wrinkles are produced here when frowning.

38A,3

110,1

The temporal region is situated on the lateral aspect of the face, in front of the external ear and above the zygomatic arch of the cheek. It is demarcated superiorly by the temporal crest. This crest indicates the upper limit of the temporalis muscle. The muscle becomes obvious when the teeth are clenched. The skin of the temple closest to the ear is particularly hirsute and is sometimes referred to as the 'beard part of the temple'. The region in front of this may consequently be called the 'non-beard part of the temple'. The superficial temporal vessels are often visible here, especially in the elderly.

110,15; 14,19
14,4; 14,5
114,1; 114,2

The orbital region is bounded by the bony rim of the orbit, which can be readily palpated round its entire extent. A supra-orbital notch may be located on the upper or supra-orbital margin, marking the site of exit of the supra-orbital nerve and vessels on to the forehead. A frontal notch may also be found towards the bridge of the nose. This is associated with the supratrochlear nerve and vessels. Above the supra-orbital margin is the eyebrow.

110,11; 110,13
110,10; 10,6
110,9; 10,5

Protecting the front of the eye are the eyelids (superior and inferior palpebrae), the upper eyelid being larger and more mobile. The interval between the eyelids is called the palpebral fissure. Extending from the margins of the eyelids are the eyelashes (cilia). The eyelashes are arranged in two or three rows. Near the attachment of the eyelashes lie the openings of a series of glands. The lower eyelid may be everted to reveal its inner surface. This surface is normally red and vascular. The upper eyelid cannot be everted sufficiently to reveal the superior fornix. Tarsal glands may be visualised as yellowish streaks on the inner surfaces of both eyelids. The lateral and medial angles of the eyelids are referred to as the lateral (or outer) and medial (or inner) canthi. The lateral canthus is relatively featureless.

118A,1; 118A,10

118A,11; 118A,3

The medial canthus shows a number of features. The canthus is separated from the eyeball by a small triangular space called the lacrimal lake (lacus lacrimalis). Here also lies a small, reddish body containing sebaceous and sweat glands (the lacrimal caruncle).

118A,4

118A,5	Lateral to the caruncle is the plica semilunaris, a fold of conjunctiva believed by some to be a vestige of the nictitating membrane of other animals. The medial end of each lid margin has no eyelashes. In this region lies a small elevation (the lacrimal papilla) on which is found a fine opening called the punctum lacrimale. The punctum opens into the lacrimal canaliculus, through which tears from the lacrimal lake pass into the lacrimal sac. The anterior surface of the eyeball shows many features. The peripheral part is the relatively avascular sclera, forming the 'white of the eye'. The sclera is seen through the thin, mucous membrane layer (bulbar conjunctiva) which contains a fine vascular network. The sclera becomes transparent over the central portion of the eye as the cornea. The boundary between the sclera and cornea is called the sclerocorneal junction or limbus. The dark, circular, central opening is the pupil. Surrounding the pupil is the iris. The iris acts as a diaphragm that controls the size of the pupil. Its colour depends on the amount and distribution of pigment.
118A,2	
118C,42	
118C,41	
110,12; 118C,36	
118A,6	
118A,7	
118A,9	
118A,8	
110,2; 10,33	The external nose is essentially pyramidal in shape. Superiorly, it is confluent with the forehead at the root of the nose, where it is supported by the nasal bones at the bridge. This part of the nose is immobile. Inferiorly, the tip of the nose is called the apex. The ridge connecting the root and apex is referred to as the dorsum of the nose. The apex and the dorsum are supported by cartilage and are mobile. The nostrils (anterior or external nares) are separated by a septum which joins the apex of the nose to the philtrum of the upper lip. The nostrils lead into the nasal vestibule. Here, the skin is characterised by having coarse hairs. The flared, lateral margins of the nose are known as the alae. The nasolabial grooves of the upper lip continue around the alae to form alar grooves.
110,4	
110,3	
126A,22	
110,7; 110,5	
110,27; 126C,36	
110,6	
110,8	
	The cheek is an extensive area of the face below the temporal region. It can be subdivided into two. There is an upper and anterior region overlying the bone of the zygomatic arch and body of the maxilla, and a lower and posterior region overlying the buccinator muscle, the ramus of the mandible and the parotid gland. The masseter muscle, parotid duct and condyle of the mandible may be visualised and/or palpated.
110,22; 110,18	
110,19; 110,16	
	The oral region presents the red zone of the lip (the vermilion) which is a characteristic of man. The sharp junction of the red zone and the skin is called the vermilion border (see Figure 95, page 273). In the upper lip, the red zone protrudes in the midline to form the tubercle. The lower lip shows a slight depression in the midline corresponding to the tubercle. In passing from the midline to the corners of the mouth, the lips widen and then narrow. Laterally, the upper lip is separated from the cheeks by nasolabial grooves. With age, similar grooves appear at the corners of the mouth delineating the lower lip from the cheeks (labiomarginal sulci). A labiomental groove separates the lower lip from the chin. In the midline, running from the upper lip to the septum of the nose, runs the philtrum. The corners of the lips (labial commissures) are usually located at the level of the maxillary canine and mandibular first premolar teeth. The lips exhibit sexual dimorphism; as a general rule the skin of the male is thicker, firmer and more hirsute, but is less mobile.
110,27	
110,25	
	The lips are said to be 'competent' when they are lightly closed at rest. The term 'incompetent' lips describes a lip posture that, at rest and with the facial muscles relaxed, does not produce an anterior seal (Figure 45). The position and activity of the lips are important in controlling the degree of protrusion of the incisors. With competent lips, the tips of the maxillary incisors lie below the upper border of the lip, this arrangement controlling the inclination of the incisors. With incompetent lips, the tips of the maxillary incisors may not be so controlled and the lower lip may even lie behind them, thus producing an exaggerated proclination of the maxillary incisors. During swallowing, it may not be possible to produce an anterior oral seal without the aid of the tongue. This behaviour (called tongue thrusting) may produce further forces tending to protrude the incisors. A tight or over-active lip musculature may produce sufficient forces to retrocline the incisors.
34,12	The chin is also a facial feature characteristic of man. It forms a distinct protuberance in the midline of the lower jaw. Passing laterally from the chin, the inferior border of
110,21; 110,23	

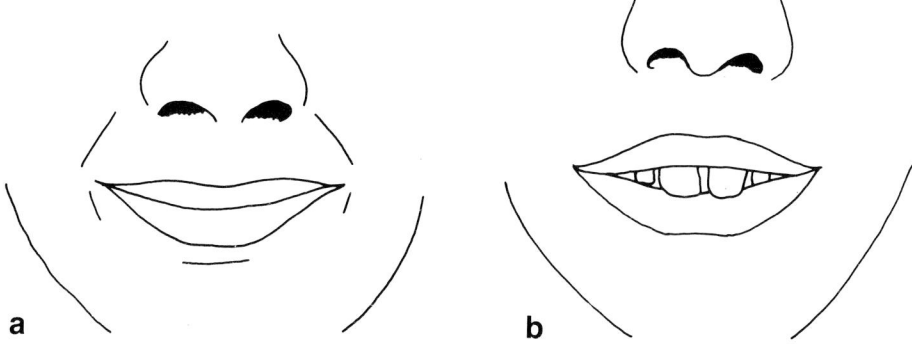

Figure 45 a) *Competent lips,* b) *incompetent lips.*

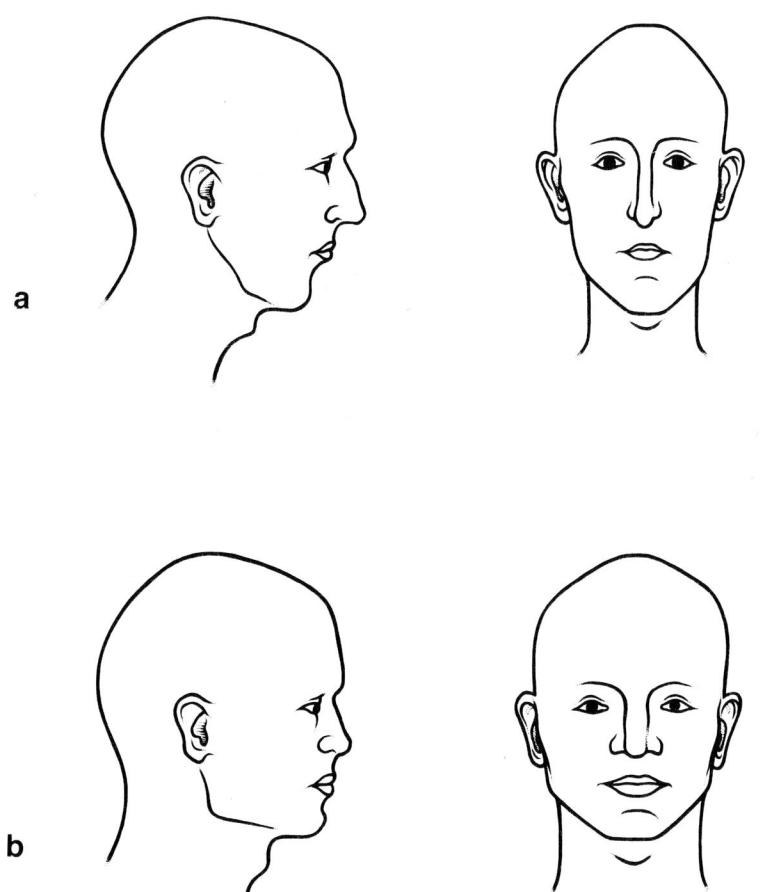

Figure 46 *Facial types:* a) *leptoprosopic, and* b) *euryprosopic.*

the mandible is readily discernible. Where the anterior border of the masseter meets the lower border of the mandible, the facial vessels are found. The junction of the posterior and inferior borders of the mandible is called the angle of the mandible.

Facial types

There are essentially two types of face: leptoprosopic and euryprosopic (Figure 46).

The leptoprosopic face is long and narrow. It tends to have a markedly convex facial profile with a protruding maxilla and a retruding mandible. The overall effect is of retrognathia. In addition, the forehead has a distinct slope, the supra-orbital ridges are prominent, the frontal sinuses are large, and the bridge of the nose is high. The eyes are close-set with a thin, long, prominent nose (often referred to as 'Roman').

The euryprosopic face is broad. The upper part of the face is less prominent than that of the leptoprosopic face. The forehead is upright and bulbous, the supra-orbital ridges are indistinct, and the frontal sinuses are small. The nose is short and the eyes are wide-set. The cheek bones are usually more prominent than those of the leptoprosopic face. The jaws are orthognathic, being relatively straight when viewed in profile.

The face shows ethnic variations. In this context, the term 'race' is used to denote a group of individuals that form a recognisable subdivision of the species. There is, of course, no such thing as a 'pure race' and indeed there may be more variation for some characteristics within an ethnic group than between groups. Nevertheless, four ethnic groups are usually considered: negroids, mongoloids, caucasoids and australoids. Apart from differences in skin colour, the groups show some distinctive facial characteristics.

Negroids generally have leptoprosopic faces. The cheek bones are prominent and the teeth are positioned within the jaws such that there is bimaxillary protrusion. In addition, the nose is short and does not protrude far from the face, and the nasal index (the ratio of breadth to height) is high. The lips are generally thick and everted.

Although mongoloids have euryprosopic faces, this tends to be associated with prognathism. Mongoloids have a vertical forehead, a low nasal bridge, short nasal protrusion and prominent cheek bones. There is a reduction in the supra-orbital ridges and the frontal sinuses are small. The cheek and orbital regions may be well padded with fat. The medial canthus of the eye is covered by a fold of skin called the epicanthal fold.

Caucasoids and australoids have many features in common. Leptoprosopic and euryprosopic facial types are found. There is a tendency to retrognathia. The lips are thin and the nasal index is usually low for caucasoids but high for australoids.

The face exhibits some sexual dimorphism, although facial types are similar until the age of 12 years. The face of the female attains its mature form earlier than the male. It is generally flatter and more gracile. The male face tends to be more protuberant and more 'knobbly', bulky and coarse. The nose of the female often has a concave to straight profile; the nose of the male is larger, wider, and longer, having a straight to convex profile. Whereas the supra-orbital ridges of the male overhang the face, those of the female are at the same level as the inferior orbital margins and cheek bones. Consequently, the cheek bones and the upper jaws of females appear relatively prominent.

The main pattern of development of the face is an enlargement of vertical facial form in excess of lateral facial growth. There is considerable growth of the nose (which is flat at birth). The essential juvenile facial type is the euryprosopic face. This is exaggerated in the presence of considerable subcutaneous fat. The most obvious feature of the aged face is the appearance of skin creases. For example, suborbital creases produce the appearance of 'bags under the eyes' and creases at the lateral corners of the eyes give rise to 'crows' feet'. Marked creases are also found on the upper lip and at the glabella. Creases appear with age because the skin becomes less firmly attached to the underlying bone and facial muscles. Should the individual become edentulous, there is loss of alveolar bone and a consequent decrease in vertical facial dimensions.

Superficial structures of the face

On reflecting the skin of the face, the following main structures are revealed:

- The muscles of facial expression
- The facial nerve
- The cutaneous branches of the trigeminal nerve and the great auricular nerve
- The facial and superficial temporal vessels
- The parotid gland and duct
- The buccal pad of fat
- The facial lymph nodes

THE MUSCLES OF FACIAL EXPRESSION (Figure 47)

The face differs from most regions of the body in not having a deep membranous fascia beneath the skin. Instead, many small slips of muscle are attached to the facial skeleton and insert into the skin. These muscles cause movement of the facial skin to reflect emotions. The muscles are grouped mainly around the orifices of the face. Indeed, it can be argued that their primary function is to act as sphincters and dilators of the facial orifices and that the function of facial expression has developed secondarily.

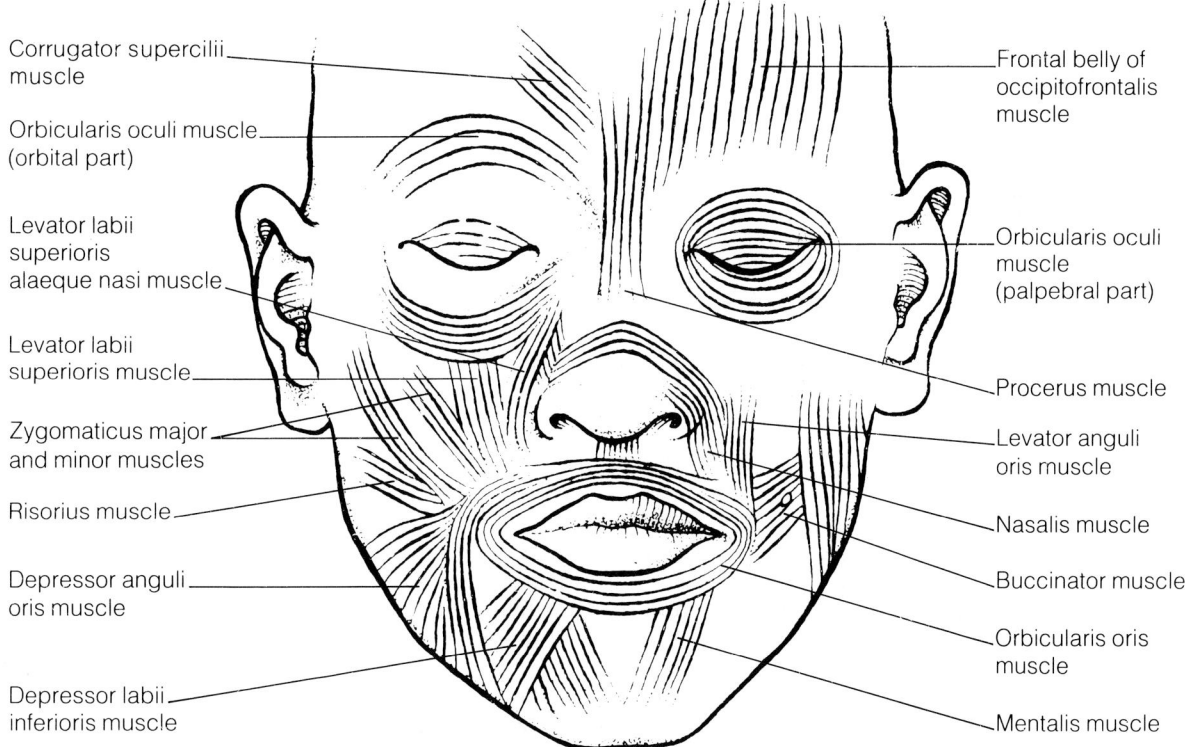

Figure 47 *The muscles of facial expression.*

The muscles of facial expression vary considerably between individuals in terms of size, shape and strength. In many instances, the names given to the muscles describe their actions. Muscles of facial expression not only lie in the face, but also in the scalp (occipitofrontalis, see page 188) and in the neck (platysma, see page 85).

Embryologically, the muscles are derived from the mesenchyme of the second branchial arch; hence they are innervated by the facial nerve.

THE ORBITAL GROUP

The chief muscle is orbicularis oculi; its fibres run circumferentially around the orbit and within the eyelids. In addition, there is the corrugator supercilii muscle.

The orbicularis oculi muscle

Attachments. This muscle is composed of three parts: the orbital, palpebral and lacrimal parts.

The orbital part is the largest and extends on to the face some distance beyond the orbital rim. It arises from three sites: from the nasal part of the frontal bone, from the frontal process of the maxilla and, between these two sites, from the medial palpebral ligament. The fibres then pass around the orbit in concentric loops.

The palpebral part is the central part and is confined to the eyelids. It arises mainly from the medial palpebral ligament and runs across the eyelids (in front of the tarsal plates) to insert into the lateral palpebral raphe.

The lacrimal part is a muscular slip that arises from the lacrimal bone. It passes behind the lacrimal sac, where some fibres insert into the lacrimal fascia. Other fibres insert into the tarsi of the eyelids near the lacrimal canaliculi and into the lateral palpebral raphe.

Innervation. The muscle is supplied by temporal and zygomatic branches of the facial nerve.

Vasculature. The arterial supply is derived from branches of the superficial temporal, facial, maxillary (infra-orbital branch) and ophthalmic arteries.

Actions. The orbicularis oculi muscle may be regarded as a sphincter of the eyelids. The orbital part is involved in forced closure (i.e. 'screwing up' the eye). The palpebral part of orbicularis oculi is involved in closing the eyelids without effort (i.e. involuntary closure during blinking). By pulling on the lacrimal fascia, the lacrimal part of the muscle is said to dilate the lacrimal sac and so aid the flow of tears into the sac.

Some fibres from the upper part of the orbicularis oculi muscle are inserted into the eyebrow and have been called the **depressor supercilii muscle**. As the name suggests, this muscle depresses the eyebrow.

The corrugator supercilii muscle

Attachments. This muscle originates from the medial end of the supra-orbital ridge on the frontal bone, deep to the orbicularis oculi muscle. It passes upwards and outwards through orbicularis oculi to insert into the skin of the middle of the eyebrow.

Innervation. This is derived from the temporal and (upper) zygomatic branches of the facial nerve.

Vasculature. Branches from the superficial temporal artery are the muscle's chief source of blood supply.

Actions. The corrugator supercilii is the principal muscle in the expression of suffering. It produces vertical ridges above the bridge of the nose when frowning by drawing the eyebrows downwards and inwards.

THE NASAL GROUP

Four muscles can be included within this group: procerus, compressor naris, dilator naris, and depressor septi. The compressor naris and the dilator naris muscles are often considered to be components of the nasalis muscle. The levator labii superioris alaeque nasi muscle can also be included in this group, although it is usually considered with the oral group of muscles.

The procerus muscle

Attachments. It arises from the nasal bone and the lateral nasal cartilage. Its fibres pass upwards to insert into the skin overlying the bridge of the nose.

Innervation. The procerus muscle is supplied by the temporal and (lower) zygomatic branches of the facial nerve. However, it has been reported that it is innervated only by the buccal branch of the facial nerve.

Vasculature. Its arteries are derived mainly from the facial artery (angular branch).

Actions. It produces transverse wrinkles over the bridge of the nose. Consequently, some consider procerus to be a member of the orbital group of facial muscles.

The compressor naris muscle (transverse part of nasalis)

Attachments. This muscle arises from the maxilla in the region overlying the root of the canine tooth. Its fibres pass over the dorsum of the nose to join the muscle from the opposite side.

Innervation. It is supplied by the buccal branch of the facial nerve, although there may also be a contribution from the zygomatic branch.

Vasculature. The arterial supply is derived from branches of the facial and maxillary (infra-orbital branch) arteries.

Actions. As indicated by the name, it is a compressor or sphincter of the nostril.

The dilator naris muscle (alar part of nasalis)

Attachments. This muscle originates from the maxilla in the region of the lateral incisor. It inserts into the greater nasal cartilage.

Innervation and vasculature. As for compressor naris.

Actions. It draws the ala of the nose downwards and laterally to dilate the nostril.

The depressor septi muscle

Attachments. This muscle arises from the incisor region of the maxilla and passes upwards from beneath the orbicularis oris muscle to insert into the cartilaginous nasal septum.

Innervation. This is provided by the buccal branch (sometimes the zygomatic branch) of the facial nerve.

Vasculature. The facial artery (superior labial branch) supplies the depressor septi muscle.

Actions. It pulls the nasal septum downwards and constricts the nostril.

THE AURICULAR GROUP

This group comprises the anterior, superior and posterior auricular muscles. The muscles are often rudimentary and show considerable variation. The most constant is the superior auricular muscle.

The anterior, superior and posterior auricular muscles

Attachments. The anterior auricular muscle arises from the epicranial aponeurosis. It passes downwards and backwards to insert into the helix of the auricle. The superior auricular muscle also arises from the epicranial aponeurosis. It passes downwards to be inserted into the upper part of the cranial surface of the auricle. The posterior auricular muscle arises from the base of the mastoid process of the temporal bone. It passes upwards to be inserted into the convexity of the concha.

Innervation. These muscles are not usually under voluntary control. The anterior and superior auricular muscles are supplied by the temporal branches of the facial nerve, the posterior auricular muscle by the posterior auricular branch.

Vasculature. The arteries are derived from the posterior auricular and superficial temporal arteries.

Actions. The auricular muscles usually show little activity. However, they may cause elevation and forward and backward movements of the ear.

THE CIRCUM-ORAL GROUP

The muscles that comprise this group show a complex arrangement. They may be subdivided into the orbicularis oris muscle itself and the muscles that radiate around it.

The orbicularis oris muscle

This muscle is the sphincter of the orifice of the mouth. Its fibres are said to encircle the orifice, lying within the upper and lower lips. However, it is possible that many fibres are derived from the other facial muscles that pass into the lips, so that relatively few fibres are truly intrinsic. The muscle may appear stratified, in which case the deep layer contains the buccinator muscle and the superficial layer comprises the other muscles merging into the lips. The arrangement of the intrinsic fibres is complex. There are oblique fibres that pass through the full thickness of the lips to connect the skin and mucous membrane. Other intrinsic fibres pass outwards from the alveolar bone in the incisor region to the angle of the mouth. It is believed that the philtrum of the upper lip is produced by the interlacing and crossing over of the intrinsic fibres of the orbicularis oris muscle in the midline.

Innervation. The muscle is innervated by the buccal and marginal mandibular branches of the facial nerve.

Vasculature. The main blood supply comes from the facial artery (superior and inferior labial branches), from the maxillary artery (mental and infra-orbital branches), and from the superficial temporal artery.

Actions. The orbicularis oris muscle is capable of various movements, including closure, protrusion and pursing of the lips.

The muscles of the lips, which are arranged around orbicularis oris, can be subdivided into superficial and deep muscles. The superficial muscles of the upper lip are the levator labii superioris and the zygomaticus major and minor muscles. The levator anguli oris muscle is the deep muscle of the upper lip. The superficial muscle of the lower lip is the depressor anguli oris muscle. The depressor labii inferioris and the mentalis muscles are the deep muscles of the lower lip. At the corners of the mouth lie the buccinator and risorius muscles.

The levator labii superioris muscle

Attachments. It arises from the maxilla at the inferior margin of the orbit, above the infra-orbital foramen. Here, the muscle is deep to the orbicularis oculi muscle. Some of its fibres pass downwards to be inserted into the skin overlying the lateral side of the upper lip. Other fibres merge with those of orbicularis oris. A small slip of muscle arises from the frontal process of the maxilla, close to the side of the nose. This is usually called the **levator labii superioris alaeque nasi** (also referred to as the angular head of the levator labii superioris). This inserts into the skin and the greater nasal cartilage of the nose, and into the skin and musculature of the upper lip.

Innervation. The zygomatic and buccal branches of the facial nerve innervate this muscle.
112,31; 112,26.

Vasculature. The muscle is supplied by branches from the facial and maxillary (infra-orbital branch) arteries.

Actions. The primary function of the levator labii superioris muscle is to elevate the upper lip. The levator labii superioris alaeque nasi muscle also dilates the nostril.

The zygomaticus major muscle
112,14; 118B,26
12,8

Attachments. This muscle takes origin from the lateral surface of the zygomatic bone, just in front of the zygomaticotemporal suture. It passes obliquely downwards to the corner of the mouth where it mingles with the orbicularis oris muscle.

Innervation. The muscle is supplied by the zygomatic and buccal branches of the facial nerve.
112,31; 112,26

Vasculature. The arterial supply is derived from the facial artery (superior labial branch).

Actions. The muscle pulls the corner of the mouth upwards and outwards, as in laughing.

The zygomaticus minor muscle
112,13; 118B,25
12,7

Attachments. From its attachment on the zygomatic bone (in front of the origin of zygomaticus major), zygomaticus minor runs downwards and forwards to insert into the lateral part of the upper lip.

Innervation. It receives its nerve supply from the zygomatic and buccal branches of the facial nerve.
112,31; 112,26

Vasculature. Its blood supply is derived from the facial artery (superior labial branch).

Actions. The muscle elevates the upper lip. Acting with other facial muscles, it produces the expression of disdain. The nasolabial furrow is associated with this muscle.

The levator anguli oris muscle
112,15; 118B,22
12,9

Attachments. It arises from the canine fossa of the maxilla (immediately below the infra-orbital foramen) and passes downwards towards the corner of the mouth. Some fibres continue around the corner of the mouth to mingle with orbicularis oris in the lower lip.

Innervation. This is derived from the zygomatic and buccal branches of the facial nerve.
112,31; 112,26

Vasculature. The blood supply arises from the facial (superior labial branch) and maxillary (infra-orbital branch) arteries.

Actions. The levator anguli oris muscle elevates the corner of the mouth.

The depressor anguli oris muscle
112,22
12,15

Attachments. It arises from an extensive area around the external oblique line of the mandible. Its fibres pass upwards to the corner of the mouth. It is partly inserted here and partly into the orbicularis oris in the upper lip.

Innervation. The muscle is innervated by the buccal and marginal mandibular branches of the facial nerve.
112,26; 112,36

Vasculature. The facial (inferior labial branch) and maxillary (mental branch) arteries supply the muscle.

Actions. The muscle depresses the corner of the mouth. It is said to be associated with the expression of grief.

The depressor labii inferioris muscle

112,21
12,14

Attachments. This muscle arises from the mandible just in front of the mental foramen (here it is covered by the anterior fibres of depressor anguli oris). The fibres pass upwards and medially to converge with the orbicularis oris muscle in the lower lip.

112,36

Innervation. It is supplied by the marginal mandibular branch of the facial nerve.

Vasculature. The blood supply is derived from the facial (inferior labial branch) and maxillary (mental branch) arteries.

Actions. It depresses the lower lip and draws it laterally. The muscle is associated with the expression of irony.

The mentalis muscle

112,20
12,17

Attachments. The muscle originates from the incisive fossa of the mandible. Its fibres descend to insert into the skin of the chin.

112,36

Innervation. This is provided by the marginal mandibular branch of the facial nerve.

Vasculature. It receives blood from the facial (inferior labial branch) and maxillary (mental branch) arteries.

Actions. The mentalis muscle raises and protrudes the lower lip (as during the expression of doubt or disdain).

The risorius muscle

112,24

Attachments. This muscle is usually poorly developed. Unlike the other facial muscles, it does not arise from bone but originates from the connective tissue overlying the parotid gland. The muscle runs horizontally across the face to insert into skin at the corner of the mouth. In some cases, risorius is indistinguishable from the facial portion of the platysma muscle.

112,26; 112,31

Innervation. The muscle is supplied by the buccal and zygomatic branches of the facial nerve.

Vasculature. This is derived from the facial artery (superior labial branch).

Actions. Risorius pulls the corner of the mouth laterally as in grinning.

The buccinator muscle

116,8; 142,44
144A,16
144A,15

This forms the musculature of the cheek. Although classified with the muscles of facial expression, the buccinator muscle functions principally during mastication. An important relationship is with the parotid duct. This pierces the muscle opposite the maxillary third molar and then runs forwards to open into the oral cavity opposite the maxillary second molar.

116C,33; 142A,43
16,12

Attachments. The buccinator muscle has two main origins. First, it arises from the anterior margin of the pterygomandibular raphe (the superior constrictor muscle of the pharynx arising from the posterior margin). Second, it is attached to the alveolar margins of the maxilla and mandible in the region of the molar teeth. A few fibres also arise from a tendinous band bridging the interval between the pterygoid hamulus and the maxillary tuberosity. Initially lying deep to the ramus of the mandible, the muscle emerges into the cheek by crossing the retromolar fossa of the mandible (behind the third molar). The fibres eventually run into the orbicularis oris muscle (where they form the deep layer of this muscle). The upper and lower fibres of the buccinator muscle pass into the substance of the corresponding lip, whereas the more centrally positioned fibres (arising mainly from the pterygomandibular raphe) decussate at the corner of the mouth to pass into the opposite lip.

Innervation. This is derived from the buccal branch of the facial nerve.

Vasculature. The arterial supply is from branches of the facial artery and from the maxillary (buccal branch) artery.

Actions. The main function of the buccinator muscle is to aid in mastication by maintaining the bolus of food between the molar teeth. In addition, it is involved in sucking and in expelling air forcibly (e.g. in whistling or playing a wind instrument).

It may be appreciated from the descriptions of the circum-oral musculature that several muscles converge and interlace at the corners of the mouth. This is said to produce a nodular region called the modiolus, which is palpable. The modiolus may be fixed by the action of some of the muscles to provide a base for the action of other muscles.

THE NERVES OF THE FACE

Both motor and sensory nerves are found on the face. The motor nerves are derived from the facial nerve (cranial nerve VII). The sensory innervation is primarily from the three divisions of the trigeminal nerve, with a small contribution from the cervical plexus via the great auricular nerve.

The facial nerve (Figure 48)

After emerging from the base of the skull at the stylomastoid foramen, the facial nerve gains access to the face by passing through the substance of the parotid gland.

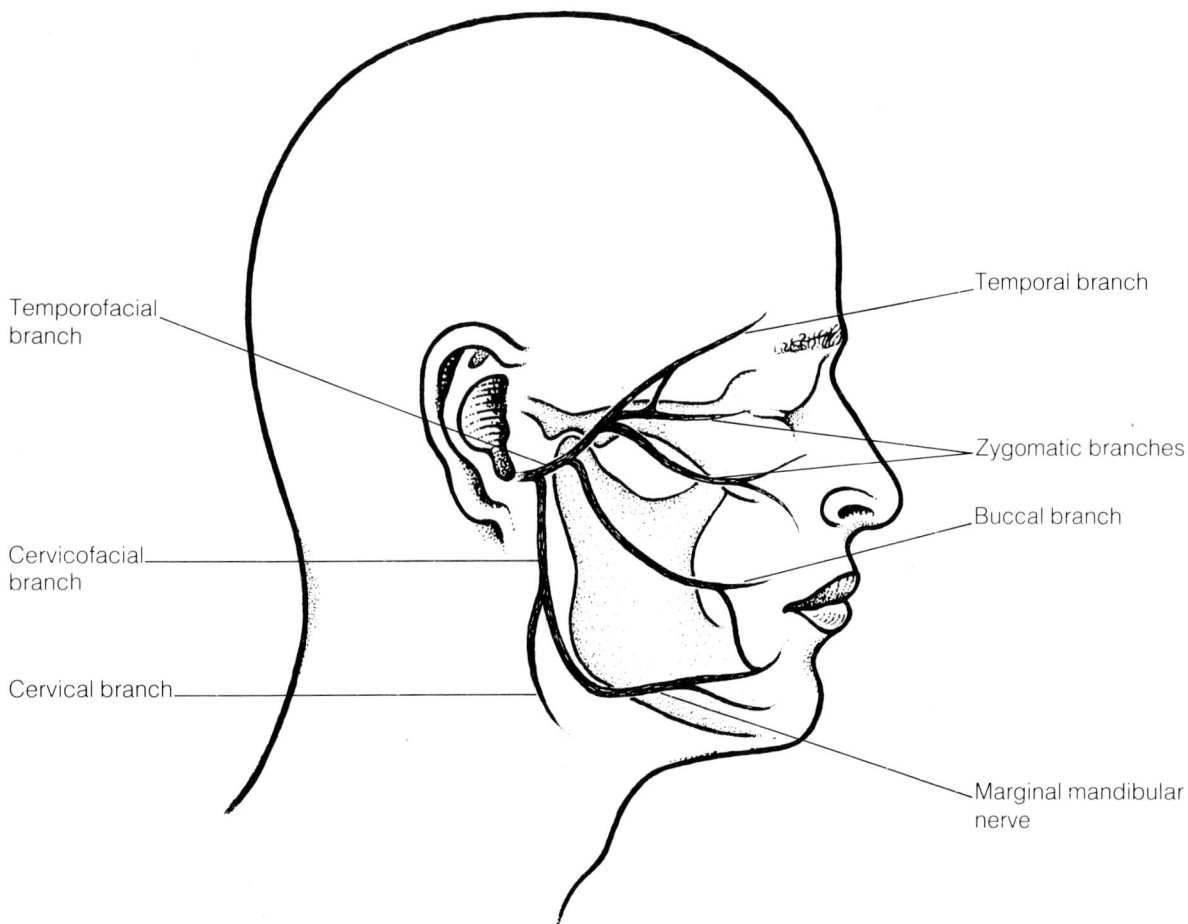

Figure 48 *The facial nerve.*

Branches in the face:
- Posterior auricular nerve
- Nerves to digastric and stylohyoid muscles
- Temporal branches
- Zygomatic branches
- Buccal branches
- Marginal mandibular nerve
- Cervical branch

The posterior auricular nerve arises close to the stylomastoid foramen. It supplies the occipital belly of the occipitofrontalis muscle and some of the ear muscles.

The nerves to the posterior belly of the digastric and the stylohyoid muscles also arise close to the stylomastoid foramen.

Just before entering the parotid gland, the facial nerve divides into temporofacial and cervicofacial trunks (Figure 48). Within the substance of the parotid gland, the trunks show a variable pattern of branching to form a parotid plexus. From this plexus usually emerge five distinct sets of branches from the anteromedial surface of the gland:

The temporal branches leave the superior surface of the gland and cross the zygomatic arch to reach the forehead. They supply the auricular muscles, the frontal belly of the occipitofrontalis muscle in the scalp, and part of the orbicularis oculi muscle.

The zygomatic branches are usually two in number. The upper branch passes above the orbit to supply the frontal belly of the occipitofrontalis muscle and the orbicularis oculi. The lower branch passes below the orbit to supply the lower part of the orbicularis oculi and to contribute to the innervation of muscles in the upper lip and nose.

The buccal branches (usually two in number) often pass across the face on either side of the parotid duct. They supply the buccinator muscle and contribute to the innervation of the muscles of the upper lip and nose.

The marginal mandibular nerve emerges from the lower border of the parotid gland and runs near the inferior border of the mandible. Initially, it may pass into the neck below the angle of the mandible, but at the anterior border of the masseter muscle it crosses back on to the face. Its function is to supply the muscles of the lower lip.

The cervical branch passes downwards from the lower border of the parotid gland to supply the platysma muscle in the neck.

It is not known to what extent sensory fibres are present in the facial nerve after it leaves the stylomastoid foramen. In some animals, 15% of the sensory fibres of the nerve may emerge at the stylomastoid foramen. Many of these are thought to be distributed to the concha of the auricle and sometimes to an area behind the ear. Others claim that some sensory fibres travel with the motor fibres to the face. Indeed, some believe the facial nerve may be involved in the appreciation of deep pain in the face as well as proprioceptive functions for the facial musculature.

The course and distribution of the facial nerve is complicated by the presence of communicating branches with other nerves. On the face, these nerves are branches from the trigeminal nerve and the cervical plexus. Though the significance of these communications is not fully known, it is thought that they may represent proprioceptive contributions from the trigeminal nerve to the facial muscles.

THE CUTANEOUS INNERVATION OF THE FACE (Figure 49)

The cutaneous innervation of the body is mapped out by the pattern of dermatomes, a dermatome being defined as an area of skin supplied by a single segment of the spinal cord. In the face, however, most of the skin is innervated by branches of the trigeminal nerve. Thus, the use of the term dermatome here is inappropriate and we must talk of peripheral nerve fields.

Three large areas of the face can be mapped out to indicate the peripheral nerve fields associated with the three divisions of the trigeminal nerve. The fields are not horizontal but curve upwards. It is claimed that the reason for this is that the skin moves upwards with growth of the brain. Embryologically, each division of the trigeminal nerve is associated with a developing facial process that gives rise in the adult to a particular area of the face. The ophthalmic nerve is associated with the frontonasal process, the maxillary nerve with the maxillary process and the mandibular nerve with the mandibular process (see pages 168 to 169).

The ophthalmic nerve

This nerve has five branches which are distributed to the face. The area supplied is extensive, including the forehead, the upper eyelid and much of the external surface of the nose. Five nerves are concerned with the cutaneous innervation:

The supra-orbital nerve is the largest ophthalmic branch on the face. It is one of the two terminal branches of the frontal nerve. It emerges from the orbit through the

118C,31; 122,4

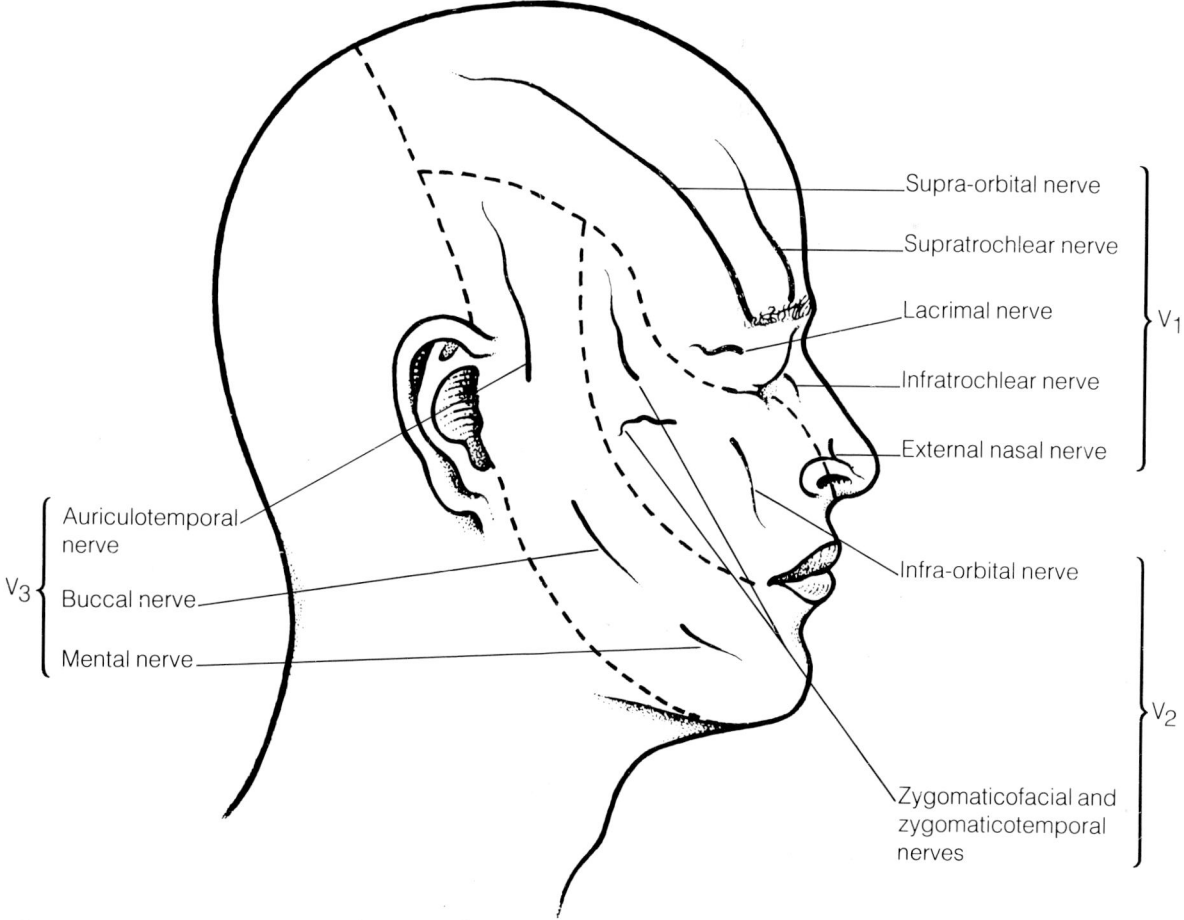

Figure 49 *The cutaneous innervation of the face.*

	supra-orbital notch (or foramen) and supplies much of the forehead and most of the upper eyelid.
122A,8	**The supratrochlear nerve** supplies a small area of skin over the medial part of the forehead and over the medial part of the upper eyelid. It is also a terminal branch of the frontal nerve and emerges from the orbit medial to the supra-orbital nerve. Its name indicates that it runs above the trochlea associated with the superior oblique muscle of the eye.
10,5	
118C,34; 122,7	
122,10	**The infratrochlear nerve** is one of the two terminal branches of the nasociliary nerve (the other being the anterior ethmoidal nerve). It supplies skin over the bridge of the nose and at the medial corner of the upper eyelid. As its name suggests, it leaves the orbit below the trochlea associated with the superior oblique muscle.
118C,34; 122,7	
	The external nasal nerve is the terminal part of the anterior ethmoidal nerve. It supplies the skin of the nose below the nasal bones (excluding the ala portion around the external nares).
120,15; 122,30	**The lacrimal nerve** is the smallest branch of the ophthalmic nerve. It emerges from the upper lateral margin of the orbit to supply the lateral part of the upper eyelid.
124C,43	

10,6

The maxillary nerve

The maxillary nerve supplies the skin of the lower eyelid, the prominence of the cheek, the ala part of the nose, part of the temple, and the upper lip. It has three cutaneous branches:

118C,40	**The infra-orbital nerve,** the largest cutaneous branch of the maxillary nerve, emerges on to the face through the infra-orbital foramen. As well as supplying skin overlying the maxilla, it gives off palpebral branches to the lower eyelid, nasal branches to the ala of the nose and labial branches to the upper lip.
10,12; 110,14	
	The zygomaticofacial nerve is one of the two branches of the zygomatic nerve. It emerges from the orbit on to the face at the zygomaticofacial foramen. It supplies skin overlying the prominence of the cheek.
14,27	
112,4	**The zygomaticotemporal nerve** is the remaining branch of the zygomatic nerve. It enters the temporal fossa via the zygomaticotemporal foramen on the deep surface of the zygomatic bone. It supplies skin over the anterior part of the temple (non-beard part of temple).
53D,26	

The mandibular nerve

The area supplied by the mandibular nerve includes the skin overlying the mandible, the lower lip, the fleshy part of the cheek, part of the auricle, and part of the temple. It has three cutaneous branches:

10,16; 110,24	**The mental nerve** is a branch of the inferior alveolar nerve. It emerges on to the face through the mental foramen of the mandible. It supplies the skin of the lower lip and the skin overlying the mandible (except around the angle of the mandible).
116A,9; 166B,46	**The buccal branch of the mandibular nerve** is sometimes referred to as the long buccal nerve to distinguish it from the buccal branch of the facial nerve. It emerges on to the face from behind the ramus of the mandible to supply the skin overlying the fleshy part of the cheek.
112,2	**The auriculotemporal nerve** appears on the face behind the temporomandibular joint, and ascends over the zygomatic arch. It supplies the tragus, concha, external acoustic meatus and tympanic membrane of the ear, and the posterior part of the temple (beard part of the temple).
110,17	

The great auricular nerve

One part of the face that does not receive its cutaneous innervation from the trigeminal nerve is the angle of the mandible. The skin in this region is supplied by the great auricular nerve (see page 85). This nerve is derived from the cervical plexus (anterior primary rami of the second and third cervical nerves). Appearing at the posterior border of the sternocleidomastoid muscle, it passes forwards and upwards across the muscle to reach the angle of the mandible.

THE ARTERIES OF THE FACE (Figure 50)

The main arterial supply to the face comes from the facial and superficial temporal arteries. These are both branches of the external carotid artery. Blood also reaches the face from some of the branches of the maxillary and ophthalmic arteries. The various branches are joined by numerous anastomoses.

The facial artery

After its origin from the external carotid artery in the neck, the facial artery passes upwards and forwards towards the inferior border of the mandible. This cervical course is described further on page 101.

The facial artery first makes its appearance on the external surface of the mandible at the anterior border of the masseter muscle. At this point, it pierces the deep fascia to pass on to the face. It then continues upwards and medially, following a tortuous

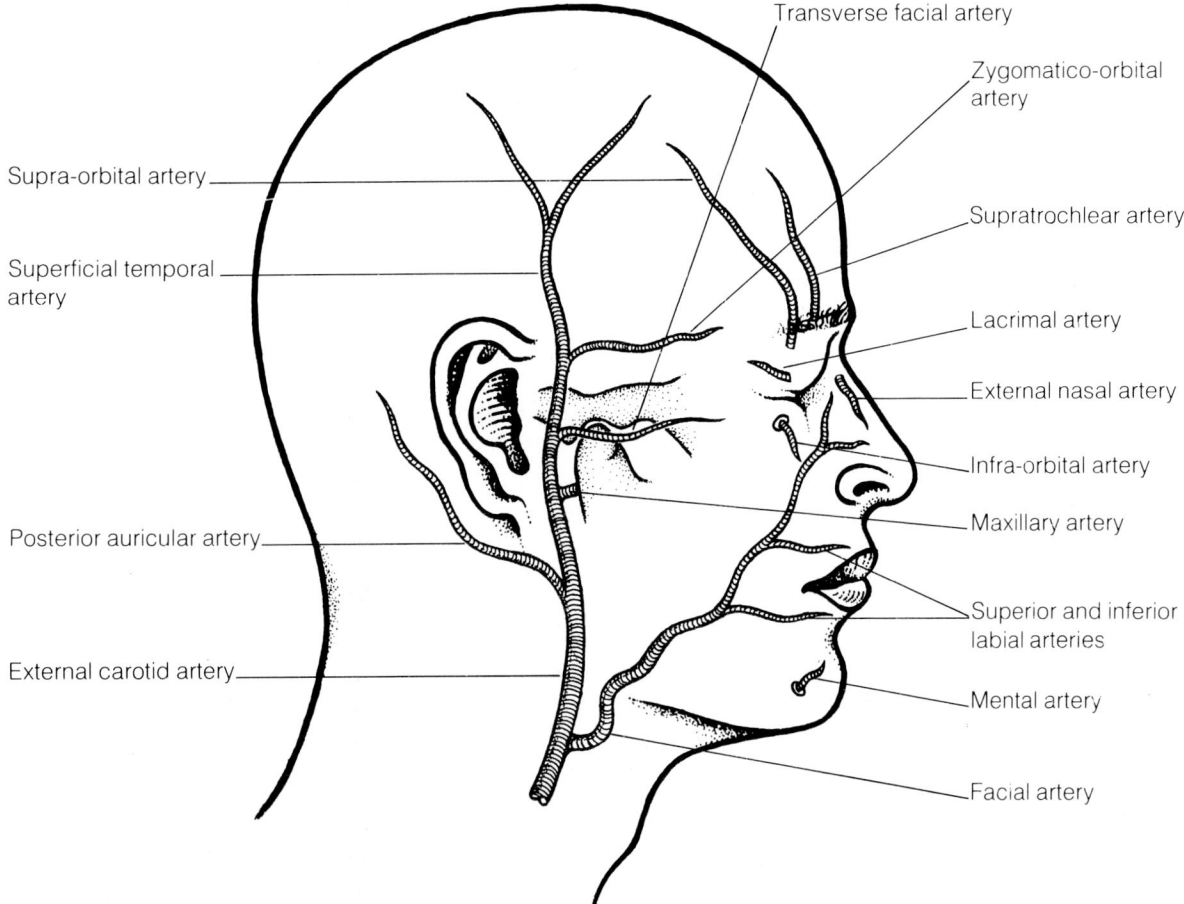

Figure 50 *The arteries of the face.*

112,17
118B,23

course towards the bridge of the nose. It lies deep to the zygomatic and risorius muscles, but superficial to the buccinator and the levator anguli oris muscles. It passes either superficially or deep to levator labii superioris. Thus, it is generally only visible on the cheek between the zygomaticus and risorius muscles. Alongside the nose, it is closely related to the levator labii superioris alaeque nasi muscle. Here, it is termed the angular artery. Throughout its course on the face, it usually lies anterior to the facial vein.

112,18

The chief branches of the facial artery on the face are the inferior labial, superior labial and angular arteries. The inferior and superior labial arteries have a tortuous course in the lower and upper lip respectively. If the margin of each lip is compressed between the thumb and index finger, the pulsations of the labial arteries may be felt. These arteries are situated deep to the orbicularis oris muscle, lying just beneath the labial mucosa. The superior labial artery has a branch supplying the nasal septum. The angular artery gives off small branches to the nose but, where there is a large branch, it has been called the lateral nasal artery. In addition, the submental artery (which arises in the neck) runs along the lower border of the mandible to supply the chin and lower lip.

102A,7

112,3

The superficial temporal artery

This is one of the terminal branches of the external carotid artery within the parotid gland. Emerging from the superior surface of the gland, the artery passes upwards towards the scalp, crossing the zygomatic process of the temporal bone. It then divides into anterior and posterior branches (these may also be called frontal and parietal branches, indicating their region of supply on the scalp). Glandular branches are given off to the parotid gland.

112,30

A transverse facial artery arises from the superficial temporal artery within the parotid gland. It crosses the masseter muscle above the parotid duct. The transverse facial artery gives branches to both the gland and the duct, to the masseter muscle, and to adjacent skin. The superficial temporal artery also gives off a middle temporal artery (piercing the temporal fascia to supply the temporalis muscle), auricular branches supplying the external lower part of the ear, and a zygomatico-orbital branch to the orbicularis oris muscle.

Facial branches from the maxillary artery

The maxillary artery is a terminal branch of the external carotid artery. It provides three branches to the face.

10,16

The mental artery arises from the first part of the maxillary artery as a terminal branch of the inferior alveolar artery. It emerges on to the face from the mandibular canal at the mental foramen. The mental artery supplies muscles and skin in the chin region and anastomoses with the inferior labial and submental arteries.

The buccal artery is a branch of the second part of the maxillary artery. It emerges on to the face from the infratemporal fossa and crosses the buccinator muscle to supply the cheek. The buccal artery anastomoses with the infra-orbital artery and with branches of the facial artery.

10,12

The infra-orbital artery arises from the third part of the maxillary artery. It runs through the infra-orbital foramen and on to the face, supplying the lower eyelid, the lateral aspect of the nose and the upper lip. The infra-orbital artery has extensive anastomoses with the transverse facial, ophthalmic and buccal arteries, and with branches of the facial artery.

Facial branches from the ophthalmic artery

The ophthalmic artery is a branch of the internal carotid artery. Five branches arise from the ophthalmic artery within the orbit to supply the face.

118C,32; 10,6

The supra-orbital artery leaves the orbit through the supra-orbital notch (or foramen). It supplies much of the upper eyelid, forehead and scalp. It anastomoses with the supratrochlear and superficial temporal arteries.

The supratrochlear artery emerges from the orbit on to the face at the frontal notch. It supplies the medial parts of the upper eyelid, forehead and scalp.

The lacrimal artery appears on the face at the upper lateral corner of the orbit to supply the lateral part of the eyelids. Within the orbit, the lacrimal artery gives off a zygomatic artery which subdivides into zygomaticofacial and zygomaticotemporal arteries. The zygomaticofacial artery then passes through the lateral wall of the orbit to emerge on to the face at the zygomaticofacial foramen, supplying the region overlying the prominence of the cheek. The zygomaticotemporal artery also passes through the lateral wall of the orbit, via the zygomaticotemporal foramen, to supply the skin over the non-beard part of the temple. The lacrimal artery anastomoses with the deep temporal branch of the maxillary artery and the transverse facial branch of the superficial temporal artery.

The dorsal nasal artery accompanies the infratrochlear nerve. It leaves the orbit at the upper medial corner. The artery supplies a region of skin around the bridge of the nose, anastomosing with the angular branch of the facial artery.

The external nasal artery is the terminal branch of the anterior ethmoidal artery from the ophthalmic artery. It supplies skin on the external nose, emerging at the junction of the nasal bone and the lateral nasal cartilage.

THE VEINS OF THE FACE (Figure 51)

The veins show a similar distribution to the arteries of the face, although there is greater variability. The chief vein is the facial vein.

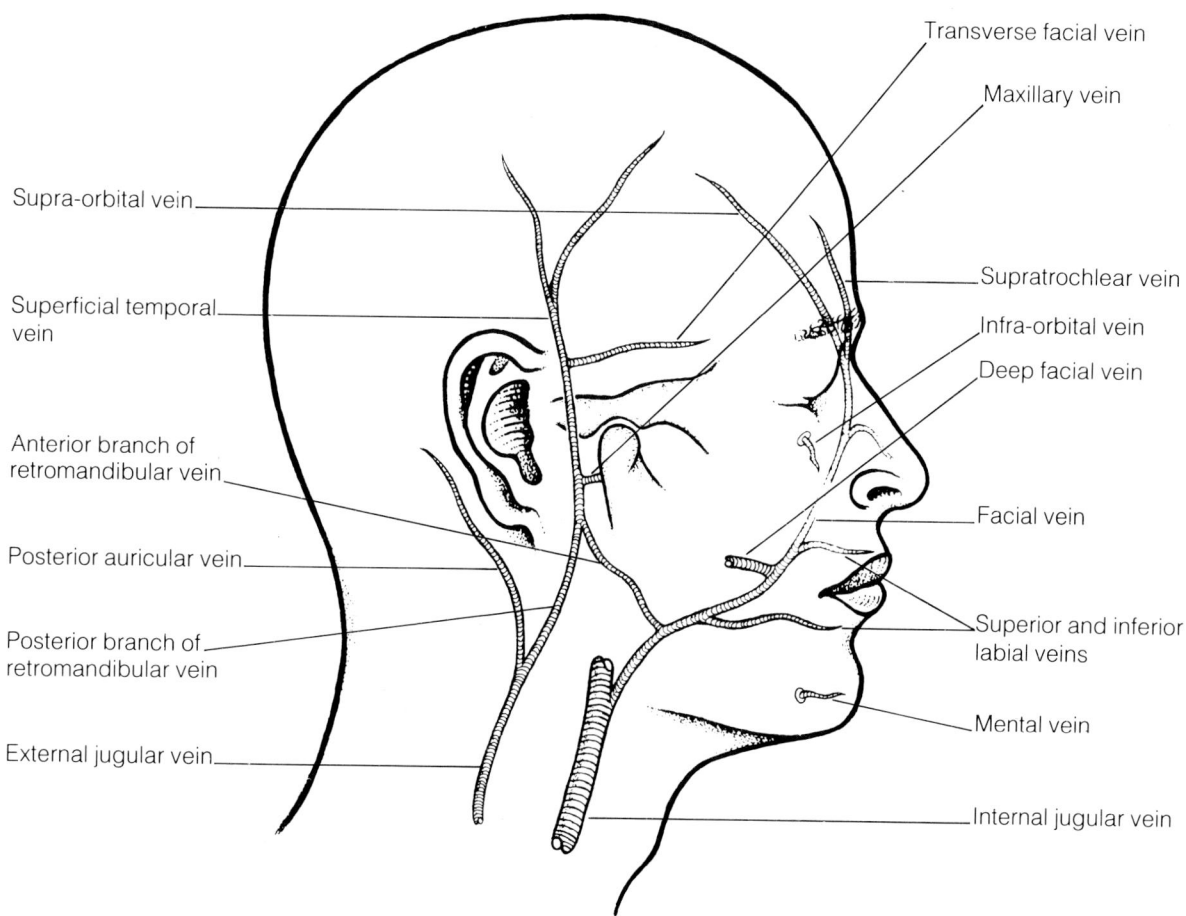

Figure 51 *The veins of the face.*

The facial vein

This begins as the angular vein at the medial corner of the eye, the angular vein being formed by the confluence of the supra-orbital and supratrochlear veins. The facial vein soon divides into a part that communicates with the superior ophthalmic vein in the orbit, and a part that remains superficial to pass downwards close behind the facial artery. The facial vein passes over the inferior border of the mandible near the anterior attachment of the masseter muscle. In the submandibular region, it receives the anterior branch of the retromandibular vein (forming what was once called the common facial vein) and, crossing the submandibular gland and the facial artery, drains into the internal jugular vein. The tributaries of the facial vein include: nasal, deep facial, and superior and inferior labial veins. The deep facial vein connects the facial vein with the pterygoid venous plexus in the infratemporal fossa.

The superior ophthalmic and infra-orbital veins are also important when considering the connections of the facial vein. The superior ophthalmic vein links the angular vein with the cavernous sinus and can receive veins from the scalp, forehead, and upper eyelid. As the facial vein has no valves, pressure or blockage may result in blood flowing into the cavernous sinus. The infra-orbital vein links the facial vein with the pterygoid venous plexus in the infratemporal fossa. It can receive veins from the lower eyelid, lateral part of the nose, and upper lip.

The superficial temporal vein

This vein is formed above the zygomatic arch by the union of anterior and posterior tributaries. The superficial temporal vein then enters the substance of the parotid gland. Here, it unites with the maxillary vein to form the retromandibular vein (see page 166). It receives the transverse facial vein and veins from the parotid gland and ear.

Buccal and mental veins drain veins from the cheek and the chin into the pterygoid venous plexus.

THE LYMPHATICS OF THE FACE

Both the distribution of the lymph nodes and the arrangement of the lymphatic vessels in the face show considerable variation. Figure 52 illustrates the main areas of lymphatic drainage.

THE PAROTID GLAND

This is the largest of the major paired salivary glands. It occupies the region between the ramus of the mandible and the mastoid process, extending upwards to the external acoustic meatus. The gland is surrounded by an unyielding capsule, the parotid capsule, derived from the investing layer of deep cervical fascia.

Within the gland lies the facial nerve, the external carotid artery (and its terminal branches), and the retromandibular vein (and its formative vessels). Lymph nodes are present both within, and on the surface of the gland. The parotid gland is a serous gland (though a small mucous component is present in the neonate) and the secretions travel in the main parotid duct to discharge into the oral cavity close to the crown of the upper second molar tooth.

In addition to the main gland, a small portion of glandular tissue may be situated above the parotid duct. This part is called the accessory parotid gland.

The parotid gland is essentially pyramidal in shape. The base (superior surface) is closely moulded around the external acoustic meatus and is called the glenoid lobe. The superficial temporal vessels and the auriculotemporal nerve appear at the superior surface and pass upwards to be distributed to the temple. The apex (inferior surface) of the gland extends beyond the angle of the mandible to overlap the digastric triangle. The lateral surface (superficial surface) has an irregular, lobulated appear-

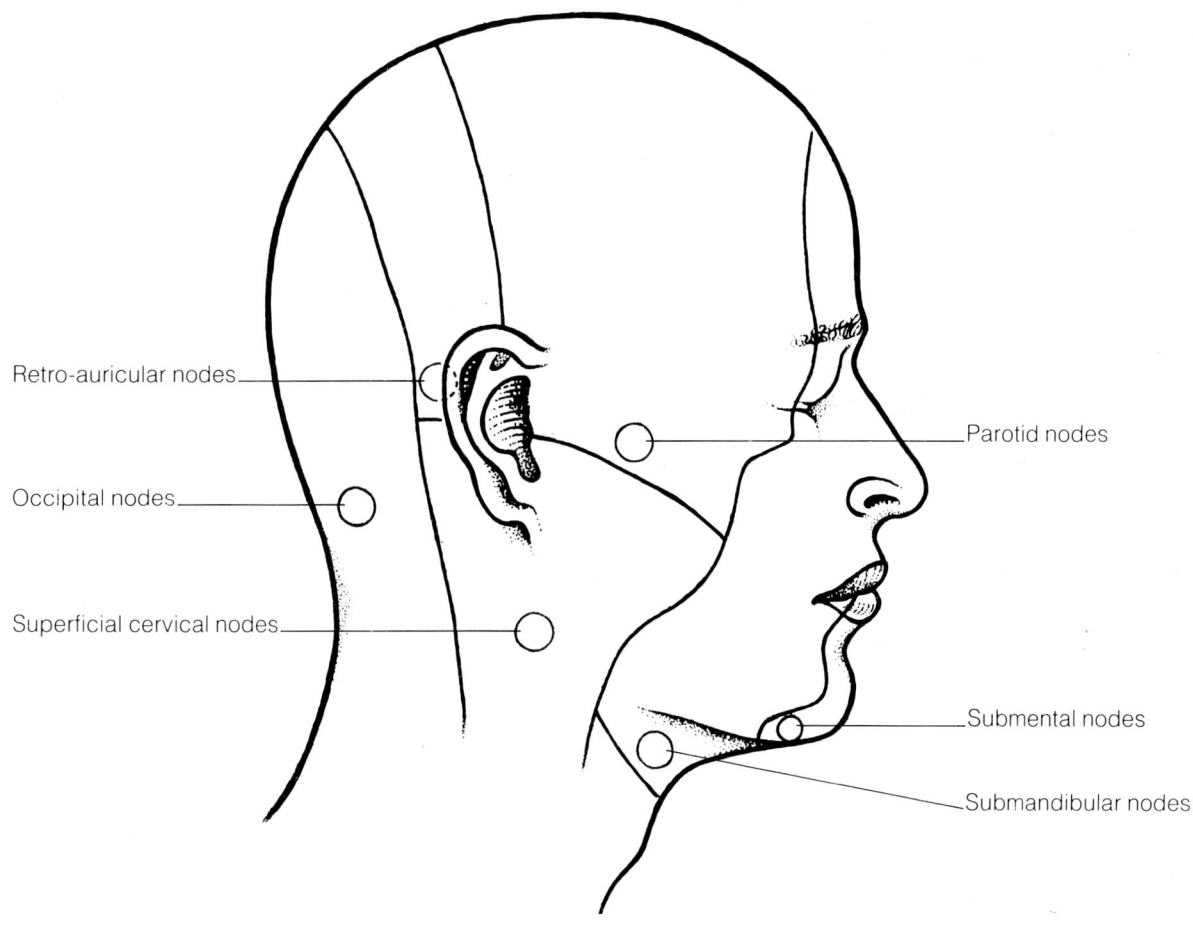

Figure 52 *The lymphatics of the face.*

ance. It is covered only by skin, superficial fascia, and part of the platysma muscle. The great auricular nerve from the cervical plexus crosses this part of the gland to be distributed to the surrounding skin and the parotid fascia. The medial surface (deep surface) rests on the 'parotid bed' (Figure 53). It can be divided into two parts. The anterior part is applied to the masseter muscle, the posterior border of the ramus, the medial pterygoid muscle, and the condyle of the mandible. The posterior part rests on the sternocleidomastoid muscle, the mastoid process, and the styloid group of muscles. At the junction between anteromedial and posteromedial surfaces, a flange of glandular tissue may extend beyond the posterior border of the mandible to reach the wall of the pharynx.

112,39

114,6; 140A,19
140A,23; 114,8
114,12; 114,10

Structures within the parotid gland

The deepest of the three main structures within the gland is the external carotid artery. The artery enters the posteromedial surface of the gland from the neck. Here, it divides into the maxillary and superficial temporal arteries. The maxillary artery emerges from the anteromedial surface of the gland and passes forwards between the neck of the condyle of the mandible and the sphenomandibular ligament to enter the infratemporal fossa (see page 180). The superficial temporal artery passes upwards to leave the gland at its superior surface. Within the substance of the gland, it gives off the transverse facial artery. This emerges at the anterior border of the parotid gland and runs above the parotid duct. The posterior auricular artery may also arise within the parotid gland.

140A,37
140C,62

142A,19

112,3

112,30
140A,36

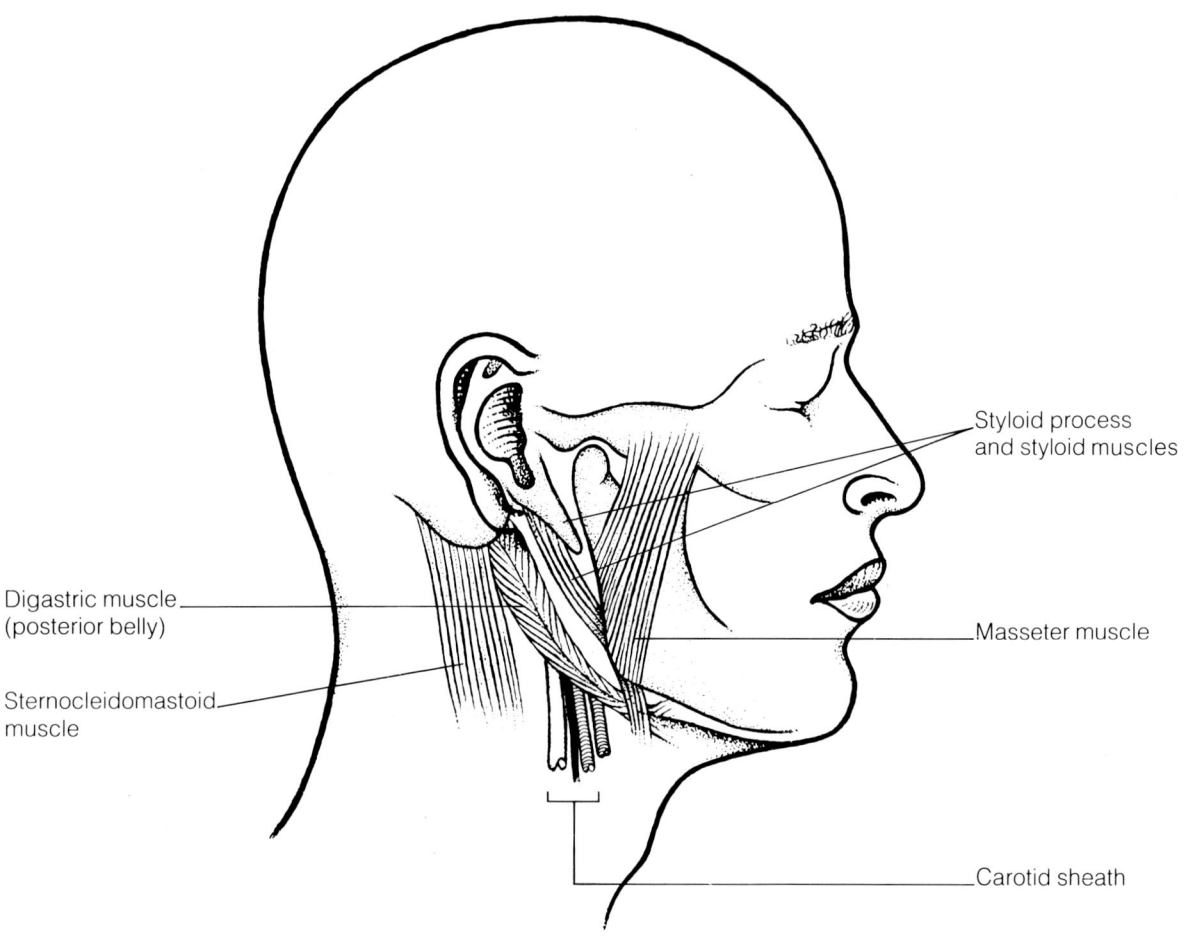

Figure 53 *Structures forming the bed of the parotid gland.*

140A,38
140C,64; 140C,65
98,5; 98,6
98,7

The veins within the parotid gland are superficial to the arteries. The superficial temporal and maxillary veins join within the upper part of the gland to form the retromandibular vein. The retromandibular vein near the apex of the gland usually divides into anterior and posterior branches. Outside the gland, the anterior branch joins the facial vein; the posterior branch unites with the posterior auricular vein to form the external jugular vein.

112

The facial nerve is the most superficial of the structures within the parotid gland. Emerging from the stylomastoid foramen, the nerve enters the parotid gland in the upper part of its posteromedial surface. It crosses the styloid process, the retromandibular vein, and the external carotid artery before dividing into a plexus in the region of the temporomandibular joint. Five main branches (which may be single or multiple) arise from the plexus and emerge from the anterior and inferior margin of

112,33,; 112,31
112,26; 112,36
112,37

the parotid gland (see Figure 48). These are temporal, zygomatic, buccal, marginal mandibular and cervical branches (see page 158). Some anatomists subdivide the parotid gland into superficial and deep parts with reference to the facial nerve.

116C,55

The auriculotemporal nerve is a branch of the mandibular division of the trigeminal nerve. It appears within the superior surface of the parotid gland, having passed from

142A,17
112,2

behind the temporomandibular joint from the infratemporal fossa (see page 178). The auriculotemporal nerve then ascends over the posterior part of the zygomatic process of the temporal bone behind the superficial temporal vessels.

The parotid capsule

The parotid gland is surrounded by a fibrous capsule which is a continuation of the investing layer of deep cervical fascia. This fascia splits to enclose the gland within a superficial and a deep layer. The superficial layer is attached above to the zygomatic process of the temporal bone, the cartilaginous part of the external acoustic meatus, and the mastoid process. The deep layer is attached to the mandible, and to the tympanic plate and the styloid and mastoid processes of the temporal bone. Between the styloid process and the angle of the mandible, the fascia is thickened to form the stylomandibular ligament. This separates the parotid and submandibular glands.

The parotid duct

This duct appears at the anterior border of the parotid gland. It passes horizontally across the masseter muscle, approximately at a level midway between the angle of the mouth and the zygomatic arch. Here, the duct lies below the transverse facial artery and receives one or more ducts from the accessory parotid gland. The parotid duct bends sharply around the anterior border of the masseter to pierce the buccal pad of fat and the buccinator muscle at the level of the upper third molar tooth. A further bend in the duct is found as it passes forwards beneath the oral mucosa before opening into the vestibule. The duct usually opens into the oral cavity opposite the crown of the upper second molar tooth. Adjacent to the opening is a small elevation of the mucosa (the parotid papilla).

The innervation of the parotid gland (Figure 54)

The innervation is related to the otic parasympathetic ganglion.

The parasympathetic secretomotor supply is from the inferior salivatory nucleus of the brainstem (see page 477). Passing with the glossopharyngeal nerve, the fibres run in the tympanic branch which contributes to the tympanic plexus on the promontory of the middle ear (see page 372). From this plexus arises the lesser petrosal nerve. This leaves the middle ear, runs in a groove on the petrous portion of the temporal bone, and then passes through the foramen ovale (or the petrosal/innominate foramen) to the otic ganglion. Synapsing at the ganglion, postganglionic fibres leave to join the

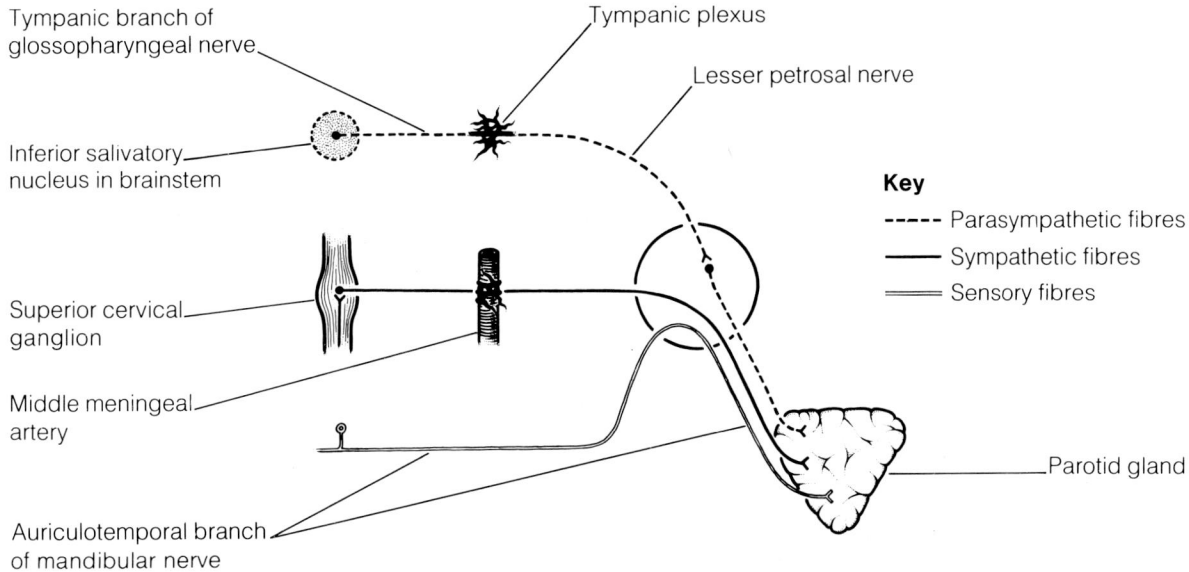

Figure 54 *The otic parasympathetic ganglion and innervation of the parotid gland.*

nearby auriculotemporal nerve, which distributes the fibres to the parotid gland. There is some evidence that secretomotor fibres from the chorda tympani branch of the facial nerve may also supply the parotid gland.

The sympathetic supply to the parotid gland is derived initially from the superior cervical sympathetic ganglion. From this ganglion, the innervation reaches the gland via the plexus around the middle meningeal artery, the otic ganglion (without synapsing) and eventually the auriculotemporal nerve. It seems likely that an alternative source of sympathetic fibres is derived directly from the sympathetic plexuses accompanying the vessels supplying the parotid gland.

Sensory fibres to the connective tissue within the parotid gland are derived directly from the auriculotemporal nerve. In passing back to the parent mandibular nerve, they run through the otic ganglion (without synapsing) via a connecting branch.

The parotid fascia is innervated by the great auricular nerve from the cervical plexus of the neck.

The vasculature of the parotid gland

The parotid gland receives its blood supply from branches of the external carotid artery which lie within the gland. Blood drains into the veins found locally. Lymphatics pass into the parotid (pre-auricular) lymph nodes. The apical part of the gland drains into the superficial cervical lymph nodes.

THE EMBRYOLOGY OF THE FACE

Facial development begins during the fourth week of intra-uterine life. Five prominences or facial processes appear around a shallow depression called the stomodeum (primitive oral cavity). The central swelling above the stomodeum is the frontonasal process. Situated laterally and below the stomodeum are the two mandibular processes. At the corners of the stomodeum are the maxillary processes (Figure 55). The mandibular and maxillary processes are components of the first branchial arch (see page 134).

The facial processes arise from mesenchymal cells accumulating beneath the surface epithelium. This mesenchyme is of neural crest origin, having migrated from the region of the neural tube. Initially, the stomodeum (lined by ectoderm) is separated posteriorly from the pharynx (lined by endoderm) by a bilaminar (ectodermal/endodermal) membrane called the buccopharyngeal membrane. This membrane degenerates by the end of the fourth week to establish continuity between the oral cavity and the pharynx (see Figure 127, page 320).

The frontonasal process contributes to the development of the nose (and possibly upper lip), the mandibular processes to the development of the lower jaw and lip, and the maxillary processes to the upper jaw and lip. With continued proliferation and migration of the underlying mesenchyme, the grooves initially separating the facial processes gradually become flattened out, and the processes merge with each other. Thus, the two mandibular processes meet in the midline, forming the lower boundary of the oral cavity. The upper boundary is formed by the maxillary processes growing in on either side and merging with the medial nasal processes of the frontonasal process (Figure 56). This merging is complete by the seventh week of intra-uterine life.

Some controversy exists concerning the precise origin of the middle third of the upper lip. One view is that the region is derived from the medial nasal processes, the lateral thirds being formed from the two maxillary processes. An alternative view is that the maxillary processes overgrow the medial nasal processes to meet in the midline, thereby contributing all the tissue of the upper lip. The latter view is largely based upon an appreciation of the innervation of the upper lip. The innervation of the fully formed upper lip is derived entirely from the maxillary division of the trigeminal nerve. This nerve is associated embryologically with the maxillary process. The nerve associated with the frontonasal process is the ophthalmic division of the trigeminal

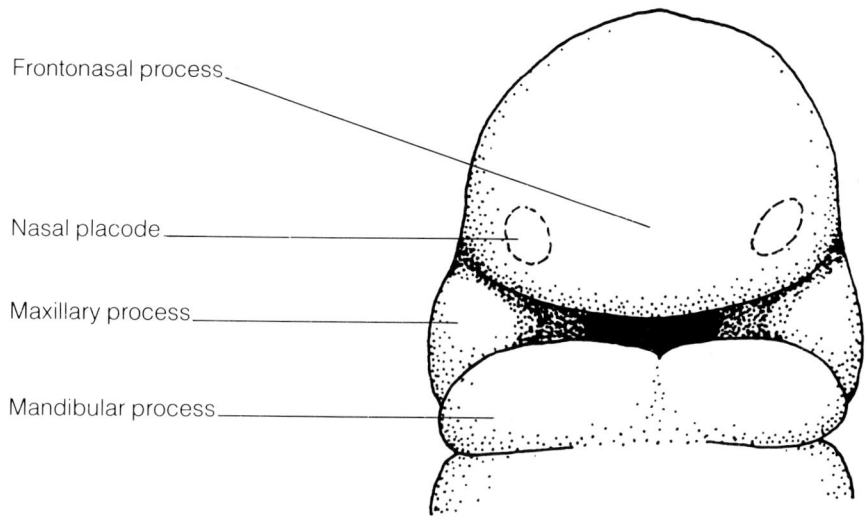

Figure 55 *The frontal aspect of the face during the fourth week of development.*

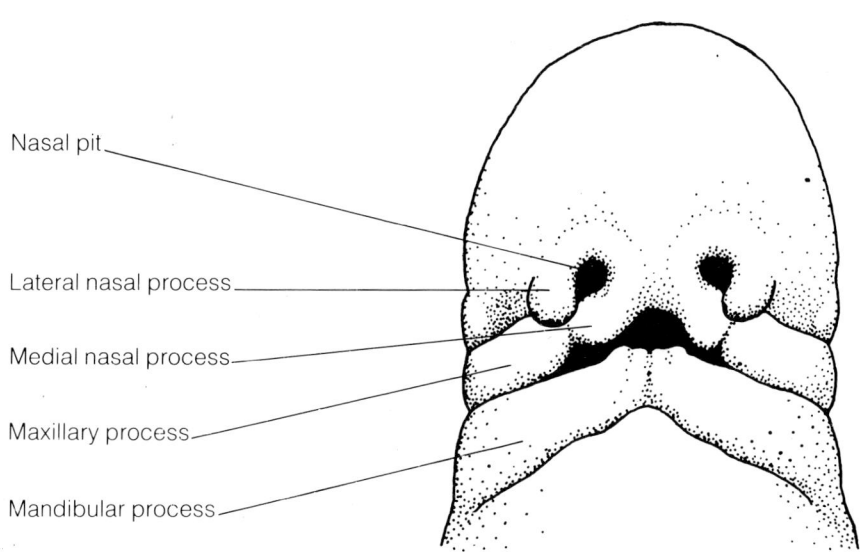

Figure 56 *The frontal aspect of the face during the fifth week of development.*

nerve, a nerve which supplies the middle third of the upper lip only if there is a bilateral cleft.

A sheet of epithelium called the primary epithelial band grows from the surface of the oral cavity into the underlying mesenchyme and divides into a labially positioned vestibular lamina and a lingually situated dental lamina. Continued proliferation of the vestibular lamina, combined with degeneration of its centrally situated cells, eventually produces a cleft that forms the vestibule. The tissue in front of the vestibule gives rise to the lips and cheeks. The tissue behind the vestibule contributes to the development of the teeth and jaws. These events are illustrated in Figure 57. The lip musculature develops from myotomes in the second branchial arch and is therefore supplied by the facial nerve.

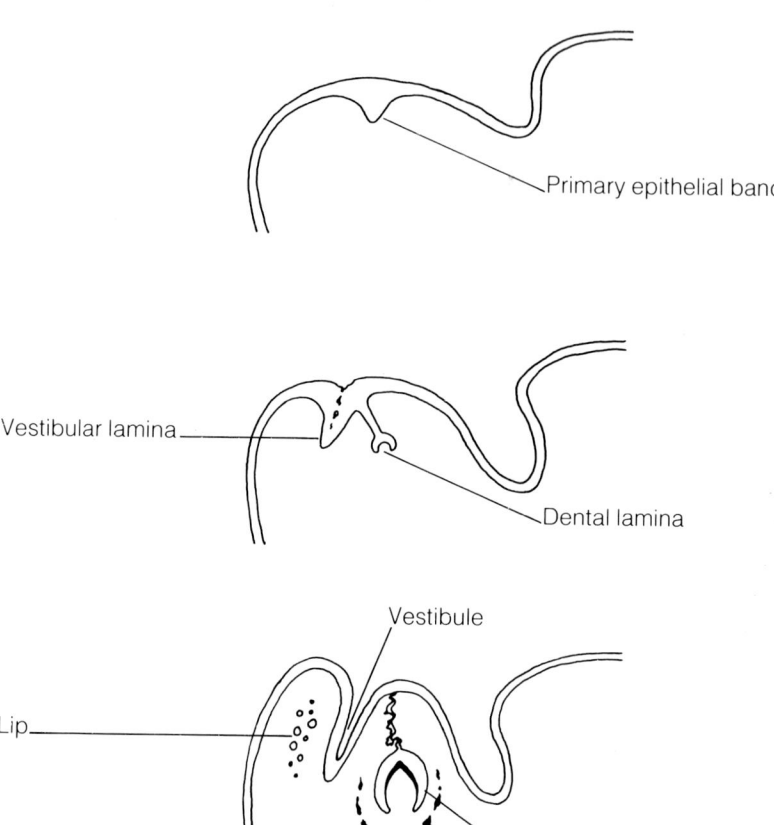

Figure 57 *The development of the vestibule.*

Cleft lip

This congenital abnormality has a multifactorial mode of inheritance.

Clefts of the lips are thought to arise primarily because the mesenchyme of the facial processes fails to proliferate or migrate. Thus, the grooves that demarcate the facial processes fail to flatten out and eventually break down. It is not known whether the epithelium has more than a passive role to play in the aetiology of the cleft lip. Depending upon the severity of the defect, the cleft may vary from a slight groove to a complete cleft. If the defect develops between the maxillary and medial nasal processes, the cleft is situated to one side of the upper lip (either unilaterally or bilaterally). More rarely, defects between the medial nasal processes or between the mandibular processes give rise to median clefts.

Deep structures of the face

Removal of the parotid gland and the superficial structures in the face reveals the ramus of the mandible and structures associated with it. Thus, a description of the deep dissection of the face includes consideration of the muscles of mastication, the temporomandibular joint and the infratemporal fossa deep to the ramus.

THE MUSCLES OF MASTICATION

Although many muscles in the head and neck are involved in mastication, the term 'muscles of mastication' is usually reserved for the four pairs of muscles (the masseter, temporalis, the lateral and medial pterygoid muscles) that are primarily responsible for moving the mandible during the comminution and processing of food before swallowing.

The masseter and temporalis muscles lie relatively superficial in the face. The pterygoid muscles are situated deep to the ramus of the mandible and are thus considered later with the infratemporal fossa.

114,6; 114,2
116A,15; 116A,10

All four muscles of mastication develop from the mesenchyme of the first branchial arch. They therefore derive their innervation from the mandibular branch of the trigeminal nerve (see Table 2, page 134).

The masseter muscle

114,6; 140A,18

Attachments. The muscle consists of three overlapping layers. A superficial part arises from the zygomatic process of the maxilla and from the anterior two-thirds of the lower border of the zygomatic arch. A middle part originates from the deep surface of the anterior two-thirds of the zygomatic arch and from the lower border of the posterior one-third of the arch. A deep part arises from the deep surface of the zygomatic arch. The three layers merge as the fibres pass downwards and backwards to insert into the lateral surface of the angle, ramus and coronoid process of the mandible.

114,6; 24,7
114,5

16,17

Innervation. The muscle is innervated by the masseteric nerve from the anterior division of the mandibular nerve.

116,17

Vasculature. The arterial supply is derived from the superficial temporal (transverse facial branch), the maxillary (masseteric branch) and the facial arteries.

Actions. The masseter muscle elevates the mandible.

The temporalis muscle

114,2; 150A,7

Attachments. The muscle arises from the floor of the temporal fossa and from the overlying temporal fascia. The attachment of the muscle is limited above by the inferior temporal line. The fibres converge towards their insertion on to the apex, the anterior and posterior borders, and the medial surface of the coronoid process. Indeed, the insertion extends down the anterior border of the ramus almost as far as the third molar tooth. The posterior fibres pass horizontally forwards, the anterior fibres pass vertically down on to the coronoid process. In order to reach the coronoid process, the temporalis muscle runs beneath the zygomatic arch. Many of the fibres (but not all) have a tendinous insertion.

16,18
114,14
114,1
115,16; 36,2

114,4
114,3; 115,3

Innervation. The muscle receives its nerve supply from the anterior division of the mandibular nerve.

116A,1

Vasculature. This is derived from the superficial temporal artery (middle temporal branch) and maxillary artery (deep temporal branches).

116A,2

Actions. The anterior (vertical) fibres of the temporalis muscle elevate the mandible, the posterior (horizontal) fibres retract it.

The temporal fascia is a strong membrane that covers the temporal fossa and the temporalis muscle. It is attached above to the superior temporal line on the cranium. Near the zygomatic arch, it splits into two layers. The superficial layer is attached to the superior margin of the zygomatic arch. The deep layer merges with connective tissue beneath the masseter muscle. As mentioned above, the temporal fascia contributes to the origin of the temporalis muscle, giving a bipennate form to the muscle. Pennate muscles are found where considerable power is required within a limited space.

THE TEMPOROMANDIBULAR JOINT

The temporomandibular joint is a synovial joint. It is formed by the condyle of the mandible articulating in the mandibular fossa (glenoid fossa) of the temporal bone. Unlike most other synovial joints, the joint cavity is divided into two by an articular disc. Although basically a hinge joint, the temporomandibular joint also allows for some gliding movements. Movement of the condylar head occurs within the mandibular fossa and down a bony prominence, called the articular tubercle, which is located immediately anterior to the mandibular fossa.

The articular surfaces of the temporomandibular joint are lined by fibrous tissue. This reflects the development of the joint. Unlike all other synovial joints, whose articular surfaces develop endochondrally and are therefore lined by hyaline cartilage, the temporomandibular joint develops in membrane.

The mandibular fossa

The mandibular fossa is an oval depression in the temporal bone lying immediately anterior to the external acoustic meatus. It is bounded anteriorly by the articular tubercle, laterally by the zygomatic process, and posteriorly by the tympanic plate. The petrotympanic fissure separates the mandibular fossa from the petrous part of the temporal bone. Occasionally, a ridge of bone (the postglenoid process) forms a prominence at the posterior boundary of the fossa.

The shape of the mandibular fossa does not exactly conform to the shape of the mandibular condyle, the articular disc moulding together the joint surfaces. The bone of the central part of the fossa is thin. This indicates that masticatory loads are not dissipated through the mandibular fossa but through the teeth and thence the facial bones and base of the cranium.

The mandibular condyle

The mandibular condyle varies considerably both in size and shape. The anteroposterior dimension of the condyle is approximately half the mediolateral dimension. The long axis of the condyle is not, however, at right angles to the ramus, but diverges posteriorly from a strictly coronal plane; if the long axes of the two condyles were extended, they would meet at an obtuse angle (approximately 150°) at the anterior border of the foramen magnum.

The articular surfaces of the condyle are the anterior and superior surfaces. These surfaces are convex. The posterior surface of the condyle is broad and flat.

The condyle is composed of a core of cancellous bone covered by a thin layer of compact bone. During the period of growth, however, a layer of hyaline cartilage lies immediately beneath the fibrous articulating surface of the condyle.

The articular head of the condyle joins the ramus through a thin bony projection called the neck of the condyle. A small depression (the pterygoid fovea) marks part of the attachment of the lateral pterygoid muscle. This fovea is situated on the anterior surface of the neck, below the articular surface.

The articular disc

The articular disc is fibrous and is moulded to the bony joint surfaces. When viewed in sagittal section, the upper surface of the disc is concavo-convex from before

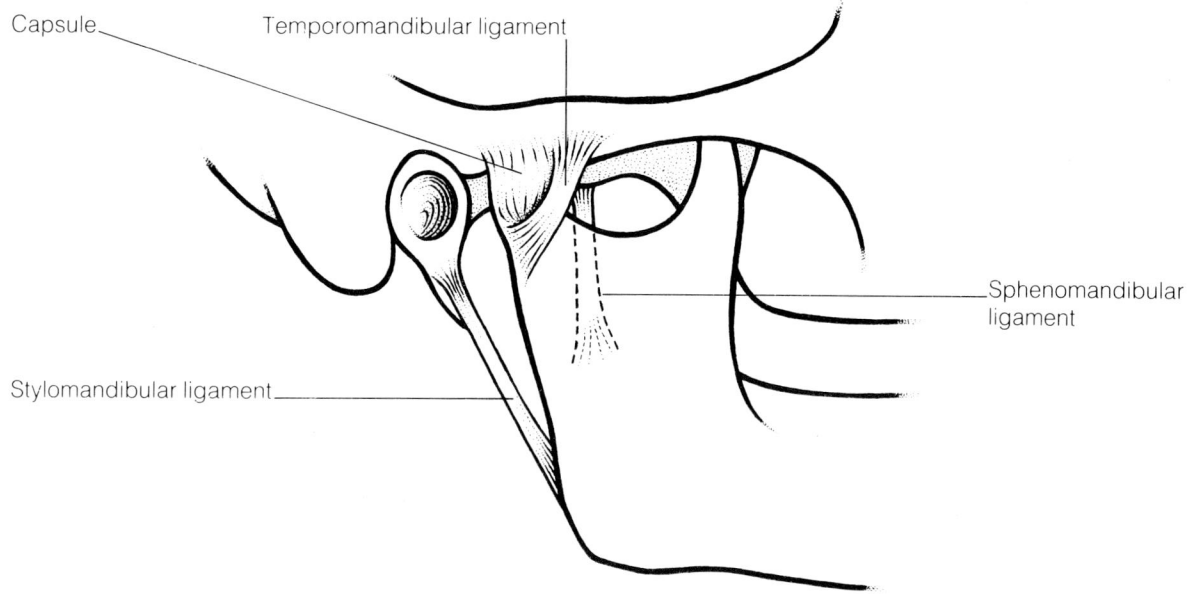

Figure 58 *The temporomandibular joint and ligaments.*

backwards, and the lower surface is concave. The disc is of variable thickness, being thinnest centrally. Its margins merge with the joint capsule. Anteriorly, it is attached to the lateral pterygoid muscle, and fibrous bands connect it to the anterior margin of the articular tubercle above, and to the anterior margin of the condyle below. Posteriorly, the disc becomes bilaminar, the upper part attaching to the anterior margin of the squamotympanic fissure, the lower part attaching to the posterior margin of the condyle.

116,19

The capsule and ligaments of the temporomandibular joint (Figure 58)

114–116

The joint capsule is attached to the neck of the condyle and to the margins of the mandibular fossa. The capsule is thin, although posteriorly it forms a thick, vascular, but loosely arranged connective tissue (the retrodiscal pad).

116,19; 36B,1
24,28

Synovial membrane lines the joint capsule, but not the articular disc.

The joint capsule is strengthened by the temporomandibular (lateral ligament). From the articular tubercle, the ligament passes downwards and backwards to attach on to the lateral surface and posterior border of the neck of the condyle.

114,9; 115,9

The accessory ligaments of the temporomandibular joint are the stylomandibular ligament, the sphenomandibular ligament, and the pterygomandibular raphe. However, none has any significant influence upon mandibular movements. The stylomandibular ligament is a reinforced lamina of the deep cervical fascia as it passes medial to the parotid salivary gland. It extends from the tip of the styloid process and from the stylohyoid ligament to the angle of the mandible. The sphenomandibular ligament is a remnant of the perichondrium of cartilage of the embryonic first branchial arch. It runs from the spine of the sphenoid bone to the lingula near the mandibular foramen. The pterygomandibular raphe passes from the pterygoid hamulus to the posterior end of the mylohyoid line in the retromolar region of the mandible.

14,18
36C,17

22,42; 36,16
116C,33
22,11; 36,11

The innervation and vasculature of the temporomandibular joint

The innervation of the joint is provided mainly from the auriculotemporal branch of the mandibular division of the trigeminal nerve. Additional fibres are supplied by the masseteric branch of the mandibular nerve.

116B,27

116,17

The vascular supply is derived from the superficial temporal artery and the maxillary artery (deep auricular branch).

Movements of the mandible at the temporomandibular joint

Mandibular movements may be classified either as bilaterally symmetrical or bilaterally asymmetrical. Because the mandible is a single bone, movement through one temporomandibular joint cannot occur without a similar co-ordinating or dissimilar reactive movement in the other joint. Depression, elevation, protrusion and retrusion of the mandible are bilaterally symmetrical movements as they require similar co-ordinated movements through both temporomandibular joints. Lateral excursions (side-to-side movements) are bilaterally asymmetrical as there are dissimilar activities at the joints.

It is said that the joint space above the articular disc is associated with anterior gliding movements, whereas the joint space below the disc is associated with hinge movements. Retrusion of the mandible takes the condyle from a position on the articular tubercle back into the mandibular fossa. Little backward movement is possible from within the fossa.

Movements of the mandible are produced principally by the four muscles of mastication, although depression is aided by the activity of other muscles.

Depression of the mandible is produced mainly by the lateral pterygoid muscles, aided by the suprahyoid musculature. With slow and conscious activity, there are initially hinge movements (in the lower joint spaces) followed by sliding of the condylar processes and the articular discs forwards and downwards along the articular tubercles (involving the upper joint spaces).

Elevation of the mandible is produced by the masseter and the medial pterygoid muscles and by the anterior fibres of the temporalis muscles. The condylar processes and the articular discs are pulled upwards and backwards along the articular tubercles into the mandibular fossae. This is accompanied by hinge movements. Hingeing may precede retrusion in slow, conscious closure.

Protrusion of the mandible is produced by the activity of the lateral and medial pterygoid muscles on both sides. The lateral pterygoids draw the condyles and the articular discs forwards along the slopes of the articular tubercles, while the medial pterygoids maintain the teeth in contact.

Retrusion of the mandible is produced by the posterior (horizontal) fibres of the temporalis muscles. These draw the condyles and the articular discs backwards and upwards along the articular tubercles, the masseter muscles maintaining the teeth in contact.

Lateral movement of the mandible is produced by the activity of the medial and lateral pterygoid muscles on one side only. Thus, for the jaw to move to the left, the right pterygoid muscles cause protrusion of the condyle down the articular tubercle on the right side, with the condyle on the left remaining within its mandibular fossa (although showing some rotation around a laterally shifting axis).

THE INFRATEMPORAL FOSSA

116; 142A

The boundaries of the infratemporal fossa (Figure 59)

26A; 140A
140A,19

The infratemporal fossa is the space located deep to the ramus of the mandible.

26A,5; 26A,19
140A,25; 140A,27
140A,39; 26A,2
140A,7; 140A,19

The fossa is bounded anteriorly by the posterior surface of the maxilla and posteriorly by the styloid apparatus, carotid sheath and deep part of the parotid gland. Medially lies the lateral pterygoid plate and the superior constrictor muscle of the pharynx. Laterally lies the ramus of the mandible. The roof is formed by the infratemporal

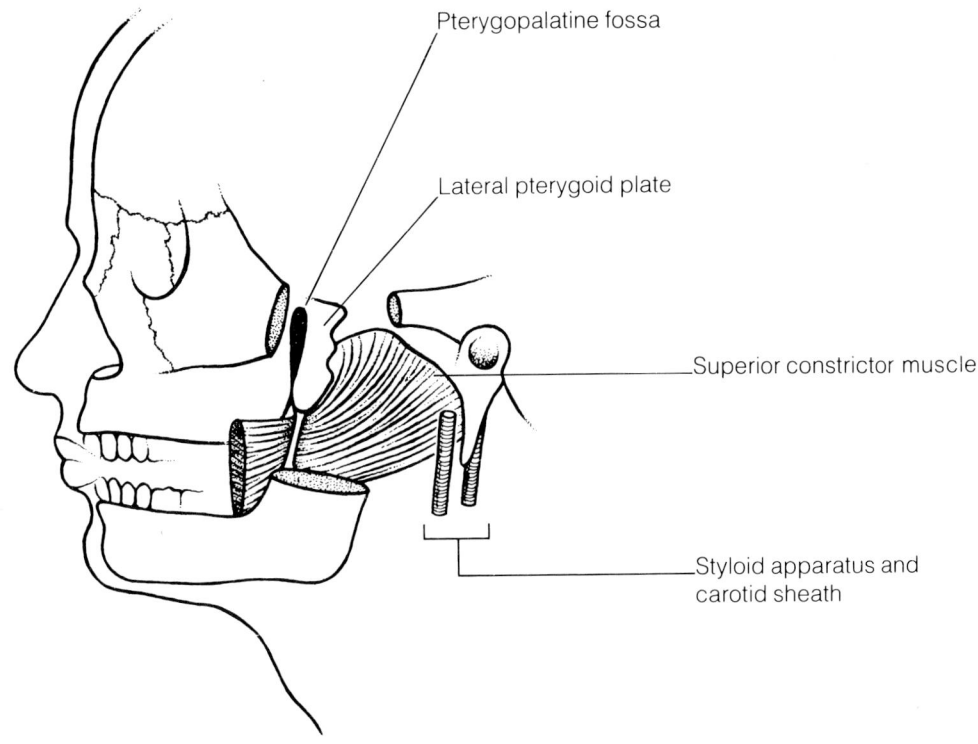

Figure 59 *The boundaries of the infratemporal fossa.* (*Note that the ramus of the mandible has been removed.*)

surface of the greater wing of the sphenoid. The infratemporal fossa has no anatomical floor, being continuous with tissue spaces in the neck.

The infratemporal fossa communicates with the temporal fossa deep to the zygomatic arch. It also communicates with the pterygopalatine fossa through the pterygo-maxillary fissure. At the base of the cranium, the foramen ovale and foramen spinosum enter the fossa through the sphenoid bone. The foramen lacerum and the petrotympanic, squamotympanic and petrosquamous fissures are also found close to the infratemporal fossa. On the medial surface of the ramus of the mandible is the mandibular foramen.

22,15

26A,1
26A,4
26A,11; 22,44
22,43; 22,46
22,38; 22,39
22,41

The contents of the infratemporal fossa

The major structures that occupy the infratemporal fossa are:
- The lateral and medial pterygoid muscles.
- The mandibular division of the trigeminal nerve.
- The chorda tympani branch of the facial nerve.
- The otic parasympathetic ganglion.
- The maxillary artery and branches.
- The pterygoid venous plexus.

The key to understanding the relationships of structures within the infratemporal fossa is the lateral pterygoid muscle. This lies in the roof of the fossa, running anteroposteriorly in a horizontal plane from the region of the pterygoid plates to the mandibular condyle. It consists of two heads, an upper head and a lower head. Deep to the muscle arise the branches of the mandibular nerve and the main origin of the medial pterygoid muscle. Superficially lies the maxillary artery. The buccal branch of

116A,3; 116A,15

116,25; 116B,10
116,16

116A,9
116A,10
116A,11; 116A,12
116A,1; 116A,2

the mandibular nerve passes between the two heads of the lateral pterygoid muscle. Emerging below the inferior border of the muscle are the medial pterygoid muscle and the lingual and inferior alveolar nerves. At the upper border emerge the deep temporal nerves and vessels. Concentrated around and within the lateral pterygoid muscle lies a venous network, the pterygoid venous plexus.

116A

The lateral pterygoid muscle

116A,15
26A,2; 116C,21
116A,3; 24,6

Attachments. The bulk of the muscle is formed by its lower head. This arises from the lateral surface of the lateral pterygoid plate of the sphenoid bone. The smaller upper head takes origin from the infratemporal surface of the greater wing of the sphenoid in the roof of the fossa. The two heads converge near the point of insertion. The fibres of

116A,18
142C,66; 36,4

the upper head insert primarily into the capsule and articular disc of the temporomandibular joint. The fibres from the lower head insert into the pterygoid fovea on the mandibular condyle.

116B,26

Innervation. The nerves to the lateral pterygoid (one for each head) arise from the anterior trunk of the mandibular nerve, deep to the muscle.

116A,16

Vasculature. The arterial supply is derived from the maxillary artery (pterygoid branches) as it crosses the lateral pterygoid muscle.

Actions. The main action of the muscle is to assist in opening the jaws by pulling forwards the mandibular condyle and the articular disc of the temporomandibular joint. In addition, the muscle is involved in protrusion and in lateral movements of the mandible. (The superior head may also help to elevate the mandible.)

116; 140; 142

The medial pterygoid muscle

116,10

Attachments. This muscle is the deepest of the four muscles of mastication. It consists of two heads. The bulk of the muscle arises as a deep head from the medial surface of

24,4

the lateral pterygoid plate. Thus, the lateral pterygoid plate of the sphenoid bone gives rise to both pterygoid muscles. A common mistake is the belief that the medial pterygoid muscle arises from the medial pterygoid plate. However, the medial pterygoid plate gives origin only to a small portion of the superior constrictor muscle of the pharynx. The smaller superficial head of the medial pterygoid muscle originates

24,5

from the maxillary tuberosity and the neighbouring part of the palatine bone (pyramidal process). From these sites, the fibres pass downwards and backwards to

36,10

insert into the roughened surface of the angle of the mandible on its medial aspect.

116B,20; 142A,9

Innervation. The nerve to the medial pterygoid muscle arises from the mandibular nerve (deep to the lateral pterygoid muscle), before the nerve divides into anterior and posterior trunks.

Vasculature. Like the lateral pterygoid muscle, the medial pterygoid derives its arterial supply from the maxillary artery.

Actions. The medial pterygoid muscle is an elevator of the mandible. Additionally, it assists in lateral and protrusive movements.

Both the lateral and medial pterygoid muscles are derived embryologically from the mesenchyme of the first branchial arch.

116; 142

The mandibular nerve (Figure 60)

This is the largest division of the trigeminal nerve and is the only one to contain motor as well as sensory fibres. Developmentally, it is the nerve of the first branchial arch and is thus responsible for supplying structures derived from it. Its sensory fibres supply the mandibular teeth and their supporting structures, the mucosa of the anterior two-thirds of the tongue and the floor of the mouth, the skin of the lower part of the face (including the lower lip) and parts of the temporal region and auricle. Its motor fibres supply the four 'muscles of mastication' and the mylohyoid, anterior belly of digastric, tensor veli palatini and tensor tympani muscles.

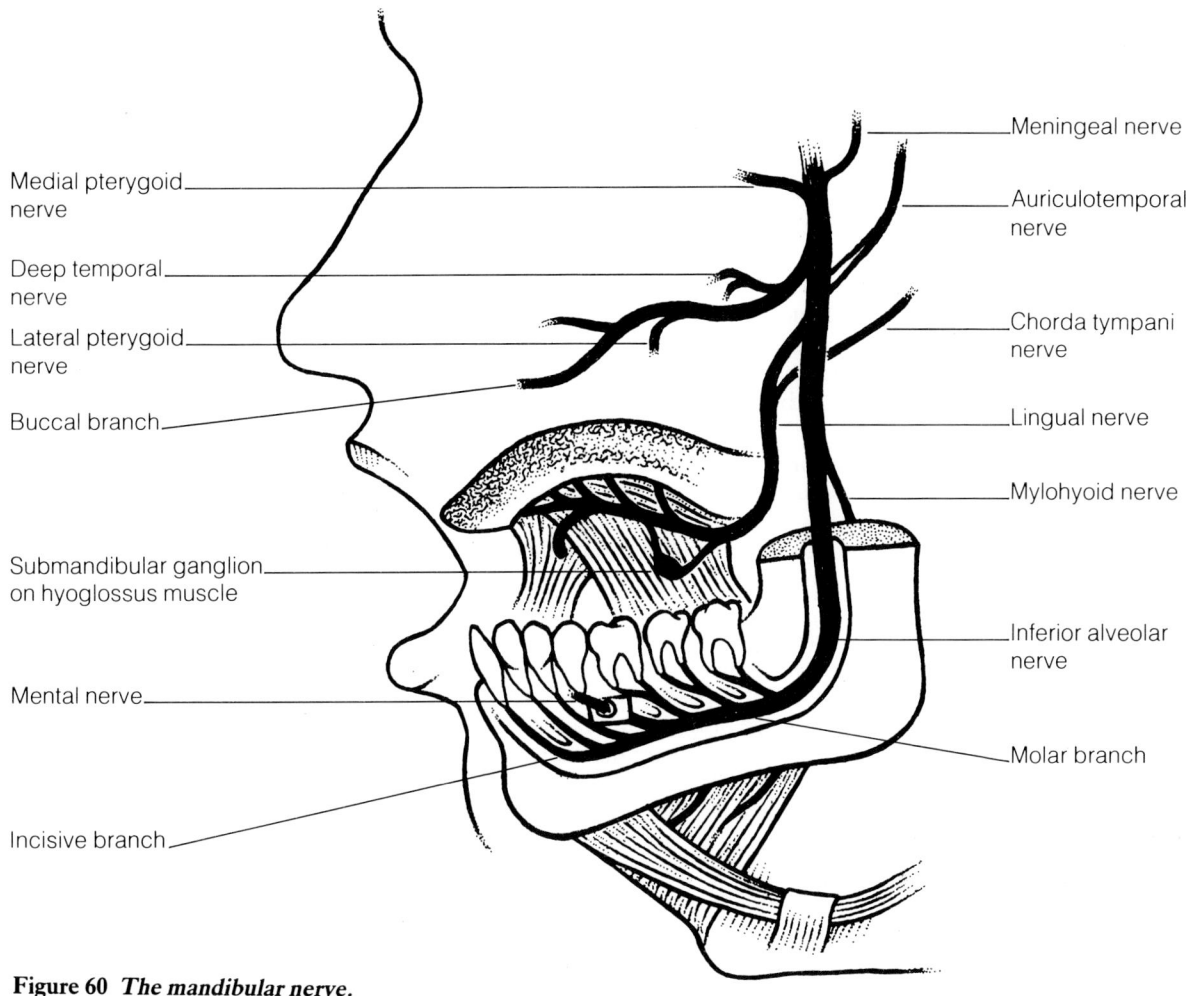

Figure 60 *The mandibular nerve.*

The mandibular nerve is formed in the infratemporal fossa by the union of the sensory and motor roots immediately after they leave the skull at the foramen ovale. At this point, the nerve lies on the tensor veli palatini muscle and is covered by the lateral pterygoid muscle. After a short course, the nerve divides into a small anterior trunk and a larger posterior trunk. Before this division, the main trunk gives off two branches — the meningeal branch and the nerve to medial pterygoid. The anterior trunk of the mandibular nerve is mainly motor, the posterior trunk mainly sensory.

Branches:

- Meningeal branch (nervus spinosus)
- Nerve to medial pterygoid
- Anterior trunk:
  Masseteric nerve
  Deep temporal nerves
  Nerve to lateral pterygoid
  Buccal nerve
- Posterior trunk:
  Auriculotemporal nerve
  Lingual nerve
  Inferior alveolar nerve

166B,47; 26,11
116C,28; 116D

28,47	**The meningeal branch of the mandibular nerve** (nervus spinosus) arises from the main trunk of the mandibular nerve. It is a 'recurrent nerve' as it runs back into the middle cranial fossa through the foramen spinosum. It supplies the dura mater lining the middle cranial fossa and the mucosa of the mastoid antrum and mastoid air cells.
116B,20; 142A,9	**The nerve to the medial pterygoid muscle** enters the deep surface of the muscle and also gives slender branches which pass uninterrupted through the otic ganglion to supply the tensor tympani and tensor veli palatini muscles.
116A,17; 116D,17	**The masseteric nerve** is usually the first branch of the anterior trunk of the mandibular nerve. It passes above the upper border of the lateral pterygoid muscle and crosses the mandibular notch (between the condylar and coronoid processes) to be distributed into the masseter muscle. It also gives an articular branch to the temporomandibular joint.
116A,1; 116D,1	**The anterior and posterior deep temporal nerves** also pass above the lateral pterygoid muscle. They subsequently enter the deep surface of the temporalis muscle.
116,26	**The nerve to the lateral pterygoid muscle** may arise separately or may run with the buccal nerve before entering the deep surface of the lateral pterygoid muscle.
166B,46; 116D,9 116A,9	**The buccal branch of the mandibular nerve** is the only sensory branch of the anterior trunk of the mandibular nerve. On emerging between the heads of the lateral pterygoid muscle, it passes downwards and forwards across the lower head to contact the medial surface of the temporalis muscle as it inserts on to the coronoid process of the mandible. It then clears the ramus of the mandible to lie on the lateral surface of
116A,8	the buccinator muscle in the cheek. At this point, it is close to the retromolar fossa of the mandible. It now gives branches to the skin of the cheek (see Figure 49) before piercing buccinator to supply its lining mucosa, the buccal sulcus, and the buccal gingiva related to the mandibular molar and premolar teeth. It may also carry secretomotor fibres to minor salivary glands in the buccal mucosa, these being postganglionic fibres from the otic ganglion.
166B,51; 116D,27 144,1 116C,55; 116C,23 142A,17 166B,51 112,2 112,3	**The auriculotemporal nerve** is the first branch of the posterior trunk of the mandibular nerve. It is essentially sensory but it also distributes autonomic fibres to the parotid gland derived from the otic ganglion. It arises as two roots that encircle the middle meningeal artery and unite behind the artery. The nerve then runs backwards under the lateral pterygoid muscle to lie beneath the mandibular condyle (between the condyle and the sphenomandibular ligament). On entering the parotid region, it turns to emerge superficially between the temporomandibular joint and the external acoustic meatus. From the upper surface of the parotid gland, the auriculotemporal nerve ascends on the side of the head with the superficial temporal vessels, passing over the posterior part of the zygomatic arch. It gives several branches along its course:

Ganglionic branches which communicate with the otic ganglion.

Articular branches which enter the posterior part of the temporomandibular joint; these carry proprioceptive information important in mastication.

Parotid branches which convey parasympathetic secretomotor fibres and sympathetic fibres to the parotid gland; these fibres are related to the otic ganglion. Sensory fibres from the auriculotemporal nerve supply the gland (with the exception of the capsule, which is innervated by the great auricular nerve).

Auricular branches (usually two) which supply the tragus and crus of the helix of the auricle, part of the external acoustic meatus, and the outer (lateral) surface of the tympanic membrane.

Superficial temporal branches which are cutaneous nerves supplying part of the skin of the temple (see Figure 49).

The lingual nerve is the second branch of the posterior trunk of the mandibular nerve. It is essentially a sensory nerve but, following union with the chorda tympani branch of the facial nerve, it also contains parasympathetic fibres. Initially, the nerve lies on the tensor veli palatini muscle deep to the lateral pterygoid muscle. Here, the chorda tympani nerve (which has entered the infratemporal fossa via the petrotympanic fissure) joins the posterior surface of the lingual nerve. Emerging from the inferior border of the lateral pterygoid muscle, the lingual nerve curves downwards and forwards in the space between the ramus of the mandible and the medial pterygoid muscle (pterygomandibular space). At this level, it lies anterior to, and slightly deeper than, the inferior alveolar nerve. This completes the course of the nerve in the infratemporal fossa. It then passes towards the floor of the mouth (see pages 301 to 302).

The chorda tympani branch of the facial nerve is distributed through the lingual nerve and has two types of fibres. Sensory fibres are associated with taste to the anterior two-thirds of the tongue. Parasympathetic fibres are preganglionic to the submandibular ganglion. Postganglionic fibres are secretomotor to the submandibular and sublingual glands.

The inferior alveolar nerve is the third branch of the posterior trunk of the mandibular nerve. Although it is essentially a sensory nerve, it also carries motor fibres which are given off as the mylohyoid nerve. Indeed, the mylohyoid nerve contains all the motor fibres of the posterior trunk of the mandibular nerve. The inferior alveolar nerve descends deep to the lateral pterygoid muscle, posterior to the lingual nerve. Here, it is crossed by the maxillary artery. On emerging at the inferior border of the muscle, it passes between the sphenomandibular ligament and the ramus of the mandible to enter the mandibular foramen.

The mylohyoid nerve is given off just before the mandibular foramen. It pierces the sphenomandibular ligament and runs in a groove (the mylohyoid groove) which lies immediately below the mandibular foramen. The mylohyoid nerve supplies the mylohyoid muscle and the anterior belly of the digastric.

The main distribution of the inferior alveolar nerve is to the mandibular teeth and their supporting structures, there being molar and incisive branches. The mental nerve is a cutaneous branch that supplies the skin of the chin and the lower lip (see Figure 49). It arises within the mandible in the premolar region, but soon exits on to the face via the mental foramen. For details of the course and distribution of the inferior alveolar nerve within the mandible see page 299.

The otic ganglion (see Figure 54, page 167)

This parasympathetic ganglion lies immediately below the foramen ovale on the medial surface of the main trunk of the mandibular nerve. It is concerned primarily with supplying the parotid gland. Like other parasympathetic ganglia in the head, three types of fibres are associated with it: parasympathetic, sympathetic and sensory fibres. However, only the parasympathetic fibres synapse in the ganglion. The preganglionic parasympathetic fibres originate from the inferior salivatory nucleus in the brainstem (see page 477). The fibres pass out in the glossopharyngeal nerve, appearing as the lesser (superficial) petrosal nerve from the tympanic plexus in the middle ear cavity (see page 372). The lesser petrosal nerve reaches the otic ganglion by a complex course. Passing through the petrous part of the temporal bone, the lesser petrosal nerve comes to lie in the floor of the middle cranial fossa. Here, it is lateral to the greater (superficial) petrosal branch of the facial nerve. The lesser petrosal nerve usually enters the infratemporal fossa through the foramen ovale to join the otic ganglion. The sympathetic root of the otic ganglion is derived from postganglionic fibres from the superior cervical ganglion. They are said to reach the otic ganglion from the plexus on the middle meningeal artery. The sensory root is derived from the auriculotemporal nerve. The postganglionic parasympathetic fibres (with sympathetic and sensory components) reach the parotid gland by way of the auriculotemporal nerve. Parasympathetic fibres may also innervate the minor salivary glands in the cheek, passing with the buccal branch of the mandibular nerve.

The innervation of tensor veli palatini and tensor tympani is derived from the nerve to medial pterygoid by a branch that passes through the otic ganglion.

The maxillary artery (Figure 61)

The maxillary artery is a terminal branch of the external carotid artery. It arises within the parotid gland at the level of the neck of the condyle of the mandible. It enters the infratemporal fossa between the deep surface of the condyle and the sphenomandibular ligament. At this point, it lies below the auriculotemporal nerve and above the maxillary vein. In the infratemporal fossa, it is closely related to the lateral pterygoid muscle. Initially, it lies near the inferior border of the muscle, crossing the inferior alveolar nerve. Its subsequent course is variable, although it usually passes superficial to the lower head of the lateral pterygoid before entering the pterygopalatine fossa through the pterygomaxillary fissure.

The maxillary artery has many branches. It is convenient to subdivide the artery into three parts: before the lateral pterygoid muscle (first or mandibular part), on the lateral pterygoid muscle (second or pterygoid part), and in the pterygopalatine fossa (third or pterygopalatine part).

The first part of the maxillary artery has five branches and all enter bone. The first branch is the deep auricular artery, supplying the skin of the external acoustic meatus and part of the tympanic membrane. A small branch contributes to the arterial supply of the temporomandibular joint. The second branch, the anterior tympanic artery, passes through the petrotympanic fissure to supply part of the lining of the middle ear. This is the companion artery to the chorda tympani nerve. The middle meningeal artery is the main source of blood to the meninges and to the bones of the vault of the skull. It ascends between the two roots of the auriculotemporal nerve and leaves the infratemporal fossa through the foramen spinosum. Its course within the cranium is described further on page 400. An accessory meningeal artery runs through the foramen ovale into the middle cranial fossa to supply the trigeminal ganglion and the dura lining the floor of the middle cranial fossa. The inferior alveolar artery accompanies the inferior alveolar nerve and has a similar distribution. Immediately before the artery enters the mandible (at the mandibular foramen), it gives off a mylohyoid branch. In the mandibular canal, it gives off branches supplying the cheek teeth before terminating in mental and incisive branches. The mental artery passes through the mental foramen on to the face to supply the lower lip, the chin and the labial mucosa related to the anterior teeth. The incisive branch continues along the incisive canal to supply the anterior teeth.

The second part of the maxillary artery also has five branches, but they differ from those of the first part in not entering bone. Muscular branches include deep temporal arteries (anterior and posterior), pterygoid arteries and masseteric arteries. The masseteric arteries also supply the temporomandibular joint. A buccal artery accompanies the buccal nerve to supply structures in the cheek. A small lingual branch may be given off to accompany the lingual nerve and supply structures in the floor of the mouth.

Branches of the third part of the maxillary artery are described with the pterygopalatine fossa (page 258).

The pterygoid venous plexus (see Figure 31)

This is situated around and within the lateral pterygoid muscle and it surrounds the maxillary artery. Its tributaries correspond to the various branches of the maxillary artery. Although it is difficult to demonstrate in the cadaver, it is very prominent in life.

The plexus communicates with the cavernous sinus, the facial vein, the inferior ophthalmic vein and the pharyngeal plexus. The connections with the cavernous sinus

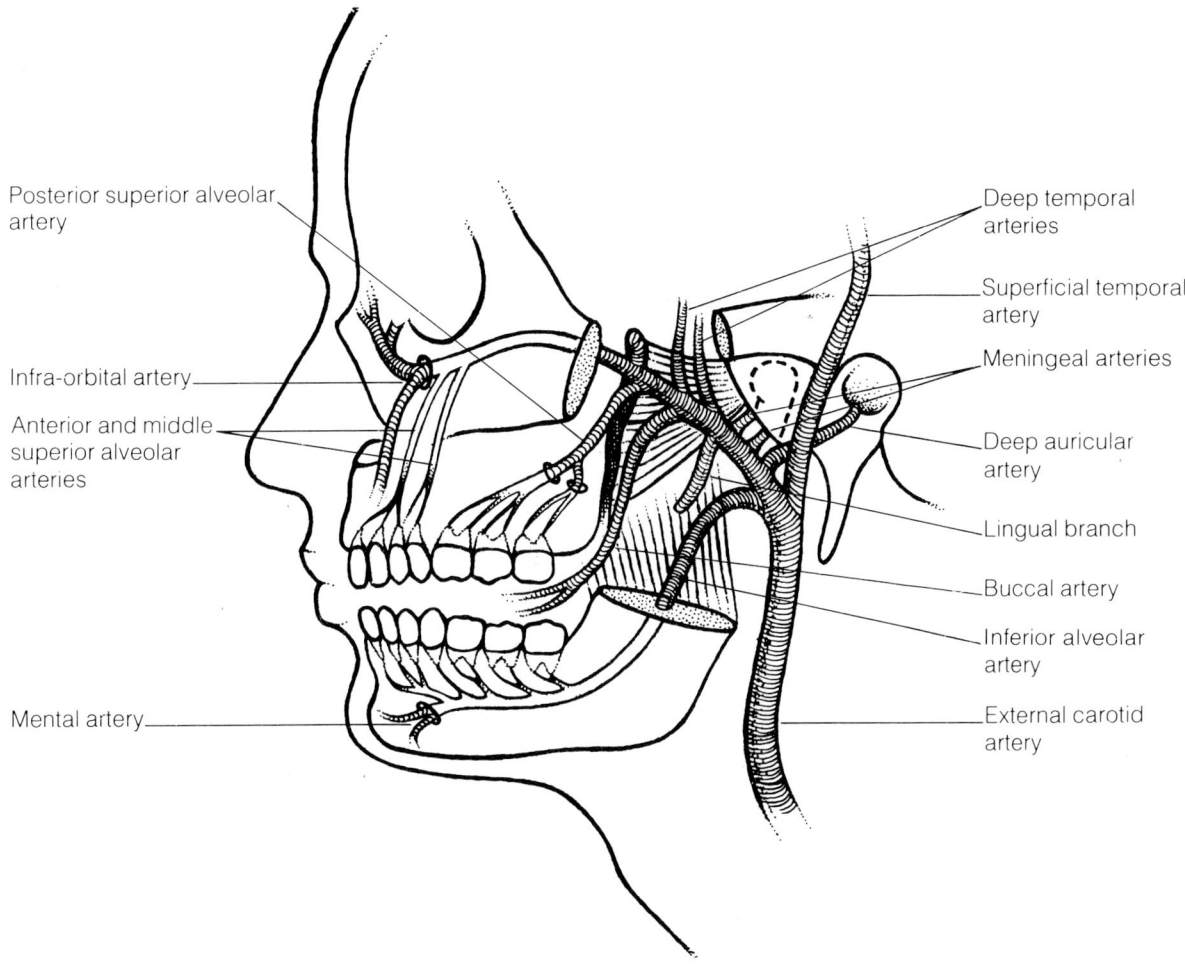

Figure 61 *The maxillary artery.*

are via emissary veins passing through the foramen ovale, foramen lacerum and, where present, the emissary sphenoidal foramen. The communication with the facial vein is via the deep facial vein which accompanies the buccal nerve. The inferior ophthalmic vein communicates with the pterygoid plexus through a branch passing through the inferior orbital fissure.

The pterygoid venous plexus drains posteriorly into the maxillary vein. The maxillary vein runs with the first part of the maxillary artery, passing deep to the neck of the condyle of the mandible to enter the parotid gland. Here, it joins the superficial temporal vein to form the retromandibular vein.

140A,38

Other features of the infratemporal fossa

In addition to the major contents described above, the infratemporal fossa also contains the sphenomandibular ligament, the tensor veli palatini muscle, the insertion of the temporalis muscle on to the coronoid process of the mandible, the maxillary nerve as it passes from the pterygopalatine fossa into the inferior orbital fissure, the posterior superior alveolar nerve(s), and a loop of the facial artery (together with its ascending palatine and tonsillar branches). These structures are considered in detail elsewhere.

116C,28
115,3
116B,4
116B,5
116C,31

SUMMARY SHEET: SUPERFICIAL STRUCTURES OF THE FACE 1

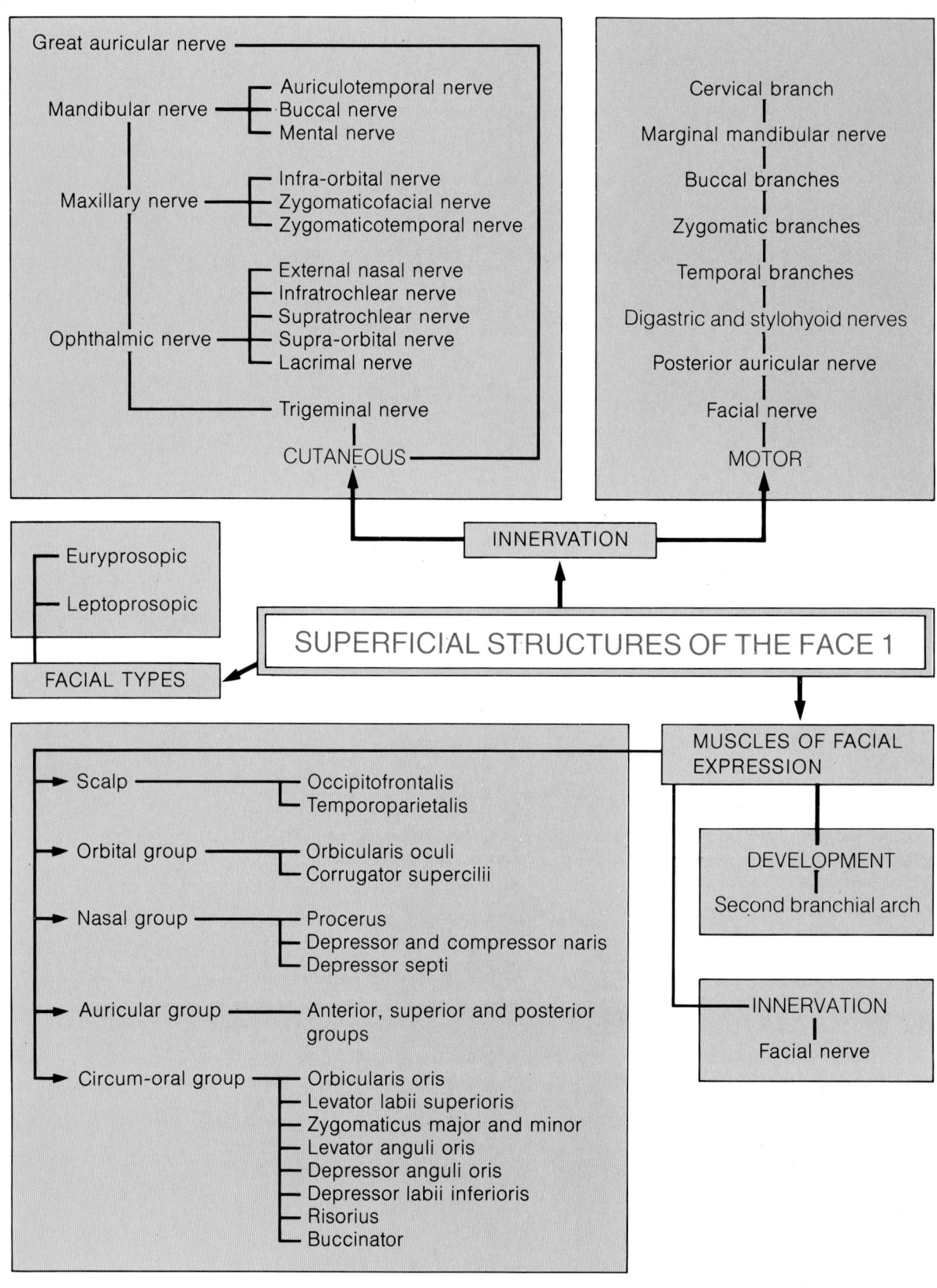

SUMMARY SHEET: SUPERFICIAL STRUCTURES OF THE FACE 2

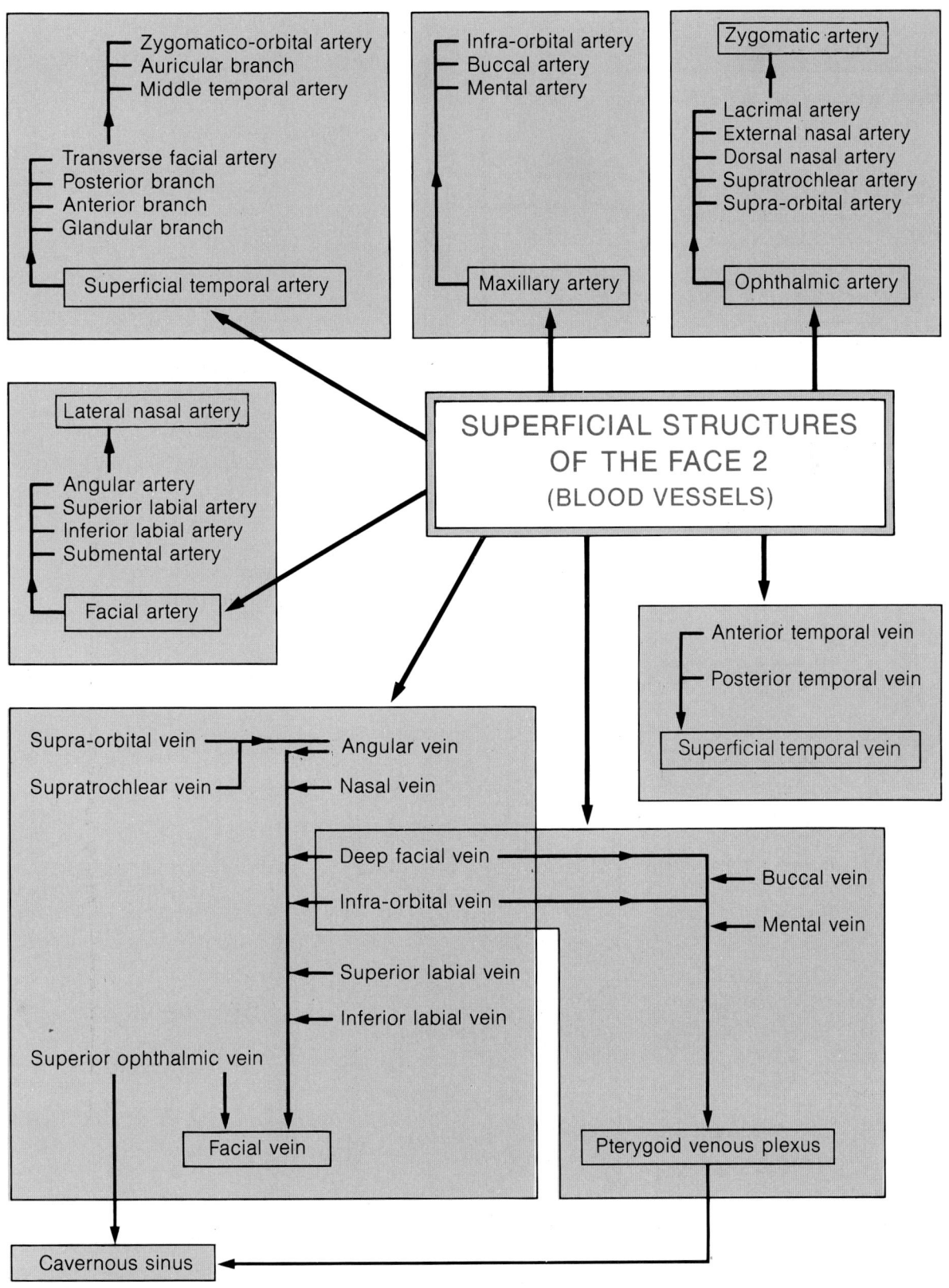

SUMMARY SHEET: SUPERFICIAL STRUCTURES OF THE FACE 3

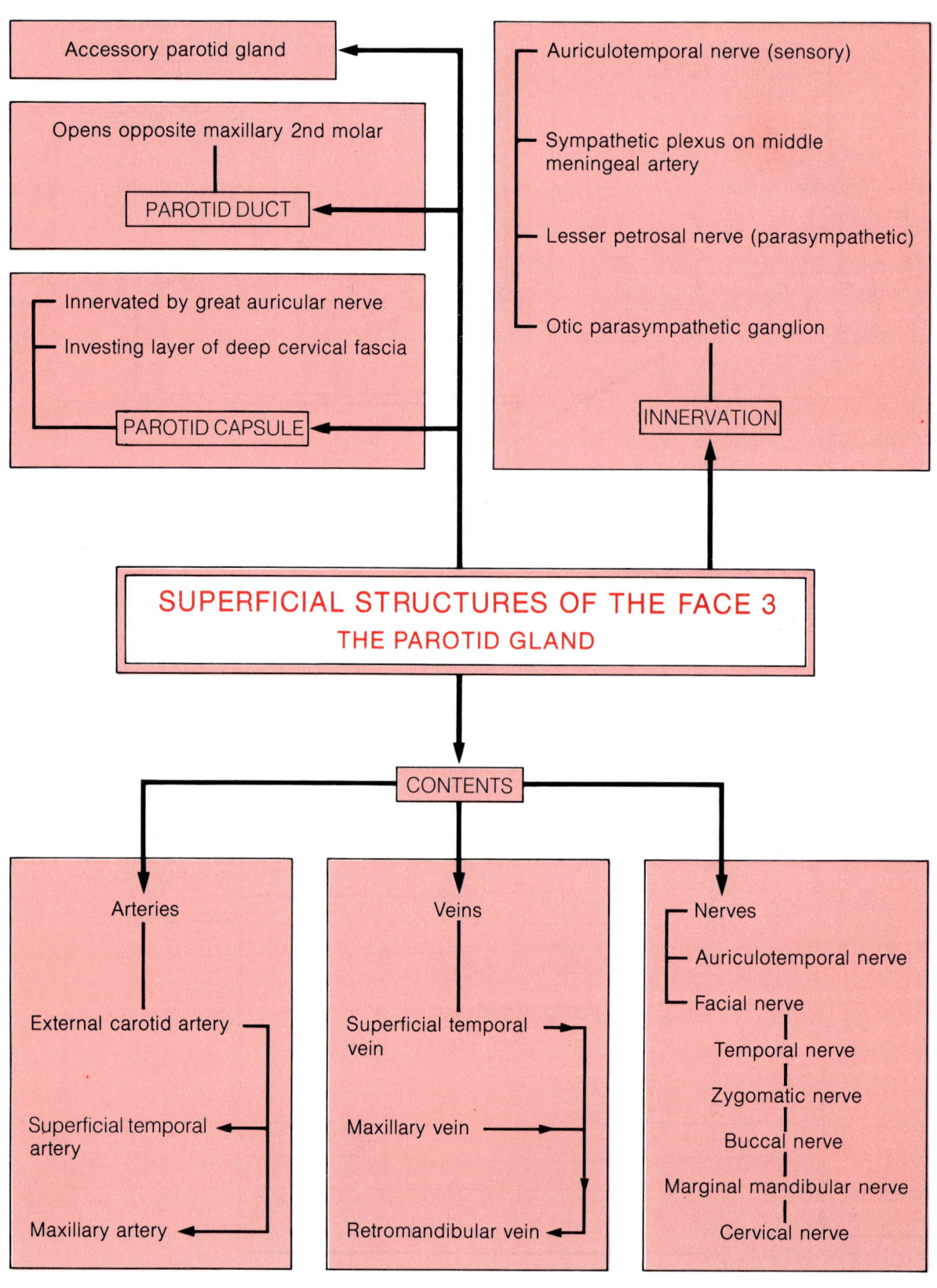

SUMMARY SHEET: DEEP STRUCTURES OF THE FACE 1

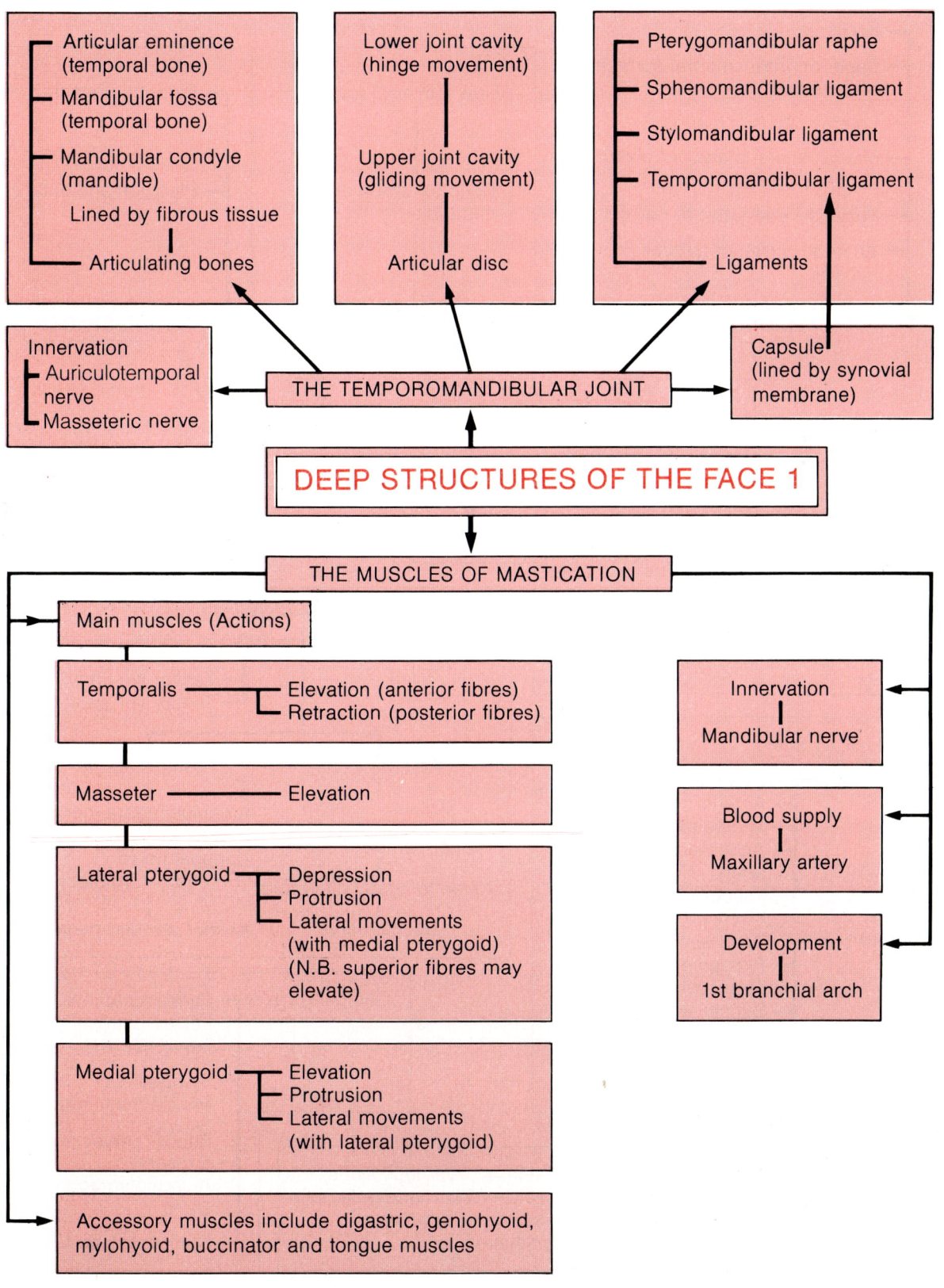

SUMMARY SHEET: DEEP STRUCTURES OF THE FACE 2

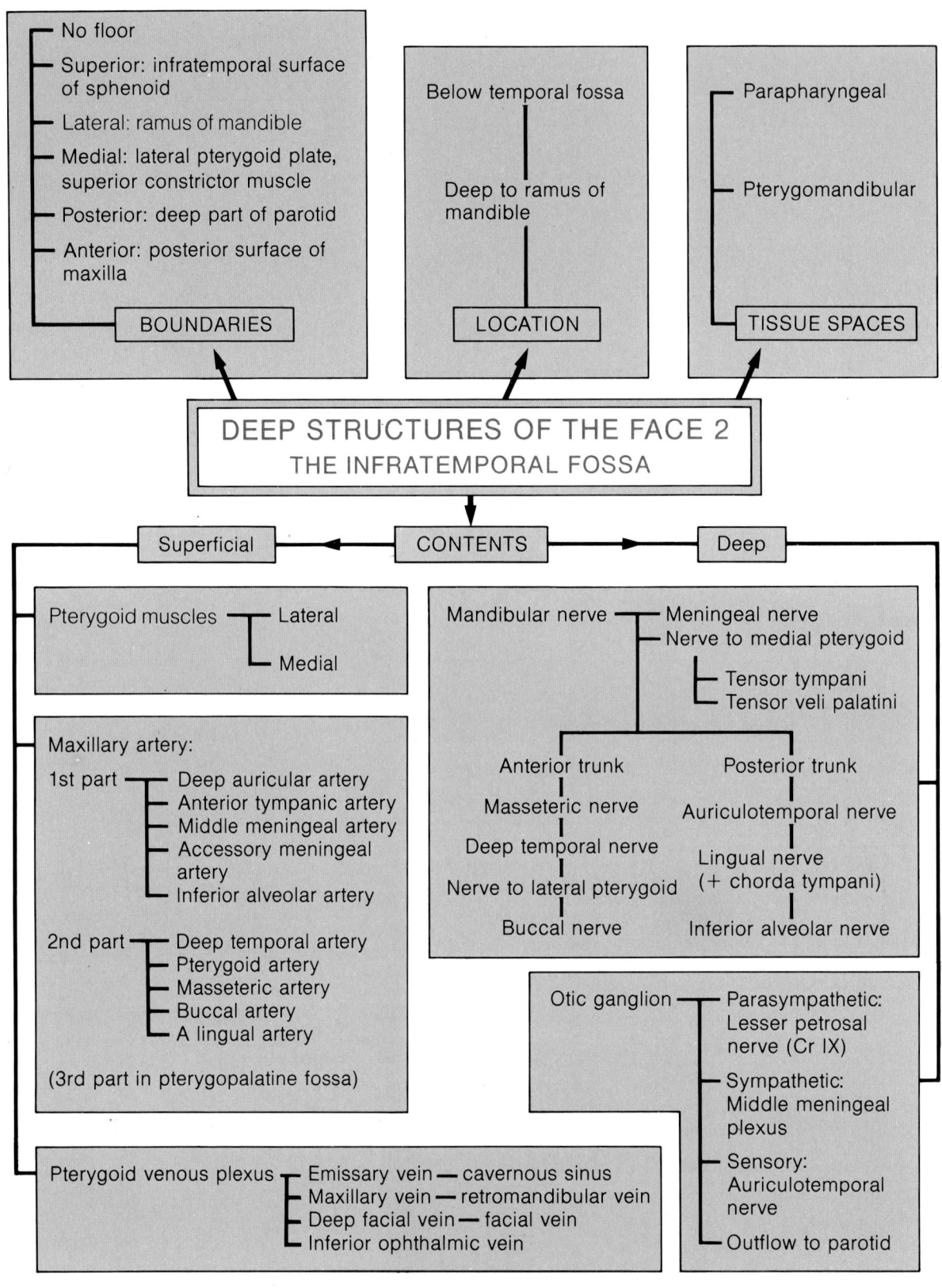

The scalp

161,18   The skin, connective tissue and muscles covering the calvaria collectively comprise the scalp. The scalp extends from the eyebrows and forehead to the superior nuchal line in the occipital region. Laterally, the scalp extends down to the zygomatic arches. Thus, the forehead and the temple are areas common to both the scalp and the face.

160   **THE LAYERS OF THE SCALP**

The scalp consists of five layers. Listed from the surface inwards, these are:

160,1   • Skin, which is thick and contains the hair
160,1   • Dense connective tissue which is richly vascularised
160,2   • The epicranial aponeurosis with the occipitofrontalis muscle
160,7   • Loose connective tissue, forming a potential subaponeurotic space
160,7   • Pericranium, the periosteum on the outer surface of the cranial vault

The three outer layers are closely adherent and can be readily separated from the underlying tissues. Consequently, they have been called by some the scalp proper.

160,2   **The epicranial aponeurosis**

18,10; 18,11   The epicranial aponeurosis is also called the galea aponeurotica. It is a continuous musculomembranous sheet which extends from the external occipital protuberance and the supreme nuchal lines to the eyebrows. The aponeurosis is continuous laterally
114,14; 114,2   with the temporal fascia overlying the temporalis muscle. Although mainly membranous, it also contains the occipitofrontalis muscles.

160; 112   **The occipitofrontalis muscle**

160,3; 112,44   *Attachments.* Each occipital belly arises from the lateral two-thirds of the supreme
16,22   nuchal line of the occipital bone and from the mastoid process of the temporal bone. It
160,2   extends forwards to become continuous with the epicranial aponeurosis. The two occipital bellies are separated in the midline by the aponeurosis as it attaches to the
160,4; 112,6   external occipital protuberance. Each frontal belly arises from the anterior margin of
160,2   the epicranial aponeurosis and passes forwards to merge with the orbital part of the
118B,12; 118B,13   orbicularis oculi muscle and with the muscle and connective tissue over the bridge of the nose.

112,33; 112,31   *Innervation.* The occipitofrontalis muscle is innervated by the facial nerve, the occipital belly by the posterior auricular branch, and the frontal belly by the temporal and zygomatic branches.

Vasculature. The muscle is supplied by branches from the superficial temporal, ophthalmic, posterior auricular, and occipital arteries.

Actions. The main function of the occipitofrontalis muscle is to elevate the eyebrows to express surprise or horror, or to produce transverse furrows of the forehead.

112,1   **The temporoparietalis muscle** is the name given to occasional muscle fibres present at the side of the scalp between the frontal belly of occipitofrontalis and the auricular muscles.

The subaponeurotic space

160,7   This potential tissue space is limited by the attachments of the epicranial aponeurosis and the occipitofrontalis muscles. Posteriorly, therefore, it is limited by the supreme nuchal lines and laterally by the zygomatic arches. Anteriorly, however, the subaponeurotic space is continuous with the eyelids and the bridge of the nose.

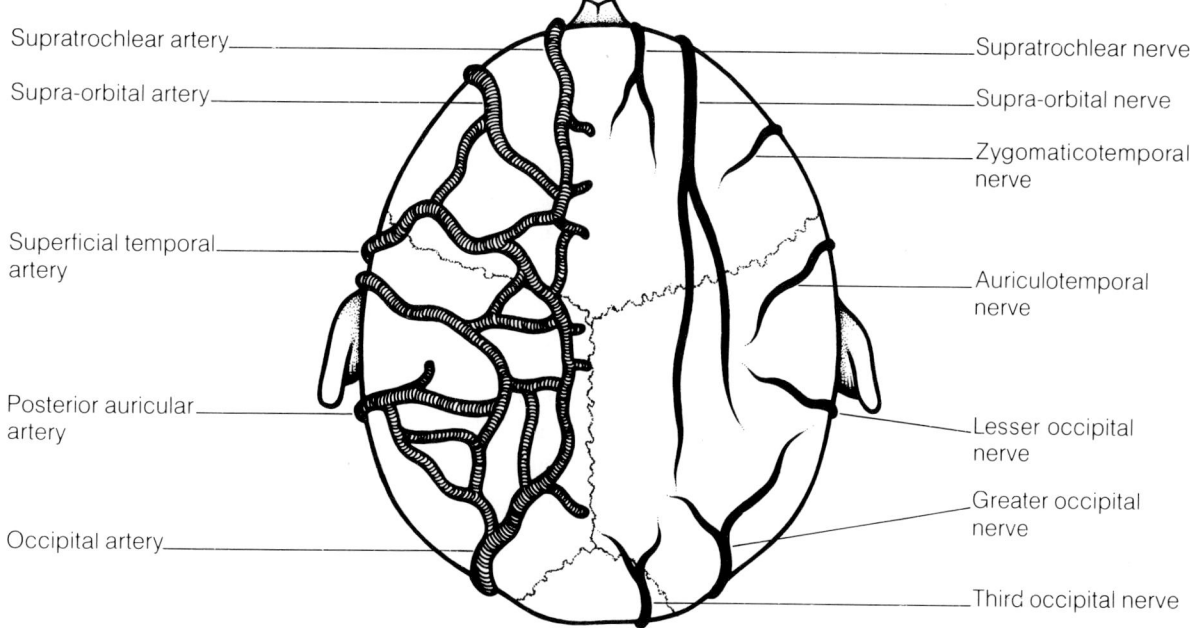

Figure 62 *The innervation and arterial supply of the scalp.*

THE CUTANEOUS INNERVATION OF THE SCALP (Figure 62)

Anteriorly and medially, the innervation is derived from the supratrochlear and supra-orbital branches of the ophthalmic nerve. Posteriorly, the skin is supplied by the dorsal rami of the second cervical nerve (greater occipital nerve) and of the third cervical nerve (third occipital nerve). Laterally, the zygomaticotemporal branch of the maxillary nerve, the auriculotemporal branch of the mandibular nerve, and the lesser occipital nerve from the cervical plexus, all contribute to the sensory innervation.

122A,8
122,4; 160,6
112,42; 99,36
99,39; 112,4
112,2; 99,10

THE ARTERIES OF THE SCALP (Figure 62)

The arteries are derived from branches of the external and internal carotid arteries. Both the arteries and the nerves enter the scalp from below. The branches from the external carotid artery are: the superficial temporal and posterior auricular arteries (which supply the scalp laterally), and the occipital artery (which supplies the scalp posteriorly). The scalp anteriorly is supplied by branches from the internal carotid, namely the supratrochlear and supra-orbital branches of the ophthalmic artery. All of the arteries in the scalp freely anastomose.

112,3
200A,2
122,5

THE VEINS OF THE SCALP

The veins have a similar distribution to the arteries, each artery in the scalp being usually accompanied by a pair of veins (venae comitantes). In addition, the veins of the scalp may communicate with the venous sinuses inside the skull via emissary veins which pass through the mastoid and parietal foramina.

The lymphatic drainage of the scalp is illustrated in Figure 52.

Case histories 3
The face and scalp

A trumpet player visited his doctor complaining that the right side of his face periodically became painfully swollen, especially at meal times. Both the pain and the swelling could be relieved by massaging the side of his face. The patient also complained of a bad taste in his mouth. On examination, he was seen to have poor oral hygiene and the region around the right parotid papilla was red. What might be the cause of the problem?

The patient had inflammation of the parotid duct (sialodochitis), which had spread from the oral region and which was associated with the poor oral hygiene. Sialodochitis is facilitated by factors causing a build-up of pressure in the mouth (in this case trumpet blowing) and can prevent the normal release of saliva, causing the gland to swell. This leads to tension within the unyielding parotid fascia, stimulating pain receptors associated with the great auricular nerve. Discharge of inflammatory products into the oral cavity is responsible for the bad taste. Massaging the face helps to discharge the accumulated saliva. Similar symptoms are caused by a stone in a salivary duct, although stones in the parotid gland are rare compared with the submandibular gland.

A patient visited her doctor shortly after returning from a winter expedition to the Antarctic. She had noticed that one side of her face appeared distorted and was expressionless. On the affected side, she could not close her eye and was unable to smile properly as that corner of her mouth drooped. She had difficulty in whistling and could not easily retain food in her mouth while eating. What is responsible for these symptoms?

The patient has Bell's palsy, which affects the facial nerve and consequently the muscles of facial expression. In the absence of obvious trauma, the aetiology of this condition is unknown, although it is commonly associated with being in a cold or draughty environment. A hemiparalysis of the muscles of facial expression leads to a loss of the normal creases and folds in the face on the affected side. This gives the face an almost expressionless (death mask) appearance. Paralysis of the muscles associated with the cheek and lips was responsible for the inability to smile, whistle and eat properly. The loss of action of the orbicularis oculi muscle and the absence of blinking can render the conjunctiva and cornea susceptible to inflammation and may eventually lead to blindness if untreated.

A facial palsy can also result from a brain lesion. Where only the lower half of the face is affected, a lesion in the contralateral side of the brain above the facial nucleus in the pons is indicated, as the upper half of the face has some ipsilateral as well as contralateral representation.

A teenager had a habit of picking the spots on his face. One of the spots above his upper lip on the left side became infected, but he continued to pick it. Some days later, he became very ill and was admitted to hospital. He was drowsy and had a high temperature. He complained of headache and nausea and had been vomiting. He had a painful, swollen right eye and double vision (diplopia) when looking to the right. A cavernous sinus thrombosis on the right side was diagnosed. How has the infection spread from the face to the cavernous sinus?

The facial vein drains the upper lip and the side of the nose. Near the medial corner of the eye, the facial vein (the angular vein) communicates with the superior ophthalmic vein. The ophthalmic vein drains directly into the cavernous sinus. As the facial vein has no valves, infection is able to spread along the ophthalmic veins, producing a swollen, inflamed eye and causing thrombosis in the cavernous sinus. The general signs of illness are the result of inflammation of the meninges. Within the cavernous sinus, the abducent nerve has been affected, with paralysis of the lateral rectus muscle and subsequent diplopia on looking to the right.

A patient presents with an abscess in a lower molar tooth. There is a large, painful swelling at the lower border of the mandible that requires immediate drainage. What important nerve and artery are liable to be damaged when making an incision in this region?

The marginal mandibular branch of the facial nerve lies near the angle of the mandible. In many cases, it may pass down into the neck for a short distance before swinging up on to the face about halfway along the lower border of the mandible. It supplies the muscles of the lip. Its normal position may be distorted because of inflammation of the facial tissues. Section of the nerve could result in permanent disfigurement of the mouth and impairment of the normal function of the lips. To reduce the possibility of sectioning the branches of the facial nerve, incisions should be made parallel, rather than at right angles, to the nerves.

The artery associated with the inferior border of the mandible is the facial artery. This passes through the submandibular salivary gland and thence weaves in and out from underneath the inferior border of the mandible before crossing on to the face at the anterior margin of the masseter muscle.

A 65-year-old woman visited her doctor in a very distressed condition. She complained of paroxysms of acute pain on one side of her face. These attacks occurred during the day and had gradually become more frequent. Each attack lasted for a few minutes and was often accompanied by salivation, lacrimation, dilation of the pupil and flushing of the face. The pain was felt particularly over the upper lip and an attack could be set off simply by touching the region. Consequently, the patient had ceased to wash her face. She also ate very little, with the result that she had lost weight. She maintained an expressionless appearance (frozen face) in the hope of reducing the number of attacks.

The patient is suffering from trigeminal neuralgia, a condition of unknown aetiology. The pain is usually limited to the area of distribution of one of the branches of the trigeminal nerve. In this case, the maxillary branch has been affected. The sensitive region related to the onset of the attack is referred to as the 'trigger area'. The right side of the face is affected more commonly than the left. The autonomic nervous system is sometimes stimulated; hence the salivation, lacrimation, dilation of the pupil and flushing of the face. The pain may be controlled by drugs or by surgical procedures aimed at producing disruption of the sensory pathway of the affected branch of the trigeminal nerve.

A young motorist, not having fastened his seat belt, received severe lacerations around the parotid region after being thrown through the windscreen of his car. He made a good recovery but some months later observed that, when he ate, he sweated profusely from the skin overlying the parotid gland. From your knowledge of the secretomotor supply to the parotid gland and the cutaneous innervation of the overlying skin, can you explain the cause of the condition?

When he received his facial injuries, many of the nerves supplying the face were severed. These included the auriculotemporal nerve (carrying secretomotor fibres to the parotid gland), and the great auricular nerve (supplying the skin overlying the parotid gland). During nerve regeneration, some of the secretomotor fibres crossed over to join the great auricular nerve. Subsequently, stimuli that would normally produce salivation now pass along the great auricular nerve and cause stimulation of the sweat glands over the distribution of the nerve (gustatory sweating).

A woman visited her doctor about a swelling in her upper lip on the left side, level with her canine tooth. She had first noticed the swelling a few years previously and it had gradually increased in size. The swelling was easily palpated, was firm and had a well-circumscribed outline. There was little discomfort. The upper teeth were all sound and there was no enlargement of lymph nodes. The swelling was diagnosed as a cyst (an epithelial-lined sac containing fluid) whose site can be accounted for by knowledge of the embryological development of the upper lip.

The patient has a naso-alveolar cyst. This is thought to be derived from epithelial remnants at the site where the embryonic maxillary, medial nasal and lateral nasal processes merge during the formation of the upper lip; hence its situation to one side of the midline of the upper lip.

A patient was given a general anaesthetic in the dental chair for extraction of a tooth. During the operative procedure, the anaesthetist wished to monitor the arterial pulse. From what site(s) on the face may the anaesthetist obtain an arterial pulse?

The superficial temporal artery may be palpated as it crosses the zygomatic arch in front of, and above, the tragus. (The facial artery may also be palpated at the anterior border of the masseter as it passes on to the face at the lower border of the mandible.)

A 50-year-old man complained of a dry, sore right eye. An ophthalmologist noted that the patient had a corneal ulcer and was unable to close his eyelids. He also found a hard, irregular mass within the substance of the right parotid gland. How might these signs be connected?

The mass in this patient's parotid gland was diagnosed as a malignant tumour (adenocarcinoma). The tumour cells had invaded and damaged the temporal and zygomatic branches of the facial nerve with resultant paralysis of the orbicularis oculi muscle. Paralysis of this muscle prevents blinking. Consequently, the anterior surface of the eye is not protected and, as the precorneal film of tears is no longer spread, the dry cornea ulcerates.

An elderly man known to be suffering from chronic myeloid leukaemia presented with a pustular, crusting rash limited to the left side of his forehead, upper eyelid and left side of his nose. He also had a painful left eye. On examination, he was found to have conjunctivitis and a small corneal ulcer. What might this rash be and can you explain its distribution?

This man has developed a herpes zoster infection (shingles). This is a common complication in immunosuppressed patients (e.g. those being treated for chronic leukaemia). The virus passes along the sensory fibres of the infected nerve. In the present case, the nerve was the ophthalmic division of the trigeminal. Thus, the rash is limited to that area of skin supplied by this nerve. There has also been involvement of the conjunctiva which is supplied in part by the lacrimal, supra-orbital and supratrochlear branches of the ophthalmic nerve. The cornea has become infected by virus particles which passed down the nasociliary nerve, through the ciliary ganglion and eventually reached the cornea via the short ciliary nerves.

A 70-year-old man visited his doctor after noticing that he was unable to elevate his right upper eyelid. In fact, the eyelid drooped (ptosis), leading to some degree of facial disfigurement. The patient also mentioned that the right side of his face was flushed and felt warm and that, when he got hot, he did not sweat noticeably on the affected side (anhidrosis). The doctor observed that the pupil in the patient's right eye was constricted (miosis) and did not dilate when exposed to light. The other eye was unaffected. On further questioning, the patient admitted to being a heavy smoker and having had a persistent cough over the past few years. The doctor referred the patient for a chest radiograph, and an opacity was discovered at the apex of the right lung, indicative of a cancer. Could this explain the patient's facial symptoms?

The facial symptoms indicate that there has been involvement of the right sympathetic chain. The sympathetic outflow to the head and neck has its origin in the thoracic region. The thoracic sympathetic trunk lies on the necks of the ribs and continues upwards over the neck of the first rib as the cervical sympathetic chain. The radiograph revealed that the patient had an apical carcinoma of the right lung. The carcinoma had spread locally to involve the sympathetic trunk at the thoracic inlet, interrupting the passage of sympathetic efferents before they reach the superior cervical ganglion. The levator palpebrae superioris muscle and the dilator pupillae muscle have a sympathetic innervation. The sweat glands on the face are also innervated by the sympathetic system. Interference with the sympathetic innervation of blood vessels results in their vasodilation, making the face red and warm. Hence, all the facial signs can be explained in terms of sympathetic denervation. The ptosis, miosis, and anhidrosis caused by paralysis of the cervical sympathetic system is known as Horner's syndrome. An additional sign associated with this syndrome is enophthalmos — a recession of the eyeball into the orbit. When the tumour also affects the brachial plexus, the syndrome is called Pancoast's syndrome.

A 55-year-old woman visited her doctor because she felt constantly cold and depressed. The doctor was shocked at the dramatic change in her facial appearance since he had last seen her two years previously. Her hair and skin were dry, her lips and tongue were thickened, and her features had become generally coarse. Her eyelids were puffy and she wore a dull, lethargic expression. The doctor also noted the patient's slow speech and slow mental processes. The patient remarked that her voice appeared to have become deeper.

This lady has myxoedema, a condition that is the manifestation of hypothyroidism in the adult. The bloated appearance is caused by the deposition of glycosaminoglycans in the dermis. The enlarged tongue may result in difficulties with speech, mastication and swallowing. The change in the tone of the patient's voice is also related to the deposition of glycosaminoglycans in the vocal cords.

A patient complained of considerable pain emanating from the right side of the mouth. Examination revealed that the second mandibular molar had a large carious cavity which had progressed to involve the pulp. Despite the patient's desire to conserve the tooth, it was necessary to extract it. The patient asked for this treatment to be carried out under local anaesthesia. Which nerves must be anaesthetised and where?

The infratemporal fossa contains four nerves that may have to be anaesthetised to undertake dental treatment. For the mandibular dentition, there are the inferior alveolar, lingual and buccal branches of the mandibular nerve. Running on the maxillary tuberosity, at the boundary of the infratemporal fossa, are the posterior superior alveolar branches of the maxillary nerve. These supply the maxillary molars. To extract a tooth it is not only necessary to anaesthetise the nerve specifically supplying the tooth, but also the nerves supplying its supporting tissues. For a mandibular molar, the tooth is innervated by the inferior alveolar nerve, its lingual gingiva by the lingual nerve, and its buccal gingiva by the buccal nerve. The inferior alveolar and lingual nerves are anaesthetised by injection into the infratemporal fossa. The buccal nerve may be 'blocked' as it passes out of the infratemporal fossa across the retromolar region of the mandible.

Whilst anaesthetising the inferior alveolar nerve in the infratemporal fossa, the patient experienced a sudden, sharp pain which she described as "like an electric shock". The onset of anaesthesia was also more rapid than usual. A week later, the patient returned complaining of persistent anaesthesia in the lower lip which recently had changed to a 'pins and needles' sensation. What has happened?

The 'electric shock' sensation and the rapid onset of analgesia indicates that the injection was so accurate that the needle had hit the inferior alveolar nerve. The prolonged impairment of sensation that followed is caused by nerve damage. The affected area is the lower lip, an area supplied by the mental branch of the inferior alveolar nerve. To test for the extent of damage, the reactions to pin-pricks or to the passing of cotton wool over the affected area may be used. This kind of injury to a nerve is usually temporary, but occasionally may be permanent. The 'pins and needles' sensation is called paraesthesia and suggests that sensation is returning. Fortunately, whilst it is common to hit the nerve during the administration of local anaesthetic, subsequent injury is rare.

A patient was given an inferior alveolar nerve block on the right side preparatory to dental treatment. Soon afterwards, the patient complained of blurred vision in the right eye. On examination a squint was noted. Can you provide an explanation?

The most likely explanation is that the local anaesthetic solution had diffused into the orbit from the infratemporal fossa via the inferior orbital fissure. The squint indicates that the anaesthetic had affected the nerves innervating the extra-ocular muscles. Very rarely, blindness may follow the administration of an inferior alveolar nerve block. This might be the result of vascular spasm in a situation where there is an uncommon vascular pattern. For example, the orbit in some people may be supplied by the middle meningeal artery. Fortunately, these visual complications are transient, passing off with the disappearance of anaesthesia.

A 45-year-old woman required a large gold inlay in her mandibular first molar tooth. To accomplish this, the inferior alveolar nerve was anaesthetised in the infratemporal fossa. The patient was discharged, but returned a few days later complaining of considerable difficulty in opening her mouth. There were no other signs or symptoms.

Difficulty in opening the jaws because of muscle spasm is called trismus. In this case, the medial pterygoid muscle is affected as a result of damage associated with the injection. The onset of trismus often occurs some time after the injection and thus can be distinguished from the general soreness and discomfort that may be experienced immediately after dental treatment. Trismus is usually caused by bleeding into the muscle (as a result of damage to blood vessels) or infection. As the patient did not complain of fever, pain or malaise, the trismus in this instance does not result from infection. The term haematoma is used to describe a localised mass of extravasated blood. In view of the highly vascular nature of the infratemporal fossa, particularly because of the many vessels of the pterygoid venous plexus, it is not unusual for haematomas to form in this region. Bleeding from the pterygoid venous plexus may produce a rapid and dramatic swelling in the region of the cheek and discoloration or bruising of the skin. However, bleeding into the pterygomandibular space may not be immediately apparent. Accidental damage to the blood vessels is nowadays more easily discerned because of the common use of aspirating syringes for delivery of the local anaesthetic.

A schoolboy attended a dental clinic for the extraction of a mandibular premolar for orthodontic reasons. During the injection of a local anaesthetic solution containing adrenaline into the infratemporal fossa, the patient complained of feeling unwell. The dentist initially thought that he was feeling faint. However, he noticed that the patient had palpitations. The dentist therefore suspected that he had injected the local anaesthetic solution into a blood vessel, especially as the anaesthetic had not worked. What blood vessels in the infratemporal fossa could be so involved?

Approximately one in 50 patients faint during dental treatment. Factors that predispose to fainting include anxiety, tiredness and lack of food. The symptoms of fainting are dizziness, nausea, slowing of the heart rate and hypotension leading to loss of consciousness. Recovery from a faint is usually rapid once the patient is laid flat and the head lowered.

Concerning the possibility of an intravascular injection, both veins and arteries within the infratemporal fossa may be involved. Most probably, the injection is into the pterygoid venous plexus. Indeed, as the maxillary artery and most of its branches are in the roof of the infratemporal fossa, it is unlikely that the injection will be intra-arterial during an inferior alveolar nerve block. However, there is a rare possibility of injecting into the inferior alveolar artery as it runs with the inferior alveolar nerve. Intravascular injection of an adrenaline-containing solution is associated with tachycardia and hypertension. There may also be ischaemia at a site remote from that of the injection (e.g. blanching of the skin of the face). These symptoms are transitory as the adrenaline is rapidly metabolised. If an aspirating syringe is used, the risk of intravascular injection is greatly reduced.

A young man visited his dental surgeon complaining of a painful swelling of the soft tissues around an erupting mandibular third molar. An infection arising from a flap of soft tissue surrounding the crown of the erupting tooth (pericoronitis) was diagnosed, and the region was syringed out. A week later, the patient returned with the same problem and the treatment was repeated. Two weeks later, he was admitted to hospital in considerable distress and, despite emergency treatment, died shortly afterwards. The patient had had trismus, difficulty in swallowing and breathing, and a high temperature (40°C). There was also considerable swelling of the soft tissues in the face, neck and at the back of the mouth. What had happened?

The infection spread from around the tooth to the tissue spaces in the infratemporal fossa. The infratemporal fossa contains two potential tissue spaces, the pterygomandibular space and the superior parts of the pharyngeal tissue spaces. The pterygomandibular space is located between the medial pterygoid muscle and the ramus of the mandible. Infection in this site is restricted inferiorly by the attachment of the muscle into the angle of the mandible. Thus, infection is prevented from spreading directly into the neck. Should the infection pass posteriorly from this region, it will spread into the parotid region. On the other hand, the parapharyngeal space between the medial pterygoid muscle and the superior constrictor of the pharynx will allow inflammatory products to spread directly into the neck. In the case described here, the trismus indicates that the infection had spread into the infratemporal fossa. The infection had then passed into the neck, as evidenced by the difficulty in swallowing and breathing (swelling impinging on the airway and oesophagus). The swelling at the back of the mouth suggests that the infection had also spread between the superior constrictor and the mucosa of the mouth into the peritonsillar region. Death was probably caused by asphyxia.

A patient had her second and third mandibular molar teeth extracted under local anaesthesia. She returned to her dentist three weeks later feeling very unwell. She complained of a stiff neck and had a high temperature. The right side of her face was tender and swollen and she had trismus. There was a purulent discharge into the mouth from the sockets of the extracted teeth, which contained some small hard fragments. The patient complained of numbness on the right side of her lower lip. In addition, her right eyelids were swollen and she had blurred vision. How might these symptoms be explained?

The patient has an acute inflammation of the bone of the mandible (osteomyelitis), which has spread to the cavernous sinus. This is most commonly related to the ingress of organisms at the site of the wound, though more rarely it can arise from an infected needle or a contaminated local anaesthetic solution. The presence of an acute inflammation is responsible for the oral signs and for the high temperature. The small, hard fragments discharged into the mouth may represent sequestrated pieces of necrosed bone. The numbness of the lip is caused by the inflammatory oedema affecting the inferior alveolar nerve and its mental branch. The involvement of the eye suggests that the infection has spread from the infratemporal fossa to the cavernous sinus in the middle cranial fossa. The pathway for this is along emissary veins connecting the pterygoid venous plexus with the cavernous sinus via openings in the sphenoid bone (i.e. the foramen ovale, the foramen lacerum, and the emissary sphenoidal foramen). The presence of infection within the cavernous sinus affects venous return from the orbit and results in oedema of the eyelids. The nerves supplying the orbital muscles will also be affected, producing the blurred vision. The spread of infection to the middle cranial fossa is also indicated by the stiffness of the neck, a sign of meningeal inflammation. The stiffness is caused by reflex spasm in the muscles controlling neck movements to prevent irritation of the inflamed meninges, which would result from neck flexion.

A young woman was referred to a dental hospital by her doctor because she appeared to have problems with her temporomandibular joints. She complained of some pain in the joints, but was most disturbed by a clicking noise when opening her jaw. Occasionally, she experienced transient locking of the jaw, almost invariably when closing her mouth. She claimed that her problems started one morning after opening her mouth especially wide during a yawn. Radiographs of her temporomandibular joints revealed that, on opening the jaw, the condyles of the mandible remained within the mandibular fossae of the temporal bones. Is this the normal pattern of movement?

The temporomandibular joint dysfunction may be related to trauma of the articular discs. It is said that the most common cause of this is a malocclusion that results in a bizarre pattern of mandibular excursions during mastication. However, a single episode, such as occurred when the young woman yawned, may also be responsible. The locking of the jaw when the mouth is being closed is symptomatic of this condition; locking when the mouth is being opened is associated with dislocation of the jaw. Normal mandibular movements during opening involve a combination of hinge movements and gliding movements down the articular tubercles. Thus, the restriction of the mandibular condyles within the mandibular fossae is abnormal.

A middle-aged man was seen in the casualty department of a hospital after having been hit over the head with a bottle. He had a large, dough-like swelling over the back of his head. The skin was intact and the swelling fluctuated on palpation.

Radiographs of the skull revealed no fracture. A haematoma of the scalp beneath the epicranial aponeurosis was diagnosed. The doctor told the man to expect to develop 'black eyes'. How is this possible?

The scalp has a rich anastomotic blood supply and bleeds easily and profusely; thus a large haematoma can readily develop. Blood passing beneath the epicranial aponeurosis enters the subaponeurotic space. This space is limited posteriorly by the attachments of the occipitofrontalis muscles on to the supreme nuchal lines. Laterally, the space is limited by the blending of the aponeurosis with the temporal fascia. Anteriorly, the subaponeurotic space extends beneath the orbicularis oculi muscles into the eyelids. Blood may therefore track downwards to produce 'black eyes'.

Following a blow to the head, blood may occasionally collect beneath the periosteum of the vault of the skull, forming a subperiosteal haematoma. Such a haematoma 'outlines' the bone as the pericranium is very firmly attached at the sutures.

Because the actions of the occipital and frontal parts of the occipitofrontalis muscles are in opposite directions, wounds involving the epicranial aponeurosis can have their margins pulled wide apart. Furthermore, bleeding from damaged arteries is considerable, as the adjacent dense connective tissue restricts retraction of the walls of the vessels.

Subaponeurotic haematomas are not infrequent in the neonate following a difficult forceps delivery. Such haematomas are more serious than in the adult. The relatively large amount of blood that can accumulate beneath the scalp can markedly deplete the circulating blood volume of the neonate. This can induce shock and a rapid deterioration in the baby's condition. Treatment would initially be aimed at restoring blood volume by replacement of body fluids. Because of these difficulties, skull dimensions are checked in babies who have difficult births.

Infections in the scalp are potentially dangerous because of its venous connections. Why?

The veins of the scalp freely anastomose with one another and are connected to the diploic veins of the skull bones and to the intracranial venous sinuses via emissary veins. Thrombosis of the scalp veins secondary to infection may result in spread of the infection and thrombosis of the venous sinuses with resultant cerebral oedema and possible death.

The orbit

The upper part of the facial skeleton shows two pyramid-shaped cavities called the orbital cavities (or orbits). They house and protect the eyes and thus should be regarded as part of the neurocranium. The eyes are the organs of special sense concerned with vision and can be thought of as extensions of the brain. The location of the orbits in the human differs from that in most other animals. In the human, the orbits are situated at the front of the face and not at the sides of the head. This facilitates binocular vision. The close relationship of the roof of the orbit and the floor of the cranial cavity is another characteristic of the human skull. This arrangement is associated with the evolutionary development of a large brain.

THE OSTEOLOGY OF THE ORBIT

The orbital aperture (aditus of the orbit) is bounded above by the supra-orbital margin of the frontal bone, laterally by the zygomatic bone and the zygomatic process of the frontal bone, below by the zygomatic bone and the maxilla, and medially by the frontal bone and the anterior lacrimal crest of the frontal process of the maxilla.

The orbital cavity is pyramidal in shape, with the apex pointing posteriorly. It has a roof, a floor, and medial and lateral walls. The bones that comprise the orbital cavity are the ethmoid, frontal, lacrimal, maxillary, palatine, sphenoid, and zygomatic bones.

The roof of the orbit is formed mainly by the orbital part of the frontal bone. The lesser wing of the sphenoid bone is situated posteriorly at the apex.

The floor of the orbit is made up of the orbital surfaces of the maxillary and the zygomatic bones, with a small contribution from the palatine bone near the apex (the orbital process of the palatine bone).

The medial wall of the orbit begins at the anterior lacrimal crest of the maxilla. Behind this lies the lacrimal bone. Most of the medial wall behind the lacrimal bone is formed by the orbital plate of the ethmoid bone. The body of the sphenoid bone contributes to the medial wall posteriorly. Between the anterior lacrimal crest of the maxilla and the posterior lacrimal crest of the lacrimal bone is the fossa for the lacrimal sac.

The lateral wall of the orbit is formed anteriorly by the orbital surface of the zygomatic bone, and posteriorly by the orbital surface of the greater wing of the sphenoid bone.

There are several foramina and fissures within the orbit:

At the superior orbital margin are the supra-orbital notch (or foramen) and the frontal notch (or foramen). These transmit respectively the supra-orbital and the supra-trochlear nerves and vessels.

In the floor of the orbit is found the infra-orbital groove and the infra-orbital canal for the infra-orbital nerve and vessels.

At the medial wall is situated the opening of the nasolacrimal canal, and the anterior and posterior ethmoidal foramina for the anterior and posterior ethmoidal nerves and vessels. The nasolacrimal canal is located antero-inferiorly, close to the orbital margin. The ethmoidal foramina lie at the junction with the roof of the orbital cavity.

At the lateral wall is the zygomatico-orbital foramen (occasionally foramina). Through this foramen pass the zygomatic branches of the maxillary division of the trigeminal nerve (with accompanying vessels).

Near the apex of the orbital cavity are the optic canal, the superior orbital fissure and the inferior orbital fissure. The optic canal lies within the lesser wing of the sphenoid bone. It transmits the optic nerve and the ophthalmic artery. The superior orbital fissure lies between the greater and lesser wings of the sphenoid, at the junction of the roof and lateral wall of the orbit. It transmits the oculomotor, trochlear, ophthalmic and abducent nerves, together with the ophthalmic veins. The inferior orbital fissure lies at the junction of the lateral wall and floor of the orbit, between the greater wing of the sphenoid and the maxilla. Through this fissure pass the infra-orbital and zygomatic branches of the maxillary division of the trigeminal nerve (with accompanying vessels).

32A,7; 57D,11

32A,11

THE CONTENTS OF THE ORBIT

The major structures that occupy the orbit are:

- The eye and optic nerve
- The extra-ocular muscles
- The oculomotor, trochlear and abducent nerves
- The ophthalmic and maxillary divisions of the trigeminal nerve
- The ciliary parasympathetic ganglion
- The ophthalmic vessels
- The lacrimal apparatus

Concerning the disposition of structures, the anterior part of the orbit is occupied chiefly by the eye and by the lacrimal apparatus which is responsible for the production and drainage of tears. The eyelids protect the exposed (anterior) surface of the eye. Running from the back of the eye to the brain is the optic nerve.

124B,35; 124C
118C,46; 118C,36
124B,24; 124B,13

Many important structures pass through the apex of the orbit. Their disposition is best understood with respect to the origin of the four recti muscles from a fibrous ring called the common tendinous ring (Figure 63). This ring surrounds the optic canal and encloses part of the superior orbital fissure. Entering the orbit through the optic canal, and thus within the common tendinous ring, are the optic nerve and the ophthalmic artery. Also entering the orbit within the common tendinous ring, but via the superior orbital fissure, are the superior and inferior divisions of the oculomotor nerve, the nasociliary branch of the ophthalmic nerve, and the abducent nerve. Entering the orbit through the superior orbital fissure outside the common tendinous ring are the trochlear nerve, and the frontal and lacrimal branches of the ophthalmic nerve.

120; 122
120B,28
32A,6
32A,7
120B,27; 122,20
120B,7; 122,21
122B,24; 122A,19
122,13; 122,29

120,20; 120B,19; 120,15

The extra-ocular muscles that move the eyeball include the recti muscles and the superior and inferior oblique muscles. The superior oblique muscle is situated above the medial rectus muscle. The inferior oblique muscle lies on the floor of the orbit, beneath the inferior rectus muscle. The orbit also contains the levator palpebrae superioris muscle. This lies immediately above the superior rectus muscle. It runs against the roof of the orbit and is attached to the upper eyelid.

122C,37
122C,14; 122C,32
122C,16; 120B,13
122C,3

The orbit contains both motor and sensory nerves. The oculomotor, trochlear and abducent nerves supply the extra-ocular muscles. The optic nerve is the sensory nerve conveying visual information from the retina of the eye. Additional sensory branches pass with the ophthalmic and maxillary nerves through the orbit to supply skin of the face and scalp as well as some orbital structures. Another neural structure in the orbit is the ciliary parasympathetic ganglion. This ganglion is located at the apex of the orbit. It is concerned with the innervation of structures within the eye, including the intra-ocular muscles associated with the iris and the lens.

122A,23; 120B,20
122D,29; 122,20
120,15; 120B,19
120,30; 122B,36

122,27

The main artery supplying the orbit is the ophthalmic artery, a branch of the internal carotid artery. Superior and inferior ophthalmic veins drain the orbit and pass back into the cavernous sinus.

122A,21
120B,18

Orbital fat fills the areas between the structures within the orbit, being particularly prominent around the optic nerve. The fat is said to play a role in stabilising the position of the eye.

a

b

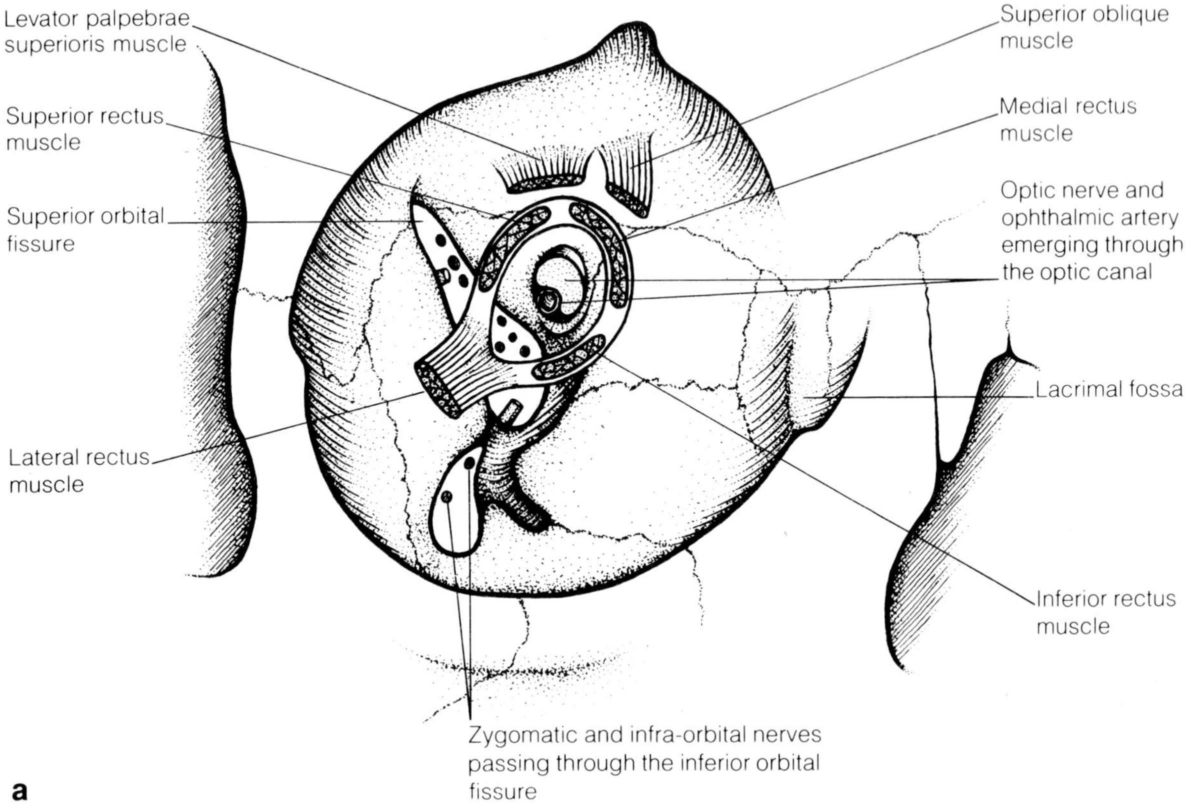

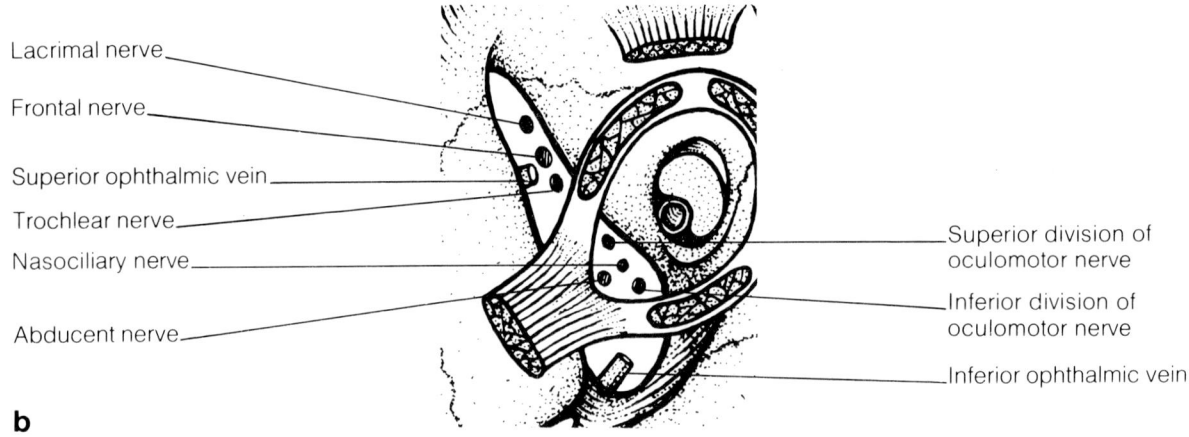

Figure 63a) ***Relationships of structures at the apex of the orbit.*** *Note that the four recti muscles originate from a common tendinous ring.* **b)** *The enlargement below shows the structures passing through the superior orbital fissure.*

118; 124; 125 **THE EYE** (Figure 64)

124 The eye is approximately spherical in shape, with a diameter of about 2.5 cm. It is not centrally positioned within the orbit but lies anteriorly. The eye is also located closer to the roof than to the floor of the orbit, and is nearer the lateral wall than the medial wall.

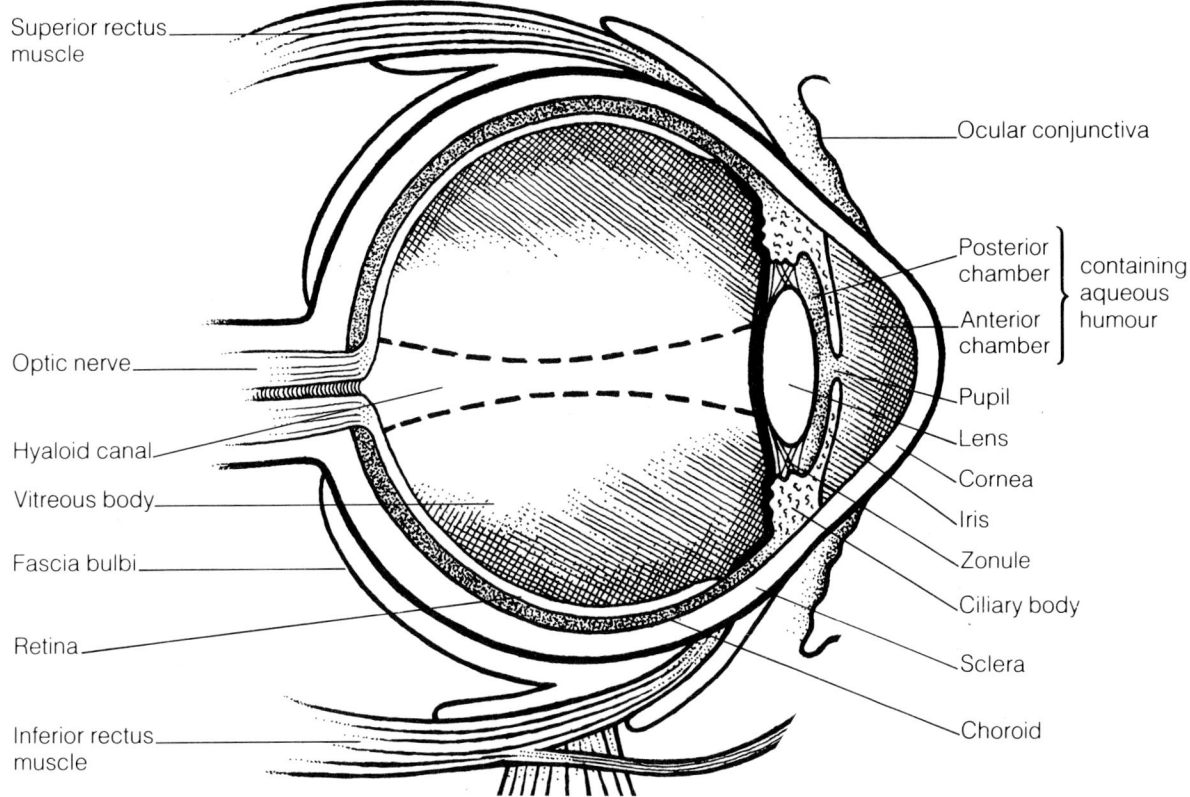

Figure 64 *The structure of the eye.*

The eye has three coats. The outer, fibrous coat is the sclera. This continues anteriorly as the cornea. The middle coat is pigmented and is called the uveal tract. This layer is highly vascular and itself consists of three parts: the choroid, the ciliary body, and the iris. The inner coat is the retina. This layer contains the photoreceptors and neurones.

125D

The exposed surface (anterior surface) of the eye shows many features. The peripheral part forming the 'white of the eye' is the relatively avascular sclera. The sclera is seen through a thin mucous membrane layer called the bulbar conjunctiva. This layer contains a fine vascular network. The sclera becomes transparent over the central portion of the eye, and is here called the cornea. The boundary between the sclera and the cornea is called the sclerocorneal junction or limbus. The dark, circular, central opening is the pupil. Surrounding the pupil is the iris.

118A,6

124B,38
118A,7
118A,9; 118A,8

The interior of the eye is divided into two segments by the lens. The area in front of the lens is called the anterior segment. This is divided into an anterior and a posterior chamber by the iris, and is filled with the clear aqueous humour. The region behind the lens is the posterior segment. This contains a jelly-like substance referred to as the vitreous body.

124B,36

124B,37

124B,35

Light entering the eye is refracted by the transparent cornea and the lens so that an image is brought to focus on the retina (the refractive indices of the aqueous humour and the vitreous body are similar to the refractive index of water). The iris acts as a diaphragm, controlling the amount of light passing into the eye. The ciliary body secretes the aqueous humour and contains muscle that controls the curvature of the lens.

The sclera and the cornea

The sclera comprises the posterior five-sixths of the external coat of the eye. It has a radius of curvature of approximately 12 mm. The sclera is white, opaque and relatively avascular. It meets the cornea at the limbus. The sclera is thick posteriorly, but is especially thin immediately behind the sites of insertion of the extra-ocular muscles and at the entrance of the optic nerve. Where it is pierced by the optic nerve, the sclera is called the lamina cribrosa. This region is so named because of the numerous perforations produced by nerve bundles.

Circumscribing the limbus is a sinus called the sinus venosus sclerae, or the canal of Schlemm (Figure 65). The sinus is essentially an endothelial-lined vessel indenting the deep surface of the sclera. The inner wall of the sinus is separated from the anterior chamber of the eye by a loose trabecular tissue called the trabecular meshwork. This tissue contains a series of endothelial-lined spaces. Aqueous fluid filters into the sinus through the trabecular meshwork. The fluid subsequently enters the bloodstream via an adjacent episcleral plexus of veins. This plexus drains into the anterior ciliary veins.

A particular feature of the sclera at the posterior end of the sinus venosus sclerae is the scleral spur. This flange of tissue gives attachment to the outer group of fibres of the ciliary muscle (see Figure 65).

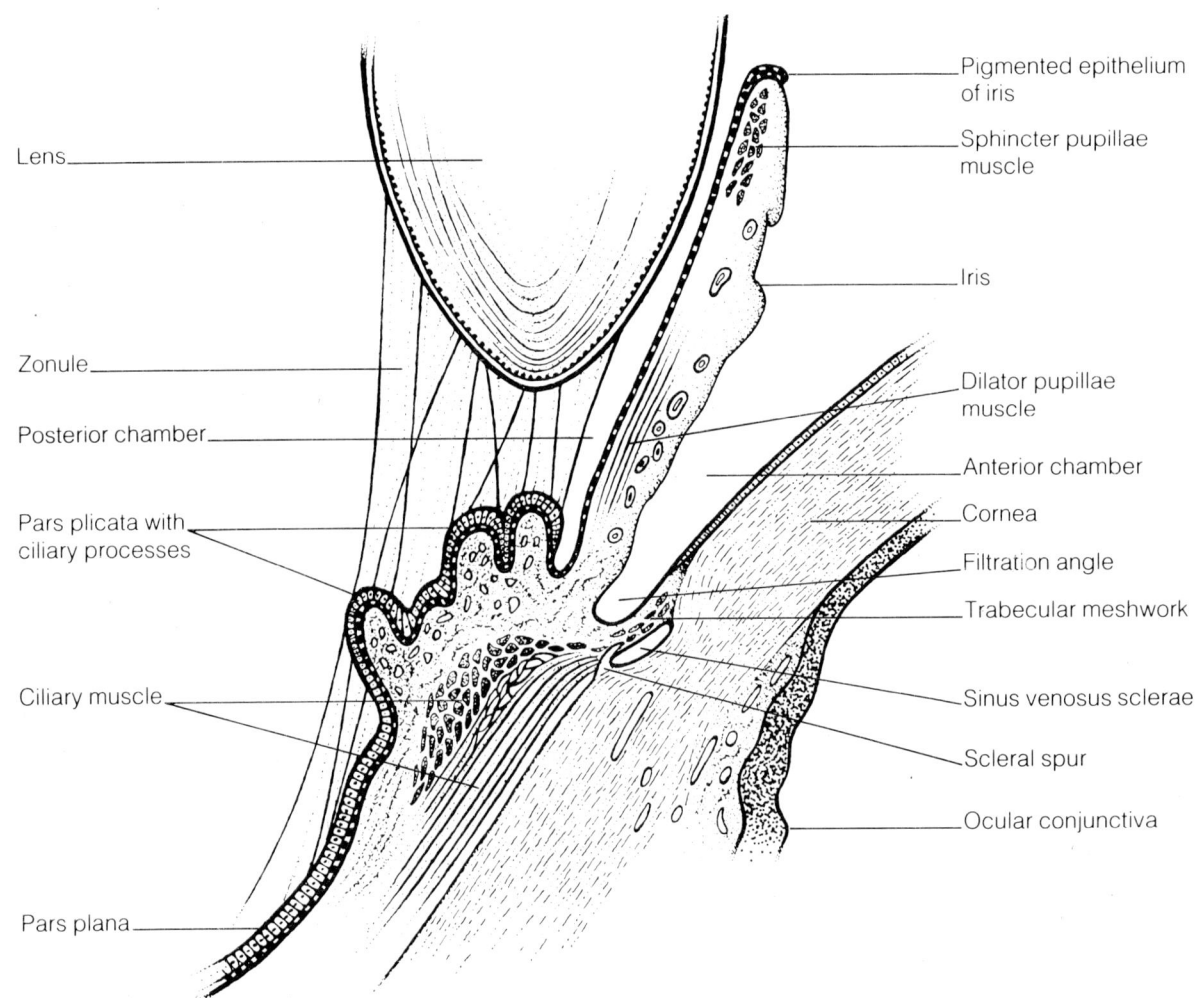

Figure 65 *Section through the iris and ciliary body.*

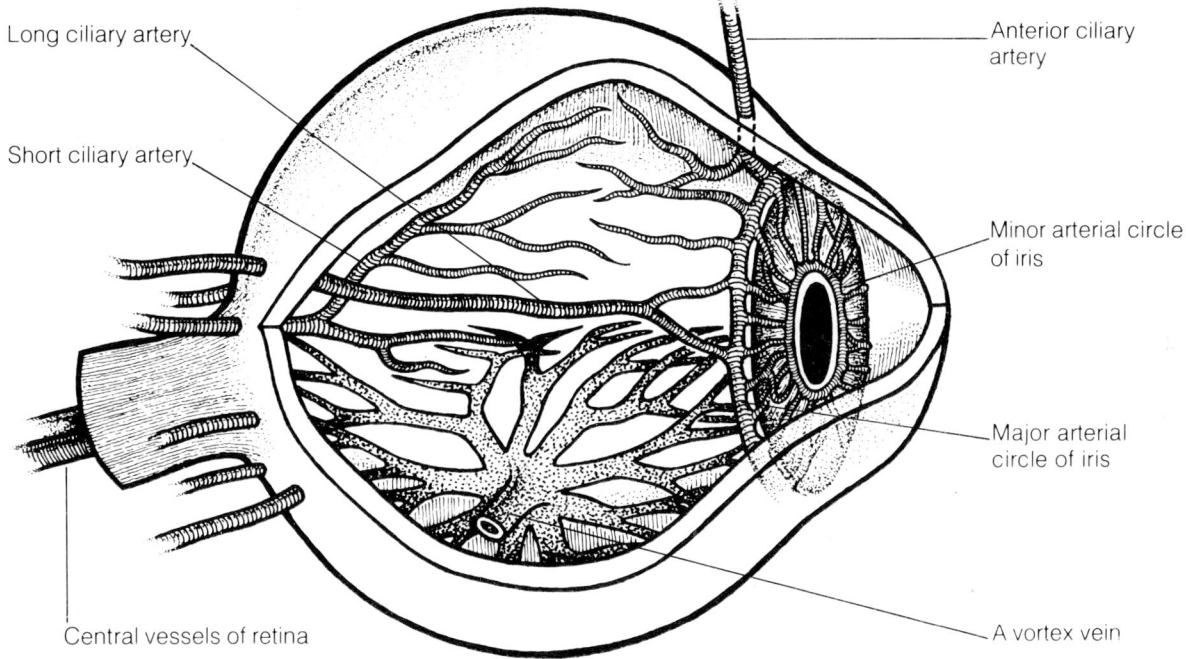

Figure 66 *The vasculature of the eye. The upper part of the diagram illustrates some ciliary arteries and the arterial circle of the iris. The lower part of the diagram illustrates one of the vortex veins.*

The sclera is pierced posteriorly around the optic nerve by the long and short ciliary nerves and vessels, anteriorly near the limbus by the anterior ciliary vessels, and just behind the equator of the eyeball by three or four vortex veins (Figure 66).

The cornea constitutes the anterior one-sixth of the external coat of the eye. Its radius of curvature is less than that of the sclera, being approximately 8 mm. Unlike the sclera, the cornea is transparent. This is because of its highly ordered structure, its lack of blood vessels, and its relative lack of fluid. The lack of a blood supply partly explains the reduced risk of early rejection of corneal grafts. The cornea is innervated by the long and short ciliary nerves (see pages 223 and 225).

124B,38; 125E,52

122,28

The uveal tract
This comprises three parts: the choroid, the ciliary body and the iris.

The choroid is the thin, vascular, pigmented layer between the sclera and the retina (Figure 64). It extends from the optic nerve to the anterior serrated margin of the retina (the ora serrata) where it is continuous with the ciliary body. The choroid is readily separated from the sclera, except around the optic nerve.

125D,45

125D,47

The numerous vessels in the choroid are arranged so that the larger ones are located externally and the smaller ones internally. Indeed, a distinct capillary layer called the choriocapillaris is found internally, which supplies the outer portion of the retina.

The vessels in the choroid are shown in Figure 66. The posterior part of the choroid is supplied by short posterior ciliary arteries (between 10 and 20 branches). These arise from the ophthalmic artery (beneath the optic nerve) and penetrate the sclera at the back of the eye. The anterior part of the choroid is supplied from three sources: the two long posterior ciliary arteries (which share a common origin from the ophthalmic artery with the short posterior ciliary arteries); the recurrent ciliary arteries (which are branches from the major arterial circle of the iris — see pages 205 and 227); and the short posterior ciliary arteries. Many anastomoses exist between the vessels supplying the two parts of the choroid.

Venous blood from the choroid (indeed for most of the eyeball) passes into four large vortex veins (venae vorticosae). The vortex vein is so named because the veins that converge to form it display a characteristic whorled pattern. The vortex veins lie equidistant from each other and pierce the sclera midway between the optic nerve and the limbus. They lie on either side of the superior and inferior rectus muscles. The two superior vortex veins generally drain into the superior ophthalmic vein. The two inferior vortex veins usually drain into the inferior ophthalmic vein.

The ciliary body lies between the iris and the choroid (Figures 64 and 65). It is ring-like and is located about 6 mm behind the limbus. In cross-section, the ciliary body is approximately triangular, being thick in front and narrow behind where it blends with the choroid at the ora serrata. The iris is attached to the middle of the anterior surface of the ciliary body.

The posterior surface of the ciliary body can be divided into two regions: the pars plana and the pars plicata. The pars plana has a smooth, black appearance with faint lines (striae ciliaris) which radiate across from the elevations of the ora serrata. The pars plicata exhibits about 70 radial ridges (the ciliary processes). These ridges are lighter in colour than the grooves between them. Arising close to the surface of the epithelium of the pars plana, and having attachments throughout the entire posterior surface of the ciliary body, are a series of transparent, fine fibres that pass inwards to insert on to the lens. These form the suspensory ligament (zonule) of the lens. The muscle within the ciliary body will affect the degree of tension within the suspensory ligament and thereby regulate the convexity of the lens, allowing the eye to accommodate for near vision.

The ciliary body contains smooth muscle. The muscle is concentrated in the pars plicata and there are two distinct groups. There is an inner circular group that forms a sphincter around the ciliary body. An outer group has radially and longitudinally orientated fibres which are attached to the scleral spur. Contraction of the ciliary muscle relaxes the suspensory ligament with the result that the anterior surface of the lens bulges (thus focusing near objects on the retina). The muscle has no antagonist. Consequently, the suspensory ligament is tensed when the muscle relaxes. This results in the lens becoming flatter (thus focusing distant objects on the retina). The ciliary muscle is supplied by parasympathetic fibres that run with the oculomotor nerve. These relay in the ciliary ganglion and enter the eye via the short ciliary nerves.

The arteries to the ciliary body (Figure 66) are derived from the anterior ciliary arteries and the long posterior ciliary arteries before they unite to form the major arterial circle of the iris (see page 205). The anterior ciliary arteries originate from the muscular branches of the ophthalmic artery and enter the front of the eye. The long posterior ciliary arteries originate from branches of the ophthalmic artery at the back of the eye. The ciliary body receives additional branches from the major arterial circle of the iris. The arteries to the ciliary processes arise from the major arterial circle of the iris. Veins from the ciliary body drain into the vortex veins.

The iris is the most anterior part of the uveal tract (Figures 64 and 65). It is a thin disc, which is perforated near the centre by the pupil. The iris meets the anterior surface of the ciliary body slightly behind the limbus.

The iris is separated from the cornea in front by the anterior chamber. Between the posterior surface of the iris and the lens is the shallow space of the posterior chamber. The anterior and posterior chambers are filled with aqueous humour and they communicate at the pupil.

The anterior surface of the iris is divided into two zones: the peripheral ring (the ciliary zone) and the central ring (pupillary zone). At the junction between the two zones is a circular ridge (the collarette). This ridge is a remnant of an embryonic vascular circle. Whereas the pupillary zone is relatively flat, the ciliary zone exhibits many radially aligned, interlacing ridges. These ridges are wavy when the pupil is dilated and straight when the pupil is constricted. In contrast to the kaleidoscopic relief of the anterior surface of the iris, the posterior surface is smooth and dark (because of the presence of a deeply pigmented epithelium).

The connective tissue within the iris is called the stroma. Between the stroma and the posterior epithelium of the iris lies some smooth muscle called the dilator pupillae muscle. This thin layer of muscle extends from the base of the iris near the ciliary body to the margin of the pupil. Here, it meets the sphincter pupillae muscle. Contraction of the dilator pupillae muscle causes the pupillary zone of the iris to slide beneath the anterior portion of the ciliary zone, thereby dilating the pupil. The muscle is innervated by sympathetic fibres running with the long ciliary nerves. The sphincter pupillae muscle forms a ring around the margin of the pupil. Contraction of the muscle constricts the pupil. In common with the ciliary muscle, the sphincter pupillae is supplied by parasympathetic fibres that run with the oculomotor nerve. These enter the eye via the short ciliary nerves after relaying in the ciliary ganglion.

122,27; 122,28

The iris is very vascular. Its blood supply (see Figure 66) is derived from the major arterial circle of the iris, which, in spite of its name, lies in the ciliary body. This circle is formed by union of the long posterior ciliary and anterior ciliary arteries. The long posterior ciliary arteries arise from the ophthalmic artery at the back of the eye. The anterior ciliary arteries arise from muscular branches towards the front of the eye. Vessels pass radially from the major arterial circle through the iris, producing striations in the ciliary zone of the iris. At the collarette, near the margin of the pupil, these vessels anastomose to form the minor arterial circle of the iris. The veins accompany the arteries. They pass to the ciliary body and eventually drain into the vortex veins.

The variation in colour of the iris is related to the amount and distribution of pigment (melanin) in its stroma. Where there is little pigment the iris appears blue. Indeed, the blue coloration comes from light scattered from the deeply pigmented posterior epithelium of the iris. Where pigment is present in the stroma, this superimposes varying degrees of brown and mottled coloration. Most babies of ethnic groups with white skin initially have blue/cloudy grey eyes, the pigment in the stroma not appearing for some months.

The retina

The retina is the inner coat of the eyeball (Figure 64). It is a thin layer and is light-sensitive. The retina extends forwards to about halfway between the limbus and the equator of the eye. This position corresponds to a serrated margin called the ora serrata. The retina at the ora serrata is firmly attached to both the choroid and the vitreous body. The retina is transparent during life but becomes white and opaque soon after death.

125D,44

125D,47

Using an ophthalmoscope, three specialised areas can be identified at the back of the retina (the fundus of the retina): the optic disc, the macula lutea, and the fovea centralis (Figure 67).

The optic disc is a pale circular area about 1.5 mm in diameter. It represents the site at which the optic nerve enters the retina. The retinal artery and vein also enter and leave at the optic disc. The depression within the optic disc is called the physiological cup. The disc is insensitive to light. Its projection in the lateral (temporal) visual field is consequently known as the blind spot.

The macula lutea is located about 3 mm lateral to the optic disc. It appears as a shallow depression, and is approximately the same size as the optic disc. Unlike the disc, however, it has a yellowish coloration and lacks a well-defined border. The macula lutea has a capillary-free zone at its centre. In the very centre is a small pit called the fovea centralis.

The central artery of the retina is seen within the optic disc dividing into superior and inferior branches (Figure 67). Each branch gives off medial (nasal) and lateral (temporal) branches, which continue to divide dichotomously. The central artery of the retina is

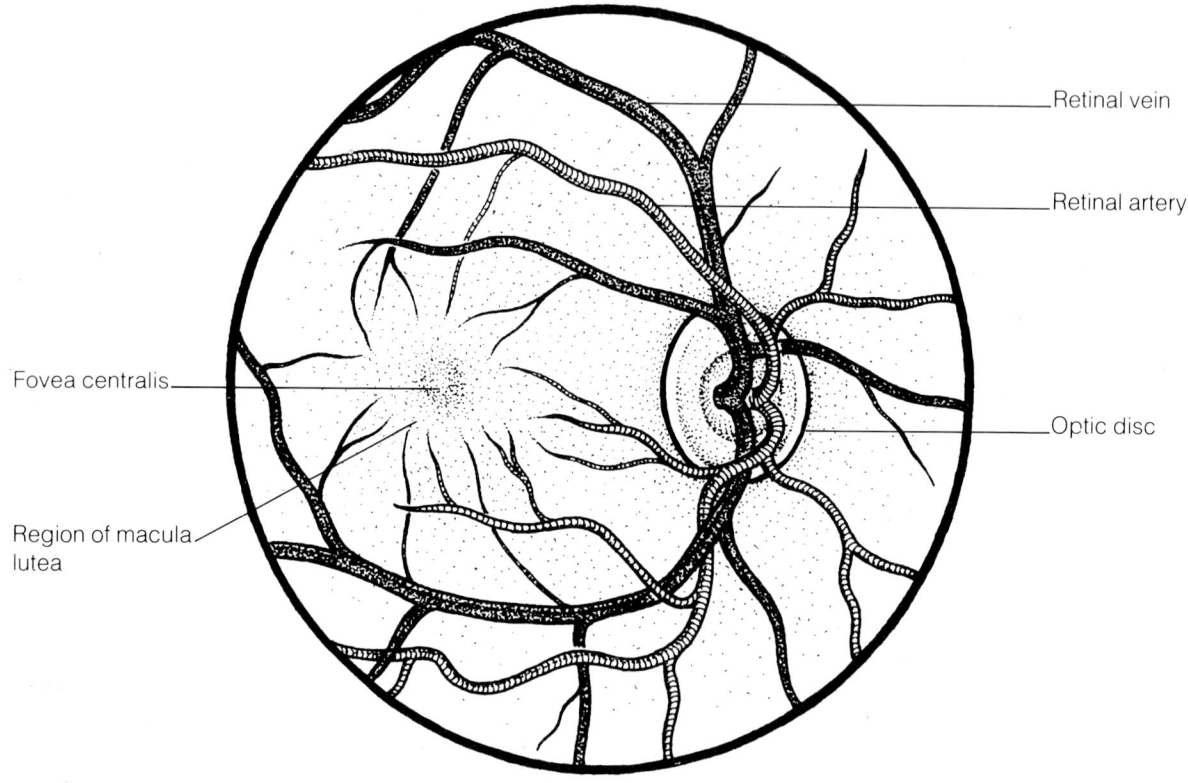

Figure 67 *The fundus of the right eye as seen through an ophthalmoscope.*

an anatomical end artery, as its capillaries do not anastomose with those of any other vessel. The artery supplies the inner region of the retina comprising the nerve fibres. The outer region containing the photoreceptors receives its nutrients by diffusion from the choroidal vessels. The retinal veins accompany the retinal arteries. Superior and inferior retinal veins join to form the central vein of the retina. This union occurs slightly proximal and lateral to the division of the companion artery. Through the ophthalmoscope, the retinal arteries characteristically appear narrower than the veins. They also appear a brighter red and, because of their convex walls, they usually reflect light and show a longitudinal streak.

The retina in most areas has a complex histology. It generally consists of 10 layers. The photoreceptors lie in the outer region (adjacent to the choroid) and the nerve fibres in the inner region. At the optic disc, however, are found only nerve fibres. At the macula lutea, the retinal vessels are absent and the elements comprising the inner layers are heaped up at the margin. This allows for the uninterrupted passage of light to the fovea centralis. The fovea is the most sensitive (and thinnest) part of the retina. It consists almost entirely of colour-sensitive photoreceptors (i.e. cones).

The lens

124B,36; 125D,45

The lens lies immediately behind the iris and the pupil (Figure 64). It is separated from the iris by the posterior chamber. The back of the lens is separated from the anterior surface of the vitreous body by a small fluid-filled space called the retrolenticular space.

The lens is a transparent, biconvex body surrounded by a highly elastic capsule. It has a gelatinous consistency, although the central part (the nucleus) is harder than the peripheral part (the cortex). The posterior surface of the lens is more convex than the anterior surface (radii of curvature 6 mm and 10 mm respectively). The most convex points on the anterior and posterior surfaces are called the anterior and posterior poles. The anterior and posterior surfaces meet at the equator of the lens.

The lens is connected to the ciliary body by the suspensory ligament. The fibres of the suspensory ligament form a layer that extends about 1.5 mm on to the anterior surface of the lens capsule and about 1.3 mm on to the posterior surface of the capsule. The fibres that insert at the equator are usually smaller than those inserting anteriorly and posteriorly.

The chambers of the eye (Figure 64)

There are three spaces within the eye: the anterior chamber, the posterior chamber, and the vitreous cavity. The anterior and posterior chambers (the anterior segment) lie in front of the lens. The vitreous cavity (the posterior segment) is located behind the lens.

The anterior chamber is bounded in front by the cornea and the limbus. Behind lies the iris and the ciliary body. Centrally, the anterior surface of the lens is seen through the pupil. The periphery of the anterior chamber, lying between the base of the iris and the limbus, is referred to as the filtration angle (Figure 65). This site is related to the trabecular meshwork and the sinus venosus sclerae at the limbus (see page 202).

124B,37

The posterior chamber is smaller than the anterior chamber. It is bounded in front by the posterior surface of the iris and behind by the lens and the suspensory ligament.

The aqueous humour is the fluid that occupies the anterior and posterior chambers. It is a colourless, transparent, protein-free fluid which is secreted continuously from the ciliary body. The anterior chamber has a capacity of about 0.2 ml and the posterior chamber of about 0.06 ml. The aqueous humour provides nutrients for the cornea and the lens. The fluid is drained from the eye through the trabecular meshwork at the limbus. It then passes through the sinus venosus sclerae into the adjacent episcleral venous plexus and eventually reaches the superior ophthalmic vein. The intra-ocular pressure is normally between 10 and 20 mm Hg. Changes in the balance between the rate of formation and the rate of drainage of the aqueous humous will affect this pressure.

The vitreous cavity is the largest cavity, occupying the posterior four-fifths of the eye. It is bounded anteriorly by the lens, the suspensory ligament and the ciliary body. Behind and laterally, it is bounded by the retina. The vitreous cavity is filled by the vitreous body which is moulded to the shape of the cavity. Thus, the vitreous body is roughly spherical in outline, but is slightly concave anteriorly where it meets the posterior surface of the lens. This concavity is called the lenticular fossa. The vitreous body is composed of a transparent, colourless, gel-like substance. It can be readily separated from the retina, except at the ora serrata and the optic disc. Condensation of some of the components of the vitreous body can give the impression of a surrounding membrane. A narrow canal (1 to 2 mm wide) may run from the lenticular fossa to the optic disc. This canal is called the hyaloid canal. It is slightly expanded at its ends and is filled with a more watery material. It represents the site of an embryonic artery (the hyaloid artery).

124B,35

The fascia bulbi (Figures 64 and 68)

The fascia bulbi (Tenon's capsule) is the thin, fibrous capsule that loosely covers the eye from the margin of the cornea anteriorly to the optic nerve posteriorly. Functionally, the eye fits into the fascia bulbi in a manner analogous to a ball and socket joint. Thus, the fascia supports the eye and allows for movement in all directions.

The inner surface of the fascia bulbi is smooth and is separated from the sclera by a fluid-filled space called the episcleral space. However, strands of the fascia pass across the space to blend with the sclera.

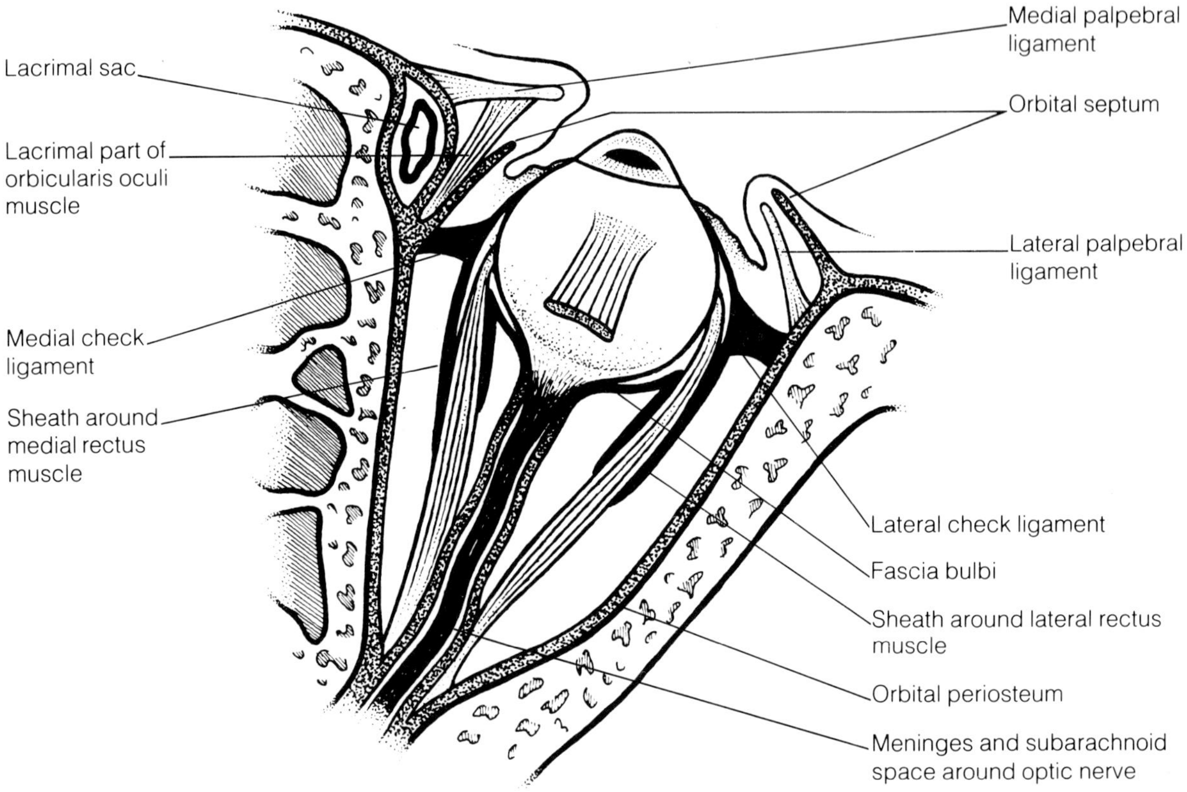

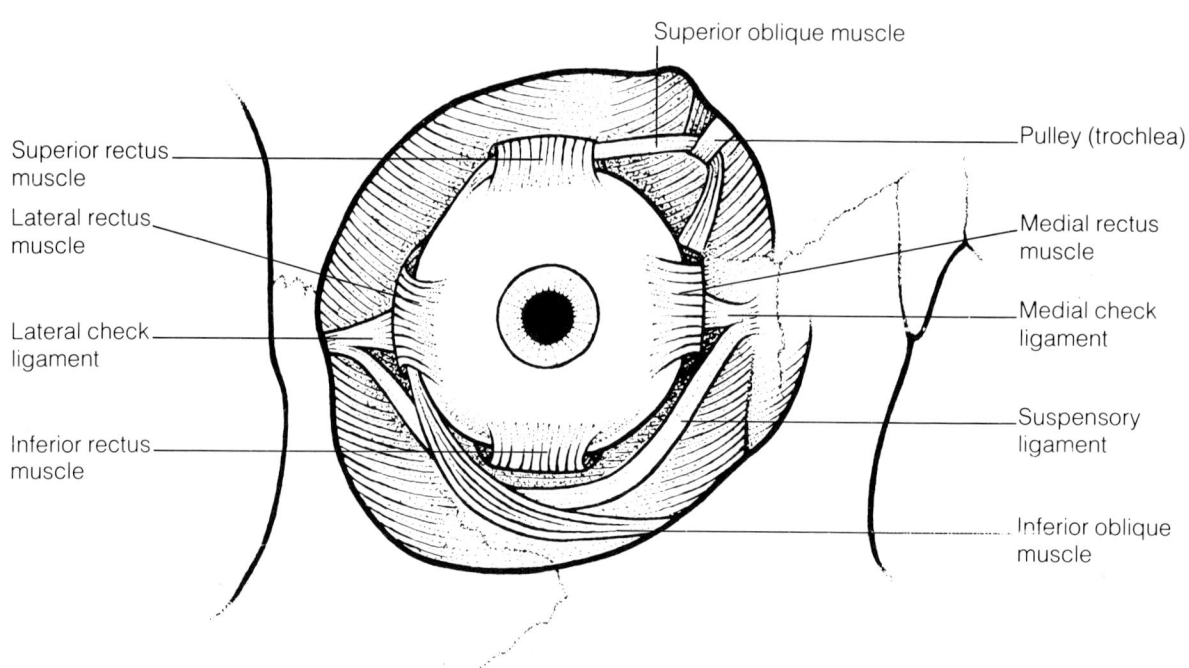

Figure 68 *Horizontal section through the orbit to show the fascia bulbi.*

Figure 69 *The eye viewed anteriorly showing ligaments and the extra-ocular muscles.*

The fascia bulbi is pierced by the tendons of the extra-ocular muscles and is reflected backwards around each muscle to provide a sheath. The fascia also projects to adjacent structures in the orbit to form check ligaments. These limit the actions of the extra-ocular muscles.

The most prominent check ligaments are the medial and lateral check ligaments (Figures 68 and 69). The medial check ligament extends from the sheath around the medial rectus muscle to the posterior lacrimal crest of the lacrimal bone on the medial wall of the orbit. The lateral check ligament extends from the sheath of the lateral rectus muscle to the marginal tubercle of the zygomatic bone on the lateral wall of the orbit.

124A,20; 14,40
124A,23; 124A,22
57,16

A check ligament from the sheath of the superior rectus muscle passes to the levator palpebrae superioris muscle. This ligament is responsible for the elevation of the upper eyelid when the gaze of the eye is directed upwards. From the sheath of the inferior rectus muscle comes a ligament that projects forwards to insert into the lower eyelid between the orbicularis oculi muscle and the inferior tarsal plate. This ligament is thought to be responsible for depression of the lower eyelid when the gaze of the eye is directed downwards.

124B,27; 124B,26

124B,33; 124B,40

Check ligaments also project from the superior and inferior oblique muscles. The ligament from the superior oblique muscle passes to the trochlea; the ligament from the inferior oblique muscle extends to the lateral part of the floor of the orbit.

122,9; 122,7
124B,34

The fascia bulbi is thickened beneath the eye to form a suspensory ligament (the suspensory ligament of Lockwood). This is connected to the medial and lateral check ligaments (Figure 69). The suspensory ligament provides sufficient support for the eye so that, even when the maxilla forming the floor of the orbit is removed, the eye will retain its position.

The fascia bulbi is pierced towards the back of the eye by the optic nerve, the ciliary nerves and vessels, and the vortex veins. Some smooth muscle is associated with the fascia bulbi (see page 220).

THE EYELIDS (Figure 70)

The eyelids (palpebrae) are two movable folds that cover the anterior surface of the eye. They protect the eye from trauma or from excessive light. The act of blinking maintains a thin film of tears over the cornea.

118A; 124B

The upper eyelid is larger and more mobile than the lower eyelid. When the eyelids are open, the upper lid just overlaps the upper part of the cornea, whereas the lower lid lies just below the cornea. The elliptical space between the eyelids is called the palpebral fissure. When the eyelids are closed, the upper lid moves down to cover the whole of the cornea.

The eyelids are covered by skin externally and by conjunctiva internally. The skeletal framework of each eyelid is formed by fibrous tissue which is arranged as a tarsal plate and an orbital septum. The chief muscle within the eyelids is the orbicularis oculi muscle. This is a muscle of facial expression.

118B,14

The skin of the eyelids is thin, elastic and almost translucent. The eyelids are demarcated from the adjacent facial skin by the superior and inferior palpebral furrows. Additional furrows appear with age just beyond the inferior orbital margins (e.g. a nasojugal furrow medially and a malar furrow laterally).

Each lid margin exhibits a small elevation approximately one-sixth of the way along from the medial canthus of the eye. This is called the lacrimal papilla. In the centre of the papilla is a small opening, the punctum lacrimale. The margin of each eyelid lateral to the lacrimal papilla is designated the ciliary part of the eyelid. From this part arise the eyelashes. These stiff hairs are arranged in two or three rows. The margin of each eyelid medial to the lacrimal papilla is called the lacrimal part of the eyelid. It lacks eyelashes.

118A,2
118C,42

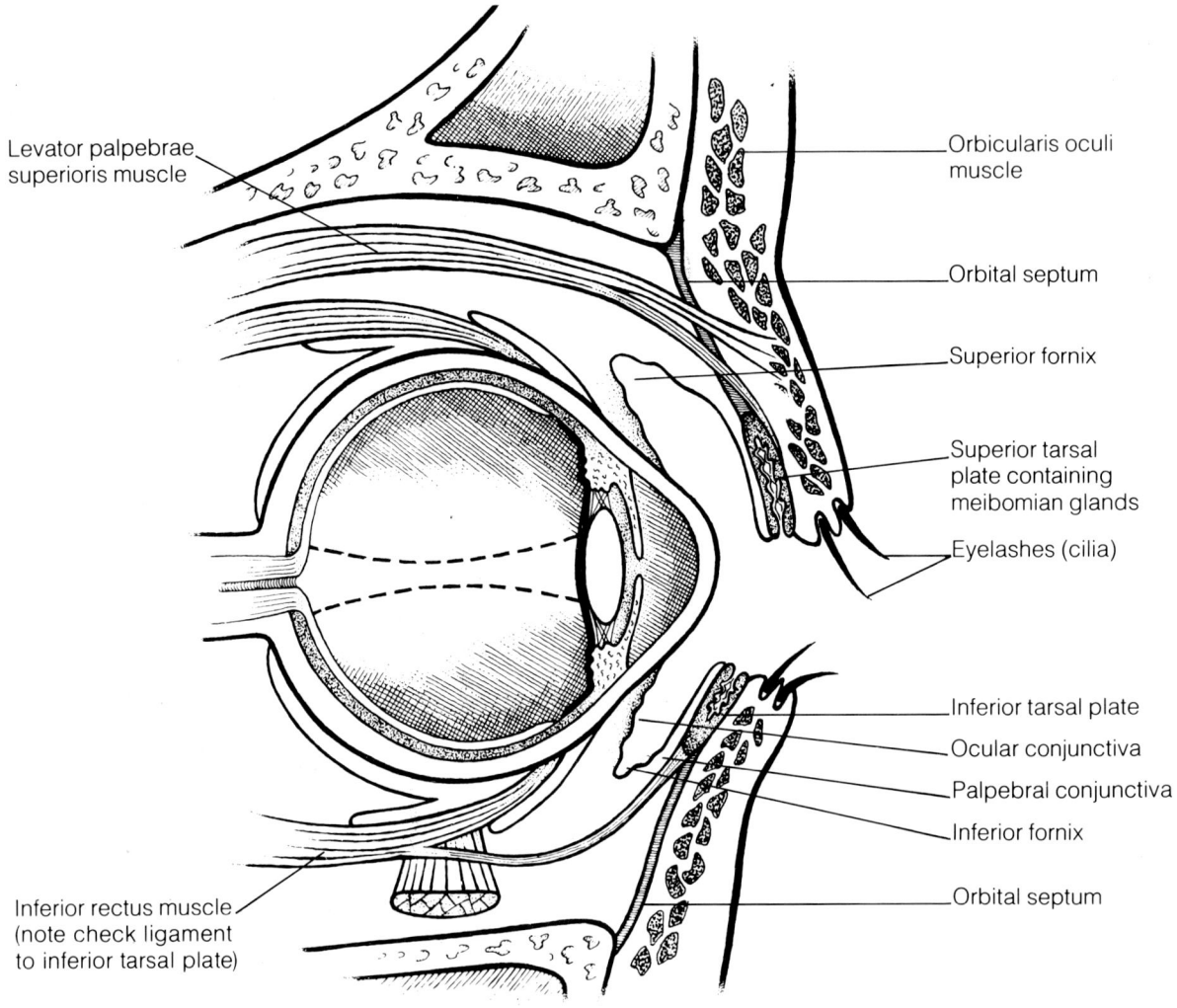

Figure 70 *Sagittal section through the orbit showing the structure of the eyelids.*

The lid margin for both the upper and lower eyelids exhibits a 'grey line' which corresponds to the mucocutaneous junction. In front of this line are the eyelashes; behind it are the openings of the tarsal glands (meibomian glands). The tarsal glands are seen as a series of parallel, faint yellow lines.

118A,3; 118A,11

The medial and lateral angles of the eye are referred to as the medial (inner) and lateral (outer) canthi. The medial canthus is about 2 mm lower than the lateral canthus, this distance being increased in orientals. The lateral canthus is relatively featureless. The medial canthus shows several features. It is separated from the eyeball by a small triangular space, the lacrimal lake (lacus lacrimalis). Here also lies a small, reddish

118A,4

body, called the lacrimal caruncle, which contains sebaceous and sweat glands, and sometimes accessory lacrimal glands. The lacrimal caruncle represents an area of modified skin and contains some fine hairs. Lateral to the caruncle is the plica

118A,5

semilunaris, a fold of conjunctiva that is believed by some to be a vestige of the nictitating membrane of other animals. In orientals, a semilunar fold of skin called the epicanthus passes from the medial end of the upper eyelid to the lower eyelid and obscures the caruncle.

124B,39

The lower eyelid can be everted to reveal its conjunctiva up to the point where it is reflected from the eyelid on to the sclera (i.e. at the inferior fornix). The upper eyelid is not easily everted.

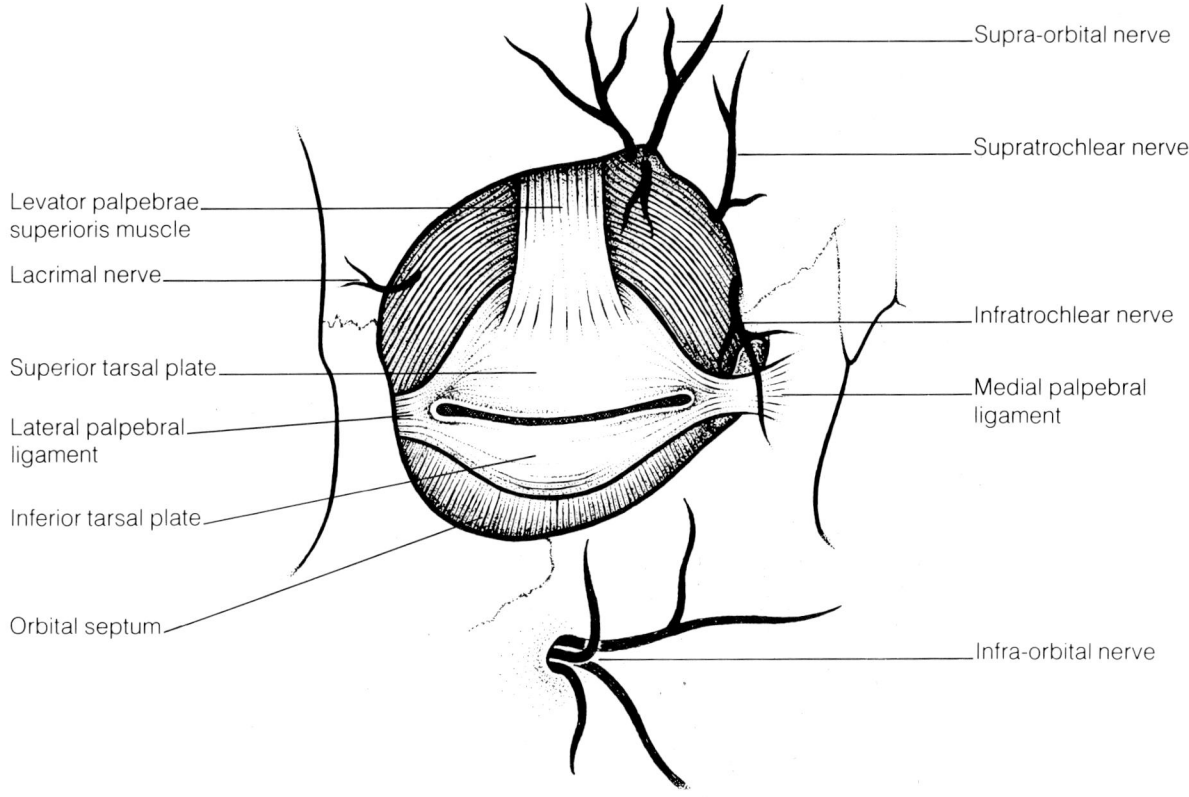

Figure 71 *Anterior view of the fibrous layer of the eyelids and the associated sensory nerves.*

The conjunctiva (Figure 70)

The conjunctiva is the thin, translucent, mucous membrane that lines the inner surface of the eyelids (palpebral conjunctiva) and which is reflected at the fornices to cover the anterior surface of the eye (ocular conjunctiva). The conjunctiva encloses a space, called the conjunctival sac, which opens to the exterior at the palpebral fissure. The superior fornix is level with the superior orbital margin, whereas the inferior fornix is just above the inferior orbital margin. The lacrimal gland drains into the lateral portion of the superior fornix.

118A,6

124B,25
124B,39; 118C,46

The palpebral conjunctiva is firmly attached to the underlying tarsal plates. Because it is translucent, the tarsal glands may be seen through it as yellowish streaks. The palpebral conjunctiva is highly vascular and provides a readily accessible site for assessing signs of anaemia.

The ocular conjunctiva is loosely attached to the sclera. Because it is translucent and relatively avascular, the white of the sclera is visible. Over the cornea, the ocular conjunctiva consists of only a layer of epithelium.

118A,6

The fibrous layer of the eyelids (Figures 70 and 71)

This comprises the orbital septum and two tarsal plates.

The orbital septum is a thin membrane that arises from the periosteum along the entire rim of the orbit. The septum passes inwards into each eyelid to merge with the tarsal plates, and is thickest laterally. It is located in front of the lateral palpebral ligament but passes behind the medial palpebral ligament and lacrimal sac (but in front of the pulley of the superior oblique muscle). The palpebral ligaments are extensions of the tarsal plates (see description that follows) and their relationships with the orbital septum are illustrated in Figure 68.

The orbital septum is pierced above by the levator palpebrae superioris muscle and below by the check ligament from the inferior rectus muscle. The lacrimal, supratrochlear, infratrochlear and supra-orbital nerves and vessels also pass through the septum.

The tarsal plates (one for each eyelid) are crescent-shaped, and about 3 cm long and 1 mm thick. The tarsal plate in the upper eyelid has a maximum height of about 1 cm, whereas the maximum height of the plate in the lower eyelid is 0.5 cm. Each plate is convex forwards, conforming to the configuration of the anterior surface of the eye. The free border is called the ciliary border, as it is adjacent to the eyelashes. The ciliary border is thick and relatively straight. The attached border is called the orbital border, as it is attached to the orbital septum. The orbital border is thin and convex.

The levator palpebrae superioris muscle is attached to both the anterior surface and to the orbital border of the tarsal plate of the upper eyelid. Where it joins the orbital border, the fibres are composed of smooth muscle. A band of smooth muscle also passes between the tarsal plate of the lower eyelid and the fascial sheath of the inferior rectus muscle.

Within the tarsal plates are found tarsal glands (meibomian glands). There are about 30 in the upper eyelid and about 20 in the lower eyelid. The glands can be seen as faint yellow streaks beneath the conjunctiva. Their ducts open on to the free margin of the eyelids behind the grey line. The oily secretions of these modified sebaceous glands lubricate the lid margins and prevent the tears from overflowing on to the face. They may also retard the evaporation of the tear-film on the cornea.

The tarsal plates are connected to the margins of the orbit by the orbital septum and by medial and lateral palpebral ligaments.

The medial palpebral ligament passes from the medial ends of the two tarsal plates to the anterior lacrimal crest and the frontal process of the maxilla. It splits at its insertion into the tarsal plates to surround the lacrimal canaliculi. The medial palpebral ligament lies in front of the lacrimal sac and the orbital septum (Figure 68).

The lateral palpebral ligament passes from the lateral ends of the tarsal plates to a small tubercle on the zygomatic bone within the orbital margin. The ligament is relatively poorly developed. It is more deeply situated than the medial palpebral ligament, lying beneath the orbital septum (Figure 68) and the lateral palpebral raphe of the orbicularis oculi muscle.

The orbicularis oculi muscle has a palpebral part, which is closely associated with the tarsal plates. It is fully described on page 152 with the muscles of facial expression. Between the muscle and overlying skin, the subcutaneous tissue is loose and easily distensible. Between the muscle and underlying tarsal plates there is a zone of loose connective tissue in which are found the main nerves to the eyelids. This region is continuous above with the subaponeurotic layer of the scalp.

The blood supply and innervation of the eyelids

The blood supply to the eyelids is derived mainly from vascular arcades near the lid margins. The arcades are formed by the medial and lateral palpebral arteries, which are branches of the ophthalmic artery (see page 227). Additional branches supplying the eyelids come from the infra-orbital, facial, transverse facial and superficial temporal arteries. The veins are larger and more numerous than the arteries. They pass either superficially to veins on the face and forehead, or deeply to the ophthalmic veins within the orbit.

The cutaneous innervation of the eyelids comes from both the ophthalmic and maxillary divisions of the trigeminal nerve (Figure 71, see also Figure 49). The upper eyelid is supplied mainly by the supra-orbital branch of the frontal nerve. Additional contributions come from the lacrimal nerve, the supratrochlear branch of the frontal nerve, and the infratrochlear branch of the nasociliary nerve. The nerve supply to the

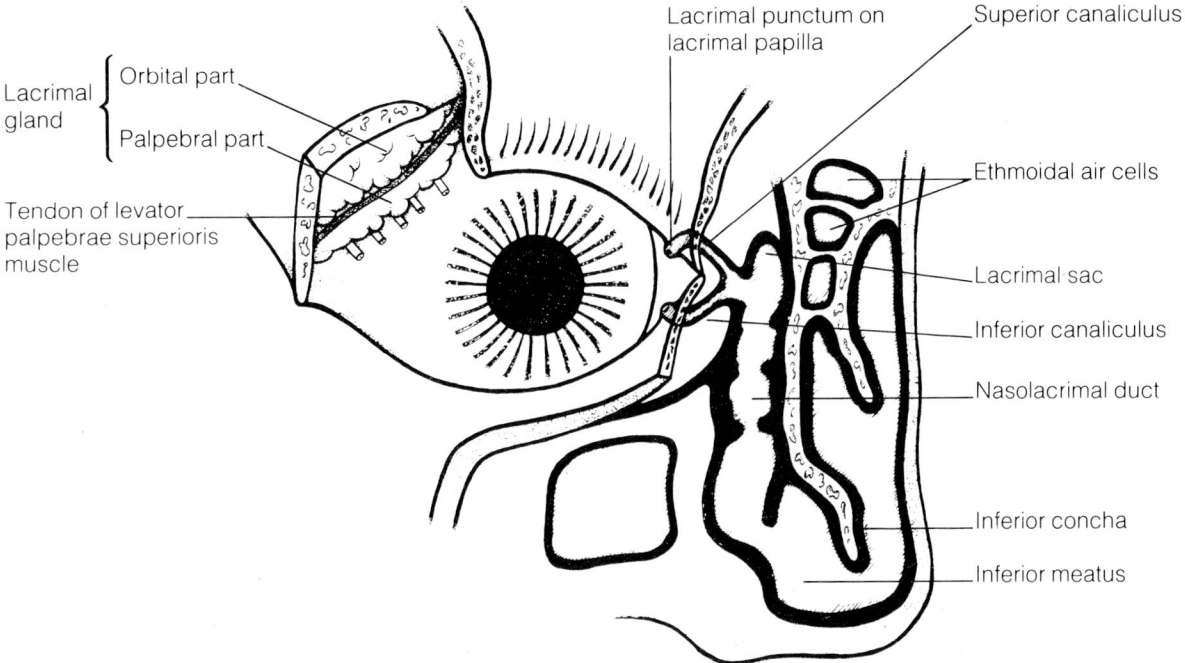

Figure 72 *The lacrimal apparatus.*

lower eyelid is principally from the infra-orbital branch of the maxillary nerve, with small contributions from the lacrimal and infratrochlear nerves.

THE LACRIMAL APPARATUS (Figure 72)

This comprises the lacrimal gland, the lacrimal canaliculi, the lacrimal sac, and the nasolacrimal duct.

118; 124C

The lacrimal apparatus is responsible for the secretion and drainage of tears. The tears form a thin film of fluid covering the exposed surface of the eye. This film has several functions. First, it prevents drying of the cornea and the conjunctiva. Second, it acts as a lubricant between the eyelids and eyeballs. Third, it improves the optical quality of the superficial part of the cornea. Fourth, tears contain a bactericidal enzyme (lysozyme). Excessive production of tears results in crying. Crying is an important means of expressing emotions.

Tears are secreted mainly by the lacrimal gland which lies at the anterolateral corner of the roof of the orbit. Small accessory lacrimal glands may also be found in both eyelids. Tears are carried from the superior fornix over the front of the eye during blinking. The fluid collects at the medial canthus of the eye where it passes through the lacrimal puncta into the lacrimal canaliculi of the eyelids. The canaliculi convey the tears into the lacrimal sac which is located at the anteromedial corner of the floor of the orbit. A nasolacrimal canal runs from the lacrimal sac into the inferior meatus of the lateral wall of the nose. Thus, the tears eventually drain into the nose.

118C,46

124B,25

118C,42; 118C,41
118C,36
118C,38
128B,17

The lacrimal gland

The lacrimal gland folds around the lateral border of the tendon of the levator palpebrae superioris muscle. As a consequence, it is divided into two parts: the orbital and palpebral parts (Figure 72). The orbital part is the larger and lies in a fossa in the frontal bone (the lacrimal fossa); the fossa is situated anterolaterally in the roof of the orbit. The palpebral part lies below the orbital part, just above the lateral aspect of the superior fornix. Thus, the palpebral part may be seen through the conjunctiva when the upper eyelid is everted.

120C,1; 120C,13

124C,41
38C,13
124C,42; 118C,46

213

The lacrimal gland is a compound, tubulo-alveolar gland. Its serous secretions pass into the lateral part of the superior fornix via several distinct lacrimal ducts. The ducts from the orbital part of the lacrimal gland pass through the palpebral part before draining into the fornix.

The innervation of the lacrimal gland is associated with the pterygopalatine parasympathetic ganglion (see page 257 and Figure 88). Parasympathetic, secretomotor fibres arise in the superior salivatory nucleus of the brainstem and pass from the brain with the nervus intermedius of the facial nerve. These preganglionic fibres leave the facial nerve at the geniculate ganglion (without synapsing) as the greater petrosal nerve. This nerve then passes along the floor of the middle cranial fossa, beneath the trigeminal ganglion, to reach the foramen lacerum. It crosses this foramen to enter the pterygoid canal where it forms the nerve of the pterygoid canal (with the deep petrosal nerve). The greater petrosal nerve subsequently runs into the pterygopalatine fossa to relay in the pterygopalatine ganglion. Postganglionic fibres pass into the maxillary nerve and run with its zygomatic branch into the orbit. They then join the lacrimal branch of the ophthalmic nerve to reach the lacrimal gland.

Sympathetic fibres innervate the blood vessels of the lacrimal gland. These fibres originate as the deep petrosal nerve from the sympathetic plexus surrounding the internal carotid artery. The deep petrosal nerve contributes to the nerve of the pterygoid canal and passes to the pterygopalatine ganglion in the pterygopalatine fossa. The fibres do not synapse at the ganglion and pass to the lacrimal gland along the same course as the parasympathetic innervation (i.e. the zygomatic and lacrimal nerves).

Sensory fibres associated with the gland are derived from the lacrimal nerve.

There may be alternative sources of innervation for the lacrimal gland. Parasympathetic fibres may be derived from orbital branches of the pterygopalatine ganglion. Sympathetic fibres may arise directly from the sympathetic plexus surrounding the lacrimal artery.

The vasculature of the lacrimal gland is derived from the lacrimal branch of the ophthalmic artery. The gland may also receive blood from the transverse facial branch of the superficial temporal artery. Venous drainage is into the superior ophthalmic vein.

The lacrimal canaliculi

The lacrimal canaliculi are fine canals (one in each eyelid) which convey tears to the lacrimal sac.

Each canaliculus opens at the margin of the eyelid, about 6 mm from the medial canthus. The canaliculus here shows a small elevation called the lacrimal papilla. The papilla has a fine opening called the lacrimal punctum. The punctum faces backwards and is seen when the eyelid is everted. A ring of dense fibrous tissue helps to maintain the patency of the punctum.

The canaliculus in the upper eyelid initially ascends before bending to pass downwards and medially to the lacrimal sac. The canaliculus in the lower eyelid initially descends and then runs medially and slightly upwards to the lacrimal sac. The canaliculi may open either separately or together in a diverticulum near the middle of the lateral surface of the upper part of the sac.

The lacrimal sac

The lacrimal sac is the blind-ended upper extremity of the nasolacrimal duct. It lies adjacent to the lacrimal groove in the anterior part of the medial wall of the orbit. The sac is bounded in front by the anterior lacrimal crest of the maxilla and behind by the posterior lacrimal crest of the lacrimal bone.

The sac is about 12 mm long and is surrounded by an extension of the orbital periosteum to form the lacrimal fascia. This fascia also extends along the nasolacrimal canal. The medial palpebral ligament crosses in front of the upper part of the lacrimal sac (Figure 68). The lacrimal part of the orbicularis oculi muscle arises behind the lacrimal sac from the posterior lacrimal crest and from the lacrimal fascia. Behind this muscle lie the orbital septum and the check ligament of the medial rectus muscle (Figure 68).

118C,37

The nasolacrimal duct

The nasolacrimal duct is the membranous canal that passes downwards from the lacrimal sac to the anterior portion of the inferior meatus on the lateral wall of the nose. The duct lies in a bony canal (the nasolacrimal canal) which is formed by the maxilla, the lacrimal bone, and the lacrimal process of the inferior nasal concha. The duct is about 15 mm long. It is directed backwards, downwards and laterally, and produces a ridge in the wall of the maxillary sinus. Several folds of mucous membrane may project into the lumen of the nasolacrimal duct (the most constant projection lying at the lower end of the duct). The shape and position of the opening of the nasolacrimal duct into the inferior meatus varies considerably and may not be easy to locate.

118C,38
118C,36; 128B,17
59E,21
62

The innervation of the lacrimal sac and the nasolacrimal duct is from the infratrochlear branch of the ophthalmic nerve and the anterior superior alveolar branch of the maxillary nerve.

122,10

The vasculature of the lacrimal sac and the nasolacrimal duct is derived from the ophthalmic (superior and inferior palpebral branches), maxillary (infra-orbital and sphenopalatine branches), and facial arteries. The venous drainage has a similar pattern to the arteries.

THE EXTRA-OCULAR MUSCLES (Figures 69, and 73 to 76)

Six muscles are involved in producing bodily movements of the eye—four recti and two oblique muscles. The term extra-ocular is used to differentiate these striated muscles from the smooth muscles inside the eye (i.e. the ciliary muscle and the dilator and sphincter pupillae muscles).

The four recti muscles are the superior, inferior, medial and lateral recti muscles. Their names indicate their positions relative to the eyeball. They all arise at the apex of the orbit from a short tendinous ring attached to the lesser wing and body of the sphenoid bone (see Figure 63). This common tendinous ring surrounds the optic canal and the medial end of the superior orbital fissure. The recti muscles run forwards as a cone of muscles and pierce the fascia bulbi to insert into the sclera over the anterior half of the eye.

122C

120B,28

32A,6; 32A,7
122B

The oblique muscles are the superior and inferior oblique muscles. The superior oblique muscle arises from the apex of the orbit and runs above the eye. The inferior oblique muscle is the only muscle that arises from the front of the orbit. It runs below the eye. As indicated by their names, both muscles have an oblique course. They pierce the fascia bulbi to insert into the sclera over the posterior half of the eye.

122C,37
122C,32

The extra-ocular muscles exhibit a number of specialised features which probably relate to the need for very fine control of eye movements. Among these features are: the small size of the muscle fibres, the high ratio of nerve to muscle fibres, and the high firing rate of the motor units.

The descriptions of the extra-ocular muscles that follow include actions that assume the eye is in the primary position (i.e. the gaze is directed straight ahead). However, the muscles can have quite different actions if the eye is already displaced from the primary position. For example, the superior and inferior recti muscles act as elevators or depressors alone only when the eye is already abducted.

The superior rectus muscle

120,17; 122,3

Attachments. The superior rectus muscle is slightly larger than the other recti muscles.

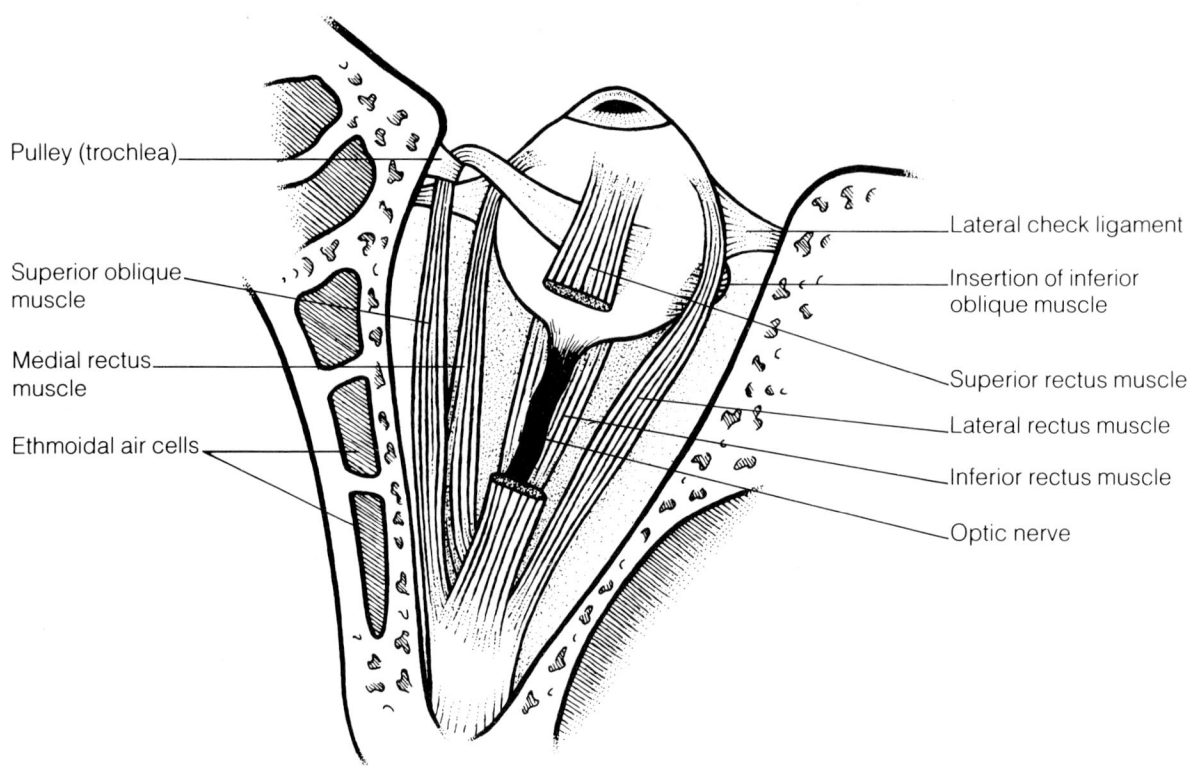

Figure 73 *The extra-ocular muscles viewed from above.* A portion of the superior rectus muscle has been removed to display the optic nerve and inferior rectus muscle.

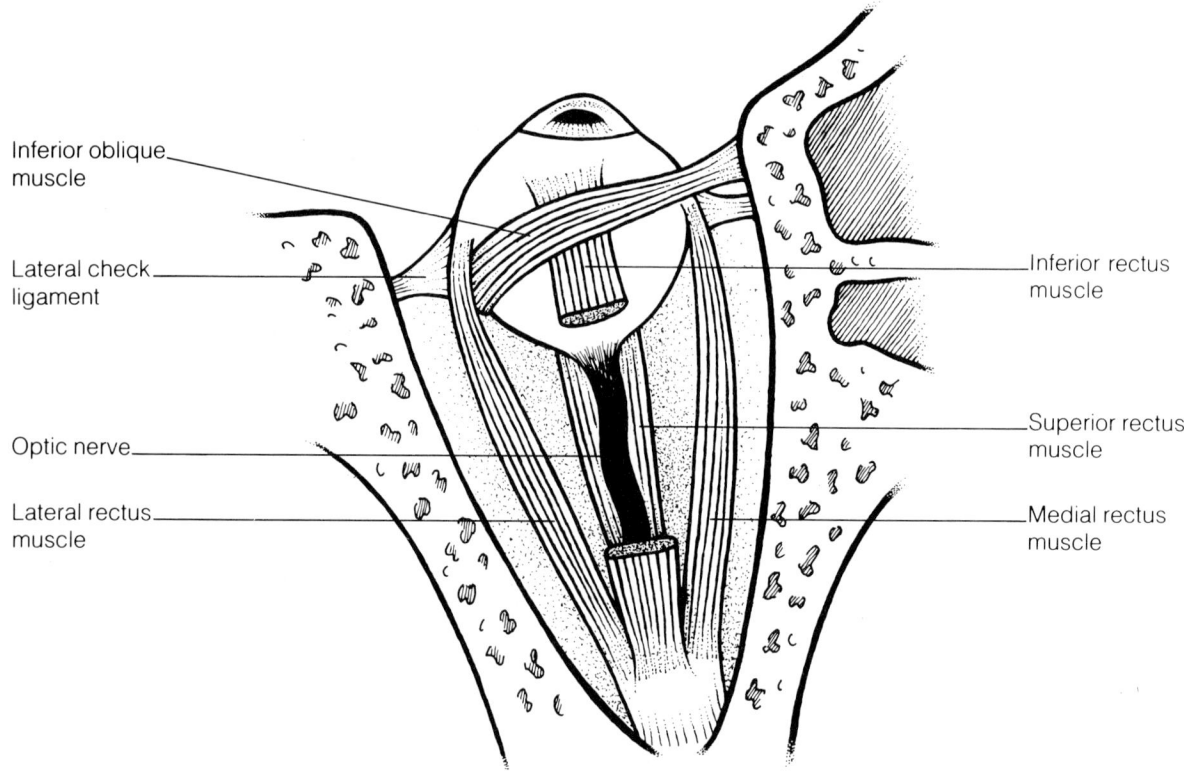

Figure 74 *The extra-ocular muscles viewed from below.* A portion of the inferior rectus muscle has been removed to display the optic nerve and superior rectus muscle.

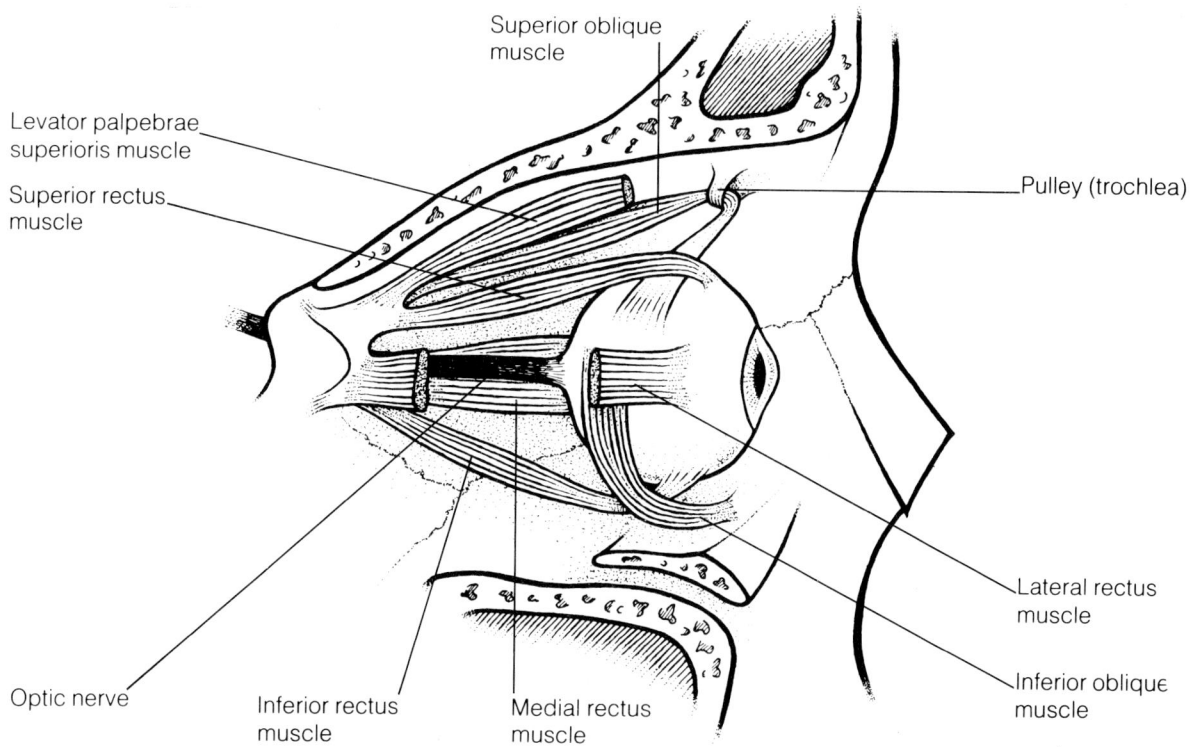

Figure 75 *The extra-ocular muscles viewed from the lateral side.* A portion of the lateral rectus muscle has been removed to display the optic nerve and medial rectus muscle.

It arises from the upper part of the common tendinous ring, above and lateral to the optic canal (see Figure 63). Some fibres also arise from the dural sheath of the optic nerve. The fibres pass forwards and laterally (at an angle of about 25° to the median plane of the eye in the primary position) to insert into the upper part of the sclera about 8 mm from the limbus (Figures 73, 75 and 76). The insertion of the superior rectus muscle is slightly oblique, the medial margin being more anterior than the lateral margin.

Innervation. The muscle is supplied by the superior division of the oculomotor nerve. This nerve enters the inferior surface of the muscle.

Vasculature. The arterial supply is derived both directly from the ophthalmic artery and indirectly from its supra-orbital branch.

Actions. Superior rectus moves the eye so that the cornea is directed upwards (elevation) and medially (adduction). To obtain upward movement alone, the muscle must function with the inferior oblique muscle. The superior rectus muscle also causes intorsion of the eye (i.e. medial rotation). Because a check ligament extends from the muscle to the levator palpebrae superioris muscle, elevation of the cornea also results in elevation of the upper eyelid.

The inferior rectus muscle

Attachments. This muscle arises from the common tendinous ring, below the optic canal (see Figure 63). It runs along the orbital floor in a similar direction to superior rectus (i.e. forwards and laterally). The muscle inserts obliquely into the sclera below the cornea, 6 mm from the limbus (Figures 74 to 76).

Innervation. The inferior rectus muscle is innervated by a branch of the inferior division of the oculomotor nerve. This branch enters the superior surface of the muscle.

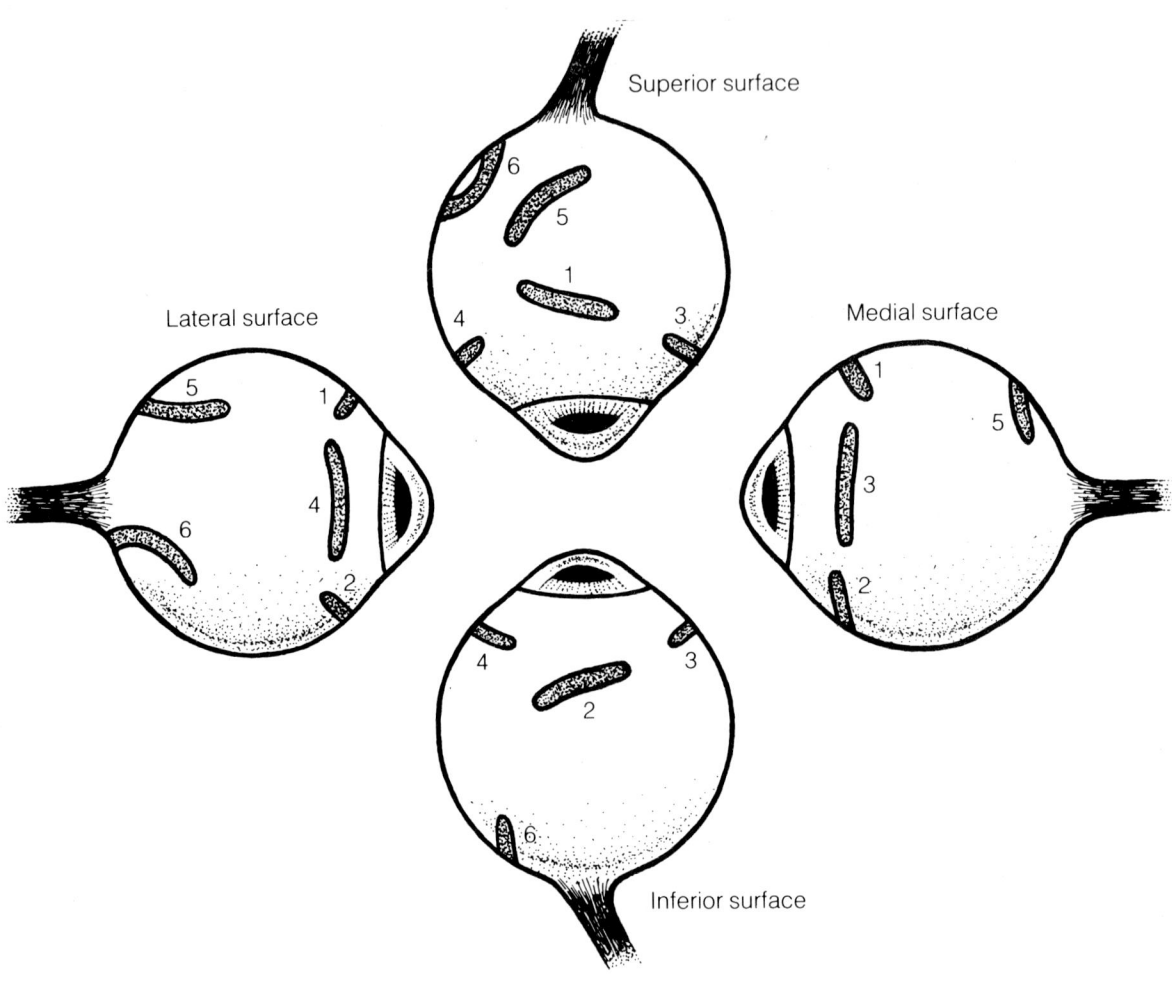

Figure 76 *The surfaces of the eye showing the sites of attachment of the extra-ocular muscles.* 1 = *superior rectus, 2 inferior rectus, 3 = medial rectus, 4 = lateral rectus, 5 = superior oblique, 6 = inferior oblique.*

Vasculature. The arterial supply is derived from the ophthalmic artery and from the infra-orbital branch of the maxillary artery.

Actions. The principal activity of the muscle is to move the eye so that the cornea is directed downwards (depression). The muscle also causes the cornea to deviate medially. To obtain downward movement alone, the muscle must function with the superior oblique muscle. The inferior rectus muscle is responsible for extorsion of the eye (i.e. lateral rotation). A check ligament passes from the inferior rectus muscle to the inferior tarsal plate of the eyelid (Figure 70). This causes the lower eyelid to be slightly depressed when the inferior rectus muscle contracts.

The medial rectus muscle

Attachments. The medial rectus muscle is slightly shorter than the other recti muscles, but is said to be the strongest. It arises from the medial part of the common tendinous ring (see Figure 63). In addition, some fibres arise from the dural sheath of the optic nerve. The muscle passes horizontally forwards along the medial wall of the orbit, below the superior oblique muscle. It inserts into the medial surface of the sclera, approximately 5.5 mm from the limbus and slightly anterior to the other recti muscles (Figures 73 and 76).

Innervation. The muscle is supplied by a branch from the inferior division of the oculomotor nerve. This branch enters the lateral surface of the muscle.

Vasculature. The arterial supply is derived from the ophthalmic artery.

Actions. The medial rectus muscle moves the eye so that the cornea is directed medially (adducted). The two medial recti muscles acting together are responsible for convergence.

The lateral rectus muscle

Attachments. This muscle arises from the lateral part of the common tendinous ring (see Figure 63). It bridges the superior orbital fissure. Some fibres also arise from a spine on the greater wing of the sphenoid. The fibres pass horizontally forward along the lateral wall of the orbit to insert into the lateral surface of the sclera, about 7 mm from the limbus (Figures 74 to 76).

Innervation. Lateral rectus receives its nerve supply from the abducent nerve. This nerve enters the medial surface of the muscle.

Vasculature. The artery to this muscle arises from the ophthalmic artery directly, and/or from its lacrimal branch.

Actions. Contraction of the lateral rectus muscle moves the eye so that the cornea is directed laterally (abducted).

The superior oblique muscle

Attachments. This muscle arises from the body of the sphenoid bone, above and medial to the common tendinous ring (see Figure 63). It runs forwards, above the medial rectus muscle, in the angle between the roof and medial wall of the orbit. The superior oblique muscle becomes narrower and tendinous as it approaches the orbital margin. The tendon passes through a fibrocartilaginous pulley (the trochlea) on the medial wall of the orbit to re-emerge at a completely different angle, being now directed backwards, outwards and downwards. The reflected tendon approaches the eye in the primary position at an angle of about 50°, passing beneath the superior rectus muscle to fan out into a broad insertion over the posterolateral aspect of the sclera (Figures 73, 75 and 76).

Innervation. The superior oblique muscle is supplied by the trochlear nerve. This nerve enters the superior surface of the muscle.

Vasculature. The muscle is supplied both directly from the ophthalmic artery and indirectly from its supra-orbital branch.

Actions. Because of its insertion into the posterior part of the eye, contraction of the superior oblique muscle will elevate the back of the eye, resulting in depression of the cornea (particularly with the eye in the adducted position). The superior oblique muscle moves the eye laterally and also causes intorsion.

The inferior oblique muscle

Attachments. Unlike the other extra-ocular muscles, the inferior oblique muscle arises anteriorly from the floor of the orbit. Its origin is from the maxilla, immediately lateral to the lacrimal groove. The muscle passes backwards, outwards and upwards along the floor of the orbit and below the inferior rectus muscle (see Figure 69). The direction of the inferior oblique muscle parallels that of the tendon of the superior oblique muscle. The inferior oblique muscle passes beneath the lateral rectus muscle to insert into the postero-inferior aspect of the sclera (Figures 74 and 76).

Innervation. The muscle receives its nerve supply from a branch of the inferior division of the oculomotor nerve. This branch enters the superior surface of the muscle.

Vasculature. The arteries supplying the muscle are derived from the ophthalmic artery and from the infra-orbital branch of the maxillary artery.

Actions. Because of its insertion into the posterior part of the eye, contraction of the inferior oblique muscle depresses the back of the eye, resulting in elevation of the cornea (particularly in the adducted position). The muscle moves the eye laterally and also causes extorsion.

The levator palpebrae superioris muscle

Although this muscle is not responsible for movements of the eye, it is considered here because of its close anatomical association with the extra-ocular muscles, and because of its functional relationship with the superior rectus muscle.

Attachments. The levator palpebrae superioris muscle arises from the lesser wing of the sphenoid bone, above the superior rectus muscle and lateral to the superior oblique muscle (Figure 63). Levator palpebrae superioris runs horizontally forwards between superior rectus and the roof of the orbit. Near the superior fornix of the conjunctival sac, the muscle fans out into a wide aponeurosis which passes into the skin and tarsal plate of the upper eyelid (Figures 70 and 71). The muscle is attached to the anterior surface and to the orbital border of the tarsal plate. At the orbital border, the fibres are composed primarily of smooth muscle. There is also an attachment to the superior fornix, via the fascial sheath of the muscle.

Innervation. The muscle is supplied by a branch of the superior division of the oculomotor nerve. This branch enters the inferior surface of the muscle. Sympathetic fibres to the smooth muscle component of levator palpebrae superioris are derived from the plexus surrounding the internal carotid artery. These nerve fibres join the oculomotor nerve in the cavernous sinus.

Vasculature. The arterial supply is derived both directly from the ophthalmic artery and indirectly from its supra-orbital branch.

Actions. As its name suggests, the levator palpebrae superioris muscle elevates the upper eyelid. (This action is opposed by the palpebral part of the orbicularis oculi muscle.) The muscle is linked to the superior rectus muscle by a check ligament, thus there is elevation of the upper eyelid when the gaze of the eye is directed upwards.

Although for descriptive convenience the actions of the extra-ocular muscles have been described for each individual muscle, it is important to appreciate that all eye movements usually require the co-ordinated activity of several muscles. Muscles that aid each other in a particular movement are called synergists; muscles that oppose each other are called antagonists. Muscles that are synergistic for one type of movement may be antagonistic for another type of movement. These concepts are further complicated in the case of the eye, where co-ordinated movements of both eyes are required in order that corresponding points of each retina will fixate on the same object. For example, when looking to the left, contraction of the left lateral rectus muscle and the right medial rectus muscle occurs together with relaxation of the left medial rectus muscle and the right lateral rectus muscle. Left and right pairs of synergistic eye muscles are called yoke muscles.

Smooth muscle in the orbit

As well as the main striated extra-ocular muscles, several aggregations of smooth (nonstriated) muscle are found within the orbit:

• The levator palpebrae superioris muscle has a smooth muscle component near its insertion on to the orbital border of the tarsal plate in the upper eyelid.

• **The orbitalis muscle** is a thin layer of smooth muscle located in the periosteum covering the floor of the orbit. It crosses the inferior orbital fissure and extends as far backwards as the cavernous sinus.

• Other groups of smooth muscle lie in the fascia bulbi, where they almost encircle the eye (being deficient laterally). There is also a band of smooth muscle that passes between the levator palpebrae superioris muscle and the tarsal plate in the upper eyelid.

Another band runs between the inferior rectus muscle and the tarsal plate in the lower eyelid.

The functional significance of smooth muscle in the orbit is not fully understood. The muscle is innervated by sympathetic nerves.

THE NERVES WITHIN THE ORBIT

Both motor and sensory nerves are found in the orbit. The motor nerves are the oculomotor, trochlear and abducent nerves. They supply the extra-ocular muscles. There are also motor nerves derived from the autonomic nervous system. Parasympathetic fibres from the oculomotor nerve (via the ciliary ganglion) supply the sphincter pupillae and ciliary muscles. Parasympathetic fibres from the facial nerve (via the pterygopalatine ganglion) supply the lacrimal gland. Sympathetic fibres supply the dilator pupillae muscle. The sensory nerves within the orbit are the optic, ophthalmic and maxillary nerves. The ophthalmic and maxillary nerves are essentially only passing through the orbit to supply the face and jaws.

The oculomotor nerve

120; 122

This is the third cranial nerve. It is the main source of innervation of the extra-ocular muscles. The oculomotor nerve also contains parasympathetic fibres which relay in the ciliary ganglion.

The oculomotor nerve emerges at the midbrain, on the medial side of the crus of the cerebral peduncle. It passes along the lateral dural wall of the cavernous sinus (see Figure 86). The oculomotor nerve then divides into superior and inferior divisions and runs beneath the trochlear and ophthalmic nerves. The two divisions of the oculomotor nerve enter the orbit through the superior orbital fissure, within the common tendinous ring of the recti muscles (see Figure 63). Here, the nasociliary branch of the ophthalmic nerve lies between the divisions of the oculomotor nerve.

124A,9; 188A,7
168,7; 166,37

166B,31; 166B,40
122A,24; 122A,19
32A,7
122A,13

The superior division of the oculomotor nerve passes above the optic nerve to enter the inferior surface of the superior rectus muscle. It supplies this muscle and provides a branch that runs to the levator palpebrae superioris muscle.

122,24

The inferior division of the oculomotor nerve divides into three branches: medial, central and lateral. The medial branch passes beneath the optic nerve to enter the lateral surface of the medial rectus muscle. The central branch runs downwards and forwards to enter the superior surface of the inferior rectus muscle. The lateral branch travels forwards on the lateral side of the inferior rectus muscle to enter the superior surface of the inferior oblique muscle. The lateral branch also communicates with the ciliary ganglion to distribute parasympathetic fibres to the sphincter pupillae and ciliary muscles (see Figure 79).

122A,19
122A,15
122C,15; 122A,17
122C,17; 122A,18

122B + C,18
122,26; 122,27

The trochlear nerve

120

This is the fourth cranial nerve. It is the only cranial nerve that emerges from the dorsal surface of the brain. The trochlear nerve passes from the midbrain on to the lateral surface of the crus of the cerebral peduncle. It runs through the lateral dural wall of the cavernous sinus (see Figure 86). The nerve then crosses the oculomotor nerve and enters the orbit through the superior orbital fissure, above the common tendinous ring of the recti muscles. Here, it lies above the levator palpebrae superioris muscle and medial to the frontal and lacrimal nerves (see Figure 63). The trochlear nerve travels but a short distance to enter the superior surface of the superior oblique muscle. Indeed, the innervation of the superior oblique muscle is the sole function of the trochlear nerve.

124A,3; 188A,8
168,8; 166,31

166,37; 32A,7
120B,20

122C,38

The abducent nerve

122

The abducent nerve is the sixth cranial nerve. It emerges from the brainstem, between the pons and the medulla oblongata. The abducent nerve is related to the cavernous

134A,6
188A,11; 168,13

sinus (see Figure 86) but, unlike the oculomotor, trochlear, ophthalmic and maxillary nerves, which merely invaginate the lateral dural wall, it passes through the sinus itself. The abducent nerve enters the orbit through the superior orbital fissure (Figure 63). Here it is situated within the common tendinous ring of the recti muscles, first below and then between the two divisions of the oculomotor nerve, and lateral to the nasociliary nerve. The abducent nerve passes forwards to enter the medial surface of the lateral rectus muscle. The innervation of this muscle is the sole function of the abducent nerve.

The ophthalmic nerve and its branches

The ophthalmic nerve is a division of the trigeminal nerve (the fifth cranial nerve) and is a sensory nerve that travels through the orbit to supply primarily the upper part of the face. Developmentally, it is the nerve of the frontonasal process (see page 168).

The trigeminal ganglion in the floor of the middle cranial fossa is the site where the ophthalmic nerve arises. The nerve passes along the lateral dural wall of the cavernous sinus (see Figure 86) and gives off three main branches just before the superior orbital fissure.

Branches of the ophthalmic nerve:

- Lacrimal nerve
- Frontal nerve:
    Supra-orbital nerve
    Supratrochlear nerve
- Nasociliary nerve:
    Sensory branches to the ciliary ganglion
    Long ciliary nerves
    Posterior ethmoidal nerve
    Anterior ethmoidal nerve (and external nasal nerve)
    Infratrochlear nerve

The lacrimal nerve enters the orbit through the superior orbital fissure, above the common tendinous ring of the recti muscles (Figure 63). Here, it is situated lateral to the frontal and trochlear nerves. The lacrimal nerve passes forwards along the lateral wall of the orbit on the superior border of the lateral rectus muscle. It passes through the lacrimal gland and the orbital septum to supply conjunctiva and skin covering the lateral part of the upper eyelid (see page 160; Figure 49). The lacrimal nerve communicates with the zygomatic branch of the maxillary nerve. By this means, parasympathetic fibres associated with the pterygopalatine ganglion are conveyed to the lacrimal gland.

The frontal nerve is the largest branch of the ophthalmic nerve. It enters the orbit through the superior orbital fissure, above the common tendinous ring of the recti muscles (see Figure 63), and lies between the lacrimal nerve laterally and the trochlear nerve medially. The frontal nerve passes forwards on the levator palpebrae superioris muscle, towards the rim of the orbit. About halfway along this course, it divides into the supra-orbital and supratrochlear nerves.

The supra-orbital nerve is the larger of the terminal branches of the frontal nerve. It continues forwards along the levator palpebrae superioris muscle and leaves the orbit through the supra-orbital notch (or foramen) to emerge on to the forehead. The supra-orbital nerve supplies mucous membrane lining the frontal sinus, skin and conjunctiva covering the upper eyelid, and skin over the forehead and scalp (see page 159; Figure 49).

The supratrochlear nerve runs medially above the pulley for the superior oblique muscle. It gives a descending branch to the infratrochlear nerve and ascends on to the forehead through the frontal notch. It supplies skin and conjunctiva covering the upper eyelid, and skin over the forehead (see page 160; Figure 49).

The nasociliary nerve passes into the orbit through the superior orbital fissure, within the common tendinous ring of the recti muscles (see Figure 63). The course of

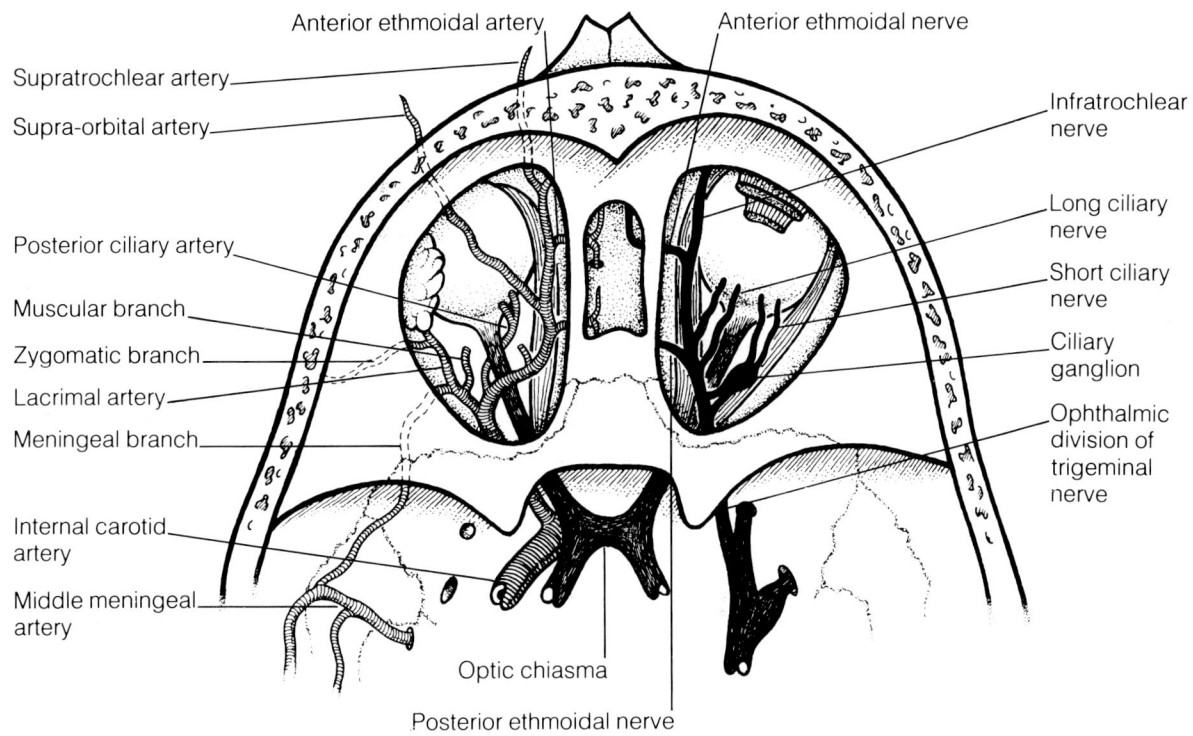

Figure 77 *The ophthalmic artery (left) and the nasociliary nerve (right).*

the nasociliary nerve within the orbit is illustrated in Figure 77. Initially, the nerve lies lateral to the optic nerve. It then runs forwards and medially across the optic nerve and, coursing between the superior oblique and medial rectus muscles, comes to lie close to the medial wall of the orbit. Near the anterior ethmoidal foramen, the nasociliary nerve divides into its terminal branches: the anterior ethmoidal and infratrochlear nerves.

The first branches of the nasociliary nerve are sensory branches to the ciliary ganglion. They leave the ganglion in the short ciliary nerves, running to the eyeball to supply the cornea, the ciliary body and the iris.

Two or three long ciliary branches arise from the nasociliary nerve as it crosses the optic nerve. These ciliary nerves pierce the sclera at the back of the eye (Figure 78) and pass forwards to provide sensory innervation to the cornea and iris. They also distribute sympathetic fibres to the dilator pupillae muscle. The sympathetic fibres originate from the superior cervical ganglion. They are postganglionic fibres which travel in the plexus surrounding the internal carotid artery. They join the ophthalmic nerve in the cavernous sinus.

The posterior ethmoidal nerve passes beneath the superior oblique muscle and leaves the orbit through the posterior ethmoidal foramen to enter the nose. It supplies the sphenoidal sinus and the posterior ethmoidal air cells.

The anterior ethmoidal nerve exits the orbit through the anterior ethmoidal foramen. It enters the anterior cranial fossa where the cribriform plate of the ethmoid bone meets the orbital part of the frontal bone. It then runs into the roof of the nose through a small foramen at the side of the crista galli. The anterior ethmoidal nerve supplies the anterior and middle ethmoidal air cells and some of the mucosa covering the nasal septum and the lateral wall of the nose (see page 247; Figures 80 and 81). It terminates on the face as the external nasal nerve (see page 160; Figure 49).

120C,33; 122,10 122C,7	The infratrochlear nerve passes forwards along the medial wall of the orbit below the pulley of the superior oblique muscle. It passes above the medial palpebral ligament to reach the side of the nose. It supplies the lacrimal sac, the caruncle, the conjunctiva at the medial canthus, and the skin on the medial aspect of the upper eyelid (see page 160; Figure 49).
122; 166B	**The maxillary nerve and its branches** The maxillary nerve is a sensory division of the trigeminal nerve (fifth cranial nerve). Most of the branches from the maxillary nerve arise in the pterygopalatine fossa (see pages 253 to 256). The maxillary nerve gives rise directly to two nerves that pass into the orbit—the zygomatic and infra-orbital nerves—and indirectly to an orbital branch from the pterygopalatine ganglion. All three nerves enter the orbit through the inferior orbital fissure.
122D,42 53E,24	**The zygomatic nerve** in the orbit is located close to the base of the lateral wall. It soon divides into two branches, the zygomaticotemporal and the zygomaticofacial nerves. These nerves run for only a short distance in the orbit before passing on to the face through the lateral wall of the orbit. They may either enter separate canals within the zygomatic bone, or the zygomatic nerve itself may enter the bone before dividing.
53,26; 23,56 122D,33	The zygomaticotemporal nerve exits the zygomatic bone at its temporal (medial) surface. It pierces the temporal fascia to supply skin over the temple (see page 160; Figure 49). The zygomaticotemporal nerve also gives a branch to the lacrimal nerve, which carries parasympathetic fibres to the lacrimal gland (see page 257; Figure 88).
53C,21; 14,27	The zygomaticofacial nerve leaves the zygomatic bone on its lateral surface to supply skin overlying the prominence of the cheek (see page 160; Figure 49).
122B,34; 166B,44 32A,15; 58,2 46E,27; 10,12	**The infra-orbital nerve** initially lies in a groove (the infra-orbital groove) on the floor of the orbit. As it approaches the rim of the orbit, it runs into a canal (the infra-orbital canal) and passes on to the face at the infra-orbital foramen. The infra-orbital nerve supplies the conjunctiva and skin of the lower eyelid. It also innervates the skin over the upper jaw (see page 160; Figure 49) and provides the middle and anterior superior alveolar nerves (see page 256).
	The orbital branch from the pterygopalatine ganglion supplies the orbitalis muscle and some periosteum. It may connect with the ciliary ganglion. The orbital branch may also leave the orbit through the posterior ethmoidal foramen to supply the posterior ethmoidal air cells and the sphenoidal air sinus.
120; 122	**The optic nerve**
188B,34 32A,6; 120B,27 120B,7 120C,27	The optic nerve is the second cranial nerve. It arises from the optic chiasma on the floor of the diencephalon (see page 524). It enters the orbit through the optic canal (see Figure 63), accompanied by the ophthalmic artery. The shape of the optic nerve changes from being flattened at the chiasma to being rounded as it passes through the optic canal. The optic nerve in the orbit passes forwards, laterally and downwards. It pierces the sclera at the lamina cribrosa, slightly medial to the posterior pole (Figures 68 and 78). The optic nerve has a slightly wavy course which allows for movements of the eye.
	Within the orbit, the optic nerve is surrounded by extensions of the three meninges (Figure 68). This reflects the fact that the nerve is really an 'outgrowth' of the brain.
120B,7; 122A,24 122A,13; 122A,29 122C	The optic nerve has important relationships with other orbital structures. As the nerve leaves the optic canal, it lies superomedial to the ophthalmic artery. The oculomotor, nasociliary and abducent nerves (and sometimes the ophthalmic veins) are situated between the optic nerve and the lateral rectus muscle (Figure 63). The optic nerve is also closely related to the origins of the four recti muscles. More anteriorly, however, the muscles diverge and the nerve becomes separated from them by a substantial amount of orbital fat. Just beyond the optic canal, the ophthalmic artery and

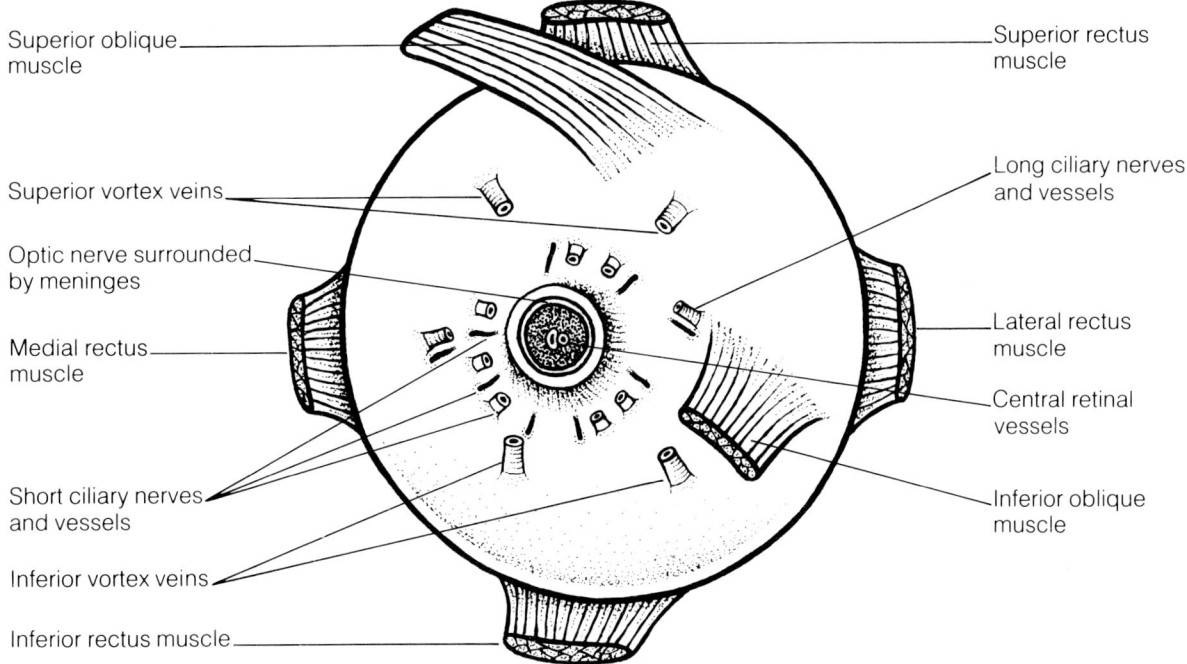

Figure 78 *The back of the eye showing the nerves and vessels.*

the nasociliary nerve cross the optic nerve to reach the medial wall of the orbit (Figure 77). The central artery of the retina enters the substance of the optic nerve about halfway along its length. Near the back of the eye, the optic nerve becomes surrounded by long and short ciliary nerves and vessels (Figure 78).

120C,7; 120C,30

The ciliary ganglion (Figure 79)

122,27

The ciliary ganglion is a parasympathetic ganglion, and is located near the apex of the orbit. It lies in front of the optic canal, between the lateral rectus muscle and the optic nerve, and close to the ophthalmic artery. The ganglion appears as a small swelling connected to the nasociliary nerve. Short ciliary nerves pass from the ganglion to the eyeball (Figure 77). Functionally, the ciliary ganglion is related to the eye, in particular the motor supply of intra-ocular muscles.

122A,13; 122,25
122,28

The parasympathetic fibres to the ciliary ganglion arise from the Edinger–Westphal nucleus of the oculomotor nerve (see page 483). The preganglionic fibres run with the oculomotor nerve into the orbit, leaving in the branch to the inferior oblique muscle. The fibres then pass to the ciliary ganglion, where they synapse. Postganglionic fibres travel to the back of the eye in the short ciliary nerves.

122,18

122,28

The sympathetic fibres to the ciliary ganglion arise from the plexus around the internal carotid artery within the cavernous sinus. These postganglionic fibres form a fine branch that enters the orbit through the superior orbital fissure, inside the common tendinous ring of the recti muscles. This branch then travels through the ganglion without synapsing, and into the short ciliary nerves.

166,38

The sensory fibres to the ciliary ganglion are derived from the nasociliary nerve. They also pass through the ganglion to the short ciliary nerves without synapsing.

The short ciliary nerves convey parasympathetic, sympathetic and sensory fibres between the eyeball and the ciliary ganglion. The nerves pierce the sclera at the back of the eye (Figure 78) and run forwards between the sclera and the choroid. The parasympathetic fibres are distributed to the sphincter pupillae and ciliary muscles.

122,28

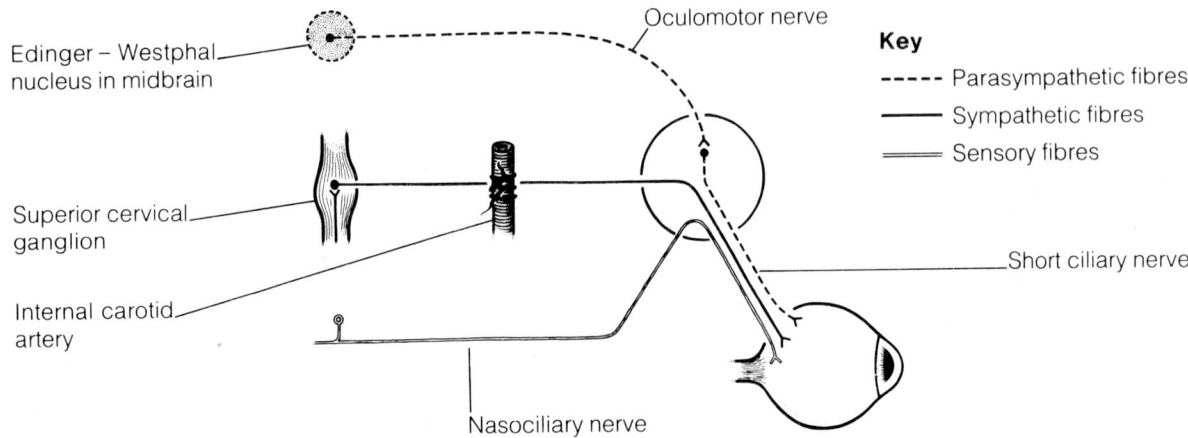

Figure 79 *The ciliary parasympathetic ganglion.*

Contraction of the ciliary muscles is associated with the accommodation reflex. The sympathetic fibres supply arteries within the eye. (The sympathetic fibres supplying the dilator pupillae muscle are thought to run in the long ciliary nerves.) The sensory fibres carry sensation from the cornea, the ciliary body and the iris.

THE ARTERIES WITHIN THE ORBIT

The main vessel supplying orbital structures is the ophthalmic artery. Its terminal branches anastomose on the face and scalp with those of the facial, maxillary and superficial temporal arteries, thus establishing connections between the external and internal carotid arteries. In addition to the ophthalmic artery, the infra-orbital branch of the maxillary artery supplies orbital structures.

The ophthalmic artery (Figure 77)

The ophthalmic artery arises from the internal carotid artery as it emerges from the roof of the cavernous sinus. It traverses the optic canal below the optic nerve and within the dural sheath.

Within the orbit, the ophthalmic artery winds across the optic nerve, passing from the lateral side to the medial side of the orbit. In this position, the artery lies immediately beneath the superior rectus muscle. It then runs a tortuous course with the nasociliary nerve and, passing between the superior oblique and medial rectus muscles, terminates near the medial canthus of the eye by dividing into the dorsal nasal and supratrochlear arteries.

Branches of the ophthalmic artery:

- Central artery of the retina
- Lacrimal artery
- Muscular branches
- Ciliary arteries
- Supra-orbital artery
- Posterior ethmoidal artery
- Anterior ethmoidal artery
- Meningeal branch
- Medial palpebral arteries
- Supratrochlear artery
- Dorsal nasal artery

Many of the branches of the ophthalmic artery accompany sensory nerves of the same name and have a similar distribution: i.e. the lacrimal, supra-orbital, posterior ethmoidal, anterior ethmoidal, and supratrochlear arteries.

The central artery of the retina is the first branch of the ophthalmic artery and is given off close to the optic canal. It initially runs beneath the optic nerve and then within the accompanying dural sheath. Before reaching the back of the eye, the artery enters the substance of the optic nerve itself (through the inferomedial surface). It then runs forwards (accompanied by the central retinal vein) in the central part of the nerve (Figure 78) to the retina. Passing through the lamina cribrosa of the sclera, the artery divides into superior and inferior branches (see Figure 67). Each branch then divides into nasal and temporal branches, usually in the region of the margin of the optic disc. These vessels supply the inner, neural part of the retina. (The photoreceptors, including the fovea centralis, are supplied by the choriocapillaris of the choroid.) The central artery of the retina is an end artery (its capillaries not anastomosing with those of any other blood vessel); therefore damage or blockage will result in varying degrees of blindness.

The ophthalmic artery gives off two main **muscular branches** (within the common tendinous ring of the recti muscles) to supply all the extra-ocular muscles. A lateral muscular branch supplies the levator palpebrae superioris, superior oblique, and lateral and superior recti muscles. A medial muscular branch supplies the inferior oblique muscle and the inferior and medial recti muscles.

The ophthalmic artery supplies the eye via the central artery of the retina (see above), and **the ciliary arteries.**

Anterior ciliary arteries arise from the muscular branches of the ophthalmic artery. They accompany the tendons of the four recti muscles (usually in pairs, although only one artery runs with the lateral rectus muscle). After supplying the conjunctiva and sclera, the anterior ciliary arteries pierce the sclera (close to the attachments of the tendons of the recti muscles) to enter the ciliary body. Here, they anastomose with the long posterior ciliary arteries to form the major arterial circle of the iris (see Figure 66).

Posterior ciliary arteries (usually two) arise from the ophthalmic artery as it runs below the optic nerve. From these arteries arise a series of short and long posterior ciliary branches. There are usually beween 10 and 20 short posterior ciliary arteries. They closely surround the optic nerve and, accompanied by the short ciliary nerves, pierce the sclera close to the optic nerve (see Figure 78). The short posterior ciliary arteries run in the choroid and supply the greater part of this layer. There are two long posterior ciliary arteries: one lies medial and one lies lateral to the optic nerve (see Figure 78). These vessels pierce the sclera close to the attachments of the medial and lateral recti muscles. The long posterior ciliary arteries run forwards in the choroid to the ciliary body. They supply the anterior part of the choroid and the ciliary body. They anastamose with the anterior ciliary arteries at the ciliary body to form the major arterial circle of the iris (Figure 66).

The ophthalmic artery also provides a **meningeal branch**, either directly, or as a branch of the lacrimal artery. The meningeal branch runs back through the superior orbital fissure to anastomose with the middle meningeal artery.

The eyelids derive their blood supply from two sources. Two **medial palpebral arteries** (superior and inferior) arise directly from the ophthalmic artery near the medial canthus. Two **lateral palpebral arteries** (superior and inferior) arise as branches of the lacrimal artery. The superior arteries supply the upper eyelid; the inferior arteries supply the lower eyelid.

The dorsal nasal artery is a terminal branch of the ophthalmic artery. It pierces the orbital septum above the medial palpebral ligament to supply the lacrimal sac and adjacent skin on the nose. It anastomoses with adjacent branches of the facial artery.

118C,35
118C,37

The infra-orbital branch of the maxillary artery enters the orbit through the posterior part of the inferior orbital fissure. It passes along the infra-orbital groove of the maxillary bone in the floor of the orbit before entering the infra-orbital canal. Eventually, it comes out on to the face through the infra-orbital foramen. As the infra-orbital artery traverses the infra-orbital canal, it provides branches that supply the inferior rectus and inferior oblique muscles and the lacrimal sac.

THE VEINS WITHIN THE ORBIT

The veins draining the orbit are the superior and inferior ophthalmic veins and the infra-orbital vein. The veins of the eyeball mainly drain into the vortex veins (see page 204).

The superior ophthalmic vein originates just above the medial palpebral ligament, where it communicates with the supra-orbital vein and with the first part of the facial vein (the angular vein). The superior ophthalmic vein passes backwards alongside the ophthalmic artery, lying between the optic nerve and the superior rectus muscle. It leaves the orbit through the superior orbital fissure to drain into the anterior part of the cavernous sinus. The superior ophthalmic vein usually passes through the superior orbital fissure above the common tendinous ring of the recti muscles. Occasionally, it can pass within or below the common tendinous ring.

The superior ophthalmic vein receives tributaries that accompany branches from the ophthalmic artery (i.e. lacrimal, muscular, anterior and posterior ethmoidal, anterior ciliary veins). It also receives the two superior vortex veins of the eyeball and the central vein of the retina. Alternatively, the central vein of the retina drains directly into the cavernous sinus (although still having a communicating branch to the superior ophthalmic vein). The superior ophthalmic vein may also receive the inferior ophthalmic vein.

The inferior ophthalmic vein originates as a plexus in the anterior part of the floor of the orbit. It runs backwards on the inferior rectus muscle and across the inferior orbital fissure. It then either joins the superior ophthalmic vein, or passes through the superior orbital fissure (within or below the common tendinous ring of the recti muscles) to drain directly into the cavernous sinus.

The inferior ophthalmic vein receives tributaries from the inferior rectus and inferior oblique muscles, and from the lacrimal sac and the eyelids. It also receives the two inferior vortex veins of the eyeball. The inferior ophthalmic vein communicates with the pterygoid venous plexus by a branch passing through the inferior orbital fissure. It may also communicate with the facial vein across the inferior margin of the orbit.

The infra-orbital vein runs with the infra-orbital nerve and artery in the floor of the orbit. It passes backwards through the inferior orbital fissure into the pterygoid venous plexus. It drains structures in the floor of the orbit and communicates with the inferior ophthalmic vein. On the face, the infra-orbital vein may communicate with the facial vein.

It is evident that the veins of the orbit link the cavernous sinus with veins on the face and in the infratemporal fossa. All of these veins lack valves. A pathway therefore exists for the intracranial spread of infection.

THE EMBRYOLOGY OF THE EYE

The eye develops from two main components. There is an outgrowth from the side of the forebrain (the optic vesicle) and a region of epithelium on the face of the embryo (the lens placode). These initially appear at the beginning of the fourth week of intra-uterine life.

The distal end of the optic vesicle invaginates to form a double-layered structure called the optic cup. The outer layer of the optic cup gives rise to the pigmented layer of the retina. The inner layer gives rise to the rest of the retina, part of the ciliary body and the inner pigmented layer of the iris. The sphincter and dilator pupillae muscles are derived from the optic cup and are therefore ectodermal in origin. The ciliary muscle, however, is derived from adjacent mesenchyme.

The proximal end of the optic vesicle is called the optic stalk. This subsequently becomes the optic nerve. That the optic nerve is surrounded by meninges is explained by its derivation from an outgrowth of the brain.

Under the inductive influence of the optic vesicle, the lens placode sinks into the underlying mesenchyme and comes to lie within the margin of the optic cup. Here it develops into the lens of the eye.

The lens placode and the optic vesicle are surrounded both externally and internally by mesenchyme. The choroid and sclera develop from the external layer of mesenchyme. The mesenchyme in front of the lens forms the connective tissue component of the cornea. Spaces appear within the mesenchyme beneath the cornea and coalesce to form the anterior chamber of the eye. The mesenchyme within the optic cup forms the vitreous body.

The extra-ocular muscles are thought to arise from three pre-otic myotomes that develop from mesenchyme around the prochordal plate. They are supplied by cranial nerves from that region (i.e. the oculomotor, trochlear and abducent nerves). The orbicularis oculi muscle originates from mesenchyme of the second branchial arch. It is therefore supplied by the facial nerve.

SUMMARY SHEET: THE EYE 1

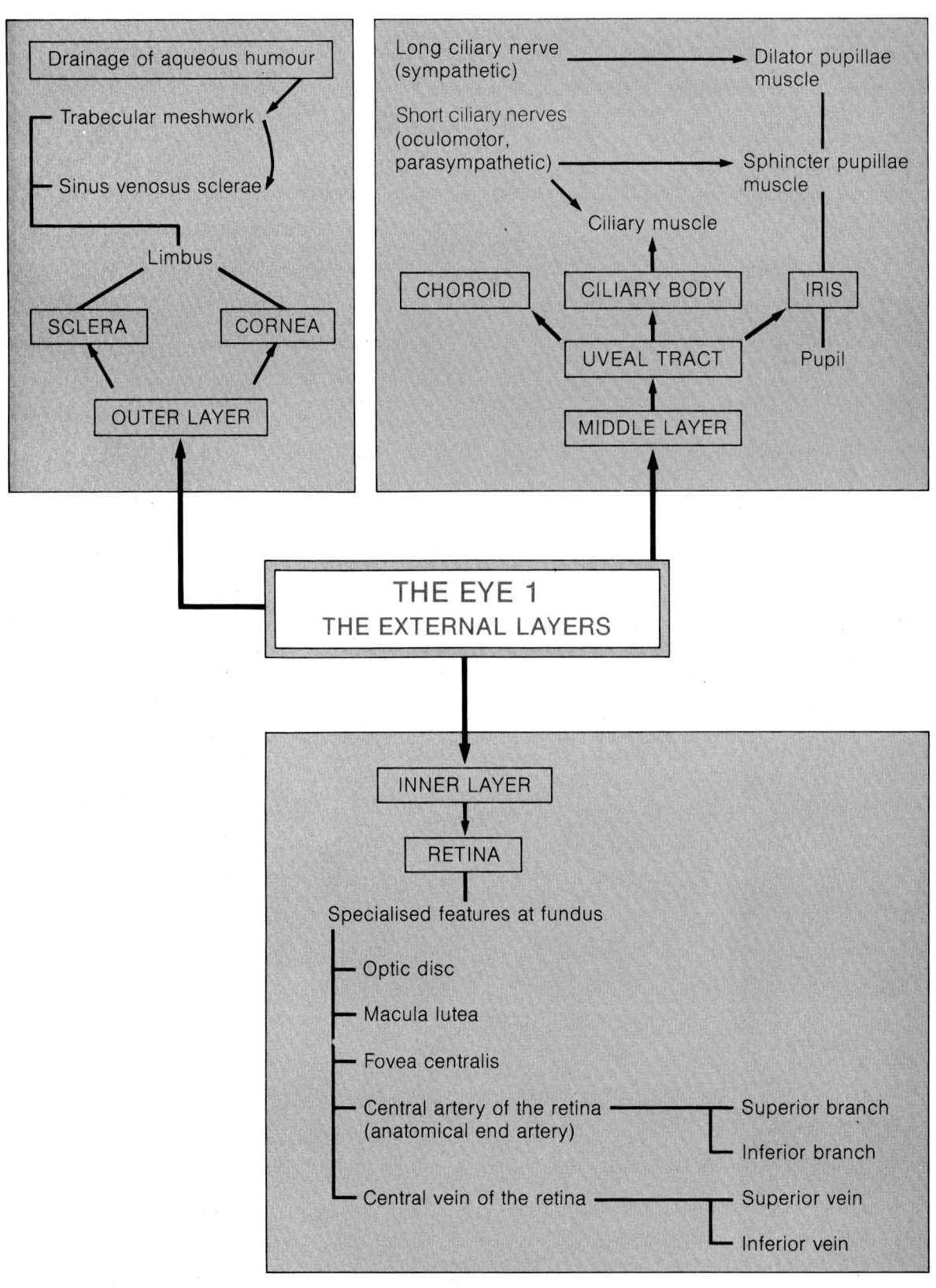

SUMMARY SHEET: THE EYE 2

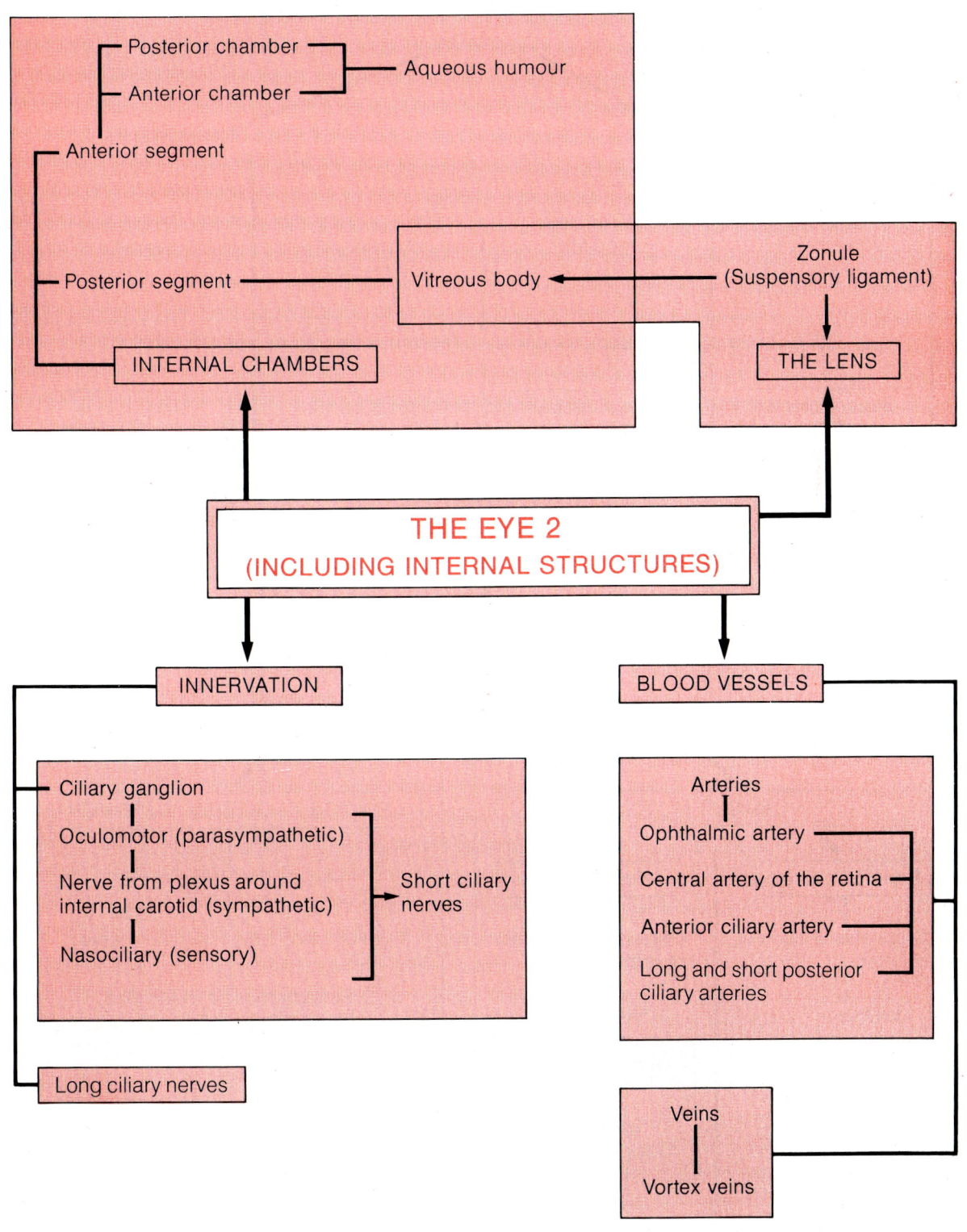

SUMMARY SHEET: THE ORBIT 1

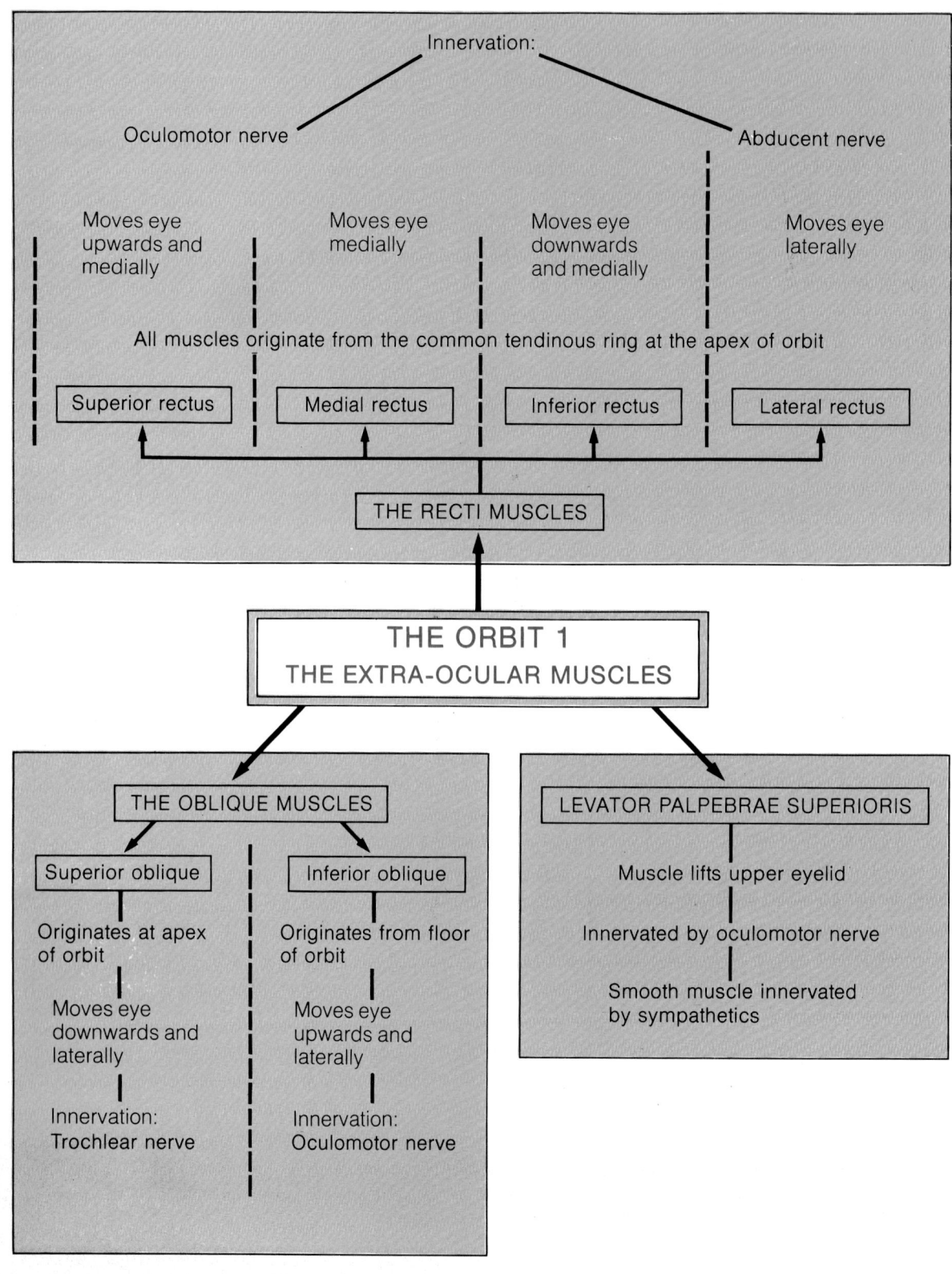

SUMMARY SHEET: THE ORBIT 2

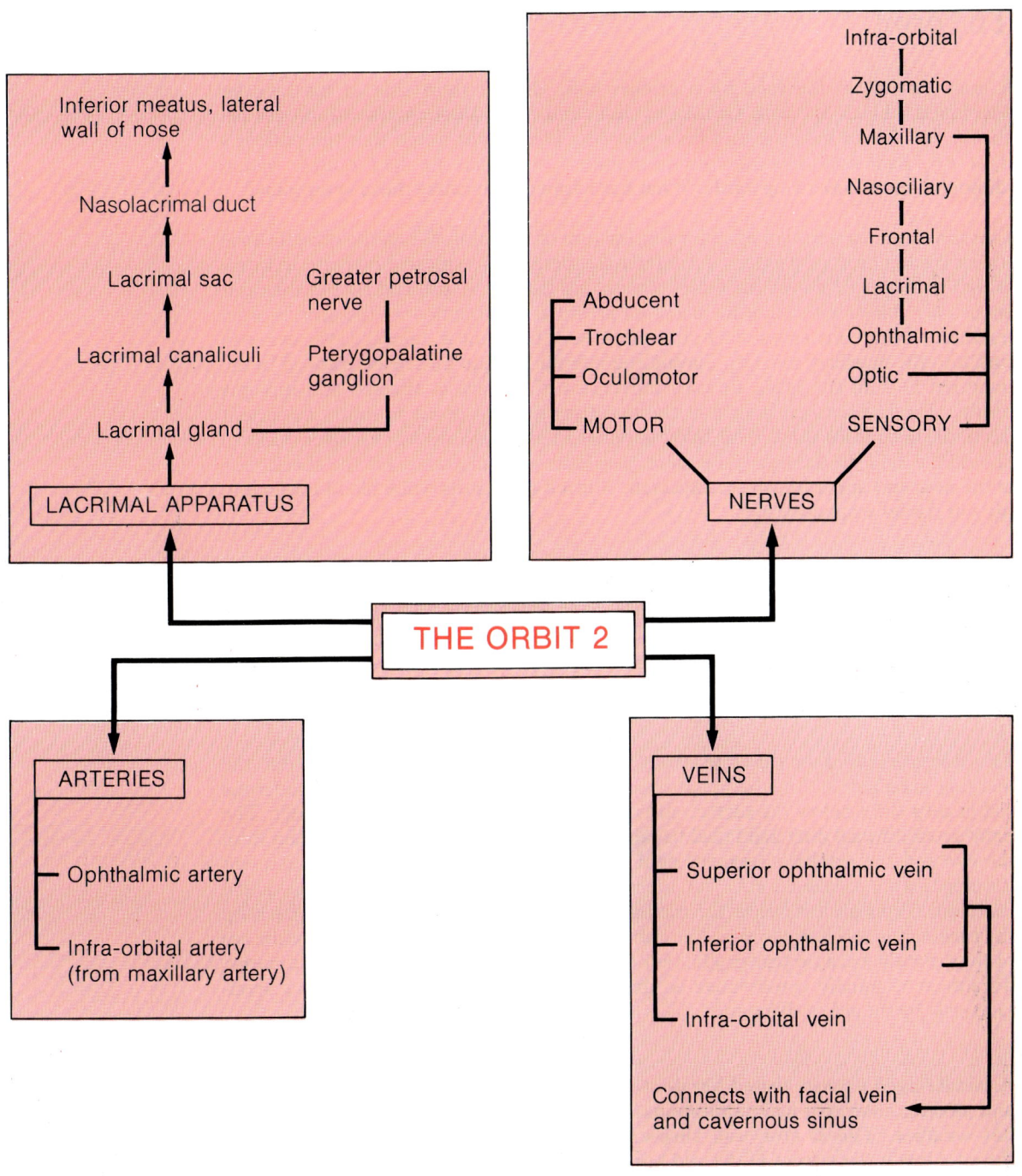

Case histories 4
The orbit

A man was hit in the face by a cricket ball. On the right side of his face he developed a black eye with swollen eyelids, and had a tingling sensation over his cheek, the side of his nose and his upper lip. The doctor noted that the patient's right eye had sunk inwards (enophthalmos), and that he had double vision (diplopia) when looking upwards. What has happened to cause these symptoms?

This is a classic description of a blowout fracture of the orbit. When the orbital rim is struck by an object of greater circumference, there is an explosive increase in intra-orbital pressure. Although the bones comprising the lateral wall and roof of the orbit are usually able to withstand the increased pressure, the thin floor of the orbit is often fractured. In rare instances where the suspensory ligament is damaged, the eye may sink into the maxillary antrum. Should the inferior oblique muscle be trapped in the fractured floor of the orbit, upward movement of the eye is affected and diplopia results. The infra-orbital branch of the maxillary nerve is prone to injury because it lies in the floor of the orbit. This nerve supplies an extensive area of skin over the face. Should the nerve be damaged, therefore, paraesthesia or anaesthesia of the cheek, side of the nose and upper lip will be expected. A black eye results when blood tracks into the soft tissues around the eye.

A mother visited her doctor with her baby who had a wet, sticky right eye. The doctor noted that there was a stagnant pool of tears in the affected eye and infection of the conjunctiva (conjunctivitis). When the doctor pressed just below the medial canthus of the eye, purulent material was expelled from the lacrimal punctum. The doctor suspected that the baby had a blocked nasolacrimal duct because of a developmental abnormality.

The nasolacrimal duct develops from the naso-optic furrow lying between the merging maxillary and medial nasal processes. From the furrow, a solid ectodermal rod of cells sinks below the surface and canalises to form the nasolacrimal duct. The baby's nasolacrimal duct had failed to canalise. Thus, tears were unable to drain from the eye and the resulting stagnant pool of tears became infected, with subsequent development of conjunctivitis. The infection tracked down to the lacrimal sac via the punctum. Pressure just below the medial canthus forced purulent material back up the canaliculus and into the eye. The baby would not develop symptoms before the age of four months because tear production does not start until the age of three months. Treatment is usually conservative as the nasolacrimal duct usually canalises spontaneously by the age of six months.

A middle-aged woman visited her doctor complaining that she had become irritable and anxious. She also had palpitations, tremor, weight loss, and an intolerance of hot weather. The doctor was suspicious that the patient was suffering from thyrotoxicosis (an excess of thyroid hormone), particularly because her eyes were markedly protruded (bilateral proptosis).

Patients with thyrotoxicosis may exhibit infiltration of lymphocytes and plasma cells into the fat, muscle and connective tissue within the orbit. There may also be inflammation and enlargement of the extra-ocular muscles. Thus, the intra-orbital contents increase in size and the globe of the eye is protruded. In extreme cases, the eyelids may be prevented from blinking and this leads to corneal ulceration.

A teenager received a cut near the corner of his right eye while playing football. He ignored it, and the injury became infected. Ten days later, he became seriously ill. He was drowsy and complained of headache, nausea and vomiting. His right eye was swollen and painful, and he had double vision when looking to the right. Can you explain the symptoms with reference to the original injury?

The patient has developed a right-sided cavernous sinus thrombosis. The angular vein at the side of the nose communicates with the superior ophthalmic vein in the orbit. The ophthalmic vein drains into the cavernous sinus. Infection has spread along these veins, producing a swollen, inflamed eye, and has caused thrombosis of blood in the cavernous sinus. The general symptoms indicate that the infection has spread into the middle cranial cavity. The abducent nerve is often affected as it is the only cranial nerve actually running through the sinus (the others invaginating the dural wall). The patient, therefore, has developed a lateral rectus palsy with subsequent diplopia on looking right.

A young girl was thrown from her pony and concussed. She was taken to hospital, where radiographs revealed a fracture of the base of the skull. The examining doctor observed that, although the girl had not received a blow to the face, there was pulsatile protrusion of the right eye. The conjunctival vessels were also grossly dilated and there was conjunctival oedema (chemosis). Using a stethoscope, the doctor heard a 'whooshing' sound (bruit) over the eye, which was synchronous with the arterial pulse. The sound was abolished by compression of the right common carotid artery in the neck. How might the clinical signs be explained in terms of the vascular connections of the orbit?

The fracture of the base of the skull was associated with rupture of the internal carotid artery as it passed through the foramen lacerum and the cavernous sinus. The end result was a carotico-cavernous fistula. The sudden increase of pressure within the cavernous sinus leads to massive dilation of the superior ophthalmic veins, and hence of the conjunctival veins. The arteriovenous communication also explains the pulsatile protrusion of the eye, which was abolished by compressing the common carotid artery in the neck.

A 'do-it-yourself' enthusiast fell from a ladder while replacing a window frame. He sustained a severe blow to his head and was rushed to hospital. During the examination, the doctor saw that the patient's level of consciousness was gradually decreasing. The patient's right upper eyelid was drooping (ptosis), the right eye had become abducted, and its pupil was fixed and dilated. A skull radiograph revealed that the squamous part of the right temporal bone was fractured. How might this be related to the clinical signs?

The fracture of the temporal bone was associated with damage to the middle meningeal artery. Extravasated blood had filled the potential space between the periosteal and meningeal layers of the dura mater, forming an extradural haematoma. This had resulted in an increase in intracranial pressure, which was responsible for the decreasing level of consciousness.

The symptoms associated with the eye were the consequence of damage to the oculomotor nerve. The ptosis was the result of partial paralysis of the levator palpebrae superioris muscle. The eye was in an abducted position because of the unopposed action of the lateral rectus muscle (supplied by the abducent nerve). Although it might be expected that the unopposed activity of the superior oblique muscle (supplied by the trochlear nerve) might cause the eye to be depressed, the muscle acts as a depressor only when the eye is in an adducted position. The pupil was fixed and dilated because of interference with the parasympathetic fibres travelling within the oculomotor nerve to the sphincter pupillae and ciliary muscles. The fracture of the temporal bone affected the oculomotor nerve because the increased intracranial pressure forced part of the temporal lobe of the brain through the opening in the tentorium cerebelli (the tentorial incisure), thereby compressing the oculomotor nerve as it passed through.

Malfunction of the oculomotor nerve can also follow from the development of an aneurysm of the left posterior communicating artery. This artery is found at the base of the brain and forms part of the vascular circle of Willis. As the oculomotor nerve is closely related to the inferior aspect of the posterior communicating artery, it can be compressed by the enlarging aneurysm.

A 60-year-old ex-marine walked into his doctor's surgery with a peculiar high-stepping gait. He complained that, when he had been washing his face earlier that morning, he had fallen backwards and hit his head. The doctor observed that the patient's pupils became constricted when he viewed a near object, but not when a light was shone in them. What neural pathways have been affected?

This man has an abnormality of the pupil known as an Argyll Robertson pupil, which is the result of syphilitic damage to the pretectal nuclei in the midbrain. Although the accommodation reflex is unaffected and reflex pupillary constriction occurs when the patient focuses on a near object, the normal reflex constriction of the pupils produced by shining a light into either eye is absent (see pages 487 to 488). The peculiar gait results from damage of the dorsal columns in the spinal cord and subsequent loss of proprioception (joint position sense). Loss of proprioception results in loss of balance when the eyes are closed. This explains why the patient fell backwards when washing his face.

A woman visited her optician complaining of progressive loss of sight in her left eye over the past year. The optician discovered that the patient was virtually blind in that eye, and that the eye was slightly protruded. The ophthalmologist to whom she was then referred further observed that her optic disc was pale and swollen. There was also some restriction of movement when the gaze was directed upwards. Subsequent radiological examination demonstrated the presence of a swelling of the optic nerve close to the apex of the orbit, and an enlargement of the optic canal. The swelling was found to be a tumour of the optic nerve sheath called a meningioma.

The optic nerve can be regarded as an extension of the brain and possesses an outer sheath of pia, arachnoid and dura mater. A meningioma is thought to arise in the arachnoid layer. The tumour causes blindness as a result of compression of the optic nerve. The pale, swollen optic disc is caused by the tumour occluding the central vessels of the retina. The optic nerve takes a sinuous course as it approaches the eye, and there is normally approximately 5 mm of slack. The loss of this slack resulting from the protrusion of the eye and the fibrosis caused by the tumour limits eye movements.

A 25-year-old man known to suffer from sarcoidosis (a nonspecific granulomatous disease affecting exocrine glands), complained of a lump in his right upper eyelid and a gritty sensation in the right eye. The swelling was painless and had been enlarging over the past three months. What might the swelling be caused by?

The patient has an enlarged lacrimal gland, a fairly common occurrence with sarcoidosis. Should the secretion of tears be reduced, a dry eye results and the patient complains of a gritty sensation. It is possible to distinguish between involvement of the orbital and palpebral parts. If the orbital part of the gland is affected, the swelling will be apparent on the face at the lateral aspect of the eyelid. If the palpebral portion is preferentially involved, this will be evident when the upper eyelid is everted. Before attempting eversion of the upper lid, the doctor will instruct the patient to look downwards. This places the sensitive cornea in a less vulnerable position and relaxes the superior rectus and the levator palpebrae superioris muscles so that the superior edge of the tarsus is not tethered during the eversion.

A 60-year-old woman, accustomed to smoking 20 cigarettes a day, had several episodes of brief (lasting one to two minutes) total loss of vision in her right eye (amaurosis fugax). She presented in the surgery during an episode that had already lasted for six hours. Examination of the fundus revealed a pale, creamy retina with a small cherry-red spot at the centre of the macula lutea.

The amaurosis fugax is caused by intermittent interference in the blood flow to the retina. This has culminated in a more prolonged episode of retinal ischaemia caused by occlusion of the central artery of the retina. Smoking predisposes to arterial degeneration and thrombosis. The cherry-red spot represents the normal appearance of the underlying choriocapillaris seen through the thin, capillary-free zone of the fovea centralis. This is now more apparent because of the contrast with the relatively ischaemic retina.

A steel-worker was taken to hospital after a fragment of metal flew into his eye. The eye was painful and watery. Following the administration of a topical anaesthetic, removal of the foreign body was attempted using a small needle. During the procedure, however, the doctor suspected that she had perforated the cornea. Why was the anaesthetic required and what sign might help the doctor assess the possibility of a perforated cornea?

The cornea has a rich sensory nerve supply. The innervation is derived from the ciliary branches of the ophthalmic division of the trigeminal nerve, particularly the long ciliary nerves. It was therefore necessary to apply topical anaesthetic to the cornea before attempting to remove the metal fragment. The reflex lacrimation that occurs following the introduction of a foreign body into the eye is mediated via parasympathetic fibres from the greater petrosal nerve, which relay in the pterygopalatine ganglion.

The central portion of the cornea is only about 0.5 mm thick and is therefore at risk of being perforated when a foreign body is removed. Small perforations are difficult to detect. As the anterior chamber of the eye is filled with aqueous humour at an intraocular pressure of 10 to 20 mm Hg greater than atmospheric pressure, fluid will quickly seep through any breach in the cornea. If dye (e.g. fluorescein) is dropped over the suspected site of perforation, it will wash away and dilute in the aqueous flow. This test is called the Seidel test.

A teenager, known to suffer from chronic sinusitis, visited her doctor complaining of a four-week history of gradual deterioration in the sight of her right eye. During the past week she had suffered with headaches, malaise and a purulent discharge from her right nostril. Although the visual field in her right eye was not grossly altered, there was a marked decrease in central visual acuity. Radiographs revealed a thickening of the mucosa of the right ethmoidal air sinuses resulting from inflammation, and erosion of the posteromedial aspect of the orbital wall caused by the presence of a mucocele (a cyst containing mucus) in the right sphenoidal sinus. How might this cyst be responsible for the symptoms reported?

The optic nerve runs in the optic canal within the sphenoid bone. The canal is closely related to the sphenoidal sinus, lying at the junction between the roof and lateral wall of the sinus. A cyst within the sphenoidal sinus can therefore erode the bone and compress the optic nerve. Early compression of the optic nerve affects primarily the centrally placed fibres that subserve central vision. The sphenoidal sinus drains into the spheno-ethmoidal recess of the nose. A sphenoidal cyst may consequently discharge into the nasal cavity and from there infection may spread to the other sinuses (in the present case the ethmoidal sinuses). Pathologies can spread easily from the paranasal sinuses into the orbit. Infection spreading to the orbit may then pass to the cavernous sinus by means of the ophthalmic veins.

An elderly man complained of distorted vision in one eye. His central vision was slightly blurred and straight lines seemed to be bowed and irregular. Ophthalmoscopy revealed a small, greyish-green area close to the fovea centralis. Several days later, there was a sudden, severe drop in visual acuity. Ophthalmoscopy now revealed a circular, elevated area, dark red in colour, which was clearly a subretinal haemorrhage at the posterior pole of the affected eye.

The underlying cause of the visual defects was small deficiencies in the basal membrane (Bruch's membrane) that separates the capillary layer of the choroid (choriocapillaris) from the retina. These deficiencies allowed vessels from the choroid to invade the pigmented epithelium of the retina, forming a subretinal, neurovascular membrane. This membrane produced the small, greyish-green area initially seen on the retina. The vessels elevated the retinal epithelium away from Bruch's membrane, causing distortion of the photoreceptors, with the resultant visual distortion. These new vessels are prone to haemorrhage, and this led to the more dramatic lesion that occurred later.

A 55-year-old woman presented to her doctor with a history of sudden blurring of central vision in one eye. This was described as a 'cobweb' in front of the eye, which the patient kept wanting to brush away. After a few days, this symptom disappeared but the woman was now aware of a large 'floater'—an opacity that seemed to obstruct her central vision. This seemed to dart away when she attempted to look at it. In addition, she experienced 'flashes of light' (photopsia) which were most apparent in the dark and on sudden movement of the eye. Her doctor noted that visual acuity was good and that the visual field was normal. Ophthalmoscopy revealed that, although the optic disc and macular region appeared unaffected, a slightly irregular ring-shaped opacity was seen 'floating' close to the disc. This was casting a shadow on the underlying retina.

This patient has the typical symptoms of a posterior vitreous detachment. With age, the vitreous body liquefies (syneresis). The internal limiting membrane of the retina may split, and the surface of the vitreous body may be pulled away from the retina. Frequently, the residual mass of solid vitreous sinks down, only maintaining a firm attachment at three sites—the anterior peripheral retina, the ora serrata and the pars plana. The ring-shaped opacity is the detached rim of the vitreous body at the optic disc (a Weiss ring). The photopsia is caused either by areas of vitreoretinal adhesion and traction, or by the residual vitreous body moving against the retina and stimulating the sensation of light.

A man was arrested by the police for unruly behaviour, including urinating in public. Realising that the man had a medical problem, the police took him to hospital. On questioning the man's wife, she reported that for the past six months her husband had been bothered by headaches and that he frequently bumped into obstacles situated on his right-hand side. Within the past three months, there had been a marked change in the man's behaviour. He had become very bad-tempered and rude, and would act in an unpredictable and bizarre manner. During examination, the casualty officer found that the man was almost blind in his right eye. Ophthalmoscopy revealed a pale optic disc that was indicative of optic nerve

atrophy. In the left eye, the optic disc was red and engorged, and had a blurred margin (papilloedema). The doctor diagnosed a space-occupying lesion in the orbital surface of the frontal lobe of the brain (Kennedy's syndrome). How could the pathology in the frontal lobe of the brain explain the signs and symptoms?

The orbital surfaces of the frontal lobes of the brain are thought to be concerned with the elaboration of social and emotional behaviour. Therefore, the presence of a lesion within this part of the brain may explain the strange behavioural patterns seen in the patient. As the optic nerve lies immediately beneath the frontal lobe, enlargement of the lobe can compress the nerve and occlude the central artery of the retina. This results in destruction and ultimately fibrosis of the optic nerve (gliosis), signified by a pale optic disc. In the present case, the right frontal lobe was enlarged, accounting for the blindness in the right eye.

The papilloedema in the left eye is an indirect effect of the brain lesion. As the brain is enclosed in a rigid cavity, any space-occupying lesion will increase the intracranial pressure. This will cause an increase in the pressure of the cerebrospinal fluid. As the optic nerves are surrounded by projections of the meninges containing cerebrospinal fluid, the vessels within the nerves would be compressed, giving rise to papilloedema in both eyes. In the case described here, however, the vessels in the right eye have already been occluded, and papilloedema is therefore evident only in the left eye.

A patient who had received a blow to the head visited his doctor with a blood-red patch on his conjunctiva. Certain anatomical features allow a differentiation between conjunctival and subconjunctival haemorrhage. What are they?

Conjunctival haemorrhages follow local trauma to the eyeball. The capillaries in the conjunctiva are ruptured and a circumscribed haematoma is formed. All its boundaries are clearly visible.

Subconjunctival haemorrhages differ in having no visible posterior boundary. This can be explained on anatomical grounds. The conjunctiva covers the sclera anteriorly and is reflected over the eyelids at the fornix. Consequently, subconjunctival haemorrhages must originate in the orbit and track underneath the conjunctiva from the side of the eyeball.

Head injuries with disruption of the anterior or the middle cranial fossa may result in bleeding into the orbit and the formation of a subconjunctival haemorrhage. For the patient, therefore, the clinical diagnosis of a subconjunctival haemorrhage will have graver consequences than that of a conjunctival haemorrhage.

A man was kicked in the eye. Because of gross lid swelling and bruising, an immediate examination of the eye was difficult. When the patient was seen a few days later, he complained of blurred and distorted vision in the affected eye. The external globe had a red elevated patch with defined margins, indicative of a conjunctival haemorrhage resulting from the trauma. The cornea appeared undamaged and no blood was seen in the anterior chamber of the eye. Ophthalmoscopy revealed a normal fundus. However, when the ophthalmoscope was focused on the region of the iris, a curved line which 'caught' the light was observed within the pupil. Displacement of the lens (subluxation) was diagnosed.

Trauma to the eye can displace the lens because of disruption of all, or part of, the zonule. This affects accommodation, and results in partial or complete loss of focusing. The curved line seen in the pupil indicates the margin of the displaced lens.

Subluxation of the lens may occur without trauma in certain medical conditions (e.g. Marfan's syndrome and homocystinuria). One complication of subluxation is 'pupil-block' glaucoma. Aqueous humour is produced in the posterior chamber of the eye and its natural passage is forwards through the pupil into the anterior chamber. If this passage is blocked by the lens, intra-ocular pressure will rise.

A far-sighted (hypermetropic) woman of short stature visited her doctor. She had been developing photographs in her dark-room three hours earlier when she had felt some discomfort in her right eye. Her vision had become 'misty' in that eye and she was aware of rainbow-like haloes in her field of vision. This was followed by severe frontal headache and nausea. On examining her right eye, the doctor noted that the pupil was partly dilated and that the cornea was slightly opaque. The patient was diagnosed as having an acute condition called angle-closure glaucoma, caused by blockage of the sinus venosus sclerae. Can you explain the underlying reason for the condition?

The axial length of the eyeball is correlated with the height of an individual. Hence, tall people are often short-sighted (myopic), as light from distant objects is brought to focus in front of the retina. Conversely, most far-sighted individuals are of short stature; their eyes may also have a shallow anterior chamber. In a shallow anterior chamber, the base of the iris may occasionally block off the filtration angle during pupillary dilation. This occurred in the case described here when the patient entered her dark-room. Dilation of the pupil blocked off the trabecular meshwork and prevented filtration of aqueous humour. The resulting accumulation of aqueous humour led to a rapid rise in intra-ocular pressure with oedema of the cornea. The altered hydration of the cornea affected its translucency and therefore its optical characteristics. This accounted for the visual symptoms and the opacity of the cornea which may appear a hazy greyish-green ('glaucos'—the colour of the sea).

Other symptoms associated with angle-closure glaucoma include stabbing eye pain, photophobia, lacrimation and vomiting. The pupil may be mid-dilated and vertically oval (perhaps in association with ischaemia of the iris). Rapid treatment with a miotic drug such as pilocarpine can bring about immediate relief by allowing the pupil to constrict and thereby pull away from the filtration angle. A surgical cut in the peripheral iris (peripheral iridectomy) may be subsequently performed to prevent further attacks.

A man was struck in the front of the eye by a direct blow from a squash ball. The force of impact was not deflected by the bony rim of the orbit. He visited his doctor shortly afterwards, by which time the vision in his affected eye was reduced to the perception of light and movement only. His iris was misshapen and, at the superior aspect of the limbus, some pigmented tissue resembling the iris was pouting through. The iris was obscured inferiorly by a crescent of blood. What may have happened to give rise to the symptoms?

Most direct blunt blows strike the inferolateral aspect of the eyeball. Force is then transmitted across the front of the eye, resulting in a rupture at its weakest point superomedially. This occurs at the limbus which is a weak point because of the proximity of the trabecular meshwork and the sinus venosus sclerae. In the present case, the ciliary body has herniated through the ruptured limbus. The later loss of vision and the presence of blood in front of the iris is the result of bleeding into the anterior chamber of the eye (hyphaema).

The nose

The nose is the upper part of the respiratory tract. It occupies the midline region of the head between the oral cavity and the cranial cavity. The nose opens on the face at the nostrils (anterior nares). Behind, it opens into the nasopharynx through the choanae (posterior nares).

The part of the nose that projects from the face is called the external nose. The nasal cavity internally is divided into two nasal fossae by the nasal septum. Prominent processes project from the lateral walls of the nasal cavity (the superior, middle and inferior nasal conchae). The lateral walls are also significant for being the sites of drainage of the ethmoidal, frontal, maxillary and sphenoidal air sinuses and of the nasolacrimal ducts.

The functions of the nose are primarily those concerned with ventilation and olfaction. Inspired air is filtered and debris is removed in a film of mucus. The air is also warmed and humidified at the nose. Another function of the nose is to act as a resonating chamber and thus to play a role in speech.

At the back of the nose, and below the apex of each orbit, lie the pterygopalatine fossae.

THE EXTERNAL NOSE

110; 126

110,2

110,4; 110,3
110,7
110,5; 110,27

110,6
110,8

The external nose is essentially pyramidal in shape. Superiorly, it is confluent with the forehead at the root of the nose where it is supported by the nasal bone at the bridge. This part of the external nose is immobile. Inferiorly, the tip of the nose is called the apex. The ridge connecting the root and apex is referred to as the dorsum of the nose. The apex and the dosum are supported by cartilage and are mobile. The anterior nares are separated by a septum which joins the apex of the nose to the philtrum of the upper lip. The anterior nares lead into the nasal vestibule. Here, the skin is characterised by having coarse hairs. The flared lateral margins of the external nose are known as the alae. The nasolabial grooves of the upper lip continue around the alae to form alar grooves.

126B,27; 54,13
54,1; 126B,26
54,14
126A,22
126A,5; 126A,6
126A,21

126B,29; 126B,30

126B,31

126B,32

The skeleton of the external nose comprises both bony and cartilaginous elements. The bony elements lie near the bridge of the nose and consist of the frontal processes of the maxillae, the nasal part of the frontal bone and the nasal bones. The cartilaginous elements form the lower part of the external nose. In the midline is the septal cartilage. This is quadrangular and fits posteriorly into the notch between the perpendicular plate of the ethmoid and the vomer. The septal cartilage is also attached to the nasal crest of the maxilla and to the anterior nasal spine. The septal cartilage at the apex of the external nose is called the columella. The columella can also be defined as the tip of the mobile part of the septum. The lateral surfaces of the external nose are supported by the lateral (upper) nasal cartilages and the greater (lower) nasal cartilages. The lateral nasal cartilages are triangular plates. They are united in the midline with each other and with the septal cartilage. The greater nasal cartilages are thin and C-shaped; the medial and lateral edges are called crura. These cartilages extend from the apex of the nose into the alae to maintain the patency of the anterior nares. The greater nasal cartilages are joined to each other, and to the septal and lateral nasal cartilages, by fibrous tissue. Fibrous tissue also joins the back of the greater nasal cartilage to the maxilla. Some lesser alar cartilages may be found here.

112,9; 112,10
112,11

The muscles of the external nose belong to the muscles of facial expression. They comprise procerus, compressor naris, dilator naris, and depressor septi. The levator labii superioris alaeque nasi muscle can also be included in this group, although it is usually considered with the oral group of facial muscles. The nasal group of facial muscles is described fully on page 153.

The cutaneous innervation of the external nose (see Figure 49; page 160) is derived from branches of the ophthalmic and maxillary divisions of the trigeminal nerve. Supratrochlear and infratrochlear branches of the ophthalmic nerve supply skin in the region of the root, bridge and upper part of the side of the nose. The external nasal branch of the ophthalmic nerve runs in a groove on the internal surface of the nasal bone to supply skin in the lower part of the nose around the midline (including the apex). The infra-orbital branch of the maxillary nerve (nasal branch) supplies the side of the lower part of the external nose.

The blood supply to the external nose is derived from three sources: the facial, ophthalmic and maxillary arteries. The facial artery gives off a lateral nasal branch (from the angular artery), and a septal branch (from the superior labial artery). The ophthalmic artery provides a dorsal nasal artery which, appearing above the medial palpebral ligament of the eyelid, supplies skin on the dorsum of the nose. An additional branch from the ophthalmic artery is the anterior ethmoidal artery. This terminates as the external nasal artery. The maxillary artery gives off an infra-orbital artery which provides nasal branches. Venous blood drains into the distal part of the facial vein (the angular vein) and into the ophthalmic veins. The lymphatics of the external nose drain into the submandibular group of lymph nodes.

118C,35

THE NASAL CAVITY

The nasal cavity is divided in the midline by the nasal septum, forming two nasal fossae. Because the septum often deviates to one side at the vomero-ethmoidal suture, the nasal fossae are frequently of unequal size. The floor of the nasal fossa is wider than its roof, giving the fossa a pear-shaped outline in cross-section.

10,19

The osteology of the nasal cavity

55; 60–62

The anterior nasal aperture (piriform aperture) is bounded mainly by the maxillary bones. The nasal bones form the superior margin of the aperture.

55,12
55,14

The nasal fossa must remain patent for ventilation. The patency is maintained by the rigidity of the bony walls. Each nasal fossa has a roof, a floor, a lateral wall and a medial wall.

The medial wall is formed by the nasal septum. It comprises the septal cartilage anteriorly, the perpendicular plate of the ethmoid bone posterosuperiorly, and the vomer postero-inferiorly. At the base of the nasal septum is the nasal crest formed by the maxillary and palatine bones.

30; 126A
126A,22; 30,27
30,26; 30,34
30,35

The bones that comprise the remaining walls of the nasal fossa are the ethmoid, frontal, lacrimal, maxillary, nasal, palatine, sphenoid, and vomer bones, and the bone of the inferior concha.

The roof of the nasal fossa is formed centrally by the cribriform plate of the ethmoid bone. Anteriorly lies the nasal bone and the nasal spine of the frontal bone. Posteriorly is located the body of the sphenoid, overlapped by the ala of the vomer and the sphenoidal process of the palatine bone. The cribriform plate transmits olfactory nerves into the roof of the nasal cavity and also the anterior ethmoidal nerves and vessels.

30,28; 32,31
30,32; 32B,50
30,21
30,28

The floor of the nasal fossa is formed by the palatine process of the maxilla anteriorly and the horizontal plate of the palatine bone posteriorly.

32,41
32,40

The lateral wall of the nasal fossa consists mainly of the medial surface of the maxilla, with the large maxillary hiatus being reduced in size by the overlapping of the lacrimal and ethmoid bones above, the palatine bone behind and the inferior concha below. From the lateral wall project the three nasal conchae. The superior and middle conchae are part of the labyrinth of the ethmoid bone. The inferior

60
62A; 62B

60A,1
60A,2; 60B,13

60A,3	concha is a separate bone. The region above and behind the superior nasal concha is called the spheno-ethmoidal recess. The region between the superior and middle nasal conchae is the superior meatus. Between the middle and inferior conchae is located the middle meatus. Below the inferior nasal concha is the inferior meatus.
60A,4	
32B,35	
32B,47	
32B,43	
62C–E	The nasolacrimal canal on the lateral wall of the nasal fossa is formed by the lacrimal groove of the maxilla articulating with the descending process of the lacrimal bone and the lacrimal process of the inferior concha.
60A,5	The sphenopalatine foramen, which links the nasal cavity with the pterygopalatine fossa, lies high up in the posterior part of the lateral wall. It is formed by the notch between the orbital and sphenoidal processes of the perpendicular plate of the palatine bone articulating with the body of the sphenoid bone. This foramen transmits the nasopalatine and posterior superior nasal nerves and the spheno-palatine vessels.
32D,54	The lateral wall of the nose is noted for being the site of drainage of the paranasal air sinuses. The sphenoidal sinus drains into the spheno-ethmoidal recess. The posterior ethmoidal air cells drain into the superior meatus. The frontal and maxillary air sinuses and the anterior and middle ethmoidal air cells drain into the middle meatus. The opening of the nasolacrimal canal lies in the inferior meatus.
32D,55; 32D,63	
32D,45; 32D,62	
32C,53	
72	The posterior nasal apertures (choanae) link the nasal fossae with the nasopharynx. The bones that contribute to this region are the palatine, sphenoid and vomer bones. The posterior border of the vomer separates the two posterior nasal apertures. Each aperture is bounded below by the posterior border of the horizontal plate of the palatine bone, laterally by the medial pterygoid plate, and above by the body and vaginal process of the sphenoid bone and the ala of the vomer. A groove on the lower surface of the vaginal process of the sphenoid bone is converted into the palatovaginal canal by articulation with the upper surface of the sphenoidal process of the palatine bone. This canal transmits the pharyngeal branch of the pterygopalatine ganglion and the pharyngeal branch of the maxillary artery. The vomerovaginal canal lies between the upper surface of the vaginal process of the sphenoid bone and the ala of the vomer. The pharyngeal branch of the sphenopalatine artery passes through this occasional canal.
72,1	
72,12; 72,7	
72,4; 72,2	
72A,5	
72B,15	
72B,14	
126–129	**The appearance of the lateral wall of the nasal fossa**
	Many of the important features of the nose are found on the lateral walls of the nasal fossae.
	The inferior margin of the lateral wall is horizontal. The upper margin is curved, being highest at its middle (beneath the cribriform plate of the ethmoid) and sloping downwards anteriorly and posteriorly.
126C,36	Corresponding to the ala of the external nose is a small recess called the vestibule. This area is lined with skin which has coarse hairs (the vibrissae). The vestibule is bounded above by a ridge (the limen nasi) which represents the upper border of the greater nasal cartilage. Beyond the ridge, the mucosa of the nose is of the respiratory type. Above the vestibule, level with the middle meatus, is a shallow depression called the atrium. A curved ridge called the agger nasi may be found above the atrium. This ridge overlies the anterior part of the ethmoidal crest of the maxilla.
126C,35	
60B,9	
126C,44; 134B,20	Behind the vestibule and atrium are the nasal conchae. The superior nasal concha is the smallest and lies posteriorly in the upper part of the lateral wall. The middle nasal concha is joined anteriorly to the superior nasal concha and is approximately twice its length. The inferior nasal concha is the longest, extending forwards in front of the middle nasal concha.
126C,42; 134B,21	
126C,40; 134B,22	
	Below the conchae lie regions called the superior, middle and inferior meatuses. The area above and behind the superior nasal concha is called the spheno-ethmoidal recess.

The superior meatus is the narrow region below the superior nasal concha and above the posterior part of the middle nasal concha. Posterior ethmoidal air cells drain by one or more openings into the anterior part of the superior meatus.

126C,43
128C,24

The middle meatus lies between the middle and inferior conchae. There is a prominent bulge in its upper part called the ethmoidal bulla. The bulla represents the site of the middle ethmoidal air cells, their opening being found on or above the bulla. The bulla forms the upper boundary of a groove, the hiatus semilunaris. The groove is bounded below by a ridge related to the uncinate process of the ethmoid bone. The hiatus semilunaris runs forwards and upwards as a curved channel with a funnel-shaped extremity (the ethmoidal infundibulum). The frontal sinus may drain through a simple opening or into a more obvious channel called the frontonasal duct. The point of drainage may then be either in front of, above, or directly into the ethmoidal infundibulum. The anterior ethmoidal air cells drain separately into the ethmoidal infundibulum or into the frontonasal duct. The maxillary sinus drains further back in the hiatus semilunaris and may have one or two openings.

126,41; 128B,15
128B,11; 32D,61
128A,2
128B,12
128B,14; 32D,46
128C,19
128C,18
128B,13; 128C,21
128A,4

The inferior meatus lies below the inferior concha. Because the concha is arched, the meatus is highest in its middle. The anterior portion of the inferior meatus is the site of drainage of the nasolacrimal duct. The opening may vary from being rounded to slit-like. When slit-like, the opening is bounded above and in front by a fold of mucosa called the plica lacrimalis.

126C,39; 128B,16
128B,17

The spheno-ethmoidal recess is a small triangular region above the superior concha. It is closely related to the cribriform plate of the ethmoid and thus the olfactory nerves. The recess receives the opening of the sphenoidal air sinus.

126C,45; 128A,7
129D,25
128,8; 128,9

A small fold called the supreme nasal concha is found above the superior nasal concha in about 60% of cases. The area between this fold and the superior nasal concha is called the supreme meatus. It receives an opening from a posterior ethmoidal air cell.

127D,46
127D,47

The appearance of the medial wall of the nasal fossa

126A

This wall is formed by the nasal septum. The septum is relatively featureless, though a small opening may be visible anteriorly on each side close to the inferior border. This opening leads backwards into a small recess called the vomeronasal organ. This is vestigeal in man but has an olfactory function in some animals. A small cartilage (the vomeronasal cartilage) may sometimes be found in association with the vomeronasal organ.

126A,23

The appearance of the roof and floor of the nasal fossa

126A

The roof is highest centrally, where it is related to the cribriform plate of the ethmoid. The roof slopes downwards anteriorly beneath the frontal bone and posteriorly beneath the sphenoid bone.

126A,4

The floor of the nasal cavity is related to the hard palate and is relatively flat and horizontal. A slight depression of the mucosa may be seen overlying the incisive canal.

126A,19

The nasal mucosa

Above the superior concha, the mucosa is called olfactory mucosa because it contains the olfactory cells. In vivo, this mucosa has a distinct yellowish colour. In contrast, the remaining nasal mucosa is of the respiratory type (ciliated columnar) and appears pink. This is particularly thick and well vascularised at the free margins of the nasal conchae.

129D,25; 135,28

The innervation of the nasal cavity (Figures 80 and 81)

Special sensation related to olfaction is associated with the olfactory nerves (i.e. the first cranial nerves). General sensation to the nasal mucosa is related to branches from the ophthalmic and maxillary divisions of the trigeminal nerves (i.e. the fifth cranial nerves).

The olfactory epithelium is located in the roof of the nasal cavity, extending on to the lateral walls of the nasal fossae (above the superior nasal conchae) and the uppermost

129D,25; 135,28

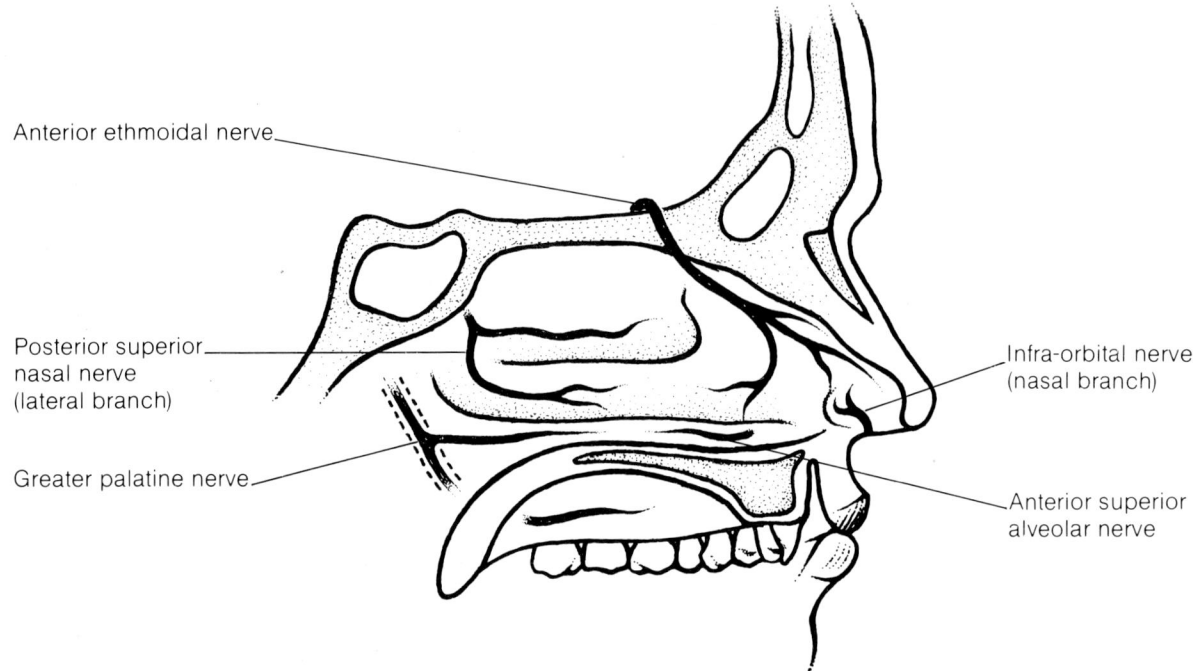

Figure 80 *The innervation of the lateral wall of the nose (excluding the olfactory nerve).*

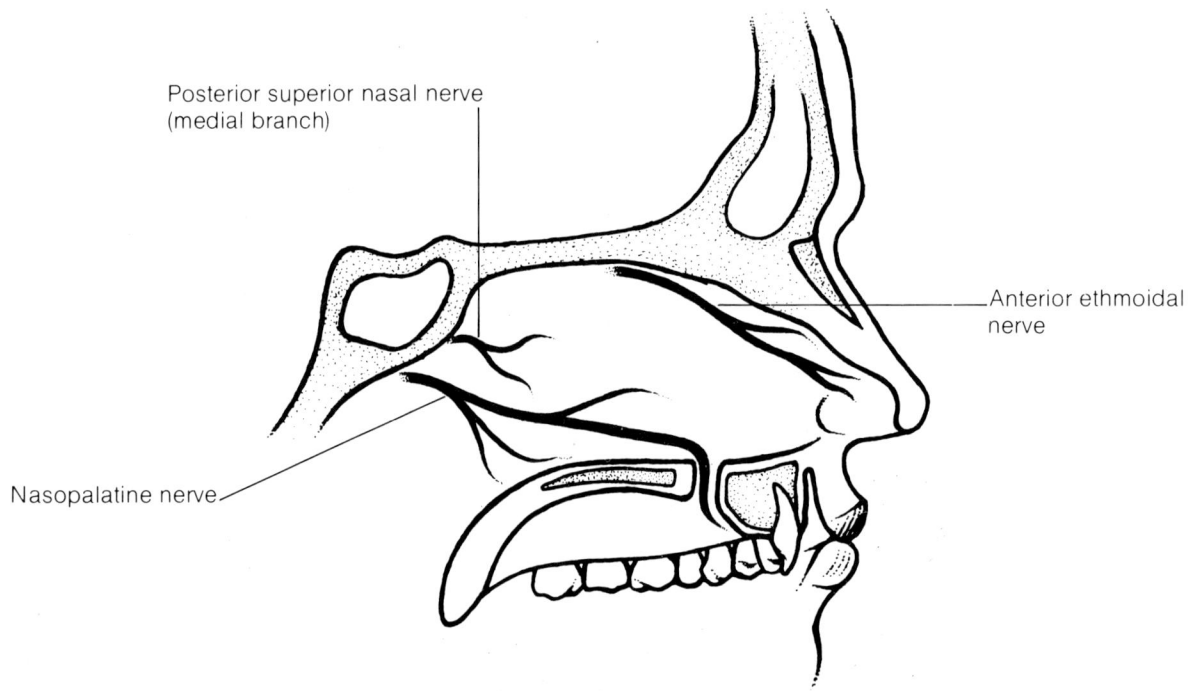

Figure 81 *The innervation of the nasal septum (excluding the olfactory nerve).*

part of the nasal septum. Filaments of the olfactory nerves (about 20 on each side) pass upwards through the cribriform plate of the ethmoid bone into the cranial cavity to synapse in the olfactory bulbs. Each filament is ensheathed by the meninges. Thus, a potential pathway exists for the spread of infection from the nose to the cranial cavity.

The anterior ethmoidal nerve is the only branch of the ophthalmic nerve that supplies the nasal mucosa. It arises from the nasociliary nerve and mainly supplies an area in front of the nasal conchae. (It also innervates the anterior extremities of the middle and inferior conchae.) After leaving the orbit through the anterior ethmoidal foramen, the anterior ethmoidal nerve enters the cranial cavity on to the cribriform plate of the ethmoid. It leaves the cranial cavity through a small slit near the crista galli and enters the roof of the nasal cavity. Here, the nerve runs in a groove on the inner surface of the nasal bone. The anterior ethmoidal nerve passes downwards and forwards and gives rise to lateral and medial internal nasal branches. The lateral internal nasal branches pass to the lateral wall of the nose, whereas the medial internal nasal branches run to the nasal septum. When the anterior ethmoidal nerve emerges at the inferior margin of the nasal bone it becomes the external nasal nerve (see page 160).

The maxillary nerve contributes many branches that supply the nasal mucosa. The infra-orbital and posterior superior alveolar nerves arise directly from the maxillary nerve. The posterior superior nasal, greater palatine and nasopalatine nerves arise indirectly by way of the pterygopalatine ganglion.

The infra-orbital nerve is the terminal branch of the maxillary nerve. After passing on to the face at the infra-orbital foramen, it provides a nasal branch that supplies the skin of the vestibule and the mobile part of the nasal septum. The anterior superior alveolar branch of the infra-orbital nerve also supplies nasal mucosa. Its nasal branch passes through a small canal in the lateral wall of the nose (below the level of the inferior concha) to innervate the anterior part of the inferior meatus and the adjacent part of the floor of the nose and adjoining nasal septum.

The posterior superior nasal nerve originates at the pterygopalatine ganglion. It enters the back of the nasal cavity through the sphenopalatine foramen and gives off lateral and medial branches. The lateral branches supply the posterosuperior part of the lateral wall of the nose around the superior and middle nasal conchae. The medial branches cross the roof of the nasal cavity to supply the septum overlying the posterior part of the perpendicular plate of the ethmoid bone.

The greater (anterior) palatine nerve also arises from the pterygopalatine ganglion. It descends in the greater palatine canal where it gives off posterior inferior nasal branches. These branches pass through small openings in the perpendicular plate of the palatine bone to supply the postero-inferior portion of the lateral wall of the nose (below and including the middle meatus).

The nasopalatine nerve passes from the pterygopalatine ganglion into the nasal cavity through the sphenopalatine foramen. It runs across the roof of the nasal cavity to reach the back of the nasal septum. It then passes downwards and forwards, lying in a groove on the vomer, to supply the postero-inferior part of the septum.

The floor of the nose is supplied anteriorly by the nasal branch of the anterior superior alveolar nerve and posteriorly by the nasal branches of the greater (anterior) palatine and by the nasopalatine nerves.

Autonomic fibres to glands and vessels in the nose are distributed with the above-mentioned branches of the maxillary nerve via the pterygopalatine ganglion. In addition, autonomic fibres are presumed to be distributed with the anterior ethmoidal nerve via the ciliary ganglion.

The arteries of the nasal cavity (Figures 82 and 83)

The general distribution of the arteries is similar to that of the nerves. The main vessels arise from the ophthalmic and maxillary arteries, with small contributions from the facial artery.

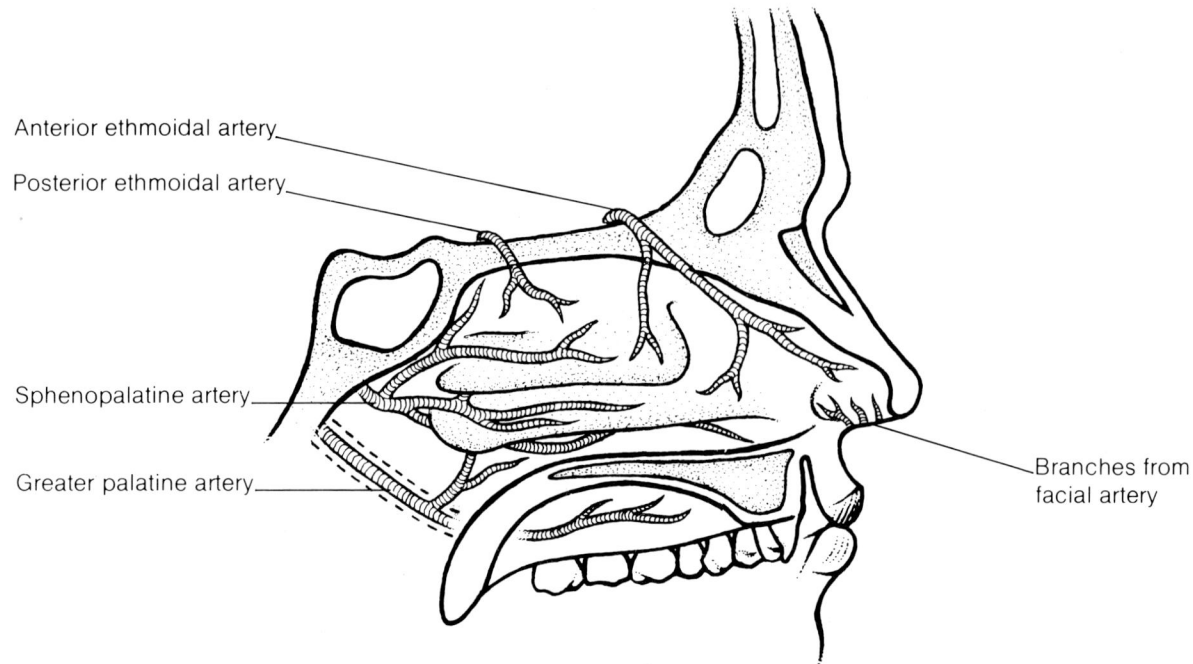

Figure 82 *The vasculature of the lateral wall of the nose.*

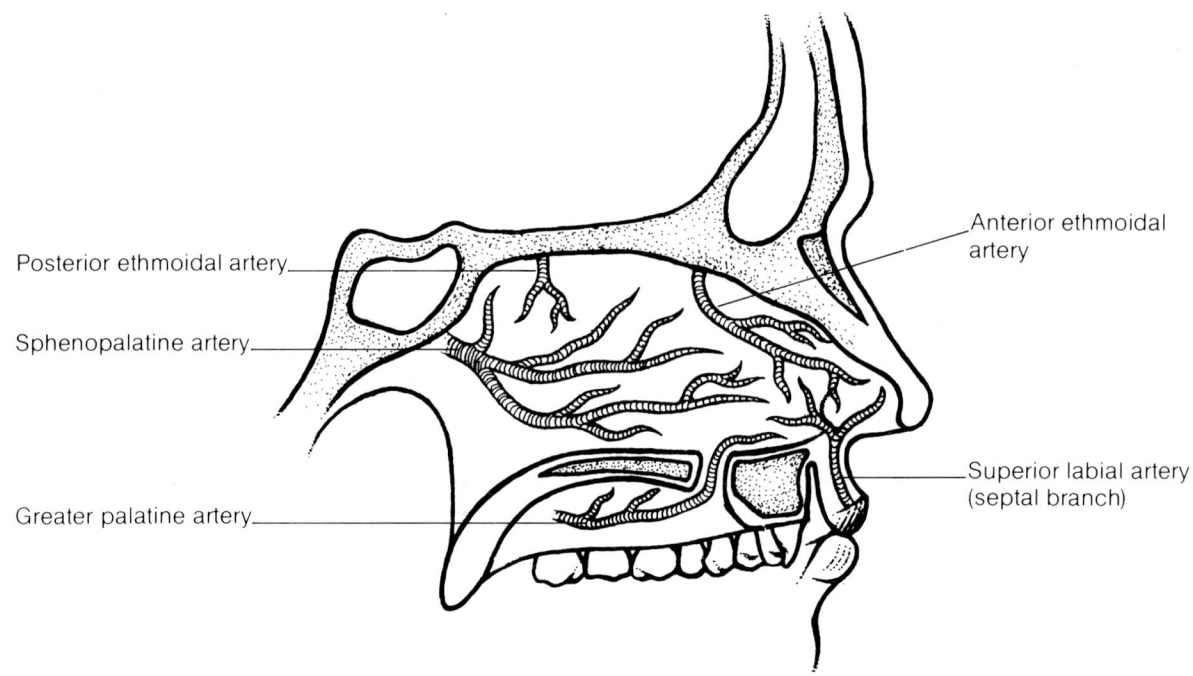

Figure 83 *The vasculature of the nasal septum.*

The ophthalmic artery provides anterior and posterior ethmoidal branches to the nasal cavity. The anterior ethmoidal artery accompanies the anterior ethmoidal nerve and supplies the anterior parts of the lateral wall and the nasal septum. Its terminal branch is the external nasal artery (see Figure 50). The posterior ethmoidal artery supplies a small part of the lateral wall of the nose around the superior nasal concha and the posterosuperior region of the nasal septum.

120C,35; 135,31

120B,29

The sphenopalatine artery is the terminal branch of the maxillary artery in the pterygopalatine fossa. It enters the lateral wall of the nose through the sphenopalatine foramen. The artery accompanies the posterior superior nasal nerve and gives off branches to supply much of the posterior part of the lateral wall of the nose. These branches lie within the middle and inferior nasal conchae for part of their course. The sphenopalatine artery crosses the roof of the nasal cavity to supply the postero-inferior part of the nasal septum.

129,26; 142A,3
60A,5
129,28

The greater palatine artery is also a branch of the maxillary artery in the pterygopalatine fossa. It accompanies the greater palatine nerve in the greater palatine canal and gives branches supplying the inferior meatus. After passing on to the hard palate through the greater palatine foramen, the greater palatine artery gives a branch that passes up through the incisive canal to supply the anterior part of the nasal septum.

142A,5

135,30

The superior labial branch of the facial artery is the main source of supply of the anterior part of the nasal septum.

112,18

The veins and lymphatics of the nasal cavity

There are conspicuous venous plexuses in the lateral walls of the nasal fossae and in the nasal septum (particularly inferiorly). Indeed, the plexuses are said to resemble plexuses in erectile tissue. Consequently, the nasal cavity is susceptible to blockage should the plexuses become engorged.

The veins draining the nose essentially correspond to the arteries. Veins from the posterior part of the nose generally pass to the sphenopalatine vein which runs back through the sphenopalatine foramen to drain into the pterygoid venous plexus. The anterior part of the nose is drained mainly by veins accompanying the anterior ethmoidal arteries, these veins passing to the ophthalmic or the facial veins.

60A,5

Lymphatics in the anterior part of the nose drain into the submandibular lymph nodes. The lymphatics posteriorly drain into the upper deep cervical nodes.

The paranasal air sinuses

The paranasal air sinuses are invaginations from the lateral wall of the nose extending into the surrounding bones. There are four sets of paired sinuses: frontal, ethmoidal, sphenoidal and maxillary. The sinuses are lined by respiratory epithelium. There is considerable variation in the morphology of the sinuses from individual to individual, and between the sinuses of each side. The precise function of the paranasal sinuses is unknown, although some believe that the sinuses lighten the skull and add resonance to the voice. It is conceivable, however, that they simply reflect the considerable growth of the bones in which they are situated.

THE FRONTAL AIR SINUSES

130A–C

The frontal sinuses lie in the frontal bone above and behind the superciliary arches. A sinus can extend into the medial part of the roof of the orbit. In such cases, a thin layer of bone often separates the sinus from the floor of the cranial cavity and from the roof of the orbit. The frontal sinuses are frequently of unequal size, the larger sinus sometimes extending across the midline. The bony septum separating the frontal sinuses

30,31; 38C,18
126,1; 128,1
134B,17

is also often asymmetrically positioned. Each sinus may be partially subdivided by additional septa. The anterior ethmoidal air cells may encroach into the frontal sinuses.

128C,19; 32D,63

The frontal sinus drains into the middle meatus of the lateral wall of the nasal fossa, either through a simple opening or as a channel called the frontonasal duct. The point of drainage varies. The frontal sinus can drain directly into the ethmoidal infundibulum of the hiatus semilunaris. More commonly, however, the ethmoidal infundibulum is blind-ended and the frontal sinus drains in front of or above the infundibulum.

The frontal sinus is innervated by the supra-orbital branch of the ophthalmic nerve. The arteries supplying the sinus are the supra-orbital and anterior ethmoidal branches of the ophthalmic artery. Venous drainage is into adjacent veins, including an anastomotic vein joining the supra-orbital and superior ophthalmic veins. Lymph vessels drain into the submandibular group of nodes.

THE ETHMOIDAL AIR SINUSES

130D–F

40,1; 134A,12

The ethmoidal air sinuses occupy the two lateral masses (labyrinths) of the ethmoid bone between the lateral wall of the nose and the medial wall of the orbit. The orbital wall is particularly thin.

The walls of the ethmoidal air sinuses are incomplete and are covered by adjacent bones (i.e. frontal, lacrimal, maxillary, sphenoidal and palatine bones). Furthermore, the sinuses may not be restricted to the ethmoid bone, but may encroach into the frontal, maxillary and sphenoidal air sinuses and into the middle nasal concha, uncinate process and agger nasi of the nose.

Each sinus is subdivided into a number of air cells (between three and 18). The air cells are separated from each other by thin, incomplete, bony septa. Three groups of air cells are usually found: anterior, middle and posterior. However, some anatomists combine the anterior and middle air cells as a single anterior group.

128C,18

The anterior ethmoidal cells occupy the anterior portion of the lateral mass of the ethmoid bone. There are approximately 11 air cells in this group. They drain into the middle meatus of the lateral wall of the nose via the frontonasal duct or the ethmoidal infundibulum.

128A,2; 32D,62

The middle ethmoidal cells produce the ethmoidal bulla in the middle meatus of the lateral wall of the nose. There are usually three middle ethmoidal air cells. They drain on or above the bulla.

128C,24; 32D,55

There are about six posterior ethmoidal cells. They lie in the posterior portion of the lateral mass of the ethmoid bone and drain into the superior meatus of the lateral wall of the nose.

The anterior and middle ethmoidal cells are innervated by the anterior ethmoidal branch of the ophthalmic nerve. The posterior ethmoidal cells are supplied by lateral branches of the posterior superior nasal nerve from the pterygopalatine ganglion and by the posterior ethmoidal branch of the ophthalmic nerve.

The ethmoidal air sinus is supplied by the lateral posterior nasal branches of the sphenopalatine artery. In addition, the anterior and posterior ethmoidal air cells are supplied respectively by the anterior and posterior ethmoidal branches of the ophthalmic artery. The veins that drain the sinuses correspond to the arteries. The sphenopalatine vein passes to the pterygoid venous plexus. The anterior and posterior ethmoidal veins join the superior ophthalmic vein. Lymphatics from the anterior and middle air cells drain into the submandibular nodes, whereas those from the posterior cells drain into the retropharyngeal nodes.

The ethmoidal vessels and nerves cross the roof of the ethmoidal air sinus; the posterior ethmoidal cells are closely related to the optic canal.

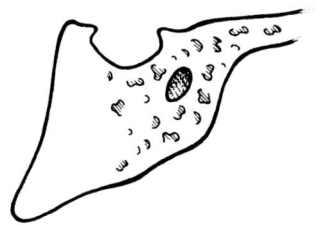 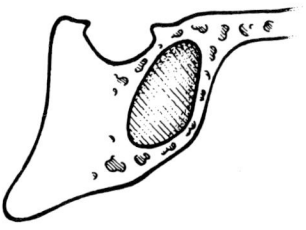 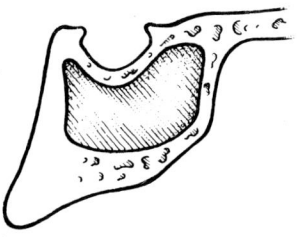

Figure 84 *Variations in the size of the sphenoidal air sinus.*

THE SPHENOIDAL AIR SINUSES

132A–E

These two sinuses lie within the body of the sphenoid bone at the back of the nasal cavity. They are separated by a bony septum which is usually positioned asymmetrically. The size of the sinuses varies considerably (Figure 84). Although usually limited to the body of the sphenoid, the sinuses can be found within the greater wings and pterygoid processes of the sphenoid, and even within the basilar part of the occipital bone. Occasionally, a posterior ethmoidal air cell may be found extending into the body of the sphenoid bone.

42A,21; 30,21
126,7; 128,8
134A,11

The sphenoidal sinuses drain into the posterior walls of the spheno-ethmoidal recesses.

128B,9; 32D,54

The sphenoidal sinuses are situated almost centrally within the cranium and have important relationships (Figures 85 and 86). The optic chiasma and the pituitary gland in the sella turcica lie above the sinuses. Laterally is the cavernous sinus. The oculomotor, trochlear, ophthalmic and maxillary nerves invaginate the lateral dural wall of the cavernous sinus. The internal carotid artery and the abducent nerve lie within the cavernous sinus itself. The basilar artery and the pons lie posteriorly and the nasopharynx inferiorly. The pterygoid canals are situated just beneath the sphenoidal sinuses.

158,49
158,58; 158,51
134A,11; 134A,9
124A,9; 134A,7
134A,8; 134A,6
134A,4; 134A,3
136,50; 132E,9
142A,6

The sphenoidal sinus is innervated by the posterior ethmoidal branch of the ophthalmic nerve and by the orbital branch from the pterygopalatine ganglion. The

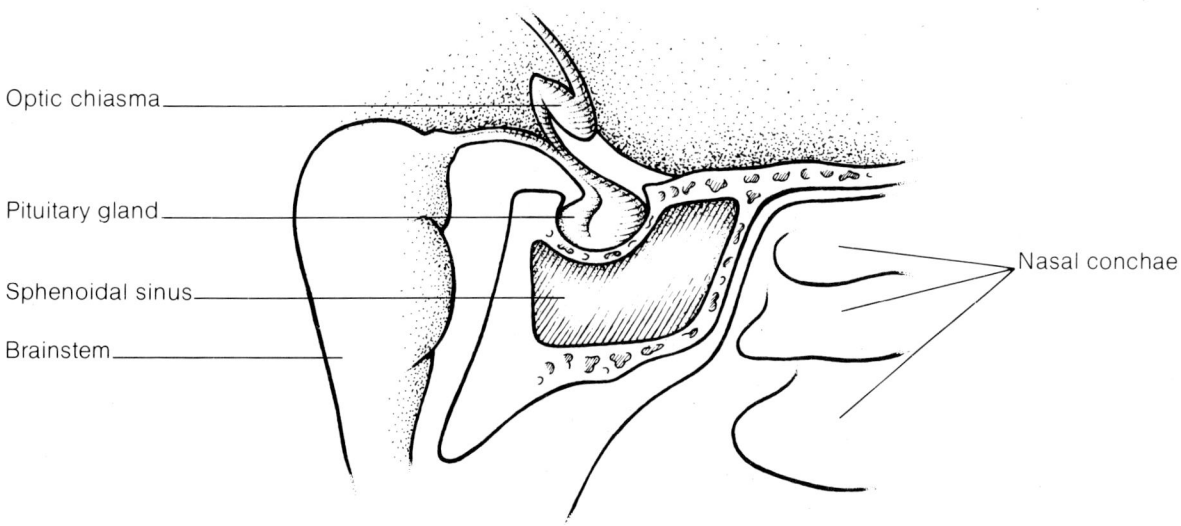

Figure 85 *Sagittal section showing relationships of the sphenoidal air sinus.*

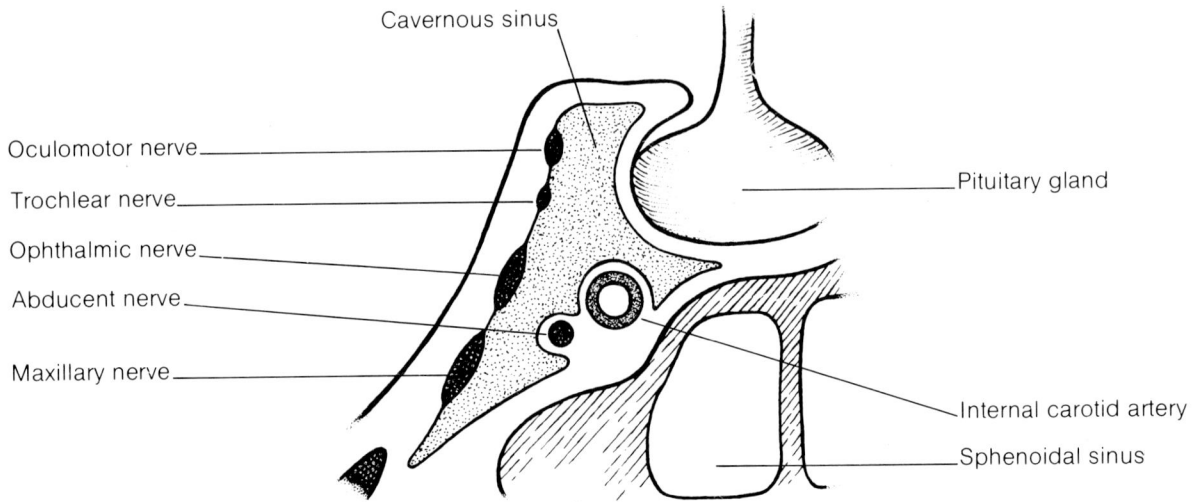

Figure 86 *Coronal section showing relationships of the sphenoidal air sinus and the cavernous venous sinus.*

arterial supply is derived from the posterior ethmoidal branch of the ophthalmic artery and from nasal branches of the sphenopalatine artery. Veins pass into those draining the nasal cavity or into the superior ophthalmic vein (via the posterior ethmoidal vein). Lymphatics from the sphenoidal sinus pass to the retropharyngeal nodes.

132F–G; 133H–J	**THE MAXILLARY AIR SINUSES**
46C,25	These are the largest of the paranasal sinuses. They are situated in the bodies of the maxillary bones.
134B	The maxillary sinus is pyramidal in shape. The base (medial wall) forms part of the lateral wall of the nose. The apex extends into the zygomatic process of the maxilla. The roof of the sinus is part of the floor of the orbit. The floor of the sinus is formed by the alveolar process and part of the palatine process of the maxilla. The anterior wall of the maxillary sinus is the facial surface of the maxilla and the posterior wall is the infratemporal surface of the maxilla. The sinus may be partially divided by incomplete bony septa.
134B,23 134B,24; 133J,11 132F,12	The medial wall of the maxillary sinus has the opening (ostium) of the sinus. The roof has the infra-orbital nerve and vessels within the infra-orbital canal. The floor of the sinus lies below the level of the floor of the nose and is related to the roots of the cheek teeth. As the size of the sinus varies considerably, this relationship will also vary. Usually, at least the second premolar and first molar are related to the floor of the sinus. However, the sinus may extend anteriorly to the first premolar (and sometimes even to the canine) and posteriorly to the third molar tooth. The anterior superior alveolar nerve and vessels (which arise from the infra-orbital nerve and vessels near the midpoint of the infra-orbital canal) pass downwards in a fine canal (canalis sinuosus) in the anterior wall of the maxillary sinus, to be distributed to the anterior teeth. The posterior superior alveolar nerve and vessels pass through canals in the posterior surface of the sinus.
62A; 62B 62B,4; 60B,15 62B,6; 62B,3 134B,23	In an isolated maxillary bone, the ostium of the maxillary sinus is large. However, the ostium in an intact specimen is considerably reduced by portions of the adjacent bones (namely the perpendicular plate of the palatine bone, the uncinate process of the ethmoid bone, the inferior nasal concha and the lacrimal bone) and by the overlying nasal mucosa. The ostium lies high up at the back of the medial wall of the maxillary sinus, being unfavourably situated for drainage. It usually opens into the posterior

part of the ethmoidal infundibulum, and hence into the hiatus semilunaris of the middle meatus of the lateral wall of the nose. An accessory ostium is sometimes present behind the major ostium.

The innervation of the maxillary sinus is derived from the maxillary nerve via its infra-orbital and posterior, middle and anterior superior alveolar branches. The arterial supply to the sinus is derived chiefly from the maxillary artery via its posterior superior alveolar, anterior superior alveolar, infra-orbital and greater palatine branches. The veins draining the sinus correspond to the arteries and pass to the facial vein or the pterygoid venous plexus. The lymphatics pass to the submandibular nodes.

The pterygopalatine fossa

The pterygopalatine fossa is situated on the lateral side of the skull, between the infratemporal (posterior) surface of the maxilla and the pterygoid process of the sphenoid bone. It is an important region because it contains blood vessels and nerves supplying the nose, palate and upper jaw: the maxillary division of the trigeminal nerve, the pterygopalatine parasympathetic ganglion, and the terminal part of the maxillary artery.

The pterygopalatine fossa communicates with the infratemporal fossa through the pterygomaxillary fissure. The anterior wall of the fossa is the infratemporal surface of the maxilla. The posterior wall of the fossa is the pterygoid process. The medial wall is formed by the perpendicular plate of the palatine bone. The lateral wall shows the pterygomaxillary fissure. The pyramidal process of the palatine bone is situated inferiorly and articulates with the tuberosity of the maxilla. It fills the triangular gap between the lower ends of the medial and lateral pterygoid plates.

The pterygomaxillary fissure transmits the maxillary artery from the infratemporal fossa, the posterior superior alveolar branches of the maxillary division of the trigeminal nerve, and the sphenopalatine veins. The fissure continues above with the posterior end of the inferior orbital fissure in the floor of the orbit. Passing through the inferior orbital fissure from the pterygopalatine fossa are the infra-orbital and zygomatic branches of the maxillary nerve, the orbital branches of the pterygopalatine ganglion and the infra-orbital blood vessels.

Opening into the pterygopalatine fossa posteriorly are the foramen rotundum from the middle cranial fossa, and the pterygoid canal from the region of the foramen lacerum at the base of the skull. The maxillary division of the trigeminal nerve passes through the foramen rotundum. The pterygoid canal transmits the nerve of the pterygoid canal (greater petrosal plus deep petrosal nerves, see page 257) and an accompanying artery.

On the medial wall of the pterygopalatine fossa lies the sphenopalatine foramen. This foramen communicates with the nasal cavity. It transmits the nasopalatine and posterior superior nasal nerves (from the pterygopalatine ganglion) and the spheno-palatine vessels.

At the base of the pterygopalatine fossa, the greater and lesser palatine nerves (from the pterygopalatine ganglion), together with accompanying vessels, pass into the hard palate to emerge at the greater and lesser palatine foramina.

The maxillary nerve (Figure 87)

This division of the trigeminal nerve (the fifth cranial nerve) contains only sensory fibres. Developmentally, it is the nerve of the maxillary process on the embryonic face (see page 168). Functionally, it supplies the maxillary teeth and their supporting structures, the hard and soft palate, the maxillary air sinus, much of the nasal cavity, and skin overlying the middle part of the face (see Figure 49).

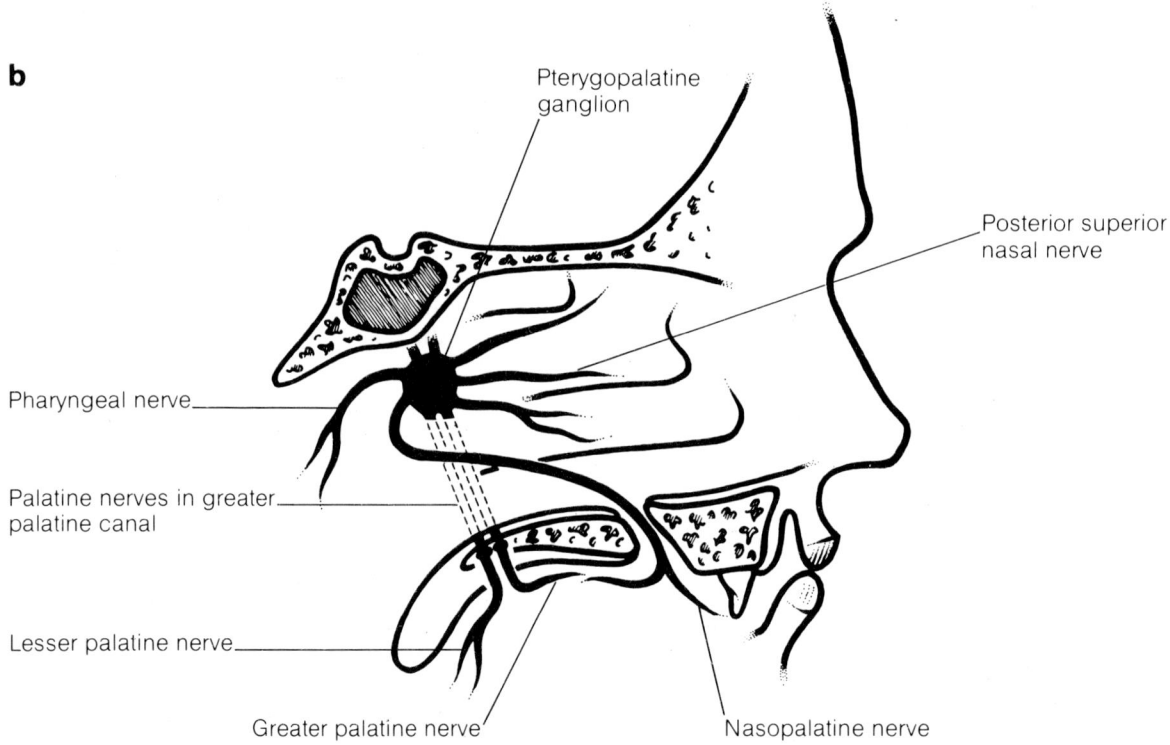

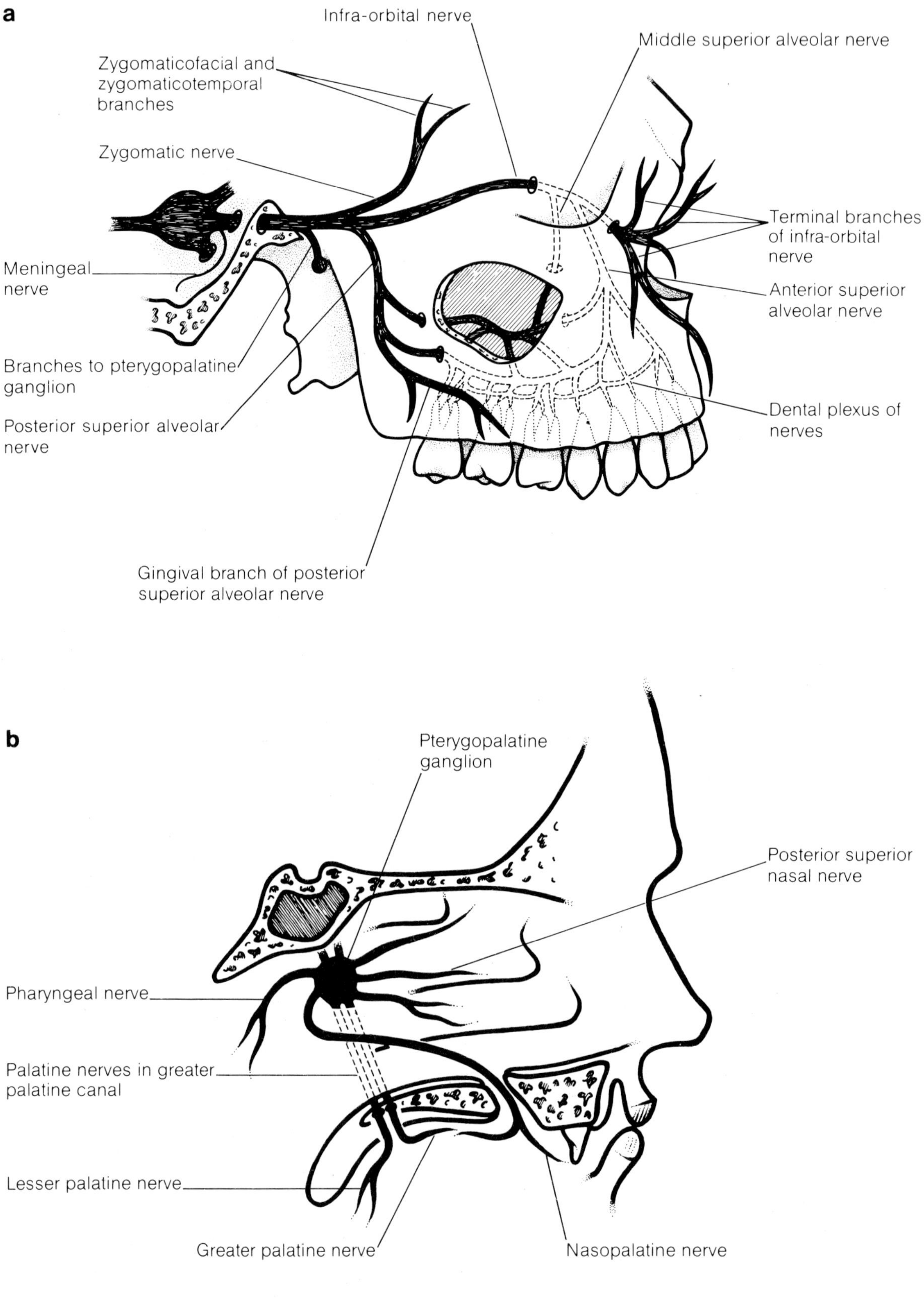

Figure 87 a) *The maxillary nerve.* b) *Branches of the pterygopalatine ganglion.*

The maxillary nerve arises from the trigeminal ganglion on the floor of the middle cranial fossa. The nerve passes along the lateral dural wall of the cavernous sinus (see Figure 86) to exit the cranial cavity at the foramen rotundum. It emerges into the upper part of the pterygopalatine fossa, where most of its branches are derived. These branches can be classified into those that come directly from the maxillary nerve, and those that are associated with the pterygopalatine parasympathetic ganglion.

Branches of the maxillary nerve:

- Branches from the main nerve trunk:

    Meningeal nerve

    Ganglionic branches

    Zygomatic nerve
    Zygomaticotemporal nerve
    Zygomaticofacial nerve

    Posterior superior alveolar nerve

    Infra-orbital nerve
    Middle superior alveolar nerve
    Anterior superior alveolar nerve

- Branches from the pterygopalatine ganglion:

    Orbital nerve

    Nasopalatine nerve

    Posterior superior nasal branches

    Greater (anterior) palatine nerve

    Lesser (posterior) palatine nerve

    Pharyngeal branch

The meningeal nerve is the only branch from the main trunk of the maxillary nerve that does not originate in the pterygopalatine fossa: it arises within the middle cranial fossa before the foramen rotundum. It runs with the middle meningeal artery and innervates the dura mater lining the middle cranial fossa.

The ganglionic branches, usually two in number, connect the maxillary nerve to the pterygopalatine ganglion. These branches contain sensory fibres, which pass through the pterygopalatine ganglion without synapsing (see page 257), and postganglionic autonomic fibres from the ganglion, which are destined for the lacrimal gland in the orbit.

The zygomatic nerve leaves the pterygopalatine fossa through the inferior orbital fissure. It passes along the lateral wall of the orbit before dividing into zygomaticotemporal and zygomaticofacial branches. These pass through the zygomatic bone to supply overlying skin (see Figure 49). The zygomaticotemporal nerve also gives a branch to the lacrimal nerve, which carries autonomic fibres to the lacrimal gland.

The posterior superior alveolar nerve(s) is one of three superior alveolar nerves that supply the maxillary teeth. The middle and anterior superior alveolar nerves are branches of the infra-orbital nerve (see below). The posterior superior alveolar nerve(s) leaves the pterygopalatine fossa through the pterygomaxillary fissure. Thence, it runs on the tuberosity of the maxilla and eventually pierces the bone to supply the maxillary molar teeth and the maxillary sinus. Before entering the maxilla, the nerve provides a gingival branch which innervates the buccal gingivae around the maxillary molars. The extra-bony course of the posterior superior alveolar nerve is variable. The nerve can subdivide into several branches just before or just after it enters the maxilla. Alternatively, it may arise as several distinct branches at the main trunk of the maxillary nerve (see Figure 116).

The infra-orbital nerve can be regarded as the terminal branch of the maxillary nerve proper. It leaves the pterygopalatine fossa to enter the orbit at the inferior orbital fissure. Initially lying in a groove in the floor of the orbit (the infra-orbital groove), the infra-orbital nerve runs into a canal (the infra-orbital canal) and passes on to the face at the infra-orbital foramen.

The middle and anterior superior alveolar nerves arise from the infra-orbital nerve in the orbit. The middle superior alveolar nerve is found in about 70% of subjects. Occasionally, it arises from the maxillary nerve in the pterygopalatine fossa. The nerve may run in the posterior, lateral or anterior walls of the maxillary air sinus, to terminate at the premolar teeth. The anterior superior alveolar nerve arises within the infra-orbital canal, generally as a single nerve but sometimes as two or three small branches. The nerve runs down the anterior wall of the maxillary sinus in a narrow, sinuous canal (the canalis sinuosus) to reach the maxillary incisor teeth. Near the anterior nasal spine, the anterior superior alveolar nerve gives off a small nasal branch which innervates the nasal mucosa around the anterior naris.

The terminal branches of the infra-orbital nerve arise as the nerve emerges through the infra-orbital foramen on to the face. Palpebral branches supply skin of the lower eyelid. Nasal branches innervate skin overlying the side of the external nose (also part of the septum between the anterior nares). Labial branches supply the skin and oral mucosa of the upper lip, the labial gingivae of the anterior teeth in the upper jaw, and the skin overlying the anterior part of the cheek covering the body of the maxilla. The peripheral nerve field for the infra-orbital nerve on the face is shown in Figure 49.

The branches of the maxillary nerve that arise with the pterygopalatine ganglion contain not only sensory fibres from the maxillary nerve, but also autonomic fibres from the ganglion, which are mainly distributed to glands and blood vessels.

The orbital nerve passes from the pterygopalatine ganglion into the orbit through the inferior orbital fissure. It supplies periosteum and, via sympathetic fibres, the orbitalis muscle. The orbital nerve can also pass through the posterior ethmoidal foramen to innervate posterior ethmoidal air cells and the sphenoidal air sinus.

The nasopalatine nerve runs from the pterygopalatine ganglion into the nasal cavity through the sphenopalatine foramen. It passes across the roof of the nasal cavity to reach the back of the nasal septum. The nasopalatine nerve then passes downwards and forwards within a groove on the vomer to supply the postero-inferior part of the nasal septum (see Figure 81). It terminates by passing through the incisive canal on to the hard palate (see Figure 118) to supply oral mucosa around the incisive papilla.

The posterior superior nasal nerve enters the back of the nasal cavity through the sphenopalatine foramen. It divides into lateral and medial branches. The lateral branches supply the posterosuperior part of the lateral wall of the nasal fossa. The medial branches cross the roof of the nasal fossa to supply the nasal septum overlying the posterior part of the perpendicular plate of the ethmoid.

The greater (anterior) palatine nerve passes downwards from the pterygopalatine ganglion, through the greater palatine canal, and on to the hard palate at the greater palatine foramen. Within the greater palatine canal, it gives off nasal branches that innervate the postero-inferior part of the lateral wall of the nasal fossa. On the palate, it runs forwards at the interface between the palatine process and the alveolar process of the maxilla (see Figure 118) to supply much of the mucosa of the hard palate and palatal gingivae (except around the incisive papilla).

The lesser (posterior) palatine nerve(s) passes downwards from the pterygopalatine ganglion, through the greater palatine canal, and on to the palate at the lesser palatine foramen (or foramina) (see Figure 118). It runs backwards to supply the soft palate.

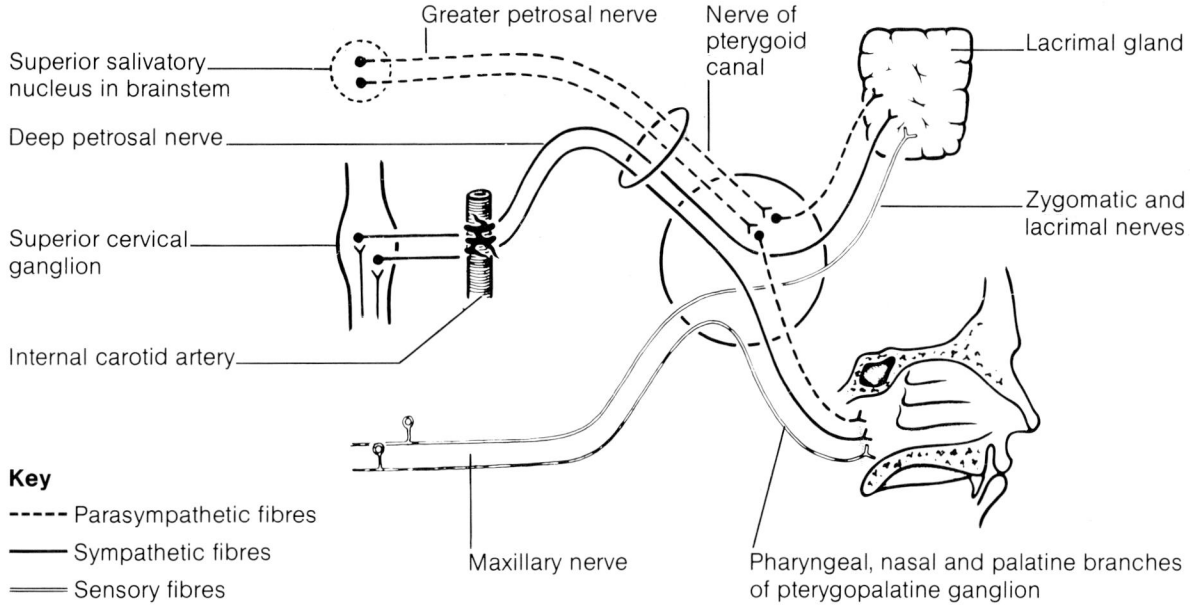

Figure 88 *The pterygopalatine parasympathetic ganglion.*

The pharyngeal branch from the pterygopalatine ganglion passes through the palatovaginal canal to supply mucosa of the nasopharynx.

23,48

The pterygopalatine ganglion (Figure 88)

129,27; 142A,4

This parasympathetic ganglion is situated below the maxillary nerve in the pterygopalatine fossa, connected by two ganglionic branches. It is concerned primarily with supplying the nose, palate, and lacrimal gland. Indeed, it can be thought of as 'the hay fever ganglion', as it is responsible for the symptoms of 'running nose and eyes'.

Like other parasympathetic ganglia in the head, three types of fibres enter the pterygopalatine ganglion: parasympathetic, sympathetic, and sensory fibres. However, only the parasympathetic fibres synapse in the ganglion. The preganglionic parasympathetic fibres originate from the superior salivatory nucleus in the brainstem. The fibres pass with the nervus intermedius of the facial nerve. They subsequently emerge as the greater (superficial) petrosal nerve (see Figure 149, page 373). This occurs within the facial canal of the temporal bone, close to the geniculate ganglion of the facial nerve. The greater petrosal nerve then passes through the bone to appear on the floor of the middle cranial fossa. It then runs medially in a shallow groove to the foramen lacerum. Passing within the foramen lacerum, the greater petrosal nerve enters the pterygoid canal which lies at the base of the pterygoid process. On leaving the pterygoid canal, the nerve emerges into the pterygopalatine fossa and joins the pterygopalatine ganglion. Postganglionic sympathetic fibres run to the pterygopalatine ganglion by a complex course. From the superior cervical ganglion, sympathetic fibres pass to the internal carotid plexus and appear as the deep petrosal nerve. This enters the pterygoid canal to reach the pterygopalatine ganglion. The greater petrosal nerve and the deep petrosal nerve join within the pterygoid canal to become known as the nerve of the pterygoid canal. The sensory fibres to the ganglion run in the ganglionic branches of the maxillary nerve.

166B,54
166B,56

28,22; 168,11
116D,66
142A,13; 42,18
142A,6
142A,4

142A,6
142A,2

The nerves leaving the pterygopalatine ganglion are the orbital nerve, the nasopalatine nerve, the greater and lesser palatine nerves, the posterior superior nasal nerves, and the pharyngeal nerve (Figure 87b). These are described above with the maxillary nerve. **The fibres supplying the lacrimal gland** first pass from the ganglion in one of the ganglionic branches to the maxillary nerve. They then travel with the zygomatic

135,29
142A,5; 129,28

and zygomaticotemporal branches. Within the orbit, they pass from the zygomaticotemporal nerve to the lacrimal nerve (of the ophthalmic nerve) to attain the lacrimal gland.

The maxillary artery

The maxillary artery continues from the infratemporal fossa into the pterygopalatine fossa through the pterygomaxillary fissure. It terminates within the pterygopalatine fossa, where it is called the third part of the maxillary artery. The first and second parts of the maxillary artery in the infratemporal fossa are described on page 180.

The third part of the maxillary artery gives branches that accompany the branches of the maxillary nerve (including those associated with the pterygopalatine ganglion).

The posterior superior alveolar artery runs through the pterygomaxillary fissure on to the maxillary tuberosity. It supplies the maxillary molar and premolar teeth, their buccal gingivae, and the maxillary air sinus.

The infra-orbital artery enters the orbit through the inferior orbital fissure. It runs on the floor of the orbit in the infra-orbital groove and infra-orbital canal to emerge on to the face at the infra-orbital foramen. The infra-orbital artery gives off the anterior superior alveolar artery within the infra-orbital canal. This branch runs downwards to supply the anterior teeth and the anterior part of the maxillary sinus. The infra-orbital artery on the face supplies the lower eyelid, part of the cheek, the side of the external nose, and the upper lip.

The artery of the pterygoid canal passes through the canal to provide branches to part of the auditory tube and tympanic cavity of the ear, and the upper part of the pharynx. The maxillary artery also provides a pharyngeal branch which passes through the vomerovaginal canal to the nasopharynx.

The descending palatine artery leaves the pterygopalatine fossa through the greater palatine canal. Within this canal, it divides into the greater and lesser palatine arteries. The greater palatine artery supplies the inferior meatus of the lateral wall of the nose before passing on to the roof of the palate at the greater palatine foramen. It runs forwards to supply the hard palate and the palatal gingivae of the maxillary teeth. It also provides a branch that runs up into the incisive canal to anastomose with the sphenopalatine artery, thereby contributing to the supply of the nasal septum. The lesser palatine artery (or arteries) emerges on to the palate at the lesser palatine foramen (or foramina). It supplies the soft palate.

The sphenopalatine artery enters the lateral wall of the nose through the sphenopalatine foramen. The artery initially accompanies the posterior superior nasal nerve and gives off branches to supply much of the posterior part of the lateral wall of the nose. The sphenopalatine artery then crosses the roof of the nose to accompany the nasopalatine nerve and to supply the postero-inferior part of the nasal septum.

Veins of the pterygopalatine fossa

The veins are extremely variable but generally correspond to the branches of the maxillary artery. The main site of drainage is into the pterygoid venous plexus in the infratemporal fossa. For example, the sphenopalatine vein draining the back of the nose passes into the pterygopalatine fossa through the sphenopalatine foramen. It reaches the pterygoid venous plexus via the pterygomaxillary fissure.

The inferior ophthalmic vein in the floor of the orbit provides a connecting branch to the pterygoid venous plexus, which passes through the inferior orbital fissure in the region of the pterygopalatine fossa.

THE EMBRYOLOGY OF THE NASAL CAVITY AND THE PARANASAL AIR SINUSES

Two zones of epithelium on the frontonasal process of the embryonic face (see page 168) become delineated to form nasal placodes (Figure 89). These placodes will

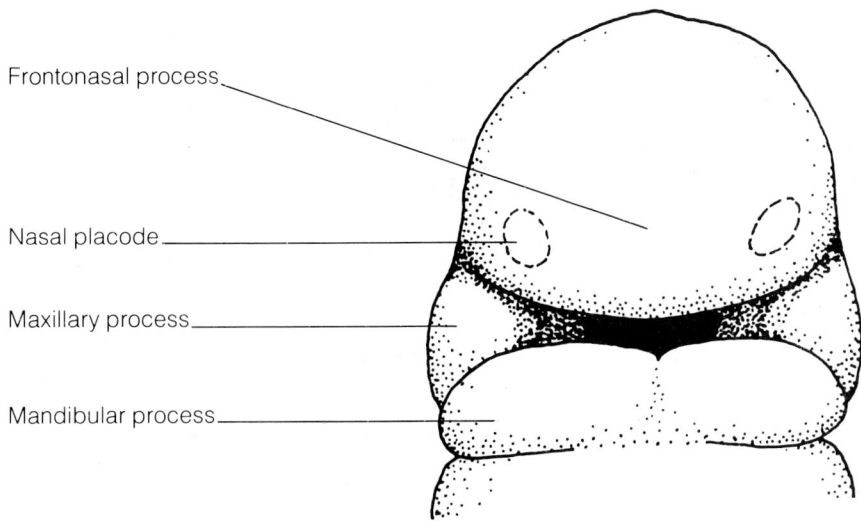

Figure 89 *The frontal aspect of the face during the fourth week of development.*

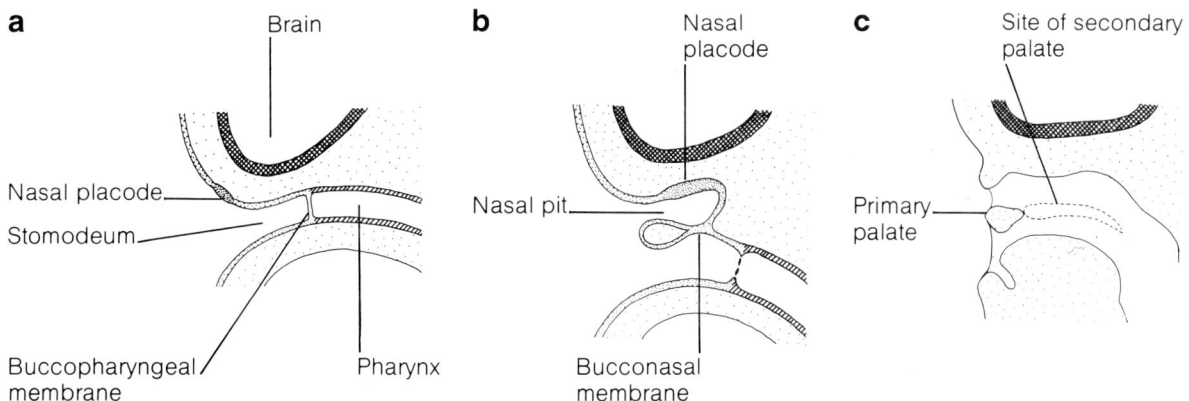

Figure 90 *Sagittal sections showing the development of the primary nasal cavity at* a) *four weeks,* b) *five weeks, and* c) *six weeks.*

eventually form the specialised olfactory epithelium found in the roof of the nasal cavity. During the fifth week, each nasal placode sinks into the underlying mesenchyme and lies in the roof of an epithelial-lined invagination called a nasal pit (Figure 90). The margins of each nasal pit become enlarged both laterally and medially (as a result of proliferation of underlying mesenchymal cells) to form lateral and medial nasal processes (Figure 91). The nasal pits extend deeper and gradually approach the epithelium lining the roof of the primitive mouth (stomodeum). Eventually, only an epithelial membrane (the bucconasal membrane) separates the oral and nasal cavities. Breakdown of this membrane at the end of the fifth week establishes continuity between oral and nasal cavities (Figure 90).

The nasal cavity so far described is called the primary nasal cavity. The tissue of the frontonasal process that separates the two primary nasal fossae is called the primary nasal septum; the tissue that separates the primary nasal cavity from the primary oral cavity is called the primary palate. The use of the term 'primary' can be deduced with the aid of Figure 90. Compared to the definitive nasal cavity, the primary nasal cavity just reaches the level of the future incisive foramen. It is only with further development of a secondary palate following elevation of two palatal shelves (from the maxillary processes of the face) that the common oro-nasal cavity is divided into oral and nasal

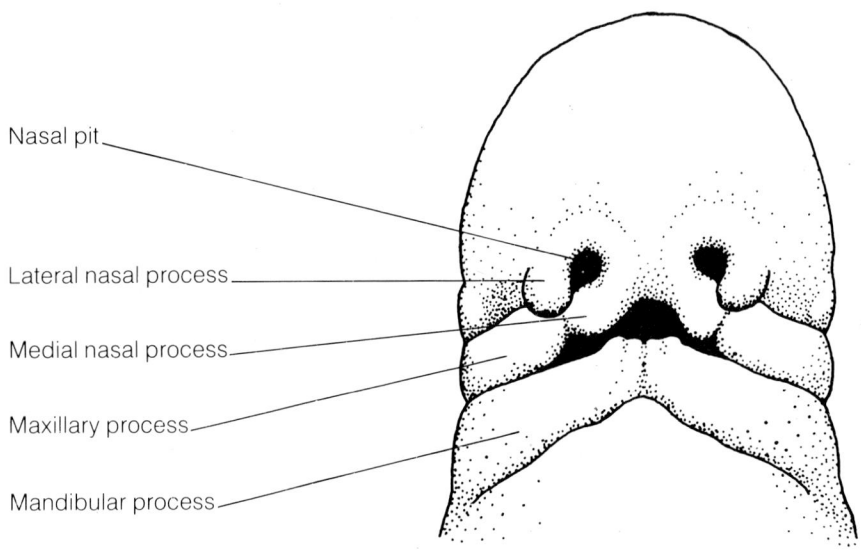

Figure 91 *The frontal aspect of the face during the fifth week of development.*

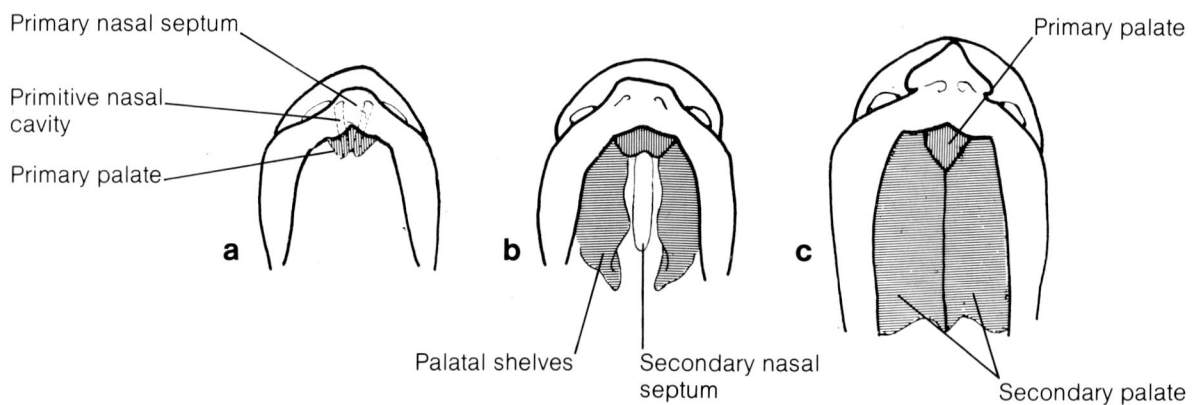

Figure 92 *Development of the nasal cavity at* **a)** *five weeks,* **b)** *six weeks, and* **c)** *eight weeks.*

chambers and the full extent of the definitive nasal cavity becomes apparent (see pages 324 to 325 for details of the embryology of the secondary palate).

A secondary nasal septum develops behind the primary nasal septum from the roof of the oro-nasal cavity. It eventually fuses in the midline with the elevated palatal shelves, dividing the common nasal cavity into right and left nasal fossae (Figure 92).

The paranasal air sinuses develop as outpocketings from the lateral walls of the nasal cavity. The maxillary sinus, the frontal sinus, and the anterior and middle ethmoidal air cells develop from the middle meatuses. The posterior ethmoidal air cells develop from the superior meatuses. The sphenoidal sinuses extend from the spheno-ethmoidal recesses. At birth, the maxillary, frontal and ethmoidal sinuses are small, but may be clinically significant. The frontal sinus, however, does not invaginate into the frontal bone until the second year.

SUMMARY SHEET: THE EXTERNAL NOSE

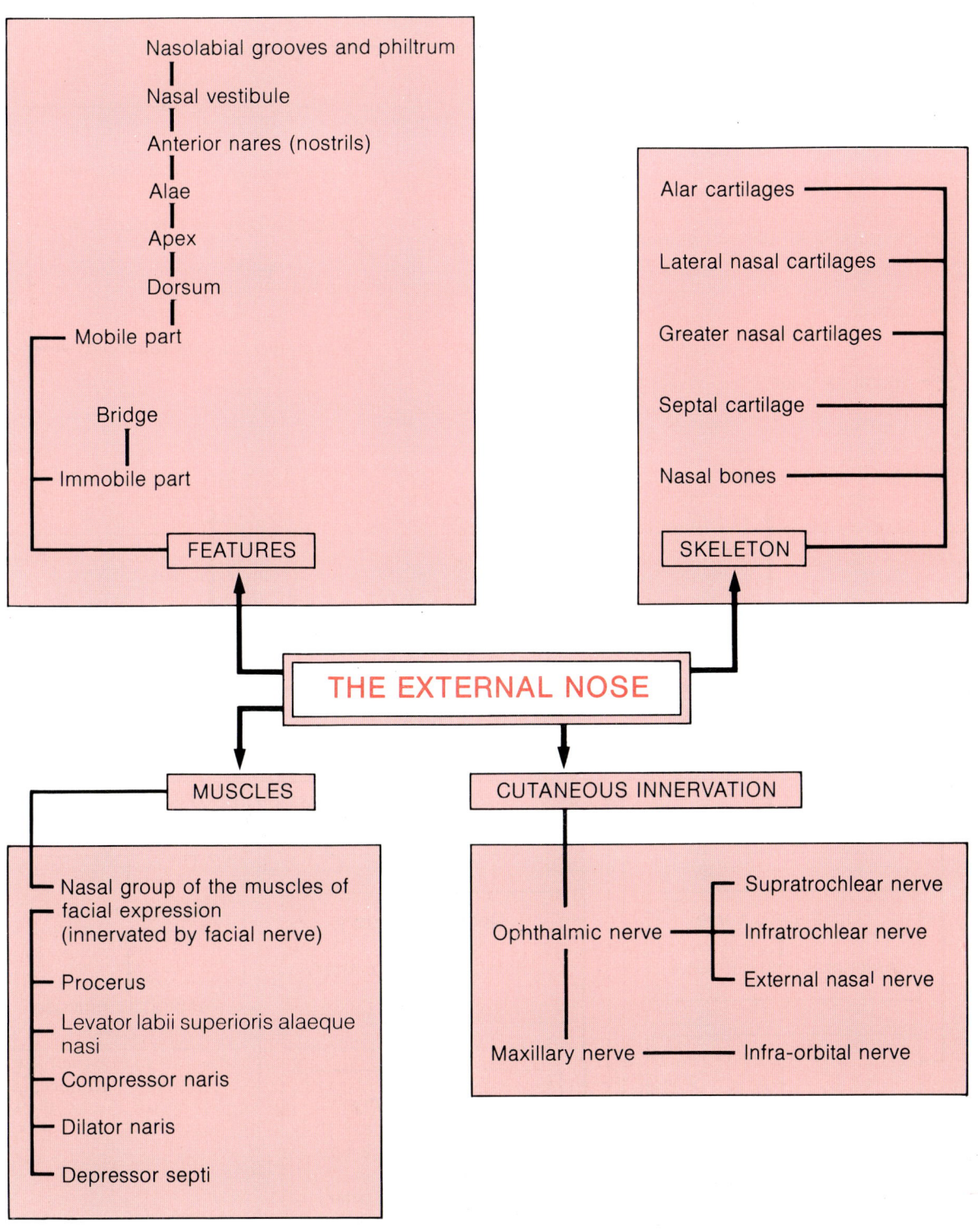

SUMMARY SHEET: THE NASAL CAVITY

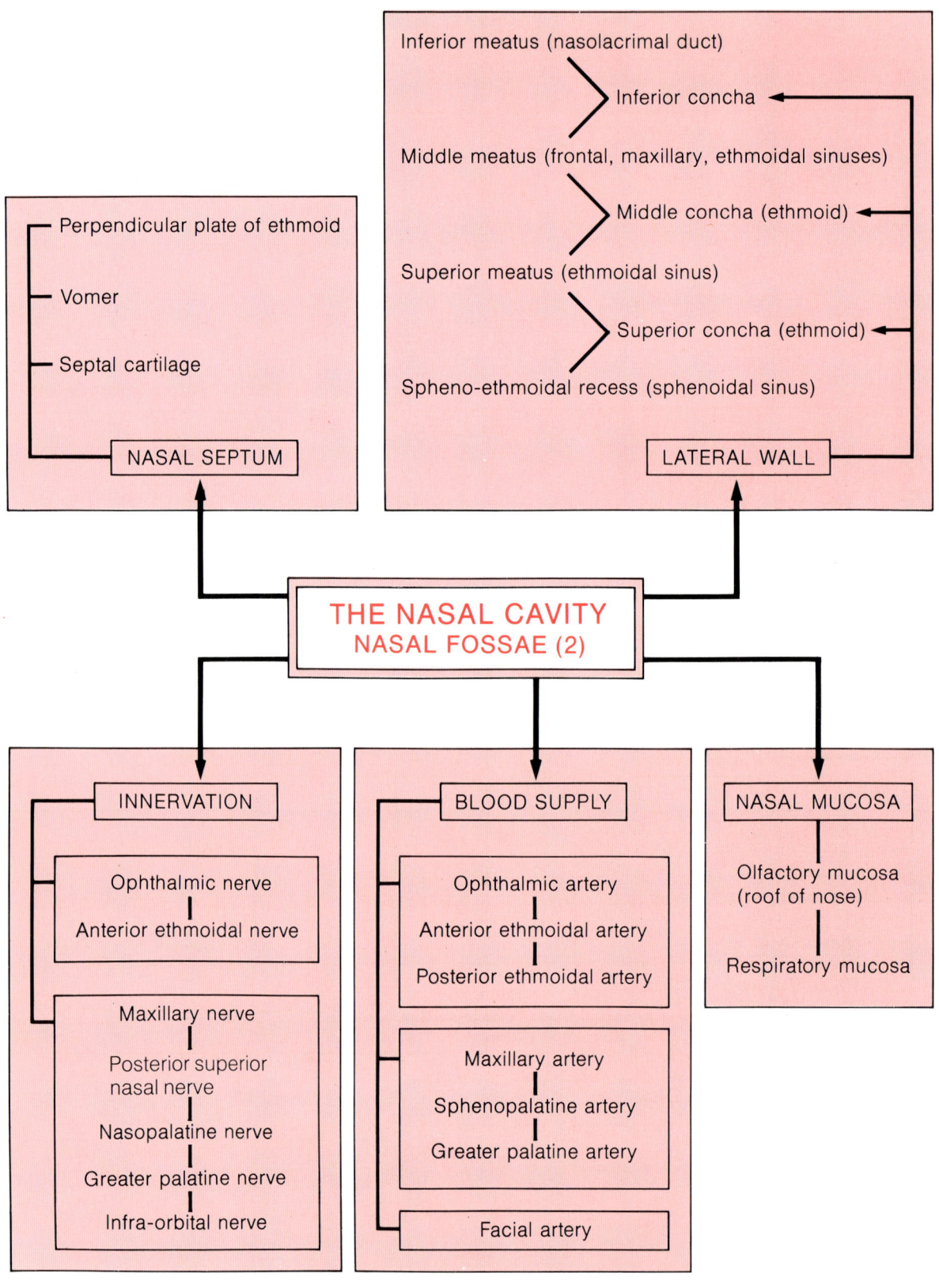

SUMMARY SHEET: THE PARANASAL AIR SINUSES

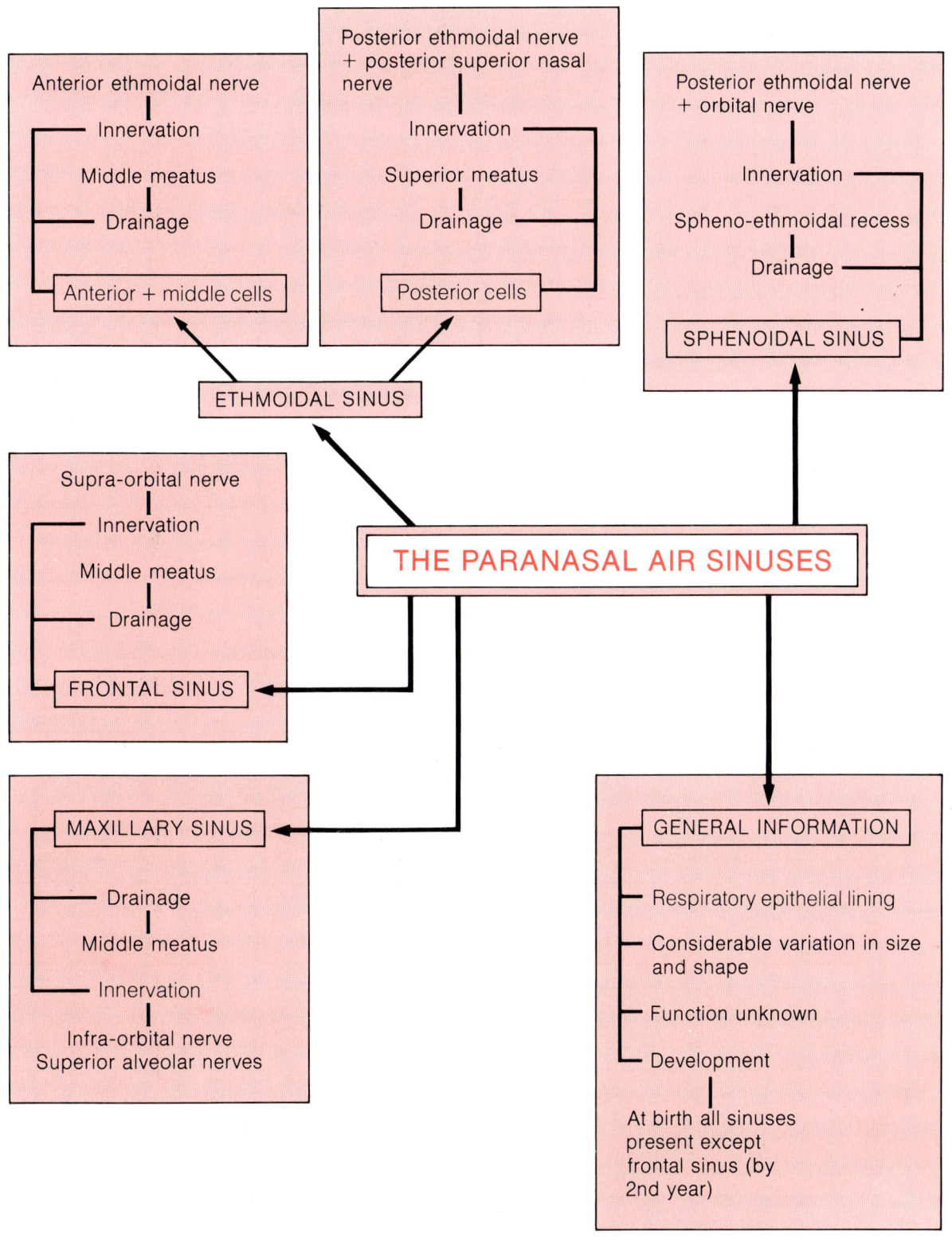

SUMMARY SHEET: THE PTERYGOPALATINE FOSSA 1

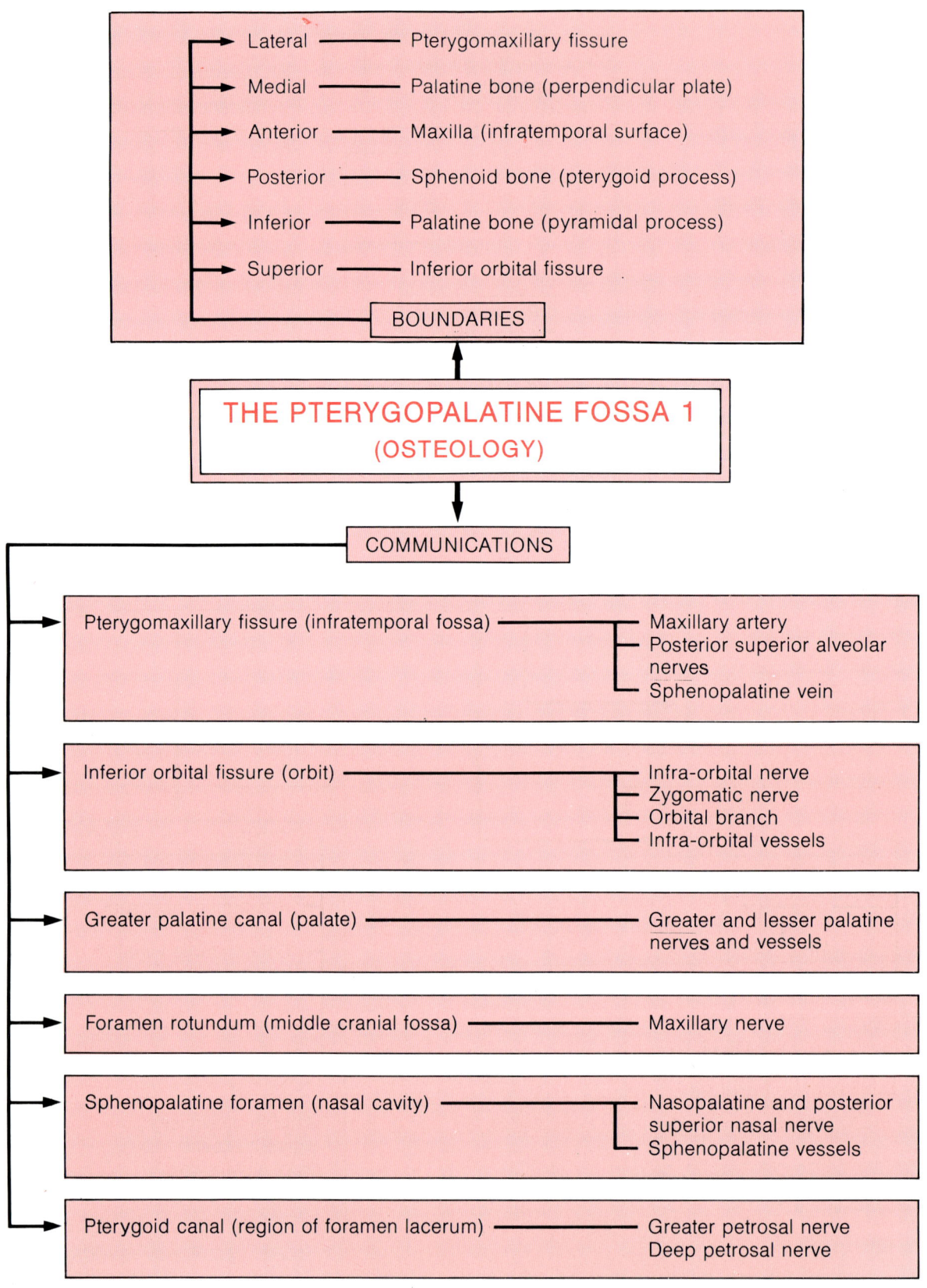

SUMMARY SHEET: THE PTERYGOPALATINE FOSSA 2

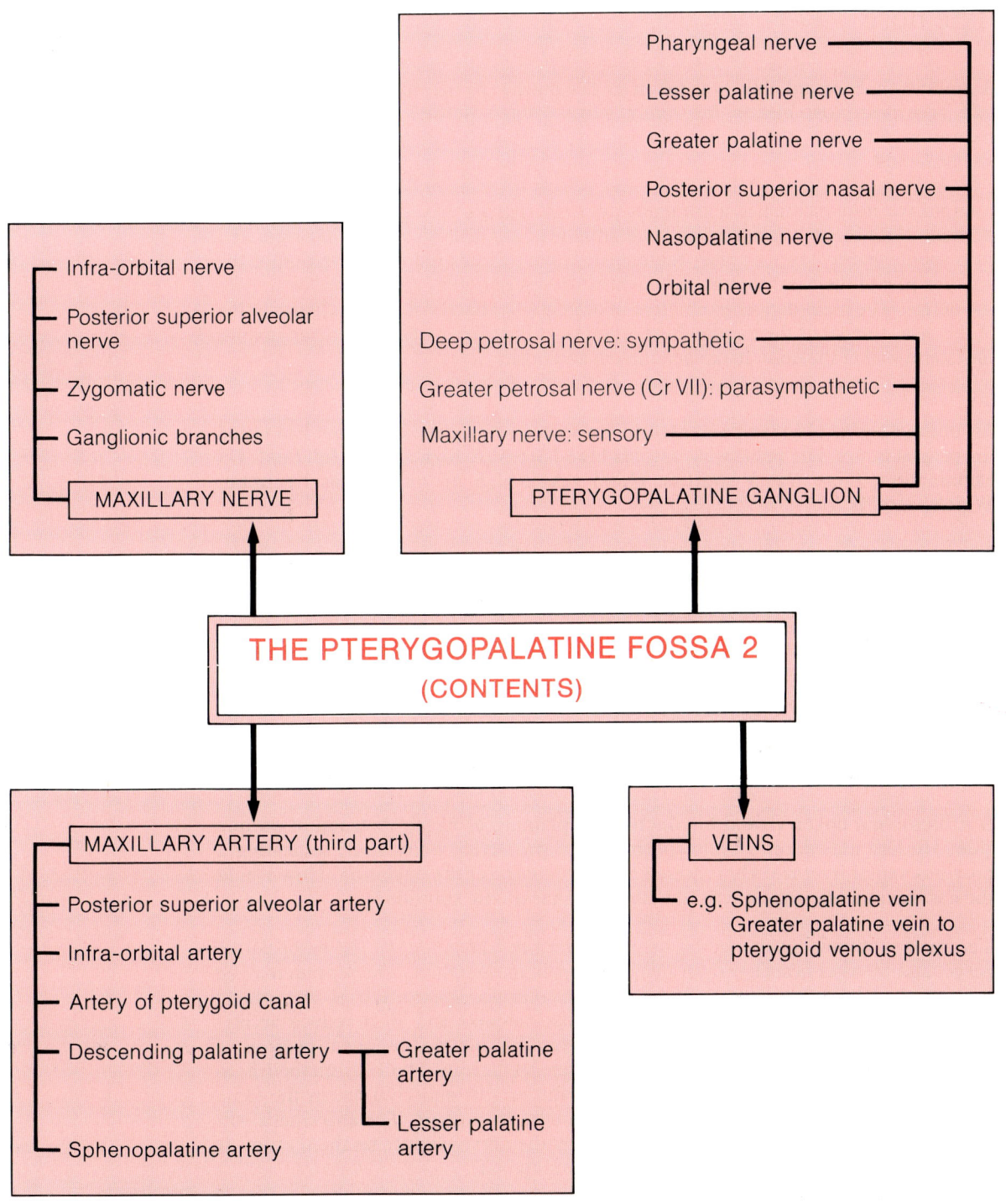

Case histories 5
The nose

A mentally subnormal child was taken by her mother to the doctor. The child had had recurrent nose-bleeds for many years, which settled quickly with little treatment. Recently, the mother had noticed that the child's nose 'whistled' when she breathed.

Compulsive nose-picking is a common problem, particularly in the mentally subnormal. Nose-bleeds (epistaxis) in such situations are usually venous in origin and stop spontaneously. In this case, the recurrent nose-bleeds led to perforation of the anterior part of the nasal septum. The whistling was caused by the passage of air over the defect. The main danger in children is of infection, nasal collapse and retardation of the development of the middle third of the face. In severe cases, defects can involve the lateral wall of the nose, maxillary antrum and orbit.

An obese 62-year-old man presented to his doctor with a nose-bleed that had lasted for several hours. His pulse was fast, his blood pressure high, and he felt faint. On examination, there was a spurting blood vessel high up on the nasal septum. Which vessel is it?

The blood supply to the nasal septum is made up of an anastomosis between several vessels. The chief area of anastomosis lies antero-inferiorly and is called Kiesselbach's plexus (or Little's area). In the case described here, the arteries that supply the upper part of the nasal septum are the ethmoidal branches of the ophthalmic artery. In the elderly, whilst venous bleeds occur as a result of trauma, bleeds from arteries are also common (especially in association with untreated hypertension). Bleeding may be so severe as to make transfusion necessary. The bleeding can be stopped by local cautery but, in severe cases, ligation of the ethmoidal artery, the maxillary artery, or the external carotid artery may be necessary.

A professional boxer visited his doctor complaining of nasal obstruction on one side. He had a deviated nasal septum with a projecting bony spur, indicating a previous septal fracture. He required surgery to correct the deviation. How would the nasal septum be anaesthetised with local anaesthetic solution?

The nasal septum is innervated by the anterior ethmoidal nerve, by the nasopalatine nerve, and by nasal branches from the pterygopalatine ganglion. The septum may be anaesthetised by submucosal injection and by surface application of anaesthetic solution. Submucosal injections may cause haematomas and thus reduce access. Injection may also cause stripping of the perichondrium and subsequent necrosis of the septal cartilage. Surface application is often inefficient. The best method is to hyperextend the neck until the external acoustic meatus and chin are in the same vertical plane and to instil anaesthetic solution into the nostrils. The anaesthetic will gravitate towards the superior meatus and block the nerves supplying the septum. It may also track back as far as the pterygopalatine ganglion.

A teenager with a history of sniffing volatile solvents presented with a series of unpleasant boils in the hair follicles of his nasal vestibule. One of the boils was large, tense and painful, and in most other anatomical sites would be considered

suitable for surgical incision and drainage of the pus. Why is this procedure not suitable in this case?

The principal venous drainage of the external nose is via the facial vein. Surgical incision of boils around the anterior nares may promote septic thrombosis and a spreading thrombophlebitis of the facial vein. Because the facial vein also receives blood from the ophthalmic veins, its obstruction may cause retrograde flow and a fatal cavernous sinus sepsis.

A newborn baby began to make respiratory efforts promptly on delivery. However, she did not become pink but turned blue-grey and lost consciousness: her respiratory efforts became progressively more feeble. The attending paediatrician immediately inserted an endotracheal tube into the baby and ventilated her. The baby quickly regained consciousness and started to breathe spontaneously. Encouraged by this, the paediatrician removed the endotracheal tube, but the baby once more exhibited respiratory distress. After reintubation, her nostrils were noted to be filled with mucus. Removal of this mucus did not improve the situation. A diagnosis of choanal atresia was made.

Choanal atresia is a congenital malformation in which there is failure of one or both of the posterior nares to canalise. The obstruction is formed by a bony or membranous septum, and mucus cannot, therefore, drain properly. Because neonates can breathe only via their noses (because of the relatively large size of the tongue in relation to the size of the oral cavity), there is partial or complete obstruction of the upper airway, and survival is dependent on insertion of an artificial airway to bypass the blockage until it can be removed surgically.

A young man was admitted to hospital having been knocked unconscious by a blow to the head. He was discharged with no apparent ill-effects, but some days later returned complaining of an inability to smell and taste. How might this be explained in terms of his original injury?

A severe blow to the head will cause the brain to move within the skull. The olfactory bulbs, lying beneath the frontal lobes, are attached to the skull by the flimsy olfactory nerves which pass through the cribriform plate of the ethmoid bone to reach the upper nasal cavity. Movements of the brain may shear these nerves at this site and cause permanent impairment of olfaction and loss of the olfactory component of taste.

A motorcyclist was involved in a crash. She could not remember what had happened, but she had two black eyes, and there was a large crack in the front of her helmet. When examined by a doctor, she complained of a headache and had a runny nose. The doctor suspected that the patient had a fractured skull, but this was not evident on a radiograph. However, the radiograph demonstrated air inside the vault of the skull. What has happened?

The motorcyclist had indeed fractured her skull, somewhere in the anterior cranial fossa. The fracture involved the nasal cavity and/or the paranasal sinuses. This allowed the passage of air into the cranial cavity. In addition, there was a leak of cerebrospinal fluid (CSF) from the cranial cavity into the nose, accounting for the runny nose. (It is possible to differentiate between CSF and the watery mucus of rhinitis by testing for the presence of glucose and the absence of mucin in the CSF.) A CSF leak has important clinical implications, because of the potential for spread of nasal bacteria to the meninges.

A patient underwent surgery to correct a blocked nostril associated with a deviated nasal septum. He complained postoperatively of discomfort in his previously patent nostril during inspiration.

Septal deviation is often associated with hypertrophy of the erectile tissue covering the inferior nasal concha on the side of the patent nostril. This in itself may cause nasal obstruction and is often a problem in that it may prevent centralisation of the nasal septum at operation. Removal of the bone of the concha solves this, but may cause problems because it reduces the surface area of the nasal cavity and may reduce the humidification of inspired air. Inspiration of dry air may subsequently cause considerable discomfort.

How would you visualise the superior nasal concha in a patient?

On viewing the nasal cavity through the nostrils after dilation with a speculum (anterior rhinoscopy), only the inferior and middle nasal conchae are visible. However, if a small mirror is placed at the back of the mouth and the patient is instructed to breathe through the nose to bring the soft palate forward, all three nasal conchae can be seen (Figure 93).

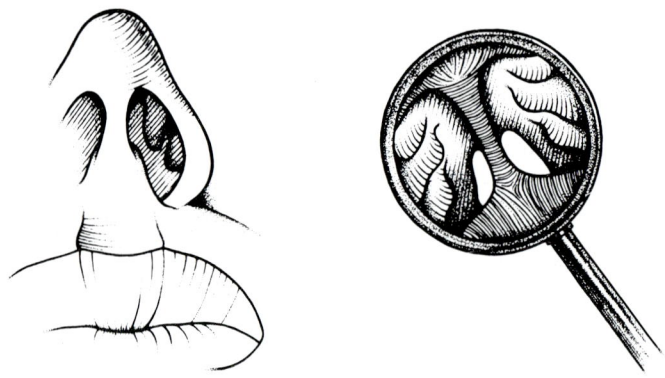

Figure 93 *Visualisation of the superior nasal concha.* Viewed directly from in front, only the middle and inferior conchae are visible. However, viewed with a mirror placed at the back of the mouth, all three conchae are seen.

A man presented with aching pain in his right maxillary teeth and a sensation of fullness and pulsation in his right cheek. He was feeling generally unwell. Examination revealed tenderness lateral to the right ala of the external nose, a deviated nasal septum, and pus in the right middle meatus. In the absence of any dental problems, how might the deviated nasal septum be responsible for producing the other symptoms?

Deviation of the nasal septum may be severe enough to impede aeration and drainage of the maxillary antrum, especially when there is concurrent inflammation of the nasal mucosa (rhinitis). In the case described here, this has led to acute maxillary sinusitis, with inflammation of the superior alveolar nerves causing referred pain in the maxillary teeth. The presence of pus in the middle meatus is further suggestive of maxillary sinusitis, this being the site of the maxillary ostium.

A patient presented to her dentist after one week's discomfort in a right maxillary molar tooth. The dentist diagnosed an apical dental abscess, extracted the tooth and prescribed a course of antibiotics. Three weeks later, the patient returned with halitosis and an unpleasant taste in her mouth. She also noticed that when she drank, her nose ran intermittently for several hours afterwards. The dentist noticed that when the patient blew her nose, there was a frothy discharge at the site of the extraction. **What is the underlying cause of this postoperative problem?**

The extraction of the tooth had either breached the bone in the floor of the maxillary antrum (which may be very thin), or exposed a pre-existing erosion secondary to infection. The result was a large oro-antral fistula, which had epithelialised and remained patent. This allowed the oral contents (especially liquids) to be forced into the sinus, causing a chronic purulent maxillary sinusitis. The leakage of purulent fluid out of the sinus (via the fistula and the ostium of the sinus) and into the mouth was responsible for the unpleasant taste and the halitosis. The passage of liquids into the nasal cavity via the fistula and ostium accounted for the running nose after drinking.

A student went scuba-diving in a cold, dirty gravel pit. Over a period of 24 hours she developed a deep-seated pain behind her eyes and some tenderness on the sides of her nose. Household analgesics having proved ineffective, she asked her doctor to call. By the time he arrived, she was feeling very unwell and in great pain. Her eyelids were puffy and her vision had deteriorated sharply. What condition might be responsible for the patient's symptoms?

Scuba-diving exposes the nose and paranasal sinuses to differential pressures and can cause sinusitis as a result of barotrauma. The situation is likely to be compounded by entry of nasal organisms into the sinuses as a result of raised intranasal pressure. In this case, the posterior ethmoidal air cells which drain into the superior meatuses, have become infected. The pain at the sides of the nose is the result of inflammation of the posterior ethmoidal branch of the nasociliary nerve, causing referred pain in the areas supplied by other branches of the ophthalmic nerve. The thin ethmoidal plate on the medial wall of the orbit has been breached, resulting in a subperiosteal abscess which has tracked forwards to cause the swelling of the eyelids. There is limitation of movement of the eyeball resulting from inflammation of the orbital contents, and a decrease in visual acuity caused by inflammation of the optic nerve which lies in close proximity to the posterior ethmoidal cells.

A child contracted scarlet fever and developed a dull ache above his eyes that got progressively worse. He also found moving and opening his eyes uncomfortable and painful. His parents noticed his condition deteriorating. He became drowsy and irritable and started to wet his bed. They also noticed that the skin over his forehead became puffy. Assuming the child had developed inflammation of the frontal sinuses, how might this be related to the symptoms?

Infective sinusitis is a common sequel to upper respiratory tract infection and to the exanthemata (rashes associated with fevers). The oedema of the forehead and the painful eye movements suggest spread of the infection through the anterior and inferior walls of the frontal sinuses. The inflammation has involved the eye musculature in the roof of the orbit, including levator palpebrae superioris. This explains the pain experienced during movements of both the eyelid and the eyeball. The personality change and the incontinence suggest that there has also been spread of infection through the posterior wall of the frontal sinuses into the anterior cranial fossa. This has caused an extradural or intracerebral abscess which has affected the function of the frontal lobes of the brain.

A 62-year-old man went to his dentist complaining of discomfort of his right maxillary molar teeth, especially during eating. The symptoms had been present for the past six months. The patient mentioned that during the same period his

right nostril had been blocked and that occasionally he had coughed up blood-flecked sputum. The dentist noticed that all the right maxillary molars were mobile without signs of inflammatory periodontal disease, and that there was enlargement of the surrounding bone. Furthermore, there was slight displacement of the right eyeball. Indeed, the patient admitted to having double vision which was gradually getting worse. The dentist suspected that there was a tumour of the mucosa lining the right maxillary sinus. How might the symptoms be explained by this diagnosis?

The tumour had spread to involve the superior alveolar nerves (causing pain in the teeth they supply) and the maxillary alveolar bone (loosening the teeth). It had also involved the lateral nasal wall (displacing the conchae and blocking the nostril). It had spread through the orbital floor to displace the globe and cause double vision. The sputum was stained by bleeding from the friable secretory surface of the tumour.

A woman was admitted to hospital after becoming severely ill with a high fever and loss of consciousness. Her relatives informed the doctor that for the preceding week she had had headaches affecting both temporal regions, which radiated to the neck and behind the ears. That morning she had complained of double vision and diminished visual acuity. Examination revealed pus in the superior meatus and spheno-ethmoidal recess, gross papilloedema (swelling of the retina and optic disc), peri-orbital oedema and neck stiffness. Infection of which paranasal air sinus might produce these symptoms?

This lady had acute sphenoidal sinusitis which had spread to the posterior ethmoidal cells; hence the presence of pus in the superior meatus as well as in the spheno-ethmoidal recess. The infection had spread laterally to cause septic thrombosis of the retinal vein and retinal oedema. The double vision can be attributed to the selective and sequential involvement of the third, fourth and sixth cranial nerves during the development of thrombosis in the cavernous sinus. The neck stiffness is an indicator of meningeal inflammation, suggesting that the infection had spread to the middle cranial fossa.

A young man complained of pain and tenderness on the right side of his nose near the tip. Clinical examination revealed no abnormality, but a radiograph showed an opacity in the anterior ethmoidal air cells. This was diagnosed as a benign tumour of the anterior ethmoidal air cells. Why should this produce pain on the external nose?

The tumour was compressing the anterior ethmoidal nerve as it passes from the nasal slit of the cribriform plate across the roof of the anterior ethmoidal air cells to the anterior ethmoidal foramen. The pressure on the nerve caused referred pain in the area supplied by its terminal external nasal branch.

A 15-year-old boy of tall stature underwent an operation to remove a tumour of the pituitary gland. The gland was approached by splitting the nasal septum and opening the sphenoidal sinus. Postoperatively, he complained of a dry right eye and crusting of his right nostril. Why?

In about 50% of people the nerve of the pterygoid canal lies in a bony canal that indents the floor of the sphenoidal sinus. It is thus at risk during transnasal hypophysectomy, especially if the sinus is opened 'off-centre'. Should the nerve of the pterygoid canal be damaged, then branches derived from the pterygopalatine ganglion will be deprived of their parasympathetic innervation. This will affect the secretion of the lacrimal gland and glands in the nasal mucosa, causing a dry nostril and eye on the affected side.

The mouth, palate and pharynx

The mouth and palate

136,41
136,30
136,40; 136,39
136,43; 136,45
136,35
136,33

The mouth or oral cavity (Figure 94) extends from the lips and cheeks externally, to the anterior pillars of the fauces internally, where it continues into the oropharynx. The mouth can be subdivided into the vestibule external to the teeth and the oral cavity proper internal to the teeth. The palate forms the roof of the mouth, separating the oral and nasal cavities. The floor of the mouth is formed by the mylohyoid muscles and is occupied mainly by the tongue. The lateral walls of the mouth are defined by the cheeks and retromolar regions.

The mouth is concerned primarily with the ingestion and mastication of food, and secondarily with phonation and ventilation.

112,19

The lips (Figure 95) have a muscular skeleton, the orbicularis oris muscle, and are covered externally by skin and internally by mucous membrane. The orbicularis oris muscle is a muscle of facial expression (see page 154) and passes circumferentially

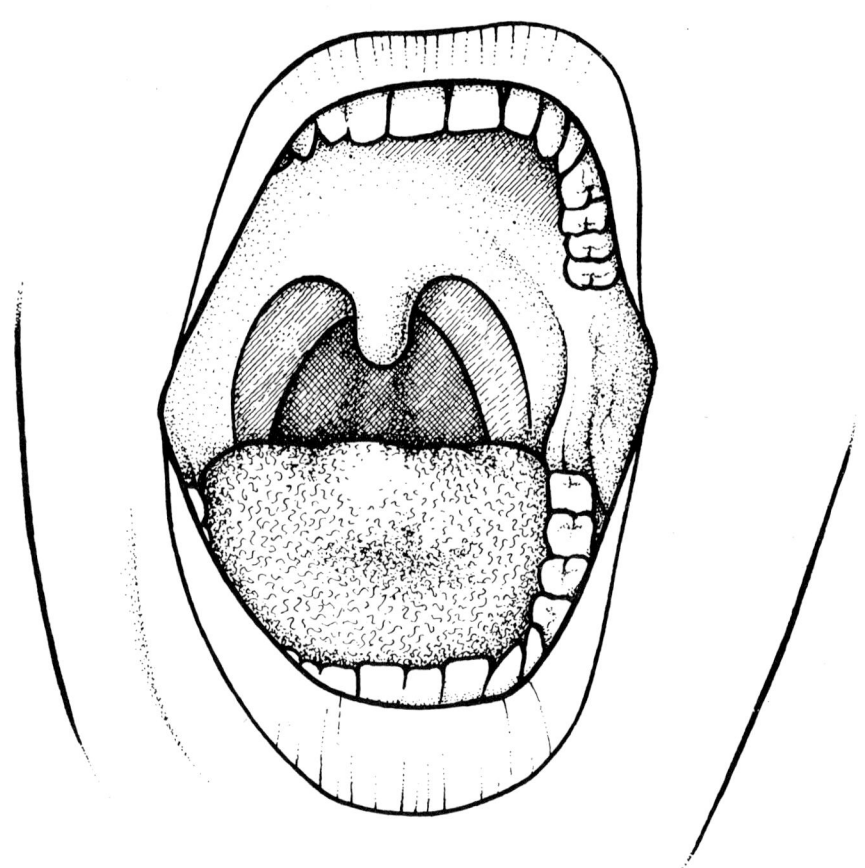

Figure 94 *Boundaries of the oral cavity.*

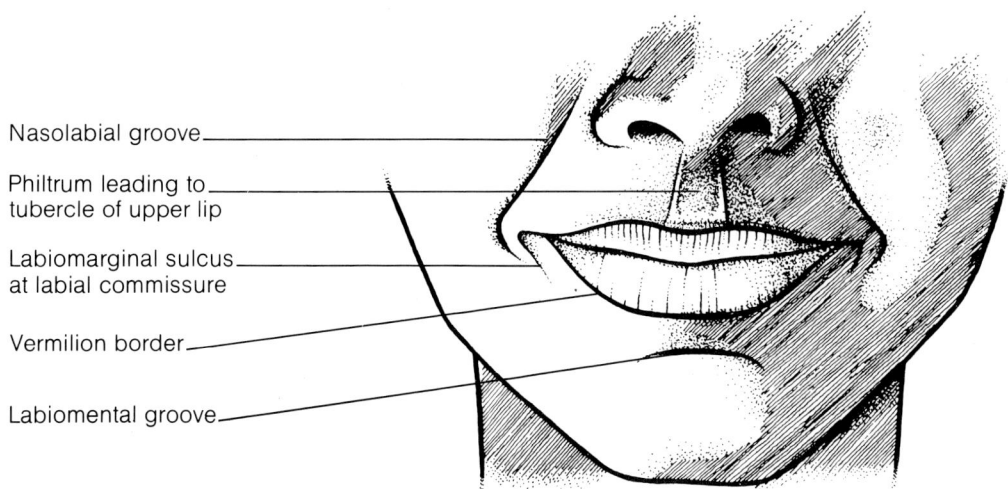

Figure 95 *Appearance of the lips and associated grooves.*

through both the upper and lower lips. The red zone of the lip (the vermilion) is a feature characteristic of man. The sharp junction between the vermilion and the skin is called the vermilion border. The vermilion of the upper lip protrudes in the midline to form the tubercle. The lower lip shows a slight depression corresponding to the tubercle. Both the upper and lower lips widen and then narrow as they pass from the midline to the corners of the lips. The upper lip is separated from the cheeks by nasolabial grooves. Similar grooves appear with age to delineate the lower lip. These are called the labiomarginal sulci. A labiomental groove separates the lower lip from the chin. In the midline of the upper lip is situated the philtrum. The corners of the lips (the labial commissures) are usually located adjacent to the maxillary canine and mandibular first premolar teeth. The lips exhibit sexual dimorphism. The skin of the male is hirsute and is generally thicker and firmer, but less mobile.

The oral vestibule (Figure 96) is a slit-like space between the lips or cheeks and the teeth. When the teeth occlude, the vestibule is a closed space that communicates with the oral cavity proper only in the retromolar regions. Where the mucosa covering the alveolus of the jaw is reflected on to the lips and cheeks, a trough or sulcus is formed which is called the fornix vestibuli. A variable number of sickle-shaped folds containing loose connective tissue run across the fornix vestibuli. The upper and lower labial frena (or frenula) are such folds in the midline. Other folds may traverse the fornix near the canines or premolars. The folds in the lower fornix are said to be more pronounced than those in the upper fornix.

The cheeks (Figure 97) extend from the labial commissures anteriorly to the ridges of mucosa overlying the ascending rami of the mandible posteriorly. They are bounded superiorly and inferiorly by the upper and the lower fornix vestibuli.

Within the cheek is the buccinator muscle which is a muscle of facial expression (see page 156). The mucosa of the cheek is tightly adherent to the buccinator muscle and is thus stretched when the mouth is opened and wrinkled when it is closed. Ectopic sebaceous glands may be evident as yellow patches (Fordyce's spots). Few structural landmarks are visible: the parotid duct drains into the cheek opposite the maxillary second molar tooth, and a hyperkeratinised line (the linea alba) may be seen at a position related to the occlusal plane of the teeth. In the retromolar region, a fold of mucosa containing the pterygomandibular raphe extends from the upper to the lower alveolus. The pterygomandibular space (in which the lingual and inferior alveolar nerves run; see page 179) lies lateral to this fold and medial to the ridge produced by the mandibular ramus.

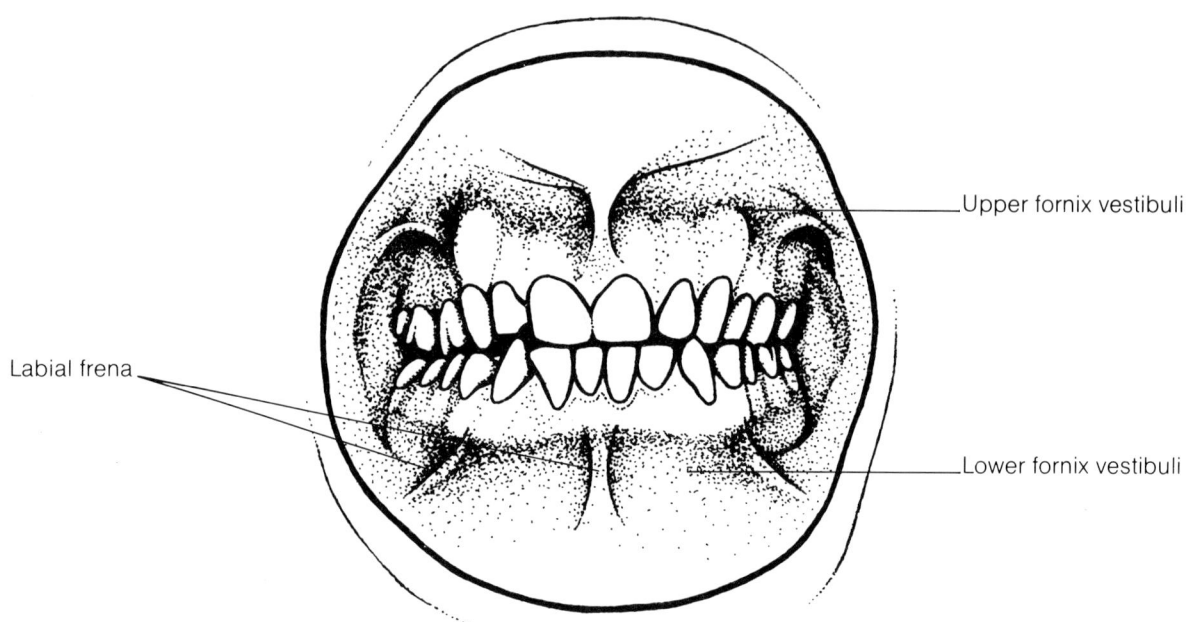

Figure 96 *The oral vestibule.*

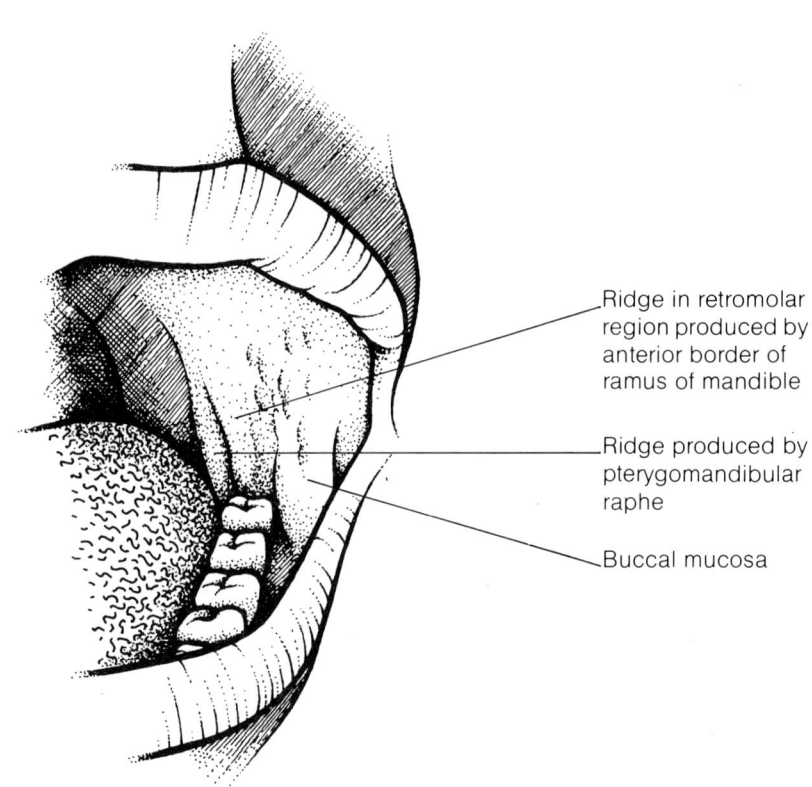

Figure 97 *Appearance of the cheek and the retromandibular region.*

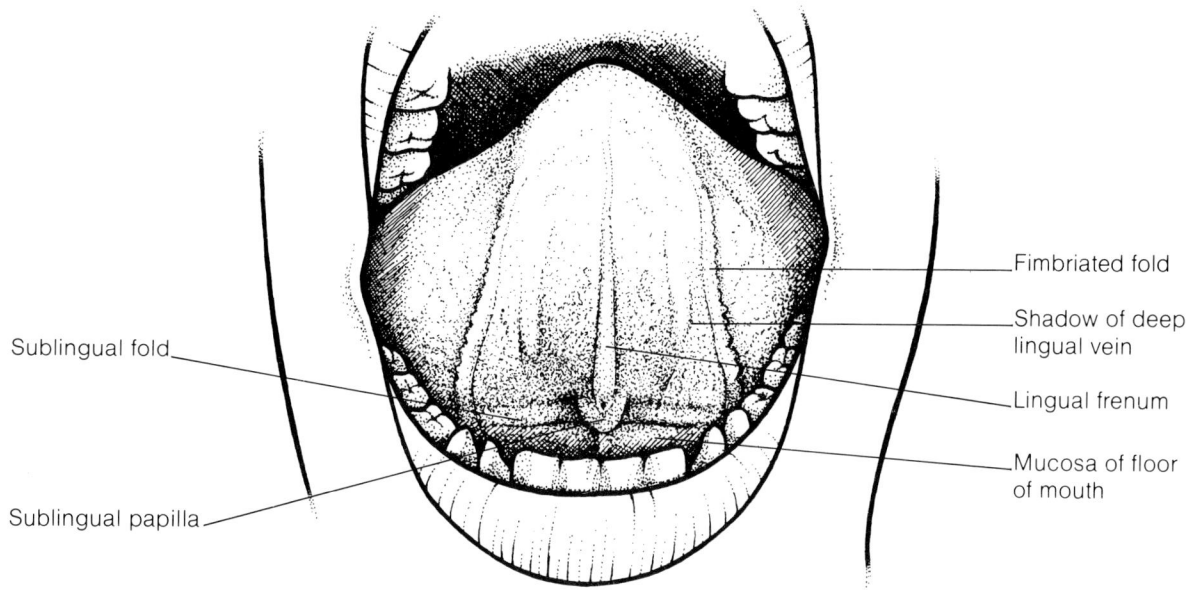

Figure 98 *Inferior surface of the tongue and the floor of the mouth.*

The floor of the mouth (Figure 98) is a small horseshoe-shaped region beneath the movable part of the tongue and above the muscular diaphragm produced by the mylohyoid muscles. In the midline, near the base of the tongue, a fold of tissue called the lingual frenum extends on to the inferior surface of the tongue. Rarely, the lingual frenum extends across the floor of the mouth and attaches on to the mandibular alveolus. The sublingual papilla is a large, centrally positioned protuberance at the base of the tongue. The submandibular salivary ducts open into the mouth at this papilla. On either side of the sublingual papilla are the sublingual folds. Beneath these folds lie the submandibular ducts and sublingual salivary glands.

The inferior surface of the tongue (Figure 98) is covered by a thin lining of non-keratinised mucous membrane which is tightly bound to the underlying muscles. In the midline, extending on to the floor of the mouth, lies the lingual frenum. Lateral to the frenum lie irregular, fringed folds of mucous membrane called the fimbriated folds. Visible through the mucosa are the deep lingual veins.

The dorsum of the tongue (Figure 99) may be subdivided into an anterior two-thirds (palatal part) and a posterior third (pharyngeal part). The junction between these two parts is marked by a shallow V-shaped groove called the sulcus terminalis. The angle (or V) of the sulcus terminalis is directed posteriorly. Near the angle may be seen a small pit called the foramen caecum. This is the primordial site of development of the thyroid gland. The mucosa of the palatal part of the tongue is partly keratinized and is characterised by the presence of numerous papillae. The most conspicuous papillae are the circumvallate papillae which lie immediately in front of the sulcus terminalis. Also found are filiform, fungiform and foliate papillae. The pharyngeal surface of the tongue is covered with large, rounded nodules called lingual follicles. These are composed of lymphatic tissue (the lingual tonsil). The posterior part of the tongue slopes towards the epiglottis, where folds of mucous membrane (the glosso-epiglottic folds) join the two. The anterior pillars of the fauces (the palatoglossal arches) extend from the soft palate to the sides of the tongue near the circumvallate papillae.

The palate (Figure 100) is divided into the immovable hard palate anteriorly and the movable soft palate posteriorly. As their names indicate, the skeleton of the hard palate is bony, whereas that of the soft palate is fibrous.

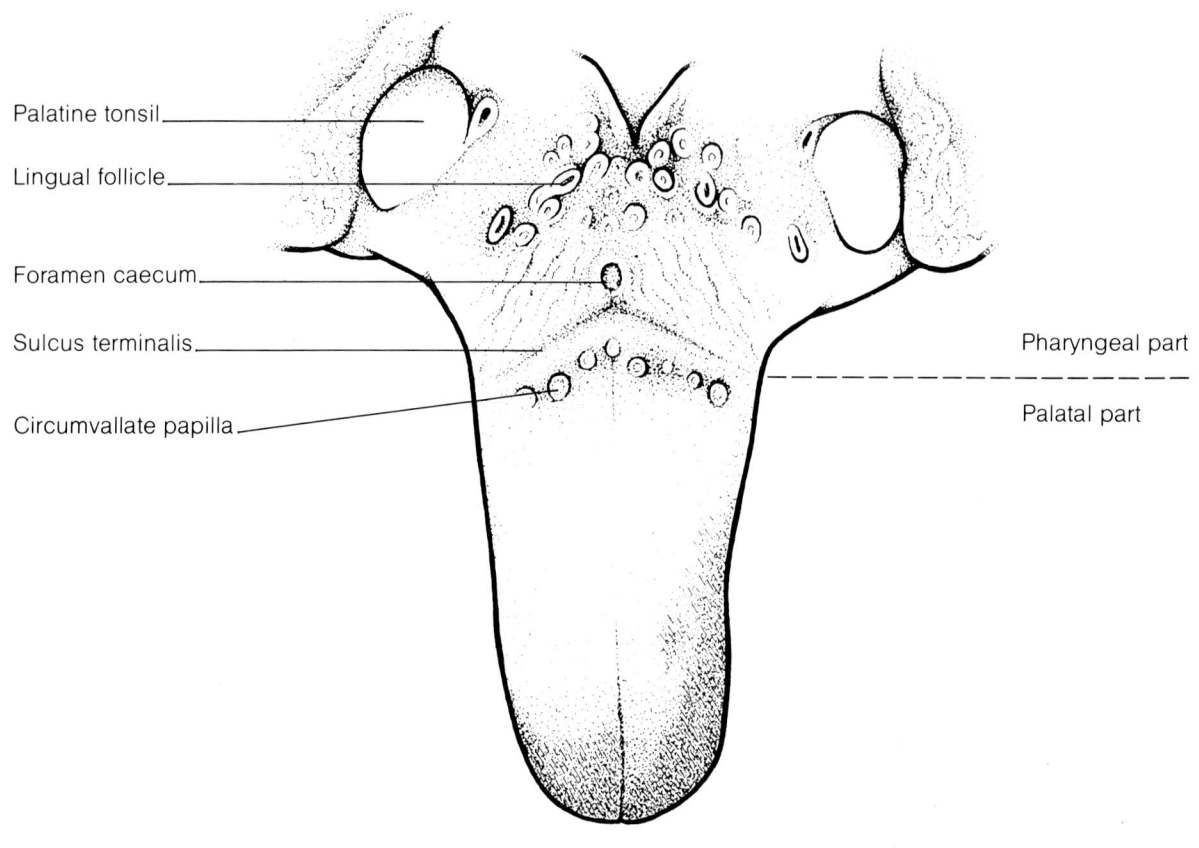

Figure 99 *Appearance of the dorsum of the tongue.*

22,2
22,6

22,3

22,1

22,5
22,7
22,8

22,52

135,29; 135,30

The hard palate is formed by the two palatine processes of the maxillary bones and the two horizontal plates of the palatine bones. It is bounded anteriorly and laterally by the alveolus of the upper jaw, which supports the teeth. A cross-shaped set of sutures traverses the palate. Running anteroposteriorly and dividing the palate into right and left halves is the median palatine suture. This suture is continuous with the intermaxillary suture between the maxillary central incisor teeth. Behind the central incisors, the junction between the palatine processes of the maxillary bones is incomplete, thus forming the incisive fossa. Incisive foramina (two lateral or one anterior and one posterior) pass into this fossa and transmit the nasopalatine nerves and the terminal parts of the greater palatine vessels. Running transversely across the palate between the maxillary and the palatine bones is the transverse palatine suture. This suture is incomplete on each side at the greater palatine foramina. Through the greater palatine foramen pass the greater (anterior) palatine nerve and vessels. Behind the foramen lie one or more lesser palatine foramina through which pass the lesser (posterior) palatine nerves and vessels.

The posterior borders of the horizontal plates of the palatine bones are concave and in the midline form a sharp ridge of bone, the posterior nasal spine. To the posterior edge of the hard palate is attached the fibrous aponeurosis of the soft palate, which is formed by the tendons of the tensor veli palatini muscles (see page 295).

The shape and size of the dome of the palate varies considerably. The mucosa covering the hard palate shows an incisive papilla immediately behind the maxillary central incisors. This covers the nasopalatine nerves as they emerge from the incisive fossa. Extending back from the incisive papilla is a ridge called the palatine raphe. The palatine rugae are irregular folds that radiate transversely from the incisive

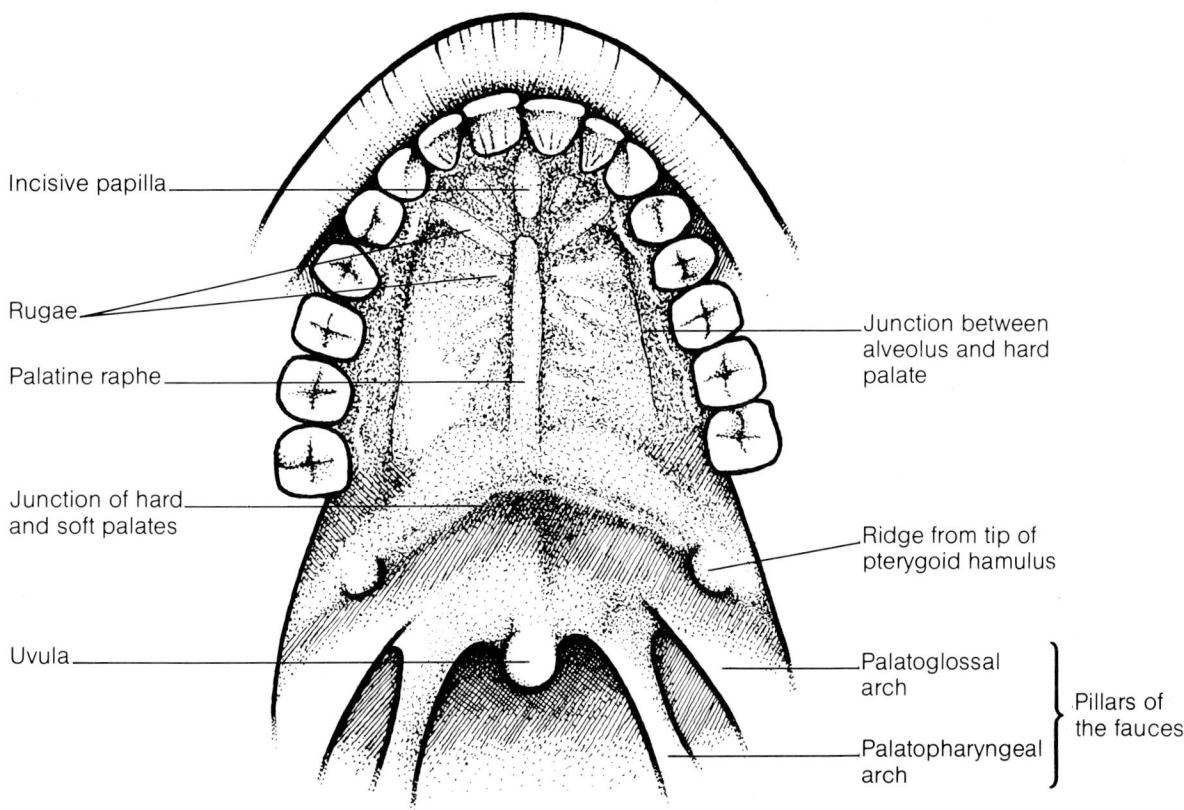

Figure 100 *Appearance of the hard and soft palates.*

papilla and the anterior part of the palatine raphe. The hard palate is covered by a mucoperiosteum (except at the junction with the alveolus, where a submucosa is present in which run the greater palatine nerves and vessels).

The boundary between the hard and soft palate is readily palpable and may be distinguished by a contrast in colour, the soft palate being a darker red with a yellowish tint. Extending from each side of the free border of the soft palate are two prominent folds called the pillars of the fauces. The anterior pillar is the palatoglossal arch; the posterior pillar is the palatopharyngeal arch. These arches cover the palatoglossus and palatopharyngeus muscles. Between the anterior and posterior pillars of the fauces is located the tonsillar fossa which houses the palatine tonsil. Extending backwards and downwards from the free edge of the soft palate in the midline is the uvula.

THE TEETH

A strict definition of a tooth is hard to construct because of the great diversity in its structure and function. A suitable definition for a human tooth might be: a hard body in the mouth, attached to, but not forming part of the jaws, which is primarily concerned with the comminution of food. Figure 101 shows the disposition of the dental tissues.

Man has two generations of teeth, the deciduous (primary) dentition, and the permanent (secondary) dentition. There are no teeth in the mouth at birth but by the age of three years the deciduous dentition is complete. By six years, the first permanent teeth appear and thence the deciduous teeth are exfoliated one by one to be

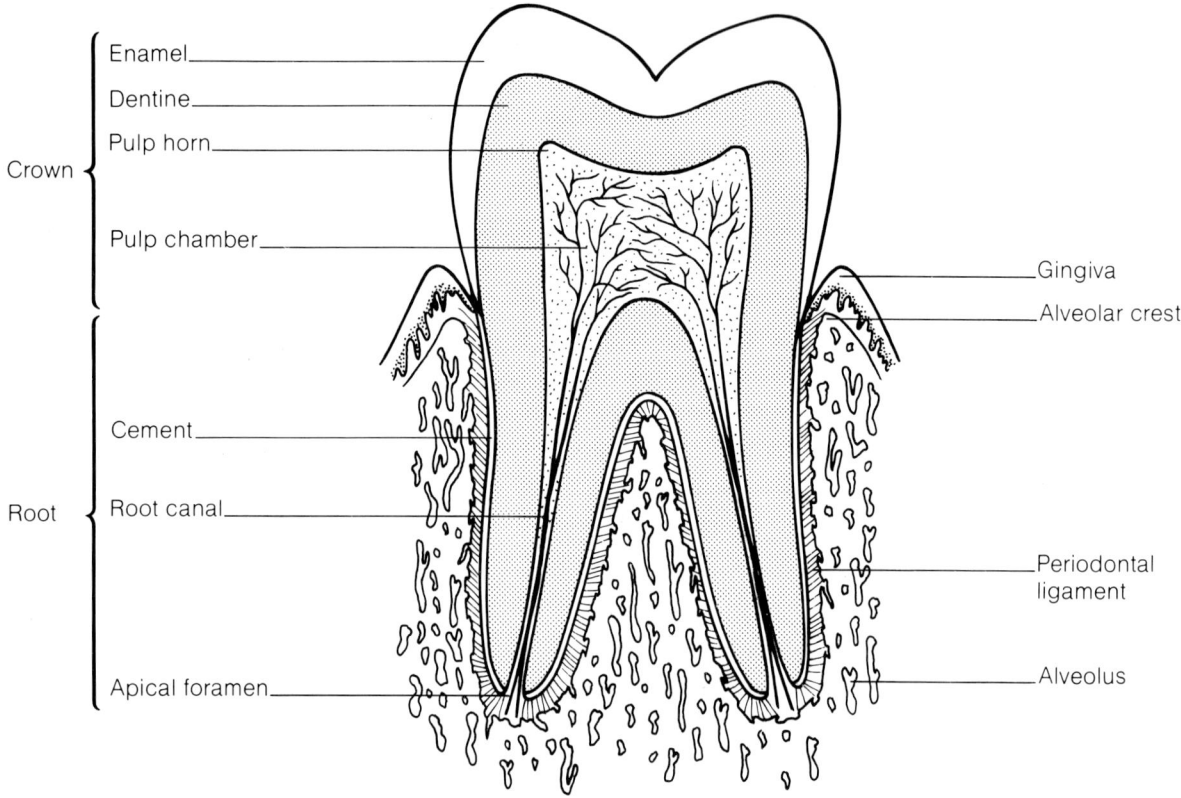

Figure 101 *Section through a molar tooth to show the distribution of the dental and supportive tissues.*

replaced by their permanent successors. There is thus a period where there is a mixed dentition. A complete permanent dentition is present at about the age of 18 years.

There are 20 teeth in the deciduous dentition — 10 in each jaw. There are 32 teeth in the permanent dentition — 16 in each jaw. In both dentitions, there are three basic tooth forms: incisiform, caniniform and molariform. Incisiform teeth (incisors) are cutting teeth, having thin, blade-like crowns. Caniniform teeth (canines) are piercing or tearing teeth, having stout, pointed, cone-shaped crowns. Molariform teeth (molars and premolars) are grinding teeth, possessing several cusps on otherwise flattened biting surfaces. Premolars are bicuspid teeth which are peculiar to the permanent dentition and which replace the deciduous molars.

The types and numbers of teeth can be expressed using a dental formula. The type of tooth is represented by its initial letter, i.e. I for incisors, C for canines, P for premolars and M for molars. The deciduous dentition is indicated by the letter D. The formula for the deciduous dentition is $DI\frac{2}{2} \, DC\frac{1}{1} \, DM\frac{2}{2} = 10$, and for the permanent dentition $I\frac{2}{2} \, C\frac{1}{1} \, P\frac{2}{2} \, M\frac{3}{3} = 16$. The number following each letter refers to the number of teeth of each type in each half of the upper and lower jaws.

Identification of teeth is made not only according to the dentition to which they belong and to the basic tooth form, but also according to their anatomical location within the jaws. The tooth-bearing regions can be divided into four quadrants, the right and left maxillary and mandibular quadrants. A tooth may therefore be identified according to the quadrant in which it is located (e.g. a right maxillary deciduous incisor or a left mandibular permanent molar). In both the permanent and deciduous dentitions, the incisors may be distinguished according to their relationship to the midline. The incisor nearest the midline is the central or first incisor, the incisor

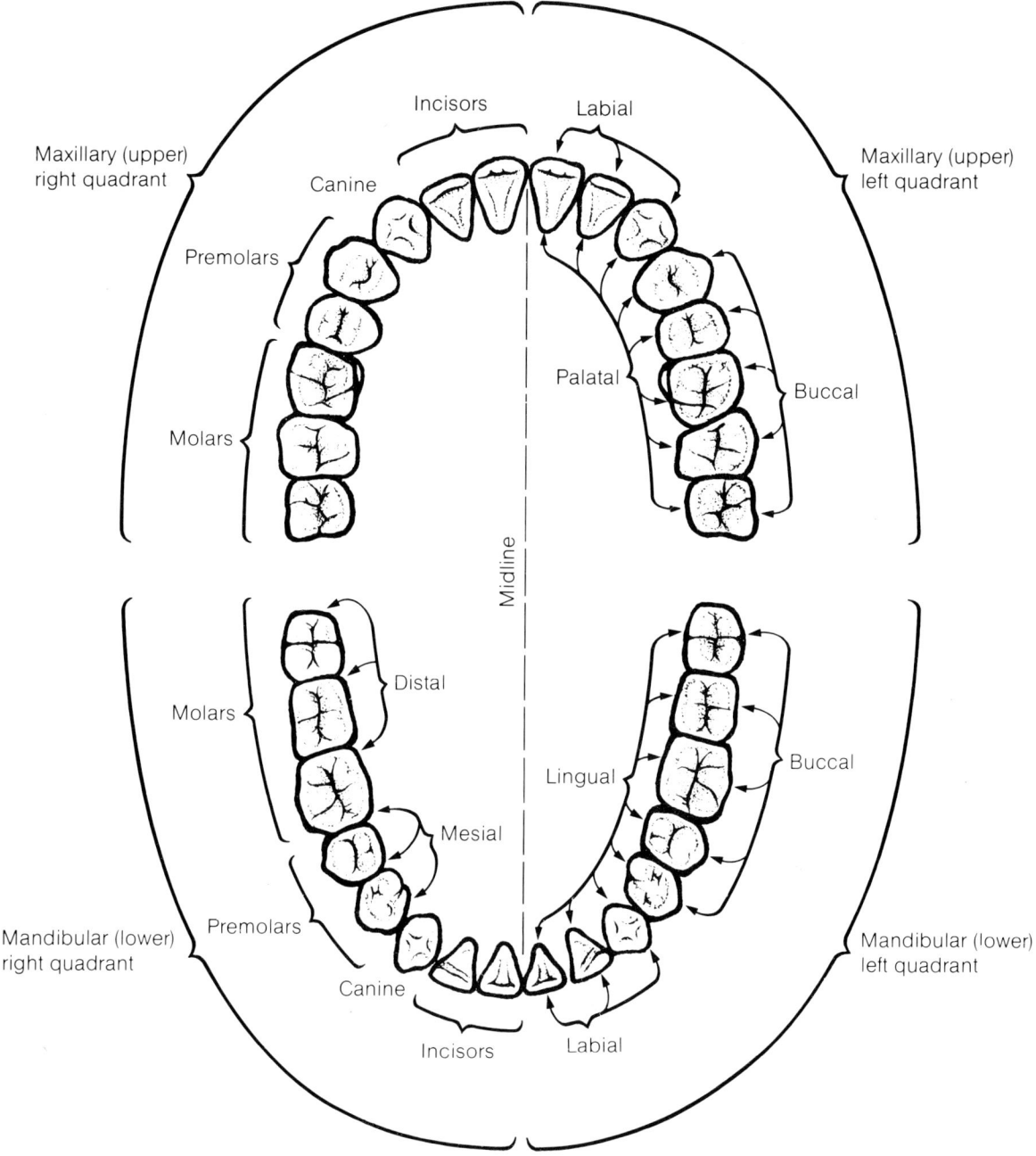

Figure 102 *Occlusal view of the permanent dentition to illustrate some dental terminology.*

that is more laterally positioned being called the lateral or second incisor. The permanent premolars and the permanent and deciduous molars can also be distinguished according to their anteroposterior relationships. The anterior premolar is the first premolar, the premolar behind it being the second premolar. Likewise, the molar most anteriorly positioned is designated the first molar, the one behind it being the second molar. In the permanent dentition, the tooth most posteriorly positioned is the third molar.

The surfaces and ridges of the teeth have a precise terminology in order to facilitate description (Figure 102). The outer surfaces of the anterior teeth (i.e. incisors and

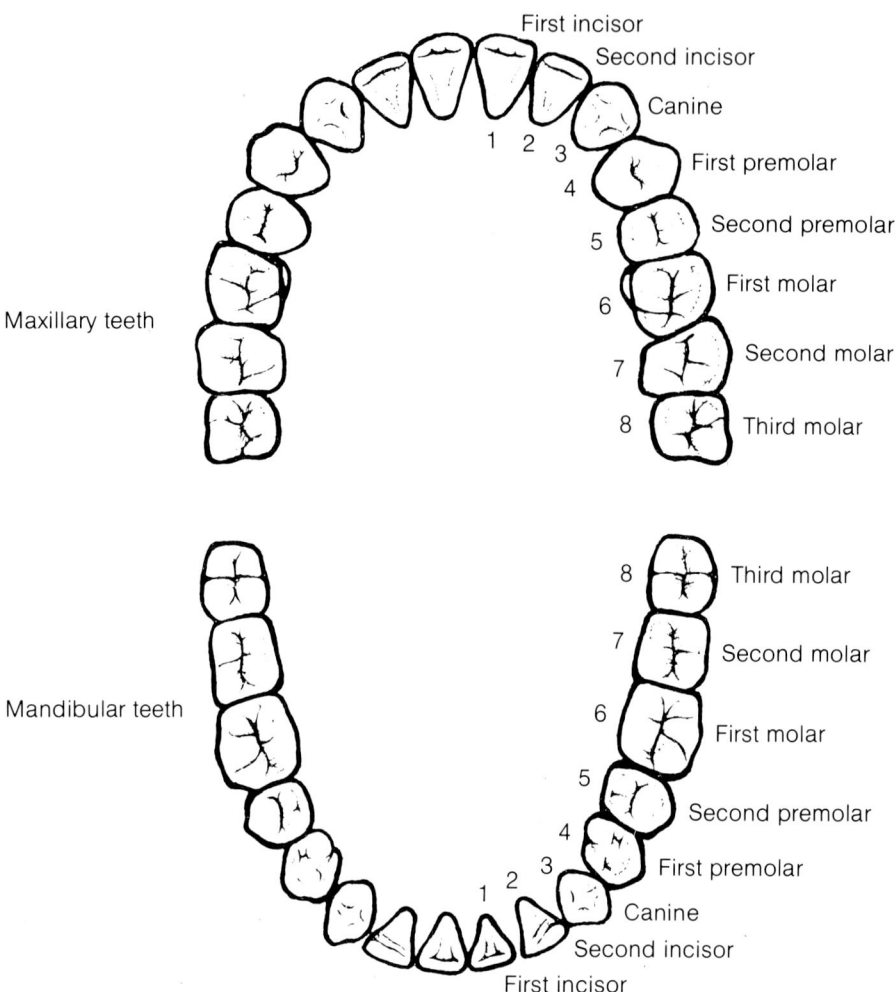

Figure 103 *Appearance of the occlusal surfaces of the permanent dentition. The numbers 1 to 8 identify the teeth according to the Zsigmondy system.*

canines) face forwards towards the lips and are consequently called labial surfaces. The corresponding surfaces of the posterior or cheek teeth (i.e. premolars and molars) face the cheeks and are consequently called buccal surfaces. The inner surfaces of all the teeth in the lower jaw face the tongue and are called lingual surfaces. The corresponding surfaces in the upper jaw are called palatal surfaces. The anterior teeth have medial and lateral neighbours; the posterior teeth have anterior and posterior neighbours. The surfaces of neighbouring teeth that face one another are called contact surfaces or proximal surfaces. Dentists usually call the proximal surface that faces forward in the arch the mesial surface; the proximal surface that faces towards the back of the arch is called the distal surface. The cutting edges of the anterior teeth are called the incisal edges. The chewing surfaces of the posterior teeth are called the occlusal surfaces.

26B; 24; 35

THE PERMANENT DENTITION (Figures 103 and 104)

The incisors

The maxillary first incisor has the greatest mesiodistal dimension of all the permanent anterior teeth. Indeed, the crown is almost as wide as it is long. The crown has a distinct wedge or chisel shape. The incisal edge presents as a narrow flattened

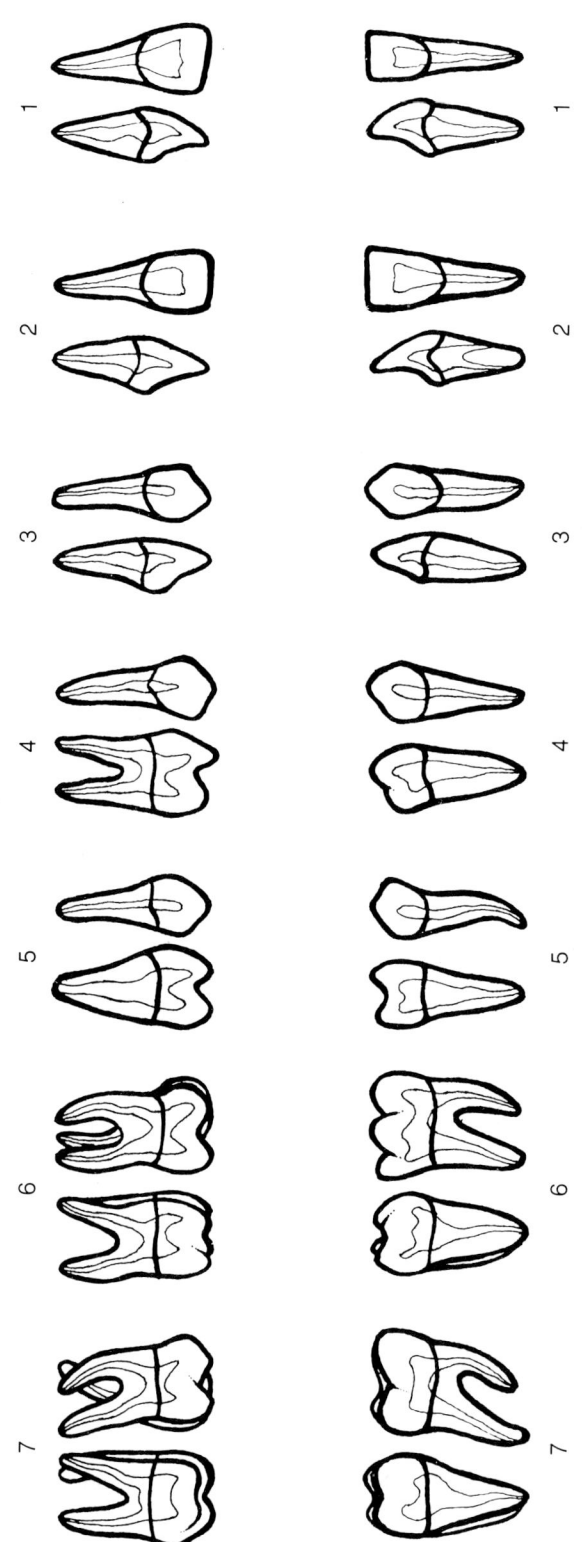

Figure 104 Buccal and distal views of the maxillary (upper) and mandibular (lower) permanent teeth. *Also shown is the outline of the pulp cavity in the young tooth (with age the pulp cavity becomes smaller). The teeth are numbered according to the Zsigmondy system.*

ridge which, before wear, may exhibit three rounded elevations called mammelons. The crown has a smooth, convex labial surface. Palatally, the middle and incisal regions of the crown are concave, giving a shovel-shaped appearance to the incisor. The palatal surface of the crown is bordered by mesial and distal marginal ridges. Near the root there is a prominent bulge called a cingulum. The mesial surface of the crown is straight and approximately at right angles to the incisal edge. However, the disto-incisal angle is more rounded and the distal outline more convex. The incisor has a single root, which tapers towards the apex. The root is conical and appears narrower from the palatal than the labial aspect.

The maxillary second (lateral) incisor is one of the most variable teeth in the dentition. Usually, it is a diminutive form of the maxillary central incisor. However, its crown is narrower and shorter than that of the central incisor, and the crown:root length ratio is considerably decreased. A pit often lies in front of the cingulum on the palatal surface. It is called the foramen cacumen and may extend some distance into the root.

The mandibular permanent incisors are the narrowest teeth in the permanent dentition. The first incisor is smaller than the second. They have similar shapes and can be distinguished from the upper incisors not only by their size but also by their mesiodistally flattened roots, and by their inconspicuous marginal ridges and cingula.

The crown of the first incisor is almost twice as long as it is broad. The incisal edge is straight and at right angles to the long axis of the tooth. Both the mesio-incisal and the disto-incisal angles are right angles. Three mammelons are present on the incisal edge of the newly erupted tooth. The single root is flattened mesiodistally and is frequently grooved on the mesial and distal surfaces.

The second (lateral) incisor is similar to the first incisor but it can be distinguished because of some asymmetry. The distal surface of the crown diverges at a greater angle from the long axis of the tooth, giving a fan-shaped appearance. The disto-incisal angle is more acute and rounded. When viewed from above, the distal aspect of the incisal edge deviates lingually.

The canines

The maxillary canine is a stout tooth. Its prominent cusp occupies at least one-third of the crown height. The mesial slope of the cusp is shorter than the distal slope. Prominent ridges pass from the tip of the cusp down the middle of the labial and palatal surfaces. The palatal surface has distinct mesial and distal marginal ridges and a well-defined cingulum. The longitudinal ridge passing down the palatal surface meets the cingulum and is separated from the marginal ridges by distinct grooves or fossae. The root is the largest in the dentition, and is triangular in cross-section. Its mesial and distal surfaces are often grooved longitudinally.

The mandibular canine is similar to the maxillary canine but is smaller, more slender and more symmetrical. The cusp is less prominent, occupying one-fifth of the crown. Indeed, the height of the crown is greater than the width. No longitudinal ridges run from the tip of the cusp on to the labial and lingual surfaces. The cingulum, marginal ridges and fossae on the lingual surface are indistinct. The labial and mesial surfaces of the crown are clearly defined, there being a distinct acute angle between them. However, the labial surface merges gradually into the distal surface.

The premolars

The maxillary first premolar appears ovoid when viewed occlusally, the buccopalatal dimension being greater than the mesiodistal. It is also wider buccally than palatally. The mesial and distal borders of the occlusal surface are marked by distinct marginal ridges. The palatal cusp is shorter than the buccal cusp, and its tip lies more mesially. The buccal and palatal cusps are separated by a central occlusal fissure. This fissure crosses the mesial marginal ridge on to the mesial surface. The cervical third of the mesial surface is marked by a pronounced depression called the canine fossa. There are usually two roots: the buccal and palatal roots.

The maxillary second premolar is similar to the first premolar. Its occlusal surface is, however, more rounded. The central fissure appears shorter and does not cross the mesial marginal ridge. The two cusps are smaller and more equal in size. There is no canine fossa and the root is usually single.

The mandibular premolars differ from the maxillary premolars in that the occlusal surfaces appear rounder. Furthermore, the cusps are of unequal size, with the buccal cusp being more prominent.

The mandibular first premolar is the smallest premolar. The buccal cusp is broad with its tip overlying the midpoint of the crown. The lingual cusp is less than half the size of the buccal cusp. The cusps are connected by a transverse ridge which divides the occlusal surface into mesial and distal fossae. The distal fossa tends to be larger than the mesial fossa. The root is single, conical and oval in cross-section. It is grooved longitudinally both mesially and distally, the mesial groove being more prominent.

The mandibular second premolar differs markedly from the mandibular first premolar in both size and morphology. The tooth is larger than the first premolar. The cusps are separated by a well-defined mesiodistal occlusal fissure, and they are not joined by a transverse ridge. The lingual cusp is prominent and is usually subdivided into mesiolingual and distolingual cusps, the mesiolingual cusp being larger. The root is thicker than that of the first premolar.

The molars

The maxillary first molar is usually the largest molar. Viewed occlusally, the crown appears rhombic, the mesiopalatal and distobuccal angles being obtuse. The longest diameter of the crown runs from the mesiobuccal to the distopalatal corners. The crown has four cusps: the mesiobuccal, mesiopalatal, distobuccal and distopalatal cusps. The mesiopalatal cusp is the largest; the distopalatal cusp is the smallest. An accessory cusplet of variable size is seen on the palatal surface of the large mesiopalatal cusp in about 60% of people. This cusplet is called the tubercle of Carabelli. The four main cusps are separated by an H-shaped fissure. An oblique ridge crosses the occlusal table diagonally from the mesiopalatal to the distobuccal cusp through the bar of the 'H'. Mesial and distal marginal ridges are present, with the mesial ridge frequently showing several tubercles. On either side of the oblique ridge lies a fossa. The mesial fossa is larger than the distal fossa. A fissure runs from the distal fossa parallel to the oblique ridge and on to the palatal surface between the two palatal cusps. There are three roots: two buccal and one palatal. The palatal root is the largest and is circular in cross-section. The buccal roots are slender and are flattened mesiodistally. The mesiobuccal root is usually larger and wider than the distobuccal root.

The maxillary second molar resembles the first molar, but is smaller, with a more pronounced rhomboid form. The occlusal fissure pattern is more variable. The distopalatal cusp is reduced in size. A tubercle of Carabelli is not usually found on the mesiopalatal cusp. The second molar has three roots, two buccal and one palatal. However, they are shorter and less divergent than those of the maxillary first permanent molar. The roots may be partly fused.

The maxillary third molar is the most variable tooth in the dentition. Its shape may range from that of a maxillary first permanent molar to a simple peg-like tooth. The roots are often fused and irregular.

The mandibular molars differ from the maxillary molars in that their crowns are oblong, their fissure pattern is cross-shaped, and their lingual cusps are more equal in size. Whereas the buccal surface of the crown of a mandibular molar is more convex than its lingual surface, this relationship is reversed for a maxillary molar. Mandibular molars have only two roots.

The mandibular first molar has a crown that is pentagonal in outline. It is broader mesiodistally than buccolingually. The occlusal surface is divided into buccal and lingual parts by a mesiodistal fissure which arises from a deep central fossa. The mandibular first molar is the only permanent molar with five cusps. There are three cusps on the buccal side: mesiobuccal, distobuccal and distal. There are two lingual cusps: mesiolingual and distolingual. The mesiolingual cusp is the most prominent; the distal cusp is the smallest. The buccal cusps are shorter than the lingual cusps. There are mesial and distal roots. They are both markedly flattened mesiodistally and the mesial root is usually deeply grooved.

The mandibular second molar has a crown that is smaller than that of the first molar and is rectangular in shape. It has only four cusps: the mesiobuccal, mesiolingual, distobuccal and distolingual cusps. The mesial cusps are slightly larger than the distal cusps. The cusps are separated by a cross-shaped occlusal fissure pattern, although there may be numerous supplemental grooves. The mesial and distal roots are flattened mesiodistally and are smaller and less divergent than those of the first molar; indeed, they may be partly fused.

The mandibular third molar has a variable morphology, although not as variable as that of the maxillary third permanent molar. The crown is usually the smallest of the mandibular molars, but occasionally it may be as large as the mandibular first molar. The roots are greatly reduced in size and are often fused.

The pulp cavities of the permanent teeth (Figure 104)

What the layman refers to as 'the nerve of a tooth' is in reality an inner core of loose connective tissue called the dental pulp. This tissue occupies a central cavity within the tooth, called the pulp cavity. The pulp cavity can be subdivided into a pulp chamber in the crown, and root canals that pass into the root(s). The pulp cavity generally follows the contours of the tooth. Pulp horns (or cornua) extend from the pulp chambers to the mesial and distal angles of the incisor teeth and towards the cusps of the cheek teeth. Each root usually contains one root canal. However, the mandibular molars have two root canals in their mesial roots. Where roots are fused, the tooth still maintains the usual number of root canals. Each root canal opens by a foramen (or foramina) at the apex of the root. The pulp chambers of the anterior teeth merge imperceptibly into the root canals. In the cheek teeth, however, the pulp chambers and root canals are distinct. The size of the pulp cavity decreases significantly with age. The apical foramen is wide when the tooth first erupts into the oral cavity, but narrows with subsequent development of the root.

THE DECIDUOUS DENTITION (Figures 105 and 106)

The deciduous teeth differ from the permanent teeth in the following respects:

1 The deciduous teeth are smaller than their permanent successors, although the mesiodistal dimensions of the premolars are generally less than those for the equivalent deciduous molars.

2 Deciduous teeth have more consistent shapes.

3 The crowns of deciduous teeth appear bulbous, often having pronounced labial or buccal cingula.

4 The cusps of newly erupted deciduous teeth are more pointed.

5 The crowns of deciduous teeth are whiter, less mineralised and become very worn.

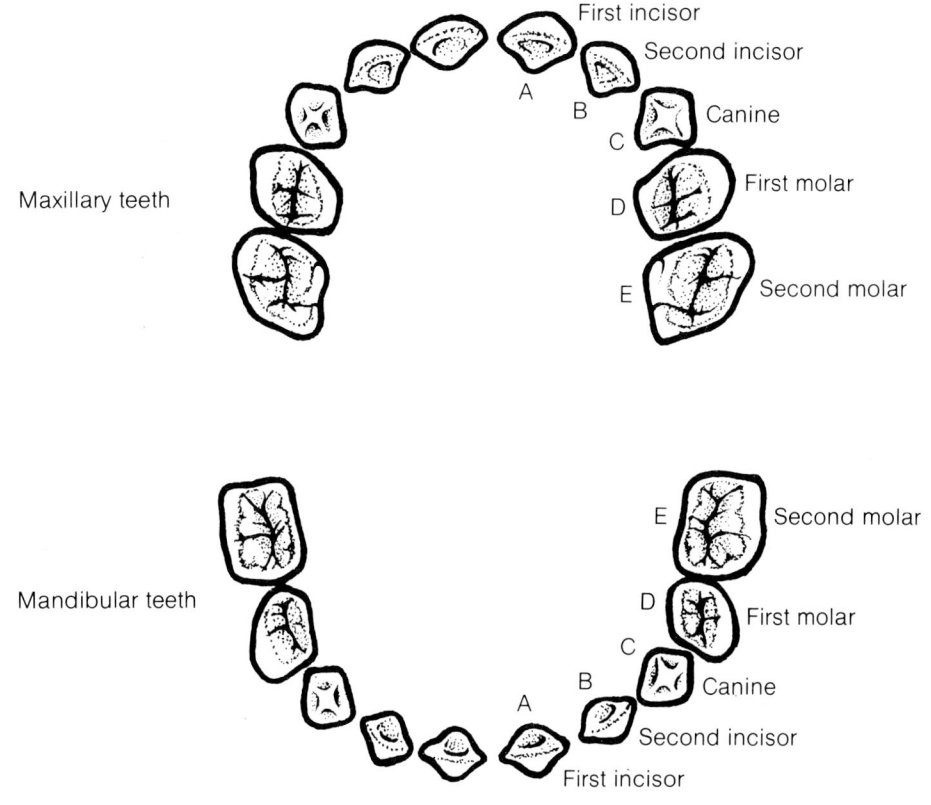

Figure 105 *Appearance of the occlusal surfaces of the deciduous dentition.* The letters A to E identify the teeth according to the Zsigmondy system.

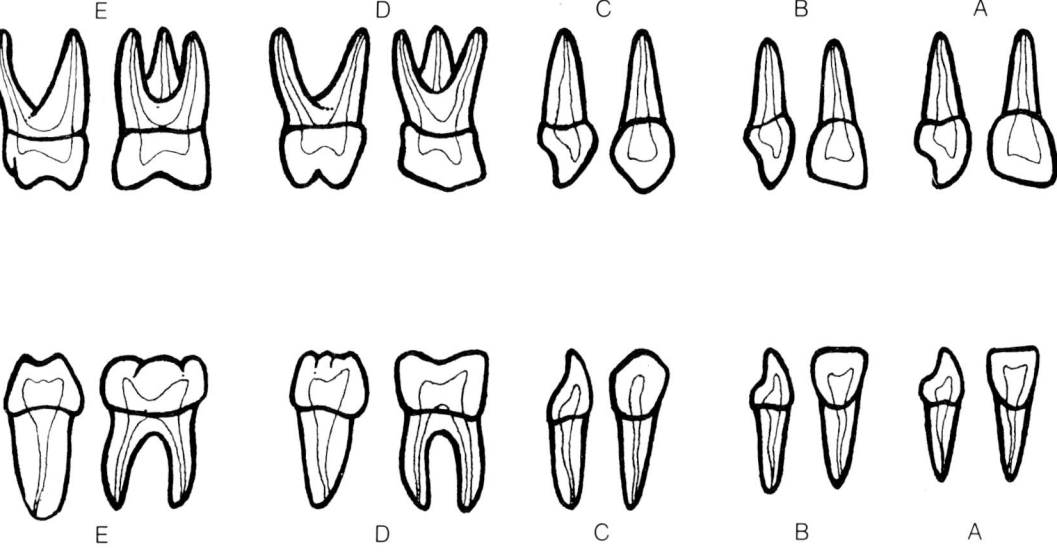

Figure 106 *Buccal and distal views of the maxillary (upper) and mandibular (lower) deciduous teeth.* Also shown is the outline of the pulp cavity. The teeth are lettered according to the Zsigmondy system.

6 The junctions between the crowns and the roots of deciduous teeth are more sharply demarcated than those of the permanent teeth, the enamel bulging at the cervical margins rather than gently tapering.

7 The roots of deciduous teeth are shorter and less robust than those of the permanent teeth.

8 The roots of the deciduous incisors and canines are longer in proportion to the crown than those of their permanent successors.

9 The roots of the deciduous molars are widely divergent, extending beyond the dimensions of the crown. This provides space for the initial development of the underlying permanent premolars.

10 The pulp cavities of deciduous teeth are proportionally larger than those of the permanent teeth.

11 The roots of the deciduous teeth undergo physiological resorption in order to allow for the eruption of the permanent teeth.

The morphology of a deciduous incisor or canine resembles that of the permanent successor. The second deciduous molars are similar to the permanent first molars. Thus, only the first deciduous molars have distinctive morphologies.

The maxillary first deciduous molar is intermediate in form between a premolar and a molar. The tooth is generally bicuspid. The occlusal surface is an irregular quadrilateral, the buccal and palatal margins being parallel. The mesiopalatal angle is markedly obtuse. The buccal and palatal cusps are separated by a mesiodistal fissure. The buccal cusp is more prominent. Marginal ridges link the cusps. A prominent bulge called the molar tubercle lies at the mesiobuccal corner near the root. The tooth has three roots: two buccal and one palatal. The palatal root is the largest root and is round in cross-section. The mesiobuccal root is flattened mesiodistally; the distobuccal root is smaller and more circular.

The mandibular first deciduous molar is molariform but has a number of unique features. The occlusal surface is elongated mesiodistally and is an irregular quadrilateral, the buccal and lingual surfaces being parallel. The mesiolingual angle is markedly obtuse. The occlusal table can be divided into buccal and lingual parts by a mesiodistal fissure. The buccal part consists of two cusps: the mesiobuccal and the distobuccal cusp. The mesiobuccal cusp is the larger of the two. The lingual part of the tooth is narrower than the buccal part and has two cusps: the mesiolingual and the distolingual cusps. The lingual cusps are separated by a lingual fissure and the mesiolingual cusp is larger than the distolingual cusp. The buccal cusps are slightly larger than the lingual cusps. A transverse ridge may connect the mesial cusps, dividing the mesiodistal fissure into a distal fissure and a mesial pit. The mandibular first deciduous molar has two divergent roots (mesial and distal). These are flattened mesiodistally. The mesial root is often grooved.

THE OCCLUSION OF THE PERMANENT TEETH

Occlusion is the relationship of the dental arches when tooth contact is made. The relationships of the jaws in function are so complex and various that our understanding of the functional articulation of teeth remains poor. To simplify analysis, several static occlusal positions have been strictly defined. These are classified into symmetric and asymmetric occlusal positions, corresponding with the classification of mandibular movements into symmetric and asymmetric movements (see page 174). The symmetric occlusal positions include centric occlusion and bilaterally protrusive occlusion. The asymmetric occlusal positions are those associated with lateral gliding movements. The most commonly used position is centric occlusion.

Centric position has been defined as the terminal position of physiological jaw movements. It is the relationship between the two arches when the teeth are brought

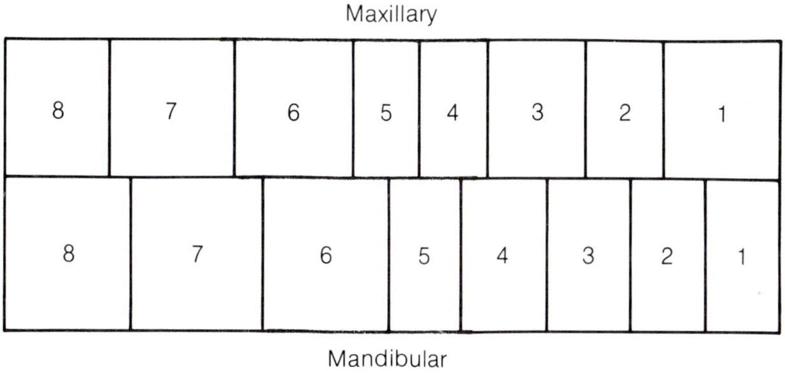

8	7	6	5	4	3	2	1
8	7	6	5	4	3	2	1

Maxillary (top) / Mandibular (bottom)

Figure 107 *The relationship between maxillary and mandibular permanent teeth in normal centric occlusion.* The teeth are identified according to the Zsigmondy numerical system.

into contact with the mandibular condyles centrally positioned at rest within the mandibular fossae. The key to the intercuspal relationships between the teeth in centric position is to be found in the relative positions of the maxillary and mandibular first permanent molars. In normal occlusion, each arch is bilaterally symmetrical. Since the anterior maxillary segment is slightly larger than the anterior mandibular segment, each maxillary tooth will contact its corresponding mandibular antagonist and its distal neighbour. This relationship continues around the dental arches so that the maxillary first permanent molar will contact the distal part of the mandibular first permanent molar and the mesial part of the mandibular second permanent molar (Figure 107).

The maxillary arch is a little larger and broader than the mandibular arch. Consequently, the maxillary arch slightly overlaps the mandibular arch, such that the buccal cusps of the maxillary teeth extend a few millimetres beyond the buccal occlusal edge of the mandibular teeth. This overlap is called overjet. Thus, the buccal cusps in the lower jaw fit into the fissures between the buccal and palatal cusps in the upper jaw. The lingual cusps in the lower jaw lie on the lingual side of the palatal cusps in the upper jaw (Figure 108).

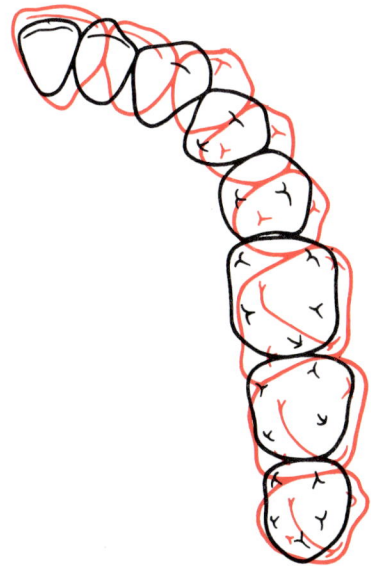

Figure 108 *The relationships between the maxillary and mandibular permanent teeth in normal centric occlusion* shown by the superimposition of the occlusal surfaces of the maxillary teeth (red) on those of the mandibular teeth (black).

The maxillary incisors overlap the mandibular incisors in two planes. Not only is there overjet (approximately 1 to 2 mm) but there is also a vertical overlap called overbite. The overbite in centric occlusion is such that the palatal surfaces of the maxillary incisors overlap the incised third of the labial surfaces of the mandibular incisors.

During mastication, the cusps and associated ridges of the mandibular teeth are carried across those of the maxillary teeth, producing shearing forces that comminute the food.

THE ORAL MUCOSA

The oral mucosa is composed of a layer of stratified squamous epithelium, which lies upon a connective tissue of varying thickness called the lamina propria. An additional layer of connective tissue, the submucosa, may or may not be present. Indeed, the oral mucosa shows a number of regional variations which depend upon functional demands.

There are three types of oral mucosa: masticatory, lining and specialised mucosa. Masticatory mucosa is found in regions that are particularly exposed to stresses associated with mastication. Among its characteristic features are a keratinized

TABLE 3 Principal features and regional variations of the oral mucosa.

REGION	EPITHELIUM		LAMINA PROPRIA		SUBMUCOSA		TYPE OF MUCOSA
	Thickness	Keratinization	Papillae	Fibre types	Density	Attachment	
Labial and buccal mucosa	Thick	Non-keratinized	Short and irregular	Collagen and some elastic	Dense	Firmly to underlying muscle	Lining
Transitional (red) zone of lip	Thin	Keratinized	Long and narrow	Collagen and some elastic	Dense	Firmly to underlying muscle	Specialised
Alveolar mucosa	Thin	Non-keratinized	Short or absent	Many elastic fibres	Loose	Loose attachment to periosteum	Lining
Attached gingiva	Thick	Keratinized and parakeratinized	Long and narrow	Dense collagen firmly attached to underlying periosteum	No distinct submucosa		Masticatory
Floor of mouth	Thin	Non-keratinized	Short and broad	Collagen and some elastic	Loose	Loose attachment to underlying muscle	Lining
Ventral surface of tongue	Thin	Non-keratinized	Short and numerous	Collagen and some elastic	Not very distinct layer; attached to underlying muscle		Lining
Dorsum of tongue (anterior two-thirds	Thick	Primarily keratinized	Long	Collagen and some elastic	Not very distinct layer; attached to underlying muscle		Specialised gustatory
Dorsum of tongue (posterior one-third)	Variable	Generally non-keratinized	Short or absent	Collagen and some elastic	Not very distinct layer; attached to underlying muscle		Lining gustatory
Hard palate	Thick	Keratinized	Long	Dense collagen in submucosa laterally, but lamina propria firmly bound to periosteum without submucosa in midline			Masticatory
Soft palate	Thick	Non-keratinized	Short	Many elastic fibres	Loose	Loose attachment to underlying tissues	Lining

epithelium and a thick lamina propria which is tightly bound to underlying periosteum. Lining mucosa is less exposed to masticatory loads and has a non-keratinized epithelium which lines a thin elastic lamina propria and a submucosa. Specialised mucosa has the characteristics neither of lining mucosa nor of masticatory mucosa. The mucosa covering the gingivae and palate is masticatory mucosa. The mucosa covering the lips, cheeks, alveolus and vestibule, floor of mouth, ventral surface of tongue, and soft palate is lining mucosa. The mucosa of the dorsum of the tongue is a specialised gustatory mucosa, which exhibits a considerable number of papillae. Some of the papillae are keratinized (the filiform papillae), others are non-keratinized (the fungiform and circumvallate papillae). Part of the mucosa of the lip (the vermilion) may also be regarded as specialised.

Table 3 summarises the regional variations of the oral mucosa.

THE SALIVARY GLANDS

Salivary glands are compound, tubular, acinous, merocrine glands whose ducts open into the oral cavity. They secrete a fluid, the saliva, which among its many functions aids in the mastication, digestion and deglutition of food.

There are three pairs of major salivary glands: the parotid, the submandibular and the sublingual glands. Numerous minor salivary glands are also scattered throughout the oral mucosa.

The parotid gland

The parotid gland is the largest salivary gland. It is almost entirely serous. The parotid duct runs through the cheek and drains into the mouth opposite the maxillary second permanent molar tooth. The parotid gland is situated in front of the external ear and consequently is described in detail in relation to the face (see pages 164 to 168).

112,35

112,29; 144A,15

The submandibular gland

This gland is intermediate in size between the parotid and sublingual glands. The submandibular gland is a mixed gland, serous cells predominating over the mucous cells in the ratio of approximately 3:2.

96,46; 140D

The submandibular gland is found in the floor of the mouth and in the suprahyoid region of the neck. A large part of the gland is visible just beneath the inferior border of the mandible. There is a prominent depression on the inner surface of the mandible, below the mylohyoid line, for the gland (the submandibular fossa). The gland has an important relationship with the mylohyoid muscle, wrapping around the posterior free edge of the muscle (not unlike a letter C). Thus, the gland is divided into a superficial and a deep part.

114,7

140D; 34C,22
142A,26; 142A,38

140D,73; 140D,74

The superficial part of the submandibular gland is located in the digastric triangle of the neck. Indeed, it occupies most of this region and often extends to overlap the digastric muscle. Posteriorly, it comes to lie close to the apex of the parotid gland, with only the stylomandibular ligament intervening. Superiorly, the superficial part of the submandibular gland lies under the inferior border of the mandible. Inferiorly, it is partially enclosed by the investing layer of deep cervical fascia. The medial surface lies on the mylohyoid muscle and hooks round the posterior free edge of this muscle to link with the deep part of the submandibular gland.

140D,73; 96,46

96,44; 96,5

96,47

142A,38; 142A,26

The superficial part of the submandibular gland is related to the facial artery, the facial vein, the cervical branch of the facial nerve, the mylohyoid nerve, and the submandibular lymph nodes. The facial artery passes across the upper surface of the gland and often produces a distinct groove. The artery then runs between the gland and the inferior border of the mandible before crossing on to the face. The facial vein and the cervical branch of the facial nerve (sometimes also the marginal mandibular branch of the facial nerve) cross the gland superficially. The submandibular lymph nodes are usually located close to the superficial part of the gland, beneath the inferior border of the mandible. Indeed, some lymph nodes are embedded in the substance of

96,8; 140D,75
96,11; 96,6

96,3

the gland. The mylohyoid nerve and the mylohyoid and submental vessels are found between the gland and the mylohyoid muscle.

140D,74; 142A,26
142A,35; 138B,32
138B,8

The deep part of the submandibular gland is situated on the superficial surface of the hyoglossus muscle in the floor of the mouth. It extends forward to reach the posterior end of the sublingual salivary gland. Indeed, the two glands may merge to form a sublingual-submandibular complex. It is from the deep part of the submandibular gland that the submandibular duct emerges. The submandibular gland on the hyoglossus muscle is closely related to the lingual and hypoglossal nerves (see Figure 113).

138B,6
138B,5; 138B,13

Where the submandibular gland folds around the posterior border of the mylohyoid muscle, it lies close to the styloglossus muscle, stylohyoid ligament and the glossopharyngeal nerve. These structures separate the gland from the wall of the pharynx.

138,6; 140D,71
142A,39
142A,38
142A,35
138B,5

138B,7

140B,58; 140B,56

The submandibular duct (Wharton's duct) commences in the superficial part of the submandibular gland but emerges from the deep part. The duct is approximately 5 cm long. It runs forwards towards the floor of the mouth between the mylohyoid and the hyoglossus muscles (see Figures 112 and 113). In the floor of the mouth, the duct initially lies below the lingual nerve. As these structures pass behind the sublingual gland, the duct is crossed twice by the lingual nerve, first on its lateral side then on its medial side. The duct now continues forward to empty at the sublingual papilla in the floor of the mouth (see Figure 98). The terminal part of the submandibular duct lies just beneath the oral mucosa on the sublingual fold. Just before it opens into the mouth, the submandibular duct may receive a duct(s) from the anterior part of the sublingual gland.

138B,8; 140D,70
144A,20
140B,57; 140B,51
140B,56

The sublingual gland

The sublingual salivary gland is the smallest of the major salivary glands. It is located in the floor of the mouth, above the mylohyoid muscle and below the oral mucosa. The sublingual gland produces a distinct ridge under the tongue, the sublingual fold (see Figure 98). The gland is narrow and flat and can be described as almond-shaped.

The sublingual gland is a mixed gland. It differs from the submandibular gland, however, because the mucous elements predominate over the serous in a ratio of approximately 3:1.

140D,7; 34C,23
140B,52
140B,58; 138B,6
138B,5; 138B,13
138B,9

The sublingual gland lies adjacent to the sublingual fossa of the mandible. The genioglossus muscle runs medial to the gland. The two sublingual glands almost meet in the midline. Between the sublingual gland and the genioglossus muscle lie the submandibular duct (near the upper border of the gland), the lingual and hypoglossal nerves (as they run into the substance of the tongue) and the sublingual artery (see Figures 112 and 113).

140B,56

140B,58
140B,55

The sublingual gland can be subdivided into anterior and posterior parts. Each part differs according to the method of drainage. Unlike the other major salivary glands, there is not a single main salivary duct, but as many as 20 small ducts. The ducts from the posterior part of the sublingual gland empty at the sublingual fold in the floor of the mouth. The ducts from the anterior part of the gland may unite to form a larger duct which either joins the submandibular duct or drains directly into the floor of the mouth at the sublingual papilla. This larger duct is sometimes referred to as Bartholin's duct.

The minor salivary glands

140A,14; 142B,61

The minor salivary glands include the labial, buccal, palatoglossal, palatal and lingual glands. The labial and buccal glands contain both mucous and serous elements. The palatoglossal glands are mucous glands and are located around the pharyngeal isthmus. The palatal glands are also mucous glands. They are located in both the soft and hard palate. The anterior and posterior lingual glands are mainly mucous. The anterior glands are embedded within muscle near the ventral surface of the tongue and open by means of four or five ducts near the lingual frenum. The posterior glands are located in the root of the tongue. Around the circumvallate papillae are serous glands (of von Ebner).

The innervation of the salivary glands

The salivary glands are innervated by way of cranial parasympathetic ganglia. The secretomotor supply for the parotid gland is derived from the glossopharyngeal nerve via the otic ganglion (see 167 to 168 and Figure 54). The secretomotor supply for the submandibular and sublingual glands is provided by the facial nerve via the submandibular ganglion (see page 304 and Figure 119). The minor salivary glands are innervated by parasympathetic fibres travelling with sensory branches of the trigeminal nerve (e.g. palatine glands by the greater and lesser palatine nerves via the pterygopalatine ganglion).

THE ORAL MUSCULATURE

THE MUSCLES OF THE LIPS AND CHEEK

Within the lips is the orbicularis oris muscle, and within the cheek is the buccinator muscle. These belong to the group called the muscles of facial expression, and are consequently described with the face (see pages 151 to 157). The orbicularis oris muscle is innervated by the buccal and marginal mandibular branches of the facial nerve. The buccinator muscle is supplied by the buccal branch of the facial nerve.

THE MUSCLES OF THE TONGUE

The tongue is composed of intrinsic and extrinsic muscles. The intrinsic muscles are restricted to the substance of the tongue, whereas the extrinsic muscles arise outside the tongue.

The intrinsic muscles of the tongue can be divided into three fibre groups: the transverse, longitudinal and vertical groups (Figure 109). However, these groups cannot readily be distinguished because their fibres intercalate. The transverse fibres pass laterally from a sheet of connective tissue that runs longitudinally through the midline of the tongue (the lingual septum). The longitudinal fibres are subdivided into the superior and inferior longitudinal muscles of the tongue. The vertical fibres pass directly between the upper and lower surfaces of the tongue. They are particularly prominent at the lateral borders of the tongue. The intrinsic musculature

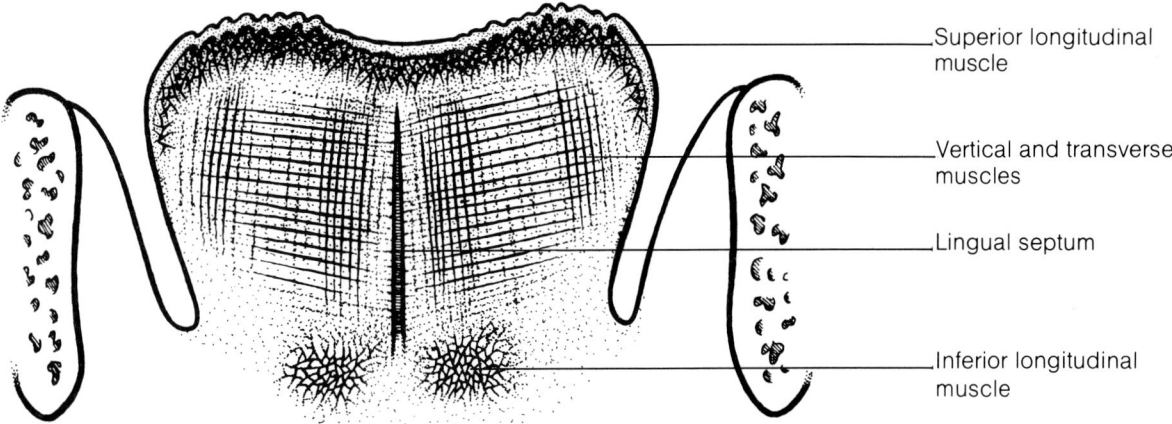

Figure 109 *Coronal section of the tongue to show the intrinsic musculature.*

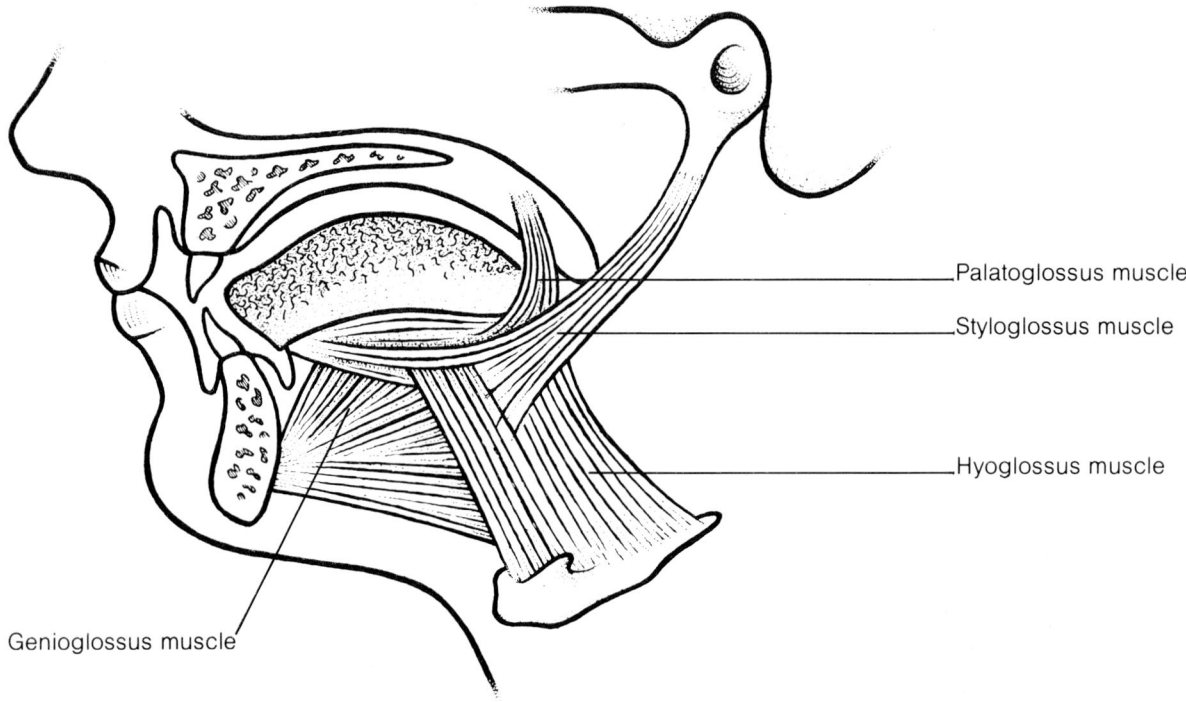

Figure 110 *Extrinsic musculature of the tongue (lateral view).*

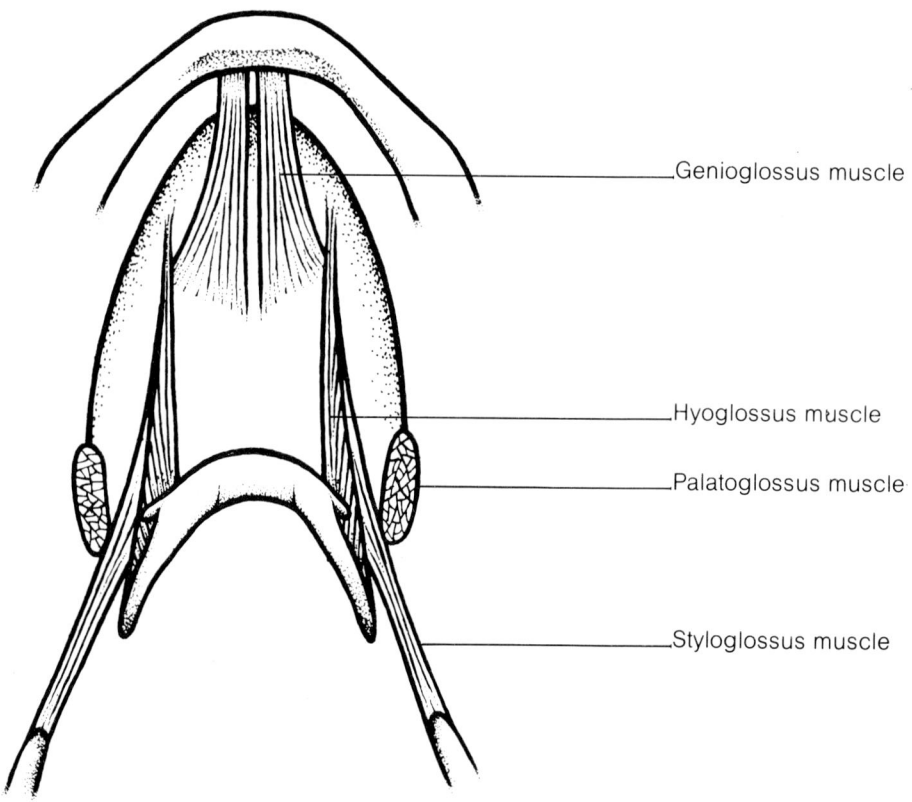

Figure 111 *Extrinsic musculature of the tongue (inferior view).*

is responsible for changing the shape of the tongue. The muscles receive their motor innervation from the hypoglossal nerve.

The extrinsic muscles of the tongue arise from the skull and hyoid bone and thence spread into the body of the tongue. The extrinsic musculature is composed of four pairs of muscles: genioglossus, hyoglossus, styloglossus and palatoglossus (Figures 110 and 111).

The genioglossus muscle

Attachments. This muscle lies anteriorly, near the median plane. It originates from the superior genial tubercle (mental spine) on the medial surface of the body of the mandible. The genioglossus muscle fans out into the substance of the tongue. The two genioglossus muscles cannot easily be separated near their origins. As the muscles enter the tongue, however, a thin strip of connective tissue intervenes. The superior fibres of the genioglossus muscle pass upwards and anteriorly towards the tip of the tongue. Some of the inferior fibres insert on to the body of the hyoid bone.

Innervation. Genioglossus receives its motor innervation from the hypoglossal nerve.

Vasculature. The lingual artery (sublingual branch) and the facial artery (submental branch) supply this muscle.

Actions. The genioglossus muscle is a protractor and depressor of the tongue.

The hyoglossus muscle

Attachments. This is a thin, quadrilateral muscle which provides an important landmark in the floor of the mouth (see Figures 112 and 113). It originates from the superior border of the greater horn of the hyoid bone. It passes vertically upwards to insert into the side of the tongue. A part of the muscle is attached to the base of the lesser horn of the hyoid bone and has been called chondroglossus. At its origin, the hyoglossus muscle is separated from the attachment of the middle constrictor muscle of the pharynx by the lingual artery.

Innervation. The hypoglossal nerve supplies the hyoglossus muscle.

Vasculature. The lingual artery (sublingual branch) and the facial artery (submental branch) supply hyoglossus.

Actions. The muscle depresses the tongue.

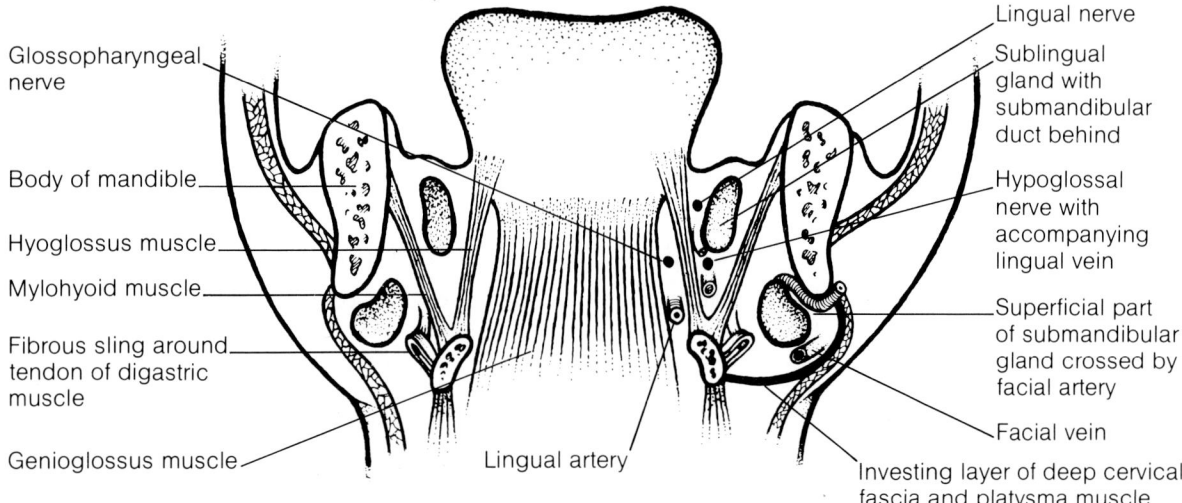

Figure 112 *Coronal section through the floor of the mouth* to show the relationships of structures to the mylohyoid and genioglossus muscles.

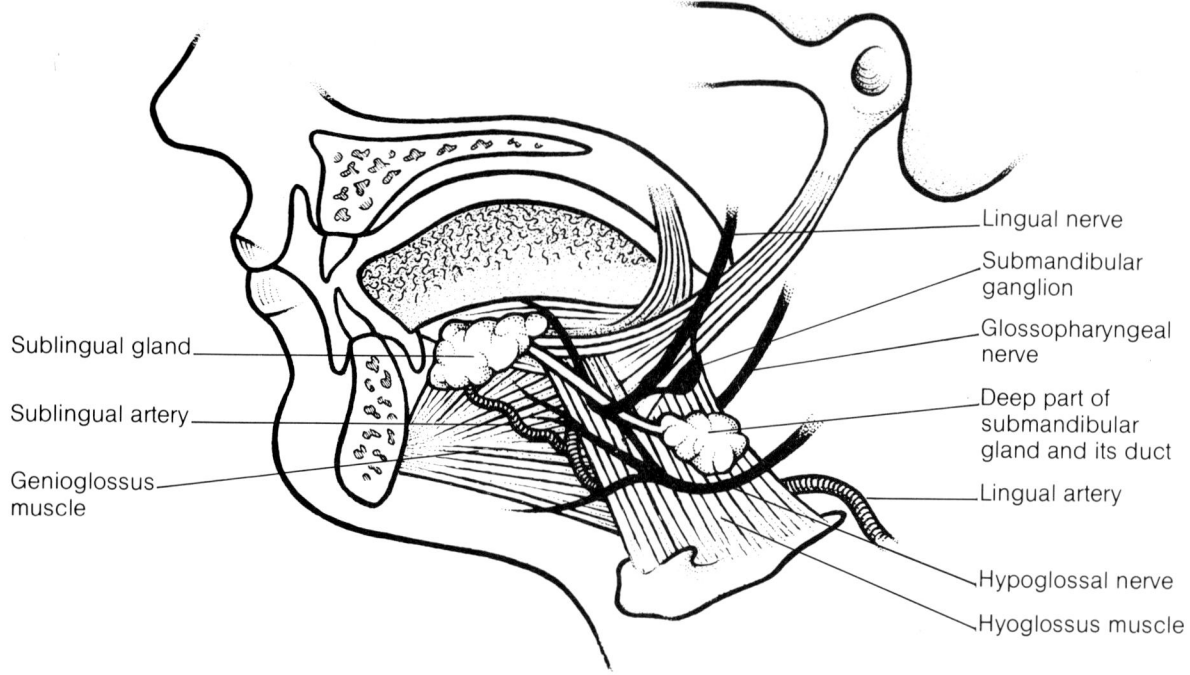

Figure 113 *Floor of the mouth showing the relationships of structures to the hyoglossus muscle.*

The styloglossus muscle

Attachments. Styloglossus arises from the tip of the styloid process of the temporal bone and from the upper end of the stylomandibular ligament. It passes downwards and forwards to enter the tongue below the insertion of the palatoglossus muscle. At this point, its fibres intercalate with the fibres of the hyoglossus muscle. The muscle then continues forwards towards the tip of the tongue.

Innervation. The muscle is supplied by the hypoglossal nerve.

Vasculature. The sublingual branch of the lingual artery supplies the styloglossus muscle.

Actions. Styloglossus retracts and elevates the tongue.

The palatoglossus muscle

Palatoglossus is more closely associated with the palate than the tongue. It is therefore described with the muscles of the palate (see page 297).

THE MUSCLES IN THE FLOOR OF THE MOUTH

The floor of the mouth is the region located between the medial surface of the mandible, the inferior surface of the tongue, and the mylohyoid muscles. The mylohyoid muscles are attached to the mylohyoid lines of the mandible and consequently structures above these lines are related to the floor of the mouth, whereas structures below the lines are related to the upper part of the neck.

The two mylohyoid muscles form a muscular diaphragm for the floor of the mouth. Above this diaphragm are found the genioglossus and geniohyoid muscles medially and the hyoglossus muscle laterally. Below the diaphragm lie the digastric and stylohyoid muscles.

The mylohyoid muscle

Attachments. The mylohyoid muscle arises from the mylohyoid line (internal oblique line) on the medial surface of the body of the mandible. Its fibres slope downwards, forwards and inwards. The anterior fibres of the mylohyoid muscle interdigitate with the corresponding fibres on the opposite side to form a median raphe. This raphe is attached above to the symphysis menti of the mandible, and below to the hyoid bone. The posterior fibres of the mylohyoid muscle are inserted on to the anterior surface of the body of the hyoid bone (near its lower border).

Innervation. The muscle is supplied by the mylohyoid branch of the mandibular division of the trigeminal nerve.

Vasculature. The mylohyoid muscle receives its arterial supply from three sources: the lingual artery (sublingual branch), the maxillary artery (the mylohyoid branch of the inferior alveolar artery), and the facial artery (submental branch).

Actions. The muscle raises the floor of the mouth during the early stages of swallowing. It also helps to depress the mandible when the hyoid bone is fixed. Conversely, it aids in elevation of the hyoid bone.

The geniohyoid muscle

Attachments. This muscle originates from the inferior genial tubercle (mental spine). It passes backwards and slightly downwards to insert on to the body of the hyoid bone (anterior surface).

Innervation. The innervation is shared with the thyrohyoid muscle, namely the first cervical spinal nerve travelling with the hypoglossal nerve.

Vasculature. The blood supply is derived from the lingual artery (sublingual branch).

Actions. The geniohyoid muscle elevates the hyoid bone and is a weak depressor of the mandible.

The main structures found above the mylohyoid muscle (i.e. in the floor of the mouth) are: the lingual, glossopharyngeal and hypoglossal nerves, the sublingual gland and the deep part of the submandibular gland, the lingual artery and the submandibular parasympathetic ganglion. The relationships of these structures to the hyoglossus and mylohyoid muscles are illustrated in Figures 112 and 113.

The structures related to the inferior surface of the mylohyoid muscle (i.e. the digastric and submental triangles of the neck; see pages 86 and 87) include: the superficial part of the submandibular gland, the facial, submental and mylohyoid blood vessels, the submandibular and submental lymph nodes, the mylohyoid nerve and the veins that form the anterior jugular veins in the submental triangle.

THE MUSCLES OF THE PALATE (Figure 114)

The soft palate has a fibrous aponeurosis whose shape and position is altered by the tensor veli palatini, the levator veli palatini, the palatoglossus, and the palatopharyngeus muscles. The musculature of the soft palate also includes the musculus uvulae.

The tensor veli palatini muscle

Attachments. This muscle arises from the scaphoid fossa of the sphenoid bone (at the root of the pterygoid plates) and from the lateral side of the cartilaginous part of the auditory tube. The fibres converge towards the pterygoid hamulus, where the muscle becomes tendinous. The tendon bends at right angles around the hamulus to become the palatine aponeurosis. The aponeurosis is attached to the posterior border of the hard palate. Medially, it merges with the aponeurosis of the other side. Posteriorly, it becomes indistinct and merges with the submucosa at the posterior edge of the soft palate.

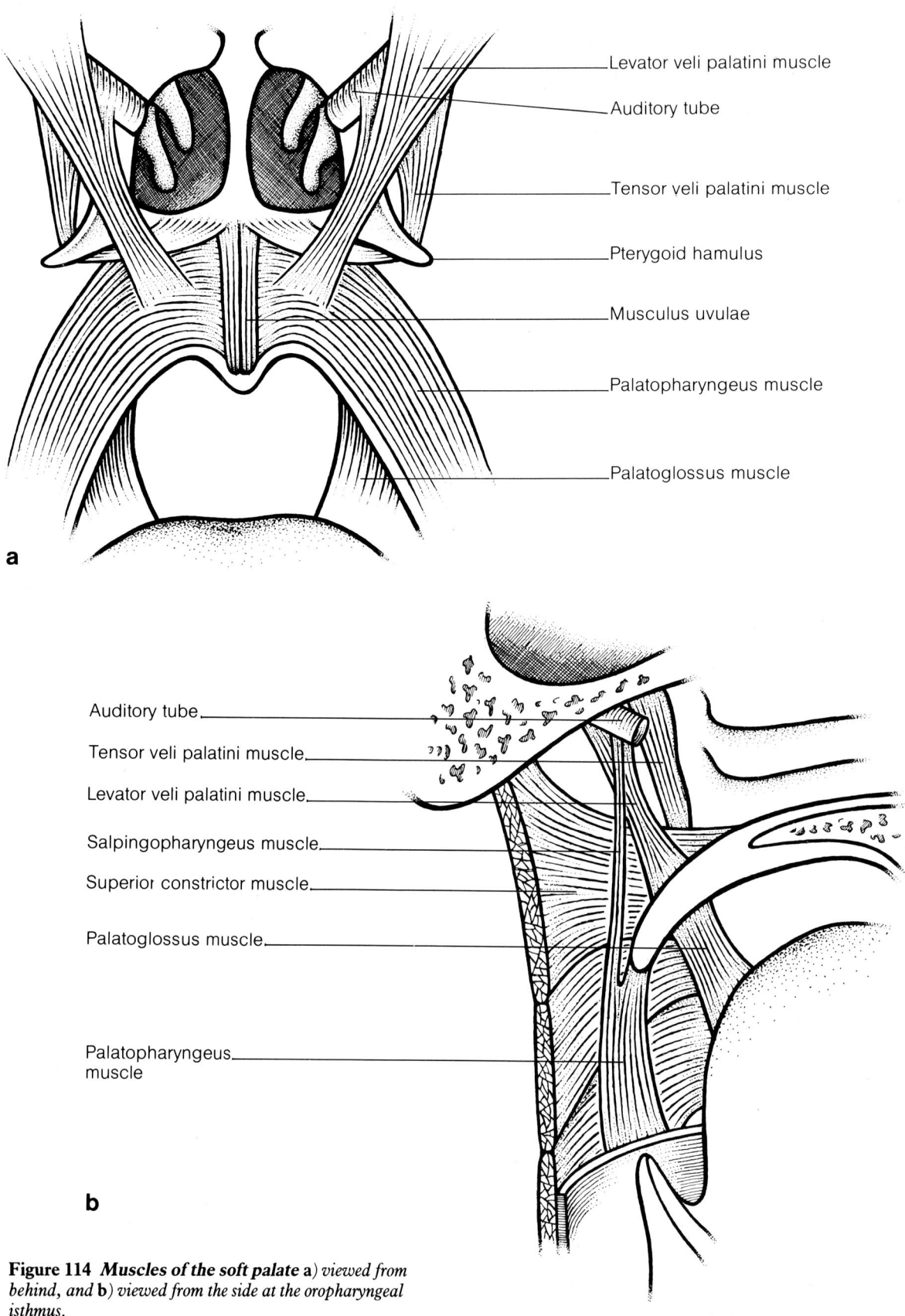

Figure 114 *Muscles of the soft palate* a) *viewed from behind, and* b) *viewed from the side at the oropharyngeal isthmus.*

Innervation. The motor innervation of the muscle is derived from the mandibular nerve via the nerve to the medial pterygoid muscle.

142A,9

Vasculature. The arterial blood supply is derived from the facial artery (ascending palatine branch) and the maxillary artery (descending palatine branch).

Actions. When the tensor veli palatini muscles act together, the palatine aponeurosis becomes taut and horizontal and provides a platform upon which other palatine muscles may act to change the position of the soft palate. Acting singly, the muscle pulls the soft palate laterally.

The levator veli palatini muscle

142C,81; 144A,9
144B,68; 146B,34
24,26

Attachments. This muscle originates from the base of the skull at the apex of the petrous part of the temporal bone, and from the medial side of the cartilaginous part of the auditory tube. The muscle curves downwards, medially and forwards to enter the palate immediately below the opening of the auditory tube. In its course, the muscle passes between the base of the skull and the superior margin of the superior constrictor muscle of the pharynx (see Figure 123).

146B,34; 146B,32

Innervation. The nerve supply to the muscle is derived from the cranial part of the accessory nerve via the pharyngeal plexus.

Vasculature. The blood supply is derived from the facial artery (ascending palatine branch) and the maxillary artery (descending palatine branch).

Actions. The levator muscles of the palate form a U-shaped muscular sling. When the palatine aponeurosis is stiffened by the tensor muscles, contraction of the levator muscles produces an upwards and backwards movement of the soft palate. In this way, the nasopharynx is shut off from the oropharynx by the apposition of the soft palate on to the posterior wall of the pharynx.

The palatoglossus muscle

138B,22; 144B,86

This muscle can be classified either as a muscle of the palate or as an extrinsic muscle of the tongue.

Attachments. The palatoglossus muscle arises from the palatine aponeurosis and runs to the tongue as the anterior pillar of the fauces. Its fibres intercalate with the transverse muscle fibres of the tongue.

Innervation. Unlike the extrinsic muscles of the tongue, which are innervated by the hypoglossal nerve, the palatoglossus muscle is supplied by the cranial part of the accessory nerve via the pharyngeal plexus.

Vasculature. The muscle receives its blood supply from the facial artery (ascending palatine branch) and the ascending pharyngeal artery.

Actions. The palatoglossus muscles acting together raise the tongue and narrow the oropharyngeal isthmus.

The palatopharyngeus muscle

144B,71

Although this muscle arises from the soft palate, it is described as a longitudinal muscle of the pharynx (see page 316).

Musculus uvulae arises from the posterior nasal spine at the back of the hard palate and from the palatine aponeurosis. It passes backwards and downwards to insert into the mucosa of the uvula. Its innervation and vasculature is similar to that of the levator veli palatini muscle. It moves the uvula upwards and laterally.

24,1

136,46

THE INNERVATION OF THE ORO-DENTAL TISSUES

The oral mucosa receives its sensory innervation primarily from the maxillary and mandibular divisions of the trigeminal nerve. The trigeminal nerve also supplies the teeth and their supporting tissues. The salivary glands are supplied by secretomotor parasympathetic fibres from the facial and glossopharyngeal nerves. The motor innervation of the oral musculature is derived mainly from the mandibular, facial, accessory and hypoglossal nerves.

THE INNERVATION OF THE TEETH AND GINGIVAE

The dentition in the lower jaw is innervated by the mandibular nerve. The teeth receive their nerve supply from the molar and incisive branches of the inferior alveolar nerve. The lingual gingivae are supplied mainly by the lingual branch of the mandibular nerve. The labial gingivae are innervated by the mental branch of the inferior alveolar nerve and the buccal gingivae by the buccal branch of the mandibular nerve.

The innervation of the dentition in the upper jaw is derived almost entirely from the maxillary nerve. The teeth are supplied by the anterior, middle and posterior superior alveolar branches. The palatal gingivae are innervated by the nasopalatine and greater (anterior) palatine branches via the pterygopalatine ganglion. The labial and buccal gingivae are supplied by the infra-orbital and the posterior superior alveolar branches.

Table 4 summarises the nerve supply to the teeth and gingivae.

The inferior alveolar nerve

The inferior alveolar nerve is the terminal branch of the posterior trunk of the mandibular nerve (see Figure 60). It arises in the infratemporal fossa, deep to the lower head of the lateral pterygoid muscle. On emerging from beneath this muscle, the inferior alveolar nerve lies within the pterygomandibular space. Here, it gives off a mylohyoid branch which is a motor nerve to the anterior belly of the digastric muscle and the mylohyoid muscle (it may also have sensory fibres which enter the mandible in the mental region to participate in the nerve supply to the lower incisors). The inferior alveolar nerve enters the mandible through the mandibular foramen.

TABLE 4 Sensory innervation of the teeth and gingivae. The teeth are numbered according to their position along the tooth row. (See Figure 103.)

Maxilla	Nasopalatine nerve	Greater palatine nerve				Palatal gingiva
	Anterior superior dental nerve	Middle superior alveolar nerve		Posterior superior alveolar nerve		Teeth
	Infra-orbital nerve	Posterior superior alveolar nerve and buccal nerve				Buccal gingiva
	1 2 3	4	5	6	7 8	
Mandible	Mental nerve	Buccal nerve and perforating branches of inferior alveolar nerve				Buccal gingiva
	Incisive nerve	Inferior alveolar nerve				Teeth
	Lingual nerve and perforating branches of inferior alveolar nerve					Lingual gingiva

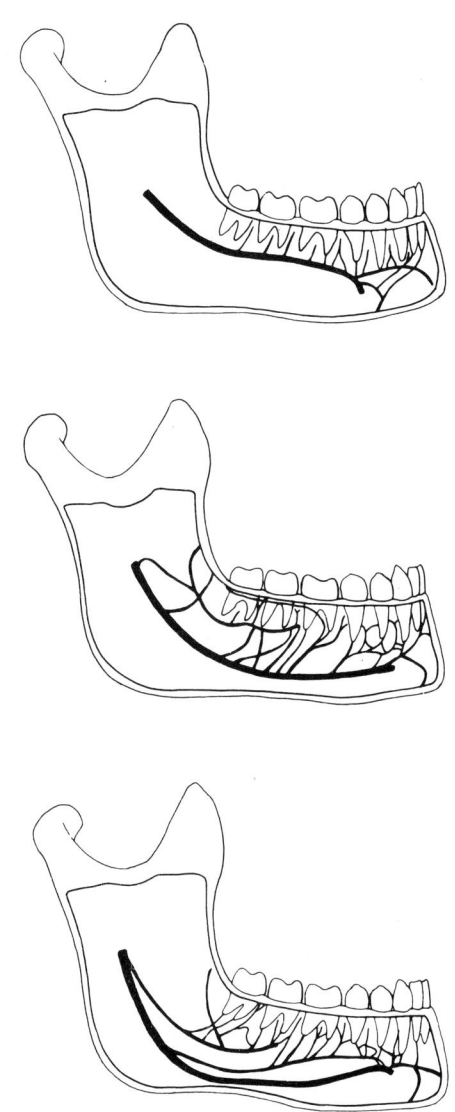

Figure 115 *Variations in the course of the inferior alveolar nerve* (*modified after Carter RB, Keen EN. J Anat 1971; 103: 433*).

The course of the inferior alveolar nerve through the mandible is variable (Figure 115). The molar branches to the premolar and molar teeth come either directly from the inferior alveolar nerve in the mandibular canal by short or long branches, or indirectly from the nerve outside the mandibular canal by a series of alveolar branches. The mandibular canal may be closely related to the roots of the mandibular molars, even to the extent of occasionally perforating a root.

The main trunk of the inferior alveolar nerve divides near the premolars into mental and incisive nerves. The mental nerve runs for a short distance in a mental canal before leaving the body of the mandible at the mental foramen to emerge on to the face. It supplies the skin and mucosa of the lower lip, and the labial gingivae of the mandibular anterior teeth. The incisive nerve runs forwards in an incisive canal. This nerve usually innervates only the incisor and canine teeth, but occasionally it supplies the first premolar.

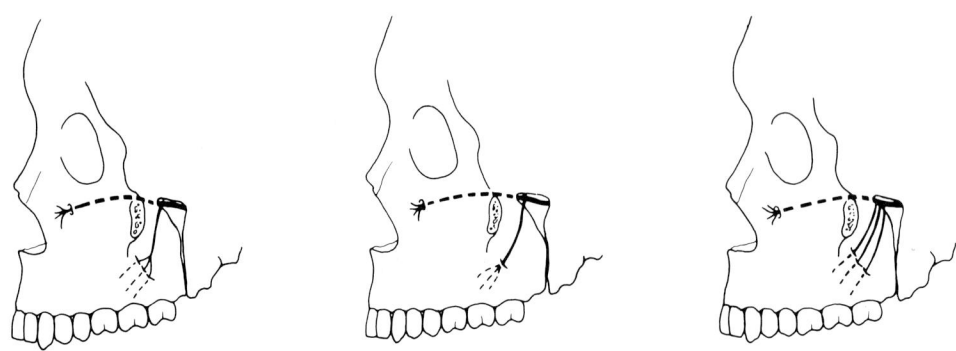

Figure 116 *Variations in the extra-bony course of the posterior superior alveolar nerve.*

The superior alveolar nerves

There are usually three superior alveolar nerves supplying the maxillary dentition: the posterior, middle and anterior superior alveolar nerves (see Figure 87a).

The posterior superior alveolar nerve arises from the maxillary nerve in the pterygopalatine fossa (see page 255). It descends on to the posterior wall of the maxilla, passing through the pterygomaxillary fissure, and divides into dental and gingival branches. The dental branches enter the maxilla and run in narrow canals above the roots of the molar teeth. The gingival branch does not enter the bone but runs along the outer surface of the maxillary tuberosity to supply the buccal gingivae of the maxillary molar teeth. Figure 116 illustrates some variations in the course of the posterior superior alveolar nerve.

The middle superior alveolar nerve is found in about 70% of individuals. The nerve generally arises in the floor of the orbit from the infra-orbital branch of the maxillary nerve. It can also arise directly from the maxillary nerve in the pterygopalatine fossa. The middle superior alveolar nerve runs in the posterior, lateral or anterior wall of the maxillary air sinus. It terminates above the roots of the premolar teeth.

The anterior superior alveolar nerve arises from the infra-orbital nerve within the infra-orbital canal. It generally appears as a single nerve, but occasionally as two or three branches. The nerve runs in the anterior wall of the maxillary sinus and terminates near the anterior nasal spine after giving off a small nasal branch.

The three superior alveolar nerves form a plexus just above the roots of the maxillary teeth. Indeed, it is difficult to trace the precise innervation of the teeth from a specific superior alveolar nerve. As a general rule, however, the incisors and canine are supplied by the anterior nerve, the molars by the posterior nerve and the premolars by the middle nerve.

THE SENSORY INNERVATION OF THE LIPS AND CHEEKS

The mucosa of the upper lip is supplied by the infra-orbital branch of the maxillary nerve. The lower lip is innervated by the mental branch of the mandibular nerve.

The mucosa of the cheek is innervated by the buccal branch of the mandibular nerve.

The buccal nerve

The buccal nerve is the terminal branch of the anterior trunk of the mandibular nerve. It arises in the infratemporal fossa (see page 178 and Figure 60), behind the upper head of the lateral pterygoid muscle. The buccal nerve passes between the two heads of the lateral pterygoid muscle and crosses the infratemporal fossa. It runs into the

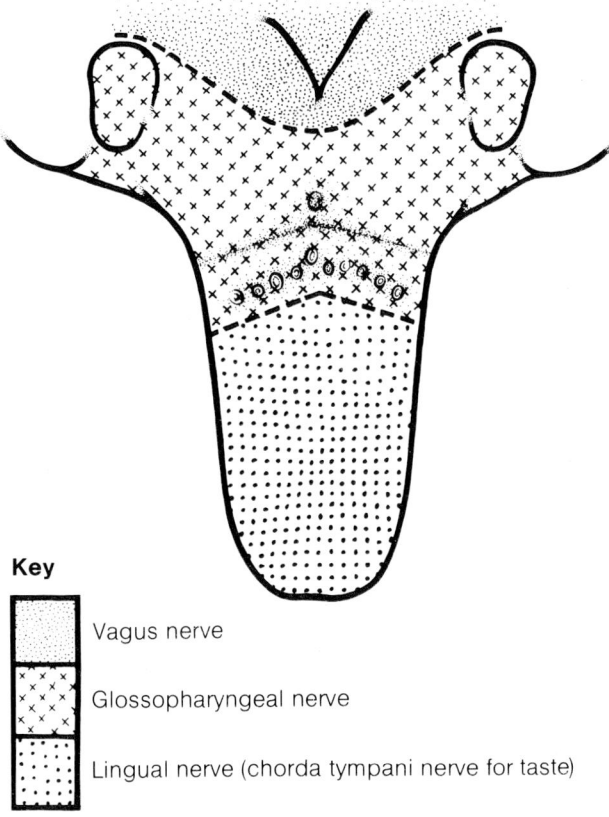

Figure 117 *Sensory innervation of the dorsum of the tongue.*

upper part of the retromolar fossa at the anterior border of the ramus of the mandible. The buccal nerve breaks up into several branches within the buccinator muscle. It innervates both the mucosa and the skin of the cheek (see Figure 49 for the cutaneous innervation) and the buccal gingivae of the mandibular cheek teeth (perhaps even of the maxillary cheek teeth).

THE SENSORY INNERVATION OF THE TONGUE AND THE FLOOR OF THE MOUTH

Concerning general sensation (i.e. excluding taste), three distinct nerve fields can be recognised on the dorsum of the tongue (Figure 117). The anterior part of the tongue in front of the circumvallate papillae is supplied by the lingual branches of the mandibular nerves. Behind, and including, the circumvallate papillae, the tongue is innervated primarily by the glossopharyngeal nerves. Small areas on the posterior part of the tongue around the epiglottis are supplied by the superior laryngeal branches (internal branches) of the vagus nerves. Concerning taste, the anterior part of the tongue is innervated by the chorda tympani branches of the facial nerves. These are distributed through the lingual nerves. The posterior part of the tongue, including the circumvallate papillae, has a similar innervation for taste as that for general sensation.

138C,49
138C,47
138C,44

138C,17

The mucosa on the ventral surface of the tongue and on the floor of the mouth is supplied by the lingual branches of the mandibular nerves.

The lingual nerve

The lingual nerve is derived from the posterior trunk of the mandibular nerve within the infratemporal fossa (see page 179 and Figure 60). It receives the chorda tympani

116,11; 142A,10
116B,25
116,22; 142A,16

301

116A,15
116,10; 142A,21
116,12
138B,5; 138B,24

142A,10; 142A,38
142A,35; 144A,26
138B,6

138B,8
142A,40

branch of the facial nerve beneath the lateral pterygoid muscle. At the level of the mandibular foramen, the lingual nerve lies on the medial pterygoid muscle and is anterior to the inferior alveolar nerve. The lingual nerve then leaves the infratemporal fossa, passing downwards and forwards to lie close to the lingual alveolar plate of the mandibular third molar tooth. Before curving forwards into the tongue, the nerve is found above the origin of the mylohyoid muscle and lateral to the hyoglossus muscle (see Figures 112 and 113). On the superficial surface of the hyoglossus muscle, the lingual nerve twists twice around the submandibular salivary duct, first on the lateral side of the duct and then on the medial side. It enters the tongue behind the sublingual salivary gland. Suspended from the lingual nerve as it runs across the hyoglossus muscle is the submandibular parasympathetic ganglion.

The lingual nerve itself supplies the mucosa covering the anterior two-thirds of the dorsum of the tongue, the ventral surface of the tongue, the floor of the mouth, and the lingual gingivae of the mandibular teeth.

116,22; 142A,16

142A,40

The chorda tympani fibres travelling with the lingual nerve are of two types: sensory and parasympathetic. The sensory fibres are associated with taste for the anterior two-thirds of the dorsum of the tongue. The parasympathetic fibres are preganglionic fibres that pass to the submandibular ganglion. Postganglionic fibres are distributed to the submandibular and sublingual salivary glands (see page 304 and Figure 119).

THE SENSORY INNERVATION OF THE PALATE (Figure 118; see also Figure 87)

142A,2
142A,4
135,29
142A,5

The sensory supply to the palate is derived mainly from branches of the maxillary nerve via the pterygopalatine ganglion. A small area behind the incisor teeth is supplied by the nasopalatine nerves. The remainder of the hard palate is innervated by the greater palatine nerves. The soft palate is supplied by the lesser palatine nerves. There is evidence to suggest that some areas supplied by the lesser palatine nerves may also be innervated from the facial nerves. The posterior part of the soft palate and the uvula may be supplied by the glossopharyngeal nerves.

The nasopalatine nerve

135,29; 135,30
22,1

This nerve runs along the nasal septum from the pterygopalatine ganglion and emerges on to the hard palate at the incisive fossa behind the maxillary first incisor teeth. The nasopalatine nerve innervates the gingivae behind the maxillary incisor teeth.

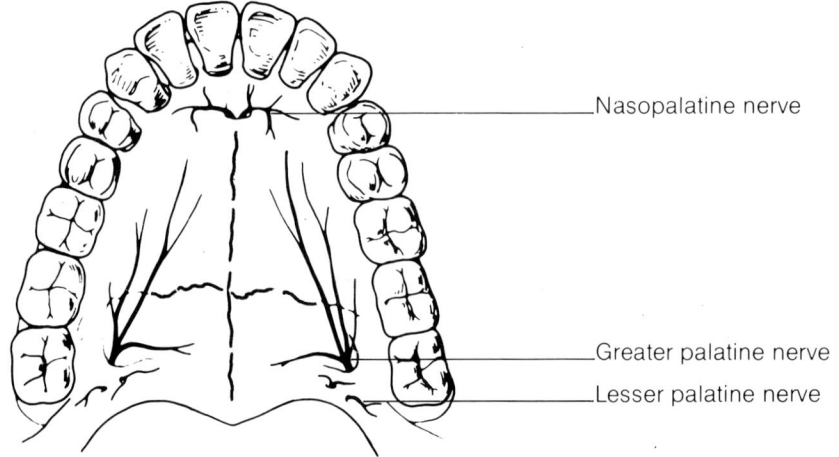

Figure 118 *Sensory innervation of the palate.*

The greater and lesser palatine nerves

These nerves pass from the pterygopalatine ganglion, down the greater palatine canal at the back of the lateral wall of the nose. The greater palatine nerve runs through the greater palatine foramen and on to the back of the hard palate. It passes towards the front of the hard palate at the interface between the palatine and alveolar processes of the maxilla. In addition to supplying the mucosa of the palate, the greater palatine nerve innervates the palatal gingivae for the maxillary cheek teeth. The lesser palatine nerve emerges on to the palate at the lesser palatine foramen. It runs backwards into the soft palate.

142A,4; 142A,5
22,7
22,4

22,8

THE SENSORY INNERVATION AT THE OROPHARYNGEAL ISTHMUS

The mucosa over the pillars of the fauces is supplied by the glossopharyngeal nerve.

144B,84

THE INNERVATION OF THE ORAL MUSCULATURE

The innervation of the various muscles associated with the mouth is derived from the mandibular division of the trigeminal, the facial, the cranial part of the accessory and the hypoglossal cranial nerves, and the first cervical spinal nerves. The innervation is summarised in Table 5.

THE INNERVATION OF THE SALIVARY GLANDS

The lesser petrosal branch of the glossopharyngeal nerve supplies the parotid gland via the otic parasympathetic ganglion (see page 167 and Figure 54). Postganglionic fibres pass to the gland through the auriculotemporal branch of the mandibular nerve.

166B,53
142A,11
142A,17

The greater petrosal branch of the facial nerve probably supplies palatal and pharyngeal glands via the pterygopalatine parasympathetic ganglion (see page 257 and Figure 88). Postganglionic fibres reach the palate with the nasopalatine, greater palatine and lesser palatine branches of the maxillary nerve.

166,54; 167,11
142A,4
135,29; 142A,5

TABLE 5 Innervation of the oral musculature.

REGION	MUSCLE	NERVE
LIPS	Orbicularis oris	Facial
CHEEKS	Buccinator	Facial
TONGUE (intrinsic musculature)	Transverse Longitudinal Vertical	Hypoglossal
TONGUE (extrinsic musculature)	Genioglossus Hyoglossus Styloglossus	Hypoglossal
	Palatoglossus	Accessory (cranial part)
FLOOR OF MOUTH	Mylohyoid	Mandibular division of trigeminal
	Geniohyoid	Hypoglossal (C1 fibres)
PALATE	Tensor veli palatini	Mandibular division of trigeminal
	Levator veli palatini Palatoglossus Palatopharyngeus Salpingopharyngeus Musculus uvulae	Accessory (cranial part)

142A,16
142A,40

The chorda tympani branch of the facial nerve provides secretomotor fibres via the submandibular parasympathetic ganglion to the submandibular and sublingual salivary glands. It probably also provides the innervation of minor salivary glands in the lips, cheeks and tongue.

142A,40

The submandibular ganglion (Figure 119)

142A,35; 142A,38
142A,10; 142A,26

This parasympathetic ganglion is found in the floor of the mouth, on the superficial surface of the hyoglossus muscle and under cover of the mylohyoid muscle (see Figure 113). The ganglion lies between the lingual nerve and the deep part of the submandibular gland. Indeed, it is suspended by two roots from the lingual nerve. The prime function of the submandibular ganglion is to supply the submandibular and sublingual salivary glands.

In common with the other parasympathetic ganglia in the head, the submandibular ganglion has a parasympathetic, a sympathetic and a sensory supply. Only the parasympathetic fibres synapse within the ganglion.

22,38
142A,16; 142A,10

Preganglionic parasympathetic fibres originate from the superior salivatory nucleus in the brain stem. The fibres pass with the nervus intermedius of the facial nerve into the internal acoustic meatus and exit the skull with the chorda tympani nerve at the petrotympanic fissure. The chorda tympani nerve joins the lingual nerve in the infratemporal fossa and by this route the parasympathetic fibres reach the submandibular ganglion. It is claimed that the preganglionic parasympathetic fibres pass via the posterior root linking the ganglion to the lingual nerve.

142A,25

The sympathetic supply to the ganglion is derived from the superior cervical ganglion. It reaches the submandibular ganglion via the sympathetic nerve plexus surrounding the facial artery.

142A,10

The sensory supply arises from the adjacent lingual nerve.

Branches from the ganglion pass directly to the submandibular gland. The sublingual gland, however, is supplied by fibres that re-enter the lingual nerve by the anterior connecting root.

THE BLOOD SUPPLY OF THE ORO-DENTAL TISSUES

THE BLOOD SUPPLY OF THE TEETH

116A,16

The main arteries to the teeth and jaws are derived from the maxillary artery, a terminal branch of the external carotid. The alveolar arteries (the superior and inferior alveolar arteries) follow approximately the same course as the alveolar nerves.

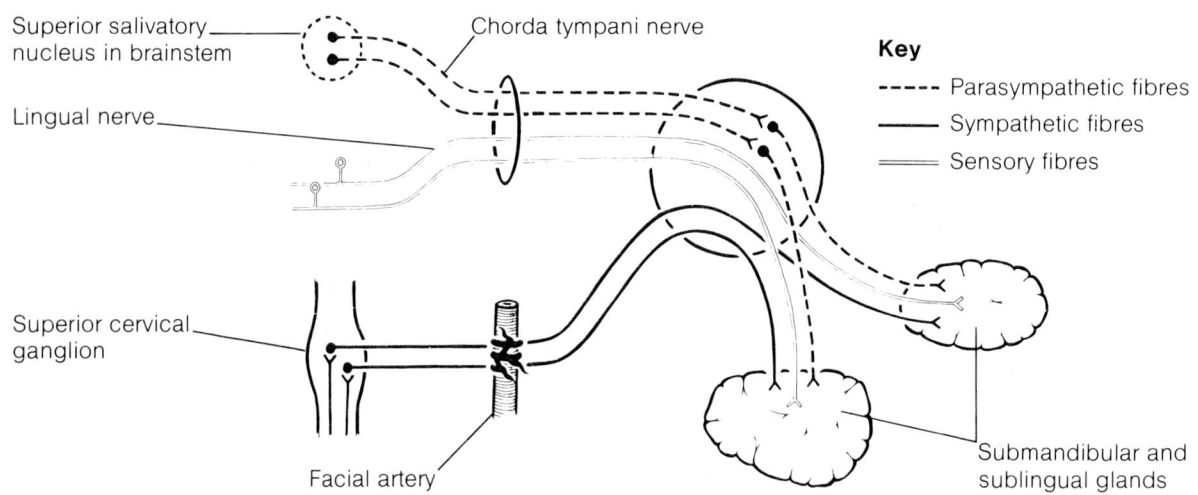

Figure 119 *The submandibular parasympathetic ganglion.*

The inferior alveolar artery supplies the mandibular teeth. It is derived from the maxillary artery before it crosses the lateral pterygoid muscle in the infratemporal fossa (see page 180 and Figure 61). A mylohyoid branch is given off before the inferior alveolar artery passes through the mandibular foramen to enter the mandibular canal. The inferior alveolar artery terminates within the body of the mandible as the mental and incisive arteries. The mental artery emerges on to the face at the mental foramen (see Figure 50). The incisive artery continues forwards within the incisive canal to supply the mandibular anterior teeth.

There are three superior alveolar arteries that supply the maxillary dentition: the posterior, middle and anterior superior alveolar arteries. The superior alveolar arteries form plexuses just above the root apices of the teeth.

The posterior superior alveolar artery usually arises from the maxillary artery in the pterygopalatine fossa. It courses tortuously over the maxillary tuberosity before entering the upper jaw to supply the molar and premolar teeth.

The middle superior alveolar artery, when present, arises from the infra-orbital artery (a branch of the maxillary artery in the pterygopalatine fossa). The middle superior alveolar artery runs down the lateral wall of the maxillary sinus, terminating near the canine tooth.

The anterior superior alveolar artery also arises from the infra-orbital artery. It runs down the anterior wall of the maxillary sinus towards the anterior teeth.

The gingival tissues derive their blood supply from the maxillary and lingual arteries.

The buccal gingivae associated with the mandibular cheek teeth are supplied by the buccal branch of the maxillary artery and by perforating branches from the inferior alveolar artery. The labial gingivae around the anterior teeth are supplied by the mental artery and by perforating branches of the incisive artery. The lingual gingivae are supplied by perforating branches from the inferior alveolar artery and by the lingual artery of the external carotid.

The buccal gingivae around the maxillary cheek teeth are supplied by gingival and perforating branches from the posterior superior alveolar artery and by the buccal artery. The labial gingivae of the anterior teeth are supplied by labial branches of the infra-orbital artery and by perforating branches of the anterior superior alveolar artery. The palatal gingivae are supplied primarily by branches of the greater palatine artery.

THE BLOOD SUPPLY OF OTHER ORAL STRUCTURES

The palate derives its blood supply from the greater and lesser palatine branches of the maxillary artery. (The nasopalatine artery does not reach the palate, the greater palatine artery passing through the incisive foramen; see Figure 83.)

The cheek is supplied by the buccal branch of the maxillary artery, the floor of the mouth and tongue by the lingual arteries. The lips are mainly supplied by superior and inferior labial branches of the facial arteries.

The lingual artery is the third branch of the external carotid artery. It reaches the floor of the mouth by passing between the hyoglossus muscle and the middle constrictor of the pharynx (Figure 113). The artery here crosses the stylohyoid ligament and is accompanied by the lingual veins and the glossopharyngeal nerve. At the anterior border of the hyoglossus muscle, the lingual artery (with the lingual nerve) bends sharply upwards towards the genioglossus muscle of the tongue. The branches of the lingual artery in the floor of the mouth are: the dorsal lingual branches, the sublingual artery and the deep lingual artery. The dorsal lingual branches supply the back of the dorsum of the tongue and the region around the pillars of the fauces. The sublingual artery supplies the sublingual gland and other structures in the floor of the mouth. The deep lingual artery is the terminal part of the lingual artery, which is found on the inferior surface of the tongue near the lingual frenum.

THE VENOUS DRAINAGE OF THE ORO-DENTAL TISSUES

The veins from the teeth are collected into inferior alveolar veins for the lower jaw, and superior alveolar veins for the upper jaw. These may drain anteriorly to the facial veins (through the mental foramina for the inferior alveolar veins) or posteriorly to the pterygoid plexuses in the infratemporal fossae (through the mandibular foramina for the inferior alveolar veins).

No accurate description is available concerning the venous drainage of the gingivae, although it may be assumed that buccal, lingual, greater palatine and nasopalatine veins are involved. These veins run into the pterygoid plexuses, apart from the lingual veins which pass directly into the internal jugular veins.

The lingual veins are variable, but they usually follow two routes. The dorsal lingual vein drains the dorsum of the tongue. It passes with the lingual artery deep to the hyoglossus muscle, where it becomes known as the lingual vein. The lingual vein joins the internal jugular vein at the level of the greater horn of the hyoid bone. The deep lingual vein is visible through the mucosa of the ventral surface of the tongue (see Figure 98). It joins the sublingual vein (from the sublingual salivary gland) to become the vein accompanying the hypoglossal nerve on the hyoglossus muscle. This terminates by either joining the lingual vein or by draining directly into the internal jugular vein.

The veins of the hard palate generally pass into the pterygoid plexus, those of the soft palate into the pharyngeal plexus.

Venous blood from the lips drains into the facial veins via superior and inferior labial veins.

THE LYMPHATIC DRAINAGE OF THE MOUTH (Figure 120)

The principal sites of drainage of lymphatic vessels from the oro-dental tissues are: the submental, submandibular, and jugulodigastric lymph nodes. The manner of drainage is, however, so variable that there is no precise and accurate description.

The lymph vessels from the teeth usually run directly into the submandibular lymph nodes on the same side. However, lymph from the mandibular incisors drains into the submental lymph nodes. Occasionally, lymph from the molars may pass directly into the jugulodigastric group of nodes.

The lymph vessels of the labial and buccal gingivae of the maxillary and mandibular teeth unite to drain into the submandibular nodes, though in the labial region of the mandibular incisors they may drain into the submental lymph nodes. The lingual and palatal gingivae drain into the jugulodigastric group of nodes either directly, or indirectly through the submandibular nodes.

Lymphatics from the bulk of the palate terminate in the jugulodigastric group of nodes. Vessels from the posterior part of the soft palate terminate in pharyngeal lymph nodes.

Lymphatics from the anterior two-thirds of the tongue may be subdivided into two groups of vessels: marginal and central vessels. The marginal lymphatics drain the lateral third of the upper surface of the tongue and the lateral margin of its lower surface. The remaining regions drain into the central vessels. The marginal vessels pass to the submandibular lymph nodes of the same side. The vessels at the tip of the tongue pass to the submental lymph nodes. Central vessels behind the tip drain into ipsilateral and contralateral submandibular lymph nodes. Some marginal and central lymph vessels pass directly to the jugulodigastric group of nodes (or even the jugulo-

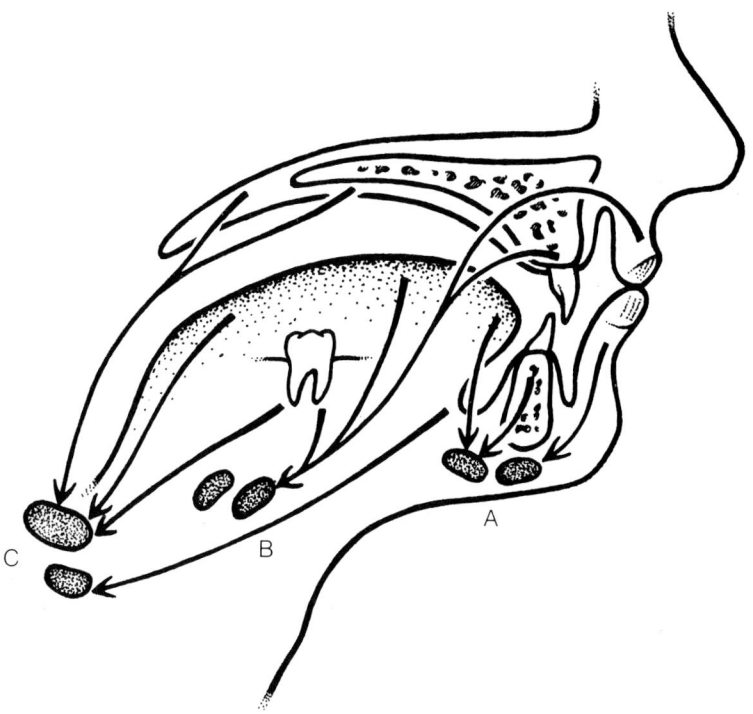

Figure 120 *The lymphatic drainage of the oro-dental tissues.* A = submental lymph nodes. B = submandibular lymph nodes. C = jugulodigastric lymph nodes.

omohyoid nodes). Lymphatics from the posterior third of the tongue drain into the deep cervical group of nodes, vessels centrally draining both ipsilaterally and contralaterally.

At the oropharyngeal isthmus lie the palatine tonsils between the pillars of the fauces, and the lingual tonsils on the pharyngeal surface of the tongue. These tonsils form part of a ring of lymphoid tissue known as Waldeyer's tonsillar ring. The other components are the tubal tonsils and adenoid tissue (pharyngeal tonsils) in the nasopharynx (see pages 310 and 311).

142B,52
138C,44
126A,15; 126A,12

THE TISSUE SPACES AROUND THE JAWS (Figures 121 and 122)

The dissemination of infection in soft tissues is influenced by the natural barriers presented by bone, muscle and fascia. Around the jaws, however, the tissue spaces are primarily defined by muscles, principally the mylohyoid, buccinator, masseter, medial pterygoid, superior constrictor and orbicularis oris muscles. None of the 'spaces' are actually empty and they should be regarded merely as potential spaces which are normally occupied by loose connective tissue. It is only when inflammatory products destroy the loose connective tissue that a definable space is produced.

The important potential tissue spaces are:

Lower jaw:
- Submental
- Submandibular
- Sublingual
- Buccal
- Submasseteric
- Parotid
- Pterygomandibular
- Parapharyngeal
- Peritonsillar

Upper jaw:
- Palatal
- Facial
- Infratemporal

(The spaces are paired except for the submental, sublingual and palatal spaces.)

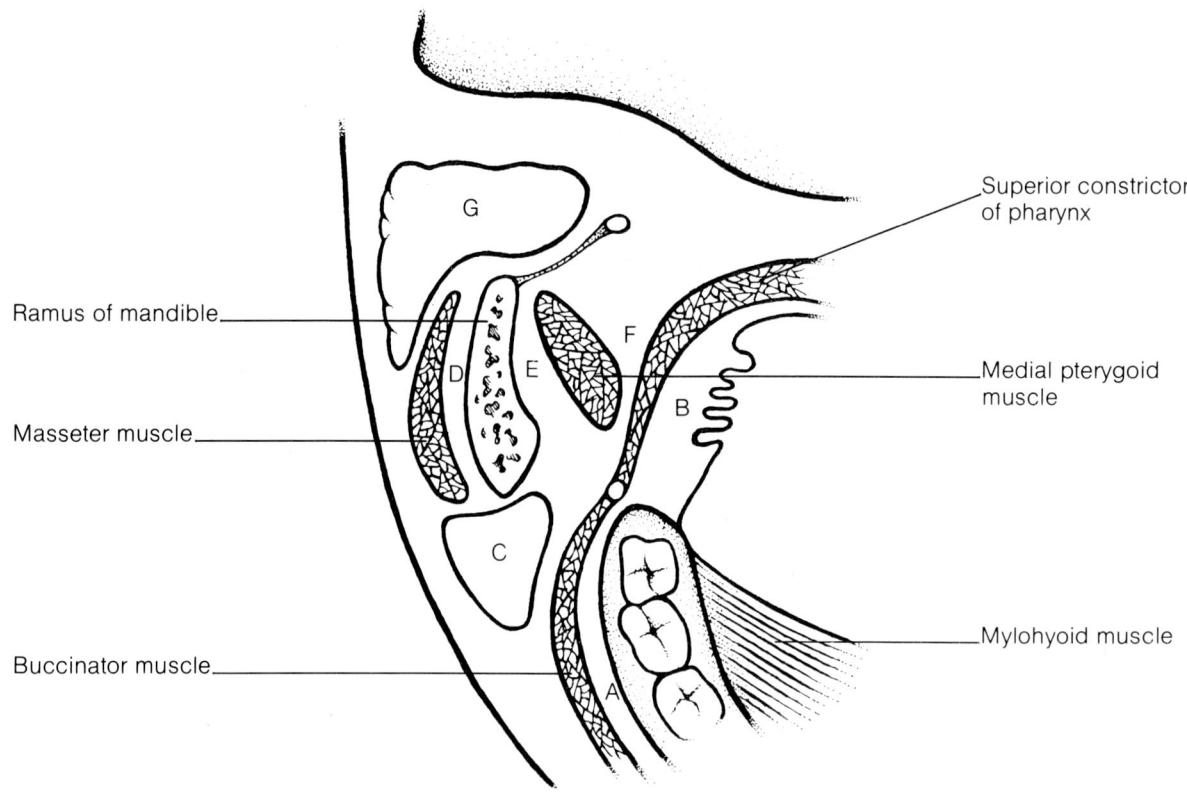

Figure 121 *Tissue spaces in the retromandibular region.* A = *fornix vestibuli.* B = *peritonsillar space.* C = *buccal space (filled with buccal pad of fat).* D = *submasseteric spaces.* E = *pterygomandibular space.* F = *parapharyngeal space.* G = *parotid space.*

102A	The submental and submandibular spaces are located below the inferior border of the mandible, beneath the mylohyoid muscle, in the suprahyoid region of the neck. The submental space lies beneath the chin in the midline, between the mylohyoid muscles and the investing layer of deep cervical fascia. It is bounded laterally by the two anterior bellies of the digastric muscles. The submental space communicates posteriorly with the two submandibular spaces. The submandibular space is situated between the anterior and posterior bellies of the digastric muscle. It communicates with the sublingual space around the posterior free border of the mylohyoid muscle.
102A,9	
102A	
102A,9; 102A,44	
102A,8	
140B,51	The sublingual space lies in the floor of the mouth, above the mylohyoid muscles. It is continuous across the midline and communicates with the submandibular spaces over the posterior free borders of the mylohyoid muscles.
116C,8; 140A,15	The buccal space is located in the cheek, on the lateral side of the buccinator muscle.
140A,19; 140A,18	Between the lateral surface of the ramus of the mandible and the masseter muscle is a series of spaces called the submasseteric spaces. These spaces are formed because the fibres of the masseter muscle have multiple insertions on to most of the lateral surface of the ramus.
140A,19; 140A,23	Between the medial surface of the ramus of the mandible and the medial pterygoid muscle lies the pterygomandibular space.
140A,39	Behind the ramus of the mandible is located the parotid space, in and around the parotid gland.
140A,7; 140A,23	The parapharyngeal space is bounded by the superior constrictor of the pharynx and the medial surface of the medial pterygoid muscle. This space is restricted to the

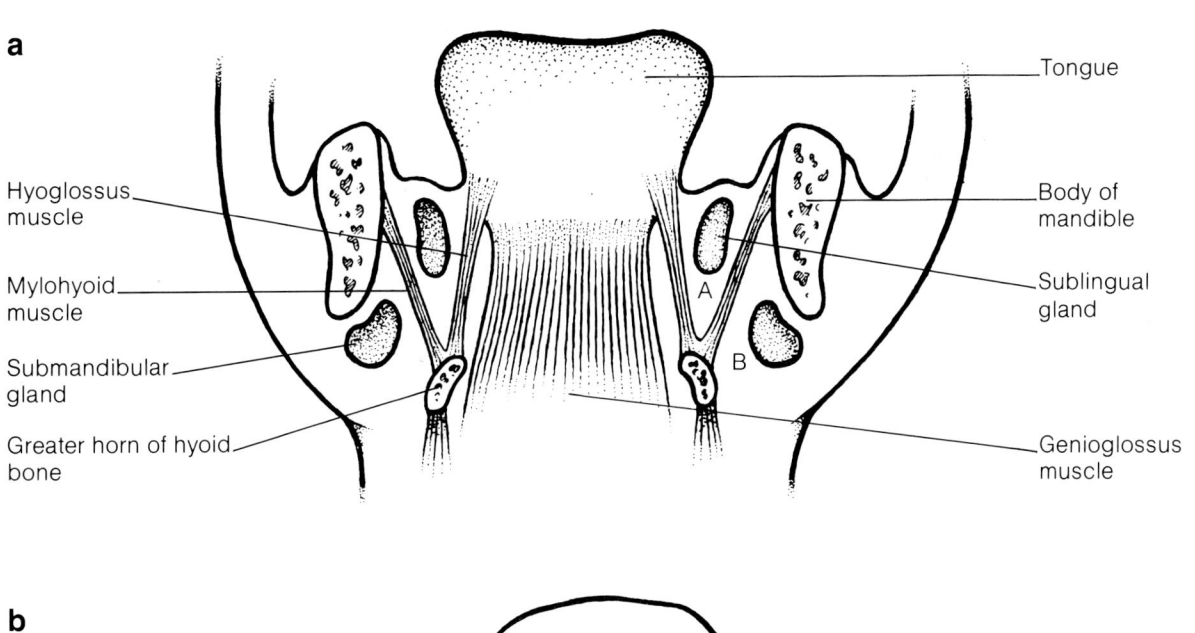
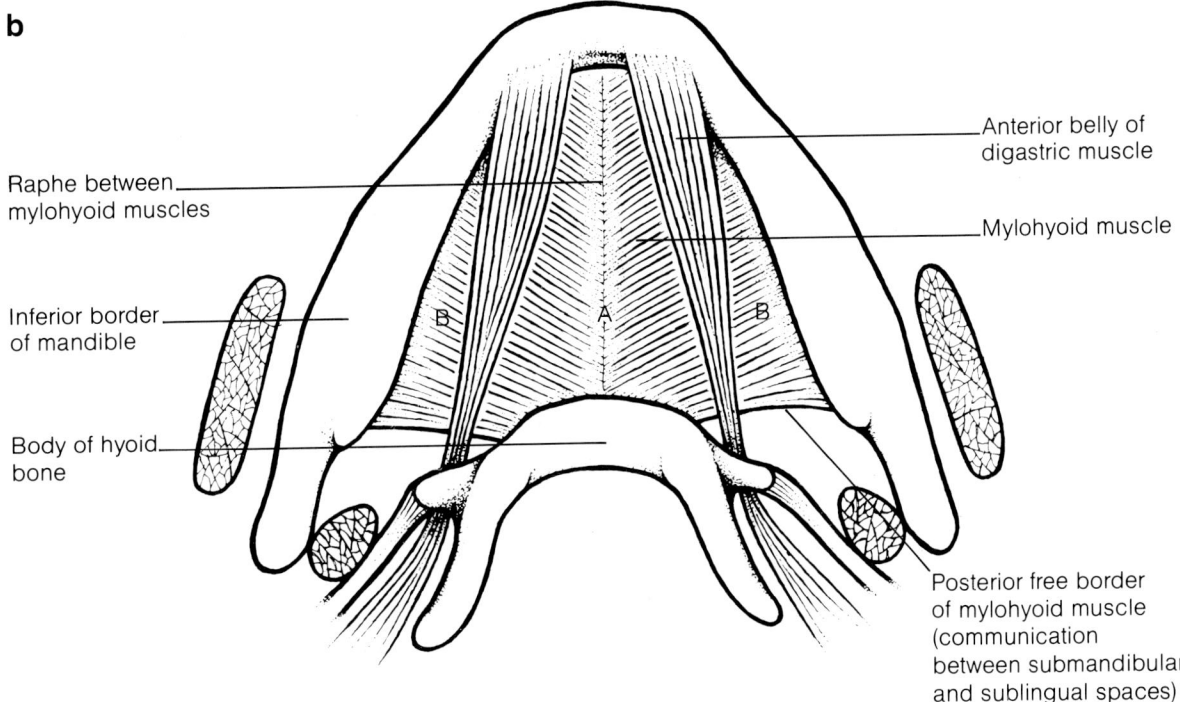

Figure 122 *Tissue spaces in the floor of the mouth.* **a)** *Coronal section through floor of mouth. A = sublingual tissue space. B = submandibular tissue space.* **b)** *Inferior view of floor of mouth (suprahyoid region of neck). A = submental tissue space. B = submandibular tissue spaces.*

infratemporal region of the head and the suprahyoid region of the neck. It communicates with the retropharyngeal space which itself extends into the retrovisceral space in the lower part of the neck (see pages 93 to 99 for a description of the tissue spaces in the neck).

The peritonsillar space lies around the palatine tonsil between the pillars of the fauces. It is part of the intrapharyngeal space and is bounded by the medial surface of the superior constrictor of the pharynx and its mucosa.

142B,52

142B,42

112	In the upper jaw, the muscles of facial expression define a number of very small tissue spaces in the face. One such space is the canine fossa, between the levator labii superioris and zygomaticus muscles.
112,12; 112,14	

There is no true tissue space in the hard palate, as the mucosa there is firmly bound to the periosteum. However, inflammation can strip away some of this periosteum to produce a well-circumscribed abscess.

140A — The infratemporal space is the upper extremity of the pterygomandibular space. It is closely related to the maxillary tuberosity and therefore the upper molars.

The pharynx

136 — The pharynx is a common passage for the alimentary and respiratory systems. It links the oral cavity to the oesophagus, and the posterior apertures of the nasal cavity to the inlet of the larynx. During swallowing, the airway is temporarily closed by elevation of the soft palate and of the inlet of the larynx beneath the base of the tongue.

136,45; 136,28

136,21; 136,11
136,4; 136,20
136,13
146A,31; 146A,29
146A,25; 146A,1

136,51
136,32; 136,19
136,47
136,30; 136,21

The pharynx is a fibromuscular tube situated in front of the vertebral column. It extends from the base of the skull to the lower border of the cricoid cartilage of the larynx. It is widest above and narrows at its junction with the oesophagus below. The principal muscles of the pharynx (the constrictor muscles) arise from structures at the sides of the head and neck and pass posteriorly to insert into a midline fibrous band called the pharyngeal raphe. The pharynx is thus not a complete tube, being semicircular in cross-section, with communications anteriorly with the nasal cavity, the oral cavity and the laryngeal cavity. Indeed, the pharynx is subdivided into three regions according to the cavity with which it is continuous, i.e. the nasopharynx, the oropharynx and the laryngopharynx.

136,50 — The pharynx also communicates with the middle ear by way of an auditory tube.

THE NASOPHARYNX

136,47

136,45
136,51

The nasopharynx is the uppermost part of the pharynx and lies above the soft palate. Anteriorly, it begins at the posterior nasal apertures (choanae). Posteriorly and inferiorly, it ends at the pharyngeal isthmus (i.e. the opening between the back of the soft palate and the posterior wall of the pharynx).

136,4
136,45

The roof and posterior wall of the nasopharynx lie against the base of the skull (primarily the basilar part of the occipital bone). Its sloping floor is the upper surface of the soft palate. With the exception of the soft palate, the nasopharynx is rigid and thus contributes to the patency of the airway.

126,15

126,16
126C,37

126C,14

126,13

The most prominent feature on each side of the nasopharynx is a triangular elevation with rounded margins called **the tubal elevation.** This is related to the underlying cartilaginous end of the auditory tube. Within the margins of the tubal elevation lies the opening of the auditory tube. Below this opening, the mucosa bulges because of the underlying levator veli palatini muscle. A fold of mucosa runs vertically downwards from the posterior margin of the tubal elevation. This is called the salpingopharyngeal fold because it overlies the salpingopharyngeus muscle. A smaller fold named the salpingopalatine fold may be present at the anterior margin of the tubal elevation. Lymphatic material comprising the tubal tonsil is found around the opening of the auditory tube. Behind the auditory tube is a small depression called the pharyngeal recess.

148H
148H,67; 150,15

The auditory tube has also been called the eustachian tube and the pharyngotympanic tube. It links the lateral wall of the nasopharynx to the anterior wall of the tympanic cavity (see Figure 146). In its course from the ear to the pharynx, it passes downwards, forwards and medially (at approximately 30° to the horizontal plane and 45° to the sagittal plane). There is a bony part near the ear and a cartilaginous part near the pharynx. The cartilaginous part is twice as long as the bony part. (This relationship is the converse of that pertaining to the external acoustic meatus: see page 364.) The bony part of the auditory tube gradually narrows as it passes from the tympanic cavity. It ends as an isthmus, which is the junction of the squamous and petrous portions of the temporal bone. The isthmus has a jagged margin for the attachment of the cartilaginous part of the auditory tube. The bony and cartilaginous parts meet at an obtuse angle. The cartilaginous part is formed by a triangular plate of cartilage which is fixed to the base of the skull in the groove between the petrous part of the temporal bone and the greater wing of the sphenoid bone. The cartilaginous part widens as it passes from the isthmus to the pharyngeal orifice. The pharyngeal orifice produces the tubal elevation in the nasopharynx and is directed downwards and backwards. The auditory tube has two functions. First, it permits the passage of air into the tympanic cavity to enable equalisation of pressure on either side of the ear drum (tympanic membrane). Second, it allows drainage of mucus from the middle ear and mastoid air cells. The mucosa lining the auditory tube has a ciliated columnar epithelium.

The cartilaginous part of the auditory tube gives attachment to the tensor veli palatini, the levator veli palatini and the salpingopharyngeus muscles. The tensor veli palatini intervenes between the tube and the mandibular nerve, the otic ganglion, the chorda tympani nerve and the middle meningeal artery. The bony part of the tube lies below the canal for the tensor tympani muscle. Below and medially lies the carotid canal.

The innervation of the auditory tube is derived from the pharyngeal branch of the maxillary nerve (via the pterygopalatine ganglion) and the tympanic plexus of the middle ear (see pages 257 and 372). The arteries supplying the tube arise from the ascending pharyngeal branch of the external carotid artery and from the maxillary artery (middle meningeal branch, artery of the pterygoid canal). The veins drain into the pterygoid venous plexus of the infratemporal fossa.

Variable amounts of lymphatic tissue are scattered within the posterior wall of the nasopharynx. Such tissue is particularly evident in children where it forms a mass called **the pharyngeal tonsil (adenoids)**. The prominence of the pharyngeal tonsil lies close to the nasal septum and displays a median recess called the pharyngeal bursa. The tonsil may become so enlarged that it interferes with nasal respiration.

In the roof of the nasopharynx may be found a small collection of glandular tissue related to the adenohypophysis of the pituitary gland. This is called **the pharyngeal hypophysis.** It is thought to be a remnant of Rathke's pouch, and can secrete hormones.

Where the nasopharynx meets the soft palate, a ridge called **Passavant's ridge** becomes evident during swallowing. This contains Passavant's muscle (see page 316).

THE OROPHARYNX

The oropharynx is the middle part of the pharynx and lies below the soft palate. It is delineated from the oral cavity proper by the palatoglossal arches (the anterior pillars of the fauces). The region between the palatoglossal arches is called the oropharyngeal isthmus.

The roof of the oropharynx is the under-surface of the soft palate. The floor of the oropharynx is formed by the root of the tongue and extends back to the tip of the epiglottis. The posterior wall of the oropharynx lies adjacent to the second and third cervical vertebrae.

The root of the tongue in the floor of the oropharynx shows numerous lingual follicles. These contain lymphatic tissue and collectively form the lingual tonsil. The anterior surface of the epiglottis is joined to the tongue by three folds of mucosa called the median and lateral glosso-epiglottic folds. Between these folds are two depressions called valleculae.

The lateral wall of the oropharynx presents two prominent folds designated the pillars of the fauces. These folds diverge inferiorly and bound a triangular area called the tonsillar fossa or sinus. The anterior fold or palatoglossal arch runs from the soft palate to the side of the tongue and contains the palatoglossus muscle (see Figure 114b). The posterior fold or palatopharyngeal arch passes from the soft palate to merge with the lateral wall of the pharynx. It contains the palatopharyngeus muscle (see Figure 114b).

The palatine tonsil is a collection of lymphatic material that lies beneath the mucosa in the tonsillar fossa. Its size is variable, tending to be large in children and small in adults. It is a frequent site of infection. The medial surface of the palatine tonsil is the visible surface within the oropharynx. It exhibits several slit-like invaginations, the tonsillar crypts. One of the crypts is particularly deep and is called the intratonsillar cleft. It is a remnant of the second branchial pouch (see Figure 44). The overall size of the palatine tonsil cannot be appreciated from a consideration of its medial surface alone. Indeed, it can extend some distance beyond the tonsillar fossa (e.g. upwards into the soft palate). The tonsillar bed that lies adjacent to the lateral surface of the palatine tonsil is formed by the palatoglossus and superior constrictor muscles. More deeply are found the styloglossus muscle and the glossopharyngeal nerve.

THE LARYNGOPHARYNX

The laryngopharynx is the lowest part of the pharynx. It extends from the upper border of the epiglottis to the lower border of the cricoid cartilage of the larynx. The laryngopharynx continues into the oesophagus. It is delineated from the oropharynx by the lateral glosso-epiglottic folds.

The obliquely-sloping inlet of the larynx lies in the anterior part of the laryngopharynx. This inlet is bounded above by the epiglottis, below by the arytenoid cartilages of the larynx, and laterally by the aryepiglottic folds (see Figure 133). Below the inlet, the anterior wall of the laryngopharynx is formed by the posterior surface of the cricoid cartilage. Situated in the lateral wall of the laryngopharynx is the lamina of the thyroid cartilage of the larynx. Between the inner surface of this lamina and the outer surface of the cricoid and arytenoid cartilages is found a recess called the piriform fossa. The posterior wall of the laryngopharynx lies adjacent to the bodies of the third to the sixth cervical vertebrae.

THE WALL OF THE PHARYNX

The wall of the pharynx consists mainly of muscle. Internally, there is also a layer of mucosa and a membrane called the pharyngobasilar fascia. Externally, there is a loose connective tissue layer which is sometimes referred to as the buccopharyngeal fascia.

THE MUSCLES OF THE PHARYNX (Figures 123 to 125)

The six pairs of muscles that comprise the pharynx can be divided into two groups. One group comprises three pairs of constrictor muscles that run transversely across the pharynx: the superior, middle, and inferior constrictors. The other group comprises three pairs of muscles that run longitudinally down the pharynx: the salpingopharyngeus, stylopharyngeus and palatopharyngeus muscles.

The constrictor muscles arise at the sides of the head and neck and insert posteriorly into the pharyngeal raphe. This raphe passes longitudinally down the back of the pharynx from the pharyngeal tubercle on the basilar part of the occipital bone. The

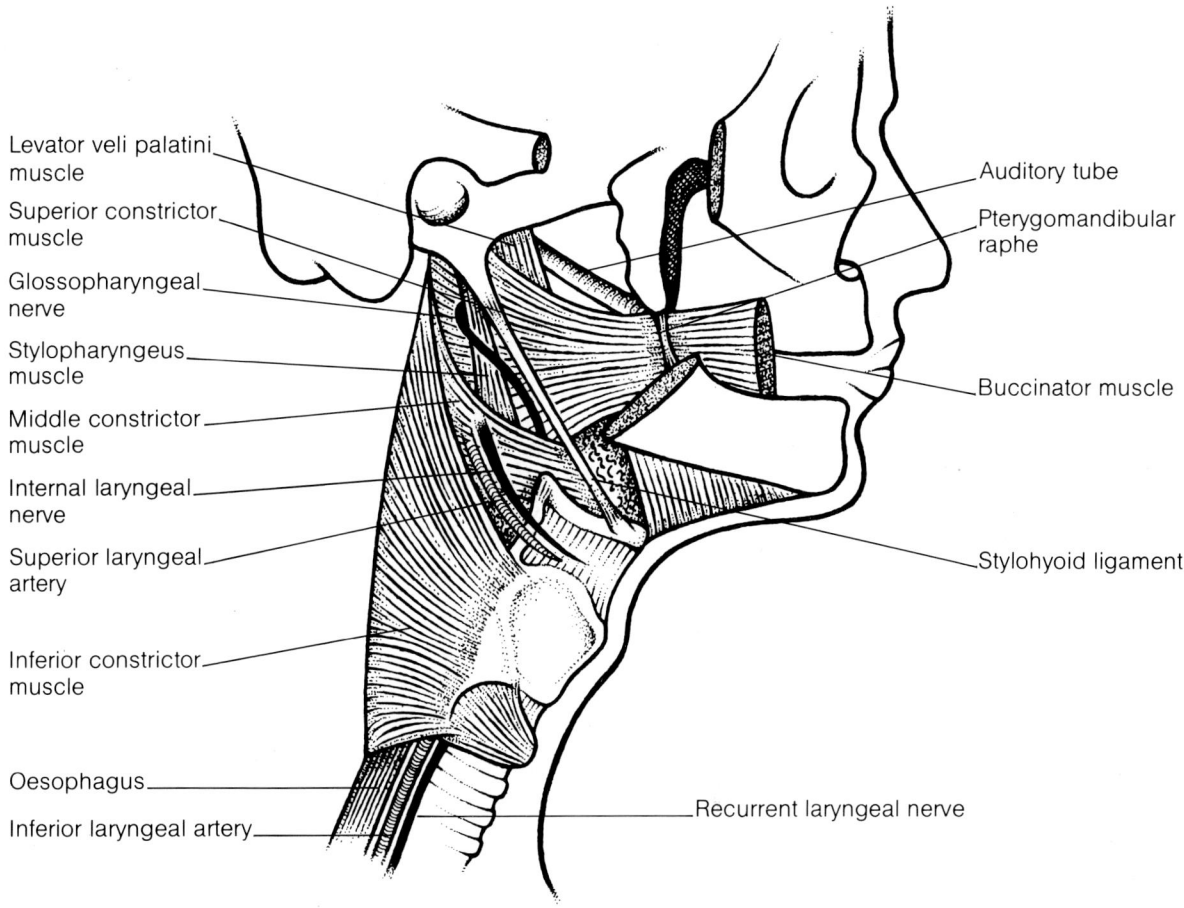

Figure 123 *The pharynx viewed from the side.*

constrictor muscles generate an ordered wave of contraction which carries the food bolus into the oesophagus. The constrictor muscles overlap each other such that the superior constrictor 'sits' within the middle constrictor and the middle constrictor 'sits' within the inferior constrictor (Figures 123 to 125). Important structures enter the pharynx in the intervals between the constrictors. The longitudinal muscles of the pharynx are attached on to the larynx. They elevate the pharynx and larynx during swallowing.

The superior constrictor muscle

Attachments. The muscle originates mainly from the posterior border of the pterygomandibular raphe. This raphe provides the sites of origin for both the superior constrictor and buccinator muscles. It runs between the pterygoid hamulus of the sphenoid bone and the back of the mylohyoid line of the mandible. The superior constrictor muscle also arises from the bone adjacent to each end of the pterygomandibular raphe and from the side of the tongue. The fibres of the muscle fan out to be attached posteriorly to the pharyngeal raphe and to the pharyngeal tubercle on the occipital bone.

Innervation. The superior constrictor muscle is innervated by the cranial part of the accessory nerve.

Vasculature. The blood supply of this muscle is derived mainly from the ascending pharyngeal artery (pharyngeal branch) and facial artery (tonsillar branch).

Actions. The muscle constricts the upper part of the pharynx.

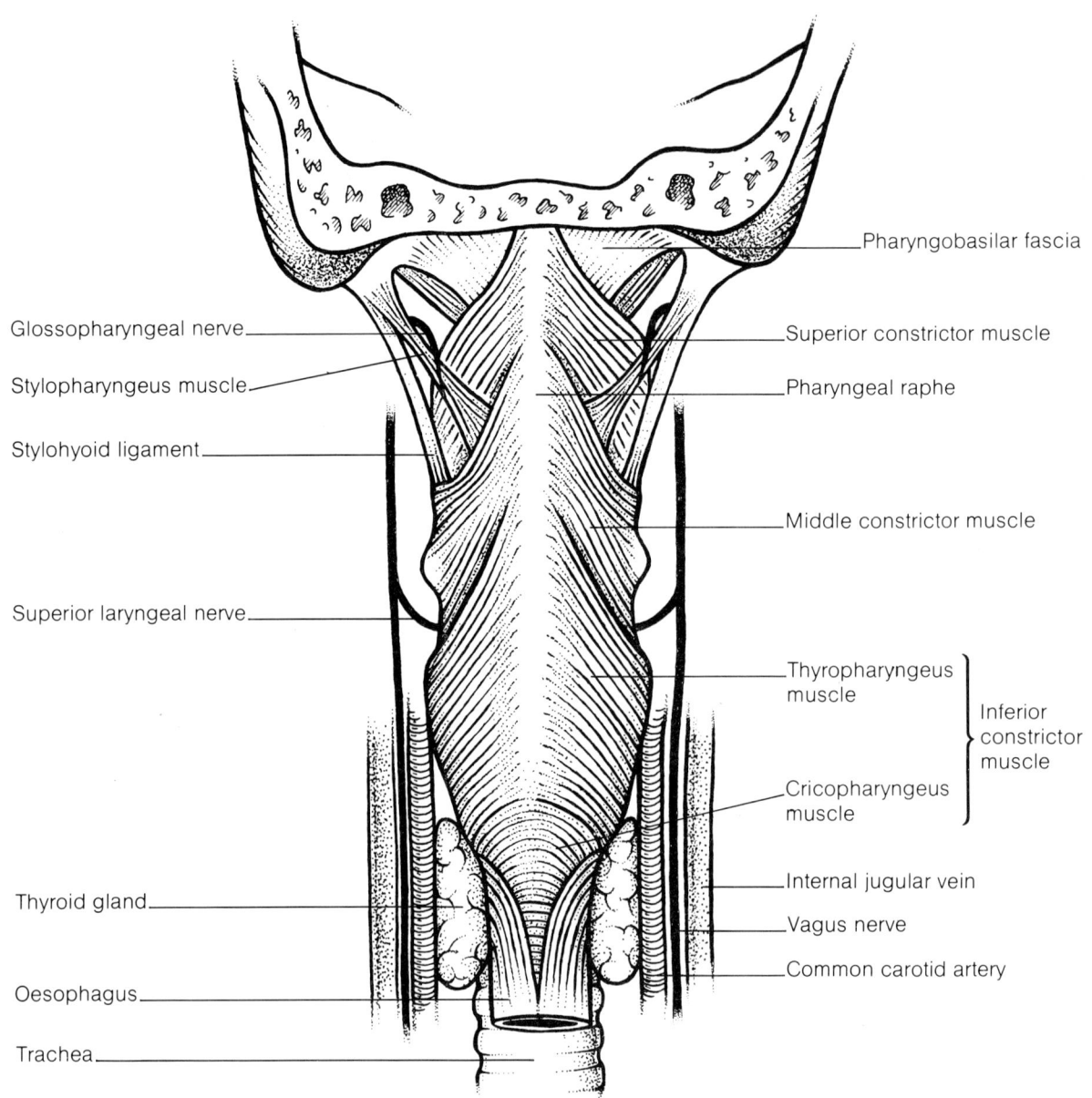

Figure 124 *The pharynx viewed from behind.*

The middle constrictor muscle

Attachments. This muscle arises from the lower part of the stylohyoid ligament and from the hyoid bone (the lesser horn and the whole of the upper border of the greater horn). The fibres fan out to be inserted posteriorly into the pharyngeal raphe.

Innervation. The middle constrictor muscle is supplied by the cranial part of the accessory nerve.

Vasculature. Branches from the ascending pharyngeal artery (pharyngeal branch) and facial artery (tonsillar branch) are the chief sources of blood supply for the middle constrictor muscle.

Actions. The muscle constricts the pharynx during swallowing.

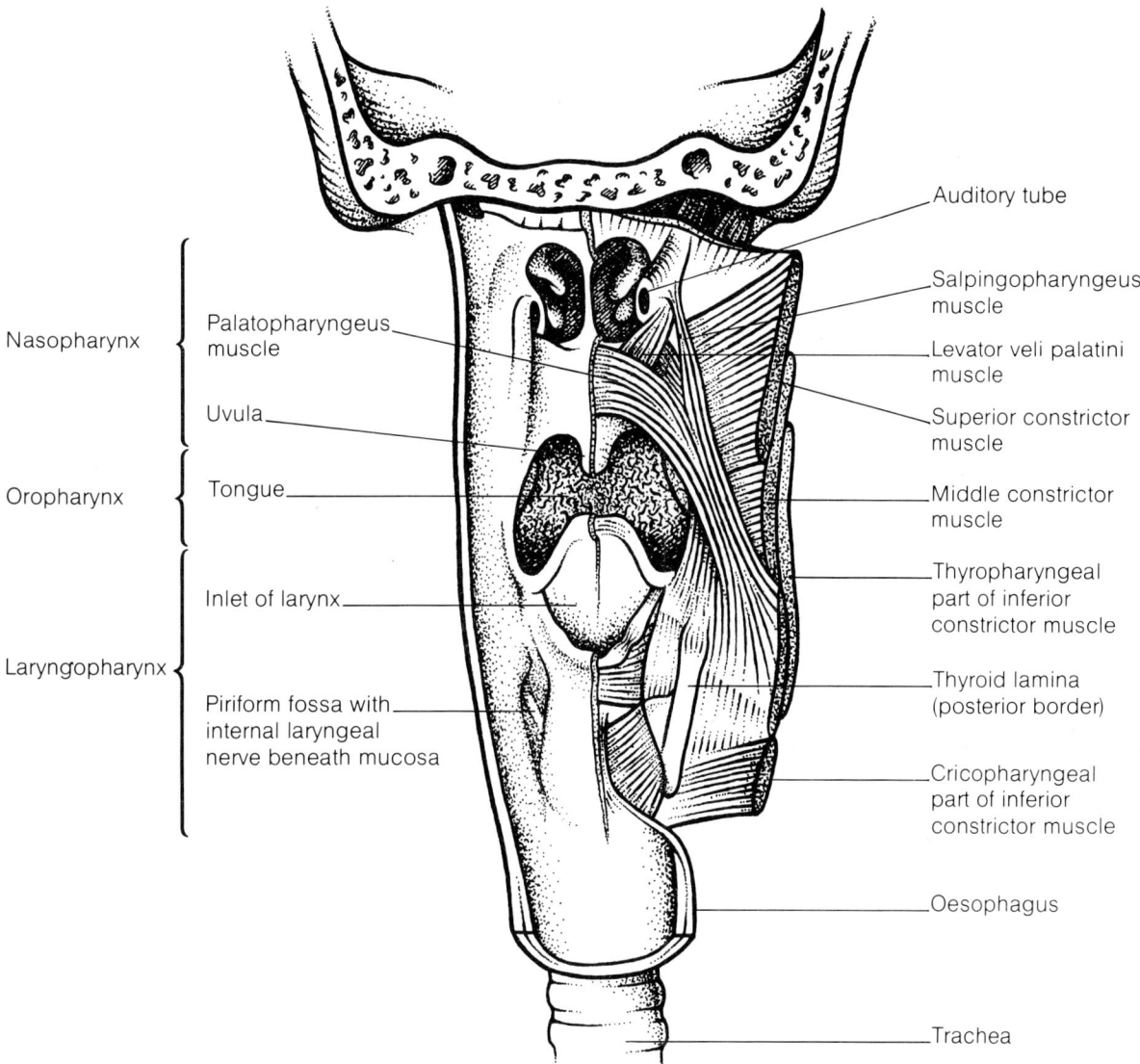

Figure 125 *Internal view of the anterior wall of the pharynx.*

The inferior constrictor muscle

Attachments. The inferior constrictor muscle has two main sites of origin: the thyroid and cricoid cartilages of the larynx. Indeed, the muscle consists of two parts — the thyropharyngeus and cricopharyngeus muscles.

144A,40; 144B,77
146A,25; 154,17
152D,26; 152G,26

The thyropharyngeus muscle arises from the oblique line on the lamina of the thyroid cartilage. Its fibres pass backwards and upwards to insert into the pharyngeal raphe.

146A,25
152D,26
146,1

The cricopharyngeus muscle arises from the lateral surface of the cricoid cartilage, just behind the origin of the cricothyroid muscle. Whereas the other constrictor muscles run backwards and upwards to insert into the pharyngeal raphe, cricopharyngeus passes only backwards to encircle the pharynx in a region lacking a pharyngeal raphe.

146A,24
152G,26

Between the two parts of the inferior constrictor muscle, some additional fibres arise from a tendinous cord that loops over the cricothyroid muscle.

Innervation. The muscle is supplied by the cranial part of the accessory nerve. Cricopharyngeus is also supplied by the recurrent laryngeal nerve and the external branch of the superior laryngeal nerve.

Vasculature. The arterial supply to the muscle is derived from the ascending pharyngeal artery (pharyngeal branch) and the inferior thyroid artery (muscular branches).

Actions. The thyropharyngeus muscle constricts the lower part of the pharynx. The cricopharyngeus muscles act as a sphincter at the junction of the laryngopharynx and the oesophagus.

The palatopharyngeus muscle

Attachments. This muscle has two heads of origin which enclose the levator veli palatini muscle. An anterior head arises from the back of the hard palate. A posterior head arises from the upper surface of the palatine aponeurosis. The palatopharyngeus muscles on each side meet in the midline at their origins in the palate (see Figure 114).

The two heads of the palatopharyngeus muscle merge to pass down within the palatopharyngeal arch at the back of the mouth. The muscle lies on the internal surface of the constrictor muscles of the pharynx and inserts into the posterior border of the thyroid cartilage of the larynx.

Innervation. The nerve supply to the palatopharyngeus muscle is the cranial part of the accessory nerve.

Vasculature. The muscle derives its arterial supply from the facial artery (ascending palatine branch), the maxillary artery (descending palatine branch) and the ascending pharyngeal artery (pharyngeal branch).

Actions. From its superior attachments, palatopharyngeus elevates the pharynx and larynx. It can also elevate the side of the tongue, and draw together the palatopharyngeal arches to close the oropharyngeal isthmus. The muscles may depress the soft palate when acting from their inferior attachments.

Passavant's muscle is a sphincter-like muscle that encircles the pharynx at the level of the palate. Contraction of this muscle forms a ridge (Passavant's ridge) against which the soft palate is elevated. Controversy exists concerning the derivation of the muscle. Some anatomists consider Passavant's muscle to be derived from the superior constrictor and palatopharyngeus muscle. Others claim that it is a distinct palatine muscle that arises from the anterior and lateral parts of the upper surface of the palatine aponeurosis.

The salpingopharyngeus muscle

Attachments. This muscle arises from the cartilage of the auditory tube, close to the opening in the nasopharynx. The salpingopharyngeus muscle runs down the internal surface of the pharynx, producing the salpingopharyngeal fold. It merges with the palatopharyngeus muscle.

Innervation. The nerve supply to this muscle is the cranial part of the accessory nerve.

Vasculature. The muscle derives its arterial supply from the same sources as the palatopharyngeus muscle.

Actions. The salpingopharyngeus muscle elevates the pharynx. It may also open the cartilaginous end of the auditory tube during swallowing.

The stylopharyngeus muscle

Attachments. This muscle arises from the medial surface of the base of the styloid process (temporal bone). It passes down into the pharynx between the superior and middle constrictor muscles. Some fibres merge with the constrictor muscles, others insert into the posterior border of the thyroid cartilage of the larynx.

152E,35

Innervation. Unlike the other pharyngeal muscles, the stylopharyngeus muscle receives its innervation from the glossopharyngeal nerve and not from the cranial part of the accessory nerve.

146A,6

Vasculature. Its arterial supply is derived from the ascending pharyngeal artery (pharyngeal branch).

146A,3

Actions. The muscle elevates the pharynx and larynx.

THE MUCOSA AND FASCIA OF THE PHARYNX

The mucosa varies in different parts of the pharynx. The epithelium covering the nasopharynx is of the ciliated columnar type and therefore resembles the epithelium of the nose. The epithelium of the oropharynx and laryngopharynx is of the stratified squamous type and resembles that found in the mouth. This epithelium is tightly bound to the underlying pharyngobasilar fascia.

The mucosa of the pharynx contains aggregations of lymphatic material forming the pharyngeal and tubal tonsils in the nasopharynx, and the palatine and lingual tonsils in the oropharynx. This ring of lymphoid material is referred to as Waldeyer's ring.

The pharyngobasilar fascia is a distinct membranous fascia that lies between the mucosa and the muscle of the wall of the pharynx. It is particularly well developed superiorly, where it extends above the free margin of the superior constrictor muscle to the base of the skull. It is attached to the basilar part of the occipital bone, the petrous part of the temporal bone in front of the carotid canal, the border of the medial pterygoid plate and to the pterygomandibular raphe. The pharyngobasilar fascia above the superior constrictor muscle is pierced by the levator veli palatini muscle, the cartilaginous end of the auditory tube and the ascending palatine artery.

144A,10; 146A,2

144A,9; 144B,68
144B,67

The external surface of the wall of the pharynx is covered by a thin connective tissue layer, the buccopharyngeal fascia. The use of the term fascia to describe this connective tissue is debatable, as it is not a membranous layer. Indeed, a membranous layer in this site would be disadvantageous because considerable mobility of the pharynx is necessary during swallowing. On the buccopharyngeal connective tissue is found the pharyngeal venous plexus and the pharyngeal nerve plexus. The posterior part of the wall of the pharynx is separated from the prevertebral musculature and the prevertebral fascia by loose connective tissue occupying a potential tissue space called the retropharyngeal space (see page 96).

STRUCTURES PASSING THROUGH THE WALL OF THE PHARYNX
(Figure 123)

In the interval between the superior constrictor muscle and the base of the skull pass the levator veli palatini muscle, the cartilaginous end of the auditory tube and the ascending palatine artery.

144A,9; 144B,68
144B,67; 116C,31

Between the superior and middle constrictor muscles run the stylopharyngeus muscle, the glossopharyngeal nerve, the styloglossus muscle, the lingual nerve and artery, and the hypoglossal nerve.

116C,49; 146A,11
144A,13; 144A,7
144A,29; 144A,25

Passing between the middle and inferior constrictor muscles are the internal branch of the superior laryngeal nerve (from the vagus nerve) and the superior laryngeal artery (a branch of the superior thyroid artery).

144A,31; 154A,30
154A,28

Underneath the lower border of the inferior constrictor muscle run the recurrent laryngeal nerve (from the vagus nerve) and the inferior laryngeal artery (a branch of the inferior thyroid artery).

154,23; 144A,45
144A,46

THE INNERVATION OF THE PHARYNX

Most of the pharynx derives its sensory nerve supply from the glossopharyngeal nerve through its tonsillar and pharyngeal branches (see Figure 32). The pharyngeal branch arises just before the glossopharyngeal nerve passes on to the posterior surface of the stylopharyngeus muscle. This branch then joins the pharyngeal branch of the vagus to form the pharyngeal plexus. The tonsillar branch of the glossopharyngeal nerve supplies the region around the oropharyngeal isthmus.

The anterior part of the nasopharynx is not supplied by the glossopharyngeal nerve but by the pharyngeal branch of the maxillary nerve. Furthermore, the soft palate is innervated by the lesser palatine branch of the maxillary nerve. Both the pharyngeal and lesser palatine nerves are branches of the maxillary division of the trigeminal nerve via the pterygopalatine ganglion (see pages 253 to 258).

The lower part of the pharynx is innervated by the superior laryngeal branch of the vagus nerve.

The muscles of the pharynx derive their innervation from the nucleus ambiguus in the brainstem (see pages 469 to 470). Fibres pass from this nucleus within the glossopharyngeal, vagus and cranial accessory nerves. The cranial accessorynerve, however, joins the vagus nerve shortly after emerging through the jugular foramen of the skull. These fibres, together with those already in the vagus, reach the pharyngeal plexus via the pharyngeal branch of the vagus to supply most of the muscles of the pharynx. The stylopharyngeus muscle, however, is supplied by fibres from the nucleus ambiguus that run with the glossopharyngeal nerve.

The pharyngeal plexus lies on the external surface of the middle constrictor muscle. It is formed by the pharyngeal branches of the glossopharyngeal and vagus nerves, with contributions from the superior cervical sympathetic ganglion. The glossopharyngeal nerve supplies only sensory fibres to the plexus. The vagus contains motor fibres associated with the cranial part of the accessory nerve which, in addition to supplying the muscles of the pharynx, also supply the muscles of the soft palate (see page 108).

THE VASCULATURE OF THE PHARYNX

The pharynx receives its blood supply from many sources, including the ascending pharyngeal artery (pharyngeal branch), the inferior thyroid artery, the facial artery (ascending palatine and tonsillar branches), the maxillary artery (pharyngeal, greater palatine branches and the artery of the pterygoid canal), and the lingual artery (dorsal lingual branch).

The veins of the pharynx drain into the pharyngeal plexus. This plexus is situated on the posterior wall of the pharynx. Pharyngeal veins drain the plexus into the internal jugular or into the brachiocephalic vein (via the inferior thyroid vein). The pharyngeal plexus may communicate with other veins, including the facial vein and the pterygoid venous plexus.

The lymphatic vessels from the pharynx drain into the deep cervical lymph nodes either directly or indirectly via the paratracheal or retropharyngeal nodes. In addition, lymph from the area around the epiglottis passes into the infrahyoid nodes.

SWALLOWING

Swallowing involves an ordered sequence of reflex events that carry food (or saliva) from the mouth into the stomach. During swallowing, the nasopharynx is isolated from the oropharynx by elevation of the soft palate. As this process occupies a total of only a few minutes in every day, the pharynx is normally maintained in the respiratory position.

The first stage of swallowing is voluntary and involves the passage of the bolus of food on to the tongue and towards the oropharyngeal isthmus. The airway remains patent during this phase.

TABLE 6 Principal events during swallowing.

STAGE	MECHANISMS ASSOCIATED WITH PASSAGE OF BOLUS	MECHANISMS ASSOCIATED WITH PROTECTING AIRWAY
1. VOLUNTARY		
Bolus in mouth	Mouth closed (temporalis, masseter, medial pterygoid)	**Airway open**
	Lips closed (orbicularis oris)	Pillars of fauces contracted against posterior surface of tongue (palatoglossus, palatopharyngeus)
	Tongue grooved, anterior part raised against palate (intrinsic tongue muscles, genioglossus)	
2. INVOLUNTARY		
Bolus passes into oropharynx	Posterior part of tongue moves upwards and backwards (styloglossus, mylohyoid)	**Nasopharynx closed off**
	Groove in tongue flattened out (intrinsic tongue muscles)	Soft palate tensed and elevated (tensor veli palatini, levator veli palatini, Passavant's muscle)
	Pillars of fauces contract behind bolus	
Bolus passes over epiglottis to lateral food channels	Pharynx elevated (stylopharyngeus, salpingopharyngeus, palatopharyngeus)	**Inlet of larynx closed off**
		Larynx elevated beneath epiglottis and posterior part of tongue (stylopharyngeus, salpingopharyngeus, palatopharyngeus, thyrohyoid)
		Laryngeal inlet reduced by approximation (interarytenoid, thyro-arytenoid) and tension (lateral crico-arytenoid, interarytenoid) of aryepiglottic folds
Bolus passes into oesophagus	Relaxation of cricopharyngeus	**Airway re-established**
		Soft palate and larynx return to original positions

When the bolus reaches the oropharyngeal isthmus during the second stage of swallowing, the process becomes involuntary. The soft palate and larynx are elevated and a wave of muscular activity of the pharyngeal constrictor muscles carries the bolus through the pharynx. This stage ends when the bolus passes into the oesophagus.

During the third stage of swallowing, the bolus passes down the oesophagus and into the stomach.

The main events during swallowing are summarised in Table 6.

THE EMBRYOLOGY OF THE MOUTH, PALATE AND PHARYNX

The mouth develops as a shallow depression called the stomodeum. During the fourth week of intra-uterine life, the stomodeum becomes surrounded by five facial processes. Below the stomodeum lie the two mandibular processes that constitute the first branchial arch. Above the stomodeum and in the midline is situated the frontonasal process; laterally are located the maxillary processes (Figure 126).

Initially, the stomodeum (lined by ectoderm) is separated posteriorly from the developing pharynx (lined by endoderm) by a bilaminar (ectodermal/endodermal) membrane called the buccopharyngeal membrane. This membrane degenerates by the end of the fourth week to establish continuity between the oral cavity and the pharynx (Figure 127).

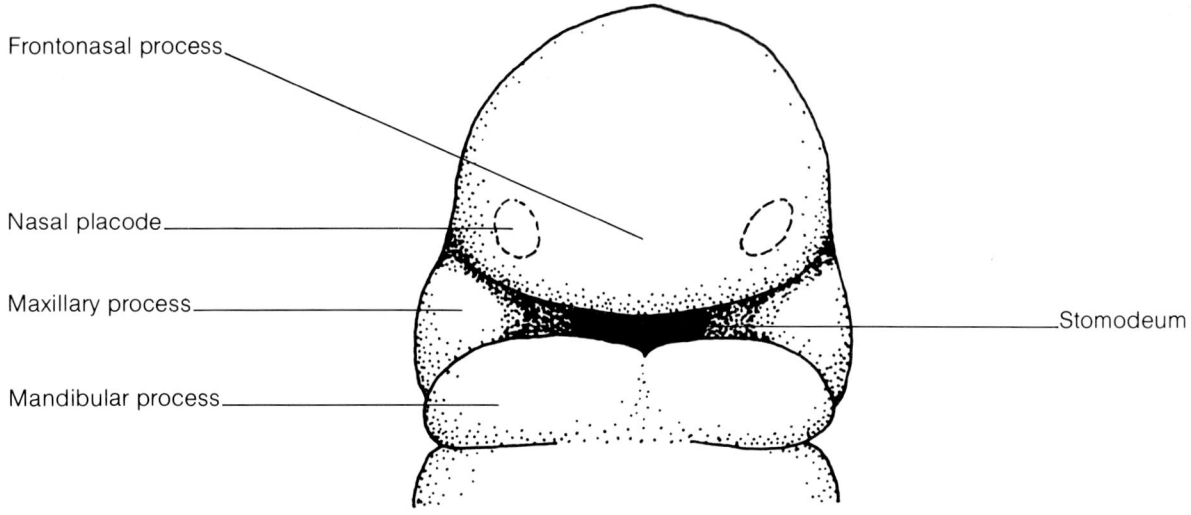

Figure 126 *Frontal aspect of the face during the fourth week of development.*

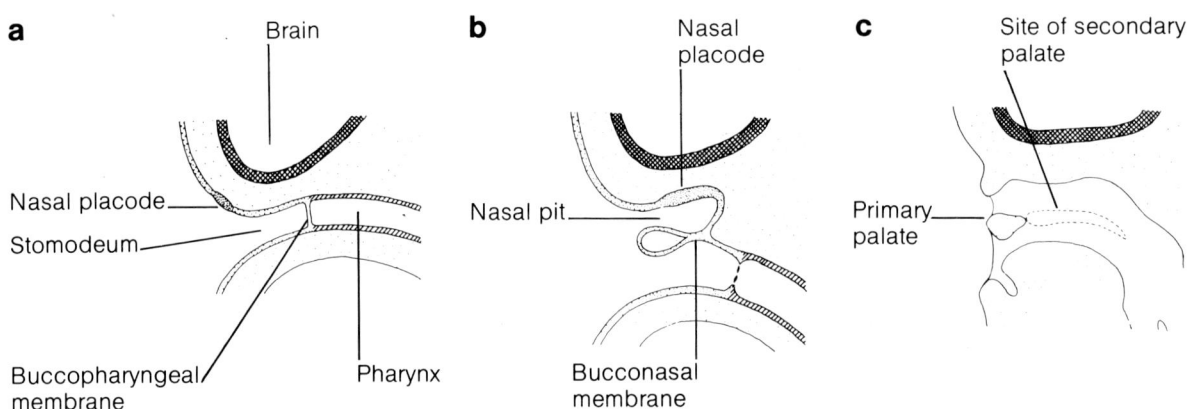

Figure 127 *Sagittal sections showing the development of the oral cavity, a) fourth week, b) fifth week, c) sixth week.*

The teeth develop from a structure called the dental lamina (Figure 128). This lamina is derived from a sheet of epithelium called the primary epithelial band, which grows from the surface of the oral cavity into the underlying mesenchyme. A series of enamel organs forms along the dental lamina, one for each deciduous tooth. The enamel organ is responsible for outlining the shape of the tooth and for the production of dental enamel. The mesenchyme around the enamel organ (the dental papilla and dental follicle) is thought to be of neural crest origin. The dental papilla forms the dentine and pulp; the dental follicle forms the dental cement, periodontal ligament, and possibly also some of the alveolar bone. The permanent teeth develop as downgrowths from the deciduous teeth, except for the permanent molars which arise directly from the dental lamina.

The chronology of tooth development and the order of eruption are given in Table 7 and Figure 129. Because no individuals are exactly alike in their development, the dates given are only approximate. Variations of six months either way are not unusual, but variations of a year or more usually indicate an abnormality. The development of the permanent dentition is generally more advanced in girls. There do not appear to be any sex differences in the deciduous dentition. As chronological age is an unreliable

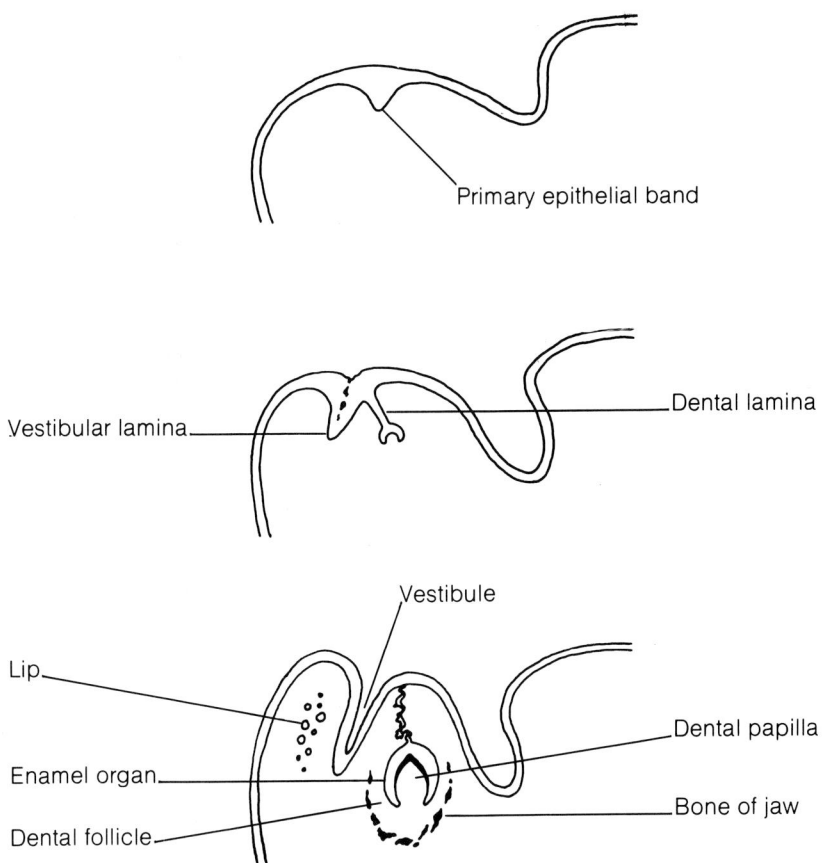

Figure 128 *Development of the dental lamina and a tooth germ.*

guide to the progress of development of an individual child, dental age used in conjunction with skeletal age is a useful index of maturity.

The tongue develops from the ventral wall of the pharynx (Figure 130). The anterior two-thirds arises from three swellings associated with the first branchial arch: two lateral lingual swellings and a median swelling called the tuberculum impar. The posterior one-third of the tongue is derived mainly from a swelling of the third branchial arch called the copula (with a small contribution from the fourth arch in the region of the epiglottis). These swellings appear during the fourth week of intra-uterine life. The copula overgrows the second branchial arch to merge with the swellings of the first arch.

The diverse embryological origin of the tongue helps to explain its general sensory supply. General sensation to the anterior two-thirds is from the lingual branch of the mandibular nerve, the nerve of the first branchial arch. General sensation to the posterior third is supplied by the glossopharyngeal and the superior laryngeal nerve, the nerves of the third and fourth branchial arches.

The muscles of the tongue arise from occipital myotomes. These migrate into the developing tongue, carrying with them their nerve supply (the hypoglossal nerve).

The thyroid gland develops on the tongue between the tuberculum impar and the copula (see page 135). This site is marked on the adult tongue by the foramen caecum.

The palate develops from two sources. By the sixth week, a primary palate is present anteriorly, which is the inferior surface of the frontonasal process (Figures 127 and

TABLE 7 The chronology of tooth development and eruption.

	DECIDUOUS DENTITION		
Tooth	First evidence of calcification (months in utero)	Crown completed (months)	Eruption (months)
Maxillary			
A	3 to 4	4	7
B	4½	5	8
C	5	9	16 to 20
D	5	6	12 to 16
E	6 to 7	10 to 12	21 to 30
Mandibular			
A	4½	4	6½
B	4½	4½	7
C	5	9	16 to 20
D	5	6	12 to 16
E	6	10 to 12	21 to 30

	PERMANENT DENTITION		
Tooth	First evidence of calcification	Crown completed (years)	Eruption (years)
Maxillary			
1	3 to 4 months	4 to 5	7 to 8
2	10 to 12 months	4 to 5	8 to 9
3	4 to 5 months	6 to 7	11 to 12
4	1½ to 1¾ years	5 to 6	10 to 11
5	2 to 2½ years	6 to 7	10 to 12
6	Birth	2½ to 3	6 to 7
7	2½ to 3 years	7 to 8	12 to 13
8	7 to 9 years	12 to 16	17 to 21
Mandibular			
1	3 to 4 months	4 to 5	6 to 7
2	3 to 4 months	4 to 5	7 to 8
3	4 to 5 months	6 to 7	9 to 10
4	1¾ to 2 years	5 to 6	10 to 12
5	2¼ to 2½ years	6 to 7	11 to 12
6	Birth	2½ to 3	6 to 7
7	2½ to 3 years	7 to 8	12 to 13
8	8 to 10 years	12 to 16	17 to 21

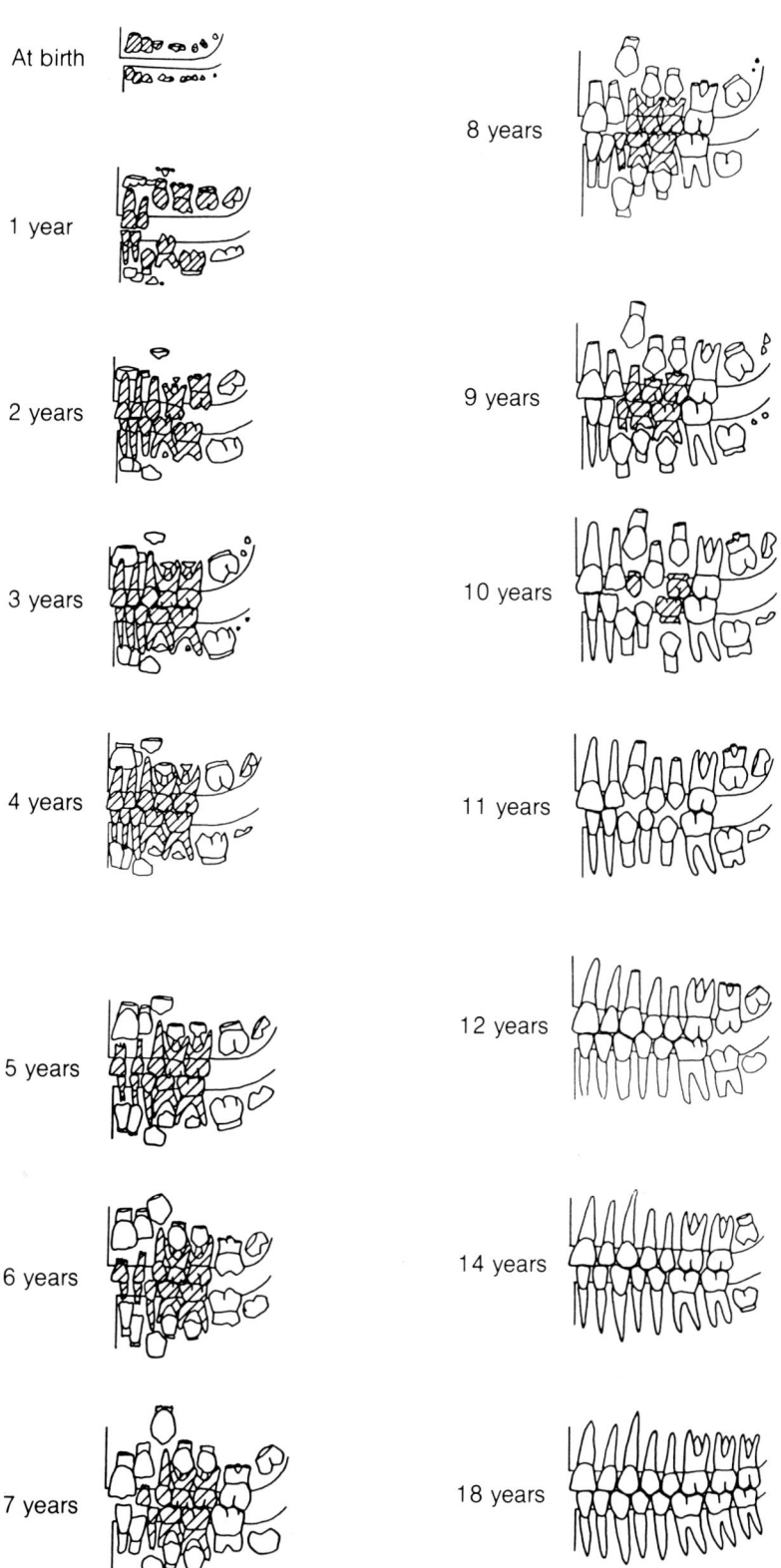

Figure 129 *Chronology of tooth development and order of eruption.* Shading indicates the deciduous teeth.

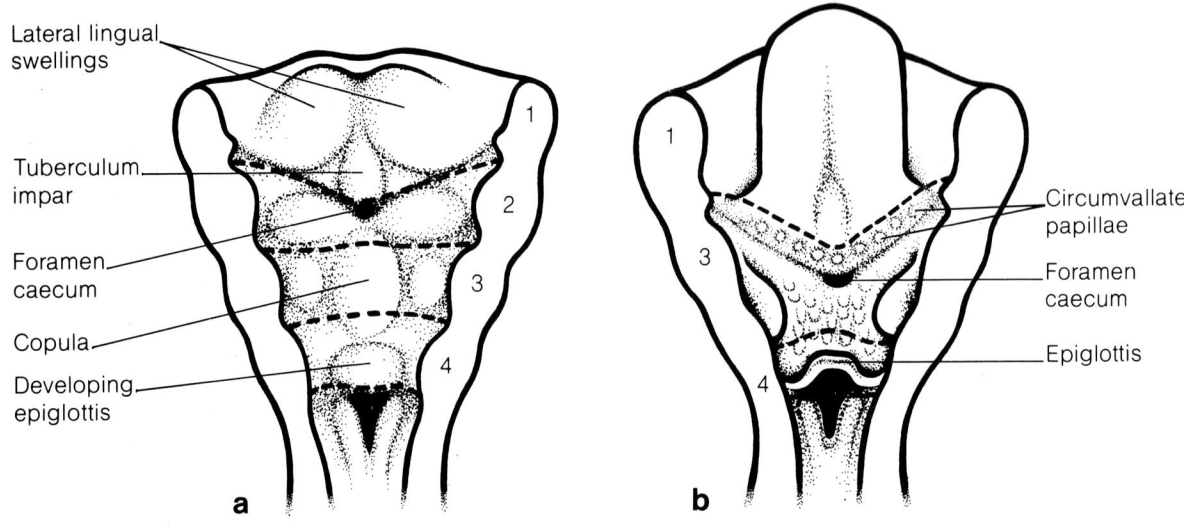

Figure 130 *Ventral aspect of the pharynx to display the developing tongue,* **a**) *fifth week of development,* **b**) *fifth month of development. The numbers 1 to 4 indicate the positions of the branchial arches.*

131). Compared to the definitive palate, the primary palate extends back only as far as the future incisive foramen. It is only with the development of a secondary palate that the common oro-nasal cavity is divided into oral and nasal chambers, and that the full extent of the definitive palate becomes apparent.

The secondary palate appears during the sixth week of development (Figures 127 and 131). A palatal shelf grows from each maxillary process into the common oro-nasal cavity, behind the primary palate. By the seventh week, the shelves lie vertically by the lateral margins of the tongue. Two crucial events occur during the eighth week. First, the tongue moves from its position between the palatal shelves. Second, the palatal shelves change from being aligned vertically to being horizontal. The shelves now contact and meet the primary palate. Fusion of the palate is complete by about the twelfth week of development. In the adult, the site at which the three palatal components meet is represented by the incisive papilla which overlies the incisive canal.

Considerable research has been undertaken to elucidate the mechanism of palatal shelf elevation. The many theories can be broadly divided into two categories. Some claim that the shelves play only a passive role, being elevated by the activity of extrinsic

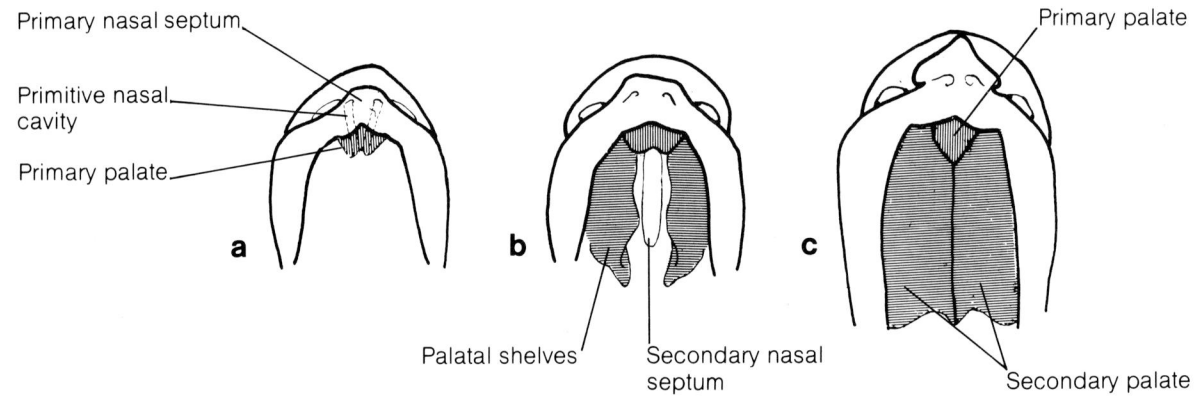

Figure 131 *The development of the palate,* **a**) *fifth week,* **b**) *sixth week,* **c**) *eighth week.*

factors, such as movement of the tongue, growth of the mandible or changes in the angulation of the cranial base. Alternatively, there is evidence that shelf elevation results from some intrinsic property of the shelves themselves, perhaps being related to differential growth, vascular changes, contraction associated with connective tissue elements, or binding of water to ground substance.

Ossification of the hard palate occurs intramembranously from four centres: one in each maxilla and one in each palatine bone. Incomplete ossification of the palate from these centres defines the median and transverse palatine sutures. The posterior one-third of the palate behind the nasal septum remains unossified as the soft palate. It is invaded by muscles originating from the branchial arches. These are the tensor veli palatini muscles from the first arch, and the levator veli palatini, palatoglossus, palatopharyngeus muscles and the musculus uvulae from the lower branchial arches.

Cleft palate is a congenital abnormality with a multifactorial mode of inheritance. It arises either because of interference with normal growth, elevation, adherence and fusion of the secondary palatal shelves and/or because of interference with fusion between primary and secondary palates. Clefts can be induced in rodents under a number of experimental conditions (e.g. cortisone injection, excess vitamin A administration), though it has yet to be shown whether similar aetiological factors can produce clefts in humans. Clefts of the palate take various forms. The mildest form of cleft is a cleft of the uvula. In more severe cases, it affects the hard palate, often passing laterally at the junction between the primary and secondary palates to involve the alveolar processes.

The pharynx is the uppermost part of the embryological foregut and is lined by endoderm. It is initially separated from the ectodermally-lined stomodeum by the buccopharyngeal membrane. Continuity between the two regions is established during the fourth week when the buccopharyngeal membrane degenerates. The branchial arch system develops from the pharynx and gives rise to many important structures in the head and neck (see Table 2; page 134).

SUMMARY SHEET: THE MOUTH 1

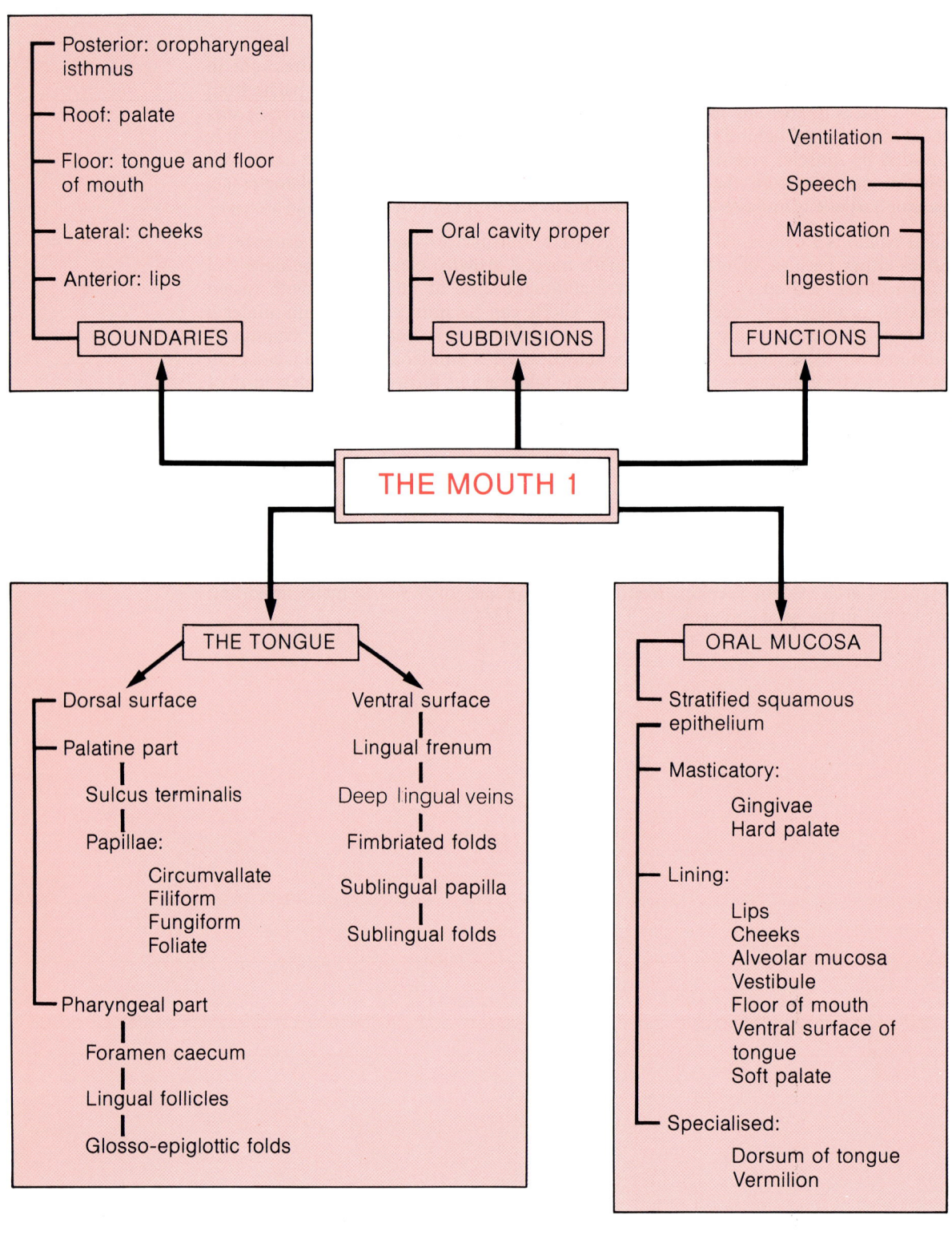

SUMMARY SHEET: THE MOUTH 2

	CUSPS	ROOTS	ERUPTION (months)
INCISORS	–	1	≃7
CANINES	1	1	≃16 to 20
MOLARS (UPPER) 1st	2	3	≃12 to 16
2nd	4	3	≃20 to 30
MOLARS (LOWER) 1st	4	2	≃12 to 16
2nd	5	2	≃20 to 30

THE DECIDUOUS DENTITION → $DI\frac{2}{2} DC\frac{1}{1} DM\frac{2}{2} = 10$

THE MOUTH 2

THE PERMANENT DENTITION → $I\frac{2}{2} C\frac{1}{1} P\frac{2}{2} M\frac{3}{3} = 16$

	CUSPS	ROOTS	ERUPTION (years)
INCISORS	–	1	≃6 to 9
CANINES	1	1	≃9 to 12
PREMOLARS	2	1 or 2	≃10 to 12
MOLARS (UPPER)	4	3	1st 6 to 7 2nd 12 to 13 3rd 17 to 21
MOLARS (LOWER)	4 (1st 5)	2	1st 6 to 7 2nd 12 to 13 3rd 17 to 21

SUMMARY SHEET: THE MOUTH 3

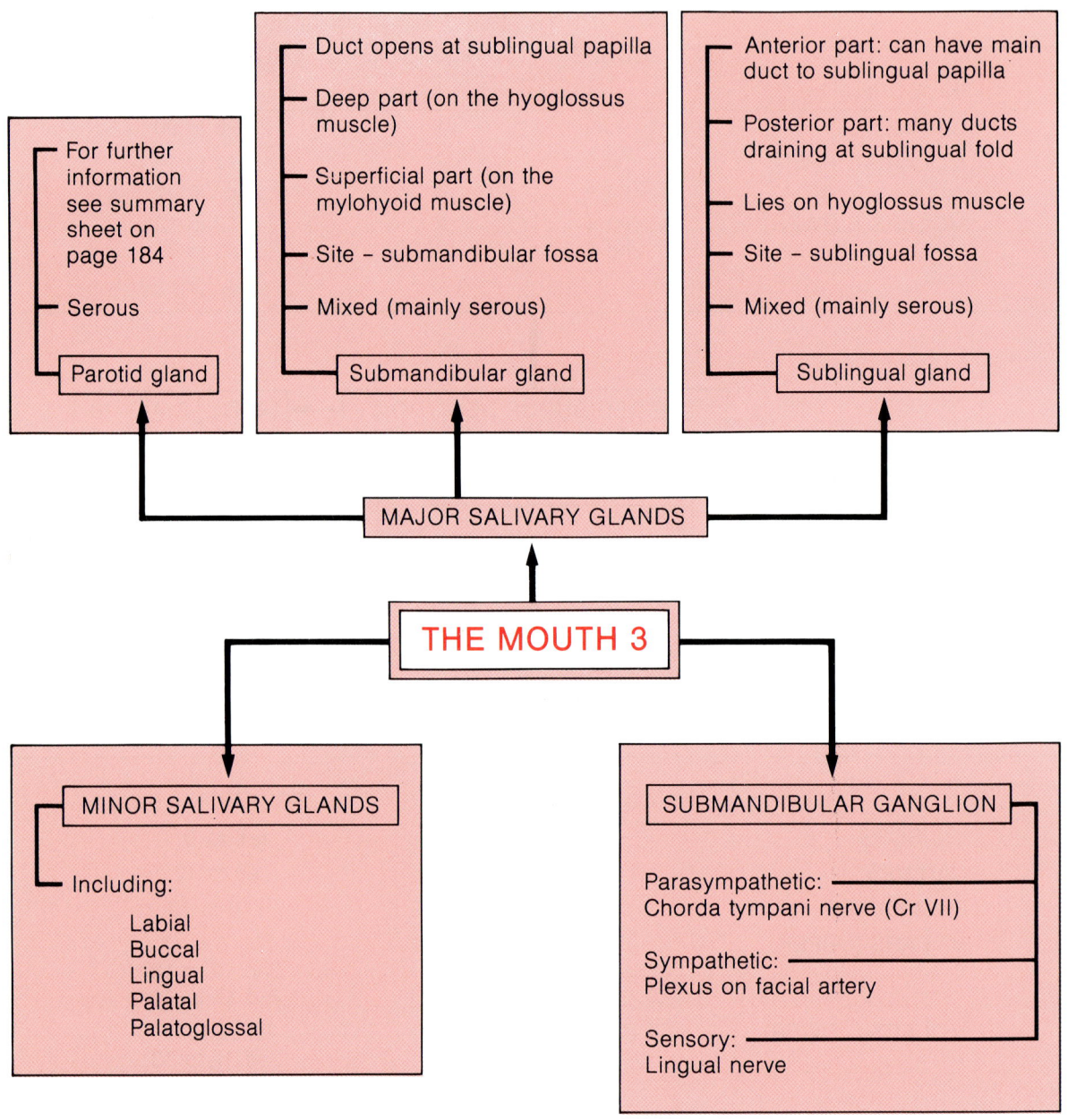

SUMMARY SHEET: THE MOUTH 4

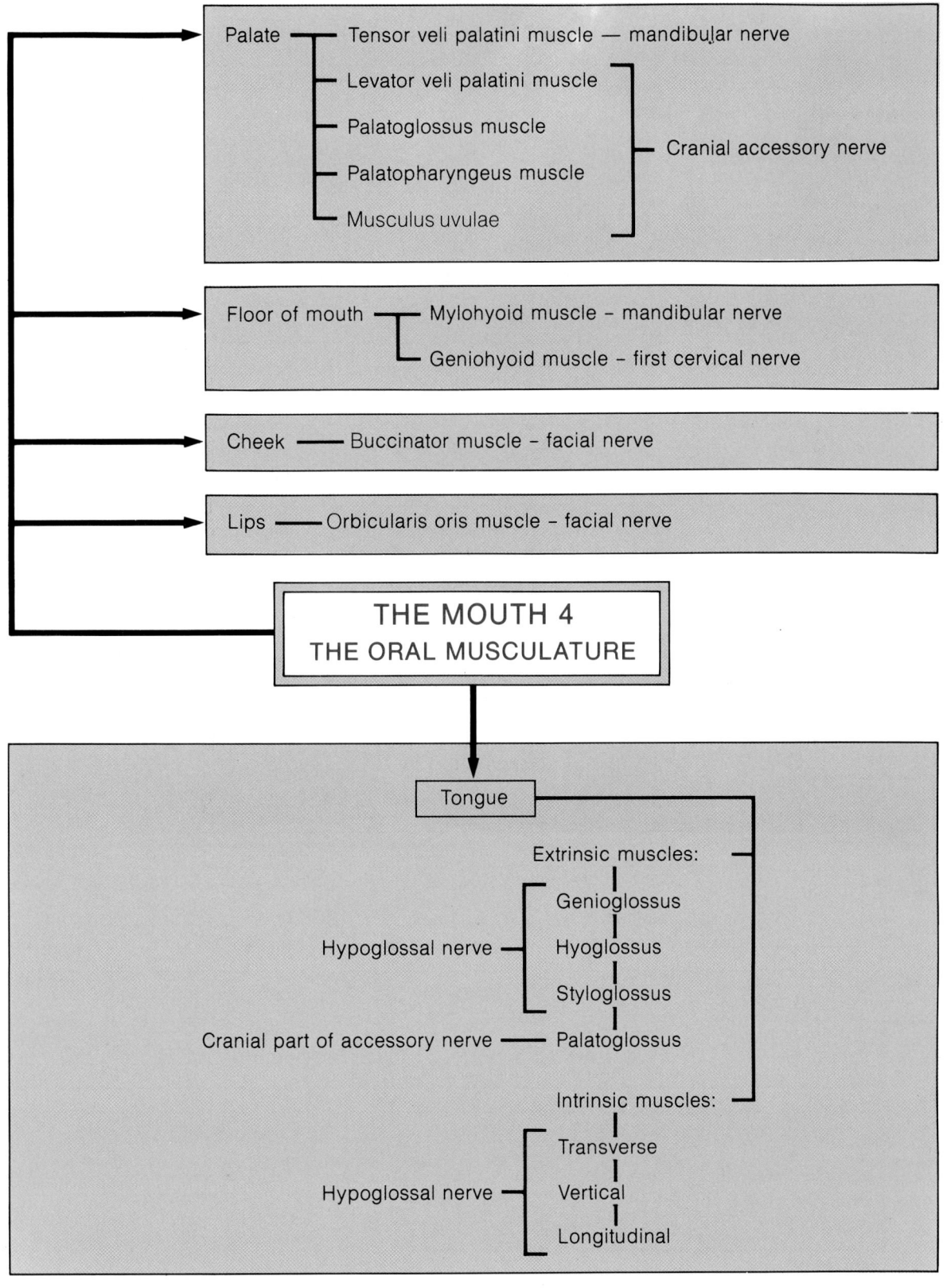

SUMMARY SHEET: THE MOUTH 5

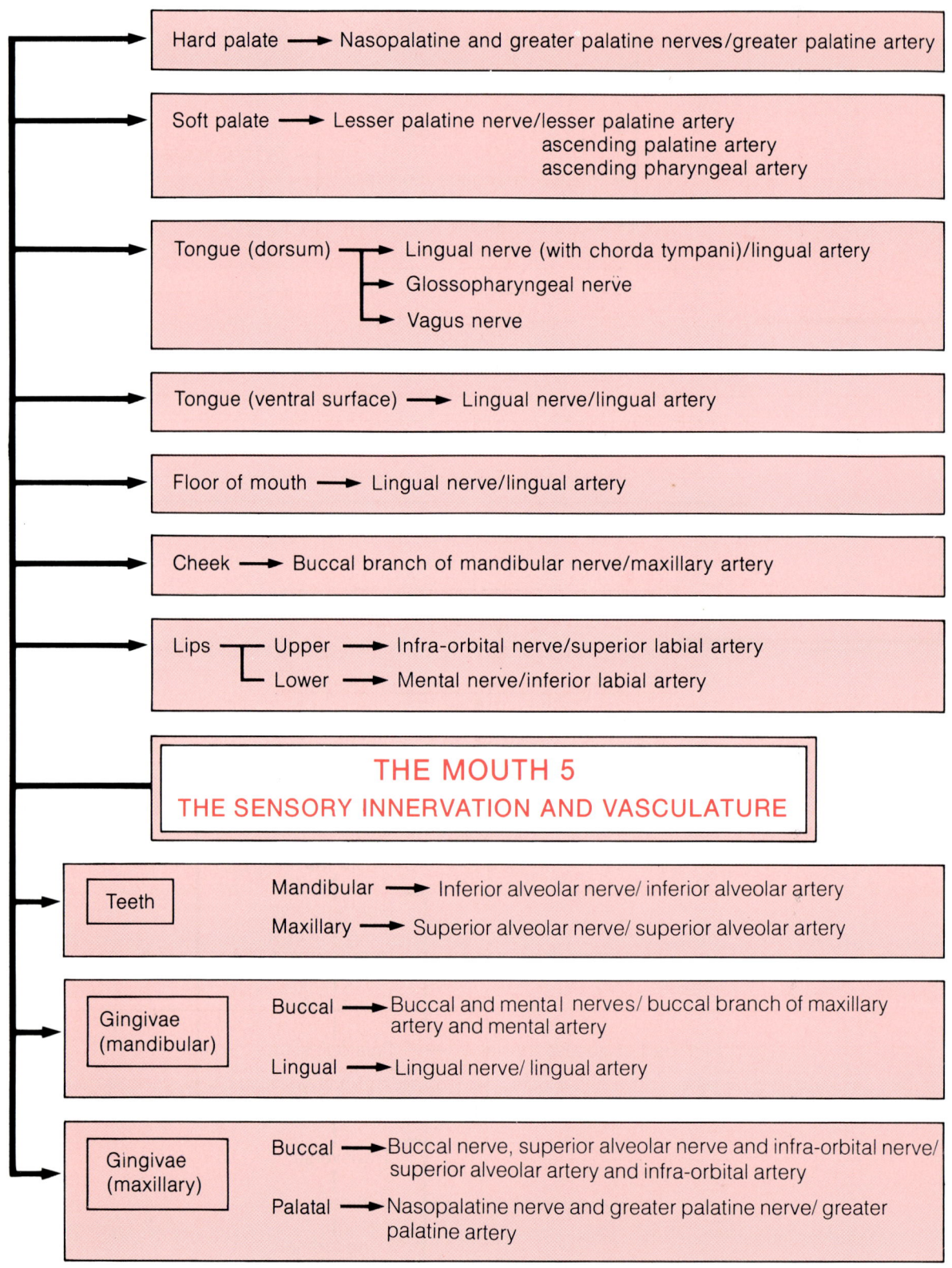

SUMMARY SHEET: THE PHARYNX

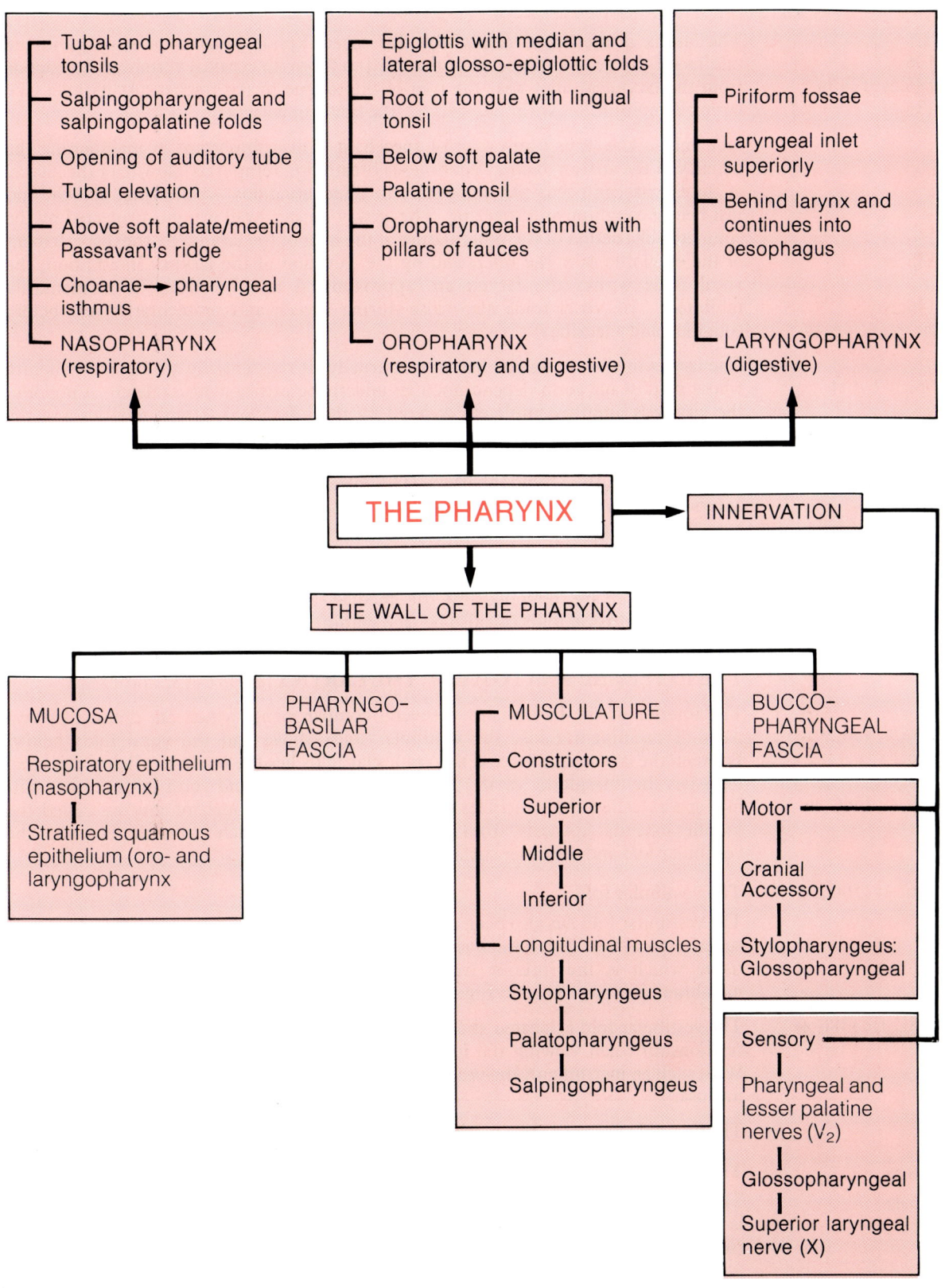

The larynx

136
136,28; 136,14
136,21

The larynx is the organ responsible for speech (phonation). It is situated in the midline of the neck at the level of the third to the sixth cervical vertebrae. It extends from the laryngeal inlet near the root of the tongue to the trachea. At its inlet, the larynx communicates with the pharynx (the laryngopharynx).

Not only is the larynx an organ of speech, it is also important in maintaining the patency of the airway to allow continuous breathing. Temporary closure of the airway at the larynx can occur physiologically in three situations: swallowing, speech, and just before coughing and sneezing. The airway during swallowing is protected by the

136,28
136,29

136,25

sphincter-like action of the musculature at the inlet of the larynx, by the displacement of the epiglottis over the inlet, and by the elevation of the larynx. In addition, the vocal folds within the larynx are approximated and breathing is momentarily inhibited. The vocal folds also close during speech and momentarily just before coughing and sneezing.

96; 102; 104

The larynx in situ shows few features externally. It is essentially a tube-like structure whose rigidity and form depend upon an underlying cartilaginous skeleton. Anteriorly, the larynx is almost completely covered by the infrahyoid (strap) muscles and the thyroid gland. The only feature usually visible is the laryngeal prominence of the

136
136,28; 156A
156A,1
156A,3
156A,4
156A,16; 138C,16
138C,42; 138C,43
138C,41; 138C,40

thyroid cartilage (Adam's apple). From the posterior aspect, the larynx forms the anterior wall of the laryngopharynx. The inlet of the larynx (or aditus) is bounded anteriorly and superiorly by the epiglottis, posteriorly and inferiorly by the mucosa over the arytenoid cartilages, and laterally by the aryepiglottic folds. The pharynx extends along the sides of the inlet to form the piriform fossae. From above, the epiglottis and the root of the tongue are separated by depressions called the valleculae. The valleculae are bounded by the median and lateral glosso-epiglottic folds. Vestibular and vocal folds can also be seen within the larynx.

THE INTERNAL ANATOMY OF THE LARYNX (Figures 132 and 133)

136,23; 136,25
156,2
136,19
136,24; 156C,29

The interior of the larynx shows several compartments which are defined by two pairs of prominent folds, the vestibular folds above and the vocal folds below. Between the laryngeal inlet and the vestibular folds lies the vestibule. Below the vocal folds lies the infraglottic cavity. Between the vestibular and vocal folds are two slit-like spaces called the ventricles (or sinuses). A small pouch of mucosa called the saccule extends upwards from the anterior end of each ventricle between the vestibular fold and the inner surface of the thyroid cartilage.

136,23; 156C,28

The vestibular fold

156D,27

The vestibular fold has also been called the false vocal fold, the ventricular fold, or the superior vocal fold. It is a thick ridge of mucosa with a thin central layer of connective tissue which is the inferior free edge of a membrane called the quadrangular membrane (see page 337 and Figure 135).

138C,41; 138C,40

The vestibular fold is located above and lateral to the vocal fold. This is an important relationship when viewing the internal anatomy of the larynx with a laryngoscope. With such an instrument, the vestibular fold appears simply as an inward bulge of the mucosa.

The fissure between the two vestibular folds is called the rima vestibuli.

136,25
156,32
156D,41

The vocal fold

The anterior three-fifths of the vocal fold is formed by the vocal cord or ligament. This is the thickened free edge of a membrane of the larynx called the cricovocal membrane (see page 337 and Figure 135). Because the mucosa covering the vocal fold in this

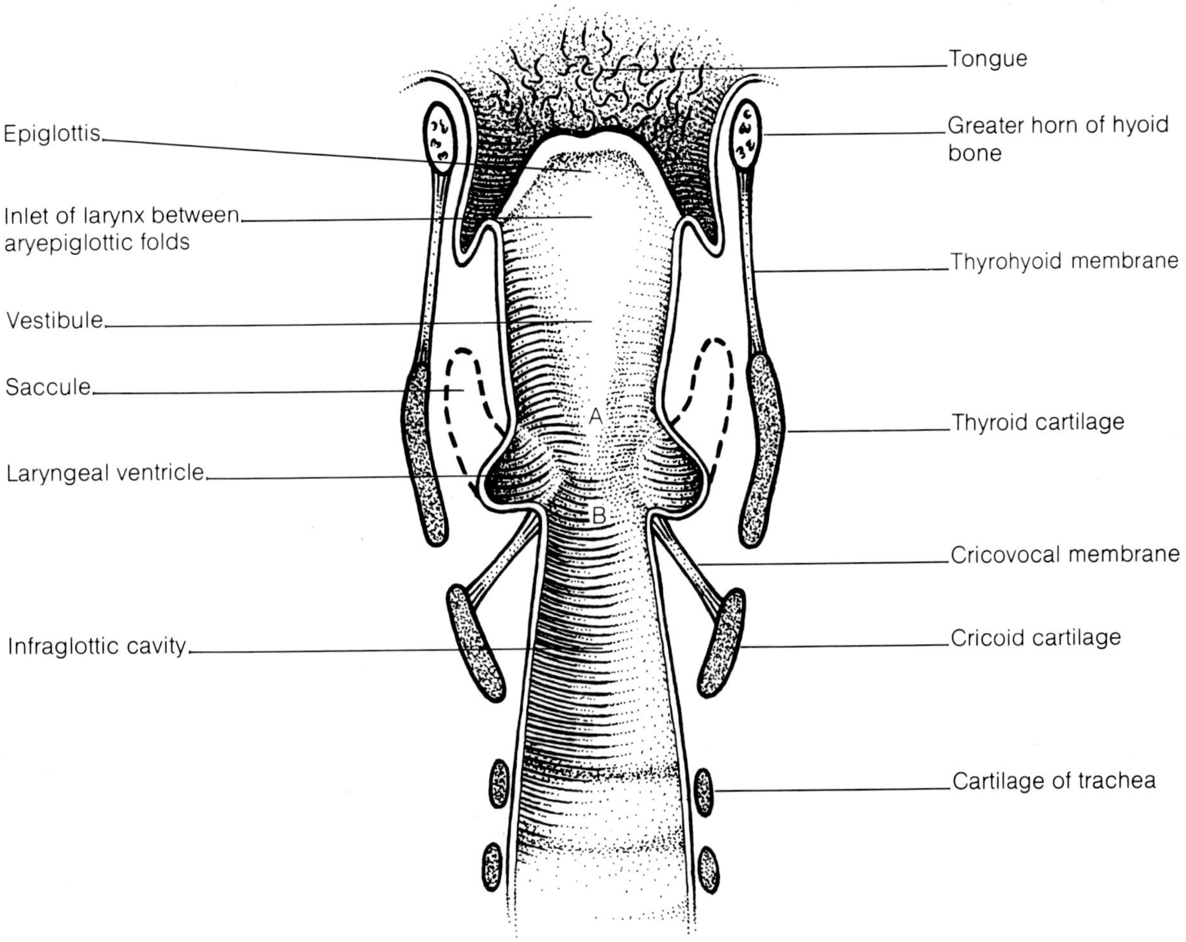

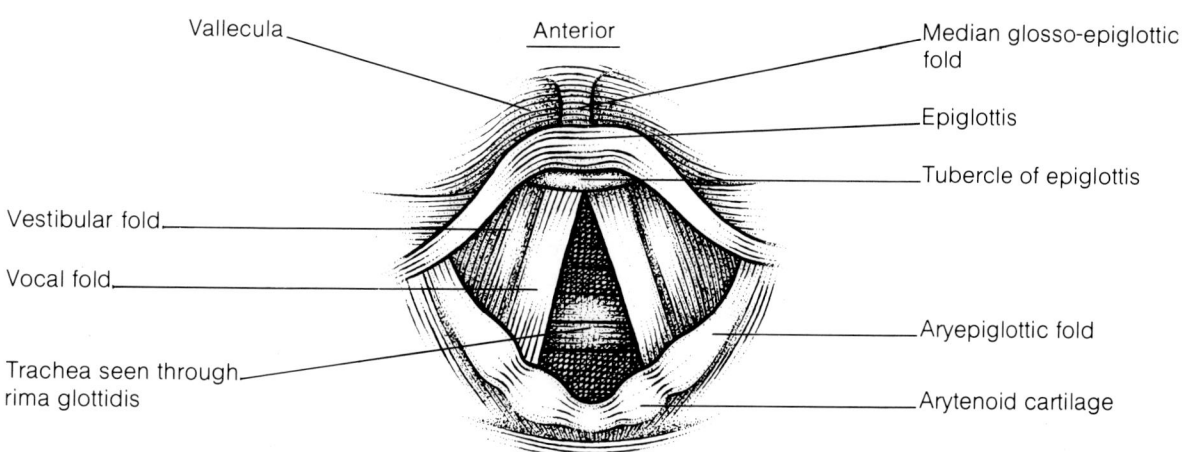

Figure 132 *Internal anatomy of the larynx as displayed in a coronal section, seen from behind.* A = *rima vestibuli between vestibular folds.* B = *rima glottidis between vocal folds.*

Figure 133 *Interior of the larynx as viewed with a laryngoscope.*

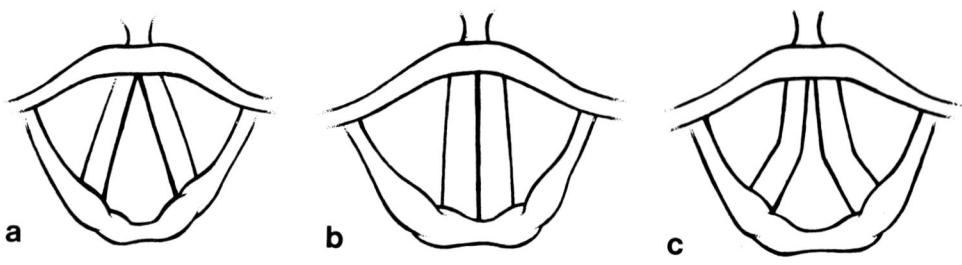

Figure 134 *The rima glottidis* a) *during quiet respiration,* b) *during speech, and* c) *during whispering.*

156C,30   region is firmly bound down to the vocal ligament, the fold appears pearly white in the living. The posterior two-fifths of the vocal fold is formed by the vocal process of the arytenoid cartilage.

138C,39   The fissure between the two vocal folds is called the rima glottidis (or glottis). That part between the vocal ligaments is called the intermembranous part. That part between the arytenoid cartilages is named the intercartilaginous part. The shape and size of the rima glottidis vary greatly during respiration and phonation (Figure 134). In quiet respiration, it has a triangular shape. During speech, the vocal folds are brought together. In whispering, the vocal folds are slightly separated at the intermembranous part, whereas a triangular space remains at the intercartilaginous part.

THE SKELETON OF THE LARYNX (Figure 135)

The skeletal framework of the larynx consists of cartilages and membranes. Its function is to prevent collapse of the air passages and to give attachment to a series of muscles.

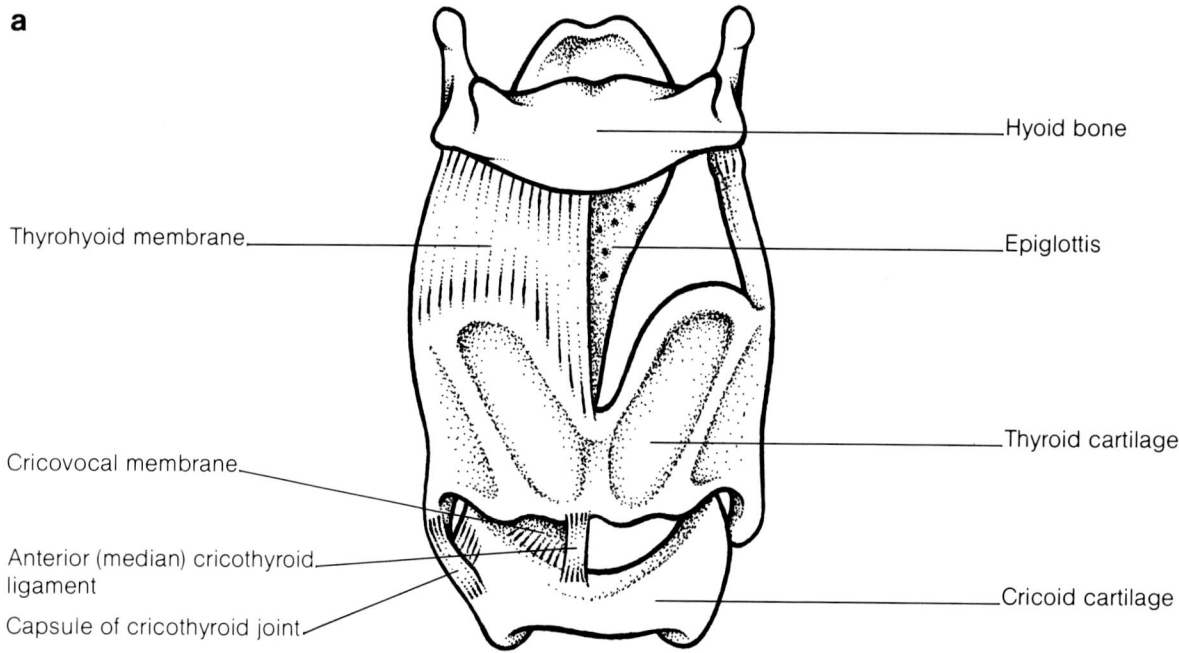

Figure 135 *Skeleton of the larynx (cartilages and membranes),* a) *viewed from the front,* b) *viewed from behind.* c) *The larynx in sagittal section. (Note that the quadrangular membrane is shown only for the larynx in sagittal section.)*

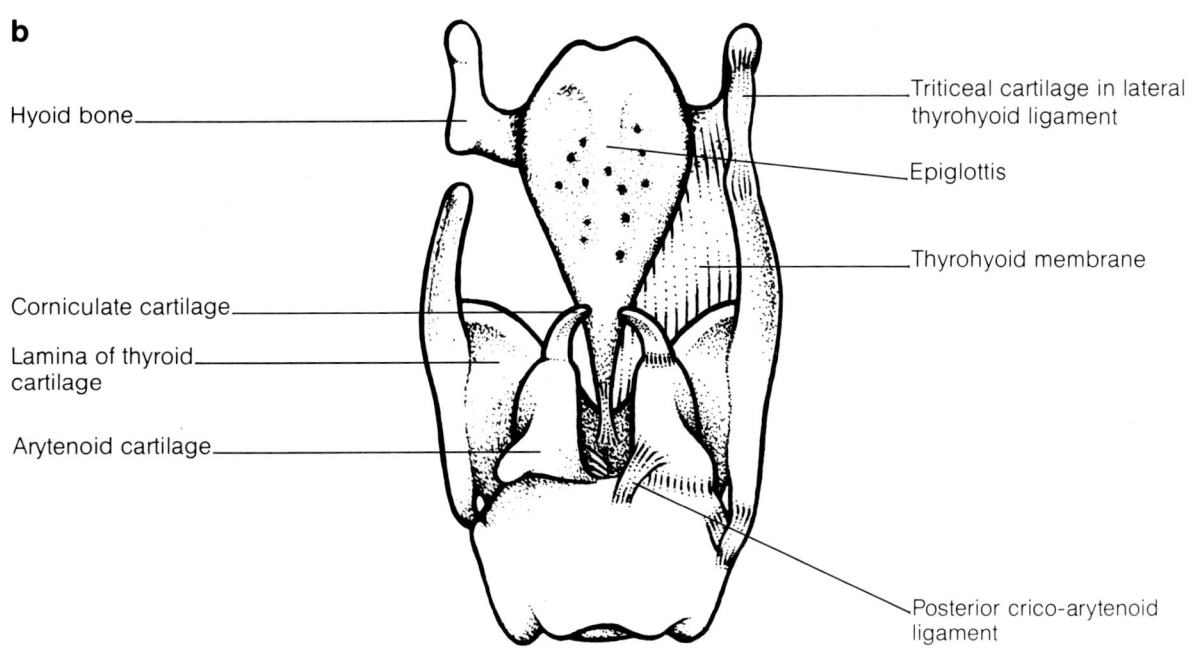

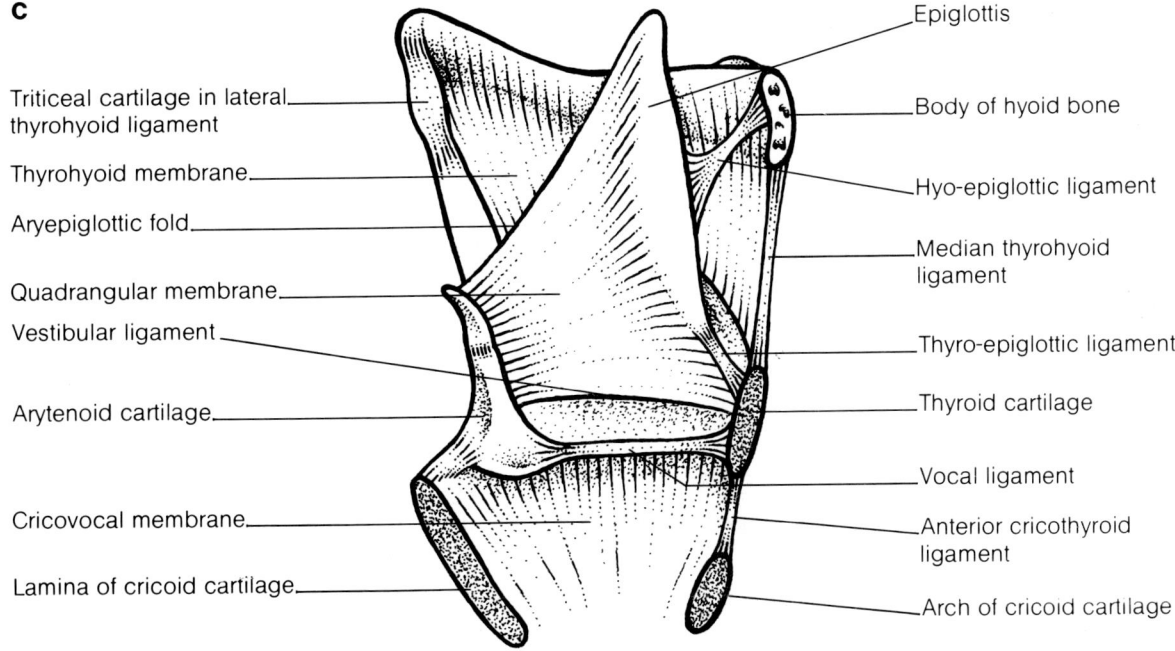

THE LARYNGEAL CARTILAGES

152; 156E

The major cartilages of the larynx are the thyroid, cricoid and arytenoid cartilages. The minor cartilages are the cuneiform and corniculate cartilages. The arytenoid, cuneiform and corniculate cartilages are paired. Associated with the larynx is the epiglottic cartilage.

The epiglottic cartilage, the cuneiform cartilage and the corniculate cartilages are composed of elastic cartilage. The other laryngeal cartilages are hyaline cartilages and may ossify in old age.

The thyroid cartilage

152C–E; 154,13

This is the largest and most prominent cartilage, forming most of the anterior and lateral walls of the larynx.

152C,19; 154C,13
152C,22; 154C,14
152C,21
152C,18; 152C,20
152C,25

The overall shape of the thyroid cartilage takes the form of a shield. It consists of two flattened, quadrilateral laminae which are joined anteriorly to form the laryngeal prominence (Adam's apple). Above this prominence, the laminae are separated by a deep V-shaped notch called the thyroid notch. Posteriorly, the laminae project upwards and downwards as the superior and inferior horns. On the external surface of each lamina lies an oblique ridge which is the site for muscle attachments. The ridge runs downwards and forwards from the superior horn towards the lower border of the cartilage. It is bounded above and below by a tubercle. The thyroid cartilage shows sexual dimorphism: in the male, it considerably increases in size at puberty and the thyroid prominence becomes very distinct.

152C,23; 152C,24

The cricoid cartilage

152F,G; 156D,E

Unlike the thyroid cartilage, the cricoid cartilage forms a complete ring. Indeed, it is the only complete cartilaginous ring in the air passages. It comprises the most inferior and posterior part of the larynx and supports the entrance to the trachea. The shape of the cricoid cartilage resembles that of a signet ring, showing a narrow arch anteriorly and a flat, quadrangular lamina posteriorly. Where the arch meets the lamina are small articular facets for the inferior horns of the thyroid cartilage. The superior edge of the lamina has sloping shoulders, and articular facets for the arytenoid cartilages. The cricoid cartilage may appear more prominent in the female.

152G,49; 156,24
152F,45; 156,8
152G,48; 156,11
152F,46; 156,42

The arytenoid cartilages

152F,G; 156E

156E

The arytenoid cartilages lie in the postero-inferior part of the larynx, on the superior edge of the lamina of the cricoid cartilage. They contribute to the margin of the inlet of the larynx. Each cartilage is pyramidal in shape, although the superior process or apex of the pyramid is really the corniculate cartilage. The base of the arytenoid cartilage presents the articulating surface with the cricoid. The arytenoid cartilage has a process anteriorly called the vocal process (for attachment of the vocal ligament), and a process laterally named the muscular process (for the attachment of some of the muscles of the larynx).

152F,36; 152,43
152F,38; 156E,42
152F,37; 156C,30
156C,32; 152F,37
156E,40

The minor cartilages of the larynx

152F,43; 156D,39

The corniculate cartilages surmount the arytenoid cartilages, thus completing their pyramidal shapes.

156D,38

The cuneiform cartilages lie within the aryepiglottic folds at the inlet of the larynx.

Small triticeal cartilages are found in the ligaments joining the tips of the superior horns of the thyroid cartilage to the tips of the greater horns of the hyoid bone.

Articulations of the laryngeal cartilages

154,34; 156,11

The inferior horn of the thyroid cartilage articulates with the cricoid cartilage by way of a synovial joint. This joint has a well-developed capsule which is strengthened posteriorly by fibrous bands. The joint permits a rotary movement with activity of the cricothyroid muscle (see page 341), such that the thyroid cartilage tilts forwards and

154B,19; 154B,31

downwards with upward movement of the arch of the cricoid cartilage (see Figure 137).

The joints between the bases of the arytenoid cartilages and the lamina of the cricoid cartilage are also synovial. The capsules of the joints are strengthened by posterior crico-arytenoid ligaments which are said to limit forward movements of the arytenoids (see Figure 135b). Rotation and gliding movements of the arytenoids occur at these joints, both types of movement being responsible for opening and closing the rima glottidis.

156E,42

Synovial or cartilaginous joints link the corniculate cartilages to the arytenoids.

156D,39

The epiglottis

152H; 156,1

The epiglottis consists of a thin lamina of elastic cartilage covered on all sides with mucous membrane. It is leaf-shaped, the 'stalk' providing the means of attachment to the larynx via a thyro-epiglottic ligament. A depression for this ligament lies just below the thyroid notch on the inner surface of the thyroid cartilage. The epiglottis is also anchored to the posterior surface of the body of the hyoid bone by a hyo-epiglottic ligament. The sides of the epiglottis are attached to the arytenoid cartilages by the aryepiglottic folds. Median and lateral glosso-epiglottic folds pass from the root of the tongue to the anterior surface of the epiglottis. The epiglottis projects upwards and backwards over the vestibule of the larynx and gives the appearance of a 'lid'. However, it does not seem to function as such, as its surgical removal has no adverse affects. The posterior surface of the cartilage of the epiglottis shows numerous small indentations or perforations in which lie mucous glands.

156C,33
152E,30
156C,35; 152B,15

156A,3
138C,42; 138C,43

152H

THE LARYNGEAL MEMBRANES

156

The larynx has thyrohyoid, quadrangular and cricovocal membranes. The thyrohyoid membrane is external to the larynx, whereas the paired quadrangular and cricovocal membranes are internal. All the membranes are composed of fibro-elastic tissue. There are also two ligaments, the anterior cricothyroid ligament and the cricotracheal ligament.

The thyrohyoid membrane

154C,29; 156D,36

This membrane extends from the upper border of the thyroid cartilage to the upper border of the inner surface of the hyoid bone (both body and greater horns). Between the membrane and the hyoid bone lies a bursa.

152B,14
152B,16

The thyrohyoid membrane is thickened in three places to form ligament-like structures. In the midline is found the median thyrohyoid ligament. At the lateral margins are found the lateral thyrohyoid ligaments, connecting the tips of the superior horns of the thyroid cartilage to those of the greater horns of the hyoid bone. The lateral ligaments may contain triticeal cartilages.

The thyrohyoid membrane is pierced by the superior laryngeal vessels and the internal laryngeal nerves as they course into the larynx (see Figure 138).

154A,28; 154A,30
154C,37

The quadrangular membrane

156,27; 156D,37

Each quadrangular membrane passes from the lateral margin of the epiglottis to the arytenoid cartilage on its own side. It is often poorly defined. The membrane shows two free borders. The upper and posterior border forms the aryepiglottic fold. The lower border forms the ventricular fold. Within the aryepiglottic folds lie the cuneiform cartilages.

152H,52
152G,52
156A,3
156C,28
156D,38

The cricovocal membrane

154C,39; 156D,41

This membrane is more pronounced than the quadrangular membrane, and arises from the side of the larynx at the upper border of the arch of the cricoid cartilage. It passes internally, deep to the lamina of the thyroid cartilage, to become attached anteriorly to the inner surface of the thyroid cartilage close to the midline, and posteriorly to the vocal process of the arytenoid cartilage.

152G,51

152E,33
152G,33

The cricovocal membrane has an upper free margin which passes across the larynx. This is thickened to form the vocal ligament.

The anterior (median) cricothyroid ligament

This is considered by some anatomists to be a superficial part of the cricovocal membrane. It is situated anteriorly in the midline, passing from the upper border of the cricoid cartilage to the lower border of the thyroid cartilage.

The cricotracheal ligament

This ligament joins the lower border of the cricoid cartilage to the first ring of the trachea.

Unfortunately, considerable differences in terminology are found in different textbooks with respect to the laryngeal membranes. In some, the cricovocal membrane and the anterior cricothyroid ligament are collectively called the cricothyroid ligament, the cricovocal membrane being designated the lateral cricothyroid ligament. Such terminology ignores the fact that the cricovocal membrane shows a thickened ligament only where it becomes the vocal ligament. Furthermore, it is attached not only to the thyroid cartilage but also to the arytenoid cartilage. Another collective term found in the literature is conus elasticus. To add further to the confusion, some anatomists restrict the term conus elasticus to the anterior cricothyroid ligament.

THE MUCOSA OF THE LARYNX

The larynx is lined internally and on its outer, posterior surface by mucous membrane. The mucous membrane on the posterior surface also forms the anterior wall of the laryngopharynx. Internally, the mucosa lines the quadrangular and cricovocal membranes and contributes to the vestibular and vocal folds. Under cover of the vestibular folds, there is a diverticulum called the laryngeal saccule. The size of this saccule varies greatly.

Stratified squamous epithelium is found on the anterior surface of the epiglottis, the upper half of the posterior surface of the epiglottis, the upper part of the aryepiglottic folds, the outer and posterior surface of the larynx, and over the vocal folds. The remaining regions have ciliated columnar epithelium.

Numerous mucous glands are scattered throughout the mucosa, but particularly on the epiglottis and in the saccule. It is thought that those in the saccule are responsible for lubricating the vocal folds. The mucous glands in the margins of the aryepiglottic folds are referred to as the arytenoid glands. Taste buds are also found in the laryngeal mucosa.

THE MUSCLES OF THE LARYNX

The muscles can be categorised as extrinsic or intrinsic.

THE EXTRINSIC MUSCLES OF THE LARYNX

These muscles have an attachment outside the larynx and include the infrahyoid (strap) muscles of the neck (see pages 88 to 90), and the stylopharyngeus, palatopharyngeus and inferior constrictor muscles of the pharynx (see pages 315 to 317).

The extrinsic muscles are responsible for movements of the whole larynx, i.e. elevation and depression during swallowing, respiration and phonation. The thyrohyoid, stylopharyngeus and palatopharyngeus muscles elevate the larynx. The omohyoid, sternohyoid and sternothyroid muscles depress the larynx. Of these three muscles, sternothyroid is the only one that has an attachment on to the larynx and which therefore depresses the larynx by direct action. The omohyoid and sternohyoid muscles can cause depression only indirectly by pressing on the larynx.

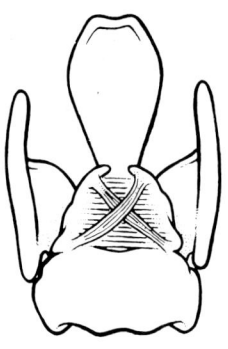
a Interarytenoid muscle (transverse and oblique fibres)

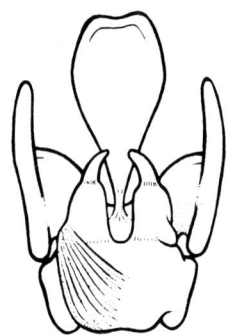
b Posterior crico-arytenoid muscle

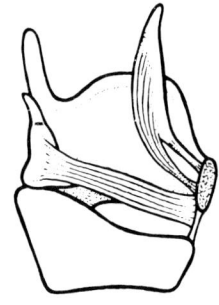
c Thyro-arytenoid and thyro-epiglottic muscles

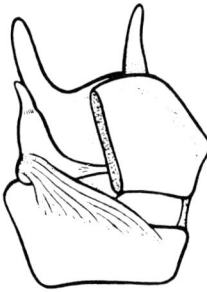
d Lateral crico-arytenoid muscle

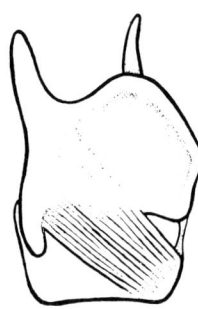
e Cricothyroid muscle

Figure 136 *Intrinsic muscles of the larynx.*

Because the larynx and the hyoid bone are connected by the thyrohyoid membrane, elevation of the larynx can occur by the actions of the suprahyoid musculature (mylohyoid, digastric, stylohyoid and geniohyoid muscles).

THE INTRINSIC MUSCLES OF THE LARYNX (Figure 136)

These are confined to the larynx. Within this group are the posterior crico-arytenoid, lateral crico-arytenoid, interarytenoid, thyro-arytenoid and cricothyroid muscles. With the exception of the interarytenoid muscle, the muscles are paired.

The intrinsic muscles of the larynx are mainly concerned with the activities of the vocal folds. Subsidiary parts of the interarytenoid and thyro-arytenoid muscles (the aryepiglottic and thyro-epiglottic muscles) modify the inlet of the larynx.

Whereas most of the intrinsic muscles lie internally (under cover of the thyroid cartilage or the mucosa), the cricothyroid muscles appear on the outer aspect of the larynx.

The posterior crico-arytenoid muscle

156A,7

Attachments. This muscle arises from a broad depression on the posterior surface of the lamina of the cricoid cartilage. Passing upwards and laterally, it is inserted into the muscular process of the arytenoid cartilage.

152F,G,42

152F,G,42

Innervation. The recurrent laryngeal branch of the vagus nerve provides the motor supply.

154,23

Vasculature. The posterior crico-arytenoid muscle receives its blood supply from the laryngeal branches of the superior and inferior thyroid arteries.

Actions. This is the only muscle that opens the rima glottidis and it does so in two ways (Figure 137). First, the upper fibres, being almost horizontal, rotate the arytenoid cartilage. Second, the lower fibres, being more vertical, cause sliding of the arytenoid cartilage down the sloping superior margin of the cricoid cartilage.

The lateral crico-arytenoid muscle

Attachments. This muscle originates from the lateral side of the upper border of the arch of the cricoid cartilage. It extends upwards and backwards beneath the thyroid cartilage to insert on to the muscular process of the arytenoid cartilage.

Innervation. The recurrent laryngeal nerve supplies the lateral crico-arytenoid muscle.

Vasculature. It receives its blood supply from the laryngeal branches of the superior and inferior thyroid arteries.

Actions. The lateral crico-arytenoid muscle rotates the arytenoid cartilage in a direction opposite to that of the posterior crico-arytenoid muscle, thereby closing the rima glottidis (Figure 137).

The interarytenoid muscle

Attachments. This is a single muscle in two parts, which runs posteriorly between the muscular processes of the arytenoid cartilages. Many of its fibres run transversely across the posterior surfaces of the arytenoids (the transverse arytenoid part), but some run obliquely from the muscular process of one arytenoid to the apex of the opposite cartilage (the oblique arytenoid part). The oblique fibres form two thin

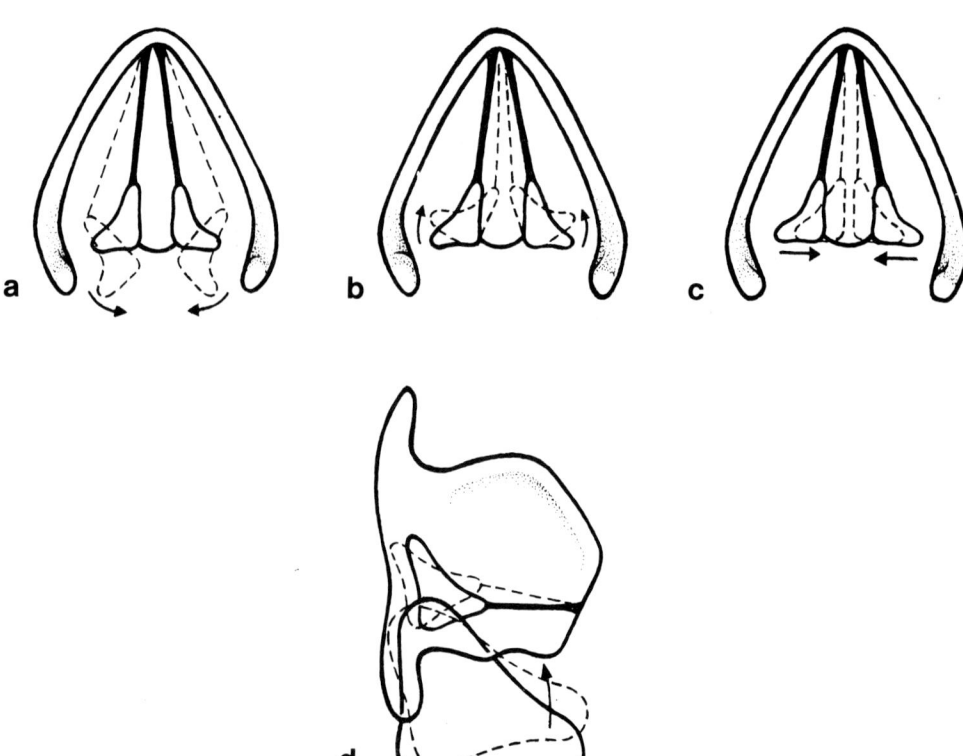

Figure 137 *Movements of the vocal folds.* **a** = *opening of the rima glottidis by rotation of the arytenoids.* **b** = *closure of the rima glottidis by rotation of the arytenoids.* **c** = *closure of the rima glottidis by approximation of the arytenoids without rotation.* **d** = *tensing of the vocal folds by tilting of the anterior part of the cricoid cartilage.*

bands which cross to produce a distinctive X shape. Some of the oblique fibres continue into the aryepiglottic folds as the aryepiglottic muscles.

156B,17

Innervation. The muscle is innervated by the recurrent laryngeal nerve.

154,23

Vasculature. The blood supply is derived from the laryngeal branches of the superior and inferior thyroid arteries.

Actions. The interarytenoid muscle closes the rima glottidis by approximating the arytenoid cartilages. This is accomplished by drawing the arytenoids upwards along the sloping shoulders of the cricoid lamina, without rotation (Figure 137). The aryepiglottic muscles modify the inlet of the larynx. However, their poor development limits their action as sphincters of the inlet.

The thyro-arytenoid muscle

156,20

Attachments. This muscle lies lateral to the vocal fold. It arises on the inner surface of the thyroid cartilage in the midline. It also arises from the cricovocal membrane. The thyro-arytenoid muscle passes backwards, upwards and outwards to be inserted into the base and anterior surface of the arytenoid cartilage. The lower and deeper fibres form a distinct bundle that runs parallel with, and lateral to, the vocal ligament. This bundle is sometimes referred to as the vocalis muscle and is attached to the vocal process of the arytenoid cartilage. There is doubt as to whether the fibres of vocalis are also attached to the vocal ligament. The upper fibres of the thyro-arytenoid muscle may extend into the aryepiglottic fold to form the thyro-epiglottic muscle.

152E,32
152G,32
156C,31
152F,39
156B,18; 152E,31

Innervation. All parts of the thyro-arytenoid muscle are supplied by the recurrent laryngeal nerve.

154,23

Vasculature. The arterial blood supply is derived from the laryngeal branches of the superior and inferior thyroid arteries.

Actions. The primary function of the thyro-arytenoid muscle is to shorten the vocal ligament and adjust the tension within it during phonation. In addition, it can rotate the arytenoid cartilage medially and so aid closure of the rima glottidis. Relaxation of the posterior parts of the vocal ligaments by the vocalis muscles, with tension in the anterior parts of the ligaments, is responsible for raising the pitch of the voice. The thyro-epiglottic muscles widen the inlet of the larynx.

The cricothyroid muscle

154,19; 154,31

Attachments. The cricothyroid muscle arises from the anterior and anterolateral parts of the external surface of the arch of the cricoid cartilage. Its fibres pass upwards and backwards to insert into the thyroid cartilage. Two distinct parts can be recognised. The anterior and superior fibres constitute the straight part of the cricothyroid muscle. This inserts into the lower border of the thyroid lamina. The posterior and inferior fibres constitute the oblique part of the cricothyroid muscle. This inserts into the inferior horn of the thyroid cartilage.

152G,29
152D+E,29
154,19

154B,31

Innervation. Unlike the other intrinsic muscles of the larynx, the cricothyroid muscle is innervated not by the recurrent laryngeal nerve but by the external branch of the superior laryngeal nerve.

154B,16

Vasculature. The muscle is supplied by the cricothyroid branch of the superior thyroid artery and by the inferior laryngeal branch of the inferior thyroid artery.

Actions. The cricothyroid muscle tenses and elongates the vocal ligaments. This is accomplished by elevating the arch of the cricoid cartilage and tilting back the upper border of its lamina (Figure 137). As a result, the distance between the angle of the thyroid cartilage and the vocal processes of the arytenoids is increased. A similar activity results if the muscles pull the thyroid cartilage forward. Indeed, this is thought to be the principal activity during phonation, as the lamina of the cricoid cartilage is held in position against the vertebral column by the cricopharyngeus muscles.

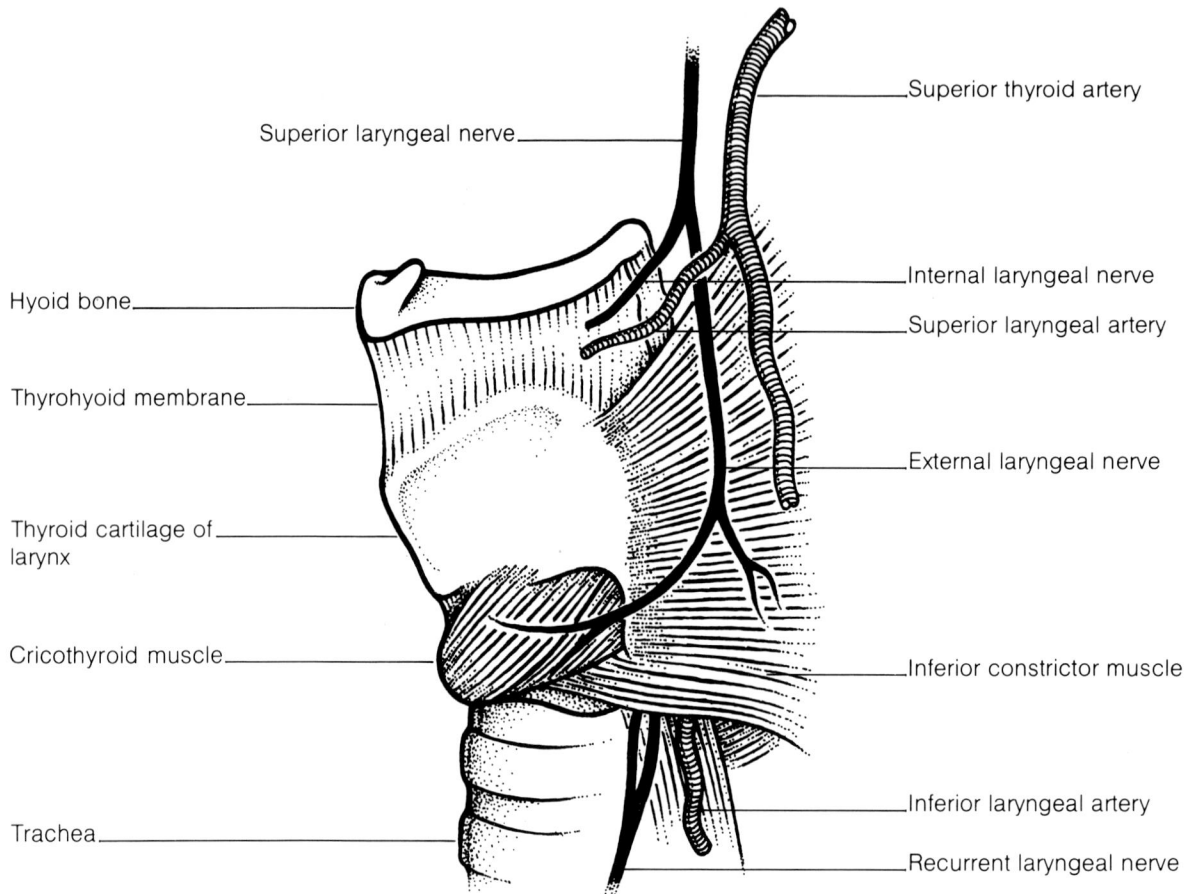

Figure 138 *Arterial and nerve supply to the larynx.*

THE BLOOD SUPPLY OF THE LARYNX (Figure 138)

154A,28
154A,27

154A,30
154C,37
154A,22; 154A,25
144A,46
144A,40
144A,45

The blood supply of the larynx is derived mainly from two pairs of arteries: the superior and inferior laryngeal arteries. The superior laryngeal artery supplies the larynx above the vocal folds. It is derived from the superior thyroid artery, a branch of the external carotid artery. The superior laryngeal artery runs down towards the larynx with the internal branch of the superior laryngeal nerve. It enters the larynx by penetrating the thyrohyoid membrane. The inferior laryngeal artery supplies the larynx below the vocal folds. It is a branch of the inferior thyroid artery which itself is derived from the thyrocervical trunk of the subclavian artery. The inferior laryngeal artery runs up and into the larynx deep to the lower border of the inferior constrictor muscle. It is accompanied in its course by the recurrent laryngeal nerve. The cricothyroid branch of the superior thyroid artery may also supply the larynx.

Venous return from the larynx occurs via superior and inferior laryngeal veins. These run parallel to the laryngeal arteries. They are tributaries of the superior and inferior thyroid veins respectively.

The lymph vessels draining the larynx above the vocal folds accompany the superior laryngeal artery, pierce the thyrohyoid membrane, and end in the upper deep cervical lymph nodes. Below the vocal folds, some of the lymph vessels pass through the cricovocal membrane to reach the prelaryngeal and/or pretracheal lymph nodes. Others run with the inferior laryngeal artery to join the lower deep cervical nodes.

THE INNERVATION OF THE LARYNX (Figure 138)

The vocal folds form a dividing line for both the sensory and the secretomotor innervation of the mucosa within the larynx. Above the vocal folds, the mucosa is innervated by the internal laryngeal branch of the superior laryngeal nerve. Below the vocal folds, the mucosa is supplied by the recurrent laryngeal nerve. The vocal folds themselves are innervated by both nerves.

136,25

The motor supply to the intrinsic muscles of the larynx is derived mainly from the recurrent laryngeal nerve. The cricothyroid muscle, however, is supplied by the external branch of the superior laryngeal nerve.

Thus, the chief nerves supplying the larynx are the superior laryngeal and recurrent laryngeal branches of the vagus, both of which contain sensory and motor fibres.

The superior laryngeal nerve leaves the trunk of the vagus at its inferior ganglion. It curves downwards and forwards by the side of the pharynx, medial to the internal carotid artery. It divides into external and internal branches. The external branch continues downwards and forwards on the lateral surface of the inferior constrictor muscle to end in the cricothyroid muscle. The internal branch is larger and enters the larynx after piercing the thyrohyoid membrane.

144A,;54

144A,39
144A,40; 144A,41
144A,31
154A,30; 154C,37

The origins of the recurrent laryngeal nerves differ according to side. The right recurrent laryngeal nerve issues from the vagus nerve in front of the subclavian artery. It then passes below and behind the artery. The left recurrent laryngeal nerve arises in the thorax (on the arch of the aorta). Both nerves run up the neck towards the larynx in 'grooves' between the oesophagus and trachea. The upper part of the recurrent laryngeal nerve is sometimes referred to as the inferior laryngeal nerve. It enters the larynx deep to the lower border of the inferior constrictor muscle. Here, it usually divides into two branches: an anterior motor branch and a posterior sensory branch.

106B,27; 106B,2
106B,11
154A,23
154A,24; 154A,21

SPEECH

The aquisition of language is probably the most complex sensorimotor development in the individual's life. Sounds are produced initially in the larynx by the co-ordinated movements of abdominal, thoracic and laryngeal muscles. Subsequent modification of laryngeal sound to produce meaningful speech occurs principally within the pharyngeal, oral and nasal cavities.

Phonation is the term used to describe the mechanism of voice production at the larynx. It involves vibration of the vocal folds, mainly in the horizontal plane. The vocal folds are separated during quiet respiration, but are approximated during speech. The apposed vocal folds provide a barrier to the passage of expired air. The air pressure increases until it overcomes the resistance. Consequently, air flows momentarily between the vocal folds into the pharynx. The folds return to their apposed position as a result of their elasticity and the negative pressure created by the rapid flow of air through the constricted rima glottidis. The cycle is repeated with a periodicity of the order of milliseconds. In this manner, expired air escapes as a series of rapid 'puffs' that form sound waves.

The character of a sound has three properties: intensity, pitch, and timbre. The intensity depends upon the pressure of the expired air. The pitch depends on many factors, including the length, shape and degree of tension of the vocal folds. The quality of the voice or timbre depends primarily on a series of resonators (see below).

The deeper pitch of the adult male voice is related to the greater length of the vocal folds (approximately 15 mm in males compared with 11 mm in females). The more rapid growth of the male larynx at puberty is responsible for the 'breaking of the voice.'

Articulation is the term used to describe the mechanism whereby laryngeal sound is modified within resonating chambers by the activity of organs such as the lips, tongue and soft palate to produce speech. This is necessary as the sound generated at the larynx carries a limited amount of speech information. Indeed, the fundamental laryngeal note has a thin and reedy quality.

The resonators of the human voice are those air-filled spaces to which sound waves have access. By a process of sympathetic vibration, the resonators act as acoustic filters, amplifying selected frequencies and attenuating others. The supraglottic resonators include the laryngeal chambers above the vocal folds, the pharynx, the oral and nasal cavities and the paranasal sinuses. It is because of the considerable alterations in both size and shape that can occur within the supraglottic resonators that the diversity of sound can be produced. For most sounds, the nasal cavity and nasopharynx are closed off by elevation of the soft palate against the posterior wall of the pharynx. The degree of elevation varies between sounds. However, for the sounds m, n and ng, resonance is produced in the nasopharynx and nasal cavities by depression of the soft palate and closure of the mouth.

The classification of sounds

Sounds may be voiced (i.e. the vocal folds vibrate in their production) or breathed (i.e. the vocal folds do not vibrate in their production). The two main groups of speech sounds are vowels and consonants.

A vowel sound is produced when the air flow is uninterrupted. In producing different vowel sounds, air is channelled or restricted by the position of the lips and tongue. All vowels are voiced.

A consonant is produced when the air flow is impeded before it is released. Consonants may be voiced (e.g. b, d, z) or breathed (e.g. p, t, s).

Consonants may be classified in two ways: according to the place of articulation (bilabials, labiodentals, linguodentals, linguopalatals, glottals), or the manner of articulation (plosives, fricatives, affricatives, nasals, laterals, semi-vowels).

In bilabial sounds, the two lips are used. In labiodental sounds, the lower lip meets the maxillary incisors. Linguodental sounds involve the tip of the tongue contacting the maxillary incisors (and adjacent hard palate). For linguopalatal sounds, the tongue meets the palate.

In plosive sounds, there is a complete stoppage of air, while fricatives require only a partial stoppage of air. Affricatives also require a partial stoppage of air, but there is a subsequent rapid release of air. In nasal sounds, the mouth is obstructed but the nasal passages remain open. Lateral sounds require air to leave the sides of the mouth. Semi-vowels are brief vowel articulations which are followed by a true vowel articulation of longer duration.

Table 8 provides examples of the various categories of consonant articulations. Figure 139 shows the various positions of the tongue and palate.

Although one may describe the position of articulators for a particular vowel or consonant, it must be remembered that there are no fixed positions during speech, only continuous movement. Hearing is an important monitoring system in controlling speech, but feed-back from sensory receptors in the oral cavity (particularly the tongue) also plays an important role.

THE EMBRYOLOGY OF THE LARYNX

The respiratory system develops at the end of the third week of intra-uterine life from a diverticulum (the respiratory diverticulum) of the ventral surface of the foregut. The diverticulum grows downwards in the midline to give rise to the trachea and lungs. It becomes separated from the foregut by a partition called the oesophagotracheal septum, except at its uppermost part which forms the inlet of the larynx. The laryngeal cartilages are derived from the cartilages of the branchial arches. The

TABLE 8 Examples of the categories of consonant articulations.

			Place of articulation					
			Linguodentals		Linguopalatals			
	Bilabial	Labiodental	(Dental)	(Alveolar)	(Alveolar)	(Palatal)	Glottal	
Voicing	− +	− +	− +	− +	− +	− +	− +	
Plosives	p b			t d		k g		
Fricatives		f v	θ ð	s z	ʃ ʒ		h	
Affricatives				tʃ				
				tr dr		j		
Nasals	m			n		ng		
Laterals				l				
Semi-vowels	w							

(Manner of articulation)

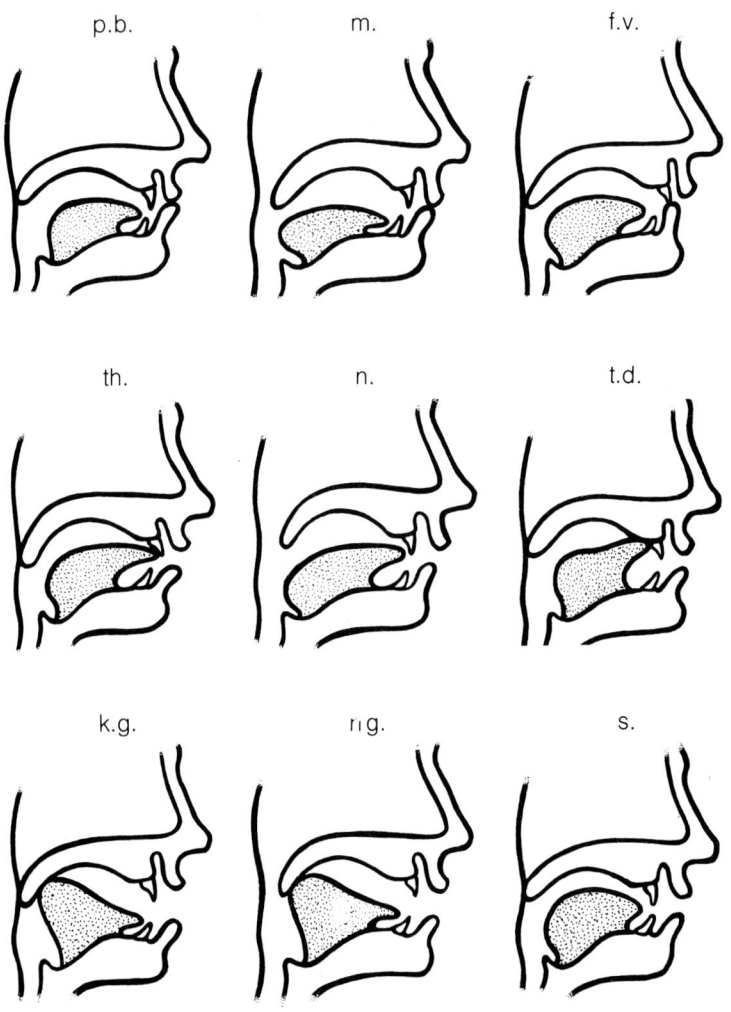

Figure 139 *Configuration of oral structures during consonant articulations.*

thyroid cartilage probably arises from the fourth arch. The cricoid and arytenoid cartilages probably develop from the sixth arch. The epiglottis is derived from a midline swelling on the posterior aspect of the fourth branchial arch (see Figure 130). The cricothyroid muscle is derived from the fourth branchial arch and is therefore supplied by the external branch of the superior laryngeal nerve. The remaining intrinsic muscles of the larynx develop from the sixth branchial arch and are thus supplied by the recurrent laryngeal nerve (see Table 2, page 134).

SUMMARY SHEET: THE LARYNX

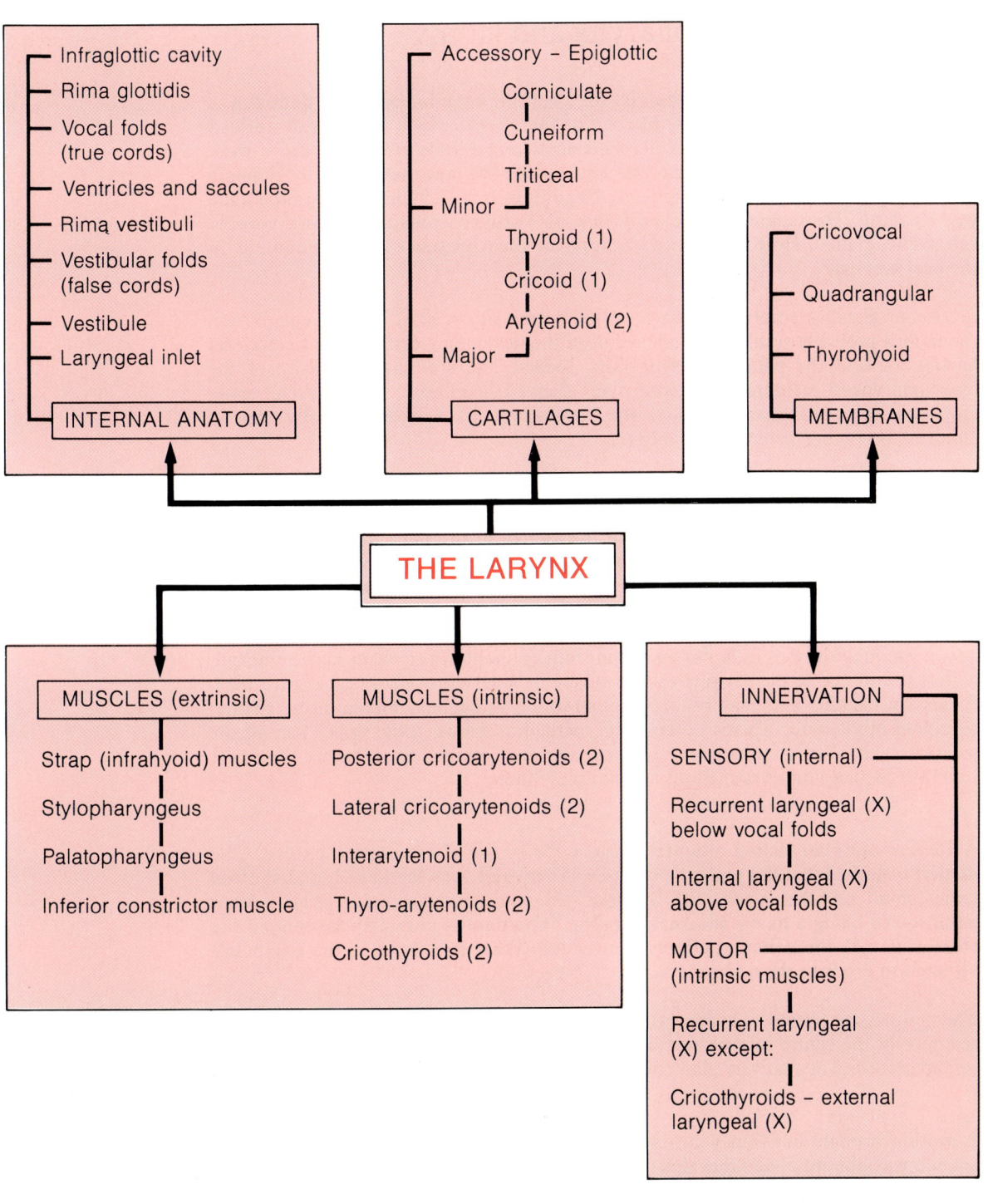

Case histories 6
The mouth, palate, pharynx and larynx

An 18-year-old youth visited his dentist for a routine examination. The dentist was surprised to see a change in the mucosa of the cheek, which he had not noticed during previous examinations. Numerous small, light-yellow macular areas were seen over the entire mucosa, the mucosa having the appearance of a 'chamois leather'. The dentist observed some duct openings from which a greasy substance was exuding. He diagnosed the presence of Fordyce spots, which are ectopic sebaceous glands. How would you account for their presence developmentally in the oral mucosa?

Sebaceous glands are normally located in the dermis of the skin. Their occurrence in the mouth probably results from inclusion in the developing oral cavity of ectoderm having some of the potentialities of skin. Indeed, the oral cavity in front of the buccopharyngeal membrane is ectodermally derived (see Figure 127). Fordyce spots are found in approximately 80% of the population. They are not usually seen in children, becoming prominent at puberty.

On examining a maxillary occlusal radiograph, an orthodontist noticed a large, well-circumscribed, radiolucent area in the centre of the hard palate. No abnormality was evident in the mouth, and all the maxillary teeth were vital. The clinician suspected that the lesion was a fissural cyst with an embryological origin. How might the cyst have arisen?

The position of the cyst indicates an origin from epithelium persistent in the embryological fusion line of the palatine shelves of the maxillary processes (see Figure 131). A median palatine cyst may be lined by stratified squamous epithelium or ciliated columnar epithelium. This is because the epithelium between the upper part of the palatine shelves differentiates into ciliated columnar epithelium, whereas the lower part differentiates into stratified squamous epithelium.

A 70-year-old man visited a dental surgery for repair of his upper denture. The dentist observed that the patient's palate was very red, was thickened, and showed many small nodules (each with a small red spot in the centre). The patient admitted to being a heavy smoker of a pipe. The dentist correctly associated the stomatitis with the irritating effects of tobacco. What structures in the palate are affected to produce the nodules?

The nodules are related to hyperplasia of the minor salivary glands of the palate. The red spot in the centre of each nodule represents the dilated (or sometimes partially occluded) orifice of a salivary gland duct.

A mother brought her four-year-old son to the doctor's surgery believing that his speech was abnormal and that this might be the result of a hearing impairment. On questioning the mother, the doctor established that the child's vocabulary was well developed, but his pronunciation was often very 'nasal' in character. Examination of the child's ears with an auriscope revealed one normal tympanic membrane and one showing evidence of fluid in the middle ear (the tympanic membrane being retracted, its colour dull compared with the usual pearly-grey, and its mobility reduced). On examination of the mouth, a groove was evident in the midline of the soft palate and the uvula was bifid. A submucous cleft palate was diagnosed.

In a submucous cleft palate, the tensor veli palatini muscles are deficient in the midline and therefore do not interdigitate to form a complete palatine aponeurosis. It is therefore not possible for the patient to tense the soft palate, with the result that palatal closure is at times inefficient. There is 'nasal' speech associated with the nasal escape of air. As the tensor veli palatini may be involved in opening the auditory tube, the presence of an abnormality in the muscle can be associated with middle ear dysfunction. This predisposes to the accumulation of fluid in the middle ear and may result in deafness (which compounds language problems).

A 54-year-old dentist visited his doctor saying that, while cleaning his teeth in front of the mirror, he had noticed that his uvula was not centrally positioned, but was displaced to the right. He also admitted to having had mild difficulty for several months in swallowing large lumps of food (dysphagia). He had no other symptoms. The doctor suspected that the patient might have a benign tumour. In what organ might this be situated?

This patient had a tumour in the deep part of the left parotid gland. These tumours often grow to a large size before they are diagnosed because they rarely cause any symptoms. The deep part of the parotid gland lies adjacent to the wall of the pharynx, lateral to the tonsillar fossa, and its enlargement can cause medial displacement of the tonsil and soft palate. It will also cause dysphagia. Tumours of the superficial part of the parotid gland enlarge laterally and will produce an obvious facial swelling when relatively small. If malignant, they may also cause facial palsy by damaging the facial nerve in its course through the gland.

A young man noticed a painless swelling at the back of his tongue in the midline. Thinking it might be a tumour, he visited his local hospital. The examining doctor believed that the lesion was harmless and had an embryological explanation. To confirm this, the doctor conducted a test which showed that the swelling took up radioactive iodine. What might the swelling be?

The uptake of radioactive iodine confirms that the lesion is an aggregation of thyroid tissue. The thyroid gland forms with the developing tongue between the tuberculum impar and the copula (see Figure 130). Although the embryonic gland usually migrates into the neck, fragments of thyroid tissue may be found anywhere along its pathway. When found on the tongue, it is referred to as a lingual thyroid. On the adult tongue, the site of development of the thyroid gland is normally marked by a pit called the foramen caecum.

A young man suffered from an upper respiratory infection. Some days later, one side of his tongue became enlarged, red and tender, and he found it difficult to avoid traumatising the region. His doctor noted that the lesion was associated with a set of parallel folds of mucosa on the lateral margin of the tongue posteriorly. What lingual structures have been affected?

Foliate papillae may sometimes be seen as vertical folds of mucosa on the lateral borders of the tongue posteriorly. In most people these are rudimentary. In some, however, they may be large and contain significant amounts of lymphoid tissue. Thus, an upper respiratory infection may result in lymphoid hyperplasia and the symptoms described above.

A 70-year-old woman became worried about the appearance of her tongue. Purplish-blue nodular areas were seen on its ventral surface, on either side of the lingual frenum. What might these be?

The nodules are lingual varicosities resulting from dilatations and increased tortuosity of the deep lingual veins. These are not uncommon in the elderly, but are of no clinical significance because complications such as thrombosis, ulceration and haemorrhage rarely occur.

A 55-year-old man visited his dentist complaining of an ulcer on the back of his tongue near the midline. He had first noticed it about two years before but, as it had not caused him any problems, he had not bothered to seek treatment. Recently, he had experienced some pain from the ulcer with some bleeding. He also complained of earache. The patient admitted to being a heavy smoker and drinker. On examination, there was no evidence of any local factor causing irritation. The upper deep cervical lymph nodes on both sides of the neck were enlarged and rock hard. A malignant tumour (carcinoma) of the tongue was diagnosed. How would you account for the symptoms of earache and the spread to the upper deep cervical lymph nodes?

The earache is the result of referred pain, the glossopharyngeal nerve supplying most of the posterior part of the tongue as well as the ear via the tympanic plexus. Lymph nodes on both sides of the neck should always be examined where tumours around the mouth are suspected. Lymph vessels from the posterior part of the tongue pass to the upper deep cervical lymph nodes. Lateral vessels from the side of the tongue pass to ipsilateral nodes. However, the central vessels can pass both ipsilaterally and contralaterally.

A 50-year-old man visited his dentist because of discomfort and pain in the floor of his mouth. At meal-times, the left side of his mouth became swollen. He also complained of a bad taste. On examination, there was evidence of inflammation around the sublingual papilla. What was responsible for these symptoms?

The patient has a stone (a sialolith) within the left submandibular duct, which is partially obstructing the flow of saliva. This causes the gland to swell as a result of the increased secretion at meal-times. The stone predisposes to infection and this is responsible for the bad taste. A sialolith may be bimanually palpated and may be diagnosed on radiographs.

A dental student who had just started his anatomy course suddenly became aware one morning of two hard, smooth lumps on the inside of his mandible, which projected into the floor of the mouth. Not knowing of any normal anatomical structure in this region, he approached his anatomy instructor for an explanation.

The student has a well-recognised anatomical variation on the lingual aspect of the mandible in the canine-premolar region, called tori mandibulares. The lumps are hyperostoses (localised, non-neoplastic, external bony overgrowths), which are usually bilateral. They are found in approximately 7% of the population. The mucosa overlying the tori is often extremely thin and is easily ulcerated. An additional problem is that food debris may collect beneath mandibular tori. A further example of a hyperostosis occurs in the midline of the hard palate (in about 20% of the population). This is called a torus palatinus. It is usually necessary to remove mandibular or palatine tori only when constructing a denture.

A 5-year-old girl was taken to her dentist for a routine examination. The dentist noticed that she had a speech defect. The lingual frenum was seen to extend across the floor of the mouth to be attached on to the alveolus behind the central incisor teeth. This condition is known as ankyloglossia or tongue-tie. Where does the lingual frenum normally terminate?

The lingual frenum runs down the midline of the ventral surface of the tongue and usually ends at the sublingual papilla. Ankyloglossia often requires correction by surgery, although in some circumstances it can correct itself. Ankyloglossia is a hereditary condition, having a dominant mode of inheritance.

A young woman was admitted to hospital for the surgical removal of a mandibular permanent third molar tooth. The operation was performed under general anaesthesia but proved to be very difficult. During the procedure, a sharp surgical instrument slipped and a nerve running close to the inner side of the tooth was severed. What was this nerve and what are the likely consequences of such damage?

The lingual branch of the mandibular division of the trigeminal nerve runs on the lingual alveolus of the mandibular third molar. Oral surgeons usually protect this nerve by means of a retractor. Should the nerve be damaged, the effects on that side are: loss of general sensation of the mucosa covering the anterior two-thirds of the tongue, the ventral surface of the tongue, the floor of the mouth and the lingual gingivae for the mandibular dentition. Accompanying the lingual nerve is the chorda tympani branch of the facial nerve. These merge in the infratemporal fossa beneath the lateral pterygoid muscle. Damage to the lingual nerve therefore also results in loss of taste to the anterior two-thirds of the tongue and loss of the parasympathetic innervation to the submandibular and sublingual salivary glands.

For several weeks, a 20-year-old woman had suffered pain and discomfort around a partly impacted mandibular third molar tooth. Her condition suddenly deteriorated and she became feverish (pyrexial). She experienced difficulty in swallowing and breathing and in opening her mouth. There was considerable swelling of the tissues in the floor of her mouth and in the upper part of the neck. Can you explain these symptoms?

This patient has a condition known as Ludwig's angina. The symptoms are the result of infection in the region of the third molar tooth spreading to produce a diffuse inflammation (cellulitis) in the tissue spaces in the floor of the mouth (the submandibular and sublingual spaces) and around the oesophagus and larynx. Oedema around the larynx and the swollen and displaced tongue can cause asphyxiation, and a tracheostomy may be required.

A 10-year-old boy was taken to the dentist by his parents who were worried that one of his maxillary first permanent incisors had still not erupted. Could this simply be an example of normal variation in eruption times?

The maxillary first permanent incisors erupt between the ages of seven and eight years. This can vary by a few months either way, but a two-year delay is indicative of an abnormality. Delayed eruption of a permanent tooth may be caused by the congenital absence of the tooth. This particularly affects third molars, mandibular second premolars and maxillary lateral incisors. Alternatively, the tooth may have formed but not erupted as a result of crowding or malpositioning. Teeth particularly affected by this are the mandibular third molars and the maxillary canines. In the case of the maxillary central incisor, the commonest cause of delayed eruption is the presence of a small supernumerary tooth (a mesiodens) along the path of eruption.

This can be identified radiographically. Following surgical removal of this obstruction, the maxillary central incisor will erupt. In cases of delayed eruption, the deciduous predecessor may be retained for a prolonged period.

Two basic techniques are used to anaesthetise a tooth for dental treatment using local anaesthetic solution: infiltration and block techniques. Infiltration involves depositing the solution on the alveolar bone immediately above the root apex. The solution seeps through the bone to anaesthetise the nerves as they pass through the apical foramen. Block technique requires that the solution is deposited around the main nerve trunk some distance away from the tooth to be treated. The choice between infiltration and block techniques depends upon whether the region to be anaesthetised is very localised or covers the whole nerve field supplied by the main nerve trunk. In addition, various anatomical features of the jaws dictate the technique that can be used. What are these features?

During infiltration, the bone covering the root of the selected tooth must not be so thick as to prevent the solution seeping through. Also, the number of vascular openings has an influence. Thus, the labial alveolar plates of the anterior teeth for both jaws are suitable for infiltration, being thin with many vascular openings. On the other hand, the buccal alveolar plates for both jaws are thick and have prominent ridges crossing them (the base of the zygomatic process for the maxilla and the external oblique ridge for the mandible). Blocking of the main nerve trunk is possible only if the nerve has an extra-bony course. Thus, the inferior alveolar and lingual nerves can be blocked as they course through the pterygomandibular space in the infratemporal fossa. The posterior superior alveolar nerve can be blocked as it passes across the maxillary tuberosity from the pterygomaxillary fissure. The nasopalatine and greater palatine nerves can be blocked as they emerge on to the hard palate. Blocking of the infra-orbital nerve (perhaps reaching the anterior superior alveolar nerve) and the mental nerve (perhaps reaching the incisive nerve) may be possible by depositing local anaesthetic solution at the infra-orbital foramen and the mental foramen respectively.

A patient required the extraction of a maxillary second permanent molar tooth. What nerve(s) had to be anaesthetised for this treatment?

For extraction, the nerves supplying both the tooth and the gingivae need to be anaesthetised. The maxillary second permanent molar and its buccal gingiva are supplied by the posterior superior alveolar branches of the maxillary nerve. The palatal gingiva is innervated by the greater palatine nerve from the pterygopalatine ganglion. The posterior superior alveolar nerve is 'blocked' as it courses from the pterygomaxillary fissure on to the tuberosity of the maxilla. The greater palatine nerve is 'blocked' as it runs in the submucosa at the junction of the hard palate with the maxillary alveolus.

In order to anaesthetise the posterior superior alveolar nerve, local anaesthetic solution is deposited along the posterior surface of the maxilla, just below the pterygomaxillary fissure. From a consideration of the anatomy of the region, what are the possible hazards?

If the needle is not kept close to the maxilla, the pterygoid venous plexus and the lateral pterygoid muscle may be entered. Damage to the pterygoid venous plexus can result in a haematoma, which usually shows itself by a rapid swelling of the face. Because the inferior orbital fissure lies close to this region, the haematoma may spread through the orbit to produce a 'black eye'. Injection into the muscle may cause spasm.

Diffusion of anaesthetic solution into the orbit through the inferior orbital fissure could result in transient squints and double vision (if the oculomotor, trochlear or abducent nerves are anaesthetised), or temporary blindness (if the optic nerve is anaesthetised). Anaesthetic solution may also be injected intra-arterially into the maxillary artery or its branches.

The root of a mandibular first premolar tooth fractured whilst the tooth was being extracted. The retained root fragment was removed surgically. The patient returned to her dentist several days later complaining that her lower lip on the side of the extracted tooth had remained numb. What might be the reason for this?

The mucosa and skin of the lower lip are supplied by the mental branch of the inferior alveolar nerve. This emerges on to the face at the mental foramen of the mandible, which lies close to the root apex of the mandibular first premolar. The nerve is therefore susceptible to damage during surgical procedures in this region. The inferior alveolar nerve itself may also be damaged during the extraction of mandibular molars, the nerve sometimes lying close to the root apices (occasionally grooving the roots).

An elderly patient visited her dentist to have her dentures modified. Only the left maxillary permanent third molar remained. The dentist decided to extract this tooth. A radiograph of the tooth was taken and the tooth was subsequently extracted without difficulty. The dentist then asked the patient to blow through her nose with the nostrils occluded. Why?

When extracting maxillary cheek teeth there is the possibility of damaging the floor of the maxillary air sinus, creating a pathway between the sinus and the oral cavity (an oro-antral fistula). Indeed, in older patients whose teeth may have been non-functional for a considerable time, the root may be fused to the bone forming the floor of the maxillary sinus. Thus, an oro-antral fistula will result when bone comes away with the tooth during extraction. This can be confirmed by increasing the air pressure within the nasal cavity, resulting in the appearance of bubbles within the blood clot at the site of the fistula. A precautionary radiograph helps to assess the likelihood of such an event.

A 65-year-old woman went to her dentist complaining of pain associated with her left maxillary molar teeth during the past nine months. The symptoms were particularly noticeable during eating. She also complained that during the same period her left nostril had been blocked and that occasionally she had coughed up blood-flecked sputum. The dentist noticed that all the left maxillary molars were mobile. However, the oral hygiene was good and there were no signs of inflammatory periodontal disease. Radiography revealed that there was enlargement of the jaw. The dentist suspected that there was a tumour of the mucosa lining the left maxillary sinus. How might the symptoms be explained by this diagnosis?

The tumour had spread to involve the superior alveolar nerves (causing pain in the teeth they supply) and the maxillary alveolar bone (loosening the teeth). It had also involved the lateral nasal wall (displacing the conchae and blocking the nostril). The sputum was stained by bleeding from the friable secretory surface of the tumour.

Infections of the teeth are the most frequent cause of swellings around the face and jaws. What anatomical features determine the spread of infection from the teeth?

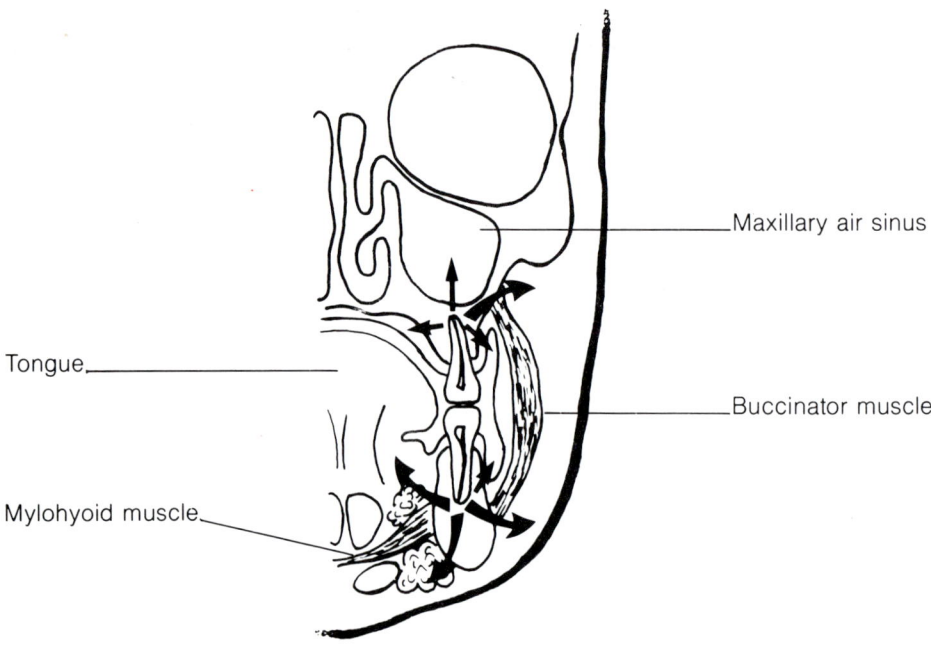

Figure 140 *Coronal section through the jaws to illustrate possible routes for the spread of infection from periapical dental abscesses.*

The relationship of the root apex to the adjacent musculature is the most important factor in determining the spread of infection (Figure 140).

For maxillary teeth, an abscess may drain outwards below the orbicularis oris muscle or inwards into the palate. Where the abscess drains above the orbicularis oris or buccinator muscles, it may enter other tissue spaces in the face. Abscesses from the maxillary incisor teeth may involve the nasal cavity, while those from the cheek teeth may drain into the maxillary sinus.

Abscesses from the mandibular anterior teeth and premolars may drain either outwards into the vestibule (above the orbicularis oris and buccinator muscles) or inwards into the floor of the mouth (above the mylohyoid muscles). Unlike the premolars, the root apices of the mandibular molars generally lie below the origin of the mylohyoid muscle and are situated more lingually. Abscesses from these molar teeth therefore usually drain into the submandibular tissue space.

Infections from around both the maxillary and mandibular teeth can also spread behind the ramus of the mandible into the spaces associated with the infratemporal fossa and the upper part of the pharynx (see pages 307 to 310).

A 65-year-old man was diagnosed as having a tumour in the left submandibular salivary gland. It was necessary to remove the gland surgically. What hazards does the surgeon face in removing the gland by means of an extra-oral approach, and what safeguards are necessary to preserve important anatomical structures in the region?

The initial incision is made in the skin about 5 cm below and parallel to the lower border of the mandible, beginning at the angle and passing forwards. The incision includes the platysma muscle so that a skin flap including this muscle is raised. This location ensures that damage to the marginal mandibular branch of the facial nerve is avoided, because the nerve runs down from the face below the inferior border of the mandible in approximately 20% of cases. Both the marginal mandibular and the cervical branch of the facial nerve are protected if platysma is raised with the skin flap.

The superficial part of the submandibular gland contains facial vessels, and segments of these vessels are removed with the gland as it is difficult to dissect them free. When removing the gland, care must be taken to protect the lingual and hypoglossal nerves which pass respectively above and below the deep part of the gland on the hyoglossus muscle.

A teenager presented with a severe nosebleed (epistaxis). The bleeding was centred around the back of the nasopharynx, with some blood running down the patient's throat. The epistaxis was controlled by anterior and posterior nasal packs. After a few days the packs were removed. Examination of the roof of the nasopharynx (by posterior rhinoscopy) showed the presence of a cherry-red swelling. Could this be the pharyngeal tonsil (adenoids) and the cause of the bleeding?

The pharyngeal tonsil is a collection of lymphoid material in the roof of the nasopharynx. However, despite the similarity of location, the tonsil is unlikely to be responsible for severe epistaxis. The cherry-red swelling is more likely to be an angiofibroma — a benign tumour that is very vascular. An inadvertent biopsy of this type of tumour will therefore result in severe haemorrhage. The use of radiographical techniques (i.e. tomography, CT scan and arteriography) can help to confirm the diagnosis of this tumour.

A boy went to his doctor complaining of a sore throat. The doctor diagnosed tonsillitis but did not give the boy any antibiotics as she did not feel the symptoms were severe enough. The boy returned one week later feeling very unwell. He was unable to open his mouth and was dribbling saliva. He had acute pain on one side of his throat which was made worse by swallowing. He also spoke in a muffled voice. The symptoms are indicative of 'quinsy' (an abscess in the peritonsillar tissue space). How would you account for the symptoms and how might the abscess now spread?

An abscess in the tonsillar region accounts for the extreme tenderness, especially when the patient tries to swallow (hence the dribbling) and tries to speak (hence the muffled quality of the voice). The inability to open the mouth (trismus) is caused by spasm of the muscles of mastication secondary to inflammation (usually associated with the pterygoid muscles). The peritonsillar space is an important part of the intrapharyngeal tissue space. Infections can spread down the inside of the pharynx or through the pharynx into the parapharyngeal space.

A young girl complained of earache soon after having a palatine tonsillectomy. Examination of the affected ear with an auriscope showed a normal tympanic membrane with no sign of an ear infection. Can you provide an explanation for the pain?

The likely reason for the symptom is referred pain secondary to tonsillectomy. Pain in the region of the palatine tonsil is carried by the glossopharyngeal nerve. This nerve also supplies the middle ear cavity, including the medial wall of the tympanic membrane.

A 50-year-old man visited his doctor with a history of pain and discomfort on the right side of his neck each time he turned his head towards the right. On examining the right side of the patient's neck, the doctor could feel a hard structure behind the angle of the mandible which extended down towards the hyoid bone. What might this structure be?

The patient has a calcified stylohyoid ligament. This could be confirmed by a radiograph. Symptoms are produced when the neck is moved, causing compression of structures against the ligament.

A patient was referred to a hospital consultant after his doctor had noticed some enlarged lymph nodes on the left side of the neck. On indirect laryngoscopy, an ulcerating (probably malignant) lesion of the left piriform fossa was found. On further questioning, the patient also mentioned that he had had some left-sided earache. How might this symptom be explained with reference to the tumour?

This would appear to be a case of referred pain from the primary lesion. The piriform fossa is supplied by the internal branch of the superior laryngeal nerve of the vagus. The vagus nerve also supplies the posterior aspect of the tympanic membrane and part of the external acoustic meatus.

A 50-year-old woman presented with pain in the right side of her neck, behind the thyroid cartilage. The pain was made worse when she swallowed. She admitted to having eaten fish two days previously, when she thought a bone had stuck in her throat. The pain was mild initially but had worsened over the next two days. On indirect laryngoscopy, the doctor noticed pooling of saliva and inflammation of the surrounding mucosa. What has happened?

The patient has indeed swallowed a fish bone. This has lodged in the right piriform fossa (a common area for this to happen). Subsequent inflammation of the mucosa around the bone made it painful to swallow. Reflex spasm of the cricopharyngeus muscle leads to difficulty in swallowing saliva. The saliva consequently pools in the oesophageal inlet and is spat out by the patient to prevent inhalation. Denture-wearers may be more prone to swallowing foreign bodies as they have diminished palatal sensation secondary to shielding of the mucosa by the denture, and the tendency to chew food boluses inefficiently. Should there be rupture of the pharyngeal wall, infection of the mediastinum (mediastinitis) may ensue because of tracking of pharyngeal contents along the retropharyngeal space and into the posterior mediastinum.

A 65-year-old lady was admitted to hospital because of a chest infection and general debilitation. On taking a history, the doctor discovered that the patient had for some time experienced difficulty in swallowing. This had progressed to such a degree that she could only swallow liquids. The patient noticed that swallowing became more difficult as the day progressed. On examination, the lady was found to have a swelling on the left side of her neck. The swelling was soft and fluctuant. Further questioning revealed that recently the patient had begun to regurgitate undigested food and suffered from halitosis. The doctor diagnosed a pharyngeal pouch — a protrusion or diverticulum of pharyngeal mucosa and submucosa through a weakness in the pharyngeal wall.

The weak spot in the pharyngeal wall usually occurs between the two components of the inferior constrictor muscle — i.e. between the transversely arranged fibres of the cricopharyngeus and the more oblique fibres of the thyropharyngeus muscles. At this site, the diverticulum is known as Killian's dehiscence. The diverticulum is free to enlarge laterally, producing a swelling of the side of the neck. It may compress the oesophagus, producing difficulty in swallowing. The opening of the diverticulum can come to lie directly below the pharynx as it descends towards the oesophagus. Undigested food fills the pouch and can remain there for days before being regurgitated, causing halitosis. When the patient lies down at night, the contents may empty into the oesophagus. Some of the contents may be aspirated into the lungs, producing chest infections.

During general anaesthesia, it is often necessary to pass an endotracheal tube through the mouth and into the trachea to facilitate ventilation of the patient. A medical student was asked to insert such a tube under the supervision of a consultant anaesthetist. The patient was already sedated and lying on his back with his head propped up by three pillows. The student opened the patient's mouth and inserted a laryngoscope. Briefly visualising the vocal cords lying just behind the epiglottis, he then inserted the endotracheal tube. On connecting the tube to the oxygen supply, the student placed his stethoscope over the chest but was unable to hear any breath sounds. The consultant immediately noticed that the stomach had started to expand. Why?

The pharynx is a common chamber leading from the oral and nasal cavities above to the trachea and oesophagus below. The trachea is situated in front of the oesophagus. In the present case, the student had incorrectly placed the endotracheal tube into the oesophagus and the pressure of the anaesthetic gases had caused the stomach to expand. Because the oesophagus lies posteriorly, it is more or less in line with the oral cavity, the trachea being at more of an angle. Therefore, a tube passed blindly down the pharynx is more likely to enter the oesophagus unless precautions are taken. One of the most important of these is to position the inlet of the larynx in line with the oral cavity. This can be achieved by extending the head at the atlanto-occipital joint and flexing the lower part of the cervical spine (into the position used when 'sniffing the morning air') (Figure 141). In the case described here, the presence of three pillows had actually flexed the head and neck. In addition, the vocal cords should be clearly visualised.

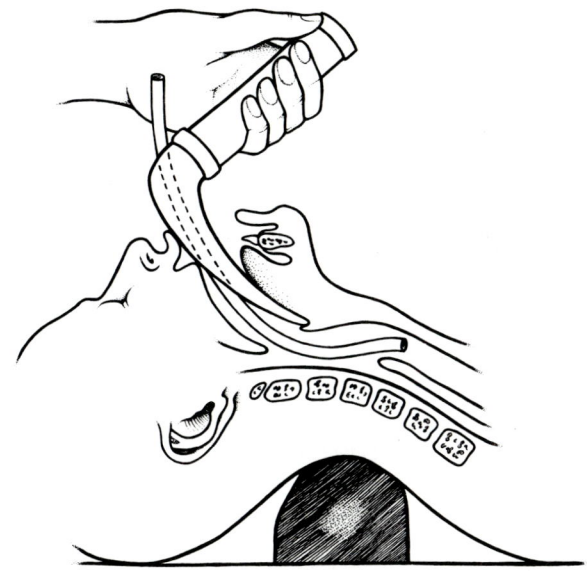

Figure 141 *Diagram to show the position of the head for the successful introduction of an endotracheal tube.*

A young baby was noted to have considerable difficulty inhaling, producing grunting noises with each intake of breath (stridor). She had no signs of infection and had a normal cry. The stridor was not noticeable during sleep. Examination of the airway under general anaesthesia did not indicate any pathological problems.

As the baby can cry normally, the problem is associated with inspiration rather than expiration. In the absence of pathology, the most likely explanation is laryngomalacia

— a condition in which the unossified cartilaginous skeleton of the juvenile larynx is prone to collapse and acts as a valve, partially or completely obstructing the airway on inspiration. In most cases, little treatment is necessary as the baby will outgrow the problem by two years of age.

A 30-year-old woman with a malignant tumour of the thyroid gland, underwent near-total removal of the gland. During recovery from the anaesthetic, she became cyanosed and exhibited stridor. Why?

When the nerve supply to the intrinsic muscles of the larynx is damaged, the paralysed vocal folds usually take up one of two positions: the cadaveric or paramedian position (Figure 142). In the cadaveric position, all the muscles are affected and the paralysed fold lies wide of the midline. This occurs when there is a high lesion of the vagus nerve, affecting both the recurrent laryngeal and external laryngeal nerves. In the paramedian position, the paralysed fold lies just off the midline. This occurs where the recurrent laryngeal nerve only is affected, the unaffected cricothyroid muscle (innervated by the external laryngeal nerve) drawing the vocal fold towards the midline.

Following near-total removal of the thyroid gland, bilateral damage to the recurrent laryngeal nerves is a frequent occurrence. This leads to bilateral incomplete paralysis of the vocal folds. However, the cricothyroid muscles are unaffected, as their innervation is via the external laryngeal branches of the superior laryngeal nerves. In the case described here, the unopposed action of the cricothyroid muscles has resulted in both vocal folds being in the paramedian position, thus compromising the airway. An immediate tracheostomy was needed.

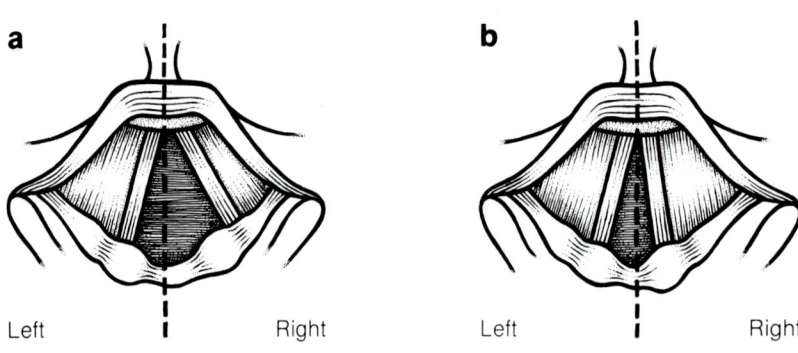

Figure 142 *Diagrams to show the main positions taken up by a paralysed vocal fold.* a) *Left vocal fold normal, right vocal fold in cadaveric position.* b) *Left vocal fold normal, right vocal fold in paramedian position.*

A 54-year-old schoolmaster presented with severe angina. During examination, a bruit (murmur) was heard over the right internal carotid artery. Detailed investigations showed stenosis (narrowing) of the main coronary arteries and of the right internal carotid artery. Treatment required a heart–lung bypass. However, the lesion in the carotid artery increases the possibility of stroke during this procedure. Therefore, a combined simultaneous coronary artery bypass and a carotid endarterectomy (widening of the diseased arterial lumen) was performed by two teams of surgeons. The operation was successful, but postoperatively the patient had hoarseness of his voice. Laryngoscopy revealed that the right vocal fold remained in the paramedian position during vocalisation. Which team was to blame for the damage?

The hoarseness is caused by paralysis of the muscles that move the vocal folds as a result of damage to one or both of the recurrent laryngeal nerves or the parent vagus nerves. On the left side, the recurrent laryngeal nerve arises in the thorax as the vagus nerve passes over the arch of the aorta. On the right, the recurrent laryngeal has its origin from the vagus nerve as it crosses the first part of the subclavian artery. Thus, the right vagus nerve would be susceptible to damage during the endarterectomy surgery, whereas the left recurrent laryngeal nerve would be at risk during the heart–lung bypass. As the right vocal fold remained in the paramedian position during both breathing and vocalisation then it must have been the right vagus nerve that was damaged (during the endarterectomy).

A patient presented with a hoarse voice caused by paralysis of one of his vocal folds. Which fold is more likely to be affected?

The left vocal fold is most frequently involved because of the greater length of the left recurrent laryngeal nerve. Because this nerve passes underneath the arch of the aorta in the thorax, it can be affected in conditions such as bronchial carcinoma.

A student was out having a meal with her parents. She made a funny remark just as her father was eating a large mouthful of food. He started to laugh but then began to choke. He rapidly lost consciousness and fainted. What do you think has happened?

This is a typical case of 'café coronary' (although it has nothing to do with a coronary thrombosis). When foreign material lodges in the region of the larynx it can completely obstruct the airway. Foreign bodies may become impacted in the inlet of the larynx or in the rima glottidis and cause suffocation by mechanical obstruction. Alternatively, small objects may lodge in the ventricles of the larynx. Here, they may irritate the mucous membrane and cause reflex glottic spasm, with consequent suffocation. The loss of consciousness and the faint are the result of reflex vasovagal stimulation, which causes massive vasodilation of the peripheral vessels and reduced cerebral blood flow.

The mother of a 3-year-old boy was worried because he had been unwell for 48 hours. He had a hoarse voice and a high temperature (39°C). On examination, the boy's neck was not particularly tender, but when the doctor examined his larynx with the aid of a mirror (indirect laryngoscopy), the area above the vocal folds appeared red and swollen. The child had laryngitis and there is an anatomical explanation for the inflammation being restricted to the supraglottic region.

The mucous membrane of the larynx is loosely applied to underlying structures, except at the vocal folds where it is tightly bound to the upper border of the cricovocal membrane. The vocal folds have no submucosa and they therefore remain less affected, even when the rest of the larynx is oedematous and red. Swelling of the larynx above the vocal folds can lead to suffocation as a result of oedema of the supraglottic region, as the tissue fluid cannot drain past the vocal folds.

A 40-year-old woman with a history of rheumatoid arthritis presented at the clinic with pain in her throat and hoarseness of her voice. On indirect laryngoscopy, she was found to have red and swollen arytenoid areas and the mobility of her vocal cords was reduced. What is the likely cause of the signs and symptoms?

Rheumatoid arthritis is a condition in which the synovial joints are inflamed. The crico-arytenoid joints are synovial joints and can become involved in the disease. The inflammation results in pain and restricted movement of the vocal cords, which is associated with hoarseness of the voice.

A 60-year-old glass-blower decided to learn to play the trumpet when he retired. Two years later, he was alarmed to notice a lump in the side of his neck. This varied in size, becoming larger after he had been playing the trumpet. Fearing that it might be a malignant growth, he consulted his doctor. During the examination, the doctor discovered that the enlargement was soft and non-pulsatile, and increased in size with Valsalva's manoeuvre (muscular contractions of the chest and abdominal wall against a closed rima glottidis to raise the air pressure within the respiratory system). There were no enlarged lymph nodes. What might the lump be?

The signs and symptoms indicate the presence of a laryngocele. This develops as a diverticulum from the ventricle of the larynx. It may be associated with the presence of persistently raised pressure in the larynx (in this case glass-blowing and later trumpet-playing). Initially, the laryngocele may cause a bulging of the vestibular fold on the affected side. It may gradually enlarge along the path of the superior laryngeal nerve and vessels to present as a compressible mass in the side of the neck. The laryngocele may be confined entirely to the larynx and presents with the symptoms of a mass in the area of the aryepiglottic or the vestibular folds (e.g. hoarseness, stridor). If a laryngocele contains air, it can be diagnosed radiographically.

The ear

The ear is the sense organ concerned with hearing and balance. It can be subdivided into three parts: the external, middle and internal ear (Figure 143). All parts are associated with, or lie within, the temporal bone on the lateral aspect of the skull.

The external ear consists of the auricle, external acoustic meatus and tympanic membrane. It is the part that funnels air vibrations towards the middle ear.

The middle ear is a cavity within the temporal bone (the tympanic cavity) which contains the ear ossicles. These small bones conduct vibrations from the tympanic membrane, across the tympanic cavity, to the internal ear.

The internal ear consists essentially of a convoluted tube (the membranous labyrinth) in which lie the receptors for hearing and balance. The membranous labyrinth itself is housed in a bony chamber called the osseous labyrinth. Movements of fluids within the internal ear (perilymph and endolymph) are responsible for stimulating the various receptors. The vestibulocochlear nerve reaches the internal ear through the internal acoustic meatus.

THE EXTERNAL EAR

148A–C

THE AURICLE

The auricle is composed of a thin, elastic, fibrocartilaginous plate covered by adherent, hairy skin.

148A,1
148A,6
148A,16
148A,15
148A,3; 148A,4
148A,5; 148A,2
148A,8
148A,10
148A,11
148A,13
148A,12
148A,14

Laterally, the auricle appears irregular in shape, showing a number of more or less consistent folds and depressions. The curved rim of the auricle is known as the helix. The anterior end of the helix is called the crus of the helix. A small tubercle (the auricular tubercle of Darwin) may be seen where the curve turns postero-inferiorly. Running within and alongside the helix is the antihelix. The antihelix divides anteriorly into two crura, between which is situated a depression called the triangular fossa. The depression between the helix and the antihelix is called the scaphoid fossa. The antihelix circumscribes the concha of the auricle, a prominent depression that leads into the external acoustic meatus. Projecting into the concha are two folds of tissue, the tragus and the antitragus. The tragus lies anteriorly and projects from the face. The antitragus lies inferiorly and above the lobule of the auricle. The tragus and the antitragus are separated by the intertragic notch. Unlike other parts of the auricle, the lobule is composed of soft tissues and does not have a cartilaginous skeleton.

148B + C
148B,1
148,17
148,19; 148B,13
148B,20
148C,25
148C,26

The cartilage of the auricle has a shape that conforms closely with the folds and depressions described above. The part of the cartilage corresponding to the helix shows a small process anteriorly, which is called the spine of the helix. Separating the tail of the helix posteriorly from the antitragus is a fissure (the antitragohelicine notch). The cranial or medial surface of the cartilage shows an eminence that corresponds to the depression of the concha. This conchal eminence is crossed by an oblique ridge (the ponticulus) which marks the site of attachment of the posterior auricular muscle. The cartilage of the auricle is continuous with the cartilaginous part of the external acoustic meatus.

The muscles of the auricle can be subdivided into extrinsic and intrinsic sets of muscles. The extrinsic muscles are the posterior, superior and anterior auricular muscles, which belong to the muscles of facial expression (see pages 153 to 154). The intrinsic muscles are rudimentary. The muscles are innervated by the facial nerve.

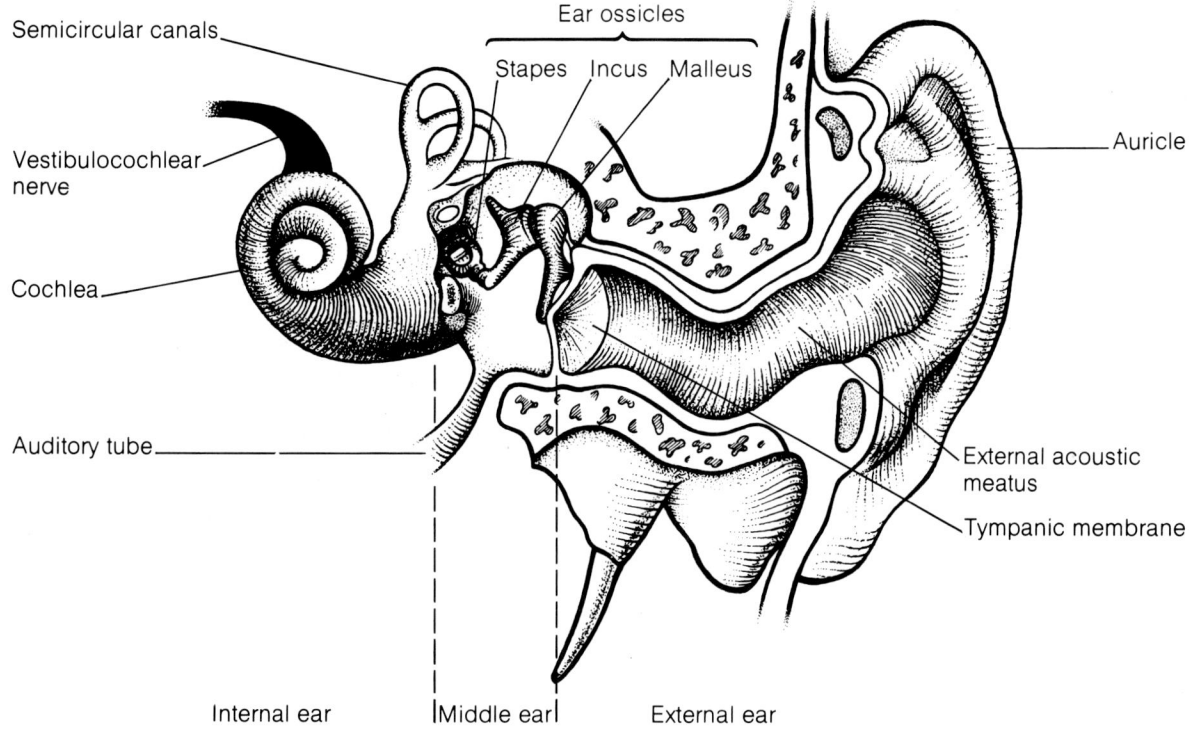

Figure 143 *The subdivisions of the ear*: the external, middle and internal ear.

The cutaneous innervation of the auricle is complex and is not fully determined. This is perhaps because the external ear represents an area where skin originally derived from the branchial region meets skin from the postbranchial region. Most of the skin is innervated by the great auricular nerve from the cervical plexus (second and third cervical nerves). The lesser occipital nerve from the cervical plexus (the second and sometimes the third cervical nerves) may supply the upper part of the cranial surface of the auricle. The auriculotemporal branch of the mandibular division of the trigeminal nerve innervates skin around the tragus and the crus of the helix. The concavity of the concha and a small area of the cranial surface near the mastoid is supplied by the auricular branch of the vagus and by the facial nerve.

112,39
112,41
112,2

The great auricular nerve, the lesser occipital nerve and the auriculotemporal nerve are described on pages 85 and 178 respectively. The auricular branch of the vagus nerve arises from the superior ganglion below the jugular foramen. It then enters the temporal bone through the mastoid canaliculus near the base of the styloid process. After passing through the temporal bone, the auricular branch exits through the tympanomastoid fissure on the posterior aspect of the opening of the external acoustic meatus. It now divides into two branches, one supplying the skin of the auricle and the other joining the posterior auricular branch of the facial nerve. Fibres from the facial nerve supplying the auricle may be derived from either its posterior auricular branch or from fibres joining the auricular branch of the vagus during its passage through the temporal bone.

50D,48
50A,15

The vasculature of the auricle is rich and there are many arteriovenous anastomoses. The vessels are derived mainly from the posterior auricular and the superficial temporal branches of the external carotid artery. The occipital artery may also contribute. The venous drainage is via the posterior auricular and superficial temporal

144A,59; 112,3
112,43
98,6

veins. The lymphatics of the auricle empty into parotid, retro-auricular, and superficial cervical nodes.

THE EXTERNAL ACOUSTIC MEATUS

150A,3; 114,13

150A,4; 114,9
148D,31; 148D,34

The external acoustic meatus is a tunnel into the temporal bone that extends medially from the concha of the auricle to the tympanic membrane. It does not run a straight course, but is S-shaped. In front of the meatus lies the temporomandibular joint and a portion of the parotid salivary gland. Behind lie the mastoid air cells.

148B + C,21

14,14; 50A,11

The external acoustic meatus can be subdivided into two portions. The outer third of the meatus is cartilaginous; the inner two-thirds is osseous. The cartilaginous portion is continuous with the cartilage of the auricle. It is fixed by fibrous tissue to the circumference of the osseous portion. The osseous portion is formed chiefly by the C-shaped tympanic part of the temporal bone. The wall is completed above by the squamous and petrous parts of the temporal bone. The osseous portion is narrower than the cartilaginous portion.

Both the cartilaginous and osseous portions of the external acoustic meatus are covered by firmly adherent skin. The skin over the cartilaginous portion possesses numerous hairs and sebaceous glands and the wax-secreting ceruminous glands.

The innervation of the skin of the meatus has a dual origin (see Figure 145). The auriculotemporal branch of the mandibular division of the trigeminal nerve supplies the anterior and superior walls. The auricular branch of the vagus nerve supplies the posterior and inferior walls. The facial nerve may also contribute via its communication with the vagus nerve.

The blood supply of the meatus is derived from the posterior auricular artery, the superficial temporal artery and, near the tympanic membrane, the deep auricular branch of the maxillary artery. The veins drain into the external jugular and maxillary veins and into the pterygoid venous plexus. The lymphatics of the external acoustic meatus drain into the parotid, retro-auricular and superficial cervical nodes.

148F,39; 150,17

THE TYMPANIC MEMBRANE (Figure 144)

150A,3; 150A,16

The tympanic membrane is commonly known as the ear drum. It is a thin, semi-transparent membrane that lies at the medial end of the external acoustic meatus, separating the external ear from the middle ear.

148H,63; 150C,39

The tympanic membrane consists of three layers. The outer layer is derived from the skin overlying the external acoustic meatus and consists of a stratified squamous epithelium. The inner layer is part of the mucous membrane lining the middle ear and consists of a ciliated columnar epithelium. The intermediate layer is fibrous. The fibres are attached to the handle of the malleus ear ossicle and radiate to the periphery. At the periphery, many circular fibres are also found. The margin of the tympanic membrane is thickened and forms a fibrocartilaginous ring.

150,28

The tympanic membrane lies obliquely at the end of the external acoustic meatus such that it slopes downwards and forwards as well as laterally. The fibrocartilaginous ring slots into a groove (the tympanic sulcus) in the tympanic part of the temporal bone. The groove is, however, deficient superiorly where two bands called the anterior and posterior malleolar folds pass across the membrane to the lateral process of the malleus. The membrane is lax above these folds. The terms pars flaccida and pars tensa are used to distinguish the lax and tense parts of the tympanic membrane. The membrane is tensed by the activity of the tensor tympani muscle within the middle ear (see page 371).

When viewed with an auriscope, the tympanic membrane appears as a concave, pearly-grey disc. In the depth of the concavity, a reddish-yellow streak indicates the site of the handle of the malleus. This is referred to as the umbo. A bright cone of light

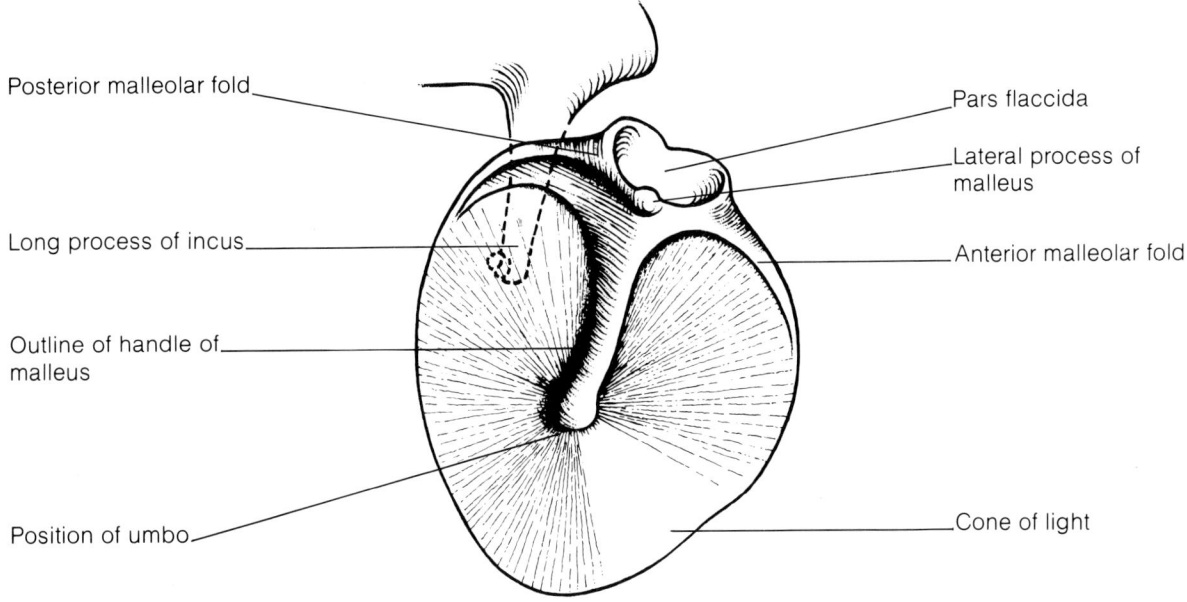

Figure 144 *Right tympanic membrane (external surface).*

is seen in the antero-inferior quadrant of the membrane when it is illuminated for inspection. Above the umbo and close to the roof of the external acoustic meatus lies a white spot which represents the lateral process of the malleus. The anterior and posterior malleolar folds and the pars flaccida can also be observed in this region. The long process of the incus ear ossicle may be discerned as a whitish streak lying posterior and parallel to the upper part of the handle of the malleus.

The innervation of the tympanic membrane is illustrated in Figure 145. The outer surface of the membrane is supplied by the auriculotemporal branch of the mandibular nerve and the vagus nerve. The facial nerve may also contribute. The tympanic branch of the glossopharyngeal nerve supplies the inner surface of the tympanic membrane.

The arterial supply of the tympanic membrane is derived from the deep auricular and anterior tympanic arteries and from the stylomastoid artery. The deep auricular artery arises from the maxillary artery before the lateral pterygoid muscle in the infratemporal fossa and pierces the wall of the external acoustic meatus. The anterior tympanic artery also arises from the maxillary artery before the lateral pterygoid. It enters the tympanic cavity through the petrotympanic fissure. The stylomastoid artery is a branch of the posterior auricular artery. It passes through the stylomastoid foramen. The deep auricular artery supplies the outer surface of the tympanic membrane; the others supply the inner surface. For both surfaces, a vascular ring is found around the margin of the membrane. However, although numerous small branches enter the membrane around the margins, significant branches also descend across the pars flaccida and the umbo.

The venous drainage of the tympanic membrane is complex. The veins from its outer surface drain into the external jugular vein, whereas those from the deep surface drain into the transverse sinus, the dural veins and the venous plexus of the auditory tube.

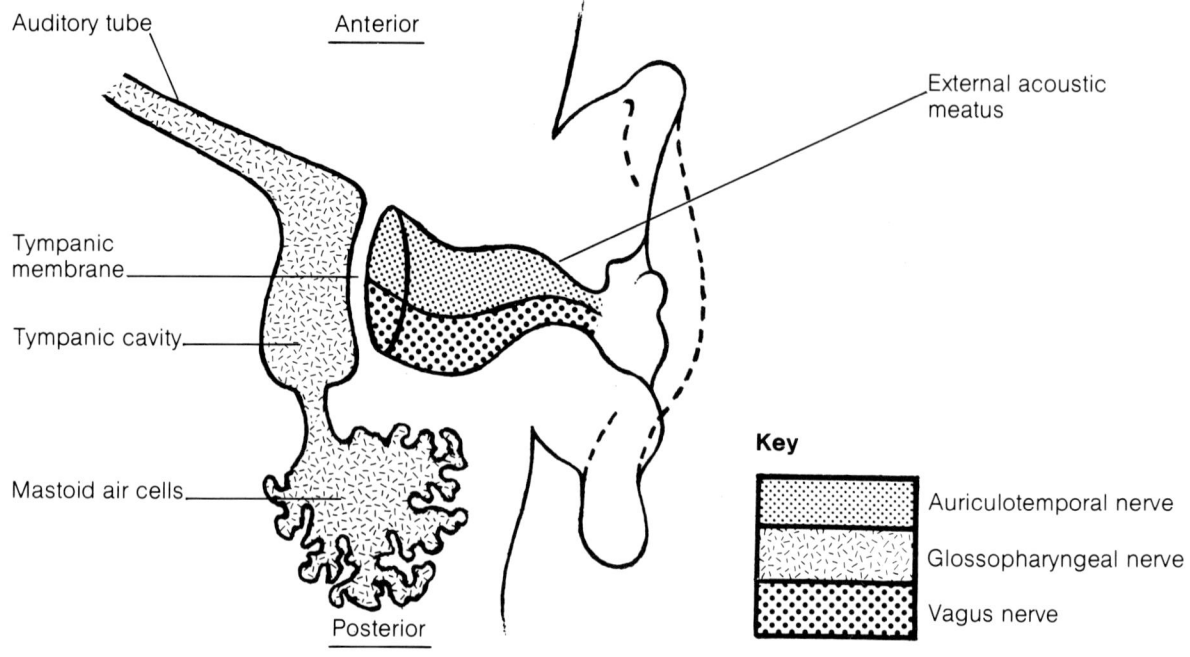

Figure 145 *Sensory innervation of the external and middle ears. Horizontal section.*

THE MIDDLE EAR

THE TYMPANIC CAVITY

The tympanic cavity lies mainly within the petrous part of the temporal bone. It can be thought of as a diverticulum of the pharynx as it is connected to the nasopharynx by the auditory tube and is derived embryologically from the first (and possibly second) pharyngeal pouch.

The tympanic cavity is irregular in shape but resembles a biconcave lens (Figure 146). It is compressed laterally, the anteroposterior diameter (≈ 15 mm) being about four times greater than the transverse diameter (≈ 4 mm). It is narrowest opposite the centre of the tympanic membrane (≈ 2 mm). The cavity can be subdivided into the tympanic cavity proper opposite the tympanic membrane, and the epitympanic recess above the membrane.

Figure 147 shows the main features of the cavity.

The roof of the tympanic cavity

This is formed by a thin plate of bone called the tegmen tympani. This plate projects anteriorly to form the roof of the canal for the tensor tympani muscle, posteriorly into the mastoid antrum and laterally into the epitympanic recess. The tegmen tympani separates the middle ear from the middle cranial fossa and the temporal lobe of the brain.

The floor of the tympanic cavity

A thin plate of bone separates the floor of the tympanic cavity from the internal jugular vein. Where the vein is small, the bone is thick and may contain accessory mastoid air cells. In places, however, the bone may be deficient such that only a thin membrane separates the vein from the tympanic cavity. A small aperture lies near the medial wall of the tympanic cavity, the (inferior) tympanic canaliculus for the tympanic branch of the glossopharyngeal nerve.

50D,45

The lateral wall of the tympanic cavity

This wall is largely occupied by the tympanic membrane with the attached handle of the malleus. In the angle between the lateral and posterior walls lies a small aperture called the posterior canaliculus for the chorda tympani nerve. From this canaliculus, the chorda tympani branch of the facial nerve and the posterior tympanic branch of the stylomastoid artery enter the tympanic cavity. The chorda tympani then passes across the upper part of the tympanic membrane beneath the mucosa and over the neck of the malleus (see Figure 148). It leaves the tympanic cavity through the anterior canaliculus for the chorda tympani nerve. The anterior canaliculus demarcates the medial end of the petrotympanic fissure through which the chorda tympani nerve emerges into the infratemporal fossa. The fissure also allows the passage of the anterior tympanic branch of the maxillary artery into the tympanic cavity.

148F,39; 148F,37

148F; 148E,38
150,29

22,38

The epitympanic recess is a small portion of the tympanic cavity that extends upwards and beyond the lateral wall. It can be likened to the peak of a cap extending laterally from the middle ear above the external acoustic meatus. It contains the head of the malleus and the body and short process of the incus.

148F,48

148F,37; 148F,47

The anterior wall of the tympanic cavity

This wall is occupied mainly by the opening of the auditory tube. This tube communicates with the nasopharynx and has a dual purpose. First, it allows pressures on either side of the tympanic membrane to be equalised. Second, it allows drainage of the secretions from the middle ear, including the mastoid air cells. The auditory tube is fully described on page 310. Above the auditory tube lies a canal which

148H,52; 150,15
148H,67; 126,16

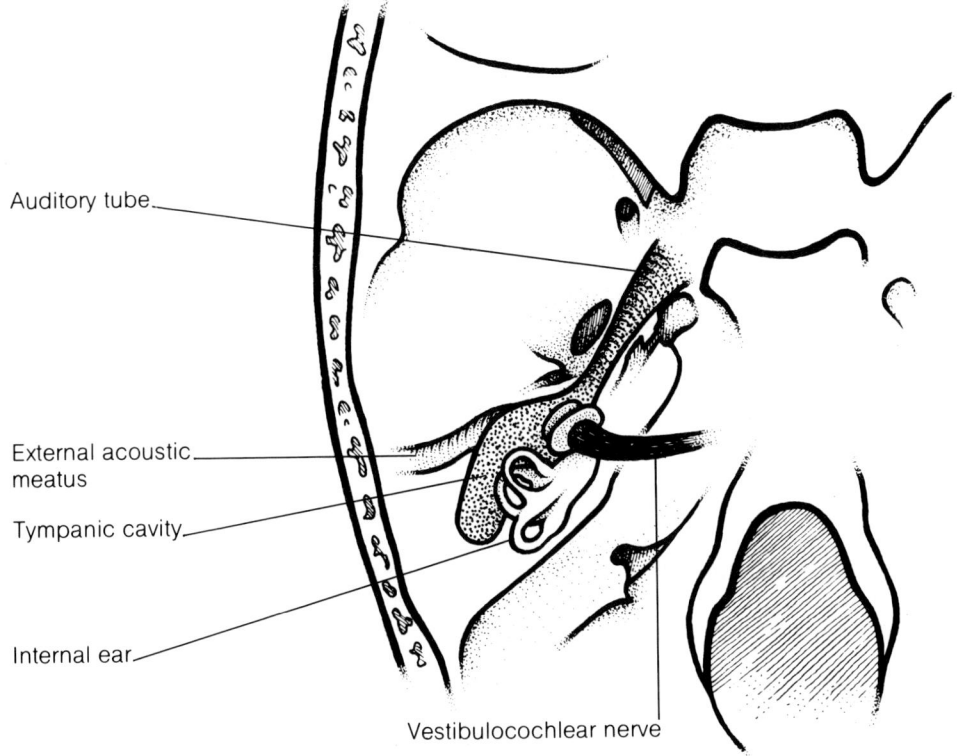

Figure 146 *Diagram to show the shape of the middle ear and its position within the temporal bone.*

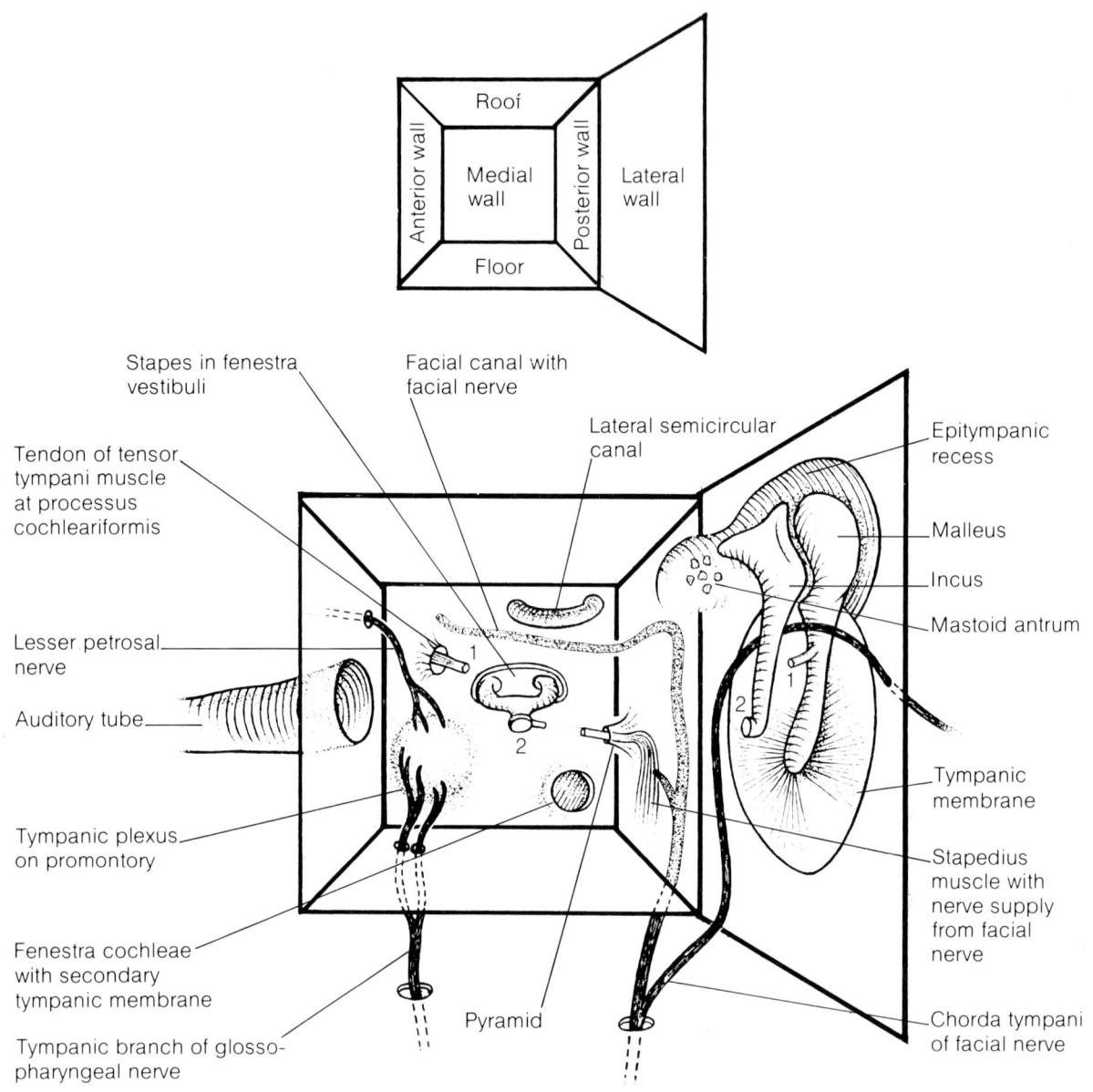

Figure 147 *Diagram representing the walls of the tympanic cavity to show the contents of the middle ear. Note that the lateral wall is displaced to display the remaining parts of the tympanic cavity. 1 = tendon of tensor tympani muscle, 2 = articulation of incus with stapes.*

148,53; 150,13

148,59; 150A,14

contains the tensor tympani muscle. The canal runs on to the medial wall of the tympanic cavity where the tendon of the muscle emerges. Below the auditory tube, the anterior wall is often thin where it forms the posterior wall of the carotid canal. This part is perforated by the caroticotympanic nerves from the sympathetic plexus on the internal carotid artery and by tympanic branches of the internal carotid artery.

The posterior wall of the tympanic cavity

148,30; 150,21

22,22

148F,49

Running vertically from the roof to the floor is the facial canal, containing the facial nerve. This canal is a continuation of the canal found in the medial wall and terminates as the stylomastoid foramen. Close to the roof of the posterior wall of the tympanic cavity lies the opening or aditus of the mastoid antrum. Below the aditus is a small depression (the fossa incudis) where the short process of the incus is attached by ligamentous fibres. Below the fossa incudis projects a hollow cone called the pyramid, through which emerges the tendon of the stapedius muscle. A canal for the muscle runs down within the posterior wall in front of the facial canal. Behind the posterior wall of the tympanic cavity is the posterior cranial fossa and the sigmoid sinus.

150,31
150,32

148E,44

The mastoid antrum can be regarded as the air sinus of the petrous part of the temporal bone. Unlike the other cranial air sinuses, the mastoid antrum is almost adult size at birth. It communicates with the middle ear through the aditus. This aperture leads forwards from the posterior wall of the tympanic cavity into the epitympanic recess.

148F,50

148F,49
148F,48

In view of the surgical importance of the mastoid antrum (viz. mastoidotomy, mastoidectomy), its topographical relationships are of significance. The anterior wall of the antrum shows the aditus in its upper part. Here, it is closely related to the lateral semicircular canal of the internal ear. Antero-inferiorly lies the descending part of the facial canal. The posterior wall of the antrum has the sigmoid sinus and the cerebellum behind it. The medial wall of the antrum is related to the posterior semicircular canal of the internal ear. The lateral wall of the antrum relates to the suprameatal triangle on the outer surface of the temporal bone. The roof of the antrum is part of the tegmen tympani and is related to the middle cranial fossa and the temporal lobe of the brain. The floor of the antrum is the site for communication with the mastoid air cells.

148F,49
148D,28
148D,30; 148E,44
148D,29

14,15
148F,45; 28,26
148F,34

The mastoid air cells lie mainly, but not exclusively, within the mastoid process of the temporal bone. At birth, the air cells are only just beginning to appear.

148,34; 150A,2
14,13

The mastoid air cells generally comprise a series of intercommunicating chains of thin-walled cavities. However, there is considerable variation in both number and arrangement. Indeed, three types of mastoid process can be identified: sclerotic (where there are few air cells), pneumatic (where there are many air cells), and mixed (where there are some regions with air cells and others with bone marrow). The air cells within the mastoid process are sometimes separated into two distinct groups by a plate of bone (the false bottom). This reflects the fact that the mastoid arises from both the squamous and petrous parts of the temporal bone divided by an extension of the petrosquamous fissure.

The medial wall of the tympanic cavity

This separates the cavities of the middle and internal ears.

150C,16; 150C,23

The most prominent feature of the medial wall is a bulge (called the promontory) formed by the cochlea of the internal ear. Above the promontory and close to the roof of the middle ear is a ridge related to the lateral semicircular canal of the internal ear. Between this ridge and the promontory lie the horizontal part of the facial canal, the fenestra vestibuli and the processus cochleariformis.

148G,56; 150,20
150,18
148G,28
148G,30
148G,55; 148G,57

The horizontal part of the facial canal transmits the facial nerve towards the posterior wall of the tympanic cavity. The fenestra vestibuli is one of the two sites of communication between the middle and internal ears (the other being the fenestra cochleae). The fenestra vestibuli lies above and behind the promontory and just below the facial canal. It is an oval opening that houses the footplate of the stapes ear ossicle. The processus cochleariformis is situated anterior to the fenestra vestibuli. It is a hollow bony process around which the tendon of the tensor tympani muscle bends. Although present on the medial wall, the processus cochleariformis is derived from the partition between the auditory tube and the canal for the tensor tympani on the anterior wall of the tympanic cavity.

148G,30
148G,55

150,33; 150E,53
148G,57; 150,28
148H,64; 150,28

The fenestra cochleae lies below and behind the promontory. It is a round opening covered by a membrane, the secondary tympanic membrane. Also behind the promontory is a depression called the sinus tympani.

148G,61

150E	**THE BONES OF THE MIDDLE EAR**
148H; 150D	The tympanic cavity is bridged by a chain of three ossicles (the malleus, incus and stapes) which passes from the lateral to the medial wall.
	The malleus (Figure 148)
148H,63	This is the largest ossicle. It is situated most laterally, alongside the tympanic membrane.
	The name is derived from its shape, being like a hammer or mallet in appearance. Five parts can be identified: the head, neck, handle and anterior and lateral processes.
150E,41; 148F,37 148F,46	The head of the malleus extends above the tympanic membrane into the epitympanic recess. It is the part that articulates with the incus. The articulation is demarcated by a facet on the posterior surface.
150E,42 148F	The narrow neck of the malleus lies adjacent to the pars flaccida of the tympanic membrane. The chorda tympani nerve crosses it medially.
150E,45 148H,63	The handle of the malleus may also be referred to as the manubrium. It is attached to the tympanic membrane.
150E,44	The lateral process of the malleus projects from the upper end of the handle. Anterior and posterior malleolar folds pass from the process to connect with the tympanic sulcus.
150E,43	The anterior process of the malleus also projects from the upper end of the handle. In the fetus, it is continuous with the cartilage of the first branchial arch. It is much less distinct in the adult and it connects with the petrotympanic fissure by some ligamentous fibres.

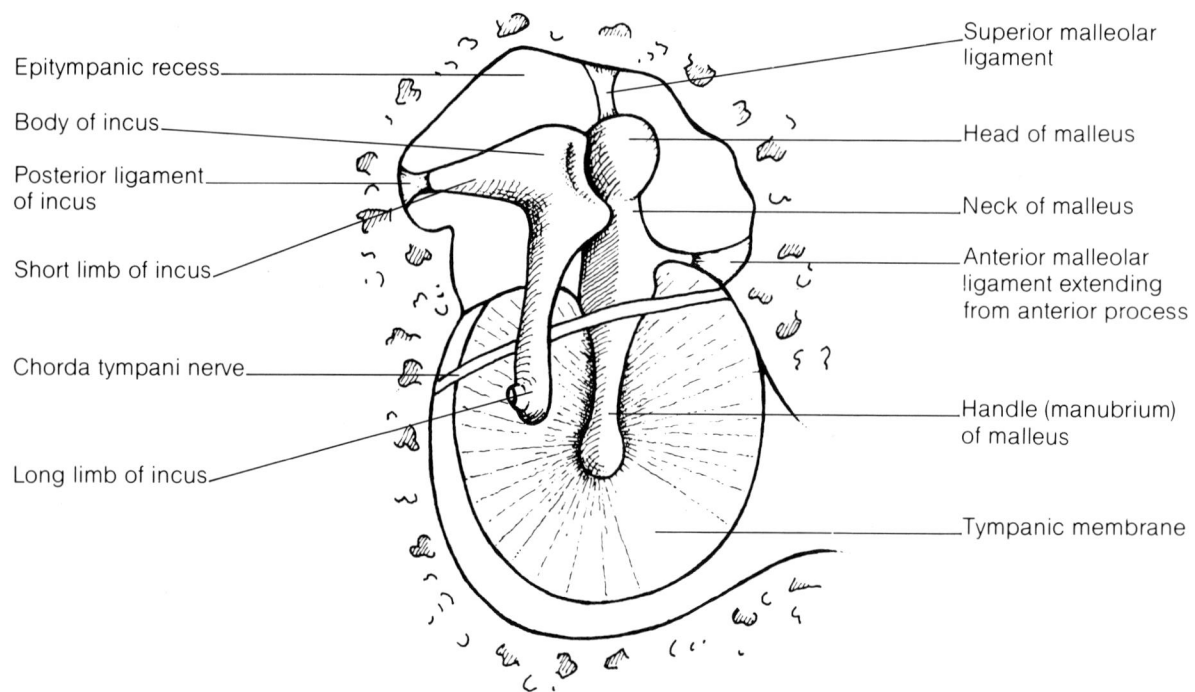

Figure 148 *Diagram of the lateral wall of the tympanic cavity (internal surface) illustrating the shapes and positions of the malleus and incus ear ossicles.*

The incus (Figure 148)

The incus is situated in the epitympanic recess and articulates with both the malleus and stapes. It resembles an anvil or a bicuspid tooth. It shows a body and two slender, root-like processes (the long and short limbs of the incus).

The body of the incus articulates with the malleus. The long limb of the incus projects vertically downwards into the tympanic cavity. It terminates as the lenticular process, which articulates with the head of the stapes. The short limb of the incus passes posteriorly and is attached by ligamentous fibres to the fossa incudis on the posterior wall of the tympanic cavity.

The stapes

This is the smallest ear ossicle. It takes its name from its resemblance to a stirrup.

The head of the stapes articulates with the incus. At the neck of the stapes lies the attachment for the stapedius muscle. Anterior and posterior limbs connect the neck to the base (footplate) of the stapes. The footplate fits into the fenestra vestibuli on the medial wall of the tympanic cavity. A ring of ligamentous fibres (annular ligament) attaches the margins of the footplate and the fenestra.

The joints between the ear ossicles are synovial. Their articular capsules possess much elastic tissue. The following ligaments attach the ossicles to the walls of the tympanic cavity:

- The anterior ligament of the malleus connects the malleus to the petrotympanic fissure and then passes to the spine of the sphenoid where it is continuous with the sphenomandibular ligament.
- The lateral ligament of the malleus passes from the head of the malleus to the posterior part of the tympanic notch.
- The superior ligament of the malleus joins the head to the roof of the epitympanic recess.
- The superior ligament of the incus attaches the body of the incus to the roof of the epitympanic recess.
- The posterior ligament of the incus passes from the short process of the incus to the fossa incudis.
- The annular ligament of the stapes attaches the baseplate to the margin of the fenestra vestibuli.

THE MUSCLES OF THE MIDDLE EAR

Two muscles are found in the tympanic cavity: the tensor tympani and the stapedius muscles.

The tensor tympani muscle

Attachments. This muscle lies within a bony canal situated above the auditory tube in the anterior wall of the tympanic cavity. It originates from this canal, from the cartilaginous part of the auditory tube and from an adjacent area of the greater wing of the sphenoid. It emerges into the tympanic cavity as a slender, tendinous structure. This turns sharply around the processus cochleariformis and crosses the tympanic cavity to insert into the upper part of the handle of the malleus.

Innervation. The nerve to the tensor tympani muscle is derived from the mandibular division of the trigeminal nerve. It initially forms part of the nerve to the medial pterygoid muscle but then passes through the otic ganglion to enter the tensor tympani muscle. The glossopharyngeal nerve (via the tympanic plexus) may also contribute to the motor innervation of the muscle.

Vasculature. The arterial supply is derived from the superior tympanic branch of the middle meningeal artery.

Actions. By drawing the handle of the malleus medially, the muscle tenses the tympanic membrane and helps to damp sound vibrations.

The stapedius muscle

Attachments. The muscle arises from within the hollow pyramid on the posterior wall of the tympanic cavity. It emerges at the apex of the pyramid as a tendinous structure and then inserts into the neck of the stapes.

Innervation. It is supplied by a branch of the facial nerve that is given off in the facial canal.

Vasculature. The muscle receives its blood supply from branches of the posterior auricular, anterior tympanic and middle meningeal arteries.

Actions. It helps to damp excessive sound vibrations, and functions when the sound is too loud.

THE MUCOSA OF THE MIDDLE EAR

The mucosa covers not only the walls of the tympanic cavity but also the structures contained within it, including the ossicles and their musculature. It is a ciliated columnar epithelium, except for regions posteriorly and within the mastoid antrum and air cells, where it is non-ciliated. Goblet cells are limited to an area around the orifice of the auditory tube. The sensory innervation of the mucosa is derived from the tympanic branch of the glossopharyngeal nerve via the tympanic plexus.

THE NERVES OF THE MIDDLE EAR

The tympanic plexus (Figure 149)

This plexus lies on the promontory of the medial wall of the tympanic cavity. It receives contributions from two main sources:

- Sensory fibres from the tympanic branch of the glossopharyngeal nerve (with possible contributions from the facial nerve).
- Vasomotor, sympathetic fibres from the internal carotid plexus.

The tympanic branch of the glossopharyngeal nerve passes into the temporal bone through the ridge between the carotid canal and the jugular foramen.

The following branches emanate from the plexus:

- Branches supplying the mucosa of the tympanic cavity and the auditory tube. The mucosa of the mastoid air cells is innervated from both the tympanic plexus and the nervus spinosus of the mandibular nerve.
- The lesser petrosal nerve destined to supply the parotid gland (see page 167).
- Communicating branches to the greater petrosal nerve.

The facial nerve within the temporal bone (Figure 149)

The facial nerve enters the temporal bone through the internal acoustic meatus (accompanied by the vestibulocochlear nerve). Initially, there are two separate components: the motor root supplying the muscles of the face, and the nervus intermedius which contains sensory fibres concerned with the perception of taste and parasympathetic (secretomotor) fibres to various glands. The two components merge within the meatus. At the end of the meatus, the facial nerve enters its own canal, the facial canal, which runs across the medial wall and down the posterior wall of the tympanic cavity to the stylomastoid foramen. As the nerve enters the facial canal, there is a bend in which lies the geniculate ganglion. The branches that arise from the facial nerve within the temporal bone can be subdivided into those that come from the geniculate ganglion and those that arise within the facial canal.

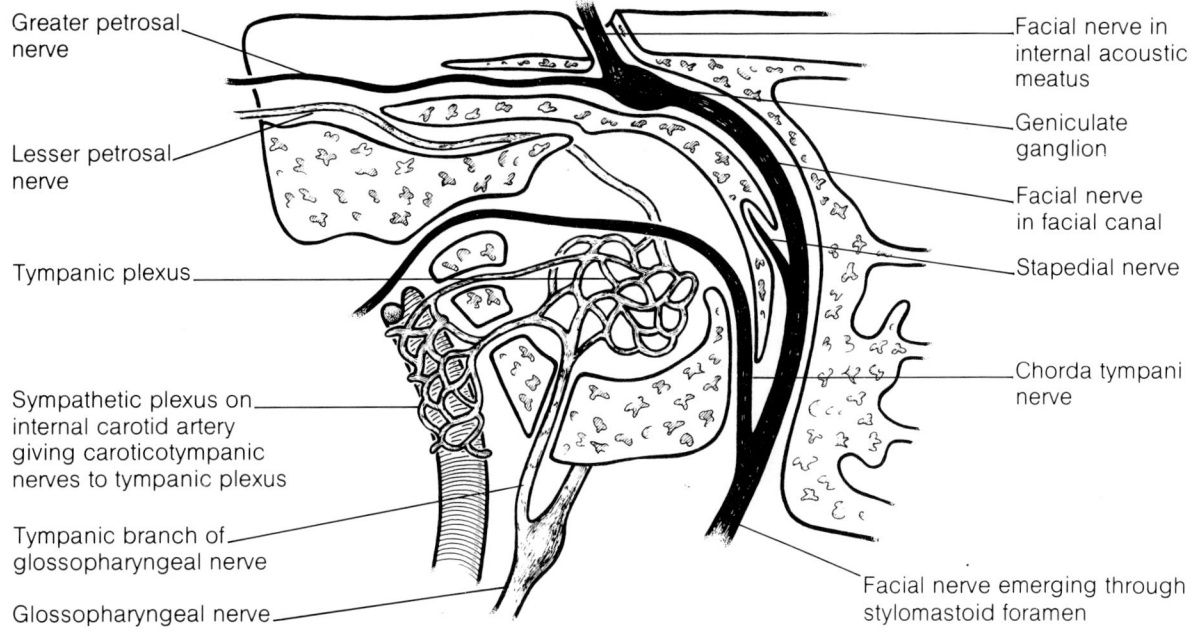

Figure 149 *The tympanic plexus and the course of the facial nerve within the temporal bone.*

The main branch from the geniculate ganglion is the greater (superficial) petrosal nerve, a branch of the nervus intermedius. The nerve passes anteriorly to exit the petrous bone and enter the middle cranial fossa. It contains parasympathetic fibres going to the pterygopalatine ganglion, and taste fibres from the palate (see page 257). The geniculate ganglion also communicates with the lesser petrosal nerve.

Within the facial canal, close to the pyramid, arises the nerve to the stapedius muscle. The chorda tympani nerve is given off just before the stylomastoid foramen. This is the second, and last, branch from the nervus intermedius. It contains parasympathetic fibres going to the submandibular ganglion and taste fibres from the anterior two-thirds of the tongue. The chorda tympani initially runs within its own canal before entering the tympanic cavity to cross the malleus (see Figure 148). It then enters another canal before leaving the temporal bone through the petrotympanic fissure (see page 179 for the subsequent course). Another branch of the facial nerve is the branch communicating with the auricular branch of the vagus.

THE BLOOD SUPPLY OF THE MIDDLE EAR

The inner surface of the tympanic membrane and the anterior part of the tympanic cavity are supplied chiefly by the anterior tympanic branch of the maxillary artery. The posterior part, including the mastoid air cells, is supplied by the stylomastoid branch of the posterior auricular (or of the occipital artery). The anterior tympanic artery enters the tympanic bone through the petrotympanic fissure, and the stylomastoid artery enters via the stylomastoid foramen. Additional sources of supply come from the inferior tympanic branch of the ascending pharyngeal artery, the middle meningeal artery, the caroticotympanic branches from the internal carotid, and branches from the artery of the pterygoid canal.

The veins of the tympanic cavity drain either externally into the pterygoid venous plexus or internally into the superior petrosal sinus. The lymphatics drain into the upper deep cervical or parotid nodes and the retropharyngeal nodes.

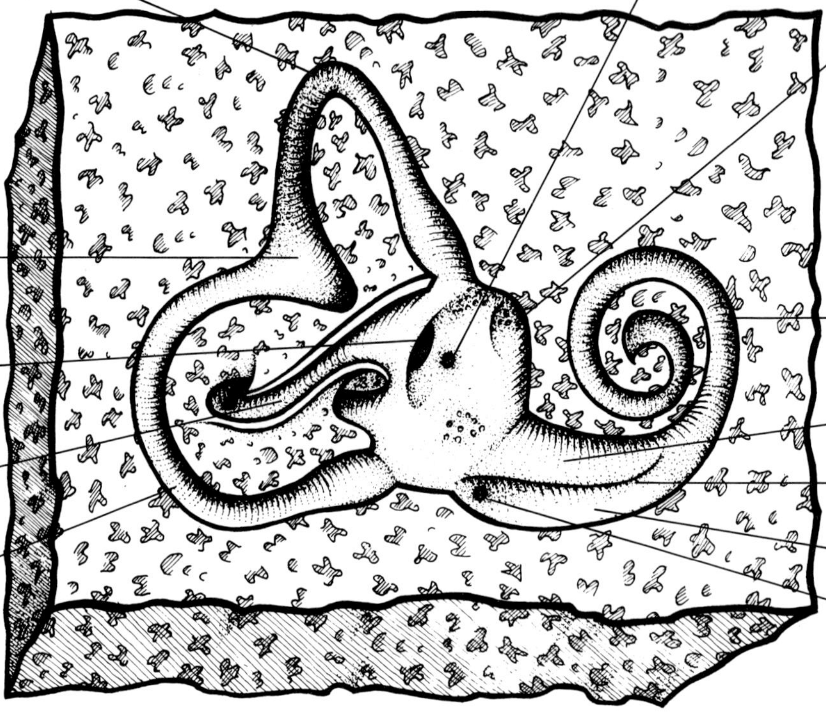

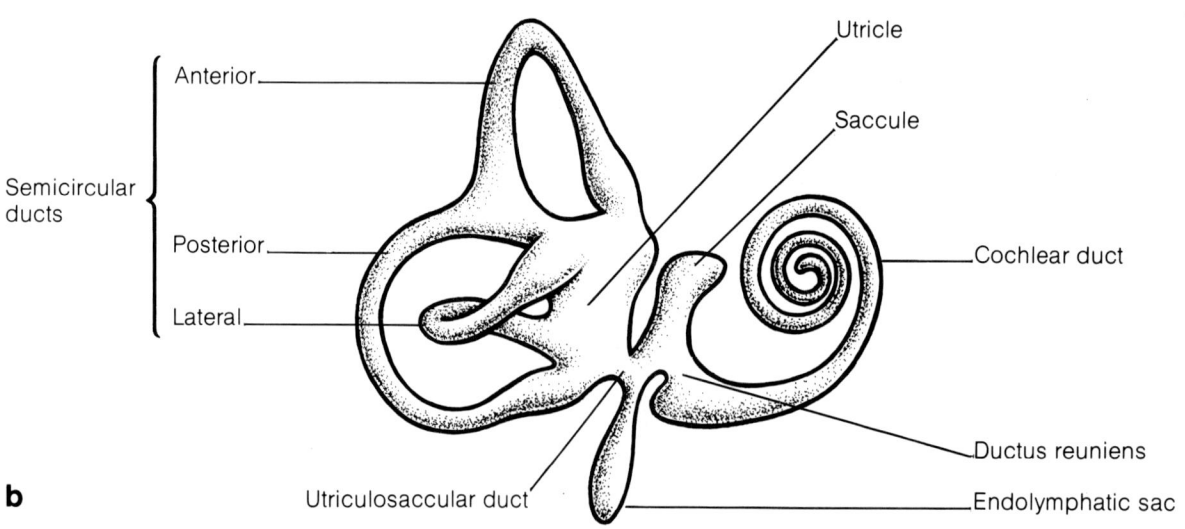

Figure 150 *The internal ear.* **a**) *The osseous labyrinth within the petrous part of the temporal bone.* **b**) *The membranous labyrinth.*

THE INTERNAL EAR (Figure 150)

THE OSSEOUS LABYRINTH

The osseous labyrinth consists of a series of chambers within the petrous part of the temporal bone. It contains the membranous labyrinth and some clear fluid called perilymph, and is lined with a thin layer of internal periosteum. The osseous labyrinth communicates with the middle ear by means of the fenestra vestibuli and the fenestra cochleae. It opens into the posterior cranial fossa via the aqueducts of the vestibule and the cochlea. The osseous labyrinth can be subdivided into three parts — the vestibule, the cochlea and the semicircular canals – although it is crossed by a sponge-like arrangement of bony trabeculae.

150C,23

148G,55; 148G,61
50B,24

The vestibule is the central portion of the osseous labyrinth. It contains the saccule and utricle of the membranous labyrinth. Indeed, the medial wall of the vestibule shows two distinct depressions for these structures. Anteriorly lies a spherical recess for the saccule. Posterosuperiorly lies an elliptical recess for the utricle. At the lower margin of the elliptical recess is the opening of the aqueduct of the vestibule. In this aqueduct lies the ductus endolymphaticus of the membranous labyrinth. On the lateral wall of the vestibule is found the fenestra vestibuli which is occupied by the base of the stapes. On the medial wall and the floor of the vestibule are minute foramina for the passage of branches of the vestibular nerve.

150C,23

148G,55; 150C,33

The cochlea is situated anterior to the vestibule and contains the cochlear duct of the membranous labyrinth. It is the part of the internal ear concerned with sound perception. In appearance, the cochlea resembles the shell of a snail. It presents as a hollowed canal of two and three-quarter turns, which diminishes in size from its base to its apex (the cupola). The cochlea spirals around a bony central pillar called the modiolus (Figure 151). A shelf of bone (the spiral lamina) projects from the modiolus in a manner likened to the thread of a screw. The lamina thus partially divides the canal of the cochlea. The division is completed by the cochlear duct. The cochlear duct occupies only the central portion of the canal (Figure 152), two passageways being seen on either side — the scala vestibuli and the scala tympani. The scala vestibuli is functionally the ascending spiral and the scala tympani the descending spiral. The scalae are continuous at the cupola around the apical extremity of the

150,18

150C,38; 150C,34

150C,37; 150C,36

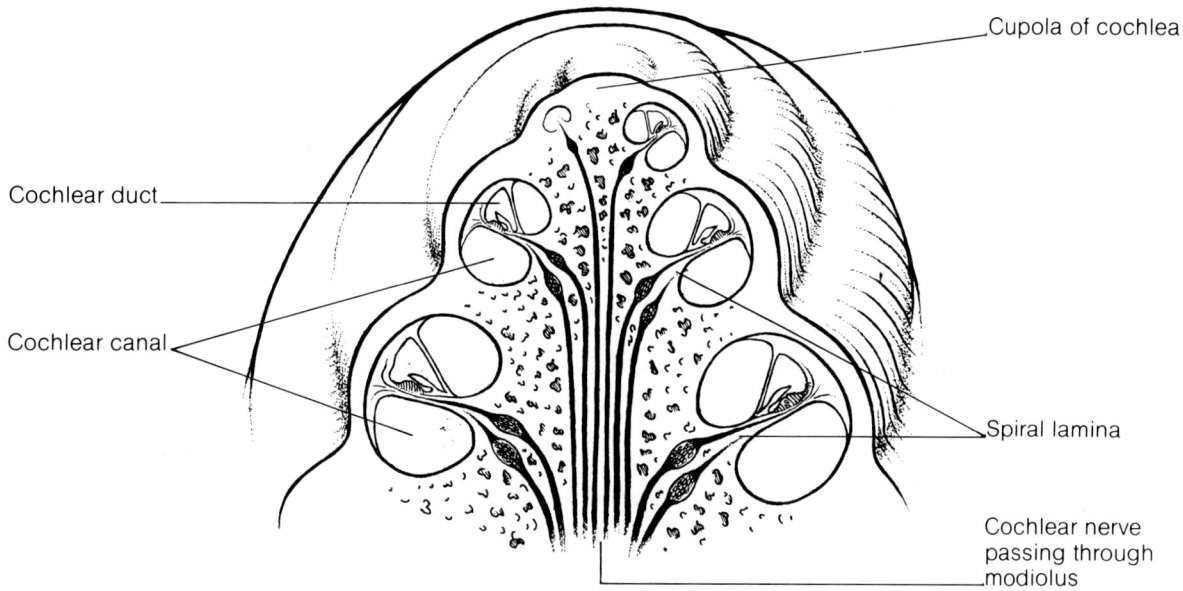

Figure 151 *Section through the cochlea of the internal ear.*

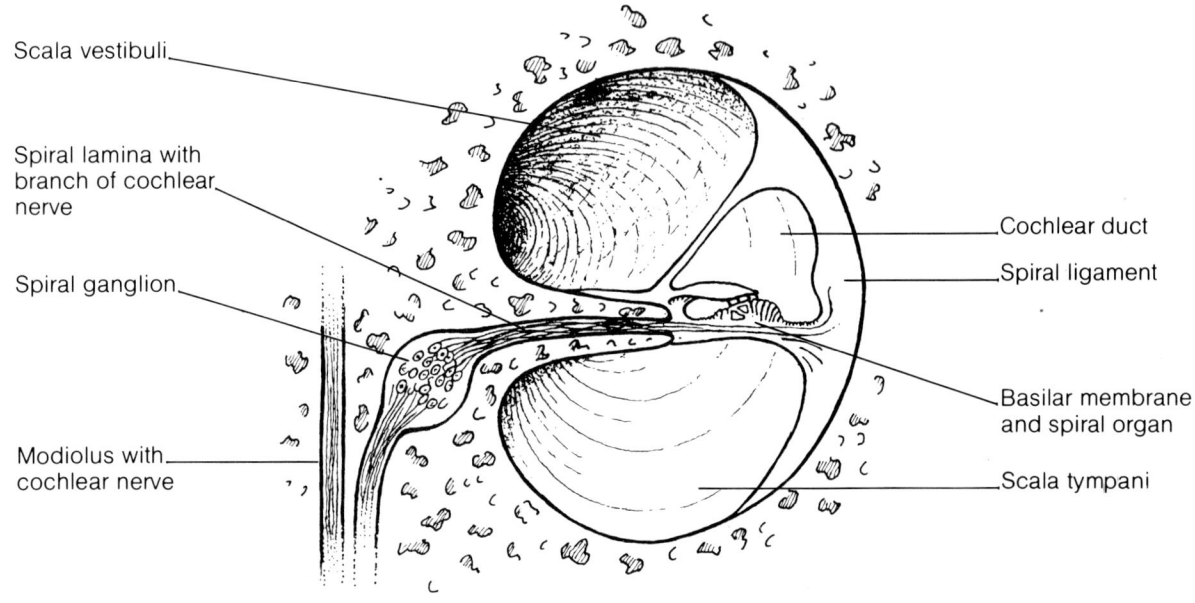

Figure 152 *Diagram to show the appearance of a single turn of the cochlea.*

150C,25

membranous cochlear duct. This region is called the helicotrema. The cochlear nerve passes centrally through the modiolus. Branches from the nerve run out along the spiral lamina to reach the sound receptors in the cochlear duct. Three openings lie near the base of the cochlea:

- The fenestra vestibuli leads into the scala vestibuli.
- The fenestra cochleae marks the termination of the scala tympani and is closed in life by the secondary tympanic membrane. This membrane allows for the dissipation of pressure from the internal ear into the tympanic cavity.
- The aqueduct of the cochlea also lies at the end of the scala tympani. It passes through the petrous bone as the cochlear canaliculus to open below the internal acoustic meatus where, because the arachnoid mater is attached to the margins of the opening, it allows the drainage of perilymph into the cerebrospinal fluid in the subarachnoid space.

148D,27–29

The semicircular canals are concerned with balance, particularly by maintaining a stable retinal image with head movement. They lie posterosuperiorly to the vestibule and are three in number: anterior (superior), posterior and lateral. Each canal occupies about two-thirds of a circle and thus opens into the vestibule at two ends. However, as the posterior end of the anterior canal and the anterosuperior end of the posterior canal join to form a common crus, there are in total five orifices for the semicircular canals into the vestibule (see Figure 150a). One end of each canal is dilated to form a structure called the ampulla.

148D,27

148G,54; 28,27

The anterior semicircular canal lies in a vertical plane across the long axis of the petrous bone. Its ends are posterior (at the common crus) and anterolateral (the ampullated end). The canal may be responsible for the ridge known as the arcuate eminence on the floor of the middle cranial fossa. However, the ridge may correspond to the occipitotemporal sulcus of the temporal lobe of the brain.

148D,29; 150D,22

The posterior semicircular canal also lies in a vertical plane but along the long axis of the petrous bone. Thus, it lies approximately at right angles to the anterior canal. Its ends are anterosuperior (at the common crus) and inferior (the ampullated end).

The lateral semicircular canal is 30° off the horizontal plane. Its ends are posterior and anterior (the ampullated end). Its ampulla produces a bulge in the middle ear, on the medial wall of the aditus and epitympanic recess above the facial canal.

The relationship between the semicircular canals is such that, whereas the lateral canals lie in the same plane, the posterior canal of one side lies nearly parallel with the anterior canal of the opposite side.

THE MEMBRANOUS LABYRINTH

The shape of the membranous labyrinth gives form to the osseous labyrinth. Essentially, the membranous labyrinth consists of a convoluted tube containing a fluid called endolymph. It is subdivided into four parts: the cochlear duct, the saccule, the utricle, and the semicircular ducts.

The cochlear duct is the spirally arranged tube that comprises the anterior part of the membranous labyrinth. It lies within the cochlea of the osseous labyrinth and is attached to the outer wall and the spiral lamina (see Figure 152). The duct commences at a blind extremity below the cupola and spirals downwards to end by joining the saccule through a minute canal (the ductus reuniens).

Figure 153 shows the constituent parts of the cochlear duct. It appears triangular in cross-section. The base of the triangle is the endosteum lining the bony canal. The endosteum is greatly thickened to form the spiral ligament of the cochlea. A specialised zone, the stria vascularis, lies on the outer wall of the cochlear duct. It is a stratified epithelium carrying a rich plexus of intra-epithelial capillaries. The floor of the cochlear duct is formed by the spiral lamina from the modiolus and by a fibrous band called the basilar membrane. This membrane runs from the lamina to the spiral ligament of the cochlea. It supports the spiral organ of Corti which contains the sound receptors. The roof of the cochlear duct is formed by a delicate membrane called the vestibular membrane.

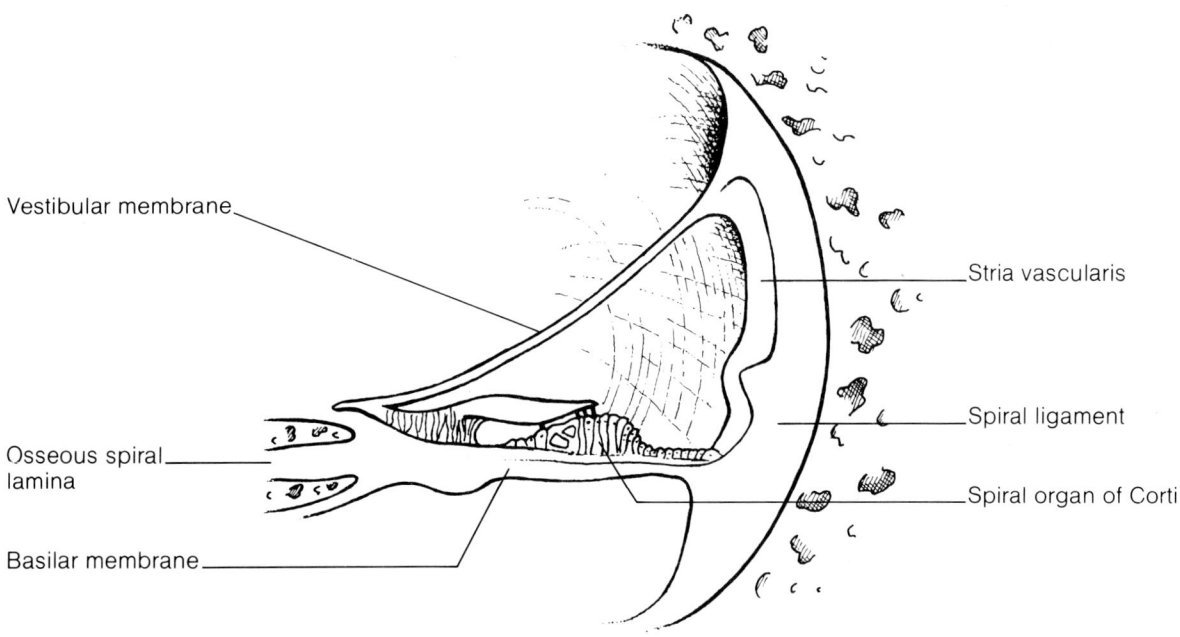

Figure 153 *Section through the cochlear duct.*

The saccule is a small sac of the membranous labyrinth which lies in the spherical recess of the vestibule of the osseous labyrinth. It is connected anteriorly with the cochlear duct by the ductus reuniens and posteriorly with the utricle by the ductus endolymphaticus and the utriculosaccular duct. The anterior wall of the saccule shows a discrete, oval thickening called the macula of the saccule. This contains sensory nerve endings concerned with balance.

The utricle is a fibrous sac that is larger and less round than the saccule. It lies within the elliptical recess of the osseous labyrinth. Within the floor of the utricle is a thickening called the macula of the utricle. This is in a plane approximately at right angles to the macula of the saccule. It also contains sensory endings concerned with balance. The five openings of the semicircular ducts appear posteriorly. Anteromedially, the utriculosaccular duct passes towards the saccule and ductus endolymphaticus.

The ductus endolymphaticus arises between the saccule and the utricle. It passes along the aqueduct of the vestibule of the osseous labyrinth and, on emerging through an aperture close to the internal acoustic meatus, ends in the endolymphatic sac beneath the dura on the posterior surface of the petrous bone.

The semicircular ducts are three in number (anterior, posterior and lateral), corresponding to the semicircular canals of the osseous labyrinth. The ducts are only one-quarter of the width of the bony canals but are similar in form. At the ampullae, they almost fill the canals. Each duct lies against the outer surface of the lining periosteum of its canal. The ducts open into five orifices in the posterior part of the utricle. The sensory receptors are concerned with balance and are situated in the ampulla at a transverse crest called the crista.

PERILYMPH AND ENDOLYMPH

Perilymph is the fluid that occupies the perilymphatic spaces between the osseous and the membranous labyrinths. In composition, perilymph resembles cerebrospinal fluid or extracellular tissue fluid. The source of the fluid and its precise mode of drainage is uncertain, though the connection between the perilymphatic space and the subarachnoid space through the cochlear canaliculus has been implicated.

Endolymph is the fluid within the membranous labyrinth. It closely resembles intracellular fluid, being rich in potassium. The source of the fluid is unknown, but it is thought to drain via a vascular plexus associated with the endolymphatic sac.

The perilymph and endolymph have important roles in both auditory and vestibular functions. When the stapes moves within the fenestra vestibuli, there is movement of perilymph within the scala tympani. This stimulates the auditory receptors in the spiral organ of Corti. The fluid movements continue down the scala tympani and pressure is dissipated by displacement of the secondary tympanic membrane which overlies the fenestra cochleae. Concerning the vestibular apparatus, the semicircular canals give information about **movements,** the saccule and utricle about the **position** of the head. Movement of the head results in movement of the endolymph in the semicircular canals. This stimulates receptors in the ampullae. The receptors in the maculae of the saccule and utricle are essentially stretch receptors. The saccule is affected when there has been lateral tilting of the head, while the utricle is affected following anterior or posterior flexion.

THE INNERVATION OF THE INTERNAL EAR (Figure 154)

The nerve to the internal ear is the vestibulocochlear nerve (eighth cranial nerve). It arises alongside the facial nerve on the lateral side of the brain between the pons and the medulla oblongata. It passes into the petrous part of the temporal bone through

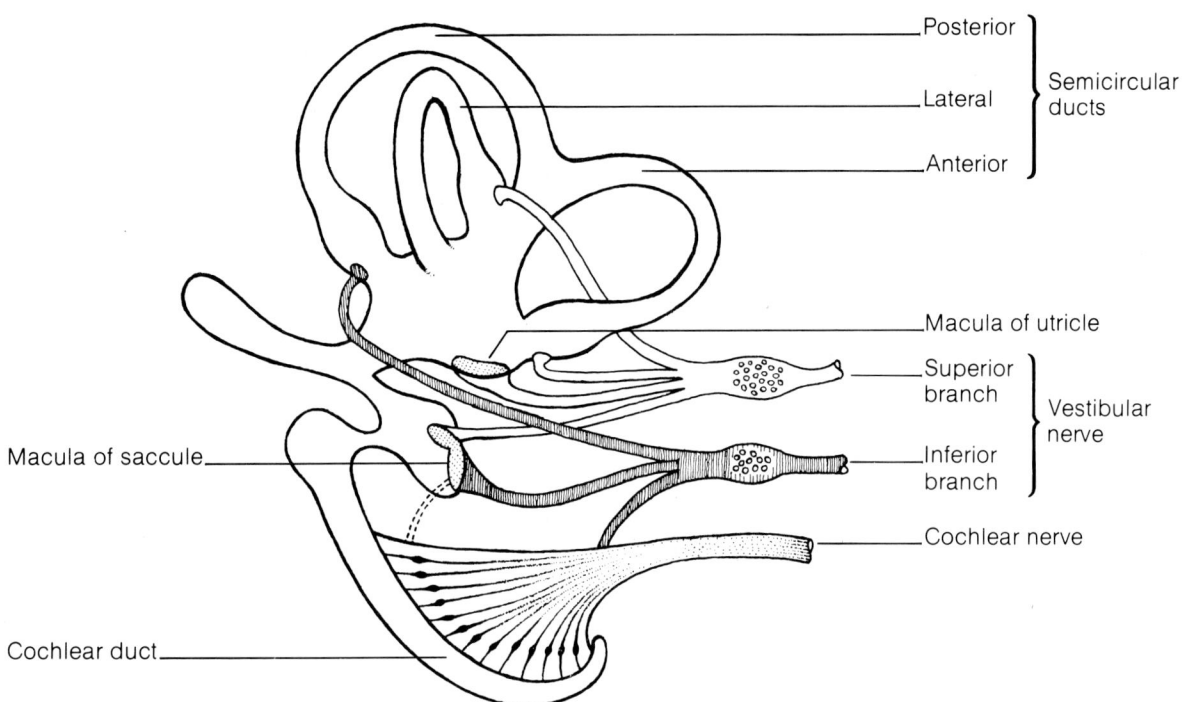

Figure 154 *The distribution of the vestibulocochlear nerve* to the membranous labyrinth of the internal ear.

the internal acoustic meatus. Within the meatus, the nerve divides into two vestibular branches (superior and inferior) and a cochlear branch.

168,16; 28,43
150,24; 150,25

The vestibular nerves arise from a ganglion (Scarpa's ganglion). The superior vestibular nerve supplies the maculae of both the saccule and utricle, and the ampullae of the anterior and lateral semicircular ducts. The inferior vestibular nerve supplies the ampulla of the posterior semicircular duct and the macula of the saccule. A small branch of the nerve communicates with the cochlear nerve.

150,24

The cochlear nerve is associated with a ganglion (the spiral ganglion), which is situated within the modiolus along the spiral lamina. Fibres pass through the lamina to end in the spiral organ of Corti. A branch of the cochlear nerve may also contribute to the innervation of the macula of the saccule.

150,25
150C,38, 150C,34

THE BLOOD SUPPLY OF THE INTERNAL EAR

The arteries of the internal ear are derived from two sources: the labyrinthine artery and the stylomastoid artery.

The labyrinthine artery arises from the anterior inferior cerebellar artery, or directly from the basilar artery. It passes through the internal acoustic meatus and divides into vestibular and cochlear branches.

150,26

A labyrinthine vein accompanies the artery and drains into the superior petrosal sinus or the transverse sinus. In addition, a small vein from the cochlea passes through the cochlear canaliculus to drain into the internal jugular vein.

THE EMBRYOLOGY OF THE EAR

The external acoustic meatus is derived from the first branchial cleft (see Figure 44). The ectodermal cells that lie towards the bottom of the meatus (i.e. adjacent to the future tympanic membrane) proliferate to form a solid clump of cells called the meatal plug. During the seventh month of intra-uterine life, the meatal plug breaks down, thereby enlarging the inner part of the external meatus and defining the lateral surface of the tympanic membrane. The auricle develops from six swellings called auricular hillocks. Three swellings lie in front of the first branchial cleft and three behind the cleft. They are produced by proliferation of the underlying mesenchyme from the first and second branchial arches. The swellings fuse to form the definitive auricle.

The tympanic cavity and the auditory tube are derived from the first (and possibly second) branchial pouch (see Figure 44). The proximal part of the first branchial pouch forms the auditory tube. The distal part of the pouch enlarges to form the tubotympanic recess from which arises the primitive tympanic cavity. Further expansion of the tympanic cavity in late fetal life produces the mastoid antrum. Pneumatisation of the mastoid region of the temporal bone in the child results in the appearance of the mastoid air cells which establish continuity via the mastoid antrum with the tympanic cavity.

The tympanic membrane is derived from the branchial membrane of the first arch interposed between the first branchial cleft and the first branchial pouch. Its lateral surface is therefore lined by ectoderm and its medial surface by endoderm. There is only a thin intervening layer of connective tissue. Thus, whereas for the remaining branchial membranes the close relationship between ectoderm and endoderm is subsequently lost as a result of intervening mesenchyme, the embryonic relationship persists for the first branchial membrane in the adult tympanic membrane.

The malleus and incus are derived from the dorsal aspect of the cartilage of the first branchial arch. The stapes arises from the dorsal part of the cartilage of the second branchial arch. The muscle associated with the malleus, the tensor tympani muscle, is derived from mesenchyme from the first branchial arch and is therefore innervated by the mandibular division of the trigeminal nerve. The muscle attached to the stapes, the stapedius muscle, arises from the second branchial arch and is supplied by the facial nerve (see page 134; Table 2).

The internal ear is first evident during the fourth week of intra-uterine life as a thickening of the surface ectoderm by the side of the developing hindbrain. This thickening is called the otic placode. It sinks into the underlying mesenchyme to form the otic vesicle. The otic vesicle gives rise to the membranous labyrinth. The mesenchyme around the otic vesicle initially chondrifies to form the otic capsule. This capsule subsequently ossifies, forming the osseous labyrinth. That part of the otic capsule adjacent to the otic vesicle breaks down to form the space (containing perilymph) between the osseous and membranous labyrinths.

SUMMARY SHEET: THE EXTERNAL EAR

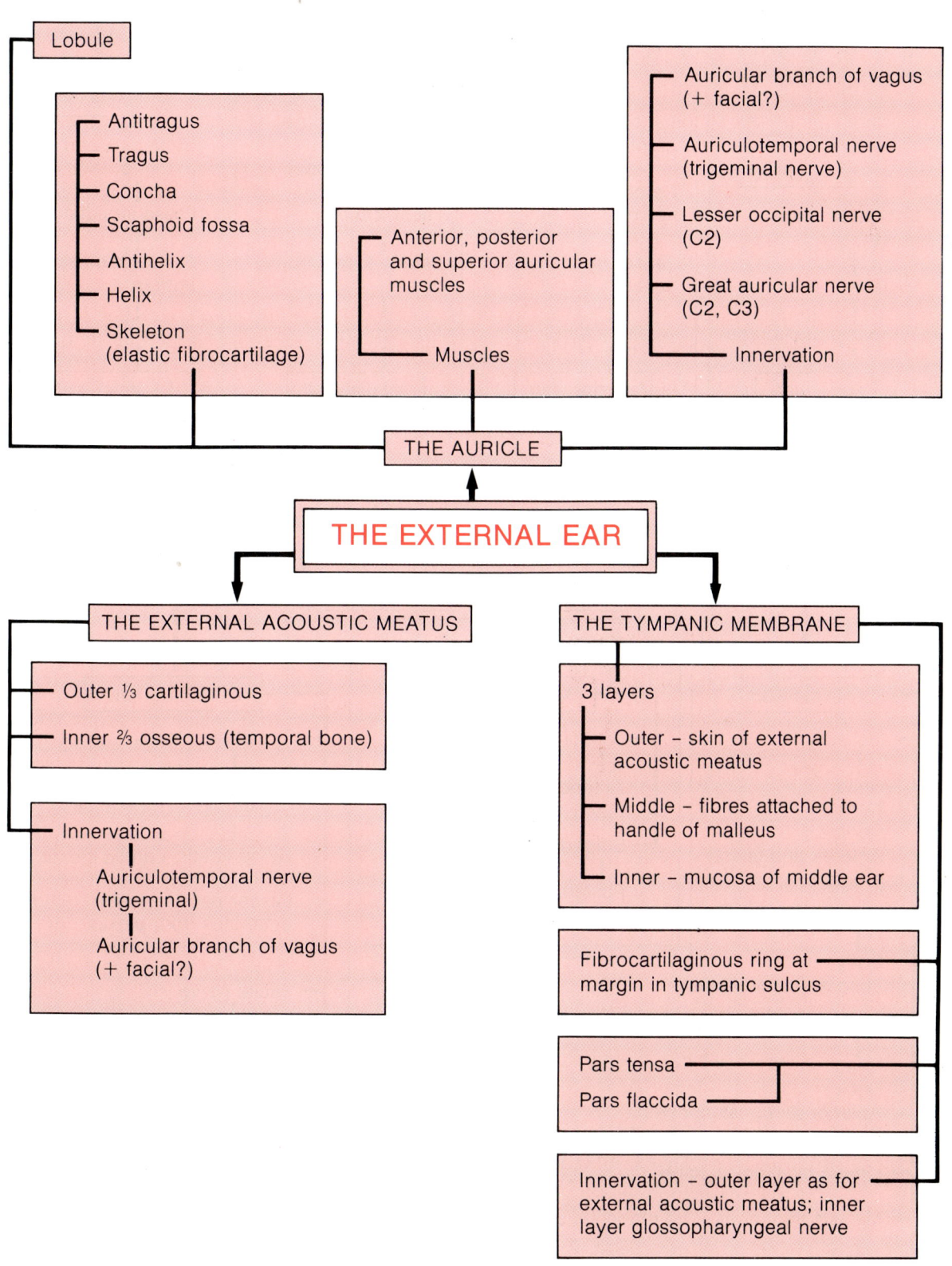

SUMMARY SHEET: THE MIDDLE EAR

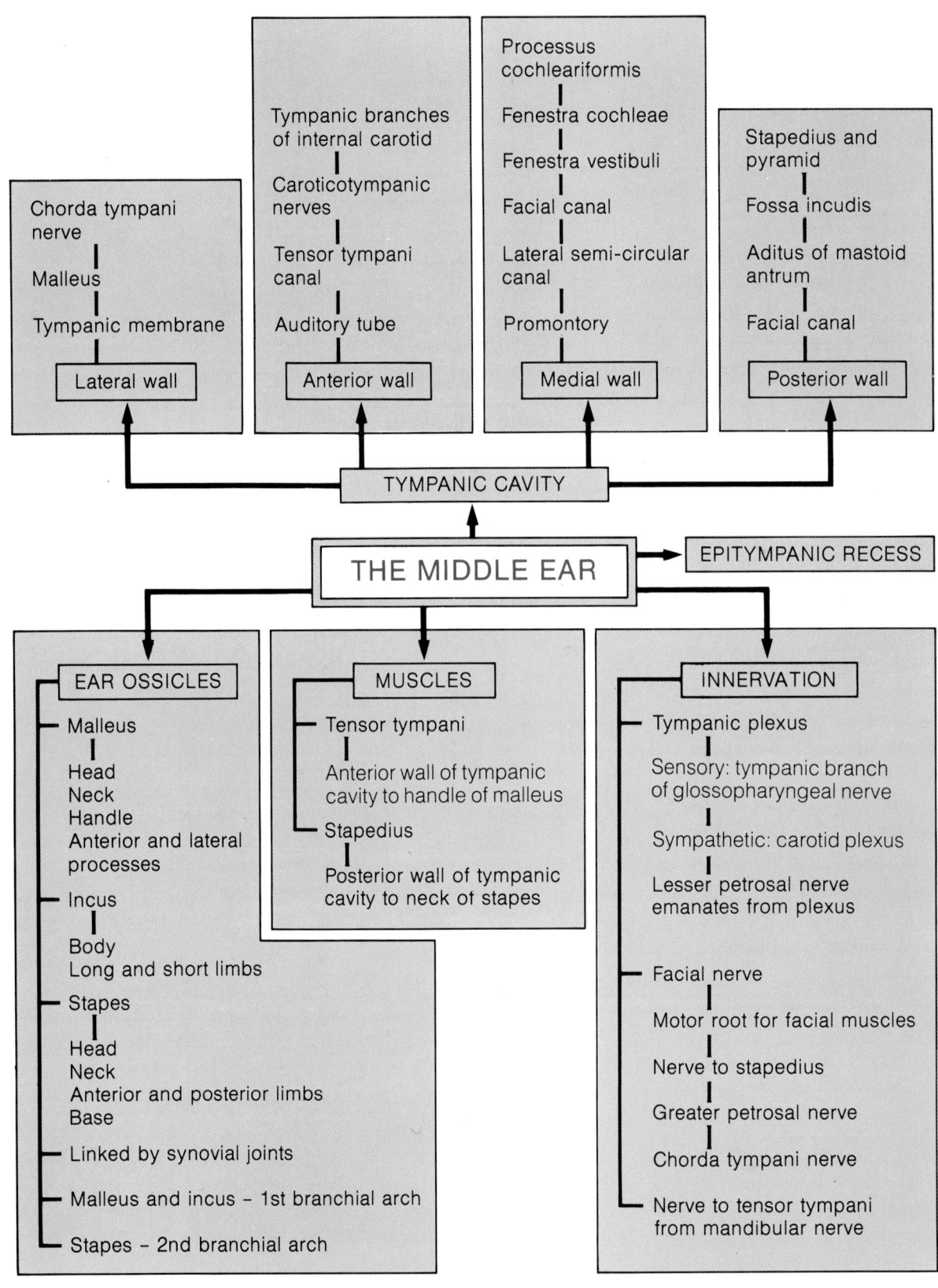

SUMMARY SHEET: THE INTERNAL EAR

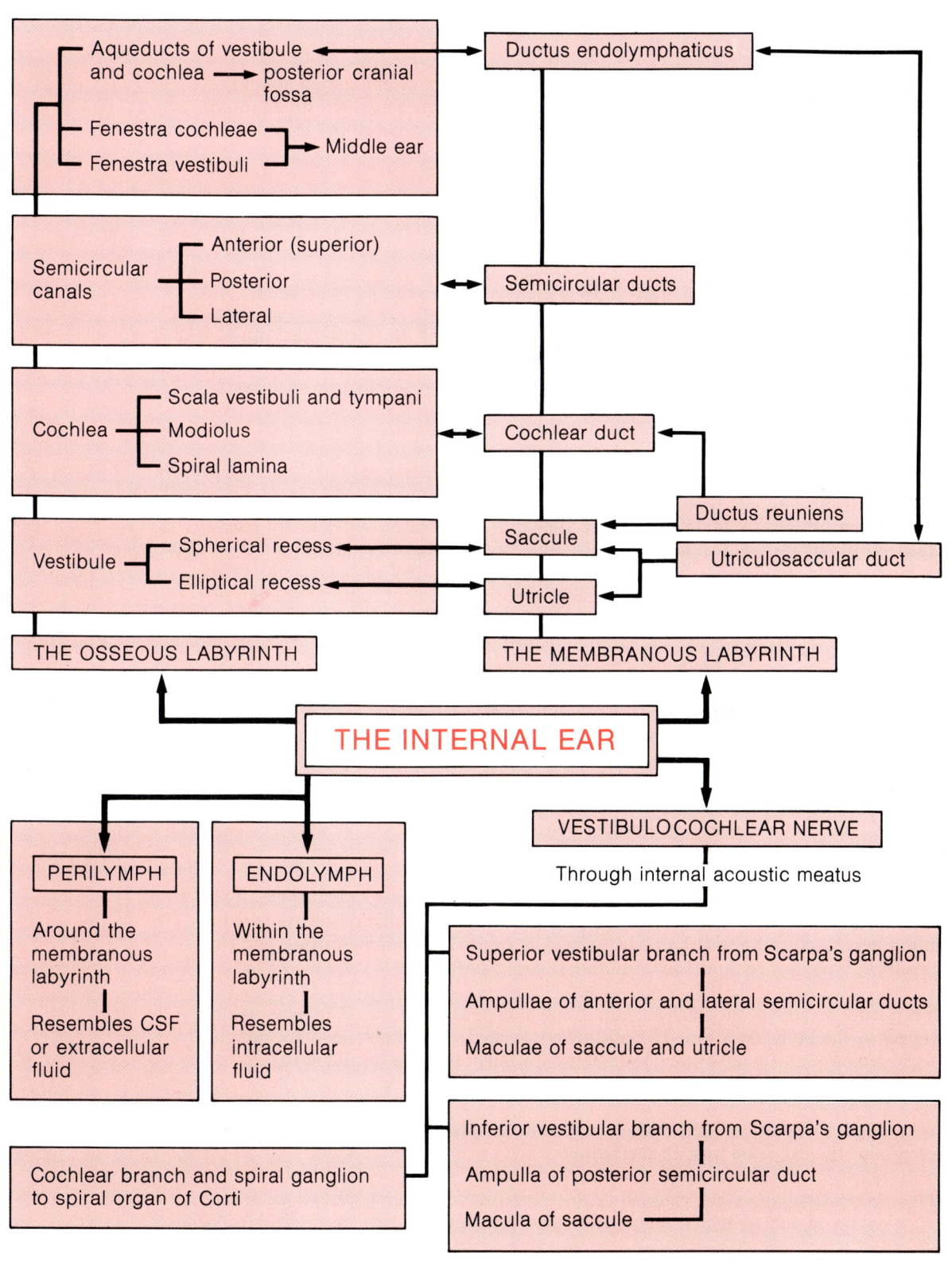

Case histories 7

The ear

A mother took her four-year-old son to the doctor as she was concerned about the appearance of his ears, which were abnormally prominent. The doctor diagnosed a congenital abnormality known as 'bat ears' which affects one of the folds of the auricle.

In this condition, the main defect is the poor definition of the antihelix. Corrective surgical treatment aims at the restoration of the antihelix.

A worker had the upper half of his auricle cleanly amputated in an accident. He was rushed to hospital clutching the amputated part of his ear in a handkerchief. The casualty officer was confident that there was a chance of a successful regrafting.

The reason for such confidence is that the auricle has a blood supply greatly in excess of its metabolic requirements.

A young baby presented with a fleshy lump in front of her right ear. The ear itself was slightly mis-shapen. A red sore area, which was moist and wept intermittently, was evident beside the lump. The baby appeared very docile and was not easily startled by loud noises. The lump was diagnosed as an accessory auricle.

The auricle is normally formed by the fusion of six enlargements (auricular hillocks) from the first and second branchial arches. The first branchial cleft lies between these arches and eventually forms the external acoustic meatus. This cleft is initially separated from the first branchial pouch by a meatal plug and eventually by the tympanic membrane. In the case described here, the maldevelopment of the auricle has been accompanied by maldevelopment of the cleft such that it was bifid, one arm forming the external acoustic meatus and the other a pre-auricular sinus that communicated with the middle ear cavity and the auditory tube. The sinus allowed saliva to pass from the oropharynx and reach the skin where it caused a localised area of inflammation. The docile nature of the baby suggests that there may be other defects in the middle ear causing deafness (at least in one ear).

A rugby player visited his doctor with a painful swelling on his right auricle. This had appeared after a rugby match the previous day. The patient had ruptured a blood vessel and a haematoma had formed between the perichondrium and the cartilage. Why do you think this condition is very painful?

It is painful because of the small amount of subcutaneous tissue, the skin being tightly bound to the perichondrium. If inadequately treated, the blood clot may become organised into scar tissue, leading to an ugly deformity known as 'cauliflower ear'.

Why does a doctor pull the auricle upwards, backwards and outwards before examining the ear drum with an auriscope?

This action straightens the external acoustic meatus and thereby facilitates direct examination of the tympanic membrane. The cartilaginous part of the meatus is slightly concave anteriorly, thus being straightened by drawing the auricle posteriorly and upwards.

A man received a blow on the chin during a fight. When seen in the casualty department, blood was coming out of one of his ears. Why?

The blow has driven the condyle of the mandible into the tympanic part of the temporal bone, disrupting the anterior wall of the external acoustic meatus and causing bleeding.

A child was taken by her father to see the doctor after she had got a plastic bead stuck in her external acoustic meatus. Where is the most likely site for such a foreign body to come to rest?

Although the site will partly depend on the size of the bead, in practice foreign bodies tend to lodge at the narrowest point of the meatus, which is at the junction of the cartilaginous and bony parts.

A professional diver working on an offshore oil-rig went to his doctor complaining of sudden onset of right-sided deafness and discharge from his right ear. He had had an itching ear for several weeks. His tympanic membrane was not visible, as the swollen external acoustic meatus was blocked with blood-stained pus and was also very tender.

Recurrent infection of the external ear (otitis externa) is common amongst divers. This does not usually cause deafness, unless associated with anatomical narrowing of the external acoustic meatus. Such narrowing is most commonly caused by infection of the hair follicles of the external acoustic meatus. In the case described here, it is possible that the meatus has been narrowed by the development of a bony outgrowth (osteoma) from the anterior or posterior walls of the meatus. This may result from repeated exposure to cold water, but the mechanism of production of the tumour is not understood. A similar deafness may be seen in congenital atresia of the external ear, in which the auricle is missing and the external acoustic meatus is stenosed.

A 35-year-old lady visited her doctor with acute pain in her left external acoustic meatus and drooping of the left side of her face. Examination revealed blistering vesicles over the surface of the external acoustic meatus and tympanic membrane. Herpes zoster oticus (Ramsay Hunt syndrome) was diagnosed, caused by infection of the geniculate ganglion of the facial nerve by the chickenpox virus. How would you explain the symptoms?

Both the motor and the sensory components of the facial nerve are affected. Damage to the motor component explains the facial palsy. Involvement of the sensory component gives rise to a shingles-type rash in areas where the nerve contributes to the cutaneous innervation (i.e. the external acoustic meatus and tympanic membrane). Impairment of parasympathetic function can also occur in severe cases. Consequently, there may be a dry eye because of interruption of the secretomotor supply to the lacrimal gland via the greater petrosal nerve. Interference with the chorda tympani nerve may also occur, although this is usually asymptomatic as contralateral secretomotor and gustatory function is intact and there may be alternative pathways. Occasionally, the vestibulocochlear nerve is also involved, resulting in deafness, dysequilibrium and nausea.

An elderly patient with a history of heart trouble visited her doctor for the removal of some wax from her ears. On syringing the external acoustic meatus, the patient suddenly collapsed. Could there be any connection between the treatment and her collapse?

The external acoustic meatus is supplied in part by the auricular branch of the vagus nerve and stimulation of this nerve may reflexly stimulate cardiac branches of the vagus, precipitating cardiac arrhythmia. Syringing of the ear may produce coughing, nausea and vomiting by similar means.

While using an auriscope to examine the tympanic membrane of a patient, a doctor noted a crescentic blue structure visible through the lower part of the membrane. What normal anatomical structure might this be?

This is an abnormally high superior bulb of the internal jugular vein. It is usually located below the floor of the tympanic cavity, but in places the bone may be deficient and the bulb may bulge upwards into the middle ear. In these situations, the bulb may be mistaken for a middle ear effusion. If the tympanic membrane is then incised for drainage, severe haemorrhage may result.

A child, well known to his doctor as a frequent sufferer from earache, returned to the surgery with his mother. She was concerned because the child was now not doing so well at school and had become disobedient and inattentive, often having to be asked several times to do things. She wondered if there might be some psychiatric problems. On examination, the child was found to have a sore throat.

This is a typical history of a child with deafness caused by a complaint called 'glue ear'. The middle ear cavity becomes filled with fluid which damps the movements of the tympanic membrane and the ear ossicles. The auditory tube is important in the functioning of the middle ear, as air can pass from the nasopharynx to the middle ear to replace air that is being removed from the middle ear by diffusion into nearby capillaries. Consequently, pressure in the middle ear is equalised with the external air pressure. In children, the tubal tonsils and adenoids are large in relation to the nasopharynx. Further enlargement with upper respiratory tract infections may therefore easily obstruct the auditory tube. The pressure in the ear falls and causes retraction of the tympanic membrane, damping its movement in response to sound. Furthermore, the mucus secreted into the middle ear cavity does not drain. The condition may be cured by insertion of a drainage grommet into the tympanic membrane and by removal of the adenoids. The safest part of the membrane in which to insert the grommet is antero-inferiorly, as this avoids the ear ossicles and the chorda tympani nerve. This area is also less vascular and, if the tympanic membrane is sucked in, it avoids the fenestra cochleae on the medial wall of the tympanic cavity.

One management problem associated with patients suffering from motoneurone disease is difficulty in swallowing saliva. One method of alleviating this involves an ear operation. What is the connection?

As the secretion of saliva from the parotid gland is controlled by the lesser petrosal nerve, the amount of secretion can be reduced (thereby reducing the frequency of swallowing) by interrupting the fibres entering the nerve. This can be achieved by sectioning the nerve at its origin from the tympanic plexus on the promontory on the medial wall of the tympanic cavity.

An inexperienced surgeon performed a mastoidectomy on a two-year-old child in order to drain an infected mastoid. He made the usual skin incision behind the ear to expose the mastoid region. Postoperatively, the surgeon was upset to see that the child's face was not moving on the operated side. What has happened?

The surgeon has cut the facial nerve. In very young children, the mastoid process is poorly developed and the facial nerve lies in a superficial position as it exits from the base of the skull at the stylomastoid foramen. Thus, the initial incision should take account of this and be made at a higher level.

A nine-year-old girl presented with deafness and pain in her left ear. She felt hot and unwell and had a temperature of 38°C. Examination of her ear revealed a red, bulging, inflamed tympanic membrane. A middle ear infection (otitis media) was diagnosed and some antibiotics and analgesics prescribed. The girl was sent home to bed. Two days later, the pain had spread over the side of her head and jaw. Her mastoid process was swollen and extremely tender and her auricle was displaced anteriorly. She had a profuse, smelly discharge from her external acoustic meatus, her deafness had worsened and she was febrile with a fast pulse. What has happened?

Her otitis media has not been controlled by antibiotics and has spread to the mastoid air cells to cause acute mastoiditis. It has also ruptured the tympanic membrane, causing the purulent discharge. The perception of pain in the jaw and the side of the head is the result of referred pain, the mastoid air cells and the side of the head and jaw sharing a common nerve supply (from the mandibular division of the trigeminal nerve).

A child presented to the doctor with earache in her left ear. The doctor correctly diagnosed acute otitis media and prescribed a course of antibiotic therapy. The following morning, the parents noted that the child was dribbling her food at breakfast and that, when attempting to smile, the left corner of her mouth did not move. What complication has occurred to explain these symptoms?

Although the facial nerve normally runs through the tympanic cavity in a bony canal, in about 10 to 20% of people this bony covering may be absent. In these people, infection in the middle ear cavity may affect the nerve and this may be manifested as a transient facial palsy.

A 76-year-old woman had been unwell for a week. She was found lying in her bed unconscious and could not be roused. On admission to hospital, she was found to have a high fever (40°C), a pulse of 120 per minute and her neck was notably stiff. Her left tympanic membrane was found to be red and bulging. Meningitis was diagnosed, the infection having spread from the middle ear. What is the likely route by which this has occurred?

The infection has probably spread through the relatively thin tegmen tympani which separates the cranial cavity from the middle ear. The stiff neck and depressed level of consciousness are clear indications of meningeal infection.

A 52-year-old woman had complained of pulsatile ringing in her right ear (tinnitus) and dizziness for eight years. A series of tests had revealed that she had slowly progressing deafness in the right ear but little else. She had lost hope of a cure and decided to put up with her deafness. However, she developed further symptoms. Her face started to droop on the right side and her right eye became sore. She had difficulty in swallowing food, especially liquids, and often choked. She noticed that she could not comb her hair with her right hand, being unable to lift her arm high enough. Examination of her right external acoustic meatus revealed a red, irregular mass that distorted the tympanic membrane and bled profusely when touched gently. Indeed, this patient had a slowly growing tumour that arose from the adventitia of the jugular bulb (which lies just below the middle ear cavity). What is the explanation for these signs and symptoms?

The tumour has eroded the temporal bone, affecting the internal ear and producing tinnitus and dizziness. The additional symptoms are caused by compression of the seventh, ninth, tenth and eleventh cranial nerves by the slowly expanding tumour. The facial nerve has been damaged in the middle ear cavity, causing the palsy on the right side of the face. The glossopharyngeal and vagus nerves have been damaged in the jugular foramen, disrupting the function of the pharyngeal plexus and interfering with swallowing. The accessory nerve has also been compressed in the jugular foramen, causing weakness of the right trapezius muscle. This prevents the arm being raised to head height. Interference with the cranial accessory nerve impedes the function of the constrictor muscles of the pharynx and disrupts swallowing.

A 55-year-old woman had for some years been unable to read road signs while in a moving car, although she had perfect vision when the car was stationary. Which part of the ear is likely to be malfunctioning?

The semicircular canals are likely to have been damaged, as they are responsible for maintaining a stable retinal image with head movements. The internal ear can be damaged following the administration of the antibiotic streptomycin, a drug that can affect both the vestibular and auditory functions of the internal ear.

A young woman with a history of neurofibromatosis (multiple tumour formation affecting the Schwann cells) visited her doctor complaining of slowly progressive deafness in her right ear over a period of five years. She was also aware of rushing and ringing noises (tinnitus) in the same ear. Recently, her right forehead had started to feel numb. What has happened to produce these symptoms?

This patient has a neurofibroma affecting her vestibulocochlear nerve in the region of her right internal acoustic meatus. This has destroyed the nerve, producing the deafness and the tinnitus. The numbness of her forehead indicates that the tumour has started to press on the trigeminal nerve (which is medially related to the vestibulocochlear nerve).

The cranial cavity

<small>28; 30; 162
168

162</small>

The cranial cavity is the interior of the skull that accommodates the brain and associated structures. Many of the contents are located near the floor of the cranial cavity, and this region can be divided into three distinct fossae: the anterior, middle and posterior cranial fossae. The fossae have a marked step-like appearance, such that the floor of the anterior cranial fossa is at the highest level and the floor of the posterior cranial fossa is lowest (see pages 21 to 24 for a detailed description of the osteology).

<small>64,1; 64,5
64,4; 64,8
64,7; 184,2</small>

The anterior cranial fossa is formed by the orbital parts of the frontal bone, the cribriform plates of the ethmoid bone with the crista galli, and the lesser wings and jugum of the sphenoid bone. It is occupied mainly by the frontal lobes of the cerebral hemispheres of the brain.

<small>67,1
64,3
67,6; 67,7
158,50; 184,28</small>

The middle cranial fossa consists of a central part formed by the body of the sphenoid bone, and right and left lateral parts each formed by the greater wing of the sphenoid bone and the squamous and petrous parts of the temporal bone. The central part is occupied mainly by the pituitary gland. The lateral parts contain the temporal lobes of the cerebral hemispheres of the brain.

<small>28,42; 28,35
28,28
28,34; 28,15

158,18; 166,27
158,10; 158,25
158,22; 158,26
184,22</small>

The posterior cranial fossa is formed by the basilar, lateral and lower squamous parts of the occipital bone, the petrous and mastoid parts of the temporal bones, a small part of the mastoid angles of the parietal bones, and the dorsum sellae and posterior part of the body of the sphenoid bone. Unlike the other cranial fossae, the posterior cranial fossa has a well-defined roof. This roof is formed by a fold or septum of dura mater called the tentorium cerebelli. The posterior cranial fossa contains the lowest part of the midbrain and the pons, cerebellum and medulla oblongata. The region of the cranial cavity immediately above the tentorium cerebelli contains the occipital lobes of the cerebral hemispheres of the brain.

<small>160,12–14</small>

Intervening between the brain and the bones of the cranial cavity are three layers called the meninges.

<small>160; 170</small>

THE MENINGES (Figure 155)

The whole of the brain and spinal cord are enveloped by three membranes or meninges: dura mater, arachnoid mater and pia mater. The meninges line the cranial cavity and the vertebral canal, providing support and protection for the neural tissue within.

The dura mater is sometimes called the pachymeninx (pachy meaning thick). The arachnoid and pia mater together constitute the leptomeninges (lepto meaning thin).

<small>160,12; 170,1</small>

The dura mater

The dura mater is the outermost and thickest meninx. It is inelastic with a high content of collagen fibres.

The dura mater in the cranial cavity is fused for much of its extent with the internal periosteum of the skull bones. The dura is a combined membrane comprising outer and inner layers. The outer layer is called the endosteal layer; the inner layer is called the meningeal layer. These layers are not easily separable, except where venous sinuses occur between them.

The dura is particularly adherent to the bones of the cranium at the base of the skull and along the sutures (most notably in childhood and old age). The attachment is mediated by fibrous tissue and the connective tissue of blood vessels. The endosteal layer of the dura is continuous with the pericranium through sutures and foramina, and also with the periosteal lining of the orbital cavity via the superior orbital fissure.

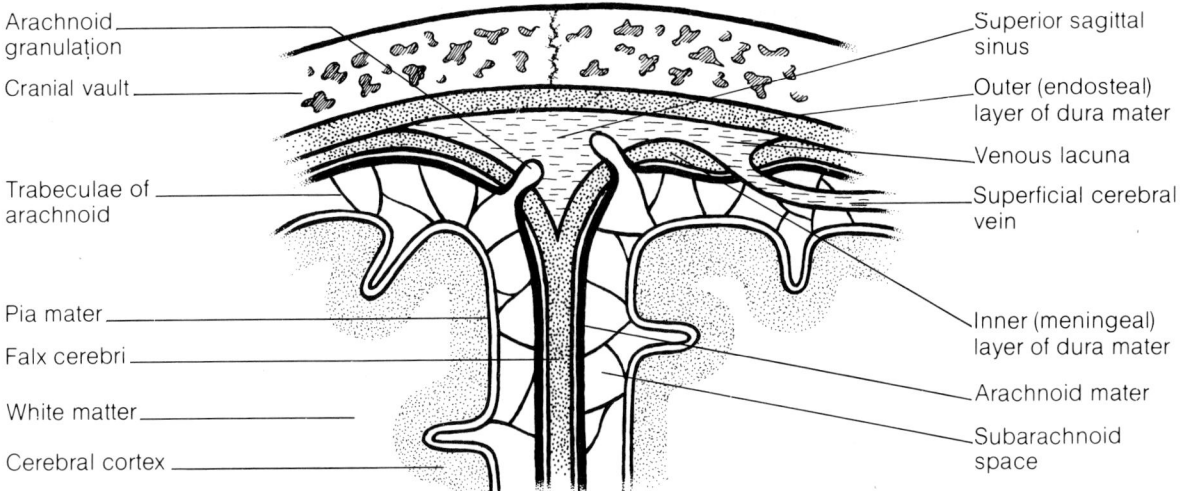

Figure 155 *Coronal section through the superior sagittal sinus showing the arrangement of the meninges.*

The dura forms sleeves around the cranial nerves, being continuous with their epineuria. The dura surrounds the whole of the optic nerve, becoming continuous with the sclera of the eyeball.

The inner surface of the dura is smooth. It has a ubiquitous lining of arachnoid mater from which it is separated by only a thin film of serous fluid.

Folds of the meningeal layer of the dura form four fibrous partitions or septa that broadly divide the cranial cavity. These are the large falx cerebri and tentorium cerebelli, and the smaller falx cerebelli and diaphragma sellae.

The falx cerebri is a sickle-shaped fold of dura that lies along the median sagittal plane in the longitudinal fissure between the two cerebral hemispheres (i.e. above the corpus callosum). Anteriorly, it is attached to the crista galli. Its attachment to the skull continues superiorly and posteriorly along the margins of the superior sagittal venous sinus to the internal occipital protuberance. At its posterior and inferior limit, the falx cerebri is continuous with the tentorium cerebelli. The inferior sagittal venous sinus runs in its free inferior border, and the straight sinus along its junction with the tentorium cerebelli.

158,4; 164,2
158,5
158,61; 164,14
158,2; 20,14
158,18; 164,25
158,62; 164,3
158,17; 164,27

The tentorium cerebelli lies between the cerebellum and the occipital lobes of the cerebral hemispheres. It forms the crescent-shaped roof to the posterior cranial fossa. The tentorium cerebelli is notched anteriorly so that the roof of the posterior cranial fossa is incomplete behind the dorsum sellae of the sphenoid bone. This notch is called the tentorial incisure, its margin being named the free border of the tentorium. The midbrain lies in this incisure, together with part of the adjacent cerebellar vermis. Closely related to the free border of the tentorium are the great cerebral vein (of Galen) and medial regions of the temporal lobes of the cerebral hemispheres (the parahippocampal gyri). The tentorium cerebelli is attached at its periphery to the margins of the transverse and superior petrosal venous sinuses, and to the posterior clinoid processes of the dorsum sellae. However, the free border of the tentorium continues anteriorly to gain attachment to the anterior clinoid processes. A cavernous venous sinus is located on each side of the sella turcica between two layers of dura mater (i.e. the layer lining the sella turcica and the layer extending from the free border of the tentorium).

158,18; 164,25
158,22
164,5; 168,27
164,23
158,15; 168,24
182A,10
168,22; 168,12
168,29
168,27; 168,32
168,33; 166,26

A recess is present in the tentorium cerebelli near its attachment to the apex of the petrous part of the temporal bone and beneath the superior petrosal sinus. This recess is called the trigeminal cave and is occupied by the roots of the trigeminal nerve connecting with the pons, and by part of the trigeminal ganglion.

158,19; 168,23

196A,6
28,39

The falx cerebelli is a small median fold at the back of the cranial cavity, which extends from the inferior surface of the tentorium cerebelli. It passes into the cerebellar notch between the cerebellar hemispheres. Posteriorly, the falx cerebelli is attached to the internal occipital crest and to the margins of the occipital sinus.

166A,17; 168,31
28,14; 158,50

The diaphragma sellae roofs the sella turcica. The pituitary gland (hypophysis) lies beneath it, with the pituitary stalk (infundibulum) passing through a central aperture. Within the sella turcica, the dura (together with the arachnoid and pia mater) blends with the capsule of the pituitary gland.

160,13; 170,2

The arachnoid mater

The arachnoid mater closely lines the dura mater, being separated from it by a potential space called the subdural space. This space contains a thin film of serous fluid and is probably in continuity with the lymph spaces of the cranial and spinal nerves. The arachnoid mater does not follow the sulci and fissures of the brain (except where these are occupied by the septa of the dura mater). In this respect, it resembles the dura mater but not the pia mater.

160,15

Between the arachnoid and pia mater is a variable space called the subarachnoid space. This is filled with cerebrospinal fluid and it contains the major arteries and veins supplying the central nervous system. The arachnoid in the cranial cavity is connected to the pia mater by a close meshwork of fine trabeculae. Indeed, it lies so close to the pia mater that virtually a single membrane is formed (the 'pia-arachnoid').

173,15; 158,27

In some regions, however, named cisternae containing cerebrospinal fluid occur where separation of the pia and arachnoid is greater. The subarachnoid space is in communication with the brain ventricles via the three apertures of the fourth

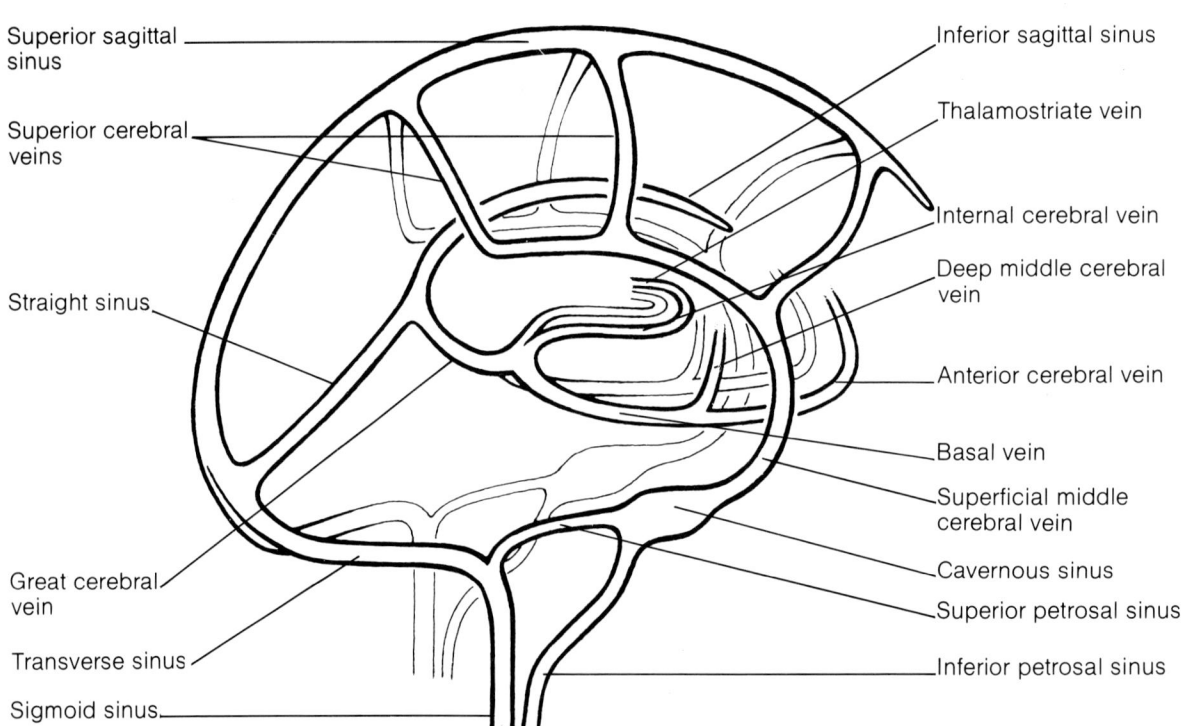

Figure 156 *The dural venous sinuses and the main cerebral veins.*

ventricle, and with the venous system via arachnoid villi and granulations (Figure 155). The space also continues into the roof of the nasal cavity along the olfactory nerves.

The pia mater

The pia mater is a vascularised, areolar membrane. The pia mater covers almost the whole external surface of the brain. Indeed, the pia is in intimate contact with the brain, extending into every sulcus and fissure.

The pia mater forms the tela choroidea of the third ventricle (roof) and of the fourth ventricle (inferior part) and the choroid plexuses of all the ventricles of the brain. It ensheaths precapillary blood vessels entering the nervous tissue, forming perivascular cuffs (with the arachnoid mater). The ensheathment is particularly marked for the minute blood vessels that pass perpendicularly into the cerebral cortex. The interface between the pia and the nervous tissue is provided by a basement membrane in contact with the end-feet of astrocytic glial cells (see page 415). Breaches in the pia roofing the fourth ventricle allow passage of cerebrospinal fluid into the subarachnoid space via a median aperture (foramen of Magendie) and two lateral apertures (foramina of Luschka).

The dural venous sinuses (Figures 155 and 156)

Between the meningeal and endosteal layers of the dura mater are found a series of venous sinuses. (The inferior sagittal and straight sinuses are exceptional in being located between two layers of meningeal dura.) The venous sinuses drain blood from the brain and from the bones of the cranium. Many of the sinuses can be easily located on the dry skull because they lie in prominent grooves along the cranium. The sinuses may be subdivided into two groups, depending upon location. There is an antero-inferior group at the base of the cranium and a posterosuperior group.

Antero-inferior sinuses
- Cavernous (paired)
- Intercavernous
- Basilar plexus
- Sphenoparietal (paired)
- Superior petrosal (paired)
- Inferior petrosal (paired)
- Middle meningeal (paired)

Posterosuperior sinuses
- Superior sagittal
- Inferior sagittal
- Straight
- Transverse (paired)
- Sigmoid (paired)
- Occipital

(The unpaired sinuses are situated in the midline)

The cavernous sinus lies adjacent to the body of the sphenoid bone. Although most of the dural venous sinuses are open channels, the cavernous sinus is crossed by many trabeculae.

Communication between the two cavernous sinuses occurs in front of and behind the pituitary gland. These are called the anterior and posterior intercavernous sinuses. The two cavernous sinuses and the intercavernous sinuses together are sometimes referred to as the circular sinus.

The tributaries of the cavernous sinus are: the sphenoparietal sinus, the superficial middle cerebral vein, the inferior cerebral veins, veins from the orbit (superior and inferior ophthalmic veins, and sometimes the central vein of the retina).

The cavernous sinus drains via the superior petrosal sinus into the transverse sinus close to its junction with the sigmoid sinus. It also drains into the internal jugular vein via the inferior petrosal sinus.

Emissary veins link the cavernous sinuses with the pterygoid venous plexuses in the infratemporal fossa, passing through the foramen lacerum, foramen ovale and the emissary sphenoidal foramen.

	Several important structures lie within the walls of the cavernous sinus (see Figure 158). The oculomotor and trochlear nerves, and the ophthalmic and maxillary divisions of the trigeminal nerve invaginate the lateral dural wall of the sinus. The internal carotid artery (with its sympathetic plexus) and the abducent nerve lie inside the cavernous sinus.
166,37; 166,31 166,40; 166,42 166,38; 166,39	

166,37; 166,31
166,40; 166,42
166,38; 166,39

Several important structures lie within the walls of the cavernous sinus (see Figure 158). The oculomotor and trochlear nerves, and the ophthalmic and maxillary divisions of the trigeminal nerve invaginate the lateral dural wall of the sinus. The internal carotid artery (with its sympathetic plexus) and the abducent nerve lie inside the cavernous sinus.

168,28

The basilar venous plexus is located in the dura covering the clivus of the posterior cranial fossa. It connects with a number of other venous sinuses, including the superior and inferior petrosal sinuses and the cavernous sinus. Further connections are with the marginal sinuses of the occipital sinus (around the foramen magnum) and with the internal vertebral plexus (within the vertebral canal).

168,34

The sphenoparietal sinus arises from meningeal veins, particularly the middle meningeal vein. It is located near the posterior rim of the lesser wing of the sphenoid bone and empties into the cavernous sinus. It usually receives anterior temporal diploic veins and often the superficial middle cerebral vein.

168,12; 28,29

168,22

The superior petrosal sinus, as its name suggests, occupies a groove on the upper border of the petrous part of the temporal bone. It is situated in the anterior part of the attached margin of the tentorium cerebelli. The superior petrosal sinus passes backwards and outwards from the cavernous sinus to the transverse sinus at the junction with the sigmoid sinus. It may pass above, below or even surround the trigeminal cave. Tributaries of the superior petrosal sinus include superior cerebellar and inferior cerebral veins, and veins from the brainstem.

168,17
28,30

The inferior petrosal sinus lies at the lower border of the petrous bone, occupying a groove in the region of the petro-occipital suture. It passes from the back of the cavernous sinus to terminate at the jugular foramen in the superior bulb of the internal jugular vein. The inferior petrosal sinus receives veins from the internal ear, the brainstem and the lower surface of the cerebellum.

20B,15

The middle meningeal veins, despite their name, are two small venous sinuses that run with the middle meningeal artery. These sinuses may be designated the frontal and parietal branches. They lie in meningeal grooves on the parietal bone, and pass backwards to drain into the pterygoid venous plexus (via the foramen spinosum and/or the foramen ovale). The middle meningeal veins may have a direct connection with the cavernous sinus or an indirect one through the sphenoparietal sinus. In addition to draining the meninges, they may receive as tributaries some diploic veins and the superficial cerebral and inferior cerebral veins.

158,2; 170,5
20B,14

The superior sagittal sinus runs along the median sagittal plane of the calvaria. It is situated in the attached margin of the falx cerebri. Small at its origin near the crista galli of the ethmoid bone, the sinus enlarges as it passes posteriorly. At its termination in the region of the internal occipital protuberance, the superior sagittal sinus becomes dilated to form the confluence of sinuses. Here, the superior sagittal sinus is usually displaced to the right side to join the right transverse sinus.

158,3

The superior sagittal sinus drains the superior cerebral veins. These veins pass into the sinus at an oblique angle. Along the sides of the sinus are situated the openings of some irregularly shaped dilations called venous lacunae. There are usually three lacunae on each side, and they connect with the meningeal and diploic veins. The lacunae contain arachnoid villi and are therefore thought to be a site at which cerebrospinal fluid passes into the bloodstream.

28,6

The superior sagittal sinus is connected to the veins of the face and the scalp through emissary veins. Rarely, an emissary vein passes through the foramen caecum lying immediately in front of the crista galli, linking the anterior extremity of the sinus with nasal veins. Emissary veins may also join the facial veins. Parietal emissary veins pass through the parietal foramina to link the superior sagittal sinus with the superficial temporal and occipital veins.

The inferior sagittal sinus is considerably smaller than the superior sagittal sinus. It lies in the free edge of the falx cerebri, above the corpus callosum. The inferior sagittal sinus drains the falx cerebri and receives some small veins from the medial surface of the cerebral hemispheres. At the anterior end of the junction between the falx cerebri and the tentorium cerebelli, the inferior sagittal sinus unites with the great cerebral vein to form the straight sinus.

The straight sinus is found at the site of attachment between the falx cerebri and the tentorium cerebelli. It is formed by the union of the inferior sagittal sinus and the great cerebral vein. The straight sinus passes downwards and backwards to enter the confluence of sinuses. It continues as the transverse sinus opposite to that receiving blood from the superior sagittal sinus (i.e. usually the left transverse sinus). The straight sinus receives some superior cerebellar veins.

The transverse sinuses begin at the confluence of sinuses. They are large sinuses, the right being generally larger. The right sinus is usually a continuation of the superior sagittal sinus, the left usually a continuation of the straight sinus. The transverse sinus passes laterally across the occipital bone in the attached margin of the tentorium cerebelli. It drains into the sigmoid sinus. The transverse sinus receives inferior cerebellar and inferior cerebral veins, and posterior temporal and occipital diploic veins. It also receives the superior petrosal sinus where it continues into the sigmoid sinus (at the postero-inferior corner of the parietal bone).

The sigmoid sinus is the main site of drainage for the dural venous sinus system. The sinus runs from the transverse sinus into the internal jugular vein at the jugular foramen. It has an S-shaped course, grooving the temporal and occipital bones. The sigmoid sinus communicates with the occipital veins extracranially via mastoid and condylar emissary veins. The mastoid emissary vein passes through the mastoid foramen near the posterior border of the mastoid part of the temporal bone. The condylar emissary vein passes through the condylar canal (posterior condylar canal) of the occipital bone. Emissary veins through this canal may also link with the vertebral veins.

The occipital sinus has two parts. First, there are two marginal sinuses that pass around the margins of the foramen magnum. Second, there is a single sinus that links the marginal sinuses with the confluence of sinuses. This part of the occipital sinus runs upwards from the foramen magnum in the attached margin of the falx cerebelli. The marginal sinuses communicate with the internal vertebral venous plexus (see page 67). There are also communications with the terminal part of the sigmoid sinus and with the basilar venous plexus.

The dural venous sinuses eventually drain into the extracranial venous system via the sigmoid sinuses and the internal jugular veins. Additional communications exist through the diploic veins of the skull and the emissary veins.

The diploic vessels run in the diploë of the cranial vault. The diploic arteries are small and numerous, and arise from the arteries of the scalp and/or of the dura mater. The diploic veins show considerable anastomoses and eventually drain into the meningeal veins, the dural venous sinuses, and the pericranial veins. Although the anastomotic pattern is complex, five main diploic veins can usually be recognised on each side: the frontal, anterior temporal (two), posterior temporal, and occipital diploic veins.

The frontal diploic vein is found at the front of the cranium. It drains extracranially into the supra-orbital vein of the forehead, and intracranially into the superior sagittal

20,4

158,62

158,15
158,17

158,17; 164,27
158,62; 158,15

162,8; 168,22

162,10; 168,21

168,12

162,10; 168,21
28,32; 28,31

22,26

22,32; 45,9

sinus. The anterior temporal diploic veins run either side of the coronal suture. They drain extracranially into the superficial temporal vein and/or intracranially into the sphenoparietal sinus. The posterior temporal vein runs in the parietal bone. It drains intracranially into the transverse sinus. The occipital diploic vein is the largest of the diploic veins. It is confined to the occipital bone. This diploic vein either drains extracranially into the occipital vein (at the suboccipital region of the neck) or intracranially into the transverse sinus (near the confluence of sinuses). It may also drain into the occipital emissary vein.

Emissary veins are veins that pass through foramina in the cranium to link the dural venous sinuses intracranially with veins extracranially. They show considerable variability. Emissary veins may be found associated with both the vault of the skull and the cranial base.

20,4

20,26

For the cranial vault, there are three named emissary veins: the parietal, mastoid and occipital emissary veins. The parietal emissary vein passes through the parietal foramen to link the superior sagittal sinus with veins on the scalp. The mastoid emissary vein passes through the mastoid foramen to connect the sigmoid sinus and the posterior auricular or occipital veins. The occipital emissary vein passes through the region of the occipital protuberances to link the confluence of sinuses and the occipital vein. It may receive the occipital diploic vein.

29C,48; 29C,49
29C,52; 22,36

28,41; 45,11

45,9

51,51

The emissary veins at the cranial base are particularly variable and are often not named. Several emissary veins link the cavernous sinus and the pterygoid venous plexus. These may pass through the foramen ovale, the foramen lacerum and/or the emissary sphenoidal foramen. An unnamed emissary vein passes through the carotid canal to connect the cavernous sinus with the internal jugular vein. Another unnamed emissary vein traverses the hypoglossal canal to link the basilar plexus or sigmoid sinus with the internal jugular vein. A condylar emissary vein runs through the condylar canal to join the sigmoid sinus and the occipital vein. An unnamed emissary vein may pass through the squamous part of the temporal bone to link the transverse sinus to the external jugular vein via a petrosquamous sinus.

28,6

An emissary vein may traverse the foramen caecum in the region of the crista galli of the ethmoid bone to join the superior sagittal sinus to veins in the nose.

The ophthalmic veins are not usually categorised as emissary veins, but they link the veins on the face extracranially with the cavernous sinus intracranially.

The vasculature of the meninges

The meningeal arteries are numerous and are derived from several sources: the ascending pharyngeal, occipital and maxillary branches of the external carotid artery, the internal carotid artery and its ophthalmic branch, and the vertebral branch of the subclavian artery. Despite being called meningeal vessels, they principally supply the bones of the skull. Indeed, the cranial dura mater is relatively avascular. Furthermore, the arachnoid mater does not appear to have a direct blood supply and the pia mater receives numerous vessels derived from arteries supplying the brain.

161,16–17
116C,23; 142A,18

22,43; 28,47
168,10
161,16; 161,17
20B,15; 28,46

The largest and principal meningeal artery is the middle meningeal artery. This is a branch of the first part of the maxillary artery (i.e. before the lateral pterygoid muscle in the infratemporal fossa; see page 180). It enters the middle cranial fossa of the skull through the foramen spinosum of the greater wing of the sphenoid bone. Passing up the squamous part of the temporal bone, the middle meningeal artery divides into frontal (anterior) and parietal (posterior) branches which ramify over the anterior and middle cranial fossa (with some small branches to the posterior cranial fossa). The frontal branch passes close to the pterion, where it frequently lies in a small canal in the greater wing of the sphenoid bone. A branch of the middle meningeal artery passes through the superior orbital fissure to anastomose with the lacrimal artery. Although mainly supplying the cranial bones, the middle meningeal artery also provides branches to the trigeminal ganglion and to the tympanic cavity (including a branch to the tensor tympani muscle). In addition to its intracranial branches, the middle meningeal artery gives off small branches that pass through the greater wing of the sphenoid bone to anastomose with the deep temporal branches of the maxillary artery in the temporal fossa.

Other arteries supplying the middle cranial fossa include the accessory meningeal artery and small branches from the internal carotid. The accessory meningeal artery arises from the first part of the maxillary artery and enters the middle cranial fossa through the foramen ovale. A meningeal branch from the ophthalmic branch of the internal carotid artery enters the middle cranial fossa through the superior orbital fissure.

116B,24
22,44; 28,48

Other arteries supplying the anterior cranial fossa are derived from the anterior and posterior ethmoidal branches of the ophthalmic artery. These enter the anterior cranial fossa from the orbit via the anterior and posterior ethmoidal foramina.

120,32; 120,29
32A,27; 32A,28

The posterior cranial fossa is provided with meningeal arteries which arise from the occipital artery, the vertebral artery, and the ascending pharyngeal artery. The meningeal branches from the occipital artery enter through the jugular and mastoid foramina, and the condylar canal. There is also an occasional branch of the occipital artery which passes through the parietal foramen. The meningeal branches from the ascending pharyngeal artery enter through the foramen lacerum and the hypoglossal canal.

22,35; 22,26
22,32
20,4
22,46; 22,34

The meningeal veins accompany the meningeal arteries and should be more correctly considered as small venous sinuses. They connect with some of the large venous sinuses and with the diploic veins. The most prominent are the middle meningeal veins accompanying the middle meningeal artery and these are described with the venous sinuses on page 394.

The innervation of the meninges

The dura mater is supplied by branches derived mainly from the trigeminal cranial nerve and from the upper cervical spinal nerves. The leptomeninges do not appear to have a sensory nerve supply.

The meningeal nerves of the anterior cranial fossa are derived chiefly from the anterior and posterior ethmoidal branches of the ophthalmic division of the trigeminal nerve. The ethmoidal nerves enter the anterior cranial fossa through the anterior and posterior ethmoidal foramina of the orbit. Their meningeal branches also innervate the anterior part of the falx cerebri. The meningeal branch of the mandibular nerve also contributes to the supply of the anterior cranial fossa.

122A,11

32A,27; 32A,28

The middle cranial fossa is innervated by the meningeal branch (nervus spinosus) of the mandibular division of the trigeminal nerve. This enters the cranial cavity through the foramen spinosum. An additional supply is derived via the meningeal branch of the maxillary nerve, and some twigs arise directly from the trigeminal ganglion.

22,43; 28,47

The posterior cranial fossa is supplied by ascending meningeal branches derived from the upper three cervical spinal nerves. Branches from the second and third cervical nerves pass through the foramen magnum. Those of the first and second cervical nerves pass through the hypoglossal canal (perhaps as the meningeal branch of the hypoglossal nerve) and through the jugular foramen (perhaps as the meningeal branch of the vagus nerve). The tentorium cerebelli and the posterior part of the falx cerebri are innervated by the recurrent tentorial branch of the ophthalmic division of the trigeminal. The supratentorial dura is supplied by branches from all three divisions of the trigeminal nerve and also directly from its ganglion.

22,31; 28,40
22,34; 28,41
28,31

Sensory and sympathetic fibres pass to the cranial meninges with branches of the internal carotid and vertebral arteries. Meningeal blood vessels, including vessels in the pia mater, are also innervated by sympathetic fibres that arise directly from the superior cervical ganglion as lateral branches.

THE CONTENTS OF THE ANTERIOR CRANIAL FOSSA

This fossa contains mainly the frontal lobes of the cerebral hemispheres. Located in the floor of the fossa are the olfactory nerves and the anterior ethmoidal nerves and vessels.

The olfactory nerve is the first cranial nerve. Approximately 15 to 20 delicate olfactory nerves pass on each side from the roof of the nasal fossa through the cribriform plate of the ethmoid bone to the olfactory bulb. These nerves are accompanied by minute sheaths of the meningeal layer of the dura mater and of the leptomeninges. From the olfactory bulbs, olfactory tracts can be seen passing along the base of the frontal lobes of the brain. Running with the olfactory system on each side is a small nerve called the nervus terminalis. It passes from the nose and alongside the olfactory bulb. However, its functional significance is unknown.

The anterior ethmoidal nerve and vessels pass across the cribriform plate and enter the nasal cavity through a fissure situated close to the crista galli. The anterior ethmoidal nerve is derived from the nasociliary branch of the ophthalmic division of the trigeminal nerve.

There are few venous sinuses in the anterior cranial fossa. The superior and inferior sagittal sinuses commence in this region.

THE CONTENTS OF THE MIDDLE CRANIAL FOSSA

Laterally, the middle cranial fossa supports the temporal lobes of the cerebral hemispheres. Centrally, the pituitary gland is retained in the sella turcica by the overlying diaphragma sellae.

The pituitary gland or hypophysis cerebri (Figure 157) is an endocrine gland connected to the hypothalamus of the brain by a pituitary stalk or infundibulum. This stalk passes through a central aperture in the diaphragma sellae. The gland is surrounded by a capsule that merges with the meninges accompanying the stalk

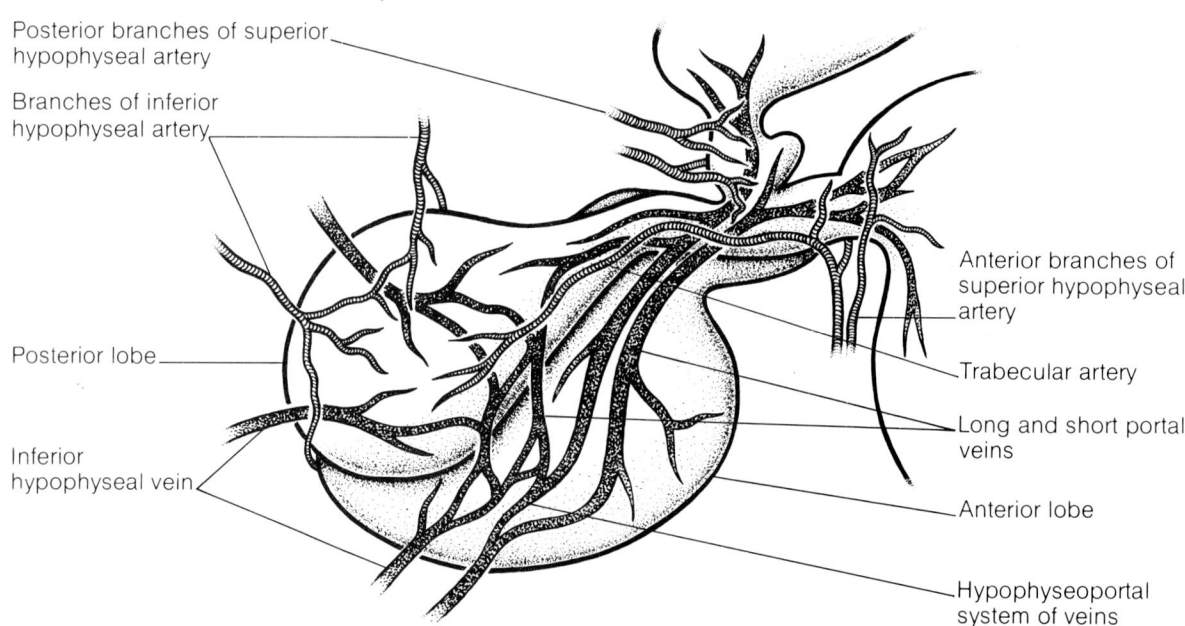

Figure 157 *Blood supply of the pituitary gland.*

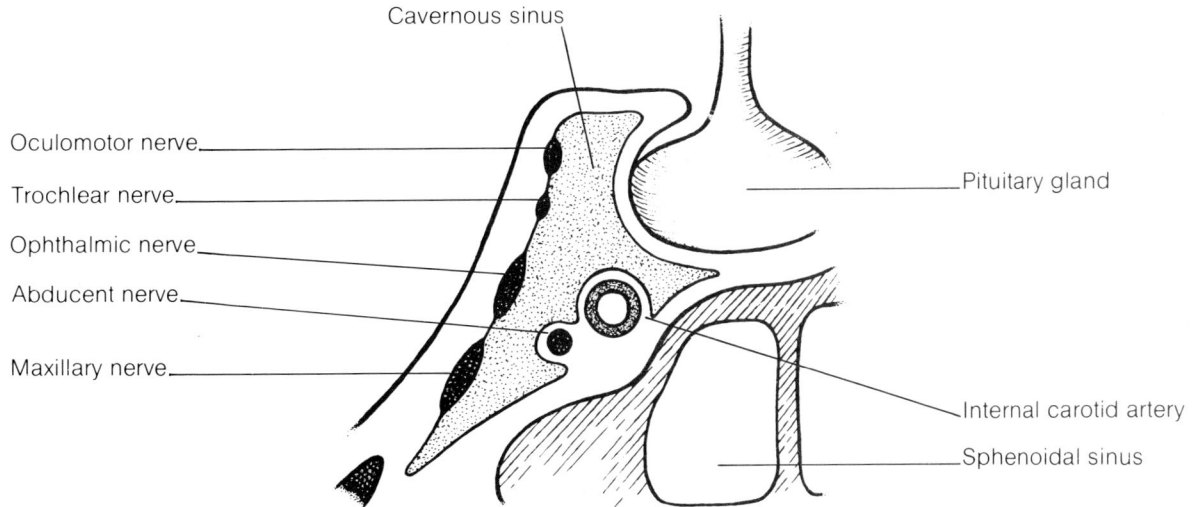

Figure 158 *Relationships of the pituitary gland.*

through the central aperture. The gland can be subdivided into an anterior lobe and a posterior lobe. This subdivision indicates different embryological and functional characteristics. The anterior lobe develops from Rathke's pouch, whereas the posterior lobe develops as a downgrowth of the brain (see page 426). The anterior lobe can also be referred to as the adenohypophysis. Although some endocrinologists regard the terms posterior lobe and neurohypophysis as being synonymous, strictly speaking the neurohypophysis includes the posterior lobe, the pituitary stalk, and the median eminence of the tuber cinereum at the base of the brain.

The hormones of the anterior lobe of the pituitary gland are controlled by hormone-releasing factors from the hypothalamus which pass to the gland by means of a hypophyseoportal venous system. The hormones of the posterior lobe are transported in precursor form directly from the hypothalamus by axons within the pituitary stalk. These axons arise from cells of the supra-optic and paraventricular nuclei in the hypothalamus. Accessory or ectopic pituitary tissue can be found between the pituitary fossa and the nasopharynx. This reflects the embryonic path of Rathke's pouch.

The pituitary gland has a number of important relationships (Figure 158). Below are the sphenoidal air sinuses in the body of the sphenoid bone. The optic chiasma is situated above the pituitary gland and in front of the pituitary stalk. However, in only a small percentage of individuals does the chiasma lie in the chiasmatic groove between the optic canals. Lying lateral to the pituitary gland are the cavernous sinuses and their contents. Indeed, the internal carotid arteries within the sinuses often lie immediately adjacent to the pituitary gland.

The blood supply of the pituitary gland (Figure 157) is derived from inferior and superior hypophyseal arteries which originate from the internal carotid arteries. The inferior hypophyseal artery comes from the internal carotid as it passes through the cavernous sinus. The superior hypophyseal artery may arise as a series of arteries that appear either directly from the internal carotid artery as it emerges from the cavernous sinus or from the anterior and posterior cerebral arteries. The inferior hypophyseal artery divides into lateral and medial branches. These anastomose with their fellows of the opposite side to form a complete arterial ring around the posterior lobe of the pituitary gland. The superior hypophyseal arteries are arranged into anterior and posterior branches which pass to the region of the pituitary stalk. Trabecular arteries are branches of the superior hypophyseal arteries which pass down to the posterior lobe. Thus, the posterior lobe and the pituitary stalk receive

158,49
158,58

168,33
168,6

their blood supply directly from the hypophyseal arteries. However, the anterior lobe of the pituitary gland has no arterial branches supplying it and receives its blood from a hypophyseoportal system of veins. The portal system begins around the pituitary stalk as a series of long and short portal vessels. These veins pass into vascular sinusoids between clumps of secretory cells in the anterior lobe. A region of the pituitary gland between the anterior and posterior lobes called the pars intermedia appears to be avascular. From the hypophyseoportal system, blood drains into inferior hypophyseal veins. These veins drain into the dural venous sinuses. The venous drainage of the posterior lobe is by three routes: directly into superior hypophyseal veins, indirectly via the hypophyseoportal system, and into veins in the hypothalamic region. The hypophyseoportal system is important functionally because it is the route by which hormone-releasing factors pass from the hypothalamus to the anterior lobe of the pituitary gland.

180,36; 180,30

The innervation of the pituitary gland can be subdivided into a constituent intrinsic innervation and a vasomotor innervation. The constituent innervation is composed of the numerous nerve fibres that pass from the hypothalamus, down the pituitary stalk to the posterior lobe. The vasomotor supply is derived from the sympathetic plexus associated with the internal carotid artery.

The cranial nerves found in the floor of the middle cranial fossa are the optic, oculomotor, trochlear, trigeminal and abducent nerves. Other nerves passing across the floor of this fossa are the greater and lesser petrosal nerves, meningeal branches of the trigeminal nerve and sympathetic nerves.

168,4
28,19; 158,58

The optic nerve is the second cranial nerve. It is seen just above the internal carotid artery, entering the optic canal. The nerve is connected to the optic chiasma in the central part of the middle cranial fossa. On entering the cranial cavity, the meningeal sheaths of the optic nerve become continuous with the cranial meninges. Although it is usually stated that the optic chiasma lies within the chiasmatic groove of the sphenoid bone, this appears to occur in very few cases. Furthermore, in most instances the chiasma lies above the diaphragma sellae, which roofs the central part of the fossa. At the optic chiasma, the nerve fibres are so arranged that those associated with the nasal parts of the retinae of the eyeballs decussate (see pages 590 to 591). From the optic chiasma two optic tracts appear, but soon merge into the substance of the base of the brain.

28,12

184,7; 188A,4

188A,6

168,7; 168,8
168,13; 166B,C

The oculomotor, trochlear and abducent nerves pass through the cavernous sinus towards the superior orbital fissure. The oculomotor nerve is the third cranial nerve. It enters the cavernous sinus in the area where the free and attached margins of the tentorium cerebelli cross over. Immediately behind the oculomotor nerve lies the very fine root of the trochlear nerve. This is the fourth cranial nerve. The oculomotor and trochlear nerves emerge from the midbrain. The abducent nerve is the sixth cranial nerve. It has a long course, arising from the brain stem in the posterior cranial fossa (see below) and then in the cavernous sinus in the middle cranial fossa. The abducent nerve is situated inside the cavernous sinus with the internal carotid artery.

186,10; 186,11
184,19

168,14; 166B,33

The trigeminal nerve is the fifth cranial nerve. It appears on the floor of the middle cranial fossa, first as nerve roots crossing the petrous ridge of the temporal bone, then as a trigeminal ganglion, and finally as three major divisions (the ophthalmic, maxillary and mandibular divisions). The trigeminal ganglion is situated in a recess of the dura (the trigeminal cave) close to the apex of the petrous part of the temporal bone. When the dura is removed, the ophthalmic, maxillary and mandibular nerves may be identified. The ophthalmic and maxillary nerves pass through the lateral wall of the cavernous sinus (Figure 158). The mandibular nerve passes directly to the foramen ovale, through which it leaves the skull to reach the infratemporal fossa. The ophthalmic nerve passes into the orbit through the superior orbital fissure, but before doing so divides into lacrimal, frontal and nasociliary branches in the middle cranial fossa. The maxillary nerve leaves the middle cranial fossa through the foramen

166B,34; 166B,40
166B,42; 166B,47
168,9; 28,45

28,48
29B,51

rotundum and eventually emerges into the pterygopalatine fossa. All three divisions of the trigeminal nerve (and the ganglion itself) have meningeal branches. The branches associated with the ophthalmic and maxillary nerves arise within the middle cranial fossa. The meningeal branch from the mandibular nerve is given off in the infratemporal fossa and enters the cranium through the foramen spinosum.

29B,50

28,47

The greater petrosal nerve is a branch of the facial nerve and carries parasympathetic fibres to the pterygopalatine fossa. It emerges from the region of the middle ear on to the floor of the middle cranial fossa at a hiatus, and then grooves the bone as it passes forwards to the foramen lacerum.

166B,54; 168,11

28,22
28,49

The lesser petrosal nerve is a branch of the glossopharyngeal nerve and carries parasympathetic fibres to the otic ganglion. Like the greater petrosal nerve, the lesser petrosal nerve arises in the middle ear and passes on to the floor of the middle cranial fossa through a hiatus in the petrous part of the temporal bone. The nerve lies in a groove located lateral to that of the greater petrosal nerve and runs towards the foramen ovale.

166B,53

28,23
28,48

The venous sinuses within the middle cranial fossa are the cavernous, intercavernous, sphenoparietal and superior petrosal. The internal carotid artery is seen by the side of the body of the sphenoid bone, associated with the cavernous sinus. Just behind the optic canal, it gives off the ophthalmic artery. The meningeal arteries are the middle meningeal and the accessory meningeal branches from the maxillary artery, and branches from the ascending pharyngeal, ophthalmic and lacrimal arteries.

168,33
168,34; 168,12
168,6
168,5
168,10

THE CONTENTS OF THE POSTERIOR CRANIAL FOSSA

This fossa contains the lowest part of the midbrain and the pons, cerebellum and medulla oblongata. Located in its floor are the roots of the trigeminal nerve and the abducent, facial, vestibulocochlear, glossopharyngeal, vagus, accessory (both cranial and spinal parts) and hypoglossal nerves. Also found are the meningeal branches of the upper cervical spinal nerves.

158

The roots of the trigeminal nerve are seen passing between the trigeminal cave at the back of the middle cranial fossa and the lateral side of the pons. They cross the petrous part of the temporal bone close to the apex.

168,14

The root of the abducent nerve is seen lying medial to and slightly below the roots of the trigeminal nerve, on the clivus of the posterior cranial fossa. It penetrates the dura mater between the apex of the petrous part of the temporal bone and the dorsum sellae to reach the cavernous sinus.

168,13

Both the trigeminal and abducent nerves run much of their intracranial course in the middle cranial fossa and have already been described.

The facial and vestibulocochlear nerves are respectively the seventh and eighth cranial nerves. They both pass into the opening of the internal acoustic meatus, lying below and lateral to the roots of the trigeminal nerve.

168,15; 168,16
28,43; 198F,44

The glossopharyngeal, vagus and accessory nerves are the ninth, tenth and eleventh cranial nerves. All pass into the jugular foramen which lies just below the internal acoustic meatus. The spinal accessory nerve is seen passing up from the cervical region of the spinal cord, through the foramen magnum, to join the cranial part of the accessory nerve shortly before the jugular foramen.

168,18; 198F,45
28,31
168,19; 198F,47

The hypoglossal nerve is the twelfth cranial nerve and is situated below the jugular foramen and behind the vertebral artery. It leaves the posterior cranial fossa through the hypoglossal canal.

168,20

28,41

The meningeal branches from the upper three cervical nerves pass through the foramen magnum to supply dura in the posterior cranial fossa.

168,21; 168,17
168,28; 168,23
168,22; 168,12
168,25; 168,26

The dural venous sinuses in the posterior cranial fossa are: the sigmoid, inferior petrosal, basilar, occipital and straight sinuses, with the transverse and superior petrosal sinuses in the attached margins of the tentorium cerebelli. The vertebral arteries pass up through the foramen magnum and unite to form the basilar artery on the clivus. Meningeal branches from the ascending pharyngeal and occipital arteries are also present.

The central nervous system

The morphology and characteristics of nervous tissue

The nervous system is the information-handling and executive system of the body. Its integrity and optimal operation are vital to the wellbeing of the organism. It is essentially composed of neurones and their supportive cells. Neurones are electrically excitable cells specialised for the reception, integration and transmission of information. The supportive cells have a variety of forms, but they are collectively known as the neuroglia. The nervous system contains little connective tissue, but it has an extensive network of blood vessels.

NEURONES

All pathways carrying information in the nervous system involve simple or complex chains of neurones. Such information must be transmitted through neurones (intracellular transmission) and between neurones (intercellular transmission). The information is carried by transient electrochemical changes affecting the plasma membranes of neurones. Accordingly, the plasma membranes and the general morphology of neurones demonstrate features facilitating the flow of such information. Neurones presumably also have specialisations allowing for the storage of information, but these are as yet poorly understood.

The typical neurone (Figure 159) consists of a cell body (the soma) from which extend several fine cytoplasmic processes (the neurites). One of these processes conveys information away from the soma and is known as the axon. The processes that convey information to the soma are called dendrites (or dendrons). The dendrites normally branch profusely and are responsible for the large surface area and wide diversity of shapes displayed by neurones.

Neurones are classified as being multipolar, bipolar and unipolar (Figure 160). Most neurones are multipolar, having multiple neurites. The shape of the soma and

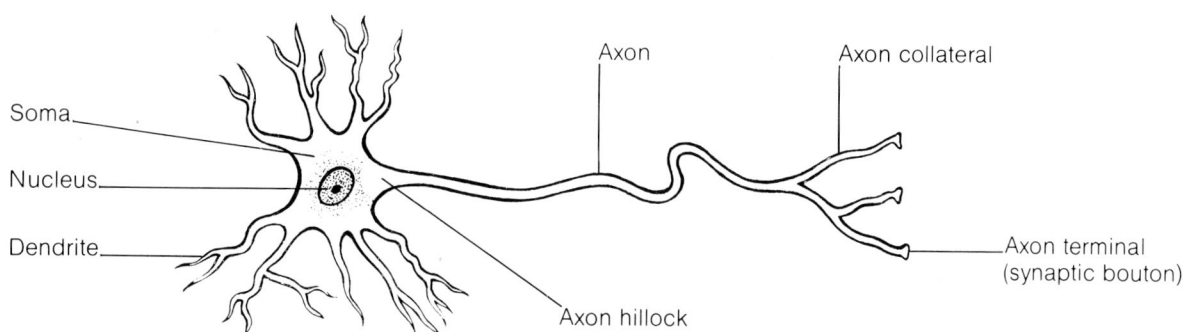

Figure 159 *A typical neurone.*

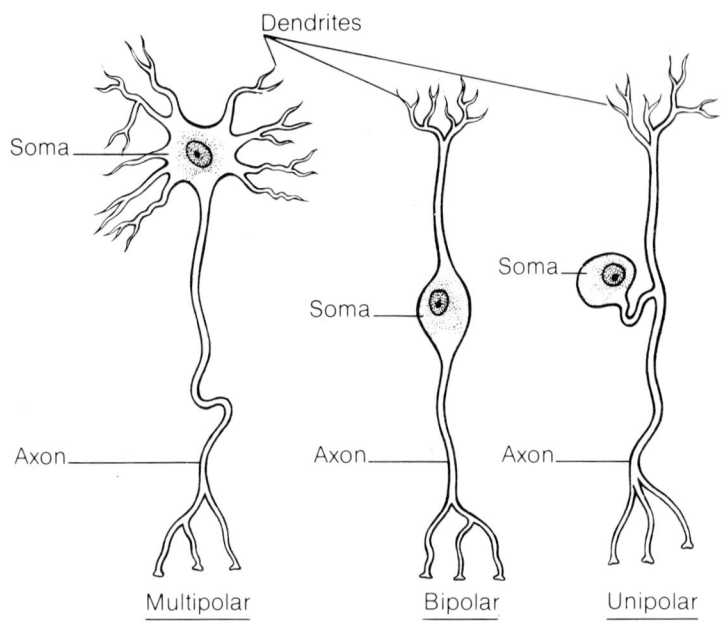

Figure 160 *The classification of neurones* according to the number of their neurites.

dendrites gives rise in many cases to their names (e.g. pyramidal, stellate, granular, fusiform or spindle-shaped). Bipolar cells have an axon and one dendrite, which emerge at opposite ends of the soma. These cells occur as the primary sensory neurones of the special sensory pathways: in the olfactory mucosa, in the retina and in the vestibulocochlear nerve. Unipolar (pseudo-unipolar) cells have a spherical body with a single neurite that bifurcates, giving rise to a central (axonal) process and a peripheral (specialised dendritic) process. These cells are found in ganglia of spinal and cranial nerves. Some neurones have dendrites but no axon (e.g. amacrine cells in the retina and granule cells in the olfactory bulb).

The soma is the site of the nucleus and is therefore the metabolic and biosynthetic centre of the neurone. Virtually all protein synthesis occurs in the soma. The cytoplasm around the nucleus is the perikaryon (although the term perikaryon is commonly used synonymously with soma). Neuronal somas vary in size, ranging from less than 5 μm to more than 120 μm across. The dimensions vary with the extent of the neurites and the metabolic requirements of the cell. Any neurite cut off from its soma undergoes degeneration.

The dendrites provide the major receiving or receptor surface of a neurone. They are usually less than 1 mm long. Most of the specialised contacts through which neurones communicate with other neurones (i.e. synapses) are found on dendrites. The possession of a large number of synapses allows the influence of many neurones to converge on a single cell. Dendrites often show small excrescences called dendritic spines (gemmules). These are typically about 2 μm long and are a common site of synapses.

The axon is the major outflow or effector portion of a neurone. It provides the means by which neurones transmit information over long distances. Axons vary from less than 1 mm to more than 1 m in length. The cytoplasm of an axon is called the axoplasm; its plasma membrane is called the axolemma. The protrusion of the soma leading to the axon is known as the axon hillock. The nearest part of the axon to the soma is called its initial segment. The branches of axons are known as their collaterals. Distally, axons usually divide into many fine terminal branches (telodendria) which end in apposition to other neurones (or sometimes muscular or glandular cells). Their terminations are usually bulbous, and are known as synaptic boutons or knobs. The possession of many axon terminals allows the influence of one neurone to diverge to many other cells.

Substances are transported through the axoplasm to and from the soma by a process called axoplasmic transport. A similar mechanism occurs in dendrites. Transport from the soma to the axon terminal is referred to as being anterograde, transport in the reverse direction being retrograde. A considerable variety of substances is transported (e.g. mitochondria, proteins, peptides, neurotransmitters), the rate of travel varying from less than 1 mm to more than 400 mm per day. Rates above 100 mm per day are considered fast, below 10 mm per day slow. Essential trophic substances pass in both anterograde and retrograde directions. If an axon with all its collaterals is severed close to the soma, both the axon and the remainder of the neurone will degenerate. Changes may also be seen in the neurones or other cells upon which the cut axon terminated. Viruses (e.g. that which causes rabies) and toxic substances (e.g. tetanus toxin) can also be transported. Axoplasmic transport of substances that can be localised and stained has provided a valuable means of tracing neuronal connections.

Myelin consists of multiple layers of the plasma membrane of certain neuroglial cells that have been tightly wound round an axon (Figure 161a). Axons possessing such a sheath are described as myelinated (or medullated); axons lacking this sheath are referred to as unmyelinated. In the brain and spinal cord, myelin sheaths are provided by neuroglia called oligodendrocytes. In the cranial and spinal nerves, they are provided by Schwann cells (sometimes called neurolemma cells).

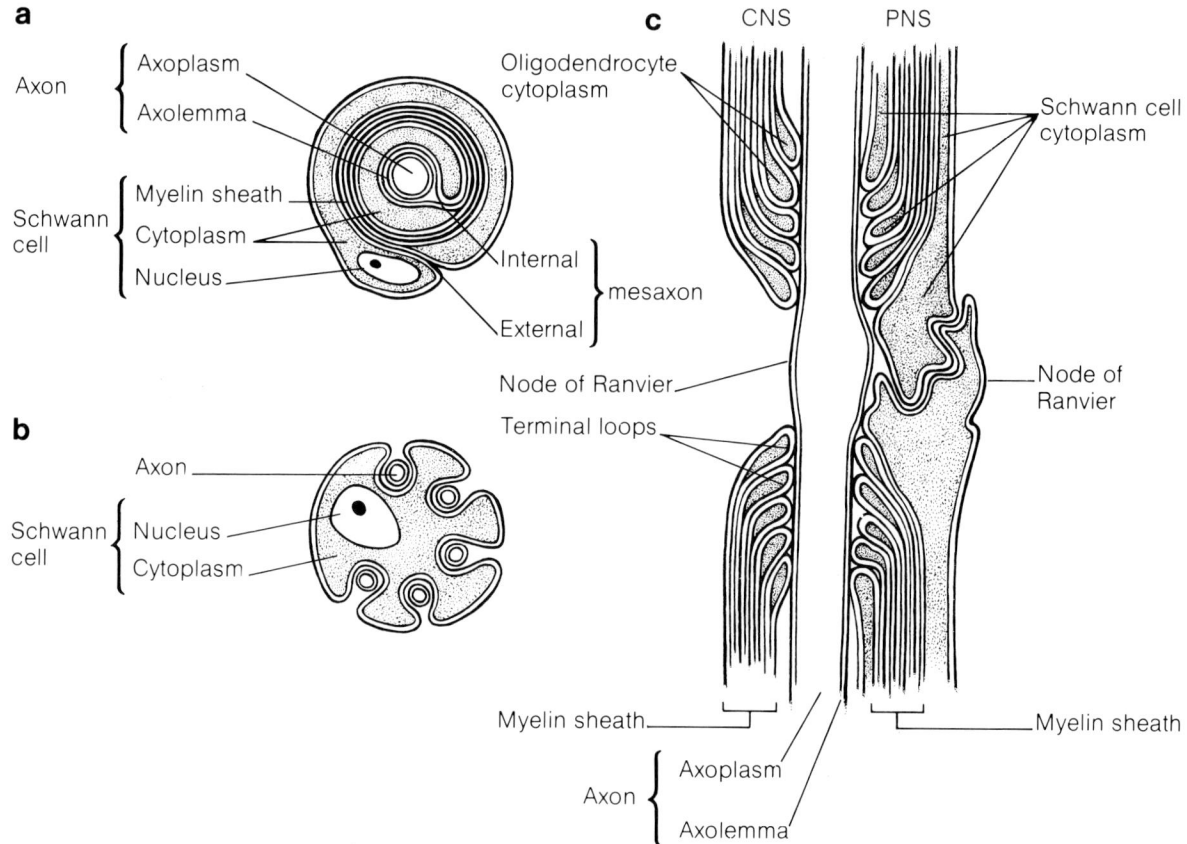

Figure 161 *Axonal myelination.* a) *An axon and its myelin sheath.* b) *Unmyelinated axons in relation to a Schwann cell.* c) *A node of Ranvier in the central nervous system (CNS), and in the peripheral nervous system (PNS).*

A myelin sheath is formed by a tongue of an oligodendrocyte or a Schwann cell winding itself around an axon to form a spiral in which the leading edge of the tongue is innermost. As the spiral tightens, the neuroglial cytoplasm is squeezed out, allowing the internal surfaces of the plasma membrane to become apposed. A double layer of plasma membrane is thus added to the sheath for each rotation. Interstitial fluid is also gradually excluded, resulting in the external surfaces of the plasma membrane of adjacent layers becoming closely apposed. The neuroglial cell nucleus and most of its cytoplasm remain peripheral. The thickness of a myelin sheath may vary from a few to over 100 double layers. Incisures (or clefts of Schmidt–Lantermann) are places where not all the cytoplasm of a Schwann cell has been squeezed out of the myelin spiral; they have not been observed for oligodendrocytes.

One oligodendrocyte may provide myelin sheaths for dozens of axons. A Schwann cell provides a sheath for only a single axon. Along its length, an axon will be ensheathed by a succession of neuroglia. Both the length of axon myelinated by a single neuroglial cell, and the thickness of its myelin sheath, are roughly proportional to the axon's diameter. Axons under 1 μm in diameter are rarely myelinated, whereas those over 2 μm nearly always are. The initial and terminal portions of axons are usually unmyelinated.

The induction, eventual size and continued presence of a myelin sheath appears to be determined by the neurone. A sheath breaks down once its axon has degenerated.

The unmyelinated axons of peripheral nerves are still associated with Schwann cells, but the axons lie in simple invaginations (Figure 161b). There may be 20 or more such axons associated with a single Schwann cell, with one axon per invagination. In a few instances (most notably the olfactory nerve), multiple axons occur in each invagination. Unmyelinated axons in the central nervous system are also invaginated in neuroglia.

Along the axon, between adjacent ensheathing neuroglial cells, are gaps in the myelin called nodes of Ranvier (Figure 161c). These nodes are spaced at intervals (internodal distances) which, for different axons, vary from about 200 μm to more than 1 mm. Both the gaps at the nodes and the internodal distances tend to be shorter for oligodendrocytes than for Schwann cells. Furthermore, the nodes between oligodendrocytes are freely exposed to the interstitial fluid. In peripheral nerves, cytoplasmic processes of adjacent Schwann cells interdigitate across the nodes.

The nodes of Ranvier are where axons branch. More importantly, they are the basis of a specialised, rapid form of discontinuous axonal conduction known as saltatory conduction (see page 408). Such conduction is made possible because, between the nodes, the myelin sheath greatly increases the effective insulation of the axon and restricts the diffusion and movement of tissue fluid in the immediate vicinity of the axolemma. For fibres with a diameter of more than 2 μm, saltatory conduction is more energy-efficient. More significantly, the discontinuous conduction along the axon is much faster than the continuous conduction that would occur in the absence of myelin. For small-diameter fibres, myelination does not confer a net advantage. Therefore, small-diameter fibres are unmyelinated and display continuous conduction.

Myelination begins in the fetus. The process continues rapidly postnatally, but is not completed for some axonal pathways until early adulthood. This is especially so for phylogenetically recent pathways (e.g. the corticospinal pathway). Although myelin is constantly renewed and replaced throughout life, malnutrition in infancy may result in permanently deficient myelination. Demyelination seriously impairs axonal conduction. This is the basis of the neurological problems of demyelinating diseases such as multiple sclerosis. In this condition, oligodendrocytes are destroyed and inadequately replaced in scattered regions throughout the brain and spinal cord where there are concentrations of myelinated fibres (i.e. white matter).

Neuronal transmission — intracellular transmission (Figure 162)

Neurones transmit information by means of the release of chemicals, and by both slow (graded) and fast (all-or-none) electrical signals. Most intracellular transmission is electrical, whereas most intercellular transmission is chemical.

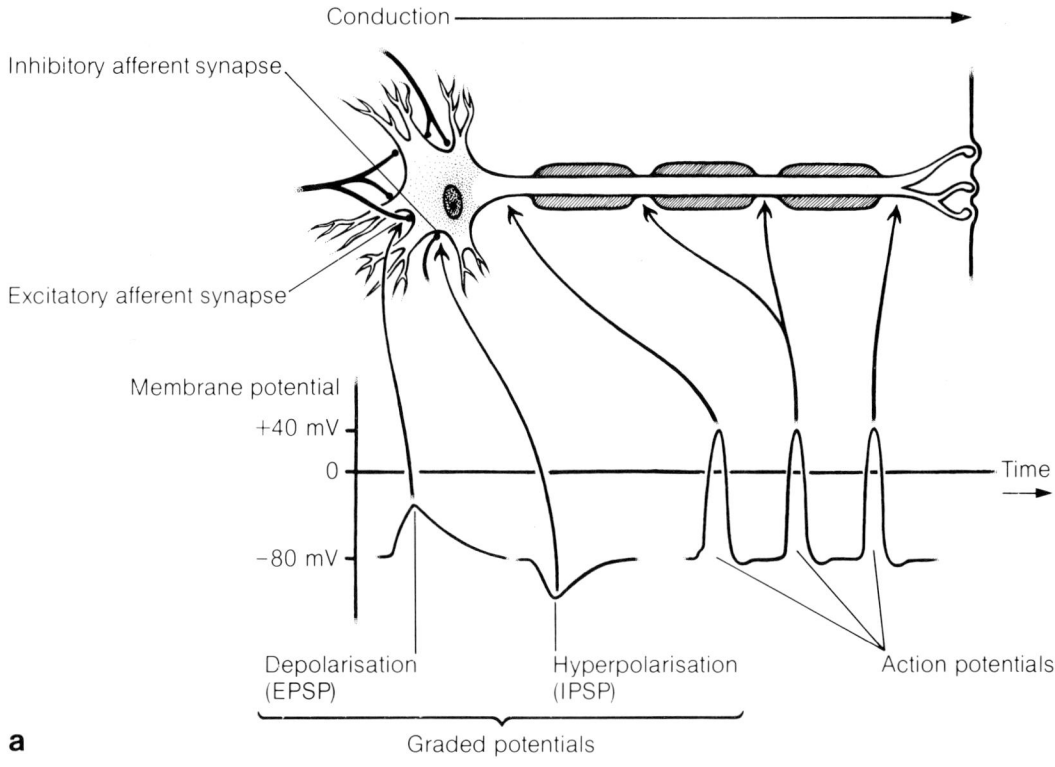
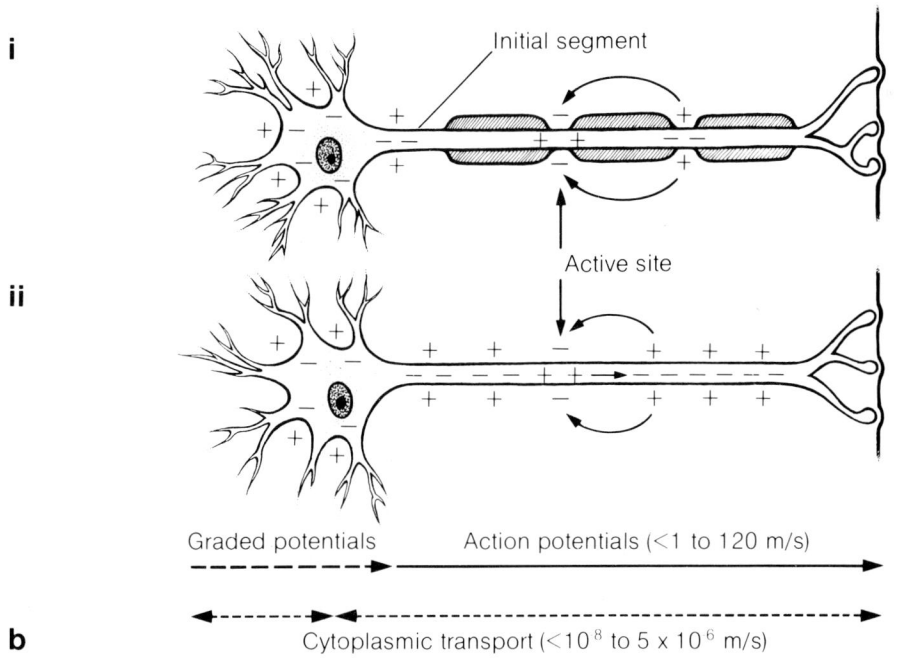

Figure 162 *Intracellular transmission of information.* a) *Membrane potential changes during information transmission.* (EPSP = excitatory postsynaptic potential. IPSP = inhibitory postsynaptic potential.) b) *Ionic changes during action potential propagation in* i) *a myelinated axon, and* ii) *an unmyelinated axon. The active site is where the membrane is briefly depolarised—see action potentials in Figure 162a.*

Neurones, in common with most cells, maintain their cytoplasm at an electrical potential about 80 mV below that of the surrounding interstitial fluid. This potential difference across the plasma membrane (called the resting potential) is maintained by cellular metabolism and by the properties of the membrane. It is produced by the selective permeability and active transport of substances through the plasma membrane (particularly of sodium and potassium ions by the sodium pump). The result is that, compared to the interstitial fluid, the cytoplasm has high concentrations of potassium and macromolecular ions, and low concentrations of sodium, chloride and calcium ions.

Chemicals released at synapses (called neurotransmitters) by one neurone bind to receptors in the plasma membrane of the other neurone participating in the synapse, causing local, selective changes in the ionic permeability of the membrane (see page 410). The effect of the altered influx or efflux of ions is to change the potential difference across the plasma membrane in the region of the synapse (i.e. to generate a postsynaptic potential). If the potential difference is increased, the cell is said to be hyperpolarised and is made less excitable. If it is decreased, the action is one of depolarisation. Such electrical changes typically last for several milliseconds. They may summate algebraically with similar potentials generated in other regions of the membrane (spatial summation) and/or with preceding or succeeding potentials (temporal summation). Summated potentials vary over a continuous range of sizes and durations, and hence are essentially analogue in form. They are known as graded or slow potentials. Importantly, they allow the integration of the inputs from different synapses.

When the summated potential exceeds a threshold level of depolarisation, further ion channels are opened in electrically excitable parts of the plasma membrane. These channels allow the rapid influx of sodium ions, followed by an outflow of potassium ions into the interstitial fluid. The result is rapid reversal in the polarity of the potential difference across the membrane, followed by its recovery to the normal resting potential (usually within a millisecond or so). Such an event is called an action potential (spike potential or impulse). For most purposes, it may be considered an all-or-none event because, once the threshold depolarisation is exceeded, the action potential generated is of virtually the same amplitude and duration on each occasion. Accordingly, the information transmitted by neuronal signals coded as action potentials is carried in a digital form; the times of occurrence of the impulses (interspike intervals or frequency) encode the message.

Unlike subthreshold graded potentials, an action potential generates sufficient depolarisation in adjacent regions of the membrane to become self-propagating. Thus, once initiated, an action potential propagates without decrement and with an amplitude, duration and speed determined by the membrane properties of the cell and the size of its processes. Because the plasma membrane becomes refractory (inexcitable) while the outflow of potassium ions associated with the action potential continues, the production of one action potential is prevented from automatically generating others. The refractory period (usually about 1 ms) also limits the maximum frequency at which information may be transmitted to less than 1 kHz.

It is normal for axons to conduct action potentials, although not all do so. Some large dendrites also conduct action potentials, but most dendrites probably do not. Conduction over short distances (less than 1 mm), notably by dendrites, is usually by graded (slow) potentials. Conduction over long distances is by axons and by action potentials. Action potentials are generated at a trigger zone. This is usually at the axonal hillock or initial segment of the axon, but such zones may occur in dendrites. Propagation of an action potential away from the cell body is known as orthodromic conduction. Conduction in the opposite direction is called antidromic conduction, but is not found under normal conditions.

In myelinated fibres, action potential conduction occurs in a discontinuous form (saltatory conduction) in which the action potential jumps (or saltates) from one node of Ranvier to the next. The increased insulation of the axon provided by the myelin sheath, and the restricted diffusion of interstitial fluid, prevent action potential generation at the internodes. However, the conditions allow an action potential at one node to generate sufficient current to cause depolarisation above the threshold for action potential production at the next node. Action potentials are thus propagated along the axon from node to node, rather than continously along the axon as occurs in unmyelinated axons. Conduction velocities in large myelinated fibres may reach 120 m/s; in unmyelinated fibres they are less than 4 m/s. For large myelinated fibres (greater than 2 μm), saltatory conduction is faster than conduction would be in an unmyelinated axon with a diameter equal to that of the myelin sheath. Myelination of large fibres is therefore economical of space, as well as producing savings in energy requirements. These advantages do not hold for small fibres, which accordingly remain unmyelinated.

Action potential propagation can be blocked by chemicals that interfere with the passage of ions across the membrane. Local anaesthetics are examples of such chemicals. Another example is tetrodotoxin (a poison found in tissues of the puffer fish), a substance that prevents action potential generation by blocking the passage of sodium ions across neuronal membranes.

Neuronal transmission — intercellular transmission (Figure 163)

Intercellular communication in the nervous system takes place at specialised regions of cellular apposition called synapses. Signals pass across the synapse from what is known as the presynaptic cell to the postsynaptic cell. Depending on the means by which signals are passed, synapses are classified as either chemical or electrical.

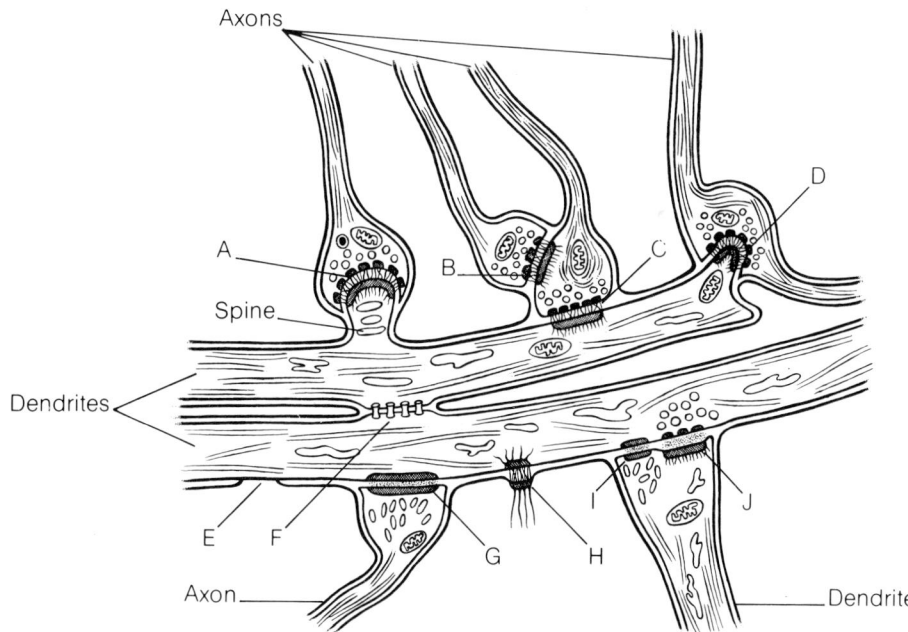

Figure 163 *Intercellular transmission of information: types of synapses.* A = Axospinous or axodendritic synapse with round synaptic vesicles and asymmetric pre- and post-synaptic membrane densities typical of an excitatory synapse. B = Axo-axonal excitatory synapse causing presynaptic inhibition of the effects of synapse C. C = Axodendritic excitatory synapse between the terminal bouton of an axon and a dendritic shaft. D = Axodendritic excitatory synapse formed 'en passage' by the non-terminal part of an axon. E = Non-synaptic tight junction (zonula occludens). F = Gap junction (communicating junction) or electrical synapse. G = Axodendritic synapse with flattened vesicles and symmetric pre- and post-synaptic membrane densities typical of an inhibitory synapse. H = Non-synaptic desmosome (macula adherens). I, J = Reciprocal dendrodendritic synapse: I = inhibitory to the horizontal dendrite; J = excitatory to the vertical dendrite.

Most synapses occur between neurones, although they are also found between neurones and other types of cell (e.g. muscle and glandular cells). Most neurones receive, and give rise to, well over 1,000 synapses; the number varies widely, some of the largest cells receiving more than 200,000 synapses.

Synapses take a variety of forms. The most common is between an axon and a dendrite (axodendritic synapses), but they may also be found between an axon and either a soma (axosomatic synapses) or another axon (axo-axonic synapses). Synapses between an axon and a dendritic spine are sometimes called axospinous synapses. Where a single branch of an axon makes repeated synapses along its length, it is said to form 'synapses en passage'; only the last is the terminal synapse. In addition, synapses may form between dendrites (dendrodendritic synapses) or in other combinations (e.g. dendrosomatic, somatosomatic synapses). Dendrodendritic synapses may set up local circuits between adjacent neurones, bypassing the somas and axons. Some of these synapses are reciprocal, indicating that an area of the membrane allows transmission from one cell to a second, with an adjacent area allowing transmission from the second to the first. Sometimes, synapses between more than two neurites occur in close proximity, so allowing complex interactions between the various participating neurones. In certain parts of the brain (e.g. the cerebellum and the olfactory bulb), there are small regions where a large number of neurites participate in complex synaptic interactions. Each region is isolated from the surrounding tissue by a neuroglial sheath. Such a structure is known as a synaptic glomerulus.

Most synapses in mammals are chemical, the effect of the presynaptic neurone on the postsynaptic neurone being mediated by a chemical transmitter substance (neurotransmitter). This type of synapse is usually 1 or 2 μm in length (sometimes considerably larger), has a narrow synaptic cleft (10 to 30 nm), and shows specialisations of the plasma membranes of both pre- and post-synaptic cells.

Close to the plasma membrane, in the cytoplasm of the presynaptic ending, are small membrane-limited packets called synaptic vesicles. These vesicles are usually between 20 and 200 nm in diameter, and store neurotransmitter. Electrical signals (either local graded potentials or action potentials) reaching the plasma membrane of the presynaptic ending cause the release (exocytosis) of the contents of the synaptic vesicles. For any one vesicle, the release is all-or-none, so that the content of a vesicle is known as a quantum of neurotransmitter (typically in the region of a few thousand molecules). Many vesicles will normally be released during neurotransmission, their number being related to the magnitude of the presynaptic electrical signal and to the resultant concentration of intracellular calcium ions. The neurotransmitter diffuses across the narrow synaptic cleft between the two cells and binds to specific molecular receptors in the postsynaptic membrane. Such binding affects the permeability of the postsynaptic membrane to certain ions and so changes the potential difference across the membrane. Thus, a postsynaptic electrical signal (the postsynaptic potential) is generated. In certain cases, complex chains of biochemical events may also be initiated in the postsynaptic cell. This system ensures that chemical synapses essentially transfer signals in one direction only. Such signals are subject to a synaptic delay (typically of a fraction of a millisecond).

Neurones (and sometimes surrounding neuroglia) have techniques for the recovery (uptake) and/or inactivation (degradation) of transmitter substances, so enabling a synapse to be rapidly and repeatedly used. In some cases, neurotransmitters may have receptors on the presynaptic terminal as well as on the postsynaptic neurone. Such receptors are called autoreceptors and allow control of further release by the neurotransmitter itself.

The electrical effects produced by the neurotransmitter in the postsynaptic neurone may be to decrease the potential difference across its membrane (i.e. to depolarise or excite) or to increase it (i.e. to hyperpolarise or inhibit). The resultant excitation or inhibition depends on the transmitter substance itself, on the particular receptor at which a given transmitter acts, and on the ionic channel controlled by that receptor.

TABLE 9 Neurotransmitters. The table should be read as an introductory guide to neurotransmitters, as there is no definitive list in a field that is undergoing rapid development. Similarly, the listed receptor types are not exhaustive. It should be appreciated that neurones may co-release neurotransmitters whose effects may interact. Effects may also be presynaptic (autoregulatory) as well as postsynaptic, or modulatory rather than simply depolarising or hyperpolarising. Neurones releasing acetylcholine are termed cholinergic, those releasing adrenaline are adrenergic, etc.

NEUROTRANSMITTER (and abbreviation)	COMMON RECEPTOR TYPES	NEUROTRANSMITTER (and abbreviation)
Acetylcholine (ACh)	Nicotinic, muscarinic ($M_{1,2}$)	**Adenohypophyseal-related peptides**
Monoamines		Gonadotropin releasing hormone (GnRH) or luteinizing hormone releasing hormone (LHRH)
Adrenaline (A) or epinephrine (E)	$\alpha_{1,2}$; $\beta_{1,2}$	
Noradrenaline (NA) or norepinephrine (NE)		Corticotropin releasing hormone (CRH)
Dopamine (DA)	$D_{1,2,3}$	Somatostatin (SS)
5-Hydroxytryptamine (5-HT) or serotonin	$S_{1,2,3}$	Thyrotropin releasing hormone (TRH)
Histamine	$H_{1,2}$	**Tachykinins**
Amino acids		Neuromedin K (NK)
Glycine (Gly)		Substance K (SK)
γ-Amino-butyric acid (GABA)	$GABA_{A,B}$	Substance P (SP)
Glutamate (Glu)	NMDA, kainate, quisqualate	**Other peptides**
Aspartate (Asp)		Angiotensin II (AII)
Opioid peptides		Bombesin
Methionine-enkephalin (Met-Enk)		Cholecystokinin (CCK)
Leucine-enkephalin (Leu-Enk)	κ, μ, δ	Neuropeptide Y (NY)
β-Endorphin (β-End)		Vasoactive intestinal peptide (VIP)
Dynorphin (Dyn)		**Purines**
Neurohypophyseal peptides		Adenosine triphosphate (ATP)
Oxytocin (Oxy)		Guanosine triphosphate (GTP)
Vasopressin (VP) or antidiuretic hormone (ADH)	$V_{1,2}$	

Furthermore, axons terminating on the synaptic boutons of other neurones may inhibit the release of neurotransmitter at that bouton by causing its depolarisation. This presynaptic inhibition is thus caused by an excitatory neurotransmitter/receptor combination. Neurones may release more than one transmitter substance, but are believed to release the same substance(s) at all their terminals. As indicated above, the action of the neurotransmitters on the postsynaptic elements need not be the same for all its terminals. Many substances have been identified as probable neurotransmitters (Table 9). Some of these chemicals (particularly certain peptides, e.g. β-endorphin) seem to have a more diffuse (less specific) action, with a slower onset and longer-lasting effects than that expected for a classical neurotransmitter. Indeed, such substances may act primarily by potentiating or diminishing actions of other transmitters (neuromodulation), or via more generalised, not even necessarily synaptic, effects (neurohumoral actions). Thus, neurotransmitter actions may be complex and subject to modulation, making possible a further level of neuronal interaction and integration.

Chemical neurotransmission in many instances may be manipulated pharmacologically by substances that mimic, potentiate, antagonise or block a neurotransmitter's action. Such pharmacological agents may affect neurotransmitter synthesis, storage, release, or its inactivation, or may bind to its receptors. To mention two of the many examples, poisons such as curare (which blocks the receptors) and botulinus toxin (which prevents release) prevent acetylcholine-mediated transmission from neurone to muscle. Disease processes may also affect such neuromuscular transmission; the muscular weakness of myasthenia gravis is a result of the loss of acetylcholine receptors in muscle. The actions of tranquillisers, antidepressants, and many psychoactive drugs (such as heroin and LSD) are produced through their effects at central synapses (particularly those at which monoamine or opioid neurotransmitters are used).

Recently, research has shown that electrical synapses (gap or communicating junctions, with a separation of 2 nm) exist between either the dendrites or somas of many contiguous neurones. These gap junctions are like those found in cardiac and smooth muscle. Such junctions place cells in direct ionic communication with each other, allowing electrotonic coupling between neurones. They may be bidirectional and are rapid, being without synaptic delay as no neurotransmitter diffusion is involved. Electrical synapses may be important where synchronous activation of a group of neurones is required (e.g. for neurones within the medulla oblongata that control respiration).

Neuronal dependence

The neurone doctrine as originally developed held that the individual nerve cell was the genetic, anatomical, trophic and functional unit of the nervous system. The lack of cytoplasmic continuity at synapses and trophic independence of neurones were emphasised. However, it is now clear that this view can no longer be upheld and that the viability of a neurone is strongly dependent upon its contacts with other neurones (as well as upon neuroglia). Three pieces of evidence may be cited to support this view. First, damage in one region may cause anterograde or retrograde changes that pass transneuronally (i.e. that affect cells that are postsynaptic or presynaptic to the damaged neurone). Second, substances are transported in both directions through the cytoplasm of axons and dendrites, and it has been demonstrated that a variety of molecules (in addition to transmitter substances) are trans-synaptically transferred. Indeed, many chemicals probably pass in both directions across all synapses. Third, cytoplasmic continuity has been demonstrated between neurones at gap junctions. Neurones can therefore no longer be considered either as metabolically independent or as unidirectional transmitters of information; their inter-relationships are now known to be considerably more complex.

Neuronal plasticity and recovery following injury

The responses of neurones to injury and disease are of considerable clinical importance. The functional losses that occur shortly after a cerebrovascular accident (i.e. a stroke) may eventually be subject to a remarkable degree of recovery. The stages of recovery may be separated into short-term and long-term responses.

Immediately following major damage, the nervous system typically goes into a state of shock, there being suppression of many functions involving the damaged region or sites receiving its projections. The mechanisms responsible are uncertain, although they probably include: signals from neurones in the damaged or denervated regions, local vascular changes, and other changes as part of a stress response to the injury. This state can last for several days and functional losses cannot be fully assessed until it resolves. Short-term recovery in the following days and weeks probably depends to a large degree on factors such as: a) reduction of oedema and resorption of blood (with the consequent removal of compression blocks on axonal conduction); b) removal of local vasoconstriction and its resultant hypoxic block of neuronal function; and c) repair and restoration of function of cells that have not been severely damaged.

Long-term recovery depends upon the nervous system's adaptive capabilities. One form of adaptation is called vicarious functioning. This involves the training of a different set of neuronal pathways, which previously may have played only a minor role, to substitute for those lost or damaged. Thus, different solutions are sought to a neurological problem. Another form of adaptation is behavioural substitution. This involves the adoption of behavioural strategems to avoid functional loss (e.g. using a different set of muscles to perform an action). However, such adaptability still depends on the plasticity of nervous tissue (i.e. on the modifiability of interneuronal connections).

The changes that follow damage to neurones and their connections are:

Denervation hypersensitivity

Reactive synaptogenesis

Axonal regeneration

Synaptic hypereffectiveness

Denervation hypersensitivity occurs when a neurone loses synapses because of degeneration of other neurones. Such a neurone develops an increased responsiveness to transmitter substances, and in addition probably releases chemicals that encourage re-innervation by other axons (e.g. nerve growth factor released by neurones of the sympathetic nervous system).

Reactive synaptogenesis involves the formation of new functional synapses by undamaged axons. Such axons react by sprouting new collaterals. However, there is variation between systems. For some axons, notably those of cells that use monoamine transmitter substances, this capacity is highly developed, but for many others a given type of collateral sprout will replace only certain types of lost synapses on adjacent neurones. It is noteworthy that such neuronal reorganisation is not necessarily behaviourally beneficial; it is not even necessarily true that the new presynaptic terminals release the same transmitter substance, far less signal the same information as the old.

Regeneration of damaged axons provides an important process for functional recovery in the peripheral nervous system (see page 443), but is very much more restricted in the brain and spinal cord. Sprouting from damaged axons in the central nervous system normally occurs only over very short distances (longer for neurones using monoamine neurotransmitters). Accordingly, regeneration following transection of central tracts is very limited and typically ineffective functionally. It seems likely that differences in glial cells may, at least in part, underly the difference in central and peripheral regenerative capacity. The lack of appropriate nerve growth factors in the adult central nervous system may also provide an explanation. Experiments have shown that when embryonic nervous tissue is transplanted into the brains of adult animals, its neurones readily grow and establish functional connections with the adult tissue.

Synaptic hypereffectiveness occurs when the terminals of any surviving collaterals of damaged axons receive and release more transmitter substance than normal. This is related to the fact that the neurone's supply of transmitter substance is no longer divided between so many synapses.

Partial denervation of a neurone may result in the unmasking of previously functionally ineffectual inputs. Indeed, it has been shown that, following severance of a major input to a neurone, subsidiary inputs greatly increase their influence. This effect appears to result either from removal of inhibition or from hypersensitivity, rather than because of collateral sprouting.

Although plastic changes in the nervous system need not be functionally advantageous in themselves, they may be made beneficially adaptive if harnessed by learning. Indeed, retraining following injury is especially important for the long-term recovery and rehabilitation of patients who have suffered brain damage. There is considerable evidence that adaptive changes following injury are particularly well developed in the young and become less so with age.

The ability to modify neuronal connections in the absence of damage is also well established. One of the most important properties of the nervous system is that of allowing the organism to adapt to experience and the environment. Such neuronal plasticity is particularly noteworthy in fetal and neonatal nervous tissue, although the phenomenon of learning strongly implies that certain systems are plastic throughout life. It appears that in development the composition of the major neuronal cell groups and their gross axonal connections are determined genetically. However, the precision and details of the connections are determined environmentally (i.e. as a result of learning and experience). Environmental influences are strongest in early life. Indeed, it is well established that the effects of abnormal visual experience in childhood, e.g. as a result of a squint (strabismus), produce changes in connections within the visual system with resultant effects on visual perception. These effects can be fully corrected only if action is taken during childhood, the adult seeming no longer to have the necessary neuronal plasticity.

It has been shown that during development many synaptic connections that are not functionally active are lost. In particular, there is loss of the synapses of presynaptic fibres that do not participate in causing activity in the postsynaptic neurones. Indeed, during development there is an overproduction of neurones and connections. The subsequent cell death (and loss of synapses) is considerable and is probably a method of achieving specificity of neuronal connections in the presence of imprecision in the formation of the initial, genetically determined contacts. It is probably also a protective mechanism, as persistence of early hyperinnervation following damage or loss of function in another input is likely to be adaptively beneficial.

There is accumulating evidence that synaptic changes resulting in altered neuronal connectivity underly all learning. It further seems plausible that many synapses may be labile, being dependent on continual usage or on other influences for their maintenance and efficacy. However, the precise nature of such changes, and the conditions necessary for their occurrence, are not yet established.

NEUROGLIA

Neuroglia is a collective term for the cells (glial cells) that provide metabolic, structural and functional support for neurones. Glial cells are usually small and are at least 10 times more numerous than neurones. Unlike neurones, they do not have axons, do not conduct action potentials, and do not form chemical synapses. Glial cells do, however, form gap junctions with other glial cells and with neurones.

In contrast to most neurones, neuroglia retain a capacity for cell division. Indeed, glioblasts (developmental stem cells capable of division and differentiation into macroglial cells) are found in the central nervous system throughout life. In accordance with their proliferative capacity, neuroglia are responsible for more than half the number of all intracranial tumours.

Neuroglia are essential for normal neuronal function:

They regulate the ionic composition of the interstitial fluid, thereby controlling the chemical milieu of neurones (a critical factor for axonal conduction and synaptic transmission).

They provide a route for the passage of nutrients, oxygen and waste products between neurones and the vascular system or the cerebrospinal fluid within the ventricles of the brain.

They form and maintain the myelin sheaths surrounding axons, thereby providing electrical insulation, and enabling rapid (saltatory) conduction of action potentials.

They have a role in combating infection and in reacting to injury — by swelling (so contributing to oedema), by proliferation, and by phagocytosis.

They have specialised functions, particularly in the guidance of neurones and their connections, during development.

They supply metabolites and trophic factors to neurones.

The neuroglia are categorised into two major groups. In the brain and spinal cord the glial cells are referred to as the central neuroglia. In the peripheral nervous system they are known as peripheral neuroglia.

Central neuroglia (Figure 164)

In the central nervous system the main glial cell types are: astrocytes, oligodendrocytes, ependymal cells, and microglia. The first three types are collectively called macroglia.

Astrocytes are 'star-shaped' with widely radiating processes. Two types can be distinguished. Those with longer, thinner processes are called fibrous astrocytes. Those with shorter, more profusely branched processes are called protoplasmic astrocytes (mossy cells). Fibrous astrocytes are found chiefly in the regions of the brain and spinal cord occupied by aggregations of axons (i.e. within fibre tracts or white matter). Protoplasmic astrocytes are mainly found where there are aggregations of neuronal cell bodies (i.e. within grey matter).

Astrocytes encapsulate the parts of neurones not covered by oligodendrocyte processes. They also cover the outer and inner surfaces of the brain and spinal cord by: a) forming a layer beneath the pia mater, the pia-glial membrane (gliosal or external limiting membrane); b) forming a layer next to the ependymal lining of the ventricles of the brain, the ependymal-glial membrane (internal limiting membrane); and c) investing blood capillaries with processes known as vascular feet or end-feet (perivascular limiting membrane).

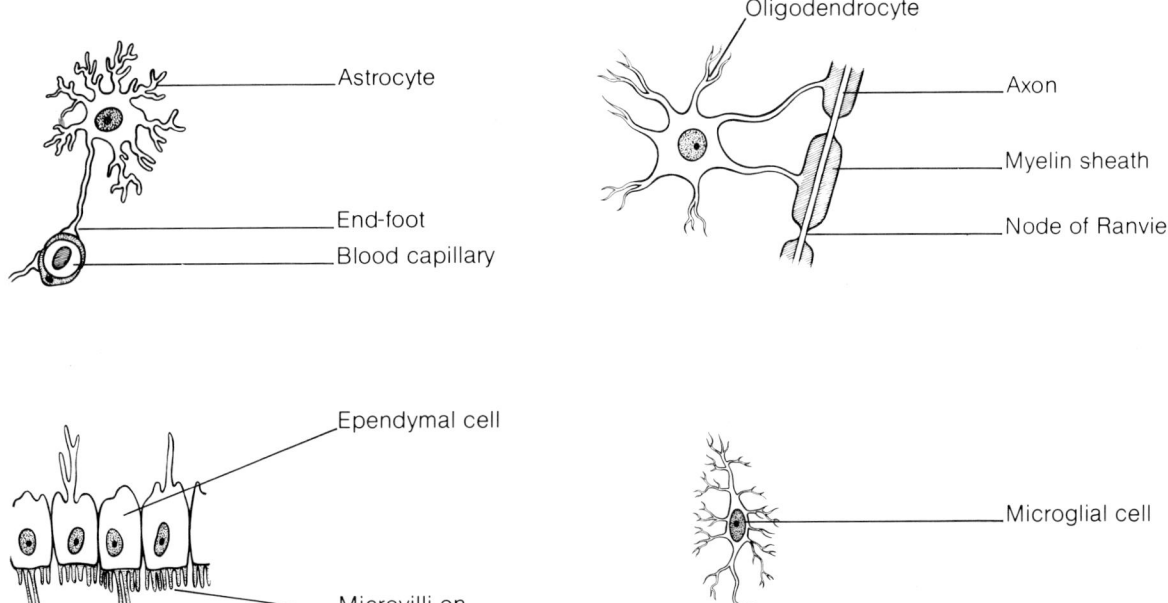

Figure 164 *Types of central neuroglial cells.*

Astrocytes play a particular role in neuronal metabolism and function through: control of the extracellular potassium ion concentration (essential for normal neuronal electrical transmission); provision of a primary glycogen store for the energy needs of the neurones; involvement in the uptake of certain neurotransmitters after their release from synapses.

Astrocytes perform a scavenging, phagocytic function. Following injury, they proliferate and produce a form of scar tissue (gliosis).

Oligodendrocytes have, as their name suggests, relatively few processes. They occur as perineuronal satellite cells surrounding neuronal cell bodies, probably serving a metabolic function. They occur also as interfascicular oligodendrocytes, providing the insulating myelin sheaths of axons within the brain and spinal cord (see page 405). One oligodendrocyte may provide myelin for many axons.

Ependymal cells form an epithelial layer (one cell thick) that lines the ventricles of the brain and the central canal of the spinal cord. The ventricular surface of the cells shows microvilli and/or cilia. Besides assisting the flow of cerebrospinal fluid, ependymal cells take up and transport substances from the ventricles, and secrete substances into them. The cells are linked by gap junctions, desmosomes, and by tight junctions in the choroid plexuses (see pages 572 to 574).

Special ependymal cells called tanycytes are found in the walls of the third ventricle. They are specialised for the transport of particular hormones and hormonal releasing factors, probably in both directions, between cerebrospinal fluid and hypothalamic neurones and/or blood vessels.

Microglia are small cells that correspond to the macrophages of connective tissue. They migrate from the blood during embryological development, and also in response to inflammation or degeneration within the central nervous system. Following injury, microglial cells proliferate rapidly and act as phagocytes to remove cellular debris.

Apart from the main types of central neuroglia, three minor types of glial cells are also found. Pituicytes (within the neurohypophysis) and Müller cells (within the retina) have similarities with astrocytes. Radial glial cells (notably the Bergmann cells of the cerebellum) provide a guidance mechanism for neuronal migration during development.

Peripheral neuroglia

The main glial cell types associated with the peripheral nerves are: capsular cells, Schwann cells, lemmal cells, teloglial cells, and supporting cells of the sensory epithelia.

Schwann cells (sometimes called neurolemma cells) are the main type of glial cell found outside the brain and spinal cord. All peripheral axons are closely associated with Schwann cells (see Figure 161). They provide the insulating myelin sheaths of myelinated fibres. Unmyelinated fibres are contained within invaginations of the cell body of the Schwann cell. As with central glial cells, the Schwann cells are important in axonal conduction, in maintaining the extracellular milieu, in neuronal metabolism, and in neurotransmitter action. They have a phagocytic role, and are important in neuronal growth and regrowth both chemically (through the production of trophic factors) and anatomically (by providing channels along which axonal sprouts may grow).

Capsular cells (inner satellite cells) surround the neuronal cell bodies found in the peripheral ganglia. They are cytologically similar to Schwann cells, and are of connective tissue origin.

Terminal gliocytes include **lemmal cells** and **teloglial cells.** Lemmal cells contribute to certain specialised sensory nerve endings. Teloglial cells encapsulate the terminals of motor axons. Both are similar in cytology to Schwann cells.

Supporting cells of the sensory epithelia may also be regarded as peripheral glial cells.

Introduction to the topography of the central nervous system

SUBDIVISIONS OF THE NERVOUS SYSTEM

The nervous system is divided for descriptive convenience into two main parts: the central nervous system (CNS) and the peripheral nervous system (PNS). The central nervous system comprises the brain and spinal cord. The peripheral nervous system comprises the cranial and spinal nerves and their associated ganglia (aggregations of nerve cell bodies).

A further subdivision of the nervous system relates to those parts of the peripheral and central nervous systems that regulate and control visceral functions. This is called the autonomic nervous system (ANS). The autonomic nervous system innervates viscera, glands, blood vessels, and smooth and cardiac muscle. The system should include both sensory and motor pathways, although some neurologists incorporate only the efferent or motor pathways.

The tissues comprising the central nervous system can be categorised as either grey matter or white matter.

Grey matter contains primarily neuronal cell bodies and dendrites, together with the mainly poorly myelinated initial/terminal portions of axons. It is a region of neuronal (synaptic) interaction. The complex of dendrites, axon terminals and neuroglial processes is known as neuropil. A nucleus is an aggregation of neuronal cell bodies in the central nervous system. In the peripheral nervous system, such an aggregation is called a ganglion. The laminated grey matter at the surface of the cerebellum and the cerebral hemispheres is known as cortex.

White matter contains primarily well-myelinated axons. It is therefore a region of neuronal transmission (from one area of grey matter to another). Gatherings of aligned axons are known by a wide variety of terms, depending upon their size, shape and location. Most commonly, they are called tracts, but they may be known as funiculi, fasciculi, lemnisci, bundles, striae, roots, radiations, peduncles, brachia, crura or capsules. The white matter beneath the cortex of the cerebellum and cerebral hemispheres is known as a medullary centre or core.

Before proceeding to a description of the gross topography of the central nervous system, some definitions should be provided concerning types of neuronal and axonal function.

Sensory receptors are specialised for the detection of stimuli. Accordingly, they respond to, and signal changes in, the internal or external environment. Essentially, the receptors effect the conversion of physical or chemical energy to electrical signals. Primary sensory neurones are the first neurones of a sensory pathway (i.e. those that either contact or are themselves the sensory receptors). Their cell bodies are mainly found in the sensory ganglia of the cranial and spinal nerves. Effector neurones are those that terminate in relation to muscles or glands (i.e. in relation to the effector tissues that produce responses). It is only through these neurones that the nervous system influences other tissues and produces behaviour. Motor neurones (motoneurones) are effector neurones that send their axons to skeletal muscle fibres. Their cell bodies are found in the motor cranial nerve nuclei and the ventral horns of the spinal cord. Strictly speaking, interneurones (internuncial or intercalated cells) are any neurones forming part of a pathway between the primary sensory neurones and the effector neurones but, in practice, the term usually means cells whose axons remain within the specific area of grey matter under consideration and so have localised effects.

Neurones with long axons that make connections between different parts of the central nervous system are known as Golgi type I neurones. Neurones with short axons that terminate close to their own soma are called Golgi type II neurones. However, there are also intermediate types which are not readily classified.

An axon with its attendant sheath is called a fibre. Fibres may be classified according to function in a number of different ways. Efferent fibres are those that run away from a nucleus or the central nervous system. Afferent fibres pass towards a nucleus or the central nervous system.

The fibres within the central nervous system are categorised as association, commissural, decussating, or projection fibres. Association fibres are those that connect one part of a structure with another part. Commissural fibres connect corresponding regions in the right and left halves of the central nervous system. Decussating fibres are those that cross the midline to connect between different structures on each side. Projection fibres are those passing between a region of grey matter and other structures as either afferents or efferents.

GROSS TOPOGRAPHY OF THE CENTRAL NERVOUS SYSTEM

The spinal cord is the part of the central nervous system that occupies most of the vertebral canal of the vertebral column. It is in continuity with the brain through the foramen magnum of the skull (the junction with the medulla oblongata of the brainstem being just caudal to the foramen magnum). In the adult, the spinal cord ends at approximately the level of the intervertebral disc between the first and second lumbar vertebrae. Thus, it lies chiefly in the cervical and thoracic parts of the vertebral column. The spinal cord is almost cylindrical in shape. It is about 450 mm long and weighs about 30 g. The sharply tapered terminal part of the spinal cord in the upper lumbar region is called the conus medullaris.

A typical transverse section of the spinal cord consists of an H-shaped core of grey matter, with white matter surrounding it. The ventral and dorsal limbs of the H-shaped grey matter are known as the ventral and dorsal horns. The surrounding white matter of the spinal cord can be subdivided into ventral, dorsal and lateral funiculi or columns.

The brain is a delicate, semi-fluid structure with the consistency of stiff porridge. It lies within the cranial cavity of the skull and has a volume of approximately 1300 ml. The adult brain weighs about 1500 g in the male; that of the female weighs about 100 g less. There is, of course, a considerable variation between individuals ($\pm$ 200 g). Comparing species, the ratio of brain size to body surface area is pre-eminent for the human. There are approximately 10^{10} to 10^{11} neurones in the adult brain.

The brain can be subdivided into three parts: the hindbrain (rhombencephalon), the midbrain (mesencephalon), and the forebrain (prosencephalon or cerebrum). Another subdivision consists of the forebrain, the cerebellum, and the brainstem which makes connections between them. The brainstem is the stalk-like continuation of the spinal cord, and includes the medulla oblongata, pons and midbrain. These subdivisions are shown in Table 10. The hindbrain lies in the posterior cranial fossa below the

TABLE 10 Major divisions of the brain.

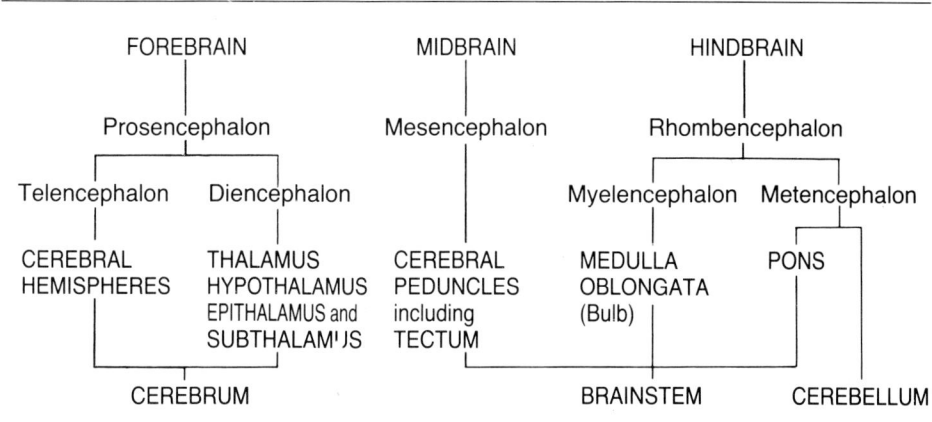

tentorium cerebelli. The forebrain is situated above the tentorium. The midbrain passes through the tentorial incisure.

The hindbrain consists of three parts: the medulla oblongata, the pons and the cerebellum. The medulla oblongata is a continuation of the spinal cord, the junction between them being located just caudal to the foramen magnum and cranial to the rootlets of the first cervical nerve. On the ventral surface of the brainstem, two prominent transverse grooves demarcate the boundaries of the medulla oblongata and the pons, and the pons and the midbrain. The cerebellum is ovoid in form and is highly folded with its characteristic appearance resulting from the multitude of parallel ridges (folia) separated by narrow fissures. It exhibits two cerebellar hemispheres connected by a central region called the vermis. The cerebellum is connected to each part of the brainstem by cerebellar peduncles. The superior cerebellar peduncles connect with the midbrain, the middle cerebellar peduncles with the pons, and the inferior cerebellar peduncles with the medulla oblongata.

158,26; 158,25
158,22; 158,30
158,21
188A

176A,25
196A,4
196A,3; 196A,2
198C,24
198C,25; 188A,23
198C,26

The midbrain consists of a right and a left cerebral peduncle. The cerebral aqueduct runs through the midbrain. The part of the midbrain dorsal to the cerebral aqueduct is called the tectum. The tectum is characterised by the presence of two pairs of swellings (the superior and inferior colliculi). The boundary between the midbrain and the diencephalon of the forebrain passes just cranial to the superior colliculi dorsally, and just caudal to the mamillary bodies ventrally.

158,10; 188,26+39
198A,3
198A,2
158,13; 158,12
158,56; 188A,25

The forebrain comprises the diencephalon and the two cerebral hemispheres (telencephalon).

The diencephalon has two major parts (the thalamus and the hypothalamus) and two minor parts (the epithalamus and the subthalamus). The anterior boundary of the diencephalon is marked by the optic chiasma and the lamina terminalis. The boundary between the thalamus and the hypothalamus is marked by shallow grooves passing from the interventricular foramina to the aqueduct of the midbrain. The pituitary gland (hypophysis) is connected to the ventral surface of the hypothalamus by the infundibulum (pituitary stalk). The pineal body is found in the roof of the diencephalon, close to the junction with the midbrain.

158,9; 158,57
180,31; 180,32
180,9; 180,22
158,50
180,30; 180,13
180,21

The cerebral hemispheres comprise by far the largest part of the human brain. The hemispheres are covered by the folded cerebral cortex, displaying a striking pattern of convolutions called gyri. The gyri are separated by clefts called sulci. The two cerebral hemispheres are separated by a longitudinal fissure whose floor is formed by the massive fibre bundle of the corpus callosum. For topographical convenience, each

172; 176; 182

172,11
180,3–6

162
28,10
28,21+24

158,18

192; 193

hemisphere is divided into four lobes: frontal, parietal, temporal and occipital (Figure 165). The lobes are named from their proximity to the corresponding bones of the vault of the skull. The frontal lobe lies above the orbital plate of the frontal bone in the anterior cranial fossa. The temporal lobe lies within the middle cranial fossa, related to the temporal bone. The occipital lobe is the most posterior lobe. It rests on the superior surface of the tentorium cerebelli, above the posterior cranial fossa. The parietal lobe is intermediate between the other lobes. Deep within the medullary core of each cerebral hemisphere are nuclei known as the basal nuclei (basal ganglia).

Although Table 10 summarises the generally accepted subdivisions of the brain, some anatomists choose to use slightly different schemata. For example, some include the diencephalon with the brainstem. Also, because the embryological division between the telencephalon and diencephalon is usually taken to pass through the interventricular foramina, parts of the hypothalamus could be thought of as belonging to the telencephalon.

190; 191

The ventricles of the brain (Figure 166) are cavities that are filled with liquid called cerebrospinal fluid (CSF). There is a lateral ventricle within each cerebral hemisphere.

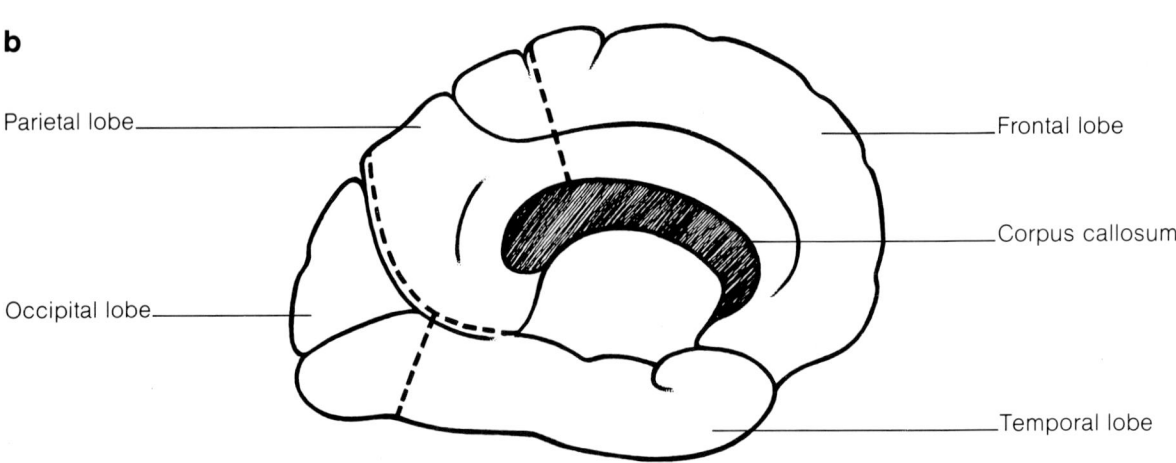

Figure 165 *The lobes of the cerebral hemisphere,* a) *lateral view, and* b) *medial view.*

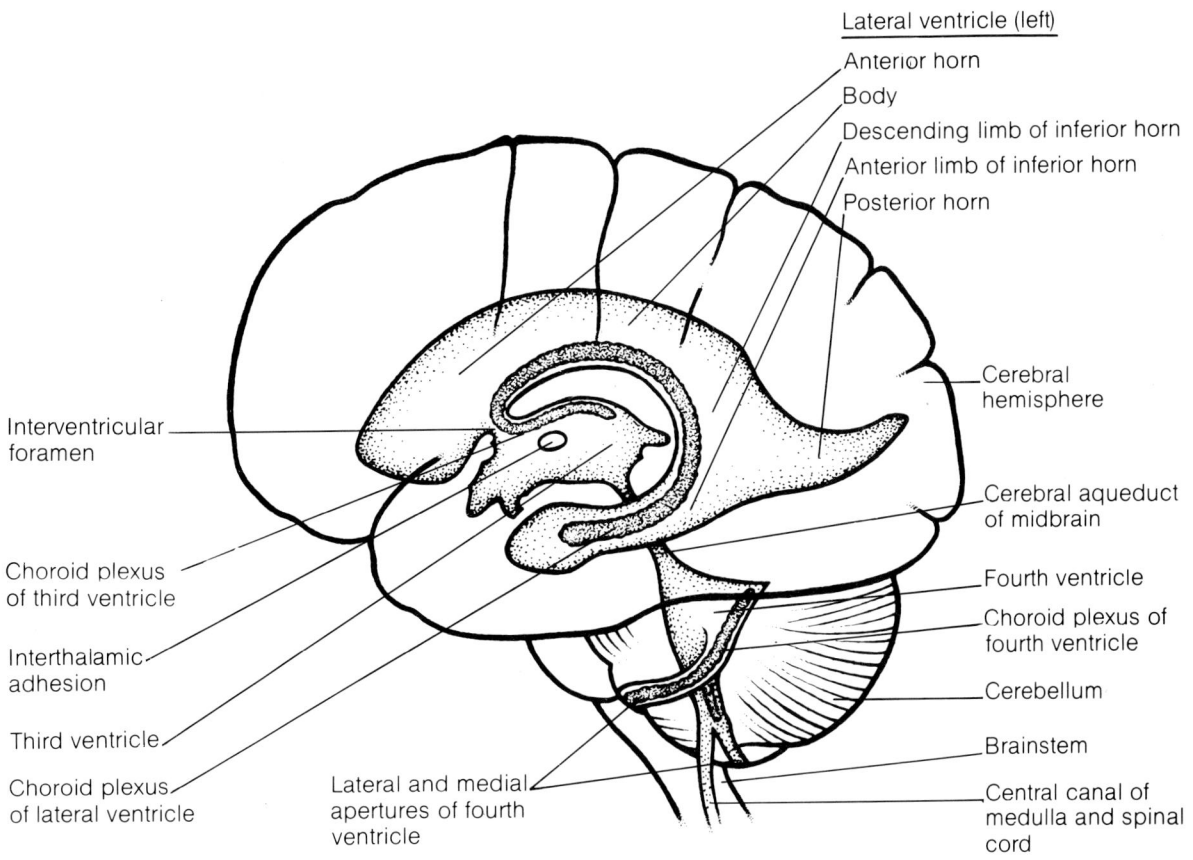

Figure 166 *The ventricular system of the brain,* also showing the location of the choroid plexuses, as if viewed in the left side of a transparent brain.

The third ventricle is a narrow median cavity within the diencephalon. It is in communication with each lateral ventricle via an interventricular foramen (of Munro). The third ventricle communicates with the fourth ventricle (located in the hindbrain beneath the cerebellum) through the narrow channel of the cerebral aqueduct (of Sylvius) of the midbrain. A central canal runs through the caudal medulla and spinal cord.

180,11
180,9
180,19
180,15; 180,22

The cerebrospinal fluid in the ventricles is produced by specialised tissue known as choroid plexuses. These plexuses are located within all four ventricles (Figure 166). The formation and the circulation of cerebrospinal fluid is further described on pages 572 to 574.

190,6; 190,30
180,12; 180,18

A note on descriptive nomenclature

Many of the nuclei and fibre tracts within the central nervous system have been primarily studied in subhuman species. Their names have then been applied to the human nervous system without translation, failing to account for our upright posture or the resultant flexure of the nervous system's axis at the mesencephalic-diencephalic junction. Thus, the commonly used and long-established names of certain nuclei and tracts are inconsistent with nomenclature based upon the conventions associated with the standard anatomical position. Neuroanatomical nomenclature is therefore inevitably problematic, particularly for the diencephalon, the brainstem and the cerebellum. Authorities differ on spinal cord terminology, but anterior may be taken as being synonymous with ventral, and posterior with dorsal. For the cerebral hemispheres, where correspondence between the human and subhuman brain is less obvious, descriptive terminology follows that of the anatomical position.

Consistency is maximised and confusion minimised by following a convention that ignores flexures and imagines that the axis of the central nervous system passes in a straight line from the frontal lobe of the cerebrum to the tip of the spinal cord (conus medullaris). Accordingly, the frontal lobe is cranial or anterior. The conus medullaris is caudal or posterior. The surface of the spinal cord nearest the viscera is ventral (in the anatomical position it is anterior), its opposite surface dorsal (rather than posterior). The same conventions then apply for the medulla, pons and midbrain. It is therefore consistent that the cranial part of the cerebellum is its anterior lobe. Structures adjacent to the base of the skull and therefore seen on the base of the brain (e.g. pyramids, cerebral peduncles, hypothalamus) are described as being ventral rather than anterior. Within the cranial cavity, structures nearer to the top of the head are superior in position to those that are inferior and nearer the base of the skull. Thus, the more cranial of the colliculi of the midbrain are designated the superior colliculi, and it is possible to speak of the surface of the cerebellum adjacent to the tentorium cerebelli as the superior surface. As far as nomenclature within the cerebral hemispheres is concerned, structures are anterior (frontal) or posterior (occipital), superior or inferior. The corresponding directions in the diencephalon are anterior (frontal)/posterior (caudal) and dorsal/ventral. The base of the cerebrum is therefore inferior and also ventral.

THE EMBRYOLOGY OF THE NERVOUS SYSTEM (Figures 167 to 170)

An understanding of the contorted topography of the human brain is facilitated by consideration of its development. Only an outline account of that development is presented here.

The nervous system develops from the ectoderm of the embryonic disc. A neural plate appears as a thickening of the ectoderm at about two and a half weeks after fertilisation. At the edges of this plate are the presumptive neural crest cells. A groove then appears running longitudinally down the neural plate, and the edges of the plate fold dorsally to form a neural tube (Figure 167). This tube is located in the midline above the notochord and extends between the primitive node and the buccopharyngel membrane. The central nervous system develops from the neural tube. The peripheral nervous system develops mainly from the cells of the neural crest.

Initially, the neural tube is open at both ends (anterior and posterior neuropores). The neuropores close during the fourth week of intra-uterine life. Failure of closure of the posterior neuropore results in a defect of the spinal cord and/or vertebral column in the lower lumbar region, called spina bifida. Failure of closure of the anterior neuropore is associated with maldevelopment of the brain, called anencephaly.

Before the neural tube is complete, the anterior (cephalic) end already shows enlargement. Two transverse constrictions divide this enlargement into three vesicles. Thus, from an early stage, the eventual subdivisions of the central nervous system (i.e. forebrain, midbrain and hindbrain, together with the spinal cord) are apparent. The lumen of the neural tube eventually develops into the ventricular system of the brain. The anterior neuropore represents the most anterior extent of the lumen of the neural tube; it is closed by development of the lamina terminalis (Figure 168).

As the neural tube closes, the neural crest cells at each side of the neural plate are brought together to form a temporary cap between the dorsal surface of the neural tube and the surface ectoderm. The cells divide and migrate laterally to form segmented clusters. These clusters give rise to cells of the dorsal root ganglia of the spinal nerves and of the ganglia of the cranial nerves. In total, neural crest cells give rise to:

All the primary sensory neurones of the ganglia of the spinal nerves and most of the primary sensory neurones of the ganglia of the cranial nerves (some neurones, however, arise from other ectodermal tissue, i.e. from the ectodermal placodes).

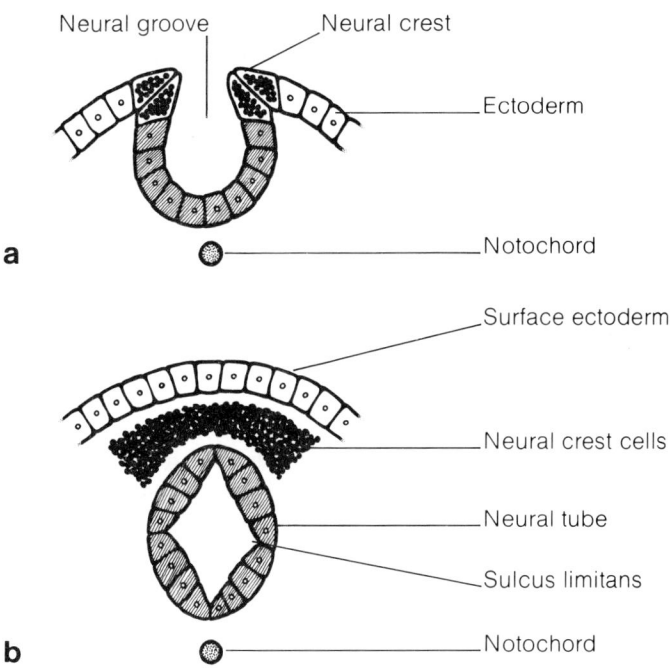

Figure 167 *Formation of the neural tube from ectoderm.* Transverse sections of part of a developing embryo to show the formation of a) the neural groove, and b) the neural tube.

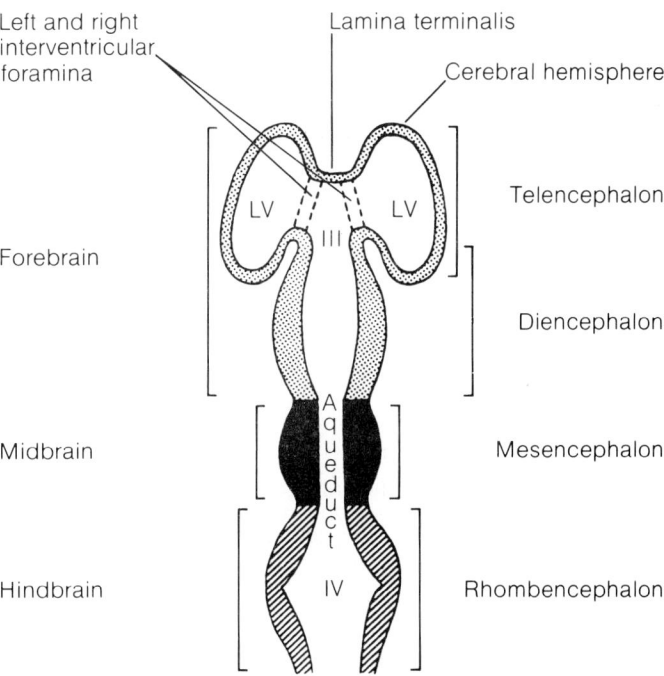

Figure 168 *The major divisions and the ventricular system of the developing brain.* (LV = lateral ventricle. III, IV = third and fourth ventricles.)

Neurones of the autonomic nervous system that have their cell bodies in peripheral (autonomic) ganglia and that innervate cardiac and smooth muscle, and glands (together with cells of the adrenal medulla).

The neuroglia of the peripheral nervous system and certain sensory receptor cells.

Some of the cells of the two inner meninges (pia mater and arachnoid mater) surrounding the brain and spinal cord.

A wide variety of non-nervous cells (e.g. the odontogenic mesenchyme).

Specialised patches of ectodermal tissue called ectodermal placodes give rise to: some of the neurones of the cranial nerve ganglia; the receptor neurones (and probably associated epithelial cells) of the olfactory mucosa; and the receptor cells of the inner ear (i.e. vestibular and cochlear hair cells).

As the neural tube develops, it is surrounded by mesodermal cells. Several supporting structures of the nervous system are derived from these cells, including: parts of the skull; the vertebrae, intervertebral discs and ligaments; the outer meninges and sheaths of peripheral nerves; the blood vessels and the microglial cells.

The development of the spinal cord

In transverse section, the neural tube is seen to consist of three layers: the ependymal (ventricular or matrix), the mantle (intermediate) and the marginal layers. The ependymal layer is innermost and contains many rapidly dividing neuro-epithelial cells. Its central cells eventually form the ependymal lining of the spinal canal (and of the ventricles in the brain). The mitosing cells produce neuroblasts and spongioblasts. The neuroblasts develop into neurones; the spongioblasts develop into neuroglial cells (astrocytes and oligodendrocytes). The neuroblasts migrate outwards to form the mantle layer. The mantle layer develops into the grey matter of the spinal cord (and of much of the brain, though not the cerebellar or cerebral cortices). The axons of the neurones in the mantle layer form the marginal layer. The marginal layer is the outermost layer of the neural tube. It gives rise to the white matter of the spinal cord (Figure 169a).

Dorsally and ventrally, the walls of the neural tube remain thin and form the roof and floor plates. On either side of the central canal, a groove called the sulcus limitans marks the division between the dorsal (alar) lamina and the ventral (basal) lamina. Cells of the dorsal laminae form the sensory, dorsal horns which receive the afferent fibres of the spinal nerves. Cells of the ventral laminae become the motor, ventral horns which give rise to the efferent fibres passing to skeletal muscle. They also form the autonomic, intermediolateral cell columns (lateral horns) which give rise to preganglionic autonomic efferents passing to the autonomic ganglia (Figure 169a). Eventually, the dorsal laminae become apposed above the spinal canal and are separated only by the dorsal median septum in the adult. An invagination of the floor plate results in development of a ventral median fissure, and the central canal becomes small.

The development of the brain

Three flexures develop in the enlarged anterior end of the neural tube from which the brain will be formed (Figure 170). At the cervical or neck flexure, between the spinal cord and hindbrain, the neural axis bends downwards (ventrally). Within the hindbrain, the pontine flexure bends the neural axis acutely upwards (dorsally). Within the midbrain, the midbrain (cephalic) flexure bends the neural axis acutely downwards (ventrally) once more. The part of the hindbrain caudal to the pontine flexure is the myelencephalon and develops into the medulla oblongata. The part of the hindbrain cranial to the pontine flexure is the metencephalon; this becomes the pons and cerebellum.

The pontine flexure buckles the walls of the neural tube outwards, causing the floor of the cavity in the hindbrain (and the floor of the fourth ventricle into which it develops)

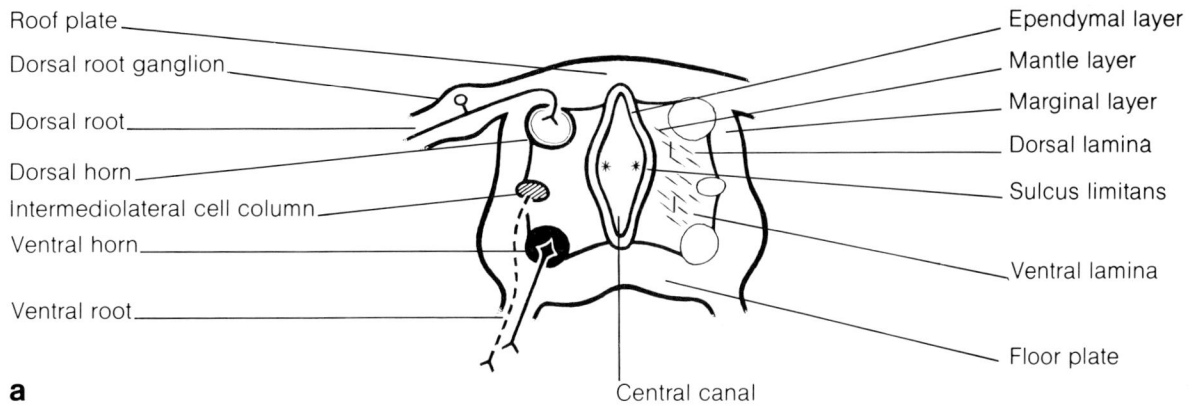
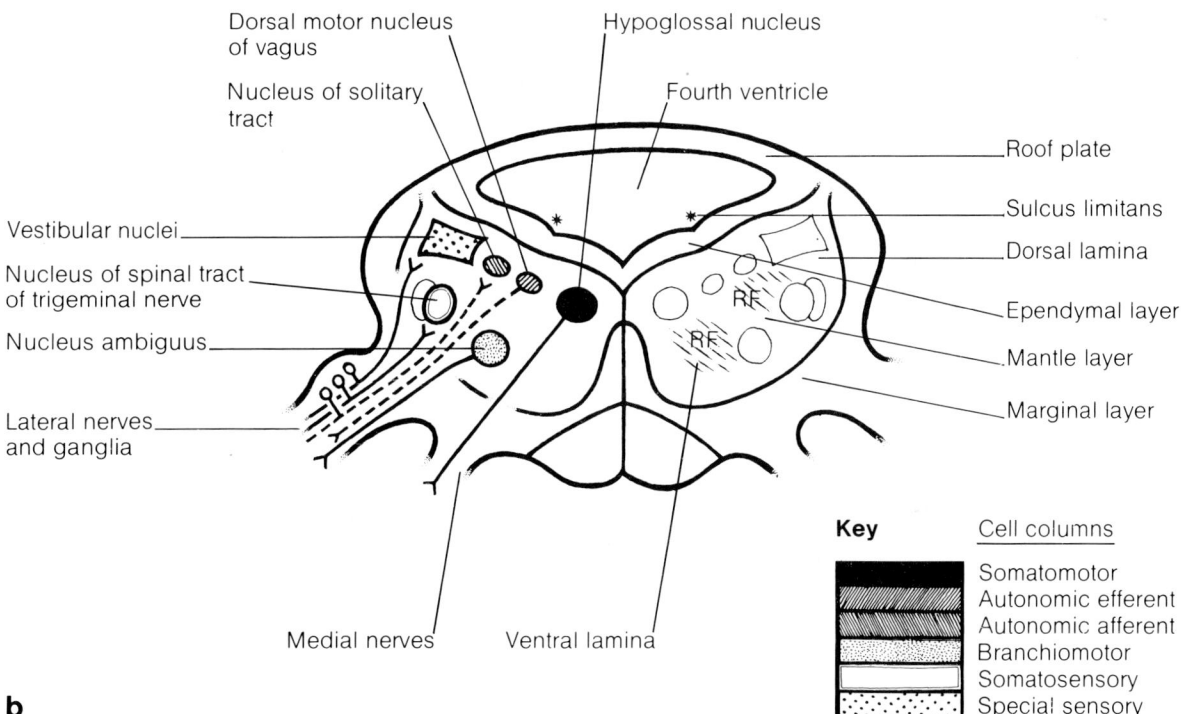

Figure 169 *Transverse sections of the developing neural tube* to show the relative positions of the sensory, motor and autonomic cell groups. a) *Spinal cord*. b) *Cranial part of the medulla oblongata*. (I = interneurones, RF = reticular formation.)

to become rhomboid. This feature of the hindbrain accounts for the term rhombencephalon. The buckling causes the dorsal laminae to lie lateral (rather than dorsal) to the ventral laminae, the roof plate becoming much extended (Figure 169b). Thus, in the region of the fourth ventricle (in the cranial part of the medulla oblongata and in the pons), sensory nuclei of the cranial nerves lie lateral (rather than dorsal) to nuclei with autonomic (parasympathetic) or somatomotor function. Other groups of nuclei dealing with specialised sensory functions, with the innervation of muscles developmentally associated with the branchial arches, and with the cerebellum are also found within the grey matter of the medulla and pons. Note that the major, myelinated fibre tracts (white matter) of the medulla are found mainly externally, as in the spinal cord (hence the derivation of the terms medulla spinalis and medulla oblongata or myelencephalon).

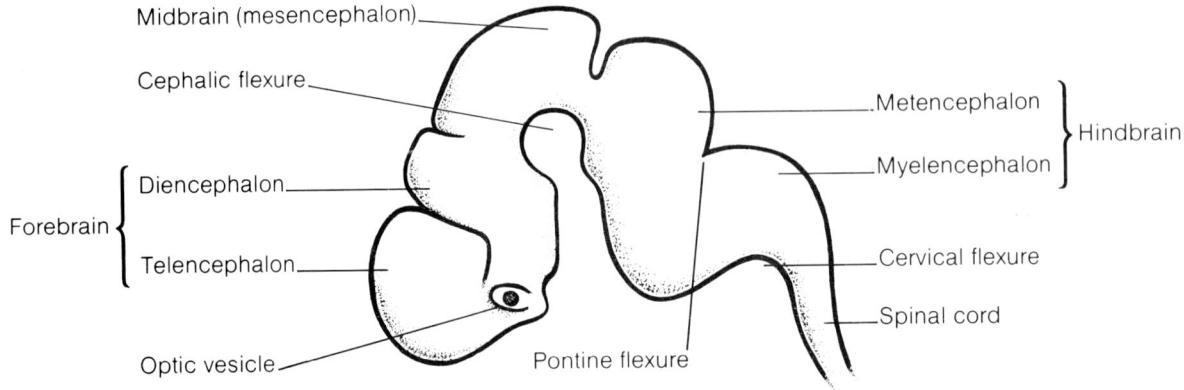

Figure 170 *Flexures and major divisions of the developing nervous system* (lateral view).

The cerebellum grows from dorsal extensions of the dorsal laminae of the metencephalon (the rhombic lips). As in the developing spinal cord, cell proliferation occurs in an inner zone, close to the ventricle. However, two regions develop within this zone. Neuroblasts from one region migrate directly outwards to the developing cerebellar plate, eventually forming the neurones of the intracerebellar nuclei and the large neurones (Purkinje and Golgi cells) of the cerebellar cortex. From the second region, cell migration occurs along the outer surface of the cerebellar plate, subsequently passing inwards and giving rise to the small cells of the cerebellar cortex (notably the granule cells of its innermost layer). The events of these migrations must occur in a precisely ordered sequence for normal cerebellar connections to be made. The cells that send axons out of the cortex (the Purkinje cells) direct these axons centrally, so that the cerebellum develops white matter internally (i.e. a medullary core) rather than externally. Correspondingly, the grey matter of the cerebellar cortex is superficial. It is closely folded to maximise the ratio of cortical surface area to cerebellar volume.

The midbrain (mesencephalon) remains essentially cylindrical in transverse section. The sensory areas (superior and inferior colliculi) of the tectum, above the cavity of the midbrain, develop from the dorsal laminae. The midbrain cavity becomes the cerebral aqueduct.

The forebrain cavity develops several out-pouchings. Two optic vesicles extend anteriorly towards the surface ectoderm. From their walls develop the retinae and optic nerves. The neurohypophysis develops in the floor of the infundibular recess. (The remainder of the pituitary gland, the adenohypophysis, develops from a dorsal process of the stomodeum, Rathke's pouch.) A cerebral vesicle develops on each side of the lamina terminalis at the anterior end of the neural tube. The cavities of the cerebral vesicles become the two lateral ventricles; their walls become the cerebral hemispheres (telencephalon). The original forebrain cavity becomes the third ventricle, its walls forming the diencephalon.

Each cerebral hemisphere enlarges laterally, dorsally, anteriorly and posteriorly around the diencephalon. Essentially, growth of the hemisphere causes it to bend back upon itself (so forming the lateral sulcus). Consequently, the cerebral hemisphere shows a curved axis that runs posteriorly through the frontal lobe, posteroinferiorly through the parietal and occipital lobes, and anteriorly into the temporal lobe. Both the lateral ventricle and the structures within its walls (e.g. caudate nucleus, hippocampus and fornix) follow this curvature.

As elsewhere, the walls of the developing cerebral hemispheres may be divided into three zones: ependymal or ventricular, mantle or intermediate, and marginal layers.

Dividing cells near the ventricle of each cerebral hemisphere migrate through the intermediate zone to form the cortical plate just below the external, marginal layer. The migrations occur in waves, the later cells always passing through the layers of earlier cells to take an external (superficial) position. This is described as 'inside-out' migration; it is very precisely ordered and is essential for the formation of intracortical connections. The projection fibres of the cortical neurones pass centrally, rather than peripherally as in the spinal cord. Thus, as in the cerebellum, the grey matter of the cerebral hemisphere lies superficially, whereas the white matter is located deeply, forming the medullary core of the hemisphere. Folding of the cortical mantle to form gyri and sulci allows a large surface area of cortex to be accommodated within a minimal volume. The marginal layer becomes the outermost, molecular layer of the cerebral cortex.

To the side of the diencephalon, masses of grey matter (which give rise to the basal nuclei) develop from the mantle layer at the base of each cerebral hemisphere. Eventually, projection fibres running to and from the grey matter of the cortex at the surface of the cerebral hemisphere pass through this region, separating its nuclei and forming the internal capsule.

Commissural fibres connecting the two cerebral hemispheres must at first pass through the lamina terminalis. As the hemispheres develop, their major commissure, the corpus callosum, expands dorsally, anteriorly and posteriorly above the diencephalon. Thus, the roofs of all three of the original brain vesicles become overgrown by nervous tissue: that of the hindbrain by the cerebellum, that of the midbrain by the colliculi, and that of the forebrain by the cerebral hemispheres and corpus callosum.

Most of the brain's neurones are generated and differentiate during early prenatal life. Formation of new neurones has largely ceased by birth, although some neurones of the cerebellum and of the hippocampal formation of the cerebral cortex are produced postnatally. Because neurones generally do not divide and are not replaced, the neonate already has its lifetime's supply of neurones.

Growth in size of neurones, and the formation of synapses, occur particularly in late prenatal and early postnatal life. At birth, the volume of the brain is about 25% of that of an adult. By the end of the first year, it is 75% of that of the adult, adult size being reached by about the eighteenth year. Much of the postnatal growth is of the neuroglial cells, and particularly of myelin sheaths. There is no good evidence of a general decline in the number of neurones with age. However, there is loss of neuropil so that the brains of elderly people may reduce to only 70% of their early adult size.

The spinal cord and spinal nerves

The spinal cord (medulla spinalis) is that part of the central nervous system that is located extracranially; it is contained within the vertebral canal of the vertebral column. Although the spinal cord constitutes only 2% of the central nervous system, it has considerable functional significance. First, it is linked to the periphery of the body through 31 pairs of spinal nerves. Second, the spinal cord organises many somatic and autonomic reflexes. Third, it passes sensory information to the brain (including proprioceptive information about the activities of neurones in the spinal cord). Fourth, the spinal cord conveys instructions from the brain for the control of both somatic and autonomic activities.

136,10; 200B
80A,12

THE SPINAL NERVES

The spinal nerves constitute part of the peripheral nervous system, the other component being the cranial nerves. However, much of what follows concerning the spinal nerves also applies to the cranial nerves (particularly receptors, effectors, reflexes and neural control mechanisms).

The 31 pairs of spinal nerves are named according to the intervertebral foramen through which they emerge. On each side of the body there are eight cervical nerves, twelve thoracic nerves, five lumbar nerves, five sacral nerves, and one coccygeal nerve. The first cervical spinal nerve emerges between the occiput of the vault of the skull and the atlas, the eighth cervical nerve through the intervertebral foramen between the seventh cervical and the first thoracic vertebrae. For the other spinal nerves, each nerve emerges through the interverterbal foramen below the vertebra which corresponds numerically (e.g. the second thoracic nerve emerges between the second and third thoracic vertebrae).

Each spinal nerve is formed from, and is connected to, the spinal cord by a dorsal (posterior) and a ventral (anterior) root. In some individuals, however, the first cervical nerve and/or the coccygeal nerve lack dorsal roots. The dorsal roots contain afferent (sensory) fibres from muscles, skin and other tissues. Dorsal roots of the first thoracic to the second lumbar nerves, and the second to the fourth sacral nerves, typically carry autonomic afferents. The ventral roots contain efferent (motor) fibres: of somatomotor function in all nerves; and additionally of autonomic function for the first thoracic to the second lumbar nerves, and the second to the fourth sacral nerves. The ventral roots also contain some afferent fibres, probably mainly autonomic afferents or nociceptive afferents. Thus, should the dorsal roots of spinal nerves be severed, not all sensory fibres will be interrupted.

On each dorsal root, there is a fusiform swelling called the dorsal root ganglion. This ganglion contains the cell bodies of the primary afferent neurones. It is located in, or close to, the intervertebral foramen through which the corresponding spinal nerve emerges. The spinal nerve is formed by the union of the dorsal and ventral roots, just distal to the ganglion. Dorsal root ganglia contain neuroglia, but no neurones other than the unipolar primary sensory neurones. Each ganglion is encapsulated by connective tissue that is continuous with the spinal dura mater.

Each spinal nerve consists of an aggregation of axons and axon-like dendrites, with neuroglia and connective tissue sheaths. Myelinated and unmyelinated fibres are found, both being associated with Schwann cells (see page 416). The plasma membrane of the Schwann cell is itself surrounded by a basement membrane and then by a fine connective tissue sheath called the endoneurium. Together the basement membrane and the endoneurium are called the neurolemma. The whole nerve trunk is surrounded by a dense, collagenous sheath called the epineurium. The epineurium is continuous with the dura mater surrounding the spinal cord. Connective tissue septa, called the perineurium, divide the nerve into distinct fasciculi. The epineurium and perineurium support and protect the nerve fibres and provide the route by which blood and lymph vessels gain access to the nerve.

The typical spinal nerve contains an autonomic component and a somatic component.

The autonomic component innervates blood vessels, smooth muscle and cardiac muscle, glands, and viscera. It is responsible for subconscious monitoring and maintenance of the normal functioning of these structures. It also gives rise to visceral sensations (including pain) and contributes to feelings such as hunger, thirst, sexual arousal and general wellbeing/malaise.

The somatic component comprises fibres primarily innervating skin and muscle in a specific area of its half of the body. The area of skin innervated is called a dermatome; the volume of skeletal muscle innervated is called a myotome. Fibres of the somatic component also pass to other deep tissues, such as meninges and joints. The somatic component is responsible for the sensations of touch, pressure, temperature and pain, as well as for more complex appreciations such as tickle, itch, texture and vibration. This component is also responsible for conscious and unconscious information concerning body position, movement and muscle tension (i.e. proprioceptive information). Furthermore, by activating the skeletal musculature, it produces movement and maintains posture.

Because a dermatome is an area of skin supplied by a specific spinal nerve, much of the skin of the head cannot be mapped as dermatomes as it is supplied by branches of the trigeminal cranial nerve (see Figure 49, page 159). Dermatomes are, however, found on the back of the head and around the neck. Posteriorly, an orderly, segmental pattern of innervation is found, although there is no dermatome for the first cervical nerve. Anteriorly, the skin of the neck shows parts of the dermatomes of the third and fourth cervical nerves. The dermatomes for the fourth and fifth cervical nerves usually extend on to the chest to the second intercostal space. For the remaining parts of the body:

Posteriorly — the skin over the shoulders shows parts of the dermatomes of the fourth to the seventh cervical nerves; the dermatomes of the eighth cervical to the second lumbar nerves cover the back; the dermatomes of the third lumbar to the third sacral nerves are associated with the buttocks; and the dermatomes of the fourth sacral nerve to the coccygeal nerve cover the perianal region.

Anteriorly — the dermatomes of the upper limb are those of the fifth cervical to the first thoracic nerves. The chest (below the second intercostal space) and the abdomen show dermatomes for the first thoracic to the first lumbar nerves. The skin of the lower limbs is supplied by the first lumbar to the second sacral nerves.

There is considerable overlap between dermatomes of adjacent spinal nerves. In general, each area of skin is innervated by at least two successive spinal nerves, the overlap being greater for the sense of touch than for painful or thermal sensations. Such overlap makes it difficult to determine the precise boundaries of the dermatomes. More importantly, it implies that transection of a single spinal nerve does not normally result in a region of anaesthesia. However, overlap is much less extensive where dermatomes of non-adjacent spinal nerves are apposed, i.e. along the ventral axial lines of the upper and lower limbs. (The ventral axial line of the upper limb is found between the dermatome of the first thoracic nerve and those of the fourth to the seventh cervical nerves; for the lower limb, the line lies between the dermatome of the second sacral nerve and those of the lumbar nerves.) Moreover, little overlap is found for the cutaneous tissue of the head. The areas innervated by the three principal divisions of the trigeminal nerve (see Figure 49) and the dermatome of the second cervical nerve do not demonstrate any significant overlap in innervation. The part of the dermatome of the second cervical nerve distal to that of the third, is not overlapped by it. Accordingly, transection of any of the principal divisions of the trigeminal nerve, or of the second cervical nerve, will produce a local, but total, sensory loss. Transection of a peripheral cutaneous nerve will produce anaesthesia within the territory it supplies. However, such territories will not normally correspond to dermatomes.

The sequence of myotomes is less clearly ordered than that of the dermatomes. This is related to the fact that adjacent myotomes commonly fuse during development. Most trunk and limb muscles receive innervation from more than one spinal nerve; possibly from as many as four. Correspondingly, a lesion must involve more than one spinal nerve before a muscle is completely paralysed. However, there are muscles that are innervated by only a single spinal nerve (the muscles in the head are each innervated by a single nerve, albeit a cranial nerve).

The cervical spinal nerves are responsible for movements of: the head by the neck muscles (the first to the fourth cervical nerves via the cervical plexus); the diaphragm (the third to the fifth cervical nerves via the phrenic nerve); and the upper limb (the fourth cervical to the first thoracic nerves via the brachial plexus). The thoracic spinal nerves control movements of the trunk by the muscles of the thorax and abdomen. The lumbar and upper sacral nerves (i.e. the first lumbar to the third sacral nerves) control movements of the lower limb via the lumbosacral plexus. Sacral and coccygeal nerves also supply muscles of the perineal region.

200C,46; 200C,47

The typical spinal nerve divides shortly after leaving the intervertebral foramen into four branches or rami: dorsal (posterior), ventral (anterior), meningeal, and autonomic (rami communicantes).

The dorsal rami supply the muscles and skin of the back and the posterior regions of the head and neck. The large ventral rami provide innervation to the superficial and deep tissues of the lateral and anterior parts of the body wall and the neck. They also innervate all the tissues of the limbs. The small meningeal branches supply the dura mater and vertebral column. The autonomic branches are described later (see pages 437 to 443).

100,29–31; 100,34

Both the dorsal and ventral rami subdivide into superficial (cutaneous) and deep (muscular) peripheral nerves. Excepting most of the thoracic spinal nerves (specifically the second to the eleventh thoracic nerves), the ventral rami anastomose to form plexuses (the cervical, brachial and lumbosacral plexuses). The peripheral nerves emerging from these plexuses contain contributions from a number of consecutive spinal nerves. Some smaller-scale anastomoses occur between dorsal rami of adjacent cervical and sacral spinal nerves. It should be noted that peripheral nerves are often described as 'mixed' because they contain both afferents and efferents. Most nerves are also mixed in that they supply both muscular and cutaneous tissue and contain contributions from more than one spinal nerve.

There are two systems for classifying the types of fibres in peripheral nerves. First, sensory afferents have been subdivided into categories (I to IV) according to function and axonal diameter. Second, fibres have been categorised (A to C) according to the speed with which they conduct action potentials (conduction velocity) and function. Unmyelinated fibres belong to class IV or C; the fastest, largest fibres to class Ia or A. Details of the classifications are given in Table 11.

At the terminations of peripheral nerve fibres are found receptors or effectors.

Receptors are sensory organs specialised for the detection of physical or chemical stimuli, and for the transduction of the incident energy into electrical signals. They are the means by which information about the internal and external environment is converted into neuronal signals.

A sensory receptor may be part of a neurone (e.g. olfactory receptors, free nerve endings) or formed from specialised, non-neuronal elements that are contacted by a neurone (e.g. retinal rods or cones, muscle spindle organs). Primary sensory neurones are those that either contact, or themselves form, the sensory receptors. Accordingly, they are the first neurone of any sensory pathway and their axons are primary sensory afferents. A primary sensory neurone with its receptors forms a sensory unit.

A sensory unit is responsive to only a restricted range of stimuli (e.g. mechanical deformation, or heat, or hydrogen ion concentration) occurring within a localised part of the body. A receptive field is defined as that area within which an appropriate stimulus of sufficient magnitude will result in a detectable response in the primary sensory neurone.

The various types of sensation are called modalities (e.g. tactile, visual, nociceptive modalities). The specificity of the receptors determines the type of stimuli detected by a sensory unit. The central connections of the primary sensory neurone are the main determinant of the sensation that will be experienced (e.g. either incident light or pressure on the eyeball will produce visual sensations).

Sensory receptors may be classified into three groups: exteroceptors, interoceptors, and proprioceptors.

Exteroceptors respond to stimuli in the external environment and lie at, or close to, the body surface. Examples are receptors of the special senses, such as the retinal rods

TABLE 11 Fibre classes in peripheral nerves.

AFFERENTS

GROUP	AXONAL DIAMETER μm (conduction velocity m/s)	ORIGIN (receptor)	SENSORY MODALITY (stimulus)	REFLEX and EFFECT ON MUSCLES
IV (C) (unmyelinated)	< 2 (< 3)	Free endings in viscera, glands, blood vessels, skin, muscle	Autonomic Pain (dull, slow) Touch Temperature	Autonomic (≥ trisynaptic) reflex
III (Aδ) (thinly myelinated)	1 to 5 (5 to 30)	Free endings in skin, muscle, viscera, etc.	Pain (sharp, fast) Temperature Touch Autonomic	Withdrawal (flexor) and crossed extensor (multisynaptic) reflexes Flexor + Extensor −
II (Aβ, Aγ)	5 to 12 30 to 70	Encapsulated endings in skin and joints. Secondary endings of muscle spindles	Descriminative touch and pressure. Proprioception: joint position, dynamic stretch of muscle	Stretch-evoked withdrawal reflex
Ib (Aα) (thickly myelinated)	12 to 17 (70 to 100)	Golgi tendon organs of muscles and Ruffini endings of joints	Proprioception: active contraction of muscle	Lengthening reaction (disynaptic) reflex Agonist − Synergist − Extensor +
Ia (Aα) (thickly myelinated)	17 to 20 (100 to 120)	Primary endings of muscle spindles	Proprioception: stretch of muscle	Myotatic or stretch (monosynaptic) reflex Agonist + (Synergist +) Antagonist −

EFFERENTS

GROUP	AXONAL DIAMETER μm (conduction velocity m/s)	TYPE	EFFECTOR
C (unmyelinated)	< 2 (< 3)	Postganglionic autonomic	Cardiac and smooth muscle, and glands
B (thinly myelinated)	< 3 (3 to 20)	Preganglionic autonomic	Autonomic ganglia
Aγ (thinly myelinated)	2 to 8 (10 to 40)	Fusimotor	Intrafusal skeletal muscle fibres
Aβ	3 to 15 (20 to 90)	Skeleto- and fusi-motor	Both extra- and intra-fusal skeletal muscle fibres
Aα (thickly myelinated)	10 to 20 (60 to 120)	Skeletomotor	Extrafusal muscle fibres

and cones responsible for vision, and cutaneous receptors, such as the various encapsulated and free nerve endings underlying tactile sensations.

Interoceptors are also known as visceroreceptors because they are found within the walls of viscera, glands and blood vessels. They are responsible for subconscious monitoring of bodily homeostasis. However, they may also give rise to conscious visceral sensations (e.g. of bladder fullness).

Proprioceptors are found in muscles, joints, ligaments and fasciae. They include muscle spindle organs and tendon organs (of Golgi). There are also proprioceptors within the labyrinth of the inner ear. Proprioceptors monitor body position and movement. The part of proprioceptive information that reaches consciousness is responsible for the awareness of body position and movement (kinaesthesia).

Table 12 describes some types of sensory receptors.

TABLE 12 Sensory receptors.

RECEPTOR TYPE	RECEPTOR	LOCATION	STIMULUS
CUTANEOUS EXTEROCEPTORS			
Nociceptors:			
Polymodal	Free ending (C fibre)	All skin	Tissue damage from chemicals, heat, etc.
High threshold	Free ending (A fibre)	All skin	Tissue distortion
Thermoreceptors:			
Hot or cold	Free ending (A or C fibre)	All skin	Change of local skin temperature
Mechanoreceptors:			
Touch	Free ending (A or C fibre)	All skin	Low intensity changes in pressure
	Associated with hair follicle	Hairy skin	Movement of hair
	Tactile corpuscle (of Meissner) or end bulb (of Krause)	Glabrous skin Lips, digits	Changes in pressure
	Tactile meniscus (Merkel disc; Type I)	All skin	Sustained pressure change (slowly adapting)
Stretch	Ruffini capsule or terminal (Type II)	All skin	Sustained tension (slowly adapting)
Vibration	Lamellated corpuscle (Pacinian corpuscle)	Glabrous skin	Rapid changes in pressure
PROPRIOCEPTORS			
Mechanoreceptors:			
Muscle length	Primary (annulospiral) ending of spindle organ	Skeletal muscle	Rate of stretching and stretch
	Secondary (flower-spray) ending of spindle organ	Skeletal muscle	Sustained stretch (slowly adapting)
Muscle tension	Neurotendinous ending	Skeletal muscle	Tension (slowly adapting)
Vibration (?)	Lamellated corpuscle	Skeletal muscle	Rapid tension changes?
Nociceptors:	Free endings	Skeletal muscle	Tissue distortion or damage
Mechanoreceptors:			
Joint position	Type I (Ruffini?)	Synovial joints	Tension (slowly adapting)
Joint acceleration	Type II (Pacinian?)	Synovial joints	Change of tension
Joint strain (?)	Type III (Golgi?)	Synovial joints	Tension (high threshold, slowly adapting)
Nociceptors:	Type IV (free endings)	Synovial joints	Tissue distortion or damage

TABLE 12 (continued).

RECEPTOR TYPE	RECEPTOR	LOCATION	STIMULUS
INTEROCEPTORS			
Baroreceptors:	Spatulate endings	Arterial adventitia	Arterial blood pressure
Chemoreceptors:	Glomus cells/free endings	Carotid and aortic bodies	P_{CO_2}, P_{O_2}, pH
	Free endings	Intima of blood vessels and heart	P_{O_2}, pH
	Free endings	Gut	pH, glucose, amino acids, osmolarity
Mechanoreceptors:	Free endings, Pacinian corpuscles and others	Smooth muscle	Stretch or tension (slowly and rapidly adapting)
Irritant receptors:	Free endings	Respiratory and alimentary tracts	Irritant particles or chemicals
Nociceptors:	Free endings	Smooth muscle, glands, connective tissue	Tissue distortion or damage
SPECIAL SENSORY RECEPTORS			
Visual photoreceptors:	Rods	Retina	Photons (λ = 400 to 700 nm)
	Cones	Retina	Red/green/blue light
Auditory mechano-receptors:	Inner and outer hair cells	Cochlea	Sound pressure waves (20 Hz to 20 kHz)
Vestibular mechano-receptors:	Type I and II hair cells	Cristae of semicircular canals	Angular acceleration of head
		Maculae of utricle and saccule	Linear acceleration (position) of head
Olfactory chemo-receptors:	Olfactory rods	Olfactory mucosa	Odoriferous substances
Gustatory chemo-receptors:	Taste bud receptor cells	Tongue and oral cavity	Hydrogen and certain metal ions, sweet and bitter tasting chemicals

Effectors are the contractile or secreting tissues of the body (i.e. muscles or glands). The effector neurones are those that control these tissues through their axonal terminations (effector junctions). These neurones may be divided into two classes: somatic and visceral (autonomic).

Somatic effector neurones are motor neurones (motoneurones). Their terminals end in relation to skeletal muscle (striated muscle). The axons typically terminate at specialised synapses known variously as neuromuscular junctions, myoneural junctions, or motor end plates. The cell bodies of motoneurones are found in the ventral horns of the spinal cord (in the case of the cranial nerves, in motor nuclei of the brainstem; see Figure 176). There is thus a direct axonal link from the central nervous system to the effector organs. Furthermore, the axons are myelinated so that transmission is rapid (see Table 11).

At the neuromuscular junction, the synapse is enclosed; the endoneurium is continuous with the endomysium of the muscle fibre and the axon terminal is still covered by a Schwann cell (teloglial cell) although all myelin is lost in the preterminal region. Both pre- and post-junctional regions show specialisations and the synaptic cleft is only 20 to 50 nm wide. The neurotransmitter is acetylcholine, and the receptor has nicotinic rather than muscarinic properties. The result of activation is always depolarisation with resulting contraction of the muscle fibre.

Motoneurones may be subdivided according to whether they terminate in relation to muscle spindle organs (i.e. end on intrafusal muscle fibres) or in relation to extrafusal muscle fibres. Fusimotor neurones (γ-motoneurones) control the setting of muscle spindle organs. Skeletomotor neurones (α-motoneurones) control the main muscle fibres that produce the contractile power of the muscle. A very small proportion of motoneurones send collaterals to both intra- and extra-fusal fibres; these are called β-motoneurones.

Fusimotor neurones determine the proprioceptive information sent to the central nervous system from muscle spindle organs. They therefore play a vital part in the control of motor activity. In particular, they govern the tension developed in a resting muscle (i.e. muscular tonus). Fusimotor neurones fall into two categories: dynamic (γ1) and static (γ2). Dynamic γ-motoneurones, which terminate in the discrete, plate endings (true motor end plates) of intrafusal fibres, act mainly to increase the transient (phasic) response of the primary (Ia) spindle afferents to rapid changes of muscle length. Static γ-motoneurones terminate by more distributed, trail endings (which do not form proper motor end plates) on the intrafusal fibres. They act chiefly to augment the sustained (tonic) response of primary (Ia) and secondary (II) spindle afferents to a maintained muscle length. Fusimotor neurones are smaller than α-motoneurones and their axons are less heavily myelinated. They constitute about 30% of all motoneurones, being particularly common for postural and antigravity muscles of the lower limbs, trunk and neck.

A single α-motoneurone innervates many extrafusal muscle fibres, but each muscle fibre is innervated by only one α-motoneurone. An α-motoneurone with the muscle fibres it innervates is known as a motor unit. It is the basic efferent unit of the locomotor system. For muscles which are required to make very precise movements (e.g. the extra-ocular muscles), motor units may contain as few as six muscle fibres. For muscles where little precision is required (e.g. latissimus dorsi), motor units may contain a thousand muscle fibres. The strength of contraction of a muscle depends on the number of active motor units and the rate of production of action potentials by their neurones. The α-motoneurones are the final common output pathway (of Sherrington) by which the activity of the central nervous system produces behaviour through the skeletal musculature.

Visceral (autonomic) effector neurones are categorised according to the effector tissue: cardiomotor neurones supply heart muscle; visceromotor neurones innervate the smooth muscles of the viscera; vasomotor neurones supply the smooth muscle of blood vessels; pilomotor neurones supply the arrectores pilorum muscles of cutaneous hairs; sudomotor neurones are associated with sweat glands; and secretomotor neurones innervate salivary and digestive glands. The cell bodies of these neurones are found in the autonomic ganglia and visceral plexuses of the peripheral nervous system. Accordingly, transmission from the central nervous system to the effector tissues is indirect, involving at least one intermediate synapse. Furthermore, the axons of the effector neurones are usually unmyelinated. Thus, transmission is slower than in the somatic efferents (see Table 11).

The neuroeffector junctions of the visceral neurones are relatively unspecialised in comparison to motor end plates, and they are often widely separated and unenclosed. Thus, there are typically few specialisations of the pre- or post-junctional membranes. The junctional cleft is very variable in width. In the rapidly responding smooth muscle of the iris, the cleft is only 20 nm. In visceral and vascular muscle, however, it may be more than 1 μm across. At such junctions, the axon is usually no longer invested by a Schwann cell, but is fully exposed to the interstitial fluid. It often lies in a shallow groove on the surface of the effector cell and is marked by a small swelling or varicosity. The neurotransmitters are various, including: acetylcholine (with muscarinic receptors), noradrenaline, 5-hydroxytryptamine, peptides and purines. Their effects may be depolarising or hyperpolarising, or modulatory on other neurotransmitters. The structure of such junctions allows the action of their neurotransmitters to be readily influenced by circulating hormones, locally released chemicals, and other neurotransmitters.

Spinal reflexes are an important mechanism by which the spinal cord controls various bodily functions. Reflexes associated with the cranial nerves of the brainstem have similar features.

A reflex pathway is the simplest functional unit of the nervous system which uses both receptor and effector neurones. A reflex is an involuntary, stereotyped response to a stimulus and is usually adjustive or protective in purpose.

The simplest reflex pathway involves only two neurones, a primary sensory neurone synapsing directly on an effector neurone. Such a reflex pathway is called monosynaptic. An example is the pathway of the myotatic (stretch) reflex that may be evoked by tapping the patellar tendon; the primary (Ia) afferents of the muscle spindles synapse directly on α-motoneurones innervating the same muscles.

Most reflexes are not monosynaptic but involve at least one interneurone. These are accordingly called di- or multi-synaptic reflexes. An example is provided by limb withdrawal (flexor reflex) in response to noxious stimuli. Reflexes may involve complex, integrated actions. For example, if one leg is withdrawn in response to a noxious stimulus, the other leg is reflexly extended (crossed extensor reflex).

Neuronal control mechanisms have been found widely, both within the spinal cord and within the brain.

Man-made engineering devices commonly have self-regulating control mechanisms (such as the regulators or servo-mechanisms of refrigerators, cars and electronic amplifiers). In the nervous system, control mechanisms occur at both the biochemical and neuronal level, and may be of considerable subtlety. Indeed, the nervous system may eventually be understood as a complex set of interlocking control mechanisms.

An example of a neuronal control mechanism is provided by the pathway underlying the myotatic reflex. This pathway acts to maintain the muscle at a constant length in the presence of perturbations or unwanted variations (noise). Thus, an increase in the length of the muscle, and hence of the fibres of the muscle spindle organ, causes excitation of the spindle afferents. These in turn excite the motoneurones, whose activity brings about reflex contraction of the muscle, returning it towards its original length. Such a system confers constancy and stability. It also allows the initial signals to the α-motoneurones to be less precisely determined. The myotatic reflex provides an example of a control mechanism involving negative feedback (i.e. the effect of the control loop is to reverse any change from the set or desired level). Such a system needs a means of comparing the actual with the wanted response (in this case the comparator is the spindle organ), and allowing this difference to alter the future response. The recurrent pathway feeding back this difference to the input (in this case the primary afferent) completes the control loop, i.e. forms a closed loop control mechanism.

Control systems having feedback usually involve negative feedback loops, as these confer stability. Positive feedback is used only where an explosive change is required in response to a small input; it is used in the generation of action potentials (see page 408).

A simple example of a negative feedback loop occurring within the central nervous system is provided by the action of Renshaw cells. Renshaw cells are inhibitory interneurones within the ventral horn of the spinal cord. They receive an excitatory input from collaterals of α-motoneurones. Among the synapses their axons make are those upon the same group of α-motoneurones. This circuit causes α-motoneuronal activity to be self-limiting, as activity of the α-motoneurones excites the Renshaw cells, which in turn inhibit the motoneurones. Besides controlling their own activity, α-motoneurones that cause contraction of the agonist muscle will also cause, via interneurones, inhibition of the motoneurones controlling antagonist muscles and excitation of motoneurones controlling synergist muscles. Thus, activity within the neurones of one system may cause enhancement or reduction of activity in other systems, thereby providing local mechanisms of co-ordination and co-operation.

A further type of control mechanism involves feed-forward inhibition. For example, when the eyes are moved, instructions pass not only to the motoneurones controlling the extra-ocular muscles, but also to the visual system so that it may correctly interpret or ignore visual signals generated as a result of the movement of the eyes. Signals to the sensory system of this type are known as corollary discharges or efference copy. The mechanism underlying such signals is often feed-forward inhibition (i.e. at the same time as instructions are passed to the motoneurones, inhibitory signals are sent to neurones of the sensory system). Feed-forward inhibition is also widely used in the control of movement. Because there is no direct neuronal pathway from the output back to the input, feed-forward mechanisms are open-loop control systems. For such a system to provide an accurate control mechanism, the circuitry must normally demonstrate plasticity, so that the system may adapt (learn) to provide adequate error correction as a result of repeated trials. The advantage of feed-forward control over feedback control is speed; it enables correction in advance of, rather than after, the event.

Control mechanisms within themselves make possible further means of regulation. Thus, both the set point and the gain of a control system may be varied (i.e. the point about which the system operates and how well it resists or follows change). In muscle, such functions are achieved through the activity of γ-motoneurones, which control both muscle spindle length and its sensitivity to change. By this means, a muscle may be held at any desired length, and the resistance to change of that length varied.

Considerable efficiency is achieved by the use of local control mechanisms. Both the afferent and efferent information necessary for operation of the system may remain local, so avoiding the need for large fibre tracts travelling over long distances (with consequent increases in reaction times). Only the reduced amount of information relating to the current state and future operation of the control mechanism itself needs to be so transmitted. It follows that the mechanism providing the higher control can be made correspondingly smaller and more computationally efficient (being freed from consideration of the finer details of operations). Accordingly, the nervous system has, in part, a hierarchical organisation, with higher levels controlling those below. For example, in the control of movement, many of the basic components of a movement are organised within the spinal cord (or the brainstem in the case of cranial nerves). Information from the brain selects, modifies, and concatenates these components into voluntary actions.

Neuronal pathways consist of both serial and parallel connections. Serial connections are made from one neurone to each successive neurone. Parallel connections are duplicates of these serial connections. Note that in the nervous system, connections are rarely exactly duplicated. Usually, similar but overlapping (i.e. only partially duplicated) connections are made between neurones of one group and those of another. Each neurone of one group makes synapses on a number of neurones of the second group. Each neurone of the second group receives synapses from a number of neurones of the first group. Thus, the connections between the two groups have both serial and parallel properties.

Without serial connections, information could not pass from one part to another part of the system, and be successively (heirarchically) processed. Parallel pathways allow for the provision of redundancy or spare capacity, and for protection against the effects of damage. If any link in a serial circuit is broken, the circuit ceases to operate. With parallel connections, however, a second link may allow continued function. Moreover, parallel circuits allow contemporaneous rather than sequential processing.

It is this property that gives the nervous system its surprisingly rapid computational capacity in many tasks (compared with serially calculating computers in use today).

In summary, then, the nervous system displays a mixture of parallel and hierarchical organisation, with both local and distal control circuits.

THE AUTONOMIC NERVOUS SYSTEM

The autonomic nervous system comprises those parts of the peripheral and central nervous systems innervating smooth and cardiac muscle, blood vessels, glands and viscera. It includes both afferent and efferent components.

The autonomic nervous system plays an important role in bodily homeostasis, feeding and reproduction. It is particularly concerned with maintaining and controlling the body's internal environment, whereas the somatic system may be said to be primarily concerned with the external environment. The autonomic nervous system also has an important role in emotional reactions.

The autonomic nervous system has also been called the visceral, vegetative, or involuntary nervous system. None of these names is entirely appropriate. The system does not operate autonomously without reference to other, somatic activity in the central nervous system. It does not control merely viscera, nor solely growth and metabolism. Furthermore, although its actions are mainly involuntary and lacking in conscious control, so are many other activities of the nervous system. Moreover, through manipulation of the emotions, or with techniques such as yoga or biofeedback, the activities of the system may be at least partly brought under conscious control. In addition, the afferents of the autonomic nervous system can give rise to conscious sensations.

The autonomic and somatic nervous systems overlap and are interdependent. In the periphery, autonomic fibres share the same peripheral nerves as the somatic fibres. However, the autonomic fibres have their own ganglia and they can form their own nerves (e.g. splanchnic nerves) and plexuses (e.g. carotid, cardiac, pulmonary, mesenteric, and pelvic plexuses). Centrally, autonomic neurones are gathered into specific autonomic nuclei, but they receive somatic as well as autonomic afferents and they influence somatic as well as autonomic efferents. Spinal and peripheral parts of the autonomic nervous system are capable in certain instances of operating in the absence of supraspinal influences. However, such regions are usually under the control of the brain via descending spinal tracts, and lesions of these tracts usually cause at least temporary disruption of 'autonomic' function. Such higher central control of the autonomic nervous system is chiefly organised through the hypothalamus and the reticular formation of the brainstem. It is noteworthy that the reticular formation itself has both somatic and autonomic functions.

Autonomic afferents

Most activity in autonomic afferents never reaches consciousness, being mainly concerned with subconscious monitoring of the internal milieu of the organism. Such information is used to effect reflex adjustments through the autonomic efferents. When sensations do reach consciousness, they are often vague, poorly localised and affective in character (e.g. feelings of general malaise, hunger and sexual arousal). Stimulation of visceral nociceptors or over-stimulation of other visceroreceptors (e.g. by distension or muscular spasm of a viscus) can lead to severe, distressing pain.

Autonomic afferents (Aδ and C fibres), which pass via sympathetic and sacral parasympathetic nerve trunks and the spinal nerves, have their cell bodies in the dorsal root ganglia. Fibres may reach the spinal cord by either dorsal or ventral roots. They mainly terminate in laminae I, V and VII of the dorsal horn. Interneurones make connections with the efferent autonomic neurones of the intermediolateral cell column of the spinal grey matter. Multisynaptic autonomic reflex pathways passing through the spinal cord are thereby made possible. Such subconscious, reflex control of autonomic activity is the main use that is made of autonomic afferent input.

At least some of the autonomic afferents contain substance P and vasoactive intestinal peptide (VIP). They may release these substances from peripheral as well as central terminals. Some of these afferents are nociceptive and terminate on dorsal horn neurones which also respond to pain of somatic origin. This finding provides an anatomical basis for referred pain in which pain of visceral origin is referred to the dermatomes supplied by the spinal nerves within which the visceral afferents reached the spinal cord. Thus, pain from cardiac ischaemia (angina pectoris) is felt in the upper part of the chest and the medial aspect of the left arm. It should be noted that supraspinal (possibly thalamic) interactions may also contribute to the phenomenon of referred pain.

Sacral autonomic afferents convey sensations of bladder and rectal distention and sexual arousal, as well as of pain of pelvic visceral origin. They are chiefly concerned with information needed for the autonomic control of the pelvic viscera.

Autonomic afferents also pass via the ninth and tenth cranial nerves. Their cell bodies are found in the inferior ganglia of these nerves. Their central processes terminate in the caudal part of the nucleus of the solitary tract in the medulla oblongata. These afferents do not seem to convey nociceptive information, though they may convey sensations such as hunger and nausea as well as vague tactile and thermal sensibilities from the upper part of the alimentary tract. They mainly transmit information required for subconscious autonomic regulation (notably of heart rate, blood pressure, breathing and digestion) from the heart, the aorta and the region of bifurcation of the common carotid arteries, the lungs, and the greater part of the alimentary tract. They are further mentioned with the appropriate cranial nerves.

Autonomic efferents

Autonomic efferent pathways are characterised by the presence of ganglia. Information must therefore pass via at least two neurones from the central nervous system to the effector tissue (smooth muscle, heart muscle or glandular epithelium). Preganglionic neurones are those whose cell bodies are within the central nervous system and whose axons terminate within the autonomic ganglia. Their cell bodies are found within the parasympathetic efferent nuclei of the brainstem (see pages 464 to 465) or the intermediolateral cell column of the spinal cord. Most of their axons are myelinated (see Table 11). They all release acetylcholine, the postganglionic cells being excited via nicotinic receptors.

Postganglionic neurones are those whose cell bodies are within the autonomic ganglia and whose axons pass to the effector tissues. Their axons are normally unmyelinated. Unlike skeletal muscle, autonomic effector tissues do not normally atrophy or degenerate following denervation, although they do develop denervation hypersensitivity (particularly to circulating adrenaline in the case of tissues which have lost noradrenergic innervation). Regenerative capacity is well developed in peripheral autonomic fibres.

The autonomic ganglia allow modulation, integration and divergence of signals. Besides synapses of preganglionic axons, they may also receive synapses from autonomic afferents. Further, many contain interneurones. The ganglia are found in three locations: in front of the vertebral column (prevertebral), to the side of the vertebral column (paravertebral), and in, or near, the walls of the target tissues (terminal).

The efferent part of the autonomic nervous system may conveniently be divided into two systems: sympathetic and parasympathetic. Their main actions are summarised in Table 13.

The sympathetic system gives rise to efferents in the ventral roots of the segments of the spinal cord between T1 and L2 (or L3), i.e. the portion of the spinal cord that

TABLE 13 Major features of the operation of the sympathetic and parasympathetic systems.

	SYMPATHETIC	**PARASYMPATHETIC**
Primary effects	Energy expenditure and emergency system (fight and flight)	Protective and restorative system (feeding and resting)
	Generalised and long-lasting Hormonal (adrenaline) supplementation	Discrete and of short duration
CNS outflow	Thoracolumbar: T1 to L2/3	Craniosacral: III, VII, IX, X and S2 to 4
Ganglia	Paravertebral and prevertebral, close to spinal cord	Terminal, close to organ
Typical divergence	~ 1 : 100	~ 1 : 2
Distribution	Throughout body	Restricted
Transmitter at effectors*	Noradrenaline (usually)	Acetylcholine
Eye	Pupillary dilation	Pupillary constriction and accommodation
Lacrimal gland	—	Secretion ↑ , vasodilation
Salivary glands	Secretion ↓	Secretion ↑ , vasodilation
Heart	Excitor	Depressor
Bronchi and lungs	Dilation	Constriction, secretion ↑ ?
Alimentary tract	Peristalsis ↓ Secretion ↓ Sphincters closed	Peristalsis ↑ Secretion ↑ Sphincters open
Liver	Glycogenolysis ↑	—
Adipose tissue	Release of fatty acids	—
Bladder	Retention	Emptying
Reproductive organs	Ejaculation/uterine contractions ↑	Vasodilation (erection)
Adrenal medulla	Secretion ↑	—
Sweat glands	Secretion ↑	—
Blood vessels†	Vasoconstriction (usually)	—
Arrectores pilorum muscles	Piloarrection	—

* The postganglionic sympathetic neurotransmitter is acetylcholine in the blood vessels of skeletal muscle and at exocrine sweat glands of hairy skin (see text for further comments on neurotransmitters within both systems).

† The sympathetic system may produce vasodilation in skeletal musculature. The parasympathetic system may produce local vasodilation where it innervates glandular epithelium.

gives rise to the first thoracic to the second or third lumbar spinal nerves. Accordingly, it is alternatively known as the thoracolumbar division of the autonomic nervous system. Most of its postganglionic neurones release the neurotransmitter noradrenaline.

The parasympathetic system gives rise to efferents in the ventral roots of some or all of the sacral segments of the spinal cord between S2 and S4, and also in the third, seventh, ninth, and tenth cranial nerves. It is therefore sometimes called the craniosacral division of the autonomic nervous system. The postganglionic neurones mainly release acetylcholine.

The sympathetic system, in broad terms, facilitates the use of energy reserves and the expenditure of energy by the body (e.g. energy consumed through increased skeletomotor activity). The system fulfils this role at all times, but is most strongly activated by emotional excitement, fear, cold or pain. It is therefore the system that prepares the body to react to emergencies or respond to stress (the 'fight or flight reaction'). In this capacity, it acts on the heart to increase blood flow, and directs blood to skeletal muscles and the brain (rather than the gut and skin). It reduces activity of the alimentary tract, but causes elevation of blood sugar levels and dilates the bronchioles. The sympathetic system is also responsible for sweating, piloarrection, dilation of the pupils, and ejaculation.

For many reactions (particularly the 'fight or flight reaction'), the sympathetic system is broadly activated and has widespread and long-lasting effects. Several features of the organisation of the sympathetic system facilitate such prolonged and integrated action. First, there are numerous interconnections between its neurones. Second, there is considerable divergence from its preganglionic neurones to the much larger number of postganglionic neurones. Third, the main postganglionic neurotransmitter (noradrenaline) is comparatively slowly inactivated. Fourth, noradrenaline is readily potentiated in its actions by circulating adrenaline, this potentiation being made possible by the open, unspecialised morphology of the effector junctions. However, it is an exaggeration to imagine that all the neurones of the sympathetic system are always, or even commonly, activated together.

It is important to appreciate that, because all arteries (and many veins) receive sympathetic innervation, the sympathetic system supplies nearly all parts of the body. Indeed, the sympathetic fibres acting chiefly upon arterioles are the main neural means of controlling both overall blood pressure and the distribution of blood to different tissues.

Despite its widespread distribution and its importance in the body's reactions to emergencies, the sympathetic system is not essential to life. Deliberate lesioning of parts of the sympathetic system (sympathectomy) is used for the relief of some clinical conditions. Such lesions have to take account of the alternative routes that may be taken and the intermixing of fibres in the sympathetic chain, the regenerative capacities of the peripheral nervous system, and the possible occurrence of aberrant (intermediate) ganglia.

Sympathetic ganglia are either prevertebral or paravertebral in position. The prevertebral ganglia are found distributed within the plexuses surrounding the origins of visceral branches of the abdominal aorta (notably the coeliac, and superior and inferior mesenteric ganglia). The paravertebral ganglia are ovoid structures found anterolaterally on either side of the bodies of the vertebrae. On each side, they are linked by longitudinal fibres to form a sympathetic trunk or chain. These trunks extend from the second or third cervical vertebra to the coccyx. Paravertebral ganglia are found along the entire extent of the chain. In the neck, there are two or three cervical ganglia: superior, middle (sometimes absent), and inferior ganglia (often the inferior cervical ganglion is fused with the first thoracic ganglion to form a stellate or cervicothoracic ganglion). There are typically eleven thoracic ganglia, four lumbar ganglia, and four or five sacral ganglia in each chain. Additionally, the two chains fuse caudally, and here show an unpaired coccygeal ganglion (ganglion impar). Sometimes, groups of ganglion cells are found within the grey rami and hence beyond the sympathetic chain. These are called intermediate ganglia.

Sympathetic ganglia contain interneurones. In particular, there are interneurones called SIF cells (small intensely fluorescent cells) which release dopamine and produce inhibition of the postganglionic neurones.

Sympathetic preganglionic neurones have their cell bodies in the intermediolateral cell column of the lateral horn of the thoracic part of the spinal cord (segments T1 to L2/L3). There have been suggestions that some sympathetic preganglionic neurones may be found in other segments also. Their axons pass via the ventral roots (often having ascended or descended by a number of segments in the white matter of the spinal cord) to the spinal nerves. In the thoracic region, all such fibres appear to enter the ipsilateral ventral root, but in the lumbar region some pass to the contralateral side. After a short distance, each spinal nerve between T1 and L2 (or L3) gives rise to a white ramus communicans which passes to a paravertebral sympathetic ganglion. The preganglionic fibres pass through the white rami communicantes, and at the ganglia they may:

Terminate in the first ganglion they reach.

Pass up or down the sympathetic trunk, synapsing in one or more ganglia.

Pass straight through the trunk without synapsing to form splanchnic nerves.

The splanchnic nerves terminate in prevertebral ganglia. Some preganglionic fibres end on cells of the adrenal medulla, causing them to release mainly adrenaline, but also some noradrenaline and a small amount of dopamine.

Fibres of postganglionic neurones of the paravertebral ganglia may ascend or descend the sympathetic chain before leaving it. They may leave it either in a grey ramus communicans, or as a separate nerve (e.g. splanchnic nerves), or by passing along blood vessels. They reach their target tissues either via the branches of peripheral nerves or via blood vessels (sometimes passing through peripheral plexuses). One or more grey rami communicantes pass to each spinal nerve. Additionally, most of the cranial nerves receive sympathetic fibres. In some instances, the white and grey rami fuse. In any case, the division of function is not complete; both may contain myelinated (white) and unmyelinated (grey) fibres, pre- and post-ganglionic fibres, and some autonomic afferent fibres. Where a spinal nerve shows both types of rami communicantes, the white ramus is located distal to the grey ramus.

The neurotransmitter of most postganglionic sympathetic neurones is noradrenaline. The actions of noradrenaline may be excitatory or inhibitory depending on the receptors at which it acts (all four recognised receptors are found: α_1, α_2, β_1, β_2). In particular, it can act on neighbouring postganglionic parasympathetic fibres to inhibit their release of acetylcholine. However, exceptions are provided by those postganglionic sympathetic neurones supplying the exocrine sweat glands of hairy skin, and blood vessels of skeletal muscle (including those of the tongue). These exceptions use acetylcholine. In the case of the sweat glands, vasoactive intestinal peptide (VIP) is also released; it produces local vasodilation. It has also been suggested that some neurones within the walls of blood vessels and the bronchial tubes may release other neurotransmitters or neuromodulators, such as 5-hydroxytryptamine, adenosine triphosphate or somatostatin.

On each side of the body there is a long pathway from the hypothalamus, through the lateral part of the brainstem, to the upper two thoracic segments of the spinal cord, through the white rami communicantes and cervical sympathetic chain to the superior cervical ganglion, and then via postganglionic fibres to the ipsilateral orbit. A lesion anywhere along this pathway produces a series of signs known as Horner's syndrome. There is dilation of the pupil (miosis), drooping of the upper eyelid (ptosis), and an apparent sinking of the eye into the orbit (enophthalmos). The ptosis occurs because the upper eyelid contains some smooth muscle innervated by sympathetic fibres. Horner's syndrome also involves flushing (vasodilation) and loss of sweating on the same side of the face.

The parasympathetic system has functions that are generally more discrete, localised, and often shorter in duration than those of the sympathetic system. It has actions complementary to, and often opposite to, those of the sympathetic system. It promotes changes that restore and conserve energy stores within the body. The parasympathetic system is most active during quiescence and contentment. It slows the heart and increases activity of the alimentary tract (increasing peristalsis and secretion, and causing relaxation of sphincters). It promotes salivation, digestion, and removal of waste by emptying the bladder and rectum. It produces engorgement of erectile tissue in the genital organs, and secretion from their accessory glands. In the eye, the parasympathetic system causes pupillary constriction, accommodation of the lens, and the production of tears.

Except in the alimentary tract, divergence between preganglionic and postganglionic neurones is less in the parasympathetic system than in the sympathetic system. Furthermore, the ganglia are not commonly interconnected. The postganglionic neurotransmitter is mainly acetylcholine, acting at muscarinic receptors. It is rapidly inactivated by the enzyme acetylcholinesterase. Salivation, lacrimation and gastric secretion are promoted by the co-release of vasoactive intestinal peptide which produces local vasodilation.

The spinal preganglionic neurones are found in the intermediolateral cell column of the sacral part of the spinal cord (S2 to S4 segments). Their axons pass via the ventral roots to the pelvic splanchnic nerves (nervi erigentes). The cranial preganglionic neurones are found in the parasympathetic efferent nuclei of the brainstem (i.e. the Edinger–Westphal or accessory oculomotor nucleus in the midbrain, the salivatory nucleus in the cranial medulla oblongata and caudal pons, and the dorsal motor nucleus of the vagus in the medulla oblongata). Their axons pass in the third, seventh, ninth and tenth cranial nerves. The preganglionic fibres are all myelinated. They are also long because parasympathetic ganglia are terminal in position.

The parasympathetic ganglia lie within, or close to, the walls of the structures they innervate. Within the head, they form discrete ovoid masses. Within the rest of the body, they are typically more distributed and may take the form of widely scattered groups of cells located in intramural plexuses.

The postganglionic fibres are fairly short, unmyelinated, and primarily release acetylcholine. Besides its action on the target tissues, acetylcholine acts on adjacent postganglionic sympathetic fibres to inhibit the release of noradrenaline. The reciprocal inhibitory actions of acetylcholine and noradrenaline upon parasympathetic and sympathetic fibres in part explain the opposing actions of the two systems upon organs that receive dual innervation.

Further information on the parasympathetic components of the cranial nerves may be found on pages 464 to 465.

The enteric nervous system was once considered as part of the parasympathetic system. However, it is now recognised as a system in its own right, although it is under autonomic control.

The enteric nervous system is formed by the neurones in the wall of the alimentary tract from the middle third of the oesophagus to the rectum. The neurones include sensory, effector, and interneurones. They are found in the myenteric plexus (of Auerbach), between the layers of smooth muscle, and in the submucosal plexus (of Meissner). Within the submucosal plexus, there appear to be sensory neurones that project to other neurones in the gut wall. Other sensory neurones project to the prevertebral sympathetic ganglia as well as to the spinal cord and, via the vagus, to the brainstem. The myenteric plexus includes the effector neurones. Among its intrinsic neurones are postganglionic parasympathetic neurones, which use acetylcholine as their neurotransmitter. These have an excitatory action. Inhibition seems to be produced by other intrinsic neurones that release adenosine triphosphate. Yet other neurones of the enteric nervous system contain and release substance P, cholecystokinin, vasoactive intestinal peptide, 5-hydroxytryptamine, enkephalins and a considerable number of other possible neurotransmitters. It has been estimated that the system has

10^8 neurones (as many as the spinal cord). They control peristalsis, secretion and the operation of sphincters within the gut. Their extrinsic control is provided by postganglionic sympathetic fibres, and by mainly preganglionic parasympathetic fibres originating in the vagus and pelvic splanchnic nerves. The sympathetic fibres appear to act by inhibiting the parasympathetic neurones. However, much of the control of the enteric nervous system is local, and essentially independent of central nervous system influences. Indeed, the system is capable of operating in the absence of such input.

Applied anatomy of the peripheral nervous system

Both spinal and cranial nerves are often anaesthetised by the injection of local anaesthetic solutions. A further clinical feature of considerable importance relates to the fact that peripheral nerves recover from injury in a manner different from that of fibres within the central nervous system (see page 413). Regeneration of peripheral nerves has significance for both accidental and surgical nerve damage.

The conduction of action potentials in peripheral nerves may be blocked in a variety of ways, resulting in differential anaesthesia and/or paralysis. Local anaesthetic agents (e.g. procaine) impair ionic movements across the plasma membrane. They first block conduction of action potentials at the nodes of Ranvier of small myelinated (Aδ) fibres. Subsequently, conduction is blocked in unmyelinated fibres. Conduction in large myelinated fibres is blocked only with high concentrations of anaesthetic agents. Thus, painful sensations that travel via small-calibre fibres may be blocked, whereas tactile sensations and muscle contraction is preserved. However, when blood flow is obstructed (e.g. by a tourniquet or by local compression) so that the nerve is deprived of oxygen and becomes ischaemic, conduction is first blocked in large myelinated fibres. Thus, 'pins and needles' may be felt, although a limb is partially paralysed.

Recovery of function following damage to a peripheral nerve is dependent on the type of injury:

Compression of fibres, without degeneration, impairs function but complete recovery (which may not be immediate) follows relief of the pressure.

When a nerve is crushed, axons and myelin are damaged and distally undergo degeneration. However, the endoneurial sheaths are likely to be spared. Regeneration occurs by sprouts from the ends of the proximal parts of the axons growing along the endoneurial tubes. As long as the appropriate sprout enters the correct endoneurial tube, full recovery is eventually possible.

If the nerve is severed, the distal parts of the axons and their myelin sheaths will again degenerate. Following severance, particularly if there is a gap (which may fill with scar tissue) between the proximal and distal parts, recovery will be slow and is unlikely to be complete. Microsurgical apposition of fasciculi may be necessary if misalignment or a gap of more than a few millimetres is present. If the fibres are misaligned, a tangled mass of nerve fibres, Schwann cells and connective tissue (a traumatic neuroma) may form. This may cause a variety of unpleasant sensations (paraesthesiae) such as the phantom limb pain that can follow amputation.

Once an axonal sprout has entered the distal endoneurial tube (assisted by Schwann cell guidance), it may grow at rates of 2 to 5 mm/day. Following re-establishment of the connections with an appropriate target tissue, remyelination will occur (although often with reduced internodal distances). A further problem that may arise is that the degeneration or disintegration of muscles or other end organs may prevent full re-innervation. Sensory recovery occurs first for sensations of deep pressure, then for poorly localised cutaneous pain, followed by appreciation of temperature and lastly, with the greatest likelihood of being incomplete, for the finer, discriminitive aspects of touch.

THE GROSS ANATOMY OF THE SPINAL CORD (Figure 171)

The spinal cord is a continuous, cylindrical structure, typically about 450 mm long and up to 14 mm in diameter. On its ventral surface is a narrow fissure called the ventral median fissure.

The spinal cord of the adult occupies most of the cranial two-thirds of the vertebral canal. It usually ends close to the intervertebral disc between first and second lumbar vertebrae, although this level varies from the lower third of the twelfth thoracic

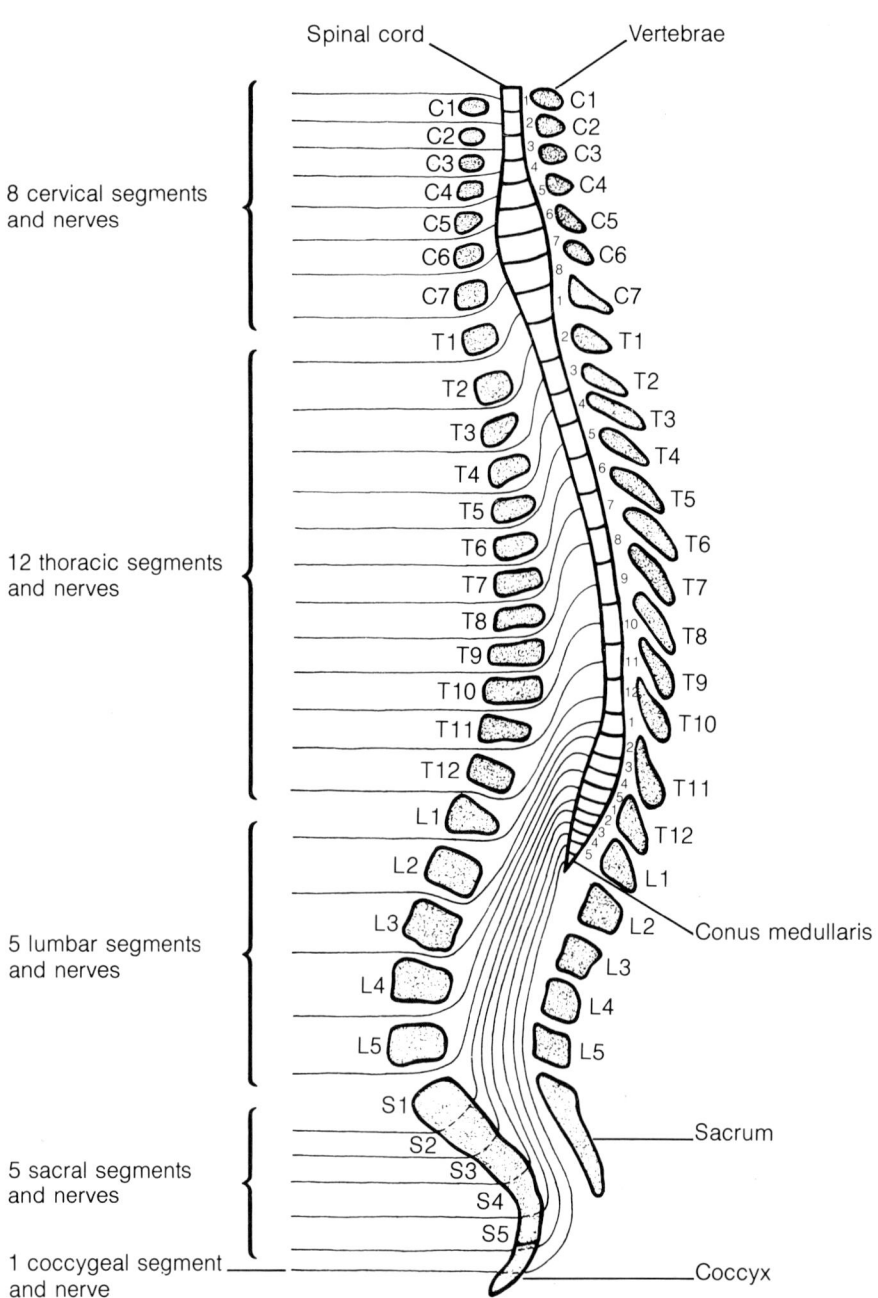

Figure 171 *Lateral view of the spinal cord within the vertebral canal,* illustrating the segments of the spinal cord and the positions of the spinal nerves.

vertebra to the disc between the second and third lumbar vertebrae. During the initial period of intra-uterine life, the spinal cord extends for the full length of the vertebral column. After the third month of intra-uterine life, the vertebral column grows in length more rapidly than the spinal cord, so that, by the start of the seventh month, the end of the spinal cord is usually close to the lower border of the third lumbar vertebra.

The spinal cord connects with the medulla oblongata of the brainstem near the cranial border of the atlas, just below the foramen magnum. 158,30; 158,26
158,43; 158,21

Through their dorsal and ventral roots, each of the 31 pairs of spinal nerves is associated with a successive region of the spinal cord. The cord may therefore conveniently be divided into 31 segments, each named after the corresponding pair of spinal nerves. By convention, the abbreviations C, T, L and S indicate respectively the cervical, thoracic, lumbar and sacral segments of the spinal cord. 200B,32; 200B,37

The spinal cord gradually increases in diameter from its caudal to its cranial end. However, superimposed on this trend are two expansions associated with segments responsible for innervation of the limbs. The cervical enlargement extends approximately from the fourth cervical segment to the first thoracic segment. The lumbosacral enlargement extends from the first lumbar segment to the third sacral segment. The conical caudal end of the spinal cord is known as the conus medullaris. From its end, the spinal meninges are continued as a thin strand of connective tissue, the filum terminale, which is attached to the coccyx.

Because of the differential growth of the vertebral column and spinal cord during development, the more caudal spinal cord segments come to lie opposite vertebrae placed more cranially in the vertebral column than the segmental name might suggest. In the average adult, segments of the upper cervical spinal cord lie approximately opposite the bodies of the same-numbered vertebrae. Lower cervical and thoracic segments are located opposite the bodies of the vertebrae numbered one to two positions more cranial (e.g. the second thoracic segment is opposite the body of the first thoracic vertebra; the twelfth thoracic segment lies opposite the tenth thoracic vertebra). Lumbar segments lie opposite the eleventh and twelfth vertebral bodies, and sacral segments opposite the first lumbar vertebra.

Because the dorsal and ventral roots of the lower lumbar, sacral and coccygeal nerves are of necessity long, they form a tassel extending caudal to the termination of the spinal cord. This resembles a horse's tail and is consequently named the cauda equina.

The important external relations of the spinal cord are with the vertebrae and the intervertebral discs of the vertebral column. The spinal cord is separated from them by the spinal meninges and extradural adipose tissue. Arteries and veins supplying the spinal cord lie within the subarachnoid space and are described on pages 582 and 584. Internal vertebral venous plexuses are located between the dura mater and the vertebral periosteum (see page 393). The spinal part of the accessory nerve ascends lateral to cervical segments of the spinal cord. 198F; 200B
136
200B,33

THE SPINAL MENINGES

As for the brain, the spinal cord is invested by three meninges: the pia mater, arachnoid mater and dura mater. These are similar to, and continuous with, their cranial counterparts. 198F,55; 198F,56

The spinal dura mater extends caudally as a cylindrical sheath over the spinal cord from the foramen magnum as far as the caudal border of the second sacral vertebra. It is separated from the vertebral periosteum by the extradural (epidural) space containing loose fat and areolar tissue together with the internal vertebral venous plexus. The spinal dura mater is attached to the circumference of the foramen magnum and to the posterior (dorsal) surfaces of the second and third cervical 198F,56; 200B,35

158,28

vertebrae. Along its length it is secured by fibrous extensions to the posterior longitudinal ligament of the vertebrae. Caudally, it invests the filum terminale and is continuous with the vertebral periosteum at the coccyx. It also forms sleeves around the spinal nerve roots, becoming continuous with the epineuria of the spinal nerves in the intervertebral foramina. Possibly in order to restrict movement of the nerve roots, the epineuria in the cervical region are closely attached to the periosteum of the transverse processes of the adjacent vertebrae.

The spinal pia mater covers the whole of the spinal cord and lines its ventral median fissure. It invests the dorsal and ventral roots and is continuous with the perineuria of the spinal nerves. At the caudal end of the spinal cord the pia continues as the filum terminale to be attached to the coccyx. Laterally, 18 to 24 triangular folds are attached by their apices to the arachnoid and dura. These folds form the ligamentum denticulatum and provide anchorage for the cord.

The spinal arachnoid mater lines the spinal dura mater, including the meningeal sleeves of the spinal nerve roots. Unlike the cranial arachnoid, it has few trabeculi linking it to the pia mater. Thus, the subarachnoid space surrounding the spinal cord is fairly regular and roughly annular in transverse section.

The lumbar cistern of cerebrospinal fluid is the subarachnoid space between the end of the spinal cord (opposite the first or second lumbar vertebra) and the end of the dural sac (at the second sacral vertebra). It contains parts of the filum terminale and cauda equina. The lumbar cistern provides a convenient place from which samples of cerebrospinal fluid may be obtained for analysis (lumbar puncture). Also, local anaesthetic may conveniently be infused into the lumbar cistern or extradural space in the region of the cauda equina to produce anaesthesia in the territories of the caudal spinal nerves (spinal block or epidural anaesthesia). This technique is very useful in childbirth, and is also used in surgery (e.g. hip replacements).

THE INTERNAL STRUCTURE OF THE SPINAL CORD

In transverse section, the spinal cord displays an H-shaped region of grey matter surrounded by white matter (Figure 172). As with other regions of the central nervous system, the spinal cord is approximately bilaterally symmetrical, so that all but midline structures are paired. The H-shaped grey matter is made up of dorsal and ventral horns on each side linked by a short bar of tissue known as the grey commissure. The spinal canal is located centrally in the grey commissure. The white matter may be subdivided broadly into: dorsal funiculi or columns (situated medial to the dorsal horns and above the grey commissure); lateral funiculi (found lateral to the dorsal and ventral horns, and between the lines of emergence of the dorsal and ventral roots of the spinal nerves); and the ventral funiculi (situated ventral and medial to the ventral horns, and medial to the lines of emergence of the ventral roots). A ventrolateral funiculus can also be located on each side, within the lateral and ventral funiculi, and ventrolateral to the ventral horn. The dorsal funiculi are separated by a thin connective tissue sheet called the dorsal median septum. The ventral funiculi are separated by the ventral median fissure.

The ascending and descending tracts of the white matter, and the major regions of the grey matter, are shown in transverse sections of the cord at sacral, lumbar, thoracic and cervical levels in Figure 173. Note that, by convention, quoted levels used to describe the spinal cord (e.g. L1 to S3) refer to segments of the cord and not to vertebrae. The main features of the grey matter and the white matter will now be described.

The dorsal horn has essentially the same shape and composition throughout the spinal cord. Within each segment, the dorsal horn has a general somatotopical organisation, neurones dealing with information from lateral regions being lateral to those neurones

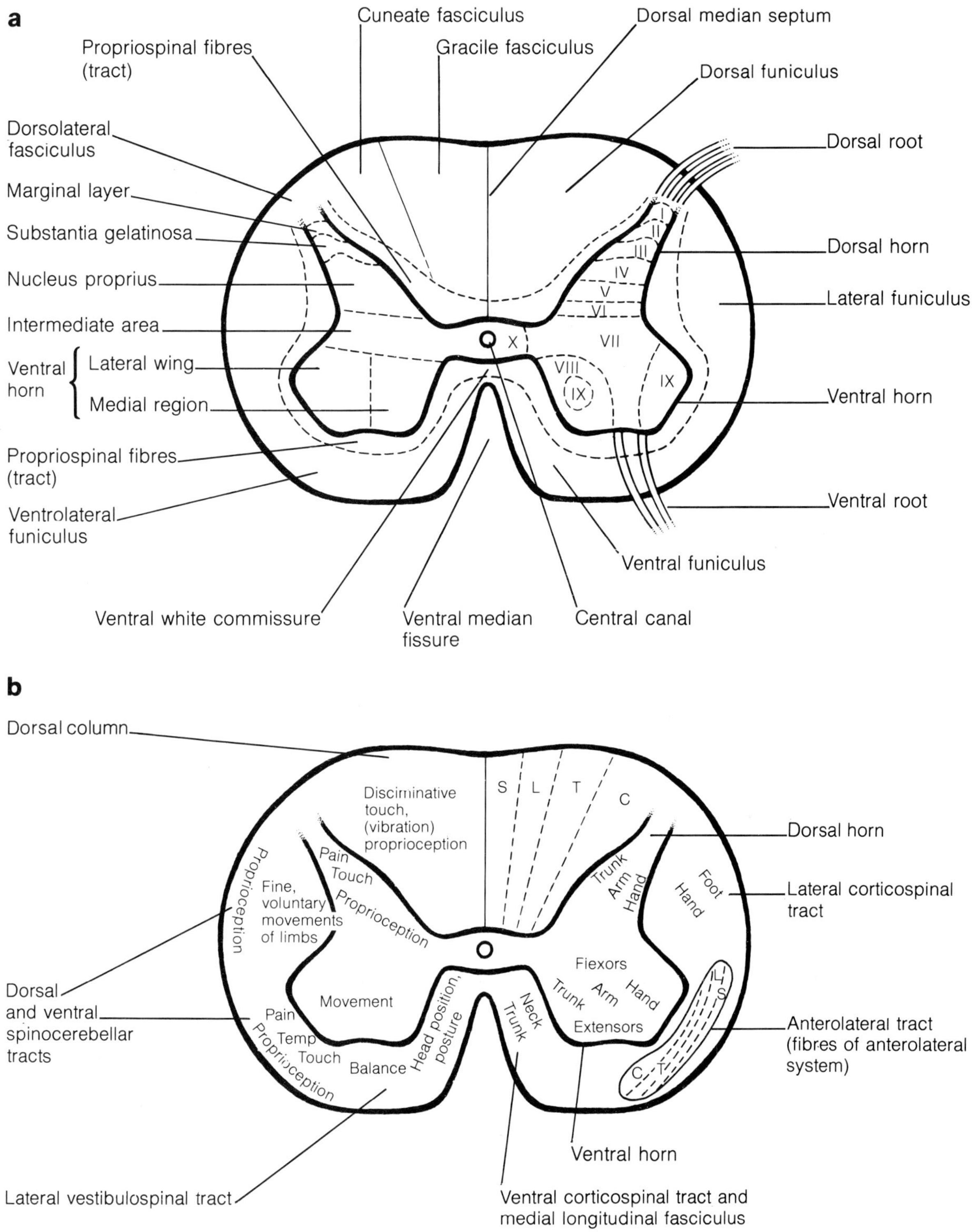

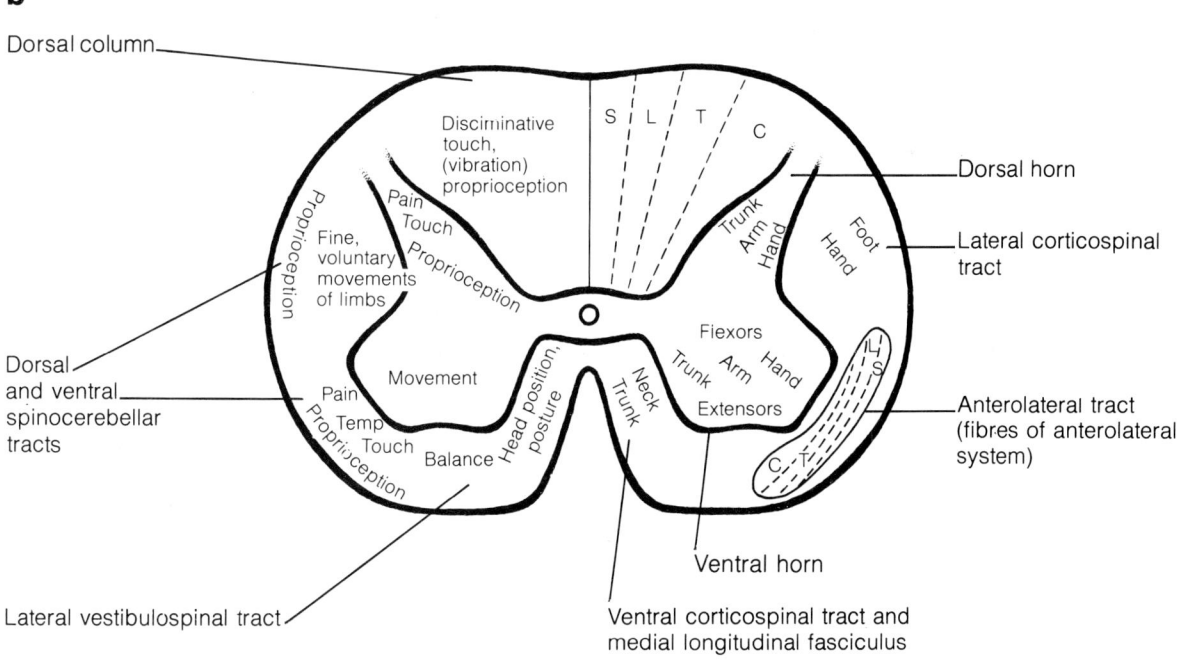

Figure 172 *The main features of the internal anatomical organisation of the spinal cord.* **a)** *Main divisions of white and grey matter. I to X = laminae of Rexed.* **b)** *Chief modalities and somatotopical organisation. (Temp. = Temperature.) Abbreviations relating to dermatomal origin of sensations: C = cervical, T = thoracic, L = lumbar, S = sacral.*

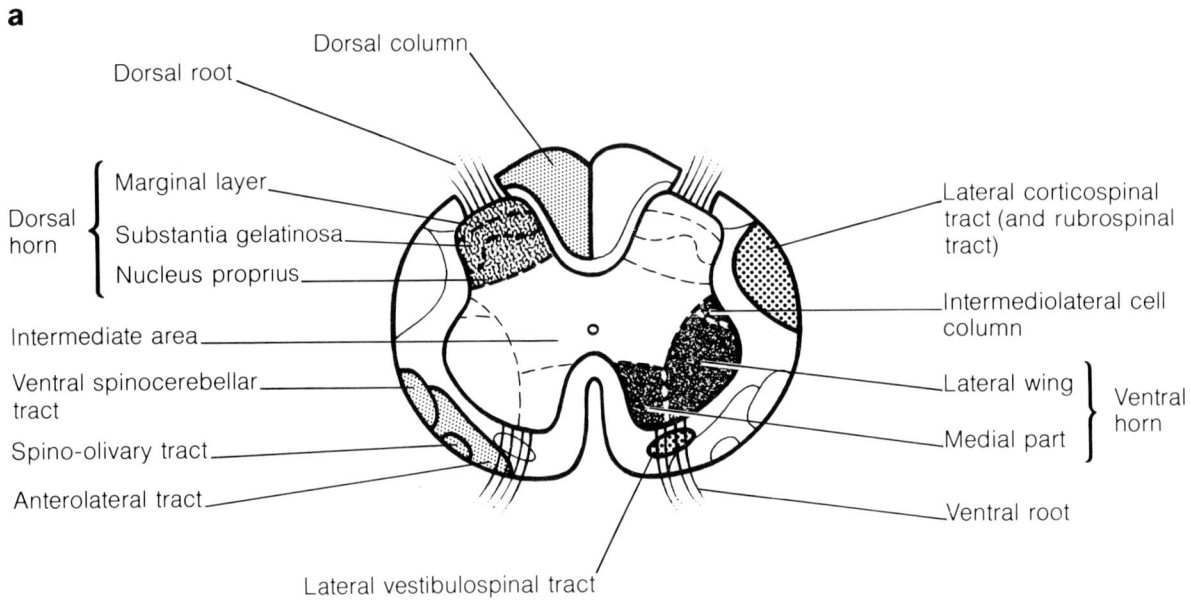

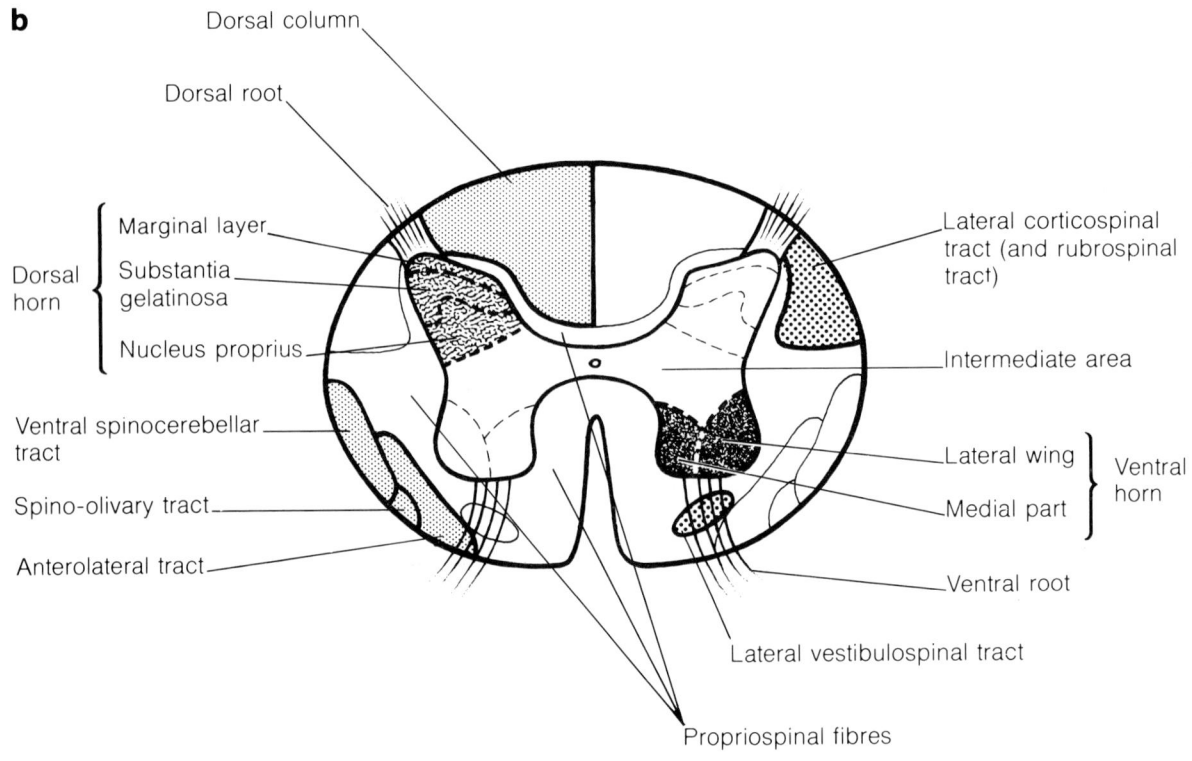

Figure 173 *Transverse sections of* a) *sacral* b) *lumbar* c) *thoracic* d) *cervical segments to show the positions of the major ascending and descending tracts. Sensory areas and tracts are labelled on the left, motor on the right. (The pontine and medullary reticulospinal tracts consist of fibres scattered in the ventral and lateral funiculi; they are not illustrated.)*

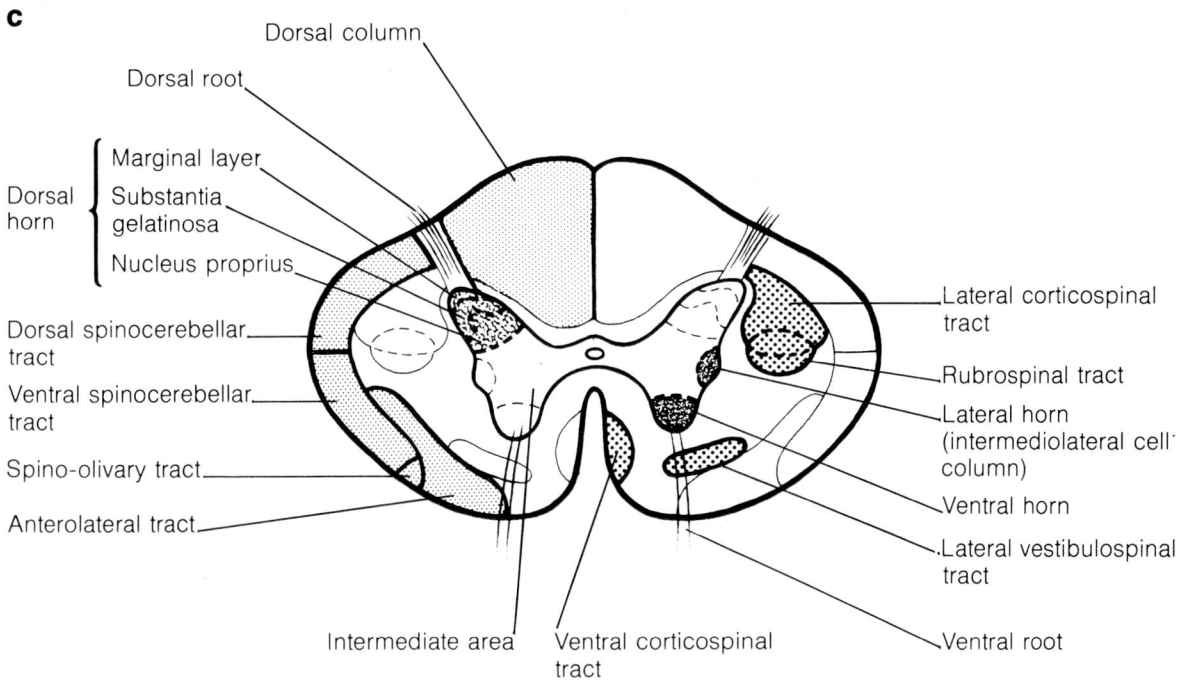

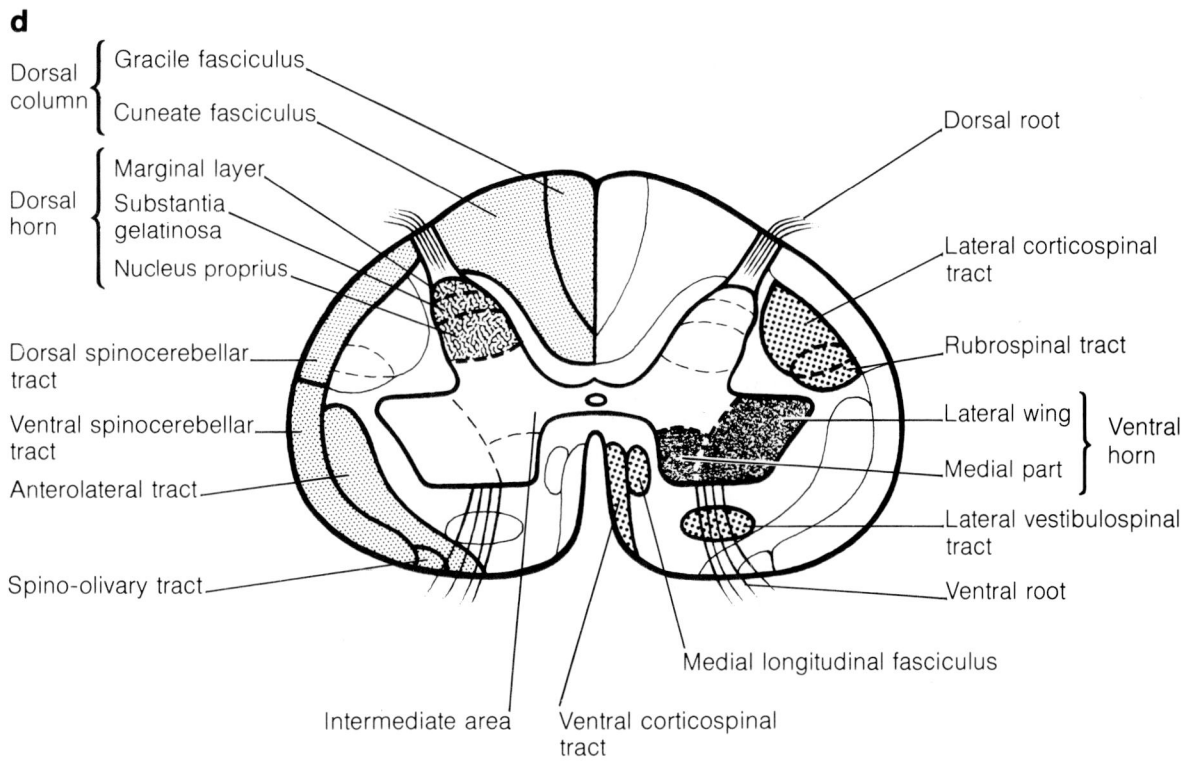

dealing with more medial parts of the body. The important functions of the dorsal horn are:

To receive terminations of primary sensory afferents.

To provide a pool of interneurones for the initial processing and modulation of sensory input (including modulation by supraspinal influences).

To distribute this information to other parts of the spinal cord for use in sensory processing and motor output (spinal reflexes).

To give origin to the ascending tracts that transmit information to the brain (including sensory information originating from peripheral receptors and information relating to what is going on in the spinal cord — i.e. proprioceptive information of peripheral and of central origin).

The ventral horn is greatly expanded by a lateral wing in the cervical (C4 to T1) and lumbosacral (L1 to S3) enlargements. The interneurones and motoneurones of the lateral wing of the ventral horn control the upper and lower limb musculature. Unlike the lateral wing, the remaining, more medially placed, portion of the ventral horn is present at all levels of the spinal cord. This medial region is concerned with the control of the neck and trunk musculature. The important functions of the ventral horn are thus:

To control the contraction of skeletal muscle by the activity of its motoneurones.

To provide interneurones that coordinate and control both motoneurones and spinal reflexes (some of these interneurones distribute information to the brain, as well as to other parts of the spinal cord).

To receive descending tracts transmitting information from the brain for the control of movement and posture.

The lateral horn is a small triangular extension of the grey matter found between the dorsal and ventral horns, approximately from segments T1 to L2. The lateral horn can also be called the intermediolateral cell column. Within the lateral horn are the cell bodies of the preganglionic sympathetic neurones. Their axons pass to the para- and pre-vertebral sympathetic ganglia whose neurones subsequently innervate the smooth and cardiac muscle and the glands of the body. An intermediolateral cell column is also found in the sacral region (approximately between segments S2 and S4). This column is not associated with a lateral extension of the grey matter (i.e. there is no lateral horn in the sacral region). It contains the cell bodies of the preganglionic parasympathetic neurones. Their axons pass in the pelvic splanchnic nerves (nervi erigentes) to the parasympathetic ganglia whose neurones innervate the pelvic viscera.

Different segments of the spinal cord may be recognised in transverse section by their different size and shape, the absolute and relative amounts of white and grey matter, and the shape of the grey matter (Figure 173). Segments from more cranial regions, and from the cervical and lumbosacral enlargements, are largest. Cervical segments are oval, being broader from side-to-side than dorsoventrally. The quantity of white matter decreases as the cord passes caudally (most descending fibres having already terminated before caudal levels, and there being few segments for ascending fibres to have arisen from). The area of grey matter is greatest in the cervical and lumbosacral enlargements. In these regions, the ventral horns possess prominent lateral wings. Lateral horns are found only in the thoracic and the uppermost lumbar segments.

The grey matter of the spinal cord

Besides its division into dorsal, ventral and lateral horns, the grey matter of the spinal cord may also be subdivided into 10 laminae on cyto-architectural grounds. These laminae (of Rexed) are illustrated in Figure 172a. A region called the intermediate area is found between the dorsal and ventral horns (Figure 172a). This area includes the lateral horn.

The dorsal horn may be subdivided according to the following scheme: the marginal layer (Rexed's lamina I); the substantia gelatinosa (lamina II); and the nucleus proprius (lamina III to VI).

The marginal layer (of Waldeyer), or Rexed's lamina I, forms a thin cap at the top of the dorsal horn. It contains a variety of mainly fusiform neurones with nearly planar dendritic arborisations. These neurones have important roles in nociception and thermal sensibility. They receive input from the smaller, laterally placed dorsal root fibres (particularly thinly myelinated fibres, and including both somatic and autonomic nociceptive afferents) and from lamina II cells (Figure 174). They send axons through the ventral white commissure into the contralateral ventrolateral (anterolateral) tracts and so to the brainstem and thalamus. In addition, collaterals of the axons of some cells pass to other spinal cord segments.

The substantia gelatinosa (of Rolando) or Rexed's lamina II is characterised by the presence of a large number of very small cells. It stains very palely using most common histological preparations. Some authorities include parts, or all, of lamina III within the substantia gelatinosa. As the marginal layer, the substantia gelatinosa is very important in mechanisms underlying the perception of pain. It receives inputs from both small (particularly unmyelinated) fibres and large-calibre dorsal root fibres. Its neurones send their axons into the dorsolateral fasciculus where they commonly bifurcate to give rise to both an ascending and a descending branch, so providing association connections to other spinal cord segments. The axons terminate in lamina II of these other spinal cord segments, but often in relation to dendrites of the neurones of the marginal layer or nucleus proprius. Some axons pass in other propriospinal fasciculi to similar terminations, some establish commissural connections.

Certain of the lamina II cells appear to use the opioid enkephalin as their neurotransmitter. They terminate on the synaptic endings of afferent fibres (which may use substance P as their neurotransmitter), so presynaptically inhibiting transmission of nociceptive information (Figure 174b). These connections provide an anatomical basis for opioid-induced analgesia. Inputs from the larger-diameter dorsal root fibres and from the brainstem (particularly those from the nucleus raphe magnus which use 5-hydroxytryptamine as a neurotransmitter) also excite lamina II cells and so inhibit nociceptive transmission (Figure 174b). These inputs allow selective filtering or blocking ('gating') of the passage of nociceptive information. Input from the large-diameter fibres provides an anatomical substrate for the lessening of pain which may be produced by rubbing or shaking a sore region. Input from the brainstem is under the control of neurones of the periaqueductal grey of the midbrain (Figure 174b). These links provide a basis for the relief of intractable pain which may be afforded (in some cases) by stimulation of the periaqueductal grey.

The nucleus proprius of the dorsal horn is an inconsistently defined region. Some authorities restrict it to Rexed's laminae III and IV. However, it is more useful to follow the definition adopted here and regard it as including laminae III to VI. It is the largest region of the dorsal horn and contains a variety of shapes and sizes of neurone, including many large cells. The nucleus proprius receives the intermediate and larger-sized dorsal root afferents (Figure 174a). However, through their dendrites its cells are in contact with dorsal root afferents of all sizes. The cells of laminae III and IV respond particularly to tactile stimuli, though responses to other stimuli may also be found. Most of the axons from cells of laminae III and IV remain within the spinal grey matter, but some decussate and run in the contralateral ventrolateral funiculus to the brainstem or thalamus as part of the anterolateral system. Lamina V is a thick layer where many of the neurones respond to tactile or nociceptive stimuli (including pain of visceral origin). Axons from lamina V also pass to the contralateral ventrolateral funiculus. Lamina VI is prominent only in segments within the cervical and lumbar enlargements of the spinal cord. It forms the base, and thus the last layer, of the dorsal horn. Lamina VI receives many of the large-diameter dorsal root fibres that carry proprioceptive information. With laminae IV and V, it also receives inputs from fibres descending from the brain (notably lateral cortico-, rubro-, and reticulo-spinal tracts).

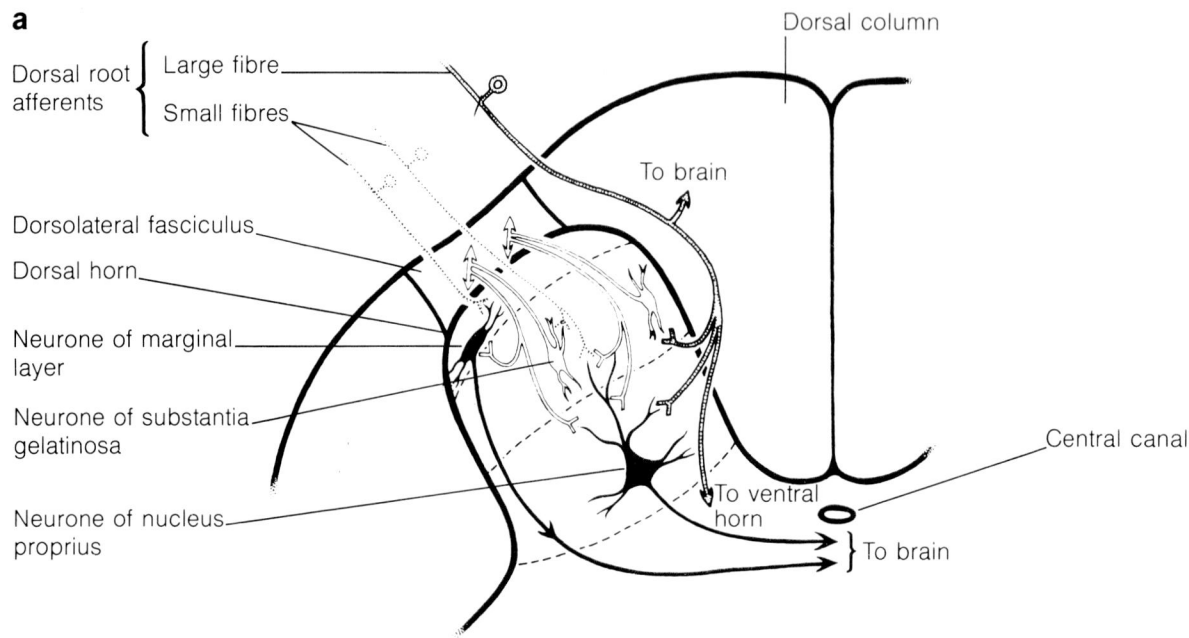

Figure 174a *Some connections of the dorsal horn of the spinal cord.* Small diameter afferents (conveying nociceptive, temperature and crude tactile signals) of the dorsal root terminate in the superficial layers of the dorsal horn. Large diameter afferents (conveying fine touch and proprioceptive signals) end in the substantia gelatinosa and nucleus proprius (as well as passing to the ipsilateral dorsal column and ventral horn). Neurones of the substantia gelatinosa send axons via the dorsolateral fasciculus to the dorsal horn of adjacent segments. These axons terminate upon various neurones of the dorsal horn and upon the terminals of small diameter afferents. Neurones (tract cells) of the marginal layer and nucleus proprius project to the brain mainly via the contralateral anterolateral tracts.

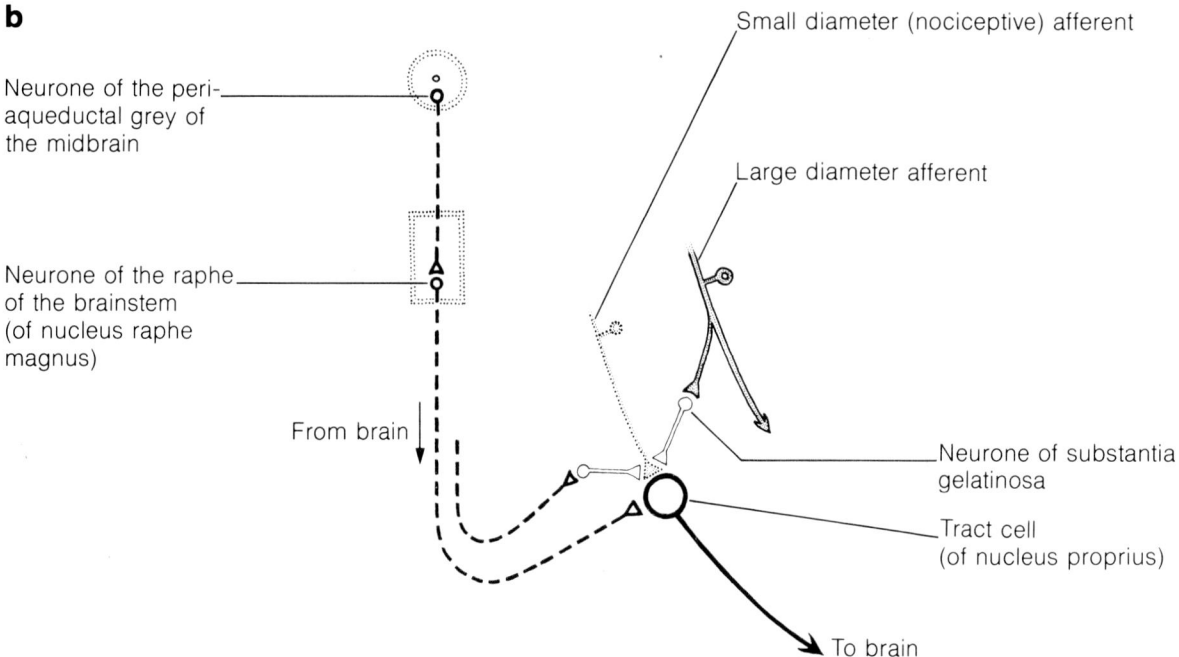

Figure 174b *The putative mechanism underlying modulation of the transmission of nociceptive information.* Signals from large diameter afferents of the dorsal root excite neurones of the substantia gelatinosa. Substantia gelatinosa cells may, in turn, presynaptically inhibit the action of nociceptive afferents, so modulating the transmission of nociceptive information. Signals from the brainstem may act similarly, either via substantia gelatinosa cells or by more direct connections upon tract cells. Some of the descending fibres from the brainstem (from the raphe) are 5-hydroxytryptaminergic. Certain substantia gelatinosa cells involved in presynaptic inhibition use enkephalin as neurotransmitter.

Some cells of lamina VI and the ventral parts of lamina V give rise to fibres of the ipsilateral ventral or contralateral rostral spinocerebellar tracts.

The subnucleus caudalis of the nucleus of the spinal tract of the trigeminal nerve extends into the dorsal horn of the first two cervical segments. Cyto-architecturally, the nucleus corresponds to the laminae I to IV of the dorsal horn, and is continuous with them. The spinal tract of the trigeminal is similarly in continuity with the dorsolateral fasciculus. The subnucleus caudalis is concerned with nociceptive, thermal and tactile stimuli (see pages 480 to 481).

The intermediate area of the spinal grey matter, which comprises Rexed's lamina X and most of lamina VII, is found between the dorsal and ventral horns. Lamina X is the small central region surrounding the spinal canal which includes the dorsal and ventral grey commissures. Lamina VII is lateral to lamina X. It contains many interneurones plus three distinct cell columns: the intermediolateral column, the intermediomedial column, and the nucleus dorsalis (of Clarke).

The intermediolateral cell column is found laterally in the intermediate area between segments C8 and L3, and segments S2 and S4. Between segments T1 (or C8) and L2 (or L3), it has a triangular, lateral extension called the lateral horn. In this region, the intermediolateral cell column contains the somas of preganglionic sympathetic neurones whose axons pass via the ventral roots of segments T1 to L2 (or L3) to sympathetic ganglia. The parasympathetic preganglionic neurones of the intermediolateral cell column of segments S2 to S4 are also laterally located, but do not produce a lateral horn. Their axons pass via some, or all, of the ventral roots of segments S2 to S4 to the parasympathetic ganglia of the pelvic viscera.

The intermediomedial cell column is found almost throughout the spinal cord, just lateral to lamina X. It receives dorsal root fibres, including visceral afferents, so that it may provide interneurones participating in autonomic reflexes. Descending pathways from the hypothalamus, autonomic nuclei, and the reticular formation of the brainstem (including those from monoamine cell groups) terminate on neurones within lamina VII to influence both sympathetic and parasympathetic efferents.

The nucleus dorsalis is alternatively known as the thoracic nucleus or Clarke's column. It is found medially, close to the base of the dorsal horn, approximately between segments T1 and L2. The cells of this nucleus receive dorsal root fibres, including those that have passed in the dorsal columns from more caudal segments and particularly those that carry proprioceptive information. Axons leave the nucleus and form the dorsal spinocerebellar tract of the ipsilateral lateral funiculus. Other cells within lamina VII (and adjacent parts of laminae V and VI) in lumbar and sacral segments give rise to the ventral spinocerebellar tract of the contralateral lateral funiculus. Similarly placed neurones of cervical segments give rise to the rostral spinocerebellar tract. Their axons ascend with those of the ipsilateral ventral spinocerebellar tract.

Neurones in the intermediate area receive fibres from descending pathways (including lateral cortico-, rubro-, and reticulo-spinal tracts and specific descending autonomic fibres), as well as from both dorsal and ventral horns. They give rise to many propriospinal fibres and to fibres joining the contralateral anterolateral system.

The ventral horn contains laminae VIII, IX and part of lamina VII. It may be divided into medial and lateral regions (Figure 172a). The medial region deals with axial musculature and is found at all levels. The lateral region is an expansion (wing) that deals with limb musculature and which is found only in the cervical and lumbosacral enlargements of the spinal cord.

Lamina VII is continued ventrally into the lateral half of the ventral horn in the cervical and lumbosacral 'limb' enlargements. Its neurones receive fibres mainly of the laterally descending pathways (lateral cortico-, rubro-, and reticulo-spinal tracts), as well as collaterals of dorsal root fibres and axons of neurones of the nucleus proprius and the intermediate area. Lamina VII provides interneurones for the laterally located motoneurones of the limb enlargements of the ventral horn, and hence plays an important role in spinal reflexes and general motoneuronal control. Renshaw cells are a variety of interneurone found in lamina VII, which receive the axon collaterals of α-motoneurones. Their axons terminate in relation to many other neurones in the same and other segments (including α-motoneurones) and are inhibitory.

Lamina VIII occupies much of the ventral horn. In the cervical and lumbosacral enlargements it continues medially but does not extend into the lateral expansions of the ventral horn. The neurones in lamina VIII receive fibres from the medial tracts descending from the brain (notably ventral cortico-, vestibulo- and reticulo-spinal tracts). The chief axonal connections of lamina VIII are with other neurones in the medial part of the ventral horn (notably medially placed motoneurones). They additionally give rise to commissural fibres terminating in the medial part of the contralateral ventral horn. In common with the interneurones of lamina VII, lamina VIII plays a major role in the organisation of motoneuronal function, specifically for the medial motoneuronal groups. Some neurones in laminae VII and VIII send axons into the contralateral anterolateral system.

Lamina IX is the lamina in which the large α-motoneuronal and smaller γ-motoneuronal somas are located. It is important to appreciate that their dendrites extend widely in the ventral horn. These neurones send axons through the ventral roots to the extra- and intra-fusal fibres of skeletal muscle.

Motoneurones of the medially located part of lamina IX are found in the ventral horn throughout the spinal cord. They form a series of overlapping, longitudinal columns of cells, each column dealing with a particular muscle group within the neck or trunk. Motoneurones of the laterally located parts of lamina IX are found only in the lateral wings of the cervical and lumbosacral enlargements. They innervate limb musculature. Motoneurone groups innervating distal limb musculature are found most laterally, whereas those innervating proximal limb muscles, and the shoulder and pelvic girdles are in an intermediate position. The motoneurones supplying the trunk musculature are located medially. Interneurones and motoneurones controlling flexor muscles are in general located more dorsally in the ventral horn than those controlling extensor muscles (Figure 172b).

Motoneurones of lamina IX receive some direct connections from descending tracts (cortico-, vestibulo- and reticulo-spinal tracts) and from the primary sensory afferents from muscle spindle organs. However, most of their inputs are provided by the interneurones of laminae VII and VIII.

Within lamina IX are found the spinal accessory and phrenic nuclei. The nucleus of the spinal part of the accessory nerve (cranial nerve XI) extends from the first to the fifth (or sixth) cervical segment. Axons which form the nerve emerge laterally between the dorsal and ventral roots of the spinal nerves in these segments, and innervate the ipsilateral sternocleidomastoid and trapezius muscles (see page 109). The nucleus of the phrenic nerve extends from the third to the fifth cervical segments. Its motoneurones innervate the diaphragm, their axons passing via the ventral roots of the third to the fifth cervical spinal nerves (see pages 113 to 114).

The white matter of the spinal cord

The white matter of the spinal cord is made up of a number of ascending and descending tracts. The positions of the main tracts are illustrated in Figure 173. Essentially, though not without exception, these tracts lie near to the areas of grey matter from which they originate or in which they terminate. Fibres travelling the shortest distance in the spinal white matter usually lie closest to the grey matter. These features minimise the need for fibres to cross and interchange with other fibres, and result in most of the tracts of the spinal cord displaying an orderly segmental (somatotopic) organisation (Figure 172b).

As elsewhere in the central nervous system, most fibre tracts are named from their main site of origin and of termination. Thus, the term spinoreticular tract indicates that the fibres pass from the spinal grey matter to the reticular formation of the brainstem. Conversely, reticulospinal fibres arise in the reticular formation of the brainstem and terminate in the spinal cord.

The propriospinal tracts (Figure 172a) contain fibres passing between different segments and the two sides of the spinal cord. Thus, they are made up of ascending and descending associational and of commissural spinospinal fibres. These fibres are adjacent to and surround the spinal grey matter, being found in the dorsal, lateral and ventral funiculi. The fibres run adjacent to the parts of the grey matter that they interconnect, with the shortest fibres being innermost.

The ventral white commissure lies ventral to the ventral grey commissure below the spinal canal. It provides a main connection between the two sides of the spinal cord. The commissural fibres it contains mainly connect medial parts of the ventral horn, so allowing co-ordinated, bilateral movements of trunk and neck musculature. It also contains decussating fibres, particularly of the anterolateral system, of the ventral spinocerebellar tract and of the ventral corticospinal tract.

The dorsolateral fasciculus or Lissauer's tract (Figure 174a) is the small zone of white matter external to the most dorsal part of the dorsal horn. Its main components are small, unmyelinated (and poorly myelinated) sensory afferents entering via the dorsal roots, and intersegmental propriospinal fibres of laminae I, II and III (particularly axons of neurones of the substantia gelatinosa). These various fibres are concerned with nociceptive, thermal and tactile information. The dorsolateral fasciculus also contains a small number of fibres from the reticular formation of the brainstem (particularly 5-hydroxytryptamine-containing fibres from the nucleus raphe magnus) which terminate in the dorsal horn. These fibres provide control of sensory processing, particularly the transmission of nociceptive information.

Also within the dorsolateral fasciculus, and within dorsolateral parts of the adjacent lateral funiculus, are found axons of dorsal horn neurones projecting ipsilaterally to the gracile or cuneate nuclei of the medulla oblongata. These fibres provide a pathway parallel to that of the dorsal columns for the passage of tactile and proprioceptive information to the gracile and cuneate nuclei.

The major ascending pathways that convey information to the brain are those of the dorsal funiculi (dorsal columns), of the anterolateral system of the ventrolateral funiculi, and of the spinocerebellar tracts of the lateral and ventrolateral funiculi. Their positions are indicated in Figure 173. The dorsal column and anterolateral system pathways are also described on pages 598 to 601. The spinocerebellar tracts are further described on pages 503 and 520.

The dorsal column lies between the dorsal horn and the dorsal median septum. In the upper thoracic and cervical regions, it is divided by a dorsal intermediate neuroglial septum into the (medial) gracile fasciculus and (lateral) cuneate fasciculus. Most of the fibres in the dorsal column are primary sensory axons of the ipsilateral dorsal root ganglion cells. These axons are the larger, well-myelinated fibres that lie medially in the dorsal roots (their collaterals passing to the dorsal horn). They convey tactile information, particularly concerning fine, discriminative touch (including vibration sensibility), and proprioceptive information (particularly that underlying conscious appreciation of body position and movement). These fibres are segmentally arranged, fibres from caudal dermatomes being medial, with fibres from more cranial segments being successively added to the lateral aspect of the tract (Figure 172b). Most of the fibres entering the spinal cord below T6 pass in the gracile fasciculus to the ipsilateral gracile nucleus of the caudal medulla oblongata. Most fibres entering above T6 pass in the cuneate fasciculus to the ipsilateral cuneate nucleus and accessory cuneate nuclei of the caudal medulla oblongata. Neurones of the accessory (lateral) cuneate nucleus give rise to the cuneocerebellar tract which conveys proprioceptive information to the cerebellum. The remaining fibres form the first part of the dorsal column/medial lemniscal pathway.

Besides the axons of primary sensory neurones, the dorsal column contains axons of cells of the nucleus proprius of the dorsal horn which also terminate in the gracile and cuneate nuclei of the medulla oblongata. In addition, some dorsal root afferents travel in the dorsal column for short distances before terminating in the spinal grey matter. Many of these are collaterals of the main axons, and provide information used in spinal reflexes. Another group is formed by those dorsal root fibres that ascend in the dorsal column before terminating in the ipsilateral nucleus dorsalis (Clarke's column) of the thoracic cord. There are also some descending fibres that provide a return pathway from the cuneate and gracile nuclei, through the dorsal column to the dorsal horn. As previously mentioned, intersegmental and commissural propriospinal fibres pass in the dorsal columns as well.

The anterolateral system comprises fibres that make up much of the ventrolateral funiculus. The fibres arise from both dorsal and ventral horns, and run chiefly contralaterally to reach the thalamus, the reticular formation, or the peri-aqueductal grey and tectum of the midbrain. They convey pain, temperature, crude or light touch, and pressure information from the contralateral side of the body to the brain. Additionally, autonomic afferent and some proprioceptive information passes to the brain via this pathway. Sensations such as tickle, itch, and libidinous feelings are also transmitted. Fibres conveying pain and temperature sensibilities typically ascend one or two segments (tactile fibres up to six segments) before decussating to join the contralateral ventrolateral funiculus. The fibres show some somatotopic organisation, with successive fibres being added medially from sacral to cervical regions (Figure 172b). However, fibres conveying the various sensory modalities are not closely grouped, and they are not segregated with respect to their points of termination. Thus, spinothalamic, spinoreticular and spinotectal fibres are intermingled. Pain and temperature fibres passing to the thalamus tend to be grouped more dorsolaterally in the tract and form the lateral spinothalamic tract. The term ventral spinothalamic tract is used to refer to fibres conveying crude touch and pressure information to the thalamus. Note that many spinoreticular fibres run ipsilaterally.

The dorsal spinocerebellar tract forms the margin of the spinal cord that lies ventral and lateral to the line of entry of the dorsal roots. Its fibres originate in the ipsilateral nucleus dorsalis (Clarke's column) of segments T1 to L2. The fibres pass to the ipsilateral cerebellum via the inferior cerebellar peduncle. The dorsal spinocerebellar tract carries detailed proprioceptive information relating chiefly to the caudal half of the body. This information is necessary for the fine co-ordination of movement and posture.

The ventral spinocerebellar tract forms the margin of the spinal cord ventral to the dorsal spinocerebellar tract, and dorsal to the line of emergence of the ventral roots. Its axons originate in the contralateral nucleus proprius and intermediate area of lumbar and sacral segments, and pass to the cerebellum mainly via the superior cerebellar peduncle. It carries integrated proprioceptive information concerning the caudal parts of the body. In cervical regions this tract is joined by ipsilaterally coursing fibres of similar origin and function but concerned with the upper limb and trunk. These fibres constitute the **rostral spinocerebellar tract**; they enter the cerebellum via the inferior and superior cerebellar peduncles. **The spino-olivary tract** is found at the ventromedial border of the ventral spinocerebellar tract. This tract conveys proprioceptive information from the contralateral spinal cord to the ipsilateral inferior olivary nuclear complex of the medulla oblongata, from where it is relayed to the cerebellum.

The major descending pathways may be divided into those of the lateral funiculi, and those of the ventral and ventrolateral funiculi. Within the lateral funiculus are: the lateral corticospinal tract, the rubrospinal tract, the medullary (lateral) reticulospinal tract, and fibres concerned with the control of autonomic reflexes. Within the ventral and ventrolateral funiculi are: the ventral corticospinal tract, the component tracts of the medial longitudinal fasciculus (the medial vestibulospinal tract, the tectospinal tract, the interstitiospinal tract and the pontine reticulospinal tract), and the lateral vestibulospinal tract. The positions of these tracts are illustrated in Figure 173. Further information concerning their origin and course is given on pages 602 to 607. Note that there are some descending fibres in both the dorsal columns and the dorsolateral fasiculi (these have been mentioned above).

The lateral corticospinal tract is the largest tract of the lateral funiculus, and lies lateral to the dorsal horn. It is somatotopically organised with fibres terminating in cervical segments lying medial to those travelling to sacral levels (Figure 172b). It contains fibres from the contralateral cerebral cortex. There are typically very few fibres from the ipsilateral cerebral cortex. The fibres terminate ipsilaterally, mainly upon neurones of lamina VII, though some fibres end directly on motoneurones (particularly those of the lateral wings of the ventral horn in the cervical and lumbosacral enlargements). More than half the fibres of the lateral corticospinal tract terminate at cervical levels. The primary function of the tract is to allow the voluntary control of fine movements of the distal limbs (although it may also influence proximal limb and axial musculature). Activity of its fibres leads chiefly to excitation of the motoneurones of flexor muscles, with complementary extensor inhibition. Some of its fibres terminate in the nucleus proprius of the dorsal horn, where they may influence the processing and transmission of sensory signals.

The rubrospinal tract is found immediately ventral to, and just overlapping, the ventral part of the lateral corticospinal tract. It arises from the contralateral red nucleus of the midbrain. The fibres are somatotopically arranged. They terminate in ventral parts of the ipsilateral nucleus proprius and in lamina VII. Via interneurones, activity of the rubrospinal tract leads to excitation of motoneurones controlling flexor muscles, and inhibition of extensor motoneurones (particularly of the proximal limb muscles). The tract's most important function seems to be control of the tonus of flexor muscles. Some neuroanatomists have suggested that this tract is rudimentary in mankind, however there is no good evidence for this view.

The ventral corticospinal tract is found medially in the ventral funiculus. It carries fibres from the ipsilateral cerebral cortex. These fibres mainly decussate just before terminating in the medial parts (lamina VIII) of the contralateral ventral horn. Their function is to allow voluntary control of muscles of the neck, upper trunk and shoulder girdle; the tract does not usually descend below upper thoracic segments.

The lateral vestibulospinal tract is found ventrolaterally in the white matter of the spinal cord. It originates in the ipsilateral lateral vestibular nucleus of the brainstem, and terminates in medial parts of the ipsilateral ventral horn at all levels. Chiefly by way of interneurones, the major action of its fibres is to excite motoneurones controlling extensor muscles and increase their tonus. Its main function is to allow the maintenance of balance and posture and to counteract unwanted influences of gravity during movement.

The medial longitudinal fasciculus descends in the ventral funiculus to cervical and upper thoracic levels, though some groups of fibres within it continue to sacral levels. Within the spinal cord, the fasciculus contains the medial vestibulospinal tract, the tectospinal tract, the interstitiospinal tract and the pontine reticulospinal tract.

The medial vestibulospinal tract arises in the ipsilateral medial vestibular nucleus of the brainstem and terminates in the medial part of the ipsilateral ventral horn of the cervical and upper thoracic segments. It provides direct, monosynaptic inhibitory inputs to motoneurones controlling neck musculature, so allowing vestibular information to influence the position and movement of the head.

The tectospinal tract arises from cranial parts of the contralateral tectum of the midbrain (deep to the superior colliculus) and passes mainly to the upper cervical segments of the spinal cord. Its function is to effect reflex turning of the head in response to salient visual or auditory stimuli.

The interstitiospinal tract runs from the ipsilateral insterstitial nucleus (of Cajal) in the midbrain to all spinal levels. It is believed to be important for rotating the body about its vertical axis.

The pontine (medial) reticulospinal tract originates in the reticular formation of the ipsilateral pons. Many of its fibres run in the medial longitudinal fasciculus at cervical levels, but below this they become more scattered in the ventral funiculus. They terminate mainly in relation to neurones of the medial part of the ipsilateral ventral horn (at all spinal levels). The fibres are not somatotopically organised. The function of these fibres is to control the general level of excitability within the spinal cord, so influencing both reflex and volitional movements. Together with other fibres of reticular origin, they play an important role in respiration.

The medullary (lateral) reticulospinal tract arises from the reticular formation of the medulla oblongata. Most fibres come from ipsilateral neurones, but some come from contralateral cells. They are not somatotopically organised. The fibres are scattered in the ventral and medial parts of the lateral funiculus, and terminate diffusely at all levels of the spinal cord. Their functions are believed to be similar to those of the pontine reticulospinal tract, although they may also influence processing of sensory signals and autonomic activity.

Other descending fibres provide small components to the spinal white matter. These include fibres from: the cerebellum (fastigiospinal fibres); the nucleus of the solitary tract within the medulla oblongata (solitariospinal fibres); the Edinger–Westphal nucleus (parasympathetic accessory nucleus of the oculomotor nerve) of the midbrain; and fibres from monoamine- (noradrenaline and 5-hydroxytryptamine) and peptide- (including oxytocin and vasopressin) containing neurones of the brainstem and hypothalamus. These fibres have important roles in the control of autonomic as well as sensory and motor functions.

Applied aspects of the internal anatomy of the spinal cord
The pattern of sensory loss following injury to the spinal cord can be illustrated by consideration of the effects of injury to one half of the spinal cord (i.e. right or left half). Such a lesion will produce a contralateral loss of pain and temperature sensibility because of interruption of the anterolateral system fibres. There will also be an ipsilateral loss of discriminative tactile abilities and of knowledge of body position (conscious proprioception) because of interruption of dorsal column plus spinocerebellar fibres. Basic tactile sensibility will be little impaired, fibres subserving this modality passing to the brain both contralaterally (in the anterolateral system) and ipsilaterally (in the dorsal columns). The discriminative tactile loss will be for all dermatomes caudal to the lesion where the dorsal column lesion is complete. The nociceptive and thermal loss with a complete lesion of the anterolateral system fibres will be for all dermatomes caudal to the lesion, excepting those of the one or two segments immediately below the lesion. Fibres of these segments will be spared because they ascend before decussating. However, an incomplete lesion that does not damage the deeper anterolateral system fibres will spare the fibres most recently added to the tract, so that only dermatomes of segments far caudal to the level of the lesion will show sensory loss (Figure 172b). Lesions involving dorsal column and/or spino-

cerebellar fibres will also result in a loss of co-ordination (ataxia) for ipsilateral movements organised by segments below the lesion.

A lesion confined to the dorsal horn which extends over several segments will produce a localised, ipsilateral loss of pain and temperature sensibilities. Dermatomal overlap masks the effects of smaller lesions. There will also be disruption of spinal reflexes. In contrast to the effects of lesions of the dorsal roots, dorsal column-mediated functions remain intact.

Large lesions of the intermediate grey area produce localised, ipsilateral losses of pain and temperature sensibility. These losses will be bilateral if the ventral white commissure is involved because there is interruption of the decussating anterolateral fibres. Lesions of the intermediate grey area will additionally interfere with spinal reflexes. Long-lasting, major disturbances of visceral function usually arise only where there is extensive bilateral damage in the region of the intermediolateral cell column. However, unilateral changes in sweating, vasodilation and piloarrection may occur as a result of lesions confined to one side. Correspondingly, damage to both lateral funiculi, producing bilateral interruption of descending autonomic fibres, results in far greater autonomic disturbance than a unilateral lesion. Note, however, that unilateral damage to the upper two thoracic segments, or their descending autonomic input, results in Horner's syndrome (see page 441).

The symptoms of damage to motor pathways are classically divided into those of the lower motoneurone type and those of the upper motoneurone type. This distinction is of clinical importance, although the terminology may be criticised from a strictly anatomical viewpoint. 'Lower motoneurones' are neurones whose axons terminate upon skeletal muscle fibres (i.e. motoneurones). 'Upper motoneurones' are neurones whose activity influences that of the lower motoneurones via descending tracts. Lower motoneurone lesions are produced when there is interruption of motoneuronal innervation of muscle as a result of lesions to the ventral horn or the ventral roots of the spinal cord (or because of peripheral disturbance of function). The symptoms are weakness (paresis) or paralysis of the affected muscles, together with diminished (or absent) tonus and reduced (or abolished) reflex contraction in response to appropriate stimuli. Further, denervated muscles rapidly become wasted. In contrast, upper motoneurone lesions are characterised by muscular weakness or paralysis in the presence of increased tonus, and exaggerated myotatic reflexes. There are also disturbances of reflexes evoked by superficial (cutaneous) stimuli. Muscular wasting is comparatively slight. If there is no paralysis, inco-ordination of movements may be the most obvious effect of the lesion. The symptoms of an upper motoneurone lesion may result from interruption of the descending tracts within the spinal cord, in which case the symptoms are ipsilateral to the lesion. However, the lesions are frequently supraspinal, often within the internal capsule of the cerebral hemisphere. For supraspinal lesions, the motor loss is contralateral to the lesion (see page 564).

The brainstem and cranial nerves

The brainstem comprises the medulla oblongata (commonly referred to as the medulla), the pons and the midbrain. It lies in the posterior cranial fossa, for the most part adjacent to the clivus. The brainstem extends from just caudal to the foramen magnum to the dorsum sellae of the sphenoid bone, and is about 75 mm long. The cerebellum and fourth ventricle are located dorsally. The midbrain passes through the tentorial incisure.

158,26
158,25; 158,10
136,4
158,21; 158,52
158,22; 158,23
158,18; 164,5

The brainstem is continuous with, and makes connection between, the spinal cord, the cerebellum and the cerebrum. It contains tracts and nuclei concerned with the interconnections between these regions, including those of ascending sensory and descending motor pathways. It gives rise to the last 10 of the 12 pairs of cranial nerves and contains their nuclei (together with the pathways originating from these nuclei).

158,30; 158,22
158,9; 158,57

The brainstem contains the brainstem reticular formation. This exerts a controlling influence over the whole of the central nervous system and is necessary for consciousness. Irreversible loss of function of the brainstem is the criterion for brain death.

THE EXTERNAL TOPOGRAPHY OF THE BRAINSTEM

Because of the complexity of the cranial nerves in terms of components, origin and diversity of function, it is first necessary to describe the basic topography of the brainstem.

The external appearance of the brainstem and the points of emergence of the cranial nerves are described initially from the ventral aspect and then from the dorsal aspect.

THE VENTRAL ASPECT OF THE BRAINSTEM

Ventrally, the caudal part of the medulla is related to the dens of the axis and the anterior margin of the foramen magnum. Occipito-axial ligaments, meninges and cerebrospinal fluid separate the brain tissue from the bone. The cranial part of the medulla and the caudal part of the pons lie above the basilar part of the occipital bone. The cranial part of the pons and the ventrolateral parts of the midbrain are related to the dorsum sellae of the sphenoid bone. The vertebral arteries run ventral to the medulla, in the subarachnoid space of the pontine cistern. They join to form the unpaired, median, basilar artery which lies ventral to the pons. (For the major branches and distribution of these arteries see pages 577 to 579.) Ventral to the midbrain is the interpeduncular cistern. Within this cistern, the basilar artery gives rise to the right and left posterior cerebral arteries which pass laterally and dorsally around the cerebral peduncles of the midbrain before continuing above the tentorium cerebelli. Soon after its origin, each posterior cerebral artery usually forms a junction with the posterior communicating artery. The oculomotor nerve passes close to this junction.

Laterally, the caudal medulla is related to the lateral margins of the foramen magnum and to the spinal part of the accessory nerve. The inferior petrosal sinus, the sigmoid sinus (close to the jugular foramen), and the petrous part of the temporal bone lie lateral to the rest of the brainstem. Parts of the cerebellum are found dorsolaterally. On each side, the free margin of the tentorium cerebelli passes close to the lateral surface of a cerebral peduncle of the midbrain.

The medulla oblongata (myelencephalon) is in continuity with the spinal cord at the upper border of the atlas, just caudal to the foramen magnum and just cranial to the emergence of the roots of the first cervical nerves. Both externally and internally, the transition from the spinal cord to the medulla is gradual. The medulla is shaped roughly like a truncated cone. It is about 30 mm in length and expands cranially. Its ventral border with the pons is marked by a salient transverse groove.

Prominent longitudinal ridges run the length of the medulla on either side of the ventral median sulcus. These are called the pyramids and they contain corticospinal fibres. Most of these fibres decussate at the caudal end of the medulla to form the lateral corticospinal tract of the spinal cord. In so doing, small bundles of fibres interrupt the caudal part of the ventral median sulcus before it develops into the ventral median fissure of the spinal cord. Dorsolateral to each pyramid, and in the more cranial part of the medulla, is a prominent oval protuberance called the olive. Deep to the olive is situated the inferior olivary nucleus. The hypoglossal (XII) nerve emerges as a series of rootlets along the ventrolateral sulcus, between the olive and the pyramid. These rootlets emerge in line with the ventral roots of the spinal nerves.

Dorsal to the cranial part of the olive lies the inferior cerebellar peduncle. The rootlets of the glossopharyngeal (IX), vagus (X) and cranial accessory (XI) nerves emerge as a craniocaudal series along the dorsolateral sulcus between the olive and the inferior cerebellar peduncle. As they emerge, the rootlets are in line with the dorsal roots of the spinal nerves.

From the transverse groove marking the junction between the medulla and pons arise the abducent (VI), facial (VII) and vestibulocochlear (VIII) nerves. The abducent nerve emerges a few millimetres from the midline, in line with the rootlets of the hypoglossal nerve. The facial and vestibulocochlear nerves emerge about 15 mm from the midline, close to the rootlets of the glossopharyngeal nerve. The vestibulocochlear nerve is larger and more lateral than the facial nerve. The facial nerve appears as two distinct parts. The smaller part is the sensory and parasympathetic division of the nerve, known as the nervus intermedius. It is usually seen as a slender bundle between the main, motor division of the facial nerve and the vestibulocochlear nerve. As the vestibulocochlear nerve leaves the brainstem, it usually passes across part of the choroid plexus of the fourth ventricle. The choroid plexus emerges through the lateral aperture of the fourth ventricle; adjacent and lateral to the plexus is the flocculus of the cerebellum. A deep recess called the cerebellopontine angle lies between the medulla, pons and cerebellum, close to the points of emergence of cranial nerves VII, VIII, IX and X.

188A,11
188A,12; 188A,14

188A,13

188A,12
184,17; 198F,44
188A,21; 198F,76
184,18

The pons (of the metencephalon) is about 30 mm long and is considerably wider than the medulla. A transverse groove on its ventral surface marks its junction with the medulla, and another marks its junction with the midbrain. A shallow, median sulcus called the basilar sulcus runs along the ventral surface of the pons. This contains the basilar artery, and explains the use of the term 'basilar part of the pons' to describe the region ventral to the more dorsal, tegmental part of the pons. The basilar part is marked by coarse, transverse striations produced by pontocerebellar fibres. These fibres converge laterally to form the middle cerebellar peduncle. Laterally, and just in the cranial half of the pons, the trigeminal (V) nerve emerges as a large sensory root and a small motor root. The point of emergence of this cranial nerve is taken as marking the border between the pons and the middle cerebellar peduncle. The motor root of the trigeminal nerve is medial to the sensory root.

158,25; 188A,9

158,47; 184,26

188A,23
188A,10

The midbrain (mesencephalon) is only about 20 mm long. Its boundary with the pons is indicated ventrally by a transverse groove. Its boundary with the diencephalon passes immediately caudal to the mamillary bodies of the hypothalamus. The midbrain is made up of a right and a left cerebral peduncle. Ventrally, between the two peduncles, is a deep recess known as the interpeduncular fossa. In the roof of this fossa is the posterior perforated substance, which is pierced by many small central branches of the posterior cerebral arteries. On each side of the posterior perforated substance, an oculomotor (III) nerve emerges through the cerebral peduncle into the interpeduncular fossa.

158,10

158,56; 188A,25
188A,26; 188B,39

188A,24; 188B,37
186,9
188A,7

Laterally, each cerebral peduncle has on its surface a shallow, lateral sulcus. The part of the cerebral peduncle ventral to the lateral sulcus is known as the crus cerebri and consists of white matter composed of efferent fibres from the cerebral cortex. Note that 'crus cerebri' (basis pedunculi) is not synonymous with 'cerebral peduncle'. The part of the midbrain dorsal to the crus cerebri and ventral to the cerebral aqueduct is called the tegmental part. The part dorsal to the cerebral aqueduct is known as the tectum. Some authors include the tectum within the term tegmentum, and some exclude the tectum from the term cerebral peduncles. Lateral to each cerebral peduncle is the free margin of the tentorium cerebelli and the parahippocampal gyrus of the temporal lobe of the cerebral hemisphere.

164,23; 198A,3
198A,4
198A,2

164,5
188A,27; 194,17

THE DORSAL ASPECT OF THE BRAINSTEM

198

The caudal part of the medulla oblongata is related dorsally to the posterior margin of the foramen magnum and to the cerebellomedullary cistern. The cranial medulla, the pons and the caudal midbrain are related to the cerebellum and the fourth ventricle. Dorsal to the cranial part of the midbrain are the pineal body, the great cerebral vein and its cistern, the splenium of the corpus callosum, and the free margin of the tentorium cerebelli.

158,26; 158,21
158,27; 158,25
158,10; 158,23
158,14; 158,15
180,6; 164,5

461

198C	**The medulla oblongata** dorsally shows the continuation of the gracile and cuneate fasciculi of the dorsal columns of the spinal cord (see pages 455 to 456). These fasciculi are bounded by five shallow longitudinal grooves, which are extensions of grooves of the cervical spinal cord. Towards the midline, the gracile fasciculus is found between the dorsal median sulcus and the dorsal intermediate sulcus. The more lateral, cuneate fasciculus lies between the continuations of the dorsal intermediate sulcus and the dorsolateral sulcus of the spinal cord. Immediately lateral to the cuneate fasciculus is the spinal tract of the trigeminal nerve; this tract is in continuity with the fibres of the dorsolateral fasciculus of the spinal cord. The nucleus of the spinal tract of the trigeminal nerve lies beneath the fibres of this tract.
198C,34; 198C,33	More cranially in the dorsal medulla, the gracile and cuneate fasciculi broaden slightly and show elevations. These elevations are called the gracile and cuneate tubercles and mark the positions of the underlying gracile and cuneate nuclei. These nuclei are known alternatively as the dorsal column nuclei. Towards the cranial end of the gracile tubercles, in the midline and above where the central canal of the caudal
198C,35	medulla opens into the fourth ventricle, is the obex. Thus, the obex marks the cranial end of the caudal, closed part of the medulla. The remaining, cranial part of the
198A,5	medulla is known as the open part, and is ventral to the fourth ventricle.
198A,5; 198A,6 198A,8; 198A,10–21	**The fourth ventricle** is the tent-shaped space lying dorsal to the pons and the cranial part of the medulla, and ventral to the cerebellum. The roof has three apices, a median (the fastigium), and two lateral, dorsal recesses. The ventricular system of the brain is described on pages 420 to 421.
198F,43 198C,27 198C,31	The floor of the fourth ventricle is diamond-shaped and is therefore referred to as the rhomboid fossa. The fossa is divided into two triangles by a median groove (sulcus). The lateral apices of the triangles are known as the lateral recesses of the fourth ventricle. The imprecisely defined dorsal boundary between the medulla and pons may be taken as passing through the lateral recesses. The caudal apices together form the calamus scriptorius. The lateral boundaries of the caudal, medullary part of the
198C,34; 198C,33 198C,26 198C,24	fossa are formed by the gracile and cuneate tubercles caudally, and by the inferior cerebellar peduncle cranially. The lateral boundaries of the cranial, pontine part of the fossa are provided by the superior cerebellar peduncles.
	The caudal part of the floor of the fourth ventricle is clearly divided into triangular areas, each named from the underlying cranial nerve nuclei. To each side of the
198C,37 198C,35 198C,36	midline is a slightly elevated region called the hypoglossal triangle, which lies above the hypoglossal nucleus. The apex of the hypoglossal triangle points to the obex. Lateral to the hypoglossal triangle is a slight depression, the vagal triangle. Beneath this is found the dorsal motor nucleus of the vagus. The apex of the vagal triangle points cranially.
198C,28 198C,27 198C,32	The median eminence is an elevation of the floor of the fourth ventricle on either side of the median groove. Lateral to it, and separated from it by the sulcus limitans, is the vestibular area. This area is related to the underlying vestibular nuclei. The acoustic tubercle in the floor of the lateral recess is a small mound produced by the underlying dorsal cochlear nucleus (part of the auditory pathway). Near the middle of the floor of the fourth ventricle, within the pontine part of the median eminence, is the facial
198C,29	colliculus. This slight elevation marks where the motor fibres of the facial nerve loop over the abducent nucleus (Figure 178g). Aberrant pontocerebellar fibres (the striae
198C,30	medullares) commonly form coarse transverse striations across the middle of the rhomboid fossa caudal to the facial colliculus.
	The roof of the caudal part of the fourth ventricle is formed from parts of the cerebellum, the inferior medullary vela (two laterally placed, thin sheets of white
198D,39; 198A,9	matter, pia and ependyma), and the tela choroidea and choroid plexus. The lateral apertures (foramina of Luschka) of the fourth ventricle are at the ends of its lateral
198C,31	recesses and open into the pontine cistern. The choroid plexus commonly protrudes
198F,76; 188A,21	through these apertures and is visible near the cerebellopontine angle. The median

aperture (foramen of Magendie) lies caudally above the obex and opens into the cerebellomedullary cistern. Through these three apertures, cerebrospinal fluid leaves the ventricular system and enters the subarachnoid space.

198D,40; 198C,35
173,15

The roof of the cranial part of the fourth ventricle is formed medially by a thin lamina of white matter, the superior medullary velum, and laterally by the superior cerebellar peduncles. The lingula, a small part of the cerebellum, is found dorsally on the superior medullary velum. Cranially, the fourth ventricle narrows towards the cerebral aqueduct of the midbrain.

198A,22; 198C,24
198A,21

198A,3

The pons beneath the cranial part of the floor of the fourth ventricle is the dorsal, tegmental part of the pons. The constricted, most cranial part of the pons is known as the isthmus of the hindbrain. It is continuous cranially with the midbrain.

198A,6

The midbrain dorsally has four conspicuous hillocks, the quadrigeminal bodies or colliculi. These colliculi form part of the tectum, the portion of the midbrain dorsal to the aqueduct. The two inferior colliculi lie caudally and form part of the auditory pathway. A trochlear (IV) nerve emerges caudal to each inferior colliculus, at the caudal limit of the midbrain. Its fibres pass laterally around the midbrain. The two superior colliculi lie cranial to the inferior colliculi and are visual in function. Each of the four colliculi has a brachium that forms a slight ridge running ventrolaterally and cranially away from its craniolateral aspect. Lateral to each superior colliculus is the medial geniculate body of the diencephalon. The pineal body, part of the epithalamus, lies between the superior colliculi. The border between the midbrain and diencephalon passes immediately cranial to the superior colliculi. It is marked in the midline by the posterior commissure, which is usually regarded as being part of the epithalamus. Beyond this point, the aqueduct of the midbrain opens out into the third ventricle of the diencephalon.

198A,2-4
198A,2
198A,3; 198,1
198C,38
164,6; 186,11
180,24; 180,23
190,14; 190,13
188B,44
180,13; 190,15

180,22; 180,11

THE INTERNAL TOPOGRAPHY OF THE BRAINSTEM

The internal topography of the brainstem is more varied and more complex than that of the spinal cord. The component systems found in the brainstem are:

The nuclei of the cranial nerves and the tracts and nuclei of the pathways to which they give origin.

The somatosensory pathways and other ascending fibres originating in the spinal cord (but not associated with the cerebellum).

Motor nuclei and descending pathways (other than those of the cranial nerves).

The fibre connections of the cerebellum and their associated precerebellar nuclei.

The reticular formation and its connections.

The broad disposition of these component systems follows the expected pattern discernible from the development of the dorsal (sensory) and ventral (motor) laminae of the embryo. In general, therefore, descending fibres from the cerebral cortex, nuclei associated with the cerebellum, and most motor nuclei are found ventrally. Other ascending and descending sensory and motor tracts are found either laterally or medially. Sensory nuclei and most of the nuclei of the cranial nerves are found dorsally. The reticular formation occupies an intermediate position.

The medulla may be divided into the pyramids ventrally and the tegmentum dorsally. The pons is divided into a basilar part ventrally and a tegmental part dorsally. The midbrain consists of the left and right cerebral peduncles. The ventral part of each cerebral peduncle is the crus cerebri plus the substantia nigra; its remaining, dorsal part is the midbrain tegmentum. That part of the midbrain tegmentum dorsal to the aqueduct is known as the tectum. Note that some anatomists define the tegmentum so as to exclude regions dorsal to the central lumen and/or define the cerebral peduncle so as to exclude the tectum.

Details of the positions of the individual nuclei and fibre tracts in the brainstem are most clearly seen by studying transverse sections (Figure 178a to j).

THE CRANIAL NERVES AND THEIR NUCLEI

The 12 pairs of nerves with direct attachments to the brain are numbered in craniocaudal order and named:

I	Olfactory	VII	Facial
II	Optic	VIII	Vestibulocochlear
III	Oculomotor	IX	Glossopharyngeal
IV	Trochlear	X	Vagus
V	Trigeminal	XI	Accessory
VI	Abducent	XII	Hypoglossal

Detailed accounts of the distributions of the cranial nerves peripherally are considered in earlier chapters.

Unlike the spinal nerves, there is no simple, combined anatomical and functional classification of the cranial nerves and their nuclei. Besides the four types of fibres found in spinal nerves (i.e. afferents and efferents of somatic and automatic varieties), cranial nerves also contain fibres of the special senses and fibres supplying muscles that originate from the branchial arches (branchiomotor fibres, see pages 135 and 425). The afferents of the special senses are conventionally divided into somatic types (nerves II and VIII) and autonomic types (smell and taste). Including this last division, seven different types of fibres may be found in the cranial nerves. An eighth type is found in cranial nerve VIII (and possibly II) – i.e. special sensory efferent fibres.

Table 14 lists the cranial nerves and provides information concerning their nuclei and component fibre types. The nuclei are shown in plan view in Figure 175, and in transverse sections in Figures 178a to j (see pages 490 to 499).

It is helpful to divide the last 10 cranial nerves, which are attached to the brainstem, into medial and lateral groups. The nerves comprising the medial group (nerves III, IV, VI and XII) emerge near the midline and innervate muscles derived from the myotomes of embryological somites. The nerves of the lateral group (nerves V, VII, VIII, IX, X and XI) innervate muscles of branchial arch origin and/or have specialised sensory and/or parasympathetic functions. (Nerve III also has a parasympathetic component.)

Whereas the sensory (dorsal horn), motor (ventral horn) and autonomic (intermediate area) regions of the spinal cord of the adult remain in positions similar to those found in the developing neural tube (see Figure 170a), the positions of these regions in the brainstem alter during development. In particular, in the open part of the medulla, their ventrodorsal arrangement becomes mediolateral (see pages 424 to 425; Figure 170b). Furthermore, the neurones belonging to the different functional categories, and the various cranial nerves, gather into discrete nuclei to form interrupted rather than continuous columns. Thus, the motor nuclei of cranial nerves III, IV, VI and XII are all found close to the midline, just ventral to the central lumen (i.e. aqueduct, fourth ventricle or central canal). These nuclei contain motoneurones which are the cranial counterpart of those of the ventral horn of the spinal cord. By comparison, the nuclei giving rise to preganglionic parasympathetic efferent fibres lie in positions slightly more lateral in the case of the dorsal motor nucleus of the vagus (X), and salivatory nucleus (of VII and IX), but more dorsal for the parasympathetic accessory oculomotor nucleus of III (the Edinger–Westphal nucleus). These nuclei form the cranial equivalent of the intermediolateral cell column of the sacral spinal cord. The parasympathetic afferent nucleus of the solitary tract is found dorsolateral to the dorsal motor nucleus of the vagus. Its caudal parts receive the visceral (autonomic) afferent fibres of cranial nerves IX and X; its cranial parts receive the gustatory afferent fibres of nerves VII and IX. The nuclei supplying muscles of branchial arch origin (i.e. motor

TABLE 14 Cranial nerve nuclei and their main functions.

Nerve	EFFERENTS (motor functions)			AFFERENTS (sensory functions)			
	Somatomotor	Parasympathetic	Branchiomotor	Autonomic reflex	Somatosensory	Special sensory	Taste and smell
	(General somatic efferents)	(General visceral efferents)	(Special visceral efferents)	(General visceral afferents)	(General somatic afferents)	(Special somatic afferents)*	(Special visceral afferents)
I							Olfactory bulb (smell)
II						Lateral geniculate body and superior colliculus (vision)	
III	III nucleus (eyeball)	Edinger–Westphal nucleus (pupil, lens)			Mesencephalic V nucleus		
IV	IV nucleus (eyeball)				Mesencephalic V nucleus		
V			Motor V nucleus (mastication)		Mesencephalic (proprioception), principal (touch), and spinal V nucleus (touch, pain, temperature)		
VI	VI nucleus (eyeball)				Mesencephalic V nucleus		
VII		(Superior) salivatory nucleus (saliva and tears)	Motor VII nucleus (facial expression)		Mesencephalic and spinal V nucleus		Nucleus of solitary tract (taste)
VIII						Dorsal and ventral cochlear nuclei (hearing) and 4 vestibular nuclei (equilibration)	
IX		(Inferior) salivatory nucleus (saliva)	Nucleus ambiguus (pharynx)	Nucleus of solitary tract (blood pressure, respiration)	Spinal V nucleus		Nucleus of solitary tract (taste)
X		Dorsal motor nucleus X (gut, heart, etc.)	Nucleus ambiguus (pharynx and larynx)	Nucleus of solitary tract (gut, heart, etc.)	Spinal V nucleus		
XI	Spinal XI: Spinal nucleus (trapezius and sternocleidomastoid)		Cranial XI: Nucleus ambiguus (larynx)				
XII	XII nucleus (tongue)						

*The eighth and possibly the second nerve contain special sensory efferents.

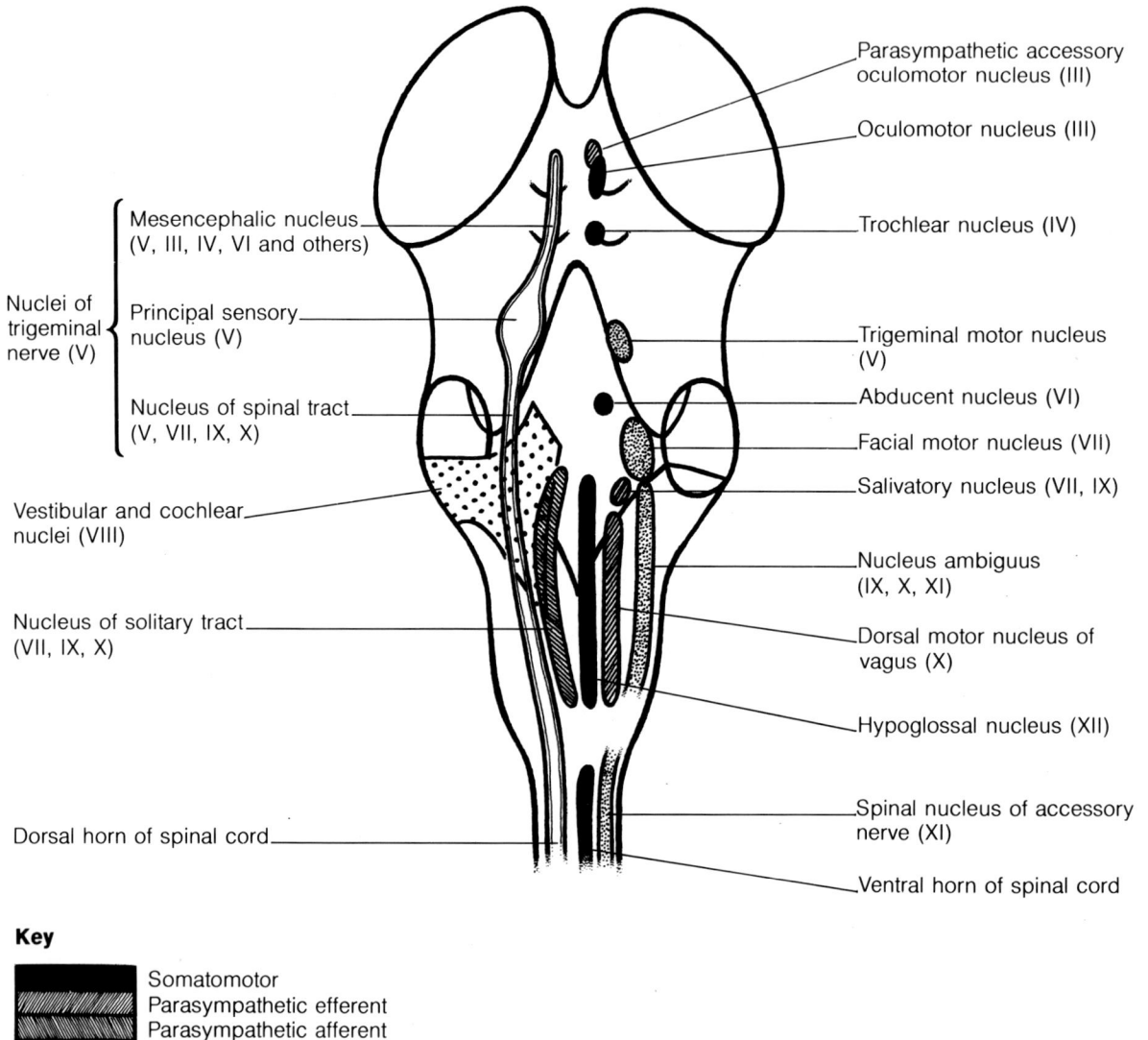

Figure 175 *Positions of the cranial nerve nuclei.* Dorsal view of the brainstem with the cerebellum removed. See also Figures 169 and 178. The cranial nerves associated with the nuclei are shown in brackets.

V, motor VII, and the nucleus ambiguus of IX, X and XI) lie more ventrally. Note that, apart from their position, these nuclei are like those of the medial group of nerves in that they contain motoneurones innervating striated muscle. The spinal accessory nucleus (of XI) may be included in this group, although its intermediate position may reflect the ambivalent origin of the muscles it supplies. The somatic sensory trigeminal (V) nuclei (mesencephalic, principal and spinal nuclei) are found in a lateral position. The special sensory cochlear and vestibular nuclei of the eighth nerve are found dorsolaterally in the upper medulla and lower pons.

The nuclei of the cranial nerves are interlinked, so providing the basis of cranial nerve reflexes. Few of these links are direct. Most connections are made through neurones of the reticular formation. Indeed, an important function of the reticular formation is as a provider of interneurones for the cranial nerve nuclei. Accordingly, the pathways underlying cranial nerve reflexes are often complex, and responses may be multiple and highly organised. Clinically, testing such reflexes is of great diagnostic value for they provide a 'window' into the brainstem. Complete loss of these reflexes is one of the important criteria for establishing irreversible loss of function of the brainstem (i.e. brain death).

Most of the cranial nerve nuclei (especially the motor nuclei) are under the direct or indirect influence (via reticular neurones) of both the cerebral cortex and the cerebellum. Motor nuclei of the cranial nerves may also be influenced indirectly from other parts of the motor system (e.g. red nucleus, basal nuclei), while the parasympathetic nuclei are subject to hypothalamic modulation. Fibres passing from the cerebral cortex to the motor nuclei of the cranial nerves are usually called corticonuclear fibres. They arise chiefly from the motor areas of the frontal cortex and are necessary for voluntary control of movements. Fibres to nuclei of the sensory pathways arise chiefly from the appropriate sensory area of the cerebral cortex (auditory, visual, etc.) and allow modulation of the transmission of afferent information. Cerebellar connections (which typically include both afferents and efferents) are concerned with the fine co-ordination and organisation of movements. As already indicated, all cranial nerve nuclei supply afferents to, and receive efferents from, the reticular formation. The afferents from sensory nuclei have arousing as well as reflex functions.

In general, the ganglia of the cranial nerves are similar to those of the spinal nerves. Cranial nerves III, VII, IX and X have peripheral parasympathetic ganglia. Nerves V, VII, IX and X have proximal ganglia that contain unipolar neurones of somatic sensory function or of afferent autonomic function (including gustatory function), and which are equivalent to the dorsal root ganglia of the spinal nerves. The special sensory ganglia of cranial nerve VIII contain bipolar cells. Cranial nerve II originates in the specialised ganglion cells of the retina, nerve I from the bipolar olfactory rods of the olfactory mucosa. Nerves IV, VI, XI and XII have no ganglia as they are largely if not entirely motor in function (although occasional ganglion cells of proprioceptive afferents may occur along their course).

The olfactory (I) and optic (II) cranial nerves are the only cranial nerves not attached to the brainstem. The olfactory nerves are attached to the cerebral hemispheres (see page 537), the optic nerves to the diencephalon (see pages 523 to 524). The cranial nerves are now described in reverse order, i.e. beginning with the hypoglossal nerve (XII) which emerges from the medulla.

The hypoglossal nerve

The hypoglossal (XII) nerve emerges from the medulla oblongata as a series of 10 to 15 rootlets between the pyramid and olive. The rootlets are in line with those of the ventral roots of the spinal nerves. The hypoglossal nerve passes laterally above the vertebral artery and leaves the cranial cavity through the hypoglossal canal to innervate all the ipsilateral intrinsic and extrinsic muscles of the tongue (excluding palatoglossus). The presence of proprioceptive fibres within the nerve is disputed. As there are spindle organs in the tongue musculature, the proprioceptive fibres may travel either with the lingual branch of the mandibular division of the trigeminal nerve or alternatively with the first cervical nerve.

188A,17
188A,18; 188A,19
158,32; 168,20
168,25; 28,41

The hypoglossal nucleus is the sole nucleus associated with the hypoglossal nerve. It extends almost the full length of the medulla oblongata, from just cranial to the termination of the ventral horn of the spinal cord to just caudal to the pons (Figure 178c to e). Throughout its length, the nucleus lies near the midline and just ventral to either the central canal or the fourth ventricle (beneath the hypoglossal triangle). The closeness of the right and left nuclei explains why damage involving one nucleus often also affects the other. The large motoneurones of the nucleus send their axons ventrally (lateral to the medial lemniscus) and then ventrolaterally (dorsal to the

198C,37

pyramid) to their point of emergence (Figure 178d and e). All muscles are innervated ipsilaterally, the motoneurones of individual muscles probably being grouped into columns.

The hypoglossal nucleus receives fibres from the reticular formation, the nucleus of the solitary tract (gustatory nucleus), and the sensory nuclei of the trigeminal nerve (conveying intra-oral and pharyngeal tactile information). The connections are involved in the reflex movements of licking, sucking, chewing, swallowing and speech. Commissural, cerebellar and presumably proprioceptive connections are made via the reticular formation. Voluntary control of tongue movement is provided by fibres from motor regions of the cerebral cortex. These corticonuclear fibres terminate contralaterally (although there is some individual variation and the more medial motoneurones may be bilaterally innervated). When a lesion weakens or paralyses the tongue muscles of one side, only the genioglossus muscle (assisted by the geniohyoid) of the opposite side will pull the tongue forward when it is protruded. Thus, a peripheral or nuclear lesion (lower motoneurone lesion) will result in deviation of the tongue to the side of the lesion on protrusion. The muscles of that side will also waste. A supranuclear (upper motoneurone) lesion will result in protrusion to the side opposite to the lesion because of the contralateral origin of the corticonuclear innervation. Unilateral lesions are usually fairly well accommodated. However, bilateral lesions have severe effects on eating, swallowing and speaking.

The accessory nerve

The accessory (XI) nerve has spinal and cranial components.

198F,47; 188A,16

The spinal part of the accessory nerve emerges as a series of rootlets from the five or six most cranial of the cervical segments of the spinal cord. The rootlets emerge laterally between the dorsal and ventral roots of the spinal nerves. They join to form a common trunk that runs cranially, dorsal to the ligamentum denticulatum. The trunk enters the cranial cavity through the foramen magnum where it lies dorsal to the vertebral artery. The spinal part of the accessory nerve continues cranially along the lateral surface of the medulla and joins with its cranial part before leaving the cranial cavity through the jugular foramen.

198F,50
198F,49; 168,19
168,25; 198F,48
188A,16; 188A,15
168,18; 28,31

Fibres of the spinal part of the accessory nerve supply the ipsilateral trapezius and sternocleidomastoid muscles. A small number of the fibres are proprioceptive afferents whose cell bodies are found along the intracranial course of the nerve. The remaining proprioceptive fibres associated with these muscles are derived from the second, third and fourth cervical spinal nerves.

The spinal nucleus of the accessory nerve forms a column of motoneurones in cervical segments 1 to 5 (or 6). Caudally, the cells lie dorsolaterally in the ventral horn (see Figure 173d). The more cranial part of the nucleus is located more centrally and approaches the caudal end of the nucleus ambiguus (Figure 178b). The motoneuronal axons pass laterally through the lateral funiculus to their points of emergence (Figure 178b). The nucleus receives reflex and descending inputs typical of those of the ventral horn. In particular, medial vestibulospinal and tectospinal afferents are concerned with the reflex control of head position in response to vestibular and visual or auditory signals. Voluntary raising of the shoulder (via the trapezius muscle) or turning of the head (via the sternocleidomastoid muscle) are effected through the corticospinal fibres. These fibres originate in the contralateral cerebral cortex in the case of the trapezius muscle, but come from the ipsilateral cortex for the sternocleidomastoid muscle. Consequently, damage to the descending pathway will result in weakness or paralysis of the trapezius contralaterally, and the sternocleidomastoid ipsilaterally. There will be an inability or weakness in turning the head towards the affected shoulder. The pattern of cortical innervation allows one cerebral hemisphere (controlling movements of the opposite limbs) to turn the face towards the side performing the movements. Unilateral lesions of the nerve or nucleus cause weakness in shrugging the ipsilateral shoulder and restrict movements of the upper limb. In addition, unless the lesion occurs distal to the sternocleidomastoid, there is difficulty in turning the head away from the affected side.

The cranial part of the accessory nerve is formed by the most caudal rootlets of the series emerging from the dorsolateral sulcus, dorsal to the olive. The fibres of these rootlets briefly join the spinal part of the accessory nerve to pass through the jugular foramen. They then pass to the vagus via the internal ramus of the accessory nerve. Accordingly, they may be classified as aberrant vagal fibres. The axons mainly arise from motoneurones of caudal parts of the nucleus ambiguus and innervate muscles of the larynx through the recurrent laryngeal branch of the vagus. Some parasympathetic fibres from the dorsal motor nucleus of the vagus may also travel in the internal ramus of the accessory nerve.

188A,15
188A,19
168,18; 168,19
28,31

The glossopharyngeal and vagus nerves

The glossopharyngeal (IX) and vagus (X) nerves emerge as a line of rootlets that passes caudally from the cerebellopontine angle along the medulla dorsal to the olive. The rootlets of the accessory nerve continue this series caudally. The most cranial fibres form the glossopharyngeal nerve, the main, middle group form the vagus nerve, and the caudal rootlets form the cranial and spinal roots of the accessory nerve. All these nerves leave the cranial cavity through the jugular foramen.

188A,15
188A,19

168,18; 28,31

The glossopharyngeal nerve supplies branchiomotor fibres to the stylopharyngeus muscle and, by way of the otic ganglion, secretomotor fibres to the parotid salivary gland. It receives sensory (touch, pain, temperature) fibres from the mucous membrane of the soft palate, pharynx, tonsils, middle ear and the posterior one-third of the tongue. It carries gustatory afferents from the taste buds of the posterior one-third of the tongue (including the vallate papillae) and afferents from the baroreceptors of the carotid sinus and chemoreceptors of the carotid body. Cell bodies of the afferents are located mainly in the inferior (petrosal) ganglion, with some somatosensory ganglion cells being in the smaller, superior ganglion.

The vagus nerve is the most widely distributed of all the cranial nerves. It supplies branchiomotor fibres to the intrinsic muscles of the larynx and pharynx (excluding stylopharyngeus), soft palate (excluding tensor veli palatini), and to the striated muscle fibres of the upper oesophagus. It also contains proprioceptive afferents from these muscles. The vagus nerve carries afferents (touch, pain, temperature) from the skin at the back of the auricle and from the external acoustic meatus, and from the mucous membrane of the epiglottis, lower pharynx and larynx. The nerve does not play an important role in the perception of taste, but innervates some taste buds in the region of the epiglottis and the vallecula. It carries afferents from the baroreceptors and chemoreceptors of the aortic arch and from specialised pulmonary receptors. It supplies preganglionic parasympathetic fibres to smooth muscle and glands, and receives visceral afferents from the lower respiratory tract and gastrointestinal tract. Similarly, it innervates the heart (cardiac inhibition). Cardiomotor fibres pass in only the caudal rootlets. Ganglion cells of the afferents are found in the superior (jugular) or inferior (nodose) ganglion, those in the superior ganglion being of somatosensory (other than proprioceptive) afferents. Four-fifths of vagal fibres are afferents.

108,7

The nuclei associated with the glossopharyngeal nerve are the nucleus ambiguus, the nucleus of the solitary tract, and the nucleus of the spinal tract of the trigeminal nerve. The parasympathetic efferent component arises in the salivatory nucleus.

The nuclei associated with the vagus nerve are the nucleus ambiguus, the nucleus of the solitary tract, and the nucleus of the spinal tract of the trigeminal nerve. The parasympathetic efferent component arises mainly from the dorsal motor nucleus of the vagus.

The nucleus ambiguus sends branchiomotor fibres to the glossopharyngeal, vagal and cranial accessory nerves for innervation of the muscles of the pharynx and larynx. It is therefore the nucleus of phonation and deglutition. Some neurones in or adjacent to it pass via the vagus to the heart. It is an indistinct nucleus, comprising a column of scattered motoneurones found ventromedial to the nucleus of the spinal tract of the trigeminal nerve and within the reticular formation of the ventrolateral medulla (Figure 178c to e). It extends most of the length of the medulla, from just cranial to the spinal nucleus of the accessory nerve to just caudal to the motor nucleus of the facial nerve. Fibres from the nucleus arch dorsomedially towards the dorsal motor nucleus of the vagus and then pass ventrolaterally to emerge dorsal to the olive. The motoneurones of the cranial part of the nucleus send axons to the glossopharyngeal nerve; those of the caudal part send axons to the accessory nerve; those of the main, intermediate part send axons to the vagus nerve.

The nucleus ambiguus receives information from receptors in the muscles and mucous membranes of the pharynx, larynx, lower respiratory tract and alimentary tract. This information is transmitted via the glossopharyngeal and vagal nerves to either the sensory nuclei of the trigeminal nerve or the nucleus of the solitary tract. There, it is relayed to the nucleus ambiguus, mainly by way of the reticular formation. Such connections form the basis of reflex movements of the pharyngeal and laryngeal muscles, including those in gagging, coughing and vomiting. Voluntary control of these muscles during swallowing and voice production is provided by corticonuclear fibres from both cerebral hemispheres. Accordingly, unilateral supranuclear lesions typically have little effect. The cerebellum and reticular formation are involved in co-ordination of the muscles.

The dorsal motor nucleus of the vagus sends preganglionic parasympathetic fibres to the vagus. A few such fibres may pass via the cranial part of the accessory nerve. The nucleus is a long column of small cells found dorsal to the hypoglossal nucleus and lateral to the central canal caudally (Figure 178c to e). More cranially, it is situated lateral to the hypoglossal nucleus and ventral to the vagal triangle (in the floor of the fourth ventricle). Fibres pass ventrolaterally to emerge dorsal to the olive. They are joined by some cardiomotor fibres from the region of the nucleus ambiguus.

The dorsal motor nucleus of the vagus has important connections with the adjacent reticular formation and with the nucleus of the solitary tract for the control of cardiovascular, respiratory and gastrointestinal function. The autonomic reflex afferents pass via the vagus and glossopharyngeal nerves, and via the spinal nerves and the anterolateral system of the spinal cord. Major descending influences come from the hypothalamus and cerebral cortex.

The parasympathetic efferents of the salivatory nucleus are described with the facial nerve (see page 477).

The nucleus of the solitary tract receives the primary autonomic afferent fibres of the glossopharyngeal and vagus nerves in its caudal and medial parts. Its expanded, cranial and more lateral part is the gustatory nucleus; this receives the primary taste fibres of the glossopharyngeal and facial (intermedius) nerves. The afferent fibres pass medially, then turn to form a prominent longitudinally running bundle of myelinated axons, the solitary tract; this is surrounded by the grey matter of the nucleus. The nucleus extends from the most caudal part of the pons, where it lies lateral to the dorsal motor nucleus of the vagus and medial to the vestibular nuclei, to the caudal medulla where it lies dorsal to the dorsal motor nucleus of the vagus (Figure 178c to e). In its most caudal part, the left and right nuclei become continuous above the central canal to form the commissural nucleus of the vagus. Most of the cells of the nucleus are small. Some of them use adrenaline or noradrenaline as neurotransmitter.

Autonomic afferents to the non-gustatory parts of the nucleus of the solitary tract originate in abdominal and thoracic viscera. They also include afferents from the baroreceptors of the carotid sinus (afferent arc of the carotid sinus reflex) and of the aortic arch, which monitor blood pressure, and the chemoreceptors of the carotid bodies and (less important) aortic bodies, which monitor blood oxygen tension and carbon dioxide levels. Afferents from different sources terminate in different parts of the nucleus. General visceral afferents terminate primarily in the most caudal third of the nucleus. Baroreceptor and chemoreceptor afferents terminate in the intermediate third of the nucleus. In addition, vagal fibres generally terminate more caudally than glossopharyngeal afferents. The connections of the gustatory part (cranial third) of the nucleus are considered with the description of the facial nerve (see pages 477 to 478).

The visceral parts of the nucleus of the solitary tract have strong connections with the adjacent reticular formation and, directly and indirectly, with the dorsal motor nucleus of the vagus and the nucleus ambiguus. Furthermore, some fibres leave the nucleus to pass (primarily contralaterally) as the solitariospinal tract to the phrenic nucleus of the cervical cord, and to the nuclei of the intercostal muscles and sympathetic intermediolateral cell column of the thoracic cord. They are joined by fibres from adjacent parts of the reticular formation. The tract includes both adrenergic and noradrenergic fibres. These connections are the basis of essential cardiovascular, alimentary and respiratory reflexes. Among the reflexes controlled are vomiting and coughing. The lateral region of the intermediate third of the nucleus forms part of the medullary respiratory (inspiratory) area.

Other connections of the visceral parts of the nucleus of the solitary tract include direct or indirect connections with more distal parts of the reticular formation, and with the spinal cord, cerebellum, hypothalamus, amygdala and cerebral cortex. Anterolateral system fibres from the spinal cord terminate in or near the nucleus. The hypothalamic connections are reciprocal and include those with neurones of its paraventricular nucleus which release vasopressin. Fibres from the cerebral cortex have a bilateral distribution. Hypothalamic and cortical afferents allow emotional and conscious experiences to influence visceral reflexes and respiration.

Somatosensory afferents of the vagus and glossopharyngeal nerves pass to the sensory trigeminal nuclei. Such afferents probably include those from receptors of the mucous membranes of the pharynx and larynx, although such fibres may additionally or alternatively pass to the nucleus of the solitary tract. Note that such fibres may be classified as either somatosensory or autonomic afferents; functionally, these categories may overlap. The central course of proprioceptive fibres is unknown.

A supranuclear lesion of vagal and glossopharyngeal nuclei usually causes only minor disturbances (possibly affecting swallowing) if it is unilateral. Lesions of the nuclei rarely occur without damage to surrounding structures.

Unilateral lesions of the glossopharyngeal nerve produce ipsilateral loss of sensation in the pharynx, the pillars of the fauces, and the posterior one-third of the tongue (including taste sensations). The ipsilateral gag (pharyngeal) and carotid sinus reflexes will be lost. Weakness of the pharynx causes deviation of the uvula towards the unaffected side.

Unilateral lesions of the vagus nerve will produce ipsilateral loss of sensation in the lower pharynx and larynx. The ipsilateral gag and carotid sinus reflexes will be lost. There will be ipsilateral weakness or paralysis of the soft palate, pharynx and larynx, causing (at least initially) difficulty with swallowing and speech. The uvula will tend to deviate to the normal side.

Bilateral lesions involving the ninth and tenth nerves are more likely to be of central than peripheral origin. Difficulties with respiration, swallowing and coughing, as well as possible over-stimulation of the heart, make such lesions potentially fatal.

The vestibulocochlear nerve

The vestibulocochlear (VIII) nerve is a large nerve that enters the brainstem laterally in the transverse groove between the pons and medulla, in the cerebellopontine angle, having passed medially from the internal acoustic meatus (Figure 178f). It contains afferent and efferent fibres innervating the cochlea and labyrinth of the internal ear.

188A,14

168,16; 28,43

Cell bodies of the afferents from the cochlea are found in the spiral (cochlear) ganglion. Their central processes pass to the dorsal and ventral cochlear nuclei. These fibres are the auditory, exteroceptive part of the nerve. The auditory receptor cells are the hair cells of the spiral organ of Corti within the cochlea (Figure 176b). Efferent fibres passing to the hair cells originate in the superior olivary nuclei and constitute the olivocochlear bundle.

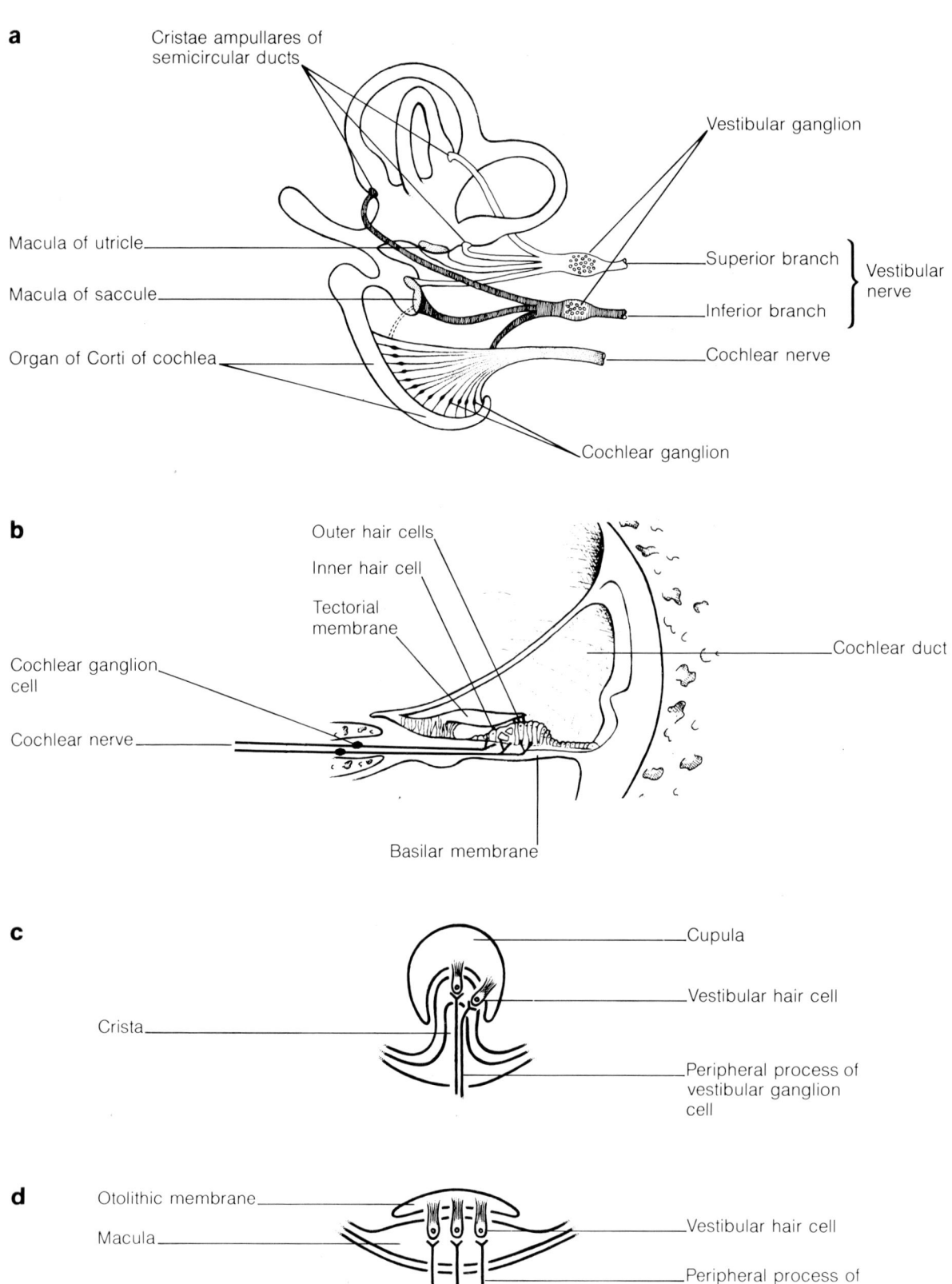

Figure 176 *Peripheral connections of the vestibulocochlear nerve.* **a)** *Innervation of the inner ear.* **b)** *The organ of Corti (transverse section).* **c)** *A crista ampullaris (transverse section).* **d)** *A macula (transverse section).*

The auditory receptor cells are found in a single row of about 3,500 inner hair cells and in three to five rows of outer hair cells (12,000 to 15,000) in each cochlea. Sound waves (20 Hz to 20 kHz) cause vibrations of the basilar membrane and bending of the auditory hairs. An inner hair cell typically responds to a very restricted range of sound frequencies. It is contacted by peripheral processes of 10 to 20 cochlear ganglion cells. There are about 30,000 cochlear ganglion cells. A single peripheral process innervates many outer hair cells; in fact 90% of these processes contact inner hair cells. The cochlea has a tonotopic map; high frequencies are detected by hair cells near the base of the cochlea, low frequencies near the apex.

Cell bodies of the afferent fibres from the labyrinth are found in the vestibular ganglion (of Scarpa) lying in the internal acoustic meatus. Their central processes pass to the vestibular nuclei and, via the inferior cerebellar peduncle, to the flocculonodular lobe and uvula of the cerebellum. These fibres constitute the proprioceptive, vestibular component of the nerve. They enter the brainstem medial to the auditory fibres. The vestibular receptor cells are found in five areas: three cristae, one within the ampulla of each semicircular canal, and two maculae, one in the utricle and one in the saccule (Figure 176a). The small number of efferent fibres passing to the vestibular receptor cells originate in the caudal pons from cells lateral to the abducent nucleus.

Vestibular receptor cells are hair cells that are either fat (type I) or thin (type II). They number about 20,000 in each ear. Type I cells receive the greater innervation. The hairs of the receptor cells of the cristae ampullares of the kinetic labyrinth project into the gelatinous cupula which extends across the semicircular duct (Figure 176c). They detect movements of the cupula primarily produced by angular acceleration of the head. The hair cells of the maculae project into a gelatinous otolithic membrane (membrana statoconiorum) which contains crystals of calcite (Figure 176d). These hair cells detect linear acceleration, particularly changes in position of the head with respect to gravity. The plane of the macula of the utricle is nearly parallel to the base of the skull and is perpendicular to that of the saccule. The utricle chiefly detects vertical acceleration, the saccule forward (or backward) acceleration.

The location and major connections of the central pathways of the vestibulocochlear nerve will be summarised here. Further information on the auditory and vestibular systems may be found on pages 592 to 595.

The dorsal and ventral cochlear nuclei are found on the lateral aspect of the inferior cerebellar peduncle, close to the junction of the medulla and pons (Figure 178f). The dorsal cochlear nucleus produces an elevation (the acoustic tubercle) in the floor of the lateral recess of the fourth ventricle. Cochlear nerve fibres subserving hearing pass dorsally on entering the brainstem and bifurcate to supply both dorsal and ventral cochlear nuclei. Fibres originating in the nuclei pass transversely across the pontine tegmentum in the dorsal, ventral and intermediate acoustic striae. The ventral acoustic stria intersects the medial lemniscus and forms the trapezoid body (Figure 178g). These fibres terminate in the superior olivary nuclei or pass into the lateral lemniscus. Some fibres end in the reticular formation; they are associated with the arousing or startling effects of sounds.

The superior olivary nuclei are found in the ventral tegmentum of the caudal pons (Figure 178g). They are immediately cranial to the inferior olivary nuclei, but the two sets of nuclei are not functionally related. Neurones in the trapezoid body (nucleus of the trapezoid body) are considered to belong to the superior olivary nuclei. The other such nuclei are the lateral and medial (accessory) superior olivary and peri-olivary nuclei. The superior olivary nuclei are important in locating the origin of sounds in space. They give rise to fibres passing to the superior olivary nuclei of the opposite side and to the lateral lemniscus of the same side. The olivocochlear bundle originates from both ipsilateral and contralateral superior olivary nuclei. The fibres from the contralateral nuclei (which are in the majority) cross in the dorsal pontine tegmentum. Both sets of fibres pass initially within the vestibular nerve before reaching the cochlea, where they terminate in relation to the hair cells. The fibres are believed to be cholinergic and to act to improve the signal to noise ratio of auditory signals. Other fibres pass from the superior olivary nuclei to the motor nucleus of the trigeminal nerve, which innervates the tensor tympani muscle, and to the motor nucleus of the facial nerve, which innervates the stapedius muscle. Reflex contraction of stapedius (and possibly of tensor tympani) during speech, swallowing, and in the presence of loud sounds protects the middle and internal ear from excessive vibration.

The lateral lemniscus commences lateral to the superior olivary nuclei and runs cranially in the lateral pontine tegmentum (Figure 178g to i). Most of its fibres terminate in the central nucleus of the inferior colliculus in the caudal midbrain. It contains fibres mainly from the contralateral cochlear nuclei and the ipsilateral superior olivary nuclei. It also contains fibres from the inferior colliculus which pass caudally. Scattered nuclei (nuclei of the lateral lemniscus) of auditory function occur along its length.

198,1

The inferior colliculus of the tectum of the midbrain is laminated and contains a large central nucleus and smaller peripheral nuclei (Figure 178i). It forms part of the main auditory pathway and is also important for auditory reflexes and locating the sources of sounds in space. Its main input is from the lateral lemniscus. Its main efferent projection is via the brachium of the inferior colliculus (inferior quadrigeminal brachium) which leaves its dorsolateral aspect and terminates in the ipsilateral

188B,44
176A,15

medial geniculate body (Figure 178j). From there, fibres pass to the auditory cortex (see Figure 188a and page 552) on the inferior bank of the lateral sulcus of the cerebral hemisphere.

The left and right inferior colliculi are interconnected by their commissure. By this means, some fibres reach the contralateral medial geniculate body. Other fibres pass from the inferior colliculus to deep layers of the superior colliculus, so allowing auditory stimuli to influence activity in the tectospinal tract (e.g. turning the head towards the source of a sudden sound). Other fibres pass caudally to other auditory nuclei (superior olivary and dorsal cochlear) and to the reticular formation. The inferior colliculus itself receives descending fibres from the auditory cortex of both cerebral hemispheres.

The bilateral connections of the auditory system beyond the cochlear nuclei mean that unilateral lesions typically produce only slight, mainly contralateral, hearing loss. Peripheral lesions result in ipsilateral deafness, or the hissing or whistling sounds of tinnitus.

198C,32
198C,35
198C,26; 198C,24

There are four vestibular nuclei: inferior (spinal or descending), medial (triangular or nucleus of Schwalbe), lateral (of Deiters) and superior (of Bechterew). They occupy a lozenge-shaped region mainly beneath the lateral, vestibular area of the floor of the fourth ventricle (Figures 175; 178e to g). The inferior nucleus extends caudally to the obex of the medulla. The superior nucleus extends into the cranial part of the pons. The medial nucleus almost reaches the midline in the caudal pons. At all levels the nuclei remain medial to the inferior and the superior cerebellar peduncles. After entering the brainstem, fibres of the vestibular nerve from the cristae (kinetic labyrinth) pass dorsally and medially beneath the inferior cerebellar peduncle; they bifurcate and chiefly terminate in the superior and medial nuclei. These nuclei give rise to the vestibular fibres of the medial longitudinal fasciculus and form part of the system subserving reflex co-ordination of head, neck and eye movements. Vestibular nerve fibres from the maculae of the static labyrinth follow a course similar to those from the cristae, but mainly end in the lateral, inferior and medial nuclei. Fibres turning caudally run through the inferior vestibular nucleus, giving it a stippled appearance in myelin-stained sections. The lateral nucleus gives rise to the lateral vestibulospinal tract which is important for the maintenance of balance and posture.

The vestibular nuclei are strongly interconnected, including by commissural fibres which link the left and right nuclei. They also have reciprocal connections with the cerebellum and with the reticular formation. These links make possible further vestibular influences upon movement (via the cerebellum and via the reticulospinal tracts). They contribute to control of eye movements and provide a route for the cardiovascular and nausea-inducing effects of overstimulation of the vestibular system. There are few fibres from the spinal cord to the vestibular nuclei and probably no direct inputs from the cerebral cortex. Some fibres from the vestibular nuclei pass to the thalamus (nuclei ventralis posterior and ventralis lateralis) for relay to the cerebral cortex (see page 550).

The lateral vestibulospinal tract leaves the lateral vestibular nucleus ventromedially (see pages 604 to 605). It passes caudally (initially through the inferior vestibular nucleus) and then through the lateral medulla (dorsal to the inferior olivary nucleus and medial to fibres of the anterolateral system) before reaching the spinal cord. These fibres are important for equilibration, and postural and righting reflexes. Their unopposed action causes rigidity of extensor muscles.

The medial longitudinal fasciculus extends the full length of the brainstem and the cervical spinal cord (see pages 594 to 595). Throughout its course it is found medially, a short distance ventral to the central lumen (Figure 178b to j). It has both a descending and an ascending vestibular component.

The descending part of the medial longitudinal fasciculus in the medulla runs ventral to the fourth ventricle and central canal, and dorsal to the tectospinal tract and medial lemniscus. It contains fibres from both right and left medial vestibular nuclei. These fibres leave each nucleus and pass medially in the caudal pons. They then turn caudally (some axons bifurcate to give rise to an ascending as well as a descending branch), to terminate in the reticular formation and inferior olivary nucleus. Other fibres, from the ipsilateral medial vestibular nucleus, pass to the cervical spinal cord as the medial vestibulospinal tract. Some of these fibres terminate in the spinal nucleus of the accessory nerve, allowing reflex vestibular influence upon head position. Other descending fibre groups are also found in the medial longitudinal fasciculus: the interstitiospinal and pontine reticulospinal tracts, and non-vestibular fibres passing to the inferior olivary nucleus (the tectospinal tract joins it in the cervical region).

Ascending fibres of the medial longitudinal fasciculus come mainly from the vestibular nuclei, and primarily from the medial and superior vestibular nuclei; some fibres also arise from the contralateral abducent nucleus. The fibres from the superior vestibular nucleus run ipsilaterally and are inhibitory (GABA-ergic). Those from the medial vestibular nucleus mainly ascend contralaterally and are both excitatory and inhibitory. In the pons, the medial longitudinal fasciculus runs cranially near the midline beneath the floor of the fourth ventricle, passing close to the abducent nucleus. In the midbrain, it lies ventral to the aqueduct and ascends close to the trochlear, oculomotor and interstitial nuclei. The fasciculus distributes vestibular fibres bilaterally to the motor nuclei of cranial nerves III, IV and VI (which control the extrinsic eye muscles), to the interstitial nucleus of Cajal, and to the rostral interstitial nucleus of the medial longitudinal fasciculus. The connections with the nuclei controlling the extra-ocular muscles are essential for the conjugate (co-ordinated) eye movements of normal binocular vision. They are the basis of the vestibulo-ocular reflex which allows maintenance of visual fixation in the presence of head movements. These connections are specific and precise in their origins and terminations. A unilateral lesion of the medial longitudinal fasciculus cranial to the abducent nucleus results in an inability to adduct the eye on the affected side (because of paralysis of the ipsilateral medial rectus muscle) on attempted gaze towards the unaffected side. Bilateral lesions may result in dissociated movements of the eyes on attempted lateral gaze.

Disturbance of vestibular connections may be detected by the presence of abnormal and involuntary rhythmic oscillatory movements of the eyeballs (nystagmus) when an attempt is made to look sideways or up or down. These oscillations have alternating slow (initial) and fast (return or recovery) components. The direction of the nystagmus is named after the fast phase (this is unfortunate as the fast phase is the compensatory and not the primary movement). Sometimes vestibular dysfunction can produce a subjective sensation of rotation (vertigo). Impulses arising in the utricle are the chief cause of motion (travel) sickness. Unilateral lesions of the vestibular nerve lead to problems of ipsilateral co-ordination and balance (falling to the ipsilateral side), nystagmus (towards the contralateral side), and other signs of vestibular dysfunction (nausea, etc.). Over a period of time, adequate adjustment to these losses is normally achieved.

The facial nerve

The facial (VII) nerve emerges laterally, into the cerebellopontine angle, from the groove between the medulla and pons (Figure 178f). It emerges medial to the vestibulocochlear nerve and has two roots: the larger root that lies more medially is the main, motor root; the smaller root is the nervus intermedius. The combined nerve is sometimes referred to as the intermediofacial nerve, but more commonly the term 'facial' is used to describe both components. The facial nerve passes laterally together with the vestibulocochlear nerve into the internal acoustic meatus.

The fibres from the major, branchiomotor root innervate muscles of facial expression (i.e. the muscles of the scalp, the muscles surrounding the apertures of the orbit, nose, mouth and ear) plus the platysma, stapedius and stylohyoid muscles and the posterior belly of the digastric muscle. The course of the small number of proprioceptive afferents from these muscles is disputed, but involves either the trigeminal nerve and/or the facial nerve.

The nervus intermedius is usually visible as a slender, separate bundle, arising between VII major and the eighth nerve. It contains gustatory afferents from taste buds of the anterior two-thirds of the tongue (excluding the vallate papillae), and somatic sensory fibres innervating the external ear; the cell bodies are in the geniculate ganglion. It also carries preganglionic parasympathetic efferents to the submandibular ganglion for relay to the sublingual and submaxillary salivary glands, and to the pterygopalatine ganglion for relay to the lacrimal, nasal, palatine and pharyngeal glands. It may contain a small number of autonomic afferents.

The nuclei associated with the facial nerve are the motor nucleus, the salivatory nucleus, the nucleus of the solitary tract, and the nucleus of the spinal tract of the trigeminal nerve.

The motor nucleus of the facial nerve is found in the ventrolateral pontine tegmentum and the most cranial part of the medulla oblongata (Figure 178f and g). It is immediately cranial to the nucleus ambiguus and caudal to the motor nucleus of the trigeminal nerve (Figure 175). It is a prominent nucleus containing perhaps 10,000 large motoneurones. These motoneurones are arranged anatomically and functionally into subnuclei sending axons into the various muscular branches of the facial nerve. These axons travel by a circuitous route through the pons from the nucleus to the main, motor root of the facial nerve (VII major). The axons pass dorsomedially and cranially before gathering into a compact bundle that runs cranially, medial to the abducent nucleus. They then turn laterally (forming the internal genu of the facial nerve) and pass over the abducent nucleus, in the facial colliculus, before passing ventrolaterally and caudally to their point of emergence from the brainstem (Figure 178g). The aberrant route is explained by migration of the facial and abducent nuclei during development.

The facial motor nucleus receives afferents either directly or, more commonly, indirectly via the reticular formation from the sensory trigeminal nucleus. These links effect the corneal reflex (blinking on touching the cornea), reflex suckling and other reflexes evoked by stimulation of the tissues of the face. Blinking and narrowing of the palpebral fissure in response to visual stimuli requires involvement of the optic nerve and the superior colliculus (which gives rise to tectonuclear fibres). Blinking in response to loud sounds is a result of signals passing via the vestibulocochlear nerve, nuclei of the auditory pathway and the reticular formation. Reflex contraction of the stapedius muscle in response to loud sounds is brought about via fibres from the adjacent (and contralateral) superior olivary nuclei of the auditory pathway. Paralysis of the muscle leads to increased sensitivity to sounds (hyperacusis).

The facial motor nucleus receives a major input from the face region of the motor cortex of the cerebral hemisphere (see Figure 188a). These corticonuclear fibres make both direct and indirect connections with the facial motoneurones. The connections are bilateral for muscles above and around the eye (occipitofrontalis and orbicularis oculi), but contralateral for muscles below the orbit. Accordingly, supranuclear lesions produce contralateral weakness or paralysis of the lower, but not the upper face. Complete lesions of the nerve or nucleus paralyse all facial muscles.

Other fibres pass to the facial motor nucleus or its environs from the nucleus of the solitary tract, the spinal cord, the contralateral red nucleus of the midbrain and, probably via nuclei of the reticular formation, from the globus pallidus of the basal nuclei. These fibres may evoke and modulate facial expressions and muscular movements during speech. The input from the basal nuclei is believed to underly emotional influences upon facial expression (e.g. in laughter), although fibres from the hypothalamus or orbitofrontal cortex may also be involved. This input may account for the spontaneous smiles that can encompass both sides of the lower half of a face otherwise paralysed on one side by a supranuclear lesion. Loss of this input is presumably responsible for the expressionless, mask-like face of Parkinson's disease.

The salivatory nucleus is a poorly defined collection of neurones scattered in the dorsolateral reticular formation of the cranial medulla and caudal pons (Figures 175 and 178f). It forms a lateral and cranial extension of the dorsal motor nucleus of the vagus. The neurones are distributed ventral to the nucleus of the solitary tract and dorsal to the motor nucleus of the facial nerve. It has been customary to group them into an inferior salivatory nucleus (caudally) that supplies preganglionic parasympathetic fibres to the glossopharyngeal nerve, and a superior salivatory nucleus (cranially) that sends fibres to the nervus intermedius part of the facial nerve. Modern studies, however, indicate that these nuclei are confluent. The action of the parasympathetic innervation of salivary glands is to produce copious, watery saliva. In contrast, sympathetic activity results in the production of viscid saliva.

The afferent connections of the salivatory nucleus are only partially known. Inputs from the gustatory part of the nucleus of the solitary tract stimulate reflex salivation. Olfactory and visual influences may reach the nucleus via the neighbouring reticular formation. Intra-oral or ocular stimulation may evoke salivation or lacrimation by way of the sensory trigeminal nuclei. Lacrimation in response to emotional stimuli is probably produced by hypothalamic afferents. Aberrant regeneration following peripheral lesions of the nervus intermedius can lead to lacrimation in addition to salivation in response to food (so-called 'crocodile tears').

The gustatory nucleus is the enlarged lateral and cranial part of the nucleus of the solitary tract (Figure 175). It lies in the upper medulla, mainly cranial to the level of entry of fibres of the vagus (Figure 178e). It receives primary taste fibres. Gustatory chemoreceptor cells are found within taste buds (gustatory caliculi). They have a half-life of just over a week and are constantly replaced. Taste buds are located in the epithelium of the tongue, the inferior surface of the soft palate, the palatoglossal arches and the posterior wall of the pharynx. In all, there are about 10,000 taste buds. They occur in the largest numbers along the sides of the vallate papillae, but are also numerous on the folia linguae and the posterior third of the tongue. They are found on the fungiform papillae but only sparsely on the soft palate and pharynx. The tip, lateral margins and posterior region of the tongue are most important for taste. Each ganglion cell typically innervates many, possibly widely distributed taste buds. Each taste bud receives peripheral processes of about 10 ganglion cells. The gustatory nucleus receives the primary gustatory afferents of the nervus intermedius part of the facial nerve, their cell bodies being in the geniculate ganglion, and of the glossopharyngeal nerve, the cell bodies being in the inferior (petrosal) ganglion. Taste fibres of the vagus are often described as terminating in the gustatory nucleus; the few taste buds in the region of the epiglottis that are supplied by the vagus typically disappear in infancy. The facial afferents terminate generally more cranially than the glossopharyngeal fibres. In sub-primate species olfactory stimuli have been shown to influence taste-responsive neurones of the cranial part of the nucleus of the solitary tract. These signals may reach the nucleus via the ophthalmic branch of the trigeminal nerve and its sensory nuclei.

The axons of the neurones of the gustatory nucleus directly or indirectly (via reticular neurones) establish contact with neurones of the salivatory nucleus, so forming a reflex pathway for producing saliva in response to the taste of food. Connections also pass to the dorsal motor nucleus of the vagus so that gastric secretions and gastrointestinal motility may be increased. Connections to the hypoglossal nucleus allow reflex movements of the tongue in response to taste. Further, pathways to the motor facial and motor trigeminal nuclei make possible reflex movements of the mouth and jaws.

Ascending fibres from the cranial part of the nucleus of the solitary tract pass, probably through the reticular formation, to reach the nucleus ventralis posterior medialis of the thalamus, from where further connections pass to the cerebral cortex (face region of SmI). The connection to the thalamus has been variously described as contralateral, ipsilateral or bilateral. There is evidence that it is supplemented by a pathway that passes indirectly via the pontine taste area. The pontine taste area corresponds to the parabrachial nuclei found adjacent to the superior cerebellar peduncle of the cranial pons (see Figure 178h). Other fibres terminate in the reticular formation and the hypothalamus. Further information on the gustatory pathway is given on pages 596 to 597.

The small number of somatic sensory fibres of the nervus intermedius pass to the spinal tract of the trigeminal nerve and terminate in its nucleus. They are further mentioned on page 480.

The abducent nerve

The abducent (VI) nerve emerges medially, just cranial and lateral to the pyramid, from the transverse groove between the pons and medulla. It runs cranially, passing through the cavernous sinus on the lateral side of the internal carotid artery, and gains the orbit through the superior orbital fissure, to innervate the lateral rectus muscle of the ipsilateral eye.

The abducent nucleus is the sole nucleus associated with the abducent nerve. It is situated medially in the dorsal pontine tegmentum beneath the facial colliculus of the floor of the fourth ventricle (Figure 178g). The motor fibres of the facial nerve run medial and dorsal to it. The medial longitudinal fasciculus is medial, and the superior vestibular nucleus lateral to it.

The abducent nucleus contains both typical motoneurones and internuclear cells. The two neuronal types are present in roughly equal numbers and, apart from the destination of their axons, are morphologically similar. The axons of the motoneurones pass ventrally and slightly caudally through the tegmentum and basilar part of the pons to form the sixth nerve (Figure 178g). Their activity causes lateral deviation (abduction) of the ipsilateral eye. The axons of the internuclear cells decussate and ascend in the contralateral medial longitudinal fasciculus to the oculomotor nucleus. The internuclear cells can by this means excite motoneurones innervating the medial rectus muscle of the contralateral eye and so cause medial deviation (adduction) of that eye. Thus, activity of neurones of the abducent nucleus results in *both* eyes turning to the ipsilateral side. Correspondingly, damage to the nucleus produces paralysis of lateral gaze towards the side ipsilateral to the lesion and involves movements of both eyes. The effect of a nuclear lesion should be contrasted with a lesion of the nerve. Lesions of the abducent nerve (which are not uncommon because of its vulnerability to stretching or compression against the superior border of the petrous part of the temporal bone, or against the superior cerebellar or internal carotid arteries) produce paralysis of lateral movements of only the ipsilateral eye. In such a case, the eye will be adducted because of the unopposed action of the medial rectus, so causing a squint (strabismus), and the patient will experience double vision (diplopia) with the two images being laterally displaced.

Neurones of the abducent nucleus are influenced by the sources to be expected of a nucleus controlling movements of the eyeball: proprioceptive, vestibular, cerebellar, reticular and visual. The nucleus also receives information from neurones of the oculomotor and trochlear nuclei. Proprioceptive feedback from the lateral rectus muscle probably reaches the abducent nucleus via the mesencephalic nucleus of the trigeminal nerve. The vestibular input has a strong direct component. Vestibular afferents are chiefly ipsilateral and are mainly from the medial vestibular nucleus (additionally, there may be a few primary afferents of the vestibular nerve which end in the abducent nucleus). These afferents underly the vestibulo-ocular reflex, maintaining gaze in the presence of head movements (see pages 594 to 595). There are both direct cerebellar influences and indirect cerebellar influences (via reticular or vestibular neurones). These are important for co-ordinating eye movements and for the fine-tuning of oculomotor reflexes. The major reticular input is provided by the paramedian pontine reticular formation. Visual information exerts indirect control over the abducent nucleus. The information is relayed from the superior colliculus and the cerebral cortex (chiefly the contralateral frontal eye field) to the paramedian pontine reticular formation. This in turn projects to the abducent nucleus. The paramedian pontine reticular formation or parabducent nucleus, lies in the medial part of the cranial pontine tegmentum between the abducent and trochlear nuclei. It is essential to the organisation of conjugate (co-ordinated) horizontal movements of the eyes (see also pages 486 and 507), and gives rise to ipsilateral excitatory and contralateral inhibitory inputs to the abducent nuclei. A further reticular region, also concerned with the integration of eye movements, projects bilaterally to the abducent nucleus. This region is the nucleus prepositus hypoglossi which is found medially between the hypoglossal and abducent nuclei. It may be regarded as a caudal extension of the paramedian pontine reticular formation.

A supranuclear lesion involving the descending cortical fibres originating in the frontal eye fields causes a temporary paralysis of voluntary lateral gaze. The eyes cannot voluntarily be moved to look to the side contralateral to the lesion, although the eyes will track an object moved in that direction, i.e. the fixation reflex remains intact. (Note that some of these descending cortical fibres follow an aberrant course by passing through the midbrain near to the medial lemniscus.)

The trigeminal nerve

The large trigeminal (V) nerve emerges laterally from the pons at the ventromedial border of the middle cerebellar peduncle (Figure 178h). It has two roots, motor and sensory.

188A,10
188A,23

The fibres from the motor root innervate the ipsilateral muscles of mastication (masseter, temporalis, medial and lateral pterygoid muscles), anterior belly of digastric, and mylohyoid muscles, and the tensor veli palatini and tensor tympani muscles. The fibres from the sensory root are responsible for somatic sensation from the ipsilateral face, top of head, oral and nasal cavities, air sinuses and meninges.

The nerve divides into its three branches distal to the trigeminal ganglion. The ganglion contains cell bodies of somatic sensory afferents, probably including those of proprioceptive fibres of cranial nerves III, IV, VI and VII. However, most if not all, of the cell bodies of the primary, proprioceptive afferents for the muscles of mastication (and probably also, at least in some instances, for the extrinsic eye muscles) are found within the central nervous system in the mesencephalic nucleus of the trigeminal nerve. The proprioceptive afferents travel in the motor root. The ophthalmic and maxillary branches of the trigeminal nerve pass along the lateral wall of the cavernous sinus. The branches of the ophthalmic nerve enter the skull through the superior orbital fissure, and the maxillary nerve comes through the foramen rotundum. The mandibular branch, which carries the motor fibres, passes through the foramen ovale.

166B,34

166,40; 166,42
166C,40
29B,51; 166C,42
29,50; 166B,47
28,48

The nuclei associated with the trigeminal nerve are the motor nucleus and three sensory nuclei: the nucleus of the spinal tract, the principal nucleus, and the mesencephalic nucleus.

The motor nucleus of the trigeminal nerve is found in the lateral tegmentum of the cranial part of the pons, medial to fibres of the trigeminal nerve and to the principal sensory nucleus of the trigeminal. It is an ovoid collection of typical motoneurones interspersed between which are some smaller neurones. The motoneurones are organised into subgroups innervating individual muscles. Their axons pass ventro-laterally, medial to the sensory trigeminal fibres, through the basilar part of the pons to their point of emergence in the motor root (portio minor) of the trigeminal nerve.

The motor trigeminal nucleus receives fibres from other trigeminal nuclei, the reticular formation, the cerebellum, and (mainly indirectly) the cerebral cortex. In addition to involvement in multisynaptic reflexes, primary proprioceptive neurones of the mesencephalic nucleus of the trigeminal nerve establish a monosynaptic myotatic reflex arc with motoneurones controlling the ipsilateral jaw muscles. Multisynaptic reflexes involving intra-oral stimuli are established via the nucleus of the spinal tract of the trigeminal and involve reticular neurones. Fibres from the superior olivary nuclei of the auditory pathway underly reflex contraction of the tensor tympani muscle. Contraction of this muscle is usually said to dampen oscillation of the ear ossicles. However, some authorities claim that its action heightens hearing. Reticular inputs allow hypothalamic, amygdalar and salivatory signals to influence mastication. Normally, inputs from the cerebral cortex are bilateral so that supranuclear lesions are without major effect. Lesions of the nucleus or nerve produce ipsilateral muscular weakness and wasting, the mandible deviating towards the affected side when the mouth is opened.

The sensory fibres of the portio major of the trigeminal nerve pass dorsomedially on entering the pons and continue through the lateral pontine tegmentum. They then turn cranially or caudally to terminate in one or more of the three sensory nuclei of the trigeminal: the mesencephalic, principal (main), and nucleus of the spinal tract (Figure 175). Many of the fibres bifurcate, giving rise to an ascending and a descending branch. The general topography and reflex connections of the sensory trigeminal nuclei are discussed below; further details of their ascending connections will be found on pages 598 to 601.

The spinal (descending) tract of the trigeminal runs caudally through the lateral pontine tegmentum and lateral medulla as far as the two most cranial of the cervical segments of the spinal cord (Figure 178b to g). It contains central processes of primary somatic sensory neurones of the trigeminal and geniculate (nervus intermedius) ganglia, and of the superior and inferior ganglia of the glossopharyngeal and vagus nerves. The fibres of cranial nerves X, IX and VII are most dorsal. Ventral to them are those of the mandibular, then the maxillary and, most ventral, the ophthalmic division of the trigeminal nerve. Some fibres of cranial nerves V, VII, IX and X leave the tract to terminate in the nucleus of the solitary tract. The trigeminal fibres may include some conveying signals from olfactory receptors in the nasal mucosa. The tract also contains fibres interconnecting different parts of the sensory trigeminal nuclei. Caudally, the tract is continuous with the dorsolateral fasciculus of the spinal cord.

The nucleus of the spinal tract lies medial to the tract throughout its length. The fibres of the tract terminate (and in some cases arise) in the nucleus. The nucleus has three subdivisions: the pars oralis, pars interpolaris and pars caudalis. Cranially, the pars (subnucleus) oralis merges with the principal sensory nucleus in the cranial pons. Caudally, the pars caudalis merges with the dorsal horn of the spinal cord. The pars interpolaris extends from just caudal to the boundary of the pons and medulla to the level of the obex. The pars caudalis is cyto-architecturally, pharmacologically, and functionally similar to laminae I to IV of the dorsal horn. Indeed, in addition to trigeminal fibres it receives afferents from the dorsal roots of the upper cervical nerves. It plays an important role in processing nociceptive information relating to the head. The mechanisms of this processing (including gating and modulation) appear similar to those of the dorsal horn (see page 451 and Figure 174). Thermal and crude tactile sensations are probably also dealt with in the subnucleus caudalis. Representation of the face is such that the anterior, peri-oral region is most cranial and the more lateral and posterior regions of the face are caudal. Thus the representation of the body surface (dermatomal map) of the spinal cord becomes smoothly continuous with that for the areas supplied by the trigeminal. Other parts of the spinal nucleus may show a similar pattern of topographical representation for the front of the head. The pars interpolaris receives both cutaneous and proprioceptive information. It sends fibres to the cerebellum via the inferior cerebellar peduncle and may be regarded as the head region's homologue of the accessory (lateral) cuneate nucleus. The pars oralis receives primarily tactile information, particularly from internal surfaces.

The nucleus of the spinal tract receives fibres (directly and via reticular neurones) from the contralateral cerebral cortex (somatosensory areas SmI and SmII), reticular formation, and red nucleus (to the pars interpolaris). Fibres leave the nucleus and pass to the contralateral thalamus (nucleus ventralis posterior medialis) and so to the somatosensory cortex. These fibres decussate and travel in the contralateral ventral trigeminothalamic tract (trigeminal lemniscus) which runs through the medulla, pons and midbrain in association with the medial lemniscus (Figure 178d to j).

Total lesions of the spinal tract of the trigeminal nerve or its nucleus result in loss of pain and temperature sensibility over the ipsilateral face and top of head. In contrast to lesions of the nerve, where all sensation is lost, tactile input is preserved because of the ascending afferents to the principal nucleus. Partial lesions of the spinal tract may be produced surgically for the relief of intractable pain. It is possible to produce anaesthesia within the territory of individual divisions of the nerve by exploiting the separation of the fibres within the tract. If the tract is lesioned caudally, only lateral parts of the territory of a division will be affected. Similarly, the somatotopic representation within the subnucleus caudalis means that lesions within the most caudal part of the subnucleus will affect only the top and sides of the head and the under-surface of the chin. As the lesion extends cranially the peri-oral region will be the last to be affected.

Fibres, including collaterals of the trigeminothalamic tract, pass to the reticular formation. These connections establish trigeminal reflexes via the hypoglossal nucleus (tongue movements), nucleus ambiguus (swallowing), motor facial nucleus (including the corneal reflex) and motor trigeminal nucleus (jaw opening). Both eyes close on touching one cornea, as impulses from one ophthalmic nerve are relayed to both motor facial nuclei. If the ophthalmic nerve or its central connections are damaged, the corneal reflex will be lost on that side. However, both eyes will close when the contralateral cornea is stimulated. With facial nerve lesions no blinking can be evoked in the eye ipsilateral to the lesion by touching either eye. Further trigeminal reflexes include: lacrimation in response to irritating or noxious stimuli, produced by parasympathetic efferents of the nervus intermedius which originate in the salivatory nucleus; and sneezing as a result of irritation of the nasal mucosa, impulses passing to neurones controlling respiration in the reticular formation, the nucleus ambiguus and the spinal cord (phrenic and intercostal nuclei).

The principal sensory nucleus of the trigeminal nerve is found in the lateral tegmentum of the cranial pons, lateral to the motor trigeminal nucleus (Figure 178h). It receives mainly the larger-diameter tactile afferents of the trigeminal nerve. Together with the mesencephalic nucleus it may be regarded as the equivalent of the gracile and cuneate nuclei for the head region. Importantly, it deals with the finer, more discriminative aspects of touch. The principal nucleus contains an inverted somatotopic map of the face, the ophthalmic afferents terminating ventrally. It receives fibres from the other sensory trigeminal nuclei, the reticular formation and the cerebral cortex (somatosensory areas). Fibres leave it to pass into the contralateral ventral trigeminothalamic tract and terminate in the contralateral thalamus (nucleus ventralis posterior medialis) for relay to somatosensory cortex. Some fibres pass ipsilaterally in the dorsal trigeminothalamic tract, through the dorsal pontine and mesencephalic tegmentum to the thalamus (nucleus ventralis posterior medialis) (Figure 178h to j).

The mesencephalic nucleus of the trigeminal nerve is a slender line of cells extending from the principal nucleus through the midbrain in the lateral part of the pontine and peri-aqueductal grey matter (Figure 178h to j). The nucleus contains large unipolar neurones which are the only examples of primary sensory neurones found in the central as opposed to the peripheral nervous system. In addition, it contains bipolar and multipolar cells. The trigeminal afferents form a narrow crescentic bundle, the mesencephalic root of the trigeminal, adjacent to the nucleus. Details of the connections and even the composition of the nucleus are not well established. It contains primary proprioceptive neurones for the muscles of mastication (though there may also be some ganglion cells in the motor root of the trigeminal nerve), the temporomandibular joint and the periodontal tissues. It receives proprioceptive input from the extrinsic eye muscles and other head musculature (probably facial, possibly lingual and laryngeal). The information it receives via the alveolar nerves concerns pressure exerted on the teeth and periodontium, presumably allowing control of the force of bite.

Fibres leave the nucleus to terminate in the motor nucleus of the trigeminal, so establishing the monosynaptic myotatic and other jaw reflexes. Other fibres pass to the principal nucleus of the trigeminal, to the cerebellum, the superior colliculus, and the thalamus (nucleus ventralis posterior medialis) for relay to the cerebral cortex (somatosensory areas SmI and SmII).

The trochlear nerve

198D,38
198D,1; 164,24
164,6; 186,11
166C,31; 29B,51

186,10; 186,9
186,12

The slender trochlear (IV) nerve is the only cranial nerve to emerge dorsally. It emerges medially, caudal to the inferior colliculus near the junction of the midbrain with the pons, passes laterally round the cerebral peduncle, along the lateral wall of the cavernous sinus and enters the orbit through the superior orbital fissure to innervate the superior oblique muscle of the ipsilateral eye. Together with the oculomotor nerve it passes between the posterior cerebral and the superior cerebellar arteries ventrolateral to the brainstem.

The trochlear nucleus is a compact group of motoneurones, intermixed with interneurones or internuclear cells (see also the abducent nucleus). It is found in the ventral peri-aqueductal grey matter of the caudal midbrain (Figure 178i). The oculomotor nuclear complex lies immediately cranially; the trochlear nucleus may be regarded as an outpost of it. The medial longitudinal fasciculus encapsulates the ventral and lateral aspects of the nucleus.

198A,22

The motoneuronal axons follow a course suitable to the bizarre characteristics of the fourth nerve. The axons curve dorsally and caudally round the margin of the peri-aqueductal grey matter and decussate in the superior medullary velum before emerging dorsomedially. Hence, the trochlear nerve is triply unique: it emerges dorsally, all its fibres arise from a contralateral nucleus, and the action of the muscle it supplies is inverted through acting over a pulley. The superior oblique muscle contributes to abduction, depression and medial rotation (intortion) of the eye. Lesions of the nerve affect the *ipsilateral* eye, with deviation and consequent diplopia being maximal when the patient looks downward and medially (an affected patient may experience particular problems when walking downstairs). Lesions of the nucleus afflict the *contralateral* eye in the same way. The central connections of the trochlear nucleus are similar to those of the oculomotor nucleus (see below).

The oculomotor nerve

188A,7; 158,55
188A,26
168,7; 166C,37
29,51

The oculomotor (III) nerve emerges ventrally and medially through the cerebral peduncle of the midbrain in the caudal part of the interpeduncular fossa (Figure 178j). It passes via the lateral wall of the cavernous sinus and enters the orbit via the superior orbital fissure to innervate the superior, inferior and medial recti, inferior oblique and levator palpebrae superioris muscles. It also supplies the ipsilateral eye with parasympathetic fibres which pass via the ciliary ganglion and the short ciliary nerves to the sphincter pupillae and ciliary muscles. Activity in these fibres results in constriction of the pupil and accommodation through relaxation (bulging) of the lens.

The course of proprioceptive afferents from the extrinsic eye muscles (supplied by nerves III, IV and VI) is uncertain. There are such afferents peripherally in all the nerves and it is possible that there are also a few ganglion cells in each of the nerves. However, it seems probable that the fibres eventually join the ophthalmic branch of the trigeminal nerve. Each extra-ocular muscle is innervated by between 2,000 and 7,000 motoneuronal axons. There are 20 to 70 spindle organs per muscle.

The oculomotor nuclear complex comprises: a) a number of subnuclei containing motoneurones, interneurones and internuclear cells collectively known as the oculomotor nucleus, and b) adjacent parasympathetic preganglionic cell groups, collectively known as the parasympathetic accessory oculomotor or Edinger–Westphal nucleus (Figures 175, 178j). A midline group of cells known as the nucleus of Perlia has been difficult to identify in primates. The interstitial nucleus of Cajal, the nucleus of Darkschewitsch and the nuclei of the posterior commissure are also known as accessory oculomotor nuclei.

The oculomotor nucleus is found ventrally in the peri-aqueductal grey matter throughout the cranial midbrain. Fibres of the medial longitudinal fasciculus lie laterally. The Edinger–Westphal nucleus is medial, dorsal and cranial to the oculomotor nucleus. Fibres leave both these nuclei and pass through the midbrain tegmentum either medial to, or through, the red nucleus and substantia nigra to reach the interpeduncular fossa.

Within the oculomotor nucleus the dorsal cell column innervates the ipsilateral inferior rectus muscle, the intermediate column innervates the ipsilateral inferior oblique, and the ventral cell column innervates the ipsilateral medial rectus; fibres from the medial cell column decussate and supply the contralateral superior rectus muscle. The caudal central nucleus is an unpaired, median cell group. It sends both crossed and uncrossed fibres to levator palpebrae superioris. Internuclear cells interconnect the oculomotor, trochlear and abducent nuclei. Some of them project to the cerebellum.

The oculomotor and trochlear nuclei receive inputs from a variety of sources: visual, vestibular and proprioceptive; cerebral cortical, superior collicular, cerebellar, and reticular. The various cell groups of the oculomotor, trochlear and abducent nuclei are also linked by internuclear cells. Direct projections from the vestibular nuclei and the abducent nucleus pass in the medial longitudinal fasciculus and innervate specific subnuclei according to a precise pattern. These connections subserve the vestibulo-ocular reflex, and co-ordination between the eye muscles during horizontal movements of the eyes (see also the abducent nerve).

Inputs to the oculomotor and trochlear nuclei come from three reticular areas implicated in the control of conjugate movements of the eyes: the nucleus prepositus hypoglossi, the rostral interstitial nucleus of the medial longitudinal fasciculus, and the interstitial nucleus of Cajal. The nucleus propositus projects ipsilaterally to the medial cell column and may instigate or influence upward movements of the eyes. The rostral interstitial nucleus of the medial longitudinal fasciculus is a nucleus cranial to the oculomotor nucleus and the interstitial nucleus of Cajal, on the border of the midbrain with the diencephalon. Its cells are found among fibres at the most cranial extent of the medial longitudinal fasciculus. They project ipsilaterally to the oculomotor nucleus and appear to be responsible for generating vertical (downward) eye movements. Fibres from the interstitial nucleus of Cajal that pass through the posterior commissure to reach the oculomotor and trochlear nuclei play a role in vertical (upward) and rotational (torsional) movements of the eyes (Figure 177b).

Visual information from the superior colliculus and cerebral cortex influences the oculomotor complex almost entirely by indirect connections to the reticular regions controlling conjugate movements of the eyes (although some axons of superior collicular origin may terminate on neurones of all the nuclei supplying the extra-ocular muscles) (Figure 177b). Inputs from the cerebellum are both direct and indirect. Additionally, information from the cervical spinal cord (concerning head and neck movements) reaches the nuclei via reticular or vestibular relays. In contrast to lateral movements of the eyes, vertical movements require symmetrical muscular activation of the two eyeballs. Vertical movements are less commonly impaired by lesions in one cerebral hemisphere than lateral movements, because there is likely to be bilateral and symmetrical involvement in their initiation.

Proprioceptive feedback from the extra-ocular muscles also seems to be indirect. The cell bodies of the afferents may be either scattered along cranial nerves III, IV or VI, or (most commonly) in the trigeminal ganglion, or in the mesencephalic nucleus of the trigeminal nerve. Information may be transmitted either to the mesencephalic nucleus or to the pars interpolaris of the nucleus of the spinal tract of the trigeminal nerve. Reticular neurones complete the circuits.

The Edinger–Westphal (parasympathetic accessory oculomotor) nucleus contains cells that are smaller than the motoneurones of the oculomotor nucleus. The right and left nuclei are confluent in the midline, cranially. Cells of the nucleus send preganglionic parasympathetic fibres to the ipsilateral ciliary ganglion. The great majority (about 95%) of the postganglionic fibres supply the ciliary muscle; the remainder supply the sphincter pupillae. The nucleus also gives rise to fibres terminating in the cerebellum, medulla and spinal cord. Some fibres terminate in the gracile and cuneate nuclei of the medulla, others in laminae I and V of the dorsal horn. Their function is not clear.

158,13

Afferents to the Edinger–Westphal nucleus come from nuclei within the pretectal area. The pretectal area extends cranial to the superior colliculus. There are both crossed and uncrossed afferents, many of them passing through the posterior commissure. These fibres mediate reflex pupillary constriction in response to light (direct and consensual light reflex). The afferent path requires the integrity of either optic nerve. Constriction of a pupil requires that the ipsilateral oculomotor nerve is intact whichever eye is stimulated. Monitoring of the light reflex during coma is important as its loss may signal a life-threatening rise in intracranial pressure.

When gaze is directed from a far to a near object there is convergence of the optic axes (contraction of both medial recti), accommodation (focussing of the lens produced by contraction of the ciliary muscle) and pupillary constriction (contraction of sphincter pupillae). This reflex (the near response) involves the visual cortex, corticotectal fibres, the superior colliculus and the pretectal area. (Note that the reflex is cortical as well as subcortical.) Accommodation may be effected via neurones of the reticular formation of the caudal midbrain. As with most movements of the eyes, fine adjustment of the reflex probably involves cerebellar and reticular inputs.

188A,27; 188A,7
164,5; 164,20

If the oculomotor nerve is damaged, then: a) the affected eye points downwards and outwards (external strabismus or squint) and is largely paralysed, b) the pupil is dilated (mydriasis) and the light reflex and accommodation reflexes are lost, and c) there is drooping of the upper eyelid (ptosis). The parasympathetic fibres are the finest and also run superficially; commonly, they are the first to be affected. In cases where raised intracranial pressure causes the temporal lobe to herniate through the tentorial notch, the ipsilateral oculomotor nerve is liable to be compressed, producing some or all of the above signs. Nuclear lesions may reproduce all or parts of these symptoms. Because of their proximity, both right and left nuclei may be involved.

The optic nerve

188B,33

The optic (II) nerve is attached to the diencephalon rather than the brainstem. From the point of view of development, it is a tract of the central nervous system rather than a peripheral nerve. It contains the axons (about 10^6 fibres) of the ganglion cells of the retina of the ipsilateral eye. It may contain a few efferent fibres from the pretectal region. The visual receptors, the retinal rods and cones, are linked to the ganglion cells by bipolar neurones, assisted by horizontal and amacrine cells (see also pages 590 to 591). The ganglion cell axons terminate in the lateral geniculate body, the superior colliculus and pretectal area of the midbrain, and the hypothalamus.

188B,41; 180,24
190,14; 180,36

The optic nerve passes from the retina through the optic canal and partially decussates in the optic chiasma. The optic chiasma is found in the midline, anterior to the infundibulum (pituitary stalk) at the anterior and ventral margin of the hypothalamus. The retinal blood supply (ophthalmic artery and vein) passes to the eye in the subarachnoid space of the meninges surrounding the optic nerve. Hence raised intracranial pressure impedes venous drainage. Such conditions produce observable oedema of the optic nerve head (optic disc or papilla), a sign of potential diagnostic value. Complete lesions of the optic nerve produce blindness of the ipsilateral eye.

The optic chiasma is where fibres from the nasal part of the retina of each eye decussate (see pages 590 to 591). Because the lens system of the eye results in an inverted image being projected upon the retina, the decussating fibres signal information from the temporal half of the field of vision of each eye. The defect of vision caused by damage to all these decussating fibres is accordingly known as a bitemporal hemianopia. Such damage can be produced by, for example, a pressure-induced conduction block as a result of an expanding tumour of the pituitary gland.

The optic tract is the continuation of the optic pathway from the optic chiasma. The partial decussation of each optic nerve in the optic chiasma results in each optic tract containing axons from the temporal half of the retina of the ipsilateral eye, and the nasal half of the retina of the contralateral eye. Taking account of the inversion of the retinal image, the optic tract thus conveys signals from both eyes relating to stimuli in the contralateral half of the visual world. Complete lesions of an optic tract result in blindness for the nasal half of the field of vision (visual field) of the ipsilateral eye and for the temporal half of the visual field of the contralateral eye. Such a defect of vision is known as a contralateral homonymous hemianopia, as the blindness is for corresponding hemifields in both eyes, the region of blindness being contralateral to the lesion in each case. A similar pattern of blindness is produced by complete lesions of the visual pathway to the cerebral cortex anywhere central to the optic chiasma.

The optic tract runs from the optic chiasma laterally, dorsally and caudally round the most cranial part of the cerebral peduncle. A few fibres leave the optic tract or chiasma to terminate bilaterally in the suprachiasmatic nucleus of the hypothalamus. These fibres are important for the control of circadian and neuro-endocrine rhythms. About 90% of optic tract fibres travel in its large, lateral root to terminate in the ipsilateral lateral geniculate body of the thalamus. The optic radiation (geniculocalcarine tract) connects the lateral geniculate body with the primary visual cortex of the occipital lobe. This part of the visual pathway is further described on pages 565 to 566 and 550. The smaller medial root of the optic tract passes into the brachium of the superior colliculus and ends in the cranial part of the ipsilateral midbrain, in both the superior colliculus and the pretectal area that is cranial to it (Figures 177a and 178j).

The superior colliculus is a laminated structure containing layers of both grey and white matter. Via the brachium of the superior colliculus, the superficial layers receive fibres from both retinae and from the ipsilateral cerebral cortex (corticotectal fibres). The corticotectal fibres arise from widespread areas of occipital, temporal and parietal cortex, and from the frontal eye field. The superficial layers send fibres to the deep layers. The deep layers of the superior colliculus are not exclusively visual in function. They receive afferents from other sensory systems: somatosensory fibres from the spinal cord (spinotectal fibres) and brainstem nuclei (trigeminal and dorsal column nuclei), and auditory fibres (chiefly from the inferior colliculus). These layers are also supplied with fibres from regions concerned with movement co-ordination (cerebellum and substantia nigra) and from the reticular formation. The two superior colliculi are linked by the commissure of the superior colliculi. They additionally receive fibres from the thalamus (nonspecific nuclei and the ventral lateral geniculate nucleus).

The efferents of the superior colliculus reach many regions and to a certain extent parallel its afferents. Ascending efferents (chiefly from the superficial, visual layers) pass to the thalamus (mainly the pulvinar, lateral geniculate body and nonspecific thalamic nuclei), to nuclei within the pretectal area, and to the interstitial nucleus of

Cajal. Descending fibres (which arise mainly from the deeper layers) have three main targets. First, efferents pass to the pontine and inferior olivary nuclei for relay to the cerebellum. Second, the tectospinal tract passes to the cervical spinal cord, allowing the head to be turned towards salient stimuli. Its fibres decussate ventral to the peri-aqueductal grey matter and descend through the brainstem in the midline just ventral to the medial longitudinal fasciculus (see pages 604 to 605 and Figure 175i to b). In the brainstem, these fibres are sometimes called the predorsal bundle; in the spinal cord, they become incorporated into the medial longitudinal fasciculus. Third, fibres accompanying the tectospinal tract pass bilaterally to the reticular formation. (Such fibres are sometimes referred to as tectobulbar rather than tectoreticular.) These fibres terminate particularly in regions of the reticular formation responsible for co-ordination of eye movements (notably the paramedian pontine reticular formation and nucleus prepositus hypoglossi). Some collicular fibres may contact dendrites of motoneurones controlling the extra-ocular muscles, but the main influence of the superior colliculus on eye movements is via the reticular formation and cerebellum (Figure 177b).

Collicular projections are important for: a) selecting salient visual stimuli, b) moving the head and eyes so that a stimulus is focused in the centre of gaze of both eyes (foveation), and c) holding a stimulus in the centre of gaze even if the subject and/or

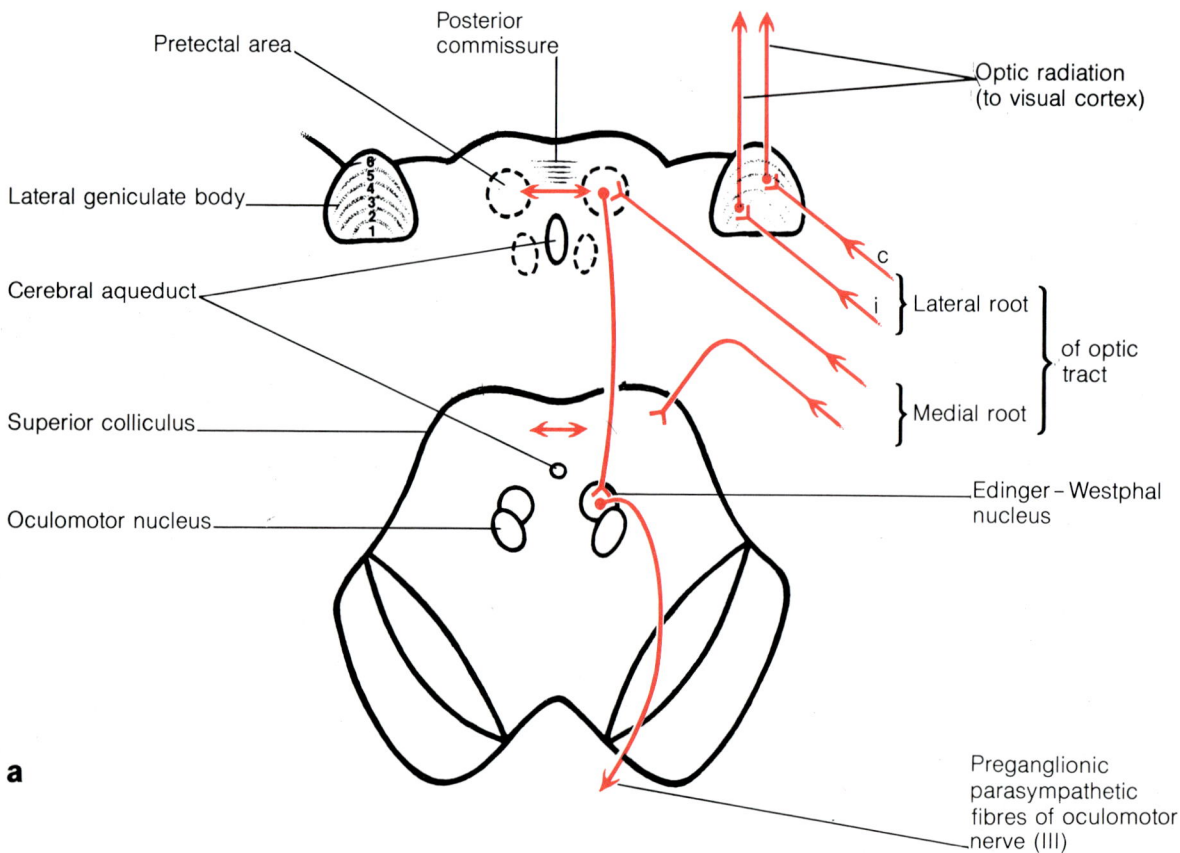

Figure 177 *Subcortical connections of the optic tract and the control of visual reflexes.* a) *Subcortical terminations of the optic tract and the control of the pupillary light reflex. The lower section is of the cranial midbrain; the upper is the dorsal part of a section through the mesencephalic-diencephalic junction. Fibres of the lateral root of the optic tract from the contralateral eye (c) terminate in layers 1, 4 and 6 of the lateral geniculate body; those from the ipsilateral eye (i) terminate in layers 2, 3 and 5. Neurones of the lateral geniculate body project via the optic radiation to the ipsilateral visual cortex. Fibres of the medial root of the optic tract terminate in the superior colliculus and pretectal area. Neurones of the pretectal area project to both parasympathetic accessory oculomotor (Edinger–Westphal) nuclei which control pupillary diameter. This pathway underlies the light reflex.*

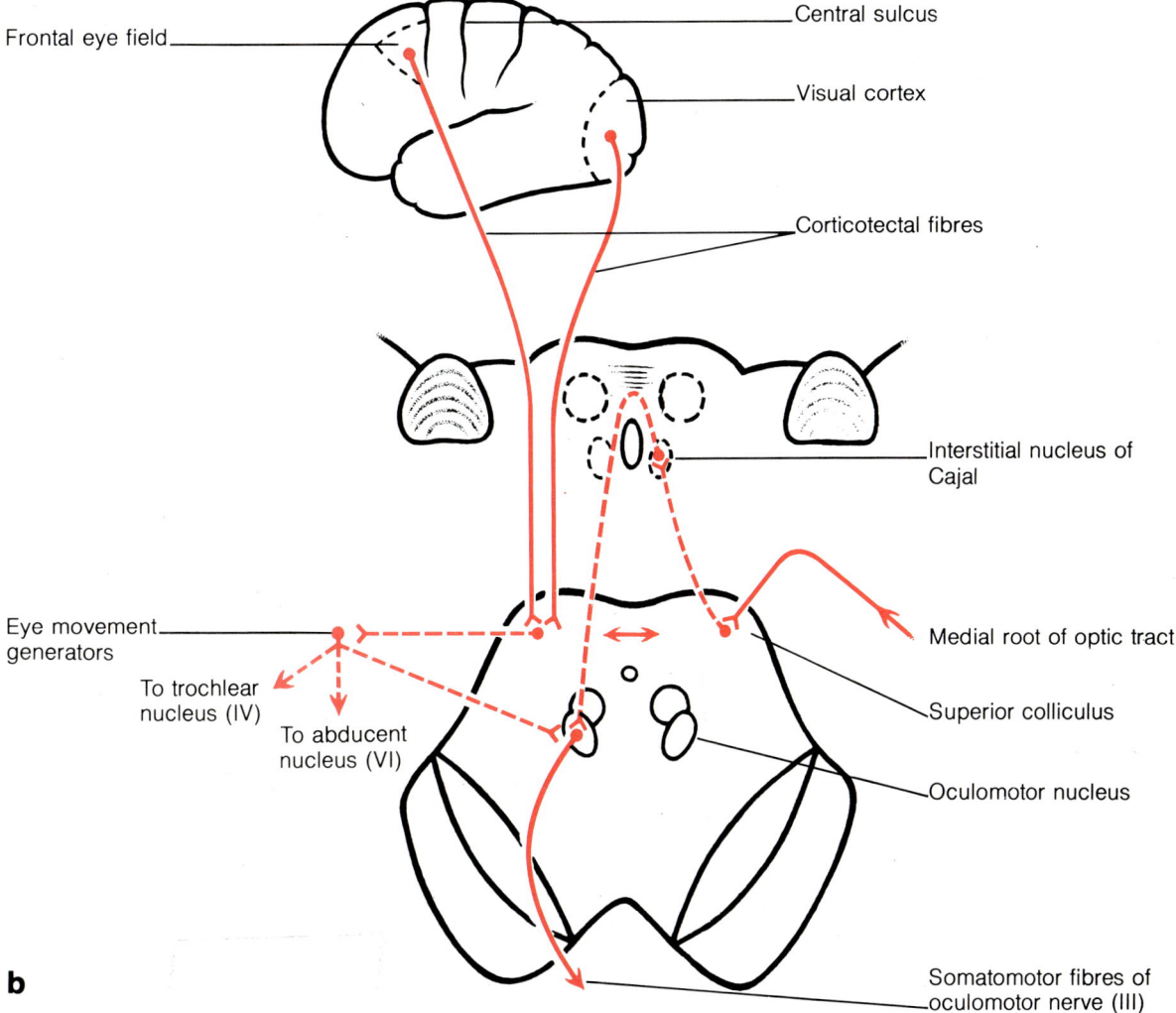

Figure 177 b) *The control of eye movements.* *Fixation and tracking of visual stimuli require inputs from the optic tract and visual cortex to the superior colliculus. Voluntary movement of the eyes is effected via fibres from the frontal eye fields. Connections to the nuclei controlling the extra-ocular muscles are via eye movement generators. The eye movement generators are: the rostral interstitial nucleus of the medial longitudinal fasciculus, the interstitial nucleus of Cajal, the paramedian pontine reticular formation, and the nucleus prepositus hypoglossi. With the exception of the interstitial nucleus of Cajal, their connections and positions are indicated only schematically. Accommodation (focusing of the lens) requires inputs from visual cortex to the superior colliculus and/or pretectal area. Parasympathetic preganglionic fibres pass from the Edinger–Westphal nucleus via the oculomotor nerve.*

stimulus are moving (fixation). Both voluntary and reflex aspects of these tasks are normally under cerebral cortical control via corticotectal and corticoreticular fibres. It has further been suggested that the superior colliculus plays a role in attentional mechanisms (see pages 509 and 534).

The pretectal area lies cranial to the superior colliculus near the border of the midbrain and diencephalon. It contains a number of small cell groups that receive retinal, superior collicular, reticular, thalamic and cortical fibres. Fibres pass between the two sides, either in the posterior commissure or ventral to the aqueduct. A few fibres may pass from the pretectal area to the retina. Neurones in the pretectal area are essential for the pupillary light reflex. Nuclei on each side receive fibres from both retinae and project to both left and right parasympathetic oculomotor nuclei (Edinger–Westphal nuclei). From each parasympathetic oculomotor nucleus, pre-

158,13
158,10; 158,9

ganglionic fibres pass through cranial nerve III to the ciliary ganglion. Stimulation of postganglionic fibres causes pupillary constriction. Lesions of an optic nerve block both direct (ipsilateral) and consensual (contralateral) pupillary constriction when light is shone in that eye. Both direct and consensual light reflexes will remain intact for the other eye.

The olfactory nerve

The olfactory (I) nerve is attached to the cerebral hemisphere. It will be described briefly here for the sake of completeness. The olfactory system is discussed further on pages 555 to 556 and 588 to 589.

The fila olfactoria, the fine axons of the olfactory chemoreceptor neurones (olfactory rods), gather into about 20 filaments that pass from the olfactory mucosa of the roof of the nasal cavity through the cribriform plate of the ethmoid bone. They terminate in the ipsilateral olfactory bulb which is located on the orbital surface of the frontal lobe. The olfactory tract passes posteriorly from the olfactory bulb to the anterior perforated substance, lateral to the optic chiasma. Axons of neurones (chiefly mitral cells) of the olfactory bulb pass through the tract to the olfactory cortex of the medial temporal lobe (see Figure 188b).

Lesions of the olfactory nerve, bulb or tract produce ipsilateral anosmia (a dysfunction that is not subjectively debilitating, but is clinically demonstrable). Depending on the lesion, the olfactory nerve is capable of regeneration. Note that primary olfactory neurones (olfactory rods) are continually replaced, having an expected life of several weeks. Inadequate replacement in later life may lead to anosmia.

Irritative lesions resulting in anomalous impulses can give rise to subjective olfactory sensations. Reflex salivation and gastric secretion are evoked via amygdalar, hypothalamic and reticular links with the salivatory nucleus (of cranial nerves VII and IX) and the dorsal motor nucleus of the vagus nerve.

The nervus terminalis is a small group of fibres that pass through the cribriform plate in association with the olfactory nerve. The nerve courses between the nasal mucosa and the anterior perforated substance and septal area. Some of its fibres may reach the hypothalamus. Amongst the unmyelinated fibres of the nerve are found both bipolar and multipolar neurones (ganglion cells). Its function is unknown; it has been suggested variously that it is either sensory or sympathetic.

The vomeronasal nerve of sub-primate species is probably absent in adult humans.

ASCENDING SENSORY PATHWAYS (see pages 590 to 601)

The ascending sensory pathways of the brainstem are the auditory and vestibular (see cranial nerve VIII, pages 471 to 476), the gustatory (see cranial nerve VII, pages 477 to 478), part of the visual (see cranial nerve II, pages 484 to 488), and the somatosensory. The somatosensory pathways comprise the dorsal and ventral trigeminothalamic tracts (see cranial nerve V, pages 480 to 482), the ascending pathways to the cerebellum (see cerebellar connections and precerebellar nuclei, pages 502 to 504), and the dorsal column/medial lemniscal pathway and anterolateral system fibres (see pages 455 to 456).

The dorsal column fibres of the medial, gracile fasciculus (of Goll) and lateral, cuneate fasciculus (of Burdach) continue dorsally into the medulla and terminate respectively in the ipsilateral gracile and cuneate nuclei of the caudal medulla (Figures 178b to d, pages 598 to 599). These nuclei lie beneath the gracile and cuneate tubercles and are collectively known as the dorsal column nuclei. The nuclei contain both projection cells and interneurones. They process information concerning touch (particularly the finer, more discriminative aspects of touch, and vibration sensibility) and proprioception (particularly that necessary for conscious appreciation of body position and movement). They are somatotopically organised, the lower limb and trunk being medial (gracile nucleus), the upper limb and trunk being lateral (cuneate nucleus).

Besides dorsal column afferents, the dorsal column nuclei also receive fibres that ascend dorsolaterally in the spinal cord. Their major descending inputs are from somatosensory and motor regions of the cerebral cortex, the red nucleus, and the reticular formation. The main efferents of the dorsal column nuclei pass via the medial lemniscus to the thalamus, but other fibres pass caudally via the dorsal columns to the dorsal horn of the spinal cord. Some fibres pass to the cerebellum and superior colliculus.

The medial lemniscus (mesial fillet) is formed by fibres originating in the dorsal column nuclei which arc transversely across the medulla (internal arcuate fibres), decussate and then turn cranially close to the midline (Figure 178c). Thus, the medial lemniscus conveys information from the contralateral side of the body. Cuneate fibres (upper body) are dorsal, and gracile fibres (lower body) are ventral. It is joined by some fibres from accessory dorsal column nuclei (the accessory cuneate nucleus, and nucleus z which is located just cranial to the gracile nucleus). The medial lemniscus ascends close to the midline and above the pyramid through the medulla, and is found medially and ventrally in the tegmental part of the pons (Figure 178c to h). Passing through the pons, it broadens out, and cuneate fibres (upper body) come to lie medially, and gracile fibres (lower body) laterally. In the midbrain it gradually becomes more lateral and dorsal in position and is joined by spinothalamic and trigeminothalamic fibres of the spinal and trigeminal lemnisci (Figure 178i and j). The somatotopical organisation of the dorsal column/medial lemniscal pathway rotates through 90° at its decussation in the medulla and a further 90° in the pons. These rotations bring into register the orientation of the somatotopical maps of the medial and spinal lemnisci. The medial lemniscus terminates ipsilaterally in the thalamus, chiefly in its nucleus ventralis posterior lateralis.

Damage to the medial lemniscus produces impaired discriminative tactile ability, vibration sensibility and proprioceptive sense contralateral to the lesion. If ventral trigeminothalamic fibres are also damaged, there will in addition be sensory loss in the contralateral head region. Neighbouring structures are likely to be involved as well, and these (particularly the nuclei and fibres of cranial nerves) provide valuable clues as to the level of the lesion.

The anterolateral system fibres (see pages 600 to 601) pass from the ventrolateral funiculus of the spinal cord into the medulla. They originate from 'tract cells' chiefly of the contralateral spinal cord. They ascend ventrolaterally through the medulla and pons (Figure 178b to h). Fibres of sacral origin course superficially, those of cervical origin run deepest. In the cranial part of the pons they ascend immediately lateral to the medial lemniscus in the spinal lemniscus. In the midbrain, the spinal lemniscus is found laterally, dorsal to the medial lemniscus (Figure 178i and j). Fibres of sacral origin are now dorsolateral, and cervical are ventromedial. The spinal lemniscus continues to its termination in the thalamus (nucleus ventralis posterior lateralis, posterior nuclear group and intralaminary nuclei). During passage through the brainstem, many fibres and collaterals leave the pathway to terminate in the reticular formation, peri-aqueductal grey and superior colliculus.

The anterolateral tract conveys pain, temperature, basic tactile, autonomic and proprioceptive information, chiefly concerning the contralateral side of the body. Spinoreticular fibres, notably those conveying nociceptive information, are the first link in an alternative sensory pathway to the thalamus via the reticular formation. Damage to the anterolateral pathway in the brainstem causes a loss of pain and temperature sensibility on the opposite side of the body. As for the medial lemniscus, neighbouring structures are also likely to be involved and provide valuable localising signs.

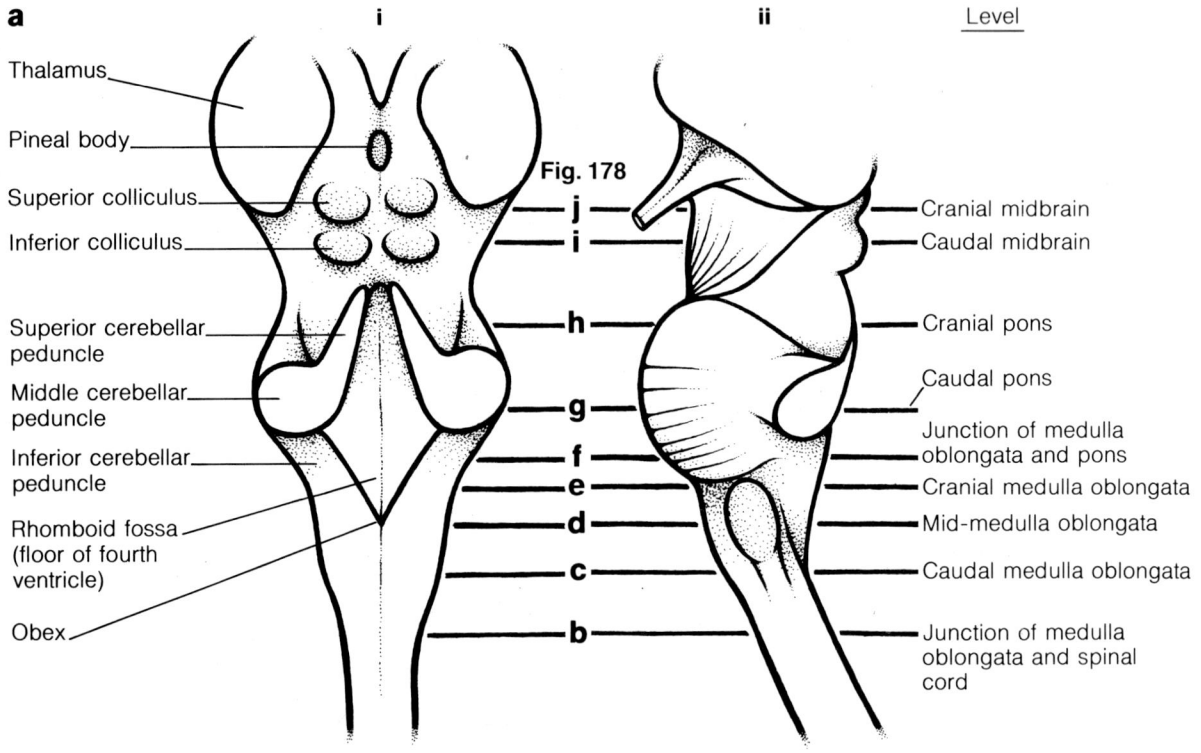

Figure 178 *Transverse sections of the brainstem.*
The levels of the transverse sections are shown in Figure 178a in relation to i) a dorsal and ii) a lateral view of the brainstem after removal of the cerebellum. In Figures 178b to 178j, sensory tracts and their associated nuclei are shaded and labelled on the left, motor on the right. Nuclei of the reticular formation are labelled in italics; their positions, but not their boundaries, are indicated. The sections are conventionally orientated with dorsal at the top.

Clues to the level of an isolated transverse section of the brainstem are provided by its outline and by the presence of either the central canal, fourth ventricle or cerebral aqueduct. The caudal medulla is approximately circular in outline, with a central canal. The cranial medulla has an increasingly scalloped outline, with the fourth ventricle extending dorsally. The pyramids are found ventrally throughout the medulla. In sections through the pons, the middle cerebellar peduncle is prominent laterally; dorsally, the fourth ventricle is enclosed by the cerebellum. Sections through the midbrain are characterised by the colliculi dorsally and the deep interpeduncular fossa ventrally; the cerebral aqueduct is surrounded by the peri-aqueductal grey. Knowledge of the external features of the brainstem is of considerable assistance in further identifying the level of a section and in understanding its internal topography.

Key for sections shown in Figures 178b to j

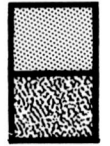 Sensory tracts

Sensory nuclei

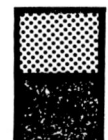 Motor tracts

Motor nuclei

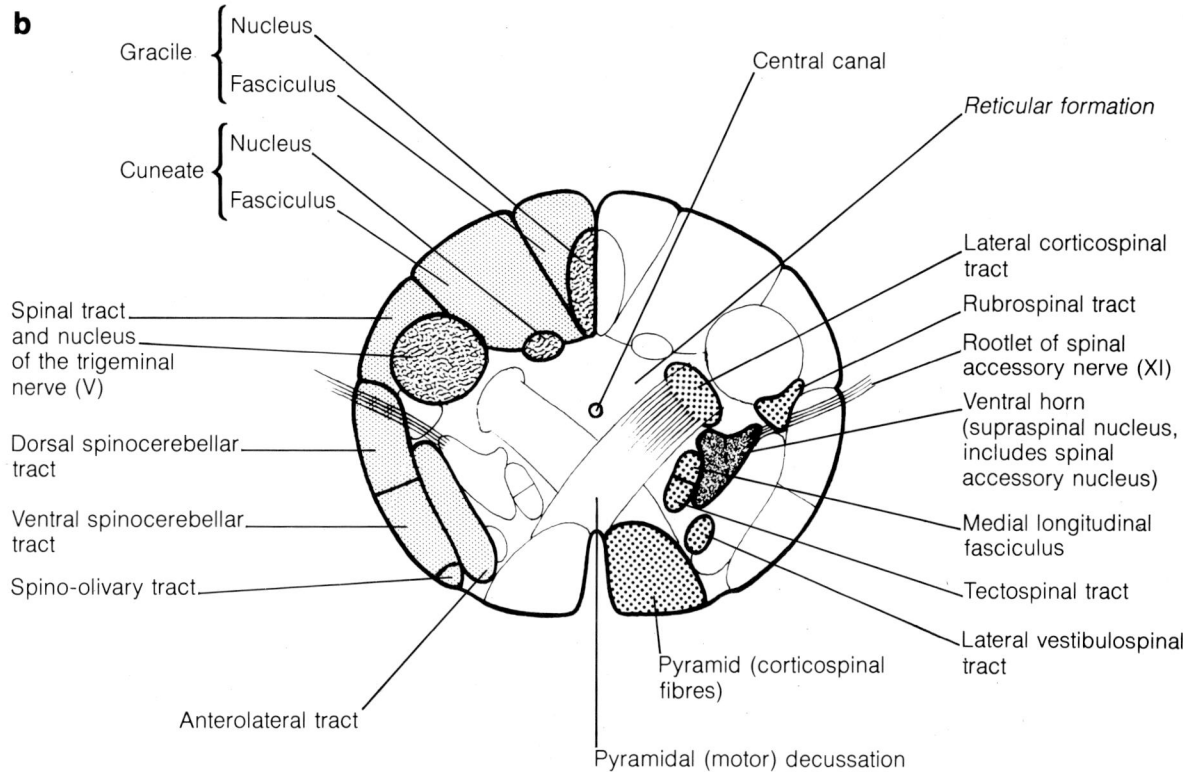

Figure 178b *Junction of spinal cord and medulla oblongata (motor decussation).*

Identifying features:

1 The outline is essentially circular. The canal is nearly central. Ventrally, small bundles of decussating fibres from the pyramids interrupt the ventral median sulcus.

2 Ventrally are the pyramids. Their decussating fibres (the motor decussation) pass obliquely across the midline to form the lateral corticospinal tract.

3 Dorsolateral to each pyramid is the cranial extension of the ventral horn of the spinal cord (supraspinal nucleus). Motoneurones found laterally in this region belong to the spinal nucleus of the accessory nerve. Fibres of the spinal part of the accessory nerve are shown emerging laterally.

4 Dorsally, the gracile and cuneate fasciculi are prominent and the caudal ends of the gracile and cuneate nuclei are seen.

5 Dorsolaterally is the spinal tract and nucleus (pars caudalis) of the trigeminal nerve. The nucleus forms a continuation of the dorsal horn of the spinal cord.

6 The other ascending tracts (dorsal and ventral spinocerebellar, spino-olivary and anterolateral) and descending tracts (rubrospinal, medial longitudinal fasciculus and associated tectospinal tract, and lateral vestibulospinal) are located ventrolaterally, close to their positions within the spinal cord.

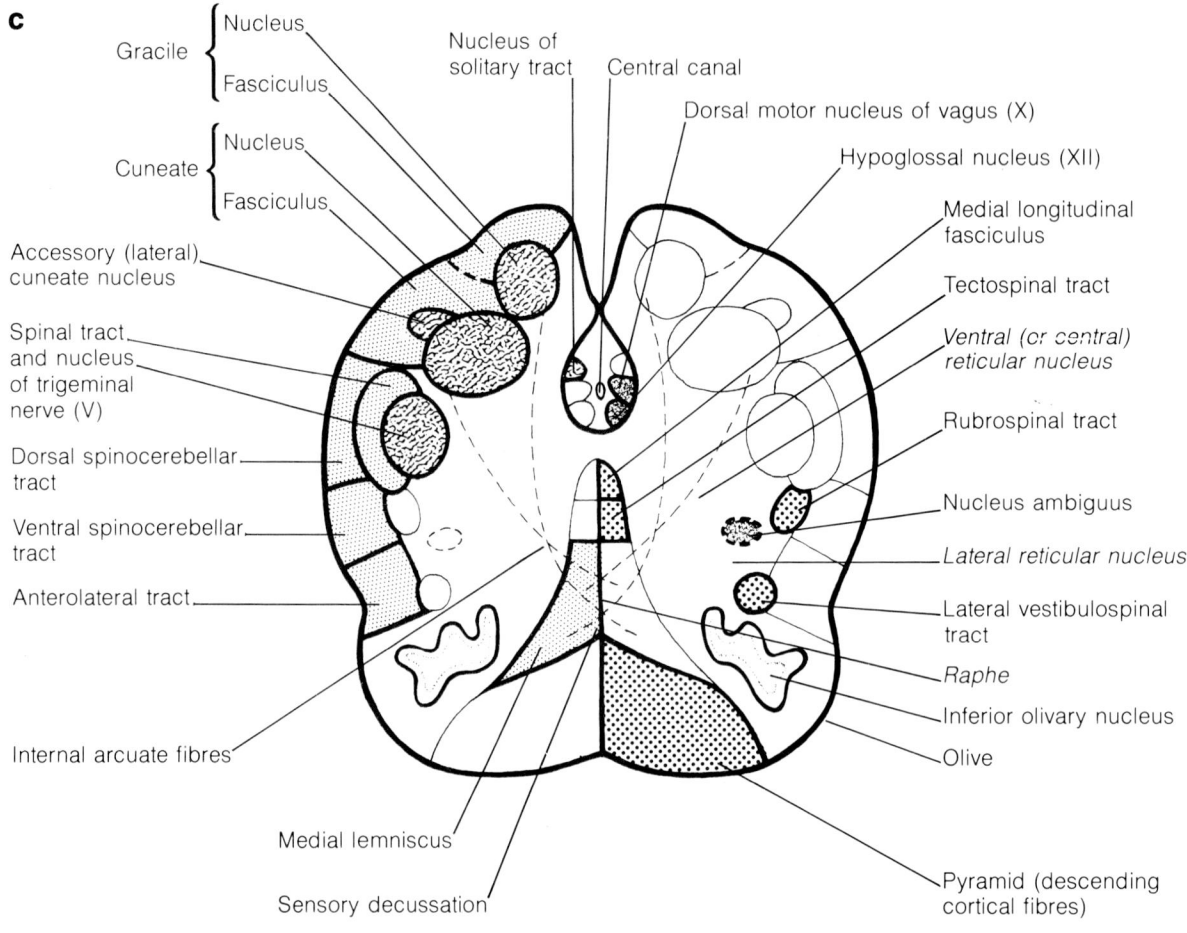

Figure 178c *Caudal medulla oblongata (sensory decussation).*

Identifying features:

1 The outline is nearly square. The central canal is more dorsal than central.

2 Ventrally, the pyramids are well formed.

3 Dorsally are the gracile and cuneate fasciculi and their nuclei. Internal arcuate fibres leave the nuclei, cross the midline as the sensory decussation, and form the medial lemniscus of the opposite side.

4 The medial lemniscus is located medially, dorsal to each pyramid. Dorsal to this lemniscus are the tectospinal tract and medial longitudinal fasciculus.

5 Immediately lateral to the main cuneate nucleus is the accessory (lateral) cuneate nucleus. Its axons form the cuneocerebellar tract.

6 Dorsolateral to each pyramid is the caudal end of the inferior olivary nucleus.

7 Surrounding the central canal is grey matter. Ventromedially within this grey area is the hypoglossal nucleus, one of the five long cranial nerve nuclei of the medulla. Dorsolateral to the hypoglossal nucleus is the dorsal motor nucleus of the vagus. Dorsal to the dorsal motor nucleus of the vagus is the solitary tract and its nucleus.

8 Laterally, within the reticular formation dorsal to the inferior olivary nucleus, is the (indistinct) nucleus ambiguus.

9 Dorsolaterally are the spinal tract and nucleus (pars caudalis) of the trigeminal nerve.

10 Other ascending and descending tracts lie laterally.

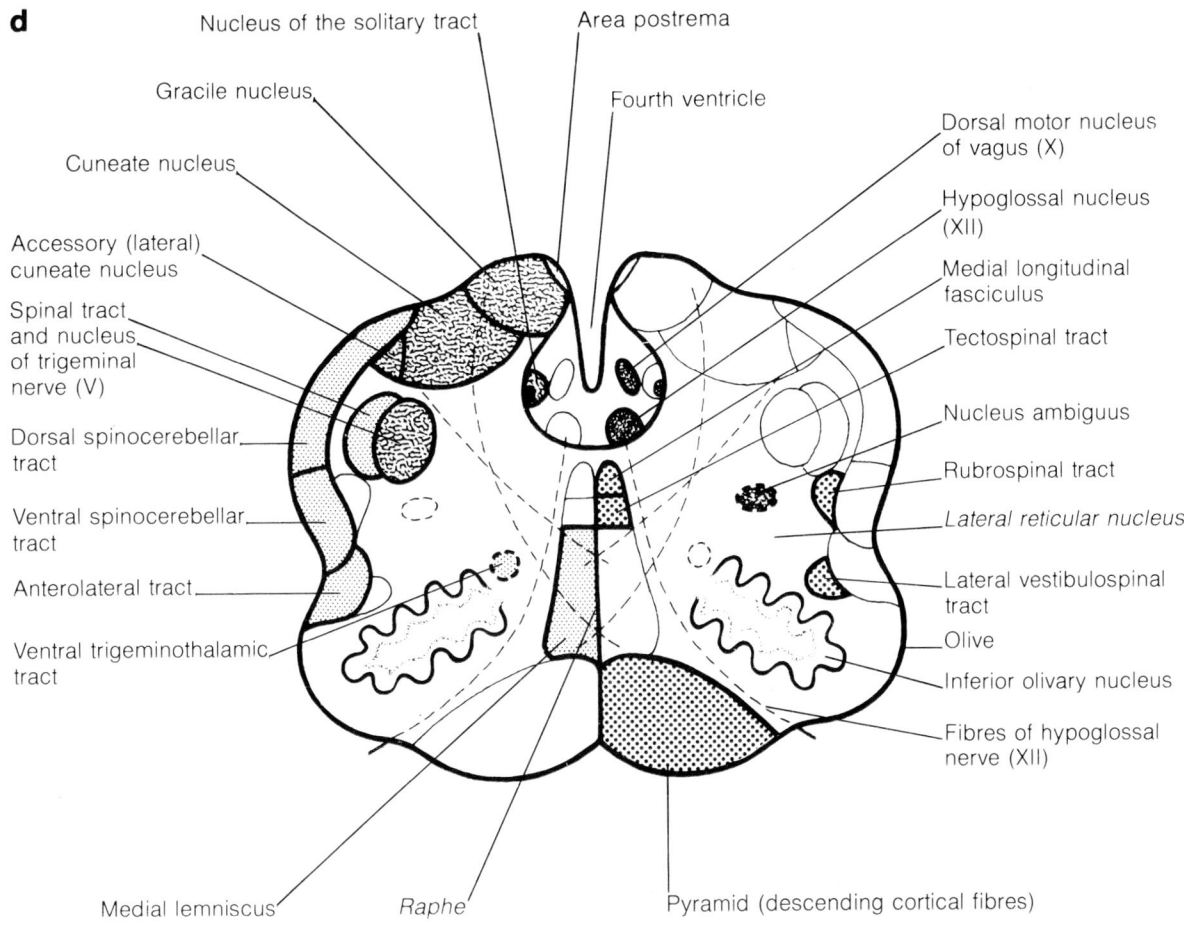

Figure 178d *Mid-medulla oblongata (obex or caudal calamus scriptorius).*

Identifying features:

1 The outline is indented and dorsally the gracile and cuneate tubercles raised by the underlying nuclei are conspicuous. In the midline dorsally, a deep groove marks the most caudal part of the fourth ventricle (at the boundary of the closed and open parts of the medulla). The median aperture of the fourth ventricle lies above this point.

2 The pyramids are prominent ventrally.

3 Dorsal to each pyramid, the medial lemniscus is now large. Associated with it, but represented by scattered bundles of fibres rather than a discrete tract, is the ventral trigeminothalamic tract. This tract remains close to the medial lemniscus throughout the brainstem and forms the trigeminal lemniscus. The tectospinal tract and the medial longitudinal fasciculus lie immediately above the medial lemniscus.

4 The inferior olivary nucleus is large and lies beneath the olive. The medial and dorsal accessory olivary nuclei are not shown.

5 The five long cranial nerve nuclei of the medulla (hypoglossal nucleus, dorsal motor nucleus of the vagus, nucleus of the solitary tract, nucleus ambiguus and nucleus [pars interpolaris] of the spinal tract of the trigeminal nerve) are all in the same relative positions as in Figure 178c. Fascicles of the hypoglossal nerve are illustrated.

6 The dorsal spinocerebellar tract is now dorsolateral. The remaining sensory and motor tracts are found laterally.

7 The reticular formation occupies large, mainly paramedian areas of the section. The area postrema is found dorsally, in the walls of the ventricle.

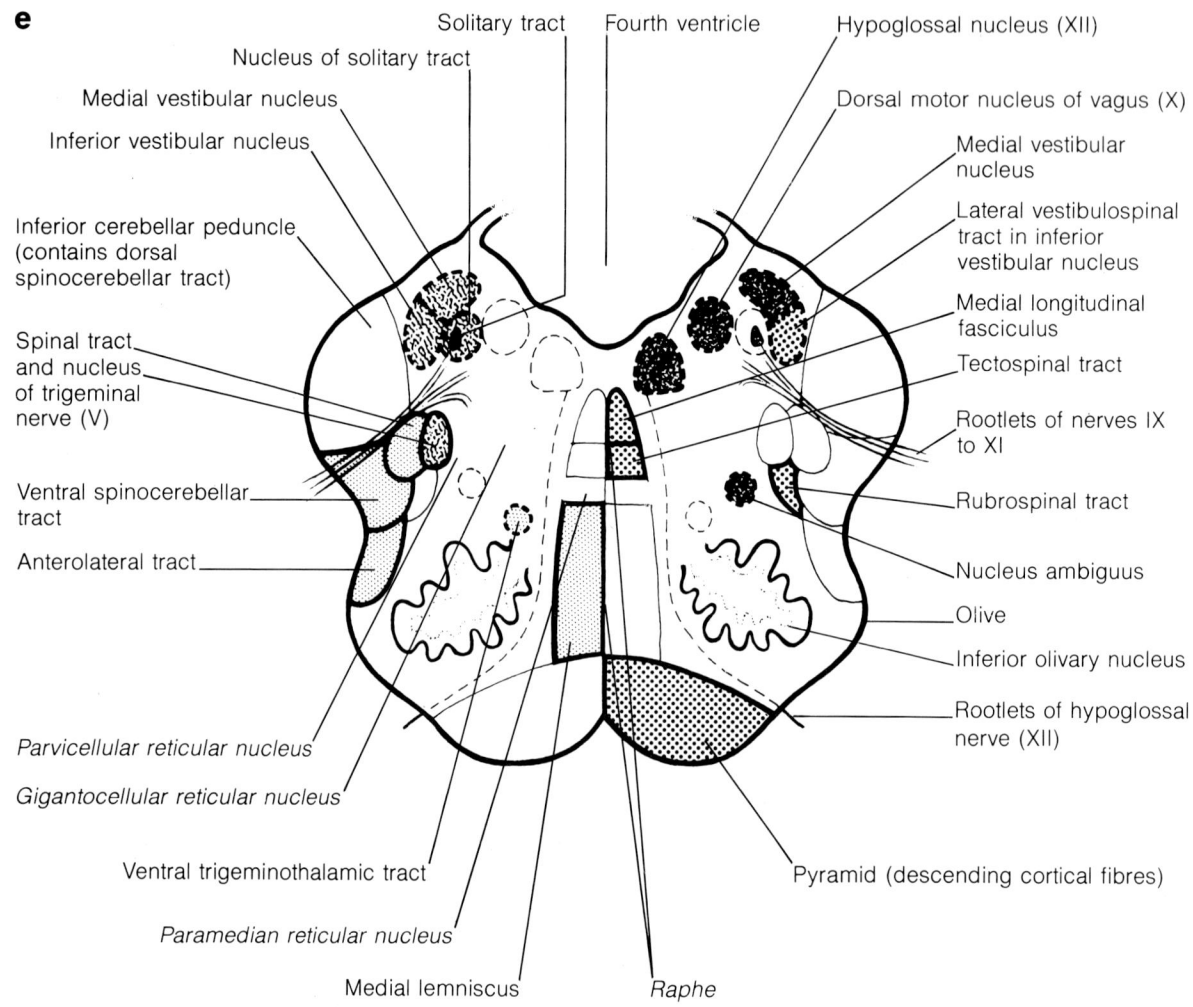

Figure 178e *Cranial medulla oblongata (open part of medulla).*

Identifying features:

1 The outline is scalloped, with bulges formed ventrally by the pyramids, ventrolaterally by the olives, and dorsolaterally by the inferior cerebellar peduncles. Dorsally, the floor of the fourth ventricle is divided into the hypoglossal and vagal triangles which lie above the respective cranial nerve nuclei.

2 Medially, the pyramids are prominent ventrally. The medial lemniscus, tectospinal tract and medial longitudinal fasciculus continue to be found dorsal to each pyramid. Other sensory and motor tracts lie laterally.

3 Ventrolaterally, the inferior olivary nucleus is large, deep to the olive. Axons leaving the nucleus pass diagonally to the contralateral inferior cerebellar peduncle.

4 The five long cranial nerve nuclei of the medulla occupy the same relative positions as they do more caudally, but are now spread out mediolaterally. The hypoglossal nucleus is dorsal and near to the midline. Fibres from it pass ventrally before emerging between the olive and pyramid. The dorsal motor nucleus of the vagus is lateral to the hypoglossal nucleus; some fibres from it are shown passing into the vagus nerve which emerges, along with the glossopharyngeal and cranial part to the accessory nerves, ventral to the inferior cerebellar peduncle. The solitary tract lies within its nucleus lateral to the dorsal motor nucleus of the vagus. Ventrally is the spinal tract of the trigeminal; it lies lateral to its nucleus (pars interpolaris). The position of the nucleus ambiguus within the reticular formation dorsal to the inferior olivary nucleus is indicated.

5 The inferior and medial vestibular nuclei are located between the nucleus of the solitary tract and the inferior cerebellar peduncle.

6 The large remaining, intermediate areas of the section are occupied by the reticular formation.

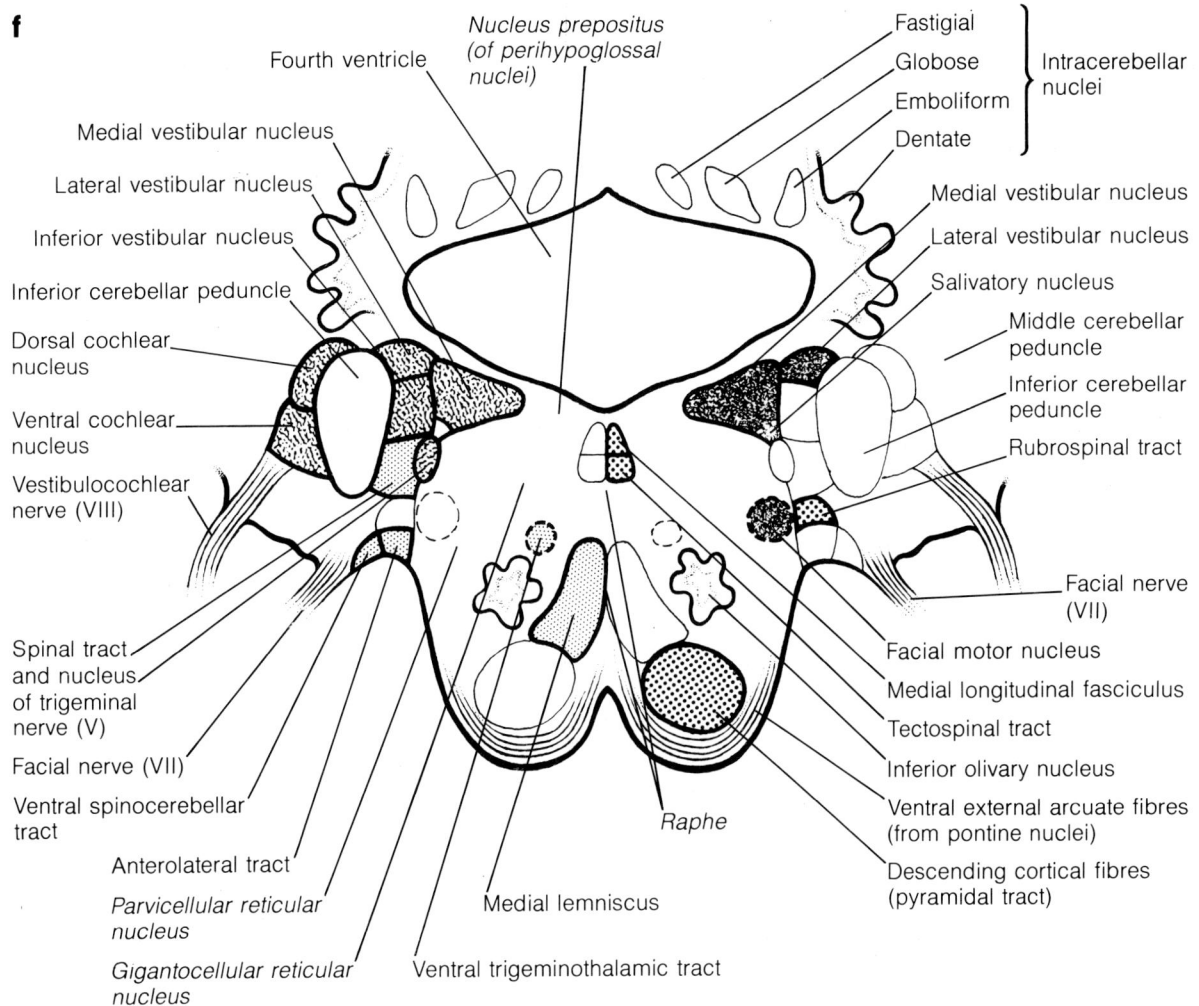

Figure 178f *Junction of medulla oblongata and pons.*

Identifying features:

1 Ventrally, the section's outline is relatively smooth, being formed by ventral external arcuate fibres from the pontine nuclei. These fibres pass dorsolaterally into the middle cerebellar peduncle. Laterally and dorsally, the brainstem is continuous with the cerebellum. Dorsal to the fourth ventricle, the four pairs of intracerebellar nuclei are visible within the medullary core of the cerebellum. The fourth ventricle is close to its widest. Just caudal to this level, the lateral apertures of the fourth ventricle are at the ends of its lateral recesses.

2 Ventrally are the corticospinal and corticonuclear fibres that form the medullary pyramids.

3 Medially, the medial lemniscus broadens out as it passes cranially, while the laterally adjacent inferior olivary nucleus diminishes in size. The medial longitudinal fasciculus and tectospinal tract are dorsal.

4 Most laterally are fibres of the middle cerebellar peduncle. The inferior cerebellar peduncle is medial to it and is surrounded by nuclei.

5 Ventrolaterally, the vestibulocochlear nerve is seen entering the brain from the cerebellopontine angle. Its nuclei are found on both sides of the inferior cerebellar peduncle. The dorsal and ventral cochlear nuclei are lateral to the peduncle. The medial, inferior and lateral vestibular nuclei are medial to it, beneath the vestibular area of the floor of the ventricle.

6 Ventromedial to the inferior cerebellar peduncle are the spinal tract of the trigeminal nerve and its nucleus (pars oralis). Other sensory and motor tracts are found more ventrally.

7 In the region just dorsal to the trigeminal nucleus are the scattered cells of the salivatory nucleus.

8 Ventromedial to the trigeminal nucleus is the motor nucleus of the facial nerve. The facial nerve may be seen ventrolateral to its motor nucleus where it enters the cerebellopontine angle, medial to the vestibulocochlear nerve.

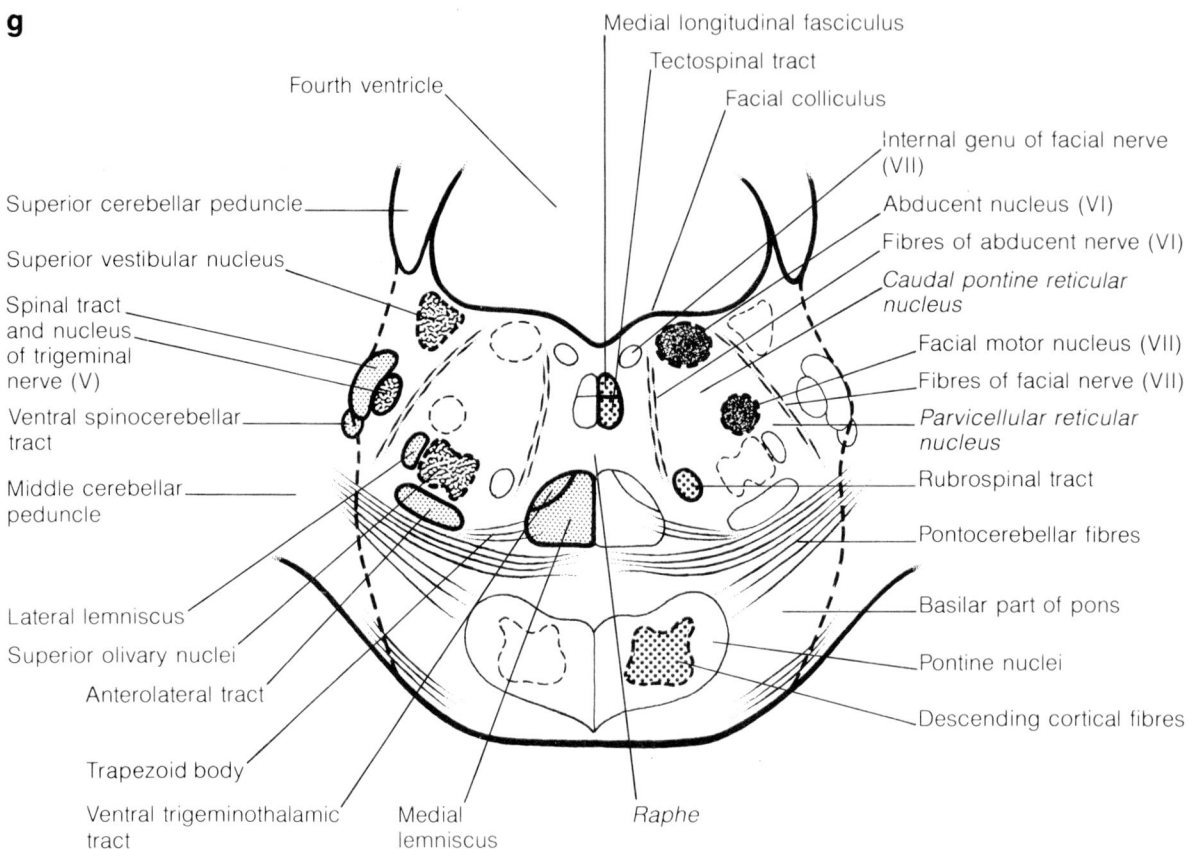

Figure 178g *Caudal pons (facial colliculus).*

Identifying features:

1 Laterally, the pons is continuous through the massive middle cerebellar peduncle with the cerebellum. Dorsally, the fourth ventricle is enclosed by the pons and cerebellum. The superior cerebellar peduncles form its dorsolateral walls. In its floor is the facial colliculus.

2 The pons is divided into a dorsal, tegmental part and a ventral, basilar part. The basilar part of the pons consists of caudally running fibres of cerebral cortical origin (cortico-pontine, -nuclear, -spinal fibres), broken up into irregular bundles by scattered pontine nuclei and by transversely running pontocerebellar fibres. Pontocerebellar fibres pass from the pontine nuclei to the contralateral middle cerebellar peduncle.

3 Each side of the pontine tegmentum is divided into three regions (medial, intermediate, lateral) by fibres of the abducent and facial nerves.

4 In the floor of the fourth ventricle is an elevation, the facial colliculus, caused by the underlying abducent nucleus. Fibres from the abducent nucleus pass ventrally, caudally and slightly laterally to their point of emergence as the abducent nerve near the pontine-medullary junction.

5 Immediately medial to the abducent nucleus is the internal genu of the facial nerve. Fibres from the motor facial nucleus pass dorsomedially and cranially, gather into a compact bundle which runs cranially (medial to the abducent nucleus), before turning laterally (dorsal to the abducent nucleus), and then passing ventrolaterally and caudally to emerge in the facial nerve (see Figure 178f). These fibres give the facial colliculus its name.

6 In the medial pontine tegmentum, the medial lemniscus lies ventrally with the ventral trigeminothalamic tract or trigeminal lemniscus. These lemnisci are crossed by decussating fibres of the auditory pathway, the fibres forming the trapezoid body. Dorsally, the tectospinal tract and medial longitudinal fasciculus are close to the abducent nucleus.

7 In the intermediate part of the tegmentum are the abducent, motor facial and superior olivary nuclei. Ventrally, above fibres of the anterolateral system (tract) and cranial to the inferior olivary nucleus, are the superior olivary nuclei of the auditory pathway. The lateral lemniscus commences just lateral to the superior olivary nuclei.

8 In the lateral tegmentum are the spinal tract of the trigeminal nerve and its nucleus (pars oralis) and, dorsally, the superior vestibular nucleus.

9 Much of the tegmentum is occupied by the pontine reticular formation.

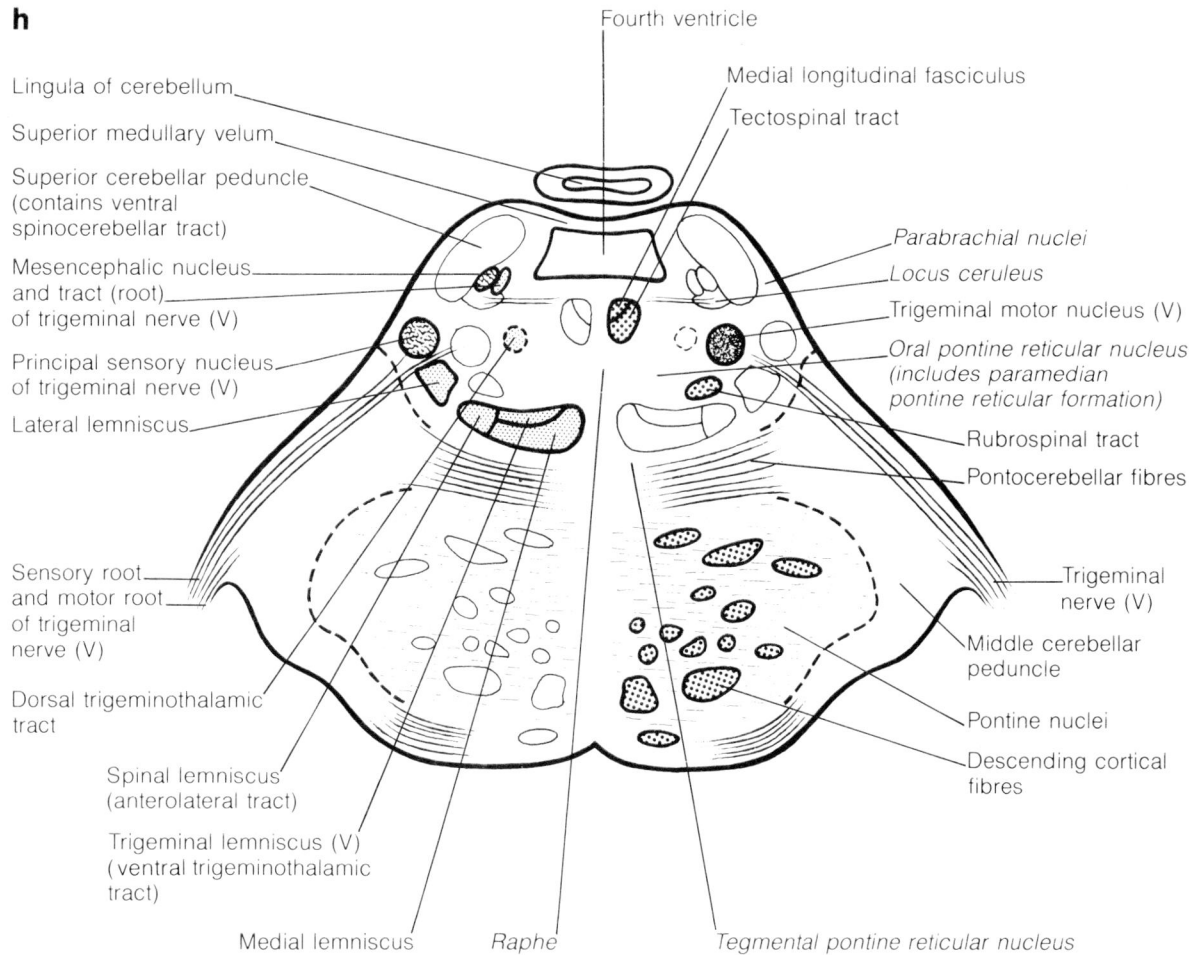

Figure 178h *Cranial pons (level of trigeminal nerve).*

Identifying features:

1 The large, ventral portion of the section is of the basilar part of the pons. Laterally, the trigeminal nerve emerges through the middle cerebellar peduncle. Dorsally, the fourth ventricle narrows as it approaches the cerebral aqueduct of the midbrain.

2 The medial, trigeminal and spinal (somatosensory), and lateral (auditory) lemnisci form a band across the ventral pontine tegmentum. The dorsal trigeminothalamic tract is found more dorsally. The medial longitudinal fasciculus and tectospinal tract are dorsomedial.

3 Fibres of the trigeminal nerve may be traced centrally to its laterally located, principal sensory nucleus and more medial, motor nucleus. The principal sensory nucleus is continuous caudally with the nucleus of the spinal tract. The mesencephalic nucleus of the trigeminal nerve is a thin column of cells lying ventrolateral to the fourth ventricle.

4 Large, melatonin-pigmented cells of the nucleus of the locus ceruleus are found just ventral to the mesencephalic nucleus of the trigeminal nerve.

5 The superior cerebellar peduncles are conspicuous lateral to the fourth ventricle. The lingula of the cerebellum is immediately dorsal to the superior medullary velum which roofs the ventricle.

6 Large areas of the tegmentum are occupied by the pontine reticular formation.

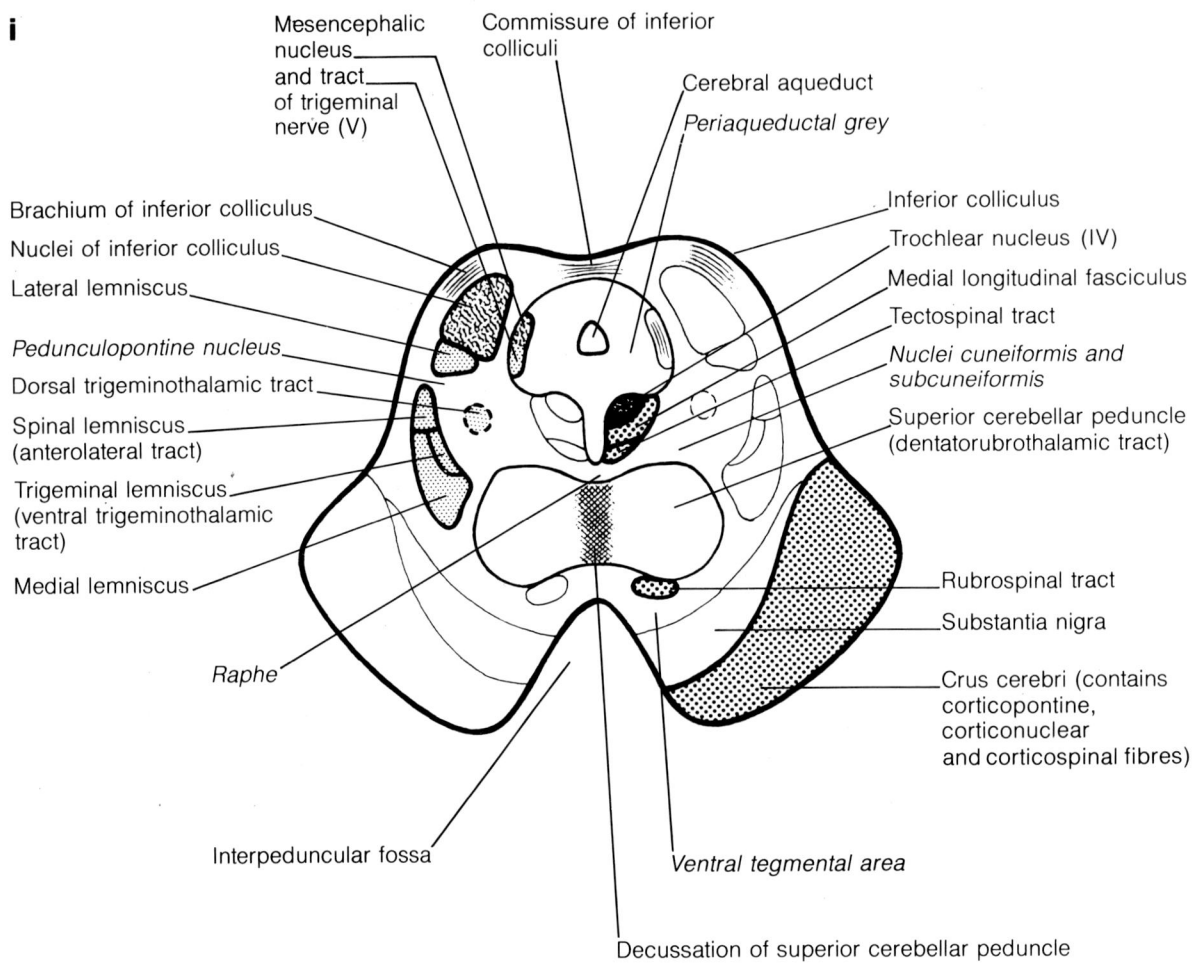

Figure 178i *Caudal midbrain (level of inferior colliculus).*

Identifying features:

1 Ventrally is the deep interpeduncular fossa. Ventrolaterally are large protuberances containing the fibre tracts of the crus cerebri. Dorsally are the eminences produced by the inferior colliculi. The cerebral aqueduct is surrounded by the peri-aqueductal grey. Above the crus cerebri on each side are darkly pigmented cells of the substantia nigra. The ventral tegmentum is dominated by the continuations of the superior cerebellar peduncles.

2 Ventrolaterally, the crus cerebri contains cortico-pontine, -nuclear and -spinal fibres of the internal capsule of the cerebral hemisphere.

3 The substantia nigra occupies the region dorsomedial to the crus cerebri. Its pars reticulata is the cell-poor part adjacent to the crus cerebri. Its darkly pigmented, dorsal region is its pars compacta.

4 Medial to the substantia nigra is the ventral tegmental area (of Tsai).

5 Ventromedially, much of the tegmentum is taken up by the superior cerebellar peduncles (dentatorubrothalamic tracts) and their decussation.

6 Dorsal to each superior cerebellar peduncle and ventromedial to the peri-aqueductal grey, the medial longitudinal fasciculus lies close to the trochlear nucleus. Axons from the trochlear nucleus pass dorsally and caudally, describing a semicircle round the edge of the peri-aqueductal grey and decussating in the superior medullary velum, before emerging just caudal to the inferior colliculus as the trochlear nerve.

7 Laterally, in the peri-aqueductal grey, is the mesencephalic nucleus of the trigeminal nerve.

8 Laterally, in the tegmentum, are grouped the somatosensory fibres of the medial, trigeminal and spinal lemnisci.

9 More dorsally, the auditory fibres of the lateral lemniscus form a capsule ventral to the inferior colliculus before terminating there. Dorsally, the inferior colliculi are linked by their commissure. Dorsolaterally, axons pass from the inferior colliculus to the medial geniculate body via the brachium of the inferior colliculus.

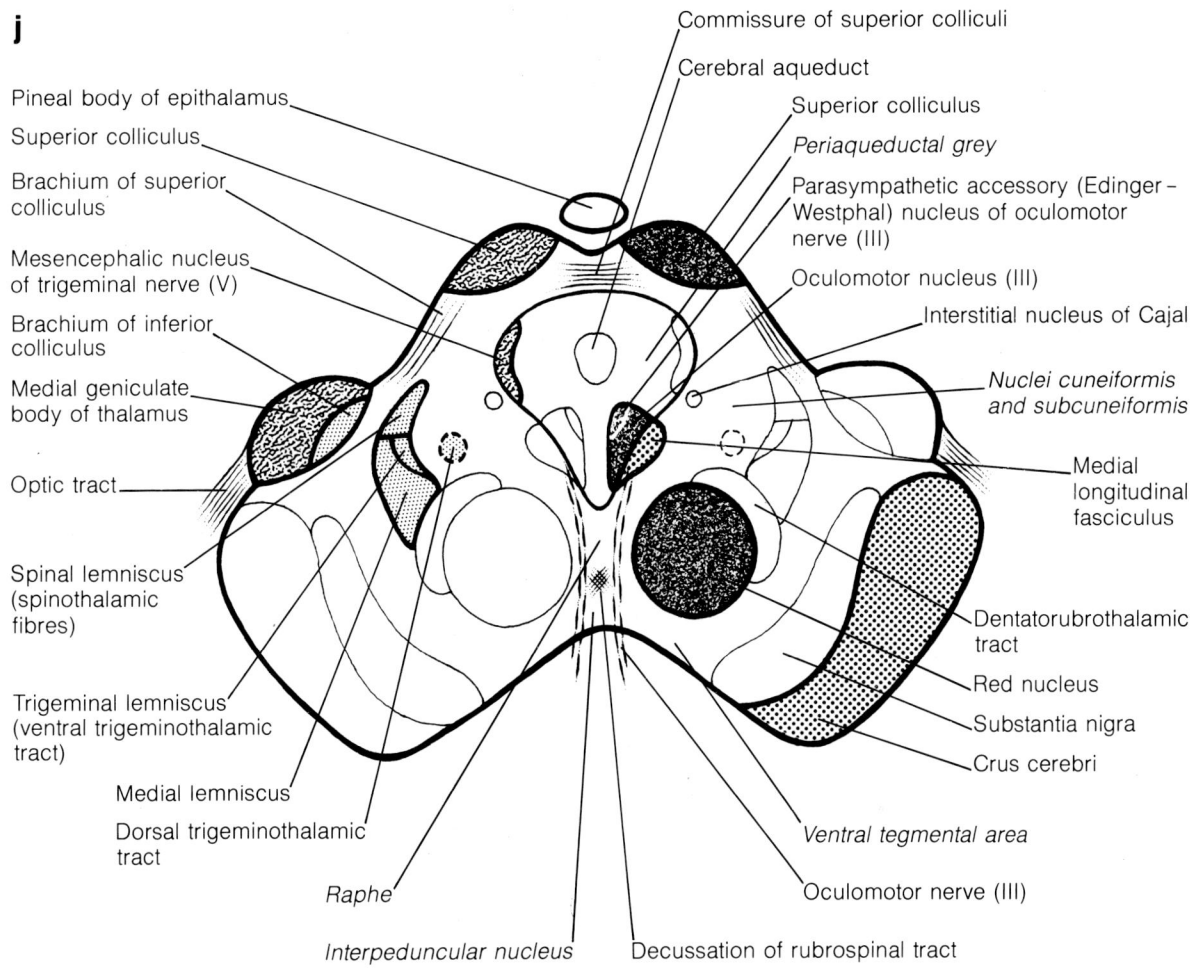

Figure 178j *Cranial midbrain (level of the superior colliculus).*

Identifying features:

1 The shape of the section is similar to that through the caudal midbrain. Many structures are continuations of those of the caudal midbrain and occupy similar positions (see Figure 178i). However, the superior colliculi are dorsal, and the red nuclei are conspicuous in the tegmentum.

2 Ventromedially on each side is the red nucleus. It gives rise to the rubrospinal tract which decussates in the ventromedial tegmentum.

3 Dorsomedial to the red nucleus and ventromedial to the peri-aqueductal grey is the medial longitudinal fasciculus.

4 Medially, adjacent to the medial longitudinal fasciculus, is the oculomotor nucleus. Dorsomedial to this nucleus is the parasympathetic accessory nucleus of the oculomotor nerve. Fibres run ventrally from the nuclei through the tegmentum to emerge in the interpeduncular fossa as the oculomotor nerve.

5 Dorsally are the superior colliculi, interconnected by their commissure. The pineal body of the epithalamus lies between them. Ventrolateral to each superior colliculus is its brachium. The brachium contains corticotectal fibres plus fibres of the medial division of the optic tract. The superior colliculus gives rise to the tectospinal tract. The tectospinal tract decussates immediately, ventral to the peri-aqueductal grey, and runs caudally, ventral to the medial longitudinal fasciculus.

6 Laterally, the medial geniculate body of the thalamus is near to the crus cerebri and the superior colliculus. It receives the auditory fibres of the brachium of the inferior colliculus. Fibres of the medial division of the optic tract pass close to it on their way to the superior colliculus.

MOTOR NUCLEI AND DESCENDING PATHWAYS (see pages 602 to 607)

All pathways descending to the spinal cord pass through (e.g. corticospinal) or originate in (e.g. rubrospinal) the brainstem. Some descending pathways terminate in the brainstem without reaching the spinal cord (e.g. corticopontine). Some of these tracts, their origins and terminations are described elsewhere: solitariospinal (page 471), vestibulospinal (pages 475 and 457 to 458), tectospinal (page 486), and reticulospinal (page 506). The pathways followed by fibres from the cerebral cortex, red nucleus and interstitial nucleus of Cajal will now be delineated.

194,10

194,22; 188B,39

180,24;
198A,4

Cerebral cortical fibres descend in the internal capsule of the cerebral hemisphere to reach the brainstem. Most cortical fibres that are to terminate in the brainstem or spinal cord enter the ipsilateral crus cerebri, the ventral part of the cerebral peduncle of the midbrain (Figure 178i and j). Other fibres pass to the midbrain more directly from the internal capsule, notably corticotectal fibres, which reach the superior colliculus via its brachium, but also some fibres that course through the midbrain tegmentum.

The fibres of the crus cerebri are conveniently divided into three categories according to their sites of termination: corticopontine, corticospinal and corticonuclear.

188A,23

The corticopontine fibres are by far the largest group (about 20 million fibres from each hemisphere). They arise from widespread regions of the cortex, but particularly from motor, somatosensory and visual areas. Those from frontal cortex occupy the medial part of the crus cerebri, those from other cortical regions are found laterally. The fibres terminate on the ipsilateral pontine nuclei in the basilar part of the pons (Figure 178g and h). The pontine nuclei give rise to transversely coursing fibres that enter the contralateral middle cerebellar peduncle and terminate in the cerebellar cortex. This corticopontocerebellar pathway (see Figure 182) provides the major route by which the cerebral cortex sends information to the cerebellum. However, the cortex additionally influences the cerebellum via the reticular formation and the inferior olivary nuclei, both of which receive corticonuclear fibres and project to the cerebellum.

176A,1-3

194,23; 198A,7
194,24; 196B,8

The corticospinal fibres (see pages 602 to 603) are found amongst fibres of the middle third of the crus cerebri. They originate from the frontal and parietal cortex on either side of the central sulcus, including motor, premotor and somatosensory cortex. Fibres supplying the cervical cord are more medial, fibres to lumbar and sacral segments more lateral. These fibres pass between the pontine nuclei and their efferents in the ventral pons, and then gather into the pyramid of the ventral medulla (Figure 178h to b). Each pyramid contains rather more than a million fibres. In the most caudal part of the ventral medulla about 85% (70 to 90%; the figure varies between individuals) of the fibres decussate to form the lateral corticospinal tract. Almost all the remaining fibres continue ipsilaterally as the ventral corticospinal tract, a few pass into the ipsilateral lateral corticospinal tract (some of which terminate in the spinal accessory nucleus). The presence of these corticospinal fibres in the medullary pyramids gives the corticospinal tract its alternative name, the pyramidal tract. Note, however, that the term pyramidal tract is sometimes used to include corticonuclear fibres. Corticospinal fibres are essential for voluntary movements, particularly fine movements of the distal limbs. Lesions cause contralateral paresis or paralysis. Damage to surrounding structures will usually give valuable clues as to the likely level of any lesion. Corticospinal fibres, chiefly originating in the parietal lobe, also allow cortical control of somatosensory transmission.

Corticonuclear fibres include all those fibres from the cerebral cortex that terminate in the brainstem, with the exception of corticopontine fibres. They are sometimes called corticobulbar fibres. However, as such fibres end at all levels of the brainstem, and 'bulb' refers to only its caudal part, the term is not entirely appropriate. These fibres terminate in relation to cranial nerve nuclei and many other nuclei of the brainstem. As for corticospinal fibres, most synapses are made with interneurones (i.e. neurones of the reticular formation), rather than directly upon projection cells such as motoneurones. Such fibres are more precisely called corticoreticular. Most corticonuclear fibres travel in close association with the corticospinal fibres for much of their course. They then travel dorsally and transversely across the brainstem to their termination. Some of these fibres are collaterals of corticospinal fibres. Indeed, in general terms the functions of corticospinal and corticonuclear fibres are allied. Some corticonuclear fibres (chiefly corticoreticular fibres) leave the crus cerebri and pass caudally close to the medial lemniscus.

Corticonuclear fibres influencing the motor nuclei of the cranial nerves form an important subgroup. They arise chiefly from motor and premotor areas of the frontal cortex. Some of these fibres terminate directly on motoneurones of the motor facial, motor trigeminal and hypoglossal nuclei. The spinal nucleus of the accessory nerve is similarly contacted by corticospinal fibres. The nuclei innervating muscle groups not normally contracted on only one side receive both ipsilateral and contralateral corticonuclear fibres; i.e. the motor trigeminal, the dorsal part of the facial motor nucleus, the nucleus ambiguus and the medial part of the hypoglossal nucleus usually receive bilateral innervation. The ventral part of the facial motor nucleus (lower facial muscles), most of the hypoglossal nucleus, and part of the spinal accessory nucleus (innervating trapezius) are supplied from the contralateral cortex. The sternocleidomastoid part of the spinal accessory nucleus is supplied ipsilaterally by corticospinal fibres. Indirectly, and according to a complex pattern, the motor nuclei of cranial nerves III, IV and VI which supply the extra-ocular muscles also receive bilateral inputs. However, note that voluntary conjugate lateral movements of the eyes normally originate in the contralateral frontal eye field. This pattern of corticonuclear innervation is subject to some individual variation, particularly for the motor trigeminal and hypoglossal nuclei. A lesion of corticonuclear fibres (most commonly in the internal capsule) will produce a pattern of disability of the upper motoneuronal type and according to the details of the innervation given above. Further details are given under the individual cranial nerves (see pages 464 to 488).

The cortex can influence other descending pathways via corticonuclear fibres terminating in the red nucleus, superior colliculus and reticular formation. Corticoreticular fibres terminate particularly in areas giving rise to reticulospinal fibres. The cortex seems to have relatively little influence over vestibular pathways. Corticonuclear fibres also end in the substantia nigra, allowing modulation of its influence upon motor activity. 194,26; 180,24

194,25

Other corticonuclear fibres pass to sensory nuclei of the brainstem: the dorsal column nuclei, sensory trigeminal nuclei, auditory nuclei (particularly the inferior colliculus), superior colliculus and nucleus of the solitary tract. These fibres arise particularly from areas of cortex having the same sensory function as those of the nuclei. They allow modulation of the transmission of sensory information and of the sensory input into local reflexes.

The interstitial nucleus of Cajal is a small collection of large neurones just ventral to the peri-aqueductal grey and lateral to the most cranial part of the oculomotor nucleus, near the junction of the midbrain with the diencephalon (Figure 177b). It receives fibres from the vestibular nuclei via the adjacent medial longitudinal fasciculus. Fibres also reach it from areas concerned with visual signals, eye movements and their co-ordination (the pretectal area, superior colliculus, frontal eye fields (probably), nucleus prepositus hypoglossi and cerebellum). Fibres leave the nucleus and pass to areas concerned with the co-ordination of eye movements. Fibres reach the contralateral oculomotor and trochlear nuclei via the posterior commissure. Other fibres enter the ipsilateral medial longitudinal fasciculus to terminate in the trochlear nucleus, the medial vestibular nucleus, the paramedian pontine reticular formation and the nucleus prepositus hypoglossi. Fibres reach the spinal cord via the interstitiospinal tract which runs within the ipsilateral medial longitudinal fasciculus (see pages 604 to 605). The fibres terminating in the brainstem are believed to be concerned with producing conjugate vertical and/or rotational movements of the eye balls. Activity in the interstitiospinal tract appears to result in rotation of the head and body about the vertical axis.

The red nucleus (nucleus ruber) is a large ovoid nucleus found in the cranial midbrain tegmentum and extending into the subthalamic region of the diencephalon (Figure 178j). It has a high concentration of iron-containing pigment and appears slightly pink in fresh material (hence its name). Fibres of cerebellar origin (the dentatorubrothalamic tract) pass through the nucleus and form a capsule round it. It contains large cells in its caudal, magnocellular part and small cells throughout, including its cranial, parvicellular part. The nucleus is somatotopically organised with representation being contralateral. The region dealing with the head is dorsomedial, the arm region intermediate, and the leg region ventrolateral.

The principal afferents of the red nucleus come from the cerebellum and cerebral cortex. The cerebellar afferents are most numerous. They arise from contralateral intracerebellar nuclei (globose, emboliform and dentate) and pass in the contralateral superior cerebellar peduncle, the decussation being in the caudal midbrain tegmentum. Corticorubral fibres arise from frontal cortical areas: from the ipsilateral motor cortex and bilaterally from premotor areas.

Efferent projections of the red nucleus pass to the spinal cord, cerebellum and brainstem. Fibres forming the rubrospinal tract (Monakow's bundle) decussate immediately in the ventral midbrain tegmentum (Figure 178j). The rubrospinal tract continues contralaterally, ventrally through the pons and laterally through the medulla, to enter the lateral funiculus of the spinal cord (Figures 178i to b, 173d to a). The rubrospinal tract may be regarded as an indirect corticospinal pathway. It acts chiefly in the control of flexor limb muscles (see also pages 602 to 603).

The connections of the red nucleus with the cerebellum are both direct and indirect. Indirect cerebellar connections are made by way of fibres passing to the inferior olivary nucleus, accessory (lateral) cuneate nucleus, reticular formation and vestibular nuclei. Rubronuclear (rubrobulbar) fibres leave the rubrospinal tract during its course through the brainstem and terminate in the contralateral motor facial nucleus, the principal sensory and spinal trigeminal nuclei and the dorsal column nuclei. These latter connections suggest a role for the red nucleus in modulation of sensory transmission.

The substantia nigra occupies the region dorsomedial to the crus cerebri throughout the midbrain and extends into the subthalamus of the diencephalon (Figures 178i and j). It is particularly well developed in the human brain, where it is the largest nucleus in the mesencephalon. The darkly pigmented, melatonin-containing cells that give the region its characteristic black appearance even in unstained tissue are in the more dorsal part of the region, its pars compacta. The cell-poor part of the substantia nigra that is adjacent to the crus cerebri is its pars reticulata. Functionally, the substantia nigra belongs with the basal nuclei (basal ganglia) and is further mentioned on pages 567 and 570.

The substantia nigra receives inputs from the basal nuclei (caudate nucleus, putamen and globus pallidus). The cells of the pars compacta give rise to nigrostriatal fibres that terminate in the caudate nucleus and putamen, and use the neurotransmitter dopamine. Loss of dopaminergic neurones is the characteristic lesion of Parkinson's disease (paralysis agitans); symptoms typically become clinically significant once dopaminergic cell loss exceeds about 80% of the initial number. The cells of the pars reticulata project to the thalamus (nuclei ventralis lateralis and ventralis anterior) and to the superior colliculus and reticular formation.

CEREBELLAR CONNECTIONS AND PRECEREBELLAR NUCLEI

Within the brainstem there are nuclei (precerebellar nuclei) and fibre tracts whose primary function is communication with the cerebellum. Their topographical contribution to the brainstem is described here; their functional contribution to the cerebellum is discussed on pages 518 to 521.

The cerebellar peduncles carry fibres to and from the cerebellum. There are three pairs: inferior, middle and superior cerebellar peduncles. The inferior cerebellar peduncle is found dorsolaterally in the cranial medulla (Figure 178e and f). It forms a lateral boundary to the floor of the caudal part of the fourth ventricle. It passes cranially and slightly dorsolaterally from the medulla into the cerebellum. It carries a number of different types of afferent and efferent cerebellar fibres (spinal, vestibular, reticular and inferior olivary). The middle cerebellar peduncle is found laterally in the pons and is formed almost entirely by transversely coursing axons which originate in the contralateral pontine nuclei (corticopontocerebellar connections) (Figure 178f to h). The superior cerebellar peduncle is found dorsolaterally in the cranial part of the pons (Figure 178g and h). It forms the walls of the cranial part of the fourth ventricle. It contains some cerebellar afferents but most of its fibres are efferents. Most of these fibres continue cranially and ventrally, and decussate in the ventral tegmentum of the caudal midbrain (Figure 178i). Fibres continue round and through the red nucleus of the cranial midbrain, forming the dentatorubrothalamic tract which continues into the diencephalon.

198C,26

198C,25

198C,24

194,26

Spinocerebellar fibres form the ventrolateral margin of the caudal medulla (Figure 178b to d). They move dorsally as they pass cranially. The dorsal spinocerebellar tract and part of the rostral spinocerebellar tract enter the ipsilateral inferior cerebellar peduncle. The remaining fibres of the rostral spinocerebellar tract and those of the ventral spinocerebellar tract pass laterally through the cranial medulla and caudal pons before moving dorsally to join the superior cerebellar peduncle (Figure 178e to h).

The accessory (lateral or external) cuneate nucleus (of Monakow) is a group of large cells similar in morphology and function to those of the nucleus dorsalis (of Clarke) in the thoracic segments of the spinal cord. It is found dorsolaterally in the caudal medulla, lateral and cranial to the cuneate nucleus (Figure 178c and d). Its main afferents are branches of well-myelinated, proprioceptive fibres of the ipsilateral dorsal roots of the cervical and upper thoracic segments. Most of its efferents pass to the cerebellum via the ipsilateral inferior cerebellar peduncle, forming the cuneocerebellar tract. Some fibres join the contralateral medial lemniscus to pass to the thalamus (nucleus ventralis posterior lateralis). In counterpart to this, some fibres from the dorsal column nuclei join the cuneocerebellar tract.

The inferior olivary nuclear complex consists of the principal olivary nucleus, and a medial and a dorsal accessory olivary nucleus. The principal nucleus has a striking, convoluted topography and is found beneath the olive of the cranial part of the medulla. It occupies a large region of the ventrolateral medulla and extends to the caudal pons (Figure 178c to f). The accessory inferior olivary nuclei are located medial and dorsal to the principal nucleus, as indicated by their names. Essentially, the whole complex has the same function. It receives afferents from all the main areas that send information to the cerebellar cortex (as mossy fibres) and gives rise to the great majority of the climbing fibres of the cerebellar cortex. The dense band of myelinated afferents round the principal inferior olivary nucleus is known as the amiculum olivae. The efferents from the inferior olivary complex stream transversely across the midline, then run dorsolaterally into the contralateral inferior cerebellar peduncle.

196B,12

The pontine nuclei are collections of cells scattered among the fibre bundles of the basilar part of the pons (Figure 178g and h). Their chief source of afferents is the ipsilateral cerebral cortex. Corticopontine fibres originate in all lobes of the cerebral hemisphere. The superior colliculus and other nuclei within the brainstem also provide afferents to specific pontine nuclei. The axons of the pontine nuclear cells (pontocerebellar fibres) run transversely, crossing the midline, and form the contralateral middle cerebellar peduncle. The arcuate nucleus, a nucleus often found ventral to the pyramid of the medulla, and the pontobulbar nucleus, found ventrolateral to the inferior cerebellar peduncle near the pontine-medullary junction, are displaced pontine nuclei.

There are other sources of cerebellar afferents within the brainstem. In particular, four regions of the reticular formation are strongly connected to the cerebellum and so may be regarded as precerebellar nuclei, i.e. the paramedian reticular, lateral reticular, reticulotegmental (tegmental pontine reticular) and perihypoglossal nuclei. The other nuclei are mentioned with the description of the cerebellum and in sections dealing with their more salient functions (see pages 520 to 521).

THE RETICULAR FORMATION OF THE BRAINSTEM

There is at present no exact definition of the reticular formation, although there is widespread agreement as to the main regions that the term includes. The term originated from the appearance of certain phylogenetically primitive regions of the brainstem which contain scattered neurones enmeshed in an irregular network (reticulum) of fibre bundles. These regions are interspersed between the major ascending and descending fibre tracts and the sensory, motor, precerebellar and cranial nerve nuclei of the brainstem. Hence, the reticular formation may also loosely be defined as the residuum left after removal of the more prominent parts of the brainstem to which some more precise functions may be ascribed.

Knowledge of the characteristics of the connections and functions of the classically defined parts of the reticular formation has led to the inclusion of other regions within modern usage of the term. Thus, the reticular formation should be understood to include: a) areas of the brainstem of classical reticular appearance; b) areas of grey matter adjacent to the aqueduct, fourth ventricle or central canal; c) monoamine cell groups found within the previous two regions; d) the nonspecific nuclei of the thalamus; and e) parts of the hypothalamus. Although not to be included within the term, the intermediate area of the spinal cord may represent a caudal extension of the reticular formation.

The functions of the reticular formation may be grouped broadly under three (overlapping) headings: a) co-ordination and regulation of sensory and motor functions; b) organisation and regulation of autonomic and vital functions; and c) a general energising and/or inactivating function. The reticular formation is responsible for regulating the level of neuronal activity within both specific regions and the whole of the central nervous system. Hence, it controls the level of behavioural arousal and is necessary for consciousness.

Many reticular neurones have long and profusely branching axons which may have local, ascending, and descending branches. Most commonly their dendrites extend primarily transverse to the brainstem, so facilitating the establishment of synaptic contact with ascending and descending fibres of a variety of different systems. These dendrites are often long, radially organised, and have a simple and regular branching pattern (isodendritic).

The connections of the brainstem reticular formation taken as a whole may be summarised as being to and from virtually all the central nervous system. The cerebral cortex, basal nuclei, thalamus, hypothalamus, cerebellum, spinal cord and non-reticular nuclei of the brainstem are so connected. The major sensory and motor pathways send fibres (often collaterals) to the reticular formation, although an exception is provided by the medial lemniscus. Accordingly, the reticular formation collects and disperses an immense amount of information. Because the number of reticular neurones is limited, there is considerable convergence and integration of the incoming information. Further, the outputs of many (though by no means all) reticular neurones must signal general, rather than detailed information.

Connections from the reticular formation to the spinal cord and cerebellum arise from medullary and pontine parts of the reticular formation. All regions of the reticular formation, but particularly the mesencephalic reticular formation, send fibres to nonspecific nuclei of the thalamus, the hypothalamus and other parts of the forebrain. Fibres pass either directly or indirectly to the whole of the cerebral cortex and the basal nuclei. The forebrain is supplied by two routes: a dorsal pathway to the nonspecific nuclei of the thalamus (chiefly midline and intralaminary nuclei), and a ventral, subthalamic pathway through the hypothalamus and basal forebrain.

All regions of the reticular formation supply both local and more distal brainstem nuclei (including other reticular nuclei). Therefore, there are many pathways through the reticular formation, both multisynaptic and oligosynaptic. Such pathways provide adjuncts to the main ascending and descending tracts for information traversing the brainstem. Notably, nociceptive information is transmitted by both types of route.

Reticular connections are predominantly ipsilateral, although there are interconnections between reticular nuclei on both sides of the brainstem. Two reticular association fibre bundles are mentioned below, but often reticular axons are scattered rather than gathered into fasciculi.

The central tegmental tract or fasciculus (of Forel) conveys many ascending and descending axons of reticular neurones. It stretches between the red nucleus and the inferior olivary nucleus (additionally, it contains rubro-olivary axons). It is found ventrolaterally in the pontine tegmentum, moving centrally and dorsomedially in the tegmentum of the midbrain.

The dorsal longitudinal fasciculus (of Schütz) is a small bundle of fibres that runs from the hypothalamus to the medulla, passing ventromedially through the peri-aqueductal grey and medially through the periventricular grey beneath the fourth ventricle. It contains mainly ascending reticular fibres (some of them monoaminergic) which pass to other reticular nuclei and the hypothalamus. Descending fibres (some of them peptidergic) include those of hypothalamic origin, and fibres that pass to autonomic nuclei of the medulla and pons. Its fibres are probably concerned with olfactory, gustatory and emotional influences on salivation, lacrimation and other oral, facial and visceral reflexes.

Individual nuclei may be delineated within the reticular formation. Reticular nuclei of the brainstem may be grouped into four columns: lateral, intermediate, medial (raphe) and periventricular/peri-aqueductal. The lateral nuclei occupy the lateral one-third of the reticular formation. They are mainly small-celled and primarily receptive in function. Their axons chiefly terminate locally and medially. The intermediate nuclei occupy most of the medial two-thirds of the reticular formation. Some of their cells are very large and the nuclei are mainly effector in function. Nuclei of the raphe are median or paramedian in position. They are described with the monoaminergic cell groups below. Other nuclei are found amongst the large numbers of small neurones that populate the grey matter adjacent to the fourth ventricle and cerebral aqueduct. The main reticular nuclei of the brainstem are listed in Table 15. Information concerning diencephalic parts of the reticular formation will be found on pages 524 to 528 and 533 to 534.

Functional aspects of the reticular formation

The involvement of the reticular formation in sensation is exemplified by nociception. Reticular afferents are responsible for the arousing (and possibly the aversive) quality of painful stimuli. Reticular connections provide an alternative pathway to that of spinothalamic fibres. Reticular efferents may modulate/gate sensory transmission. The peri-aqueductal grey contains some neurones that use enkephalin as neurotransmitter. These cells project to the nucleus raphe magnus of the medulla, from where fibres descend to the dorsal horn of the spinal cord (Figures 179b and 174b). Stimulation of the peri-aqueductal grey activates these pathways and can produce analgesia.

Influences of the reticular formation upon motor activity occur via its connections with the spinal cord, cranial nerve nuclei, cerebellum and basal nuclei.

TABLE 15 Nuclei of the reticular formation. (The letters in brackets in the position column refer to the levels of the brainstem shown in Figure 178.) Abbreviations: Ach = acetylcholine, DA = dopamine, NA = noradrenaline, 5-HT = 5-hydroxytryptamine.

NUCLEUS	POSITION	MAIN CONNECTIONS/FUNCTION
LATERAL		
Lateral reticular (of lateral funiculus)	Medulla; caudal, lateral (c, d)	Cerebellum, cord, hypothalamus
Parvicellular reticular	Medulla and pons; lateral (e, f, g)	Receptive, local reflexes
Lateral and medial parabrachial	Pons; cranial, dorsolateral (h)	Nucleus of solitary tract; gustatory, respiratory
Pedunculopontine	Midbrain; caudal, lateral (i)	Basal nuclei
INTERMEDIATE		
Ventral reticular	Medulla; caudal, intermediate (c)	
Paramedian reticular (of anterior funiculus)	Medulla; mid, medial (e)	Cerebellum
Gigantocellular reticular (magnocellular reticular)	Medulla and caudal pons; intermediate (e, f)	Cord, diencephalon; respiratory, cardiovascular
Pontine reticular: oral and caudal (includes paramedian pontine reticular formation	Pons and caudal midbrain; intermediate (g, h)	Cord, diencephalon; respiratory, eye movements, sleep
Tegmental pontine reticular	Pons; medial (h)	Cerebellum
Parabigeminal area	Midbrain; caudal, lateral	Superior colliculus; visual
Cuneiformis and subcuneiformis	Midbrain; intermediate (i, j)	Forebrain; desychronised EEG (Ach)
Interpeduncular	Midbrain; ventral, median (j)	Habenula, limbic
Ventral tegmental area (of Tsai)	Midbrain; ventral (i, j)	Forebrain (DA)
MEDIAL		
Raphe	Brainstem; median and paramedian (c to j and Figure 179b)	All CNS (5-HT); sleep, mood, nociception, autonomic
PERIVENTRICULAR/PERI-AQUEDUCTAL		
Perihypoglossal	Medulla; cranial, medial (f)	Cerebellum; eye movements
Area postrema	Medulla; obex, medial (d)	Chemoreceptor trigger zone, emesis
Locus ceruleus (pigmentosus pontis)	Pons; cranial, ventrolateral (h)	All CNS (NA); arousal
Peri-aqueductal grey (central grey)	Midbrain (i, j)	Reticular; nociception
Dorsal and deep (ventral) tegmental (of Gudden)	Midbrain; ventral peri-aqueductal grey	Hypothalamus, interpeduncular nucleus, limbic

The medullary (lateral) reticulospinal tract originates from the nucleus reticularis gigantocellularis (Table 15; Figure 178e and f). The fibres are scattered and pass caudally through the reticular formation of both sides (though predominantly ipsilaterally) to the spinal cord (see pages 606 to 607). They end diffusely, chiefly in the intermediate area of the spinal cord (see page 458). These fibres influence postural, respiratory, autonomic (cardiovascular) and other reflexes, and modulate transmission of sensory information.

The pontine (medial) reticulospinal tract originates from the pontine reticular nucleus (Table 15; Figure 178g and h). Most of the fibres run caudally in association with the medial longitudinal fasciculus (see pages 606 to 607). The fibres terminate diffusely, mainly in medial parts of the ipsilateral ventral horn (see page 458). Their activity influences spinal reflexes and voluntary movements, particularly postural adjustments, head movements and respiratory activity.

Pontine and medullary neurones giving rise to reticulospinal fibres receive afferents from the brainstem (reticular, vestibular, superior collicular, and from cranial nerve nuclei), and notably large numbers of fibres from the spinal cord (spinoreticular fibres) and the cerebral cortex (from both hemispheres and particularly from somatosensory and motor regions). Half of these neurones also have ascending branches that pass to the midbrain and/or diencephalon.

Four groups of nuclei have such strong connections with the cerebellum that they may alternatively be regarded as precerebellar nuclei; the lateral reticular, paramedian and perihypoglossal nuclei of the medulla, and the tegmental pontine reticular nucleus of the metencephalon. The pedunculopontine nucleus of the midbrain is closely linked with the substantia nigra and the globus pallidus of the basal nuclei.

Other sensory and motor functions of the reticular formation include its important role in both the reflex and the voluntary control of cranial nerve and other brainstem reflexes. Reticular neurones act as interneurones for the cranial nerve nuclei. Among the important and complex reflexes organised by the reticular formation are swallowing, coughing, sneezing and vomiting. Its role in eye movements, respiration and cardiovascular functions will now be described.

Conjugate (co-ordinated) eye movements are organised by reticular nuclei which receive fibres from, and are under the control of, the superior colliculus, vestibular system, cerebellum and cerebral cortex (visual areas and the frontal eye fields) (Figure 177b, see also pages 547 to 552). The perihypoglossal nuclei (nucleus prepositus, nucleus intercalatus of Staderini and nucleus of Roller), found between the hypoglossal and the abducent nuclei (Figure 178f), appear to be important mainly for vertical (upward) eye movements. The paramedian pontine reticular formation (parabducent nucleus) found between the abducent and trochlear nuclei (Figure 178h, see also page 479), organises lateral movements, though its more cranial part may also be concerned with vertical movements. The rostral interstitial nucleus of the medial longitudinal fasciculus is probably important for downward movements of the eyes. The interstitial nucleus of Cajal seems to be necessary for vertical (upward) and rotatory eye movements (Figure 177b, see also pages 501 to 502). These four paired, medially placed nuclei are all interconnected. They all project to the oculomotor nucleus (and perhaps the trochlear nucleus). The paramedian pontine reticular formation and the nucleus prepositus hypoglossi send fibres to the abducent nucleus. Some of the connections are made via the medial longitudinal fasciculus, but others pass through the reticular formation. Fibres going from the interstitial nucleus of Cajal to the contralateral oculomotor nucleus pass through the posterior commissure. The region that co-ordinates horizontal and vertical components to produce oblique eye movements has not been established. Reticulospinal fibres from the paramedian pontine reticular formation and interstitiospinal fibres (from the nucleus of Cajal) probably assist the tectospinal tract in turning the head towards visual stimuli.

Respiration is a motor, but also an autonomic/vital function under reticular formation control. Neurones in and around the nucleus of the solitary tract correspond to the dorsal, and those in the region of the nucleus ambiguus to the ventral, medullary respiratory region on each side. However, neurones whose activity is related to either inspiration or expiration are found more widely in the reticular formation of the medulla and pons. Their exact location in the human brainstem has not been established. Many of the respiratory neurones appear to be linked by gap junctions, so facilitating their synchronous discharge. The so-called 'pneumotaxic centre' has been identified as the medial parabrachial nucleus of the cranial pons (Figure 178h). It receives fibres from the nucleus of the solitary tract and influences the rate of breathing via connections with medullary respiratory neurones. The pontine 'apneustic centre', which is facilitatory to inspiratory neurones, has not been anatomically defined. Many, though not all, respiratory neurones project to cervical and thoracic segments of the spinal cord for control of the diaphragm and intercostal muscles. Others control the larynx and pharynx via the nucleus ambiguus. They receive inputs from the chemoreceptors (monitoring the $P{CO_2}$ and pH of the blood) of the vagus and glossopharyngeal nerves via the nucleus of the solitary tract. Other inputs come more directly from the area postrema (Table 15, Figure 178d), and possibly from local blood vessels. Respiratory neurones also receive feedback information from ascending spinoreticular fibres. Descending influences from the hypothalamus and cerebral cortex allow emotional and volitional factors to alter breathing.

Cardiovascular regulation is effected by a variety of linked neural systems including the caudal pontine and medullary reticular formation, hypothalamus, and autonomic nuclei and nerves of the brainstem and spinal cord. Reticular neurones (including adrenergic and noradrenergic cells) concerned with such functions are found in regions adjacent to the nucleus ambiguus, dorsal motor nucleus of the vagus and nucleus of the solitary tract. They appear to be intermingled with respiratory neurones. They have pressor, depressor, accelerator and decelerator functions. These neurones send fibres to the dorsal motor nucleus of the vagus for control of parasympathetic input to the heart, and to the intermediolateral cell column of the spinal cord for control of sympathetic cardiac innervation. Inputs reach these neurones from the chemoreceptors and baroreceptors of the vagus and glosso- pharyngeal nerves via the nucleus of the solitary tract. They also receive input from the area postrema and, via spinoreticular fibres, from the spinal cord. Descending inputs include connections from the peri-aqueductal grey, cerebellum, hypothalamus and cerebral cortex (particularly orbitofrontal cortex), so making possible alteration of heart rate and blood pressure in response to emotional and other influences.

Arousal levels are controlled by the reticular formation in co-operation with dien- cephalic and telencephalic regions. The activity of neurones of the reticular formation determines whether a person is awake or asleep, and hence the gross level of consciousness. Large lesions of the reticular formation, if they are not fatal, can produce coma. Smaller lesions may produce somnolence or insomnia. The cerebral cortex is additionally necessary for consciousness, particularly its content, i.e. what a person is aware of, rather than that the person has awareness. However, for its normal operation the cerebral cortex requires the reticular formation.

Changes within the reticular formation result in changes in the cortical EEG (electroencephalogram). The EEG may be categorised as desynchronised or synchronised. Desynchronised EEG (characterised by waves of low voltage and high frequency, i.e. greater than 15 Hz) is found during alertness and during the paradoxical phase of sleep, so-called rapid eye movement (REM) sleep. REM sleep is accompanied by muscular and autonomic changes and dreaming. Synchronised EEG (higher amplitude, slower waves) is found during drowsiness, relaxation and slow wave sleep. Under normal circumstances, slow wave sleep is the initial type of sleep and is followed by REM sleep; the two states then alternate during sleep.

Synchronisation of the EEG primarily involves a pathway from the reticular formation to the nonspecific nuclei of the thalamus (midline, intralaminar, and ventralis anterior nuclei) and so to the cerebral cortex. Parts of the pontine reticular nucleus seem to be important in producing EEG synchronisation. They may suppress activity of the mesencephalic reticular formation. It has been suggested that the 5-hydroxy- tryptaminergic neurones of the raphe are necessary for the production of slow wave sleep and the induction of REM sleep, but their exact involvement remains to be clarified.

Desynchronisation of the EEG requires the integrity of a pathway that passes by a subthalamic route through the hypothalamus and basal forebrain to the cortex. At least in part this pathway involves cholinergic neurones (probably of the nucleus cuneiformis of the midbrain and/or of the basal forebrain nucleus – see page 513 and 566 to 567), but other reticular fibres (including monoaminergic axons) also follow this route. Desynchronisation of the EEG, whether during arousal or during REM sleep, is correlated with increased activity of certain neurones of the midbrain reticular formation. Thus, cranial parts of the reticular formation appear to control cortical de.ynchronisation, while more caudal regions may be responsible for synchronisation. The pontine reticular formation appears to be particularly important for sleep (REM and slow wave) and, together with the vestibular nuclei, for muscular and autonomic changes during sleep. The mesencephalic and cranial pontine reticular formation seem more important for wakefulness.

There is some, as yet inconclusive, evidence that sleep may be induced as a result of a build-up of one or more peptides within the cerebrospinal fluid of the brain ventricles, this build-up possibly being detected by raphe or other reticular neurones. However, although there may prove to be neurohumoral assistance or triggering, sleep and the cortical EEG are under neuronal rather than neurohumoral control. Note that sleep is positively promoted and is not the mere absence of arousal.

Lesions producing hypersomnolence include those of the cranial midbrain tegmentum, medial and nonspecific thalamic nuclei, and the posterior hypothalamus. A lesion of the anterior hypothalamus or in the region of the cranial raphe nuclei produces insomnia (though the insomnia after raphe lesions does not extend beyond a few days). Large lesions of the midbrain, cranial pontine reticular formation or hypothalamus result in coma. Lesions of the caudal pontine reticular formation may be associated with stupor or coma, but accompanied by a desynchronised EEG. Lesions of the medullary reticular formation are usually rapidly fatal because of the disruption of respiratory and cardiovascular functions.

198A,4
158,9; 158,57

The reticular formation is probably also involved in processes of attention. It has been hypothesised that selective attention involves the neurones of the midbrain reticular formation, superior colliculus, cerebral cortex and reticular nucleus of the thalamus (see page 534).

The concept of an ascending reticular activating system (ARAS) is useful in emphasising the central role played by the reticular formation in the control of arousal. However, that role is complex and occurs in co-operation with forebrain structures (cortex, thalamus, hypothalamus, etc.). The concept is limited since not all reticular axons that may be described as having an activating function ascend (e.g. reticulospinal fibres), and some ascending fibres may inactivate rather than activate (e.g. promote slow wave sleep). Further, although they must operate in conjunction with one another, arousal appears to involve a number of subsystems (cholinergic, monoaminergic and other subsystems), and these are not entirely of classical reticular origin.

Monoaminergic and cholinergic pathways

Within the brainstem reticular formation there are groups of neurones that use a monoamine as their neurotransmitter, either a catecholamine (adrenaline, noradrenaline, or dopamine) or the indoleamine, 5-hydroxytryptamine (serotonin). These neurones, including their axonal distributions, may be demonstrated by fluorescence histochemistry.

The numbers of monoamine neurones are small (typically a few hundred or a few thousand per group) but their axonal distributions are characterised by being extremely widespread. These distributions are not without pattern, but a single neurone may innervate many different areas. Axonal connections are very largely ipsilateral. As with postganglionic sympathetic axonal terminations, the terminals of central monoaminergic fibres frequently lack normal synaptic specialisations and have large junctional gaps. Central monoaminergic neurones have strongly developed capacities for regeneration and re-innervation of denervated regions. These properties have been exploited experimentally to restore loss of function by grafting fetal brain tissue (particularly dopaminergic neurones) into adult brains.

The precise functions of the various central monoaminergic cell groups are poorly understood. They may play a role in controlling the level of neuronal excitability, in the enabling of information transfer, or in regulating metabolic or trophic activity over widespread areas of the central nervous system. They are implicated in the

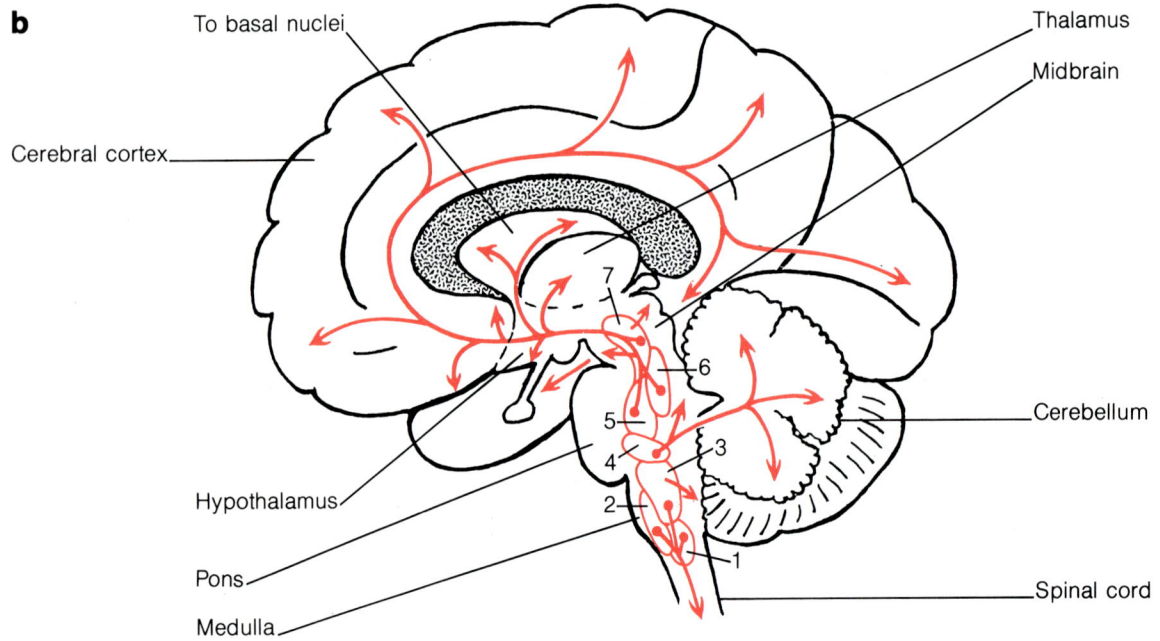

Figure 179 **Monoaminergic and cholinergic pathways of the reticular formation** (*projected on to a median sagittal section of the brain*). a) *Noradrenergic pathways.* b) *5-hydroxytryptaminergic pathways and the raphe nuclei:* 1 = *nucleus raphe obscurus,* 2 = *nucleus raphe pallidus,* 3 = *nucleus raphe magnus,* 4 = *nucleus raphe pontis,* 5 = *nucleus raphe medialis,* 6 = *nucleus raphe dorsalis,* 7 = *nucleus raphe linearis.* c) *Adrenergic and dopaminergic pathways.* d) *Cholinergic reticular pathways.*

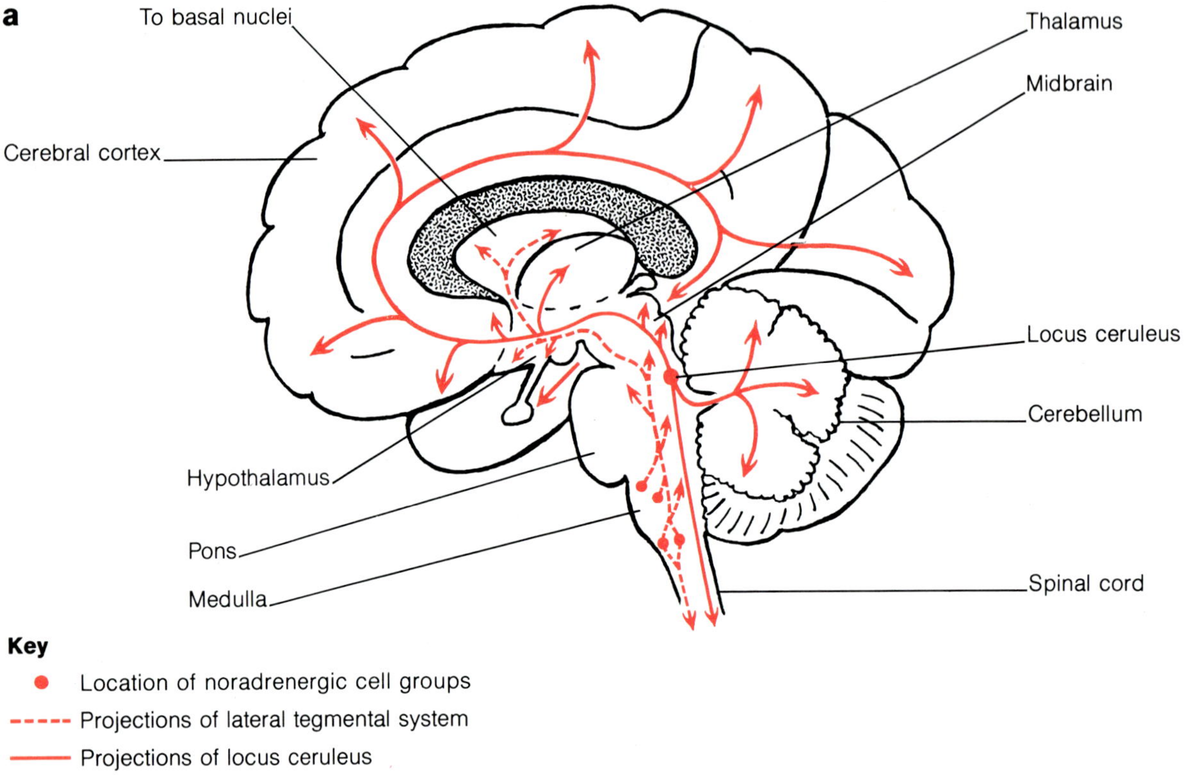

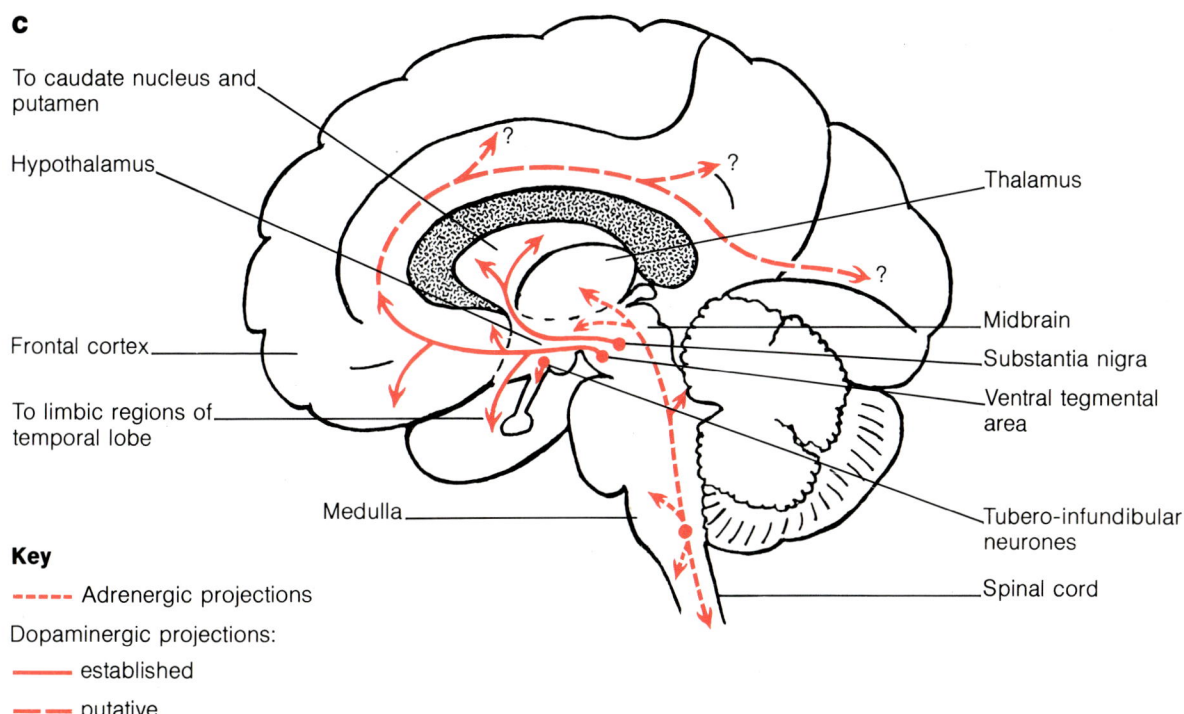

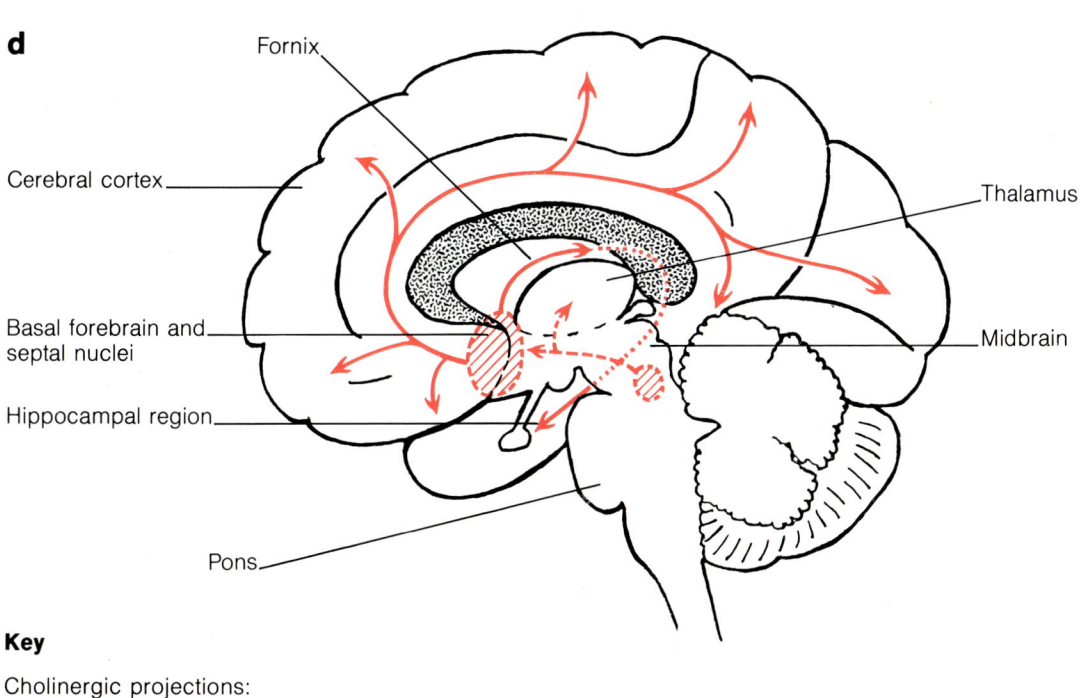

control of the level of arousal and of the sleep/wake cycle. They seem to be involved in the modulation of mood and motivation, of pleasure and anguish. They form part of central autonomic pathways. Many of the drugs used to treat psychiatric illness alter the operation of central monoaminergic pathways (e.g. monoamine oxidase inhibitors and tricyclic antidepressants).

Noradrenergic cell groups collectively innervate all regions of the central nervous system. There are two systems: the lateral tegmental system and that of the locus ceruleus (Figure 179a).

The lateral tegmental system comprises six main cell groups (designated A1 to 5 and A7). These are found ventrolateral to the nucleus of the solitary tract and the dorsal motor nucleus of the vagus in the medulla, and in the lateral pontine tegmentum. Caudal groups project to the spinal cord. All groups project cranially, via the central tegmental tract and ventral tegmentum of the midbrain, to parts of the brainstem or to restricted parts of the forebrain, notably the hypothalamus. The lateral tegmental system appears to be involved in homeostasis, particularly autonomic activities related to feeding, and endocrine functions of the hypothalamus.

The locus ceruleus (locus caeruleus, coeruleus, nucleus pigmentosus pontis) is by far the largest noradrenergic cell group (A6). It is a collection of large, darkly pigmented neurones located ventrolaterally in the periventricular grey of the cranial pons (Figure 178h). It innervates the spinal cord, parts of the brainstem (including the raphe), the cerebellum, thalamus, cerebral cortex and other forebrain regions. An individual neurone may provide axonal branches to both cerebellar and cerebral cortices. All axons ascending to the thalamus and cerebral cortex pass through the dorsal midbrain tegmentum and medial forebrain bundle. Afferents to the nucleus come from many different areas, e.g. cerebral cortex, hypothalamus, cerebellum and reticular formation (including nuclei of the raphe). Innumerable functions have been proposed for the locus ceruleus; it may play a role in behavioural arousal.

Adrenergic neurones are few in number. They are found scattered in the reticular formation of the cranial medulla close either to the dorsal motor nucleus of the vagus and nucleus of the solitary tract or to the nucleus ambiguus. Their axonal distribution is restricted. Axons pass to the intermediolateral cell column of the spinal cord, to autonomic nuclei within the brainstem and to the hypothalamus and thalamus (Figure 179c). These adrenergic neurones appear to be involved in autonomic, particularly cardiovascular, functions.

The raphe nuclei are found along the midline of the brainstem from the caudal medulla to the cranial midbrain (Figure 179b). Many, but not all, of the neurones of these nuclei use 5-hydroxytryptamine as neurotransmitter; the nine such cell groups (designated B1 to 9) are found within the nuclei. Collectively, raphe neurones innervate all regions of the central nervous system, with caudal neurones sending axons caudally and cranial neurones projecting via the medial forebrain bundle to the forebrain. Unusually, some fibres end in relation to the brain ventricles. Further, the dendrites of some neurones are found in close proximity to blood vessels passing through the nuclei. As for the noradrenergic nuclei, inputs to nuclei of the raphe come from widespread regions of the central nervous system including the cerebral cortex, hypothalamus, cerebellum, spinal cord and reticular formation (including the locus ceruleus). The raphe nuclei seem to be concerned with regulation of autonomic activities, mood and motivation, and sleep. The nucleus raphe magnus provides monosynaptic inhibitory terminals to neurones of the intermediolateral column of the spinal cord. It also plays a specific role in modulation of the transmission of nociceptive information via axons that pass bilaterally to the dorsal horn (see Figure 174b, and also page 451).

Dopaminergic neurones are found in the retina (certain amacrine cells which may be involved in changes during dark adaptation), olfactory bulb and hypothalamus (notably those of the tubero-infundibular dopamine pathway which inhibits prolactin secretion). A few such neurones are found scattered in the periventricular and peri-aqueductal grey; their connections are mainly local and their function is unknown. Most dopamine neurones are found in the ventral midbrain. In the ventral midbrain they extend from the ventral tegmental area (of Tsai) laterally into the substantia nigra on both sides (see Figure 178i and j). Dopaminergic neurones have a more specific and restricted pattern of axonal distribution than the noradrenergic or 5-hydroxy-tryptaminergic systems (Figure 179c). Neurones of the pars compacta of the substantia nigra give rise to the nigrostriatal dopamine pathway which terminates in the putamen and caudate nucleus of the basal nuclei (see pages 569 to 570). Neurones of the ventral tegmental area give rise to the mesocortical or mesolimbic dopamine systems. Their axons pass through the medial forebrain bundle to parts of the limbic system, frontal and medial cortical areas. Over-activity within this pathway may give rise to some of the symptoms of schizophrenia; major tranquillisers such as the phenothiazines, which block dopamine receptors, affect the operation of this system. It has been suggested that the dopamine neurones of the midbrain are involved in motivation and drive. Pleasurable sensations may be evoked by stimulation of what are believed to be dopaminergic axons that pass through the hypothalamus to the septal nuclei of the limbic system. The loss of dopamine neurones which occurs in Parkinson's disease (paralysis agitans) may be responsible for the poverty of movement (hypokinesia), lethargy and lack of emotion which are characteristic of the disease.

198A,4; 194,25

Acetylcholine is used as a neurotransmitter by many neurones within the central nervous system. Such cholinergic neurones include all the motoneurones and preganglionic autonomic neurones of the spinal cord and cranial nerve nuclei. There are also cholinergic cells in the cerebral cortex and basal nuclei. In addition, there are two further, interconnected groups of cholinergic neurones (Figure 179d). The first group is found in the reticular formation of the dorsal tegmentum of the caudal midbrain and cranial pons. These neurones may play a role in arousal and desynchronisation of the EEG. They project to the thalamus and to the basal forebrain nucleus (substantia innominata). The basal forebrain nucleus (of Meynert) is a scattered collection of large neurones anterior to the hypothalamus in the region of the anterior perforated substance. Many neurones in the basal forebrain nucleus and its medial extension into certain of the septal nuclei (the nucleus of the diagonal band of Broca and the medial septal nucleus) are cholinergic. These neurones form the second cholinergic group. They project diffusely to all regions of the cerebral cortex. Their loss has been hypothesised to underly memory loss in dementias of the Alzheimer type.

198A,4

188B,32

The cerebellum

The cerebellum is part of the metencephalon of the hindbrain. It occupies the greater part of the posterior cranial fossa below the tentorium cerebelli and forms about 10% of the brain by weight. It is a phylogenetically ancient structure that shows development parallel to that of the cerebral cortex in mammals.

158,22
158,18

The chief function of the cerebellum is the fine co-ordination of movement and posture. Changes in intracerebellar synaptic connections are likely to play a major role in the learning of such co-ordination. Lesions of the cerebellum produce clumsy, ataxic movements. Each half of the cerebellum is primarily concerned with the ipsilateral side of the body. Its operations are largely automatic and do not reach consciousness.

Structurally, the cerebellum consists of an outer lamina of grey matter, the cerebellar cortex, overlying white matter, the medullary core. Within the medullary core are four paired intracerebellar nuclei (fastigial, globose, emboliform and dentate nuclei) which give rise to most of the fibres leaving the cerebellum. The cerebellum communicates with the rest of the brain via three paired peduncles: the inferior cerebellar peduncles (attached to the medulla oblongata), the middle cerebellar peduncles (attached to the pons) and the superior cerebellar peduncles (attached to the midbrain).

198A; 198B

198B,23

198C,26
198C,25
198C,24

THE EXTERNAL APPEARANCE OF THE CEREBELLUM

196A,3
196A,2

The cerebellum comprises two large laterally placed lobes, the cerebellar hemispheres, linked by a median region, the vermis. Craniocaudally the cerebellum is shaped like a wedge; it has a superior, a caudal and a ventral surface.

196A,4; 196A,5

The cerebellar surface is marked by a regular pattern of mainly transverse parallel ridges (folia) and grooves (fissures). Estimates suggest that six-sevenths of the cerebellar surface is buried in the fissures; it would measure a metre across if laid flat.

158,22; 158,18
196A,2

164,5
158,17; 168,23

162,9; 162,8
168,12

Superiorly, the cerebellum is adjacent to the tentorium cerebelli; at the gross level this surface of the cerebellum is relatively flat. The vermis lies in the midline separated from the hemispheres by two shallow grooves; it is roughly 10mm wide. Cranially, the vermis extends just beyond the free margin of the tentorial notch. The straight sinus lies within the dura at the junction of the falx cerebri with the tentorium, above all but the cranial end of this superior surface of the vermis. The posterolateral margins of the cerebellar hemispheres are related to the transverse sinuses which run along the fixed margin of the tentorium. The superior petrosal sinuses are similarly placed in relation to the craniolateral margins of the hemispheres.

196A,6
158,19

44B,19; 162,9
162,10; 168,21
158,22; 158,23
158,26; 158,25

Caudally, the cerebellar hemispheres are separated by the cerebellar notch (posterior cerebellar incisure). The falx cerebelli lies in the cerebellar notch; within it runs the occipital sinus. The cerebellum is convex caudally and ventrally where it is related to the cerebellar fossa of the occipital bone. Cranially and ventrolaterally, the cerebellum is related to the sigmoid sinus and the petrous part of the temporal bone. Ventrally and medially, the cerebellum is related to the fourth ventricle and to the dorsal surface of the brainstem, from the caudal medulla to the caudal midbrain. Dorsal to the pons, the median dorsal recess of the fourth ventricle occupies a deep pyramidal fossa

TABLE 16 Subdivisions of the cerebellum. The middle and anterior lobes together form the corpus cerebelli. Note that alternative names are used in subhuman neuroanatomy.

LOBE	LOBULE		FISSURE
	VERMIS	HEMISPHERE	
Anterior	Lingula (I)		
	Central (II, III)	Alar central	
			Postcentral
	Culmen (IV)	(Anterior) Quadrangular	
			Primary
Middle (posterior)	Declive (V)	Simplex (posterior quadrangular)	
			Postlunate (posterior superior)
	Folium (VI)	Superior semilunar	
			Horizontal
	Tuber (VII)	Inferior semilunar (and gracile)	
			Prepyramidal
	Pyramid (VIII)	Biventral	
			Secondary (postpyramidal or retrotonsillar)
	Uvula (IX)	Tonsil (and paraflocculus)	
			Dorsolateral (posterolateral)
Flocculonodular	Nodule (X)	Flocculus	

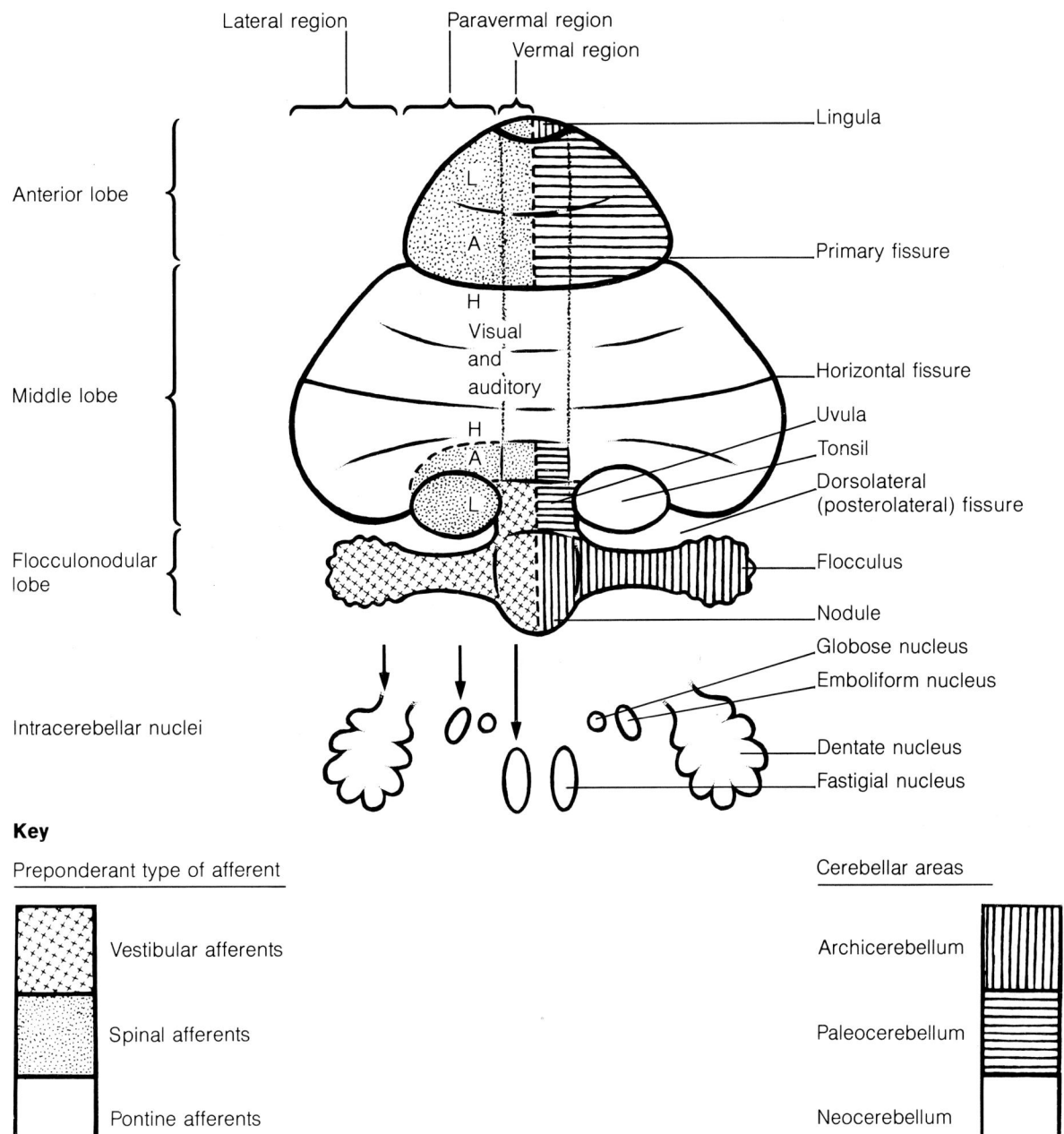

Figure 180 *Topographical divisions of the cerebellum.* The cerebellar surface has been unrolled so that the lingula is at the top of the diagram and the nodule at the bottom. The intracerebellar nuclei are shown below. Somatotopical organisation is indicated: H = head, A = arm, L = leg; the distal limbs are represented laterally, the trunk medially.

beneath the vermis. The apex of the fossa is known as the fastigium. Caudally and medially, the cerebellomedullary cistern is found dorsal to the medulla and caudal to the vermis near the foramen magnum. In this region, the cerebellar hemispheres are more widely and deeply separated by the vallecula cerebelli which is continuous dorsally with the cerebellar notch.

The deeper fissures of the cerebellum divide it into lobes and lobules (see Table 16, Figure 180), but only certain subdivisions of the cerebellum are readily discernible from its external surface. About three-quarters of the distance across the craniocaudal extent of its superior surface, the primary fissure marks the boundary between the anterior and the middle (or posterior) cerebellar lobes. Ventrally, lateral to the cerebellopontine angle and adjacent to each lateral aperture of the fourth ventricle (and usually partially covered by choroid plexus and crossed by the eighth nerve) is a small, partly detached area of the cerebellum known as the flocculus. The horizontal fissure passes laterally across the cerebellar hemisphere, cranial and lateral to the flocculus. Caudal to the flocculus is the dorsolateral (posterolateral) fissure which divides the flocculonodular lobe from the remainder of the cerebellum, the corpus cerebelli. Caudal to the flocculus a cerebellar tonsil protrudes dorsolateral to the medulla on each side and extends towards the foramen magnum. Caudally, dorsal to the medulla, the uvula is the portion of the vermis between the cerebellar tonsils.

The proximity of the tonsils to the foramen magnum is of clinical importance. In patients with raised intracranial pressure, there is a risk that release of pressure following injudicious withdrawal of cerebrospinal fluid from the lumbar cistern may lead to a sudden caudal movement of the hindbrain. The hindbrain may then be compressed in the foramen magnum ('coning'). The pressure of the tonsils upon the medulla may cause the respiratory and cardiovascular neurones of the medulla to cease functioning, resulting in the patient's death.

The divisions of the vermis may be readily determined from its medial surface once the cerebellum has been sagittally sectioned. The multibranched and foliated appearance of this cut surface has given rise to the name arbor vitae.

The small strip of cerebellum lying on the superior medullary velum near the midbrain is the lingula. Caudal to the lingula and to the median dorsal recess of the fourth ventricle is the nodule. Laterally, the nodule is connected to the flocculus; together they form the flocculonodular lobe. Caudally, the small nodular lobule is separated from the uvula by the deep dorsolateral (posterolateral) fissure. Superiorly, the deep primary fissure separates the declive of the middle lobe from the culmen of the anterior lobe. Other lobules and fissures are given in Table 16.

The cerebellar peduncles

The three pairs of cerebellar peduncles are attached to the cerebellum ventrally. They form parts of the roof and walls of the fourth ventricle, so that the surface of the cerebellum in this ventral region is formed by white rather than grey matter. They convey both cerebellar afferents and efferents.

The inferior cerebellar peduncle (restiform plus juxtarestiform body) is the most caudal. It passes cranially and slightly dorsolaterally from a dorsolateral position in the cranial medulla into the medullary core of the cerebellum (see Figure 178e and f). Most of its fibres arise, course and terminate ipsilaterally. It contains cerebellar afferent fibres: of the dorsal and rostral spinocerebellar tracts, the cuneocerebellar tract, and from the inferior olivary complex, the reticular formation, the vestibular nuclei and nerve, other cranial nerve nuclei (notably trigeminal) and aberrant pontocerebellar fibres. Cerebellar efferents passing through the inferior peduncle include those to the vestibular nuclei, inferior olivary nuclei and reticular formation of the brainstem, and a few to the spinal cord. Most of the efferents originate in the fastigial nuclei.

The middle cerebellar peduncle (brachium pontis) is the most massive. It passes laterally and dorsally into the cerebellum from the dorsolateral pons (see Figure 178f to h). It is almost entirely formed by transversely coursing axons of cells of the nuclei of the contralateral basilar part of the pons (pontocerebellar fibres from the pontine nuclei). It also contains a few reticulocerebellar fibres.

The superior cerebellar peduncle (brachium conjunctivum) is the most cranial. It leaves the cerebellum medially and passes cranially and ventrally through the dorsal part of the cranial pons (see Figure 178g and h). Its fibres decussate in the ventral tegmentum of the caudal midbrain (see Figure 178i). Most fibres in the superior cerebellar peduncle are efferents from the ipsilateral dentate, globose or emboliform nuclei. These fibres terminate in the thalamus (most importantly, nucleus ventralis lateralis), and in the red nucleus, reticular formation, inferior olivary nuclei and cranial nerve nuclei of the brainstem. Afferent fibres include the ventral, and part of the rostral, spinocerebellar tracts, and fibres from the tectum of the midbrain, the reticular formation and the hypothalamus.

198B,23
194,6
194,26

THE INTERNAL ORGANISATION OF THE CEREBELLUM

Cerebellar cortex covers the cerebellum everywhere but ventromedially. Cerebellar cortex (Figure 181) is about 1 mm thick and has three layers: molecular (most superficial), Purkinje and granular (adjacent to the white matter). Afferents to the cerebellar cortex terminate either as climbing fibres or as mossy fibres. Both types of fibre are excitatory. Purkinje cells of the cerebellar cortex project to the intracerebellar nuclei. Their axons are the only efferents of the cerebellar cortex; they are inhibitory.

198A; 198B

The medullary core of the cerebellum contains projection fibres of the cerebellar cortex and the intracerebellar nuclei. There are a few commissural and association fibres (Purkinje cell axon collaterals).

The intracerebellar nuclei are found ventrally and medially within the white matter of the cerebellum, dorsal to the cranial medulla and caudal pons (see Figure 178f). From medial to lateral they are: the fastigial, globose, emboliform and dentate nuclei. These nuclei give rise to the great majority of cerebellar efferents and send fibres to cerebellar cortex. Their axons are excitatory. The nuclei receive excitatory collaterals of climbing fibres and some mossy fibres, and inhibitory axons of the Purkinje cells of cerebellar cortex.

198B,23

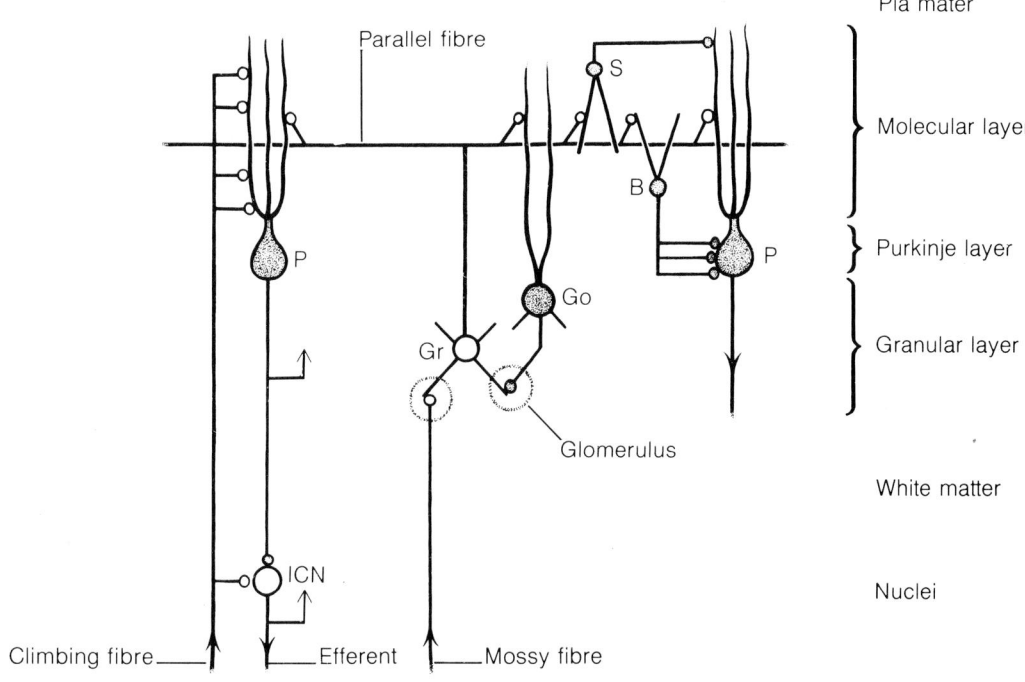

Figure 181 *The main connections of the cerebellar cortex. Neuronal types:* B = *basket,* Go = *Golgi,* Gr = *granule,* ICN = *intracerebellar nuclear,* P = *Purkinje,* S = *stellate cells. Inhibitory neurones and synapses are shaded; excitatory are unshaded. Purkinje and intracerebellar nuclear cells send collaterals to the cerebellar cortex.*

198C,24–26	The connections of the cerebellum pass in the cerebellar peduncles. Inputs come from widespread regions of the central nervous system. These inputs convey a considerable variety of information, not merely motor and proprioceptive signals. The cerebellum has outputs that influence neurones projecting to the spinal cord or cranial nerve nuclei. Thus, its main connections are directly and indirectly with: a) the spinal cord, b) cranial nerve nuclei (notably the vestibular nuclei), c) the cerebral cortex, and d) the reticular formation.

The cerebellum is topographically organised; there are two somatotopic maps in each hemisphere, representation being mainly ipsilateral (Figure 180). Because representation within the cerebral cortex is contralateral, fibres of pathways making connection between the cerebral cortex and cerebellum decussate.

Divisions of the cerebellar cortex into archi-, paleo- and neo-cerebellar regions may be made on comparative phylogenetic, anatomical and functional grounds (Figure 180). However, there is some overlap in the functions and distal connections of these regions, and there is co-operation between them in the normal working of the cerebellum.

196B,14; 198A,10 198A,21; 198A,11	**The archicerebellum** (the phylogenetically oldest part) corresponds to the flocculonodular lobe plus the lingula. Together with parts of the uvula it has primarily vestibular connections and is concerned with balance and eye movements. Dysfunction of these vestibulocerebellar areas or their connections produces nausea, nystagmus (rhythmical oscillation of the eyeballs) and a tendency to fall to the side of the lesion.

The paleocerebellum is made up of most of the anterior lobe plus much of the remainder of the vermis. It is the main area of termination of spinal afferents. Dysfunction of such spinocerebellar areas produces disturbances of posture and muscle tone, and possibly of speech.

The neocerebellum, which in man makes up about 90% of the whole, comprises most of the middle lobe. It receives its main input from the cerebral cortex via the pontine nuclei. It is particularly concerned with fast, and with skilled movements. Dysfunction of such pontocerebellar areas or their connections produces a loss of muscle tone together with asynergia and dysmetria (difficulties with the force, direction and distance covered by movements), particularly of the limbs, but also possibly affecting speech.

198B,23 198C,24	Cerebellar lesions, especially of neovermal regions, may also affect the precision of eye movements and autonomic reflexes. If the dentate nucleus or its efferents in the superior cerebellar peduncle are damaged, a coarse, arrhythmic tremor is seen during movements (intention tremor), in addition to the symptoms of pontocerebellar damage. The effects of cerebellar lesions (unless midline) are usually evident ipsilaterally. With time, considerable if not complete compensation for these defects is usually achieved.

Cerebellar afferents (Figure 182)

Afferents to the cerebellum may be divided into two main types according to their pattern of termination within the cerebellar cortex: climbing fibres and mossy fibres (Figure 181).

Climbing fibres originate almost entirely from the contralateral inferior olivary nuclear complex of the medulla oblongata. They enter the cerebellum via the inferior peduncle and terminate in the cerebellar cortex by making multiple excitatory synaptic contacts on Purkinje cells. In addition they give off collaterals to the intracerebellar nuclei and other cells of the cerebellar cortex.

Other sources of cerebellar afferents give rise to mossy fibres which terminate in the granular layer of cerebellar cortex. Some of these sources send fibres to the intracerebellar nuclei. A few monoaminergic (noradrenaline, 5-hydroxytryptamine) afferents terminate diffusely.

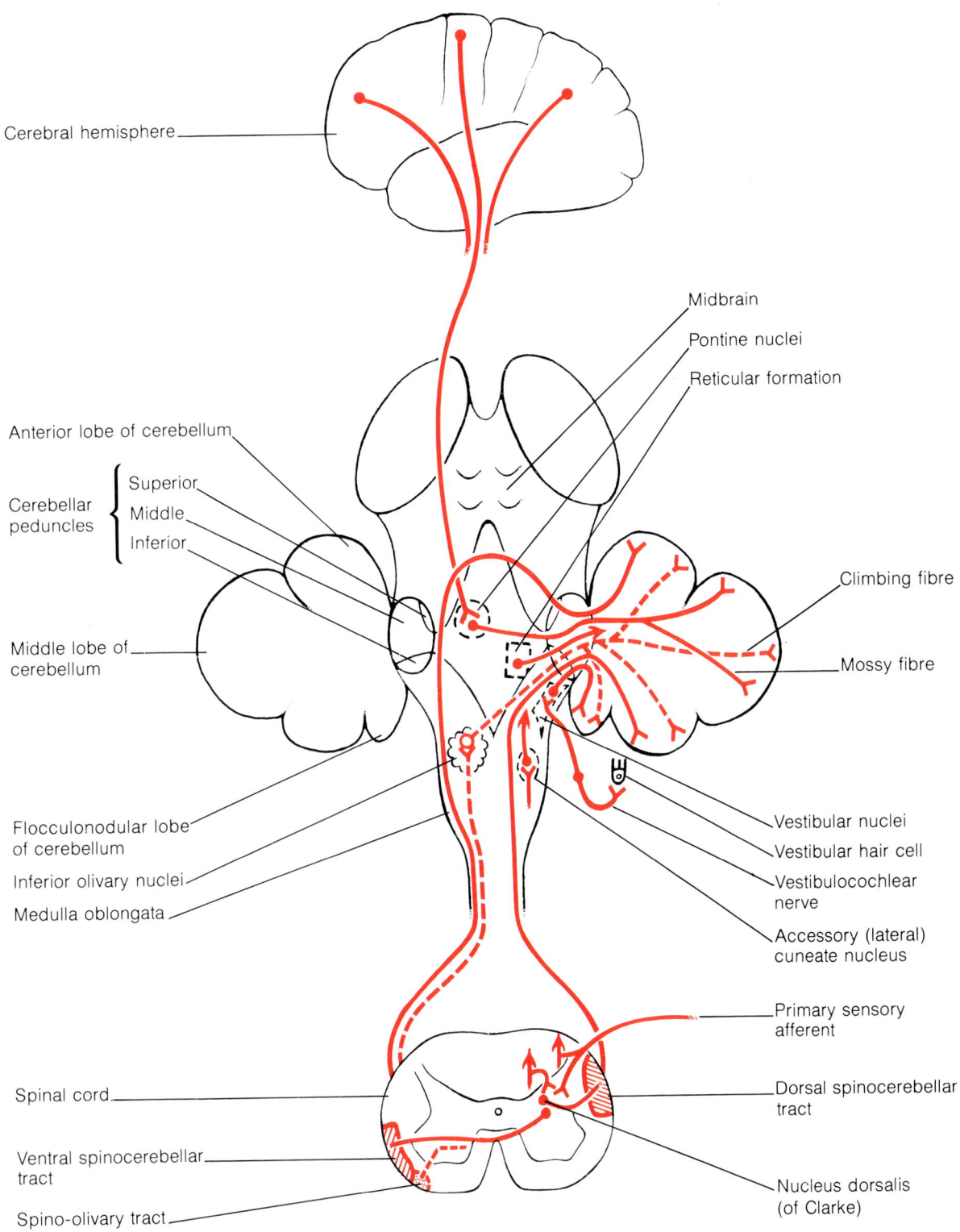

Figure 182 *The major afferent connections of the cerebellum. The diagram shows a lateral view of the left cerebral hemisphere, a dorsal view of the brainstem with the cerebellum displaced laterally, and a transverse section of the thoracic spinal cord. Fibre tracts are indicated by individual red lines. Reticular afferents come from several different nuclei (not illustrated). The rostral spinocerebellar tract and some smaller groups of afferents are not shown.*

The inferior olivary nuclear complex (principal, medial accessory and dorsal accessory olivary nuclei) of the medulla oblongata gives rise to climbing fibres. It receives inputs from all those regions that are the source of mossy fibres and/or are major sources of pathways to the spinal cord. Thus, afferents include cortical (corticonuclear), spino-olivary (both direct fibres and collaterals of spinocerebellar axons), vestibular and reticular fibres. There are also afferents from the red nucleus, the superior colliculus, and the interstitial nucleus of Cajal. In addition, the intracerebellar nuclei (which receive collaterals of the climbing fibres) project to the inferior olivary nuclei. The inferior olivary nuclei are precisely topographically organised, as are their projections to the cerebellum. Furthermore, the topographical representation of the climbing fibres is in register with that of the mossy fibres.

The pontine nuclei of the basilar part of the pons are by far the largest source of cerebellar afferents. Their axons form the middle cerebellar peduncles, almost all the axons passing to the contralateral cerebellum. However, some vermal fibres terminate bilaterally. The great bulk of the input to pontine nuclei comes from the cerebral cortex. These projections arise from widespread areas but are most dense from motor and somatosensory regions and parts of visual association cortex. Other inputs to the pontine nuclei include fibres from the superior and inferior colliculi and from the intracerebellar nuclei.

Proprioceptive information is conveyed from the spinal cord to the cerebellum by several pathways.

The dorsal spinocerebellar tract arises mainly from the nucleus dorsalis (Clarke's column) of the thoracic and lumbar segments of the spinal cord. It runs ipsilaterally and dorsolaterally in the cord and medulla to enter the cerebellum through the inferior peduncle. It carries specific modality and positional information from the lower limb and caudal trunk. The upper limb, upper trunk and neck equivalent is the cuneocerebellar tract, which originates in the accessory (lateral) cuneate nucleus of the medulla. Its fibres also remain ipsilateral and pass through the inferior cerebellar peduncle.

The ventral spinocerebellar tract arises from cells of spinal segments below mid-thoracic levels. Most of the fibres immediately decussate and run cranially just ventral to the dorsal spinocerebellar tract. Some fibres may enter the cerebellum through the inferior peduncle, but most decussate a second time and enter via the superior peduncle. Some fibres end bilaterally. The ventral spinocerebellar tract carries information relating to the lower half of the body. This information concerns the internal state of the spinal cord and the activity of afferents which may evoke withdrawal reflexes (flexor reflex afferents); the information is typically multimodal and does not contain precise somatotopic information. The equivalent for the upper half of the body is the rostral spinocerebellar tract which arises from the upper thoracic and cervical cord. It travels ipsilaterally with fibres of the ventral spinocerebellar tract and enters the cerebellum via both inferior and superior cerebellar peduncles.

Indirect spinocerebellar pathways are provided by spino-olivary fibres to the inferior olivary nucleus, by spinoreticular fibres (particularly to the lateral reticular nucleus of the medulla) and by projections from the gracile and cuneate nuclei of the medulla to the cerebellum.

Vestibular afferents enter the cerebellum medially in the inferior peduncle (in the juxtarestiform body). They mainly arise from the vestibular nuclei, but some are central processes of vestibular ganglion cells.

Reticulocerebellar fibres come chiefly from specific regions of the pontine and medullary reticular formation (paramedian and lateral reticular, reticulotegmental and perihypoglossal nuclei). These regions all receive both cerebellar and cerebral cortical inputs. Some fibres arise from the raphe and locus ceruleus.

Other brainstem sources of cerebellar afferents include the sensory nuclei of the trigeminal nerve (pars interpolaris and pars oralis of the nucleus of the spinal tract, and the mesencephalic nucleus), the nucleus of the solitary tract, and the red nucleus.

The cerebellar cortex (Figure 181)

The cerebellar cortex has the most regular anatomical structure in the brain. The granular layer, the innermost of the three layers, contains multitudinous (10^{10} to 10^{11}) small granule cells and a much smaller number of large Golgi cells. The Purkinje layer contains the somas of the Purkinje cells. The outer, molecular layer contains the dendrites of the Purkinje cells and the axons (parallel fibres) of the granule cells. It also contains basket and stellate cells. Essentially, the cerebellar cortex contains the same pattern of neuronal connections repeated across its entire extent. This feature suggests that in each individual region the operation performed on the incoming data is fundamentally the same whatever the source of the data.

The granular layer receives fibres from extracerebellar sources and from the intracerebellar nuclei. The mossy fibres end in glomeruli (small isolated regions of complex synaptic interactions) in the granular layer, where they make excitatory contact with the dendrites of the granule cells. Axons of the inhibitory Golgi cells also terminate in the glomeruli. The granule cells send their axons (parallel fibres) to the molecular layer, where they make excitatory synapses with Golgi, stellate, basket and Purkinje cells. With the exception of the granule cells, all cells of the cerebellar cortex are inhibitory.

The Purkinje cell dendritic tree is designed to maximise the number of parallel fibres with which it makes contact; it is broad across the folium at right angles to the direction of the parallel fibres, which run along the folium. A single Purkinje cell may synapse with over 10^5 parallel fibres. Each Purkinje cell also receives a massive excitatory synaptic input from a single climbing fibre. One climbing fibre excites many Purkinje cells; there are about 10^6 inferior olivary neurones and over 10^7 Purkinje cells. The Purkinje cell is the sole output cell of the cerebellar cortex. It is GABA-ergic and inhibitory. Most Purkinje cell axons terminate in the ipsilateral intracerebellar nuclei, but some leave the cerebellum to end in the vestibular nuclei or reticular formation. They also give rise to collaterals which terminate in the cerebellar cortex.

It can be shown that the divergence from 10^{7-8} mossy fibres to 10^{10-11} granule cells effects a pattern discrimination function. It is believed that plasticity of the parallel fibre–Purkinje cell synapses may allow the learning of the precise contexts or conditions within which movements or corrections to movements should be performed. The cerebellum has been shown to be the critical locus for the changes that produce the learning which underlies classical conditioning.

Cerebellar efferents and the intracerebellar nuclei (Figure 183)

Efferents from the cerebellar cortex fall into three broad categories: vermal and flocculonodular, paravermal, and lateral (see Figure 180). Thus, there is a medial to lateral organisation in cerebellar outputs.

The fastigial nucleus, which is medially located, receives the axons of the Purkinje cells of the mainly medial, vermal and flocculonodular cortex. This nucleus sends axons bilaterally via the inferior cerebellar peduncle (fibres reach the contralateral peduncle via the hook bundle or uncinate fasciculus of Russell) to the reticular formation and vestibular nuclei of the brainstem (a few fibres pass to the thalamus, superior colliculus, pretectal area and spinal cord). The primary function of this system appears to be the co-ordination and unconscious maintenance of posture and balance. It chiefly acts upon extensor muscle tone. It also forms an important part of the system controlling reflex eye movements.

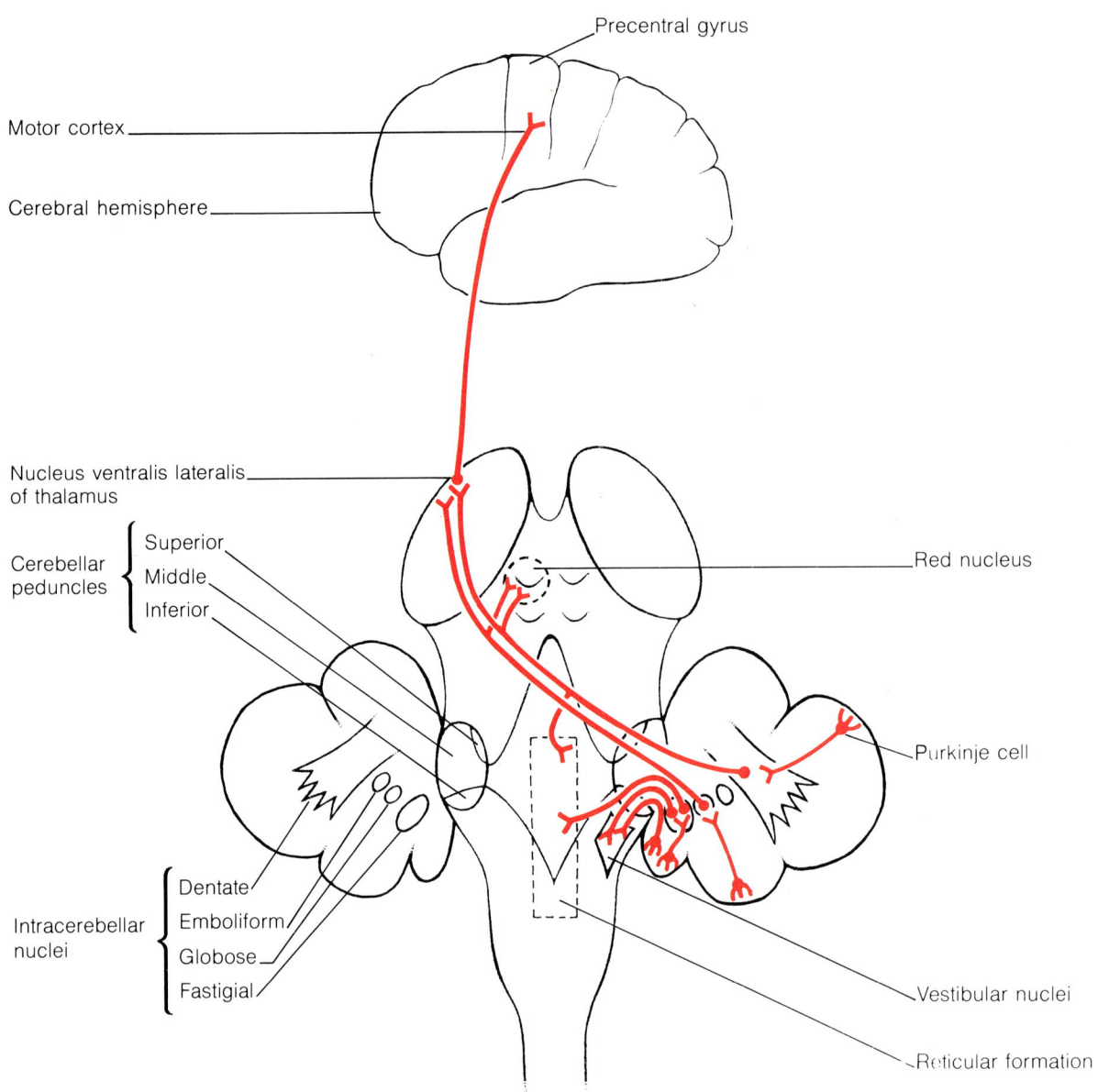

Figure 183 *The major efferent connections of the cerebellum.* The outline of structures is similar to that seen in Figure 182. Although only ipsilateral connections are shown, it is to be noted that the fastigial nucleus projects bilaterally to the reticular formation and vestibular nuclei.

194,26
198C,24

The globose and emboliform nuclei, which are both small and located in an intermediate position, receive input from the paravermal cortex. These nuclei send projections primarily to the contralateral red nucleus of the midbrain via the superior cerebellar peduncle, although some axons pass to the thalamus (mainly its nucleus ventralis lateralis). This system is believed to be important for making corrections to on-going movements. It particularly influences the tonus of flexor muscles.

198B,23

The dentate nucleus, which is large, lobulated and laterally located, receives the projections of the lateral zone of the cerebellar cortex. It is by far the biggest of the intracerebellar nuclei and gives rise to about half a million fibres passing into the superior cerebellar peduncle. This nucleus sends fibres via the superior cerebellar

198C,24

peduncle to the contralateral red nucleus and, more importantly, to the contralateral thalamus, so forming the dentatorubrothalamic tract. Some fibres also go to the reticular formation and the inferior olivary nucleus. Most of the thalamic fibres terminate in the nucleus ventralis lateralis, but some end in caudal parts of ventralis anterior, and a few in cranial parts of ventralis posterior lateralis and in nonspecific nuclei. The nucleus ventralis lateralis projects to the primary motor cortex (area 4) and ventralis anterior to the premotor cortex (area 6) of the cerebral hemisphere (see Figures 184 and 189). This system is believed to be important in the pre-programming of movement, especially for skilled and/or rapid voluntary movements.

194,26
194,6

Thus the cerebellum, which is so important for the smooth co-ordination of movement, has no major projection to the spinal cord or motor cranial nerve nuclei. However, its outputs go to motor areas of the cerebral cortex, the red nucleus, the vestibular nuclei and the reticular formation; these structures are the main sources of descending fibres to the spinal cord and motor cranial nerve nuclei. The cerebellum's role is effected via its feed-back and feed-forward influences. The anatomical substrates of these influences are the many neuronal loops formed by its own connections, together with those of its target structures. The plasticity of cerebellar synapses allows the fine tuning of these control circuits and thus the detailed regulation of motor activity.

The diencephalon

The diencephalon is the medial and ventral part of the forebrain. It is continuous with, and lies immediately cranial to, the midbrain of the brainstem. Laterally and cranially (anteriorly) it is continuous with the telencephalon (cerebral hemispheres). The diencephalon constitutes only about 2% of the brain, but it occupies a pivotal position both anatomically and functionally.

The diencephalon is divided into two major and two minor subregions. The major subregions are the thalamus and hypothalamus; the minor subregions are the epithalamus and subthalamus. The thalamus is dorsal and anterior. It is the major source of afferents for the cerebral cortex. The hypothalamus is ventral and anterior. It controls many of the homeostatic mechanisms of the body and is vital to the preservation of the individual and the species. Connected to it and controlled by it is the pituitary gland. The optic nerves are also attached to the hypothalamus. The epithalamus is a small region near the dorsomedial boundary with the midbrain; it includes the pineal body. The subthalamus is a caudal, ventral and lateral region effecting continuity with the midbrain and containing connections of the basal nuclei.

158,9; 158,57

158,50; 158,58
158,10
158,14

In the intact brain, the only externally visible part of the diencephalon is its ventral (hypothalamic) surface. The frontal lobes of the cerebral hemispheres are anterior, with the main bulk of the hemispheres being lateral and dorsal. The fornices and the corpus callosum pass dorsal to the diencephalon. The internal capsule and parts of the basal nuclei are lateral. The brainstem is caudal.

188
158,60;
158,7; 158,5
194,6; 194,10
194,29; 194,23

Ventral to the hypothalamus is the interpeduncular cistern of cerebrospinal fluid within the interpeduncular fossa. This cistern contains the cerebral arterial circle (of Willis). Beneath it lies the sella turcica of the body of the sphenoid bone, containing the pituitary gland flanked by the cavernous sinuses, through which the internal carotid arteries pass as they enter the cranial cavity (see pages 393 to 394).

180,36; 180,27
158,54
158,52
168,30; 168,33
168,6

THE GROSS TOPOGRAPHY OF THE DIENCEPHALON

The caudal boundary of the diencephalon with the midbrain passes immediately caudal to the two adjacent, hemispherical, mamillary bodies ventrally, and to the posterior commissure dorsally. The posterior commissure is found in the midline cranial to the superior colliculi of the midbrain and caudal to the point of attachment of the pineal body. The cranial boundary of the diencephalon with the cerebral

180,36; 180,21
180,26

180,24
180,13

hemispheres passes cranial to the optic chiasma ventrally, and then continues dorsally through the lamina terminalis and the left and right interventricular foramina. Laterally, the diencephalon is bounded by the internal capsule. The caudate nucleus and stria terminalis of the telencephalon lie dorsolaterally. Dorsally, the diencephalon is bounded by the lateral ventricle, the roof of the third ventricle, and, caudally above the epithalamus, the cistern of the great cerebral vein.

External surface of the diencephalon

The ventral surface, which forms part of the hypothalamus, comprises the optic chiasma, the mamillary bodies and the intervening prominence, the tuber cinereum. A median conical process, the infundibulum of the pituitary stalk, passes ventrally from the tuber cinereum and makes connection to the pituitary gland. The raised area of the tuber cinereum at the base of the infundibulum is the median eminence.

Internal surface of the diencephalon

The diencephalon forms the walls of the third ventricle and is almost completely bisected by it. Continuity is maintained ventrally in the hypothalamic floor of the third ventricle, and more dorsally by an interthalamic adhesion (when present). Note that there is a single subregion of the diencephalon known as the thalamus, but that the term thalamus is also commonly used to refer to one of the two symmetrical groups of nuclei that make up the subregion. Hence the occurrence of the phrase 'the two thalami'. A similar ambiguity exists for the term hypothalamus.

The third ventricle (see Figure 166) is a slit-like median cavity about 30 mm long and 30 mm high. Its roof is formed by the tela choroidea and the choroid plexus of the third ventricle. The hypothalamus and the thalamus form the main part of its lateral walls. Its anterior wall comprises the lamina terminalis ventrally, and the columns of the fornices and the anterior commissure more dorsally. Most of its floor is made up of parts of the hypothalamus; craniocaudally these are: the optic chiasma, the tuber cinereum and infundibulum, and the mamillary bodies. More caudally, the floor and walls are formed by the cerebral peduncles of the midbrain and their continuations into the subthalamus and epithalamus. Ventrally, the ventricle has an optic recess above the optic chiasma and a funnel-shaped infundibular recess within the pituitary stalk. Dorsally and caudally it has a pineal recess which projects into the stalk of the pineal body, and a suprapineal recess which projects caudally above it.

The third ventricle communicates with each lateral ventricle via an interventricular foramen (of Monro), which is a small, comma-shaped opening lying between the column of the fornix and the anterior extension (anterior tubercle) of the thalamus. The third ventricle communicates with the fourth ventricle via the cerebral aqueduct of the midbrain. On each side between the cerebral aqueduct and the interventricular foramen runs a shallow hypothalamic sulcus. The hypothalamic sulcus marks the division between the dorsal and ventral parts of the diencephalon. Below the sulcus are the hypothalamus and subthalamus. Above the sulcus are the thalamus and epithalamus. An interthalamic adhesion (massa intermedia) bridges the ventricle and joins the two thalami in 70 to 80% of individuals.

THE HYPOTHALAMUS

The hypothalamus is the ventral, anterior and medial part of the diencephalon. It forms the floor and parts of the lateral walls of the third ventricle below the hypothalamic sulcus. The thalamus is dorsal, the internal capsule lateral, and the subthalamus caudal and lateral to it.

The hypothalamus represents less than 0.4% of the total brain weight, yet it is essential for life as it controls many important bodily functions. It is necessary for the maintenance of homeostasis and for the expression of basic patterns of emotional behaviour (e.g. defensive, aggressive, reproductive). Thus, the hypothalamus controls or influences: growth and metabolism, appetite, thirst and water balance, cardiovascular function, temperature regulation, aspects of the sleep/wake cycle including its circadian entrainment, the reproductive organs and sexual behaviour, and emotional reactions involving fear, stress, anger, etc. This regulation is exercised both neuronally and hormonally. In particular, the hypothalamus exerts a controlling influence over the autonomic nervous system and the pituitary gland. Accordingly, the hypothalamus forms part of the reticular formation, the limbic system, the autonomic nervous system and the endocrine system.

158,51; 158,50

The hypothalamus is crossed by many fibre tracts, and cell groups are often ill-defined both anatomically and functionally. The reason why such a comparatively small number of neurones is able to control or influence so many processes may be that hypothalamic cell groups, like reticular nuclei of the brainstem, exert a gating/enabling or general activating/inactivating function on other systems. Such an action might allow the hypothalamus to select between different behavioural programs, the programs themselves being stored and executed elsewhere.

Connections of the hypothalamus may be grouped into a number of classes: sensory, autonomic and endocrine, limbic and reticular. Pathways are often complex and precise details have not always been established; general features will be emphasised here.

Inputs (sometimes direct, but more importantly indirect) to the hypothalamus come from both the general and the special sensory systems. Olfactory (via the olfactory bulb and olfactory cortex) and gustatory (via the nucleus of the solitary tract) afferents influence feeding and appetite. Visual input (via the optic pathway) to the suprachiasmatic nucleus allows light to entrain biological (including endocrine) rhythms. General sensory afferents, including nociceptive and autonomic afferents from the brainstem and spinal cord, let such signals affect emotional, autonomic and endocrine reactions and behaviour (e.g. anger in response to pain, maternal suckling).

Autonomic information from the spinal cord or cranial nerve nuclei (e.g. nucleus of the solitary tract, dorsal motor nucleus of the vagus) reaches the hypothalamus mainly indirectly via the reticular formation. In addition, there are inputs from local and more distant neurones which contact local blood vessels or cerebrospinal fluid in the third ventricle. Such inputs may be the basis of the control of blood osmolarity (water balance) and body temperature. There may also be endocrine feedback upon hypothalamic cells. These influences allow control of autonomic and endocrine functions, feeding, drinking, and sexual and other behaviours.

Limbic influences upon the hypothalamus come directly both from the amygdala and septal nuclei, and from the hippocampal region of the cerebral cortex. The major direct input from the hippocampal region passes to the mamillary body, but some fibres (of the medial corticohypothalamic tract) end in other hypothalamic nuclei (particularly near the ventromedial nucleus). There is also a strong indirect influence through hippocampal inputs to the amygdala and the septal nuclei. Other afferents include small numbers of fibres from frontal (particularly orbitofrontal) cortex and from the midline nuclei of the thalamus. Reticular afferents arise particularly, but not exclusively, from the midbrain. They include monoaminergic fibres.

Hypothalamic efferents pass to: the brainstem reticular formation (notably midbrain areas, including peri-aqueductal and tectal regions), the spinal cord and autonomic cranial nerve nuclei, limbic nuclei (the amygdala, septal nuclei and parts of the thalamus) and the pituitary gland. A few fibres from the lateral hypothalamic area pass directly to the motor area of the cerebral cortex. Vasopressinergic and oxytocinergic fibres have been traced to many limbic regions (e.g. septal nuclei, amygdala, hippocampus), to the thalamus, epithalamus, brainstem (including cranial nerve and reticular nuclei, and the substantia nigra) and spinal cord. Control of autonomic functions is exercised via both direct and indirect (reticular) pathways. The major thalamic input is via the mamillothalamic tract to the anterior nuclei, but other fibres pass to the medial thalamus.

The nuclei of the hypothalamus

The hypothalamus is divided into a number of areas or nuclei, although the precise boundaries of these regions are not always distinct.

Most anteriorly, the pre-optic area extends between the optic chiasma, the lamina terminalis and the anterior commissure (which is just ventral and anterior to the interventricular foramen). Anteriorly it blends into the basal forebrain area (substantia innominata) deep to the anterior perforated substance. Although it may be argued on embryological grounds that this area belongs to the telencephalon, functionally it forms part of the hypothalamus. Its neurones control the release of gonadotropins, and the area demonstrates sexual dimorphism.

Most caudally, the mamillary bodies consist of two rounded groups of nuclei (large, medial and smaller, intermediate and lateral nuclei). They receive the fibres of the fornix which originate in subicular cortex (see Figure 192a, and pages 556 to 559). Fibres leave the mamillary body dorsally in the fasciculus mamillaris princeps. Most of these fibres continue ipsilaterally as the mamillothalamic tract which runs dorsally to the anterior nuclei of the thalamus (see Figure 192b). Some fibres turn caudally to form the mamillotegmental tract which ends mainly in the dorsal and ventral tegmental nuclei of the midbrain reticular formation (though some fibres pass to pontine levels). These and other reticular nuclei in turn send fibres via the mamillary peduncle to the mamillary bodies. The mamillary bodies are reciprocally connected with the septal nuclei, and some fibres pass to the hippocampal region.

Between the pre-optic area and mamillary bodies, the hypothalamus may be divided into medial and lateral areas. The division between them is marked by the fornix and the mamillothalamic tract.

The medial area may be subdivided into a narrow periventricular zone and a wider intermediate zone. It may also be subdivided into a supra-optic region anteriorly (near to the optic chiasma) and a tuberal region posteriorly (near to the tuber cinereum).

The supra-optic region on each side contains the anterior hypothalamic area (nucleus), which surrounds three well-defined nuclei. The suprachiasmatic nucleus is a group of small cells immediately dorsal to the optic chiasma. Some of these cells synthesise vasopressin. The nucleus receives visual information direct from the retina and from the ventral part of the lateral geniculate body of the thalamus. Its efferents probably terminate in other hypothalamic nuclei. The paraventricular nucleus lies in the walls of the third ventricle dorsal to the suprachiasmatic nucleus. The supra-optic nucleus lies lateral and caudal to the optic chiasma. These two latter nuclei contain large as well as small cells and are the major magnocellular hypothalamic nuclei. Their neurones synthesise either oxytocin or vasopressin. Axons from the nuclei form the hypothalamohypophyseal tract (supra-opticohypophyseal tract) which ends in the posterior pituitary. However, other, small neurones of the parvicellular part of the paraventricular nucleus project (primarily ipsilaterally) to: a) the nucleus of the solitary tract and the dorsal motor nucleus of the vagus in the medulla (mainly by vasopressinergic fibres), and b) to the dorsal horn (laminae I and II) and the intermediolateral cell column of the spinal cord (mainly by oxytocinergic fibres). These connections allow control of autonomic functions. The parvicellular neurones of the paraventricular nucleus may also be the source of vasopressinergic fibres passing to limbic forebrain areas (septal nuclei, amygdala, hippocampus, habenula).

The tuberal region contains four nuclei: arcuate, ventromedial, dorsomedial and posterior. The arcuate (infundibular) nucleus is located in and just dorsal and caudal to the median eminence. Dorsal to it is found the ventromedial nucleus which, in turn, is ventral to the dorsomedial nucleus. The posterior nucleus (area) lies caudal to these nuclei and is in continuity with the peri-aqueductal grey of the midbrain.

The tubero-infundibular tract passes from the tuberal region (particularly from the arcuate nucleus) to the median eminence. Dopaminergic neurones are found within the arcuate and dorsomedial nuclei and close to the third ventricle (periventricular).

The tubero-infundibular dopaminergic neurones inhibit prolactin secretion (i.e. dopamine is a prolactin-inhibiting factor). Cell bodies containing β-endorphin are found within the ventral tuberal region. Their axons distribute to the supra-optic, paraventricular and suprachiasmatic nuclei and to periventricular regions.

The organum vasculosum of the lamina terminalis or supra-optic crest is found in the midline in the lamina terminalis. The subfornical organ is located between the left and right interventricular foramina. Both are circumventricular organs (see pages 584 to 585) and may be important in the maintenance of water balance and other endocrine functions. In addition, both the median eminence and the posterior lobe of the pituitary gland are circumventricular organs.

180,32
180,9

The tracts of the hypothalamus

Several named tracts pass through the hypothalamus.

Postcommissural fibres of the fornix on each side pass caudally and ventrally from just anterior to the interventricular foramen through the middle of the hypothalamus to the mamillary body (see Figure 192b). These fibres come from the hippocampal region of the temporal lobe (see pages 556 to 559). The mamillothalamic tract (Vicq d'Azyr's bundle) passes dorsally from the mamillary body to the anterior nuclei of the thalamus. Shortly after leaving the mamillary body, fibres leave the mamillothalamic tract to form the mamillotegmental tract, which passes caudally to reticular nuclei of the midbrain. The mamillary peduncle contains hypothalamic afferents from the midbrain reticular formation. Most of the above connections terminate ipsilaterally.

180,9
180,26

180,21

The medial forebrain bundle is a loose collection of fibres which passes through the lateral hypothalamic and preoptic areas. It extends from the septal region of the basal forebrain to the midbrain. It contains ascending monoaminergic (noradrenergic, 5-hydroxytryptaminergic and dopaminergic), cholinergic and other fibres of brainstem reticular origin and descending fibres of amygdalar, septal and other basal forebrain origin. Some fibres terminate in the hypothalamus, some originate there and many pass through. Fibres of hypothalamic origin ascend to septal nuclei (medial septal nucleus and diagonal band of Broca) and descend to the midbrain reticular formation.

The dorsal longitudinal fasciculus (see page 505) contains fibres from the medial and periventricular parts of the hypothalamus (these fibres mainly terminate in the midbrain), together with hypothalamic afferents from the brainstem.

The stria terminalis and ventral amygdalofugal pathway (see pages 568 to 569) bring forebrain afferents from the region of the amygdala to the hypothalamus and also convey hypothalamic efferents to the amygdala.

The supra-optic decussations are small bundles of fibres that cross the midline dorsal to the optic chiasma. They are not numerous and are not all of hypothalamic origin.

180,31

Two tracts allow the hypothalamus to control the pituitary gland: the tubero-infundibular tract and the hypothalamohypophyseal tract.

The tubero-infundibular tract arises mainly from the tuberal region (particularly the arcuate nucleus) and terminates in the median eminence. The axons secrete releasing or inhibiting factors which control the anterior lobe of the pituitary via the hypophyseoportal venous system (see pages 399 to 400). These factors (mainly peptides) control release of prolactin, growth hormone, gonadotropin, follicle-stimulating hormone, thyrotropin and corticotropin.

180,29

The hypothalamohypophyseal (supra-opticohypophyseal) tract arises from the neurosecretory cells of supra-optic and paraventricular nuclei. The axons (about 10^5 fibres) release oxytocin and vasopressin (antidiuretic hormone) within the posterior lobe of the pituitary gland. The axons pass through the pituitary stalk.

158,50; 158,51

The optic chiasma is described with the visual pathway (see pages 435 and 590 to 591).

180,31

The functions of the hypothalamus

In general, specific functions have been but poorly localised within the hypothalamus. Furthermore, lesions disrupt fibres of passage as well as individual nuclei. Usually, lesions must affect both sides before major signs develop. However, the closeness of the left and right nuclei increases the possibility of bilateral damage.

Bilateral lesions of the mamillary bodies may cause or contribute to the amnesia of Korsakoff's syndrome (see page 561).

Anterior and medial hypothalamic areas control parasympathetic activity of the autonomic nervous system. Sympathetic activity is controlled by posterior and lateral regions. Bilateral lesions of the anterior hypothalamus may cause hyperthermia (pyrexia) and also insomnia. Posterior lesions may produce a fall or total loss of regulation of body temperature; they also cause somnolence and emotional lethargy. The suprachiasmatic nucleus appears to be responsible for the entrainment of the circadian rhythm of bodily functions to the 24-hour light/dark cycle.

Lesions of the lateral hypothalamus can result in loss of appetite, while lesions of the ventromedial nucleus can result in hyperphagia. Lesions of the ventromedial nucleus may also result in aggressive behaviour. Electrical stimulation of this region is perceived as unpleasant, while stimulation near the median forebrain bundle in the preoptic region gives rise to pleasurable sensations.

The supra-optic nucleus appears to be particularly important for the control of water balance. Interruption of the hypothalamohypophyseal tract causes diabetes insipidus (increased urine production and compensatory water intake). Hypothalamic lesions can also cause hypogonadism, precocious puberty, or general signs of reduced pituitary function.

THE SUBTHALAMUS

The subthalamus (ventral thalamus) is the ventral, caudal and lateral part of the diencephalon. It is an ill-defined region which is continuous caudally with the midbrain tegmentum. Cranially and medially is the hypothalamus, dorsally the thalamus, and laterally the internal capsule as it merges into the crus cerebri of the midbrain.

The subthalamus contains: a) continuations of midbrain nuclei and tracts (see Figure 178j); the cranial ends of the red nucleus and substantia nigra; the medial, spinal and trigeminal lemnisci; cerebellothalamic fibres; the fasciculus retroflexus, and other fibre bundles; b) nuclei and tracts associated with the basal nuclei (see pages 567 to 571); the subthalamic nucleus, subthalamic fasciculus, lenticular fasciculus (Forel's field H2), ansa lenticularis, thalamic fasciculus (Forel's field H1), and prerubral field (Forel's field H); and c) small interposed nuclei. The largest of the interposed nuclei is the zona incerta. Scattered cells of the prerubral field (cranial to the red nucleus) are continuous with the thalamic reticular nucleus and are sometimes called the subthalamic reticular nucleus.

The zona incerta lies between the thalamic and lenticular fasciculi, ventral to the thalamus and dorsal to the subthalamic nucleus. It receives fibres from frontal cortex and sends axons to the red nucleus, superior colliculus and pretectal area of the midbrain.

The subthalamic nucleus (body of Luys) is a lens-shaped nucleus immediately dorsomedial to the internal capsule as it continues into the crus cerebri of the midbrain. Caudally, it abuts the substantia nigra. It is strongly connected to the globus pallidus of the basal nuclei (via the subthalamic fasciculus) and the substantia nigra, and receives fibres from frontal cortex and the reticular formation (see also page 570). Lesions of the nucleus cause contralateral hemiballismus.

The thalamic fasciculus (Forel's field H1) contains as its major components: the fibres of the lenticular fasciculus and ansa lenticularis (i.e. efferents of the globus pallidus), cerebellothalamic fibres, and thalamostriate fibres (i.e. fibres passing from the intralaminary thalamic nuclei to the putamen and caudate nucleus).

THE THALAMUS

The thalamus is the major dorsal and anterior subregion of the diencephalon. On each side, the thalamus comprises a large (about 4 cm in greatest diameter) ovoid mass of nuclei. It is covered by a thin layer of white matter; dorsally this is called the stratum zonale and laterally, the external medullary lamina. The two thalami may be linked by an interthalamic adhesion.

The thalamus is the major source of afferents to the cerebral cortex. All sensory pathways with the exception of the olfactory system relay in the thalamus. The thalamus also receives important motor (cerebellar and basal nuclear), limbic and reticular inputs. The nuclei of the thalamus are classified and listed with their main connections in Table 17.

The thalamus is subdivided into the metathalamus, comprising the medial and lateral geniculate bodies, and the dorsal thalamus, which forms the great bulk of the whole. A Y-shaped internal medullary lamina divides the dorsal thalamus into three main groups of nuclei: anterior, medial and lateral (Figure 184).

The nuclei of the thalamus (Figure 184)

In the dorsal thalamus, the anterior nuclei are within the anterior thalamic tubercle which forms the posterior boundary of the interventricular foramen. The large nucleus medialis dorsalis (mediodorsal or dorsomedial nucleus) is separated from the third ventricle by the small midline nuclei. The lateral nuclei have a dorsal and a ventral tier. The dorsal tier nuclei are craniocaudally: lateralis dorsalis, lateralis posterior and the pulvinar. The large, posterior, pulvinar nucleus forms a smooth projection beneath the splenium of the corpus callosum and overhangs the midbrain. The ventral tier nuclei are craniocaudally: ventralis anterior, ventralis lateralis, ventralis posterior (lateralis, medialis, and inferior) and the posterior group. The ill-defined posterior group is caudal to ventralis posterior, medial to the pulvinar and close to the midbrain.

Intralaminary nuclei are located within the internal medullary lamina, the largest being the centrum medianum. The thalamic reticular nucleus is a thin layer of neurones surrounding the anterior and lateral aspects of the thalamus, just outside the external medullary lamina.

In the metathalamus, the medial geniculate body (nucleus) is on the ventral surface of the pulvinar, lateral to the superior colliculus of the midbrain (see Figure 178j). The lateral geniculate body (nucleus) is lateral to the medial geniculate body.

Thalamic nuclei may be broadly classified into specific and nonspecific on the basis of their connections with the cerebral cortex. However, anatomically and functionally the division is not absolute. Nonspecific nuclei include: the midline, the intralaminary, and the reticular nuclei. Specific nuclei make up the remainder (as mentioned above) and form the main mass of the thalamus. Nucleus ventralis anterior and nuclei of the posterior group may be classified as both specific and nonspecific.

The specific thalamic nuclei form by far the largest part of the thalamus and are responsible for most of the information that is sent to the cerebral cortex. A specific nucleus is characterised by having a precise topographical projection to a limited region of the ipsilateral cerebral cortex. This region of cortex is the major target of its axons. These axons chiefly terminate in layer IV or III of the cortex. The cortical region projects back topographically upon the nucleus, the axons primarily arising from cells in layer VI. Such a thalamic nucleus is termed 'cortically dependent' because typically its neurones degenerate if there is a lesion of the cortical region with which it is connected. This dependence emphasises the close relationship between the cerebral cortex and thalamus; they may effectively be considered part of a single, integrated piece of neuronal machinery.

TABLE 17 Thalamic nuclei: classification and main connections. Type A = association nucleus; NS = nonspecific nucleus; R = relay nucleus. The numbers in brackets refer to Brodmann areas of the cerebral cortex.

NAME (ABBREVIATION)	SOURCES OF MAJOR SUBCORTICAL INPUTS	MAJOR OUTPUT TARGETS	TYPE	FUNCTIONAL CLASSIFICATION
Specific nuclei:				
Anterior division (Ant)				
Anterior ventralis (AV)	Mamillary body	Cingulate gyrus	R	Limbic
Anterior dorsalis (AD)		Subicular cortex		
Anterior medialis (AM)				
Medial division				
Medialis dorsalis (MD)	Amygdala	Prefrontal cortex	A	Association and limbic
	Olfactory areas	Medial temporal cortex	(R)	
	Thalamus			
Lateral division: ventral tier				
Ventralis anterior (VA)	Globus pallidus	Premotor cortex (6)	R	Motor and nonspecific
	Substantia nigra	Prefrontal cortex	(NS)	
	Cerebellum	Thalamus		
	Thalamus			
Ventralis lateralis (VL)	Intracerebellar nuclei	Motor cortex (4)	R	Motor
	Substantia nigra			
	Globus pallidus			
Ventralis posterior lateralis (VPL)	Dorsal column nuclei	Sensorimotor cortex (SmI, II)	R	Somatosensory (body)
	Spinal cord			
	Vestibular nuclei			
Ventralis posterior medialis (VPM)	Trigeminal nuclei	Sensorimotor cortex (SmI, II)	R	Somatosensory (head)
	Nucleus of solitary tract			Taste
Ventralis posterior inferior (VPI)	Vestibular nuclei	Sensorimotor cortex (2)	R	Somatosensory Vestibular
Posterior group:				
Suprageniculatus	Reticular formation	Insular cortex	NS	Nonspecific?
Limitans	Midbrain tectum			
Posterior (PO)	Spinal cord	Retro-insular cortex	R	Somatosensory
	Trigeminal nuclei	Postauditory cortex	(NS)	(Nociceptive)
Lateral division: dorsal tier				
Lateralis dorsalis (LD)	Thalamus	Parietal association cortex	A	Association
		Cingulate gyrus		Limbic
Lateralis posterior (LP)	Thalamus	Parietal association cortex	A	Association
	Superior colliculus			(Somatosensory/visual)
Pulvinar (Pul)	Superior colliculus	Occipital, parietal and	A	Association
	Pretectal area	temporal association cortex	(R)	(Visual, speech)
	Thalamus	Frontal eye field		
Metathalamus				
Lateral geniculate body (LGB)	Retina	Primary visual cortex (17)	R	Visual
Medial geniculate body (MGB)	Inferior colliculus	Primary auditory cortex	R	Auditory
	Spinal cord	Insular and opercular cortex		Association (Nociceptive?)
	Superior colliculus			

TABLE 17 (continued).

NAME (ABBREVIATION)	SOURCES OF MAJOR SUBCORTICAL INPUTS	MAJOR OUTPUT TARGETS	TYPE	FUNCTIONAL CLASSIFICATION
Nonspecific nuclei:				
Midline group				
Parataenialis	Hypothalamus	Hippocampal region	NS	Limbic
Paraventricularis (Mid)	Thalamus	Cingulate gyrus		
Reuniens	Reticular formation	Amygdala		
Rhomboidalis		Hypothalamus		
Intralaminary nuclei				
Centrum medianum (CM)	Reticular formation	Caudate nucleus	NS	Reticular
Parafascicularis (Pf)	Thalamus	Putamen		
Paracentralis (Pc)	Sensory and motor systems	Cortex		
Centralis lateralis (CL)		Thalamus		
Centralis medialis		Reticular formation		
Reticular nucleus				
Reticularis (Retic)	Thalamus	Thalamus	NS	Reticular (Attention?)
	Cortex	Reticular formation		
		Superior colliculus		

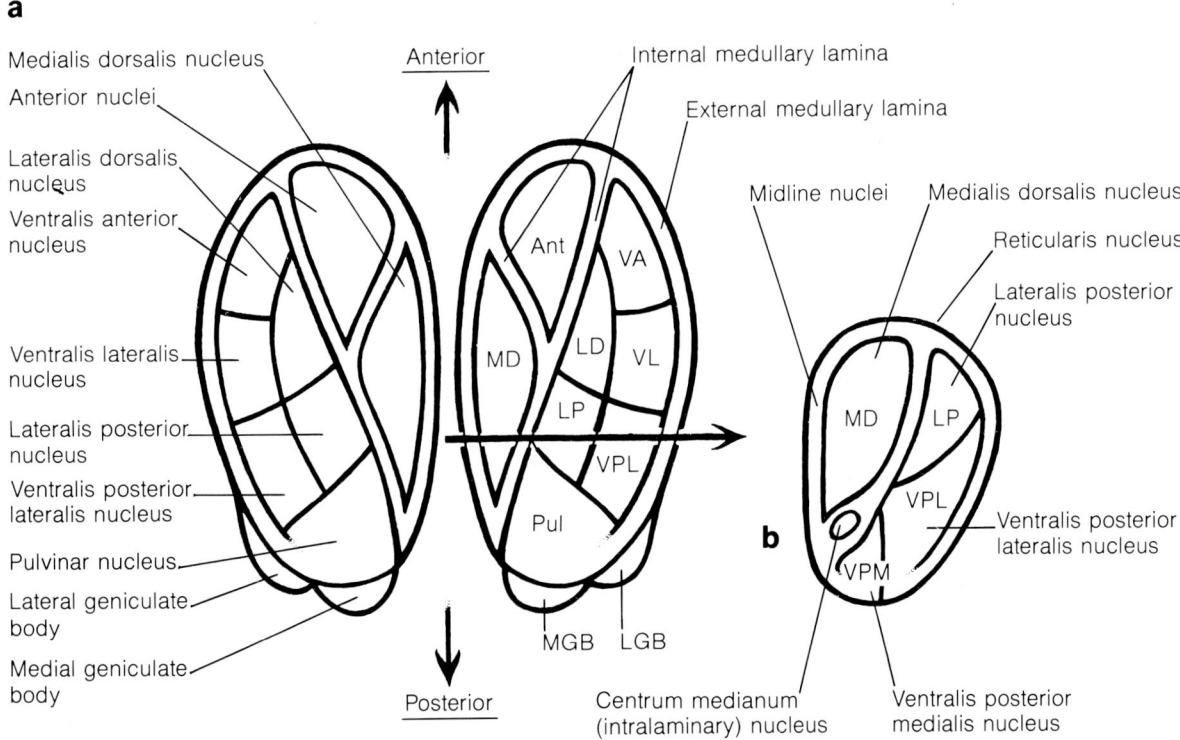

Figure 184 *The nuclei of the thalamus.* **a)** *Horizontal section of left and right thalami.* **b)** *Frontal (coronal) section of right thalamus (as indicated by arrow in Figure 184a). The common abbreviations used for the nuclei of the thalamus are shown.*

192,27–29

The connections of the specific thalamic nuclei are dominated by both afferent and efferent connections with the ipsilateral cerebral cortex. These connections travel almost entirely in four thalamic peduncles which form part of the internal capsule of the cerebral hemisphere (see Figure 194, and also pages 564 to 566). These connections are known as thalamic radiations. On the gross scale, these connections have a topographic, anterior to posterior organisation. The anterior (frontal) peduncle interconnects anterior and medial thalamic nuclei (including ventralis anterior) with the frontal lobe. The superior (centroparietal) peduncle interconnects most of the lateral nuclei with paracentral regions of the frontal and parietal lobes. The posterior (occipital) peduncle interconnects the more posterior nuclei (pulvinar and lateral geniculate body) with posterior regions of the parietal, temporal and occipital lobes. The small inferior (temporal) peduncle contains connections of the temporal lobe and insula (including those of the medial geniculate body, and of the amygdala). (The inferior peduncle plus amygdalar connections to the hypothalamus constitute the ansa peduncularis.) Some afferents from subicular cortex reach the anterior nuclei via the fornix.

The specific functions with which the individual thalamic nuclei are involved are most easily determined by reference to the function of their main afferents and/or the functions of the cortical region to which they project (Table 17, see Figures 187 and 188).

Efferents of specific thalamic nuclei end in the cortex or in other thalamic nuclei (a few fibres go to the amygdala). Connections between the right and left thalami are few or none. Non-cortical afferents include fibres from the reticular formation of the brainstem (which pass to all nuclei), inputs from other thalamic nuclei, and non-thalamic subcortical afferents. The specific thalamic nuclei are categorised as relay nuclei if they receive a major, non-thalamic subcortical input. Association thalamic nuclei receive their main subcortical input from other thalamic nuclei. However, again the categories show overlap.

The relay nuclei (and their main subcortical inputs) are: ventralis posterior (somatosensory; medial, spinal, and trigeminal lemnisci), ventralis lateralis (motor; cerebellothalamic fibres of the dentatothalamic tract), posterior group (sensory including nociceptive; spinal and trigeminal lemnisci, tectal and reticular fibres), ventralis anterior (motor; basal nuclear fibres via ansa lenticularis and the lenticular fasciculus), anterior group (limbic; mamillothalamic tract). The lateral geniculate body (visual; optic tract) and medial geniculate body (auditory; brachium of the inferior colliculus) are also relay nuclei. Nucleus medialis dorsalis (limbic; amygdalar via ansa peduncularis, and possibly olfactory and hypothalamic fibres) may also be regarded as in part a relay nucleus, as may the pulvinar (visual inputs from the superior colliculus and pretectal area).

The association nuclei (medialis dorsalis, lateralis dorsalis, lateralis posterior and the pulvinar) are particularly highly developed in the human thalamus. Their main connections are with other thalamic nuclei and with association areas of the cerebral cortex. They are concerned with higher order sensory processing, intellectual abilities and personality (see Figures 187, 188 and pages 553 to 555).

The motor and sensory functions of the thalamic relay nuclei will now be described in more detail.

Motor functions of the thalamus involve primarily nuclei ventralis anterior and ventralis lateralis. Nucleus ventralis anterior is the major thalamic termination of fibres from the ipsilateral basal nuclei (globus pallidus and substantia nigra); nucleus ventralis lateralis for fibres from the contralateral cerebellum (mainly its dentate nucleus), though these connections overlap to some extent (see Figures 183 and 195). Nucleus ventralis lateralis projects to motor cortex (area 4), nucleus ventralis anterior to premotor cortex (area 6), and diffusely (i.e. in a nonspecific manner) to prefrontal cortex. In nucleus ventralis lateralis neurones dealing with contralateral movements of head muscles are medial, those for the arm are intermediate and for the leg lateral; trunk movements are represented dorsally, distal limb movements ventrally. Lesioning of nucleus ventralis lateralis has been used to relieve movement disorders (e.g. tremor, hemiballismus) of cerebellar or basal nuclear origin.

Sensory inputs to the thalamus chiefly end in nucleus ventralis posterior and the metathalamus. The nucleus ventralis posterior (ventrobasal complex) is subdivided into lateral, medial and inferior nuclei. Nucleus ventralis posterior lateralis receives precise somatosensory input relating to the contralateral body from the ipsilateral medial and spinal lemnisci (see pages 598 to 601). Nucleus ventralis posterior medialis receives input for the head region via the dorsal and ventral trigeminothalamic tracts (see pages 598 to 601). It also receives gustatory fibres (see pages 596 to 597). Nucleus ventralis posterior inferior and nucleus ventralis posterior lateralis have vestibular connections (see pages 594 to 595). Nucleus ventralis posterior has a somatotopic map with the head being medial, the arm intermediate and the leg lateral. Proprioceptive afferents terminate mainly anteriorly, cutaneous afferents posteriorly. Lesions cause contralateral loss of deep sensations and discriminative tactile ability, while appreciation of thermal, noxious and crude tactile stimuli is less affected. Efferents pass to sensorimotor cortex (areas SmI and SmII).

The posterior group of thalamic nuclei (particularly the medial part of the posterior nucleus) which receives spinal and reticular afferents conveying somatosensory information (especially nociceptive) is believed to play an important role in pain perception. Somatotopic information is bilateral and imprecise. Efferents to the cortex end within and in regions surrounding the insula, notably in the retroinsular cortex immediately posterior to SmII on the parietal operculum (see pages 600 to 601). Lesions of the posterior thalamus can give rise to severe pain in response to innocuous tactile stimuli ('thalamic pain').

The medial geniculate body is part of the auditory pathway (see pages 592 to 593). Its ventral, laminated and parvicellular part projects to auditory cortex. Medial neurones respond to high frequencies, lateral to low. The remaining regions (including the medial, magnocellular) receive additional input from the anterolateral system and the superior colliculus; they project to areas surrounding primary auditory cortex.

188B,44

The lateral geniculate body receives most of the fibres of the optic tract. It projects to visual cortex (see pages 590 to 591). The main, dorsal part of the lateral geniculate body has six layers, lamina 1 being most ventral (see Figure 177a). Fibres from the ipsilateral eye terminate in laminae 2, 3 and 5, and fibres from the contralateral eye in laminae 1, 4 and 6. Laminae 1 and 2 are magnocellular and probably deal with monochromatic vision. Laminae 3 to 6 are parvicellular and are concerned with colour vision. Afferents from the centre of the retina end caudally, from the inferior retina laterally, and from the superior retina medially. Lesions cause a contralateral homonymous hemianopia (see pages 485 and 590 to 591). There is a small, ventral region that sends fibres to the superior colliculus and pretectal area of the midbrain, and to the suprachiasmatic nucleus of the hypothalamus.

188B,41; 188B,40

Most thalamic lesions will be likely to involve more than one nucleus and, if laterally placed, may include parts of the internal capsule. Lesions of any nucleus (not merely relay nuclei) will affect cortical function. For example, speech deficits have been reported with lesions of the left pulvinar nucleus (see pages 552 to 553) and there is evidence for anterograde amnesia following damage to the nucleus medialis dorsalis (see page 561). The interdependence of the thalamus and cerebral cortex, together with the possibilities for synaptic remodelling following lesions, makes it difficult to interpret claims that there is conscious appreciation of stimuli (e.g. of pain) at the thalamic level.

194,6; 194,10

The nonspecific thalamic nuclei include the midline and intralaminary nuclei and the reticular nucleus of the thalamus. They may be regarded as diencephalic extensions of the reticular formation of the brainstem.

The midline nuclei are relatively small in the human thalamus. Their connections are mainly with limbic structures (hypothalamus, amygdala, cingulate gyrus and hippocampal region), and other nonspecific nuclei (including nucleus ventralis anterior). It is possible that lesions of these nuclei contribute to amnesia of diencephalic origin (such as that of Korsakoff's syndrome; see page 561).

The intralaminary nuclei are well developed in the human thalamus. Their efferents go chiefly to the putamen and caudate nucleus of the basal nuclei, but they additionally project, probably by collaterals, to widespread regions of the cerebral cortex. They constitute a source of nonspecific afferents to the cortex, the fibres terminating mainly in layers V and VI. Connections with the cortex are not reciprocal. Most of their cortical input is from motor and premotor areas. The intralaminary nuclei have many interconnections with other thalamic nuclei, most notably with ventralis anterior, and send some fibres to the brainstem reticular formation and the superior colliculus. The nucleus centralis lateralis possibly has a role in nociception and sensory integration, as subcortical input from spino- and trigemino-thalamic fibres, the superior colliculus, vestibular nuclei and reticular formation terminates particularly in this nucleus. The nuclei centrum medianum and parafascicularis may be more concerned with control of motor activity, as input from the cerebellum, globus pallidus and substantia nigra ends mainly in these nuclei. The intralaminary nuclei receive many fibres from the reticular formation of the brainstem (see pages 504 to 513) and functionally may be considered to be an extension of it. They have been implicated (together with nucleus ventralis anterior) in the control of cortical EEG (particularly its synchronisation) and sleep. They probably exercise a general activating/ inactivating or modulatory role on thalamic and cortical neurones. Via the connections of these nuclei, arousing (e.g. noxious) stimuli can affect many regions of the cerebral cortex.

The reticular nucleus of the thalamus is so positioned that connections between the cortex and the rest of the thalamus pass through it. Excitatory collaterals of both thalamocortical and corticothalamic fibres terminate in the nucleus. These inputs are strictly topographic in organisation and involve projections of all the thalamus (including the midline and intralaminary nuclei) to the cortex. Furthermore, neurones of the nucleus send inhibitory axons to the thalamic region from which they receive input. Some neurones in the nucleus project to the mesencephalic reticular formation and superior colliculus. The reticular nucleus is ideally connected to control or select out ('gate') thalamocortical activity and it has been suggested that it may play a part in mechanisms of selective attention. It has also been proposed that nucleus ventralis anterior, through its widespread connections with prefrontal cortical areas, may be important for attentive processes.

THE EPITHALAMUS

The epithalamus is the small, dorsal, caudal and medial part of the diencephalon. It comprises the pineal body, the posterior commissure, the subcommissural organ and the left and right habenulae and their connections.

The pineal body (pineal gland, epiphysis cerebri) is a median, fusiform endocrine organ located between the superior colliculi of the midbrain. Dorsal to it is the great cerebral vein and the splenium of the corpus callosum. It is attached to the diencephalon by its stalk. The pineal stalk splits, either side of the pineal recess of the third ventricle, into an inferior and a superior lamina. The pineal body is highly vascularised and contains pinealocytes and neuroglial cells. It is innervated by noradrenergic postganglionic sympathetic fibres of the superior cervical ganglion. It secretes 5-hydroxytryptamine and melatonin into both cerebrospinal fluid and blood. Melatonin synthesis is under control of the light/dark cycle, being increased in the dark. A multisynaptic pathway originating in the suprachiasmatic nucleus of the hypothalamus and passing via the superior cervical ganglion controls this periodicity. The influence of the daily melatonin rhythm is noticeable in the effects of changes of time zone ('jet-lag'). Lesions of the pineal body in childhood may result in precocious puberty. Post-pubertal calcification makes the pineal body a useful reference point in brain scans.

The posterior commissure crosses the midline in the inferior lamina of the pineal stalk, immediately cranial to the superior colliculi of the midbrain. Amongst other components, it contains fibres from nuclei in or near the pretectal area of the midbrain, in particular from the interstitial nucleus of Cajal. These latter fibres are not commissural but pass to the oculomotor nucleus (see pages 483 and 501). Lesions of the posterior commissure cause impairment of vertical eye movements, but not of pupillary light reflexes.

180,24

The subcommissural organ is ventral to the posterior commissure at the entrance to the aqueduct of the midbrain. It is one of the circumventricular organs (see pages 584 and 585), as is the pineal body.

180,22

The habenula is a small, pear-shaped eminence found on each side of the midline immediately anterior to the pineal stalk. It comprises a small medial and a large lateral nucleus. The left and right nuclei are interconnected via the habenular commissure which passes in the superior lamina of the pineal stalk. The stria medullaris thalami, which runs posteriorly along the dorsomedial margin of the thalamus, close to the edge of the roof of the third ventricle, terminates in the habenula. It contains fibres of limbic origin (from the septal nuclei, hypothalamus and anterior thalamic nuclei) and from the globus pallidus of the basal nuclei. Fibres from the habenula pass ventrally and caudally through the subthalamus as the fasciculus retroflexus (habenulo-interpeduncular tract) to the interpeduncular nucleus of the ventral midbrain tegmentum. The habenula has other connections with the midbrain, including reciprocal connections with nuclei of the raphe. It has been suggested that these pathways are one means by which emotional factors may influence autonomic responses.

The cerebral hemispheres

The cerebral hemispheres or telencephalon form the main bulk of the forebrain and more than 80% of the whole brain. Together with the diencephalon they fill nearly the whole of the cranial cavity above the tentorium cerebelli.

162; 176; 182

164,25

Each cerebral hemisphere comprises: a) an outer layer of grey matter, the cerebral cortex; b) an underlying medullary core, the centrum semi-ovale; c) deeply placed masses of grey matter, chiefly the basal nuclei but also the basal forebrain nucleus and the septal region of the limbic system; d) a cavity filled with cerebrospinal fluid, the lateral ventricle; and e) the olfactory bulb (which receives the olfactory nerve) and olfactory tract.

192,1
192,13
192,31
192,5
190; 188A,1
188A,2

The cerebral cortex and its connections are responsible for conscious perception, voluntary action (including speech), personality, emotion, memory and thought. The cerebral cortex is able to influence or control all aspects of central nervous system function, both conscious and unconscious. The limbic system is important in emotional and life-supporting behaviours and plays a central role in memory functions. The basal nuclei form an adjunct to the cerebral cortex in its normal operation. Their precise function is poorly understood, but their dysfunction gives rise to disorders of movement.

THE EXTERNAL TOPOGRAPHY OF THE CEREBRAL HEMISPHERES
(Figure 185)

The two cerebral hemispheres together form roughly half a sphere, the convex surface being superior. Most anteriorly, at the frontal poles, the surface is more rounded than it is most posteriorly, at the occipital poles. Laterally, on each side, the temporal lobe extends anteriorly and inferiorly towards the temporal pole (like the thumb of a boxing glove), divided from the rest of the hemisphere by the lateral sulcus.

172,10; 176A,10
172,2; 176A,27
176A,21; 184,29
176A,15

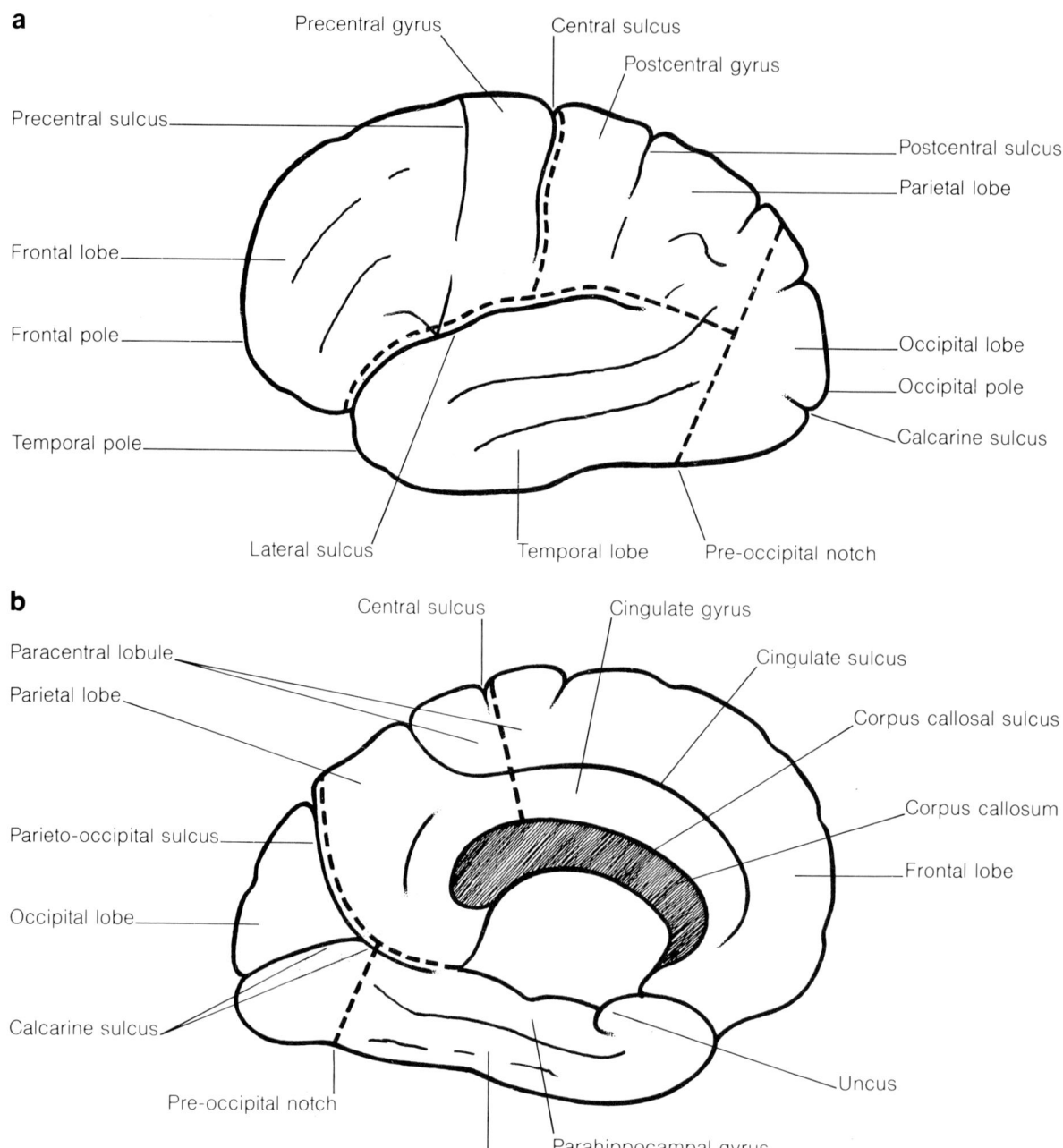

Figure 185 *Major sulci, gyri and the lobes of the cerebral hemisphere.* a) *Lateral view,* b) *medial view.*

172,11
180,3–6

158,4; 164,2
164,1; 164,3
182B,23

The right and left hemispheres are approximately symmetrical and are separated by the sagittal longitudinal fissure. They are joined by commissures, by far the largest of which is the corpus callosum. The corpus callosum is found centrally, deep within the longitudinal fissure. Above the corpus callosum the longitudinal fissure is occupied by the falx cerebri within the margins of which run the superior and inferior sagittal sinuses. The anterior cerebral arteries (and veins) and their branches run in all but posterior parts of the longitudinal fissure, the main trunks anteriorly passing round, and lying close to, the corpus callosum. Branches of the posterior cerebral artery are found posteriorly.

The adjacent, medial surfaces of the hemisphere on either side of the longitudinal fissure are vertical and flat at the gross level. The superolateral surface of each hemisphere is convex and lies adjacent to the vault of the cranium. Branches of the cerebral arteries (chiefly the middle) and superficial cerebral veins pass across this surface. Branches of the middle meningeal artery run extradurally. The superomedial border lies between the medial and superolateral surfaces and is adjacent to the superior sagittal sinus. A line drawn slightly lateral to the midline from just above the inion to just above the nasion lies superficial to this border.

The inferior or basal surface of the cerebral hemisphere is related to the base of the skull. The inferior and superolateral surfaces meet at its inferolateral border. Anteriorly, the inferolateral border is known as the superciliary border. In relation to external features, the superciliary border corresponds to a line drawn above the eyebrows that laterally passes caudally and slightly superiorly to the pterion. The inferior surface meets the medial surface at the inferomedial border anteriorly (in the medial orbital border) and posteriorly (in the medial occipital border); centrally, the inferior surface borders the diencephalon and midbrain.

Anteriorly on the inferior surface, orbitofrontal parts of the hemisphere lie in the anterior cranial fossa above the orbital part of the frontal bone. Medially in this region, the olfactory bulb is above the cribriform plate of the ethmoid bone. It receives the fascicles of the olfactory (I) nerve which pass through the bone from the nasal cavity. The olfactory tract passes posteriorly towards the anterior perforated substance (see pages 555 and 556). The anterior perforated substance is found anterior and lateral to the optic chiasma of the hypothalamus; the internal carotid artery divides into anterior and middle cerebral arteries inferior to it. The anterior perforated substance is penetrated by the striate arteries (small branches from the cerebral arterial circle).

Anterior and inferior parts of the temporal lobe are found inferior and lateral to the hypothalamus, adjacent to the sphenoid bone and the squamous part of the temporal bone in the middle cranial fossa. The most medial part of the temporal lobe, the parahippocampal gyrus, is immediately lateral to the free margin of the tentorium cerebelli at its incisura, and to the cerebral peduncles of the midbrain. Folded medial and superior to the parahippocampal gyrus, and so largely hidden by it, are the hippocampus and dentate gyrus of the hippocampal formation (see Figure 192a). More posterior parts of the inferior surface (parts of the temporal and occipital lobes) rest on the tentorium cerebelli. The straight sinus runs posteriorly along the sagittal junction of the falx cerebri and tentorium, near the medial occipital border of the hemisphere. The transverse sinus runs along the attached margin of the tentorium near the posterior inferolateral border of the hemisphere. Branches of all the cerebral arteries are found on the inferior surface of the hemisphere.

When the brain has been sagittally sectioned, the large, white and curved cut surface of the corpus callosum, together with the structures inferior to it, may be seen. The most anterior part of the corpus callosum is its genu. The genu is continuous postero-inferiorly with the rostrum of the corpus callosum. The rostrum of the corpus callosum narrows postero-inferiorly towards the lamina terminalis. Frontal areas of cortex are interconnected by the genu and rostrum of the corpus callosum. The complete arch formed by these commissural fibres is the forceps minor. The expanded, caudal part of the corpus callosum is its splenium. Posterior areas of the cortex are interconnected through the splenium, the fibres forming the forceps major. Immediately inferior to the splenium is the great cerebral vein and the pineal body. Between the genu and the splenium is the body (trunk) of the corpus callosum.

Anteriorly beneath the corpus callosum, and attached to its body, genu and rostrum, is the septum pellucidum. The septum pellucidum is composed of two thin vertical laminae which provide a partition between the anterior horns of the lateral ventricles.

Along its arched inferior edge run the fornices. Each fornix arises in the hippocampal region of the temporal lobe, the crus of the fornix following a curved course through the lateral ventricle. The body of the fornix is found just beneath the body of the corpus callosum, and superior to the thalamus. Fibres crossing the midline in this region form the hippocampal commissure (commissure of the fornix) and interconnect the hippocampal regions of the two hemispheres. Inferior to the body of the fornix is the tela choroidea and choroid plexus of the roof of the third ventricle. The fornix then arches anteriorly and inferiorly, the anterior (descending) column of the fornix forming the anterior boundary of the interventricular foramen. Immediately inferior to the foramen the column is split by fibres of the anterior commissure into a precommissural and a postcommissural component.

The transverse fissure is produced by the extension of the cerebral hemispheres posteriorly above the thalamus and midbrain. Superiorly, it is bounded by the fornices and corpus callosum and, inferiorly, by the dorsal surface of the thalamus. Posteriorly, this fissure is continuous with the cleft occupied by the tentorium cerebelli between the cerebellum and cerebrum. The tela choroidea of the third ventricle lies within the transverse fissure. The internal cerebral veins run within it.

THE SULCI, GYRI AND LOBES OF THE CEREBRAL HEMISPHERES

The surface of the cerebral hemisphere is folded into convolutions, gyri, separated by grooves, sulci. The deepest sulci are alternatively called fissures. The pattern of convolutions is very variable both between individuals and even between hemispheres within one individual. Nevertheless, the major sulci and gyri are sufficiently constant to be readily recognisable in all brains and provide important topographical and functional landmarks. Certain sulci and gyri will be identified first (Figure 185). The remaining major sulci and gyri will be described after the lobes of the hemisphere have been defined.

The lateral (sylvian) sulcus is prominent and runs almost horizontally across the central three-fifths of the superolateral surface of the hemisphere near its most lateral extent. It divides the temporal lobe (inferiorly) from the frontal and parietal lobes (superiorly). Anteriorly it has a short stem; this divides into the major, posterior ramus which forms nearly all of the sulcus, and two minor rami which groove the frontal lobe — a horizontal, anterior ramus and a vertical, ascending ramus. The posterior margin of the lesser wing of the sphenoid bone protrudes into the stem of the lateral sulcus. When unqualified, the term lateral sulcus may be taken to mean its stem plus the posterior ramus. On the side of the head, the lateral sulcus is positioned deep to a line extended posteriorly and slightly superiorly for about 70 mm from the pterion, and which then curves superiorly towards the parietal eminence. The main trunk of the middle cerebral artery runs deep within the sulcus on the surface of the insula.

The central (rolandic) sulcus runs obliquely (laterally, anteriorly and inferiorly) across the superolateral surface of the hemisphere, from near the superomedial border of the hemisphere close to its midpoint to just above the lateral sulcus. It is most easily identified by the two gyri which run parallel to it and on either side of it; the precentral gyrus anteriorly and the postcentral gyrus posteriorly (these gyri may be interrupted by small sulci). The precentral and postcentral gyri are limited by the precentral and postcentral sulci. Posteriorly and anteriorly, sulci and gyri course nearly horizontally. The superior end of the central sulcus lies beneath the skull about 10 mm behind the midpoint of the line extending from the nasion to the inion. The sulcus passes laterally, anteriorly and inferiorly at 60 to 70° to the median plane and is more than 80 mm long.

The parieto-occipital sulcus is situated on the medial surface of the hemisphere and runs obliquely, anteriorly and inferiorly from the superomedial border of the hemisphere about 40 mm from the occipital pole, to meet the calcarine sulcus approximately 15 mm behind the splenium of the corpus callosum.

The calcarine sulcus runs anteriorly and slightly superiorly just above the inferomedial border, from close to the occipital pole, to meet the parieto-occipital sulcus. The calcarine sulcus then continues anteriorly and slightly inferiorly (continuing the line of the parieto-occipital sulcus). It ends below the splenium of the corpus callosum, separated from the edge of the hemisphere by the isthmus (a superior continuation of the parahippocampal gyrus). The posterior part of the calcarine sulcus is that part between the junction with the parieto-occipital sulcus and the occipital pole. When the calcarine sulcus extends on to the lateral surface of the hemisphere, it is commonly capped by the small lunate sulcus. The posterior cerebral artery runs posteriorly in the calcarine sulcus.

182A,7
182A,5
180,6
182A,10
176A,28; 182B,34

The corpus callosal sulcus lies between the cingulate gyrus and the corpus callosum on the medial surface of the hemisphere.

182A,20; 182A,19
180,4–6

The cingulate sulcus limits the cingulate gyrus anteriorly and superiorly. The sulcus parallels the callosal sulcus, running from below the genu of the corpus callosum until it turns upwards to the superomedial border of the hemisphere (about 40 mm behind the midpoint of the part of that border between the frontal and occipital poles). At this point, the cingulate sulcus is slightly posterior to the termination of the central sulcus. This latter part of the cingulate sulcus forms the inferior and posterior boundary of the paracentral lobule. The paracentral lobule is the cortex on the medial surface of the hemisphere surrounding the end of the central sulcus. Posteriorly, the cingulate gyrus continues round the splenium of the corpus callosum to the region of the calcarine sulcus, where it is continuous through the isthmus with the parahippocampal gyrus.

182A,21; 182A,19
182A,20; 180,4
176A,10; 176A,27
182A,1
182A,2
182A,19; 180,6
182A,7; 182A,10

The lobes of the cerebral hemispheres (Figure 185) are useful topographically, although their boundaries do not necessarily mark functional divisions.

Each cerebral hemisphere has four lobes: frontal, parietal, temporal and occipital. They are adjacent to the bones of the skull after which they are named, but do not correspond precisely in extent to these bones.

The frontal lobe extends from the frontal pole to the central sulcus. It is separated from the temporal lobe by the lateral sulcus. On the medial surface, its posterior boundary is given by a vertical line joining the central sulcus to the corpus callosum.

176A,10; 176A,2
176A,15
182A,1; 180,5

The temporal lobe is separated superiorly from the frontal and parietal lobes by the lateral sulcus and a horizontal line continuing the main part of the sulcus posteriorly to meet a line joining the parieto-occipital sulcus to the pre-occipital notch. The pre-occipital notch (incisure) is the shallow indentation 40 to 50 mm in front of the occipital pole on the inferolateral border of the hemisphere. The boundary with the parietal lobe on the medial surface of the hemisphere runs along the calcarine sulcus and a line continuing the sulcus to the margin of the hemisphere near the splenium of the corpus callosum. The boundary with the occipital lobe is given by lines joining the two ends of the parieto-occipital sulcus with the pre-occipital notch.

176A,15
176A,30; 176A,26
182A,7
180,6
182A,5; 176A,30; 176A,26

The occipital lobe extends from the occipital pole to the parieto-occipital sulcus on the medial surface of the hemisphere. Its other boundaries are given by lines joining the two ends of the parieto-occipital sulcus to the pre-occipital notch.

176A,27
182A,5
176A,30; 176A,26

The parietal lobe occupies the position between the other lobes as defined above.

The frontal lobe on the superolateral surface of the hemisphere, anterior to the precentral gyrus and precentral sulcus, is divided into inferior, middle and superior frontal gyri. The almost horizontal superior and inferior frontal sulci divide these gyri from each other. The pars triangularis of the inferior frontal gyrus lies between the anterior and ascending rami of the lateral sulcus. The pars opercularis is the part of the inferior frontal gyrus posterior to the ascending ramus. On the medial surface of the hemisphere, the medial frontal gyrus is anterior and superior to the cingulate gyrus and cingulate sulcus. Posteriorly and superiorly, close to the central sulcus, is the anterior part of the paracentral lobule. Posteriorly, inferior to the rostrum of the

176A,3; 176A,4
176A,6; 176A,8
176A,12; 176A,9
176A,11,13; 176A,14
176A,13
182A,16; 182A,19
182A,21; 182A,1
182A,2; 180,3

corpus callosum and close to the lamina terminalis, are the paraterminal gyrus (including the prehippocampal rudiment) and, anterior to it, the subcallosal area (parolfactory gyrus). The subcallosal area lies between the small, vertical, anterior and posterior parolfactory sulci. Inferiorly, the gyrus rectus extends on either side of the medial orbital border of the hemisphere. The olfactory bulb and tract lie over the olfactory sulcus lateral to the gyrus rectus on the inferior surface of the cerebral hemisphere. Posteriorly, at the anterior edge of the anterior perforated substance, the olfactory tract divides into medial, lateral and intermediate olfactory striae. The orbital gyri and sulci of the orbitofrontal part of the frontal lobe are lateral to the olfactory sulcus.

The parietal lobe on the superolateral surface of the hemisphere, posterior to the postcentral gyrus and postcentral sulcus, is divided into the superior and inferior parietal lobules by the almost horizontal, intraparietal sulcus. Anteriorly within the inferior parietal lobule is the supramarginal gyrus which surrounds the upturned end of the posterior ramus of the lateral sulcus. Posterior to it in the inferior lobule is the angular gyrus which surrounds the upturned end of the superior temporal sulcus. On the medial surface, the precuneus is above the cingulate gyrus and its posterior boundary, the subparietal (supracplenial) sulcus. Anterior to the superior continuation of the cingulate sulcus is the posterior part of the paracentral lobule.

The temporal lobe laterally is divided into superior, middle and inferior temporal gyri by the roughly horizontal superior and middle temporal sulci. In the inferior bank of the posterior limb of the lateral sulcus are either one or two transverse temporal gyri (of Heschl) and, more posteriorly, the planum temporale. Inferiorly in the temporal lobe, the inferior temporal sulcus borders the horizontally running lateral occipitotemporal gyrus. The medial occipitotemporal (fusiform) gyrus is parallel to the latter, separated from it by the occipitotemporal sulcus; it is separated from the medial, parahippocampal (hippocampal) gyrus by the collateral sulcus. Anteriorly, and in line with the collateral sulcus, the lateral boundary of the parahippocampal gyrus (and the piriform lobe) is marked by the usually slight, rhinal sulcus. The anterior parahippocampal gyrus turns medially, inferior and lateral to the hypothalamus, forming the uncus. Anterior to the uncus is the anterior perforated substance.

The occipital lobe, on its superolateral surface, has superior and inferior (lateral) occipital gyri separated by the short, horizontal, lateral occipital sulcus. Superiorly, the transverse occipital sulcus runs nearly horizontally towards the superior temporal sulcus. Above it is the caudal part of the arcus (gyrus) parieto-occipitalis. Near the occipital pole, a descending gyrus and polar gyri may sometimes be identified. On the medial surface, the wedge-shaped region between the parieto-occipital sulcus and the posterior part of the calcarine sulcus is the cuneus. Inferior to the posterior part of the calcarine sulcus is the lingual gyrus.

The insula is an area of the cerebral hemisphere found in the floor of the lateral sulcus. It is covered by the banks of the lateral sulcus known as the frontal, frontoparietal and temporal opercula. The frontal operculum is the area triangularis between the anterior and ascending rami of the lateral sulcus. The posterior ramus separates the other two opercula. Within the insula are its long and short gyri. The insula is limited by the circular sulcus, which passes round its edge, and by the limen of the insula, found where it is notched anteriorly at the border between the frontal and temporal lobes. The gyrus ambiens extends from the limen of the insula towards the anterior perforated substance. The insula is also known as the central lobe or island of Reil.

THE CEREBRAL CORTEX

The cerebral cortex is the thin (1.5 to 4.5 mm) outer sheet of grey matter covering the cerebral hemispheres. Because it is highly folded, only about one-third of its total area of approximately $0.22\,m^2$ is visible on the surface of the cerebrum, the remainder being hidden in the sulci and fissures.

On cytoarchitectural and developmental grounds, the cerebral cortex may be divided into three main categories: archi-, paleo- and neo-cortex. Archicortex is mainly three-layered and is found in the hippocampal formation. Paleocortex is inconsistently layered; it is found in the piriform cortex of the olfactory lobe. Neocortex is six-layered. It comprises over 90% of the total human cortex. Because of the consistency of its layering, it is known as isocortex or homogenetic cortex. Archi- and paleo-cortex constitute the allocortex or heterogenetic cortex. The term pallium is sometimes used synonymously with cortex, although developmentally the pallium is the wall of the cerebral hemisphere.

The structure and connections of the neocortex (Figure 186)

Neocortex has a complex structure and details of its connections are not fully known. Only the main features of its organisation and connections will be described (Figure 186a).

From the superficial (pial) surface to the white matter, the six layers of the neocortex and their main component types of neuronal somas are:

I   Molecular layer: rich in fibres, few neurones.

II Outer granular layer: densely packed, small pyramidal, round and star-shaped (stellate) neurones.

III Pyramidal layer: mainly medium-sized pyramidal neurones.

IV Inner granular layer: mainly small stellate, with some small pyramidal neurones.

V   Ganglionic layer: mainly medium-sized and large pyramidal neurones.

VI Multiform layer: spindle-shaped and other neurones.

Layers superficial to layer IV are known as supragranular; layers deep to it are infragranular. Where investigation has been made, pyramidal cells, and stellate cells with many dendritic spines have been found to be excitatory; other stellate cells and basket cells are inhibitory. A variety of other cell types has been described.

Axons and dendrites are found in all layers. Fibres coursing parallel to the pial surface are particularly numerous in layers IV and V, where they form, respectively, the outer and inner bands of Baillarger. Despite the lamination, the primary arrangement of neocortical connections is perpendicular rather than parallel to the cell layers. Accordingly, the neocortex can be considered to be organised into 'columns' running perpendicular to the pial surface through all six layers. Anatomically, columns may be regarded as being based on the connections of corticocortical afferents, or of a large (layer V) pyramidal cell (Figure 186b). The exact size and shape of such columns may differ. For example, in primary visual cortex, afferents bearing information from the ipsilateral eye terminate within lamina IV in strips which alternate with those bearing information from the contralateral eye. These strips are 0.25 to 0.5 mm across and correspond well with 'ocular dominance columns' (within which neurones respond preferentially to stimuli seen by one or other eye). In contrast, 'orientation columns', within which neurones respond specifically to lines of light of a particular orientation, represent strips only 25 to 30 μm in width ('microcolumns'). A distance of 0.5 mm is equivalent to the horizontal spread of the dendritic tree of a large (layer V) pyramidal cell, 25 to 30 μm to the diameter of its soma. In many regions the cortex has been shown, either functionally or anatomically, to be organised in strips or lamellae. Many of these strips run mediolaterally across the cortex.

In a 25 μm by 30 μm block of neocortex passing perpendicular to the pial surface and including all six layers there are 110 ± 10 neurones. Two-thirds of these neurones are pyramidal and one-third are other types. These figures apply to any region of the neocortex and to all mammalian species investigated, including mankind. The sole exception is the binocular region (i.e. most) of the primary visual cortex of primates,

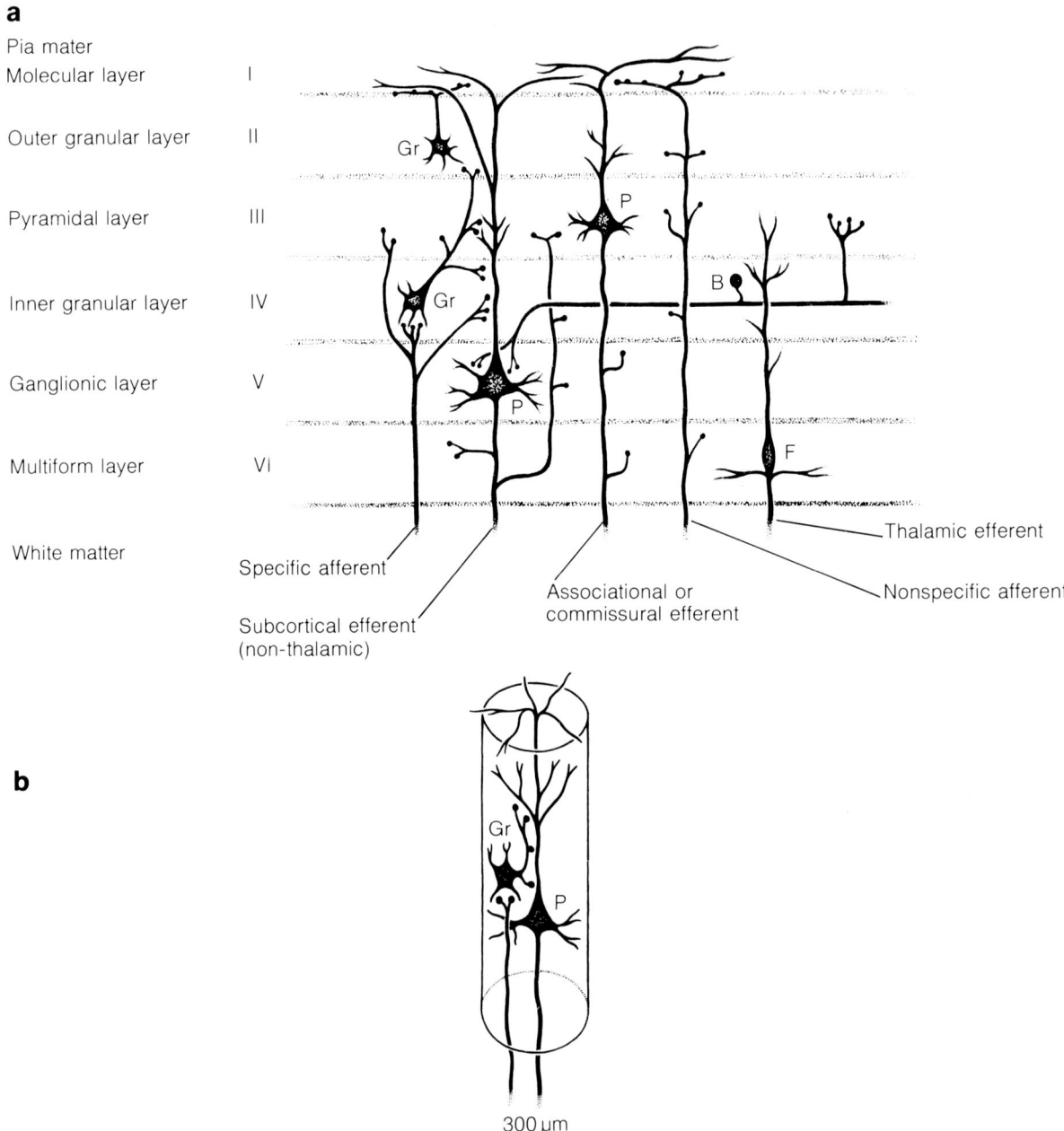

Figure 186 *Main features of the cyto-architecture and connections of the cerebral neocortex.* **a)** *The cortical layers.* **b)** *A cortical column. Neuronal types:* B = *basket,* F = *fusiform,* Gr = *granular (stellate),* P = *pyramidal cells.*

where the number of granule cells is about 2.5 times greater. In the cerebral cortex of both hemispheres there is a total of about 3×10^{10} neurones.

Afferents to the neocortex may be classified as either specific or nonspecific.

Specific afferents come from the specific thalamic nuclei or from other areas of cortex (association and commissural fibres). They terminate chiefly in layers III and IV (Figure 186a). The most numerous type of specific afferent to a given region terminates primarily in layer IV, but also in layer III. The second most numerous type of afferent terminates mainly in layer III, but also in layer IV. In primary sensory areas of cortex the most numerous afferents are thalamic in origin, followed by fibres of cortical origin. In association cortex the two roles are typically reversed. The connections demonstrate precise topographical organisation. As the number of cortical neurones is much larger than the number of thalamic neurones, the thalamocortical projection typically demonstrates considerable divergence. Thalamocortical fibres travel in the internal capsule.

192,27–29

Nonspecific afferents form a very small minority of all afferents. They come chiefly from the brainstem reticular formation, the nonspecific thalamic nuclei (intralaminary, midline and ventralis anterior) and the basal forebrain nucleus. A few fibres come from other areas, including the hypothalamus and the amygdala. Reticular afferents include noradrenergic fibres from the locus ceruleus, 5-hydroxytryptaminergic fibres from the raphe, and dopaminergic fibres from the ventral tegmental area of the midbrain. The basal forebrain nucleus and septal region give rise to cholinergic afferents. Fibres from the intralaminary nuclei of the thalamus chiefly terminate in layers V and VI. In general (though there are specific regional variations), nonspecific afferents terminate in all layers, but predominantly the upper layers (Figure 186a).

Olfactory afferents reach the paleocortex via the olfactory tract. The olfactory system is the only sensory pathway which does not pass via the thalamus and constitutes the only nonthalamic source of specific information to the cerebral cortex.

188A,2

Efferents from the neocortex are either association, commissural or projection fibres. Association and commissural fibres (corticocortical fibres) primarily originate from layer III pyramidal cells. Corticothalamic projection fibres are mainly axons of layer VI cells. Other projection fibres arise chiefly from layer V pyramidal cells. Cortical efferents are usually organised with considerable topographical precision. The great majority of the projection fibres pass through the internal capsule.

192,27–29

The cerebral cortex sends efferents directly or indirectly to all other major regions of the central nervous system. Fibres pass directly to the thalamus, the basal nuclei (although only indirectly, via the caudate nucleus and putamen, to the globus pallidus), the hypothalamus, olfactory and limbic nuclei, the brainstem and spinal cord. The largest number of fibres going to the brainstem terminate on the pontine nuclei; via these nuclei the cortex (indirectly) sends information to the cerebellum. Other fibres to the brainstem terminate in many different areas including the reticular formation, superior and inferior colliculi, red nucleus, dorsal column and cranial nerve nuclei (though input to the vestibular and some other cranial nerve nuclei is indirect). Only about 0.1% of cortical neurones send projections outside the forebrain, i.e. to the brainstem or spinal cord.

Certain efferent projections are common to widespread regions of the cortex. Thus, virtually all cortical regions project to the ipsilateral thalamus, reciprocating their specific input (an exception may be provided by parts of the anterior temporal lobe; Figure 187). Many regions (although particularly motor, somatosensory and visual areas) project to the brainstem reticular formation, and to the amygdala, claustrum, caudate nucleus and putamen of the basal nuclei. Many parts of association cortex send fibres to the ipsilateral pontine nuclei, so influencing the contralateral cerebellum.

Certain other cortical projections are particular to given regions. Thus, motor and somatosensory areas of cortex send fibres to the spinal cord, to cranial nerve and dorsal column nuclei, to the inferior olivary nuclei and to the red nucleus. Auditory cortex projects bilaterally to the inferior colliculi, visual cortex and the frontal eye field to the ipsilateral superior colliculus.

Olfactory and limbic regions of cortex project to the olfactory bulb, and the amygdaloid, septal and hypothalamic nuclei; some of these connections pass via the fornix (see pages 555 to 560).

190,16;180,8

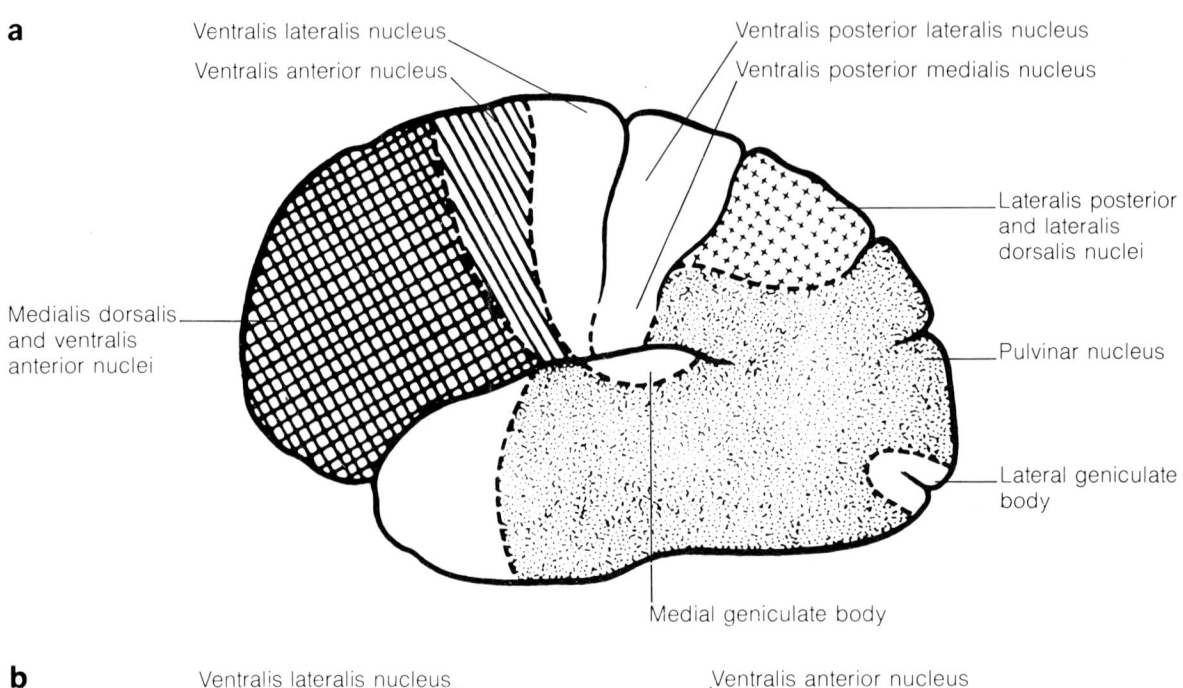
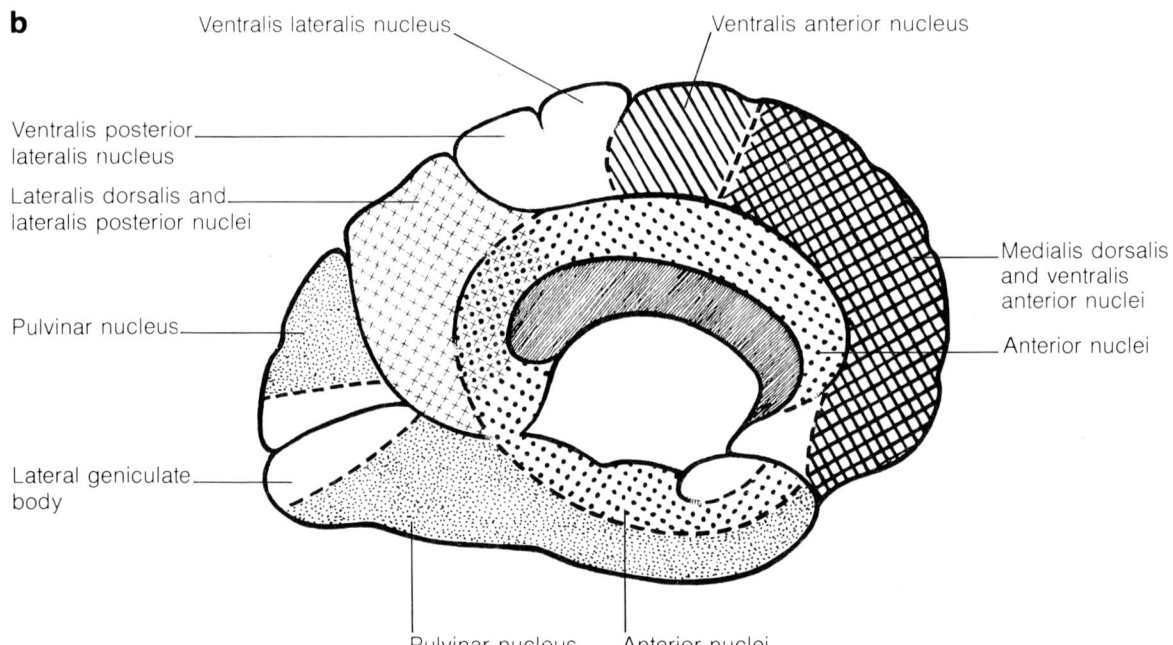

Figure 187 *The main areas of projection of the specific thalamic nuclei upon the cerebral cortex.* a) *Lateral view*, b) *medial view. Note that the projection of nucleus ventralis anterior to the prefrontal cortex is diffuse (nonspecific) rather than specific. Nucleus medialis dorsalis also projects to the medial temporal lobe and the pulvinar nucleus to the frontal eye field (not shown).*

Cortical areas (Figures 188 and 189)

The cortex may be divided both anatomically and functionally into different areas. Histologically based divisions correspond in many cases with functional areas, although current knowledge allows no complete correspondence to be made. The numbered regions of Brodmann's cyto-architectonic map of the cortex (Figure 189) provides convenient alternative designations for certain major functional areas (Figure

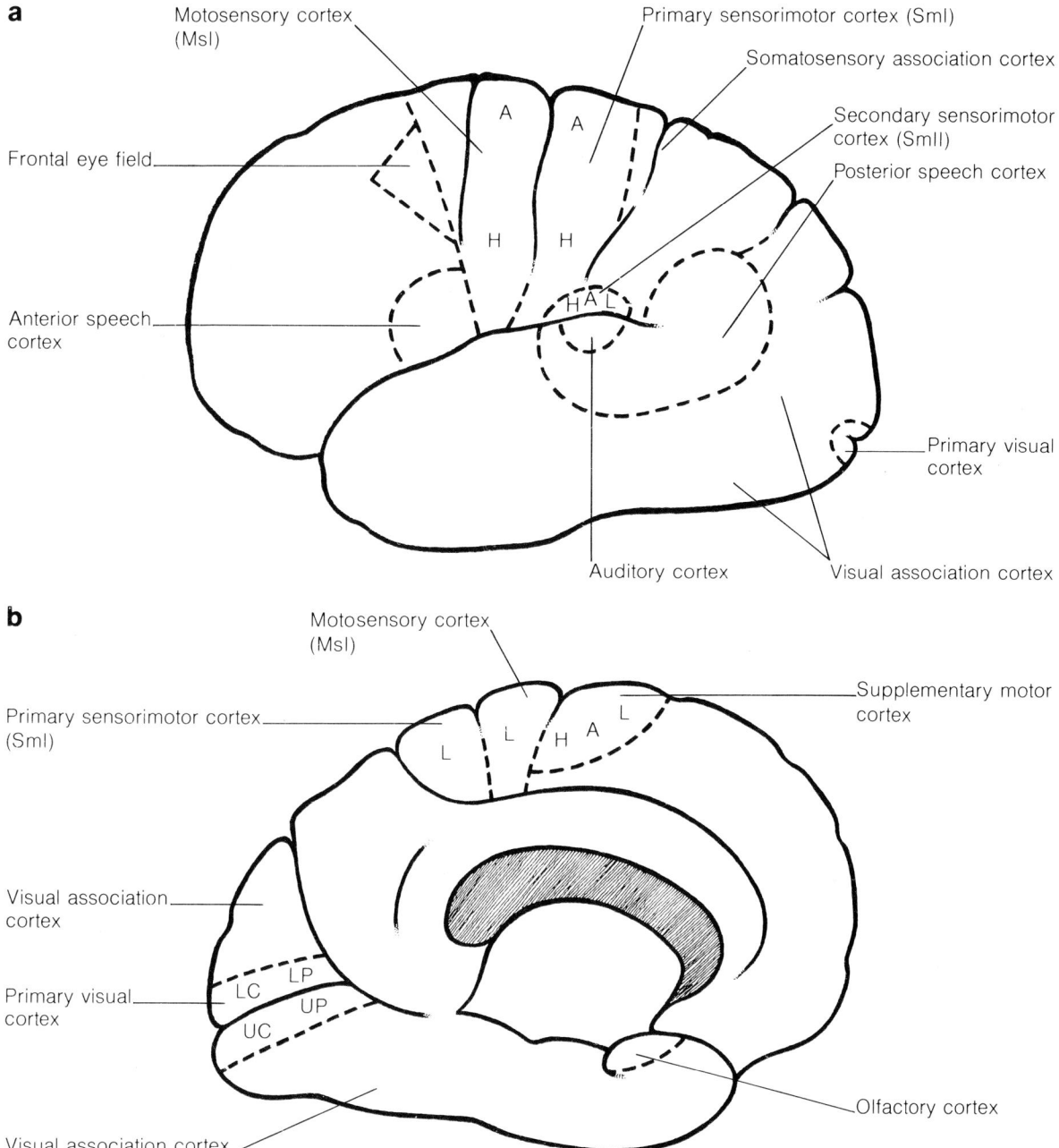

Figure 188 *The main functional areas of the cerebral cortex.* a) *Lateral view,* b) *medial view. Somatotopical representation:* H = *head,* A = *arm,* L = *leg. Representation of the contralateral halves of the visual fields:* LC = *lower central,* LP = *lower peripheral,* UC = *upper central,* UP = *upper peripheral.*

188). The sulci and gyri do not in general correspond closely with functional areas, although the major sulci bear constant relation to the primary sensory and motor areas.

The variations in cyto-architecture and function from region to region of the isocortex reflect variation in afferent and efferent connections. As mentioned above, the numbers of pyramidal and non-pyramidal cells remain constant (although this does

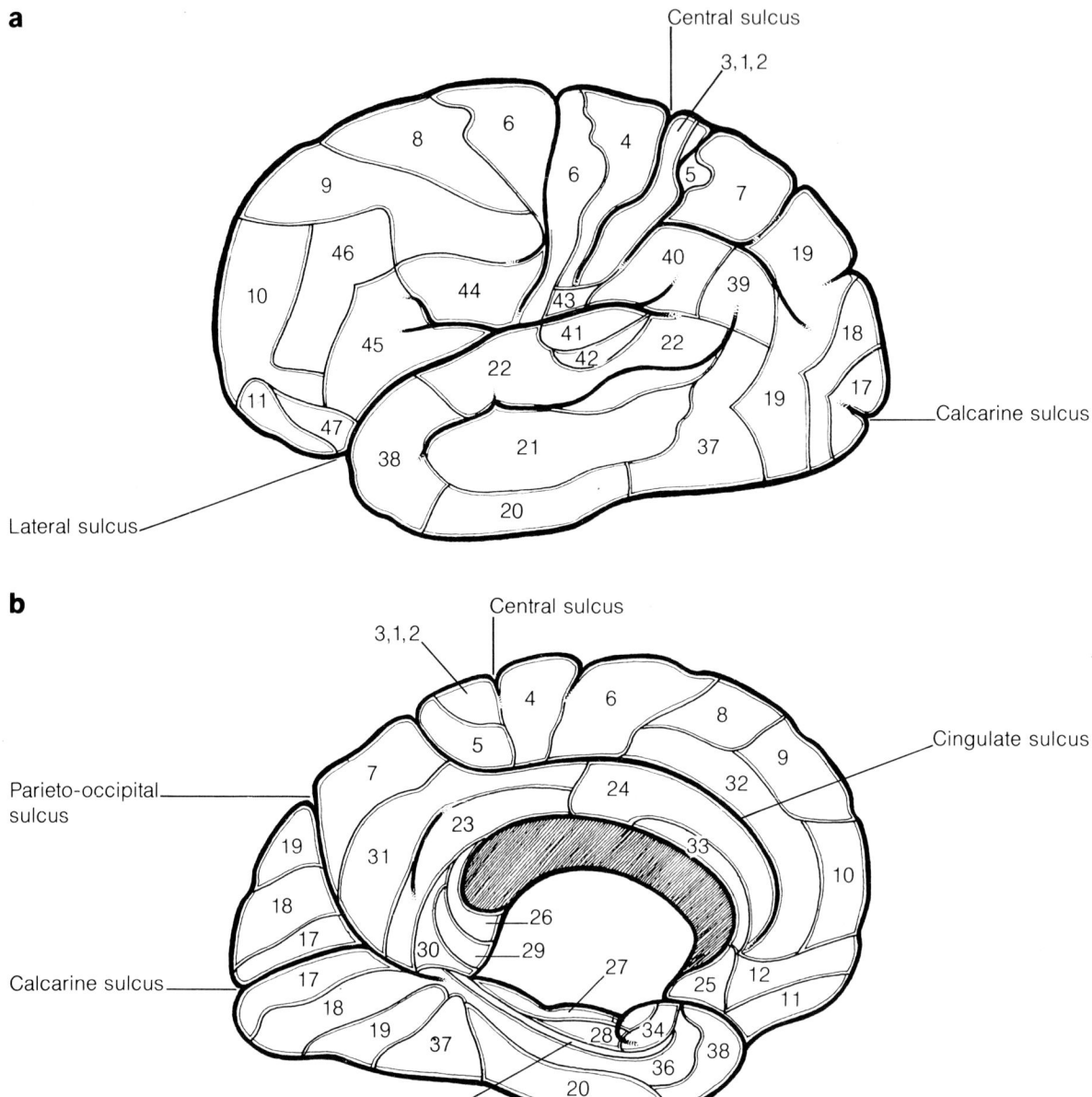

Figure 189 *Cyto-architectonic areas of the cerebral cortex as numbered by Brodmann.* a) *Lateral view*, b) *medial view*.

not preclude other differences). The primary sensory areas of the neocortex are characterised by highly developed granular layers (layers II and IV); this type of cortex is called granular or koniocortex. Motor cortex is characterised by well-developed pyramidal cell layers (layers III and V); it is (inappropriately) known as agranular cortex. Motor cortex is thicker (4.5 mm) than sensory cortex (1.5 to 2 mm). On the basis of the increasing prominence of the granular layers, five types of cortex are distinguished: agranular, frontal, parietal, polar and granular.

Motor areas of the cerebral cortex are all found in the frontal lobe, mainly in the precentral cortex; i.e. Brodmann's areas 4, 6 and 8.

Motor cortex (Brodmann's area 4) is a triangular strip of cortex occupying most of the precentral gyrus superiorly and tapering inferiorly. It extends into the anterior part of the paracentral lobule on the medial surface of the hemisphere. Its posterior limit is at the bottom of the central sulcus. Area 4 is characterised by some very large neurones (Betz cells) in layer V. These cells give rise to about 30,000 of the more than one million corticospinal fibres; the remainder arise from other layer V cells in areas 4, 6 and the parietal lobe. These areas also give rise to corticonuclear fibres.

176A,3; 172,6
182A,2
176A,2

The premotor cortex (occupying most of Brodmann's area 6) is a further triangular strip of cortex tapering inferiorly. It is immediately in front of the motor cortex on the precentral gyrus and extends into the superior and middle frontal gyri, but not on to the medial surface of the hemisphere.

176A,3; 176A,5+7

The motor and premotor areas are collectively known as motosensory area I (MsI). The name is appropriate as stimulation of this cortex may evoke sensations as well as movements (correspondingly, stimulation of sensorimotor cortex may evoke movements as well as sensations).

The precentral motor area (MsI) is concerned with voluntary movements of the opposite half of the body. Its afferents include those bearing proprioceptive and other information from somatosensory cortex. It receives cerebellar and basal nuclear input via nuclei ventralis lateralis and ventralis anterior of the thalamus (Figure 187). The precentral motor area is a major source of corticonuclear and corticospinal fibres (see pages 602 to 603). The upper and medial part of the area is concerned with movements of the lower limb and perineum. The lower and lateral parts are concerned successively with movements of the trunk, upper limb, neck and head (Figure 190a). The paracentral lobule is also involved in the voluntary control of micturition and defaecation. This topographical arrangement, the 'motor homunculus', is distorted such that parts of the body with a high density of innervation peripherally occupy a large area of cortex. There is extensive overlap rather than a sharp boundary between adjacent areas. Overall, the motor homunculus is inverted (i.e. head inferior, leg superior) and contralateral. However, the head region itself is not inverted and stimulation evokes bilateral movements of the soft palate, laryngeal, lower facial and, usually, masticatory and certain tongue muscles (see page 501). Cortex dealing with movements of distal limb muscles is found near the central sulcus; for trunk muscles it is anterior. Lesions result in impairment (either paresis or spastic paralysis of the upper motoneuronal type; see page 459) of voluntary movements, chiefly of the contralateral side and particularly for discrete, skilled movements of the distal limbs.

182A,2

The supplementary motor cortex occupies those parts of Brodmann's areas 6 and 8 found on the medial surface of the hemisphere in the medial frontal gyrus. It extends for 40 to 50 mm anterior to the paracentral lobule. This area is the second motosensory area (MsII), although other areas are sometimes erroneously given this name. Its connections parallel those of MsI, with which it is closely interconnected.

182A,18
182A,2

The functions of the supplementary motor area are poorly understood. Lesions result in forced grasping and some slowness of movements on the opposite side. They may also cause speech problems. There is probably a second motor homunculus within this area, with the posterior cortex being concerned with movements of lower limb muscles (adjacent to the corresponding lower limb region of MsI), the anterior cortex being concerned with head muscles, and the upper limb region being intermediate. Some recent evidence suggests that neurones in this area are activated during contemplated as well as during performed movements. Neurones in MsI are activated only when movements are performed.

The frontal eye field occupies parts of Brodmann's areas 6, 8 and 9 on the middle frontal gyrus adjacent to the precentral gyrus. It receives fibres from nuclei medialis dorsalis and pulvinar of the thalamus, and from visual cortex. It sends fibres to the ipsilateral superior colliculus (see Figure 177b). Voluntary and normal scanning movements of the eyes are controlled in this region, which corresponds to the area of cortex of the motor homunculus where movements of the eye muscles are represented. Lesions temporarily impair voluntary eye movements to the contralateral side, and

176A,7; 176A,3

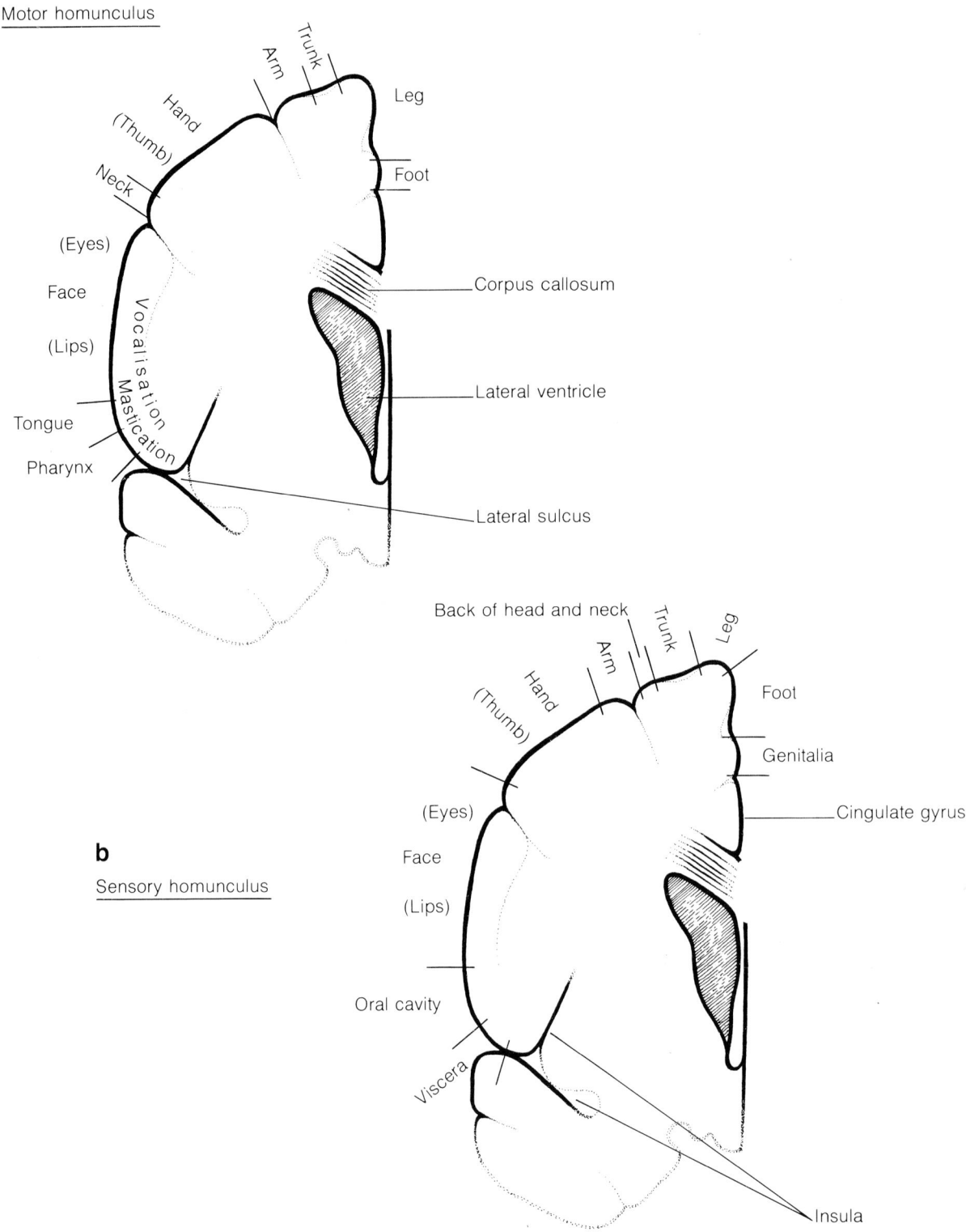

Figure 190 *Somatotopical representation within the primary motor cortex (**MsI**) and the primary somatosensory cortex (**SmI**).* **a**) *A section through the precentral gyrus illustrating the motor homunculus.* **b**) *A section through the postcentral gyrus illustrating the sensory homunculus.*

also disrupt the pattern of scanning of visual scenes; they leave reflex movements of the eyes unimpaired.

Anterior speech cortex (Broca's area) occupies Brodmann's area 44 and part of area 45 of the inferior frontal gyrus of the dominant (usually left) hemisphere, adjacent to the precentral gyrus. It thus lies immediately in front of parts of the cortex concerned with movements of the mouth, tongue and pharynx, and with vocalisation (see below under auditory and speech areas).

Somaesthetic areas are found in the parietal lobe, mainly on the postcentral gyrus.

The primary somatosensory cortex (first sensorimotor area: SmI, Brodmann's areas 3, 1, 2) occupies almost all of the postcentral gyrus and extends medially into the posterior portion of the paracentral lobule. A second area of somatosensory cortex, sensorimotor area II (SmII), is found inferior to the primary area (SmI). It occupies the buried cortex of the frontoparietal operculum, close to the insula on the superior wall of the lateral sulcus. Somatosensory association cortex is located immediately posterior to SmI in area 5 within the superior parietal lobule.

Proprioceptive and exteroceptive somatosensory information is relayed to the somatosensory areas (SmI and SmII) from the dorsal column/medial lemniscal, anterolateral and trigeminal systems via nucleus ventralis posterior of the thalamus. Fibres reach the cortex via the somatosensory radiation of the posterior limb of the internal capsule. Nucleus ventralis posterior medialis projects to the inferior postcentral gyrus (head region), nucleus ventralis posterior lateralis to superior parts (and to SmII). The sensory homunculus of SmI broadly parallels the motor homunculus of MsI and is similarly distorted, so that areas of the body of high sensitivity occupy large areas of cortex (Figure 190b). Sensations from the perineal region and lower limb are dealt with superiorly and medially, particularly in the paracentral lobule. Moving laterally and inferiorly, the cortex deals (in order) with sensations from the trunk, neck, back of head, upper limb, face, oral cavity, pharynx and abdominal cavity. Sensations originate mainly contralaterally, but come from both sides of the larynx and perineum, and from the ipsilateral oral region. There may be four separate and parallel representations of the body (i.e. four homunculi) within SmI. Area 3 receives mainly information concerning light tactile stimuli, area 1 signals from cutaneous and deep tissues, and area 2 signals from deep tissues and joints. Area 3a, at the border of areas 3 and 4 deep in the central sulcus, processes information from the primary afferents of muscle spindles. Area SmII is also somatotopically organised, the face being represented anteriorly (close to the face region of SmI) followed by the upper limb, with the lower limb being posterior. Some parts of the ipsilateral as well as the contralateral body are represented in SmII. The somatosensory pathways are summarised on pages 598 to 601.

Sensations of pain are not evoked on stimulating SmI, and neuronal responses to nociceptive stimuli are rare. Similar findings apply to SmII. However, an area immediately posterior to SmII may receive nociceptive information from the posterior thalamus. Lesions of SmI typically lead to contralateral sensory loss, although, commonly, poorly localised appreciation of pain and touch remain. Usually there is progressive recovery, especially for painful sensations, but proprioceptive, vibratory and fine, discriminative touch sensations are likely to be permanently impaired.

The four somatosensory and motor areas (SmI and SmII, and MsI and MsII) are intimately interconnected, and also with their counterparts in the opposite cerebral hemisphere. Their afferent and efferent connections overlap (although differing, in particular, in thalamic input). All areas give rise to corticonuclear and corticospinal fibres, although most arise from MsI. These fibres pass through the posterior limb or genu of the internal capsule. Somatosensory cortex projects mainly to sensory nuclei of the brainstem and the dorsal horn of the spinal cord, motor areas to motor nuclei and the ventral horn. It has been suggested that somatosensory association cortex (area 5) is important in initiating or directing purposeful movements. It receives fibres from SmI and sends fibres to MsI, as well as to the brainstem and spinal cord.

Gustatory signals originating in the contralateral half of the tongue are relayed to the cortex via nucleus ventralis posterior medialis of the thalamus. They are processed in SmI in a region (within area 43) immediately inferior and anterior to that which deals with somatic sensations from the tongue. There is probably a second taste area in SmII, near the insula (see also pages 596 to 597).

Vestibular representation is most probably in a small posterior part of the face region of SmI, on the postcentral gyrus close to the intraparietal sulcus. The region is within area 2 which receives proprioceptive input from muscle receptors. There may also be vestibular areas within area 3a, and on the superior temporal gyrus anterior and lateral to the primary auditory cortex. Vestibular information reaches the cortex from nuclei ventralis posterior inferior and lateralis, and nucleus ventralis lateralis of the thalamus. It may additionally influence the cortex via cerebellar and reticular connections (see also pages 594 to 595).

Visual areas are primarily found posteriorly and occupy most of the occipital lobe.

The primary visual cortex (Brodmann's area 17 or striate cortex) is found in parts of the cuneus and lingual gyrus surrounding, and in the walls, of the posterior part of the calcarine sulcus. Striate cortex is usually confined to the medial surface of the hemisphere but may extend laterally as far as the lunate sulcus. It is known as striate cortex because of the prominent (indeed macroscopically visible) inner band of Baillarger, the visual stria (of Gennari or Vicq d'Azyr). Striate cortex is surrounded by parastriate cortex (area 18), except anteriorly. In turn, area 18 is similarly largely surrounded by peristriate cortex (area 19). Both area 18 and area 19 have visual functions and are sometimes referred to as VII and VIII, area 17 being VI. Visual cortex (areas 17, 18 and 19) occupies virtually the whole of the occipital lobe, extending into the parietal lobe medially and superomedially, and into the temporal lobe along the lingual gyrus. Other visual areas of the cortex are found in the temporal lobe, notably in the middle and inferior temporal gyri.

Areas 17, 18 and 19 are closely interconnected. They also have commissural connections, though area 17 is unusual in having few of these. Visual information reaches area 17 primarily from the lateral geniculate body, though some fibres also come from the pulvinar. Visual signals reach areas 18 and 19 from area 17 and from the pulvinar, but there may additionally be some fibres from the lateral geniculate body. Visual input to the pulvinar comes from the superior colliculus and from areas 17, 18 and 19 (although some retinal fibres may also terminate there). The projections from areas 17, 18 and 19 (the occipital eye field) to the ipsilateral superior colliculus and pretectal area control visual reflexes such as fixation and accommodation. Connections to the frontal eye field make possible voluntary turning of the eyes towards visual stimuli (see also Figure 177b and pages 547 to 549).

A complete retinotopic map is found within primary visual cortex. Visual information is received from the ipsilateral halves of both retinae (i.e. the contralateral half of the visual field of each eye) via the lateral geniculate body and the optic radiation (geniculocalcarine tract) of the retrolenticular part of the internal capsule. The course of the optic radiation is further described on pages 565 to 566 (see also pages 590 to 591). Upper parts of the retina project above the calcarine sulcus, lower parts below, so that the cortical map of the visual world is upside down (Figure 191). The macular region of the retina (i.e. that responsible for the central 6° of vision) projects to the posterior one-third of area 17, near the occipital pole. The peripheral retina (including the monocular, nasal segment) projects anteriorly, nearer to the parieto-occipital sulcus. A disproportionately large area of cortex is devoted to central vision. The vertical meridian is represented bilaterally along the boundary between areas 17 and 18. A further retinotopic map is found in area 18. Perception of how far away visual stimuli are (i.e. stereoscopic depth perception) is accomplished in area 18. A considerable number of areas specialised for processing different aspects of visual

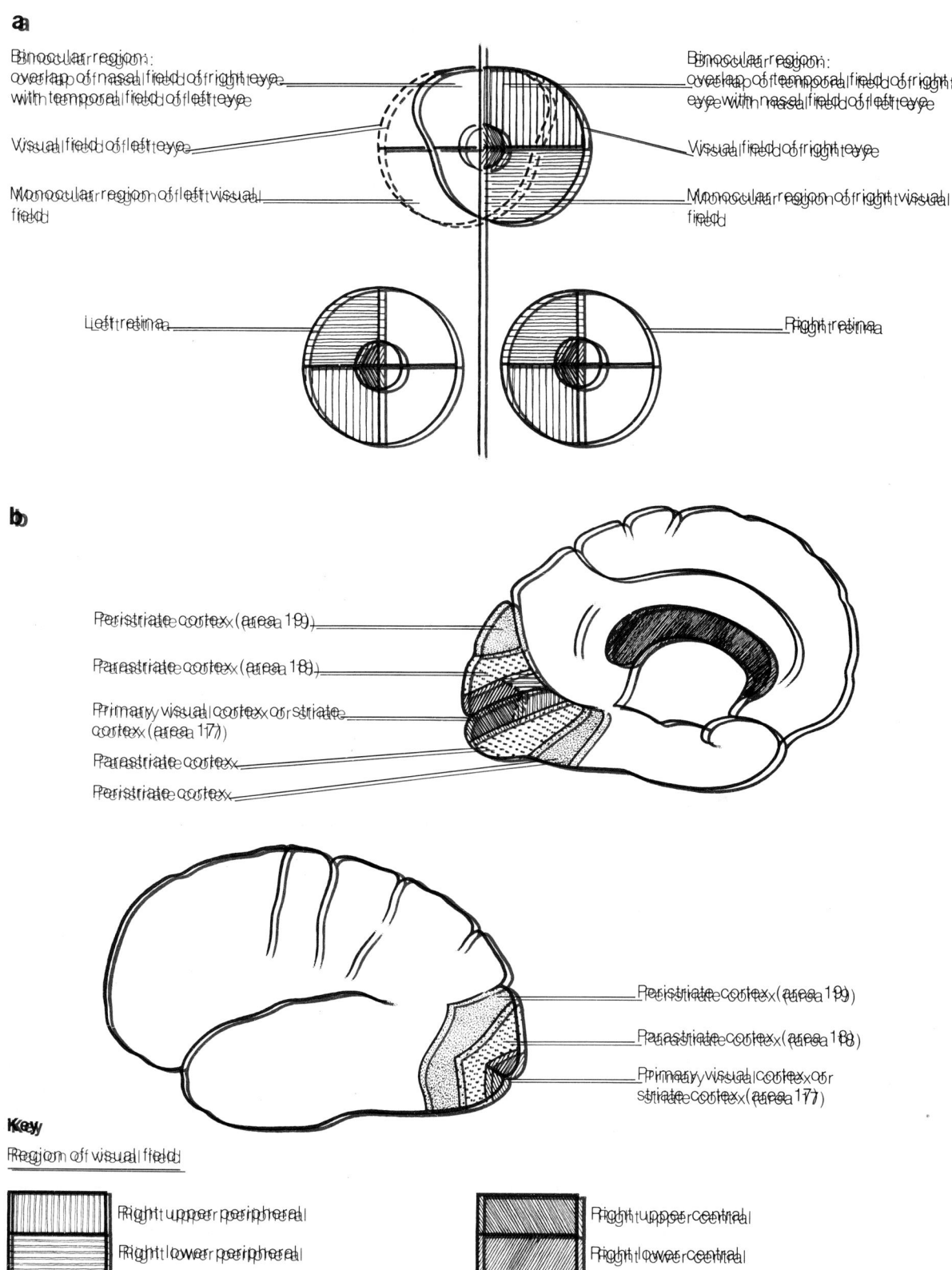

Figure 19.1 Retinotopic projections within the visual pathway. The projections shown are of different parts (see key) of the nasal field of the left eye and the temporal field of the right eye upon a) the two retinae, and b) the primary visual cortex of the left hemisphere; medial view above, lateral view below. The extent of parastriate and peristriate cortex is also shown.

stimuli (e.g. colour and motion) have been described in area 19. Visual association areas within the temporal lobe appear to be particularly important in feature detection, i.e. for establishing *what* a visual stimulus is; areas near to the parietal lobe seem to be necessary for establishing *where* a stimulus is.

In layer IV within area 17, afferents signalling information from the ipsilateral eye end in irregular strips which alternate with afferents carrying information from the contralateral eye. These strips are 0.25 to 0.5 mm across and correspond to 'ocular dominance columns' (found physiologically by determining which eye evokes stronger neuronal responses). Information from both eyes is integrated in area 17; the neurones are usually binocularly responsive. Most neurones respond selectively to bars of light of a particular orientation, shown within a restricted part of the visual field. Cells responsive to the same orientation of stimulus are found in strips of cortex ('orientation columns'). Orientation columns covering the full set of orientations (18 columns at 10° per column) occupy strips about 0.5 mm across. These strips are neither parallel nor perpendicular to the ocular dominance columns. Neurones responsive to stationary bars of light are called 'simple cells'; if movement of the bar is required they are 'complex cells'. 'Hypercomplex cells' require additional features, such as bars of a particular length or angles or corners, before they respond maximally. Most simple cells are layer IV granule cells. Complex and hypercomplex cells are usually pyramidal neurones. Most neurones within area 17 are not sensitive to the colour of stimuli (i.e. they are monochromatic contrast detectors), though there are regular, interspersed patches of colour-selective cells. In visual association cortex, neurones respond selectively to increasingly complex and specific features of visual stimuli and no simple cells are found. Nevertheless, much visual information processing occurs in parallel and not in a serially hierarchical manner.

Lesions of area 17 result in loss of conscious awareness of visual sensations (i.e. blindness). The area of loss (scotoma) is for corresponding (homonymous) parts of the contralateral visual fields. Complete lesions result in contralateral homonymous hemianopia. In some patients it has been possible to demonstrate some slight residual (though not conscious) visual function, such as the ability to point to spots of light falling within the area of scotoma (so-called 'blind-sight').

Auditory and speech areas are found in the temporal, parietal and frontal lobes. Auditory cortex is in the temporal lobe. Speech cortex usually occupies chiefly the left hemisphere.

The primary auditory cortex (AI; mainly Brodmann's area 41, but possibly including parts of areas 42 and 52) is found on the temporal operculum, in the inferior bank of the lateral sulcus opposite the end of the postcentral sulcus. It occupies a few square centimetres on the transverse temporal gyri (of Heschl) and, on the left, extends posteriorly just into the planum temporale. Within this region there is probably a tonotopic organisation with high frequencies being represented posteromedially, and low frequencies anterolaterally. Several other auditory areas lie adjacent to AI, on the temporal operculum and superior temporal gyrus. Auditory association cortex surrounds primary auditory cortex and extends on to the superior temporal gyrus, particularly into area 22. The ventral, laminated part of the medial geniculate body projects via the auditory radiation (geniculotemporal tract) in the sublenticular part of the internal capsule to primary auditory cortex (called the 'core projection'). Remaining parts of the medial geniculate body plus neighbouring posterior thalamic nuclei project to the surrounding regions (the so-called auditory 'belt projection'). All these regions project back to the thalamus and to both inferior colliculi (see also pages 592 to 593). Lesions of auditory cortex do not lead to complete deafness unless they are bilateral. Irritative lesions of this area may lead to auditory sensations.

Language areas occupy large parts of the cerebral cortex. In about nine out of every ten adults language representation is extensive in the left hemisphere and minor in the right hemisphere. For most of the remaining 10% the roles of the hemispheres are reversed, but 1 to 2% of the population have extensive bilateral representation. The left hemisphere is dominant for language in over 90% of right-handed individuals and in about 60% of left-handed or ambidextrous individuals.

Posterior speech cortex (sometimes called Wernicke's area) includes parts of the superior and middle temporal gyri, and extends superiorly into the inferior parietal lobule. This area is primarily concerned with the perception and understanding (sensory side) of speech. It receives fibres from auditory cortex and the pulvinar nucleus of the thalamus. Most characteristically, lesions give rise to receptive (sensory) aphasia, with speech itself remaining comparatively fluent.

176A,16; 176A,18
176A,31
188B,43

Anterior speech cortex (Broca's area) occupies parts of the inferior frontal gyrus (including its pars opercularis and pars triangularis) immediately in front of those parts of area MsI concerned with vocalisation. It includes area 44 and parts of area 45. It is primarily concerned with the production (motor side) of speech. It receives inputs from nucleus medialis dorsalis of the thalamus and from posterior speech cortex. Typically, lesions give rise to expressive (motor) aphasia, characterised by non-fluent speech and grammatical confusions (e.g. difficulties with prepositions, endings and passive sentences), in the absence of paralysis of the musculature involved in speech production.

176A,9
176A,14; 176A,12

Division of sensory and motor function between anterior and posterior speech areas is incomplete and in the intact brain these areas intercommunicate, so that the effects of lesions upon language function are often complex. Further, damage to subcortical white matter, the internal capsule or parts of the thalamus (e.g. left pulvinar nucleus) can also give rise to language defects.

192,27–29; 188B,43

The superior speech cortex corresponds to the supplementary motor area (see above). Damage to this area is far less devastating in its consequences for speech than that of the anterior or posterior speech areas.

Olfactory areas are found in the anteromedial temporal lobe and in the orbitofrontal cortex.

The primary olfactory cortex is a small region, the prepyriform and peri-amygdaloid areas, just anterior to the uncus of the parahippocampal gyrus. Olfactory association cortex occurs in the anterior parahippocampal gyrus and the posterolateral quadrant of orbitofrontal cortex (further details are given under the olfactory lobe and olfactory system on pages 555 to 556). Bilateral damage to primary olfactory cortex causes anosmia. Epileptic discharges in this region (uncinate fits) can give rise to imagined unpleasant smells.

182A,13; 182A,10

Association cortical areas occupy a particularly high proportion of the cerebral cortex in the human brain. The primary somatosensory, visual, auditory and olfactory and motor areas account for only 5% of the cortical surface. The remaining 95% is association cortex.

Adjacent to each primary sensory cortical area are regions of sensory association cortex which receive connections from the primary area and further process the sensory information. Lesions commonly give rise to more complex perceptual disturbances without total loss of sensory awareness. Large regions of the temporal, parietal and frontal cortex remain outside the primary sensory association areas. These regions are concerned with the integration of information from different sensory modalities (polysensory association cortex) and/or with high order and abstract mental processes. Their functions include mnemonic as well as perceptual and intellectual capacities. Understanding of the precise functions of these areas is limited and arises almost entirely from study of the effects of lesions. Some deficits arise because of interruption of the passage of information rather than damage to the site at which it should be processed. For example, if damage to the cortex or subcortical white matter prevents visual information reaching posterior speech cortex, the patient will be unable to read even if other language functions are intact. Such deficits are called disconnection syndromes.

The speech dominant hemisphere is the more important for language, arithmetic and analytical abilities. The non-dominant (usually right) side is more important in non-verbal skills such as visuospatial, synthetic and intuitive abilities. It is better at three-dimensional relationships, recognising faces, and musical talents. However, such functions are not totally confined to either the left or the right cerebral hemisphere and, in the intact brain, both hemispheres are in close communication and co-operate to produce understanding.

Prefrontal cortex is the association cortex of the frontal lobes and lies anterior to areas 4, 6 and 8. It is highly developed in the human and its functions are believed to include some of our species' highest attributes. Unilateral (usually left) lesions of anterior speech cortex produce motor aphasia and/or agraphia (inability to write). Elsewhere, the effects of unilateral lesions are typically slight compared to those of bilateral damage. During their period of popularity, prefrontal lobotomies and leucotomies (undercutting the white matter of prefrontal cortex) performed on psychotic patients produced large numbers of subjects with bilateral removals. Such removals commonly produce personality changes and a reduction in attentive capacities, initiative, foresight and appreciation of the future consequences of actions; ethical standards are lowered and social interaction compromised. Self-consciousness is typically reduced, while the patient becomes self-centred, present-centred and often puerile in behaviour. Some of these operations were performed for the relief of intractable pain. After operation, pain is still felt, but it no longer causes anxiety or distress. Cutting the association fibre bundle of the cingulum beneath the anterior cingulate gyrus has a similar effect on the appreciation of pain. Superolateral prefrontal damage results primarily in impairments of complex intellectual capabilities. Lesions of orbitofrontal cortex result chiefly in disturbances of emotional and social behaviour. Extensive bilateral lesions of inferior parts of the medial frontal lobe may produce amnesia. Bilateral lesions of the anterior cingulate gyrus may cause increased tameness.

Parietal association cortex is involved in the integration and interpretation of somatosensory, visual and auditory signals. Lesions may produce impairment of linguistic ability (aphasia) and/or skilled acts (apraxia), and/or awareness, knowledge and recognition (agnosia). These impairments occur in the absence of muscular paralysis or primary sensory losses. Although deficits vary with the site of the lesion, in general, knowledge about functional localisation is rudimentary. Lesions of the superior parietal lobule may produce contralateral inco-ordination and some muscular wasting. Lesions of the posterior speech cortex in the inferior parietal lobule of the speech dominant hemisphere (usually the left) cause receptive aphasia. Such aphasia can occur in more than one form. For example, visual (reading) and/or auditory (spoken) verbal material may be affected; there may be problems with naming objects and/or understanding the meanings of words. Lesions of the inferior parietal lobule may also cause difficulties in recognising objects by touch, or in understanding the uses of objects (such as a pen). These deficits are more commonly noticed when the damage is to the non-dominant hemisphere. Parietal lesions can also result in the neglect of sensory information from the contralateral side of the patient's own body and from the opposite half of visual space. Thus, the patient's body image is defective, there being a loss of awareness for the contralateral half of the body. The loss may extend to insight into the person's own disabilities (anosognosia), including denial of blindness or paralysis. With damage to the non-dominant hemisphere there may be deficient appreciation of spatial relationships between objects. Lesions of the white matter deep to the parietal lobe may interrupt superior fibres of the optic radiation, producing defects of the contralateral lower visual fields.

Occipital association cortex is almost entirely visual in function. Damage may produce disturbances of the visual reflexes of fixation, accommodation and convergence. Other defects, particularly noticeable if the lesion extends into the inferior parietal lobule, may include visual perceptual disturbances, loss of reading ability (alexia) and failure of visual recognition (visual agnosia).

Temporal association cortex is involved in both auditory and visual perception of complex stimuli (including voices and faces), in both written and spoken language (within parts of posterior speech cortex), and in particular types of longer-term memory function. Bilateral temporal lobectomies, including removal of large parts of the hippocampal region and the amygdala, give rise to the Klüver–Bucy syndrome: visual agnosia ('psychic blindness') in which there is no recognition of the significance of objects that are seen, examination of objects by mouth, exaggerated attention to visual stimuli, and changes in emotional, dietary and sexual habits. There is also likely to be anosmia. Such lesions, unless restricted to the anterior temporal lobe, also commonly result in profound memory disturbances (see page 561). Interference with the inferior fibres of the optic radiation within the white matter of the temporal lobe (the temporal loop of Meyer) results in visual field defects (commonly approximating a contralateral homonymous upper quadrantanopia).

The insula lies in the floor of the lateral sulcus. The functions of the insular cortex are obscure. Close to it are parts of sensory and motor cortex concerned with the viscera and taste. It has been suggested that the insula is concerned with visceral sensations, control of the alimentary tract, respiration, blood pressure and the autonomic nervous system. The limen of the insula contains olfactory cortex. Olfactory fibres also pass beneath it, as do fibres of the uncinate fasciculus which interconnects orbitofrontal and anterior temporal cortex.

The olfactory lobe and olfactory system (including paleocortex)

The olfactory lobe comprises structures related to the olfactory bulb and its connections. These structures are several, but mainly small in the human brain. Some of these structures are limbic, rather than primarily olfactory in function. The term rhinencephalon should be used synonymously with olfactory lobe, but it is inconsistently defined by different authorities, and in older texts includes much of the limbic system. The major components of the olfactory lobe are the olfactory nerve, bulb, tract and striae with their attendant grey matter, plus the pyriform lobe (the paleopallium). The pyriform lobe will now be defined before the other components of the olfactory system are described.

The pyriform (piriform) lobe mainly comprises parts of the parahippocampal gyrus medial to the rhinal sulcus. It is made up of the uncus, the entorhinal area, the peri-amygdaloid area, the prepyriform cortex with its adjacent lateral olfactory stria, and is usually taken to include the cortex of the anterior perforated substance.

The prepyriform cortex consists of the lateral olfactory gyrus (the grey matter covering the lateral olfactory stria as it forms the anterolateral boundary of the anterior perforated substance) and the gyrus ambiens. The gyrus ambiens lies lateral to the lateral olfactory stria and the peri-amygdaloid area, and extends into the limen of the insula. (The gyrus ambiens is sometimes considered to be part of the lateral olfactory gyrus.)

The peri-amygdaloid area (gyrus semilunaris or cortical amygdaloid nucleus) is at the anterior boundary of the uncus, posterior and inferior to the anterior perforated substance. It is anterior and superior to the main nuclear mass of the amygdala.

The entorhinal area (Brodmann's area 28; Figure 189b) is the main part of the anterior parahippocampal gyrus. It is medial to the rhinal sulcus and lateral to the uncus.

The uncus is subdivided into the main, uncinate gyrus and the small and posterior, band of Giacomini and intralimbic gyrus.

The rhinal sulcus marks the boundary between neocortex and the paleocortex of the pyriform lobe. Paleocortex is irregularly layered. It has six layers in the entorhinal region, but elsewhere has three or four layers or is not clearly laminated.

The olfactory system comprises the central connections of the olfactory (first cranial) nerve. The axons (fila olfactoria) of the primary olfactory neurones (olfactory rods) form the olfactory nerve which ends in the olfactory bulb.

188A,1; 168,2

The olfactory bulb is a laminated structure containing several types of intrinsic neurones. In it the fila olfactoria terminate in synaptic glomeruli within each of which many thousands of the fila olfactoria make contact with a few tens of mitral cells (see pages 588 to 589). Mitral cells provide the chief projection to olfactory cortex. This projection is ipsilateral and does not pass via the thalamus. Periglomerular and tufted cells also send processes to the glomeruli. Periglomerular cells inhibit the mitral cells of neighbouring glomeruli, so providing lateral inhibition. The granule cells, which are axonless, provide feedback inhibition to the mitral cells via dendrodendritic synapses. The olfactory bulb receives centrifugal fibres from a variety of regions including: the olfactory cortex, the nucleus of the olfactory tract, the amygdala and monoamine (noradrenaline, dopamine and 5-hydroxytryptamine) cell groups of the brainstem reticular formation.

188B,30; 168,3
188B,32

188B,31

180,33

The olfactory tract passes posteriorly from the olfactory bulb to the anterior perforated substance where it divides into three olfactory striae: medial, intermediate and lateral. The area of broadening of the olfactory tract before it divides is the olfactory trigone (olfactory pyramid). The olfactory tract contains the axons of mitral cells plus some tufted cells, centrifugal axons and scattered cells of the nucleus of the olfactory tract (anterior olfactory nucleus). The nucleus of the olfactory tract sends axons to the ipsilateral olfactory bulb and, via the medial olfactory striae and anterior commissure, to the opposite olfactory bulb (see pages 588 to 589).

188B,32
192,5; 182A,15
180,3;
180,33

The medial olfactory stria runs medially along the anterior edge of the anterior perforated substance towards the septal region beneath the rostrum of the corpus callosum. Most of its fibres originate in the nuclei of the olfactory tracts and pass through the anterior commissure. The thin layer of grey matter on its surface is the medial olfactory gyrus.

188B,32

The small intermediate olfactory stria terminates in the anterior part of the anterior perforated substance (in the olfactory tubercle).

188B,32; 176B,41
188B,27

The main, lateral olfactory stria passes posterolaterally along the anterolateral edge of the anterior perforated substance until beneath the limen of the insula. It then passes posteromedially, terminating anterior to the uncus.

188A,27
188A,1
182A,10

The prepyriform cortex covers and is lateral to the lateral olfactory stria; it comprises the lateral olfactory gyrus and the gyrus ambiens. The prepyriform cortex, and the peri-amygdaloid area (gyrus semilunaris) which is immediately anterior to the uncus, constitute the primary olfactory cortex. Some fibres from the olfactory bulb probably also pass to the entorhinal area of the anterior parahippocampal gyrus, the medial nucleus of the amygdala, the septal region and the hypothalamus.

Secondary olfactory projections (which can be difficult to distinguish from non-olfactory, limbic connections) pass to the entorhinal area, the basolateral part of the amygdala, the hypothalamus, the reticular formation, and to the posterolateral quadrant of orbitofrontal cortex. Olfactory signals probably reach this orbitofrontal area either via the stria medullaris thalami and the mediodorsal nucleus of the thalamus and/or through the uncinate fasciculus.

Connections to the hypothalamus and amygdala allow odours to trigger autonomic reflexes such as salivation and gastric secretion. The close relationship of the olfactory system with the limbic system explains the affective and evocative aspects of olfactory sensibility. A lesion of one side of the olfactory pathway produces ipsilateral anosmia, a sign which can be of diagnostic value.

The hippocampal formation (including archicortex) (Figure 192)

194,18; 182A,11
182A,13

The hippocampal formation (archipallium) comprises as its major components: the hippocampus, dentate gyrus and subicular region, and as its minor components: the indusium griseum, longitudinal striae, gyrus fasciolaris and parts of the uncus. Its cortex is archicortex. Some authorities include the entorhinal area within the hippocampal formation, while others use a more exclusive definition than the above.

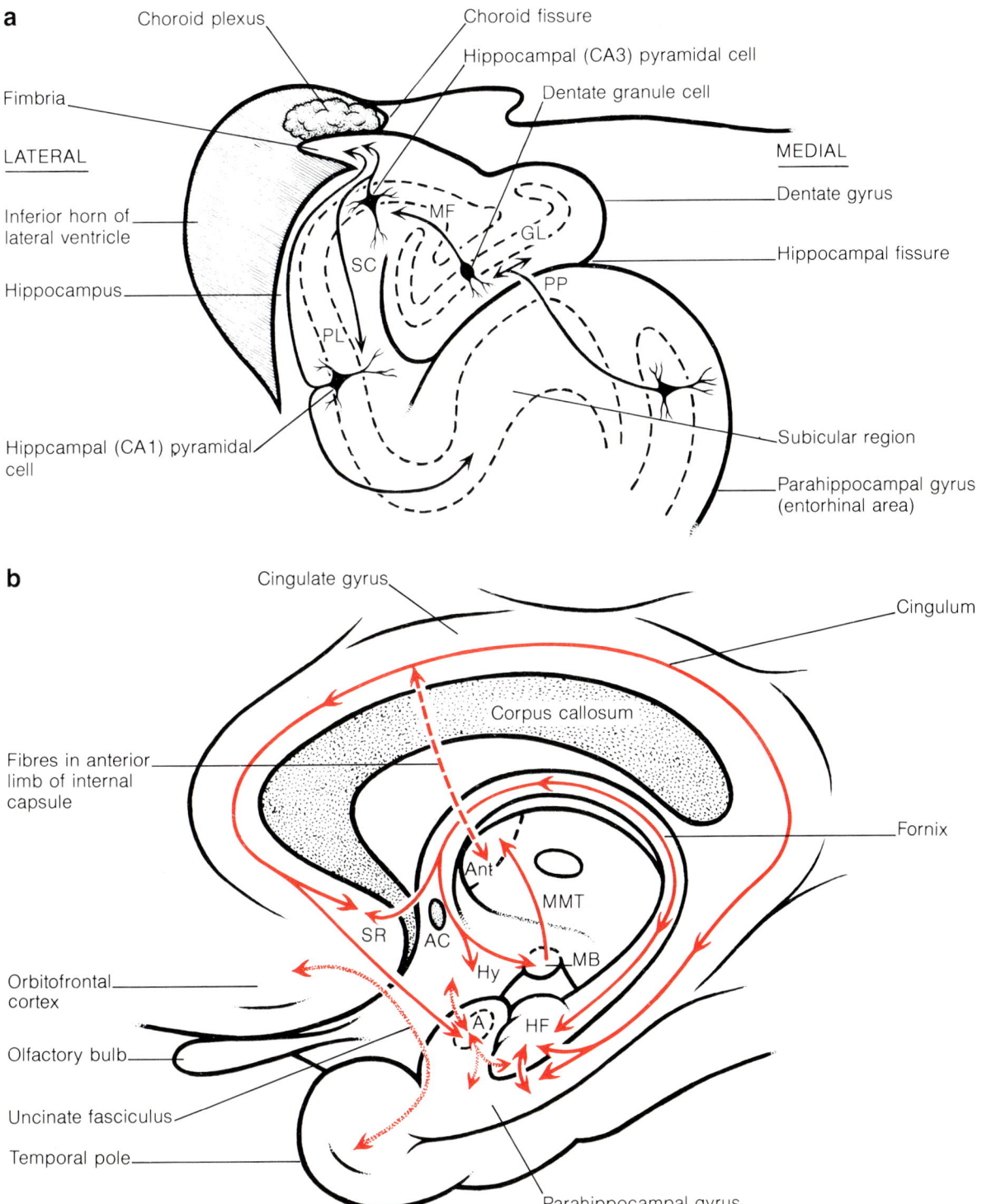

Figure 192 *Connections of the hippocampal formation.* **a)** *A frontal (coronal) section through the medial temporal lobe showing major connections of the hippocampal formation. Broken lines indicate the granular layer (GL) of the dentate gyrus, and the pyramidal layer (PL) of the hippocampus with its continuation into the six-layered cortex of the parahippocampal gyrus. MF = mossy fibre, PP = axon of perforant path, SC = Schaffer collateral.* **b)** *A lateral view of the brain to show some major connections of the hippocampal formation with other parts of the limbic system. A = amygdala, AC = anterior commissure, Ant = anterior nuclei of thalamus, HF = hippocampal formation, Hy = hypothalamus, MB = mamillary body, MMT = mamillothalamic tract, SR = septal region. Fibres pass between the amygdala and hypothalamus via the ventral amygdalofugal pathway (and also the stria terminalis which is not shown).*

In this text the hippocampal region includes the hippocampal formation and all of the parahippocampal gyrus.

The hippocampal formation is a major component of the limbic system and is closely interlinked with other limbic regions. Its precise function is unknown. Large bilateral lesions of the medial temporal lobes, including the hippocampal region and adjacent structures, cause profound amnesia (see page 561).

The hippocampus (cornu ammonis, Ammon's horn) is located superomedial to the parahippocampal gyrus. Medially, the cortex of the parahippocampal (hippocampal) gyrus turns superiorly. The cortex (of the subicular region) then folds inferomedially before curving superiorly and laterally to give the characteristic C-shaped outline of the hippocampus as seen in transverse (coronal) section (Figure 192a). The ⊐-shaped dentate gyrus (fascia dentata) caps and is interlocked with the hippocampus. The partially obliterated hippocampal fissure lies between the dentate gyrus and the hippocampus plus the subicular region. Superior to it, the dentate gyrus is notched (dentated) by transversely running blood vessels. The subicular region (composed of the prosubiculum, subiculum, presubiculum and parasubiculum) is interposed between the hippocampus and the parahippocampal gyrus. Subicular cortex is archicortex which is transitional between the three-layered archicortex of the hippocampus and dentate gyrus, and the six-layered paleocortex of the entorhinal area and neocortex of the remainder of the parahippocampal gyrus.

Anteriorly, the hippocampal formation bends medially towards the uncus. This part of the hippocampal formation forms the pes hippocampi. Within the uncus, the terminal part of the dentate gyrus turns superiorly as the band of Giacomini. Posteriorly, the hippocampus extends to below the splenium of the corpus callosum. The dentate gyrus continues posteriorly around the splenium as the thin gyrus fasciolaris and then, on the outer surface of the rest of the corpus callosum, as the rudimentary indusium griseum (supracallosal gyrus), which ends in the paraterminal gyrus. The indusium griseum is accompanied by two thin fibre bundles, the medial and lateral longitudinal striae (of Lancisi).

The hippocampus and dentate gyrus have three cortical layers. In the dentate gyrus these are the strata: polymorphe, granulosum and moleculare. The stratum polymorphe is next to the hippocampus in the hilus of the dentate gyrus. In the hippocampus the molecular layer is subdivided and the strata become: oriens (polymorphe), pyramidale, radiatum and lacunosum-moleculare. The (obliterated) hippocampal fissure separates the dentate and hippocampal molecular layers (Figure 192a).

Next to cerebellar cortex, the structure of the hippocampus and dentate gyrus is the most regular in the central nervous system. It is possible to trace a regularly repeated neuronal circuit through the region – which may be regarded as being organised as a series of overlapping lamellae, their plane being roughly transverse to the long axis of the hippocampus. Neurones of the entorhinal area (area 28) of the anterior parahippocampal gyrus project to the dentate gyrus (and hippocampus) via the perforant path (Figure 192a). The axons (mossy fibres) of the granule cells of the dentate gyrus terminate in the adjacent region (subfield CA3) of the hippocampus. Pyramidal cells of the CA3 subfield send axons (Schaffer collaterals) to the more distant part (subfield CA1) of the hippocampus. Pyramidal cells of subfield CA1 in turn project to subicular cortex. Subfield CA2 is a minor region between subfields CA1 and CA3. The hippocampus and dentate gyrus have relatively few inhibitory neurones. The hippocampus contains about seven million pyramidal cells and the dentate gyrus considerably more granule cells.

Axons of hippocampal pyramidal cells collect on the ventricular surface, deep to the stratum oriens, as the alveus. These axons of the alveus collect into a thin sheet, the fimbria. The fimbria is joined by axons of subicular neurones within the fornix; it arches posteriorly, superiorly and medially as the medial fringe of the crus of the fornix. The choroid plexus of the lateral ventricle, following the choroid fissure, forms an adjacent arch inside that of the fimbria.

Subcortical efferents of the hippocampus and subiculum pass via the precommissural part of the fornix and terminate in the septal nuclei (Figure 192b). Other subicular axons pass via the postcommissural fornix and terminate in the hypothalamus (most notably in the mamillary body), although some fibres go to the anterior and intralaminary thalamic nuclei. A small number of fibres pass via the hippocampal commissure of the fornix to the opposite hippocampal region. The other connections are predominantly ipsilateral.

Subcortical afferents to the hippocampal formation come from limbic regions (the medial septal nucleus, the amygdala, the hypothalamus, the anterior and midline nuclei of the thalamus) and the brainstem reticular formation (including mono-aminergic cell groups). However, most afferents are cortical and come from the entorhinal area, although other cortical afferents to the subicular region include fibres that travel in the cingulum from the cingulate gyrus (Figure 192b).

The entorhinal area of the anterior parahippocampal gyrus receives some olfactory input. More importantly, this area – or parts of the parahippocampal gyrus and cortex adjacent to the rhinal sulcus (pro- and peri-rhinal cortex) which project to it – receives inputs from virtually all parts of association cortex. Efferents from the hippocampus and subicular cortex (or from parts of the parahippocampal gyrus in which their fibres end) reciprocate these projections. Hence, the hippocampal formation processes information from, and influences processing in, widespread association areas of the cerebral cortex. The cingulate gyrus, orbitofrontal and inferior temporal cortical regions are strongly represented in these connections. There are also reciprocal connections between the nucleus medialis dorsalis (its medial, magnocellular part) of the thalamus and the entorhinal area and surrounding neocortex of the temporal lobe. The same part of nucleus medialis dorsalis is also connected to the inferomedial and orbitofrontal regions of the frontal lobe. These regions communicate with the medial temporal lobe via the uncinate fasciculus. Accordingly, the subcortical projections via the fornix to the septal nuclei, hypothalamus and anterior thalamic nuclei form only part of hippocampal output pathways.

The limbic system

The limbic system is a loosely defined set of structures that are closely related anatomically and functionally to the hypothalamus and hippocampus. The functional links are to emotional or visceral aspects of behaviour, and to the learning and remembering of new behaviours or experiences (especially where these are subject to confusion, or conflict with previous behaviour). Thus, the limbic system plays a particularly important role in pleasure or anxiety, aggression or amiability. However, although the constituent parts of the limbic system are closely interlinked, the structures, connections and functions of the individual components differ.

The major components of the limbic system are: the hypothalamus, the hippocampal formation, the parahippocampal and cingulate gyri, the septal region, the amygdala, the anterior nuclei of the thalamus and the fibre connections of these areas. Closely linked to these regions and usually included within the definition are: orbitofrontal cortex, parts of the medialis dorsalis, lateralis dorsalis and midline thalamic nuclei, the habenula, and parts of the midbrain reticular formation (together with their connections). Some authorities include the olfactory lobe, cortex near the temporal pole, anterior parts of the insula and the nucleus accumbens septi. The components of the limbic system are mainly found medially within the forebrain. The detailed connections and functions of these components are described under the individual regions.

The idea of a limbic system developed from the limbic lobe (of Broca). The limbic lobe comprises the cortex of: the hippocampal formation, parahippocampal gyrus, cingulate gyrus and septal area. These areas form the edge or border (limbus) of the

194,19; 194,20
190,6

182A,10

182A,14

180,36; 190,24

180,36; 190,24
182A,10; 182A,19
182A,15; 158,9

184,29; 176B,44

190,24; 182A,10
182A,19; 182A,15

cerebral cortex and surround the diencephalon. They are also linked by being of simpler structure than the rest of the cortex: archicortex (hippocampal formation), paleocortex (pyriform cortex), or meso- or juxtallo-cortex (cingulate and posterior parahippocampal gyri).

Links between structures of the limbic lobe were emphasised by Papez's proposed circuit of emotions. The modern version of Papez's circuit (Figure 192b) is: from the hippocampal formation via the fornix to the mamillary bodies of the hypothalamus, via the mamillothalamic tract to the anterior nuclei of the thalamus, via the internal capsule to the cingulate gyrus, and via the cingulum back to the hippocampal formation. The circuit is useful in emphasising the interlinkage of limbic structures. Nevertheless, it is but one of the more prominent loops that interrelate limbic structures. Furthermore, its components do not appear to be concerned primarily with emotions. It is other limbic regions (remaining parts of the hypothalamus, the amygdala, septal region, orbitofrontal cortex) which control or more strongly influence emotional behaviour and autonomic responses. (However, bilateral section of the cingulum reduces the fear and anxiety associated with chronic pain.) The functions of the components of Papez's circuit remain largely obscure, but they appear to be important in the learning of new behaviour, particularly where a discriminative motor or spatial decision is required. Bilateral lesions which include the hippocampal region or medial thalamus cause anterograde amnesia (see also page 561).

Cortical plasticity

Cortical plasticity is evident from a variety of different findings. The effects of early visual experience on the primary visual cortex provides a striking example. Clinical and experimental evidence has shown that the ocular dominance and orientation columns of primary visual cortex are genetically determined. They are present soon after birth and persist into adulthood even in the absence of visual experience. The effect of normal visual experience is to develop (tune-up) the innate response properties of the cortical neurones. However, abnormal visual experience in childhood produces irreversible changes in this usual pattern. There is a critical period, which extends from near birth to somewhere between the first 3 to 11 years of life, when the primary visual cortex has the capacity for such plastic changes. If one eye is not used (is amblyopic) during this period, ocular dominance columns for the other eye become greatly expanded and the amblyopic eye's anatomical and functional connections within primary visual cortex are greatly reduced. In the competition between afferents from the two eyes, those from the eye receiving normal visual experience prevail over those of the disadvantaged eye. Once beyond the critical period, this change is permanent and the person will always have good vision in only one eye. A squint (strabismus) uncorrected in childhood results in loss of stereoscopic vision because neurones receive effective connections from one or other, but not both eyes. Uncorrected astigmatism produces loss of visual acuity for the disadvantaged orientation; after the critical period complete compensation for this loss is not possible even with astigmatic glasses.

Similar evidence for dynamic competition between different anatomical inputs to neurones has been found for other cortical areas. However, the life-long capacity for learning makes it implausible that plasticity of all cortical areas should be subject to critical periods. Plasticity within visual cortex requires the integrity of reticular afferents and appropriate proprioceptive signals from the eye muscles. These findings suggest that there may be detailed conditions or specific systems which may enable or block plastic changes within the cortex.

In various cortical regions (the hippocampal formation has received particular attention), long-term changes in synaptic efficacy have been produced under experimental conditions: these changes involve the NMDA (N-methyl-D-aspartate) excitatory amino acid receptor. Sprouting and regeneration of cortical axons has also been demonstrated experimentally. Study of such phenomena is progressing rapidly, but

190,24; 190,16+27
158,56; 158,9
192,29; 182A,19
182A,10

their relation to memory and development, and their clinical and psychological significance have as yet only partly been explored.

Memory functions

Learning and memory are fundamental and widespread functions of the central nervous system. Certain types of memory are affected more by lesions in one area than in another, suggesting that there are different memory systems for different learning tasks. However, it is also probable that different memory systems overlap in function. The exact neural substrates of various types of memory are not yet known. Lesions usually have to be bilateral before general memory deficits become pronounced (hence the particular vulnerability of certain medial limbic structures). However, material-specific memory deficits may arise following unilateral lesions.

Lesions of the basal nuclei or cerebellum particularly affect memories for acquired skills. Changes in cerebellar circuitry underlie classical conditioning. Damage to prefrontal cortex differentially affects complex memory tasks which require changing strategies, maintained attention or elaborative encoding/decoding (such as forming images or imaginative associations). Extensive lesions of the inferomedial and orbitofrontal regions of the frontal lobes may produce larger deficits. Lesions within association cortex of the parietal, occipital or non-medial temporal lobes may result in material-specific memory losses for long-known information, as though either the site of storage of the information has been removed, or access to the stored information is no longer possible. For example, such losses may be for the meanings of words, or for well-known faces, or buildings (aphasias and agnosias); they may also be for skills (apraxia). Lesions of posterior speech cortex in the parietal lobe may also cause an inability to learn and briefly retain rapidly presented short lists of items (impaired span-type, short-term memory), e.g. telephone numbers. In such patients, learning of items presented one at a time may be normal. There are probably a number of other short-term memory stores, including stores associated with each of the sensory systems. Further, the mere occurrence of an item or event facilitates for some hours its future use or recall (priming memory), although such facilitation may occur without conscious recollection of the material. Such priming memory is believed to be another attribute of cortical association areas.

192,7+30+31
180,15

The classic amnesic syndrome follows bilateral medial temporal lobe (including hippocampal region and amygdala) or bilateral diencephalic (medial thalamic and/or possibly mamillary body) damage. It is characterised by inability to form new, consciously appreciated memories (anterograde amnesia), regardless of the type of material or sensory input. Certain learning (e.g. of new skills or for testing procedures) is still possible, but the patient shows no awareness and expresses no familiarity with this learning. There may additionally be amnesia for events which occurred quite a long time before the lesion (retrograde amnesia). Long-distant memories (e.g. for language, and of early life) remain at least relatively intact. Such patients commonly invent stories (confabulate) to hide their memory loss.

182A,10; 158,9
158,56

The precise critical loci for producing such amnesia are unclear. In particular, most patients with damage to the medial temporal lobes have extensive lesions. In any case, even confined damage would be likely to disrupt the operation of whole circuits and not only operations at the lesioned site. The sufficiency of bilateral mamillary body or fornix lesions for causing such amnesia is questionable. In the case of medial thalamic damage, it is not clear whether bilateral lesions of midline nuclei or the nucleus medialis dorsalis are sufficient to produce amnesia; both have connections with the medial temporal lobes.

158,56
158,7; 158,9

Patients with Korsakoff's psychosis following diencephalic damage display symptoms of the amnesic syndrome (with possible additional symptoms, such as general confusion, probably resulting from additional damage, commonly to frontal cortex). Early symptoms of senile dementia of the Alzheimer type include a similar pattern of memory loss. Such patients have prominent hippocampal region damage (although there may also be widespread damage to other regions), and also a possible loss of

182A,10

cholinergic input to the cerebral cortex from the septal region and basal forebrain nucleus.

THE INTERNAL STRUCTURE OF THE CEREBRAL HEMISPHERES

As the cerebral cortex comprises only the outermost few millimetres of the cerebral hemisphere, the main volume of the hemisphere consists of white matter, underlying nuclei (chiefly the basal nuclei) and the lateral ventricle.

THE WHITE MATTER

The white matter of the cerebral hemisphere is composed of association, commissural and projection fibres of the cerebral cortex.

The association fibres of the cerebral cortex may be intracortical, or run for varying lengths in the underlying white matter. Intracortical fibres are particularly evident in the inner and outer bands of Baillarger in layers IV and V of the cortex. Association fibres that pass extracortically arise mainly from layer III pyramidal cells and terminate mainly in layers III and IV. Fibres of different lengths and directions are found throughout the white matter underlying the cortex; they are known as arcuate fibres.

In certain places, the longer fibres are gathered into loose bundles (Figure 193). Each long association fibre bundle gains and loses fibres throughout its length so that fibres passing the full length of the bundle are in the minority. The superior longitudinal fasciculus runs through the central part of the white matter of the hemisphere between the occipital and frontal lobes. Amongst other areas, it interconnects the visual cortex and the frontal eye field, and posterior and anterior speech cortex. The inferior longitudinal fasciculus makes connection between occipital and temporal lobes. It carries both auditory and visual information. The cingulum runs medially, passing round the corpus callosum deep to the cingulate gyrus and through the isthmus into the parahippocampal gyrus. It particularly interconnects limbic areas, from the inferomedial frontal cortex to the entorhinal area. Cutting the cingulum in each frontal lobe has been found to be beneficial in reducing neurotic obsessions, and in relieving the anguish and emotional overtones of intractable pain. The uncinate fasciculus links the orbitofrontal cortex with that of the temporal pole; it passes deep to the limen of the insula. Amongst its fibres are again those interconnecting regions of limbic function.

The extreme capsule is found beneath the cortex of the insula; deep to it is the claustrum of the basal nuclei. The external capsule lies between the claustrum and the lentiform nucleus of the basal nuclei. Both capsules contain chiefly cortical association fibres, but also lentiform nuclear and claustral connections.

The commissural fibres of the cerebral cortex pass through the anterior commissure, the hippocampal commissure, or the corpus callosum (Figure 192b).

The anterior commissure interconnects: a) anterior parts of the temporal lobes (particularly the middle and inferior temporal gyri, but also parts of the parahippocampal gyri, and the amygdalae), and b) olfactory regions (i.e. the nucleus of the olfactory tract with the contralateral olfactory bulb; see pages 588 to 589). From the midline, the posterior, non-olfactory fibres curve laterally and posteriorly as a compact bundle through the most inferior part of the lentiform nucleus and into the inferior part of the external capsule.

The small hippocampal commissure (commissure of the fornix) interconnects parts of the left and right hippocampal regions.

The corpus callosum conveys the great majority (more than 10^8 fibres) of the commissural fibres of the cerebral cortex. Its fibres form most of the roof and parts of the walls of the lateral ventricle. They fan out/funnel in as the callosal radiation, inter-

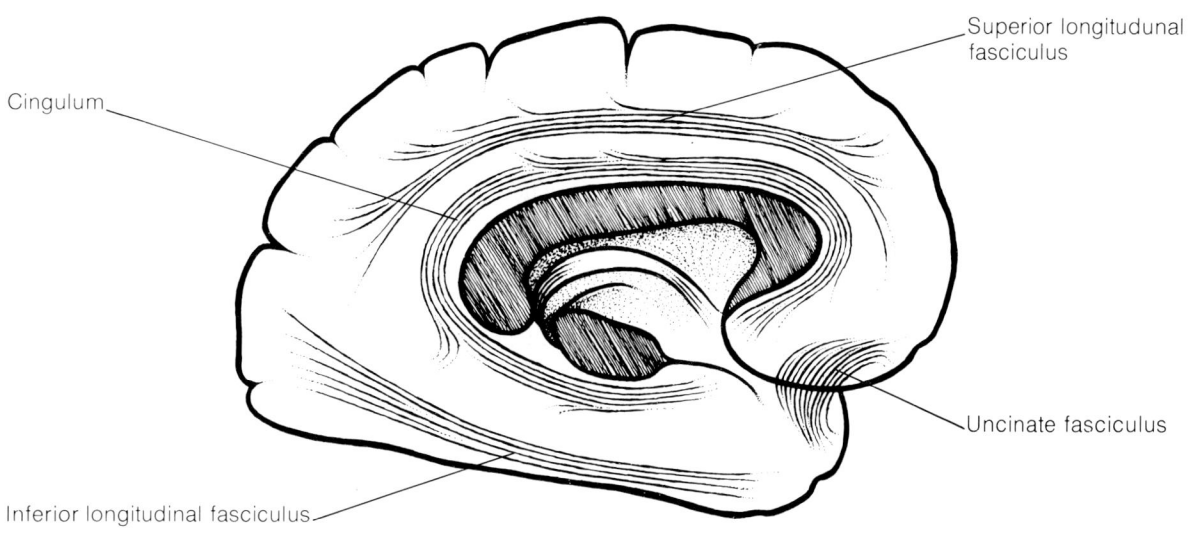

Figure 193 *Long association fibre bundles of the cerebral hemisphere (medial view).*

connecting corresponding regions of the two cerebral hemispheres. All regions are so linked except for: most of the primary visual and auditory cortex, the hand and foot regions of somatosensory and motor cortex, and the anterior temporal regions interconnected by the other commissures. Some heterotopic connections also pass through the corpus callosum.

Sectioning the cortical commissures disconnects the two hemispheres (producing a 'split-brain') so that they can no longer exchange information. In the normal brain, information, including new learning, is rapidly shared, passing through the commissural fibres from one hemisphere to the other. Although patients with divided commissures have few difficulties in every day life, the two hemispheres may independently be taught different, even conflicting, tasks in the laboratory. Certain intellectual, short-term memory and attentive impairments have also been reported. All verbal communication is with the speech-dominant hemisphere (usually the left).

The projection fibres of the cerebral cortex chiefly pass through the fornix, the internal capsule, the olfactory tract, or the substantia innominata of the basal forebrain. Some cortical connections of the basal nuclei, particularly the amygdala and claustrum, pass outside these structures. The olfactory tract contains the connections of the olfactory bulb. Some reticular and basal forebrain connections pass through the substantia innominata.

180,8; 192,12
192,27–29; 188A,2

192,33; 188A,2
188A,1

The fornix is a long arched bundle containing more than 10^6 fibres. It carries hippocampal region connections and is one of the main fibre bundles of the limbic system. The body of the fornix passes dorsal to the thalamus and beneath the body of the corpus callosum. The anterior column of the fornix forms the anterior boundary of the interventricular foramen. Posteriorly and laterally, the crus of the fornix arises in the hippocampus and the neighbouring subicular region of the medial temporal lobe. Most of the fibres of the fornix originate from these regions. Most of those from the hippocampus pass in the fimbria before gathering into the body of the fornix. Some fibres of the fornix are commissural fibres and cross the midline in the hippocampal commissure (psalterium) to distribute to the hippocampal region of the opposite hemisphere. Anteriorly, the fornix is split by the anterior commissure into a precommissural and postcommissural component. The precommissural fornix passes anteriorly and contains axons of cells of the hippocampus and some axons of subicular origin. These axons terminate chiefly in the lateral septal nucleus. The postcommissural fornix passes posteriorly and contains axons of cells of subicular cortex. These axons

180,8

180,8; 180,11
180,5; 191,28
191,5; 180,9
190,16; 190,24

190,17; 190,27

180,8; 180,33

192,5

180,26	terminate in the hypothalamus (most notably in the mamillary bodies) but some reach the anterior and intralaminary nuclei of the thalamus. The fornix also contains a small number of fibres passing to the hippocampal region from the medial septal nucleus, hypothalamus and reticular formation.
192,10	
192,27–29; 194,10	**The internal capsule** contains the vast majority of the afferent and efferent projection fibres of the cerebral cortex. Almost all the subcortical afferents to the cortex emanate from the thalamus in the thalamic radiations of the internal capsule (olfactory and some reticular afferents being two notable, but not large, exceptions). The descending fibres of the internal capsule pass to the thalamus (corticothalamic fibres also pass in the thalamic radiations), the basal nuclei (caudate and putamen), the brainstem (including the pontine nuclei, for relay to the cerebellum, and the cranial nerve nuclei) and the spinal cord. Hence, its major components are thalamocortical, corticothalamic, corticostriatal, corticopontine, corticonuclear and corticospinal fibres. Certain connections of the basal nuclei also course within it.
194,6	
194,8; 194,30	
194,22–24	
192,26	In horizontal sections of the hemisphere which pass through the thalamus, the internal capsule occupies a restricted <–shaped region (Figure 194). It has an anterior limb which lies between the head of the caudate nucleus and the lentiform nucleus. Its posterior limb lies between the thalamus and the lentiform nucleus. The anterior and posterior limbs meet at the genu between the caudate nucleus and thalamus. Posteriorly, fibres pass inferior to the lentiform nucleus and form the sublenticular part of the internal capsule. Even further posteriorly, some fibres course almost horizontally, posterior to the lentiform nucleus in the retrolenticular part of the internal capsule.
192,27–29	
192,7; 192,30–31	
192,26; 192,30–31	
192,7; 192,26	
194,29–30	
192,20	
192,29	The anterior limb of the internal capsule contains the anterior thalamic peduncle and frontal projection fibres. The posterior limb contains fibres of the superior thalamic peduncle and projection fibres of central regions of the cortex. In the posterior limb, thalamocortical fibres lie medially (close to the thalamus) and include the somatosensory radiation. More laterally, the posterior limb contains corticofugal fibres amongst which are the important corticospinal fibres. The corticospinal fibres going to lumbar and sacral segments lie posterior to those ending in cervical and thoracic regions. There is disagreement about the exact position of the corticospinal fibres in the posterior limb, and about whether corticonuclear fibres pass through the posterior limb or genu. The internal capsule is a three-dimensional shell within which the position of the genu varies relative to the capsule's component fibres. Hence, the corticonuclear fibres may pass through the genu superiorly, but not inferiorly. The sublenticular fibres contain connections of temporal cortex, including the auditory radiation of the inferior thalamic peduncle. The most posterior (retrolenticular) fibres within the internal capsule are those of the posterior thalamic peduncle plus projection fibres of occipital and temporal cortex. The optic radiation runs in this region (see below).
192,27	
192,26	
192,28	
192,20	
192,27	As so many cortical projection fibres are confined within such a small region in the internal capsule, even minor lesions can have devastating consequences. A lesion of the internal capsule of vascular origin is the most common cause of cerebral apoplexy or stroke. The major effects of a lesion of the posterior limb will be contralateral hemianaesthesia and/or contralateral spastic paralysis (of the upper motoneurone type) possibly involving the head region. On the left (or dominant) side language functions may be affected. A lesion of the retrolenticular part of the internal capsule is likely to produce a visual defect — a contralateral homonymous hemianopia, if the optic radiation is completely interrupted. Some, mainly contralateral, hearing loss will occur if the auditory radiation is involved. Smaller lesions will produce partial symptoms varying in extent and/or severity. Larger lesions will include damage to thalamic and basal nuclei.
192,20	
192,30–31; 194,29–30	Distal to the lentiform nucleus, the capsular fibres fan out anteriorly, posteriorly and superiorly, as the corona radiata, becoming mixed with association and commissural fibres of the hemisphere. Inferiorly, the internal capsule funnels into the crus cerebri
192,13; 194,9	
194,10; 194,22	

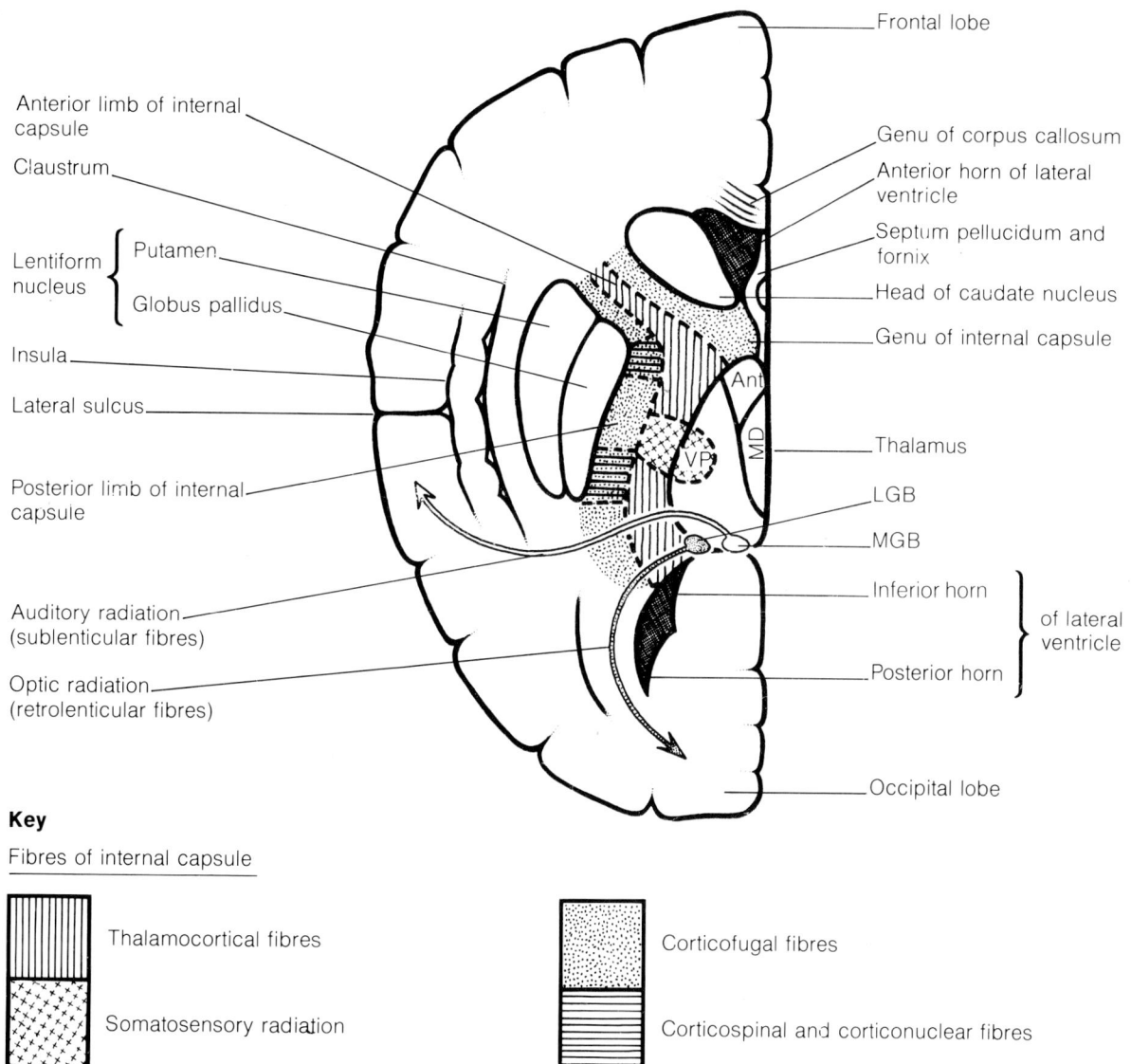

Figure 194 *The internal capsule. The position and fibre components of the internal capsule are shown on a horizontal section of the cerebral hemisphere. Thalamic nuclei: Ant = anterior, MD = medialis dorsalis, VP = ventralis posterior, LGB = lateral geniculate body, MGB = medial geniculate body.*

of the midbrain; there, only cortical fibres to the brainstem and spinal cord (corticospinal, corticonuclear and corticopontine fibres) remain.

The optic radiation follows a long course through the white matter of the temporal, parietal and occipital lobes to reach the visual cortex on either side of the posterior part of the calcarine sulcus. In so doing, its fibres pass close to the lateral ventricle. The most inferior fibres of the optic radiation, those representing the contralateral upper quadrants of the visual fields, turn anteriorly, laterally and inferiorly into the temporal lobe (forming Meyer's temporal loop) close to the inferior horn of the lateral ventricle (see Figure 166 and pages 590 to 591). They loop superiorly over the anterior part of the inferior horn before turning posteriorly and running close to the lateral walls of the inferior and posterior horns, to reach the cortex below the posterior part of the calcarine sulcus. The more superior fibres, representing the lower visual field,

193,20
193,39
182A,7; 193,40

186,30

182A,7

565

pass more directly round the descending limb of the lateral ventricle and close to its lateral wall to reach the cortex above the posterior part of the calcarine sulcus. The fibres carrying information from the macula travel in the largest, intermediate part of the optic radiation. Lateral to the descending limb and posterior horn, the visual fibres run in the internal (corticofugal, including corticotectal fibres) and external (thalamo-cortical) sagittal strata. Visual defects may be the first sign of pathological changes in the region of the ventricle. Temporal lobe lesions may produce a contralateral homonymous upper quadrantanopia. Parietal lesions may produce a contralateral homonymous lower quadrantanopia.

THE BASAL FOREBRAIN AND SEPTAL REGION

The basal forebrain and septal region consists of nuclei and overlying cortex at the base of the forebrain, anterior to the hypothalamus.

The substantia innominata is continuous posteriorly with the preoptic region of the hypothalamus and lies deep to the anterior perforated substance. Scattered, large neurones within the substantia innominata and extending superiorly into the medullary laminae of the globus pallidus of the basal nuclei constitute the basal forebrain (magnocellular) nucleus (of Meynert). The basal forebrain nucleus receives input from the reticular formation of the brainstem (including cholinergic fibres), cortical and subcortical parts of the limbic system, and the basal nuclei. Neurones of the basal forebrain nucleus, together with those of the nucleus of the diagonal band and the

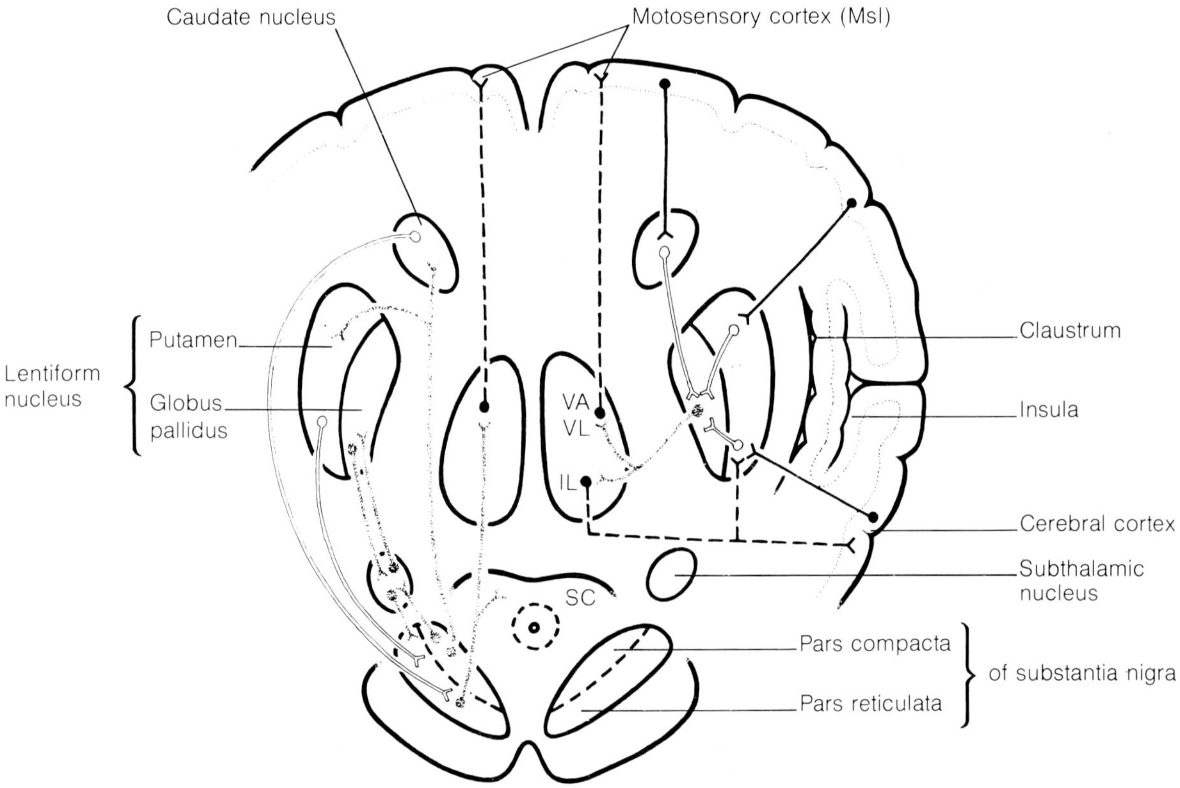

Figure 195 *Major connections of the basal nuclear system. Connections of the substantia nigra and subthalamic nucleus are shown on the left. Other connections of the caudate and lentiform nuclei are shown on the right. (Neither the nucleus accumbens septi nor reticular connections are shown.) Thalamic nuclei:* IL = *intralaminary,* VA = *ventralis anterior,* VL = *ventralis lateralis.* SC = *superior colliculus. Fibres from the cerebral cortex are indicated by solid lines; fibres from the caudate nucleus and putamen by open lines; fibres from the globus pallidus, subthalamic nucleus and substantia nigra by stippled lines; and from the thalamus by broken lines.*

medial septal nucleus send cholinergic fibres to all areas of the cerebral cortex (see Figure 179d). The precise function of these neurones is not known. They may form part of an arousal mechanism and of a system producing desynchronisation of the EEG. They also seem to be important for the normal functioning of memory.

The septal region consists of the cortical, septal area (paraterminal body) and the subcortical, septal nuclei. The septal area comprises the paraterminal gyrus and (probably) the subcallosal area, anterior to the lamina terminalis. The septal nuclei are deep to it and anterior to the anterior commissure. The parts of the septal region anterior to the anterior commissure are its precommissural portion. Some cells (of the supracommissural septum) are found superiorly, scattered in the septum pellucidum. The bed nucleus of the stria terminalis is found laterally and may be included in the septal region. The medial septal nucleus is continuous with the nucleus of the diagonal band (of Broca). The diagonal band marks the posteromedial boundary of the anterior perforated substance, close to the optic tract, and terminates in the periamygdaloid area. Its nucleus may also be included in the septal region.

192,5; 182A,15
182A,16; 180,32
180,33

180,7

192,5

188B,32; 188B,40

The septal nuclei receive fibres from the reticular formation (midbrain and mono-aminergic cell groups) via the medial forebrain bundle and a massive input from the hippocampal formation via the fornix (Figure 192b). Other inputs include fibres from other limbic regions (the hypothalamus, amygdala and cingulate gyrus).

Efferents largely reciprocate these afferents. Besides cholinergic input to the hippocampal formation (from the medial septal nucleus and nucleus of the diagonal band), fibres pass through the medial forebrain bundle to the hypothalamus and midbrain reticular formation, and through the stria medullaris thalami to the habenula (so again indirectly influencing midbrain reticular nuclei). Other fibres reach the amygdala, cingulate gyrus and thalamus (anterior and medialis dorsalis nuclei).

The septal region provides a location where limbic signals (hippocampal, hypothalamic, amygdalar and reticular) may interact and influence each other. Stimulation produces pleasurable sensations. Lesions result in behavioural over-reaction, and in emotional, learning, autonomic and endocrine disturbances. However, it is possible that some of these symptoms arise from lesions because of the involvement of fibres of passage.

THE BASAL NUCLEI (BASAL GANGLIA) (Figure 195)

The basal nuclei (basal ganglia) are areas of grey matter in the interior of the cerebral hemisphere, i.e. the caudate nucleus, lentiform nucleus, claustrum and amygdala. The anatomical terminology applicable to these nuclei is summarised in Table 18. The term corpus striatum arose from the appearance of the region, the grey matter of the nuclei being split up by interweaving white matter. The corpus striatum includes the claustrum, caudate nucleus, and putamen and globus pallidus of the lentiform nucleus. However, the striatum (and the suffix 'striate') includes only the caudate nucleus and putamen (the neostriatum).

192,7
192,30-31; 192,33

Functionally, the amygdala (the archistriatum) is part of the limbic system. The claustrum belongs neither with the limbic system nor with the remainder of the basal nuclei. The caudate nucleus, putamen and globus pallidus are closely linked both anatomically and functionally. Intimately related to these three nuclei are the subthalamic nucleus of the diencephalon and the substantia nigra of the midbrain. Accordingly, in functional terms, the basal nuclei are often implicitly understood to refer to these five nuclei (caudate nucleus, putamen, globus pallidus, subthalamic nucleus and substantia nigra). These nuclei are here referred to as the basal nuclear system. The nucleus accumbens septi has links with these nuclei as well as with the limbic system, and therefore appears also to belong to the functional grouping of the basal nuclear system.

194,31
194,8; 194,30
194,29
194,27; 194,25

The amygdala or amygdaloid complex (body or nuclei) is an almond-shaped group of nuclei about 10mm across, beneath and anterior to the uncus. It is anterior to the end

188,27

TABLE 18 The subdivisions of the basal nuclei.

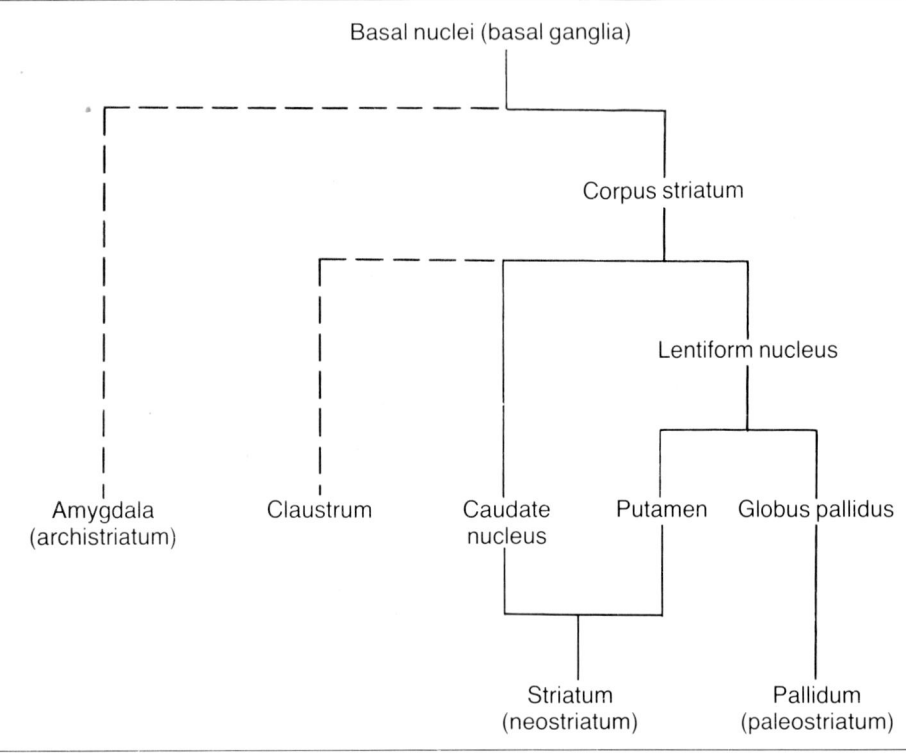

of the tail of the caudate nucleus, the pes hippocampi, and the temporal tip of the inferior horn of the lateral ventricle. Functionally, it belongs to the olfactory and limbic systems (see pages 555 to 556 and 559 to 560). It comprises a small, dorsal, corticomedial group of nuclei which includes: the paleocortex of the peri-amygdaloid area (gyrus semilunaris), the anterior amygdaloid area, the nucleus of the lateral olfactory stria, and the central and medial nuclei. This group receives fibres direct from the olfactory bulb. The larger, basolateral group contains: the basal, the lateral and the accessory basal amygdaloid nuclei. The amygdala contains high concentrations of monoamines (noradrenaline, 5-hydroxytryptamine, dopamine) and peptides (e.g. enkephalins, substance P and somatostatin). On anatomical grounds, the amygdala has been grouped with the substantia innominata as the ventral pallidum (i.e. the limbic equivalent of the globus pallidus).

The connections of the amygdala are remarkably wide and, unless otherwise stated, reciprocal. Thus, it samples and integrates information from many regions (cortical, thalamic, limbic, reticular and autonomic), so that it in turn may influence processing in these regions. The individual nuclei have specific connections, but these will not be detailed here. The amygdala has connections with: the olfactory bulb and cortex, many other cortical (including limbic) regions, the thalamus (reciprocal with nucleus medialis dorsalis, but afferents come from several specific and nonspecific nuclei), the hypothalamus, the septal region, and the brainstem (reticular formation and substantia nigra; afferents from the nucleus of the solitary tract and efferents to the dorsal motor nucleus of the vagus).

The subcortical connections of the amygdala pass via the stria terminalis or the ventral amygdalofugal pathway. Both contain a variety of afferents and efferents. The stria terminalis is a small bundle which arches, adjacent to the lateral ventricle, along the whole of the medial edge of the caudate nucleus, eventually lying between the caudate nucleus and thalamus. Many of its fibres terminate in the bed nucleus of the stria terminalis; this nucleus is lateral to the columns of the fornix and superior to the anterior commissure. Fibres from the bed nucleus pass mainly to the hypothalamus, habenula, reticular formation and nucleus accumbens septi. The ventral amygdalofugal pathway (which also contains connections of the pyriform cortex) passes through the substantia innominata, deep to the anterior perforated substance, to the preoptic region of the hypothalamus. Commissural fibres of the amygdala pass through the anterior commissure.

191,28
180,33

188B,32

180,33

The precise functions of the amygdala are little understood. It exerts a modulating influence upon the hypothalamus. Its actions affect endocrine and autonomic activity, emotionally based behaviours (fear, aggression, sexual behaviour), feeding and drinking, and possibly memory. It is thought to be involved in opiate dependency. The amygdala has a very low threshold for epileptic seizures.

The claustrum is the most laterally placed of the basal nuclei. It is a thin sheet of grey matter narrowly separated from the cortex of the insula by the white matter of the extreme capsule, and from the outer surface of the lentiform nucleus by the thin lamina of the external capsule. It has reciprocal, topographically organised connections with most of the cerebral cortex. These connections resemble those of the thalamus in that afferents arise from layer VI of the cortex and most efferents terminate in layer IV. Its functions are unknown.

192,33; 194,31
192,35; 194,13
192,34; 194,12
192,31; 192,32

The nucleus accumbens septi lies inferolateral to the septal nuclei, deep to the paraterminal gyrus and inferomedial to the head of the caudate nucleus, i.e. between the septal and basal nuclei. It is the major part of the so-called ventral striatum (the limbic-related equivalent of the caudate nucleus and putamen). It receives fibres from limbic, basal nuclear and reticular regions (notably from dopaminergic neurones of the ventral tegmental area and substantia nigra of the midbrain) and sends fibres to the globus pallidus and substantia nigra. It has been suggested that this nucleus links limbic and basal nuclear functions and is therefore important in motivation (drive).

192,5
182A,15; 190,4

The lentiform nucleus lies between the external and internal capsules. Its darker, larger and more lateral part is the putamen. Its lighter, medial part is the globus pallidus. The globus pallidus (pallidum) is separated from the putamen by a thin, lateral medullary lamina, and is itself divided into outer and inner segments by a medial medullary lamina.

192,32; 192,27–29
192,31; 194,30
192,30; 194,29

The caudate nucleus lies in the wall of the lateral ventricle and is comma-shaped. The large head of the caudate nucleus forms most of the lateral wall of the anterior horn of the lateral ventricle. It is grossly joined to the putamen inferiorly (indeed, the caudate nucleus and putamen are essentially a single mass of grey matter split by the internal capsule). The caudate nucleus is separated from the thalamus by the stria terminalis and the genu of the internal capsule. It is bounded laterally by the anterior limb of the internal capsule, and superiorly by the corpus callosum and other fibres of the hemisphere. The tapering body of the caudate nucleus arches superiorly then inferiorly in the lateral wall of the body of the lateral ventricle, remaining superolateral to the thalamus. The long thin tail of the nucleus passes inferiorly in the anterior wall of the descending limb and then anteriorly in the roof of the anterior part of the inferior horn of the lateral ventricle. It ends at the amygdala.

191,4; 192,7–8

193,7; 193,31
193,29; 192,10
192,9; 192,28
192,29; 194,8
194,1; 192,8
192,11;
192,10; 192,23
192,21; 194,14
194,15

The components of the striatum (neostriatum), the caudate nucleus and putamen, are histologically similar, having many densely packed small cells (more than 10^8) and relatively few scattered large cells (about 0.6×10^6). Certain of the intrinsic neurones are cholinergic. Most of the projection neurones are GABA-ergic. The globus pallidus (paleostriatum) contains widely separated, mainly large cells (more than 0.6×10^6).

194,8; 194,30

194,29

The connections of the basal nuclear system

The connections of the basal nuclear system are summarised in Figure 195. They are very largely ipsilateral. Because the major output of the system is via the thalamus to the ipsilateral motosensory cortex (area MsI), its main influence is upon movements of the contralateral side of the body.

194,8; 194,30   The striatum receives a topographic and mainly ipsilateral input from: the cerebral cortex, the substantia nigra, and the intralaminary nuclei of the thalamus. Afferents from the whole of the cerebral cortex end in the striatum, the putamen receiving a particularly strong projection from bilateral motor (MsI) and ipsilateral somatosensory (SmI) areas. These afferents arise from layer V pyramidal cells, are excitatory, and are probably glutaminergic. The striatum also receives fibres from the substantia nigra. These fibres are chiefly the nigrostriatal dopamine fibres arising from the neurones of the pars compacta of the substantia nigra. They pass through the subthalamus (in the prerubral field) and the internal capsule. The striatal input from the intralaminary nuclei of the thalamus (the thalamostriate fibres) also traverses the internal capsule. In addition, there are inhibitory, 5-hydroxytryptaminergic afferents from the nuclei of the raphe of the brainstem.

The striatum sends inhibitory GABA-ergic fibres to the ipsilateral globus pallidus and substantia nigra. The substantia nigra also receives a smaller, excitatory projection which uses the peptide, substance P. These projections are again topographically organised. Fibres from the caudate nucleus cross the internal capsule.

194,29   The globus pallidus receives fibres from the striatum and subthalamic nucleus. It projects to the thalamus via the fasciculus lenticularis, which crosses the internal capsule, and the ansa lenticularis, which winds inferiorly around the posterior limb of the internal capsule. The fibres are inhibitory, GABA-ergic and terminate in the nuclei ventralis anterior, ventralis lateralis and centrum medianum. The nuclei ventralis anterior and ventralis lateralis project to MsI, so allowing pallidal modulation of descending motor pathways. Other GABA-ergic fibres go to the substantia nigra, the subthalamic nucleus, the reticular formation (notably the pedunculopontine nucleus in the caudal midbrain tegmentum) and the habenula of the epithalamus.

194,25   The substantia nigra (see also page 502) receives fibres from the globus pallidus and striatum (see above). It also has inputs from the subthalamic nucleus, the raphe, the pedunculopontine nucleus of the reticular formation, the nucleus accumbens septi, the amygdala (plus the bed nucleus of the stria terminalis) and frontal cortex. Dopaminergic fibres from the darkly pigmented neurones of the pars compacta project to the striatum and, possibly, the amygdala, basal forebrain structures (including the nucleus accumbens septi) and frontal cortex. They form the nigrostriatal and part of the mesocortical or mesolimbic dopamine systems. Dopamine-containing neurones lying medially in the ventral midbrain tegmentum (ventral tegmental area of Tsai) and forming a continuous band with those in the substantia nigra make up the remainder of these projections. GABA-ergic neurones of the pars reticulata send axons to the nuclei ventralis lateralis and ventralis anterior of the thalamus, to the superior colliculus and to the reticular formation.

194,27   The subthalamic nucleus (see page 528) has reciprocal, GABA-ergic connections with the globus pallidus. It receives dopaminergic fibres and sends GABA-ergic fibres to the substantia nigra. It also receives input from the reticular formation and frontal cortex.

The striatum pallidum, substantia nigra and subthalamic nucleus therefore form a strongly interconnected system. The main input to the system is from all areas of the cerebral cortex to the striatum. There is considerable convergence from the striatum to the pallidum and the substantia nigra. (The whole system somewhat resembles a funnel.) The main output is from the pallidum and substantia nigra to nucleus ventralis anterior and nucleus ventralis lateralis of the thalamus, though smaller projections go to a variety of other regions (including the superior colliculus and intralaminary nuclei of the thalamus, and several forebrain areas via dopaminergic fibres). Nuclei ventralis anterior and ventralis lateralis in turn project upon area MsI of the cortex. By this means, influence may be exerted on (mainly contralateral) motoneurones of the brainstem and spinal cord. The projection to the superior colliculus may be concerned with influencing eye movements, though the superior colliculus also has other functions. The outputs to areas other than nuclei ventralis anterior and ventralis lateralis (notably the intralaminary nuclei) suggest that cortical activity beyond the purely motor may be influenced by the basal nuclear system. Basal nuclear and cerebellar output is integrated within nuclei ventralis lateralis and ventralis anterior, and motosensory area I (MsI). Cerebellar efferents are excitatory, pallidal and nigral inhibitory. However, although the final output of the basal nuclear system is via inhibitory axons, the presence of many feedback circuits and multiple inhibitory links make thalamic disinhibition a quite plausible consequence of striatal activation.

The functions of the basal nuclear system are far less well understood than are the symptoms of their dysfunction. Dysfunction arising from vascular or degenerative disorders gives rise to changes in muscle tone, in reflexes, and the occurrence of characteristic, abnormal involuntary movements: tremor, chorea, athetosis and ballism. Hemiballismus is commonly associated with localised, haemorrhagic lesions of the contralateral subthalamic nucleus. The violent involuntary movements of ballism most often involve the limbs, but may involve the trunk, neck and head. Parkinsonism (paralysis agitans) is associated with degeneration of the dopamine-containing neurones of the substantia nigra, although this is not usually the only locus of degenerative change. Loss of GABA-ergic and cholinergic striatal neurones is a prominent feature of Huntington's chorea. Dysfunction within any one of the components of the system must necessarily affect the performance of the whole. Like the cerebellum, the basal nuclei provide additional processing and storage capacity for the cerebral cortex. It has been suggested that basic, well-practised motor programs are held within the basal nuclear system; however it seems likely that its functions extend beyond the purely motor sphere.

THE LATERAL VENTRICLE (see Figure 166)

190–194

The lateral ventricle of each cerebral hemisphere is a C-shaped structure. From the anterior horn (cornu) in the frontal lobe, its body curves posteriorly, superior and lateral to the thalamus. The descending limb of the inferior horn passes inferiorly, posterior and lateral to the thalamus. The remainder of the inferior horn then runs anteriorly (parallel to the superior temporal sulcus) within the temporal lobe to about 30 mm from the temporal pole. The posterior horn passes posteriorly into the occipital lobe from the lower part of the descending limb. The junction of the body and the inferior and posterior horns in the descending limb is known as the atrium (trigone) of the lateral ventricle. The lateral ventricle is easily visualised by computerised tomography.

190
190,3; 190,6
190,7;
190,10
176A,17
176A,21; 190,11
193,40

The anterior (frontal) horn has as its medial wall the septum pellucidum. Its roof, anterior wall and floor are formed by the body, genu and rostrum of the corpus callosum. The lateral wall is the head of the caudate nucleus. The anterior column of the fornix lies in the medial wall just anterior to the interventricular foramen.

190,3; 190,1
192,3; 190,2
190,4; 191,28
191,5

The body (central part) of the lateral ventricle extends from the interventricular foramen to the splenium of the corpus callosum. Its floor is formed medially by the thalamus and laterally by the body of the caudate nucleus. In the groove between these two structures runs the stria terminalis and the thalamostriate vein. In the medial wall, the choroid plexus is formed by invagination through the choroid fissure above the thalamus and beneath the body of the fornix. Superior to the body of the fornix, the medial wall is completed by part of the septum pellucidum. The roof and lateral wall are formed by fibres of the body of the corpus callosum.

192,11; 194,5
192,6; 192,14
192,26; 192,8
192,9; 194,7
190,6; 190,26
190,7; 190,27
180,7; 194,2
180,5; 194,1

The posterior (occipital) horn is of variable size. Its medial wall usually contains two conspicuous ridges: the bulb and the calcar avis. The fibres of the forceps major, after

190,11; 190,19
190,20; 192,15

passing through the splenium of the corpus callosum, turn posteriorly, producing a large bulge in the upper part of the medial wall: this is the bulb of the posterior horn. Sometimes the bulb is so large as to occlude the posterior horn. Below the bulb there is normally a further prominent ridge, the calcar avis. The calcar avis is raised by the deep, anterior part of the calcarine sulcus. The posterior horn is confluent with the descending limb of the inferior horn; the same structures form the roof, lateral wall and floor of both parts of the ventricle (see below).

The descending limb (atrium) of the inferior (temporal) horn is roofed by fibres of the corpus callosum which, as they sweep over the lateral walls of the inferior and posterior (occipital) horns, are called the tapetum. The triangular area in the floor of the descending limb and posterior horn is the collateral trigone. The anterior wall of the descending limb contains the tail of the caudate nucleus immediately lateral to the stria terminalis. Medially within the descending limb are the crus of the fornix and the tail of the hippocampus, with the fimbria as a thin sheet of fibres on their lateral edge. There is usually a swelling in the choroid plexus of the descending limb: this is the glomus. The glomus calcifies in later life and provides a useful radiographic landmark.

The anterior part (limb) of the inferior (temporal) horn has in its roof continuations of the tail of the caudate nucleus and the stria terminalis. These structures end at the amygdala deep to the uncus. Other parts of the roof and lateral walls are formed by the tapetum of the corpus callosum. Parts of the optic radiation run superior and lateral to the tapetum (see pages 565 to 566). In the floor of the anterior part of the inferior horn is the collateral eminence, an elevation raised by the deep collateral sulcus. More medially is the hippocampus. The choroid plexus is formed by invagination in the choroid fissure between the stria terminalis and fimbria. Beneath the fimbria, and interlocking with the hippocampus, is the dentate gyrus (Figure 192a). In the most anterior part of the inferior horn, the hippocampus turns medially towards the uncus. Because of its characteristic, paw-like appearance, this part of the hippocampus is called the pes hippocampi.

The choroid plexuses and the cerebrospinal fluid

The integrity and optimal operation of the central nervous system is vital to the wellbeing of the organism. Nervous tissue is delicate, semi-fluid in consistency and would be easily damaged without external protection and support. The brain and spinal cord are encased in bone (the cranium and vertebral column). They are enveloped by three membranes (meninges) and made buoyant in fluid (cerebrospinal fluid). Besides mechanical protection, these structures provide barriers against infection. The brain and spinal cord are also in a closely controlled homeostatic environment. There is close regulation of their temperature and of the composition of the interstitial fluid. The blood–brain and blood–cerebrospinal fluid barriers limit the passage of many molecules into the central nervous system. The oxygen and nutritional requirements of the central nervous system receive privileged protection at times of deficiency.

THE CHOROID PLEXUSES

The choroid plexuses are formed from highly vascular pia mater and the ependymal lining of the brain ventricles. During development, the vascular pia plus a covering layer of ependyma invaginates into the ventricles in regions of the roof plate where no nervous tissue intervenes between pia and ependyma. The invaginations take the form of fringes of finger-like processes (villous processes), giving a large surface area. The ependymal cells develop many microvilli which protrude into the ventricle, further increasing the surface area. The villous processes contain blood capillaries often specialised for fluid transport by the presence of fenestrations (i.e. the endothelial cells of the capillaries have pores closed by thin diaphragms). However, the membranes of adjacent ependymal cells are linked by tight junctions (zonulae occludens), so preventing the passage of extracellular fluid and providing a basis for the blood–cerebrospinal fluid barrier. Large molecular weight substances, such as proteins and hydrophilic solutes, are prevented from passing from blood to the ventricles; other substances are actively transported by the ependymal cells. Some nerve fibres are present adjacent to the blood vessels of the choroid plexuses. They are thought to be sympathetic fibres allowing control of blood flow through the choroid plexuses.

Choroid plexuses are found in all four brain ventricles (see Figure 166).

The layer of pia mater above the roof of the third ventricle of the diencephalon becomes folded back on itself during the overgrowth of the cerebral hemispheres and the corpus callosum during development (see page 427). The double layer is known as the velum interpositum and forms the roof and floor of the transverse fissure. Where this double layer of pia fuses it is known as the tela choroidea of the third ventricle. On each side the inferior layer is attached to the thalamus, the line of its attachment following the stria medullaris thalami and being known as the taenia thalami. Two vascularised folds of the inferior layer, covered with ependyma, invaginate ventrally to form the choroid plexus of the third ventricle.

180,12; 191,30
180,3–6
191,29
191,7

On each side, a lateral invagination from the tela choroidea, above the thalamus and below the fornix produces the choroid plexus of the lateral ventricle of a cerebral hemisphere. The growth of the cerebral hemisphere during development causes the choroid plexus of the lateral ventricle to be bent in a C-shape, following the lateral aspect of the thalamus from superior, through posterior, to inferior. In the inferior horn of the lateral ventricle, the choroid plexus lies medial and anterior (or superior) to the fimbria of the fornix. It occupies the choroid fissure and extends throughout the body and inferior horn of the lateral ventricle, but not into the anterior or posterior horns. The glomus is a swelling of the choroid plexus in the descending limb of the lateral ventricle, at the junction of the anterior and posterior choroidal arteries. It may become cystic and so exert pressure on structures in the walls of the ventricle. More commonly, it partially calcifies, providing a convenient radiographic landmark. The choroid plexuses of the lateral ventricles are continuous with those of the third ventricle through the interventricular foramina.

191,29; 190,7
190,16; 191,6
190,6; 190,10
190,7
190,10
190,17; 190,26
190,3; 190,11

191,6
191,30; 190,5

The layer of pia above the roof of the caudal (inferior) part of the fourth ventricle of the hindbrain is similarly folded back on itself during development by the overgrowth of the cerebellum. The fused double layer becomes the tela choroidea of the fourth ventricle. The tela choroidea is attached along the lateral margins of the rhomboid fossa, the line of attachment on each side forming a taenia of the fourth ventricle. Within the tela choroidea are found the two choroid plexuses of the fourth ventricle. Together these form a T-shape with a double stem. The cross-piece runs transversely between the lateral recesses of the fourth ventricle, with part of the choroid plexuses emerging through the lateral apertures of the fourth ventricle (foramina of Luschka) near to the cerebellopontine angle on each side. The double stem runs caudally, either side of the midline, to the median aperture of the fourth ventricle (foramen of Magendie), immediately cranial and dorsal to the commencement of the central canal of the medulla.

198A,5
198A,10–21; 198E,42

198F,41; 198D,39
198E
198C,31
188A,21
198A,9
198D,40
198C,35

Choroid plexus tissue is specialised for the secretion of cerebrospinal fluid (CSF). About 500 ml of CSF is produced every day, approximately 70% being produced by the choroid plexuses, the remainder largely coming from brain capillaries via the extracellular fluid. There are no lymphatics to effect drainage of interstitial fluid within the brain. The total volume of CSF is approximately 150 ml, of which about 25 ml is intraventricular, the remainder being in the subarachnoid space between the pia and arachnoid, external to the brain and spinal cord. In composition, CSF resembles an ultrafiltrate of blood, but it has higher Na^+, Cl^- and Mg^{2+} concentrations and lower K^+, Ca^{2+} and glucose concentrations than blood plasma. Normal CSF contains little protein (25 mg/100 g) and less than five cells per ml. In the recumbent subject, its pressure is normally 100 to 150 mm of water. Changes in composition or pressure are valuable diagnostically.

CSF is in free communication with the extracellular fluid of the brain and spinal cord, i.e. there is no brain–CSF barrier. Accordingly, it can function as a buffer to the interstitial fluid and remove waste metabolic products into the venous system. Ependymal cells also rapidly take up certain substances (neurotransmitters, hormones) from the CSF. CSF may act to transport hormones and possibly other, neural, messengers. It also has a major structural function, as it supports, cushions, makes buoyant, and maintains a uniform, distributed pressure upon the central nervous system. It has been estimated that the effective inertial mass of the brain is reduced to about one-thirtieth (i.e. 50 g) by being floated in CSF. This buoyancy greatly reduces potential damage when the head is rapidly accelerated or decelerated. Further, the brain's weight is evenly distributed to the supporting meninges. CSF also allows some variation in the volume of the brain within the rigid cranium as CSF volume may be expanded or displaced.

THE CIRCULATION OF THE CEREBROSPINAL FLUID (see Figure 166)

Cerebrospinal fluid (CSF) is secreted by the choroid plexuses in the brain ventricles and it fills these ventricles. Fluid formed in the lateral ventricle of each cerebral hemisphere passes through an interventricular foramen (of Monro) into the median, third ventricle of the diencephalon. Fluid formed in the first three ventricles passes through the midbrain in the cerebral aqueduct (of Sylvius) to reach the fourth ventricle of the hindbrain. Fluid from all four ventricles enters the subarachnoid space via the median aperture (of Magendie) and two lateral apertures (of Luschka) of the fourth ventricle. CSF fills the spinal subarachnoid space and passes through the cisterns and over the surface of the brain to drain into the venous system.

The main site of absorption of CSF is via the arachnoid villi. These are found adjacent to the venous sinuses of the brain, most particularly along the superior sagittal sinus. Each villus is a microscopic protrusion of the pia-arachnoid (and a consequent extension of the subarachnoid space) through the meningeal layer of the dura mater into a venous sinus or nearby venous lacuna (see Figure 155). The villi act as unidirectional pressure-dependent valves, allowing CSF to pass into the venous system. Arachnoid granulations are macroscopic protrusions of the pia-arachnoid, containing many arachnoid villi; they increase in number and size throughout life, and frequently calcify. Absorption into the vertebral venous plexus, along with postural changes, produces flow of CSF over the spinal cord.

As CSF is actively secreted, blockages in its circulation will result in raised intracranial pressure (hydrocephalus or 'water on the brain'). Hydrocephalus may be internal (i.e. CSF is prevented from reaching the subarachnoid space) or external/communicating (i.e. absorption into the venous system is impeded). Blockage of the cerebral aqueduct or of the apertures from the fourth ventricle are common causes of internal hydrocephalus. Blockage of the apertures of the fourth ventricle occurs, for example, in developmental abnormalities of the posterior cranial fossa (Arnold–Chiari malformations in which there is cerebellar tonsillar herniation). Lumbar, ventricular and cisternal puncture may be used to investigate the pressure and composition of CSF for diagnostic purposes.

The subarachnoid cisterns

Cisterns are accumulations of cerebrospinal fluid within the subarachnoid space where the pia mater and arachnoid mater are widely separated. The most important are as follows:

The lumbar cistern extends from the caudal end of the spinal cord (typically at the caudal border of the first lumbar vertebra in adults) to the caudal limit of the dura mater (caudal border of the second sacral vertebra). It contains parts of the cauda equina and filum terminale.

The cerebellomedullary cistern (cisterna magna) is in the angle formed by the dorsal aspect of the medulla and the ventral (caudal) aspect of the cerebellum. Its caudal boundary is formed by a sheet of arachnoid that passes directly from the ventral surface of the cerebellum to the dorsal surface of the medulla. The cistern is continuous with the subarachnoid space surrounding the spinal cord. The median aperture of the fourth ventricle (foramen of Magendie) opens directly into this cistern.

158,27; 173,15
158,26; 158,22

158,33
198D,40

The pontine cistern (cisterna pontis) is the upward continuation of the subarachnoid space of the spinal meninges, ventral to the medulla and pons. It is just deep enough to contain the vertebral and basilar arteries. Cerebrospinal fluid enters this cistern from the fourth ventricle through its lateral apertures (foramina of Luschka) and is secreted into it directly by parts of the choroid plexus emerging through these apertures. The pontine cistern is in continuity with the cerebellomedullary cistern dorsally and interpeduncular cistern cranially.

158,47; 158,33
158,26; 158,25
158,46; 158,47
158,23; 188A,21

158,27
158,54; 180,27

The cistern of the great cerebral vein (superior cistern or cisterna ambiens) lies dorsal to the midbrain in the tentorial opening. It extends from the splenium of the corpus callosum to the superior surface of the cerebellum. It contains the great cerebral vein and the pineal body.

158,15
164,24; 164,5
180,6; 180,15
180,14; 180,13

The interpeduncular cistern (cisterna basalis) extends ventral to the midbrain and hypothalamus, and fills the interpeduncular fossa. The cistern is enclosed by a sheet of arachnoid mater that passes from one temporal lobe to the other. The frontal lobes and the optic chiasma form the anterior limit of the cistern and the caudal boundary is formed by the pons. It contains the cerebral arterial circle and is pierced by the infundibulum (pituitary stalk). Some authorities include within this cistern the cisterns that extend anteriorly and then superiorly round the corpus callosum, i.e. the cisterns of the optic chiasma and the lamina terminalis, and the supracallosal cistern. These cisterns contain the anterior cerebral arteries.

180,27; 180,21
180,36; 184,10
184,28; 184,2
184,7
184,13; 184,9
184,8
180,3–4
180,31; 180,32
182B,23; 186,3

The cistern of the lateral sulcus (of the lateral fossa) is continuous with the interpeduncular cistern and occupies the deep fissure between the temporal and the frontal lobes of the brain. It contains the middle cerebral artery.

174,3;

186,7; 186,34

The vasculature of the central nervous system

Nervous tissue requires a large and constant supply of well-oxygenated blood for its normal function. The central nervous system is highly metabolically active and there is little storage of oxygen or glucose. The brain consumes about 20% of the body's total oxygen requirement, although it makes up only about 2% of the body's weight. There is sophisticated control of the blood supply to the brain, regions of high neuronal activity receiving increased supply.

Cessation of the blood supply to the brain results in loss of consciousness within seconds. Irreversible damage (infarction) occurs within about five minutes of the deprivation of arterial supply. Considerable anatomical and functional reorganisation, involving the establishment of new synaptic connections and the learning of new conceptual and motor skills, is possible and may continue for some time following lesions of the central nervous system. Nevertheless, some permanent loss of function is likely after a major cerebrovascular accident.

186

The adventia and media of the cerebral arteries are thinner than in most other parts of the body. This feature predisposes them to aneurysm. Aneurysms occur most frequently where arteries branch: such points on the cerebral arterial circle (of Willis) are particularly common sites of aneurysms. As elsewhere, atheromatous plaques, quite common in middle age and beyond, predispose vessels to haemorrhage, thrombosis or occlusion by embolism. Cerebrovascular accidents are particularly serious because of the relative poverty of anastomotic connections, the functional

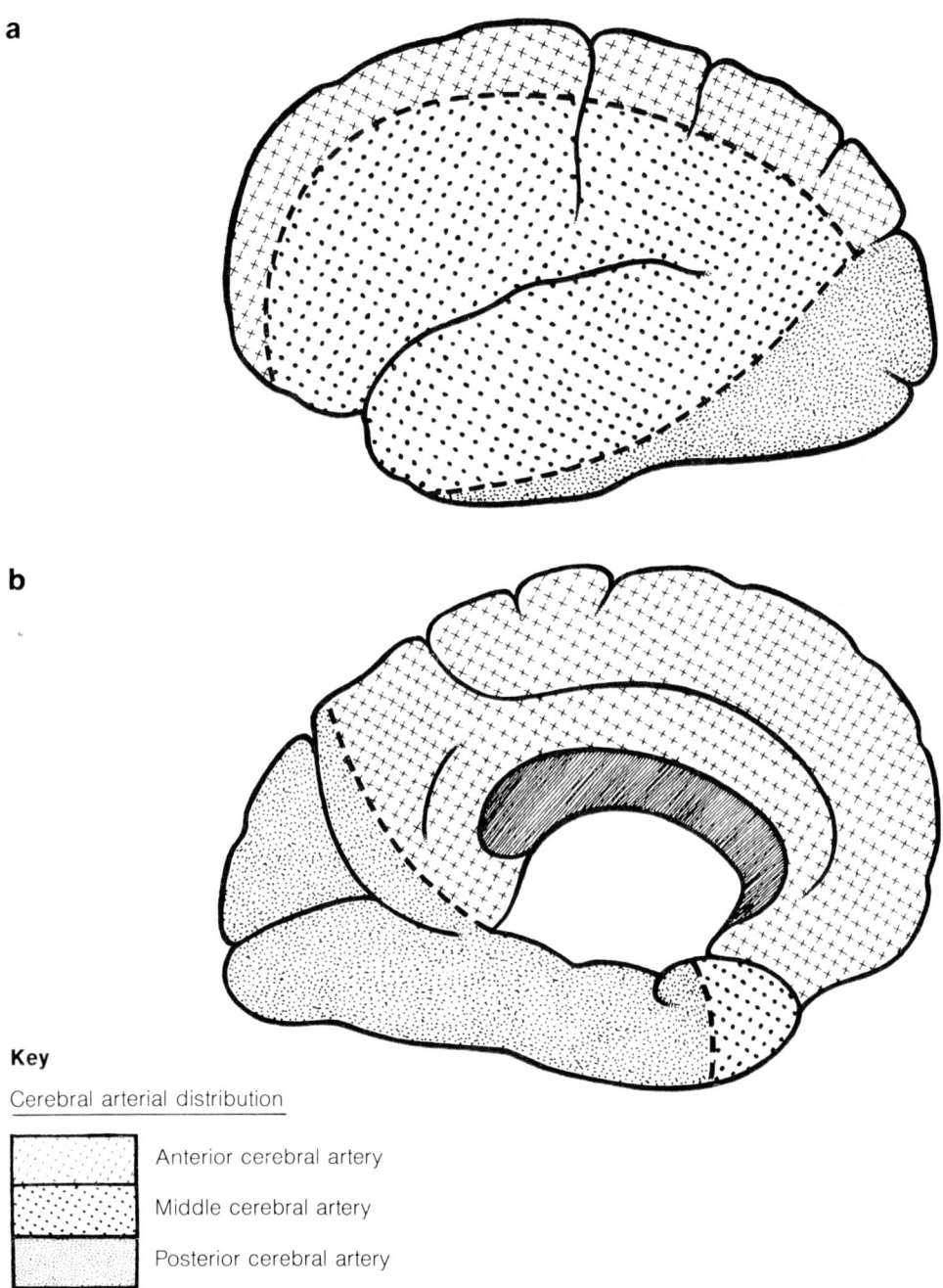

Figure 196 *The cortical territories of the cerebral arteries.* a) *Lateral view*, b) *medial view*.

importance of brain tissue, and its susceptibility to infarction. Haemorrhages also cause local neural damage. If haemorrhages are major, the resulting increase in intracranial pressure (since the cranial cavity is effectively closed) may lead to cessation of neuronal activity and death.

As already stated, control of the blood supply to the brain is closely regulated. The supply shows considerable and rapid regional changes in response to increased demand (greater neural activity within the region) or arterial insufficiency (ischaemia). This regulation appears to be controlled largely locally (by autoregulation), rather than by sympathetic nerve fibres, although non-penetrating branches of arteries receive sympathetic innervation. Cerebral arterioles respond to P_{CO_2} (pH), P_{O_2} and blood pressure to maintain constant levels of oxygen supply and carbon dioxide removal. It seems likely that central vessels are also under neuronal control (a variety of peptide or monoamine neurotransmitters effecting such a role have been proposed). The regional changes in blood supply can be demonstrated using positron emission tomography (PET scans).

THE ARTERIAL SUPPLY TO THE BRAIN

The arterial supply to the brain is normally derived entirely from the internal carotid and vertebral arteries.

168,6; 186,6
168,25; 186,21

The two vertebral arteries enter the cranial cavity through the foramen magnum. They unite to form the basilar artery which terminates by dividing into two posterior cerebral arteries. The major branches of the vertebral artery are the ventral (anterior) and dorsal (posterior) spinal arteries, and the posterior inferior cerebellar artery. Other major branches of the basilar artery are the anterior inferior cerebellar and the superior cerebellar arteries. The parts of the brain supplied by these arteries are 'vertebrobasilar territory'. Typically, this includes the whole of the brainstem, the cerebellum, most of the diencephalon, and the posterior and inferior parts of the cerebral hemispheres supplied by the posterior cerebral arteries (Figure 196).

158,46; 186,21
158,47; 186,15
158,54; 186,9
186,22; 200B,36
186,24; 186,17
182,16

The two internal carotid arteries enter the cranial cavity through the carotid canals. Each artery divides into an anterior and a middle cerebral artery after giving off an ophthalmic and an anterior choroidal branch. The territories of the anterior and middle cerebral arteries are shown in Figure 196. Small central branches of these arteries, the striate arteries, pass through the anterior perforated substance to supply parts of the diencephalon, the basal nuclei (basal ganglia) and the internal capsule. The regions supplied by these vessels are 'carotid territory', i.e. parts of the diencephalon and most of the cerebral hemispheres.

166C,38; 22,36
186,3; 186,7
166A,16; 186,33

186,36; 188B,32

The anterior cerebral arteries are linked anastomotically by the anterior communicating artery. On each side, the internal carotid artery gives rise to a posterior communicating artery which links it to the posterior cerebral artery. These connections establish anastomoses between carotid and vertebrobasilar territories, so completing the cerebral arterial circle or circulus arteriosus (of Willis). In practice, the posterior communicating arteries are variable in number and magnitude. Thus, the cerebral arterial circle is incomplete in some 10% of people, while in others one or more posterior cerebral arteries may be mainly supplied from the internal carotid artery.

186,3; 186,38
186,6; 186,8
186,9

THE VERTEBROBASILAR TERRITORY

The vertebral artery

Each vertebral artery (see also page 131) arises from a subclavian artery. It ascends through the foramina of the transverse processes of the upper six cervical vertebrae. It bends behind the lateral mass of the atlas and enters the vertebral canal by passing below the inferior border of the posterior atlanto-occipital membrane. It then passes through the dura and arachnoid. It reaches the cranial cavity through the foramen magnum and ascends in the subarachnoid space, ventral to the roots of the hypoglossal nerve, on the ventral surface of the medulla oblongata. The two vertebral arteries,

108,39; 108,48
200B,39; 80A,11
200A,16;
200A,19
198F,48;
168,25; 168,20
186,21; 186,23

which are often unequal in size, join near the caudal border of the pons to form the median (unpaired) basilar artery.

Cranial branches of the vertebral artery:
- Meningeal arteries
- Ventral (anterior) spinal artery
- Dorsal (posterior) spinal artery
- Posterior inferior cerebellar artery
- Medullary arteries

Meningeal arteries arise near the foramen magnum and supply the falx cerebelli and bones of the cerebellar fossa.

A ventral (anterior) spinal artery arises medially from each vertebral artery. It passes caudally and unites with its fellow on the ventral surface of the caudal medulla to form an unpaired, median, ventral spinal artery which descends in the ventral median fissure of the spinal cord. The ventral spinal artery supplies medial parts of the medulla oblongata, including the hypoglossal nerve and nucleus, medial longitudinal fasciculus, medial lemniscus and pyramid (corticospinal tract) (see Figure 178c). Infarction following occlusion of the artery produces loss of function of these structures, producing a constellation of signs called the medial medullary syndrome.

A dorsal (posterior) spinal artery arises from either the vertebral or the posterior inferior cerebellar artery within the cranial cavity. It supplies a small dorsal sector of the medulla (including the dorsal column nuclei and parts of the inferior cerebellar peduncle). It descends at first dorsolaterally along the medulla oblongata and then as a pair of branches lying on either side of the dorsal roots of the spinal nerves to supply the spinal cord (see page 582).

The posterior inferior cerebellar artery arises from the vertebral artery near the olive of the medulla. It is very variable in size and not uncommonly absent on one side. It passes at first dorsally and caudally, and then runs cranially, dorsal to the rootlets of the vagus and glossopharyngeal nerves. At the lower border of the pons, it turns caudally along the caudal, lateral margin of the fourth ventricle, before passing laterally into the vallecula of the cerebellum. The posterior inferior cerebellar artery gives off many small branches during its course. Some branches anastomose with branches of the anterior inferior and superior cerebellar arteries. The artery usually supplies a large dorsolateral segment of the more cranial parts of the medulla, the cerebellum (particularly inferior and caudal parts, and also the dentate nucleus) and the choroid plexus of the fourth ventricle. Its occlusion can give rise to the lateral medullary syndrome (of Wallenburg) caused by infarction of: the spinal tract and nucleus of the trigeminal nerve, vestibular nuclei and the lateral vestibulospinal tract, the nucleus ambiguus, the solitary tract and its nucleus, spinocerebellar tracts, fibres of the anterolateral system, and descending autonomic fibres (see Figure 178d). More caudally, direct medullary branches of the vertebral artery may supply many of these structures.

Medullary arteries supply intermediate parts of the medulla oblongata. They enter the medulla as small vessels adjacent to the rootlets of the glossopharyngeal, vagus and accessory nerves.

The medullary territories supplied by the various branches of the vertebral artery overlap. They also show considerable individual variation. In addition, there are anastomotic connections between the different territories. Accordingly, vertebral insufficiency is the most common cause of medullary infarction. Cranial nerve palsies may arise from compression, or intermittent stimulation, of the fibres of a nerve which is too close to a central branch of one of the vertebral (or other) arteries.

The basilar artery

The basilar artery lies in the median groove on the ventral surface of the pons and terminates near the upper border of the pons by dividing into two posterior cerebral arteries.

Branches of the basilar artery:
- Paramedian, short circumferential and long circumferential arteries
- Labyrinthine arteries
- Anterior inferior cerebellar arteries
- Superior cerebellar arteries
- Posterior cerebral arteries

The paramedian, and short and long circumferential arteries are named according to the distance they travel from the basilar artery laterally and dorsally around the pons before entering and supplying it.

A labyrinthine (internal auditory) artery usually arises from each side of the basilar artery, but may arise from one of the cerebellar arteries. It runs with the facial and vestibulocochlear nerves into the internal acoustic meatus and supplies the inner ear.

An anterior inferior cerebellar artery arises on each side from the caudal part of the basilar artery. It runs caudally and laterally towards the internal acoustic meatus, passing ventral to the abducent, facial and vestibulocochlear nerves. It reaches the craniolateral part of the ventral cerebellum where it anastomoses with the posterior inferior cerebellar artery. The anterior inferior cerebellar artery chiefly supplies the cerebellum, including lateral parts of the middle cerebellar peduncle, but also gives some supply to tegmental parts of the pons and sometimes to the cranial medulla. It is variable in size and is sometimes absent.

A superior cerebellar artery arises close to the termination of the basilar artery on each side; it may be multiple. It passes laterally around the cerebral peduncle of the midbrain, immediately caudal to the oculomotor and trochlear nerves, to reach the superior surface of the cerebellum. It anastomoses with the inferior cerebellar arteries. The superior cerebellar artery supplies anterior parts of the cerebellum, dorsal parts of the pons and midbrain, the pineal body and the choroid plexus of the third ventricle. The superior cerebellar arteries distribute mainly below the tentorium cerebelli, the posterior cerebral arteries above it.

The left and right posterior cerebral arteries are the terminal branches of the basilar artery. Each posterior cerebral artery passes round the cerebral peduncle of the midbrain, cranial to the oculomotor and trochlear nerves, and close to the optic tract. It receives the posterior communicating artery before reaching the tentorial surface of the cerebral hemisphere. There it gives off branches to the inferior surface and underlying white matter of the temporal lobe (including the uncus and hippocampal formation, but excluding the temporal pole) and the whole of the occipital lobe and neighbouring parts of the parietal lobe (Figure 196).

The cortical branches of the posterior cerebral artery include the anterior and posterior (temporo-occipital) temporal arteries and the parieto-occipital and calcarine branches that run in their respective sulci. The posterior cerebral artery supplies the visual cortex.

Small, posteromedial central branches of the posterior cerebral artery pass (together with branches of the posterior communicating artery) through the posterior perforated substance between the cerebral peduncles to supply the midbrain and (via thalamo-perforating branches) medial parts of the diencephalon. Small posterolateral central (thalamogeniculate) branches arise laterally and pass centrally to supply the midbrain (including the colliculi), the pineal body and the posterior thalamus.

Posterior choroidal arteries arise from the posterior cerebral artery to supply the choroid plexuses of the third and lateral ventricles. The medial posterior choroidal artery(ies) passes lateral to the pineal body, and medially above the thalamus in the tela choroidea of the third ventricle. It supplies dorsal parts of the midbrain and thalamus. The lateral posterior choroidal artery(ies) passes laterally into the posterior part of the choroid fissure to supply the choroid plexus of the lateral ventricle; it also supplies the fornix. The posterior choroidal artery and anterior choroidal artery (a branch of the internal carotid artery) anastomose in the glomus within the descending limb of the inferior horn of the lateral ventricle.

THE CAROTID TERRITORY

The internal carotid artery

22,36; 166C,61
166C,38
166A,9–10; 168,6
188A,28; 186,6
186,7; 186,3

The internal carotid artery is a branch of the common carotid artery. It enters the skull via the carotid canal of the temporal bone and, above the foramen lacerum, passes through the cavernous sinus and into the subarachnoid space of the interpeduncular cistern, medial to the anterior clinoid process. Inferior to the anterior perforated substance it divides into two main branches, the middle and anterior cerebral arteries.

Cranial branches:

- Cavernous arteries
- Hypophyseal arteries
- Meningeal artery
- Ophthalmic artery
- Anterior choroidal artery
- Posterior communicating artery
- Anterior cerebral artery
- Middle cerebral artery

Cavernous arteries supply the trigeminal ganglion and the dura mater.

Hypophyseal arteries are distributed to the pituitary gland and are described with the gland on pages 399 to 400.

The meningeal artery is small and supplies dura mater in the anterior cranial fossa.

168,5
28,19; 168,4

168,10

The ophthalmic artery arises in the interpeduncular cistern and passes through the optic canal (inferolateral to the optic nerve) to supply the eye (see pages 226 to 228). The artery provides a meningeal branch (either directly, or indirectly through the lacrimal artery) that runs back through the superior orbital fissure to anastomose with the middle meningeal artery.

186,33
186,31; 188A,27
194,20; 194,19
190,10

The anterior choroidal artery (which may alternatively arise from the middle cerebral artery) passes posteriorly round the cerebral peduncles close to the uncus. It passes through the choroid fissure to reach and supply the choroid plexus of the inferior horn of the lateral ventricle. It gives off small branches to the hypothalamus, cerebral peduncle, anterior medial temporal lobe (uncus, hippocampal formation and amygdala), optic tract, lateral geniculate body of the thalamus, retrolenticular and inferior parts of the internal capsule (including the optic radiation) and basal nuclei. The anterior choroidal artery anastomoses with the posterior choroidal artery in the glomus of the descending limb of the lateral ventricle. Its long subarachnoid course and small diameter make it vulnerable to thrombotic occlusion.

186,8; 164,20
186,9; 164,4

The posterior communicating artery passes caudally, superior to the oculomotor nerve, to establish a variable anastomotic connection with the posterior cerebral artery. Small central branches pass into the posterior perforated substance between the cerebral peduncles and supply parts of the midbrain and diencephalon.

The anterior cerebral artery passes medially, above the optic nerve, into the longitudinal fissure of the cerebral hemisphere. Here it immediately anastomoses with the anterior cerebral artery of the opposite side via the very short anterior communicating artery(ies). Both anterior cerebral arteries run close together around the genu of the corpus callosum and, posteriorly, continue just superior to the body of the corpus callosum (as the pericallosal arteries). Posteriorly, terminal branches of the anterior cerebral artery may anastomose with branches of the posterior cerebral artery; anteriorly, anatomoses may occur with branches of the middle cerebral artery.

Cortical branches are given off to supply the cortex and underlying white matter of the medial surface, plus a superomedial strip of the frontal and parietal lobes of the hemisphere (Figure 196). These branches include: the frontobasal (orbital) arteries, the callosomarginal artery, the frontopolar artery and the medial frontal arteries, the paracentral and the precuneal arteries. The anterior cerebral artery supplies parts of the cortex (parts of areas MsI and SmI) concerned with movements and sensations from the contralateral leg and perineal region. Occlusion results in contralateral monoplegia and anaesthesia of the lower limb. In the speech-dominant hemisphere (usually the left), it supplies fibre connections of the anterior speech cortex, so that its blockage may cause expressive aphasia. Anomalies occur in some 25% of brains; these include the occurrence of a single, unpaired anterior cerebral artery, or branches which supply the contralateral hemisphere.

Small anteromedial central branches arise from the initial part of the anterior cerebral artery and from the anterior communicating artery; these branches pass through or medial to the anterior perforated substance to supply the substantia innominata, parts of the basal nuclei, the anterior limb of the internal capsule, the hypothalamus (including the optic chiasma), the septal region and the anterior columns of the fornix. One of the more constant of these branches is called the medial striate or long central (recurrent) branch (of Heubner). It may arise proximal or distal to the anterior communicating artery.

The middle cerebral artery is the major branch of the internal carotid artery. It passes laterally and runs deep in the lateral sulcus of the cerebral hemisphere on the surface of the insula.

The middle cerebral artery gives off cortical branches which supply most of the cortex and underlying white matter of the superolateral surface of the hemisphere, the insula, and the temporal pole (Figure 196). Some of its branches run along the more constant sulci, for example the arteries of the precentral and central sulci. Other branches include the lateral frontobasal (orbitofrontal) artery, the anterior and posterior temporal arteries, the artery of the angular gyrus, and the anterior (postcentral) and posterior parietal artery. Significantly, the middle cerebral artery supplies parts of the somatosensory and motor cortex (most of SmI and MsI) dealing with the head and upper limb, the auditory cortex and, in the speech-dominant hemisphere (usually the left), the anterior and posterior speech cortex. If not fatal, major occlusion leads to: contralateral paralysis and sensory loss which is most marked in the upper limb and face, some contralateral auditory loss and, possibly, severe aphasia and/or loss of high order perceptual functions.

Small anterolateral central branches arise from near the commencement of the middle cerebral artery (and also from terminal portions of the internal carotid artery) and pass into the anterior perforated substance to supply the basal nuclei and the internal capsule.

The striate arteries (so called because they supply the corpus striatum) are small arteries that pass through the anterior perforated substance to supply the basal nuclei (caudate nucleus, putamen and globus pallidus) and internal capsule. They are sometimes divided into medial and lateral striate arteries, but the division is neither sharp nor constant. Although one of them has been called the 'artery of cerebral haemorrhage' because of its susceptibility to rupture, it is doubtful whether it is a constant branch. The striate arteries arise mainly from the middle cerebral artery.

The primary supply of the anterior and posterior limbs of the internal capsule is normally from the lateral striate arteries. The medial striate arteries supply parts of the anterior limb. The genu is usually supplied by branches from the internal carotid artery. The retrolenticular portion (including the optic radiation) and inferior parts of the posterior limb of the internal capsule are supplied by branches of the anterior choroidal artery. Occlusion or rupture of even small arteries supplying the internal capsule can have dire consequences because of the concentration of important sensory and motor connections of the cerebral cortex within this restricted region. Infarction of retrolenticular parts of the internal capsule, including the optic radiation, results in a contralateral homonymous hemianopia. Infarction of the posterior limb, including ascending and descending connections of somatosensory and motor cortex, results in a contralateral spastic paralysis (hemiplegia) and sensory loss. In some cases, the face may be spared from paralysis or, alternatively, solely involved, as corticonuclear fibres are more anteriorly placed than corticospinal fibres. Many fibres in addition to corticonuclear or corticospinal fibres are damaged in a typical lesion of the internal capsule. Aphasia may result if the lesion is in the speech-dominant hemisphere (usually the left).

The cortical branches of the cerebral arteries all supply the cerebral hemisphere by very small vessels that enter the cortex vertically from the pia mater. Short arteries terminate in the cortex while longer, medullary arteries pass through the cortex to the underlying white matter. There are no anastomoses between these penetrating vessels.

Functionally, the major cerebral arteries (anterior, middle, posterior and choroidal) are end-arteries; such anastomoses as are formed (usually near the edges of their territories) are insufficient to prevent brain damage should a major vessel be rapidly occluded. Similarly, usually the smaller branches of the cerebral arteries, and often the smaller central branches passing to the brainstem, diencephalon, basal nuclei or internal capsule are functionally end-arteries. If occlusion is gradual, adequate anastomotic connections may develop, for example, between the main cerebral arteries, or between internal and external carotid territories via meningeal vessels. Anastomoses in the subarachnoid space are common between branches that both arise from the same cerebral artery.

THE ARTERIAL SUPPLY TO THE SPINAL CORD

The arterial supply to the spinal cord is via small, central branches that arise from a single ventral (anterior) and two pairs of dorsal (posterior) spinal arteries. These spinal arteries run the length of the spinal cord. The ventral spinal artery is in the ventral median fissure. The dorsal spinal arteries lie on either side of the dorsal roots of the left and right spinal nerves. The spinal arteries are supplied by branches from the vertebral artery, and from the deep cervical, posterior intercostal and lumbar arteries via a very variable number of radicular branches which accompany the spinal nerves. There are good anastomotic connections between the dorsal and ventral spinal arteries in the subarachnoid space, but not within the spinal cord itself. Because of the variable nature of the radicular arterial supply, certain levels of the cord (C2 to C3, T1 to T4, L1) are particularly vulnerable to ischaemic damage.

THE VENOUS DRAINAGE OF THE BRAIN

The venous drainage of the cerebral hemisphere (see Figure 156)

Small veins emerging from the brain substance form a pial plexus before gathering into the main veins. The veins have no valves and are extremely thin-walled as they contain no muscular tissue. The major vessels run in the subarachnoid space before emptying into venous sinuses within the dura. The main features of the venous drainage are illustrated in Figure 156. The positions of the venous sinuses have already been described (see pages 393 to 395).

The pattern of veins on the surface of the cerebral hemisphere is variable. The superolateral surface of the hemisphere is partly drained by eight to twelve superior cerebral veins. These, joined by some small vessels from the medial surface, pass towards the superomedial border of the hemisphere and terminate in the superior sagittal sinus. The superficial middle cerebral vein runs anteriorly and superficially along the lateral sulcus. It drains adjacent parts of the superolateral surface and ends in the cavernous sinus. The inferior surface of the hemisphere is drained by small, inferior cerebral veins. Anteriorly, they drain into the anterior cerebral vein or superior cerebral veins. Posteriorly, they drain into the superficial middle cerebral or basal veins. Veins from the medial surface of the hemisphere either drain superiorly via the superior cerebral veins into the superior sagittal sinus or inferiorly into the anterior cerebral vein. There are good anastomotic connections between the superficial and deep cerebral veins.

174; 1; 172,12
158,4; 170,5
162,3-4; 174,3

168,33
174,4

The anterior cerebral vein accompanies the anterior cerebral artery in its course round the corpus callosum and drains medial and inferior regions of the hemisphere.

182,23

The deep middle cerebral vein accompanies the middle cerebral artery deep in the lateral sulcus and drains some of the deeper parts of the hemisphere (insula, basal nuclei and white matter). It unites with the anterior cerebral vein inferior to the anterior perforated substance and forms the basal vein (of Rosenthal).

186,34

188B,32

Striate veins accompany the striate arteries and emerge through the anterior perforated substance to empty into the basal vein.

186,36
188B,32

The basal vein accompanies the posterior cerebral artery, passing posteriorly around the cerebral peduncle of the midbrain, and empties into the great cerebral vein between the splenium of the corpus callosum and the pineal body.

158,16; 186,9
186,31; 158,15
158,5; 158,14

The thalamostriate (terminal) vein runs anteriorly in the groove between the caudate nucleus and thalamus. It drains deep parts of the hemisphere (including the basal nuclei and thalamus). Near the interventricular foramen, it unites with the choroidal vein of the choroid plexus of the lateral ventricle to form the internal cerebral vein.

192,9; 194,7
192,8+10; 194,6+8
192,6
191,6; 191,31

The choroidal vein drains the choroid plexus of the lateral ventricle and the hippocampal formation.

191,6
190,24

The internal cerebral vein runs posteriorly, close to the midline and above the thalamus in the tela choroidea of the third ventricle. The two internal cerebral veins unite with the two basal veins to form the single, median great cerebral vein.

191,31
158,9; 158,8
158,15; 191,32

The great cerebral vein (of Galen) runs posteriorly around the splenium of the corpus callosum. It joins with the inferior sagittal sinus (running in the inferior border of the falx cerebri deep in the longitudinal fissure) and drains into the straight sinus.

180,14; 180,6
158,62
158,17

The superficial middle cerebral vein runs superficially along the lateral sulcus. It drains much of the superolateral surface of the hemisphere. It is usually linked to the superior sagittal sinus by superior anastomotic veins, and to the transverse sinus by one or more inferior anastomotic veins. Posteriorly, these anastomotic channels may be large. The superficial middle cerebral vein ends in the cavernous sinus lying to the side of the body of the sphenoid bone.

174,3

170,5; 174,2
162,8
168,33

The venous drainage of the cerebellum and brainstem

Venous drainage of the cerebellum is via superior cerebellar veins that run superomedially to empty into the straight sinus or into the great cerebral vein, or laterally into the transverse or superior petrosal sinuses. Inferior cerebellar veins pass laterally or posteriorly to end in the inferior petrosal, occipital or sigmoid sinuses.

158,17; 158,15
168,22; 168,12
168,17; 168,21

158,16; 158,15
168,12+17; 168,22

Venous drainage of the brainstem is via a superficial venous plexus that lies deep to the arteries. Cranially, the plexus drains into the basal or great cerebral veins. Pontine veins may drain into the petrosal or transverse sinuses, the cerebellar or basal veins, or into emissary veins. Caudally, the medullary venous plexus is continuous with veins of the spinal cord. Drainage also occurs into neighbouring venous sinuses or into the internal jugular vein.

The great proportion of venous blood from the cranial cavity passes into the internal jugular vein. However, blood also leaves via a number of emissary veins, so establishing contact between venous sinuses inside the skull and veins outside (see page 396).

THE VENOUS DRAINAGE OF THE SPINAL CORD

The venous drainage of the spinal cord is into pial plexuses within which there are six longitudinally running channels. There is a ventral (anterior) and a dorsal (posterior) median longitudinal vein, and two ventrolateral and two dorsolateral longitudinal veins (running along the line of the nerve roots). These vessels drain via veins passing with the spinal nerve roots into the internal vertebral venous plexuses within the epidural fat of the vertebral column (see also page 67). These plexuses drain into lumbar, intercostal or azygos systems of veins or, cranially, into the basilar venous plexus.

THE BLOOD–BRAIN BARRIER

The nervous system operates in a protected, closely controlled environment. Part of that protection is provided by the selective permeability that exists between blood capillaries and nervous tissue. Capillaries supplying the brain, spinal cord, peripheral nerves, retina and inner ear have a continuous inner layer of endothelial cells linked together by tight junctions (zonulae occludens). At these tight junctions the plasma membranes of adjacent cells are closely apposed, so that the permeability of the capillaries is effectively restricted to that of a plasma membrane. Thus, substances can pass from the blood to nervous tissue only by lipid mediation or by active transport through the surrounding cells. Cerebral blood vessels do not in general have pores (fenestrations) nor a well-developed pinocytotic or vesicular transport system. The endothelial cells actively transport amines, amino acids and sugars. The endothelial cells are surrounded by a basement membrane. Some 85% of the surface of this membrane is covered by glial cell processes, so allowing further regulation of the transfer of chemicals. Nevertheless, the main structural counterpart of the blood–brain barrier is the seal provided by the endothelial cells.

Certain regions of the brain are not protected by a blood–brain barrier: the pineal body, the posterior lobe of the pituitary gland, the area postrema, the subfornical organ, the organum vasculosum of the lamina terminalis (supra-optic crest) and the median eminence of the hypothalamus. These regions comprise all but one of the circumventricular organs (the exception being the subcommissural organ). The regions are highly vascularised and appear to have specialised secretory and/or chemoreceptor functions. Within these regions the capillaries are fenestrated, i.e. have regions where the walls are reduced to a highly permeable, thin diaphragm, so facilitating chemical (including macromolecular) exchanges.

Capillaries within the choroid plexuses of the brain ventricles are also fenestrated. However, in these regions the ependymal cells lining the ventricles are linked by tight junctions, so producing a blood–cerebrospinal fluid barrier. Such a barrier is necessary because there is free permeability between cerebrospinal fluid and the interstitial fluid of the central nervous system.

The circumventricular organs

The circumventricular organs are: the area postrema, the subfornical organ, the organum vasculosum of the lamina terminalis (supra-optic crest), the posterior lobe of the pituitary gland, the median eminence, the pineal body and the subcommissural organ. These regions have specialised relationships with interventricular cerebrospinal fluid and/or blood capillaries. Their cells are specialised for chemoreception and/or secretion. With the exception of the subcommissural organ, blood capillaries in these regions are fenestrated so that they lack a blood–brain barrier and are therefore permeable to proteins and peptides. These regions probably take part in functions largely under hypothalamic control. They may themselves be influenced by hypothalamic and pituitary hormones and other neuropeptides found in cerebrospinal fluid. They provide a possible means of hormonal influence on brain function.

The area postrema is located dorsally in the medulla on either side of the midline close to the obex (see Figure 178d). It is known as the chemoreceptor trigger zone. It is sensitive to changes in the composition of blood plasma. The area postrema gives inputs to the surrounding reticular formation. Thus, it can influence cardiovascular functions and respiration. Activation of its neurones by substances in the arterial circulation can trigger the vomiting reflex. 198C,35

The remaining circumventricular organs are all unpaired, median structures adjacent to the third ventricle.

The subfornical organ (intercolumnar tubercle) lies between the interventricular foramina on the posterior surface of the anterior columns of the fornices. It is a chemoreceptor zone, sensitive to the osmolarity of plasma and cerebrospinal fluid. It is important, via its connections with the hypothalamus, for regulation of body fluids and water intake (thirst). It mediates the effects of angiotensin on water intake. 191,5; 180,9 191,28; 180,8

The organum vasculosum of the lamina terminalis (supra-optic crest) is found in the lamina terminalis above the optic chiasma. It appears to be a site for the release into the blood stream of gonadotropin-releasing hormone and somatostatin (growth hormone inhibiting factor). It is a third chemoreceptor zone and may play a role in the control of water balance. 180,34; 180,31

The posterior lobe of the pituitary gland (see pages 398 to 399 and 527) is connected to the ventral hypothalamus by the pituitary stalk. 158,50; 158,57 158,51

The median eminence is found in the floor of the third ventricle immediately anterior and posterior to the infundibulum. It is the site of release into the hypophyseal portal venous system of hormonal releasing factors which act on the anterior lobe of the pituitary gland (see pages 398 to 400 and 527). 180,29 180,50

The pineal body is the fourth circumventricular organ concerned with endocrine functions. It forms part of the epithalamus and is located in the midline immediately anterior to the superior colliculi of the midbrain (see page 534). 180,13; 196,1 180,24

The subcommissural organ is located just ventral and posterior to the pineal body (and to the posterior commissure) at the junction of the third ventricle with the cerebral aqueduct of the midbrain. This area is not highly vascularised but the cells are specialised for secretion. Their function is unknown though it may be related to pineal function. 158,14 158,9 158,11

The general organisation of sensory and motor pathways

This section contains some general comments concerning the organisation of the major sensory and motor pathways by way of an introduction to the diagrams that follow (the summary sheets on pages 588 to 607). The diagrams are simplified in order to highlight the main features of the pathways. Although the pathways are bilaterally symmetrical, only one side is illustrated. No distinction is made between projections established by axon collaterals and by separate neurones; a single line can represent many thousands of axons (and possibly millions of synapses).

The sensory/ascending pathways comprise both special sensory systems (the olfactory, visual, gustatory, auditory and vestibular systems) and the somatosensory or general sensory system (which includes tactile, nociceptive, temperature and proprioceptive afferent pathways).

An idealised sensory pathway consists of fibre links provided by three main neuronal stages between the receptor cell and the contralateral cerebral cortex (although the actual pathways show many variations from this scheme):

	Receptors	
Stage I	Ganglion cells	↓ Peripheral process } of spinal or cranial nerve
Stage II	Nuclear cells	↓ Central process
Stage III	Thalamic nucleus	↓ Decussation; lemniscus
	Cerebral cortex	↓ Thalamic radiation of the internal capsule

The thalamic radiations end in the primary sensory area of the cerebral cortex. From this primary area information is relayed to, and further processed in, secondary areas – sensory and polysensory association cortex. These association areas are particularly well developed in the human brain. They are responsible for higher perceptual and thought processes.

All sensory systems have descending/centrifugal fibres that make possible modulation and feedback control of the information that is being transmitted towards the cortex. Except for the olfactory system, each area of sensory cortex projects back upon the thalamic nucleus that gives rise to its sensory input. Descending connections are particularly prominent in the auditory system where they arise from the auditory cortex, the inferior colliculus and the superior olivary complex (passing from the latter as far as the receptor cells).

Sensory pathways make connections to structures other than the cerebral cortex. These include structures involved in specific reflexes (spinal cord or cranial nerve, skeletomotor or autonomic) and the reticular formation. The alerting or arousing effects of stimuli are dependent on connections with the reticular formation. Conscious awareness of a sensory stimulus probably requires information to have reached the sensory cortex, although conscious appreciation of pain might arise at the thalamic level. The reticular formation influences the passage of information at all stages of central synaptic transmission. Hence, relay nuclei are in reality regions allowing complex interactions (most contain interneurones as well as projection cells) and do not transmit unmodified all incoming information. For example, the terminals of the optic tract make up only a very small percentage of the synapses of the lateral geniculate body. Moreover, even when the number of sensory afferents is approximately equal to the number of sensory efferents (as is the case for the lateral geniculate body), pre- and post-synaptic elements are connected many-to-many, not one-to-one. Such connections preserve specificity while minimising the sensory loss when individual neurones are damaged. All systems show massive divergence on reaching the cerebral cortex.

All sensory systems demonstrate strict topographical projections; representations (maps) of the receptor surface may be found at all levels of a pathway. These maps are distorted by having large areas devoted to regions that are densely innervated at the receptor surface. Thus, much larger areas of cortex are devoted to processing information from the lips than from the back of the head, and from the centre of the retina than from its periphery. Representation in the primary sensory cortices is predominantly for stimuli from the contralateral external world (although not for the olfactory cortex). For the somatosensory and visual systems, the gross cortical representation is also vertically inverted. Hence, the body surface is represented head downwards and legs uppermost; the upper half of the visual world is represented below the lower half.

The motor/descending pathways to the spinal cord and motor nuclei of the cranial nerves originate chiefly from the cerebral cortex, the red nucleus and the superior colliculus of the midbrain, and the reticular formation and vestibular nuclei of the pons and medulla. Other areas give rise to small groups of descending fibres. The descending pathways may be considered to have two neuronal stages from the brain to skeletal muscle.

Stage I   Motor nucleus or area
Stage II   Ventral horn of spinal cord
          Muscle (effector)

↓ Tract
↓ Motoneuronal axon in spinal nerve

From the cerebral cortex or a nucleus within the brainstem, axons pass to the ventral horn of the spinal cord or to the vicinity of a motor nucleus of one of the cranial nerves. Most descending fibres terminate on interneurones (including those of the reticular formation of the brainstem) that contact α and/or γ motoneurones. The lateral corticospinal and rubrospinal tracts are crossed, run in the lateral spinal funiculus and terminate chiefly in relation to laterally placed motoneurones that innervate distal limb muscles. The uncrossed ventral corticospinal and vestibulospinal tracts and the crossed tectospinal tract run in the ventral funiculus and terminate chiefly in relation to medially placed motoneurones that innervate axial and proximal limb musculature. The connections and functions of the reticulospinal tracts are more generalised.

The descending pathways exert control over intrinsic spinal mechanisms (e.g. by modulating spinal reflexes or activating intrinsic, movement pattern generators). Their activity is regulated by the reticular formation, and co-ordinated and controlled by the cerebellum, the basal nuclear system and the cerebral cortex (most importantly, area MsI of motosensory cortex). The cerebral cortex provides massive inputs to the cerebellum and the basal nuclei. Cerebellar and basal nuclear outputs go to the nuclei ventralis lateralis and ventralis anterior of the thalamus for relay to area MsI. The cerebellum and (sometimes indirectly) the cerebral cortex send fibres to all the main sources of the descending tracts.

The clinical usage of the term 'pyramidal tract' or 'upper motoneurone' lesion to describe a spastic paralysis or paresis in which muscular tonus and deep reflexes are increased is misleading anatomically. Such a paralysis develops after major lesions of motosensory cortex, the internal capsule, or white matter of the spinal cord. However, although interruption of the fibres of the pyramidal tract (corticospinal fibres) causes the paralysis, the spasticity results from damage to non-pyramidal ('extrapyramidal') pathways involving connections of the cerebellum, basal nuclear system or other motor tracts. Similarly, 'extrapyramidal disorders' are characterised by disturbances of reflexes and muscular tonus and/or abnormal involuntary movements, without paralysis; typically these are a result of damage to the cerebellum or basal nuclear system (or their connections). However, strictly, the extrapyramidal system should include all descending pathways and regions influencing them, except for the pyramidal tract; again clinical usage is anatomically imprecise.

THE OLFACTORY SYSTEM

Although humans are microsmatic mammals, the sense of smell is not to be dismissed in normal community life. Social and commercial activities demonstrate its strong influence in everyday living. The pleasures of the dinner table and the garden would be greatly diminished without it.

The olfactory epithelium is a yellow mucosa occupying 2 to $3\,cm^2$ of the surface of the roof of the nasal cavity. It is located immediately below the cribriform plate of the ethmoid bone on the superior nasal concha and adjacent part of the nasal septum. Air is directed on to this area during sniffing. The mucosa contains supporting cells, basal cells and receptor cells. The receptor cells are bipolar neurones known as olfactory rods; they are chemoreceptors. They have cilia that are bathed in a specialised mucus secreted from the olfactory glands (of Bowman). The high lipid content of this mucus means that lipophobic substances do not dissolve and so are not detected as odours. The olfactory rods are continually replaced from stem cells within the population of basal cells. Thus, primary olfactory neurones have a limited life-span (typically several weeks); new neurones are formed throughout life to replace them. Inadequate replacement is probably a major factor in the loss of the sense of smell in later life.

The fine, unmyelinated axons of the olfactory rods, the fila olfactoria, constitute the olfactory (I) nerve. They gather into about 20 bundles to pass through the cribriform plate to the olfactory bulb. The bundles are invested in meninges, the dura being continuous with the periosteum of the nasal cavity and the pia-arachnoid continuous with the perineurium of the nerve bundles.

The olfactory bulb is an outgrowth of the telencephalon. Its structure has been likened to that of the retina or to a simplified type of cerebral cortex. The fila olfactoria terminate upon the dendrites of mitral cells within glomeruli in the olfactory bulb. Transmission of olfactory information through the olfactory bulb is modified by the actions of periglomerular cells, granule cells, neurones of the nucleus of the olfactory tract, and by centrifugal influences. As for other sensory systems, olfactory projections are topographically organised; different areas process information relating to different odours. The mitral cells provide the main projection to olfactory cortex, their axons running in the olfactory tract. They also send fibres to the nucleus of the olfactory tract (anterior olfactory nucleus); this nucleus projects back to the ipsilateral olfactory bulb and, via the anterior commissure, to the contralateral olfactory bulb.

Centrally, the olfactory tract divides into three olfactory striae: a medial stria that runs towards the septal area; a small intermediate stria that goes to the olfactory tubercle (i.e. the anterior part of the anterior perforated substance); and the main, lateral stria that runs close to the limen of the insula and ends in the primary olfactory cortex. The primary olfactory cortex lies anterior to the uncus and comprises the prepyriform cortex and the peri-amygdaloid area.

Secondary connections of the olfactory system are complex and difficult to distinguish from those of the limbic system. The close association of the olfactory and limbic systems underlies the affective and evocative aspects of the olfactory sense. Further olfactory projections (which include some primary fibres) reach parts of the parahippocampal gyrus (the entorhinal area), the amygdala, hypothalamus, reticular formation, nucleus medialis dorsalis of the thalamus and posterolateral quadrant of orbitofrontal cortex. Links with the hypothalamus allow olfactory stimuli to trigger autonomic reflexes such as salivation and gastric secretion.

The olfactory pathway is unusual in that direct connection is made to the cerebral cortex by links involving only two neurones: olfactory rods and mitral cells. Moreover, the primary neurones are the receptor cells. Connections to the cortex are ipsilateral and do not relay in the thalamus. Olfactory cortex is paleocortex.

a) *The olfactory nerve and connections within the olfactory bulb.* b) *The olfactory cortex and central connections of the olfactory bulb (a view of the base of the brain).* Neuronal types: *Gr = granule cell, M = mitral cell, P = periglomerular cell. (Tufted cells are not shown.)*

SUMMARY SHEET: THE OLFACTORY SYSTEM

a

b

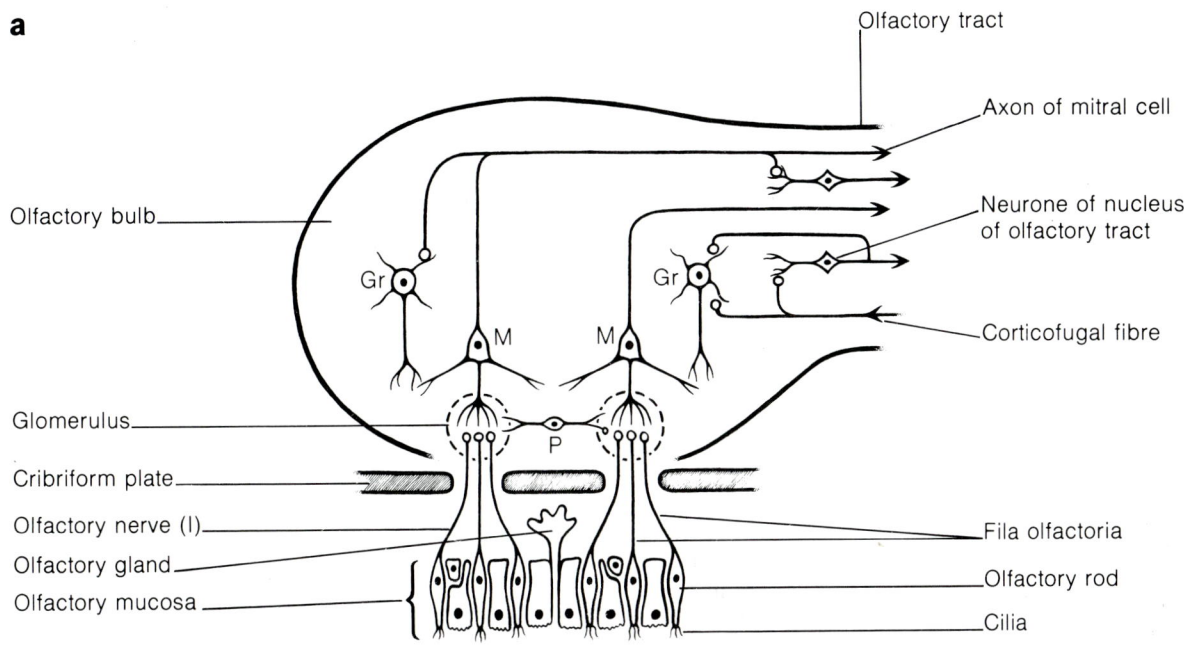

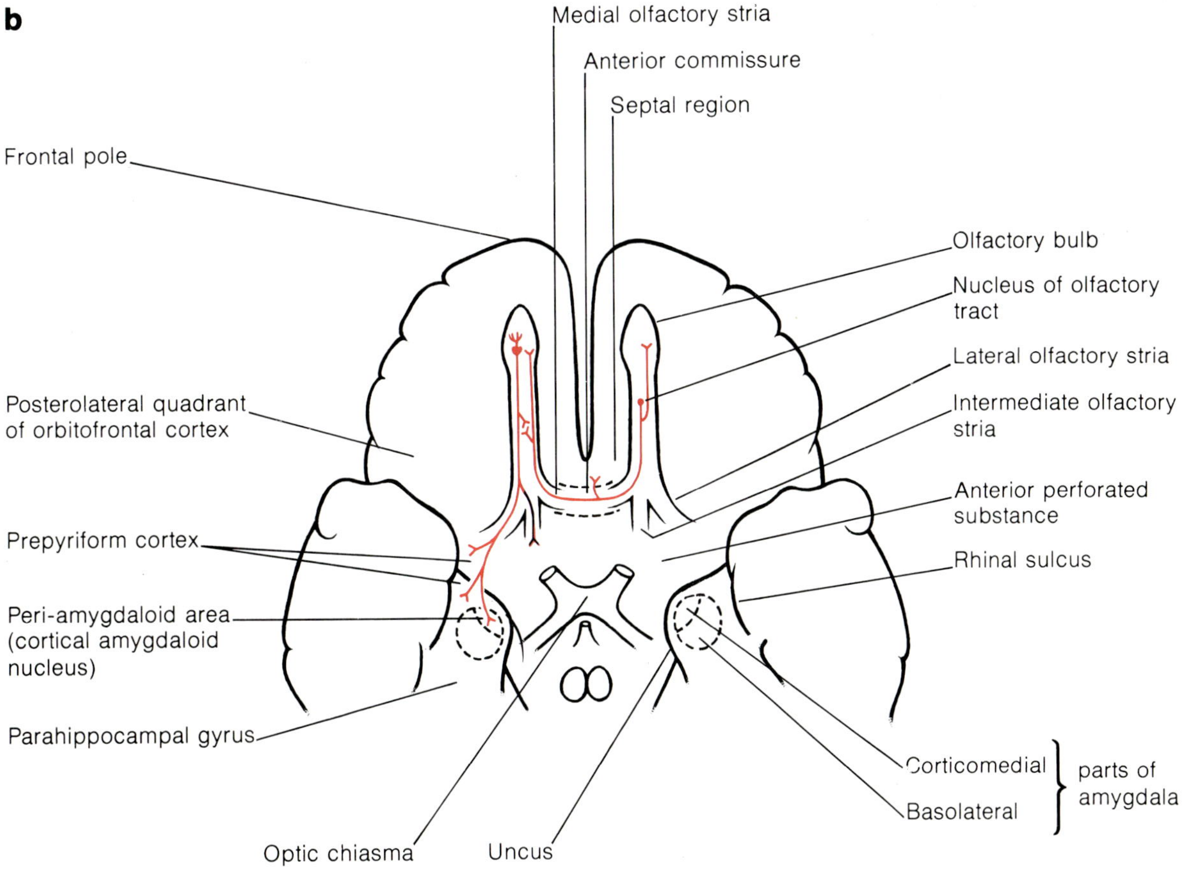

THE VISUAL SYSTEM

The human visual system is highly developed for the detection and interpretation of signals within a restricted part of the electromagnetic spectrum (wavelengths of 400 to 700 nm). Sight is the most important of the special senses and large areas of the brain are involved in processing visual information.

Developmentally, the retina and optic nerves are outgrowths from the diencephalon. The optic nerve is ensheathed by meninges, the dura being continuous with the sclera of the eyeball. Light rays are focused on the retina by the cornea and lens of the eye. The retina has a regular, laminated structure with the photoreceptor cells (rods and cones) lying closest to the pigment epithelium of the choroid. Thus, light photons pass through most of the retina before reaching the receptors. There are about 10^8 rods responsible for monochromatic vision at low light levels (scotopic vision). The 6×10^6 cones are responsible for high acuity and colour vision at moderate and high levels of illumination (photopic vision). Individual cones are maximally responsive to red, green or blue light. Most of the cones are concentrated towards the middle of the retina, in a small (2×1 mm) oval, yellow-tinted area, the macula lutea. Near the centre of the macula is a small depression about 0.3 mm in diameter, the fovea centralis. The fovea is responsible for the central 1° of vision, the macula for the central 6°. Central and peripheral components of the visual system are organised so that light from objects of visual scrutiny is focused on the fovea (foveation).

Considerable processing of visual information occurs during its passage through the retina. Further, the functional pathways vary with the level of illumination. These pathways use both electrical and chemical synapses. Rods and cones are contacted by bipolar and horizontal cells. The bipolar cells are the primary sensory neurones of the visual pathway. They synapse on to retinal ganglion cells which constitute the secondary neurones. Amacrine cells contact both bipolar and ganglion cells. Horizontal and amacrine cells have no axon and make connections running parallel as well as vertical to the retinal surface. The axons (about 10^6) of the ganglion cells run across the surface of the retina and converge on the optic disc, before becoming myelinated and leaving the eyeball as the optic (II) nerve. The optic disc is about 4 mm (approximately 15° of visual angle) medial and slightly superior to the fovea; it corresponds to the 'blind spot'.

The optic nerve fibres from the ganglion cells in the nasal half of the retina decussate in the optic chiasma; those from the temporal half remain ipsilateral. Hence, parts of the visual pathway beyond the chiasma deal with information from the contralateral halves of both visual fields. The fibres continue in the optic tract, 90% terminating in the lateral geniculate body. Neurones of the lateral geniculate body project via the optic radiation (geniculocalcarine tract) of the retrolenticular part of the internal capsule to the primary visual cortex. The primary visual cortex is found on either side of the posterior part of the calcarine sulcus in the occipital lobe. Within this cortex representation of the visual world is upside-down, with central vision being near the occipital pole. Visual association cortex occupies nearly the whole of the occipital lobe and extends into the parietal and temporal lobes.

The optic tract gives off fibres to the hypothalamus, pretectal area and superior colliculus. The pretectal area projects to the parasympathetic accessory (Edinger–Westphal) nucleus of the oculomotor nerve, which is important for light (sphincter pupillae) and accommodation (ciliary muscle) reflexes. The superior colliculus projects via the reticular formation to the nuclei of the oculomotor (III), trochlear (IV) and abducent (VI) nerves which control the extra-ocular muscles. It also projects via the pulvinar nucleus of the thalamus to visual cortex and to the cervical spinal cord via the tectospinal tract. Visual cortex and/or the frontal eye fields are important for all visual reflexes except the light reflex. Eye movement control is further mentioned in the summary sheets on pages 594 to 595 and 604 to 605.

SUMMARY SHEET: THE VISUAL SYSTEM

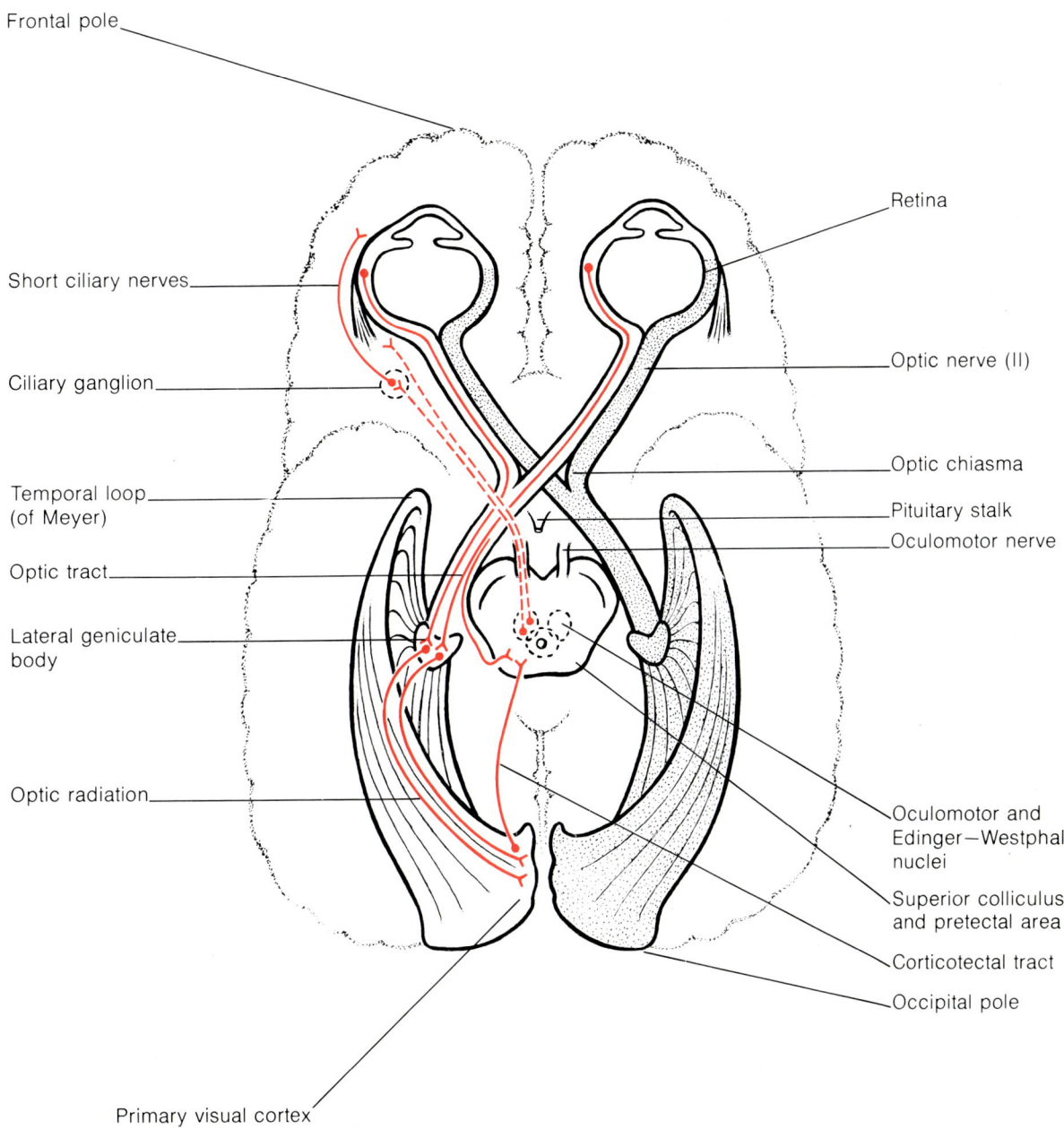

Connections of the visual system are shown as if viewed from the base of the brain.

THE AUDITORY SYSTEM

The auditory system makes possible perception of the direction, intensity and frequency of incident sound waves, together with the understanding of speech and the appreciation of music. The human ear is responsive to sound frequencies of about 20 to 20,000 Hz, speech typically using frequencies of 300 to 3,000 Hz.

The middle ear ossicles minimise the energy loss in the transferance of vibrations of the tympanic membrane to the fluid of the internal ear. Their oscillations may be controlled by the actions of the muscles tensor tympani (supplied by the trigeminal nerve) and stapedius (supplied by the facial nerve).

The auditory receptor cells are the hair cells of the organ of Corti within the cochlear duct. In each ear there are some 30,000 outer hair cells, but more important is the single row of about 3,500 inner hair cells. Vibrations of the basilar membrane cause deflections of the sensory receptor cells' hairs. The hair cells are responsive to narrow bands of sound frequencies. Cells responsive to low frequencies are found near the apex of the cochlear duct, and to high frequencies near the base. The hair cells are contacted by the peripheral processes of the cells of the spiral (cochlear) ganglion of the vestibulocochlear (VIII) nerve; the ganglion cells are located in the spiral canal of the modiolus. Their central processes pass in the VIIIth nerve through the internal acoustic meatus and bifurcate to end in both the dorsal and ventral cochlear nuclei.

Fibres from the cochlear nuclei cross the pontine tegmentum in the dorsal, ventral and intermediate acoustic striae. Fibres terminate in either the ipsilateral or contralateral superior olivary nuclei, or reach the contralateral lateral lemniscus. The superior olivary nuclei function to compare intra-aural sound differences, so contributing to the spatial localisation of sounds. The lateral lemniscus is primarily formed of axons from contralateral cochlear nuclear cells and ipsilateral superior olivary cells. Its chief termination is in the inferior colliculus, but fibres also end on neurones grouped along the lateral lemniscus. The inferior colliculus projects by way of its brachium to the ipsilateral medial geniculate body. There are commissural connections between the two inferior colliculi and some fibres pass to the contralateral medial geniculate body. Other fibres go the reticular formation and the superior colliculus. The inferior colliculus is important for the organisation of auditory reflexes, such as turning the head and eyes towards the source of a sound. The medial geniculate body projects to the ipsilateral primary auditory cortex, the auditory radiation passing in the sublenticular part of the internal capsule. The primary auditory cortex is buried in the lateral sulcus, on the temporal operculum, opposite the postcentral sulcus, in the transverse temporal gyri (of Heschl).

Auditory association cortex surrounds the primary cortex; it further processes auditory signals. Posterior speech cortex (Wernicke's area) is necessary for verbal comprehension. The left hemisphere is dominant for language in over 90% of right-handed individuals and in over 60% of others. The ability to speak requires anterior speech cortex (Broca's area) which is in inferior frontal cortex, also usually in the left hemisphere, close to the face region of motosensory cortex (area MsI).

Descending connections are well developed in the auditory system. Auditory cortex projects to the ipsilateral medial geniculate body and to both inferior colliculi. The inferior colliculus projects to the ipsilateral superior olivary complex and to both dorsal cochlear nuclei. The superior olivary nuclei of both sides send fibres to the auditory hair cells via each olivocochlear bundle.

Throughout the auditory pathway there are close topographical projections and tonotopic maps. Information is conveyed primarily contralaterally, but there is a strong ipsilateral component and there are many connections between the left and right pathways. The auditory pathway is also noteworthy for the number of its subcortical nuclei and for the prominence of its descending/centrifugal connections.

The auditory pathway is shown superimposed on a dorsal view of the brainstem and thalamus, and a lateral view of the left cerebral hemisphere.

SUMMARY SHEET: THE AUDITORY SYSTEM

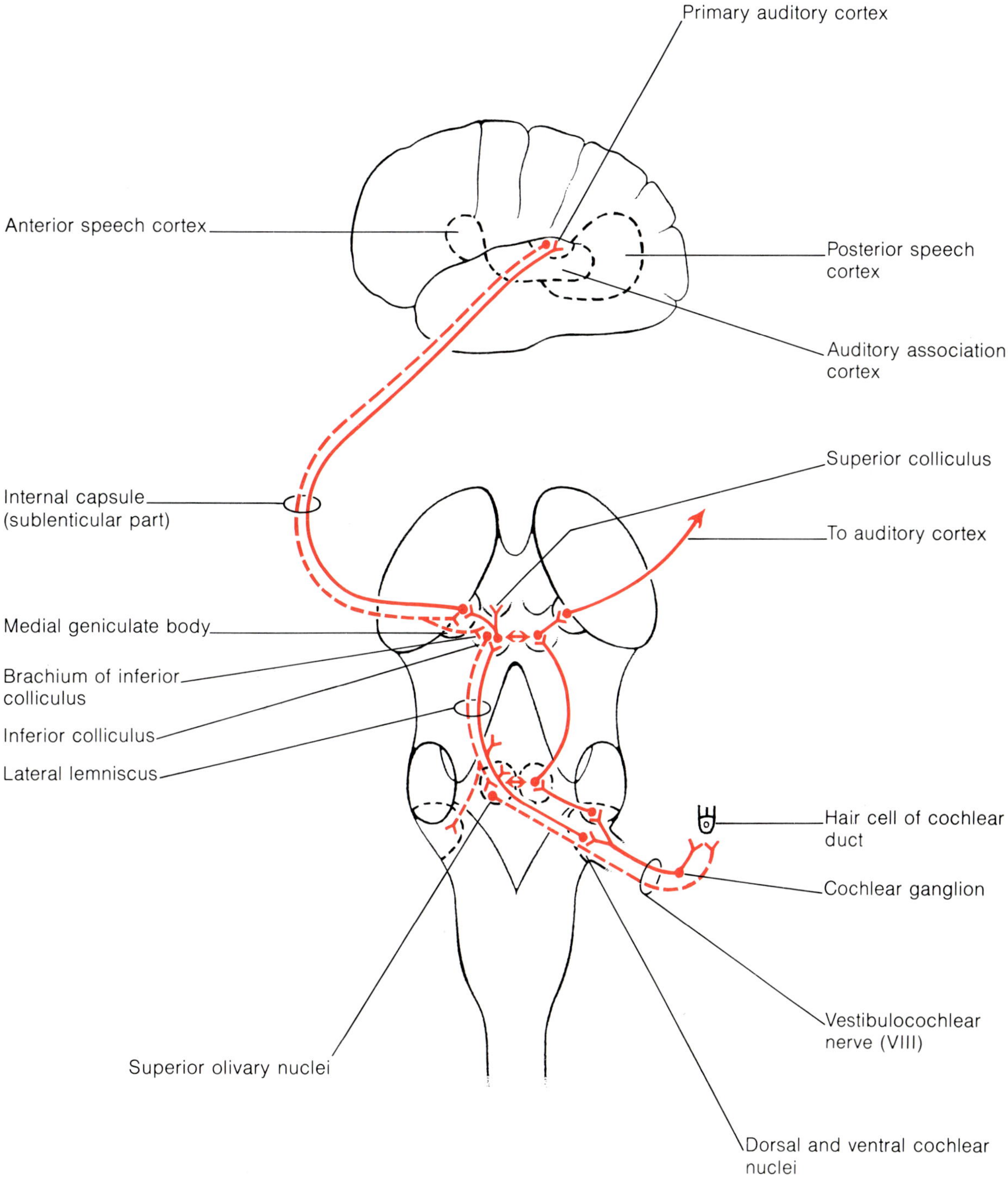

Key

Ascending fibres

Descending fibres

THE VESTIBULAR SYSTEM

The vestibular system represents the phylogenetically older, proprioceptive component of the vestibulo-cochlear (VIII) nerve. It provides information about the orientation and acceleration of the head in space, so that the body may maintain equilibrium and so that compensation may be made for the effect of head movements on the direction of gaze of the eyes. Such operations are carried out chiefly at the subconscious level. However, information about which way up the head is, and whether or not it is spinning, does reach consciousness.

The vestibular receptor cells are hair cells. They are concentrated in five regions of specialised epithelium within each membranous labyrinth: a crista in the ampulla of each of the three semicircular canals, and a macula in both the saccule and the utricle. In each crista ampullaris, hairs of the receptor cells project into a dome-shaped gelatinous cupula; angular acceleration of the head causes bending of the hairs. Hairs of the receptor cells of each macula project into a gelatinous otolithic membrane containing crystals of calcite (otoliths); linear acceleration or changes in position of the head in relation to gravity cause bending of the hairs of these receptor cells.

The hair cells are contacted by the peripheral processes of the bipolar cells of the vestibular ganglion. The vestibular ganglion is located at the distal end of the internal acoustic meatus. The hair cells are also contacted by a few efferents of the VIIIth nerve. These efferents arise as bilateral projections from cells found between the vestibular and abducent nuclei in the pons. The central processes of the cells of the vestibular ganglion pass in the VIIIth nerve to the four ipsilateral vestibular nuclei (superior, medial, lateral and inferior) and the flocculonodular lobe of the cerebellum. Fibres from the cristae ampullares mainly distribute to the superior and medial vestibular nuclei. These nuclei provide fibres to the medial longitudinal fasciculus for the co-ordination of head, neck and eye movements. Those from the maculae terminate chiefly in the lateral, inferior and medial vestibular nuclei. The lateral vestibular nucleus (Deiters' nucleus) gives rise to the lateral vestibulospinal tract which is important for posture and the maintenance of balance. (See also summary sheet on pages 604 to 605.)

The vestibular nuclei are closely interconnected. They are reciprocally connected with the ipsilateral cerebellum (via the inferior cerebellar peduncle) and with the reticular formation; these connections provide additional routes by which the vestibular system may influence eye and other movements.

Ascending fibres of the medial longitudinal fasciculus arise mainly from the ipsilateral superior and contralateral medial vestibular nuclei. They distribute bilaterally, but in a closely organised manner, to the cranial nerve nuclei controlling the extra-ocular muscles: the abducent, trochlear and oculomotor nuclei. These connections underlie vestibulo-ocular reflexes. Some ascending fibres of the medial longitudinal fasciculus pass to two nuclei cranial to the oculomotor nucleus: the interstitial nucleus of Cajal and the rostral interstitial nucleus of the medial longitudinal fasciculus.

Other ascending fibres of the vestibular nuclei reach the nuclei ventralis lateralis, and ventralis posterior lateralis and inferior of the thalamus; the projections may be bilateral. Representation in the cerebral cortex is most probably in a small area located posteriorly within the face region of the primary sensorimotor cortex (area SmI), on the postcentral gyrus close to the intraparietal sulcus. This region is within Brodmann's area 2, the area that receives proprioceptive input from muscle receptors. Claims have also been made for vestibular areas within area 3a (which lies deep in the central sulcus between area 3 of SmI and area 4 of MsI), and on the superior temporal gyrus, anterior and lateral to the primary auditory cortex. Vestibular information may also reach the cortex by other routes, including via the cerebellum or reticular formation.

SUMMARY SHEET: THE VESTIBULAR SYSTEM

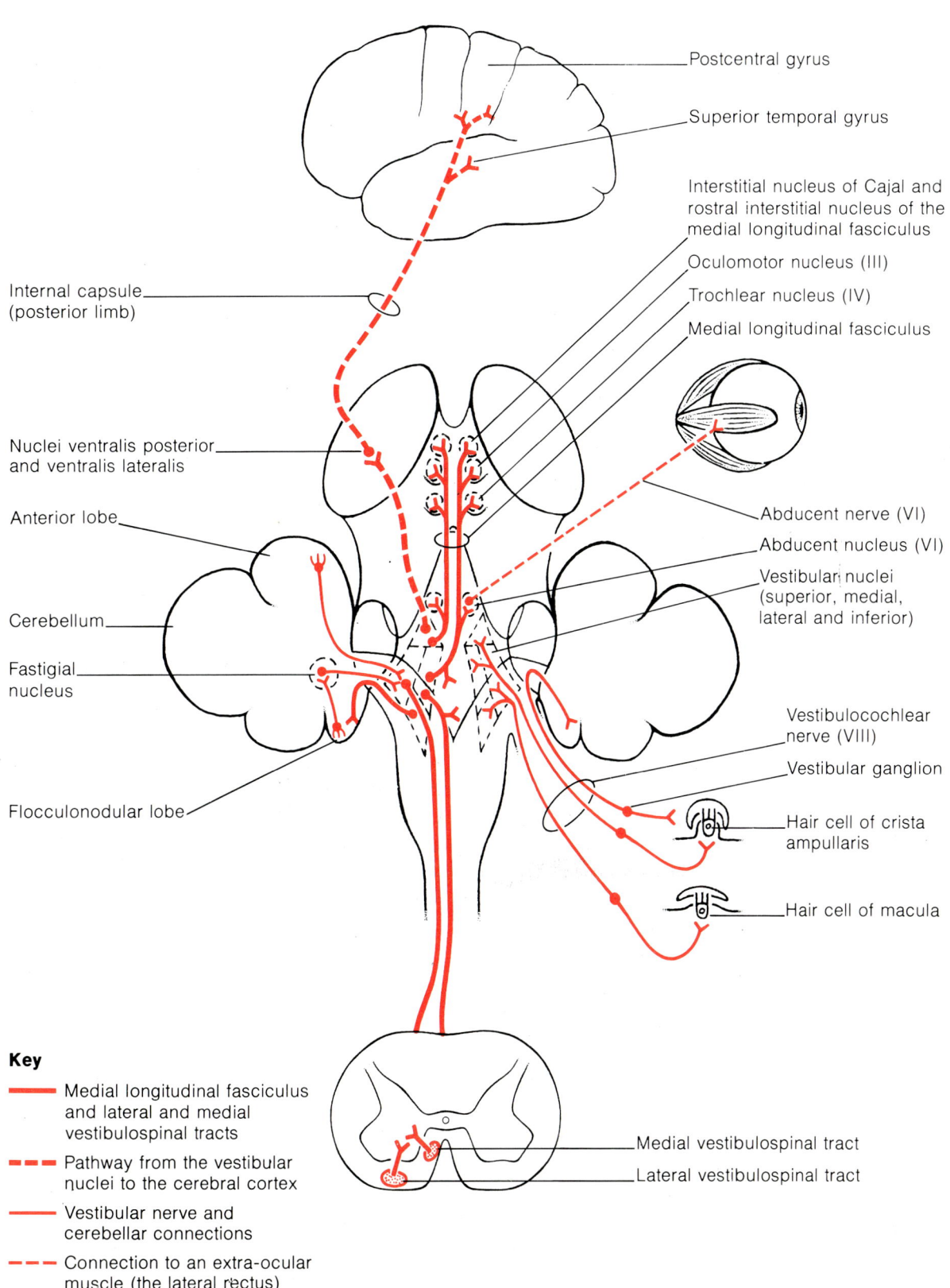

THE GUSTATORY SYSTEM

Taste is a weak sense. The delights of the gourmet depend more on stimulation of the olfactory, visual and somaesthetic pathways than on the gustatory system.

Sensory receptor cells are found within taste buds. Taste buds are found in the epithelium of the tongue, the inferior surface of the soft palate, the palatoglossal arches and the posterior wall of the pharynx. They are pear-shaped collections of about 50 cells, which include supportive and generative cells as well as gustatory receptor cells. Cells within the taste bud are continually developing and being replaced; the receptor cells have a half-life of just over a week. The receptor cells have microvilli which protrude into the saliva above the taste bud. Gustation is dependent on chemoreceptors; only substances dissolved in saliva are detectable. Four modalities of taste are recognised: salt, sweet, sour and bitter. The most sensitive regions for detecting saltiness are located anteriorly on the sides of the tongue; for sweetness the area is near the tip of the tongue; for sourness the regions are found posteriorly on the sides of the tongue; and for bitterness the area is found posteriorly on the dorsum of the tongue.

Peripheral processes of cells of the inferior ganglion of the glossopharyngeal (IX) nerve innervate taste buds on the ipsilateral half of the pharyngeal part of the tongue (i.e. those posterior to the sulcus terminalis), plus those of the circumvallate papillae. Peripheral processes of the geniculate ganglion cells of the facial (VII) nerve innervate taste buds on the ipsilateral half of the oral part of the tongue (i.e. those anterior to the sulcus terminalis), except those of the circumvallate papillae. These VIIth nerve fibres normally pass to the geniculate ganglion in the chorda tympani, but in some people they reach the geniculate ganglion via the otic ganglion and the greater petrosal nerve. The taste buds of the soft palate are usually supplied by the facial nerve through the greater petrosal nerve, the nerve of the pterygoid canal, and the greater and lesser palatine nerves. The glossopharyngeal nerve probably also innervates some taste buds of the soft palate, in addition to those of the palatoglossal arches and pharynx. It is the glossopharyngeal nerve that innervates most of the taste buds. The vagus (X) nerve is said to contact taste buds in the region of the epiglottis, but these are few in number and chiefly disappear in early infancy.

The central processes of the primary gustatory afferents pass via the glossopharyngeal nerve and nervus intermedius to the cranial and lateral part of the nucleus of the solitary tract in the cranial part of the medulla. The central gustatory pathway from the nucleus of the solitary tract to the most medial and caudal part of the nucleus ventralis posterior medialis of the thalamus is disputed. Information probably passes contralaterally but possibly also ipsilaterally. Direct fibres may be supplemented by an indirect pathway that relays in the pontine taste area (the parabrachial nuclei found close to the superior cerebellar peduncle). Additionally, fibres terminate within the reticular formation and (probably directly as well as indirectly) signals reach the hypothalamus; via these connections autonomic responses may be induced. From the nucleus of the solitary tract connections are made to the salivatory nucleus and to the dorsal motor nucleus of the vagus by interneurones within the adjacent reticular formation; these pathways underlie reflex salivation and changes in gastric secretion and gut motility.

Within the thalamus, the gustatory region is near the part of the nucleus ventralis posterior medialis subserving somaesthetic sensations from the tongue. Gustatory fibres reach the cortex via the posterior limb of the internal capsule. In the cerebral cortex, gustatory signals are processed in a separate area anterior and inferior to the area for somaesthetic sensations from the tongue within area SmI (i.e. on the frontoparietal operculum in the inferior part of the postcentral gyrus). There is probably a secondary taste area in the anterior part of SmII, also on the frontoparietal operculum, close to the insula.

SUMMARY SHEET: THE GUSTATORY SYSTEM

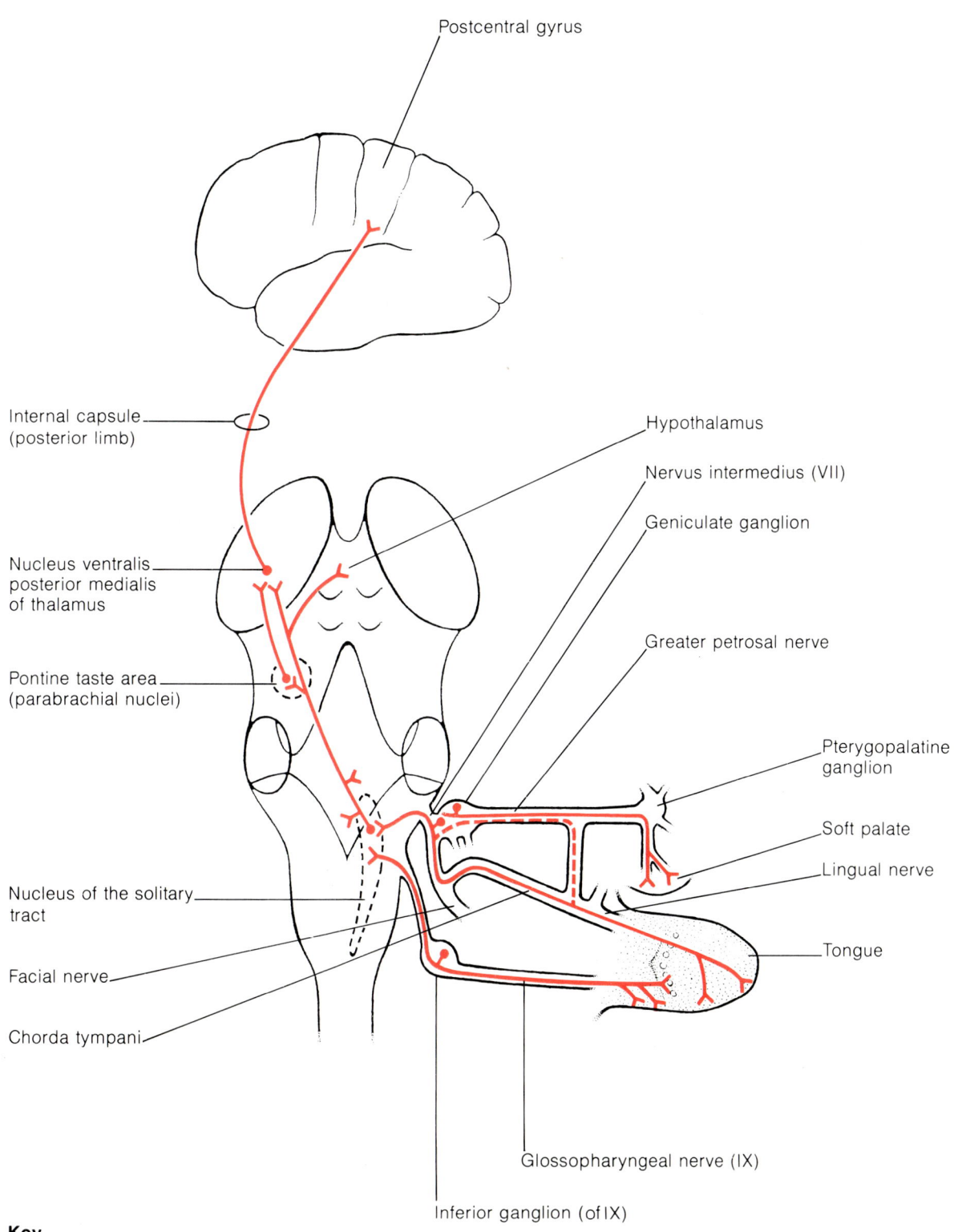

THE SOMATOSENSORY SYSTEM: the dorsal column/medial lemniscal pathway and associated central trigeminal connections

The dorsal column/medial lemniscal pathway is well developed in humans. It conveys tactile and proprioceptive signals from the skin, joints and skeletal muscles. In particular, it is responsible for the more discriminative aspects of touch, for the conscious knowledge of the position of the body in space, for stereognosis, and probably for much of the awareness of vibrations and textures. The information transmitted is detailed and precise. The receptors are chiefly encapsulated cutaneous endings (e.g. tactile corpuscles) or complex structures (e.g. muscle spindle organs). The afferents are well myelinated (group A fibres).

The cell bodies of the primary sensory neurones are found in the dorsal root ganglia of the spinal nerves. The central processes of the ganglion cells pass into the spinal cord medially in the dorsal roots. The fibres usually ascend or descend a few segments, giving off collateral branches to the dorsal and ventral horns; these collaterals are important in establishing spinal reflexes. The main axons of the ganglion cells then ascend in the ipsilateral dorsal column. These fibres (together with axons of dorsal horn neurones passing in the dorsal columns or dorsolateral fasciculus) terminate in the ipsilateral gracile and cuneate nuclei of the caudal medulla. In the dorsal columns the gracile afferents lie medially, arise from spinal nerves caudal to T6, and deal with information from the lower half of the body; the cuneate afferents lie laterally, arise from nerves cranial to T6, and serve the upper parts of the body.

Fibres from the gracile and cuneate nuclei decussate to form the contralateral medial lemniscus; this terminates in the nucleus ventralis posterior lateralis of the thalamus. Other fibres go to the cerebellum. The somatosensory radiation passes through the posterior limb of the internal capsule to the postcentral gyrus (primary sensorimotor area: SmI) and frontoparietal operculum (area SmII). The somatotopic representation is contralateral and grossly inverted: the leg and perineal areas are on the medial surface of the hemisphere in the paracentral lobule; the upper limb region is near the middle of the postcentral gyrus. Somatosensory association cortex is found immediately posteriorly (in Brodmann's area 5).

The trigeminal (V) nerve conveys discriminative tactile information for the ipsilateral half of the face and the top of the head; the axons of the trigeminal ganglion cells go to the principal sensory nucleus and pars oralis of the nucleus of the spinal tract of the trigeminal. Proprioceptive information from the ipsilateral muscles of mastication, temporomandibular joint, periodontal tissues, extra-ocular muscles and, possibly, from other muscles of the head and neck, reaches the mesencephalic nucleus of the trigeminal; this nucleus contains some primary proprioceptive neurones. Direct and indirect connections of these nuclei form the basis of cranial nerve reflexes. Signals from the nuclei are transmitted mainly via the contralateral ventral trigeminothalamic tract (trigeminal lemniscus) and the ipsilateral dorsal trigeminothalamic tract to the nucleus ventralis posterior medialis of the thalamus. Axons from this nucleus go through the posterior limb of the internal capsule to the inferior part of the postcentral gyrus (face region of SmI) and frontoparietal operculum (anterior part of SmII).

Other proprioceptive pathways (not shown opposite) transmit information from the spinal cord to the cerebellum; such information is utilised subconsciously. Detailed information from ipsilateral muscles and joints ascends via the dorsal spinocerebellar and cuneocerebellar tracts. Information concerning the state of excitation/inhibition of interneuronal groups within the spinal cord is transmitted by the contralateral ventral and the ipsilateral rostral spinocerebellar tracts. The dorsal and ventral spinocerebellar tracts send information about the lower part of the body, the rostral spinocerebellar and cuneocerebellar tracts about upper regions. Further signals reach the cerebellum indirectly via spino-olivary and spinoreticular paths.

SUMMARY SHEET: THE SOMATOSENSORY SYSTEM 1

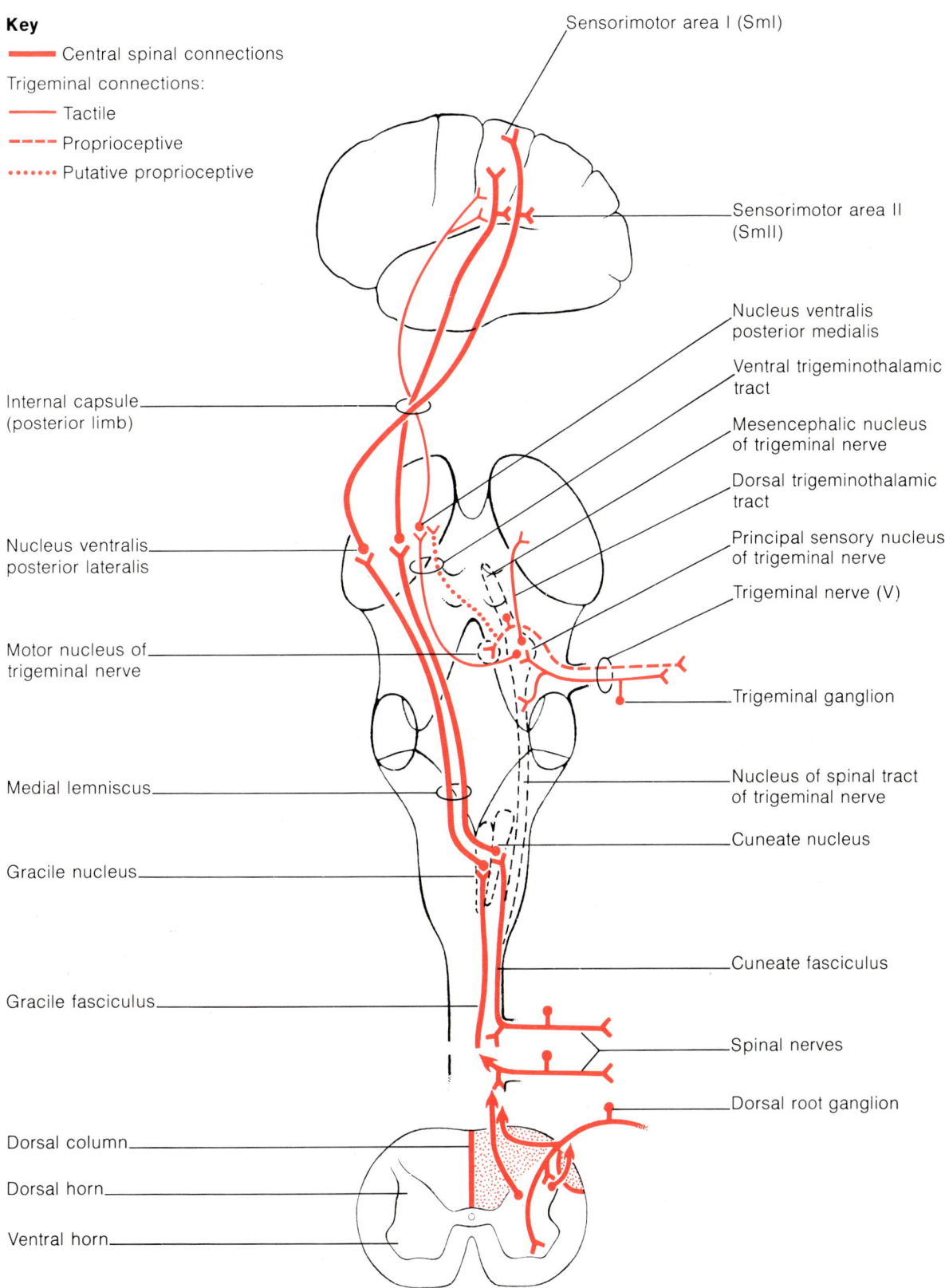

THE SOMATOSENSORY SYSTEM: the anterolateral (spinothalamic) system and associated central trigeminal connections

The anterolateral system conveys basic tactile, pressure, temperature and nociceptive information. It includes spinothalamic, spinoreticular and ascending autonomic fibres. It is particularly responsible for the alerting/arousing effects of somatosensory stimuli. The receptors are mainly free nerve endings. The fibres are chiefly unmyelinated (group C) or poorly myelinated (group Aδ). Unmyelinated afferents of polymodal nociceptors convey slow, aching pain; mainly thinly myelinated fibres from high threshold mechano-receptors convey fast, pricking pain.

Small-diameter central processes of the dorsal root ganglion cells come to lie laterally in the dorsal roots of the spinal nerves. They may ascend or descend for a segment or so before terminating, mainly ipsilaterally, in the dorsal horn. Many afferents (notably from nociceptors) end in the marginal layer (Rexed's lamina I) and substantia gelatinosa (lamina II). The substantia gelatinosa is important in the local, descending and opioid-mediated modulation of the transmission of nociceptive information. Other afferents (chiefly tactile fibres) end in the nucleus proprius (laminae III to VI). Convergence of nociceptive signals from deep and superficial tissues probably underlies referred pain.

Spinal neurones ('tract cells'), chiefly of laminae I and V, give rise to axons that form the anterolateral tracts. Most such axons cross the midline within a few segments to ascend contralaterally. Fibres of sacral origin run superficially, and cervical fibres ascend most deeply. The axons go to the reticular formation, peri-aqueductal grey, superior colliculus, and ventralis posterior lateralis, intralaminary and posterior group of thalamic nuclei. Nociceptive and other information may also reach the thalamus and cortex via the reticular formation.

Projections to the cerebral cortex pass via the posterior limb of the internal capsule. From the posterior group of thalamic nuclei (notably nociceptive signals), axons go to the posterior part of the frontoparietal operculum and neighbouring insula (retro-insular cortex). From nucleus ventralis posterior lateralis (mainly tactile signals), axons go to the primary and secondary sensorimotor areas (SmI and SmII) of the postcentral gyrus and frontoparietal operculum. Signals from the intralaminar nuclei may influence widespread areas of the cortex. The emotional overtones of pain probably arise from involvement of limbic regions.

The spinal tract of the trigeminal contains primary somatosensory afferents from regions of the head supplied by the trigeminal (V), facial (VII), glossopharyngeal (IX) and vagus (X) nerves. The few fibres of the Xth, IXth and VIIth nerves are dorsal in the tract, above (in turn) those of the mandibular, maxillary and ophthalmic divisions of the Vth nerve. The nucleus of the spinal tract of the trigeminal extends from the principal sensory nucleus in the pons to the cranial two segments of the spinal cord. It is subdivided into the pars oralis, the pars interpolaris and the pars caudalis. The pars oralis deals mainly with tactile signals. The pars interpolaris receives cutaneous and proprioceptive information and sends fibres to the cerebellum. The pars caudalis deals particularly with nociceptive signals, but also with tactile and possibly thermal information; it is continuous with, and similar to, the laminae I to IV of the dorsal horn.

Fibres from the nucleus of the spinal tract of the trigeminal pass to the reticular formation; these fibres underlie cranial nerve reflexes. Some fibres run near to the medial lemniscus in the contralateral ventral trigeminothalamic tract (trigeminal lemniscus) to reach the nucleus ventralis posterior medialis, posterior group and intralaminary thalamic nuclei. As for the anterolateral system, nociceptive information mainly reaches the posterior group and intralaminary nuclei. The nucleus ventralis posterior medialis projects via the posterior limb of the internal capsule to the inferior part of the postcentral gyrus (face region of SmI) and to the frontoparietal operculum (face region of SmII).

SUMMARY SHEET: THE SOMATOSENSORY SYSTEM 2

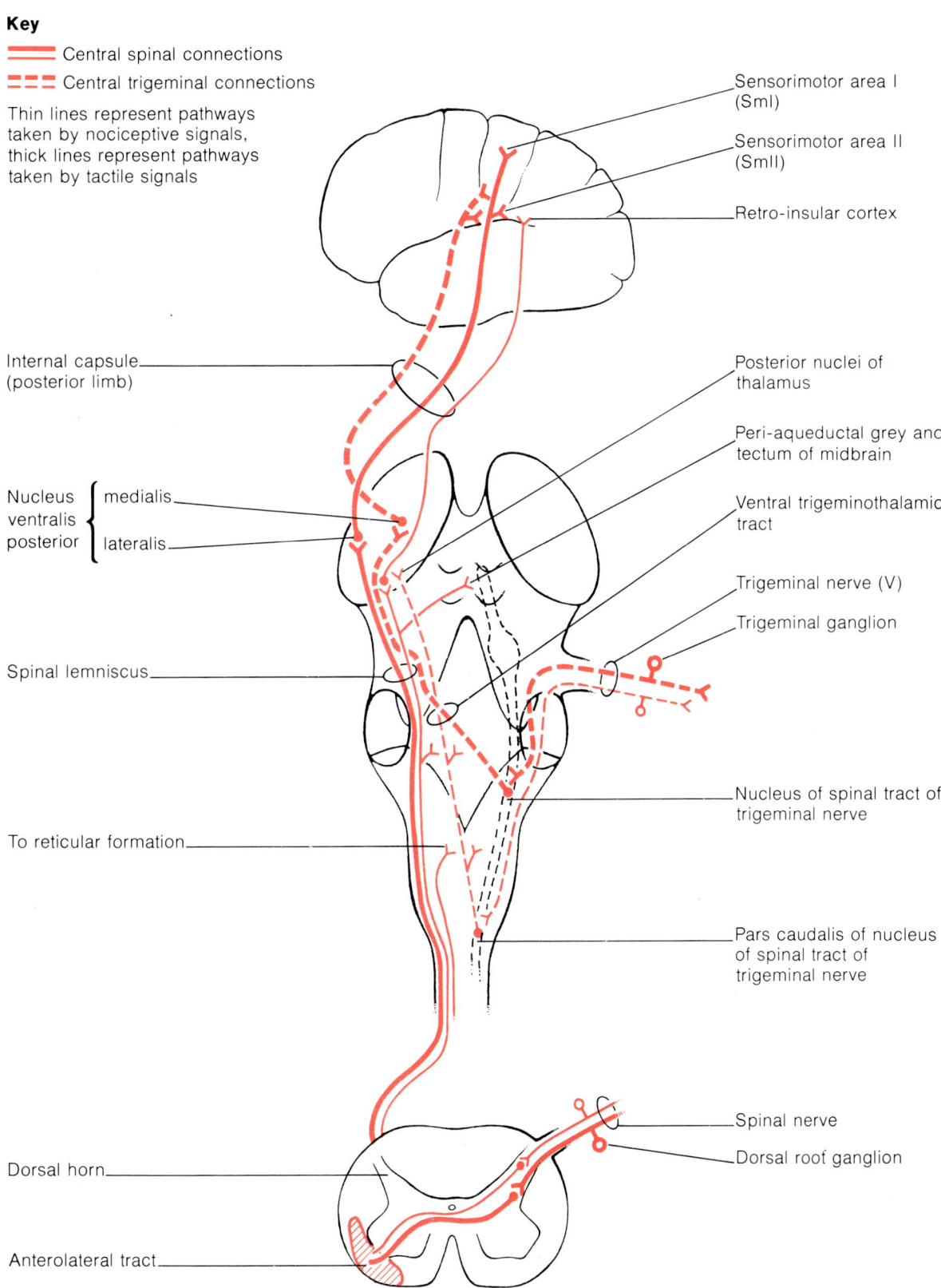

THE CORTICOSPINAL, CORTICONUCLEAR AND RUBROSPINAL TRACTS

The corticospinal (pyramidal) tract is of major significance in human motor control. Phylogenetically it develops in parallel with the cerebral cortex. Its main function is the control of discrete, voluntary movements of the distal limb muscles. Corticonuclear fibres terminate in the brainstem and subserve an equivalent function in relation to cranial nerve nuclei. In general, corticospinal fibres have a facilitatory action on motoneurones supplying flexor muscles and are inhibitory to extensors. The descending fibres originate mainly from the pyramidal cells of layer V of the cortex of the motosensory (MsI and supplementary motor) and sensorimotor (SmI and SmII) areas. Area MsI gives rise to most of the 10^6 fibres found in each medullary pyramid. In the main, precentral areas supply motor nuclei and postcentral areas supply sensory nuclei. Area MsI is somatotopically organised, the head region lying inferiorly on the precentral gyrus, with the lower limb region extending superomedially into the paracentral lobule.

Corticospinal fibres descend through the ipsilateral posterior limb of the internal capsule, fibres related to the lower limb lying posteriorly. Corticonuclear fibres lie anterior to the corticospinal fibres, possibly descending in the region of the genu. Both sets of fibres are found in the middle third of the crus cerebri, corticonuclear fibres being medial. The fibres continue through the basilar part of the pons and collect in the pyramid on the ventral surface of the medulla. Corticonuclear fibres leave the pathway along its course and distribute directly or indirectly (via reticular neurones) to the nuclei of the cranial nerves. Most nuclei receive bilateral innervation, but the part of the facial motor nucleus supplying lower facial muscles, most of the hypoglossal nucleus and the part of the spinal accessory nucleus innervating trapezius are supplied from the contralateral hemisphere. The frontal eye field controls voluntary eye movements to the opposite side. The part of the spinal accessory nucleus innervating sternocleidomastoid is supplied from the ipsilateral cortex. Other corticonuclear fibres end in the dorsal column nuclei, the red nucleus, inferior olivary nuclei and the reticular formation.

At the junction of the medulla and spinal cord most of the pyramidal fibres (75 to 90%) decussate to form the lateral corticospinal tract. The remaining fibres descend ipsilaterally, mainly as the ventral corticospinal tract, although a few form an uncrossed component within the lateral corticospinal tract. The ventral corticospinal tract typically does not descend farther than the upper thoracic segments. Most of its fibres decussate before terminating on neurones concerned with innervation of neck and shoulder musculature. The lateral corticospinal tract descends the length of the spinal cord. Its fibres chiefly end on interneurones and motoneurones lying laterally in the ventral horn, i.e. mainly those in the upper and lower limb enlargements. Some fibres terminate in the dorsal horn.

The rubrospinal tract arises from the red nucleus of the midbrain. It has similar functions to the corticospinal tract, but is of lesser importance. Rubrospinal fibres end in relation to limb rather than axial musculature; they are particularly important in control of the tonus of flexor muscles, via connections to γ-motoneurones. The red nucleus receives bilateral input from the cerebral cortex (particularly motor and somatosensory areas) and so forms part of an indirect corticospinal pathway. It also receives input from the cerebellum via the dentatorubrothalamic tract. The red nucleus is somatotopically organised. It sends efferents to cranial nerve nuclei, the dorsal column nuclei, inferior olivary nuclei and the reticular formation as well as to the spinal cord.

The rubrospinal fibres decussate immediately in the midbrain and descend laterally through the brainstem and in the lateral funiculus of the spinal cord. They terminate on interneurones (chiefly in laminae V, VI and VII) which are facilitatory to laterally lying motoneurones innervating flexor muscles.

SUMMARY SHEET: THE CORTICOSPINAL, CORTICONUCLEAR AND RUBROSPINAL TRACTS

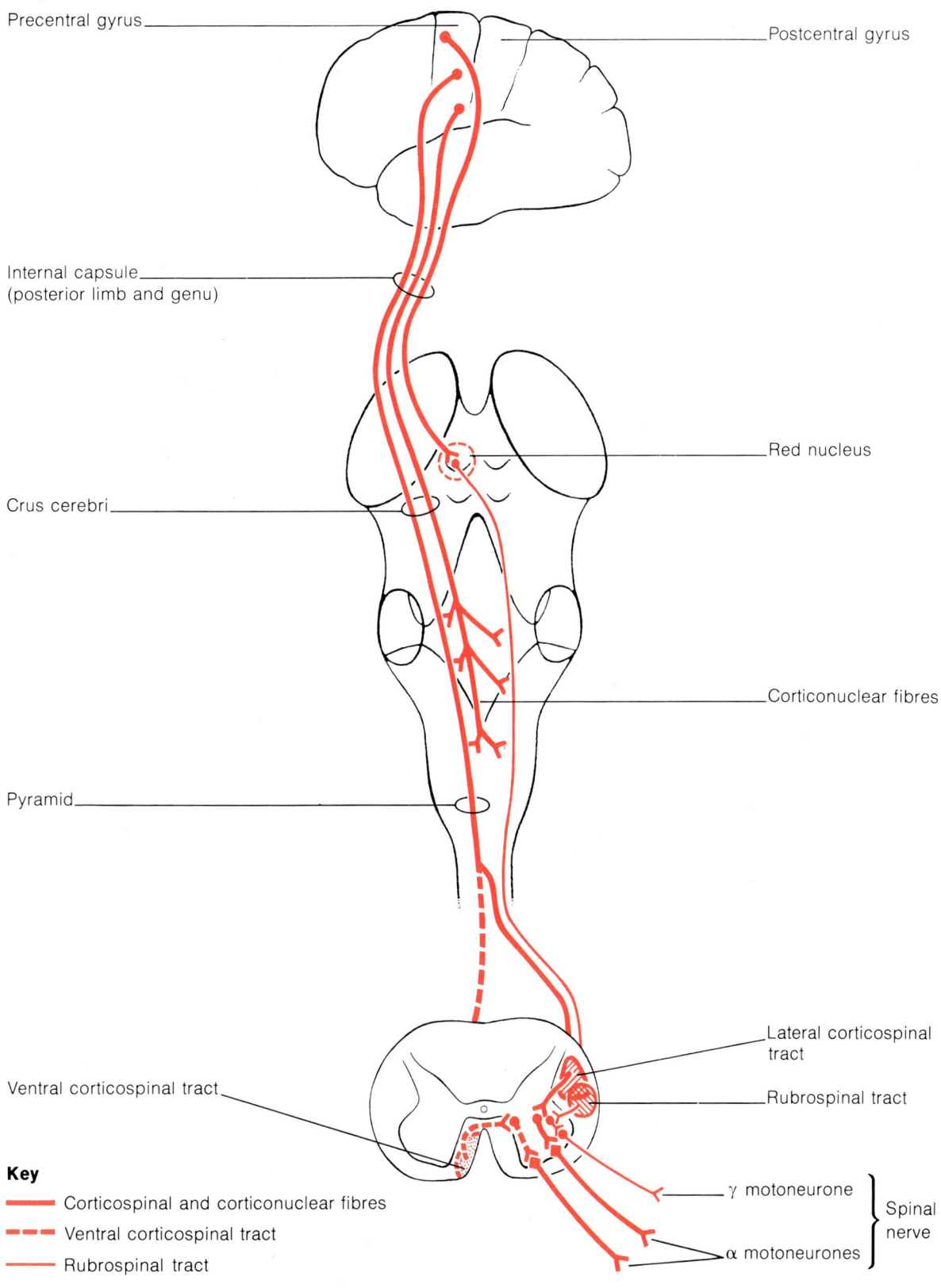

603

THE VESTIBULOSPINAL, TECTOSPINAL AND INTERSTITIOSPINAL TRACTS AND THE CONTROL OF EYE MOVEMENTS (see also summary sheet for the vestibular system on page 594 and 595)

The lateral vestibulospinal tract arises from the lateral vestibular nucleus (Deiters' nucleus). This nucleus receives inputs from the vestibulocochlear (VIII) nerve (particularly the maculae of the utricle and saccule), the cerebellum (notably parts receiving vestibular and spinal input) and the reticular formation. Both the nucleus and its connections are somatotopically organised. The lateral vestibulospinal tract descends ipsilaterally in the ventrolateral funiculus of the spinal cord. The fibres terminate chiefly in laminae VII and VIII at all levels of the cord. They end on interneurones and motoneurones that cause excitation and increase in the tone of extensor muscles (particularly antigravity muscles), and inhibit flexors. The lateral vestibulospinal tract influences the activity of both α- and γ-motoneurones, primarily the medially located motoneurones controlling axial and limb girdle musculature. It is of major importance for the maintenance of balance and posture.

The medial vestibulospinal tract descends as a component of the medial longitudinal fasciculus. It arises from the medial vestibular nucleus. This nucleus receives inputs from the vestibulocochlear (VIII) nerve (notably signals from the cristae ampullares), the parts of the cerebellum receiving vestibular input, and the reticular formation. The medial vestibulospinal tract runs ipsilaterally in the ventral funiculus as far as the upper thoracic segments of the spinal cord. Its fibres terminate mainly medially in the ventral horn. They have been shown to form monosynaptic inhibitory connections with α-motoneurones. These connections allow vestibular control over head movements.

The tectospinal tract originates from the deep layers of the superior colliculus. The superior colliculus receives a wide variety of inputs: visual (from the optic tract, visual cortex and frontal eye field), auditory (from the inferior colliculus), somatosensory (via spinotectal fibres), motor (cerebellar and substantia nigral), and reticular. Tectospinal fibres cross immediately in the midbrain and descend just ventral to the medial longitudinal fasciculus through the brainstem, to cervical segments of the spinal cord. They distribute to laminae VI, VII and VIII and allow reflex turning of the head towards salient stimuli.

The interstitiospinal tract arises from the interstitial nucleus of Cajal in the midbrain. The nucleus receives inputs from the medial longitudinal fasciculus, the tectum of the midbrain, the cerebellum and the reticular formation. The interstitiospinal tract descends mainly ipsilaterally within the medial longitudinal fasciculus, but continues to all levels of the spinal cord. The fibres terminate chiefly in lamina VIII and may be concerned with rotation of the head and body about the vertical axis.

Eye movements are produced by the extra-ocular muscles innervated by the oculomotor (III), trochlear (IV) and abducent (VI) nerves. The nuclei of these nerves are under the control of the cerebral cortex, superior colliculus, cerebellum, vestibular system and reticular formation. Certain areas of the reticular formation organise conjugate (co-ordinated) eye movements. Lateral movements (to the ipsilateral side) are organised by the paramedian pontine reticular formation. Vertical movements are controlled via the rostral interstitial nucleus of the medial longitudinal fasciculus (downward movements) and the interstitial nucleus of Cajal (upward movements), both in the cranial midbrain. However, the perihypoglossal nuclei (particularly the nucleus prepositus hypoglossi) in the cranial medulla and caudal pons, and cranial parts of the paramedian pontine reticular formation are also involved in vertical movements. The interstitial nucleus of Cajal may additionally control rotatory movements. The frontal eye fields are responsible for voluntary eye movements. Accommodation, foveation and visual fixation require the visual cortex and the superior colliculi.

SUMMARY SHEET: THE VESTIBULOSPINAL, TECTOSPINAL AND INTERSTITIOSPINAL TRACTS

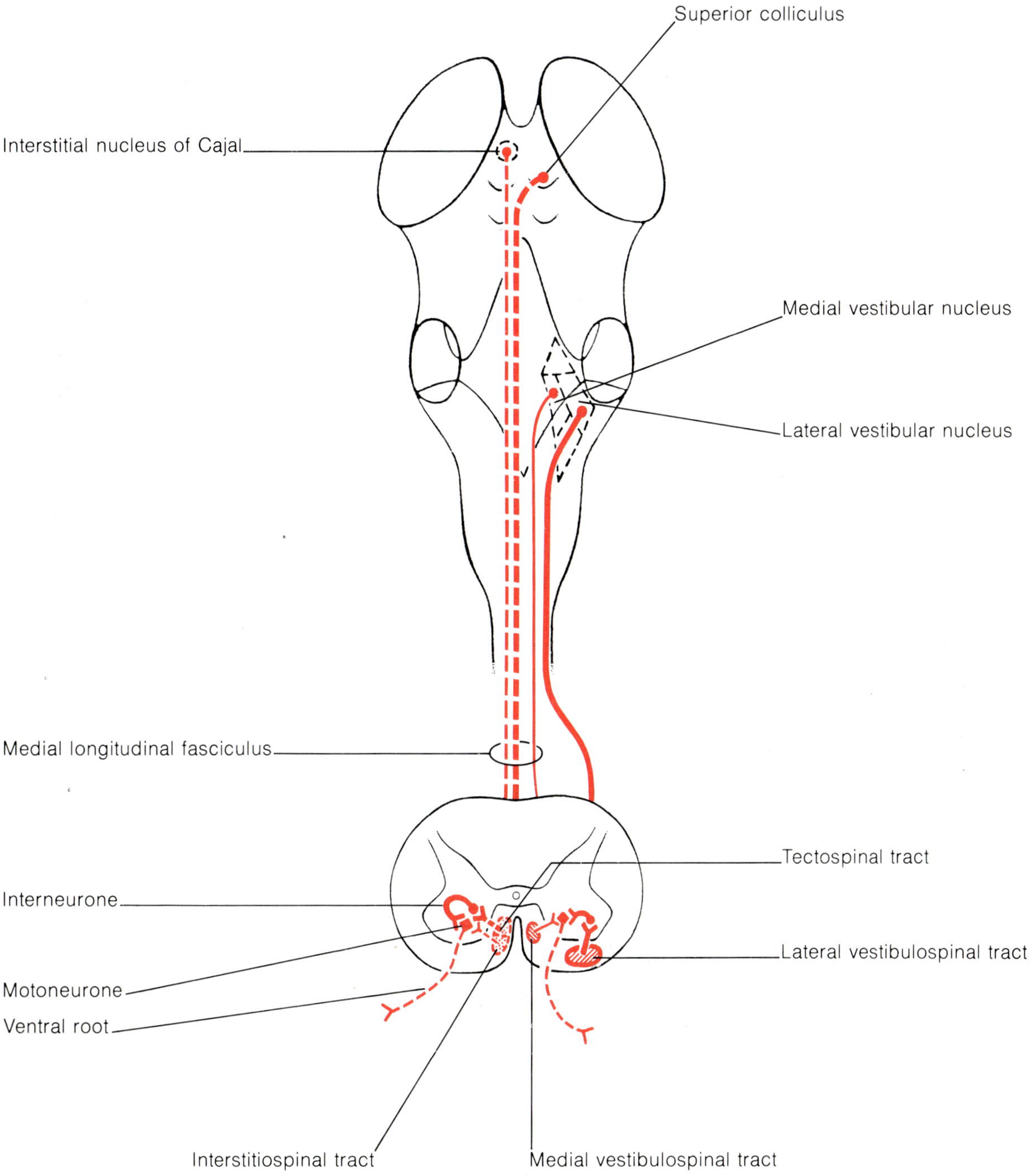

Key

— Lateral vestibulospinal tract
— Medial vestibulospinal tract
--- Tectospinal tract
--- Interstitiospinal tract

THE RETICULOSPINAL TRACTS AND OTHER DESCENDING FIBRES

The reticulospinal tracts influence the excitability of α- and γ-motoneurones and spinal reflex pathways, plus the centripetal transmission of sensory information. They may initiate activity in movement pattern generators of the spinal cord. Reticulospinal fibres have specific roles in the control of respiration, and of cardiovascular and other autonomic reflexes. The fibres are not closely somatotopically organised and are not gathered into discrete bundles. They distribute to all levels of the spinal cord, giving collaterals to widely spaced segments.

Reticulospinal fibres arise from two main regions: the gigantocellular reticular nucleus of the medulla, and the pontine reticular nucleus. These two regions occupy much of the medial two-thirds of the reticular formation of the medulla and pons. In addition, they give rise to ascending connections to the midbrain and forebrain. They receive an extremely wide variety of input, including direct fibres and collaterals from sensory and motor pathways (notably spinoreticular fibres), cranial nerve nuclei, the cerebellum and the cerebral cortex.

The pontine (medial) reticulospinal tract descends almost entirely ipsilaterally. The fibres pass mainly within the medial longitudinal fasciculus, and then in the ventral funiculus of the spinal cord. Most fibres terminate medially on interneurones in the ventral horn; some fibres decussate before terminating.

The medullary (lateral) reticulospinal tract descends mainly ipsilaterally, but a considerable number of fibres cross in the medulla to descend contralaterally. The fibres travel in the lateral funiculus and end chiefly in lamina VII of the spinal cord. The fibres end more laterally than the pontine reticulospinal fibres and overlap the terminations of the lateral corticospinal and rubrospinal tracts.

Other descending fibre groups (not illustrated) include:

Raphe–spinal fibres: most of these fibres arise from the nucleus raphe magnus of the medulla and are 5-hydroxytryptaminergic. They descend bilaterally in the dorsolateral funiculus of the spinal cord. Fibres distribute to the dorsal and ventral horns and the autonomic intermediolateral cell column. Those terminating in the substantia gelatinosa can inhibit transmission of nociceptive information.

Ceruleospinal fibres: fibres from noradrenergic cell groups, notably the locus ceruleus, run ipsilaterally in the ventral and lateral funiculus and terminate widely in the spinal cord. They influence autonomic activity and the general level of neuronal excitability in the spinal cord.

Solitariospinal fibres: the nucleus of the solitary tract receives autonomic reflex afferents via the vagus and glossopharyngeal nerves, and descending cortical and hypothalamic fibres. It projects bilaterally to neuronal groups in cervical and thoracic segments controlling the diaphragm (the phrenic nucleus) and intercostal muscles, and to the sympathetic intermediolateral cell column. These connections allow control of breathing, cardiovascular and alimentary functions.

Edinger–Westphal–spinal fibres: descending fibres from the parasympathetic accessory (Edinger–Westphal) nucleus of the oculomotor nerve end in the dorsal column nuclei and also, after passing in the lateral funiculus of the spinal cord, in laminae I and V of the dorsal horn. They may play a role in nociception.

Hypothalamospinal fibres: the paraventricular nucleus plus dorsal parts of the posterior and lateral hypothalamus send fibres (mainly vasopressinergic) to the dorsal motor nucleus of the vagus and nucleus of the solitary tract. Fibres (mainly oxytocinergic) also pass via the ipsilateral lateral funiculus to end in the sympathetic and parasympathetic intermediolateral cell columns of the spinal cord (for the control of autonomic functions) and to laminae I and II of the dorsal horn.

Fastigiospinal fibres: a few fibres from the fastigial nucleus of the cerebellum travel with the contralateral lateral vestibulospinal tract to cervical segments.

SUMMARY SHEET: THE RETICULOSPINAL TRACTS

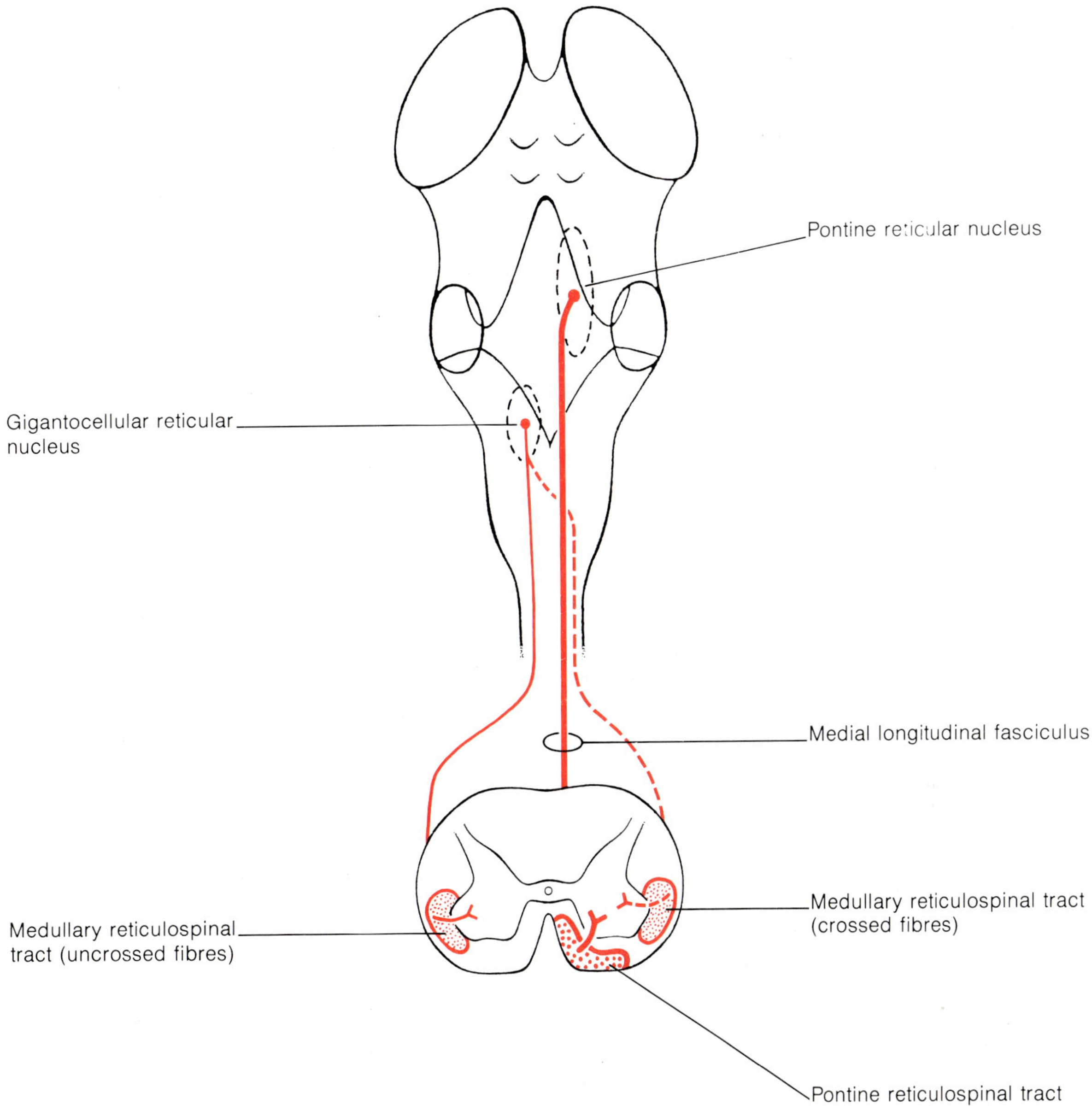

Key
— Pontine reticulospinal tract
Medullary reticulospinal tract:
— Uncrossed fibres
--- Crossed fibres

Case histories 8
The brain and cranial cavity

What anatomical features of the cavernous sinus contribute to the likelihood of thrombosis as a result of spread of infection?

The cavernous sinus is located by the side of the body of the sphenoid. Not only does it communicate with other venous sinuses (i.e. the superior and inferior petrosal sinuses, the intercavernous sinuses) but it also has connections with the extracranial venous system. First, the superior and inferior ophthalmic veins link the facial vein with the cavernous sinus. Second, emissary veins link the cavernous sinus with the pterygoid venous plexus in the infratemporal fossa, passing through the foramen lacerum, foramen ovale and the emissary sphenoidal foramen. A further feature of the cavernous sinus that may contribute to the likelihood of thrombosis relates to the considerable number of trabeculae that cross the sinus. In this respect, the cavernous sinus differs from other sinuses which present open channels. Because of the trabeculae, the flow of blood through the cavernous sinus is considerably slowed. Also, the numerous trabeculae present a large surface area of endothelium.

A 60-year-old man visited his doctor complaining of a shearing, stabbing pain on the right side of his face when he touched the skin just to the side of the corner of his mouth. Indeed, the patient claimed that he lived in constant fear of a painful attack. The doctor diagnosed trigeminal neuralgia. Initially, the condition was controlled by drugs but, after several years, the pain became uncontrollable. The man was referred to a surgeon, and it was decided to produce prolonged anaesthesia by electrocoagulation of the trigeminal ganglion in the floor of the middle cranial fossa. How might the trigeminal ganglion be approached for this procedure and what hazards and safeguards need to be borne in mind?

The trigeminal ganglion is located in the trigeminal impression behind and medial to the foramen ovale. It can be reached by passing a needle upwards through the foramen ovale either from the side of the head below the zygomatic arch and within the mandibular notch (between the condylar and coronoid processes), or more anteriorly, from the front of the face, piercing the skin just lateral to the corner of the mouth.

Concerning hazards, many of the blood vessels in the infratemporal fossa are at risk during this procedure. Intracranially, there is the danger of entering the subarachnoid space below the temporal lobe of the brain. Should local anaesthetic solution be injected into the cerebrospinal fluid, the patient will become unconscious and several of the cranial nerves will be paralysed. An important safeguard relates to the fact that the trigeminal ganglion is unusually flat for a cranial nerve ganglion. Consequently, the needle, on entering the middle cranial fossa, must be kept close to the bone of the trigeminal impression in which the ganglion sits.

A 40-year-old man suffered from a severe nasal infection. It appeared to clear up, but suddenly there was a rush of clear fluid from his nose. His doctor diagnosed a spontaneous rhinorrhoea (a watery discharge through the nasal cavity) caused by loss of cerebrospinal fluid. What anatomical considerations may account for this?

Spontaneous rhinorrhoea may result from the presence of an arachnoid diverticulum that passes into the roof of the nasal cavity through a defect in the ethmoid bone. It is also possible that the discharge may arise from the meninges which follow the olfactory nerves into the nose. Where there is a defect in the floor of the anterior cranial fossa, the protruding meninges are called a meningocele. There may also be an encephalocele, where brain tissue protrudes into the nose. These diverticula can also pass into the orbit (through the superior orbital fissure) or into the pterygopalatine fossa (having passed through the orbit) or into the nasopharynx (through the ethmoid or sphenoid bones).

A patient was taken to an operating theatre for surgical intervention into the anterior cranial fossa. The surgeon elected to enter through the forehead and made a semicircular incision with a distinct lateral curvature down the forehead. The skin flap was pulled downwards. He then made a bone flap which had a reverse C-shape with the base in the anterior temporal region. The bone flap was reflected over the temporalis muscle. What anatomical features have dictated the avoidance of a sagittal incision on the forehead, and the type of bone flap used?

The sagittal skin incision cuts across the creases of the forehead and is therefore avoided for cosmetic reasons. The semicircular incision maintains the innervation of the skin of the forehead, though resulting in some damage to the supra-orbital supply to the scalp. Care must be taken to avoid damage to the superficial temporal vessels and the auriculotemporal nerve as they pass on to the temple in front of the ear. The bone flap aims to avoid the frontal air sinus. A preoperative radiograph is taken to locate this sinus. The bone flap is attached to the temporalis muscle in order to maintain its blood supply. A possible hazard concerns the presence of the middle meningeal vessels in the region of the pterion. This region is in the vicinity of the posterior part of the bone flap. Obviously, sudden penetration through the bone and into the meninges and brain must be avoided.

A newborn baby is observed to have a raised anterior fontanelle, a raised temperature and to be in a distressed condition. The paediatrician suspects that the baby is suffering from meningitis. To test this, some cerebrospinal fluid must be obtained for analysis. How would this procedure be undertaken?

In the adult, a sample of cerebrospinal fluid can be obtained either by cisternal puncture or by lumbar puncture. The cerebellomedullary cistern (cisterna magna) is situated in the angle between the cerebellum and the dorsal surface of the medulla oblongata and is therefore accessible to a needle introduced between the external occipital protuberance and the spine of the axis forwards to the foramen magnum. Lumbar puncture is more commonly performed. For this procedure, the needle is inserted between the spines of the third and fourth lumbar vertebrae into the subarachnoid space below the termination of the spinal cord (normally at the level of the intervertebral disc between the first and second lumbar vertebrae). At birth, however, the spinal cord terminates at the third lumbar vertebra. Care must be taken, therefore, to ensure that the needle is introduced below this level.

A 50-year-old man complained of severe neuralgic pain induced by swallowing. The pain started in the region of the palatal tonsil and extended up to the ear. Which cranial nerve is producing the neuralgia and, should the cranial nerve need to be sectioned intracranially, what are the anatomical considerations?

The patient has glossopharyngeal neuralgia. This cranial nerve is distributed to the mucosa of the oropharyngeal isthmus and the tympanic cavity. The glossopharyngeal nerve emanates intracranially in the posterior cranial fossa from the lateral side of the medulla oblongata. Near the brainstem, it lies very close to the vagus nerve. Occasionally these nerves are distinctly separated as they pass through the dura mater and consequently the glossopharyngeal nerve can be operated upon without damage to the vagus nerve. However, should it not be possible to separate the two nerves, some of the rootlets of the vagus nerve can be sectioned with the glossopharyngeal nerve as they are essentially sensory. Extracranial sectioning of the glossopharyngeal nerve is possible but presents difficulties because of its close relationship to vessels within the carotid sheath. In some cases, sectioning of the glossopharyngeal nerve may not relieve the neuralgia. This is because the region normally supplied by the glossopharyngeal nerve may have a contribution from the vagus nerve.

A patient is diagnosed as having a tumour of the pituitary gland. Although it is possible for a surgeon to approach the gland via the anterior cranial fossa (see earlier case history, page 609), it is more common to take a route through the sphenoidal air sinuses.

The transphenoidal approach requires that the surgeon attains the sphenoidal air sinuses through the medial wall of the orbit, the ethmoidal air cells, and the uppermost part of the nasal cavity. The initial incision is close to the bridge of the nose, along the medial margin of one of the orbits. Structures along the medial wall of the orbit are reflected, including the trochlea of the superior oblique muscle. Care must be taken to avoid damage to orbital structures by maladroit reflection (even the optic nerve can be damaged). On entering one of the nasal fossae through the ethmoid bone, the anterior wall of the sphenoidal sinus can be seen. It is now necessary to pass centrally through the sphenoidal air sinus to reach the anterior wall of the pituitary fossa. Care must be taken not to use the sphenoidal septum as a guide for the midline as it is usually off-centre. Instead, the landmarks used are the upper posterior part of the nasal septum and the rostrum of the sphenoid bone. A complication of the procedure close to the pituitary gland is the accidental penetration of a cavernous sinus. This can cause serious damage to the internal carotid artery within the sinus and the cranial nerves passing along the walls of the sinus.

A patient was taken to hospital after collapsing suddenly. He regained consciousness after a few hours and appeared to have recovered completely by the next day. The consultant physician diagnosed a transient ischaemic attack affecting the brain (a 'stroke' in which the symptoms have disappeared within 24 hours). In making a final neurological examination before discharging the patient the doctor touched the region of the pillars of the fauces in the mouth. Why?

The doctor is testing the gag reflex. Although the patient appears to have fully recovered, there may be some residual damage to individual cranial nerves. The gag reflex assesses the integrity of the glossopharyngeal and vagus nerves. The sensory component of the reflex is the glossopharyngeal nerve which supplies much of the mucosa at the back of the mouth. The motor component is the vagus nerve whose pharyngeal branch supplies most of the muscles in the pharynx and palate. Should these nerves be damaged, the patient will experience difficulty in swallowing and may choke.

After several headaches of sudden onset during the previous six months, a 44-year-old woman was admitted to hospital following a particularly severe headache which woke her from sleep. She complained of double vision, particularly on looking to the right. On examination, her left pupil did not react directly to light nor did it react consensually (i.e. constriction of one pupil when light is shone into the other eye). The right pupil reacted normally. The left eye moved little and tended to be directed downwards and outwards wherever the patient was asked to look. She appeared to be otherwise in good health and demonstrated no other abnormalities. What is the likely cause of her symptoms?

The restricted movements of the left eye, together with its tendency to be directed downwards and outwards, indicate dysfunction of the extra-ocular muscles supplied by the oculomotor nerve (superior, medial and inferior recti and inferior oblique muscles). Downward and outward movements may be effected by the superior oblique muscle which is supplied by the trochlear nerve and by the lateral rectus muscle which is innervated by the abducent nerve.

The loss of the direct and consensual light reflexes in the left eye indicates the involvement of the parasympathetic fibres of the left oculomotor nerve. Fibres necessary for the light reflex leave each retina in the corresponding optic nerve, undergo partial decussation at the optic chiasma, and so travel via both optic tracts to both pretectal nuclei (see pages 590 to 591 and Figure 177). These nuclei are interconnected via fibres passing in the posterior commissure. Each pretectal nucleus projects to both Edinger–Westphal (parasympathetic accessory) nuclei of the oculomotor nerves. From each Edinger–Westphal nucleus, parasympathetic preganglionic fibres pass through the ipsilateral nerve to the ciliary ganglion and thence to the eyeball via the short ciliary nerves. Activity in postganglionic fibres produces pupillary constriction. Lesions of one oculomotor nerve thus prevent both direct and consensual light reflexes of the corresponding pupil, but leave the reflexes of the other eye unaffected.

The sudden onset of the symptoms in the absence of obvious infection suggests a vascular cause. The headaches indicate either small haemorrhages or enlargement of an aneurysm. Larger haemorrhages would be likely to cause disturbances of consciousness, stiffness of the neck or other signs of raised intracranial pressure. Haemorrhaging from an aneurysm will occur into the subarachnoid space, but is likely to cause local damage to nervous tissue. Oculomotor nerve palsy is not uncommonly caused by an aneurysm of the cerebral arterial circle (of Willis). Aneurysms occur particularly at the junction of arteries. In the present case, a likely site would be the junction of the left posterior communicating and internal carotid arteries (although the junction of the posterior communicating and posterior cerebral arteries provides another possible site). A left carotid (or vertebral) angiogram could be used to confirm the diagnosis.

A 66-year-old man suddenly felt a tingling sensation throughout his whole body. He also saw flashes of light. He felt faint but did not collapse and was able to call out to his wife who helped him to bed. At breakfast next morning he could not see his wife who was seated on his left. When he went out for a walk, he collided with people several times and bruised his left shoulder by walking into a lamp-post. On consulting his doctor, he was referred for neurological examination. This revealed that he was blind in the temporal half of the field of his left eye and the nasal half of the field of his right eye (left homonymous hemianopia). His optic discs were normal. There was slight thickening of his retinal arteries. His blood pressure was raised (190/110 mm Hg) and his pulse was fast (95 per minute) and irregular. What had happened?

The left homonymous hemianopia implies that the patient does not have the ability to see to his left. Homonymous hemianopias result from complete lesions of the contralateral visual pathway central to the optic chiasma (see pages 590 to 591). Fibres from the nasal retina (dealing with the temporal visual field) decussate in the chiasma, but those from the temporal retina (dealing with the nasal visual field) do not. Accordingly, the lesion may be in the optic tract, lateral geniculate body, optic radiation or primary visual cortex (striate cortex) on the patient's right.

The impairment followed an event that was sudden and brief. Such events are most commonly cerebrovascular accidents and, if very sudden, are probably caused by an embolism. Symptoms from haemorrhages are typically less sudden and are accompanied by a headache because of increased intracranial pressure. Cerebral thrombosis usually results in symptoms that develop over minutes or hours, and a history of prodromal episodes is common. The sudden increase in size of an aneurysm may produce symptoms because of pressure on surrounding structures; the optic tract, for example, is not uncommonly compressed by aneurysms. However, such an event might be expected to give rise to a headache rather than the general paraesthesia reported in this case. It should be noted that haemorrhaging from an aneurysm into the subarachnoid space would also result in severe headache. In this case, there is evidence of arterial and heart disease (thickening of the retinal arteries, raised blood pressure, fast and irregular pulse). Atrial fibrillation provides a possible source of a thrombotic embolism and might be suspected in such circumstances. The normality of the optic discs indicates that the patient is not suffering from any major and sustained rise in intracranial pressure. The optic nerves are invested in meninges and raised intracranial pressure results in oedema of the optic disc at the optic nerve head.

It is unlikely that the homonymous hemianopia resulted from blockage of an artery supplying the optic tract or lateral geniculate body. This is because there is an absence of neurological problems involving surrounding structures within the diencephalon and midbrain. It is possible to assess optic tract lesions by investigating the hemianopic pupillary reflex. This reflex involves constriction of both pupils when the intact half of each retina is illuminated, but not when the blind half is illuminated. Fibres involved in the light reflex travel in the optic tracts to the pretectal area (see pages 590 to 591). Lesions of the optic radiation or visual cortex will not affect these fibres.

It is possible that the patient has a lesion of the optic radiation within the retrolenticular part of the internal capsule (see Figure 194). However, the severity of the initial symptoms suggests the involvement of a major vessel, i.e. the right posterior cerebral artery. Besides supplying the terminal parts of the optic radiation, this artery supplies the primary visual cortex and much of the surrounding visual association cortex of the right hemisphere. A CT scan would provide the most straightforward means of confirming infarction of tissue of the right occipital lobe within the distribution of the right posterior cerebral artery.

Homonymous hemianopias resulting from posterior cerebral artery occlusion are often described as demonstrating 'sparing' of the macula (i.e. the hemianopia is incomplete at the centre of the visual field so that vision is normal within this small area). Macular sparing is difficult to demonstrate. Because it extends to an area of only one or two degrees from the midline (i.e. is 'foveal' rather than 'macular'), visual field determination of the necessary accuracy is difficult to achieve. Macular sparing may result from additional supply of the foveal region of striate cortex by the middle cerebral artery, but anatomical demonstration of such supply is lacking. Furthermore, research on monkeys suggests that there may not be an exact dividing line in the centre of vision between retinal ganglion cells projecting ipsilaterally and those projecting contralaterally. The result of such untidiness would be representation of the midline (including the foveal region) in both hemispheres. As a consequence, however, macular sparing would not be differentially diagnostic of cortical infarction. The clinical literature supports this view. Accordingly, determination of the presence of macular sparing is of limited clinical value.

A 75-year-old woman was admitted to hospital. She had felt faint over lunch and had been helped to bed, complaining of feeling weak and 'tingly' down her left side. By the time the doctor arrived half an hour later, she found she could no longer grip a mirror in her left hand. She volunteered that she had had one or two 'funny turns' over the past few weeks. Examination revealed weakness of limb and trunk muscles on the left side, with greatly reduced muscular tonus, so that the muscles felt flabby on the left compared with those on the right. Tapping the tendons of the muscles of the left limbs produced little reflex contraction of the muscles. Stroking the skin of the abdomen on the left produced no movement of the umbilicus, i.e. there was no reflex contraction of the abdominal muscles. Drawing a key across the sole of the left foot towards the big toe caused the toe to extend upwards (extensor plantar response: Babinski's sign) rather than flexing downwards as on the right. By using sharp and blunt probes, warm and cold objects, and a tuning fork, it was established that the patient experienced no sensations of pain, temperature, touch or vibration over the left half of her body, although such sensations were appreciated normally on the right and over the whole of her face and the top of her head. When her eyes were closed, the patient could not describe the positions of her left limbs when they were moved by the

examiner. The patient had a history of angina pectoris. What is the likely cause of these symptoms and what future change might be expected?

The weakness, with reduced muscular tone and diminished or absent tendon reflexes on the left side of her body, the absence of superficial abdominal reflexes on the left, and the nature of the left plantar response (Babinski's sign) are signs that would be expected shortly after an upper motor neurone lesion (i.e. a lesion damaging descending tracts to the spinal cord and in particular involving the corticospinal fibres).

The loss of appreciation of nociceptive, thermal and tactile stimuli over the left half of her body indicates a lesion affecting the anterolateral system (see pages 600 to 601). Also the patient no longer feels vibrations and has lost proprioceptive awareness for the left side of her body. These sensibilities depend upon the dorsal column/medial lemniscal system (see pages 598 to 599). Sensory appreciation is normal over the face and the top of the head, which are supplied by the trigeminal nerves. Because the sensory and motor loss affects all segments of the spinal cord, but on only the left side, the lesion is unlikely to be peripheral and is most likely to lie above the foramen magnum. Note also that within the spinal cord the left anterolateral tract carries information concerning the right side of the body, while the left dorsal column conveys information concerning the left side. Not until the cranial part of the pons is reached do fibres of both systems, relaying information relating to the left (contralateral) side of the body, lie close together in the spinal and medial lemnisci (see Figure 178h). At this level, the corticospinal fibres, although dealing with movements of the contralateral body, are not close to the lemnisci. Any lesion involving sensory and motor fibres would be expected to produce other signs, of cranial nerve or cerebellar dysfunction. Similar arguments make a mesencephalic or diencephalic lesion implausible. However, sensory and motor fibres dealing with the left (contralateral) side of the body are found close together in the posterior limb of the right internal capsule, and the sensory and motor cortices (areas SmI and MsI) are adjacent on the postcentral and precentral gyri of the right hemisphere (see Figures 194 and 188).

In the cerebral cortex, the regions dealing with the upper limb are supplied by the middle cerebral artery, while those for the lower limb are supplied by the anterior cerebral artery. Accordingly, blockage of a cortical branch of either of these arteries would not produce symptoms in both limbs. Intracranial haemorrhages may produce such symptoms by the effects of pressure (possibly in addition to loss of blood supply), but this patient reported no headaches, neck stiffness or other signs of raised intracranial pressure that would then be expected. Thus, the most likely explanation of the symptoms is damage to the posterior limb of the right internal capsule, involving the somatic sensory and motor fibres dealing with the left half of the body, but sparing the more anteriorly placed fibres dealing with the head region.

The rapid deterioration of the patient, in the absence of signs of infection, suggests a cerebrovascular accident. The blood supply of the internal capsule is mainly from the lateral striate arteries from the cerebral arterial circle of Willis (chiefly from the middle cerebral artery). Ventral and retrolenticular parts of the internal capsule are supplied by branches from the anterior choroidal artery. In this case, the progressive course of the symptoms over half an hour without headache or loss of consciousness, together with the suggestions of prodromal episodes over the previous weeks, makes arterial thrombosis of one of the right lateral striate arteries the likely cause. The presence of coronary artery disease (angina pectoris) with probable widespread atherosclerotic changes increases the probability of this diagnosis.

The internal capsule is one of the regions most commonly affected by cerebrovascular accidents ('strokes'), being particularly vulnerable to even small haemorrhages or blockages of minor vessels because of the high density of important fibre connections. In addition to the interruption of corticospinal and thalamocortical fibres because of the loss of their blood supply and consequent infarction, there is also likely to be disruption to the functioning of other structures (i.e. cerebellum, basal nuclei, red nucleus) resulting from the lesion. Cerebellar or basal nuclear signs will be masked by the paralysis. However, the eventual development of spasticity (see below) is dependent upon involvement of connections in addition to the corticospinal tract; in the rare cases of isolated corticospinal tract lesions there is permanently reduced muscular tone and diminished reflexes. The loss of superficial reflexes (such as the abdominal reflexes) is believed to be caused by interruption of a necessary cortical component of these reflexes.

Some recovery of function may be anticipated following thrombosis, recovery in general being greater the more minor the initial symptoms. Typically, after several days have elapsed, the flaccid contralateral muscular weakness (hemiparesis) or paralysis (hemiplegia) may be expected to become spastic, with increased muscular tone and reflexes. Following fairly complete capsular or motor cortical lesions, the spinal cord goes into shock because of loss of the descending inputs, and a flaccid contralateral hemiplegia results, with complete loss of superficial and deep reflexes on the affected side. The inverted plantar response (Babinski's sign) can usually be evoked several hours after the onset of symptoms. Within the next few days, tendon reflexes gradually return and become brisker than on the unaffected side. In parallel with this change, muscular tone increases on the affected side and the muscles present an increased resistance to passive movements. This increased resistance is often known as 'clasp-knife' because, with sufficient applied force, it suddenly gives way. In the upper limb, the spasticity is typically more marked in flexors; in the lower limb, in extensor muscles. Thus, on the affected side the arm tends to remain permanently flexed, while the leg is extended. Usually, aided by physiotherapy and determined effort, some capacity for performing voluntary movements will gradually return over the ensuing months and years.

A 27-year-old motor mechanic was sent to the local hospital by the garage foreman after the mechanic had badly burned his left hand on a hot engine. The patient kept saying that there was no need to make a fuss, and it was apparent that he was in no discomfort. Consequently, a neurological disorder not associated with his injury was suspected. On examination, it was found that the patient felt pin-pricks only as slight touches to his skin, and that he could not tell whether objects were warm or cold anywhere on either upper limb. Sensation elsewhere was normal; appreciation of touch, of the vibrations of a tuning fork against his body, and of the position of his upper limbs was normal. However, tendon reflexes were diminished in the left upper limb. It was also noticed that he had a tuft of hair associated with some dimpling of the skin over the lower lumbar spinal column. On questioning, he said that he had felt some numbness and tingling of the left hand for about a year and that recently this had sometimes extended to his left forearm. What is your diagnosis?

The patient has a dissociated sensory loss. Nociceptive and thermal sensitivity has been lost in both upper limbs (innervated by spinal cord segments C5 to T1), while touch and proprioceptive sensibility is unimpaired. Pain, temperature and touch information is carried in the anterolateral tracts of the spinal cord (see pages 600 to 601 and Figure 172b). However, touch is additionally conveyed, together with vibratory sensibility and conscious proprioceptive information, by the dorsal column system (see pages 598 to 599 and Figure 172b). The dissociated sensory loss and involvement of both upper limbs without other sensory impairment makes a peripheral lesion unlikely and indicates that any lesion of the anterolateral tracts must be incomplete. Interruption of the decussating sensory fibres before their entry into the ventrolateral funiculus would produce bilateral loss of pain and temperature sensibility for those segments giving rise to the decussating fibres. The decussation occurs in the ventral commissure, just ventral to the spinal canal (see pages 600 to 601). Because these fibres typically ascend a few segments before decussating, a lesion in the lower cervical spinal cord (C3 or 4 to C7 or 8) near the central canal is indicated. That the symptoms appeared to have been progressive over the period of a year does not suggest a ventral spinal arterial occlusion or haemorrhage, but rather an intrinsic tumour or syringomyelia (the presence of cavities in the spinal cord associated with glial cell proliferation).

Differential diagnosis of a tumour or syringomyelia may be difficult before operation; indeed, the two conditions are sometimes associated. The possible presence of spina bifida occulta, indicated by the slight lower lumbar deformity, increases the possibility of syringomyelia. The proliferative gliosis or cavitation of syringomyelia occurs close to the central canal of the spinal cord, most commonly in the lower cervical region. It is usually slowly progressive and is frequently associated with developmental abnormalities of the neural tube. In addition to spina bifida, syringomyelia is particularly associated with mild variants of the Arnold–Chiari malformation (in which parts of the medulla and cerebellum extend through the foramen magnum into the vertebral canal). Arnold–Chiari malformations often cause obstruction of CSF flow from the fourth ventricle, and it is possible that the increased pressure of CSF may initiate the cavitation.

The decussating anterolateral system fibres are the first to be affected in syringomyelia, but the lesion (syrinx/cavity) gradually extends, typically asymmetrically, to include successively: a) the ventral horn (resulting in a lower motoneurone lesion that often starts with wasting of the intrinsic hand muscles); b) the intermediolateral column (giving rise to sympathetic signs that include Horner's syndrome and bladder dysfunction); c) the lateral funiculi (giving rise to the symptoms of an upper motoneurone lesion resulting from interruption of the lateral corticospinal tract). Loss of myotatic (stretch) reflexes, in this case involving the left upper limb, commonly occurs as an early sign, probably as a result of interruption of the collaterals of the muscle spindle afferents as they pass from the dorsal to the ventral horn (see Figure 174a).

A young motorist was brought into hospital unconscious after his vehicle had been hit from behind by a lorry. His arms and legs were flaccid and no muscular reflexes could be evoked from them. After a week in the intensive care unit, the patient regained consciousness. He did not have full voluntary movement in any limb and had either impaired or absent appreciation of pain, temperature, touch, vibration and joint sensation at all levels below the shoulders (with the exception of a region in the upper part of the chest above the second intercostal space). A pin-prick or light pressure was felt acutely across the top of the shoulders and upper chest; there was normal sensibility in the neck and head regions. For both lower limbs, muscle tone was increased, knee and ankle jerks were hyperactive, and the limbs were held in extension. Forward movement of a sharp object along the soles of the feet produced reflex upward movements of the big toes (Babinski's sign). The tone of the muscles of the left forearm and hand was decreased and reflexes greatly diminished or absent. Fracture of the right radius precluded testing of the right arm. How would you account for the neurological symptoms? What other major functional disturbances of neurological origin would be expected?

The patient is quadriplegic as a result of 'whiplash' injury to the spinal cord. Complete transection of the spinal cord leads to paralysis and anaesthesia of all regions supplied by spinal nerves originating below the level of transection.

Initially, paralysis caused by injury to the spinal cord is flaccid, muscular tone is reduced and reflexes (cutaneous and tendinous) are absent. This state of 'spinal shock' may persist for days or weeks and is likely to be caused by the loss of descending excitatory input (corticospinal, rubrospinal, vestibulospinal and reticulospinal fibres) to the segments of the cord below the lesion (see pages 602 to 607).

Later, there is considerable local reorganisation and probably development of types of denervation hypersensitivity within the undamaged parts of the cord. Spastic paralysis then develops during the following weeks and is usually attributed to release of spinal mechanisms from the inhibitory influences of the descending pathways, though hypersensitivity may also be a contributory factor. In complete cord transections, flexor rather than extensor spasm usually develops. In the case described here, the presence of some sensations and voluntary muscular control below the shoulder, plus the extended position of the lower limbs after their initial flaccidity, indicates that the cord transection is partial rather than complete. In such cases, some recovery of function may eventually occur.

Muscles of both lower limbs show spastic paralysis. They demonstrate the typical features of paralysis or paresis resulting from an upper motoneurone lesion: weakness with hypertonia and hyper-reflexia in the absence of significant muscle-wasting. Babinski's sign was also present. These features are accounted for by interruption of descending pathways (corticospinal, rubrospinal, reticulospinal and vestibulospinal) from the 'upper motoneurones' (i.e. those neurones within the brain whose axons make up the descending pathways to the lower motoneurones of the motor cranial nerve nuclei and ventral horns of the spinal cord). Babinski's sign is not definitive of a pyramidal tract (corticospinal) lesion. Except in infancy, its presence does usually indicate loss of function in the descending spinal pathways (however its expression is subject to some individual variation).

The muscles of the left hand and forearm show a flaccid paralysis, there being hypotonia and hyporeflexia. Paralysis or paresis in the presence of hypotonia and hyporeflexia, with marked muscular wasting and possible fasciculation, are the characteristic symptoms of a lower motoneurone lesion. These symptoms are a direct consequence of the denervation of the muscles. A lower motoneurone lesion may result from injury to the ventral horn of the spinal cord (or brainstem in the case of cranial nerves), or to interruption of motoneuronal axonal transmission to the muscle. In the patient described here, injury to the left side of the lower cervical and upper thoracic segments (C6 to T1) is probable (though damage to the left brachial plexus cannot be excluded and, as the patient may still be emerging from spinal shock, further changes are possible). Lesions involving C5 as well as lower segments will result in complete upper limb paralysis. If the fifth cervical nerves are intact, the deltoids, spinati, rhomboids, bicipites and brachiales muscles will receive innervation. However, the forearm muscles (innervated by segments C7 and C8) and the intrinsic muscles of the hand (innervated by segments C8 and T1) will be paralysed. Muscle wasting would be expected to develop in time.

Interruption of ascending sensory pathways (including both dorsal column and anterolateral system fibres) would be responsible for the loss of sensation below the shoulders. The increased cutaneous sensibility of the upper thorax and top of the shoulders is consistent with a lesion just below these dermatomal levels (i.e. below C5). Such hyperaesthesia is common within the dermatomes of segments just above a cord lesion. Thus, the patient probably has severe injury to the lower cervical spinal cord producing major, but subtotal, interruption of all ascending and descending pathways. In such a patient, where immobilisation of the neck is essential, testing of neck and shoulder musculature is impracticable and inadvisable.

In addition to probable injuries to the vertebral column and associated structures in the neck, the patient is likely to have other major, neurologically based problems. Secondary involvement of the phrenic nerves (origin segments C3 to C5) resulting from oedema or circulatory disturbances may produce respiratory difficulties caused by paralysis of the diaphragm. Interruption of descending autonomic pathways will initially result in retention of urine and faeces. Later (after one to several weeks), the patient will experience double incontinence. For similar reasons, peristalsis and sweating will show initial impairment, the patient will suffer a temporary drop in blood pressure and demonstrate Horner's syndrome (ptosis, miosis, enophthalmos) on both sides. Eventually, reflex erection and ejaculation may be evoked.

A businessman in his thirties was referred for neurological examination because of repeated and severe headaches. He explained that lying down made the headaches worse and that recently they had interrupted his sleep. He confirmed that he had had the headaches for some months, but said that they were becoming more severe and frequent. They were now often associated with nausea. Indeed, he had had

two episodes of vomiting. He also mentioned that he felt lethargic and generally run-down. His concentration seemed poor and he thought he might be suffering from eye-strain. Examination revealed that both pupils were of normal size but neither constricted when light was shone in either eye. However, the pupils did constrict when the patient was asked to look at some fine print close to his eyes. On being asked to look up, the patient simply moved his head, being unable to roll his eyes upwards. Both optic discs appeared oedematous. There were no signs of acute infection nor of vascular disease. What might you expect a CT scan to reveal?

The patient displays signs of raised intracranial pressure (i.e. severe headaches with nausea, and papilloedema). His general lethargy and reduced concentration could also be symptoms of this.

Constriction of the pupils during the direct and consensual light reflex requires the integrity of the pathway from the retina through the optic nerve, optic chiasma and optic tracts to the pretectal areas on each side of the midbrain. Fibres pass from each pretectal area to both the parasympathetic accessory nuclei of the oculomotor nerves (the Edinger–Westphal nuclei). Preganglionic fibres leave each Edinger–Westphal nucleus for the ipsilateral ciliary ganglion in the orbit, and postganglionic fibres reach the sphincter pupillae muscle of the eyeball (see pages 590 to 591). The accommodation reflex requires focusing and convergence of the eyes upon near objects, together with pupillary constriction (which reduces the spherical aberration of the highly convex lens). This reflex involves connections from the visual cortex and, where voluntary, the frontal eye fields (see Figure 177). These connections are believed to pass to the superior colliculus and pretectal nucleus. It is possible that the reflex itself is organised in the caudal midbrain reticular formation. Either the pathway to the Edinger–Westphal nucleus necessary for pupillary constriction during accommodation differs in its course from that of the light reflex, or it involves more fibres of the same pathways, because there are conditions in which pupillary constriction is lost in response to light but preserved during accommodation. The most common such condition is that of the Argyll Robertson pupil found in syphilis of the central nervous system (particularly in tabes dorsalis). However, the Argyll Robertson pupil is small and often irregular in outline (possibly as a result of local damage to the iris). Accordingly, the normal size of the pupil observed in the patient described here, together with the raised intracranial pressure, suggests that syphilis is not the most likely diagnosis.

The parasympathetic fibres of the oculomotor nerve are often the most susceptible in third nerve lesions. However, the bilateral pupillary signs suggest a central origin for the dysfunction. This suggestion is supported by the paralysis of conjugate vertical movements of the eyes (Parinaud's syndrome). Such movements are probably organised in the pretectal region immediately cranial to the superior colliculi, probably in the interstitial nuclei of Cajal, whose efferents pass through the posterior commissure.

The apparently slow onset of symptoms, and the absence of signs of infection or vascular disease, suggest a tumour, probably a pinealoma. The pineal gland lies in the midline between, and cranial to, the superior colliculi (see Figure 178j). Tumours of the pineal thus compress the cranial parts of the tectum and frequently compress the cerebral aqueduct of the midbrain. Blockage of the aqueduct, with its consequent interruption of the circulation of CSF would lead to an internal hydrocephalus with the accompanying symptoms of raised intracranial pressure. Thus, a CT scan would be expected to show enlarged lateral and third ventricles, together with a tumour in the region of the pineal gland.

Late on a Saturday night, a middle-aged man was brought into hospital by the police. He had been found semi-conscious, lying in the road. He smelled strongly of alcohol, urine and vomit. He was in a confused state, being unsure where he lived or the day of the week. He continually demanded to know how he came to be in hospital. He complained of an intense headache and a stiff neck. His blood pressure was found to be raised but his temperature was normal. Sensory testing

was difficult and inconclusive. However, his neck was stiff and painful to move. There was evidence of some weakness in both right limbs. No abdominal reflexes could be evoked on the right. When the sole of his right foot was stroked with a sharp instrument, his big toe extended (Babinski's sign). He was grossly inco-ordinated. At this stage, the patient's right hand and arm began to twitch involuntarily and he lost consciousness. The man's optic discs appeared normal but his pupils were small and reacted only slowly to light. The corneal reflexes were depressed. The CT scanner was out of action and a lumbar puncture was performed to aid further diagnosis. Just after a successful tap had been obtained, the patient suddenly stopped breathing and could not be resuscitated. Provide an explanation for the events.

The patient's confusion, inco-ordination and headache could have been caused by alcohol and/or concussion. A similar cause could underly his presumed vomiting and incontinence. However, these symptoms could also be signs of raised intracranial pressure. The abnormal reflexes in the right limbs, together with a probable epileptic attack involving the right upper limb, support such a diagnosis. The weakness, absence of abdominal reflexes, and Babinski's sign suggest upper motoneurone dysfunction (possibly caused by pressure or haemorrhagic damage). The neck stiffness could be caused by irritation of the meninges by blood. The depression of the corneal reflex (dependent on the integrity of afferents in the trigeminal and efferents in the facial nerve) and light reflex (dependent on connections from the optic nerve to the oculomotor nerve supplying parasympathetic fibres to the eye), together with loss of consciousness, are again consistent with raised intracranial pressure.

In cases where there are signs of raised intracranial pressure, lumbar puncture should be performed with great care because of the possibility of herniation through the tentorial incisure or of sudden descent of the hindbrain ('coning') on release of fluid from the lumbar cistern. With descent of the hindbrain, the cerebellar tonsils may become wedged in the foramen magnum, compressing the regions of the medulla controlling respiration and so leading to respiratory failure. The risk of such a complication of lumbar puncture is very low in most circumstances of raised intracranial pressure. However, an exception is suspected intracranial tumour, particularly of the posterior cranial fossa; in such cases, lumbar puncture is clearly contra-indicated.

The symptoms in the present case are more consistent with an intracranial haemorrhage. For an unco-operative or unconscious patient, in the absence of a past history, differential diagnosis is difficult without ancillary investigations (such as a CT scan). Nevertheless, the nuchal rigidity in the absence of a raised temperature, and the relatively rapid worsening of symptoms in the presence of hypertension in a middle-aged patient, make haemorrhage a likely supposition. Definite neurological signs (for example, as here, cranial nerve palsies and a possible hemiparesis) are to be expected in any major intracranial haemorrhage. However, where there is bleeding from vessels within the subarachnoid space (subarachnoid haemorrhage), the neurological signs may be generalised rather than focal. Epidural (extradural) haemorrhages, which arise from the rupture of meningeal vessels, do not normally result in nuchal rigidity. Other types of intracranial haemorrhage (intracerebral, subarachnoid, subdural) are all likely to lead to irritation of the meninges by blood, and grossly blood-stained CSF. The fact that the optic discs had not become oedamatous suggests a relatively rapid development of the raised intracranial pressure.

In this case, a postmortem examination revealed a large subdural haematoma over the left frontal convexity of the hemisphere, resulting from rupture of a superior cerebral vein at its point of entry into the superior sagittal sinus. Pressure on the left precentral gyrus (area MsI) results in right-sided motor deficits: most corticospinal and corticonuclear fibres decussate before their termination.

A 6-year-old boy was brought for consultation because his mother was concerned that he kept having sick headaches and seemed unable to stand upright without falling over either backwards or forwards. The headaches had started a month

before and were becoming worse. On examination, the boy's gait was found to be ataxic. On being asked to stand upright with his eyes closed, he tended to fall backwards. When asked to use his index finger to touch alternately the examiner's finger and his own nose, he had to make frequent corrections and his finger showed a coarse tremor during, and especially just before, completion of the movements. This occurred whichever hand he used. Muscular tone was low in all limb muscles. Nystagmus (rhythmical oscillations of the eyeballs) was present on looking to the left or the right. Both optic discs were grossly oedamatous and some skull sutures were palpable. There was no sign of neck stiffness or infection. Explain the symptoms.

The child has signs of cerebellar dysfunction: hypotonia, ataxia, and poorly co-ordinated movements with tremor during their performance ('intention tremor'). The nystagmus is also probably of cerebellar origin.

Dysfunction of a cerebellar hemisphere typically produces a tendency to fall to the same side as the lesion (the cerebellum deals with the ipsilateral side of the body). In this patient, there is a tendency to fall forwards or backwards; such a symptom is often indicative of midline (vermal) cerebellar damage. That the damage is not confined to one hemisphere is supported by evidence of cerebellar dysfunction on both sides of the body.

The papilloedema and palpable skull sutures suggest raised intracranial pressure. The relatively gradual development of symptoms and absence of nuchal rigidity suggest a tumour rather than haemorrhage (neither subdural nor epidural haemorrhage is likely to produce the present symptoms). Note that lumbar puncture is contra-indicated if tumours of the posterior cranial fossa are suspected because of the possibility of the hindbrain becoming wedged in the foramen magnum.

A 50-year-old woman went to her doctor complaining of attacks of giddiness and vomiting. It transpired that she had become progressively deaf in her right ear over the past few years. Recently, she had heard hissing and whistling noises in that ear (especially when trying to go to sleep at night). She had also recently been troubled by dull headaches (felt mainly at the back of her head) which sometimes woke her from sleep. Neurological examination revealed profound deafness in the right ear both for air- and bone-conducted sounds. Nystagmus (rhythmical oscillation of the eyeballs) was elicited when she looked to either side. On walking, the patient was unsteady, with a tendency to veer right. Movements of her right upper limb lacked co-ordination (i.e. were ataxic). There was impairment of pain and thermal sensibility over the whole of the right side of the patient's face, with absence of the corneal reflex in the right eye. The patient's right eye was easier to force open against her resistance than was the left. What is the probable cause of these symptoms?

The patient's deafness of the right ear even for bone-conducted sound suggests that the problem is either associated with the internal ear or is of more central origin. The giddiness, nystagmus and unsteadiness of gait all suggest involvement of the vestibular system (again either in the labyrinth or more centrally). The proximity of the auditory and vestibular receptors, of the axons of the primary neurones in the vestibulocochlear nerve in the internal acoustic meatus, and of the vestibular and cochlear nuclei in the brainstem (see Figure 178f) makes it common for there to be joint occurrence of auditory and vestibular symptoms.

The ataxia of the patient's upper limb suggests damage to the ipsilateral cerebellum or its connections. Such damage might also explain the nystagmus and unsteadiness of gait. The cerebellum deals with the ipsilateral side of the body, and deviation towards the side of the lesion is a common symptom of its dysfunction.

The loss of pain and thermal sensibility covering the whole right side of the face implies involvement of all three divisions of the right trigeminal nerve. Central processes of nociceptive (and probably thermoreceptive) trigeminal ganglion cells enter the pons (see pages 600 to 601) and mainly turn caudally in the spinal tract of the trigeminal. The tract runs laterally through the medulla (see Figures 178b to f), nociceptive fibres terminating chiefly in the pars caudalis of its nucleus (see Figures 178b and c) in the caudal medulla and upper cervical segments of the spinal cord.

The corneal reflex depends on an afferent arc involving the ophthalmic division of the trigeminal and an efferent arc involving motoneurones of the facial nerve supplying the palpebral part of the orbicularis oculi muscle. The weakness of facial muscles (shown by the ease with which the right eye could be forced open) implies involvement of the right facial nerve or possibly its nucleus. The motor nucleus of the facial nerve lies in the caudal pons (see Figure 178f and g), its fibres looping around the abducent nucleus before emerging laterally at the pontine–medullary junction and continuing with the vestibulocochlear nerve through the internal acoustic meatus.

The slow onset of symptoms is suggestive of a tumour. Headaches radiating to the back of the head are consistent with a space-occupying lesion causing raised intracranial pressure below the tentorium cerebelli in the posterior cranial fossa. These symptoms are likely to be aggravated when arterial pressure to the head is raised or venous pressure lowered, for example when lying down.

Parts of the cerebellum together with the inferior cerebellar peduncle are located laterally at the pontine–medullary junction. Here, they are adjacent to the spinal tract of the trigeminal and the points of emergence of the facial and vestibulocochlear nerves in the right cerebellopontine angle, close to the opening of the right internal acoustic meatus. A tumour enlarging in the cerebellopontine angle would compress all these structures. As the first symptom was deafness, the tumour may have originated close to the vestibulocochlear nerve.

A 23-year-old woman was taken to the doctor by her father. He had found her in distress, apparently with a severe headache, but she did not seem able to put two words together. Examination revealed that the patient understood a wide variety of instructions, although she was confused by sentences in the passive. However, she spoke in only single words or very simple sentences, often uttered with much stammering and hesitation. Her speech was consistently slurred, but phonation seemed normal. She could not name a set of common objects, although she could point to the individual objects when asked. On being told to write, she took up the pen in her right hand but had to be instructed on how to hold it properly. This she did with difficulty. Her writing was very clumsy and slow. Although she could write out some individual words or copy them from a book, she could not write out whole dictated phrases. She indicated that her headache was no longer severe. Blood pressure, pulse rate and temperature were within normal limits, but her neck was stiff and painful on flexion. The grip and movements of the fingers of her right hand were weaker than those of the left, and reflexes were diminished. Twice, it was observed that there was spontaneous twitching of the fingers of her right hand with spasm of the right arm muscles, shrugging of the right shoulder and turning of the head to the right. Arm muscle strength was normal, but weakness was found on the right when the patient was asked to shrug her shoulders against resistance. When asked to bare her teeth, her mouth opened less on the right so that she appeared to talk out of the side of her mouth. However, she could frown and wrinkle her forehead normally and, on being complimented on her dress, there was no asymmetry in her smile. When requested to protrude her tongue, it deviated to the right. On being asked to look to the right, neither eye moved, although both moved normally on being asked to look to the left and when following the examiner's finger moving slowly from side to side in front of her face. No abnormalities of sensation were found. Explain these symptoms.

The patient's weakness of the muscles of the right hand, right shoulder, right lower face and tongue is suggestive of a lesion of descending motor pathways.

Of the muscles innervated by the facial nerve, those above the orbit receive bilateral descending input, while those of the lower face are supplied by fibres from the contralateral cerebral cortex which decussate in the brainstem. Thus, a lesion of the left cerebral hemisphere would result in paresis of right lower facial muscles (indicated by the partial opening of the mouth on the right) but not of the upper facial muscles (normal ability to wrinkle the forehead and frown). An unimpaired ability to smile as part of an emotional response is common in the presence of diminished function of the corticonuclear fibres. Note that when a subject is asked to smile for a photograph the expression is often seen as not 'natural'. Conversely, in disorders of the basal nuclei (such as parkinsonism), emotional facial expressions are much reduced. The clear implication is that a different pathway is involved in conveying emotional input to the facial nucleus.

Such a supranuclear lesion on the left would be expected to result in the observed deviation of the tongue to the right as corticonuclear fibres supply the contralateral hypoglossal nucleus (though exceptions to this pattern are not uncommon). Paresis of the right genioglossus muscle will result in a reduced tendency for the tongue to protrude on the right. Note that descending cortical input to the motor nucleus of the trigeminal nerve is usually bilateral and that unilateral supranuclear lesions involving the nucleus ambiguus produce dysphagia only sometimes, and may be symptomless.

Dysfunction of descending fibres from the left cerebrum to the contralateral spinal accessory nucleus explains the weakness in shrugging the shoulders on the right (weakness of the upper part of the trapezius muscle).

An impaired ability to move both eyes horizontally may arise either from pontine lesions (the paramedian pontine reticular formation near the abducent nucleus organises conjugate horizontal eye movements) or from lesions of descending fibres to the pons from the frontal eye fields (see Figure 188a). In the latter case, the dysfunction is usually transitory and resolves within a few days. In pontine lesions, the paralysis of lateral gaze or loss of conjugate lateral movements of the eyes extends to such movements when the eyes are tracking a moving object (fixation reflex). Also, there is usually paralysis of one or both lateral recti muscles supplied by the abducent nerves. As the patient's fixation reflex is intact, the lesion must lie above the pontine level. In pontine lesions, the patient is unable to look towards the side of the lesion; in lesions of a frontal eye field or of its connections with the pons, the patient cannot voluntarily look to the side opposite the lesion. That the patient could not look right is consistent with a lesion of the left frontal eye field or of its connections.

The presence of expressive aphasia (an impaired ability to speak, beyond what might be expected from palsies of the facial and hypoglossal nerves) confirms the presence of a lesion within the forebrain. The patient was right-handed, so language representation would almost certainly be in the left hemisphere. A lesion of the left cerebral hemisphere is consistent with the motor deficits. The lesion is likely to be of the anterior speech cortex (Broca's area) or its fibre connections. Such lesions are characterised by good comprehension of most verbal material in the presence of non-fluent speech. Difficulties with sentences in the passive voice are a particular feature of such lesions. Ability to express thoughts in writing is also likely to be impaired. Aphasia, voluntary eye movement disturbance, and paresis involving muscles of the head, neck and arm could be caused by dysfunction of the frontal cortex or its fibre connections, particularly in the internal capsule (see Figures 188a and 194).

The lack of clear signs of cerebellar or basal nuclear dysfunction (tremor, athetosis) and the presence of probable epileptic activity (spontaneous movements of the upper limb, shoulders and neck) make a cortical lesion more likely. The epileptiform movements occurred in regions that are represented in adjacent parts of the motor homunculus (see Figure 190a). Turning the head to the right is associated with the left sternocleidomastoid muscle which is excited from the left motor cortex via neurones of the left spinal accessory nucleus. Therefore, the sternocleidomastoid muscle is under the control of the ipsilateral motor cortex, while the trapezius is controlled from the contralateral hemisphere. Note that this sternocleidomastoid control results in turning of the head towards the contralateral side, i.e. the side with which that hemisphere is concerned.

Signs of loss of skills in the absence of paralysis (apraxia), such as the apparent forgetting of how to hold a pen, are often associated with lesions of areas just anterior to the motor cortex.

The sudden onset of symptoms, headache and nuchal rigidity are consistent with an intracranial haemorrhage. The frontal cortical areas that show signs of dysfunction are supplied by branches of the left middle cerebral artery, although branches of the left anterior cerebral artery supply their deep fibre connections. However, in a young patient with no evidence of head injury, leakage of blood from an arteriovenous malformation must form part of the differential diagnosis. Such leakage involving the cortex of the inferior frontal convexity of the left hemisphere was subsequently confirmed in this patient.

On getting up in the morning, a man in his mid-fifties found that he could not keep his balance and kept falling to his left. He felt giddy and sick. On returning to bed, the left side of his face started to feel as if it were burning and then the right side of his body gradually began to feel numb from the neck down. When he called for help, he discovered that his voice was hoarse and, on taking a sip of water, he had difficulty in swallowing. On examination, his doctor found that his left eyelid was drooping (ptosis), while the left lower lid was higher than the right (indicating enophthalmos). Furthermore, his left pupil was smaller than the right (miosis), and there was nystagmus when the patient looked sideways. He also showed impaired appreciation of pin-pricks over all of the left side of his face and the right side of the body below the face. The knee and ankle jerks were reduced on the left compared to the right and the tonus of the limb muscles was less on the left. Performance of the finger–nose test with the patient's left hand demonstrated inco-ordination (ataxia) and tremor on movement (intention tremor). What is your diagnosis?

The patient has damage to structures of the lateral medullary region of the brainstem (Wallenberg's syndrome). Concerning the loss of pain sensibility on the left side of the face and the burning sensation, nociceptive information from the face passes ipsilaterally from all divisions of the trigeminal nerve, through the spinal tract of the trigeminal in the lateral medulla, to the caudal part of its spinal nucleus (see pages 600 to 601). Nociceptive information from regions below the face is carried cranially (primarily contralaterally) in the anterolateral tracts. The fibres decussate within a few segments of their origin in the dorsal horn of the spinal cord (see pages 600 to 601). Interruption of these fibres in the lateral medulla (see Figure 178b to e) impairs pain sensibility on the right side of the body below the face.

Spinocerebellar fibres also pass through the lateral medulla. With the exception of the ventral spinocerebellar tract, these fibres convey ipsilateral proprioceptive information and enter the cerebellum mainly by its inferior peduncle. Their interruption along with that of descending rubrospinal fibres would give rise to the hypotonia, hyporeflexia, and inco-ordination found in the left limbs. The intention tremor suggests further cerebellar damage, probably including the dentate nucleus (see Figure 178f). Interruption of the spinocerebellar fibres would also contribute to the tendency to fall to the left, but this symptom would be compounded by damage to the left inferior vestibular nucleus and to the left lateral vestibulospinal tract (see Figure 178e). The vestibular dysfunction would account for the presence of nystagmus and feelings of giddiness and nausea. The nucleus ambiguus also lies in the lateral medulla and sends fibres to the glossopharyngeal, vagus and cranial accessory nerves. Damage to this nucleus would account for the patient's hoarseness (as a result of paralysis of the left vocal fold) and difficulty in swallowing (as a result of paresis of the left side of the soft palate and pharynx).

The ptosis, enophthalmos and miosis are indicative of Horner's syndrome (disturbance of the cervical sympathetic system). Descending autonomic fibres, including those travelling to the intermediolateral cell column of the first and second thoracic segments of the spinal cord, lie laterally in the medulla. Their dysfunction leads to decreased preganglionic sympathetic activity. Consequently, decreased activity of the postganglionic neurones of the superior cervical ganglion affects the dilator muscle of the iris and the non-striated fibres of levator palpebrae superioris muscle. The enophthalmos is usually apparent rather than real, being caused by the narrowing of the palpebral fissure by ptosis. Loss of sweating and of vasodilation over the left side of the head and neck might also be expected.

The symptoms in the case described here may all be explained by dysfunction of nuclei or tracts within the territory supplied by the medullary branches of the left posterior inferior cerebellar artery. Their onset and development suggest blockage rather than haemorrhage of the artery, as there is no report of headache, neck stiffness, loss of consciousness nor signs of raised intracranial pressure. In this instance, where there are suggestions of gradual development of the symptoms, the blockage is more likely to be caused by thrombosis than by embolism.

A dentist saw a 30-year-old man who complained of toothache and soreness on the left side of his mouth. Just as the dentist was about to begin his examination, his hand was jogged, causing the sharp probe he was holding to scratch the patient's left ear. To the dentist's surprise, the patient only laughed and seemed not to react to the injury. The dentist also noticed that the patient seemed unable to move his tongue in a controlled manner. When asked to protrude his tongue, it deviated to the left and looked wasted on that side. Finding no dental abnormalities, the patient was referred to a hospital specialist. This doctor first noticed that the patient's voice was hoarse and that his speech was slurred. On examination, the tongue signs were confirmed. The uvula was displaced to the right and the patient clearly experienced difficulty in drinking. Also, his breathing at times seemed laboured and was accompanied by a harsh whistling sound. The patient said he had had these difficulties over the past month, but they seemed more noticeable recently. No impairment of taste was detectable. Thermal and nociceptive sensibilities were normal around the mouth and nose, but were impaired on the chin, cheek and forehead on the left and bilaterally at the back of the head, in the neck, and across the top of the shoulders. Appreciation of tactile and vibratory stimuli was within normal limits over these regions. There was evidence of weakness of the left hand. The patient displayed mild scoliosis (lateral curvature of the vertebral column) which he had had since infancy. Explain the symptoms.

The patient has a dissociated sensory loss involving the upper cervical segments (C1 to C4) and the left trigeminal nerve.

The loss involves bilateral impairment of thermal and nociceptive sensibility in the dermatomes of the cervical segments. Such loss may occur following interruption of the fibres of the anterolateral tract system as they decussate in the commissure ventral to the spinal canal, before joining the ventrolateral funiculi (see pages 600 to 601). Appreciation of tactile stimuli is preserved, as such information also passes via the dorsal column system. Normal appreciation of vibratory stimuli suggests that the latter system is intact.

The pattern of trigeminal loss does not correspond to the cutaneous distribution of its three main divisions (ophthalmic, maxillary, mandibular; see Figure 49). Pain and temperature information passes ipsilaterally through the spinal tract of the trigeminal to the pars caudalis of its spinal nucleus in the caudal medulla and upper cervical spinal cord (see pages 600 to 601). The most caudal regions of the subnucleus deal with the top of the head and most lateral parts of the face, with the perioral and perinasal regions being dealt with most cranially. Accordingly, damage to the more caudal parts of the nucleus or to its efferent fibres will produce ipsilateral (in this instance left) sensory loss with sparing of the 'muzzle' region of the face. Touch and vibration sensibility will be maintained through fibres passing to the principal sensory nucleus of the trigeminal.

The patient displays signs (wasting and leftward deviation of the tongue) of dysfunction of the left hypoglossal nerve or its nucleus in the medulla (see Figure 178c to e). He also has signs of laryngeal and pharyngeal paralysis: uvula displaced to the right, hoarseness of voice, difficulty in swallowing, and respiratory stridor. These signs suggest dysfunction of the fibres of the left vagal and glossopharyngeal nerves which supply the larynx and pharynx, or of the nucleus ambiguus from which the fibres originate (see Figure 178c to e). The absence of taste impairment or autonomic changes rules out complete lesions of the ninth or tenth cranial nerves. The overall pattern of impairment suggests a lesion close to the spinal canal in the upper cervical segments of the spinal cord and extending to the left caudal medulla. Given that the onset appears to have been gradual rather than abrupt, either an intrinsic tumour or syringomyelia/syringobulbia is a probable diagnosis. Syringomyelia is a proliferative gliosis or cavitation of the spinal cord; syringobulbia indicates that the site of the lesion is within the medulla. The presence of scoliosis, probably originating from a developmental defect of the spine, increases the likelihood of there being a syrinx.

In the present case, the syrinx appears to have affected the upper cervical segments (C1 and C2) near and ventral to the spinal canal, and the central tegmental regions of the left caudal medulla. Such a lesion would produce: a) dysfunction of decussating anterolateral system fibres conveying information from the C4 dermatomes and above; b) disruption of efferents from the left caudal part of the nucleus of the spinal tract of the trigeminal (and, through pressure, possibly disruption of the nucleus itself); c) damage to the left nucleus ambiguus; d) interruption of the left hypoglossal nerve fibres on their way through the tegmentum (see Figures 178 d and e). Pain in segments adjacent to those in which there is sensory loss is a common symptom. The weakness of the left hand probably results from damage to pyramidal fibres decussating to form the left lateral corticospinal tract (see Figure 178b). Although not seen in this patient, nystagmus, Horner's syndrome and loss of taste are commonly encountered in syringobulbia.

A 17-year-old girl was knocked off her bicycle by a 'hit-and-run' motorist. She briefly lost consciousness but rapidly recovered and was helped home by a passer-by. A few hours later, she complained to her mother that she had developed a headache and that she was seeing double. Her mother brought her to hospital. The daughter had several lacerations and contusions on the right side of her face and head. She had to be supported by her mother while walking and showed signs of mental confusion. It was immediately noticeable that her right eye was directed downward and outward and could be moved but little. There was drooping of the right upper eyelid (ptosis). The right pupil did not constrict when light was shone in either eye, nor did it change during accommodation. The left limbs and the lower part of the left face were weak, and the left plantar response was extensor (Babinski's sign). Occasional involuntary twitching of the facial muscles was seen. On discovering a soft, mobile swelling over the right temporalis muscle, the doctor ordered an immediate operation. Explain the symptoms.

The patient has an epidural (extradural) haematoma. A skull radiograph might be expected to show a fracture, most commonly in the antero-inferior region of the parietal bone. This fracture is associated with rupture of the middle meningeal artery, usually involving the anterior branch at its point of emergence from a tunnel in the greater wing of the sphenoid bone. If the rise in intracranial pressure is not rapidly checked, death will ensue from cessation of function of the medullary neurones controlling respiration.

The patient has a right oculomotor palsy: paralysis of the superior and inferior oblique muscles and the inferior and middle recti muscles; partial paralysis of levator palpebrae superioris resulting in ptosis; loss of direct and consensual light reflex and accommodation reflex because of interruption of parasympathetic input to the sphincter pupillae muscle. This palsy is most likely to be caused by pressure on the right oculomotor nerve by the right parahippocampal gyrus of the temporal lobe as a result of its herniation through the tentorial incisure.

The patient also shows signs of a contralateral (left) upper motoneurone lesion (muscular weakness together with a positive Babinski's sign), indicating loss of activity in descending motor tracts. The paresis affects both left limbs and the lower left face. Pressure on the right crus cerebri of the midbrain, through which the oculomotor nerve emerges, might cause loss of function in corticospinal and corticonuclear fibres concerned in the control of the left limbs and lower face. Note that cortical connections with the cerebellum (corticopontocerebellar fibres), reticular formation, and red nucleus may also be disrupted by this pressure. However, the extradural haemorrhage most often spreads superiorly, producing pressure in the region of the precentral gyrus (area MsI) from which many of these descending motor cortical fibres arise. Because most of the corticospinal fibres decussate at the caudal end of the medulla, effects of lesions anywhere above this level have contralateral expression. These effects are compounded by loss of function in corticostriatal, corticorubral, corticoreticular and corticopontocerebellar fibres, so that other descending pathways to the spinal cord are also disrupted. The corticonuclear fibres to the motor nucleus of the facial nerve are bilaterally distributed to parts of the nucleus innervating the upper facial musculature, but end contralaterally on parts innervating the lower facial muscles. Thus, upper motoneurone lesions affecting the facial nerve normally affect only muscles of the lower face, contralateral to the lesion. It should be noted that occasionally herniation through the tentorial incisure results in hemiplegia ipsilateral to a haematoma. This is believed to arise from the crus cerebri contralateral to the haemorrhage being pressed against the tentorium.

A 64-year-old company director consulted his doctor because during the past year his right hand had begun to tremble embarrassingly even when he was not doing anything with it. He was not prevented from using the hand, but was finding it increasingly difficult to do up buttons and had noticed that his handwriting had become small, cramped and difficult to read (micrographia). He complained of stiffness in his limbs, especially when he wanted to start walking, and felt that he was becoming slower in everything he did. While walking, he stooped slightly and kept his right arm still, semiflexed and close to his body. During the interview he rarely blinked, smiled or altered his facial expression and he spoke in a monotone. Examination confirmed that his right hand was affected by a coarse rhythmic tremor that decreased on movement. Muscle tone appeared in general to be slightly increased, but reflexes and strength were normal. What is your diagnosis?

This patient demonstrates several of the symptoms that may be associated with Parkinson's disease (paralysis agitans): a) tremor at rest; b) rigidity (hypertonia, which may be of the 'cog-wheel' type, i.e. increased followed by suddenly decreased muscular resistance on passive movement); c) reduced and slowed voluntary movements (hypokinesia), including loss of associated movements such as swinging of the arms during walking, micrographia, poverty of facial expression and a monotonous voice.

Degeneration of the dopaminergic neurones of the substantia nigra (see Figures 178i and j, and 195) seems to be the most critical lesion underlying Parkinson's disease. However, there are typically more widespread degenerative changes, in particular involving other dopaminergic neurones. The precise causes of the various symptoms and the cause of degeneration are uncertain. The substantia nigra projects to the putamen and caudate nucleus by the dopaminergic pathway, and to the nucleus ventralis lateralis of the thalamus. The nucleus ventralis lateralis receives input from the cerebellum and globus pallidus and projects to the motor cortex (MsI).

A 71-year-old woman collapsed in the street while trying to chase a youth who had stolen her purse. On arrival in hospital, she had regained consciousness but could not be made to understand even simple instructions. Her right arm and leg showed decreased muscular tone and reflexes, but her neck was stiff. Her right plantar response was extensor (positive Babinski's sign). Her temperature was normal, but her blood pressure was high (180/90 mm Hg). The following day, the patient was better able to co-operate, though she often just shook her head if anything more than a simple instruction was given. She had difficulty in naming a variety of common objects and frequently used inappropriate or nonsense words. On being asked, she indicated that she was right-handed. Her right arm and leg were weak and the abnormalities of tone and reflexes were confirmed. Attempted movements were weak and poorly co-ordinated. Occasional spontaneous writhing movements (athetoid movements) of these limbs were seen. There was loss of appreciation of pain and temperature on the right side of the body (excluding the face and the top of the head), together with loss of sensibility to touch or vibration in the same region. With her eyes closed, the patient could not describe the position of her right arm or leg when each was moved by the examiner. No abnormalities of the cranial nerves were found. Two days later, the strength of the right limbs was approaching normal, although voluntary movements were still clumsy. However, the patient was complaining of recurrent intense pain, sometimes involving all the right half of her body. Lightly touching the right side of the patient's body evoked such pain, although the patient could not say where she had been touched.

The patient displays a transient flaccid hemiparesis of the right limbs and a positive Babinski's sign. There is also loss of pain and temperature sensibility on the right side from all dermatomes (excluding regions supplied by the trigeminal nerve). Fibres subserving these sensations travel via the anterolateral system to the contralateral (left) posterior thalamus and cerebral cortex (see pages 600 to 601). Over the same dermatomes there is loss of touch and vibration sensibility. Furthermore, proprioceptive information from the right limbs is not available to consciousness. These findings suggest that the dorsal column/medial lemniscal pathway to the contralateral (left) thalamus and cerebral cortex is additionally interrupted (simple touch stimuli may pass by either system; see pages 598 to 599).

The patient is also suffering from receptive aphasia (difficulty in understanding speech and in naming common objects, together with substitution mistakes in her own speech). The presence of aphasia (in the absence of deafness or paralysis of laryngeal or facial muscles) means that the lesion must be in the forebrain. In the right-handed patient, speech functions are almost invariably in the left hemisphere. Also the left side of the forebrain deals with the sensory and motor functions of the paretic right limbs (see pages 598 to 599 and 602 to 603).

In the absence of infection or previous history of neurological illness, the suddenness of onset of the symptoms suggests a vascular cause. The circumstances (onset during activity and/or emotional upset), the presence of raised blood pressure, the loss of consciousness and nuchal rigidity (as a possible result of irritation of the meninges by blood) all make intracranial (probably intracerebral) haemorrhage the most likely cause.

As the arm and leg regions of the primary sensory and motor areas of the cerebral cortex are supplied by different arteries (middle and anterior cerebral respectively), a single vascular accident affecting the cortex, in the absence of evidence of a massive haemorrhage, could not explain the symptoms. However, the sensory, motor and speech disturbances are explicable by damage to the left posterior limb of the internal capsule and/or adjacent posterolateral thalamus (see Figure 194). The development of spontaneous impaired sensation (dysaesthesia) and of exaggerated subjective responses to painful stimuli (hyperpathia) is suggestive of 'thalamic pain'. Moreover, the hemiparesis is transient, suggesting that the dysfunction of the motor fibres of the internal capsule is caused by oedema rather than infarction. Corticonuclear fibres subserving the head are more anteriorly placed in the internal capsule and have thus escaped involvement in this patient. The lateral and posterior nuclei of the thalamus are supplied by posterolateral (thalamogeniculate) branches of the posterior cerebral artery. In this case, the lateral and medial geniculate bodies (subserving vision and hearing respectively) have been spared and a typical 'thalamic syndrome' has developed. Infarction of the left nucleus ventralis posterior lateralis, or of fibres entering or leaving the nucleus, is responsible for the loss of touch, vibration and proprioceptive sensibilities (see pages 598 to 599). Damage to the left posterior nuclear group underlies the contralateral loss of pain and temperature sensibility, together with the development of dysaesthesia and hyperpathia (by unexplained mechanisms). The left pulvinar nucleus is reciprocally connected with the posterior speech cortex (Wernicke's area). Damage here thus accounts for the receptive aphasia (see Figures 187a and 188a). The athetoid movements and inco-ordination are likely to result from involvement of basal nuclear and cerebellar connections with the nuclei ventralis lateralis, ventralis anterior and intralaminary nuclei of the left thalamus (see Figures 195 and 183).

A 55-year-old ballet instructor awoke one morning to find herself unable to move her left arm or leg. Over the next few days this paralysis gradually disappeared, but she was distressed to discover that she could no longer fully control her left arm or leg. Both these limbs were subject to continuous, jerky involuntary movements, sometimes resulting in wild and theatrical gestures of the arm and gross kicking or high stepping of her whole leg. Explain what has happened.

The patient is suffering from hemiballismus. This is characterised by continuous, unco-ordinated activity of the axial and proximal limb musculature that is so violent that the limbs may be forcefully and aimlessly thrown about. It is commonly the result of a lesion (usually vascular in origin) of the contralateral (in this instance right) subthalamic nucleus or its connections. In the present case, arterial blockage or haemorrhage provides a plausible explanation for the sudden onset of the transient contralateral (left) hemiplegia, which at first masks the ballism.

The subthalamic nucleus is supplied by one or more of the small arterial branches of the posterior cerebral artery (or possibly of the posterior communicating artery). These branches enter the midbrain through the posterior perforated substance between the cerebral peduncles (see Figure 178j). The subthalamic nucleus is reciprocally connected with the ipsilateral globus pallidus (see Figure 195). It appears that lesions of the subthalamic nucleus result in loss of the inhibitory control of the globus pallidus over the thalamus, specifically over the nuclei ventralis anterior and over ventralis lateralis which project to motor cortical areas (MsI). However, the precise mechanism is unclear because the subthalamic nucleus sends inhibitory GABA-ergic fibres to the globus pallidus which, in turn, sends inhibitory GABA-ergic fibres to the thalamus. Presumably, some balance between parallel inhibitory circuits is upset, resulting in eventual excitation (or disinhibition) of thalamic cells. Ballism may be controllable with either GABA-mimetic drugs (including benzodiazepines) or dopamine antagonists. Stereotaxic lesions of the thalamus (usually of the nucleus ventralis lateralis) have been used in otherwise intractable cases.

A 38-year-old computer operator went to see her doctor because she kept having headaches. These were so severe that she felt as though her forehead was going to split open. She thought that there might be a link with having given up the contraceptive pill three months before. She mentioned that she had not menstruated during that time (amenorrhoea). She said she felt thoroughly run down and constantly tired. Although it was a warm summer day, the woman was wearing a scarf and long coat. On being asked whether she had anything else wrong with her, she said that she thought her eyes were not as good as they used to be and that sometimes her breasts seemed to be leaking milk (galactorrhoea). On examination, her pulse rate was 60 per minute and her blood pressure 90/60 mm Hg. Both optic discs appeared to be somewhat pale, but not unequivocally abnormal. On plotting the visual fields, there was found to be blindness in the upper and outer

quadrant of the left field and in all but the lower part of the lateral half of the right field. Account for the symptoms.

The patient's symptoms indicate hypopituitarism. Decreased secretion of gonadotropin would explain the amenorrhoea. Decreased production of corticotropin (ACTH) and thyrotropin (thyroid stimulating hormone, or TSH) leads to lethargy, fatiguability and intolerance to cold. The low blood pressure is also suggestive of adrenal cortical deficiency.

The patient has an upper bitemporal quadrantanopia, being blind in areas served by fibres of the inferior half of the nasal half of each retina. These fibres are the inferiorly placed, decussating axons of the optic chiasma. Pressure in the midline, and acting from below, will block conduction in these fibres first. With increasing pressure, more superiorly placed decussating fibres will be compromised (as in this case), until eventually the patient presents with a bitemporal hemianopia. Subsequently, the upper nasal quadrants and then the lower nasal quadrants may also become anopic. Usually (though not invariably), the loss of sight first affects the periphery rather than the centre of the visual fields. This is because the fibres subserving this region lie most posteriorly in the chiasma. Definite paleness of the optic discs would suggest likely atrophy of optic nerve fibres. Such atrophy generally occurs only as a result of a long-standing defect. Papilloedema may be expected to develop only when a space-occupying lesion in the region has reached considerable size.

The presence of frontal headaches, with evidence of midline pressure on the optic chiasma and symptoms of hypopituitarism, strongly suggest a space-occupying lesion in the region of the sella turcica. Pressure on the diaphragma sellae often produces 'bursting' headaches felt behind and just above the eyes. Pressure on the pituitary and optic chiasma would explain the other symptoms. The absence of evidence of infection or sudden onset of symptoms makes tumour growth a plausible hypothesis. The probable presence of galactorrhoea with amenorrhoea make a prolactin-secreting adenoma the most likely diagnosis. Amenorrhoea and galactorrhoea may start only when the taking of oral contraceptives has been discontinued.

A man with a repaired cleft lip and cleft palate had a history of middle ear infections. While on holiday, he developed another ear infection. He put up with the pain until he returned home. By the time he saw his doctor he was feeling very unwell with headache, drowsiness and nausea. He was deaf in his left ear, was having difficulty talking and was extremely concerned because his left hand had become clumsy and unco-ordinated. He found he kept missing objects he was trying to pick up and also that his gait had become unsteady. His other complaint was of poor vision. Examination showed that he was febrile, drowsy, and had swelling of the optic disc (papilloedema). There were oscillatory movements of his eyes (nystagmus), especially when looking to his right. His pulse was slow (40 per minute). He was unable to name objects (nominal aphasia) and his speech was slow and slurred (scanning speech). The left side of his body had normal strength but was unco-ordinated, especially when he was asked to perform rapid, repetitive movements (dysdiadochokinesia). His right side showed signs of spasticity, with very brisk reflexes. This man had chronic otitis media with an associated cholesteatoma (a cyst-like lesion) of his left ear, which had spread intracranially.

Palatal defects are often associated with recurrent otitis media, probably because of abnormalities of the anatomy of the auditory tube. In the case described here, the infection had spread rapidly from the ear to the middle and posterior cranial fossae. This had caused septic thrombosis of the sigmoid venous sinus and development of abscesses in the temporal lobe and cerebellum of the brain. The cerebellar abscess had obstructed the outflow of cerebrospinal fluid. This caused raised intracranial pressure as indicated by the low pulse and high temperature, the papilloedema, the high blood pressure, the headache and the drowsiness. The abscess had also compressed the left side of the cerebellum causing the nystagmus, the lack of muscular co-ordination, the dysdiadochokinesia and the scanning speech. The raised intracranial pressure, exacerbated by the presence of abscesses on the left, was probably responsible for the nominal aphasia (resulting from compression of the posterior speech area or its connections) and for the contralateral spasticity (resulting from compression of the left motor cortex or its subcortical connections) (see Figures 188a and 194).

A 42-year-old engineer became ill and developed a severe headache and high temperature (41°C). He became comatose and remained so for a month during which time he had occasional seizures. Herpes simplex encephalitis was diagnosed and confirmed. Raised intracranial pressure was reduced by cannulating the inferior horn of the left lateral ventricle. On regaining consciousness, the patient had a right hemiparesis and speech problem (aphasia). Over the next two weeks, these symptoms cleared completely and he was discharged from hospital. Subsequent examination revealed the following enduring symptoms: loss of smell (anosmia), drooping of the left eyelid (ptosis), a left pupil that was slightly dilated compared to the right, a slight left external squint (strabismus), and amnesia. Psychological tests indicated above-average intelligence and normal performance of language functions. He demonstrated normal recall of material after short delays (less than 20 seconds). However, the patient performed extremely poorly when asked to recall either verbal or non-verbal material (presented either visually or aurally) at longer intervals or following distraction. Even after several sessions he denied any familiarity with any of the tests and denied having previously seen the psychologist who administered them. He could not remember any significant public events nor events in his own life that had occurred subsequent to at least one year before his hospitalisation. However, earlier events were apparently normally recalled. He gave his age as 40. During testing, he often complained of his poor memory. Comment on the likely anatomical substrates of his disabilities.

The herpes simplex virus can cause meningitis and encephalitis, which is frequently fatal. Necrotic, inflammatory or haemorrhagic lesions may be widespread and oedema may be problematic. Lesions of the frontal and temporal lobes are particularly common.

The patient's initial right hemiparesis and aphasia probably arose from pressure on the left hemisphere (see Figure 188a). The anosmia is likely to be caused by damage to the olfactory nerves (although damage to both olfactory bulbs or their central connections cannot be excluded). The ptosis, dilated pupil and external strabismus of the left eye may all be produced by a left oculomotor nerve lesion. In this case, the nerve may have been damaged because of herniation of the left parahippocampal gyrus through the tentorial incisure, as a result of raised intracranial pressure arising from oedema.

The amnesia is not restricted to a single type of material or one mode of presentation. It is restricted to long-term rather than short-term memory and to recent events. Events of the patient's earlier life are recalled but there is anterograde amnesia from the time of recovery from coma, plus a retrograde amnesia of at least a year. This type of amnesia is characteristic of bilateral damage to the medial temporal lobes, though it may occur without significant retrograde amnesia. Bilateral damage to the medial thalamus can produce a similar pattern of amnesia. The related amnesia of Korsakoff's psychosis, in which there may be widespread damage including the medial thalamus and/or mamillary bodies and/or prefrontal cortex, is usually marked by a more extensive, though subtotal, retrograde amnesia and by confabulation. Short-term memory and other intellectual impairments are also commonly present. Patients whose amnesia arises from medial temporal lobe damage usually demonstrate some awareness of their disability, Korsakoff patients rarely do. Viral or haemorrhagic damage to the medial temporal lobes may in this case have been compounded by direct or ischaemic damage caused by tentorial herniation. The extent of the lesions may be determined from CT scans.

A member of a university cricket team was hit on the side of the head by the hard cricket ball. He was not knocked out, but retired hurt. By the end of the day, he had recovered sufficiently to join in the post-match celebrations. However, on the coach home his team-mates became worried as he was becoming drowsy, and he was taken to hospital. On arrival, he was barely conscious. On examination, the casualty officer noticed bruising over his left temple and that his left pupil was dilated and did not respond to light. His right pupil was smaller and reacted normally. The patient's condition continued to deteriorate rapidly. He started to vomit and soon lost consciousness. A surgeon was called. What is happening and what will the surgeon need to do?

The patient has fractured the squamous part of his left temporal bone, which has caused the left middle meningeal artery to rupture in the middle cranial fossa. Consequently, blood has escaped between the dura mater and bone and produced an extradural haematoma. Blood has also escaped from the fracture site, leading to the visible bruising. The pressure from the haematoma intracranially has compressed the oculomotor nerve against the tentorium cerebelli on the left side, producing the visual signs. The drowsiness, vomiting and loss of consciousness are the result of raised intracranial pressure caused by the haemorrhage. The surgeon will need to drill a burr hole through the left temporal bone to remove the clot of blood. If the pressure is not so relieved, displacement of the brainstem at the opening of the tentorium cerebelli will force the crus cerebri of the opposite side against the rim of the tentorium cerebelli. This would result in a left-sided paralysis of the body. Further compression would produce decerebrate rigidity, fixed dilation of both pupils, and death as the regions of the brainstem controlling vital functions are compressed.

A 45-year-old housewife was referred from a psychiatrist because of 'eccentric behaviour', depression and headaches, continuing over some months. No signs of anaesthesia or muscular weakness were found. However, when both sides of her face were touched simultaneously, the patient noticed only the touch on the right. Two-point discrimination was mildly impaired on the left cheek. The patient could not recognise familiar objects by feeling them with her left hand, although she could with her right. It was also noticed that she had difficulty tying her shoe laces and getting dressed after the examination. The patient had great difficulty in copying even simple diagrams and often seemed to ignore features on the left side of the picture. It was noticed that she wore make-up only on the right side of her face. On testing the visual fields, it was discovered that there was blindness in the inferior temporal quadrant of the left eye and the inferior nasal quadrant of the right eye (left inferior homonymous quadrantanopia). Both optic discs were oedematous. On questioning, the patient said she did not think there was anything wrong with her: it was her husband who had made her go to the doctor. Explain the symptoms.

The patient displays signs of right parietal lobe dysfunction: a) astereognosis (inability to recognise objects by feel) using the left hand; b) apraxia (loss of skills in the absence of paralysis, such as tying shoe laces and dressing); c) apractognosia or constructional apraxia (inability to copy diagrams); d) hemineglect or contralateral inattention (failure to notice features of the left side of diagrams, noticing only the stimulus on the left when touched on both sides simultaneously, and neglecting the left side of her face when using make-up – an example of autotopagnosia or lack of awareness of parts of the patient's own body); e) anosognosia (lack of acknowledgement or insight into her own disabilities). There is also mild impairment of discriminative touch (two-point discrimination) on the left: parietal lesions not involving the primary somatosensory cortex (area SmI) may produce this symptom. Lesions of SmI result in contralateral anaesthesia.

The fibres from the right halves of the two retinae pass through the optic nerves and, at the optic chiasma, enter the right optic tract (see pages 590 to 591). From the lateral geniculate nucleus, they pass through the retrolenticular part of the right internal capsule to the primary visual cortex on either side of the posterior part of the calcarine sulcus in the occipital lobe. The fibres that terminate above the sulcus pass inferiorly through the parietal lobe, lateral to the descending limb of the lateral ventricle. Because these fibres convey information from the left lower visual fields, their interruption produces a left inferior homonymous quadrantanopia.

The report of continuing headaches and the presence of oedematous optic discs suggests raised intracranial pressure. The time course makes a vascular cause less likely than a tumour. It was established at operation, following a CT scan, that the patient had an infiltrating glioblastoma multiforme in the right inferior parietal lobe.

Index

References are to page numbers.
Page numbers in brackets refer to Case Histories.
Bold type indicates a major reference.

Abnormalities, developmental
- ankyloglossia (351)
- bat ears (384)
- choanal atresia (267)
- cleft lip 170
- cleft palate 325 (348-349)
- lingual thyroid (349)
- nasolacrimal duct blockage (234)
- neural 422,574 (615)
- tori of jaws (350)
Abscess, cold (tubercular) (142-143)
- dental (191,353-354)
- extradural (269)
- intracerebral, following sinusitis (269)
- - following ear infection (628-629)
- peritonsillar (quinsy) 97 (355)
Acceleration, detection of (see System, vestibular)
Accident, cerebrovascular **412-414**, 575-576(611,612,613,614,618, 622,623,624,625,626,627,630)
Accommodation (see Reflex)
Acetylcholine 411,**433-434**,**438**,**439**, **440**,441,442,474,508,513,527,543, 562,566,567,569
Achondroplasia (77)
Acromegaly (75)
Action potential 406,**408**,409
Adam's apple 83,**336**
Adaptation (see Learning)
Adenohypophysis **399**,527,585
Adenoids **311**
Adenosine triphosphate (ATP), as neurotransmitter 411,434,441,442
Adhesion, interthalamic **524**
Aditus, of larynx (see Inlet)
- to mastoid antrum 368,**369**,377
- of orbit 13,**46**,198
Adrenaline, in neurotransmission 411, 438,**440**,441,470,471,508,509,510, 511,512
Agger nasi **244**,250
Agnosia **554**,555,561
Agraphia **554**
Air cells, ethmoidal 26,30,**37**,40,41, 47,53,244,245,**250**,251
- - infection of (238-239,269)
- - tumour within (270)
- mastoid 101,106,178,364,366,367, **369**
- - infection of (387)
- within palatine bone **37**
Air sinuses (see Sinus)
Ala, of crista galli **36**
- of nose 148,160,242
- of vomer 38,47,48
Alexia **554**
Allocortex **541**,555-559
Alveolar bone (see Process)
Alveus **558**
Alzheimer type dementia **561**
Amaurosis fugax, retinal ischaemia (237)
Amblyopia **560**
Amenorrhoea (627,628)
Amiculum olivae **503**
Amines (as neurotransmitters) 411, **509** (see also named)
Amino acids (as neurotransmitters) **411** (see also named)
γ-Aminobutyric acid (GABA) **411**, 475,521,569,570,571

Amnesia (see also Memory) 513,528, 533,534,554,555,558,560,**561**,**562** (629)
Ampulla, of internal ear **376**,378,379, 473,494
Amygdala 525,527,530,531,532,533, 543,555,556,559,560,561,562,563, 567-569,570,572,580,588,589
- connections 527,530,531,532,533, 543,556,559,562,563,567,**568-569**, 570
- - with cortex 543,**568**
- - with hypothalamus, 525,**527**
- - olfactory 555,**556**,588,589
- - with thalamus 530,531,**532**,**533**
- - stria terminalis 527,**568-569**,571, 572
- - ventral amygdalofugal pathway 527,**568**,**569**
- functions **556**,560,561,**568**,**569**
- nuclei 555,556,567,**568**,569,588,589
Analgesia **443**,451,505
Anaesthesia, in relation to CNS 409, 429,443,446,481,564,581(630)
- spinal (epidural) **446**
Anencephaly **422**
Aneurysm, carotid (139)
- of arteries of brain (236)575(610,611, 612)
Angina pectoris 438(613)
Angiogram (611)
Angiotensin **411**,**585**
Angle, cerebellopontine (pontomedullary) 461(620)
- of eyelid (see Canthus)
- filtration, of anterior chamber of eye **207**
- of mandible 15,**43**,57,83,95,149,158, 161,164,173
- of mouth (see Commissure)
Anhidrosis (193)
Ankyloglossia (351)
Anopia (628) (see also Blindness)
Anosmia (73,76,267)**488**,553,555,556 (629)
Anosognosia **554**(630)
Anoxia, effects on neural tissue, 443, 575
Ansa, cervicalis/subclavia (see Nerves)
- lenticularis **529**,532,570
- peduncularis **532**
Antidepressants **512**
Antidiuretic hormone (see Vasopressin)
Antihelix **362**
- defect of, bat ears (384)
Antitragus **362**
Antrum, mastoid 16,34,178,366,**369**
- maxillary (see Sinus)
Anulus fibrosus of intervertebral disc **64**
Aperture, lateral of fourth ventricle (of Lushka) **462**,573,**574**,575
- median of fourth ventricle (of Magendie) **462**,**463**,573,**574**,575
- nasal, anterior (piriform) 14,**46**,243
- - posterior (choanae) 18,**47-48**,242, 244
- oral 14
- orbital 13,**46**,198
Anxiety **559**,560
Apex
- of lung 128,129,130,**133-134**

- of nose 148
- of petrous part of temporal bone **19**, 22,33,400
Aphasia **553**,**554**,555,561,581,582
- expressive (motor) **553**,554,581 (620,621)
- nominal **554**(628,629)
- receptive (sensory) **553**,554,555,561, 581(626,627)
Aponeurosis, epicranial 27,153,154, **188**
- of levator palpebrae superioris **213**, 220
- palatine 17,**295**,297
Apoplexy (see Stroke)
Apparatus, lacrimal **213-215**
Appetite (see also Feeding) 524,**528**
Apractognosia (630)
Apraxia **554**,561(620,622,630)
Aqueduct
- of cochlea 375,**376**
- cerebral, of midbrain **421**,426,463, 524,574
- - blockage of (617)
- of vestibule 24,34,**375**
Aqueous humour 201,202,**207**
- failure of drainage (240)
Arachnoid, granulations 21,26,27,394, **573**
- mater **392-393**,396,**446**,573,574
- villi **574**
Arbor vitae **516**
Arch, of atlas **62**,66,67,121,126,131
- branchial 68,**134-135** (139) 173,319, 321,346,380
- of cricoid cartilage 83,**336**,341
- dental **287**
- jugular **105**
- palatoglossal **275**,**277**
- palatopharyngeal **277**
- superciliary **25**,51,249
- vertebral **61-62**,**64-65**
- zygomatic **15-16**,94,147,148,158, 160,162,172,175,188
Archicerebellum **518**
Archicortex (archipallium) **541**, 556-558
Archistriatum (see Amygdala)
Arcus (gyrus) parieto-occipitalis **540**
Area (also Cortex, Region, or Nucleus)
- association (see Cortex)
- Broca's speech 549,552,**553**,554,562, 581,592,593(620,621)
- of Brodmann (numbered) **544**, 546-553,555
- - 1,2,3 **549-550**
- - 4 **546-547**
- - 5 **549-550**
- - 6 **546-549**
- - 17,18,19 **550-552**
- - 22 **552**
- - 28 **555**,**559**
- - 41,42 **552**
- - 43 **550**
- - 44,45 549,**553**
- entorhinal 555,556,557,558,559,562, 588
- hypothalamic (see Hypothalamus)
- intermediate, of spinal cord 450,**453**, 454,456,459,504
- motor (see Cortex)

- paraterminal **567**
- peri-amygdaloid 553,555,**556**,567, 568,588,589
- pontine taste 478,**596**,597
- postrema 506,508,584,**585**
- preoptic **526**,528
- pretectal 484,485,**487-488**,530,535, 550,590
- sensory (see named of Cortex)
- septal 559,**567**
- speech (see Cortex)
- subcallosal **540**,567
- supra-optic **526**,527
- tegmental, ventral (of Tsai) 506,**513**, 543,569,570
- tuberal **526-527**
- vestibular **462**
- Wernicke's **553**,554,561,562,581, 592,593(628,629)
Arousal 504,505,**508-509**,510,511, 512,513,534,567,586,600
Arrest, cardiac (142)
Artery
- alveolar, inferior 162,**180**,304,**305**
- - - injection of anaesthetic agent (195)
- - superior 253,258,**305**
- angular, of facial 153,**162**,163,243
- of angular gyrus 581
- anterolateral central branches of middle cerebral 581
- anteromedial central branches of anterior cerebral 581
- aorta 95,99,108,119,129,469
- auricular, deep 174,**180**,364,365
- - posterior 86,88,**101-103**,154,165, 188,189,363,364,372
- - from superficial temporal artery 162
- axillary 95,115,130
- basilar 131,251,379,402,460,461, 575,577,**579**
- - branches of 579-580
- brachiocephalic 99,119,**129**
- buccal 157,162,**180**,305
- calcarine 579
- callosomarginal 581
- caroticotympanic 99,**373**
- carotid, **99-103**
- - common 82,83,90,**99**,112,115,119, 129,131
- - external 83,95,**100-103**,111,112, 165
- - internal 19,22,23,62,82,95,**99**,111, 112,226,251,368,373,399,537,577, **580-582**
- - - aneurysm (139,611)
- - - rupture of (235,610)
- - cavernous **580**
- - central, long of anterior cerebral **581**, 582
- - - long or recurrent (of Heubner) **581**
- - of middle cerebral **581-582**
- - of posterior cerebral 579-580 (627)
- - - thalamogeniculate (posterolateral central) **579** (627)
- - - thalamoperforating **579**
- - of central sulcus **581**
- cerebellar, anterior inferior 577,**579**
- - posterior inferior 577,**578**,**579** (622,623)
- - superior 577,**579**
- cerebral, anterior 112,536,537,575,

A–B

577,580,**581**,**582**(613,622,626)
- - middle 112,537,538,575,577,580, **581**,**582**(613,622,626)
- - posterior 537,538,577,**579**-**580**,582 (611,612,626,627)
- of cerebral haemorrhage **581**-**582**
- cervical, ascending 122,123,130,**131**
- - deep 101,125,130,**132**
- - superficial 92,101,123,124,131,**132**
- choroidal, anterior 573,577,**580**,**582**
- - posterior 573,**580**,582
- ciliary, anterior 203,204,205,**227**
- - posterior 203,204,205,**227**
- - recurrent 203
- circle, cerebral arterial (of Willis) (circulus arteriosus) 537, 575,**577** (611)
- communicating, anterior 577,**581**
- - posterior 577,579,**580**-**581** (627)
- - - aneurysm (236,611)
- cortical **582**
- - of anterior cerebral **581**
- - of middle cerebral **581**
- - of posterior cerebral **579**
- costocervical trunk 124,131,**132**
- cricothyroid **101**,341
- descending branch of occipital **101**
- diploic **395**
- ethmoidal, anterior 163,227,243, **249**,250
- - posterior 227,**249**
- facial 88,90,**101**,112,152,153,155, 157,**161**-**162**,171,181,212,215,226, 227,249,250,252,289,295
- - damage to, at inferior border of mandible (191)
- - transverse 162,163,165,167,171, 212,214
- frontal, branch of anterior cerebral **581**
- frontobasal
- - branch of anterior cerebral **581**
- - branch of middle cerebral **581**
- frontopolar **581**
- hyaloid 207
- hypophyseal **399**-**400**,**580**
- incisive **180**,305
- infrahyoid of lingual **101**
- infrahyoid of superior thyroid **101**
- infra-orbital 152,153,154,155,162, 212,215,217,219,**228**,243,249,**305**
- intercostal 124,125,**132**
- labial, inferior 154,155,156,**162**,305
- - superior 153,154,155,156,**162**,243, 249,305
- labyrinthine 24,**379**,**579**
- lacrimal 22,163,214,219,**227**,580
- laryngeal, inferior **131**,317,339,341, **342**
- - superior **101**,317,339,341,**342**
- lenticulostriate (see Striate)
- lingual 83,89,90,**101**,103,293,295, **305**-**6**,317
- - deep **101**,**306**
- - dorsal **305**,318
- lingual branch of maxillary **180**
- masseteric 171,**180**
- maxillary 48,165,175,179,**180**,181, 226,**258**,304
- - branches on face **162**
- medullary, branches of vertebral **578**
- - branches of cerebral arteries **582**
- meningeal, accessory 19,22,**180**,**397**, 401
- - of ascending pharyngeal 18,23,**101**, **397**,401,402
- - of carotid, internal **580**
- - middle 15,19,22,112,178,179,**180**, 227,255,311,371,372,373,**396**-**7**, 401 (609)
- - - following skull fracture (74,235, 624,630)
- - of occipital **101**,**397**
- - of ophthalmic **227**,**397**,580
- - of vertebral **578**

- mental 154,155,156,162,**180**,305
- mylohyoid **180**,290,295,**305**
- nasal, dorsal 163,**227**,243
- - external 163,243,**249**
- - lateral **162**,243
- occipital 18,23,24,83,86,88,90,91, **101**,111,122,124,125,126,188,189, 363,373
- occipital branch of posterior auricular 103
- occipitotemporal branch of posterior cerebral **579**
- oesophageal of inferior thyroid **131**
- ophthalmic 22,49,112,152,188,199, 219,220,224,**226**-**7**,485,577,**580**
- - branches on face **162**-**163**
- orbital branch of middle meningeal 22,**396**
- palatine, ascending **101**,181,297, 316,317,318
- - descending 258,297,316
- - greater 17,249,253,**258**,305
- - lesser 17,**258**
- palpebral 212,215,**227**
- paracentral **581**
- parietal branches of middle cerebral **581**
- parieto-occipital **579**
- penetrating, of cerebral arteries **582**
- pericallosal **581**
- pharyngeal, ascending 18,23,90,**101**, 122,297,311,313,316,317,318,373
- - branch of sphenopalatine **18**
- - from maxillary 18,**258**,318
- - pontine, branches of basilar artery **579**
- of postcentral sulcus **581**
- of precentral sulcus **581**
- precuneal **581**
- pterygoid 176,**180**
- pterygoid canal 99,**258**,311,318,373
- radicular **582**
- recurrent of anterior cerebral 581
- retinal **205**-**206**,225,**227**
- - occlusion of (237)
- scapular, dorsal 92,**132**
- sphenopalatine 18,215,249,250,**258**
- spinal, anterior/ventral 18,23,577, **578**,**582**
- - posterior/dorsal 18,23,577,**578**,**582**
- sternocleidomastoid of occipital 101, 111
- striate 537,577,**581**-**582** (613)
- stylomastoid 18,**101**,**103**,**365**,367, 373
- subclavian 91,95,108,113,114,129, **130**-**132**,343
- - steal syndrome (144)
- - stenosis with cervical rib (80)
- sublingual **101**,290,293,294,295,**306**
- submental 85,88,**101**,162,290,293, 295
- supra-orbital 14,162,189,217,219, 220,226,**227**,250
- suprascapular 85,86,131,**132**
- supratrochlear 14,162-163,189,**227**
- temporal, deep 163,171,**180**
- - branches of middle cerebral artery **581**
- - branches of posterior cerebral artery **579**
- - superficial 152,154,**162**,164,165, 171,174,188,189,212,226,363,364 (609)
- - - palpation for pulse (192)
- thoracic, internal 114,120,**131**
- thyrocervical trunk 85,**131**
- thyroid, inferior 108,112,113,116, 118,119,120,122,123,**131**,316,318, 342
- - - tracing parathyroids (141)
- - superior 86,89,90,**101**,104,119, 341,342
- thyroidea ima 118,**119**
- tonsillar **101**,181,313,318

- tracheal of inferior thyroid **131**
- tympanic, anterior **180**,**365**,367,372, 373
- - inferior **101**,373
- vertebral 18,23,67,83,101,112,113, 122,123,126,129,130,**131**,402,460, 575,**577**-**578**,582
- insufficiency of **578**
- in subclavian steal syndrome (144)
- zygomaticofacial 16,163
- zygomatico-orbital 162
- zygomaticotemporal 16,163
Arthritis, rheumatoid (79,359-360)
Aspartate **411**
- N-methyl-d-aspartate (NMDA) 411, **560**
Astereognosis (630)
Asterion **16**,26,27,52
Astigmatism **560**
Astrocytes (astroglia) **415**-**416**
Asynergia **518**
Ataxia (inco-ordination) **458**,**459**,**476**, **513**,**554**,**587** (617,618,619,625,626, 627,628,629)
Athetosis **571**(626,627)
Atlas 18,**62**-**63**,64,65-67,83,126,**577**
- fracture (78)
Atrium, of lateral ventricle 572
- of nose 244
Auricle 85,108,158,160,178,**362**-**364**
- accessory (384)
- haematoma (384)
- innervation 363
- muscles of **153**-**154**
- regrafting of (384)
- vasculature **363**-**364**
Axis 18,23,**63**,64,65,66 (609) (see also Dens)
Axolemma **404**
Axon 403, **404**-**405**, 406-414 (see also Fibres)
- growth **413**,**443**
- hillock **404**,408
- initial segment **404**,408
- myelinated (medullated) **405**-**409**
- reaction after transection **405**,**413**, **443**
- regeneration of **413**-**414**,**443**
- terminal (see Synapse)
- unmyelinated **405**,**406**,409
Axoplasmic transport **405**

Balance (equilibrium) 378 (see also System, vestibular)
- water 524,525,527,**528**,585
Ballism **571** (see also Hemiballismus)
Band
- of Baillarger **541**,550,562
- diagonal, of Broca 566,**567**
- of Giacomini 555,**558**
- longitudinal of atlanto-axial joint **66**
- primary epithelial **169**,320
Banks, of lateral sulcus **540**
Bar, costotransverse **62**
Baroreceptor 99(139)
Barotrauma (269)
Barrier
- blood–brain 572,**584**
- blood–CSF 572,**573**,584
Basal forebrain **566**-**567**
Basal nuclei/ganglia 420,427,**567**-**571** (see also Nuclei, basal)
Basis pedunculi (see Crus cerebri)
Bed, of parotid gland 165,166
Beetle brows (75)
Behaviour and emotions **437**,**438**,**439**, 440,442,524-525,526,527,**528**,554, 555,559,**560**,567,**569**
- aggression 440,524,525,**528**,554, 559,560,**569**
- defensive **440**,**524**
- eccentric (238-239,630,631)
- fear 440,**525**,560,**569**
- feeding and digestion 437,438,**439**,

440,**442**-**443**,**469**,470,471,512,524, 525,**528**,555,**569**
- obsessional **562**
- pleasure 442,512,**513**,**528**,559,560, 567
- reproductive/sexual **437**,**439**,440, 442,**524**,525,526,527,**528**,534,555, **569**
- stress **440**,**525**
- tameness 554,**559**,**560**,**569**
Bifurcation, of common carotid artery 83,90,**99**
Blindness (73-74,194,237)485,552, 564,582(611,612,627,628,630,631)
- psychic 555
Blind-sight 552
Blind spot (see Disc, optic)
Blood, osmolarity, neural monitoring 525,**585**
- oxygen tension, neural monitoring 432,433,**469**,470,471,**507**-**8**,577
- pressure, neural control 432,433, **438**,**440**,**469**,**508**,**528**,555,577(616)
Body
- amygdaloid (see Amygdala)
- carotid **99**,107,108,112,**469**
- ciliary 203,**204**,**227**
- of corpus callosum **537**
- of fornix **538**
- geniculate, lateral 484,485,**529**,530, 532,**533**,586,**590**,**591**
- - medial 463,**474**,**529**,530,532,**533**, 552,**592**,**593**
- of hyoid bone **68**,83,88,89,90,94, 119,293,295,337
- juxtarestiform 516,**520**
- of lateral ventricle 569,**571**,573
- of Luys (see Nucleus, subthalamic)
- mamillary **523**,**524**,525,**526**,**527**,**528**, 530,559,**560**,**561**,564(629)
- of mandible 14,**43**,51,59,85
- of maxilla 40-41,148,252
- paraterminal **567**
- pineal 419,463,**534**,537,575,583, 584,585(617)
- - tumour (617)
- quadrigeminal (see Colliculus)
- restiform **516**
- of sphenoid bone, 18,19,22,23,24, **29**-**30**,46,47,48,53,215,251,393, 399
- trapezoid **473**
- of vertebra **61**,64,121
- - fracture (78)
- vitreous 201,205,**207**
Boils, nasal (266-267)
Bone (see individual names and Process, Plates, Spine, etc.)
Botulinus toxin **412**
Bouton, synaptic **405**
Brachium
- conjunctivum (see Peduncle, cerebellar, superior)
- of inferior colliculus 463,**474**,532, 592,593
- pontis (see Peduncle, cerebellar, middle)
- of superior colliculus 463,**485**,**500**, 590
Brain **459**-**607**
- anatomy, gross **418**-**422**
- arterial supply **575**-**582**
- ascending pathways from spinal cord **455**-**456**,**586**,**598**-**601**
- barriers 572,**573**,**584**
- death **460**,467
- descending pathways to spinal cord **457**-**458**,**487**,**602**-**607**
- development **422**-**427**
- environment (homeostatic) **572**-**574**, **584**
- protection of 572,574,**584**
- subdivision **418**-**420**
- venous drainage **582**-**584**
Brainstem 418,419,**459**-**513**

B-C

- arterial supply **577-580**(622-633)
- ascending sensory pathways **488-489**
- - cerebellar connections **502-504**
- cranial nerves 460,461,463,**464-488**
- descending pathways **500-502**
- external topography **460-463**
- - dorsal aspect **461-463**
- - ventral aspect **460-461**
- internal topography 463-464, 464-489,**490-499**,500-513
- lesions involving 489,500,501,502, 509(619,620,622,623,624,625,630) (see also Nerves, cranial, named, lesions involving)
- nuclei, cranial nerve **464-488**
- - motor **500-502**
- precerebellar 502-504
- reticular formation of **504-513**
- tegmentum of **463**
- transverse sections of 490-499
- venous drainage 584
Bregma **13**,26,52
Bronchus 118
- tumour of (359)
Buds, taste 469,476,477,596
Bulb
- of brainstem 419,**500**
- end bulb, of Krause **432**
- of internal jugular vein **103**,112,394
- - in tympanic cavity (386)
- olfactory 488,535,537,540,543,555, 556,562,563,568,**588**,**589**
- of posterior horn of lateral ventricle 572
Bulla, ethmoidal **37**,245,250
Bundle (see also Fasciculus, Fibres, Striae, Tracts)
- hook, of Russell **521**
- medial forebrain 512,513,**527**,528, 567
- olivocochlear 471,472,**473-474**,592, 593
- Monakow's (see Tract, rubrospinal)
- predorsal **486**
- Vicq d'Azyr's (see Tract, mamillothalamic)
Bursa, for thyrohyoid membrane 68,**337**
- pharyngeal **311**

Calamus scriptorius **462**
Calcar avis **572**
Calcium (in synapses) **410**
Caliculus, gustatory (see Bud, taste)
Calvaria **12-13**,21,188
Canal, carotid **19**,33,34,95,99,112, 368
- central, of medula and spinal cord **421**,424,462,573
- condylar **18**,23,28,29,396
- facial 257,368,369,**372-373**
- - abscence of (387)
- - damaged by skull fracture (74)
- hyaloid **207**
- hypoglossal **18**,**23**,28,111,396
- incisive **41**,**44**,180,245,249,256,258, 305
- infra-orbital **41**,46,256
- mandibular **44**,305
- mental **44**
- nasolacrimal 39,40,41,**46**,**47**,**215**, 244
- optic **22**,29,30,**47**,199,215,217,250, 580
- palatine, greater 38,**41**,256
- palatovaginal **18**,32,48,257
- pterygoid 19,**32**,48-49,103,214,251, 257
- of Schlemm (sinus venosus sclerae) **202**,207
- - blockage (240)
- semicircular 23,34,369,**376-377**,378, 473,**594**
- for tensor tympani 34,**366**,371
- vertebral **62**,67,427,**444**

- vomerovaginal **18**,32,38-39,48
Canaliculus, for chorda tympani, 367
- cochlear 34,**376**,378,379
- lacrimal 148,212,**214**
- mastoid **18**,34,108,363
- sinuosus **256**
- tympanic **18**,34,106,367
Canthus, of eyelids **147-148**,150,**210**
Capillaries in central nervous system 573,**584**
Capsule
- external 562,569
- extreme 562,569
- internal 427,532,543,560,563,**564-566**,568,570
- - anterior limb **564**,581
- - arterial supply of 577,580,**581-582** (613)
- - fibres of **564-565**
- - genu 549,**564**,569,582,602,603
- - lesions of 459,**564**,582,587(613, 614,621,626,627)
- - posterior limb of 549,**564**,581,594, 595,596,597,598,599,600,601,602, 603
- - retrolenticular part 550,**564**,565, 582,590,591
- - sublenticular part 552,**564**,582, 592,593
- joint, temporomandibular **173**
- of parotid gland 94,96,**167**,168(190)
- of submandibular gland 96,**289**
- Tenon's **207-8**
- of thyroid gland 95,**119**,120
Cartilage, arytenoid **336**
- of auditory tube **311**
- auricular **362**
- cauliflower ear (384)
- chondrocranium, fetal **54-57**
- condylar **54**,**58**,172
- - acromegaly (75)
- corniculate **336**
- cricoid 83,118,119(140)312,315, **336**,337,338,339,341
- cuneiform **336**,337
- epiglottic 312,**337**
- of external acoustic meatus 364
- of larynx **334-337**
- Meckel's **45** (see also Arches, branchial)
- nasal, alar **242**
- - greater 153,155,**242**,244
- - lateral 153,**242**
- - septal 47,153,**242**
- of petro-occipital fissure 24
- spheno-occipital synchondrosis 19, 30,**52-53**,**54**,57
- thyroid 68,83,90,98,99,119,312, 315,316,317,**336**,337,338,341
- tracheal **118**,338
- triticeal **336**,337
- vomeronasal **245**
Caruncle, lacrimal 147,148,**210**
Catecholamines **411**,**509** (see also named)
Cauda equina **445**,446,575
Caudate nucleus (see Nucleus, caudate)
Cave, trigeminal **391-392**,401
Cavity, cranial **21-24**,48,**390-401**
- infraglottic **332**
- of mouth (oral) **272-307**
- nasal **47-48**,**243-249**
- orbital **46-47**,**198-229**
- tympanic 258,**366-369**
- ventricles of brain (see Ventricles)
Cells (see also Neurones)
- air (see Air cells)
- amacrine 484,512,**590**
- astrocytes **415-416**
- basket **521**,**541**
- Bergmann **416**
- Betz **547**
- bipolar, of retina 484,**590**
- capsular **416**

- complex 552
- cones, of retina 433,484,**590**
- ectodermal **422-424**
- endothelial **584**
- effector 418,**433-435**,438
- ependymal **416**,572,573,574
- fusiform (see spindle-shaped)
- ganglion (general) **404**,428,438,440
- - of retina 484,**590**
- glial (see Neuroglia)
- Golgi **418**,426,521
- granule or stellate **404**,426,521,541, 542,546,552,556,558
- - cerebellar **521**
- - cerebral 541,542
- - of dentate gyrus 558
- - of olfactory bulb 556,**588**,**589**
- - of visual cortex 542,552
- hair, of organ of Corti 471,473,**592**
- - of vestibular system 473,**594**
- horizontal, of retina 484,**590**
- hypercomplex 552
- internuclear **478**,482,483
- lemmal 416,**417**
- mesodermal **424**
- microglial 415,**416**
- mitral 556,**588**,589
- nerve cells (see Neurones)
- of neural crest **422-424**
- neuroblasts **424**
- neuroglia **414-416** (see also Neuroglia)
- neurolemma (see Schwann)
- oligodendrocytes 405,415,**416**
- periglomerular 556,**588**,589
- pituicytes **416**
- Purkinje 426,517,**521**
- pyramidal **404**
- - of neocortex 541,543,552
- - of hippocampus 558
- Renshaw **435**,436,454
- rod, of retina 433,484,**590**
- satellite **416**
- Schwann 405,406,**416**,428,433,443
- - acoustic neuroma (76,388)
- simple 552
- small intensly fluorescent (SIF) **440**
- spindle-shaped (fusiform) **404**,541
- spongioblasts **424**
- stellate, of cerebellum **521**
- - of neocortex **541**
- tanycytes **416**
- tract 489,**600**
- tufted **556**
Centre (see also Region or Area)
- apneustic **507**
- medullary core **417**
Centrum semi-ovale **535**
Cerebellum 418-419,**513-523**
- arbor vitae **516**
- archicerebellum **518**
- areas **518**
- arterial supply 577,**578**,**579**
- cells of 517,**521**
- connections
- - afferent **502-504**,516,517,**518-521**, 543,598
- - efferent 516,517,**521-523**,530,532, 534,547,571,587
- - cortex 426,513,517,**521**
- - cyto-architecture **521**
- - layers 517,**521**
- - regions **521**,522
- - development **426**
- - dysfunction **518**,561,587(619,622, 628,629)
- fibres
- - climbing 503,517,**518**,520,521
- - mossy 517,**518**,520,521
- - fissures 514,516
- - folium(a) 419,**514**,521
- - functions 513,**518**,**521-523**,561,571
- - lobes **514-516**
- - - anterior **514**,518
- - flocculonodular **514**,516,521,594,595

- - middle (posterior) **514**,518
- lobules **514-516**
- neocerebellum **518**
- nuclei, intracerebellar (deep) 513, 516,517,518,**521-523**,530,532 (see also Nuclei, named)
- paleocerebellum **518**
- peduncles 419,460-462,**503**,504, 513,**516-517**,518 (see also Peduncles, cerebellar)
- pontocerebellum **518**
- relationships 514-516
- spinocerebellum **518**
- subdivisions **514-516**,521
- tonsil **516**
- topography
- - external 513,**514-517**
- - internal 513,**517-523**
- venous drainage **583**
- vermis 419,**514**,518
- vestibulocerebellum **518**
- zones **521**,522
Cerebrospinal fluid (see Fluid, cerebrospinal)
Cerebrum **418-420**,**523-572** (see also Cortex, Diencephalon, and Hemispheres, cerebral)
- development **426-427**
- vasculature 577,**579-583**
Chain (trunk), sympathetic **440**,**441** (see also Nerve)
Chamber, anterior of eye 207
- posterior of eye **207**
Cheek **147**,**273**
- innervation **300-301**
- vasculature 305
- muscles **156-157**
Chemosis (235)
Chiasma, optic 22,224,251,399,400, 485,524,527,537,575,581,**590**,**591**
- damage with tumour (75,627,628)
Chin 14,**43**,**148-149**,170
Choana (see Aperture)
Choanal atresia (267)
Chondrocranium **54**,**57**
Chorea 571
Choriocapillaris **203**,227
- in retinal ischaemia (237)
Choroid **203-204**,205,227,**590**
Cilia (see Eyelashes)
Cingulum 554,559,560,**562**
Circle, arterial (of Willis) (see Artery)
Circuit, (see also Pathway or System)
- control **435-437**,523
- local **410**
- of Papez **560**
- serial and parallel **436-437**
Cistern, cisterna **574-575**
- cerebellomedullary (magna) **575** (609)
- - puncture to sample CSF(609)
- interpeduncular (basalis) **575**,580
- of lateral sulcus **575**
- of lamina terminalis **575**
- lumbar (spinal) 446,**574-575**(609, 618)
- of optic chiasma **575**
- pontine **575**
- superior (of great cerebral vein, ambiens) **575**
- supracallosal **575**
Claustrum 543,562,563,567,568,**569**
Clavicle 83,86,88,89,90,92,94,115, 129,133(142,143)
Cleft, branchial **134**,379
- intratonsillar **312**
- lip 169,**170**(628)
- median of jaws **170**
- palate **325**(348-349,628)
- of Schmidt–Lantermann **406**
- synaptic **410**,433-435
Clivus 24,28,30,394
Cochlea 369,**375-376**,377,471-473
Collaterals, axon **404**

C

- - Schaffer 558
Colliculus, facial 462
- inferior 419,**463,474**,530,543,554, **592**,593
- - brachium of **463,474,532,592**,593
- superior 419,463,483,**485-487**,530, 531,532,533,534,543,547,550,570, 571,**590,591**,**604**
- - brachium of **463,485,500,590**
Columella 242
Column
- cortical **541**,552,560
- of fornix (anterior or posterior) 524,538,**563**,571,585
- ocular dominance **541**,552,560
- orientation **541**,552,560
- of spinal cord
- - of Clarke (see Nucleus, dorsalis)
- - dorsal (posterior) **446**,453,**455-456**, 458,459,**488-489**,**598**,599(611,612, 613,614,615,622,626)
- - intermediolateral 437,438,440,442, **450**,453,512,526,606(615)
- - intermediomedial **453**
- - vertebral, cervical **61**,82,95,98, 121,124,427,**445**
Coma **508,509**
Commissure, anterior 524,526,538, 556,**562,563**,569
- corpus callosum (see Corpus callosum)
- of fornix (hippocampal) **538**,559, 562,563
- habenular **535**
- intercollicular
- - inferior **474**
- - superior **485**
- of lips **148**,273
- posterior 463,483,487,501,523,**535**
- of spinal cord 446,451,453,**455**
- - lesion of (614,623)
Competition, dynamic, between afferents **560**
Complex
- amygdaloid (see Amygdala)
- olivary (see Nuclei, olivary)
Concha, of auricle 154,158,160,**362**
- nasal, inferior 11,**39**,47,**244**,252
- - - fetal **54**
- - - ossification **39**
- - middle 37,47,**244**,250
- - superior 37,47,**244**
- - - visualisation of (268)
- - supreme **245**
- - hypertrophy (268)
- - sphenoidal **30**
Conditioning, classical 521,**561**
Conduction
- block **443**
- decremental **408**
- orthodromic **408**
- saltatory **406-409**
- speed in nerve fibres **431**
Condyle, of mandible (see Process, condylar)
- occipital **18**,**28**,62,66
Cones, retinal 433,484,**590**
Confabulation **561**(629)
Coning, of hindbrain **516**(618)
Conjunctiva 148,**211**
- haemorrhage 71 (239)
- inflammation following damage to orbicularis oculi (190)
- inflammation following blocked nasocrimal duct (234)
- with shingles (192)
Connections
- serial and parallel **436**,437
- dynamic competition between **560**
Consciousness 504,**508-509**,533,535, 552,561,575,586
Control mechanisms, neural **435-437**
Conus elasticus **338**
Conus medullaris 418,**445**
Convergence (see Reflex)

Convolution (see Gyrus)
Copy, efference **436**
Cord
- false (see Fold, vestibular)
- spinal 418,427,428,433,437,438, 439,440,441,**444-459**
- - anchorage **445-446**
- - arterial supply **582**
- - boundaries 444-445,**460**,491
- - connections
- - - ascending **455-456**,598,599,600, 601
- - - descending **457-458**,543,564-565, 602,603,604,605,606,607
- - - intrinsic 435,446,451,457, **454-455**
- - spinal nerves **427-443**
- - development **422-425**(609)
- - grey matter **446-454**
- - functions **427**,**446-450**
- - funiculi **446**,454-458
- - horns **446-454**
- - intermediate area 450,**453**,454,504
- - laminae **450-454**
- - lesions **458-459**,587(615,616)
- - marginal layer **451**
- - nucleus, dorsalis **453**,456
- - proprius **451-453**,456
- - reflexes **435**,450,453
- - relationships **444-445**
- - sections, transverse 447-449,**450**
- - segments **445**
- - subdivisions **445**,**446-450**
- - substantia gelatinosa **451**,600
- - topography, external **444-446**
- - - internal **446-458**
- - venous drainage **584**
- - vertebral levels **445**
- - white matter **454-458**
- vocal (see Fold, vocal)
Core, medullary **417**,513,535
Cornea 147,201,**203**,209,590
- opacification (240)
- perforation (237)
- ulceration (190,192)
Cornu ammonis (see Hippocampus) **558**
Corona radiata **564**
Corpora quadrigemina (see Colliculi)
Corpus (see also Body)
- amygdaloideum (see Amygdala)
- callosum 419,427,536,**537**,**562-563**, 571,572
- - body (trunk) **537**,571,580
- - connections **537**,**562-563**
- - development **427**
- - forceps major and minor **537**
- - function **563**
- - genu of **537**,580
- - splenium of **537**,571,583
- - rostrum **537**
- - transection **563**
- - striatum **567-571** (see also named structures)
- - archistriatum **567-569**
- - neostriatum **567-571**
- - paleostriatum 568,569 (see also Globus pallidus)
Corpuscles (sensory endings) **432**,433
Corollary discharge **436**
Cortex **417**
- cerebellar 513,517,**521**
- cerebral 419,535,**540-562** (see also Gyri and Areas)
- - agranular **546**
- - allocortex **541**,555-559
- - archicortex **541**,**556-559**,560
- - areas **544-559**
- - - Broca's (see Cortex, speech, anterior)
- - - Brodmann's **544**
- - - entorhinal **555**,556,557,558,**559**, 562,588
- - - septal **559**,**567**
- - - Wernicke's (see Cortex,

speech, posterior)
- - arterial supply **577**,579-581,582
- - association 543,**553-555**,559,561 (620,621,622,629,630)
- - - auditory 474,530,533,543,**552**, 553,562,563,592,593
- - - cells **541-542**
- - - columnar organization **541**,552,560
- - - connections (see also named regions) 529-534, **542-543**,545, **562-566**
- - - afferents **542-543**
- - - - nonspecific 512,513,534,541, **543**,**566-567**
- - - - specific 529-533,**542-543**,564, 565-566
- - - - association **542-543**,562
- - - - commissural **542-543**,562-563
- - - - efferents **543**,563-564,566
- - - - intracortical **541**
- - - - projection **562-566**
- - - - thalamic 529-534,**542-543**,564, 565-566
- - cyto-architecture **541-546**
- - dimensions **540-541**,542
- - electroencephalogram (EEG) **508-509**,513,534,567
- - entorhinal **555**,556,557,558,**559**, 562,588
- - eye field
- - frontal 479,485,510,530,543, **547-549**,550,562,590,604(620,621)
- - - occipital **550** (see also Cortex, visual)
- - frontal 532,538,**539-540**,546, 553-554,559,561-562,564,570
- - granular **546**
- - gustatory **550**
- - gyri 419,427,**538-540**
- - heterogenetic (heterotypic) **541**
- - higher order activities 535, **553-555**,561
- - hippocampal **556-559** (see also Formation, hippocampal)
- - homogenetic (homotypic) **541**
- - insula 530,532,533,**540**,555
- - isocortex **541**
- - juxtallocortex **560**
- - koniocortex **546**
- - lamellae (strips) **541**
- - layers **541**
- - limbic 543 **559-560**,562,568 (see also System, limbic)
- - motor 543,**546-549**,563,570,571, 581,582,602-603
- - - lesions **587**(613,618,625,628-629)
- - - precentral (MsI) 546,**547**,549 (613)
- - - premotor 530,532,**547**
- - - primary 530,532,**547**,602-603
- - - secondary (MsII) **547**,549
- - - supplementary **547**,549,553
- - neocortex **541-543**,**544-555**
- - occipital 530,532,**539**,**540**,550,551, 552,554,561,562,564,565,566
- - olfactory 488,543,**553**,**555-556**, 568,**588**,589
- - opercular 530,533,**540**,549,550, 552
- - orbitofrontal (orbital) **537**,554,555, 559,560,561,562,588,589
- - organization **541**
- - paleocortex 541,543,**555-556**,560, 568,588
- - paracentral **539** (see also Lobule)
- - parastriate **550**,551,552
- - parietal 530,532,538,**539**,**540**, 546,549,550,553,554,561,564(630, 631)
- - perirhinal **559**
- - peristriate **550**,551,552
- - plasticity **560-561**
- - polar **546**
- - prefrontal 530,532,**554**,561,562 (629)

- - prepyriform (prepiriform) 553, **555**,556,**588**,**589**
- - prorhinal **559**
- - pyriform 540,541,**555**,556,560,569
- - retro-insular 530,533,549,**600**,601
- - somaesthetic (somatosensory, sensorimotor) 530,533,543, **549-550**,570,581,582,594,595,596, 597,598,599,600,601
- - - association **549-550**,598
- - - primary (SmI) 478,530,533,**549**, 563,596,597,598,599,600,601, (613,630)
- - - secondary (SmII) 530,533,**549**, 596,597,598,599,600,601
- - speech
- - - anterior (Broca's area) 549,552, **553**,554,562,581,592,593(620,621)
- - - posterior (Wernicke's area) **553**, 554,561,562.581,592,593(628,629)
- - - superior 547,**553**
- - striate 550,**551**,552
- - strips (columns or lamellae) **541**, 552,560
- - sulci 419,427,**538-540**
- - temporal 530,532,535,537,538, **539,540**,543,550,552-553,555-559, 560,561(629)
- - venous drainage 582-583
- - vestibular **550**
- - visual 530,533,541,543,**550-552**, 562,579,590,604
- - - association **550-552**
- - - columnar organisation 541,552, 560
- - - neuronal responses **541**,552
- - - plasticity **560**
- - - primary (V I) **541**,**550-552**,560, 563,590(612)
Corticotropin 411,527(628)
Cranium, base of 10,**17-24**,**29-35**,48, 55,57,58
- vault 10,**12-13**,21,**25-29**,51
Creases, skin of face **147**,150
Crest, conchal, of maxilla **41**
- - of palatine bone **38**
- ethmoidal of maxilla **41**
- - of palatine bone **38**
- frontal **21**,26
- infratemporal **31**
- jugal **40**
- lacrimal, anterior 41,**46**,212,214
- - posterior 40,**46**,209,214
- nasal of maxilla **41**,**47**,242
- - of palatine bone **38**,47
- neural **422-424**
- occipital, external **27**
- - internal **23**,**24**,392
- palatal **38**
- of sphenoid bone **30**
- supramastoid **15**,33
- supra-optic 527,584,585
- temporal of coronoid process **44**,147
- tympanic **33**
- vertical, of nasal bone **39**
- zygomatico-alveolar **40**
Crista galli 21,36,48,391
Crus
- cerebri (of cerebral peduncle) **461**, 500,**564-565**,602,603(613,625,630)
- of fornix 538,558-559,**563**,572
- of helix **362**
- of semicircular canals **376**
Crypts, tonsillar 312
Culmen 514,**516**
Cuneus **540**,550
Curvature, of vertebral column **61**
Cup, optic **229**
Cupula 473,**594**
Curare **412**
Curvature, of vertebral column **61**
Cycle (see Rhythm)
Cyst, branchial (139)
- mucocele in sphenoidal sinus (237-8)
- naso-alveolar (192)
- palatine, median (348)

635

C-F

- thyroglossal (140)

Deafness (76,385,386,387-388)474, 552(619,620,628)
Death, of brain 460,467
Declive 514,516
Decussation 418
- motor (pyramidal) 491,**500**,602,603
- optic chiasma 485,590,591
- sensory 489,492,598,599
Defaecation, voluntary control 547
Degeneration, of nerve fibres 405
Dehiscence, Killian's (144-145,356)
Dementia 513,**561**
Demyelination 406
Dendrites (dendrons) **403-404**
- isodendritic **504**
Denervation hypersensitivity 413,438 (615)
Dens, of axis 18,23,62,**63**,64,65,460
- - subluxation (79)
Dependence, neuronal 412,438
- opiate **569**
Depolarization 408,409,**410**,411
Depression (630,631)
Dermatomes 428,429(616)
Diabetes insipidus 528
Diaphragm 113,114
Diencephalon 419,**523-535** (see also Epithalamus, Hypothalamus, Subthalamus and Thalamus)
- - arterial supply 577,579,580,581
- - development 426
- - relationships 523-524
- - subdivisions 418,523
- - topography 523-524
- - venous drainage 583
Dimorphism
- sexual 150,273,336
- - hypothalamus 526
- - lips 273
- - thyroid cartilage 336
- - tooth eruption 320
Diploë 10,395
Diplopia (71,190-191,234,235)478, 482,484(610,624)
Disc, intervertebral **64**
- - cervical spondylosis (80)
- Merkel 432
- optic 205,207,227,**485**,590(612,627, 628)
- - papilloedema (239)485(617, 618,619,630,631)
- of temporomandibular joint, articular **172-173**,174
Discharge, corollary 436
Diverticulum, arachnoid, in nose (608-609)
- respiratory 344
- pharyngeal (Killian's dehiscence) (144-145,356)
Doctrine, neurone 412
Dominance, cerebral 549,552-553, **554**,563
Dopamine (as neurotransmitter) 411, 440,441,502,509,510,511,**512-513**, 526,527,543,556,568,569,570,571 (625)
Dorsum, of external nose 148,242
- of tongue 275
Drinking (see Thirst)
Drive (see Motivation)
Duct, Bartolin's **290**
- cochlear 375,376,**377**,378
- ductus endolymphaticus 375,**378**
- ductus reuniens 377,**378**
- frontonasal 245
- jugular lymph trunk 118,132
- of lacrimal gland 214
- nasolacrimal 215,245
- - blocked (234)
- parotid (Stenson's) 147,156,162,**167**, 273
- inflammation of (190)
- right lymphatic duct 118,120,**132**

- semicircular 378
- subclavian lymph trunk 132
- of sublingual gland (also Bartolin's duct) 290
- submandibular gland (Wharton's) 275,**290**,302
- - sialolith (stone) (350)
- thoracic 118,120,130,131,**132**
- thyroglossal 135
- utriculosaccular 378
Dura mater 430
- cerebral **390-392**,396
- spinal **445-446**
Dysaesthesia (626,627)
Dysdiadochokinesia (628,629)
Dyskinesias 518,571,**587**
Dysmetria 518
Dysphagia (141,143,144-145,349,351, 356)468,470,**471**(622,623,624)
Dyspnoea (141,351,357-358,358)

Ear 362-380
- external 362-364
- - bat ear (384)
- - cauliflower ear (384)
- - innervation 363,476
- - vasculature 363-364
- - haemorrhage (73,385)
- internal 375-379
- - antibiotic-induced damage (388)
- - innervation 378-379,471-473
- - tumour (388)
- - vasculature 379
- middle 366-373
- - blood supply 373
- - bones of (ossicles) 370-371,592
- - drainage of (386)
- - fluid in (glue ear) (386)
- - infection of (387)
- - innervation 372-373,469
- - muscles 371-372
Effectors 418,**433-435**,438
Electroencephalogram (EEG) 508, 509,513,534,567
Element, costal of vertebrae 62,64
Elevation
- tubal 310
Embolism 575(611,612)
Embryology
- brain and spinal cord **422-427**
- ear 379-380
- eye and extra-ocular muscles **228-229**
- face 168-170
- larynx 344-346
- mouth 319
- neck (branchial arch system) **134-135**
- nasal cavities 258-260
- palate 321-325
- paranasal sinuses 260
- pharynx 325
- pituitary gland 399
- skull (see also individual bones)
- - fetal (including growth) **51-60**
- teeth 320-321
- thyroid gland 135
- tongue 321
Eminence, arcuate 23,34,376
- articular of temporomandibular joint (see Tubercle, articular)
- canine 40,**43**
- collateral 572
- medial, of fourth ventricle 462
- median, of hypothalamus 399,**524**, 584,**585**
Emotional behaviour (see Behaviour)
Encephalitis (629)
End plate, motor 433,**434**
Ending (see also Synapse and Junction)
- neurotendinous (of Golgi) 431,**432**
- sensory **430-433**
- trail **434**
Endolymph 378
Endoneurium 428,443
Endorphins, enkephalins (see Opioids)

Enophthalmos (193,234)**441**(622,623)
Ependyma 416,572,573,574
Epicanthus 210
Epiglottis 68,**337**
Epilepsy 553,**569**(618,620,621)
Epinephrine (see Adrenaline)
Epineurium 428,446
Epiphysis cerebri 534 (see also Body, pineal)
Epistaxis (266,355)
Epithalamus 419,523,524,**534-535**
Epithelium, of ventricles (see Ependyma)
- olfactory 424,**588**,589
Equilibration (see System, vestibular and Movement, balance)
Ethmoid bone 11,21,**36-37**,46,47,48, 51,53,54(608-609)
- cribriform plate 21,**36**,47,48,53,243, 245,247,398,537,**588**,589
- - fracture of (73)
- - tumour at (76)
- fetal 51,53
- labyrinth (lateral mass) 37,47,51,53, 250
- ossification 37
- perpendicular plate **36**,47,53
Euryprosopic face 149-150
Excitatory postsynaptic potential (EPSP) 410
Eye 200-207
- chambers 207
- displacement with tumour (270)
- extra-ocular muscles 215-221
- fascia bulbi **207-209**,220
- fields (see Fields)
- innervation 222-226,439,440,441, 442,478,479
- intra-ocular muscles 199,**215**
- layers **201-206**,484,485,487,512,530, 590,591
- lens 206-207
- - neural control of 439,440,442,482, **484**,590
- movements (see also Extra-ocular muscles)
- - conjugate 475,479,483,484,486, 501,**507**,604(617)
- - dissociated 475
- - nystagmus 475,476,518(619,622, 624,628,629)
- - scanning 547-549
- - voluntary 547-549,550,604(621)
- optic disc (blind spot) (see Disc, optic)
- pupil (see Pupil)
- reflexes (see Reflex)
- vasculature 202,203,204,205,206, 227,485,580
- vision (see System, visual)
Eyebrows 147,152,188
Eyelashes (cilia) 147,148,210,212
Eyelid 147,199,**209-213**,217,220
- drooping (235)**441**,484(622,623,624, 625,629)
- fibrous layer 211-212
- innervation 212-213
- muscles 152
- Panda eyes, haemorrhage (73)
- vasculature 212

Face
- blood supply 161-163
- deep structures 171-181
- ethnic variations 150
- infratemporal fossa 174-181
- innervation 159-161
- lymphatics 164-165
- muscles, facial expression 151-157
- - paralysis (73,74,76,77,190)
- - mastication 171-172,176
- sexual dimorphism 150
- skeleton 13-14,35-45,46-48
- - fractures (72-73)
- superficial structures 151-170

- surface markings 147-150
- temporomandibular joint 160, **172-174**,178,180,364
- types 149-150
- veins 163-164
Facet, articular of vertebrae 61,62, 63,64-65
Factors, hormonal releasing/inhibiting 411,**527**,585
Fainting (195)
Falx, cerebelli 23,28,**392**,514,578
- cerebri 21,28,36,**391**,394,395,514, 536,**577**
Fascia, axillary sheath 95
- buccopharyngeal 317
- bulbi (of eyeball) **207-209**,220
- carotid sheath 82,86,87,93,**95-96**, 101,105,111,118,130,174
- definition of **93**
- investing layer of deep cervical 82, 90,93,**94-95**,105,116,167,308
- lacrimal 152,**214-215**
- of neck **93-95**
- parotid 94,96,**167**,168(190)
- pharyngobasilar 312,**317**
- pretracheal 82,93,94,**95**,97,119,120, (140)
- - restriction on movement of thyroid gland (140)
- prevertebral 82,90,93,94,**95**,96,98, 105,113,116,121(143)
- submandibular gland, sheath of 96, **289**
- temporal 15,16,27,171,**172**,188
- thyroid, sheath 95,**119**
Fasciculus (see also Tracts)
- cuneate (of Burdach) **455-456**,462, 488,489,595,599
- dorsolateral (of Lissauer) of spinal cord 451,453,**455**,598
- gracile (of Goll) **455-456**,462,488, 489,598,599
- lenticularis 528,529,532,**570**
- longitudinal
- - dorsal (of Schütz) 505,527
- - inferior 562
- - medial (MLF) **457**,458,474,**475**, 478,483,594,595,604,605,606,607
- - - lesion 475
- - superior 562
- mamillaris princeps 526
- proprii 455
- retroflexus 528,535
- of Schütz, dorsal longitudinal 505, 527
- subthalamic 528
- tegmental, central 505
- thalamic 528,**529**
- uncinate
- - (of Russell) of cerebellum 521
- - of cerebrum 555,556,559,**562**
Fastigium 462
Fat, buccal pad 147,167
- orbital pad 199
Fauces, pillars of 275,**277**(610)
Feedback 435,436,523,556,571,586
Feed-forward 436,523
Feeding/hunger, neural control of 433,**437**,438,442-443,468,471, **477-478**,480,482,488,500,505,506, 512,524,525,**528**,555,569,596-597 (622,623,624)
Feet, end of astroglia 415
Fenestra cochleae 369,375,376
- vestibuli 369,371,375,376
Fenestrations, of capillaries 584
Fibres, muscle 434
- nerve 418 (see also Fasciculus, Pathway, Tract)
- - A group 430,431,432,437,451,455, 598,600
- - adrenergic 470,471,508,509,510, 511,**512**
- - afferent 418
- - - autonomic 437-438,451,453,464,

F

465,469,470
- - - classification of sensory 430,**431**
- - - primary sensory 428,**430**,**431**,432, 433,434,450,451,455,464,465,586, 588-601
- - - somatosensory 428-434,451-453, 455,464-465,479,480,481,482,586, **598-601**
- - - special sensory 464-465,469, 471-473,476,477-478,484-485,488, 586,**588-597**
- - - spindle organ 431,432,**434**
- - anterolateral (see Tract, anterolateral)
- - arcuate
- - - cerebral 562
- - - internal **489**
- - association, cortical 418,541-543, **562**
- - B group **430**,431,438,441,442
- - branchiomotor **464**,465
- - C group **430**,431,432,437,438,441, 442,451,**600**
- - cardiomotor **434**,439,469,470
- - centrifugal **586**
- - cerebellothalamic **517**,522,523,528, 529.532
- - ceruleospinal **606**
- - cholinergic **433-434**,438,**439**,440, 441,442,473,508,**513**,527,543,562, 566,567,569
- - classification of nerve fibres **430-431**,**464-465**
- - climbing 503,517,**518**,520,521
- - commissural **418** (see also Commissure)
- - - cortical **562-563**
- - cortical **541-543**,562-566 (see also Cortex)
- - corticobulbar **500** (see also corticonuclear)
- - corticocortical 541,543,**562**
- - corticofugal 500-501,**543**,564
- - corticonuclear 467,**500-501**,543, 547,549,564,565,582,602,603 (620-622,624-625,626)
- - corticopontine **500**,503,543,564, 565
- - corticoreticular **501**
- - corticospinal 451,453,454,457,460, **500**, 543,547,549,564,565,582,587, **602-603**(613,614,615,616,618,621, 623,624,625,626,627)
- - corticostriate 564,**570**
- - corticotectal **485**,500,566
- - corticothalamic **529-532**,534,543, 564-566
- - decussating **418**
- - descending (motor) 457-458,500-502,505,506,543,547,549,564-565, **587**,**602-606**(613,614,615,616,618, 621,623,624,625,626,627,629)
- - dopaminergic 440,441,502,509, 510,511,**512-513**,526,527,543,556, 568,569,570,571(625)
- - efferent **418**,431,**433-434**
- - - autonomic 434,**438-443**,453,464, 465,469,476,482,484
- - - classification of motor 430,**431**
- - - somatomotor 428-431,**433-434**, 450,454,464-465,467-470,471, 476-477,478,479,480,482-483,484
- - Edinger-Westphal-spinal **606**
- - fastigiospinal 458,**606**
- - GABA-ergic 475,**521**,569,**570**,571
- - group Ia **430**,431,432,**434**,435,549
- - group Ib **430**,431,432
- - group II **430**,431,432,**434**
- - group III (Aδ) **430**,431,432,437,451, **600**
- - group IV (c) **430**,431,432,434,437, 438,441,442,451,**600**
- - hypothalamospinal 526,**606**
- - intracortical 541,**562**

- - mossy
- - - cerebellar 517,**518**,520,521
- - - of dentate gyrus 558
- - motor (efferent) 418,428-431, **433-434**,438-443,464-465
- - myelinated **405-409**,430-431
- - nigrostriatal 502,513,**570**(625)
- - noradrenergic 434,438,**439**,440, 441,442,470,471,508,509,**512**,518, 527,534,543,556,568,606
- - parallel 521
- - pontocerebellar **503-504**,516,520
- - projection **418**
- - - of cortex 542-543,**563-566**
- - propriospinal **455**
- - raphe/spinal **606**
- - reticulocerebellar **507**,520
- - reticuloreticular **505**
- - rubronuclear (rubrobulbar) **502**
- - sensory (see afferent)
- - solitariospinal 458,471,**606**
- - spino-olivary **456**,519,520
- - spinoreticular 456,489,506,**600**,601
- - spinospinal **455**
- - spinotectal 456,485,**600**,601
- - striatonigral **570**
- - tectoreticular (tectobulbar) **486**
- - thalamocortical **529-532**,534, 542-543,564-566
- - thalamostriate 529,534,**570**
- - tubero-infundibular 512,**526-527**
- - unmyelinated **405-408**,430,431
- - ventrolateral (see Tract, anterolateral)
- - vestibulocerebellar 473,474,**520**, 594,595

Field(s)
- eye
- - frontal 479,485,510,530,543, **547-549**,550,562,600,604(620,621)
- - occipital **550** (see also Cortex, visual)
- H fields of Forel 528,**529**
- prerubral **528**,570
- receptive **430**
- - visual **550-552**
- visual **485**,550-551
- - defects **485**,533,552,566(611,612, 627,628,630,631) (see also Anopia, Hemianopia, Quadrantanopia)
Fila olfactoria 488,555,556,**588**,**589**
Fillet, mesial (see Lemniscus, medial)
Filum terminale 445,446,575
Fimbria 558-559,563,572,573
Fissure (see also Sulcus) 514,**538**
- calcarine **539**,540,550,565,566,579
- cerebellar **514**,515-516
- choroid(al) 559,571,572,**573**,579,580
- dorsolateral (posterolateral) 514,516
- hippocampal 558
- horizontal 514,**516**
- lateral (of Sylvius) (see Sulcus)
- longitudinal 419,**536**,537,580
- palpebral **147**,209
- primary 514,516
- of skull
- - orbital, inferior 41,**47**,48,181,199, 224,228
- - superior 22,30,**47**,112,199,215, 219,227,580(609)
- - petro-occipital **24**
- - petrosquamous **20**,32,33,34,175, 369
- - petrotympanic **20**,33,172,175,179, 180,304,365,367,370,373
- - pterygomaxillary 15,**48**,175,180, 255,258
- - squamotympanic **20**,33,173,175
- - tympanomastoid **33**,108,363
- - transverse **538**,573
- - ventral median **444**,446,460,582
Fistula, branchial (139)
- caroticocavernous (235)
- oro-antral (269,353)
Fit, uncinate 553

Flaccidity (see Paralysis/Paresis)
Flexure, of neuraxis **424**
Flocculus 461,514,**516**
Floor
- of cranial fossae **21-24**,**48**,**398-402**
- of fourth ventricle **462**
- of lateral ventricle **571**,572
- of mouth **275**
- - muscles of **294-295**
- - of posterior triangle **90**
- - of third ventricle **524**
Flow, axoplasmic **405**
Fluid
- cerebrospinal 376,392,420,421,446, 525,**572-575**,584,585
- - anaesthetic solution within (608)
- - aspiration 446,516,**574**(609,618)
- - barriers 572,**573**,574,584
- - circulation **574-575**
- - cisternal puncture (609)
- - composition **573-574**
- - functions **574**
- - haemorrhage into (618)
- - hydrocephaly (77-78)**574**(615,617)
- - loss following fracture of cranial base (73)
- - lumbar puncture 446,516,**574**(609, 618)
- - obstruction **574**(615,617,628,629)
- - rhinorrhoea (73,267)
- - secretion **573-574**
- - extracellular (interstitial) 572,**574**, 584
Fold, aryepiglottic 312,**337**
- epicanthal **150**
- fimbriated **275**
- glosso-epiglottic 275,**312**
- malleolar **364**,365,370
- palatoglossal **277**
- palatopharyngeal **277**
- salpingopalatine **310**
- salpingopharyngeal **310**
- sublingual **275**,290
- vestibular **332**
- vocal **332-334**,339,341,343
- - examination before thyroid gland surgery (141)
- - with myxoedema (193)
- - obstruction at (140)
- - paralysis (358,359,622)
Folia (folium) 419,**514**,521
Follicles, lingual **275**
Fontanelle, anterior 13,52(77)
- - tensed (77,609)
- - mastoid **16**,52
- - palpation prenatally (77)
- - posterior 13,**52**(77)
- - sphenoidal 15,52
Foramen, foramina
- for basivertebral veins on body of vertebra **61**
- caecum of skull 21,26,36,275,394
- - of tongue 135,**275**,**321**
- ethmoidal (anterior and posterior) **46**,397
- frontal **14**,25,46
- incisive **17**,276
- infra-orbital **14**,40,41,154,155,256
- innominate **31**
- interventricular (of Monro) 524,**538**, 571,573,574,583
- intervertebral **61**,67,**428**
- jugular **18**,19,**24**,28,34,103,106,108, 109,394,395
- of Luschka (lateral aperture) **462**, 573,**574**,575
- lacerum **19**,**22-23**,33,48,112,175, 181
- of Magendie (median aperture) **462-463**,573,**574**,575
- magnum 10,12,**18**,**23**,27,28,**29**,131, 402,459,**460**,**461**,516,577,578(609)
- mandibular **44**,175,179,180,298, 305,306
- mastoid **16**,34,189,396

- mental 14,**43**,156,160,179,306
- ovale **19**,**22**,**23**,31,106,175,177,179, 181(608)
- palatine, greater **17**,49,256,276
- - lesser **17**,38,49,256
- parietal **13**,26,189,396
- petrosal **31**
- for posterior superior alveolar nerves 40,**300**
- rotundum **22**,31,32,**48-49**,253,255
- sphenoidal emissary 19,**22**,**31**,181, 396
- sphenopalatine 37,**47**,49,244,253, 256
- spinosum **19**,**22**,**31**,175,178
- stylomastoid **18**,35,51,158,368,372
- supra-orbital **14**,25,46
- transversarium **62**,83,129,130,131
- vertebral 61,**62**
- zygomaticofacial **16**,36,160,163
- zygomatico-orbital 36,**46**,198
- zygomaticotemporal **16**,36,160,163
Forceps, major **537**,571-572
- minor **537**
Forebrain (prosencephalon) 418-420, 422,**523-572**(see also Cerebrum)
- basal **566-567** (see also named components)
- development **424-427**
Forehead 147,150,188(609)
Formation
- hippocampal 537,541,**556-559**,560, 567,580 (see also Region, hippocampal)
- reticular 460,463,466,467,**504-513**, 520,521,525,530,531,532,533,534, 543,556,559,563,564,566,567,568, 569,570,586,587,600,606,607
- - connections 437,453,466-467, **504-505**,506-511,530,531,532,534, 543,556,559,563,564,566,567,568, 569,570,586,587,600,606,607
- - functions **504-513**
- - - arousal **508-509**
- - - cardiovascular **508**
- - - eye movements **507**
- - - motor **505-508**
- - - reflexes **507**
- - - respiration **507**
- - paramedian pontine 479,**507**,604 (621)
Fornix
- of brain 526,527,532,538,543,558, 559,560,561,**563-564**,567,571,572
- - body **538**,563,571
- - column 524,**538**,**563**,571,585
- - commissure **538**,559,562,563
- - crus 538,558,**563**,572
- - fimbria 558-559,563,572,573
- - postcommissural 538,559,**563-564**
- - precommissural 538,559,**563**
- - conjunctival 147,210,**211**,220
- vestibuli **273**
Fossa
- canine **40**,155
- cerebellar **28**,**514**,578
- cerebral **28**
- condylar 18,**28**
- cranial **21-24**,**48**,55,57,58
- - anterior **21-22**,**48**,**398**,**537**
- - - fracture (73-74,267)
- - - surgical intervention (609)
- - middle **22-23**,**48**,**398-401**,**537**
- - - fracture (74)
- - - infection of (196)
- - posterior **23-24**,**48**,**401-402**
- digastric **43**
- incisive of mandible **43**,156
- - of maxilla **40**
- - of palate **17**,41
- incudis 369,371
- infratemporal 15,**22**,**31**,**174-178**,258
- - anaesthesia within (194)
- - haematomas (194-195)
- - infection, spread of (195-196)

F

637

F–H

- interpeduncular **461**,575
- jugular 18,**34**,108
- for lacrimal gland **26**,213
- for lacrimal sac 40,**46**,214
- lenticular **207**
- mandibular **16**,**20**,33,34,51,**172**, 174,287
- nasal 18,**47-48**,**243-245**
- piriform **312**
- - foreign body in (356)
- - tumour of (356)
- pituitary **22**,**29**,**30**
- - enlargement (75)
- pterygoid **19-20**,**31**
- pterygopalatine 15,22,**48-49**,174, 180,242,**253-258**
- - arachnoid diverticulum within (608-609)
- - fractures at (72-73)
- retromolar **44**,156,178,301
- rhomboid **462**,573
- scaphoid, of ear **362**
- - of skull 20,**31**,295
- subarcuate **34**
- sublingual **43**,290
- submandibular **43**,289
- supraclavicular **83**
- - lesser **83**
- suprasternal **83**
- temporal **15**,22,175
- triangular **362**
Fovea, centralis 205,**206**,227,590
- pterygoid **44**,172
- trochlear **26**
Fractures of skull (71-74,234,235,267, 624,630)
- of vertebrae (78,79)
Frenulum (frenum) **273**,275(350-351)
Frontal bone 11,12,14,15,21,**25-26**, 46,47,48
- fetal **52**,**53**
- internal surface **26**
- muscles attached to **26**
- nasal part **25**,242
- orbital part **25-26**,46,48,537
- ossification **26**
- squamous part 12,25,**26**
Funiculus, of spinal cord 446,**454-458**
Furrows (see Grooves or Sulci)

Galactorrhoea (627,628)
Galea aponeurotica (see Aponeurosis, epicranial)
Ganglion, ganglia
- basal (see Nuclei)
- cervical sympathetic, inferior 112, **113**,440
- - middle 83,**112-113**,440
- - superior 108,**111-112**,179,223,257, 304,318,397,440
- cervicothoracic 112,**113**,133,440
- ciliary 112,199,220,223,224, **225-226**,247,**482**,**484**,488,591
- cochlear (see spiral)
- dorsal root 422-424,**428**,598,599, 600,601
- geniculate of facial nerve 257,**372**, 373,476,477
- glossopharyngeal, inferior (petrosal) **106**,112,**469**,477,480,596,597
- - superior 106,**469**,480
- impar **440**
- intermediate (sympathetic) **440**
- otic 106,112,167-168,178,**179-180**, 469,596
- parasympathetic 438,**442**,467 (see also named ganglia)
- paravertebral 438,**440**,441
- prevertebral 438,**440**,441
- pterygopalatine 112,**257-258**,476
- Scarpa's **379**,473 (see also vestibular)
- sensory, of cranial nerves **467** (see also named ganglia)
- spiral (cochlear) **379**,471,473,**592**, 593

- stellate **113**,440
- submandibular 112,179,295,302, **304**,476
- sympathetic 438,**440**,441 (see also cervical sympathetic)
- trigeminal 251,**392**,222,255,392,396, **400**,**479**,480,598,599,600,601(608)
- vagal **108**,112,469,480
- vertebral 113
- vestibular (of Scarpa) **473**,594,595
Gating/enabling 451,504,505,509,512, 525,534
- of pain signals **451**,512
Gemules 404
Genu, of corpus callosum **537**,581
- external 372
- of facial nerve, internal **476**
- of internal capsule 549,564,**569**,582, 602,603
Giddiness (619,622)
Gingivae
- innervation **298-300**
- vasculature 305,**306**
Glabella **14**,25,51,147,150
Gland
- adrenal, innervation of **441**
- arytenoid **338**
- buccal **290**
- innervation of 431,432,433,**434-435**, 437,438,**439**,440,441,442,443,469, 470,471,476,477,478
- labial **290**
- lacrimal 210,211,**213-214**
- - enlarged (236-237)
- - innervation 214,**257-258**,439,442, 476,477,505
- - loss of (270)
- laryngeal **338**
- lingual **290-291**
- meibomian **212**
- olfactory (of Bowman) **588**
- palatal **290**
- - hyperplasia (348)
- palatoglossal **290**
- parathyroid 119,**120**
- - tetany (141)
- parotid 83,86,87,100,103,106,147, 156,158,**164-168**,178,179,180,289, 308,364
- - accessory **164**,167
- - innervation **167-168**
- - tumour (192,349)
- pineal (see Body, pineal)
- pituitary 251,392,**398-400**,485,524, 525,526,**527**,528,580,584,585
- - tumour (75,270,610,627,628)
- of saccule of larynx **338**
- salivary, neural control 434,439,442, **469**,**476**,477
- sebaceous (348)
- sublingual 179,275,**290**,302,304
- - innervation **304**
- submandibular 83,86,101,111,116, 164,179,**289-290**
- - innervation **304**
- - tumour (354-355)
- sweat 434,439,**441**
- tarsal 147,**212**
- thymus **120**
- thyroid 82,83,101,104,113,118, **119-120**,332
- - goitre (140,141)
- - lingual (349)
- - myxoedema (193)
- - surgery (140-141,141)
- - thyroglossal cyst (140)
- - tumour (358)
Glaucoma (240)
Glioblastoma (631)
Gliosis **416**
Globus pallidus 543,566,567,568,**569**, **570**,**571** (see also System, basal nuclear)
Glomerulus **410**,521,556,588,589
Glomus 572,**573**,579

Glottis **334**,340,341
- obstruction (359)
Gonadotropin 527,585(628)
Grafting of brain tissue **509**
Granulations, arachnoid 21,26,27, 394,**573**
Grey matter **417**
- peri-aqueductal (central) 451,**505**, 506,508,526
- periventricular 505,506,526
- of spinal cord 446-450,**450-454**,459
Groove, alar **148**,242
- for auditory tube **19**,31
- carotid **22**,30
- for confluence of sinuses **23**
- ethmoidal **26**,37
- for greater petrosal nerve **23**,34
- for inferior petrosal sinus **28**,34,394
- infra-orbital 41,**46**,198,256
- labiomarginal 148,273
- labiomental **148**,273
- lacrimal **41**,47,214,219
- for lesser petrosal nerve **23**,34,106
- malar **209**
- for middle meningeal vessels **21**,**23**, 26,27,32,394
- mylohyoid **44**,179
- nasojugal **209**
- nasolabial **148**,155,273
- naso-optic (234)
- for nasopalatine nerve and vessels **39**
- neural **422**
- occipital **18**,34
- palatine **38**,41
- palpebral **209**
- for petrosquamous sinus **32**
- prechiasmatic **22**,29
- of pterygoid hamulus **31**
- for sigmoid sinus **23-24**,28,34,395
- for superficial temporal arteries **32**
- for superior petrosal sinus **23**,33,394
- for superior sagittal sinus **21**,**24**,26, 27,28
- for transverse sinus **24**,27,28
- for vertebral artery **62**
Growth (see Embryology or Development)
- neural control of bodily growth 524, **527**
Growth hormone-releasing factor **527**
Gyrus **419**,427
- ambiens 540,**555**,556
- angular **540**
- cingulate 530,531,533,**539**,**554**,559, 560,562,567
- dentate 537,**558**,**559**,572
- - layers (strata) **558**
- fasciolaris 556,**558**
- frontal, inferior **539**,549,553
- - medial **539**,547
- - middle **539**,547
- - superior **539**,547
- fusiform (see occipitotemporal)
- hippocampal (see parahippocampal)
- intralimbic **555**
- lingual **540**,550
- long of insula **540**
- occipital **540**
- occipitotemporal
- - lateral **540**
- - medial **540**
- olfactory 555,**556**
- orbital **540**
- paracentral (see Lobule, paracentral)
- parahippocampal 537,540,553,555, 556,558,**559**,560,562,588,589
- - entorhinal area 555,556,557,558, 559,562,588
- - herniation **484**,537(618,624,625, 629,630)
- - uncus **540**,555,**556**,557,558,567, 572
- paraterminal **540**,558,567,569
- parietal **540** (see also Lobule, parietal)

- parieto-occipitalis **540**
- parolfactory **540**
- polar **540**
- postcentral **538**,**540**,**549**,550,598, 599,600,601
- precentral **538**,**539**,**547**,602,603
- - lesion 587
- rectus **540**
- semilunaris (peri-amygdaloid area) **555**,**556**,568
- short of insula **540**
- subcallosal (area) **540**,567
- supracallosal **558**
- supramarginal **540**
- temporal, inferior **540**,550,562
- - middle **540**,550,553,562
- - superior **540**,550,552,553
- - transverse (of Heschl) **540**,552
- uncinate **555**

Habenula **535**,559,567,569,570
Haematoma, auricle (384)
- extradural (74,235,624-625,630)
- in infratemporal fossa (194-195)
- pterygoid venous plexus 352
- in scalp (196-197)
- in sternocleidomastoid producing torticollis (141-142)
Haemorrhage (see Haematoma)
Halitosis (144-145,356)
Hamulus, lacrimal **40**
- pterygoid **20**,31,156,173,295,313
Hanging (79)
Headaches (610,611,612,616,617,618, 619,620,621,622,627,628,629,630, 631)
Hearing (see System, Auditory)
Heart, neural control of 431,433,434, 438,**439**,440,442,**469**,470,471,508, 512,524-525,606
Heat conservation and loss **440**,525, **528**(627-628)
Helicotrema **376**
Helix 154,178,**362**
Hemianopia, bilateral (75)
- bitemporal **485**(628)
- homonymous (74) 485,533,552,564, 582(611,612)
Hemiballismus **528**,532-533,571(627)
Hemineglect **554**(630)
Hemiparesis/hemiplegia (74,76) 459, 500,547,564,**582**,587(612,613,614, 625,626,627)
Hemispheres
- cerebellar 419,**514-516**,518
- cerebral 419-420,**535-572** (see also Cortex, cerebral and named components)
- - arterial supply 577,579-582
- - connections 541-543,562-566, 566-571 (see also Cortex, cerebral and Nuclei, basal)
- - cortex **540-562** (see also Cortex, cerebral)
- - development **426-427**
- - dominance 549,552-553,**554**,563
- - functions, general **535**
- - lobes **539-540**
- - nuclei **566-571**
- - relationships **535-539**
- - subdivisions **426-427**
- - topography
- - - external **535-540**
- - - internal **562-572**
- - venous drainage **582-583**
Herniation
- cerebellar tonsillar **516**,574(618)
- tentorial 461,**484**,537(618,624,625, 629,630)
Herpes simplex, encephalitis (629)
- zoster, infection (192,395)
Hiatus for greater petrosal nerve **23**,
- for lesser petrosal nerve **23**,106
- maxillary **41**,47
- semilunaris 37,**245**

H – L

Hilus of dentate gyrus 558
Hindbrain 418-419,422,424-426,459-488,491-497,500-512,513-523 (see also Brainstem and Cerebellum)
Hippocampus 537,556,557,**558**,**559**,563,568,572 (see also Region, hippocampal)
- layers (strata) 558
Hoarseness (358-359)470,**471**(622,623,624)
Homeostasis, neural control of 432,433,**437-443**,469-471,504,508,512,523,524-525,526,527,528,569,585 (627-628) (see also System, autonomic nervous)
Homocystinuria (239)
Homunculus
- motor **547**(621,622)
- sensory 549
Hormones (see also named hormones)
- in cerebrospinal fluid 574,**585**
- as neurotransmitters 410,**411**
- releasing (hypothalamic) 527,585
Horn
- Ammon's 558 (see also Hippocampus)
- of hyoid bone, greater 68,83,87,90,101,103,293,314,337
- - lesser 68,293,314
- of lateral ventricle (see Ventricle, lateral)
- of spinal cord
- - dorsal (posterior) 437,446-450,**451-453**,600
- - lateral 450,**453**
- - ventral (anterior) 446-450,**453-454**,457,458,602,603
- of thyroid cartilage, inferior 336
- - superior **336**,337
Humour, vitreous 207
Hydrocephalus (77-78)**574**(615,617)
5-Hydroxytryptamine (as neurotransmitter) 411,434,441,442,451,508,509,**512**,518,527,534,543,556,568,570,606
Hyoid bone 68,83,135,293,295,314,317
- muscles attached to 68
- ossification 68
Hyperacusis 476
Hyperaesthesia 533(615,616,626,627)
Hyperpathia (626)
Hyperphagia 528
Hyper-reflexia (see Reflex)
Hypersensitivity, denervation 413,438(615)
Hypersomnolence 509,528
Hyperthermia (hyperpyrexia) 528
Hyphaema (240)
Hypogonadism 528
Hypokinesia 513(625)
Hypophysis, cerebri (see Gland, pituitary)
- pharyngeal 311
Hypopituitarism (627,628)
Hyporeflexia (see Reflex)
Hypothalamus 524-528
- and autonomic nervous system 437,441,453,470,471,477,478,505,507-508,525,526,**528**,606
- areas **526-527** (see also named areas and nuclei)
- arterial supply 577,579,580,581
- and behaviour 524-525,528
- - emotional 524-525,528
- - reproductive 525,526,527,528
- boundaries 523-524
- and circumventricular organs 585
- connections 437,441,453,470,471,477,478,505,506,507-508,**525-527**,531,533,535,543,556,559,564,567,568,569,588,596,606
- development 426
- and endocrine system 525,526,**527**,585

- and feeding 524,525,**528**
- functions 523,524,525,526,527,**528**
- homeostasis 523,**524-525**,528
- lesions 509,**528**(627-628)
- and limbic system 525,526,527,**559**,560,567,568,569
- median eminence 524,527,**585**
- neurotransmitters 411,512,**526-527**
- nuclei (see also named) **526-527**
- and pituitary activity 525,**527**,**528**,584,585(627-628)
- relationships **523-524**
- releasing hormones **527**,585
- and reticular formation 504,505,506,508,509,512,**525**,526,527
- and sleep-wake cycle 509,525,526,**528**,534
- and suckling 525,526
- topography, external **523-524**
- - internal **526-527**
- and temperature regulation 525,**528**
- tracts (see also named) **525**,526,527
- venous drainage 583
- and water balance 524,525,526,527,**528**,585
Hypoxia (see Anoxia, Ischaemia)

Impression, trigeminal 23,34(608)
Inattention 554(630)
Incisure
- of Schmidt-Lantermann 406
- tentorial (see Notch and Herniation) 391,419,461,484,514,537,575(618,624,625,629,630)
Incontinence (616)
Inco-ordination of movements (see Ataxia)
Incus 51,365,367,369,**371**
- conductive deafness (76)
Index, nasal 150
Indoleamines 411,509 (see also 5-Hydroxytryptamine)
Indusium griseum 556,**558**
Infarction 575,578,580,581,582(611,612,613,614,622,623,626,627)
Inferior nasal concha (see Concha)
Infundibulum, of ethmoid 37,**245**,250
- of pituitary gland 398,419,524,527,575,585
Inhibition
- feedback **435**,436,523,556,571,586
- feed-forward **436**,523
- postsynaptic potential (IPSP) 410
- presynaptic **411**,451
Inion 27,537
Inlet, of larynx 312,332,339,341
- thoracic 128
- - goitre (141)
- - subclavian vein cannulation (142)
Innervation
- of ear 363,364,372-373,378-379 **471-473**
- of face **157-161**
- of larynx 343
- of meninges 397
- of neck 84-85,105-116
- of nose 243,245-247
- of orbit 221-226
- of oro-dental structures **297-304**
- of paranasal sinuses 250,**251**,253
- of pharynx 318
- of salivary glands **167-168**,291,**303-304**
- of scalp 189
- of temporomandibular joint 173
Insomnia 508,509,528
Insula 538,**540**,555,562,581
- limen of **540**,555,556,562
- misshapen (240)
Isthmus, oropharyngeal 311
- pharyngeal 310
- of thyroid gland 119(140)
- - tracheostomy (140)
Intellect 532,**553**,554,563

Interneurones 418
- in enteric nervous system 442
- reticular **466-467**
- spinal **450**,453,454
- in sympathetic ganglia 440
Internode 406,408
Ischaemia 413,443,577,582(610)
Island of Reil (see Insula)
Isthmus
- of cerebral cortex 539
- of brainstem 463

Jaws (see also Mandible and Maxilla)
- evolution 10
- innervation **297-300**
- vasculature **304-305**
Joint
- atlanto-axial 65-66
- atlanto-occipital 18,23,**66-67**,126
- of ear ossicles 371
- of laryngeal cartilages **336-337**
- - rheumatoid arthritis of (359-360)
- mandibular symphysis 43,44,**52**,54
- spheno-occipital synchondrosis 19,30,**52-53**,54,57
- - early closure (77)
- sphenopetrosal synchondrosis 19
- sternoclavicular 83,129,132
- suture (see Suture)
- temporomandibular 160,**172-174**,178,180,364
- - clicking joint (196)
- vertebral 64-67
- - between arches **64-65**
- - between bodies 64
Jugum 21,**30**,48
Junction
- gap (communicating) 412,414,416,507
- mucocutaneous of eyelid 210
- neuro-effector **433-434**,440
- neuromuscular (myoneural) 412,**433-434**
- sclerocorneal (see Limbus)
- synaptic (see Synapse)
- tight, of capillaries 416,473,**584**
- visceral neuro-effector **434-435**

Kinaesthesia 432
Koniocortex 546

Labyrinth, ethmoidal 37,47,51,53,250
- membranous 375,**377-378**
- osseous **375-377**
Lacrimal apparatus 199,**213-215**
- bone 11,**39-40**,46,47,252
- - ossification 40
Lacrimation, neural control 214,439,442,**476**,477,481,505
Lacuna, venous **394**,574
Lacus lacrimalis **147**,148,210
Lambda **13**,26,27,52
Lamellae
- of archicortex 558
- of neocortex 541
Lamina (see also Layers)
- cribrosa **202**,224,227
- of cricoid cartilage **336**,339
- dental 169,**320**
- medullary of thalamus 529
- of neocortex **541-542**
- of neural tube **424-426**,463,464
- of spinal cord (of Rexed) **450-454**
- spiral 375,376,377,379
- terminalis 422,427,**524**,537,585
- of thyroid cartilage **336**,341
- of vertebra 61,63
- vestibular 169
Language disorders (see Speech)
Laryngitis (359)
Laryngocele (360)
Laryngomalacia (357-358)
Laryngopharynx 118,**312**
Laryngotomy (140)

Larynx 82,83,95,**332-346**
- cartilages **336-337**
- foreign body in (359)
- innervation 108,**343**
- inlet 312,332,339,341
- internal anatomy of **332-334**
- membranes **337-338**
- muscles, extrinsic **338-339**
- - intrinsic **339-341**
- vasculature 101,104,131,**342**
Layers
- of cortex
- - cerebellar **517**,521
- - cerebral
- - - archicortex 541,**558**
- - - neocortex **541-542**
- - - paleocortex 541,**555**
- - of neural tube **424-426**,463,464
- marginal (of Waldeyer) of spinal cord 451
Learning 414,436,521,559,**560-561**,562,563,567
Le Fort fractures (of face) (72-73)
Lemniscus
- lateral 473,**474**,592,593
- medial **489**,503,504,528,530,532,533,**598**,599(613)
- spinal **489**,528,530,532,533,534,**600**,601(613)
- trigeminal 481,528,530,532,533,534,**598**,599,**600**,601
Lens 201,**206-207**
- of eye, neural control 439,442,482,484
- placode 228,229
- subluxation (239-240)
Leptomeninges 390
Leptoprosopic face 149-150
Lesions 412-414,443,458-459,575,582,587(see also named structures)
- lower motoneurone 459(614,615,616,619,620,622,623,624)
- upper motoneurone (supranuclear) 459,500,501,547,564,582,587,(612-614,615,616,617,618,620-622,624,625,626,627,628,629,630)
Lethargy 513,528
Leukotomy, prefrontal 554
Levator glandulae thyroidae 119
Ligament, ligamentum
- alar of dens 63,**66**
- annular 371
- apical of dens 18,23,63,**66**
- check, of eye 209,212,215,217,218,220
- cricothyroid, anterior 116(140)**338**
- - posterior 337
- cricotracheal 338
- cruciform 66
- - with subluxation of dens (79)
- denticulatum 446
- of ear ossicles 371
- flavum 64
- hyo-epiglottic 68,**337**
- interspinous 65
- lateral of temporomandibular joint 173
- lateral of thyroid gland 119
- longitudinal of vertebrae 64,65,66,95
- membrana tectoria 18,23,**66**
- nuchae 65
- palpebral, lateral 36,211,**212**
- - medial 41,152,211,**212**,224,227,228
- pterygomandibular raphe 20,43,156,**173**,273,313
- sphenomandibular 20,31,43,165,**173**,179,180,181
- spiral 377
- stylohyoid 68,173,290,314
- - calcified (355-356)
- stylomandibular 43,95,167,**173**,289,294
- supraspinous 65
- suspensory, of eyeball (Lockwood)**209**

639

L – M

- - of lens 204,**207**
- - - damage to (234)
- temporomandibular **173**
- thyro-epiglottic **337**
- thyrohyoid **337**
- transverse of atlas 62,63,**65-66**
- vocal **332**,336,337,341
Limbus 148,**201**,202,204,207,217,218
- ruptured (240)
Limen, of insula **540**,555,556,562
- nasi **244**
Line, axial of dermatomes **429**
- mylohyoid **43**,173,295,313
- nuchal, inferior **13**,18,27,124,125
- - superior **13**,18,27,83,94,124,125, 188
- - supreme **13**,27,188
- oblique, external of mandible **43**
- - internal of mandible (see Line, mylohyoid)
- - of thyroid cartilage 90,315,**336**
- temporal, inferior **15**,25,26,171
- - superior **15**,25,26,172
Lingula, of cerebellum 463,514,**516**
- of mandible **44**,173
- of sphenoid bone **30**
Lip (of mouth) **148**,150,**272-273**
- cleft 169,**170**(628)
- competence of (anterior oral seal) **148-149**
- innervation **300**
- muscles **154-157**,291
- vasculature **305**,**306**
- vermilion **148**,273
Lip, rhombic **426**
Lobe
- of brain (see also Cortex)
- - central (insula or island of Reil) 538,**540**,555,562,581
- - frontal 420,538,**539-540**,**546-549**, 553,**554**,556,559,561,562,564,571, 581(629)
- - limbic (of Broca) **559-560**
- - occipital 420,**539**,**540**,**550-552**,554, 561,562,564,565,566,572,579
- - olfactory 541,**555**
- - parietal 420,538,**539**,**540**,**549-550**, 553,**554**,561,564,566,579,581
- - pyriform (piriform) 540,**555-556**
- - temporal 420,535,537,538,**539**,**540**, 550,**552-553**,555-559,560,561,562, 563,564,565,566,571,572,579,580, 581(629)
- cerebellar **514-516**
- cerebral 420,**539-540**
- flocculonodular **514**,516,518,521, 594,595
- glenoid of parotid gland **164**
- lateral of thyroid gland **119**
- of pituitary gland **399**
- pyramidal of thyroid gland **119**,135
- of thymus **120**
Lobotomy
- prefrontal **554**
- temporal **555**
Lobule, of ear **362**
- cerebellar 514,515,**516**
- paracentral **539**,**540**,**547**,**549**,598, 600
- parietal, inferior **540**,553,**554**
- - superior **540**,**554**
Locus ceruleus (caeruleus, coeruleus) 506,**512**,520,543,590,606
Loop
- feedback (see Feedback)
- temporal (of Meyer) 555,**565**,591
Ludwig's angina (351)
Lumbar puncture **446**,516,574(609, 618)
Lung, apex of 128,129,130,**133-134**
Lymph nodes, buccal 116
- cervical, anterior **116**
- - deep 86,**117-118**,120,249,307,318, 342,373
- - - tumour (144-145,350)

- - - tubercular lymphadenitis (142-143)
- - superficial 86,**116**,165,168
- infrahyoid **116**,318
- jugulodigastric **117**,306
- - tumour 144
- jugulo-omohyoid **117**,306-307
- occipital **165**
- paratracheal **116**,118,120,318
- parotid (pre-auricular) **165**,168
- prelaryngeal **116-117**,120,342
- pretracheal **116**,118,120,342
- retro-auricular **165**,364
- retropharyngeal **116-117**,118,252,306, 318,373
- submandibular 86,**116**,165,243,249, 250,253,289-290,295,306
- submental 87,**116**,165,306
Lymphatic, drainage from ear 364, 373
- face **164**,165
- larynx 342
- mouth 306-307
- neck **116-118**
- nose and paranasal air sinuses **243**, **249**,**250**,**252**,**253**
- pharynx 318
Lyra (commissure of fornix) 538,559, 562,563

Macroglia 415-416
Macula, lutea (of retina) 205,206, 590
- of saccule **378**,379
- of utricle **378**,379,473
Macular sparing (612)
Malformations
- arteriovenous (622)
- developmental (see Abnormalities, developmental)
Malleus 11,51,364,365,367,**370**,371
- conductive deafness (76)
Mandible 11,14,15,**43-45**,51,55-59, 83,94,101,289,290,293,295, 298-299
- with acromegaly (75)
- alveolar process **43**,51,59
- body 14,**43**,51,59,85
- condylar process 15,43,**44**,45,54, 58,59,100,148,165,**172**,180
- coronoid process 15,43,**44**,45,59
- fetal **51**,**52**
- fractures (71)
- growth and remodelling **55-59**
- mandibular canal **44**,299,305
- movements of **174**
- muscles attached to **44**
- ossification **44-45**
- osteomyelitis (196)
- ramus 15,**43-44**,51,57,59,148, 164,165,174,308
Manubrium (see Sternum)
Margin, of foramen magnum 18, 65,66,67
- frontal of greater wing of sphenoid bone **30**
- - of lesser wing of sphenoid bone **30**
- - of parietal bone **26**
- infra-orbital of maxilla **40**
- maxillary of zygomatic bone **35**
- occipital of parietal bone **26-27**
- orbital of zygomatic bone **35**
- parietal of greater wing of sphenoid bone **30**
- posterior of lesser wing of sphenoid bone **30**
- sagittal of parietal bone **26**
- sphenoidal of zygomatic bone **36**
- squamous of greater wing of sphenoid bone **30**
- - of parietal bone **27**
- superior of petrous part of temporal bone **33**
- supra-orbital of frontal bone **25**

- temporal of zygomatic bone **35**
- of tentorium cerebelli, attached **391**
- - free **391**
- zygomatic of greater wing of sphenoid bone **31**
Maps
- Brodmann's **544**
- representations **586**
- - motor 532,**547**,602(621,622)
- - retinotopic (visual) 533,**550**,590
- - somatosensory 446,455,456,533, **549**,598
- - tonotopic (auditory) 533,**552**,592
Mass, lateral of atlas **62**,66,122,131
Massa intermedia **524**
Mastication, muscles of 44, **171-172**,**176**
Mastoidectomy, facial nerve (387)
Mastoiditis (387)
Meatus, acoustic, external **16**,20, 33,51,94,108,160,164,172,178, 180,**364**
- - - damage (385)
- - - tumour (385)
- - internal 24,34,379,579
- - - constriction with Paget's disease (76)
- - - damage with skull fracture (74)
- - - tumour (388)
- - nasal, inferior 38,41,**47**,215,245
- - middle 38,41,**47**,**245**,250
- - - infection (268)
- - superior 37,38,**47**,118,**245**,250
- - supreme **245**
Mechanisms, control, neural **435-437**
Mediastinitis (356)
Mediastinum 95,98,99
Medulla oblongata (myelencephalon) 418,419,459,**460-461**,**462**,463, 464 (see also Brainstem)
- arterial supply **578**,579
- boundaries **460**,**462**
- cranial nerve rootlets at 105,107, 109,111
- development **424-425**
- topography (gross) **460-461**,**462**, **491-495**
- venous drainage **583**
Medulla spinalis (see Cord, spinal)
Membrana tectoria 18,23,**66**
- statoconiorum (otolithic) **473**,594
Melatonin **534**
Membrane (see also Meninges)
- atlanto-occipital, anterior 18,**66**
- - posterior 18,**66-67**,126,577
- basal of eye (Bruch's), deficiencies in (238)
- basilar **377**,590
- bucconasal **259**
- buccopharyngeal **168**
- cricovocal 116,332,**337-338**,341, 342
- ependymal-glial (internal limiting) **415**
- gliosal (external limiting) **415**
- nictitating **148**

- otolithic (statoconiorum) **473**,594
- perivascular **415**
- pia-glial **415**
- plasma **408**,410
- quadrangular 332,**337**
- suprapleural 62,123,**133-134**(142)
- synovial, of temporomadibular joint **173**
- thyrohyoid 68,116,**337**,342
- tympanic 51,108,160,178,180, **364-365**,367,370,592
- - damage following skull fracture (73)
- - inflammation 387
- - innervation 365
- - secondary 369,376,378
- vestibular **377**
Memory (see also Amnesia)
- and limbic system 559,560, **561-562**,567,569
- long term 535.553,555,559,560, **561-562**,563,567,569(629)
- priming 561
- short term **561**,563(629)
Meninges (see also Arachnoid mater, Dura mater and Pia mater)
- cerebral **390-393**,572
- - innervation **397**
- - vasculature **396-397**
- irritation by blood (618)
- spinal **445-446**
Meningioma (76)
- optic nerve (236)
Meningitis (190-191,270,387,609, 629)
Meningocele (608-609)
Mesencephalon (see Midbrain)
Mesiodens (351-352)
Metabolism, neural control 433, 437,439,**440**,441-442,504,**524**, 525,528,569
Metathalamus **529** (see also Body, geniculate)
Metencephalon **419** (see also Pons, Cerebellum)
- development **424**
Microcolumns **541**
Microglia 415-416
Micrographia (625)
Micturition, voluntary control 547 (616)
Midbrain (mesencephalon) 391, 418,419,422,459,460,**461**,**462**, **463**,464 (see also Brainstem)
- arterial supply **579**,**580**,**581**
- boundaries **461**,**463**
- cerebral peduncle **461**,**463**
- development 422,424,**426**
- tectum **461**
- tegmentum **461**
- topography (gross) **461**,**463**, **498-499**
- venous drainage **584**
Migration, neuronal **426**,**427**
Miosis (193)441(622,623)
Modiolus, of cochlea **375**,376,379
- of facial muscles **157**
Modulation (see Neuromodulation)
Modality, sensory **430**
Mood, control of **512**,513,**524**,528, 554,**559-560**,567,569
Monoamines 411,412,**509-513**,577 (see also named neurotransmitters)
Monoplegia **581**
Motivation **512**,513,**524**,**525**,528, 569
Motoneurones (see Neurones)
Motor unit **434**
Mouth **154**,**272-310**
- innervation **301-305**
- lymphatic drainage **306-307**
- musculature **154-157**
- teeth (see Teeth)
- salivary glands (see Glands)
- vasculature **304-307**

M–N

Mucosa, laryngeal 338
- of middle ear 106,372
- nasal 245
- olfactory 245-246,259
- oral 288-289
- pharyngeal 317
Muscles, named
- arrectores pilorum 434,439,440
- aryepiglottic 341
- auricular 153-154,158,188,362
- buccinator 14,16,20,42,44,148, 154,156-157,158,162,167,178, 301,308
- chondroglossus 68,293
- ciliary 202,204,215,225,226,482, 484
- compressor naris 153
- constrictor of pharynx, inferior 90,101,108,118,131,315-316, 317,342
- - - pharyngeal diverticulum (145, 356)
- - middle 68,86,90,106,107,108, 293,314,317
- - superior 21,29,32,44,106,174, 308,312,313,317
- corrugator supercilii 14,16,26,152
- crico-arytenoid, lateral 340
- - posterior 339-340
- cricopharyngeus 118,315,341
- cricothyroid 108,316,339,341
- depressor anguli oris 14,16,44, 155-156
- - labii inferioris 14,16,44,156
- - septi 14,16,42,153
- - supercilii 152
- diaphragm 454,507-508
- digastric 20,35,44,86,87-88,90, 96,100,101,111,116,158,179, 294,308
- dilator naris 153
- - pupillae 205,215,223,439,441 (622,623)
- - - Horner's syndrome (193,622, 623)
- erector spinae 124,125
- extra-ocular 215-221,475,478,479, 482-484,507(610,611,617,620, 621,624,625,629,630)
- of facial expression 44,85,151-157, 362,476-477
- - palsy (73,74,76,77,190)349(385, 387,620-621)
- - Chvostek's sign (141)
- of floor of mouth 294-295
- genioglossus 44,68,290,293,294, 468(621)
- geniohyoid 44,68,111,295,468
- hyoglossus 68,86,90,101,103,111, 293,295,302
- infrahyoid (strap) 82,86,88-90,95, 97,98,115,118,119(140)332
- interarytenoid 340-341
- interspinales 63,126
- intertransversarii 63,126
- of larynx 469-470,471(622,623,624)
- - extrinsic 338-339
- - intrinsic 339-341
- levator anguli oris 14,16,42,155, 162
- - glandulae thyroideae 119
- - labii superioris 14,16,36,42, 154-155,162
- - labii superioris alaeque nasi 14, 16,154-155,162
- - palpebrae superioris 199,212, 217,220,441,482,483,484(622, 623,624,625,629)
- - - inflammation of (269)
- - - Horner's syndrome (193)
- - scapulae 63,90,91,105,109,123
- - veli palatini 20,35,297,316,317
- levatores costarum 64
- longissimus capitis 13,20,35,64, 124,125,126

- - cervicis 64,124,125
- longus capitis 20,29,63,95,111, 122,130
- - cervicis 95
- - colli 64,121-122
- masseter 14,17,35,36,44,59,101, 148,149,158,161,162,164,165, 167,171,172,174,308
- of mastication 44,171-172,176
- mentalis 14,17,44,156
- of mouth 154-157,291
- multifidus 63,126
- musculus uvulae 20,38,297
- mylohyoid 44,68,86,87,96,116, 179,275,289,290,294,295,302, 308
- nasalis 14,17,42,153
- of neck 85-92,120-128
- nose, external 153,242-243
- oblique of eyeball, inferior 199, 219,482
- - - damage with skull fracture (72, 234)
- - superior 199,219,475,482,507 (610-611)
- obliquus capitis inferior 126,128
- - superior 13,21,29,63,126,128
- occipitofrontalis 13,17,20,29,35, 158,188
- oesophageal 118
- omohyoid 68,83,86,87,88-89,90, 99,111,338
- orbicularis oculi 14,17,26,42,152, 158,188,212,215
- - - lacrimal part 40,152,215
- - - paralysis (190,192)
- - oris 42,154,155,156,162
- orbitalis 220
- palatal 295-297
- palatoglossus 277,294,297,312
- palatopharyngeus 20,38,277,297, 316,388
- Passavant's 316
- of pharynx 44,312-317,469-470,471 (622,623,624)
- platysma 14,17,44,82,85,86,94, 105,156,158,165
- postvertebral 123-126
- prevertebral 121-123
- procerus 14,17,39,153
- pterygoid, lateral 20,32,44,59, 172,173,174,175,176,177,178, 179,180
- - medial 20,32,38,44,59,165,174, 175,176,178,308
- - - damage to (194-195)
- rectus capitis anterior 20,29,63, 122
- - - lateralis 20,29,63,95,122
- - - posterior major 20,29,63,126, 128
- - - - minor 20,29,63,128
- - of eyeball, inferior 199,209,212, 215,217-218,224,482
- - - lateral 219,224,475,478-479 (610-611,620-621)
- - - medial 199,209,218-219,482
- - - superior 199,215-217,224,226, 482
- rhomboideus minor 64
- risorius 156,162
- rotatores 126
- salpingopharyngeus 310,316
- scalenus anterior 63,95,104,114, 115,122-123,129,130,131
- - medius 63,90,105,115,123,133
- - minimus 64,123
- - posterior 63,90,123
- semispinalis capitis 13,20,90,91, 125-126
- - cervicis 63,125-126
- - thoracis 63
- serratus anterior 123
- - posterior superior 64
- smooth, in orbit 220-221

- spinalis capitis 124
- - cervicis 63,124
- sphincter pupillae 205,215,221, 225,482,484(625)
- splenius capitis 13,20,29,35,90, 91,124,126
- - cervicis 63,124
- stapedius 369,371,372,474,476, 592
- sternocleidomastoid 13,17,20,29, 35,82,83,85-86,91,94,105,109, 115,116,165,468(620-622)
- - torticollis (141)
- - triangles of neck 86,87,88,90, 91,126-128
- sternohyoid 68,87,89-90,338
- sternothyroid 90,338
- - limitations on expansion of thyroid goitre (141)
- strap muscles of neck (see Muscles, infrahyoid)
- styloglossus 20,35,100,290,294, 312,317
- stylohyoid 20,35,68,88,294
- stylopharyngeus 21,35,87,100, 106,107,316-317,318,338
- suprahyoid 44,174,339
- temporalis 14,17,26,27,32,35,36, 44,59,147,162,171,172,174,178, 181(609,624)
- temporoparietalis 188
- tensor tympani 34,178,368,369, 371-372,474,480,592
- - veli palatini 17,20,21,32,38,177, 178,179,181,295-297,311
- thyro-arytenoid 341
- thyro-epiglottic 341
- thyrohyoid 68,90,338
- thyropharyngeus 315
- of tongue, extrinsic 291,293-294, 567-568(620,621,623,624)
- - intrinsic 291-293
- trachealis 118
- trapezius 13,21,29,64,82,83,90, 91-92,94,109,123(143)468(620, 622)
- vocalis 341
- zygomaticus major 14,17,36,155, 162
- - minor 14,17,36,155,162
Muscles, innervation
- iliocostalis cervicis 64,124,125
- innervation (see also Pathways, descending)
- - atrophy (wasting or degeneration) 438,443,459,554(614,615,616)
- - cardiac 431,433,434,437,438, 439,440,442,469,470,471,508
- - fibres (extrafusal and intrafusal) 434
- - hypertonus 459(612-614,615-616, 625,628-629)
- - hypotonus 459(612-614,615-616, 618-619,622-623,626-627)
- - skeletal 431,432,433-434,450, 454,459,464-466,587,602-607 (see also Paralysis/paresis)
- - smooth (involuntary) 431,432, 433,434-435,437-443,464,465, 469,470,471,482,484
- - tonus 434,457,458,459,501,518, 521,522,547,564,571,587,602, 604,606(612,613,614,615,616, 619,622,625,626,628,629)
Myasthenia gravis 412
Mydriasis 484(629,630)
Myelin/myelination 405-409,416, 427,433,443
Myelencephalon (see Medulla oblongata)
Myotome 429
Myxoedema (193)

Naris, anterior 148,242
- posterior 242

Nasal bone 11,14,39,46,47,147,242
- ossification 39
Nasion 537
Nasopharynx 251,257,258, 310-311,366
- angiofibroma (355)
- arachnoid diverticulum within (608-609)
Nausea 438,475,476,518,585(617, 619,622,628)
Neck 82-145
- arteries 99-103
- bones of 61-68
- compartments and general arrangement 82,93,95,120-128
- fascia 93-95
- innervation 84-85,105-116
- lymphatics 116-118
- muscles 85-92,120-128
- root of 128-134
- surface markings 83
- tissue spaces 95-99
- triangle, anterior 86-90
- - posterior 86,87,90-92,104,109, 115
- - suboccipital 101,124,126-128, 129,130,131
- veins 103-105
- viscera 82,93,95,118-120
Neglect (inattention) 554(630)
Neocerebellum 518
Neocortex 541-543,544-555
Neopallium 541
Neostriatum 567 (see also Nucleus, caudate, Putamen)
Nerve, nervus (see also Innervation)
- abducent 22,47,112,199,219, 221-222,224,400,461,464-467, 478-479,579
- - anaesthesia (353)
- - functions 478(611)
- - lesions involving 478,479(620, 621)
- - involvement in cavernous sinus thrombosis (190-191,235)
- accessory 18,24,105,106,109
- - functions 468,469
- - lesions involving 468(620,621, 622)
- - cranial 108,109,297,313,316, 318,401,460,468,469,578
- - spinal 18,23,86,91,92,109,401, 445,454,460,468
- - - damage with aneurysm (139)
- - - damage with neck wound (143)
- - alveolar, inferior 160,176,179, 180,298-299,302
- - - anaesthesia (194,352)
- - - with fracture of mandible (71)
- - - infection (196)
- - - superior anterior 215,247,252, 253,256,300
- - - superior middle 253,256,300
- - - superior posterior 48,181,252, 253,255,300
- - - anaesthesia (352-355)
- - - tumour (270,353)
- - ansa cervicalis 86,89,90,105,111, 115,119
- - subclavia 112,113,130
- - articular, of auriculotemporal 178
- - auricular, great 84,85,91,161,165, 168,363
- - - pain associated with parotid fascia (190)
- - - damage (191-192)
- - - posterior 108,154,158,188
- - - of vagus 108,363,364,373
- - - reflex stimulation 386
- - auriculotemporal 160,164,166, 173,178,179,180,189,363(609)
- - - damage (191)
- - buccal of facial 153,154,155,156, 158
- - of mandibular 160,175-176,178,

641

N

   180,181,**300-301**
- - - anaesthesia (194)
- cardiac **109**,112,113
- caroticotympanic **112**,368
- carotid 99,**107**,108
- - internal of sympathetic system **112**
- cervical, facial 85,**158**,289
- - - tumour of submandibular gland 354
- - spinal 105,108,295,397
- - - compression with cervical spondylosis (80)
- - - connections with sympathetic trunk **113**
- - - cutaneous **84-85**,113,429
- - - damage with vertebral fracture (78)
- - - first with hypoglossal 90,111,**114-115**
- - - meningeal branches 23,397,**402**
- - - motor supply to vertebral muscles 122,123,124,125,429
- - - plexuses (see Plexus, cervical and Plexus, brachial)
- - - phrenic nerve 86,**114**,129,429
- - - for spinal accessory **109**
- - - sympathetic trunk 105,107,110,**111-113**,118,120,133(139)
- - - Horner's syndrome (193)
- - - transverse **85**,113
- cervicofacial trunk **158**
- chorda tympani 20,**179**,180,301,**304**,367,370,**373**,**596**,597
- - herpes infection (385)
- - ciliary **223**(237)
- - long 203,205,**223**
- - short 112,203,223,**225-226**,227,482
- cranial 464-488 (see also individual nerves)
- - attachment to brain 459-463,**467**
- - - components 464-466
- - ganglia **467**
- - reflexes 466-467
- descendens cervicalis **115**
- - hypoglossi 89,90,**111**,115
- digastric 88,**158**
- erigentes **442**,450
- - ethmoidal, anterior 21,22,46,160,**223**,**247**,250,397,**398**
- - - compression by tumour (270)
- - - posterior 46,**223**,250,251,397
- - - inflammation 269
- - facial 18,24,**157-158**,166,**179**,291,362,**372-373**,461,464-467,**476-478**,480,579,**596**,597,600
- - - compression with forceps delivery (77)
- - - with fracture of skull (73-74)
- - - functions **476**
- - - herpes zoster infection (385)
- - - lesions involving **477**,596 (620,621)
- - - middle ear infection (387)
- - - palsy (see Palsy, facial)
- - - within temporal bone **372-373**
- - - tumour (76,349)
- frontal 22,**159**,160,199,**222**
- - ganglionic, of auriculotemporal **178**
- - - of maxillary nerve 255,257
- - to geniohyoid **111**
- - glossopharyngeal 18,24,86,**105-107**,179,290,291,295,301,302,303,312,317,318,401,438,460,464-467,**469-471**,477,480,578,**596**,597,600
- - - damage with aneurysm (139)
- - - functions **469**
- - - gag reflex (610)
- - - lesions involving **471**(622,623, 624)
- - - neuralgia (609-610)
- - - referred pain from ear (355)

- - reflexes **470**,**471**,**477**
- hypoglossal 18,23,86,90,101,103,105,110,**111**,112,114-115,290,293,295,317,401,460,464,465,466,**467-468**,577
- - accompanying first cervical spinal nerve 90,111,**114-115**
- - damage with aneurysm (139)
- - damage with submandibular tumour (354)
- - function **467**
- - lesions involving **467-468**(620,621,623,624)
- - reflexes **468**
- incisive, of inferior alveolar **179**,299
- - infra-orbital 14,41,46,47,**160**,213,**224**,**243**,**247**,252,253,**256**
- - - anaesthesia (352)
- - with fractures of skull (73,234)
- - infratrochlear **160**,163,212,213,215,222,223,**224**,243
- - intermediofacial (see facial)
- - intermedius 257,**372**,461,**476**,480,**596** (see also Nerve, facial)
- - jugular 112
- - lacrimal 22,**160**,199,212,213,**222**,224
- - laryngeal, external **108**,341,343
- - - internal 314,343
- - - referred pain 356
- - - recurrent 83,**108**,109,116,118,119,130,131,317,339,340,341,342,343
- - - - damage during surgery (358-359)
- - - - damage with thyroidectomy (358)
- - - - tumour (145)
- - - superior 90,101,**108**,119,301,343
- - - - damage with thyroidectomy (358-359)
- - laryngopharyngeal **112**
- - to lateral pterygoid **178**
- - lingual, of glossopharyngeal **107**
- - of mandibular nerve 176,**179**,180,290,295,**301-302**,304,317,318,337,341,342
- - - anaesthesia (194,352)
- - - damage with submandibular tumour (354-355)
- - - section (351)
- - mandibular 19,22,159,**160**,175,**176-179**,297-302,400
- - - anaesthesia (193)
- - marginal mandibular of facial 154,155,156,**158**,289
- - - damage to (191,354-355)
- - masseteric 171,173,**178**
- - maxillary 22,41,159,**160**,168,181,(193)212,213,**224**,**253-257**,298,300,400,401
- - to medial pterygoid **178**,297
- - meningeal, of cervical nerves 23,**397**,**402**
- - - of hypoglossal **111**,397
- - - of mandibular (nervus spinosus) 19,22,**178**,372,**397**,401
- - - of maxillary 255,401
- - - of ophthalmic 397,**401**
- - - of vagus **108**,397
- - mental 14,**160**,179,**299**
- - - anaesthesia (352)
- - - damage during tooth extraction (353)
- - molar, of inferior alveolar **179**,**299**
- - mylohyoid 86,88,**179**,290
- - nasal, external **160**,223,243,247
- - - posterior superior 49,**247**,249,250,**256**,258
- - nasociliary 22,160,199,221,**222-223**,224,225,226
- - nasopalatine 17,49,**247**,**256**,258,**302**
- - - anaesthesia (352)

- occipital, greater 84,**126**,127,189
- - - compression with fracture of atlas (78)
- - lesser 84,**85**,91,189,363
- - third 189
- - oculomotor 22,47,112,199,217,218,220,**221**,222,224,225,**400**,460,464-467,**482-484**,579
- - - anaesthesia during dental injections (353)
- - - damage with skull fracture (74)
- - - involvement in extradural haematoma (235-236)
- - - functions **482-483**
- - - lesions involving **484**(610,611,617,624,625,629,630)
- - olfactory 21,245-247,**398**,464-467,**488**,535,537,**555-556**
- - - dural sheath **247**
- - - lesions involving **488**(629)
- - - trauma (267)
- - - tumour (76)
- - ophthalmic 22,47,112,**159-160**,168-169,212,**222-224**,243,400
- - - herpes zoster infection (192)
- - - rheumatoid arthritis, loss of corneal reflexes (79)
- - optic 22,49,202,203,223,**224-225**,**400**,464-467,**484-488**,523,580,590,591
- - - involvement during dental anaesthesia (353)
- - - involvement with sinusitis (269)
- - - compression (237-238,239)
- - - damage with skull fracture (74)
- - - dural sheath 217-218
- - - lesions involving **484**,**488**
- - - meningioma (236)
- - orbital 48,214,**224**,251,**256**
- palatine, greater (anterior) 17,**247**,249,**256**,**303**
- - - anaesthesia (352)
- - - lesser (posterior) 17,107,**256**,**303**,311,318,401
- petrosal, deep 19,23,**112**,214,257
- - greater (superficial) 19,23,112,214,257,**303**,**373**,**401**,**596**,597
- - - reflex lacrimation (317)
- - lesser (superficial) 22,23,**106**,179,303,372,373,401
- - section (386)
- - pharyngeal of glossopharyngeal **107**,108,318
- - from pterygopalatine ganglion 18,**257**,311,318
- - of vagus **108**,318
- - phrenic (and accessory phrenic) 86,**114**,129
- - - damage with vertebral fracture (78-79)
- - of pterygoid canal 19,23,**112**,**257**
- - damage during hypophysectomy (270)
- - scapular, dorsal **116**,123
- - spinal **427-443** (see also Nerve, cervical)
- - - applied anatomy **443**
- - - components **428-429**,430,431
- - - dermatomes **429**
- - - ganglia **428**,**434**,**438**,440,441,442
- - - plexuses 429,**430**
- - - rami **430**,441
- - - reflexes **435**
- - - roots **428**
- - splanchnic 437,441,442,443,450
- - to stapedius 372,**373**
- - stylopharyngeal **107**,317
- - suboccipital **126**,127-128,131
- - supraclavicular 84,**85**
- - supra-orbital 14,**159-160**,189,212,**222**,250(609)
- - suprascapular **116**
- - supratrochlear 14,85,**160**,189,212,**222**,242
- - sympathetic, cervical 105,107,

   110,**111-113**,118,120,133(139)
- - temporal, deep (of mandibular nerve) **178**
- - of facial 152,153,154,**158**,188
- - - damage with parotid tumour (192)
- - superficial (of auriculotemporal nerve) **178**
- - temporofacial trunk **158**
- - to tensor veli palatini **178**,296
- - terminalis **398**,**488**
- - to thyrohyoid **111**
- - tonsillar **107**,318
- - trigeminal (see also mandibular, maxillary and ophthalmic) 22,112,158,291,**297-298**,**400-401**,461,464-467,**479-482**,589,599,600,601
- - - functions **479**
- - - lesions involving **480-481**(619,620,623,624)
- - - neuralgia (191,608)
- - - root, motor **461**
- - - sensory **461**
- - trochlear 22,47,112,199,219,**221**,222,**400**,463-467,**482**,579
- - - functions **482**(611)
- - - involvement during dental anaesthesia (352-353)
- - - lesions involving **482**
- - tympanic, of glossopharyngeal 18,**106**,112,365,367,372
- - vagus 18,24,82,86,95,105,106,**107-109**,111,118,120,129,401,438,442,443,460,464-467,**469-471**,480,578,**596**,600(609-610)
- - - damage with aneurysm (139)
- - - damage to recurrent laryngeal (358,358-359)
- - - functions **469**
- - - lesions involving **471**(622,623,624)
- - - - reflexes of **470-471**
- - vestibulocochlear 24,372,375,378-379,401,461,464-467,**471-476**,579,592-595
- - - acoustic neuroma (76)
- - - herpes zoster infection (385)
- - - functions **471**
- - - lesions involving **474**,**475**,**476**(619,620)
- - - neurofibroma (388)
- - - reflexes (610)
- - vomeronasal **488**
- - zygomatic, of facial 152,153,154,156,**158**,188
- - - damage with tumour (192)
- - zygomatic, of maxillary 46,47,222,**224**,255,257
- - - zygomaticofacial **160**,**224**,255
- - - zygomaticotemporal **160**,189,**224**,255,257
Nervous system (see System, nervous)
Neuralgia
- glossopharyngeal (609-610)
- trigeminal (191,608)
Neuraxon (see Axon)
Neurite 403
Neuroblasts 424
Neurocranium 10,51
Neuroglia 413,**414-417**
- astrocytes **415-416**
- central **415-416**
- development **424**
- ependyma **416**
- lemmal **416**,**417**
- microglia 415,**416**
- oligodendrocytes 405,406,415,**416**
- peripheral **416-417**
- pituicytes **416**
- radial (Bergmann) **416**
- in regeneration 413,443
- Schwann cell 405,**416**,433,443
- teloglia **416**,**417**,433

Neurohormone 411-412
Neurohypophysis 399,527,585
Neurolemma 428
Neuroma, traumatic 443
Neuromodulation 411-412,435
Neurones (see also Cells) 403-414
- adrenergic (see Noradrenaline)
- aging affecting 427
- amacrine of retina 484,590
- autonomic 434-435,437-443
- axons of 403,404-405,406-414
  (see also Fibres)
- axoplasmic transport 405
- bipolar 403,404,484,590
- cardiomotor 434,439,469,470
- cell body 403,404
- cholinergic (see Acetylcholine)
- degeneration of 405,412,414,443
- dendrites 403,404
- development 422-427
- dopaminergic (see Dopamine)
- effector 418,433-434
- forms of 403-404
- fusimotor 431,434
- ganglion (see Cells)
- Golgi (types I and II) 418,426,521
- granule 404,426,521.541,542,546,
  552,556,558
- horizontal of retina 484,590
- initial segment 404,408
- injury 405,412-414,443
- intercalated/interneurones/
  internuncial 418,440,442,450,
  453,454,466-467
- isodendritic 504
- monoaminergic 509-513 (see also
  Monoamines)
- motor (motoneurones) 418,431,
  433-434,435-436,454,459,
  464-467,587,602-607
- - alpha 431,434,454,602,603,604,
  605,606,607
- - beta 431,434
- - branchiomotor 425,464-466
- - of cranial nerves 464-467
- - gamma 431,434,454,602,603,
  604,605,606,607
- - lower 459 (see also Lesions)
- - somatomotor (skeletomotor)
  425,431,434,450,454,464-466
- - spinal 450,454
- - upper 459,587 (see also
  Pathways, descending)
- multipolar 403,404
- neurosecretory 526,527
- noradrenergic (see Noradrenaline)
- perikaryon of 404
- pilomotor 434
- plasma membrane of 408,410
- postganglionic 438,439,440,441,
  442,443
- - in enteric nervous system
  442-443
- preganglionic 438,439,440,441,
  442,443,450,453
- primary sensory 417,428,430,431,
  586
- pseudo-unipolar 404
- Purkinje 426,517,521
- pyramidal (see Cells, pyramidal)
- Renshaw 435-436,454
- secretomotor 434 (see also Glands,
  innervation of)
- sensory (afferent)(see also
  Pathways)
- - autonomic 431,432,433,437-438,
  464,465,469,470-471
- - primary 417,428,430,431,586
- - receptive field of 430
- - and reflex activity 431,434,435,
  450,466-467
- - somatosensory 598-601
- - special sensory 588-597
- serotoninergic (see
  5-Hydroxytryptamine)
- SIF (small intensely fluorescent)
  440
- skeletomotor 431,434
- somatomotor 431,433-434,
  464-466 (see also motor)
- sudomotor 434 (see also Glands,
  sweat)
- synapses of (see Synapses)
- transfer of information,
  intercellular 409-412
- - intracellular 406-409
- trigger zone 408
- unipolar 403,404
- vasomotor 434
- visceromotor 434
Neuropeptides 411-412,485 (see
  also named)
Neuropil 417
Neuropore 422
Neurosis, obsessional 562
Neurotransmitters 408,410-412,
  574 (see also named)
Nexus (see Synapse, electrical)
NMDA 411,560
Nociception (see Pain)
Node, lymph (see Lymph node)
- of Ranvier 406,407,408,409,443
Noradrenaline (norepinephrine)
  411,434,438,439,440,441,442,
  470,471,508,509,510,511,512,
  513,518,527,534,543,556,568,
  606
Nose, external 148,150,242-243
  (see also Cavity, nasal)
- - haemorrhage (266)
- - infection (266-267)
- - innervation 243
- - muscles 153
- - vasculature 243
Nostril 148,242
Notch, antitragohelicine 362
- cerebellar (posterior) 514,515
- ethmoidal of frontal bone 21,26
- frontal 14,25,46,147,198
- intertragic 362
- jugular, of occipital bone 28,34
- - of sternum 83
- mandibular 15,43,178(608)
- mastoid 18,34
- pre-occipital 539
- pterygoid 20,31
- sphenopalatine 37
- supra-orbital 14,25,46,147,198
- tentorial 419,537,575
- thyroid 336,337
- vertebral 61
Nucleus, nuclei (named)
- abducent 462,464,465,475,
  478-479,501,507,590,594,595,
  604
- - lesion involving 478,479
  (620-622)
- accessory, spinal 454,465,466,
  468,475,500,501
- - lesion involving 468(621,622)
- accumbens septi 567,569,570
- ambiguus 318,464,465,466,
  469-470,501,507,508
- - lesion involving 470,471(621,
  622,624)
- amygdaloid (see Amygdala)
- anterior
- - of hypothalamus 526,528
- - of thalamus 527,529,530,532,
  535,559,560,564,567
- arcuate
- - of medulla 504
- - of hypothalamus 526,527
- basal 420,427,567-571 (see also
  named components)
- - arterial supply 577,579,580,581
- - components 567-569
- - connections 502,504,506,507,
  512-513,528-529,530,532,534,
  543,547,562,563,564,566,
  567-569,570-571
- - functions 502,513,561,569,571,
  587
- - system 570-571 (see also System,
  basal nuclear)
- - - lesions 502,513,532-533,561,
  571,587(613,614,621,625,
  626-627)
- - topography 567-569
- - venous drainage 583
- basal forebrain
  (magnocellular, of Meynert) 513,
  535,543,562,566-567
- bed, of stria terminalis 567,569,
  570
- caudate 529,531,534,543,564,
  567,568,569,570-571,583 (see
  also System, basal nuclear)
- - head 569,571
- - tail 569,572
- centralis lateralis 531,534
- centrum medianum 529,531,534,
  570
- of cerebellum 513,521-523 (see
  also named nuclei)
- cochlear 462,465,466,473,474,
  592,593
- - accessory (lateral) 456,489,503
- cuneiform 506
- commissural of posterior
  commissure 483
- commissural, of vagus 470
- cranial nerve 463,464-488,491-499
  (see also named nuclei)
- cuneate 455,456,462,488-489,598,
  599
- - accessory (lateral) 456,489,503
- cuneiform 506
- Darkschewitsch 483
- Deiters (see Nuclei, vestibular)
- dentate 513,517,518,522-523,532
- - lesion 518(622)
- of diagonal band (of Broca) 527,
  566,567
- dorsal column 456,488-489,530,
  543,598,599(see also gracile and
  cuneate)
- dorsalis (thoracic, Clarke's
  column) 453,456
- dorsomedial of hypothalamus 526
- emboliform 513,517,522
- facial, motor of 464,465,466,
  476-477,501
- - lesion involving 477(620,621,
  624,625)
- fastigial 513,516,517,521,606
- geniculate (see Body)
- globose 513,517,522
- gracile 455,456,462,488-489,598,
  599
- gustatory 477-478,596,597 (see
  also Nucleus, of solitary tract)
- habenular 535 (see also Habenula)
- hypoglossal 462,464,465,467-468,
  501
- - lesion involving 468(620,621,
  623,624)
- of hypothalamus 526-527 (see also
  named nuclei)
- of inferior colliculus 474 (see
  also Colliculus, inferior)
- infundibular (arcuate) 526,527
- intercalatus of Staderini 507
- intermediolateral of spinal cord
  437,438,440,442,450,453,512,
  526,606(615)
- intermediomedial of spinal cord
  453
- interpeduncular 506,535
- interstitial, of Cajal 475,483,
  501-502,507,535,594,595,604,
  605,617
- rostral of medial longitudinal
  fasciculus 475,483,507,594,595,
  604
- intracerebellar 513,516,517,518,
  521-523,530,532
- intralaminar 489,504,505,508,
  529,531,533,534,543,559,564,
  570,571,600
- of lateral lemniscus 474,592
- of lateral olfactory stria 568
- lateralis, dorsalis 529,530,532,559
- - posterior 529.530,532
- lentiform (lenticular) 562,564,
  567,568,569 (see also System,
  basal nuclear)
- locus ceruleus 506,512,520,543,
  606
- mamillary (see Body)
- medialis dorsalis 529,530,532,
  533,547,553,556,559,560,561,
  567,568,588(629)
- midline of thalamus 529,531,
  533-534,543,559,561
- oculomotor 464,465,478,483-484,
  501,507,590,594,595,604
- - lesions involving 484,501
  (616-617)
- - accessory of 483
- parasympathetic accessory
  nucleus (Edinger–Westphal) 225,
  438,442,464,465,483,484,487,
  488,590,606
- - lesions involving 484(610-611,
  616-617)
- of olfactory tract (anterior
  olfactory nucleus) 556,562,588,
  589
- - of lateral olfactory stria 568
- olivary, accessory 473,503
- - inferior 456,503,518,520,543
- - peri-olivary 473
- - superior 473-474,592,593
- parabducent 479,507 (see also
  Formation, reticular, paramedian
  pontine)
- parabigeminal 506
- parabrachial 478,506,507,596,597
- parafascicular 531,534
- paraventricular, of hypothalamus
  526,527,606
- - of thalamus 531
- pedunculopontine 506,507,570
- perihypoglossal 504,506,507,520,
  604
- periventricular, of hypothalamus
  526
- of Perlia 483
- phrenic 454,606
- pigmentosus pontis (see locus
  ceruleus)
- pontine 500,503-504,516,518,519,
  520
- pontobulbar 504
- posterior, of hypothalamus 526,
  528
- - of thalamus 489,529,530,532,
  533,549,552,600,601(626,627)
- precerebellar 502-504 (see also
  named nuclei)
- preoptic 526,528,566,569
- prepositus hypoglossi 479,483,
  507,604
- pretectal (236)484,485,487-488,
  (611,616-617)
- proprius 451-453,456
- pulvinar 529,530,532,533,547,
  550,553(626-627)
- raphe 451,452,505,506,508,509,
  512,520,535,543,570,606
- red 502,521,522,523,528,543,602,
  603
- relay 530,532,586
- reticular, of reticular formation
  505,506
- - - gigantocellularis 506,606,607
- - - lateral 504,506,507,520
- - - paramedian 504,506,507,520
- - - parvicellular 506
- - - pontis (caudalis and oralis) 506,
  508,606,607

643

N – P

- - - tegmental pontine 504,506, 507,520
- - - ventral 506
- - of subthalamus 528
- - of thalamus 528,529,531,533, 534,590
- of Roller 507
- ruber (see red)
- salivatory 179,442,464,465,**477**
- septal 525,**526**,527,535,543,559, 563,564,566,**567** (see also Region, septal)
- of solitary tract 438,464,465, **470-471**,**477-478**,480,507,508, 521,525,**526**,530,568,596,597, 606
- subcuneiform **506**
- subthalamic **528**,567,570-571(627) (see also System, basal nuclear)
- suprachiasmatic 485,525,**526**,527, 528,534
- supra-optic **526**,527,528,606
- supraspinal **491**
- tegmental, dorsal and deep (ventral) **506**,526
- of thalamus
- - association 530,532
- - nonspecific 488,504,505,508, 523,529,531,**533-534**,543
- - relay 530,532
- - specific 529,530,**532-533**
- - ventrobasal complex 533 (see ventralis posterior)
- thoracic (see dorsalis)
- of trapezoid body **473**
- trigeminal
- - mesencephalic 465,466,479,**482**, 520,598,599
- - motor 464,465,466,**480**,501,599, (621)
- - principal sensory 465,466,**481**, 598,599(624)
- - spinal 453,462,465,466,**480-481**, 520,598,599,600,601(618,619, 620,622,623,624)
- - - pars (subnucleus) caudalis 453, **480-481**,600
- - - pars (subnucleus) intermedius 480,**481**,520,600
- - - pars (subnucleus) oralis 480, **481**,520,598,600
- - trochlear 464,465,**482**,483,501, 507,590,594,595,604
- tuberal **526-527**
- vagus
- - dorsal motor 442,462,464,465, **470**,471,507,508,525,**526**,568, 606
- - ventralis, anterior 508,523,529, 530,**532**,**534**,543,547,570,571 (627)
- - lateralis 475,522,523,529,530, **532-533**,547,550,570,571,594 (627)
- - posterior 475,529,530,532,**533**, 549,594,595
- - - inferior 529,530,**533**,550,594
- - - lateralis 489,523,529,530,**533**, 549,550,595,598,599,600,601 (626,627)
- - - medialis 478,481,529,530,**533**, 549,550,596,597,598,599,600, 601
- ventromedial of hypothalamus 525,**526**,528
- vestibular 462,465,466,473, **474-475**,520,521,530,592,593, 594,604,605(622)
- X and Z **489**
Nucleus pulposus, of intervertebral disc **64**
Nystagmus 475,476,518(619,622, 624,628,629)

Obex 462
Occipital bone 11,12,13,18-19,
23-24,**27-29**,48,52-53
- basilar part 19,20,22,23,24,**28**,48, 122,251,460
- fetal 52-53,**54**
- fracture (74)
- lateral part 23,**28-29**,48
- muscles attached to 29
- ossification 29
- squamous part 12,18,23,**27-28**,48, 52,53
Occlusion, dental **286-288**
Odour, perception of (see System, olfactory)
Oedema, and neural function **412-413**(626,629)
Oesophagus 82,83,97,98,108, 113,**118-119**,133
- atresia (144)
- compression with goitre (141)
- infection in prevertebral tissue space (143)
- tumour (144-145)
Olfactory apparatus (see System, olfactory)
- evolution 11
- epithelium **245-246**
Oligodendroglia (oligodendrocytes) 405,406,415,**416**
Olive 105,107,109,111,460
Opening (see also Aperture, Hiatus, Incisure, or Foramen)
- of aqueducts of the cochlea and vestibule 24,34,375,**376**
- of auditory tube **310-311**
- of cochlear canaliculus 376
- of nasolacrimal duct 215
- of paranasal sinuses **250**,**251**,**252**, 253
Operculum **540**
- frontal 540
- frontoparietal 533,**540**,549
- hyoid **134**
- temporal **540**,552
Ophthalmoscopy **205-206**
Opioid neurotransmitters **411**,412, 442,451,505,527,568,600
- and pain **451**,600
Ora serrata 204,**205**,207
Orbit 22,**46-47**,147,**198-229**
- anaesthetic solution diffusing into (194)
- blowout fracture (234)
- contents **199**
- evolution **11**
- extra-ocular muscles **215-221**
- eye **200-209**
- eyelids (73)**147**,**152**,**199**,**209-213**, 217,220(235)
- lacrimal apparatus 199,**213-215**
- nerves within **221**
- osteology **46-47**,**198-199**
- smooth muscle **220-221**
- vasculature **226-228**
Organ, organum
- circumventricular 526,535,**585**
- muscle spindle 431,432,**434**,549
- neurotendinous (of Golgi) 431,**432**
- spiral, of Corti 377,**471-473**,590
- subcommissural 535,584,**585**
- subfornical 527,584,**585**
- tendon (of Golgi) 431,**432**
- vasculature, of lamina terminalis 527,584,**585**
- vomeronasal **245**
Organisation, serial, parallel, hierarchical **436-437**
Oropharynx **311-312**
Osmolarity (neural monitoring of blood and CSF) 525,**585**
Osteomyelitis, pyogenic (143)
Ossicles, auditory **370-371**
Ostium, maxillary sinus **252-253**
Otitis, externa (305)
- media (387,628)
Otoliths **473**,594

Overbite **288**
Overjet **287**,288
Oxytocin 411,525,**526**,527,606

Pachymeninx **390**
Pad of fat, buccal **147**
- orbital **199**
Paget's disease (76)
Pain (see also Neuralgia)
- anaesthesia 409,429,**443**,446,481, 564,581(630)
- cingulotomy 554,560,**562**
- descending influences on 451,455, 505,512,600,**606**
- gating **451**,505,512,600,606
- intractable pain 451,481,554,562
- - and limbic system 525,554,560, **600**
- pathways
- - central, from body 451,456,458, 459,489,505,525,533,534,549, 564,586,**600-601**(612,613,614, 615,616,622,626,627)
- - - from head 479,480-481,505, 525,533,534,549,564,586, **600-601**(619,620,622,623,624)
- - peripheral 429,431,**432**,433, 437-438,443(608,609,610)
- - perception 533,586
- - phantom limb **443**
- - prefrontal lobotomy 554
- - referred 438,**600**
- - and reticular formation 451,489, 505,512,534,600,606
- - tractotomy 458,**481**
- - thalamic 533(626-627)
- - visceral 437-438,451,600
Palate **275-277**
- cleft **325**(348-349,628)
- fetal **51**
- hard 17,**275-277**
- innervation **302-303**
- primary **259**
- secondary **259**,**321-325**
- soft 275,**277**
- - muscles **295-297**
- - paresis (622)
- - vasculature **305**,**306**
Palatine bone 11,17,20,**37-38**,46, 47,48
- horizontal plate **38**,47,48
- muscles attached to 38
- ossification 38
- perpendicular plate 18,**37-38**,47, 48,252
- pyramidal process 20,**38**,48,176
Paleocerebellum **518**
Paleocortex 541,543,**555-556**,568
Paleopallium **555** (see also Paleocortex)
Paleostriatum (see Globus pallidus)
Pallidum (see Globus pallidus)
- ventral pallidum 568
Pallium **541** (see also Cortex, cerebral)
Palpebrae (see Eyelids)
Palsy, brachial plexus (Erb–Duchenne palsy) (143)
- facial, Bell's (190)
- - following forceps delivery (77)
- - infection (385,387)
- - with skull fractures (73,74)
- - with tumour (76,192,349)
- - oculomotor (610-611)
- optic (73-74)
Papilla, lacrimal 148,**209**,214
- optic (see Disc)
- parotid **167**
- - infection (190)
- sublingual 275,**290**
- tongue 275,**477**
- - circumvallate 107,**275**,469,477, 596
- - foliate (349)**477**
Papilloedema (239,270)485(617,
618,619,628,629,630,631)
Paraesthesia(e) **443**(611)
Paraflocculus, cerebellar **514**
Paralysis/paresis, caused by neural lesion (see also named nerves and nuclei) 412,443,**459**,500,501, 547,564,582,587(612-630)
- - abducent nerve 235,**478**,**479** (620,621)
- - accessory nerve (139,143)468 (620,621,622)
- - external laryngeal nerve (358)
- - of extra-ocular muscles 478,479, 482,483,484(610-611,620-622, 624-625,629)
- - facial nerve and muscles (73,74,76, 77,190,192,349,385,387)**477**,596 (619,620,621,622,624,625)
- - flaccid 412,443,**459**(612-614, 615-616,626-627)
- - glossopharyngeal nerve (139)**471** (622,623,624)
- - of hand muscles 429(614,615, 616,620-622)
- - hypoglossal nerve (139,354-355) **467-468**(620,621,623,624)
- - of larynx and pharynx 470,**471**, (620,621,622,623,624)
- - of levator palpebrae superioris 193,**441**,**484**(622-623,624-625,629)
- - oculomotor nerve (235-236)**484** (610,611,617,624,625,629,630)
- - of pupil 441,**484**(610,611, 616-617,618,622,623,624,625, 629,630)
- - recurrent laryngeal nerve (358, 359)
- - of soft palate 470,471(622,623)
- - spastic 459,500,501,547,564, 582,587(612-614,615-616, 618-629)
- - of tongue 467,468(620,621,623, 624)
- - trigeminal nerve **480**
- - trochlear nerve **482**
Paralysis agitans (Parkinson's disease) 477,502,**513**,571 (621,625)
Parasubiculum 558
Parietal bone 11,13,15,**26-27**,48, 52(624)
- fetal 52-53
- - ossification 27
Parkinson's disease (see Paralysis agitans)
Pars (subnucleus)
- caudalis, intermedius, oralis (see Nucleus, trigeminal, spinal)
- flaccida 364,365,370
- opercularis 539
- orbitalis 539,553
- plana 204
- plicata 204
- tensa 364
- triangularis 539,540,553
Path, perforant 558
Pathways
- ascending (sensory) 430-433,450, **455-456**,458,459,504,505,**532-533**, 534,542-543,549-553,564,565-566, 586,**588-601**(611,612,613,614,615, 616,619,620,622,623,624,626,627, 628,630,631)
- auditory 471-474,532,533, 552-553,564,**592-593**(619,620)
- cholinergic 513,566-567
- corticopontocerebellar **500**,503
- descending (motor) 429,431, 433-434,450,453,454,455, **457-458**,**459**,**500-502**,504,505,506, 532-533,543,547,549,564-565,587, 602-606(613,614,615,616,618,621, 623,624,625,626,627,629)
- final common **434**
- gustatory 433,469,476,477-478,525,

P

533,550,564,586,**596-597**(624)
- of limbic system **559-560**
- monoamine 438-439,**509-513**(625)
- - dopaminergic **512-513** (see also Dopamine)
- - 5-hydroxytryptamine (serotoninergic) **512** (see also 5-Hydroxytryptamine)
- - noradrenergic **438-439**,512 (see also Noradrenaline)
- olfactory 433,488,525,553, 555-556,563,586,**588-589**(629)
- somatosensory
- - anterolateral 451,454,455,**456**, 458,459,**489**,532,533,534,549, 564,586,**600-601**(613,614,615, 616,622,623,624,626,627)
- - dorsal column **455-456**,458,459, **488-489**,532,533,549,564,586, 598-599(613,614,615,616,623, 626,627)
- - nociceptive (pain) 429,431, 432-433,443,**451**,455,**456**,458, 459,479,480-481,489,505,525, 532,533,534,**549**,564,586, 600-601(612,613,614,615,616, 619,620,622,623,624,626,627)
- - proprioceptive 432,450,451,453, **455**,**456**,458,479,482,488-489, 503-504,520,532,533,549,564,586, 598-601(611,612,613,614,615,616, 622,626,627)
- - tactile 429,431,432,443,451,453, **455-456**,458,459,479,480-481, 488-489,532,533,534,549,564, 586,**598-601**(612,613,614,615, 616,623,626,627,630)
- - temperature (thermal) 429,431, 432,443,451,453,**456**,458,459, 479,480-481,489,533,549,564, 586,**600-601**(612,613,614,615, 616,619,620,623,624)
- - trigeminal 453,479-482,532,533, 549,564,586,**598-601**(608,613, 618,619,620,622,623,624)
- - for pain (see nociceptive)
- - subthalamic reticular **505**,508
- - for taste (see gustatory)
- - ventral amygdalofugal 527,568, 569
- - vestibular 433,**457-458**,471,473, 474-476,520,533,550,564,586, **594-595**,**604-605**(619,620,622)
- - visual 433,484-488,525,526,528, 532,533,534,550-552,554-555, 564,565-566,586,**590-591**(611, 612,616,617,618,624,625,627, 628,630,631)
Pedicles, of vertebrae **61**,63
Peduncle, cerebellar
- - inferior 419,460,503,513,**516**, 520,521(620,622)
- - middle 419,461,503,504,513,**516**
- - superior 419,462,503,513, **516-517**,518,522,523
- cerebral 419,460,**461**,524,537, 579,580,583
- mamillary **526**,**527**
- thalamic **532**,564,565
Peptide neurotransmitters **411-412**, 434,438,441,442,451,458,505, 509,525,526,527,568,577,585, 600,606
Perception 533,535,553,554,555, 581,**586** (see also specific senses)
- stereoscopic depth **550**,560
- speech **553**
Pericardium, fibrous 95
Perikaryon **404**
Perilymph 376,**378**
Perineurium **428**,446
Period, critical (in development) **560**
Personality 532,535,**554**
Pes hippocampi **558**,568,**572**

Pharynx 82,83,290,**310-319**
- branchial fistula (139)
- diverticulum (145,356)
- innervation 318
- laryngeal part (see Laryngopharynx)
- mucosa and fascia **317**
- muscles 312-**317**
- oral part (see Oropharynx)
- nasal part (see Nasopharynx)
- vasculature **318**
- wall of 165,**312**
- - structures passing through 317
Philtrum **148**,154,242,**273**
Phonation (speech) **343**
Photopsia (238)
Pia mater **393**,396,572,573,574
- spinal **446**
Pineal (see Body)
Pinealocytes **534**
Pinealoma (617)
Pinna (see Auricle)
Pit, nasal **259**
- suprameatal **33**
Pituicytes **416**
Placodes, ectodermal **422-424**
- lens **228**
- nasal **258-259**
- otic **380**
Planum temporale **540**,552
Plaques, atheromatous **576**
Plasticity, neural 412-414,436,521, 523,**560-561**,575
Plate
- cerebellar **426**
- cortical **427**
- cribriform of ethmoid bone 21,22, **36**,47,48,53,243,245,247,398, 537,**588**,589
- - fracture (73)
- - tumour at (76)
- floor, of neural tube 424
- horizontal of palatine bone **38**,47, 48
- motor end **433-434**
- neural **422**
- orbital, of ethmoid bone **37**,46
- - of frontal bone 21,**25-26**,46,537
- - of maxilla **41**,46
- - of zygomatic bone **35-36**,46
- perpendicular of ethmoid bone **36**,47,53
- - of palatine bone 18,**37-38**,47,48
- pterygoid of sphenoid bone 19-20, **31-32**,48,53,174
- - fracture (72-73)
- roof, of neural tube **424**,425,572
- tarsal **212**
- tympanic 16,20,**33**,53,172,364
Plexus
- choroid 421,**572-574**,575,584
- - of fourth ventricle 461,462,**573**, 575
- - of lateral ventricle 559,571-572, **573**,583
- - of third ventricle 524,538,**573**
- nerve 429,430,437,441,442,443
- - ansa cervicalis 86,**89**,90,105,111, **115**,119
- - brachial 83,86,91,95,105, **115-116**,129,130,429,430(616)
- - - damage (Erb–Duchenne palsy) (143)
- - - Pancoast's syndrome (193)
- - cardiac **109**
- - carotid, internal **112**
- - cervical 84-85,86,91,105, 113-**115**,158,429,430
- - myenteric (of Auerbach) **442**
- - palate, soft **107**
- - parotid **158**
- - pharyngeal 103,107,108,112, 317,**318**
- - submucosal (of Meissner) **442**

- - sympathetic on blood vessels 112,113,119,179,220,225,257, 304,372,400
- - tympanic 106,311,371,**372**
- - venous
- - auditory tube **365**
- - basilar **394**,584
- - medullary **584**
- - pharyngeal 180,**318**
- - pial 582,584
- - pterygoid 164,176,**180-181**,228, 250,253,258,306,311,318,364, 373(608)
- - - anaesthetic injection into (195)
- - - bleeding from (194-195,352)
- - - superficial, of brainstem 584
- - vertebral **67**,129,394,445,584
Plica lacrimalis **245**
- semilunaris 147,**210**
Pneumothorax (142)
Pole, of cerebrum 535
Pons 251,418,419,459,460,**461**, 462,**463**,464,471-482,**495-497** (see also Brainstem)
- arterial supply **579**
- boundaries **461**,**462**,463
- development **424-425**
- subdivisions **461**,496
- topography (gross) **461**,**462**,463, **495-497**
- venous drainage **584**
Ponticulus **362**
Position sense (see Pathways, proprioceptive)
Posture (see Systems, motor)
Potential
- action 406,**408**,409
- graded 406,**408**
- postsynaptic 408,**410**
- resting **408**
- summation **408**
Pouch, branchial **134**,380
- Rathke's 311,399
Precuneus **540**
Premaxilla **42-43**
Pressure
- of cerebrospinal fluid **574** (see also Hydrocephalus)
- intracranial, raised (235,239)516, 574,576(612,613,617,618,619, 620,624,625,628,629,630,631)
- intra-ocular **207**
- sensation (see Sensation)
Presubiculum **558**
Pretectum (see Area, pretectal)
Primary position of eye **215**
Priming memory **561**
Process, processus
- alveolar, mandible **43**,51,**59**
- - maxilla **41-42**,51
- anterior of malleus **370**
- articular of vertebrae **61**,63,124, 125
- ciliary **204**
- clinoid, anterior 22,30,391
- - middle **29**
- - posterior 22,30,391
- - erosion with pituitary tumour (75)
- cochleariformis 369,**371**
- condylar of mandible 15,43,**44**,45, 54,58,59,100,148,165,**172**,180
- coronoid of mandible 15,43,**44**, 45,59
- descending of lacrimal bone **40**,47
- ethmoidal of inferior concha **39**
- facial, developmental 159, **168-170**(192)222,253,**258-259**
- frontal of maxilla 14,**41**,46
- - of zygomatic bone **15**
- jugular of occipital bone 18,**28**,122
- lacrimal of inferior concha **39**,47, **215**
- lenticular of incus **371**
- mastoid of temporal bone 13,16,

18,33,**34**,51,77,82,90,94,95,124, 164,165,369
- - Battle's sign (73)
- - facial nerve damage (387)
- maxillary of inferior concha **39**
- - of palatine bone **38**
- muscular of arytenoid cartilage **336**,339,340
- odontoid of axis (see Dens)
- orbital of palatine bone **37**,46,47, 198
- palatine of maxilla 17,**41**,47,48, 243,276
- petrous 19,22,23,24,**33-34**,48,53, 54,297,361
- posterior of malleus **370**
- postglenoid **16**,33,172
- pterygoid of sphenoid bone 19-20, **31-32**,48,53,174
- - fracture (72-73)
- pyramidal of palatine bone 20,**38**, 48,176
- - of thyroid gland **119**,135
- sphenoidal of palatine bone **37**,47, 48
- spinous of vertebra (see Spine)
- styloid of temporal bone 16,18,33, **34-35**,53,68,88,95,103,173,174, 294,316
- temporal of zygomatic bone **35**
- transverse of vertebra **62**
- uncinate of ethmoid bone **37**,245, 250,252
- vaginal of sphenoid bone **31**,48
- vocal of arytenoid cartilage 334, **336**,337,341
- zygomatic of frontal bone 14,15, 25,46
- - of maxilla 15,40,**41**
- - of temporal bone 15,**33**
Prolactin, neural control of release **527**
Prominence, laryngeal 83(140)332, **336**
Promontory 369,372
Proprioception 428,430,431,**432**, 433,**450**,**473**,594,598 (see also Pathways, proprioceptive)
Proptosis (234)
Prosencephalon 419,420 (see also Cerebrum)
Prosubiculum **558**
Protuberance, mental 14,**43**,51
- occipital, external 13,27,83,94, 188(609)
- - internal **23**,27-28
Psalterium **563** (see also Commissure, hippocampal)
Psychosis, Korsakoff's (see Syndrome)
Pterion **15**,26,30,52 537(609)
Ptosis (193,235)**441**,484(622,623, 624,625,629)
Puberty, lesions affecting **528**,**534**
Pulp, dental 284,**286**
Pulvinar (see Nucleus)
Pump, sodium-potassium **408**
Punctum, lacrimal 148,**209**,214
Puncture
- cisternal/ventricular **574**
- - lumbar/spinal **446**,516,574(609, 618)
- - - contra-indication 516(618,619)
Pupil 147,**201**,204,**205**(235)
- Argyll Robertson (236,617)
- neural control 439,**441**,482,**484** (610,611,612,616,617,618,624, 625,629,630)
Purines as neurotransmitters **411**, 434,441,442
Putamen 528,531,534,543,564,567, 568,**569**,**570**,**571**
Pyramid
- of medulla oblongata 111,**460**,500,

P – S

600,607
- of middle ear **369**,372
- olfactory **556**
- of vermis **514**
Pyrexia **528**

Quadriplegia (**615**,616)
Quadrantanopia 555,**566**(627,628, 630,631)
Quantum of neurotransmitter **410**

Radiation
- auditory 552,**564**,592,593
- callosal **562-563**
- optic 485,550,554,555,564, **565-566**,572,580,581,**590**,**591** (630,631)
- somatosensory 549,**564**,582,598, 599,600,601(626,627)
- thalamic **532**,564
Ramus
- of accessory nerve, internal **469**
- of lateral sulcus 538
- of mandible 15,**43-44**,51,57,59, 148,164,165,174,308
- of spinal nerves 430,441
- - communicans 440,**441**
Raphe
- of brainstem 506,**512** (see also Nuclei)
- mylohyoid **295**
- palatine **276**
- palpebral, lateral 36,152,**212**
- pharyngeal 20,28,**312**,313
- pterygomandibular 20,43,156, **173**,273,313
Reading 553,**554**
Receptive field (see Field)
Receptor
- baroreceptor (pressoceptor) **99** (139)433,469,508
- chemoreceptor 99,433,469,508, 584,585,588,596
- neurotransmitter 408,**410-412**
- - adrenergic 411,441
- - amino acid **411**,560
- - autoreceptors 410
- - acetylcholine **411**,412
- - muscarinic **411**,442
- - - nicotinic **411**,433-434,438
- sensory **430-433**
- - auditory (hair cell) 433,**473**,592
- - exteroceptors **430-432**,433
- - interoceptors **432**,433,600
- - mechanoreceptors **432**,433
- - muscle **432**,434,436
- - nociceptors **432**,433,600
- - olfactory rods 433,488,555,**588**, 589
- - osmoreceptors 433,525,585
- - photoreceptors (rods and cones) 430,432,433,484,**590**
- - proprioceptors 431,**432**,433,434, 473
- - thermoreceptors **432**
- - vestibular (hair cell) 433,**473**,594
Recess (see also Angle, Fossa)
- dorsal of fourth ventricle **462**, **514-515**
- elliptical **378**
- epitympanic 366,**367**,369,370
- infundibular of third ventricle **524**
- lateral of fourth ventricle **462**,473
- optic **524**
- pharyngeal **310**
- pineal **524**
- spheno-ethmoidal 47,**244**
- spherical **378**
- suprapineal **524**
Recognition 554,555,**561**(629,630, 631)
Recovery of function after neurological lesion **412-414**,443, 560(612-614,615-616,626-627, 629)

Redundancy of connections **436**
Reflex 431,**435**
- accommodation-convergence 204, **226**(239)482,**484**,550,554,590, 604(617,624,625)
- acoustic **474**,476,480
- areflexia (see hyporeflexia)
- autonomic **437-438**,453,459,470, **471**,484,504,505,508,512, 524-525,**528**,569,585
- Babinski (612-614,615-616,618, 624-625,626-627)
- blink **476**,481(618,619-620)
- carotid sinus 470,**471**
- corneal (79)476,**481**(618,619-620)
- cough **470**,471,507
- cranial nerve **466-467** (see named Nerves)
- crossed extensor 431,**435**
- cutaneous 431,435,459(612-614, 618)
- deep (muscular or tendon) 431,434, **435**,436,459,587(612-614, 615-616,619,622-623,625, 626-627,628-629,630)
- exteroceptors **430-433**
- fixation (tracking) 475,479,**486**, 550,554,604(612-614)
- flexor (withdrawal) 431,**435**
- foveation **486**,550,590,604
- gag (pharyngeal) 470,**471**(610)
- hyper-reflexia **459**,587(612-614, 615-616,628-629)
- hyporeflexia 412,443,**459** (612-614,615-616,620-623, 626-627)
- light (pupillary) 236(239)**484**,486, 487,488,590,591(610,611,617, 618,624,625)
- - hemianopic (612)
- monosynaptic 431,**435** (see also myotatic and deep)
- multisynaptic 431,**435**
- myotatic 431,**435**,459,587 (see also deep)
- near response **484**
- pharyngeal (see gag)
- righting **475**
- salivation (see Salivation)
- sneezing **481**,507
- spinal 431,**435**,450,453,454,459, 586,587,606(612-616,617-618, 620-623,624-627)
- stretch (see myotatic)
- superficial (see cutaneous)
- swallowing (see Swallowing)
- sweating (see Sweating)
- vestibulo-ocular 475,479,**594**,595, (see also Nystagmus)
- visceral **437-438**,453,470,471,504, 505,508,512,524-525,528,569
- vomiting 470,**471**,507,585
- withdrawal (flexor) 431,**435**
Refractory period **408**
Regeneration, axonal **413**,438,443, 488,509,560
Region (within brain) (see also Area, Cortex)
- basal forebrain **566-567** (see also septal, Nucleus)
- hippocampal 538,**556-559**,560, 562,563,564 (see also Formation)
- - connections of 525,530,531,533, 534,**558-560**,562,563,564,567
- - functions of 555,**560**,**561**(629)
- septal 525,526,527,535,543,556,559, 562,**567**,568,569 (see also Area, Nucleus)
- subicular 556,558,559,563
- supra-optic **526**,527
- tuberal **526-527**
Representations (see Maps)
Resonators **344**
Respiration, neural control 429, 432,433,438,439,440,454,469,

470,471,506,**507-508**,555,585, 606(616,623,624)
Response, near **484**
Reticular formation (see Formation)
Retina 201,**205-206**,484,485,487, 512,530,**590**,591
Rhinencephalon 555
Rhinorrhoea (73,267,608-609)
Rhinoscopy (268)
Rhombencephalon 418,419,425 (see Cerebellum, Medulla oblongata, Pons, Brainstem)
Rhythms, circadian and neuro-endocrine 485,**525**,528,**534**
Rib, cervical **62**(80)
- thoracic, first 88,115,122,123, 128,129,130,134
- second 123
Ridge, of Passavant 311,**316**
Rigidity (see also Paralysis/Paresis, spastic)
- clasp-knife (614)
- cog-wheel (625)
- decerebrate (74,630)
- nuchal (617,618,620,622,626)
Rima glottidis **334**,337,340,341
- vestibuli 332
Ring, common tendinous of eye muscles 199,**215**,217,218,219, 221,222,228
- tracheal **118**,338
- tympanic **364**
- Waldeyer's lymphatic **317**
Rods, retinal 433,484,**590**
- olfactory 433,488,555,**588**,589
Root(s)
- of spinal nerves 109,**428**,445,446
- - dorsal or sensory **428**,445,446, 455,456,459
- - ventral or motor **428**,445,446, 459
- of trigeminal nerve **479**
- - mesencephalic **482**
Rostrum, of corpus callosum **537**
- of sphenoid bone **30**(610)
Rudiment, prehippocampal **540**
Rugae, palatine **276**

Sac, conjunctival **211**
- endolymphatic **378**
- lacrimal 148,152,**214-215**,227
Saccule, of internal ear 211,212, 375,**378**,473,590
- of larynx **332**,338
Salivation, neural control 439,442, **477**,478,**488**,505,556,588,596
Sarcoidosis (236-237)
Scala tympani 375,**376**,378
- vestibuli **375**,376
Scalp 147,**188-189**
- cutaneous innervation 84,**189**
- haematoma (196-197)
- infection (197)
- layers of **188**
- vasculature **189**
Scan, computerised tomographic (CT) 534,571,573(612,617,618, 629,631)
- positron emission tomography (PET) **577**
Scapula 83,88,92,94,123(143)
Schizophrenia **513**
Sclera 147,201,**202-203**
Sclerosis, multiple **406**
Scoliosis (623-624)
Scotoma **552**
Secretion, neural control
- of hormones 441,**527**,534,585
- gastric 439,440,**442-443**,471,488, 525,556,569,588,596
Segment, initial of axon **404**
Seidel test for perforation of cornea (237)
Seizure (see Epilepsy)
Sella(e), dorsum **22**,23,30,48,401,

459,460
- diaphragma 22,30,**392**,398,400 (628)
- tuberculum **22**,29
- turcica **22**,29,30,398
- - with pituitary tumour (75,628)
Sensations (see also Pathways, Systems)
- classification **430-432**,464-465,586
- exteroceptive 428,**431-432**,433, 586
- hearing (see System, auditory)
- interoceptive 428,432,433, **437-438**,600,601
- modality **430**
- pain (see Pain)
- pathways
- - in brainstem **488-489**
- - cortical areas **545**,549-556
- - in internal capsule **564-566**
- - in spinal cord **447**,450-456
- - summaries **586**,589-601
- - in thalamus **530-533**
- pressure 428,**432**,600,601
- proprioceptive (position and movement) 428,430,431,**432**, 433,458,473,479,482,549,594, 598,599(612,613,614,615,616, 626,627)
- receptors **430-433**
- smell (see System, olfactory)
- tactile 428,431,**432**,443,458,459, 549,598-601(612,613,614,615, 616,623,624,626,627,630,631)
- taste (see System, gustatory)
- temperature (hot and cold) 428,431,**432**,443,458,459,500, 601(612,613,614,615,616,619, 620,623,624,626,627)
- visceral 428,432,**437-438**,555,600, 601
- vibration 428,**432**,549,598,599 (612,613,614,615,616,623,624, 626,627)
- visual (see System)
Sensory unit **430**
Septum
- of cerebrum (see Region, septal)
- nasal 18,**47**,162,242,**243**,**245**
- - anaesthesia (266)
- - deviation of (268)
- - perforation (237)
- - orbital **211-212**,215,227
- - pellucidum **537-538**,567,571
- of spinal cord
- - dorsal intermediate **455**
- - dorsal median 424,**446**
Serotonin (see 5-Hydroxytryptamine)
Sexual dimorphism **150**,273,336
Sheath, carotid 82,86,87,93,**95-96**, 99,101,105,111,118,130,174
- dural, for optic nerve **224**
- fascial of eyeball **207-209**
- myelin (see Myelin)
- of styloid process **33**
- of thyroid gland **119**,120
Shingles (192,385)
Shock, neural, following injury **412-413**(612-614)
Sight (see System, visual)
Sign
- Babinski (see Reflex)
- Battle's (73)
- Chvostek's (141)
Sinus, carotid **99**,107(139-140) 469
- - compression to produce bradycardia (139-140)
- cervical **134**(139)
- of larynx **332**
- paranasal **249-253**
- - ethmoidal **250** (see also Air cells)
- - - sinusitis (269)
- - - tumour (270)

S

- - fetal 53
- - frontal 26,47,150,245,**249-250**(609)
- - maxillary 41,47,215,245,250,**252-253**
- - - fracture (71-72)
- - - inflammation (268)
- - - Paget's disease (76)
- - - relation with teeth 252
- - - tumour (253,270)
- - sphenoidal 22,29,30,47,244,245,250,**251-252**,399
- - - cyst 237-238
- - - pituitary tumour (75,270,610)
- venous
- - basilar 394
- - cavernous 164,180,220,221,222,223,226,228,251,391,**393-394**,399,400,583(608,610)
- - - infection (196,267,270,608)
- - - thrombosis (190-191,235,267,270,608)
- - confluence 23,394,**396**
- - intercavernous 393(608)
- - occipital 392,**395**,514,583
- - petrosal, inferior 18,24,28,34,**103**,394,460,583(608)
- - - superior 23,33,373,379,391,**394**,514,583(608)
- - petrosquamous **32-33**
- - sagittal, inferior **395**,398,536,583
- - - superior 21,24,26,27,28,391,**394-395**,398,536,537,574,583
- - sigmoid 24,28,34,103,129,369,**395**,460,514,583
- - sphenoparietal **394**
- - straight **395**,514,537,583
- - transverse 24,27,28,365,379,391,**395**,514,537,583
- - venosus sclerae **202**,207
- - glaucoma (240)
- Sinusitis, maxillary (268)
- - sphenoidal (237-238)
- Skin, innervation 428-430,431,**432**,439,**440**,**441**,479,480-481,598-601
- Skull
- - appearance, extracranial
- - - norma basalis **17-21**
- - - norma frontalis **13-14**
- - - norma lateralis **15-17**
- - - norma occipitalis **13**
- - - norma verticalis **12-13**
- - appearance, intracranial
- - - anterior cranial fossa **21-22**
- - - middle cranial fossa **22-23**
- - - posterior cranial fossa **23-24**
- - bone articulations
- - - cranial base **48**
- - - nasal cavity and nasal aperture **46**,47-48
- - - orbit and orbital aperture **46-47**
- - - pterygopalatine fossa **48-49**
- - calvaria **12-13**,21
- - evolutionary development **10-12**
- - fetal skull **51-60**
- - fontanelles 13,15,16,**51-52**(77,609)
- - growth of skull **54-59**
- - individual bones **11** (see named bones)
- - - of cranial base **29-35**
- - - of face **35-40**
- - - of jaws **40-45**
- - - of vault **25-29**
- - - subdivisions of skull **10**
- Sleep **508**,**509**,512,525,528,534
- Sling, digastric **87-88**
- Smell, sense of (see System, olfactory)
- Sneezing (see Reflex)
- Soma, of neurone 403,**404**
- Somatostatin **411**,568,585
- Somnolence **508**,**509**,528

Sounds, classification of **344**
Space (see also Tissue space)
- epidural (extradural) 445,446
- episcleral **207**
- perilymphatic **378**
- retrolenticular **206**
- subarachnoid 392,446,573,574,575,(608,609,612)
- subdural **392**
Sparing, macular (612)
Spasm, carpopedal (141)
Spasticity (see Paralysis/Paresis)
Speech **343-344**
- area (cortical)
- - anterior (Broca's) 549,552,**553**,554,562,581,592,593(620,621)
- - posterior (Wernicke's) **553**,554,561,562,581,592,593(628,629)
- - superior 547,**553**
- cerebellum, involvement in **518**(628-629)
- cerebral dominance 552-553,563
- cranial nerves, involvement in **468**,**471**,477,480
- language disorders 518,533,**553**,554,561,564(620,621,622,623,626,627,628,629)
- pulvinar, involvement in 533(626,627)
- scanning (628,629)
- slurred (620,621,622,623)
Sphenoid bone 11,15,19,20,21,22,23,24,**29-32**,46,47,48,52-53
- body 18,19,22,23,24,**29-30**,46,47,48,53,215,219,251,399
- fetal **52-53**
- greater wings 19,22,**30-31**,46,47,48,53,175,219,396,397
- lesser wings 21,22,**30**,46,47,48,53,215,220,538
- muscles attached to **32**
- ossification **32**
- pterygoid processes 19-20,**31-32**,48,53,174
- - fractures (72-73,624)
Spina bifida **422**(615)
Spinal block (epidural anaesthesia) **446**
Spinal cord (see Cord)
Spine, dendritic **404**
- of helix **362**
- mental **43**,293,295
- nasal, anterior of maxilla **14**,41,297
- - of frontal bone 25,47
- - posterior of palatine bone **17**,38
- of sphenoid bone 20,**31**,173
- suprameatal **33**
- of vertebra **62**,63,83
- - muscles and fascia attached to 92,94,95,124,125
Spindles, neuromuscular 431,432,**434**,549
Splenium of corpus callosum **537**
Split-brain **563**
Spondylosis, cervical (80)
Spongioblasts **424**
Spot
- blind 485,**590** (see also Disc, optic)
- yellow of retina **590** (see also Macula lutea)
Sprouting, axonal **413**,443,560
Spur, scleral **202**,204
Squamous part, of frontal bone 12,**25**,26
- of occipital bone 12,18,23,**27-28**,48,52,53
- of temporal bone 15,22,**32-33**,48,53,537
- - fracture (235,630)
Squint (see Strabismus)
Stalk
- pineal 524,**534**,535
- pituitary 392,**398**,419,524,527,575,585

Stapes 11,51,**369**,**371**
Stem, of lateral sulcus **536**
Stereognosis **598**
Sternum, manubrium 83,86,89,90,94,105,128,129
Stomodeum **168**,319
Strabismus 478,560(629)
Stratum, strata (see also Layers)
- of hippocampal formation **558**
- sagittal **566**
- zonale **529**
Stress **440**
Striae
- acoustic **473**,592
- ciliaris **204**
- longitudinal (of Lancisi) 556,558
- medullaris thalami **573**
- medullary, of fourth ventricle **462**
- - of thalamus 535,556,567
- olfactory 540,555,**556**,588,589
- terminalis (semicircularis) 527,**568-569**,571,572
- vascularis **377**
- visual (of Gennari or Vicq d'Azyr) **550**
Striatum (neostriatum) (see also Nucleus, caudate, Putamen) **567**
- ventral **569**
Stridor, respiratory (141,267,351,623-624)
Strips, cortical **541**
Stroke **412-414**,564,575,576(610,611,612,613,614,618,622,623,624,625,626,627,630)
Subiculum 526,**558**,559,563
Substance P **411**,438,442,451,568,570
-perforated, anterior 488,513,**537**,540,555,**556**,566,567,580,**581**,583,**588**,589
- - posterior **461**,**579**,580
Substantia
- gelatinosa **451**,600
- innominata (of Reichert) 513,526,563,**566-567**,568,569
- nigra **502**,513,528,530,534,567,568,569,570,571(625) (see also System, basal nuclear)
Subthalamus 419,523,524,**528-529**,570
Suckling **525**
Sulcus, sulci
- cerebral 419,**538-540**,545 (see also Fissure)
- - calcarine **539**,540,550,565,566,579
- - central **538**,539,547,581
- - cingulate **539**
- - circular **540**
- - collateral **540**,572
- - corpus callosal **539**
- - frontal, inferior **539**
- - - superior **539**
- - hypothalamic **524**
- - intraparietal **540**
- - lateral 535,**538**,539,540,555,581,582,583
- - - rami 538,539,540
- - - stem **538**
- - lunate **539**,550
- - occipital, lateral **540**
- - occipitotemporal 376,**540**
- - olfactory **540**
- - parieto-occipital **538**,539,540,579
- - parolfactory **540**
- - postcentral **538**,540,581
- - precentral **538**,539,581
- - rhinal **540**,555,559
- - rolandic (see central)
- - subparietal **540**
- - supraspleniaI **540**
- - sylvian (see lateral)
- - temporal, inferior **540**
- - - middle **540**

- - - superior **540**
- labiomarginal, of lips 273
- limitans, of neural tube **424**,462
- of spinal cord
- - dorsal (median, intermediate, dorsolateral) **462**
- - ventral median **444**,446,460,582
- - terminalis of tongue **275**
- - tympanic **364**,370
Surface markings of ear **362**
- of eye **147-148**,150,**209-210**
- of face **147-150**
- of neck **83**
Sutural bones **13**
Suture, coronal **13**,57
- fetal skull **52**,**54**,**56**,**57**
- frontal 26,**52**,57
- frontomaxillary **14**
- frontonasal **14**
- frontosphenoid **21**
- frontozygomatic **14**,15
- intermaxillary **14**,17
- lambdoid **13**,16,57
- metopic (frontal) 26,**52**,57
- occipitomastoid **13**,16,57
- palatine, median **17**,57
- - transverse **17**,57,276
- parietomastoid **13**,16
- petro-occipital **19**,394
- sagittal **13**,57
- spheno-ethmoidal **21**
- sphenosquamosal **15**
- sphenozygomatic **15**
- squamosal **15**,57
- zygomaticomaxillary **15**
- zygomaticotemporal **15**,57
Swallowing (deglutition) **318-319**,468,**469-470**,**471**,481,507(610,621,623,624)
- dysphagia (141,143,144-145,349,351,356)
- movement of thyroid gland during (140)
Sweating, gustatory (191-192)
- neural control 434,439,440,**441**,459,528(615,616,622,623)
Sympathectomy **440**
Symphysis, menti 43,44,**52**,54,295
Synapse (see also Junction) 404,408,**409-412**,413,414,433-434,434-435,440
- bouton **405**
- chemical **409-412**
- development 414,427,460
- electrical 409,**412**,507,590
- hypereffectiveness **413**
- knob **405**
Synaptogenesis, reactive **413**
Synchondrosis, spheno-occipital 19,30,**52-53**,54,57
- sphenopetrosal **19**
Syndrome
- amnesic **561**
- disconnection **553**
- Horner's (193)**441**,459(615,616,622,623,624)
- Kennedy's (238-239)
- Klüver-Bucy **555**
- Korsakoff 528,534,**561-562**(629)
- Marfan's, subclavian steal (144)
- medullary, lateral (Wallenberg's) **578**(622,623)
- - medial **578**
- Pancoast's (193)
- Parinaud's (617)
- Ramsay Hunt (385)
- thalamic 533(626,627)
Syphilis (617)
Syringobulbia (624)
Syringomyelia (615,624)
System
- anterolateral (see Pathway, somatosensory)
- auditory **592-593**

647

S – T

- - connections, central 430,432,**433**, **473-474**,543,552-553,564,586, 592-593
- - - peripheral 432,**471-474**,592
- - - lesions 474,552,564(619,620)
- - autonomic nervous **417,437-442**
- - afferents and ascending fibres 428,431,432,433,**437-438**, 451-453,456,464,465,469-470, 500
- - cranial nerves and nuclei **464**, **465**,466,469-470,471,476,477, 482
- - efferents 428,431,434-435, **438-443**,450,453,464,465,**469**, 470,476,482,484,573
- - ganglia **438**,440,442,466,469,476, 484
- - lesions 437,**441**,443,**459**,471,477, **484**(615,616,622,623,624)
- - reflexes (see Reflexes)
- - spinal cord and nerves 428,431, 432,433,**437-443**,450,453,459
- - - supraspinal control **437**
- - - descending pathways 437,441, 458,**459**,505,512,**606**
- - - - hypothalamic 505,525,526,**528**
- - - - other 535,555,560,567,**569**,585
- - - - reticular **504**,505,507,508,**512**
- - basal nuclear 567,**570-571** (see also named components)
- - cardiovascular, neural control 431,433,434,437-438,**439,440**, 441,442,**469**,470,471,**508**,512, 524,525,526,528,585,606
- - central nervous **403-607**
- - - definition **417**
- - - gross topography **418-422**
- - dorsal column (see Pathway, somatosensory)
- - endocrine, neural control 439, 441,**525,527,528,534,569**,585 (627,628)
- - enteric nervous **442-443**
- - extrapyramidal **587**
- - gustatory 586,**596-597**
- - - in brainstem **477-478**,505,506, 507,585,596
- - - in cortex **550**
- - - in diencephalon 525,**533**
- - - in periphery 433,469,476, 477-478,**596**
- - - lesions 471,**596**(624)
- - limbic **559-560** (see also named components)
- - Papez, circuit **560**
- - motor **587**
- - - components
- - - - basal nuclear system 502,507, 513,528,561,**570-571**(613-614, 621,626,627)
- - - - brainstem nuclei
- - - - - cranial nerve **464-467**(619, 620,621,622)
- - - - - reticular 501,**505-508**
- - - - - other **501-504**,513
- - - - cerebellum 518-523,561(613, 614,618,619,621,622,623,624, 626,627)
- - - - motoneurones (see Neurones, motor)
- - - - muscles (see Muscles, innervation)
- - - - pathways 447,**587**,**602-607**
- - - - - brainstem 471,475,486, **500-504**,**505-508**,513(613,622, 623,624,625,629,630)
- - - - - cerebral cortex 543,**547-549**, 550,554(613,617,618,620,621, 624,625,629)
- - - - - diencephalon 528,529,**532-533**, 534
- - - - - internal capsule 564-565,582 (612,613,614,621,626,627)
- - - - - spinal cord and nerves 428,429, 430,431,433-434,443,**447**,450, 451,**453-454**,455,**457-458**,459 (613,614,615,616)
- - - control of
- - - - balance 454,457,471,473, 474-475,476,518,521,594,595, **604,605**
- - - - co-ordination of movement (see Cerebellum, Pathways, proprioceptive)
- - - - eye movements (see System, visual)
- - - - head movements 454,457-458, 468,485-486,501-502,506,507, 547,564,**604,605**
- - - - posture 450,502,506,518,521, **604,605**
- - - - reflexes (see also Reflex) 435, 457,**459**,506,507,571,**587**,606, 607
- - - - skills 454,475,523,547,554, 561,**602-603**
- - - - tonus (see also Muscles, innervation) 434,457,**459**,518, 521,522,571,**587**,602,603,606,607
- - - - voluntary movements 450,453, 454,455,457,458,**459**,500, 506,522,523,**547**,564,571,582, **587**,**602-603**
- - dysfunctions
- - - abnormal movements 475,**518**, 533-534,**571**(618,619,620,622, 623,625,626,627)
- - - apraxia (loss of skills) 554(620, 621,622,630,631)
- - - imbalance **476**,**518**(618,629, 622,623)
- - - inco-ordination (ataxia) **458**, 476,**518**,554(618,619,622,623, 626,627,628,629)
- - - paralysis/paresis **459**,500,501, 547,564,582,**587**(612,613,614, 615,616,619,620,621,622,623, 624,625,626,627,629,630)
- - olfactory **555-556**,**588-589**
- - - bulb 556,**588**,589
- - - connections 480,488,525,530, 535,540,543,555-556,562,568, 586,**588-589**
- - - cortex 543,555,556,568,588,589
- - - lesions 488,553,588(629)
- - - lobe 555
- - - nerve **488**,555,556,588,589
- - parasympathetic (craniosacral) 438,439-440,**441-442**,450,453, 464,465,466,469-470,476,477, 482,484,508,528
- - peripheral nervous 416,**417**, **427-443**,464-467
- - reticular activating **509**
- - somatosensory 586,**598-601** (see also Pathways, Sensations)
- - - in brainstem 464,465,479, 480-482,**488-489**(622,623,624)
- - - in cortex **549-550**,554(612,613, 614,630,631)
- - - in diencephalon 528,530,531, 532,**533**,534(626,627)
- - - in internal capsule 564(613,614, 626,627)
- - - lesions 458-459,533,549,564
- - - modalities 428,**430-432**
- - - in spinal cord 451,454-456, 458-459(614,615,616)
- - sympathetic (thoracolumbar) 438, 439,**440-441**,450,453,477,508, 528,577
- - ventricular **420-421** (see also Ventricles)
- - vestibular
- - - connections
- - - - ascending 474-475,518,520, 530,533,550,**594-595**
- - - - descending 457-458,475,587, **604-605**

- - - peripheral 471,**473**,594
- - - lesions 475-476(619,620,622, 623)
- - - vestibulo-ocular reflex **475**,479, **594**,595 (see also Nystagmus)
- - visual (see also Eye, Reflexes)
- - - control of eye movement **487**,**604**
- - - lesions 475,478,479,484,535, 547-549
- - - - nuclear **478-479**,**482**,**483-484** (610,611,620,621,622,624,625, 629)
- - - - reflex 475,**479**,**483**,484,**486**, 507,550,590(618,619,620,621, 622,623,628,629)
- - - - supranuclear 475,478-479,484, 501,**507**,518,521,535,543, 547-549,550,571,590,594,604 (620,621,622)
- - - - voluntary 486,**547-549**,550 (620,621,622)
- - - control of lens 482,**484**,486,550 (616,617,624,625)
- - - control of pupil 439,**441**,482, **484**,487-488(610,611,612,616, 617,618,622,623,624,625,626, 630)
- - - vision
- - - - colour 533,550,552,**590**
- - - - double 478,**482**,**484**,560(610, 611,624,625)
- - - - eye fields 547-549,**550**(620,621, 622)
- - - - lesions **485**,488,533,552,560, 564,566,582(611,612,627,628, 630,631)
- - - - pathway for 586,**590-591**
- - - - - cortex 541,543,**550-552**,554, 555,560,590,591
- - - - - diencephalon 524,525,526,527, 528,532,**533**,534,590,591
- - - - - internal capsule 564,**565-566**, 582,590,591
- - - - - midbrain **485-488**,590,591
- - - - - retina 433,484,485,487,512, **590**,591
- - - - photopic/scotopic **590**
- - - - plasticity **560**
- - - - receptive fields **550**
- - - - reflexes **486**,487-488,550,**590**, 594-595,**604**(616,617,618,620, 621,622,623,624,625,630)
- - - - stereoscopic **550**,560
- - - - visual fields **485**,550,551

Tachycardia (139)
Tactile sense (see System, somatosensory)
Taenia
- of fourth ventricle 573
- thalami 573
Tail of caudate nucleus **569**,572
Tanycytes **416**
Tapetum 572
Tarsus (see Plates, tarsal)
Taste (see System, gustatory)
Tears 209,**212**,213
- blockage (234)
Tectum, of midbrain 419,**461**
Teeth, tooth **277-288**
- abscesses 351(353-354)
- anaesthesia (193-194,352-353)
- canines **282**
- deciduous **284-286**
- - in fetal skull **51**
- eruption 320-321
- - delayed eruption (351-352)
- growth, influence on jaws 59
- incisors 280-282
- - control of position by lips and tongue 148,149
- innervation 298-300
- inhalation of tooth (140)
- molars 283-284
- occlusion 286-288

- - malocclusion following jaw fractures (71,72-73)
- permanent **280-284**
- premolars **282-283**
- sockets **41-42**,43
- - tumour (270)
- surfaces of 279-280
- vasculature 304,**305**,306
Tegmen tympani 20,**34**,366,369
Tegmentum, of brainstem **463**
Tela choroidea 462,524,538,**573**, 579,583
Telencephalon **419-420**,**535-572** (see also Hemispheres, cerebral)
Telodendria **404**
Teloglia 416,**417**
Temperature
- body, neural control 439,**440-441**, 524,525,528
- sense (see Sensation)
Temporal bone 11,13,15-16,18-20, 22-24,**32-35**,48,51,53,54
- fetal **51**,53,54
- fracture (73,74,235-236,630)
- mastoid process 13,16,18,33,**34**, 51,77,82,90,94,95,124,164,165, 369
- muscles attached to **35**
- ossification **35**
- petrous part 19,22,23,24,**33-34**, 48,53,54,297,361,375,394
- squamous part 15,22,**32-33**,48,53
- - fracture (235,630)
- styloid process 16,18,33,**34-35**,53, 68,**85**,95,103,173,174,294,316
- tympanic part 16,20,**33**,53,172, 364
Temple **147**,**188**(609,630)
Tentorium cerebelli 24,28,30, **391-392**,394,400,419,420,460, 514,535,537(630)
Tetraplegia (see Quadriplegia)
Tetrodotoxin (TTX) **409**
Thalamus 419,523,524,**529-534**, 542,543,564,571,572,583
- arterial supply 579,**580**,581
- connections **529-534**,542-543,564, 565-566
- functions 529.**530-534**,561
- lesions involving **532-533**(626, 627)
- nuclear groups **529-534**
- relationships **523-524**,**529**,571, 572,573
- subdivisions **529**
- topography 523-524,**529**
- venous drainage **583**
Thermal sense (see Sensation, temperature)
Thirst 524,525,527,**528**,569,585
- drinking (swallowing) 468, **469-470**,**471**,481,507,547(622, 623,624)
Thought **535**
Thrombosis, of arteries of brain (611,613,614,623)
- cavernous sinus (190-191,235, 267,270,608)
Thymus (see Gland)
Thyroidectomy (141,358)
Thyrotropin 527(627,628)
Thyrotoxicosis (234)
Tinnitus (76,388,474,619-620)
Tissue spaces
- buccal **308**
- of carotid sheath **99**
- definition **95-96**
- infratemporal **310**
- intrapharyngeal **97**,309
- parapharyngeal **96-97**,**308-309**
- - infection (195)
- parotid **308**
- peritonsillar **97**,309
- - abscess (quinsy) **97**(355)
- pharyngeal **96-97**(195)**309**

T-V

- pretracheal **97-98**
- prevertebral **98-99**(143)
- pterygomandibular 179,273,298, **308**,310
- - infection (195)
- retropharyngeal **96**,97,98,309
- retrovisceral 96,**98**,99
- subaponeurotic **188**
- - haematoma (197)
- sublingual 96,**308**
- submandibular **96**,308
- - infection (351)
- submasseteric **308**
- submental 96,**308**
- suprasternal **94**

Tomography, computerised (see Scans)
Tone (tonus) **434** (see also Muscle)
Tongue **275**
- in acromegaly (75)
- ankyloglossia (350-351)
- innervation 107,111,179,**301-302**, 467,468,469,471,476,477,596, 597
- lingual thyroid **119**(349)
- lymphatic drainage **306-307**
- movements of thyroglossal cysts on protrusion (140)
- muscles, extrinsic **293-295**
- - intrinsic **291-292**
- with myxoedema (193)
- tongue thrusting **148**
- vasculature 101,103,**305-306**

Tonsil
- of cerebellum 514,516
- lingual 275,307,311,317
- palatine 97,107,307,**312**
- brachial fistula 139
- - glossopharyngeal neuralgia (609-610)
- - inflammation (355)
- - referred pain (355)
- - tubercular infection (142)
- pharyngeal 307,**311**,317
- tubal 307,**310**,317
Torticollis (141)
Torus, mandibular (350)
- palatine (350)
Touch (see Sensation, tactile; System, somatosensory)
Trabecular meshwork at limbus **202**,207
Trachea 82,83,95,97,108,113,**118**, 129,133,338
- compression with goitre (141)
- intubation (357)
- tracheostomy (140)
Tract (see also Fasciculus, Fibre, Lemniscus, Pathway)
- amygdalofugal (see Pathway)
- anterolateral 451,453,454,456, 458,459,471,489,533,534,600, 601
- ascending (sensory) (see Pathway)
- corticohypothalamic, medial **525**
- corticopontine **500**,503,543,564, 565
- corticospinal (pyramidal) **500**,543, 547,549,564,565,582,**602,603**
- - lateral 451,453,454,**457**,460,**602**, 603
- - lesions of **459**,587(613,614,615, 616,618,621,623,624,625,626, 627)
- - ventral 454,455,**457**,460,602,603
- corticotectal **485**,500
- cuneocerebellar 456,503,**520**,598
- dentatorubrothalamic 502,503, **523**,528,529
- dentatothalamic **532**
- descending (see Pathway)
- dorsolateral (of Lissauer) 451,453, **455**,598
- geniculocalcarine 485,550,564, **565-566**,590,591 (see also Radiation, optic)
- geniculotemporal 552
- habenulo-interpeduncular **535** (see also Fasciculus, retroflexus)
- hypothalamohypophyseal 526, **527**,528
- interstitiospinal 457,**458**,502,507, 604,605
- of Lissauer (see dorsolateral) 451, 453,**455**,598
- mamillotegmental **526**,527
- mamillothalamic 525,526,**527**, 532,560
- olfactory 488,535,537,540,543, **555**,**556**,563,**588**,**589**
- optic 485,532,533,567,579,**590**, **591**
- - lesions involving **485**(611,612)
- pontocerebellar 503-504,516
- propriospinal **455**
- pyramidal (see corticospinal)
- reticulospinal 451,453,454,457, 458,**606**,**607**,608
- - medullary (lateral) **458**,506,507, 607,608
- - pontine (medial) **458**,506,507, 607,608
- rubrospinal 451,453,**457**,501,**602**, 603(622)
- solitariospinal 458,471,**606**
- solitary **470**
- spinal, of trigeminal nerve 453, 462,**480**,**481**,600,601(620,622, 623)
- of spinal cord **454-458**
- spinocerebellar 451,453,455,**456**, 458,503,520,598(622)
- spino-olivary 456,520,598
- spinoreticular **456**,489,504,505, 520,598
- spinospinal **455**
- spinotectal **456**,485
- spinothalamic **456** (see also anterolateral)
- supra-opticohypophyseal 526,**527**, 528
- tectospinal 457,**458**,474,486,**604**, 605
- tegmental, central (of Forel) 505
- trigeminothalamic, dorsal **481**, 533,**598**,599
- - ventral **481**,489,533,**598**,599, **600**,601
- tubero-infundibular **526-527**
- ventrolateral (see anterolateral)
- vestibulospinal 454,**457**,**458**,474, 475,594,595,**604**,605(622)
Tract, uveal of eye 203-205
Tractotomy 458,**481**
Tragus 160,178,**362**
Tranquillisers **513**
Transmitters (see Neurotransmitters)
Tremor, intention **518**,532 (619,622)
- - of basal nuclear origin **571**(625)
Triangle, anterior **86-90**
- carotid 87,**90**
- digastric (submandibular) 86-87, 164,289
- hypoglossal **462**
- muscular 87,**88-90**
- occipital 87,**91**
- posterior 86,87,**90-92**,104,109, 115
- - pharyngeal diverticulum (145)
- - wound (143)
- submental **87**
- suboccipital 101,124,**126-128**,129, 130,131
- supraclavicular 87,**91**
- suprameatal **16**,33,369
- vagal **462**
Trigone, collateral 572
- olfactory 556

Trismus (194-195,355)
Trochlea of superior oblique **26**, 160,209,**219**,222,224(610)
Trophic effects of neurones **405**, 412,413,414
Trunk
- of corpus callosum **537**
- costocervical 124,131,**132**
- lymphatic (see Ducts)
- sympathetic 105,107,110,**111-113**, 118,120,133(139)440,441
- thyrocervical 85,**131**
Tube, auditory 19,31,34,106,258, 295,297,**310-311**,317,366, 367-368,371(628)
- neural 442
- - layers of 424-427
Tuber, cinereum **399**,524
- of vermis 514
Tubercle
- acoustic **462**
- anterior of thalamus 524,529
- articular (of temporal bone) **16**,20, 33,51,**172**,173,**174**
- - of atlas, anterior **62**,63
- - posterior **62**,128
- auricular **362**
- carotid **62**
- cuneate **462**
- genial (see Spines, mental)
- gracile **462**
- intercolumnar (see Organ, subfornical)
- jugular **28**
- lip, upper **148**,273
- marginal (Whitnall's) 36,209
- mental **43**
- olfactory **556**,588
- pharyngeal **20**,28,312
- postglenoid 16,33,**172**
- scalene 122
- of spinous process of cervical vertebra **62**
- thyroid 336
- of transverse process of cervical vertebra **62**,63
- trochlear 26
Tuberculosis, cervical lymph nodes (142-143)
Tuberosity, frontal **25**,51,147
- of maxilla **40**,48,156,255,305
- parietal 13,**27**,51
Tumour
- bronchus (359)
- of central nervous system (239, 610,615,616,618,619,627,628, 631)
- ethmoidal air sinus (270)
- external acoustic meatus (385)
- internal jugular vein (388)
- leukaemia (192)
- lung (193)
- lymph nodes, cervical (144,145)
- maxillary sinus (270,353)
- meningioma (76)
- nasopharynx (355)
- oesophagus (144-145)
- optic nerve (236)
- parotid gland (192,349)
- piriform fossa (356)
- prostate (79)
- pituitary gland (75,270,610)
- submandibular gland (354-355)
- thyroid gland (358)
- tongue (350)
- vestibulocochlear nerve (76,388, 619,620)

Ulcer, corneal (192)
Umbo **364**
Uncus **540**,555,**556**,557,558,567, 572
- of vertebra **61**
Unit, motor **434**
- sensory **430**

Utricle 375,**378**,473,590
Uvula, of cerebellum 471,514,**516**, (623,624)
- of soft palate **277**,297
- - displacement (349)

Vallecula **312**
Varicosities, nerve **434**
Vasculature
- of central nervous system **575-584**
- - accidents involving **575-577** (see also Accident, cerebrovascular)
- - arterial supply **577-582**
- - - anastomoses 576,**577**,578,579, 580,581,582
- - - to spinal cord **582**
- - barriers 573,**584**
- - neural control 575,**577**,585
- - venous drainage **582-584**
- - - of spinal cord **584**
- of ear 363-364,**373**,**379**
- of face 161-164,180-181
- of larynx 342
- of meninges 396-397
- of neck 99-105,129-132
- of nose 243,247-249
- of orbit 226-228
- of orodental structures 304-306
- of paranasal sinuses 250,252,253
- of pharynx 318
- of scalp 189
- of temporomandibular joint **173-174**
Vaso-active intestinal peptide (VIP) **411**,438,441,442
Vasopressin 411,471,525,**526**,**527**, 606
Vault, cranial **10**,12-13,21,**25-29**,51
Vein, vena
- accompanying hypoglossal nerve 103,**306**
- alveolar, inferior **306**
- - superior **306**
- angular **164**,228,243
- - infection (190,235)
- auricular, posterior 105,**166**,363
- axillary 132
- basal **583**
- basivertebral 61,**67**
- brachiocephalic 83,103,118,120, **129**,130,318
- - injection into (142)
- buccal **164**,306
- cerebellar **583**,584
- cerebral, anterior 536,**583**
- - great (of Galen) 391,461,537, 575,**583**
- - inferior 393,**583**
- - internal 538,**583**
- - middle, deep 583
- - - superficial **583**
- - superficial 537,**583**
- - superior **583**(618)
- cervical, deep 105,**130**
- - superficial **105**
- choroidal **583**
- ciliary **202**,203
- cochlear **379**
- cricothyroid 104
- diploic 105,**395-396**
- emissary 13,16,18,19,21,22,23, 105,181,189,**393**,396,**584**(608)
- facial **103**,105,162,**164**,166,180, 228,253,306,318(608)
- - common **103**,164
- - deep 164,**181**
- - infection (190-191,266-267)
- - transverse **164**
- hypophyseal **400**,585
- infra-orbital **164**,228
- intercostal **130**
- intervertebral **67**
- jugular, anterior 85,87,99,**105**, 116,295
- - - dilation with goitre (141)

V-Z

- - communicating between external and internal 105
- - external 83,85,86,91,**104-105**, 116,132,166,364,365
- - - cannulation (142)
- - - indicator of venous pressure (142)
- - internal 18,24,82,83,95,**103-104**, 105,109,111,115,118,129,164, 306,318,366,379,394,584
- - - high superior bulb (386)
- - - tumour (388)
- - posterior external 105
- labial **164**,306
- labyrinthine 379
- lacrimal 228
- laryngeal 104,**342**
- lingual **103**,306
- - deep **103**,275,**306**
- - - varicosities (350)
- - dorsal **103**,306
- mastoid emissary 16,105,189,**396**
- maxillary 164,166,180,**181**
- meningeal, middle **394**,397(609)
- mental **164**
- nasal **164**
- occipital **105**,130
- ophth...ic, inferior 22,180,181, 204,**228**,243,258,396,485(608)
- - superior 22,**164**,204,207,212, 214,**228**,243,250,396(608)
- - - dilation with caroticocavernous fistula (235)
- - - infection (190-191,235)
- pharyngeal **103**,318
- pontine **584**
- pterygoid canal **103**
- retinal **206**
- - thrombosis (270)
- retromandibular 103,105,164,

166,180
- sphenopalatine 48,**249**,258
- striate **583**
- subclavian 83,103,105,114,118, 129,130,**132**
- - cannulation (142)
- sublingual **103**,**306**
- supra-orbital 164,**228**,250
- suprascapular **105**
- supratrochlear **164**
- temporal, superficial **164**,166,181, 363(609)
- thalamostriate (terminal) 571,**583**
- thoracic, internal 120,**130**
- thyroid, inferior 103,118,119,120, **130**,318,342
- - middle 99,**104**,119,120
- - superior **104**,119,120,342
- vertebral 66,**67**,105,**129-130**,131, 445
- - tumour (79)
- vortex 203,**204**,205,228
- Velum, medullary (inferior, superior) **462**,**463**
- interpositum **573**
- Ventral thalamus (see Subthalamus)
- Ventricles of brain **420-422**,**572-574**
- fourth 421,**462-463**,573,574,575
- - apertures of **462-463**,473,474,475
- lateral 420,421,535,537,562,565, 566,569,**571-572**,573,574
- third 421,**524**,538,574,585
Ventricle of larynx 332
Vermilion 148,273
Vermis 391,419,**514**,518
Vertebra, atlas 18,**62-63**,64, 65-67,83,126
- axis 18,23,**63**,64,65,66(609)(see

also Dens)
- cervical typical **61-62**
- fractures (78,79)
- joints **64-67**
- muscles attached to **63-64**
- ossification **64**
- osteomyelitis (143)
- prominens (seventh) **62**,83
- thoracic 83,128
- vasculature 66,**67** (see also Column, vertebral)
Vertex, of skull 12
Vertigo 475
Vesicles
- of brain **422-427**
- optic **228-229**,426
- synaptic **410**
Vessels, blood
- of brain 573,**576-577**,**583**,584 (see also Vasculature, of central nervous system)
- innervation 431,432,433,437-438, 439,**440**,**441**,**508**,512,577,585
Vestibule, of labyrinth 375,376
- of larynx 332,337
- of mouth 169,242,244,**273**,375
- of nose 148,**244**
Vestibulocerebellum (flocculonodular lobe) **518**
Vibratory sense (see Sensation)
Vibrissae **244**
Villi, arachnoid 394,**573**,**574**
Virus, herpes (385,629)
Viscera, innervation of 428,432, 433,**437-443**,450,453,**469**, 470-471
Viscerocranium **10**,51
Vocalisation, voluntary control of 549 (see also Speech)

Voice, hoarseness of (358)470,**471** (622,623,624)
- monotonous (625)
Vomer 11,18,**38-39**,47,48
- ossification **39**
Vomeronasal, cartilage **245**
- organ **245**
Vomiting 470,471,507,585 (616-617)

Water balance, neural control 524, 525,527,**528**,569,585
White matter **417-418**
- cerebral **562-566**
- spinal **454-458**
Wing
- lateral of ventral horn of spinal cord **450**,453,454
- of sphenoid bone, greater 19,22, **30-31**,46,47,48,53,175,219,396,397
- - lesser 21,22,**30**,46,47,48,53,215, 220,538
Wry neck (141-142)

Zona incerta **528**
Zone (see also Area, Layer, Region)
- chemoreceptor trigger 506,**585**
- of neural tube **426**
- periventricular, of hypothalamus **526**,527
- - of reticular formation **505**,506
Zonulae occludens (tight junctions) 416,**573**,**584**
Zonule (see Ligament, suspensory)
Zygomatic bone 11,14,15,16,**35-36**, 46
- fracture (71-72)
- muscles attached to **36**
- ossification **36**